W9-AXZ-896

Manufacturing Engineering and Technology

SECOND EDITION

Manufacturing Engineering and Technology

SECOND EDITION

SERO PE KALPAKJIAN

Illinois Institute of Technology

ADDISON-WESLEY PUBLISHING COMPANY

Reading, Massachusetts • Menlo Park, California • New York
Don Mills, Ontario • Wokingham, England • Amsterdam • Bonn
Sydney • Singapore • Tokyo • Madrid • San Juan • Milan • Paris

Sponsoring Editor: Don Fowley
Production Supervisor: Loren Hilgenhurst Stevens
Design, Editorial, and Production Services: Quadrata, Inc.
Illustrator: Capricorn Design

Cover Design: Peter M. Blaiwas
Cover Photo: Marshall Henrichs
Manufacturing Supervisor: Roy Logan

Library of Congress Cataloging-in-Publication Data

Kalpakjian, Serope
Manufacturing engineering and technology / Serope Kalpakjian—2nd ed.
p. cm.
Includes bibliographical references and index.
ISBN 0-201-56950-7
1. Production engineering. 2. Manufacturing processes.
I. Title.
TS176.K34 1992
670.42—dc20 91-37657
CIP

Reprinted with corrections May, 1992

2 3 4 5 6 7 8 9 10—DO—95949392

To
Milton C. Shaw and **M. Eugene Merchant**
in gratitude

About the author

Professor Serope Kalpakjian has been teaching and conducting research in manufacturing processes at the Illinois Institute of Technology since 1963. After graduating (with High Honors) from Robert College, Harvard University, and the Massachusetts Institute of Technology, he joined Cincinnati Milacron, Inc., where he was a research supervisor in charge of metal-forming processes and machine tools. He has published numerous papers in technical journals and is the author of several articles in the *Encyclopedia of Materials Science and Engineering*, the *McGraw-Hill Encyclopedia of Science and Technology*, and the *Marks' Standard Handbook of Mechanical Engineers*. He is a co-founder and former co-editor of the *Journal of Applied Metalworking* (now the *Journal of Materials Shaping Technology*) and an associate editor of the *Journal of Tribology*. He presently serves on the editorial boards of *Manufacturing Review*, the *Journal of Manufacturing Systems*, and the *Journal of Engineering Manufacture*.

Professor Kalpakjian is also the author of four books: *Mechanical Processing of Materials* (Van Nostrand, 1967), *Manufacturing Processes for Engineering Materials* (second edition, Addison-Wesley, 1991), which received the M. Eugene Merchant Manufacturing Textbook Award in 1985, *Lubricants and Lubrication in Metalworking Operations* (with E. S. Nachtman; Marcel Dekker, 1985), and *Manufacturing Engineering and Technology* (first edition, Addison-Wesley, 1989), which received the M. Eugene Merchant Manufacturing Textbook Award in 1990. He has also edited *Tool and Die Failures: A Source Book* (ASM, 1982). He is a Fellow of the American Society of Mechanical Engineers and ASM International, and is a full member of the International Institution for Production Engineering Research (CIRP). He is a founding member and a past president of the North American Manufacturing Research Institution of the Society of Manufacturing Engineers (NAMRI/SME).

Professor Kalpakjian has received a number of awards, including citations by the Forging Industry Educational and Research Foundation for best paper (1966) and the Society of Carbide and Tool Engineers (1977), the "Excellence in Teaching Award" from the Illinois Institute of Technology (1970), the "Centennial Medallion" from the American Society of Mechanical Engineers (1980), and the "Education Award" from the Society of Manufacturing Engineers (1989).

Preface

The proper method of teaching undergraduate courses in manufacturing engineering and technology in colleges and universities, both in the United States and abroad, continues to be a subject of considerable discussion, evaluation, and some controversy. One reason is the fact that this subject has become increasingly broad and complex, with continuing rapid progress in several applied as well as analytical aspects of manufacturing engineering. Another reason is the academic mission of a particular institution and the different backgrounds, experiences, and preferences that instructors in charge of such courses bring into the classroom. An appropriate methodology is particularly crucial for an introductory course in which most students encounter this subject for the first time.

Philosophy and goals

The rapid progress made in certain areas of manufacturing and the valuable suggestions and constructive criticism received from many colleagues at various institutions led me to prepare this second edition. Although on the one hand producing a comprehensive and state-of-the-art text has been the most important goal throughout the writing of this second edition, my ambitions continue to have two additional aims: to challenge and to motivate students. Topics are presented with a balanced coverage of the relevant fundamentals and real-world practices, so that the student develops a good understanding of the important interrelationships among the many technical and economic factors involved and how engineering science impacts on practical considerations.

Features and study aids

In general, this edition follows the format, organization, balance, and introductory nature of the first edition. It has retained such features as:

- An emphasis on the influence of materials and processing parameters as an aid to understanding manufacturing processes and operations.
- Design considerations, product quality, and manufacturing cost factors.
- The competitive context of each manufacturing process and operation, highlighted with illustrative examples.
- Development of a view of each subject that incorporates both current and potential developments.

Study aids include the following:

- A description of each topic within a larger context of manufacturing engineering and technology, using extensive schematic diagrams and flowcharts.
- Many analogies, discussions, and problems designed to stimulate the student's curiosity about products and phenomena encountered in everyday life.
- Emphasis on the practical uses of concepts and information discussed in various sections.
- Extensive amount of reference information, including a large number of tables, illustrations, graphs, and bibliography.
- Numerous illustrative examples to highlight important concepts and techniques.
- Summary tables highlighting key topics and concepts and comparing advantages and limitations of manufacturing processes.
- Summary, list of key terms, and discussion of trends at the end of each chapter.

New to this edition

In response to the many reviewers' comments and suggestions, numerous changes, both major and minor, have been made throughout the text, information has been updated, and more numerical problems have been added. Some illustrations and text have been deleted or replaced because they are no longer relevant or in common practice. Furthermore, certain new topics have been added and described in some detail, including:

- Status of advanced manufacturing technologies and strategies.

- Environmental considerations.
- Rapid prototyping.
- Technological advances in topics such as sensor fusion, neural networks, diamond coatings, and seeded-gel abrasives.

This edition also covers the following subjects in greater detail:

- Solidification mechanisms in casting; evaporative casting; vacuum casting.
- Behavior and processing of polymers and composites.
- Powder metallurgy.
- Coatings.
- Limits, fits, and tolerances.
- Inspection techniques.

In addition:

- The number of questions, problems, and examples has been increased by about 150, bringing the total to more than 1500.
- New illustrations have been added or replaced; some qualitative examples have been replaced with numerical examples.
- The "Trends" section at the end of each chapter has been updated and expanded.
- All bibliographies have been updated and expanded.

Audience

As with the first edition, this text has been written for students in manufacturing, mechanical, industrial, and metallurgical and materials engineering programs. It is hoped that by reading and studying this book, students will come to appreciate the growing importance of manufacturing engineering as an academic subject that is as exciting and challenging as any other engineering discipline.

Acknowledgments

It is a pleasure to acknowledge the assistance of many people in the preparation and publication of this second edition.

I am very grateful to Donald A. Fowley, Jr., Executive Editor, and Laurie McGuire, Associate Editor, Addison-Wesley Publishing Company, for their guidance and enthusiastic help in the preparation of this book. Many thanks also to Martha Morong and Geri Davis of Quadrata, Inc., for their work in producing this book.

I gratefully acknowledge the help of many colleagues at various universities and organizations for their constructive criticisms and suggestions for modifications for certain sections in the book: P. K. Wright (University of California, Berkeley), B. Harriger (Purdue University), G. W. Fischer (University of Iowa), C. A. Brown (Worcester Polytechnic Institute), S. G. Kapoor (University of Illinois, Urbana–

Champaign), L. Mapa (Purdue University–Calumet), J. Nazemetz (Oklahoma State University), E. M. Odom and J. M. Prince (University of Idaho), D. Bourell and C. Maziar (University of Texas), R. Jaeger (Auburn University), W. J. Riffe and D. Harry (GMI Engineering and Management Institute), R. L. French (I-RON Technologies), R. Taylor (KO Steel), L. Soisson (Welding Consultants, Inc.), and J. Widmoyer (Odermath U.S.A.). Special thanks also to Professors P. Demers (University of Cincinnati) and P. Cotnoir (Worcester Polytechnic Institute) for contributing original problems to Parts II and IV.

Many thanks to the chairman of my department, H. M. Nagib of the Mechanical and Aerospace Engineering Department at the Illinois Institute of Technology, for his constant encouragement and support. The assistance of my present and former students is also acknowledged, especially S. R. Schmid, P. Grigg, S. Chelikani, and X. Z. Li. It is again with great joy that I acknowledge the assistance of my son Kent, now a graduate student at the University of California, Berkeley, as the author of Chapter 34 on "Fabrication of Microelectronic Devices." In addition, I appreciate the help of many organizations that supplied me with numerous illustrations, photographs, and technical literature on various topics. The assistance of many other colleagues in the preparation of the first edition of this text has been gratefully acknowledged in the Preface of that book.

Finally, many thanks to my wife, Jean, for her help during the writing and production of this book.

Evanston, Illinois S.K.

Contents

7 Polymers: Structure, General Properties, and Applications 209

8 Ceramics, Graphite, and Diamond: Structure, General Properties, and Applications 235

9 Composite Materials: Structure, General Properties, and Applications 255

PART II METAL-CASTING PROCESSES AND EQUIPMENT 275

10 Fundamentals of Metal Casting 279

11 Metal-Casting Processes 304

12 Casting Design; Casting Alloys; Economics of Casting 345

PART III FORMING AND SHAPING PROCESSES AND EQUIPMENT 363

Manufacturing Engineering and Technology

SECOND EDITION

General Introduction

What Is Manufacturing?

As you begin to read this Introduction, take a few moments and inspect the different objects around you: your watch, chair, stapler, pencil, calculator, telephone, and light fixtures. You will soon realize that all these objects had a different shape at one time. You could not find them in nature as they appear in your room. They have been transformed from various raw materials and assembled into the shapes that you now see. A paper clip, for example, was once a piece of wire. The wire was once a piece of metal obtained from ores.

Some objects are made of one part, such as nails, bolts, wire or plastic coat hangers, metal brackets, and forks. However, most objects—aircraft engines (Fig. 1), ballpoint pens, toasters, bicycles, computers and thousands more—are made of a combination of several parts made from a variety of materials. A typical automobile, for example, consists of about 15,000 parts, and a C-5A transport plane is made of more than 4,000,000 parts. All are made by various processes that we call manufacturing. **Manufacturing**, in its broadest sense, is the process of converting

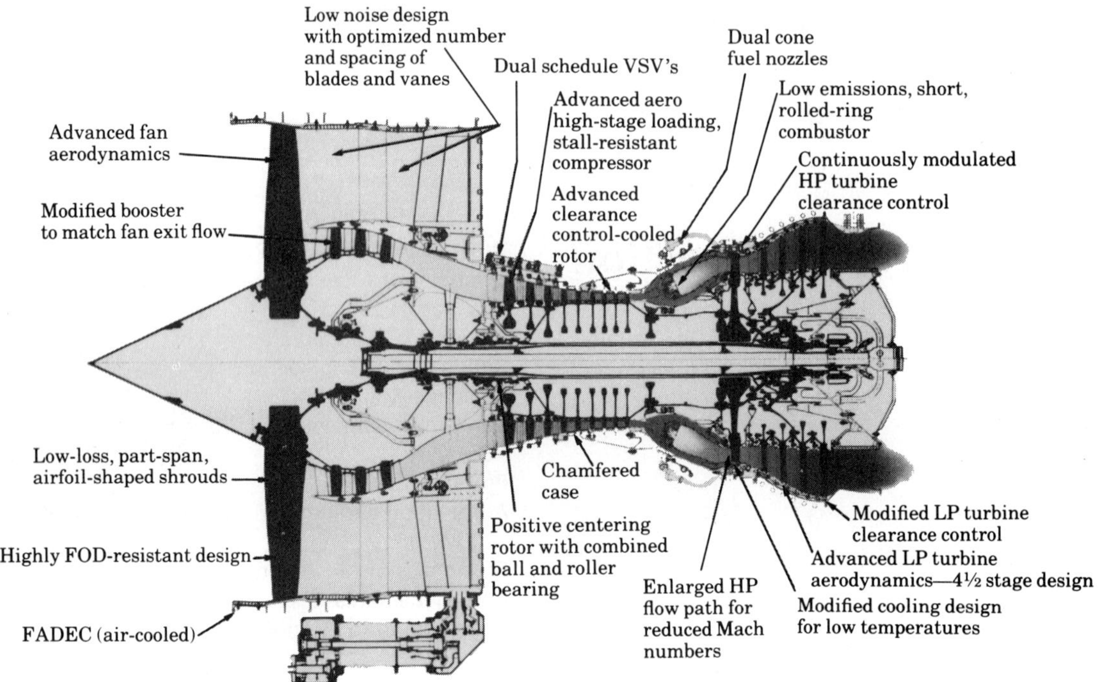

FIGURE 1
Cross-sectional view of a jet engine showing various components. Many of the materials used in this engine must maintain their strength and resist oxidation at high temperatures. *Source:* Courtesy of General Electric Company.

raw materials into products. It encompasses the design and production of goods, using various production methods and techniques.

Manufacturing is the backbone of any industrialized nation. Its importance is emphasized by the fact that, as an economic activity, it comprises approximately one-third of the value of all goods and services produced in industrialized nations. The level of manufacturing activity is directly related to the economic health of a country. Generally, the higher the level of manufacturing activity in a country, the higher is the standard of living of its people.

Manufacturing also involves activities in which the manufactured product is itself used to make other products. Examples are large presses to form sheet metal for car bodies, metalworking machinery used to make parts for other products, and sewing machines for making clothing. An equally important aspect of manufacturing activities is servicing and maintaining this machinery during its useful life.

The word *manufacturing* is derived from the Latin *manu factus*, meaning made by hand. The word manufacture first appeared in 1567, and the word manufacturing appeared in 1683. In the modern sense, manufacturing involves making products from raw materials by various processes, machinery, and operations, following a well-organized plan for each activity required. The word **product** means something that is produced, and the words product and production first appeared sometime during the fifteenth century. The word production is often used interchangeably with the word manufacturing. Whereas *manufacturing engineering* is the term used widely in the United States to describe this area of industrial activity, the equivalent term in Europe and Japan is *production engineering.*

Because a manufactured item has undergone a number of changes in which a piece of raw material has become a useful product, it has a *value*—defined as monetary worth or marketable price. For example, as the raw material for ceramics, clay has a certain value as mined. When the clay is used to make a ceramic dinner plate, cutting tool, or electrical insulator, value is added to the clay. Similarly, a wire coathanger or a nail has a value over and above the cost of a piece of wire. Thus manufacturing has the important function of adding value.

Manufacturing may produce *discrete products,* meaning individual parts or part pieces, or *continuous products.* Nails, gears, steel balls, beverage cans, and engine blocks are examples of discrete parts, even though they are mass produced at high rates. On the other hand, a spool of wire, metal or plastic sheet, tubes, hose, and pipe are continuous products, which may be cut into individual pieces and thus become discrete parts.

Manufacturing is generally a complex activity (Fig. 2), involving people who

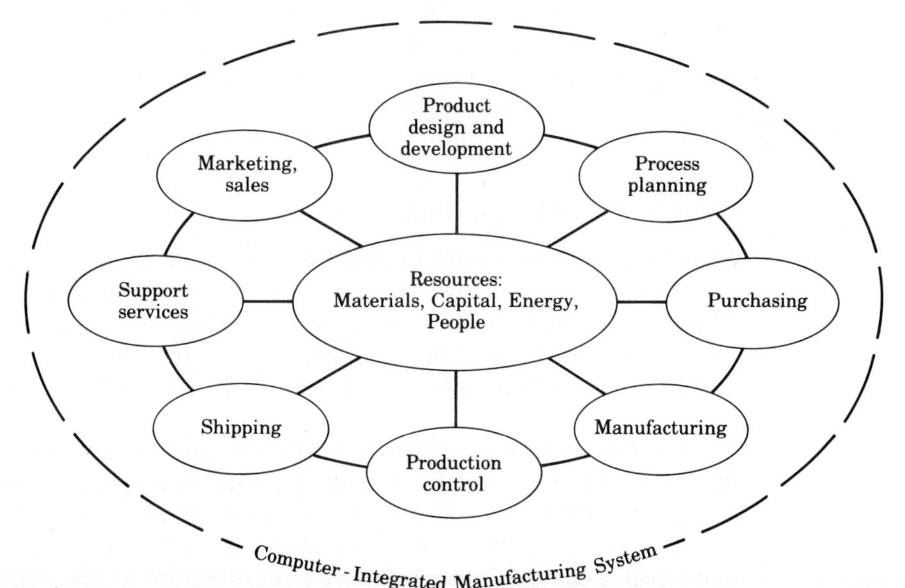

FIGURE 2
Chart showing relationships among many activities in manufacturing, involving materials, processes, machinery, and people.

have a broad range of disciplines and skills and a wide variety of machinery, equipment, and tooling with various levels of automation, including computers, robots, and material-handling equipment. Manufacturing activities must be responsive to several demands and trends:

- A product must fully meet *design requirements* and specifications.
- A product must be manufactured by the most *economical* methods in order to minimize costs.
- *Quality* must be built into the product at each stage, from design to assembly, rather than relying on quality testing after the product is made.
- In a highly competitive environment, production methods must be sufficiently *flexible* so as to respond to changing market demands, types of products, production rates, production quantities, and on-time delivery to the customer.
- New developments in *materials*, *production methods*, and *computer integration* of both technological and managerial activities in a manufacturing organization must constantly be evaluated with a view to their timely and economic implementation.
- Manufacturing activities must be viewed as a large *system*, each part of which is interrelated to others. Such systems can be modeled in order to study the effect of factors such as changes in market demands, product design, material and various other costs, and production methods on product quality and cost.
- The manufacturing organization must constantly strive for higher *productivity*, defined as the optimum use of all its resources: materials, machines, energy, capital, labor, and technology. Output per employee per hour in all phases must be maximized.

Brief history of manufacturing

Manufacturing dates back to 5000–4000 B.C. and began with the production of various articles made of wood, ceramic, stone, and metal (see Table 1 on pp. 6–7). The materials and processes first used to shape those products by casting and hammering have been gradually developed over the centuries, using new materials and more complex operations, at increasing rates of production and higher levels of quality.

The first materials used for making household utensils and ornamental objects included metals such as gold, copper, and iron. These were followed by silver, lead, tin, bronze, and brass. The production of steel in about A.D. 600–800 was a major development. Since then a wide variety of ferrous and nonferrous metals have been developed. Today, the materials used in advanced products such as computers and supersonic aircraft include engineered or tailor-made materials with unique properties, such as ceramics, reinforced plastics, composite materials, and specially alloyed metals.

Until the Industrial Revolution, which began in England in the 1750s, goods had been produced in batches, with heavy reliance on manual labor in all aspects of production. Modern mechanization began in England and Europe with the development of textile machinery and machine tools for cutting metals. This technology soon moved to the United States where it was developed further, including the important advance of designing, making, and using *interchangeable* parts. Prior to the introduction of interchangeable parts, a great deal of hand fitting was necessary because no two parts were made exactly alike. We now take for granted that we can replace a broken bolt of a certain size with an identical one purchased years later from a local hardware store.

Further developments soon followed, resulting in numerous products that we can't imagine being without because they are so common, such as typewriters, sewing machines, and engines. Since the early 1940s, major milestones have been established in all aspects of manufacturing. Note from Table 1 the progress made during the past 100 years, and especially the last two decades with the advent of the computer age, as compared to the period of 4000 to 1 B.C..

Although the Romans had factories for mass producing glassware, manufacturing methods were at first very primitive and generally very slow, with much manpower involved in handling parts and running the machinery. Today, with the help of *computer-controlled* manufacturing systems, production methods have been advanced to such an extent that, for example, aluminum beverage cans are manufactured at a rate of 10 cans per second. Screws and bolts are made at rates of thousands per minute.

Examples of Manufactured Products

In this section we review briefly the thought processes and procedures involved in designing and manufacturing some common products. Our purpose is to identify the important factors involved and, with specific examples, to show you how intimately design and manufacturing are interrelated.

Paper clips

Assume that you are asked to design and produce ordinary paper clips. What type of material would you choose to make this product? Does it have to be metallic or can it be nonmetallic, such as plastic? If you choose metal, what kind of metal? If the material that you start with is wire, what should be its diameter? Should it even be round or should it have some other cross-section? Is the wire's surface finish important and, if so, what should be its roughness? How would you shape a piece of wire into a paper clip? Would you shape it by hand on a simple fixture and if not,

TABLE 1
HISTORICAL DEVELOPMENT OF MATERIALS AND MANUFACTURING PROCESSES (DATES ARE APPROXIMATE)

	PERIOD	METALS AND CASTING	FORMING PROCESSES
Egypt: ~3100 B.C. to ~300 B.C.; Greece: ~1100 B.C. to ~146 B.C.; Roman empire: ~500 B.C. to A.D. 476; Middle ages: ~476 to 1492; Renaissance: 14th to 16th centuries	Before 4000 B.C.	Gold, copper, meteoritic iron	Hammering
	4000–3000 B.C.	Copper casting, stone and metal molds, lost wax process, silver, lead, tin, bronze	Stamping, jewelry
	3000–2000 B.C.	Bronze casting	Wire by cutting sheet and drawing; gold leaf
	2000–1000 B.C.	Wrought iron, brass	
	1000–1 B.C.	Cast iron, cast steel	Stamping of coins
	A.D. 1–1000	Zinc, steel	Armor, coining, forging, steel swords
	1000–1500	Blast furnace, type metals, casting of bells, pewter	Wire drawing, gold and silver smith work
	1500–1600	Cast iron cannon, tinplate	Water power for metalworking, rolling mill for coinage strips
Industrial revolution: ~1750 to 1850	1600–1700	Permanent mold casting, brass from copper and metallic zinc	Rolling (lead, gold, silver), shape rolling (lead)
	1700–1800	Malleable cast iron, crucible steel	Extrusion (lead pipe), deep drawing, rolling (iron bars and rods)
	1800–1900	Centrifugal casting, Bessemer process, electrolytic aluminum, nickel steels, babbitt, galvanized steel, powder metallurgy, tungsten steel, open-hearth steel	Steam hammer, steel rolling, seamless tube piercing, steel rail rolling, continuous rolling, electroplating
WWI	1900–1920		Tube rolling, hot extrusion
WWII	1920–1940	Die casting	Tungsten wire from powder
	1940–1950	Lost wax for engineering parts	Extrusion (steel), swaging, powder metals for engineering parts
Space age	1950–1960	Ceramic mold, nodular iron, semiconductors, continuous casting	Cold extrusion (steel), explosive forming, thermomechanical treatment
	1960–1970	Squeeze casting, single crystal turbine blades	Hydrostatic extrusion; electroforming
	1970–1990s	Compacted graphite, vacuum casting, organically bonded sand, automation of molding and pouring, large aluminum castings for aircraft structures, rapid solidification technology	Precision forging, isothermal forging, superplastic forming, dies made by computer-aided design and manufacturing, rapid prototyping, net-shape forming

Source: After J. A. Schey, C. S. Smith, R. F. Tylecote, T. K. Derry, T. I. Williams, and S. Kalpakjian.

(*continued*)

TABLE 1 (*continued*)

JOINING PROCESSES	TOOLS, TOOL MATERIALS, AND MACHINING	NONMETALLIC MATERIALS
	Tools of stone, flint, wood, bone, ivory, composite tools	Earthenware, glazing, natural fibers
Soldering (Cu-Au, Cu-Pb, Pb-Sn)	Corundum	
Riveting, brazing	Hoe making, hammered axes, tools for ironmaking and carpentry	Glass beads, potter's wheel, glass vessels
Forge welding of iron and steel, gluing	Improved chisels, saws, files, woodworking lathes	Glass pressing and blowing
	Etching of armor	Venetian glass
	Sandpaper, windmill driven saw	Crystal glass
	Hand lathe (wood)	Cast plate glass, flint glass
	Boring, turning, screw cutting lathe, drill press	Porcelain
	Shaping, milling, copying lathe for gunstocks; turret lathe, universal milling machine, vitrified grinding wheel	Window glass from slit cylinder, light bulb, vulcanization, rubber processing, polyester, styrene, celluloid, rubber extrusion, molding
Oxyacetylene; arc, electrical resistance, and thermit welding	Geared lathe, automatic screw machine, hobbing, high-speed steel tools, aluminum oxide and silicon carbide (synthetic)	Automatic bottle making, Bakelite, borosilicate glass
Coated electrodes	Tungsten carbide, mass production, transfer machines	Development of plastics, casting, molding, PVC, cellulose acetate, polyethylene, glass fibers
Submerged arc welding		Acrylics, synthetic, rubber, epoxies, photosensitive glass
Gas metal–arc, gas tungsten–arc, and electroslag welding, explosive welding	Electrical and chemical machining, automatic control	ABS, silicones, fluorocarbons, polyurethane, float glass, tempered glass, glass ceramics
Plasma arc and electron beam, adhesive bonding	Titanium carbide, synthetic diamond, numerical control	Acetals, polycarbonates, cold forming of plastics; reinforced plastics, filament winding
Laser beam, diffusion bonding (also combined with superplastic forming)	Cubic boron nitride, coated tools, computer integrated manufacturing, adaptive control, industrial robots, flexible manufacturing systems, untended factory	Adhesives, composite materials, optical fibers, structural ceramics, ceramic components for automotive and aerospace engines, ceramic-matrix composites

what kind of machine would you design or purchase to make paper clips? If, as the owner of a company, you were given an order of 100 clips versus a million clips, would your approach to this manufacturing problem be different?

The paper clip must meet its basic functional requirement: to hold pieces of paper together with sufficient clamping force so that the papers do not slip away from each other. It must be designed properly, including its shape and size. The design process is based partly on our knowledge of strength of materials and mechanics of solids, dealing with the stresses and strains involved in the manufacturing and normal use of the clip.

The material selected for a paper clip must have certain stiffness and strength. For example, if the stiffness (a measure of how much it deflects under a given force) is too high, a great deal of force may be required to open the clip, just as a stiff spring requires a greater force to stretch or compress it than does a softer spring. If the material is not sufficiently stiff, the clip will not exert enough clamping force on the papers. Also, if the yield stress of the material (the stress required to cause permanent deformation) is too low, the clip will bend permanently during its normal use and will be difficult to reuse. These factors also depend on the diameter of the wire and the design of the clip.

Included in the design process are considerations such as style, appearance, and surface finish or texture of the clip. Note, for example, that some clips have serrated surfaces for better clamping. After finalizing the design, a suitable material has to be selected. Material selection requires a knowledge of the *function* and *service requirements* of the product and the materials that, preferably, are available commercially to fulfill these requirements at the lowest possible cost. The selection of the material also involves considerations of its corrosion resistance, since the clip is handled often and is subjected to moisture and other environmental attack. Note, for example, the rust marks left by paper clips on documents stored in files for a long period of time.

Many questions concerning production of the clips must be asked. Will the material selected be able to undergo bending during manufacturing without cracking or breaking? Can the wire be easily cut from a long piece without causing excessive wear on the tooling? Will the cutting process produce a smooth edge on the wire, or will it leave a burr (a sharp edge)? A burr is undesirable in the use of paper clips since it may tear the paper or even cut the user's finger. Finally, what is the most economical method of manufacturing this part at the desired production rate, so that it can be competitive in the national and international marketplace and the manufacturer can make a profit? A suitable manufacturing method, tools, machinery, and related equipment must then be selected to shape the wire into a paper clip.

Transistors

Manufacturing a paper clip is relatively simple compared to manufacturing advanced products such as an integrated circuit containing hundreds of thousands of transistors. The cross-section of a typical transistor is shown in Fig. 3. Integrated

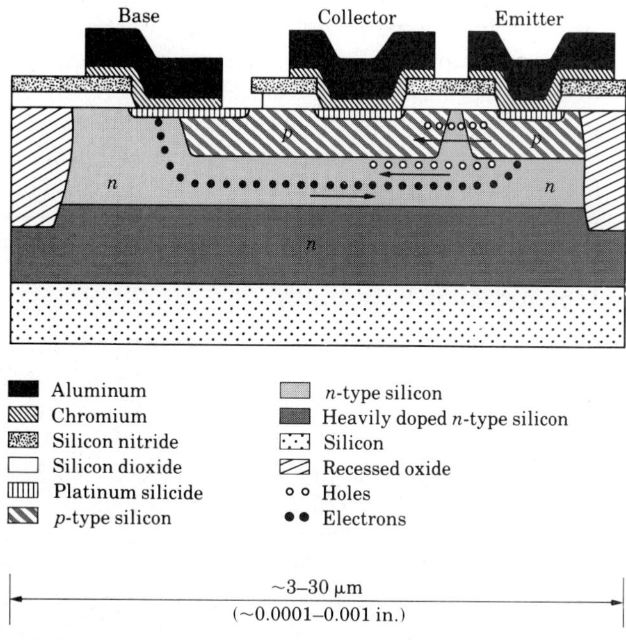

FIGURE 3

Cross-section of a bipolar transistor showing the materials used in making it. *Source:* After P. Chaudhari.

circuits (or chips) are designed and the materials and dimensions are specified by electronic and chemical engineers, material scientists, and physicists. Since the invention of the transistor in 1948, device and circuit dimensions have continued to shrink. This ongoing miniaturization of components has allowed for more complex chips with higher operating speeds and lower cost per unit. Today's technology allows for tens of millions of transistors to be fabricated onto a single 1-cm^2 chip.

Several advanced technologies have been developed to manufacture these tiny and complex parts. First, single-crystal silicon is grown, in diameters as large as 200 mm (8 in.) and lengths as much as 1 m (3 ft.), which has uniform properties and is free from defects. The crystal is then sliced into 0.5-mm (0.02-in.) thick wafers that are polished to a very smooth finish. The wafers then undergo a series of repeated film depositions and oxidations, lithographic patterning, and etching processes to construct the devices. The electrical properties of silicon are altered by introducing impurity atoms (dopants) by ion implantation and diffusion processes. Once the transistors are formed, aluminum is used to interconnect all the devices to form the functional integrated circuit. Completed chips are then placed in rugged packages for protection from the environment, such as temperature, humidity, shock, and radiation.

Because of the extremely small dimensions involved and the importance of quality control, each manufacturing step must be carried out with great care. All processing is performed in clean rooms, that is, dust- and dirt-free, since the smallest foreign particle will render a circuit inoperative. For example, some computer-chip manufacturers are now developing technologies and controls such that these products will be perfect 99.99966 percent of the time; that is, only 3.4 parts out of a million may be defective (known as six sigma quality). In addition, production techniques must be reliable and production rates must be sufficiently high to produce chips that are cost- and performance-competitive in national and international markets.

Incandescent light bulbs

The first incandescent lamp was made by T. A. Edison and lit in 1879. Many improvements have since been made in the materials and manufacturing methods for making bulbs. The components of a typical light bulb are shown in Fig. 4. The light-emitting part is the filament which, through electrical resistance, is heated to incandescence, that is, to temperatures between 2200 °C and 3000 °C (4000 °F and 5400 °F). Edison's first successful lamp had a carbon filament, although he and others had also tried various materials, such as carbonized paper, osmium, iridium, and tantalum. However, none of those materials has the high-temperature resistance, strength, and long life of tungsten, which is now the most commonly used filament material.

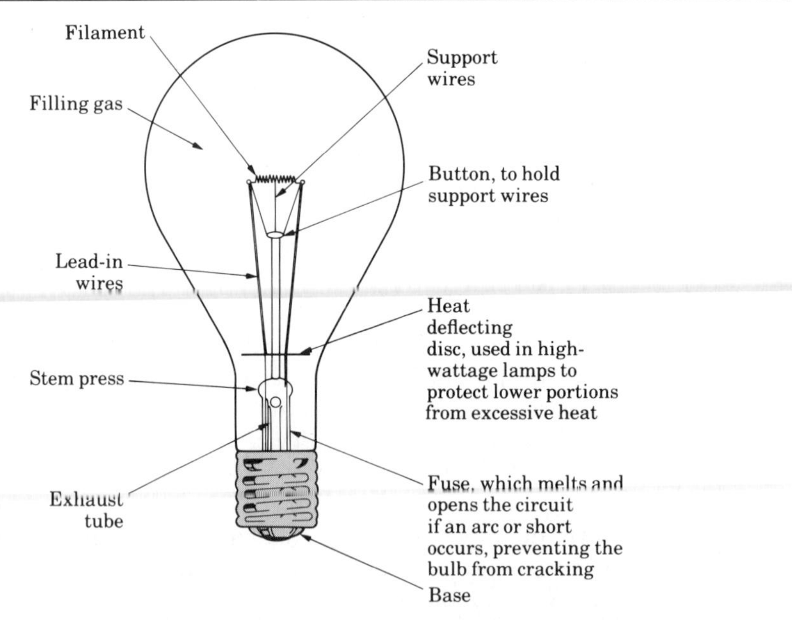

FIGURE 4
Components of a typical light bulb. *Source:* Courtesy of General Electric Company.

The first step in manufacturing a light bulb is making the glass stem which supports the lead-in wires and the filament and connects them to the base of the bulb (Fig. 5). These parts are positioned, assembled, and sealed while the glass is heated by gas flames. The filament is then attached to the lead-in wires. These and other operations in making bulbs are performed on highly automated machines.

The completed stem assembly, called the mount, is then transferred to another machine, which lowers a glass bulb over it. Gas flames are used to seal the rim of the mount to the neck of the bulb. The air in the bulb is exhausted through the exhaust tube (an integral part of the glass stem), and either evacuated or filled with inert gas. For bulbs of 40 watts and above, the gas is typically a mixture of nitrogen and argon. The filling gas should be pure; otherwise the inside surfaces of the bulb will blacken. For example, just one drop of water in the gas used for half a million lamps causes blackening in all of them. The exhaust tube is then sealed off. The last production step consists of attaching the base to the bulb, using a special cement. The machine that does the attaching also solders or welds the lead-in wires to the metal base for electrical connection.

The filament is made by first pressing tungsten powder into ingots and sintering it (heating without melting). The ingot is then shaped into round rods by swaging, and the rods are drawn through a die in a number of steps into thin wire. The wire is coiled to increase the light-producing capacity of the filament. The wire diameter for a 60-watt, 120-volt lamp is 0.045 mm (0.0018 in.) and must be controlled very accurately. If the wire diameter is only 1 percent less than that specified, the life of the lamp may be reduced by as much as 25 percent. Spacing between coils must also be very accurate to prevent heat concentration at one point and possible shorting.

Lead-in wires are usually made of nickel, copper, or molybdenum, and the support wires are made of molybdenum. The portion of the lead-in wire embedded in the stem is made of an alloy of iron and nickel, coated with copper, and has essentially the same thermal expansion as the glass. In this way, thermal stresses

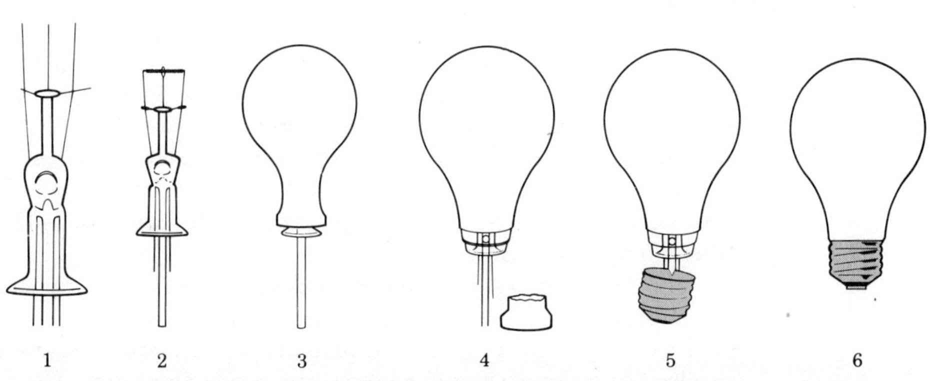

FIGURE 5
Manufacturing steps in making a light bulb (see text for details). *Source:* Courtesy of General Electric Company.

that otherwise might cause the stem to crack are not developed. The base is generally made from aluminum, specially lubricated to permit easy insertion into the socket. In the past the base was made from brass, which is both more expensive than aluminum and not as good an electrical conductor.

The glass bulbs are commonly made by blowing molten glass into a mold. Automated machinery can make bulbs at rates of 1000 a minute or higher. Several types of glasses are used depending on the kind of bulb desired. The inside of the bulb is either plain, or is frosted to reduce glare and diffuse the light better.

Jet engines

Compared to the preceding examples, designing and manufacturing a jet engine is a much more challenging and demanding task (Fig. 1). We present this figure merely to point out the complexity of this important product, which—depending on its size and capacity—costs up to a few million dollars. Selection of materials and processes, inspection and testing, and quality control are particularly critical for this engine because of its application. Failure of any of the major components in this engine can be catastrophic in terms of loss of life and property.

These diverse examples show us that each of the manufacturing operations requires thought processes that are common to making all products. In the following sections, we present an overview of the important interrelationships among product design, material selection, and manufacturing processes. The end result should be a product that meets high quality standards and expected service requirements—and is also economical to produce.

The Design Process

The design process for a product first requires a clear understanding of the functions and the performance expected of that product (Fig. 6). The product may be new, or it may be a revised version of an existing product. We all have observed, for example, how the design and style of radios, toasters, watches, automobiles, and washing machines have changed. The market for a product and its anticipated uses must be defined clearly, with the assistance of sales personnel, market analysts, and others in the organization.

The design process begins with the development of an original product concept. An innovative approach to design is highly desirable—and even essential—at this stage for the product to be successful in the marketplace. Innovative approaches can also lead to major savings in material and production costs. The design engineer or designer in charge of the product must be knowledgeable of the interrelationships among materials, design, and manufacturing, as well as the overall economics of the operation. In most cases, the majority of the decisions for selecting materials and

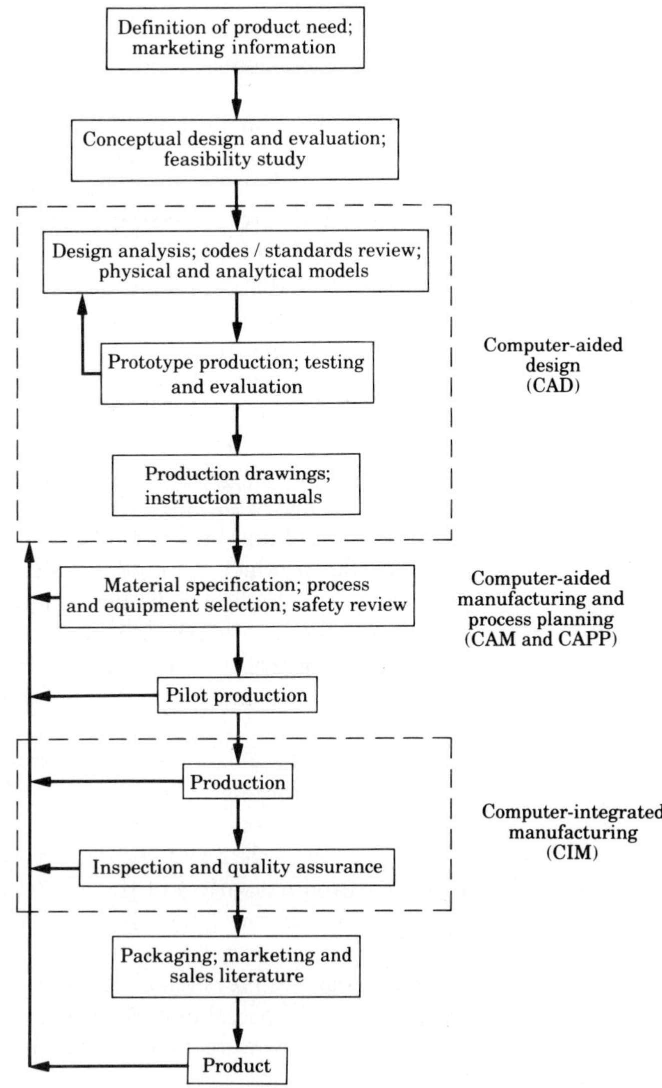

FIGURE 6 Chart showing various steps involved in designing and manufacturing a product. Depending on the complexity of the product and the type of materials used, the time span between the original concept and marketing a product may range from a few months to many years. *Simultaneous engineering* combines these stages to reduce the time span.

manufacturing processes for a product are made by the designer, with input from others in the organization.

Product design often involves preparing analytical and physical models of the product, as an aid to analyzing factors such as forces, stresses, deflections, and optimal part shape. The necessity for such models depends on product complexity. Today, constructing and studying analytical models is simplified through the use of *computer-aided design and manufacturing* techniques. On the basis of these models, the designer selects and specifies the final shape and dimensions of the product, its surface finish and dimensional accuracy, and the materials to be used. The selection of materials is often made with the advice and cooperation of materials engineers, unless the design engineer is also experienced and qualified in this area.

An important design consideration is how a particular component is to be assembled into the final product. Take apart a ballpoint pen or a toaster, or lift the hood of your car and observe how hundreds of components are put together in a limited space. Note also how difficult it is on some cars to remove a spark plug or an oil filter, much less to make repairs or perform maintenance on the engine.

The next step in the production process is to make and test a prototype, that is, an original working model of the product. Testing, either at this stage or periodically during production, is an important aspect of product manufacturing. Testing is now done statistically, and the proper interpretation of test results is crucial to maintaining the quality of a product (*statistical process control*). Total quality control of a product throughout the manufacturing process is one of the most important considerations in manufacturing engineering.

Tests must be designed to simulate as closely as possible the conditions under which the product is to be used. These include environmental conditions such as temperature and humidity, as well as the effects of vibration and repeated use and misuse of the product. Computer-aided design techniques are now capable of comprehensively and rapidly performing such simulations. During this stage, modifications in the original design, materials selected, or production methods may be necessary. Difficulties are often encountered in making the product function properly while fulfilling design, quality, and service requirements—or producing it economically.

After this phase has been completed, appropriate manufacturing methods, equipment, and tooling should be selected with the cooperation of manufacturing engineers, process planners, and all others that are to be directly involved in production. As Table 2 shows, various manufacturing processes are available for making parts with a variety of shapes, depending on the material and various other factors that we discuss later in this Introduction.

Overdesign. Surveys have indicated that many products in the past have been overdesigned. That is, they were either too bulky, were made of materials of too high a quality, or were made with unwarranted precision and quality for the intended uses.

Overdesign may result from uncertainties in design calculations or the concern of the designer and manufacturer over product safety in order to avoid user injuries or deaths and the likelihood of product liability lawsuits. Many designs are based on past experience and intuition, rather than on thorough analysis and experimentation. Overdesign can add significantly to the product's cost. We must point out, however, that this entire subject is somewhat controversial. From the consumer's standpoint, an automobile, washing machine, or lawnmower that has been operating satisfactorily for many years without needing repairs or part replacement is a good product—in other words, it is reliable and economical.

Manufacturers are sensitive to the public image of their products in the

TABLE 2
SHAPES AND SOME COMMON METHODS OF PRODUCTION

SHAPE	PRODUCTION METHOD
Flat surfaces	Rolling, planing, broaching, milling, shaping, grinding
Parts with cavities	End milling, electrical-discharge machining, electrochemical machining, ultrasonic machining, cast-in cavity
Parts with sharp features	Permanent-mold casting, machining, grinding, fabricating
Thin hollow shapes	Slush casting, electroforming, fabricating
Tubular shapes	Extrusion, drawing, roll forming, spinning, centrifugal casting
Shaping of tubular parts	Rubber forming, expanding with hydraulic pressure, explosive forming, spinning
Curvature on thin sheets	Stretch forming, peen forming, fabricating
Openings in thin sheets	Blanking, chemical blanking, photochemical blanking
Reducing cross-sections	Drawing, extruding, shaving, turning, centerless grinding
Producing square edges	Fine blanking, machining, shaving, belt grinding
Producing small holes	Laser, electrical-discharge machining, electrochemical machining
Producing surface textures	Knurling, wire brushing, grinding, belt grinding, shot blasting, etching
Detailed surface features	Coining, investment casting, permanent-mold casting
Threaded parts	Thread cutting, thread rolling, thread grinding, chasing
Very large parts	Casting, forging, fabricating
Very small parts	Investment casting, machining, etching

marketplace. In fact, some products, such as washers and driers, that require infrequent repair have been advertised as such in the public media. However, many manufacturers believe that if a product works well for an extended period of time, it may have been overdesigned. In such cases, the company may consider downgrading the materials and/or the processes used. Some industries have even been accused of following a strategy of planned obsolescence in order to generate more sales over a period of time.

Design for Manufacture

Because a particular design is eventually made into a product, design and manufacturing must be intimately interrelated. Design and manufacturing should never be viewed as separate disciplines or activities. Each part or component of a product must be designed so that it not only meets design requirements and specifications, but also can be manufactured economically and with relative ease. This approach improves productivity and allows a manufacturer to remain competitive.

This broad view has now become recognized as the area of *design for manufacture*. It is a comprehensive approach to production of goods and integrates the design process with materials, manufacturing methods, process planning, assembly, testing, and quality control. Effectively implementing design for manufacture requires that designers have a fundamental understanding of the characteristics, capabilities, and limitations of materials, manufacturing processes, and related operations, machinery, and equipment. This knowledge includes characteristics such as variability in machine performance, surface finish and dimensional accuracy of the workpiece, processing time, and the effect of processing method on part quality.

Designers must be able to assess the impact of design modifications on manufacturing process selection, assembly, inspection, tools and dies, and product cost. Establishing quantitative relationships is essential in order to optimize the design for ease of manufacturing and assembly at minimum product cost (also called *producibility*). Computer-aided design, manufacturing, and process planning techniques, using powerful computer programs, have become indispensable to such analysis. New developments include expert systems (see the section, "Automation and the Impact of Computers on Manufacturing"), which have optimization capabilities, thus expediting the traditional iterative process in design optimization.

Simultaneous Engineering

Also known as *concurrent engineering*, simultaneous engineering is a systematic approach that integrates the design and manufacturing of products with the view of optimizing all aspects involved in the life cycle of the product, including costs, quality, scheduling, and the ultimate disposal of the product. The basic goals of this approach are to minimize (a) product design and engineering changes and (b) the time and costs involved from design concept to production and introduction of the product to the market.

Traditionally, design and manufacturing have taken place sequentially rather than simultaneously or concurrently. Simultaneous engineering, which is a broader concept than design for manufacturing, (a) requires full support of a company's management, (b) requires multifunctional and interactive teamwork in an organization, including support groups, and (c) utilizes all available technologies such as systems engineering, various analytical methods, and computer technology including artificial intelligence (see below). Although this concept appears to be logical and efficient, its implementation in manufacturing organizations has taken considerable time due largely to poor teamwork and lack of appreciation of its real benefits.

There are many examples of the benefits of simultaneous engineering. As a result of the implementation of simultaneous engineering, an automotive company, for example, has reduced an engine's components by 30 percent and its weight by 25 percent, while reducing manufacturing time by 50 percent. This concept can be implemented not only in large manufacturing organizations but in smaller shops as well. This is particularly noteworthy in view of the fact that more than 80 percent of discrete parts made in the United States are manufactured in shops with 200 employees or fewer.

Selecting Materials

An ever-increasing variety of materials is available, each having its own characteristics, applications, advantages, and limitations. The following are the general types of materials used in manufacturing today:

- Ferrous metals (carbon, alloy, stainless, and tool and die steels).
- Nonferrous metals and alloys (aluminum, magnesium, copper, nickel, titanium, superalloys, refractory metals, beryllium, zirconium, low-melting alloys, and precious metals).
- Plastics (thermoplastics, thermosets, and elastomers).
- Ceramics, glass ceramics, glasses, graphite, and diamond.
- Composite materials (reinforced plastics, metal-matrix and ceramic-matrix composites, and honeycomb structures). These are also known as *engineered materials*.

As new materials are developed, there are important trends in the use of traditional materials. In the example of the new bicycle shown in Fig. 7, the spoke wheels are made of composite materials, the tubes are made of carbon-fiber reinforced plastic, and the rear dropouts, headset, and seat mast are made of titanium.

Aerospace structures, as well as products such as sporting goods, have been at the forefront of new material usage. The trend to use more titanium and composites for the airframe of commercial aircraft, such as the Boeing 757, can be seen in Table 3. Note the decline of the use of aluminum and steel in this aircraft. The trend in

TABLE 3
BOEING 757 AIRFRAME MATERIALS (WEIGHT, PERCENT)

	1980	1990	1995
Aluminum	78	70	62
Steel	12	10	8
Titanium	6	10	12
Composites	4	10	18

FIGURE 7

The Triton bicycle made with aerospace materials. A single mold is used for the whole frame. It retails for $8000 to $10,000, depending on whether it is configured as a road, racing, or mountain bike. *Source:* Courtesy of Triton R. Huffy Bicycles Company, © 1991.

materials for military aircraft is shown in Table 4. Note the extensive use of titanium and aluminum and the growing use of composite materials in these high-performance aircraft.

Properties of materials

When selecting materials for products, we first consider their mechanical properties: strength, toughness, ductility, hardness, elasticity, fatigue, and creep. The strength-to-weight and stiffness-to-weight ratios of material are also important, particularly for aerospace and automotive applications. Aluminum, titanium, and reinforced plastics, for example, have higher ratios than steels and cast irons. The mechanical properties specified for a product and its components should, of course, be for the conditions under which the product is expected to function. We then consider the physical properties of density, specific heat, thermal expansion and conductivity, melting point, and electrical and magnetic properties.

Chemical properties also play a significant role in hostile as well as normal environments. Oxidation, corrosion, general degradation of properties, toxicity, and flammability of materials are among the important factors to be considered. In

TABLE 4
MILITARY AIRFRAME MATERIALS CONTENT (WEIGHT, PERCENT)

AIRFRAME	DESIGN YEAR	ALUMINUM	TITANIUM	STEEL	COMPOSITES
F-14 Tomcat	1969	39	24	17	1
F-15 Eagle	1972	36	27	6	2
F-16 Falcon	1976	64	3	3	2
F-18 Hornet	1978	49	13	17	10
AV-8V Harrier	1982[1]	44	9	8	26
F-117A Nighthawk	1983[2]	20	25	5	40
B-1 bomber	1984[1]	41	22	15	1
C-17 transport	1986	77	9	13	7
B-2 bomber	1988[2]	19	26	6	38
ATF stealth	1989[2]	20	25	5	40
A-12 Avenger II	1989[2]	20	20	15	30

[1] *Redesigned.* [2] *Estimated.*

some commercial airline disasters, for example, many deaths have been caused by toxic fumes from burning nonmetallic materials in the aircraft cabin.

Manufacturing properties of materials determine whether they can be cast, formed, machined, welded, and heat treated with relative ease. The method(s) used to process materials to the desired shapes can adversely affect the product's final properties and service life.

Availability and cost

Availability and cost of raw and processed materials and manufactured components are major concerns in manufacturing. Competitively, the economic aspects of material selection are as important as the technological considerations of properties and characteristics of materials.

If raw or processed materials or manufactured components are not available in the desired quantities, shapes, and dimensions, substitutes and/or additional processing will be required, which can contribute significantly to product cost. For example, if we need a round bar of a certain diameter and it is not available in standard form, then we have to purchase a larger rod and reduce its diameter by some means, such as machining, drawing through a die, or grinding.

Reliability of supply, as well as demand, affects cost. Most countries import numerous raw materials that are essential for production. The United States, for example, imports the majority of raw materials such as natural rubber, diamond, cobalt, titanium, chromium, aluminum, and nickel from other countries. The broad political implications of such reliance on other countries is self-evident.

Different costs are involved in processing materials by different methods. Some methods require expensive machinery, others require extensive labor, and still others require personnel with special skills or a high level of education or specialized training.

Appearance, service life, disposal, and recycling

The appearance of materials after they have been manufactured into products influences their appeal to the consumer. Color, feel, and surface texture are characteristics that we all consider when making a decision about purchasing a product.

Time- and service-dependent phenomena such as wear, fatigue, creep, and dimensional stability are important. These phenomena can significantly affect a product's performance and, if not controlled, can lead to total failure of the product. Similarly, compatibility of materials used in a product is important. Friction and wear, corrosion, and other phenomena can shorten a product's life or cause it to fail. An example is galvanic action between mating parts made of dissimilar metals, which corrodes the parts.

Recycling or proper disposal of materials at the end of their useful service lives has become increasingly important in an age conscious of maintaining a clean and healthy environment. Note, for example, the use of biodegradable packaging materials or recyclable glass bottles and aluminum beverage cans. The proper disposal of toxic wastes and materials is also a crucial consideration.

• Example: Material selection for baseball bats.

Baseball bats for the major leagues are generally made of northern white ash because of its dimensional stability, high elastic modulus and strength-to-weight ratio, good shock resistance, and low density. The bats are made on semiautomatic lathes and then subjected to finishing operations. The straight uniform grain required for such bats has become increasingly difficult to find, particularly when the best wood is said to come from ash trees that are at least 45 years old. Consequently, consideration is being given to using other materials for bats. For the amateur market, aluminum bats (top portion of the accompanying figure) have been made for some years by various metal-forming techniques, although at first their performance was not as good as that for wooden bats. The technology has been advanced such that aluminum bats are now made from high-strength aluminum tubing and possessing various characteristics such as weight distribution, center of percussion, sound, and impact dynamics. They are usually filled with polyurethane or cork for sound damping and controlling the balance of the bat. However, aluminum bats can be sensitive to surface defects (due to use) and usually fail in fatigue (due to cyclic loading).

New developments in bat materials include composite materials, consisting of high-strength graphite and glass fibers in an epoxy resin matrix. The inner woven sleeve (lower portion of the figure) is made of Kevlar fibers, which give additional strength and dampen vibrations in the bat. These bats cost about $125 and perform and sound much like wooden bats. *Source:* Mizuno Sports, Inc.

Selecting Manufacturing Processes

Many processes are used to produce parts and shapes (see Table 2). As you can see, there is usually more than one method of manufacturing a part from a given material. The broad categories of processing methods for materials are:

- *Casting* (expendable mold and permanent mold).
- *Forming* and *shaping* (rolling, forging, extrusion, drawing, sheet forming, powder metallurgy, and molding).
- *Machining* (turning, boring, drilling, milling, planing, shaping, broaching, grinding, ultrasonic machining; and chemical, electrical, and electrochemical machining; and high-energy beam machining).

- *Joining* (welding, brazing, soldering, diffusion bonding, adhesive bonding, and mechanical joining).
- *Finishing* operations (honing, lapping, polishing, burnishing, deburring, surface treating, coating, and plating).

Selection of a particular manufacturing process depends not only on the shape to be produced but also on a large number of other factors. The type of material and its properties are basic considerations. Brittle and hard materials, for example, cannot be formed easily, whereas they can be cast or machined by several methods. The manufacturing process usually alters the properties of materials. Metals that are formed at room temperature, for example, become stronger, harder, and less ductile than they were before processing.

Manufacturing engineers are constantly being challenged for new solutions to manufacturing problems and cost reduction. For a long time, for example, sheet metal parts were cut and made by traditional tools, punches, and dies. Although they are still widely used, these operations are now being replaced by laser cutting techniques (Fig. 8). With advances in computer controls, we can automatically control the path of the laser, thus producing a wide variety of shapes accurately, repeatedly, and economically.

FIGURE 8

Cutting sheet metal with a laser beam. *Source:* Courtesy of Rofin-Sinar, Inc., and *Manufacturing Engineering Magazine*, Society of Manufacturing Engineers.

Dimensional and surface finish requirements

Size, thickness, and shape complexity of the part have a major bearing on the process selected to produce it. Flat parts with thin cross-sections, for example, cannot be cast properly. Complex parts cannot be formed easily and economically, whereas they may be cast or fabricated from individual pieces.

Tolerances and surface finish obtained in hot-working operations cannot be as good as those obtained in cold-working operations. Dimensional changes, warpage, and surface oxidation occur during processing at elevated temperatures. Some casting processes produce a better surface finish than others because of the different types of mold materials used and their surface finish.

The size and shape of manufactured products vary widely. For example, the main landing gear for the twin-engine, 400-passenger Boeing 777 jetliner will be 4.3 m (14 ft) high, with three axles and six wheels, made by forging and machining processes. At the other extreme is the generation of 0.05-mm (0.002-in.) diameter hole at one end of a 0.35-mm (0.014-in.) diameter needle (Fig. 9), using a process called electrical-discharge machining. The hole is burr-free and has a location accuracy of ± 0.003 mm (0.0001 in.).

Another small-scale manufacturing example is given in Fig. 10, which shows microscopic gears as small as 100 μm (0.004 in.) in diameter. These gears have possible applications such as powering microrobots to repair human cells, microknives in surgery, and camera shutters for precise photography. The gears are made by a special electroplating and x-ray etching technique of metal plates coated with a polymer film. The center hole in these gears is so small that a human hair cannot pass through it (see insert to figure). Such small-scale operations are called *nanotechnology* and *nanofabrication* ("nano" meaning one billionth).

Since not all manufacturing operations produce finished products, additional operations such as grinding or polishing may be necessary. These operations can add significantly to product cost. Consequently, there is now a trend, called *near-net* or *net-shape processing*, in which the part is made as close to the final dimensions and specifications as possible.

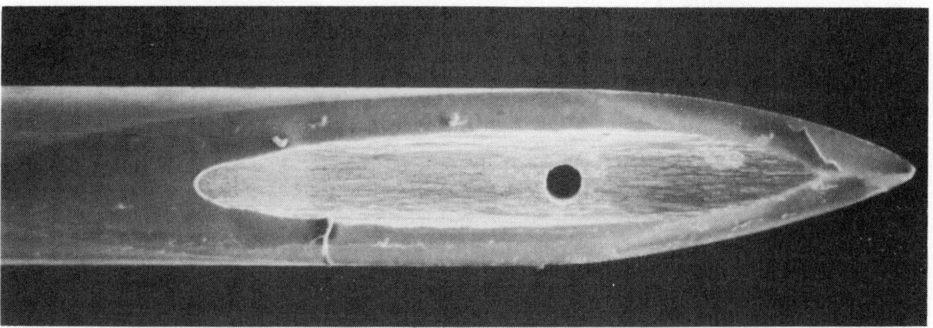

FIGURE 9

A 0.05-mm hole produced in a needle, using the electrical-discharge machining process. *Source:* Courtesy of Derata Corporation.

FIGURE 10
Microscopic gear with a diameter on the order of 100 μm, made by a special etching process. *Source:* Courtesy of Wisconsin Center for Applied Microelectronics, University of Wisconsin–Madison.

Operational and manufacturing cost considerations

The design and cost of tooling, the lead time required to begin production, and the effect of workpiece material on tool and die life are major considerations. Depending on its size, design, and expected life, the cost of tooling can be substantial. For example, a set of steel dies for stamping sheet-metal fenders for automobiles may cost about $2 million.

For parts made from expensive materials, the lower the scrap rate, the more economical the production process will be. Because it generates chips, machining may not be more economical than forming operations, all other factors being the same.

Availability of machines and equipment and operating experience within the manufacturing facility are also important. If they are not available, some parts may have to be manufactured by outside firms. Automakers, for example, purchase many parts from outside vendors, or have them made by outside firms according to their specifications.

The number of parts or products required and the desired production rate (pieces per hour) help determine the processes to be used and the economics of production. Beverage cans or transistors, for example, are consumed in numbers and at rates much higher than telescopes and propellers for ships.

The operation of machinery has significant environmental and safety implications. Depending on the type of operation and the machinery involved, some processes adversely affect the environment. Unless properly controlled, such processes can cause air, water, and noise pollution. The safe use of machinery is another important consideration, requiring precautions to eliminate hazards in the workplace.

Consequences of improperly selecting materials and processes

Numerous examples of product failure can be traced to improper selection of material or manufacturing processes or improper control of process variables. A component or a product is generally considered to have failed when:

- It stops functioning (broken shaft, gear, bolt, cable, or turbine blade).
- It does not function properly or perform within required specification limits (worn bearings, gears, tools, and dies).
- It becomes unreliable or unsafe for further use (frayed cable in a winch, crack in a shaft, poor connection in a printed-circuit board, or delamination of a reinforced plastic component).

We describe in detail in this text the types of failure of a component or product that result from design deficiencies, improper material selection, material defects, manufacturing-induced defects, improper component assembly, and improper product use.

• Example: Manufacturing a sheet-metal part by various methods.

Often there is more than one method of manufacturing a part. The selection of a particular process over others depends on a large number of factors. Consider, for instance, the following example of forming a simple dish-shaped part from sheet metal (see the accompanying figure). Such a part can be formed by placing a flat piece of sheet metal between a pair of male and female dies and closing the dies through proper application of closure force. A large number of parts can be formed at high production rates by this method, known as stamping or pressworking. The

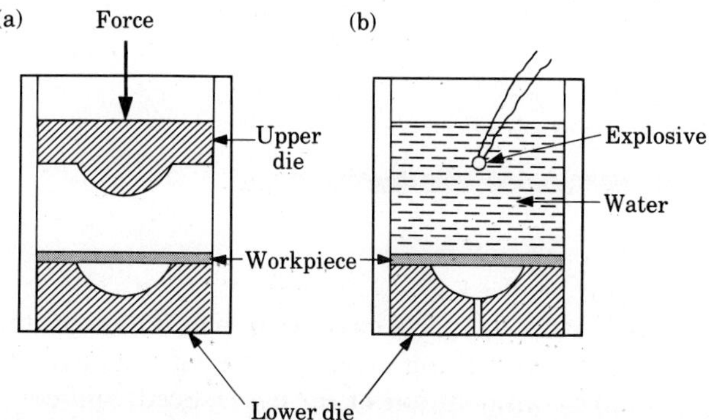

Two methods of forming a dish-shaped part from sheet metal: (a) conventional mechanical or hydraulic press, using a pair of male and female dies; and (b) explosive forming method using only one die.

dies are activated by a machine, which can be either a mechanical or a hydraulic press.

Assume now that the size of the part is very large, say, 2 m (80 in.) in diameter. Further assume that only 50 of these parts are required. In response to these constraints, we have to reconsider the total operation and ask a number of questions. Is it economical to manufacture a set of dies 2 m in diameter (which are very costly) when the production run (quantity of production) is so low? Are machines available with sufficient capacity to accommodate such large dies? What are the alternative methods of manufacturing this part? Does it have to be made in one piece?

This part can also be made by welding together smaller pieces of metal formed by other methods. (Ships and water tanks are made by this method.) Would a part manufactured by welding be acceptable for its intended purpose? Will it have the required properties and the correct shape after welding, or will it require additional processing? The same part can be made by a spinning process similar to making ceramic dishes and pottery on a potter's wheel. It can also be made by the explosive forming process shown in the figure. In this method, the male die is replaced by water, and the shock wave from the explosion generates sufficiently high pressure to form the part.

In comparing the two processes, we note that, in the explosive forming process, the deformation of the material takes place at a very high rate. In selecting this process over conventional pressworking, we have to consider various factors. For instance, is the material capable of undergoing deformation at high rates without fracture? Does the high rate have any detrimental effect on the final properties of the formed part? Can tolerances be held within acceptable limits? Is the life of the die sufficiently long under the high transient pressures generated in this process? Can this operation be performed in a manufacturing plant in a city or should it be carried out in open country? Although having only one die is an advantage, is the overall operation economical? In describing manufacturing operations throughout this text, we bring all these factors into consideration.

Assembly

After individual parts have been manufactured, they are assembled into a product. Assembly is an important phase of the overall manufacturing operation and requires consideration of the ease, speed, and cost of putting parts together. Many products are also designed so that they can be taken apart for maintenance and servicing.

There are several methods of assembly, each with its own characteristics and requiring different operations. The use of a bolt and nut, for example, requires preparation of holes that must match in location and size. Hole generation requires operations such as drilling or punching, which take additional time, require separate operations, and produce scrap. Furthermore, the presence of holes could reduce the useful life of components because of stress concentration, leading to fatigue failure of the part. On the other hand, products assembled with bolts and nuts can be taken apart and reassembled with relative ease.

Parts can also be assembled with adhesives. This method, which is being used extensively in aircraft and automobile production, does not require holes. However, surfaces to be assembled must match properly and be clean because joint strength is affected by the presence of contaminants, such as dirt, dust, oil, and moisture. Production rates in assembly with adhesives may be low because of the time required for the adhesive to be applied and to develop full strength. Unlike mechanical fastening, adhesively joined components, as well as those that are welded, are not usually designed to be taken apart and reassembled.

Parts may be assembled by hand or by automatic equipment. The choice depends on factors such as the complexity of the product and the number of parts to be assembled, the protection required to prevent damage or scratching of finished surfaces of the parts, and the relative costs of labor and machinery required for automated assembly. Variations in the dimensions of parts may be tolerated in hand assembly since the operator can make minor adjustments. In automated assembly, however, parts must have consistently uniform dimensions. Otherwise, their movement through the assembly equipment can be impeded and parts not assembled properly.

Design for assembly

Because assembly operations can contribute significantly to product cost, *design for assembly* is now recognized as an important area of manufacturing. It involves the following basic considerations:

a) Design of the product to permit assembly with relative ease. Note in Fig. 11, for example, how a design can be simplified by reducing the number of pieces required, thus making assembly of the product easier and faster—and resulting in lower product cost. Also, modifying a part's shape can allow easier assembly or make it adaptable for robotic assembly.
b) Possibility of designing multipurpose parts so that they can be used in different lines of products, thus reducing the number of different parts to be made.
c) Capabilities and limitations of each manufacturing process, particularly in regard to consistency in part shape and dimensions. The greater the variations in these characteristics, the more difficult proper assembly becomes.

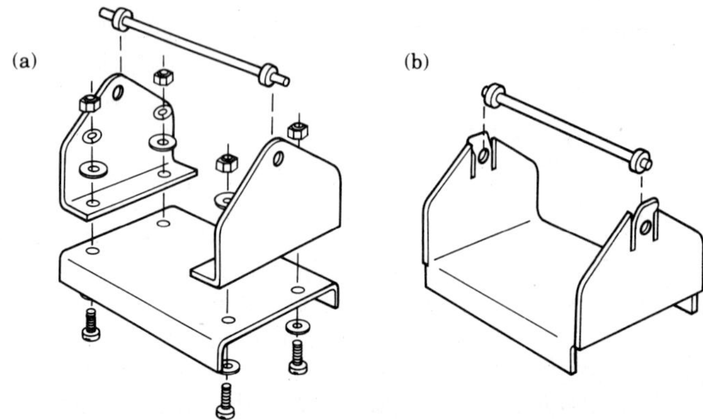

FIGURE 11
An example of design for assembly. (a) As designed, this product consists of many components and takes considerable time to assemble. (b) Redesigned product consists of only two parts and is easy to assemble, either by hand or in automated machinery. *Source:* Courtesy of G. Boothroyd.

d) Methods by which individual parts are brought together for assembly and packaged, including the use of conveyors, feeders, transfer equipment, and programmable industrial robots.

Design for assembly is an integral part of design for manufacture. Thus it should always be given due consideration as a way to improve productivity in manufacturing.

Product Quality

We all have used terms like poor quality and good quality in describing a product. What do we mean by quality? In a broad sense, *quality* is a characteristic or property consisting of several well-defined technical and aesthetic, hence subjective, considerations. The general public's perception is that a high-quality product functions reliably and as expected over a long period of time.

Product quality has always been one of the most important aspects of manufacturing, as it directly influences the marketability of a product and customer satisfaction. Traditionally, quality assurance has been obtained by inspecting parts after they have been manufactured. Parts are inspected to ensure that they conform to a detailed set of specifications and standards, such as dimensions, surface finish, and mechanical and physical properties.

However, quality cannot be inspected into a product after it is made. The practice of inspecting products after they are made is now being replaced rapidly by the broader view that *quality must be built into a product*—from the design stage through all subsequent stages of manufacture and assembly. Producing defective products can be very costly to the manufacturer, creating difficulties in assembly operations, necessitating repairs in the field, and resulting in customer dissatisfaction. Contrary to general public opinion, high-quality products do not necessarily cost more than poor-quality products.

Total quality control and quality assurance have become the responsibility of everyone involved in designing and manufacturing a product. Our awareness of the technological and economic importance of built-in product quality has been heightened further by recent pioneers in quality control, primarily Deming and Taguchi. They pointed out the importance of the management's commitment to product quality, pride of workmanship at all levels of production, and the use of modern statistical techniques to identify the causes of quality control problems.

Important developments in quality control include the implementation of *experimental design*, a technique in which the factors involved in a manufacturing process and their interactions are studied simultaneously. Thus, for example, variables affecting dimensional accuracy or surface finish in a machining operation can be readily identified and appropriate actions taken. The use of computers has greatly enhanced our capability to utilize such techniques rapidly and effectively.

Although it can be described in various terms, *product integrity* is a term that can be used to define basically the degree to which a product (a) is suitable for its intended purpose, (b) fills a real market need, (c) functions reliably during its life expectancy, and (d) can be maintained with relative ease. Product integrity has also been defined as the total product experience of the customer, or as the totality of qualities needed to conceive, produce, and market the product successfully.

Automation and the Impact of Computers on Manufacturing

The major goals of automation in manufacturing facilities are to integrate various operations to improve productivity, increase product quality and uniformity, minimize cycle times and effort, and reduce labor costs. Beginning in the 1940s, automation has accelerated because of rapid advances in control systems for machines and in computer technology.

Few developments in the history of manufacturing have had a more significant impact than computers. Beginning with computer graphics, which are rapidly replacing drafting boards, the use of computers has been extended to computer-aided design and manufacturing and, ultimately, to *computer-integrated manufacturing systems*. The use of computers covers a broad range of applications, including control and optimization of manufacturing processes, material handling, assembly, and automated inspection and testing of products.

Machine control systems

Numerical control (NC) of machines is a method of controlling the movements of machine components by direct insertion of coded instructions in the form of numerical data. Numerical control was first implemented in the early 1950s and was a major advance in automation of machines.

In *adaptive control* (AC), process parameters are adjusted automatically to optimize production rate and product quality and minimize cost. Parameters such as forces, temperatures, surface finish, and dimensions of the part are monitored constantly. If they move outside the acceptable range, the system adjusts the process variables until the parameters again fall within the acceptable range.

Major advances have been made in *automated handling of materials* in various stages of completion, such as from storage to machines, from machine to machine, inspection, inventory, and shipment. Introduced in the early 1960s, *industrial robots* are replacing humans in operations that are repetitive, boring, and dangerous, thus reducing the possibility of human error, decreasing variability in product quality, and improving productivity. Robots with sensory perception capabilities are being developed, with movements that are beginning to simulate those of humans. *Automated assembly systems* are replacing costly assembly by operators. Products are being designed or redesigned so that they can be assembled more easily by machine.

Computer technology

Computers allow us to *integrate* virtually all phases of manufacturing operations, which consist of various technical as well as managerial activities. With sophisticated software and hardware, manufacturers are now able to minimize manufacturing costs, improve product quality, reduce product development time, and maintain a competitive edge in the domestic and international marketplace.

Computer-integrated manufacturing (CIM) is particularly effective in responding to recent trends of short product life cycles, market demand for product types and quantities, and customer requirements for high-quality, low-cost products. The benefits of CIM are:

- Responsiveness to rapid changes in market demand, product modification, and shorter product life cycles.
- High-quality products at low cost.
- Better use of materials, machinery, and personnel and reduced inventory.
- Better control of production and management of the *total* manufacturing operation.

Computer-aided design (CAD) allows the designer to conceptualize objects more easily without having to make costly illustrations, models, or prototypes. These systems are now capable of rapidly and completely analyzing designs, from a simple bracket to complex structures, such as aircraft wings. The performance of structures subjected to static or fluctuating loads and temperatures can now be simulated, analyzed, and tested efficiently, accurately, and more quickly than ever on the computer. The information developed can be stored, retrieved, displayed, printed, and transferred anywhere in the organization. Designs can be optimized, and modifications can be made directly and easily at any time in the life of a product.

Computer-aided manufacturing (CAM) involves all phases of manufacturing by utilizing and processing further the large amount of information on materials and processes collected and stored in the organization's database. Computers now assist manufacturing engineers and others in organizing tasks such as programming numerical control of machines; programming robots for material handling; designing tools, dies, and fixtures; and quality control and inspection.

An important new development is *rapid prototyping*, which relies on CAD/CAM and various manufacturing techniques (using polymers or metal powders) to produce prototypes in the form of a solid physical model of a part rapidly and at low cost. For example, prototyping new automotive components (by traditional methods of shaping, forming, machining, etc.) costs hundreds of millions of dollars a year and may take a year to make. Rapid prototyping can cut these costs as well as development times quite significantly. These techniques are being advanced further so that they can be used for low-volume economical production of parts and components.

Computer-aided process planning (CAPP) is capable of improving productivity in a plant by optimizing process plans, reducing planning costs, and improving the consistency of product quality and reliability. Functions such as cost estimating and work standards (time required to perform a certain operation) can also be incorporated into the system.

The high level of automation and flexibility achieved with numerically controlled machines has led to the development of *group technology* and *cellular manufacturing*. The concept of group technology (GT) is that parts can be grouped and produced by classifying them according to similarities in design and manufacturing processes. In this way, part designs and process plans can be standardized, and families of parts can be produced efficiently and economically.

Flexible manufacturing systems (FMS) integrate manufacturing cells into a large unit, containing industrial robots serving several machines, all interfaced with a central computer. Flexible manufacturing systems have the highest level of efficiency, sophistication, and productivity in manufacturing. They are capable of producing parts randomly and changing manufacturing sequences on different parts quickly. Thus they can meet rapid changes in market demand for various types of products.

An important concept in manufacturing is *just-in-time production*, which states that supplies are purchased just in time to be used, parts are produced just in time to be made into subassemblies and assemblies, and products are finished just in time to be sold. In this way, inventory carrying costs are low, part defects are detected right away, productivity is increased, and high-quality products are made at low cost.

Artificial intelligence (AI) involves the use of machines and computers to replace human intelligence. Computer-controlled systems are becoming capable of learning from experience and making decisions so as to optimize operations and minimize costs. *Expert systems*, which are basically intelligent computer programs, are being developed rapidly with capabilities to perform tasks and solve difficult real-life problems as well as human experts would. With advances in *neural networks*, which are being designed to simulate the thought processes of the human brain, these systems have the capability of modeling and simulating production facilities, monitoring and controlling manufacturing processes, diagnosing problems in machine performance, financial planning, and managing a company's manufacturing strategy.

We can now envisage the *factory of the future* in which production takes place with little or no direct human intervention. The human role will be confined to supervision, maintenance, and upgrading machines, computers, and software. The impact of such advances on the workforce remains to be fully assessed.

A word of caution is necessary here because the implementation of some of the modern technologies we outlined above requires considerable technical and economic expertise, time, and capital investment. Largely because of lack of proper advice and incorrect assessment of the real and specific needs of a company and the market for its products, as well as poor communication among the parties involved such as vendors, suppliers, technical personnel, and company management, some of this high technology has been applied improperly. It has been implemented either too soon or at too large or ambitious a scale, involving major expenditures with questionable return on investment and disappointing results. Manufacturing engineers have the responsibility to be cognizant of all these factors in planning their operations. Company management also has the responsibility to acquire the proper technical knowledge and operational experience to help them in their decision-making process as well as in the implementation of this technology in an environment of increasingly challenging international competition.

Although large corporations can afford to implement modern technology and take risks, smaller companies generally have difficulty in doing so with their limited personnel, resources, and capital. More recently, the concept of *shared manufacturing* has been proposed. This will consist of a nationwide network of manufacturing facilities, with state-of-the-art equipment for training, prototype development, and small-scale production runs, and will be available to help small companies develop products and compete in the international marketplace.

Economics of Manufacture

The *cost* of a product is often the overriding consideration in its marketability and general customer satisfaction. Consequently, all the costs involved in manufacturing a high-quality product must be thoroughly analyzed. Typically, manufacturing costs represent about 40 percent of a product's selling price.

The concepts of design for manufacture and assembly that we described earlier also include design principles for economic production:

- The design should be as simple as possible to manufacture and assemble.
- Materials should be chosen for their appropriate manufacturing characteristics.
- Dimensional accuracy and surface finish specified should be as broad as permissible.
- Because they can add significantly to cost, secondary and finishing processing of parts should be avoided.

The total cost of manufacturing a product consists of costs of materials, tooling, and labor, as well as fixed costs and capital costs. Several factors are involved in each cost category. The cost of materials is usually the largest percentage of total manufacturing costs. This cost can be minimized by analyzing the product design to determine whether part size and shape are optimal and the materials selected are the least costly, while possessing the desired properties and characteristics. The possibility of substituting materials is an important consideration in minimizing costs.

Tooling costs depend on the complexity of part shape, the materials involved, the manufacturing process, and the number of parts to be made. Complex shapes, difficult-to-machine materials, and stringent dimensional accuracy requirements all add to the cost of tooling. Direct labor costs are usually a small proportion of the total, typically ranging from 10 percent to 15 percent. The trend toward increased automation and computer control of all aspects of manufacturing helps minimize labor involvement and direct labor costs, which continue to decline steadily. Automation also has other benefits, such as higher production rates and more consistent product quality.

Fixed costs and capital costs depend on the particular manufacturer and plant facilities. Computer-controlled machinery, which constitutes a capital cost, can be very expensive. However, economic analysis indicates that, more often than not, such an expenditure is warranted in view of the long-range benefits, such as improved productivity. In mass production of products, such as automobile engines, specialized machines are arranged in various product flow lines. These machines, too, require a major capital investment, yet the high production rate makes the cost per part competitive.

Economics of manufacturing has always been a major factor and it has become even more so as the international competition (*global competitiveness*) for high-quality products and low prices becomes a simple fact in worldwide markets (*world class manufacturing*). The markets have become multinational and dynamic, product lines have become extensive and technically complex, and the demand for quality has become commonplace. To respond to these needs while keeping costs down is a constant challenge to manufacturing companies and indeed to their very survival.

Environmental Issues

In the United States alone, nine million passenger cars are discarded a year. Six billion tons of industrial, commercial, agricultural, and domestic waste are produced each year. How do we dispose of, treat, or recycle this waste? What are the side effects? In manufacturing operations specifically, we may cite various examples and raise a number of similar questions. Metalworking fluids such as lubricants and coolants are often used in machining, grinding, and forming operations. Various fluids and solvents are used in cleaning manufactured products. Some of these fluids pollute the air and waters during their use.

Likewise, many by-products from manufacturing plants are sooner or later discarded: sand with additives used in metal-casting processes, water, oil, and other fluids from heat-treating facilities and plating operations, slag from foundries and welding operations, and a wide variety of metallic and nonmetallic scrap produced in operations such as sheet forming, casting, and molding. Further consider the fact that there are more than five million underground storage tanks in the United States containing petroleum products or hazardous chemicals, and it is estimated that two million of them may be leaking.

How do we assess the environmental impact of these activities while still maintaining their necessary functions? How do we dispose of these fluids and materials after use? Are they degradable? Can they be recycled? How do we recycle these materials? Are there toxic by-products when they are recycled? Is recycling feasible and acceptable both environmentally and economically, and what are its effects on the financial situation of a manufacturing plant?

The present and future adverse effects of these activities, their damage to our environment and to the earth's ecosystem, and ultimately their effect on the quality of human life are by now well recognized by the public as well as local and federal governments. Consider the effects of water and air pollution, acid rain, ozone depletion, the greenhouse effect, hazardous wastes, landfill seepage, and global warming. In response to these major concerns, a wide range of laws and regulations have been promulgated by local, state, and federal governments as well as professional organizations, both in the United States and in other industrial countries. These regulations are generally stringent and their implementation can have a major impact on the economic operation and financial health of industrial organizations. Regardless, however, and notwithstanding arguments by some that environmental damage from these activities is overestimated, manufacturing engineers and the management of companies have great responsibility in planning and implementing environmentally safe manufacturing operations.

Much can be gained by careful analysis of products, their design, materials used, and the manufacturing processes and practices employed in making these products. Certain guidelines can be followed in this regard, as we outline below.

a) Reducing waste of materials at their source by refinements in product design and reducing the amount of materials used.

b) Research and development in environmentally safe products and manufacturing technologies.
c) Reducing the use of hazardous materials in products and processes.
d) Ensuring proper handling and disposal of all waste.
e) Improvements in waste treatment, recycling, and reuse of materials.

Significant developments have been taking place regarding these matters. Note, for example, the recent trend in making fast-food containers from biodegradable materials and recycled paper instead of plastics such as styrofoam. In metalworking operations, the trend is to use water-based, instead of oil-based, lubricants to reduce air pollution as well as reduce health hazards to employees. In 1990, almost one third of plastic soft-drink bottles were recycled. Many metals are recovered from scrapped automobiles and reused. There are many types of plastics used in cars (about 2.5 million pounds of plastic materials are used in automobile construction per year) and numerous other products. Although they are more difficult to recycle than metals, efforts continue and manufacturers are now making products such as signposts, curbing, and park benches out of recycled plastics. Other examples include a plan by a computer company to recycle spent toner cartridges from laser printers; the aluminum drums are to be recycled and other parts reused to make new cartridges. The benefits of recycling are also evident from a study showing that aluminum from scrap, instead of bauxite ore, costs one third as much and reduces energy consumption and pollution by more than 90 percent.

From this brief introduction, it is evident that manufacturing engineers are faced with interesting challenges, both in design and production, to develop products and operations that are environmentally safe and thus contribute to the quality of human life.

Organization for Manufacture

The various manufacturing activities and functions that we have described must be organized and managed efficiently and effectively in order to maximize productivity and minimize costs, while maintaining high quality standards. Because of the complex interactions among the various factors involved in manufacturing—materials, machines, people, information, power, and capital—the proper coordination and administration of diverse functions and responsibilities is essential.

The basic functional organization for a manufacturing company can be represented by the chart in Fig. 12, which includes an outline of the functions and responsibilities of each major group in the organization.

Manufacturing engineers traditionally have had several major responsibilities:

- Plan the manufacture of a product and the processes to be utilized. This function requires a thorough knowledge of the product, its expected performance, and specifications.

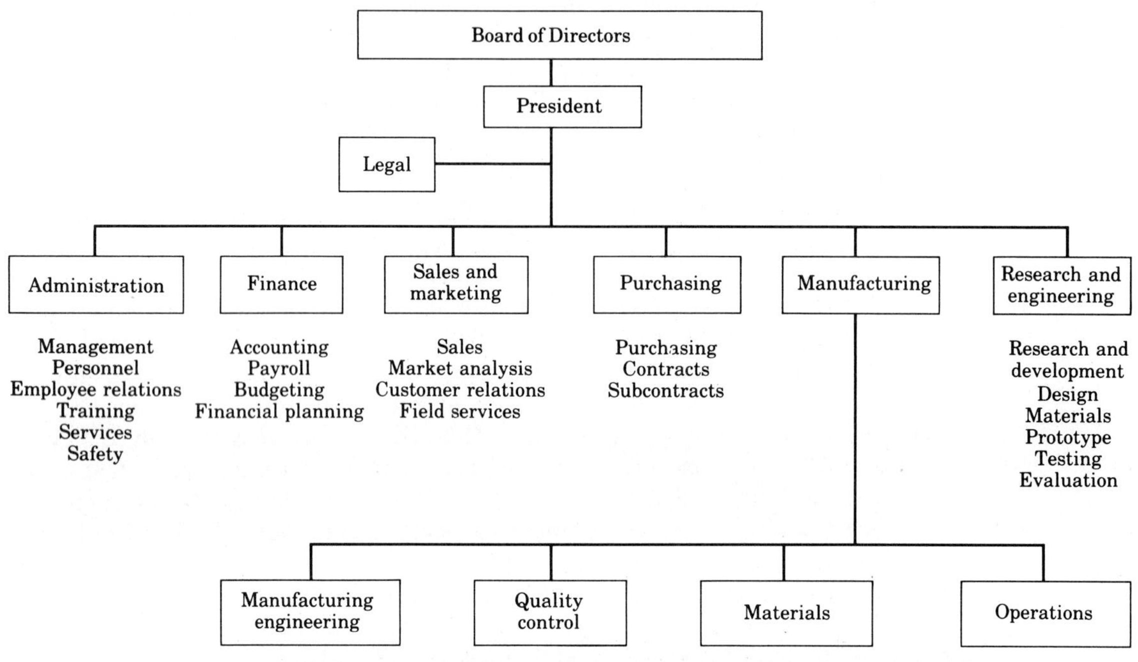

FIGURE 12
A typical organization chart for a manufacturing company. Such charts can be organized in different ways.

- Identify machines, equipment, tooling, and the personnel to carry out the plan. This function requires evaluation of the capacities and capabilities of machines, tools, and workers, so that proper functions and responsibilities can be assigned.
- Interact with design and materials engineers to optimize productivity and minimize production costs.
- Cooperate with *industrial engineers* when plant floor activities are involved, such as plant layout, machine arrangement, material handling equipment, time-and-motion study, production methods analysis, production planning and scheduling, and maintenance. Some of these activities are carried out under the name *plant engineering*, and some are interchangeably performed by both manufacturing and industrial engineers.

Manufacturing engineers, in cooperation with industrial engineers, also are responsible for evaluating new technologies, their applications, and how they can be implemented. In view of the vast amount of technical literature available, this task in itself can present a major challenge. Gaining a broad perspective of computer capabilities, applications, and integration in all phases of the manufacturing process is important. This knowledge is particularly crucial for long-range production-facility planning in view of constantly changing market demand and product mix.

There have been important trends in the operational philosophy of manufacturing organizations, particularly in the United States. Traditionally, the emphasis has been on top-down communication in the organization and strong control by management, with priorities for quick financial return (*profits first*) and growth and size (*economy of scale*). However, largely based on the experience of highly successful examples of Japanese methods of management and production technology, the trend is now toward broad-base communication across the organization, viewing the people in a company as important assets, emphasizing the need for teamwork in problem-solving, customer satisfaction, quality (*quality first*), and faster response to product demands (*economy of time*) in the marketplace.

BIBLIOGRAPHY

Selected General Textbooks on Manufacturing

Alexander, J.M., R.C. Brewer, and G.W. Rowe, *Manufacturing Technology*, 2 vols. Chichester, England: Ellis Horwood, 1987.

Alting, L., *Manufacturing Engineering Processes*. New York: Marcel Dekker, 1982.

Amstead, B.H., P.F. Ostwald, and M.L. Begeman, *Manufacturing Processes*, 8th ed. New York: Wiley, 1987.

DeGarmo, E.P., J.T. Black, and R.A. Kohser, *Materials and Processes in Manufacturing*, 7th ed. New York: Macmillan, 1988.

Dieter, G.E., *Engineering Design: A Materials and Processing Approach*, 2d ed. New York: McGraw-Hill, 1991.

Doyle, L.E., C.A. Keyser, J.L. Leach, G.F. Schrader, and M.S. Singer, *Manufacturing Processes and Materials for Engineers*, 3d ed. Englewood Cliffs, N.J.: Prentice-Hall, 1985.

Kalpakjian, S., *Manufacturing Processes for Engineering Materials*, 2d ed. Reading, Mass.: Addison-Wesley, 1991.

Lindberg, R.A., *Processes and Materials of Manufacture*, 4th ed. Boston: Allyn & Bacon, 1990.

Ludema, K.C., R.M. Caddell, and A.G. Atkins, *Manufacturing Engineering: Economics and Processes*. Englewood Cliffs, N.J.: Prentice-Hall, 1987.

Niebel, B.W., A.B. Draper, and R.A. Wysk, *Modern Manufacturing Process Engineering*. New York: McGraw-Hill, 1989.

Schey, J.A., *Introduction to Manufacturing Processes*, 2d ed. New York: McGraw-Hill, 1987.

General References

Bralla, J.G. (ed.), *Handbook of Product Design and Manufacturing*. New York: McGraw-Hill, 1986.

Encyclopedia of Chemical Technology, Kirk-Othmer, 3d ed. New York: Wiley, 1978.

The Encyclopedia of Materials Science and Engineering. Oxford: Pergamon, 1986.

Machinery's Handbook. New York: Industrial Press (revised periodically).

Metals Handbook, Desk Edition. Metals Park, Ohio: American Society for Metals, 1985.

Metals Handbook, 10th ed. Materials Park, Ohio: ASM International (various volumes and dates).

Metals Handbook, 9th ed. Metals Park, Ohio: American Society for Metals (various volumes and dates).

Standard Handbook for Mechanical Engineers, 9th ed. New York: McGraw-Hill, 1987.

Tool and Manufacturing Engineers Handbook, 4th ed., 5 vols. Dearborn, Mich.: Society of Manufacturing Engineers.

Young, J.F., and R.S. Shane (eds.), *Materials and Processes*, 3d ed., 2 vols. Schenectady, N.Y.: Genium, 1990.

PART I

Fundamentals of Materials: Their Behavior and Manufacturing Properties

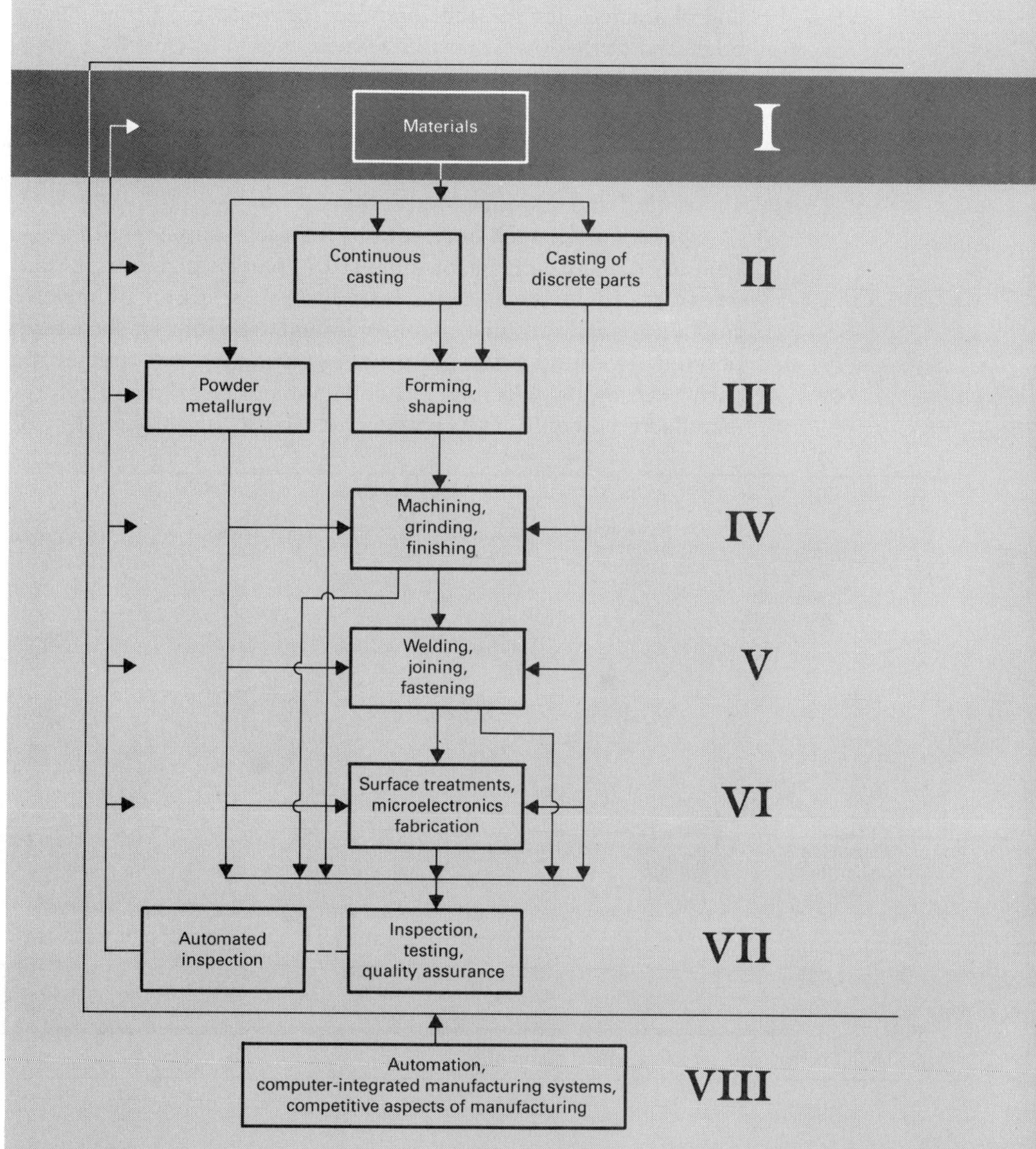

• INTRODUCTION •

We begin Part I of this text by presenting the fundamentals of materials and their behavior and applications. As we explain their behavior and engineering properties throughout the following nine chapters, we highlight the characteristics, advantages, and limitations that influence the choice of materials in the design and manufacture of products.

In order to emphasize the importance of the topics that you will study, let's use the automobile as an example of a product that contains a wide variety of materials (Fig. I.1). These materials were selected because they possess the desired properties and characteristics for the intended functions of specific parts of the automobile—and at the lowest cost. Steel was chosen for much of the body because it is strong, easy to form, and inexpensive. Plastics were used in many components because of characteristics such as a wide choice of colors, light weight, low cost, and case of manufacturing into various shapes. Glass was chosen for the windows not only because it is transparent, but also because it is easy to shape, is easy to clean, and is hard and resistant to abrasion.

We can make such observations for each component of an automobile, which typically is an assemblage of some 15,000 parts, ranging from thin wire to bumpers. As we stated in the General Introduction, selection of materials for individual components in a product requires an understanding of their technological properties, functions, and costs. Note that by saving just one cent on the average cost per part, such as by selecting a different material or manufacturing technique, the manufacturer could reduce the cost of an automobile by $150. The task of

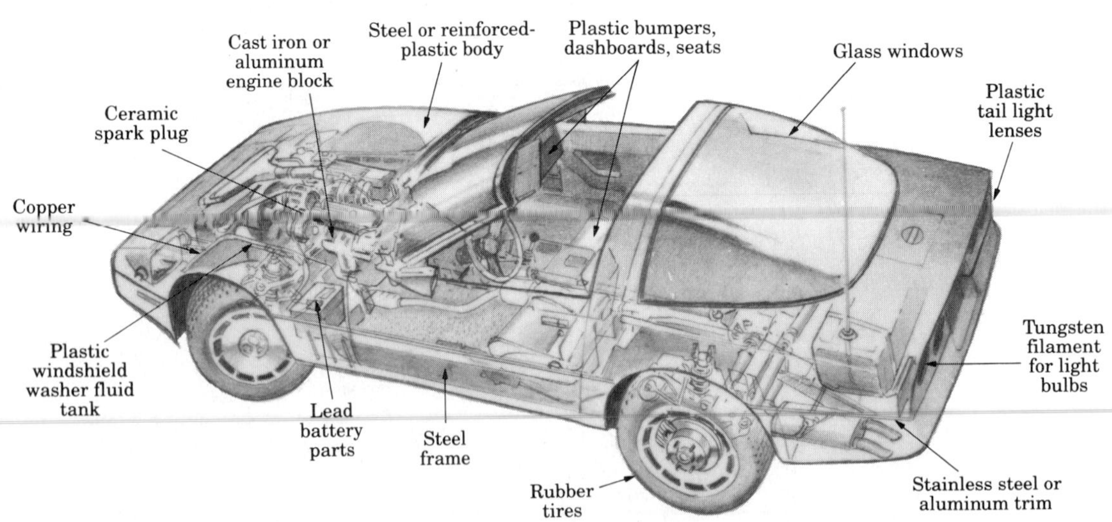

FIGURE I.1
Some of the materials used in a typical automobile.

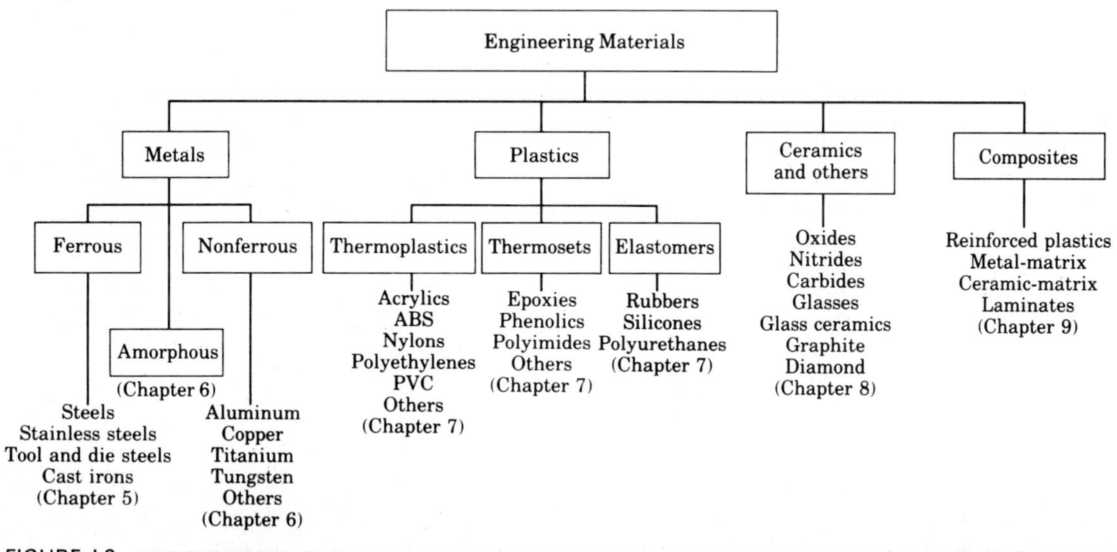

FIGURE I.2
An outline of engineering materials described in Part I.

manufacturing engineers thus becomes truly challenging, especially with the wide variety of materials now available (Fig. I.2).

In this Part, we study the fundamentals of materials so that we can understand and explain their behavior and take full advantage of their capabilities. A general outline of the various topics that we present is given in Fig. I.3 below. Studying these

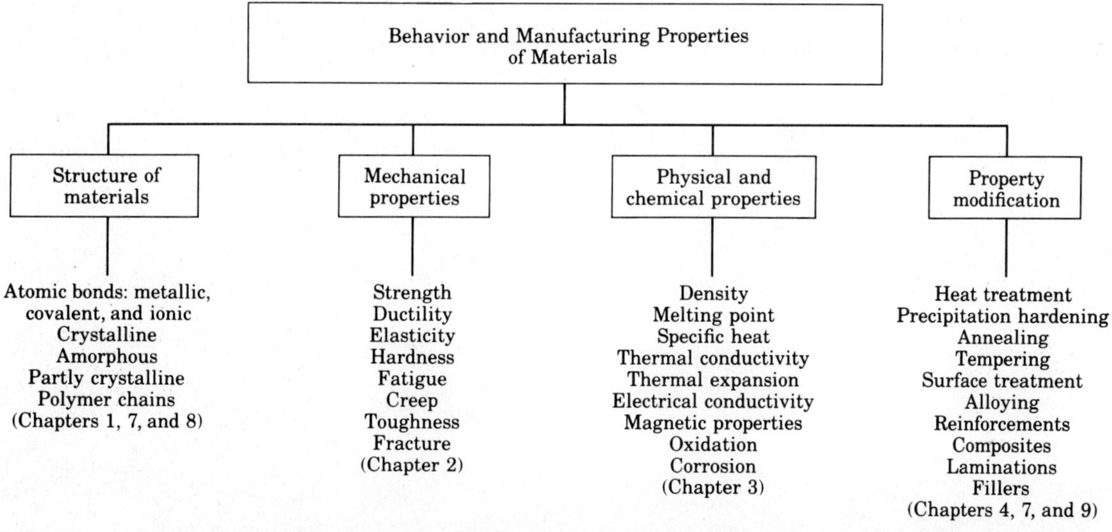

FIGURE I.3
An outline of the behavior and manufacturing properties of materials described in Part I.

topics will give you a broad view of materials and their properties that are relevant to manufacturing engineering and technology, as well as general engineering practice.

The fundamental knowledge gained in Part I on the behavior, properties, and characteristics of materials will help you to better understand the manufacturing processes described in Parts II–IV of this book. This knowledge will also enable you to better analyze the often complex relationships among materials, manufacturing processes, and machinery and tooling, as well as the economics of manufacturing operations.

1

The Structure of Metals

1.1 Introduction

Why are some metals hard and others soft? Why are some metals brittle while others are ductile and can be shaped easily without fracture? Why is it that some metals can withstand high temperatures while others cannot? We can answer these and similar questions by studying the *structure* of metals, that is, the arrangement of the atoms within metals. The structure of metals greatly influences their behavior and properties. A knowledge of structures guides us in controlling and predicting the behavior and performance of metals in various manufacturing processes. Understanding the structure of metals also allows us to predict and evaluate their *properties*. This helps us make appropriate selections for specific applications under particular force, temperature, and environmental conditions.

In addition to atomic structure, various other factors also influence the properties and behavior of metals. Among these are the composition of the metal, impurities and vacancies in the atomic structure, grain size, grain boundaries, environment, and size and surface condition of the metal, as well as the methods by which metals and alloys are made into useful products.

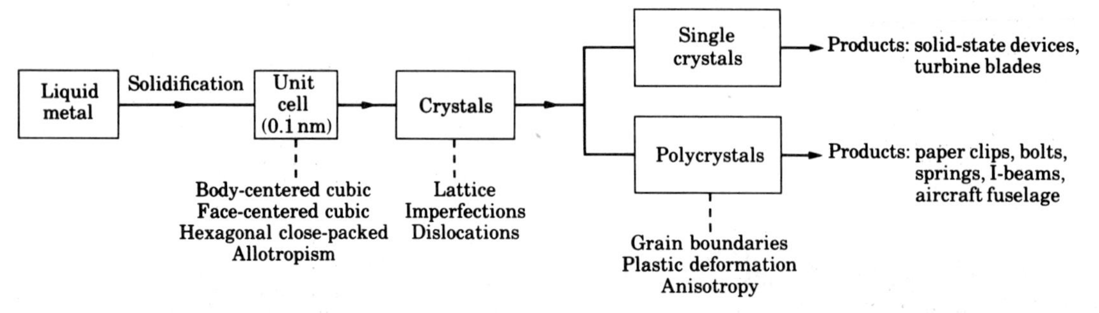

FIGURE 1.1
Outline of topics described in Chapter 1.

The topics covered in this chapter and their sequence are outlined in Fig. 1.1. The structure and general properties of materials other than metals are described in Chapters 7 (polymers), 8 (ceramics), and 9 (composite materials). We describe the structure of metal alloys, the control of their structure, and heat-treatment processes in Chapter 4.

1.2 The Crystal Structure of Metals

When metals solidify from a molten state, the atoms arrange themselves into various orderly configurations, called **crystals**. This arrangement of the atoms in the crystal is called *crystalline structure*. A crystalline structure is like metal scaffolding in front of a building under construction, with uniformly repetitive horizontal and vertical metal pipes and braces.

The smallest group of atoms showing the characteristic **lattice structure** of a particular metal is known as a **unit cell**. It is the building block of a crystal, and a single crystal can have many unit cells. Think of each brick in a wall as a unit cell. The wall has a crystalline structure; that is, it consists of an orderly arrangement of bricks. Thus the wall is like a single crystal, consisting of many unit cells.

The three basic patterns of atomic arrangement found in most metals are: (a) **body-centered cubic** (bcc), (b) **face-centered cubic** (fcc), and (c) **hexagonal close-packed** (hcp). These structures are represented by the drawings and models shown in Figs. 1.2–1.4. Each sphere in these illustrations represents an atom. The order of magnitude of the distance between the atoms in these crystal structures is

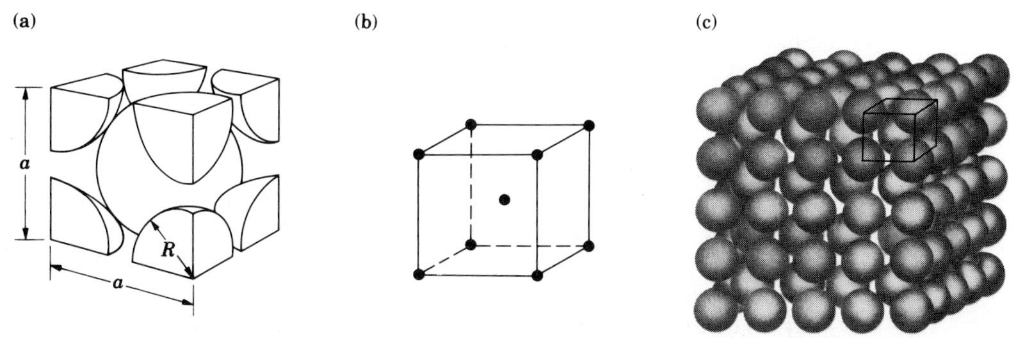

FIGURE 1.2
The body-centered cubic (bcc) crystal structure: (a) hard-ball model; (b) unit cell; and (c) single crystal with many unit cells. *Source:* W. G. Moffatt, et al., *The Structure and Properties of Materials,* Vol. I, John Wiley & Sons, 1976.

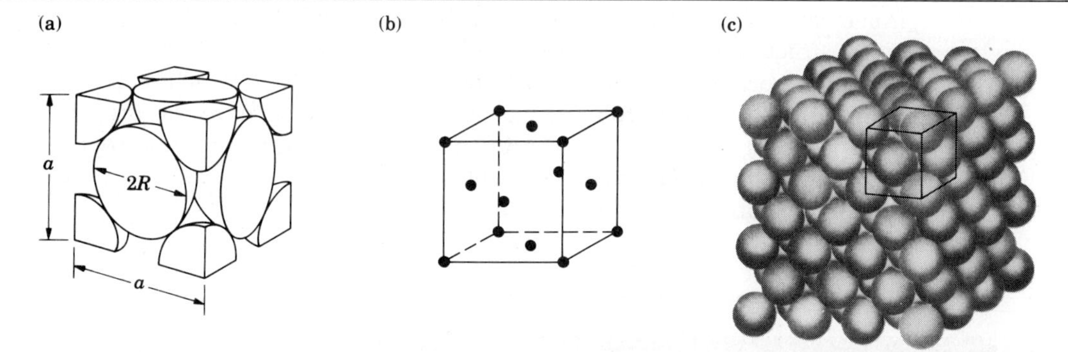

FIGURE 1.3
The face-centered cubic (fcc) crystal structure: (a) hard-ball model; (b) unit cell; and (c) single crystal with many unit cells. *Source:* W. G. Moffatt, et al., *The Structure and Properties of Materials,* Vol. I, John Wiley & Sons, 1976.

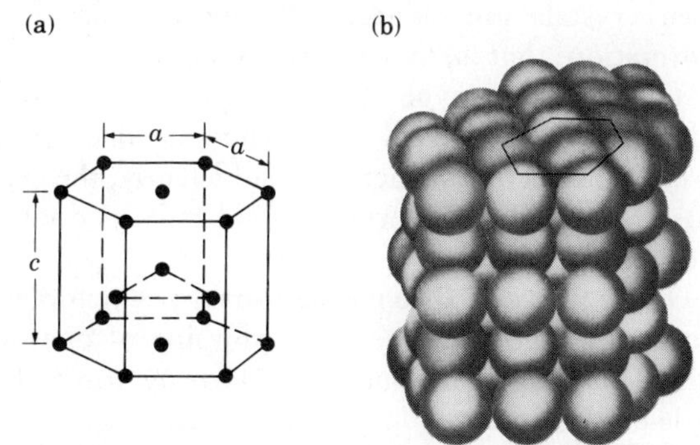

FIGURE 1.4
The hexagonal close-packed (hcp) crystal structure: (a) unit cell; and (b) single crystal with many unit cells. *Source:* W. G. Moffatt, et al., *The Structure and Properties of Materials,* Vol. I, John Wiley & Sons, 1976.

0.1 nm (10^{-8} in.). The models are known as *hard-ball*, or *hard-sphere*, models and can be likened to tennis balls arranged in various configurations in a box. As you will see, the way in which these atoms are arranged determines the properties of a particular metal. We can modify these arrangements by adding atoms of some other metal or metals. This is known as *alloying* and often improves the properties of the metal (discussed in Chapter 4).

As shown in Fig. 1.2, each atom in the bcc structure has eight neighboring atoms. Of the three structures illustrated, the fcc and hcp crystals have the most densely packed configurations. In the hcp structure, the top and bottom planes are called **basal planes**.

Why do metals form different crystal structures? This is a matter of minimizing the energy required to form these structures. Thus tungsten forms a bcc structure because it requires less energy than other structures. Similarly, aluminum forms an fcc structure. However, at different temperatures the same metal may form different structures because of the lower energy required at that temperature. For example, iron forms a bcc structure below 912 °C (1674 °F), and above 1394 °C (2541 °F), but an fcc structure between 912 °C and 1394 °C.

The appearance of more than one type of crystal structure is known as **allotropism**, or **polymorphism**, meaning many shapes. Since the properties and behavior of a metal depend greatly on its crystal structure, allotropism is an important aspect in the heat treatment of metals, as well as in metalworking and welding operations.

1.3 Deformation and Strength of Single Crystals

When crystals are subjected to an external force, they first undergo **elastic deformation**; that is, they return to their original shape when the force is removed. An analogy to this type of behavior is a helical spring that stretches when loaded and returns to its original shape when the load is removed. However, if the force on the crystal structure is increased sufficiently, the crystal undergoes **plastic deformation** (or permanent deformation); that is, it does not return to its original shape when the force is removed.

There are two basic mechanisms by which plastic deformation may take place in crystal structures. One is the slipping of one plane of atoms over an adjacent plane—the **slip plane**—under a shearing force (Fig. 1.5). The deformation of a single-crystal specimen by slip is shown schematically in Fig. 1.6. This situation is much like sliding playing cards against each other.

Just as it takes a certain amount of force to slide playing cards against each other, so a crystal requires a certain amount of shear stress (*critical shear stress*) to undergo permanent deformation. **Shear stress** is the ratio of the shearing force ap-

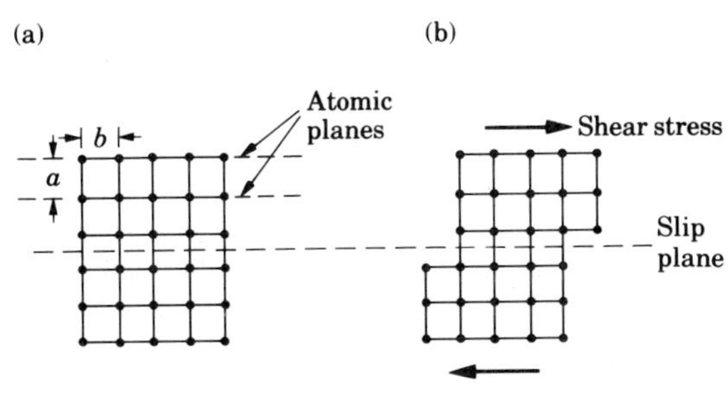

FIGURE 1.5 Permanent deformation, also called plastic deformation, of a single crystal subjected to a shear stress: (a) structure before deformation; and (b) deformation by slip. The b/a ratio influences the magnitude of shear stress required to cause slip.

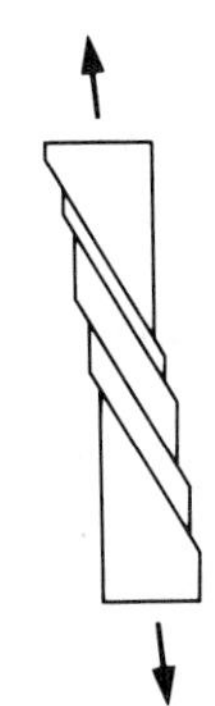

FIGURE 1.6 Permanent deformation of a single crystal under a tensile load. Note that the slip planes tend to align themselves in the direction of pulling. This behavior can be simulated using a deck of cards with a rubber band around them.

plied (Fig. 1.5b) to the cross-sectional area being sheared. Thus there must be a shear stress of sufficient magnitude within a crystal for plastic deformation to take place.

The shear stress required to cause slip in single crystals is directly proportional to the ratio b/a in Fig. 1.5(a), where a is the spacing of the atomic planes and b is inversely proportional to the atomic density in the atomic plane. As b/a decreases, the shear stress required to cause slip decreases. We can therefore state that slip in a crystal takes place along planes of maximum atomic density, or that slip takes place in closely packed planes and in closely packed directions.

Because the b/a ratio is different for different directions within the crystal, a single crystal has different properties when tested in different directions. Thus a crystal is anisotropic. A common example of **anisotropy** is woven cloth, which stretches differently when we pull it in different directions, or plywood, which is much stronger in the planar direction than along its thickness direction (it splits easily).

The second mechanism of plastic deformation is **twinning**, in which a portion of the crystal forms a mirror image of itself across the plane of twinning. Twins form abruptly and are the cause of the creaking sound (tin cry) when a tin or zinc rod is bent at room temperature. Twinning usually occurs in hcp metals.

Slip systems. The combination of a slip plane and its direction of slip is known as a **slip system**. In general, metals with slip systems of 5 or above are ductile, whereas those below 5 are not.

a) In body-centered cubic crystals, there are 48 possible slip systems. Thus the probability is high that an externally applied shear stress will operate on one of these systems and cause slip. However, because of the relatively high b/a ratio, the required shear stress is high. Metals with bcc structures, such as titanium, molybdenum, and tungsten, have good strength and moderate ductility.

b) In face-centered cubic crystals, there are 12 slip systems. The probability of slip is moderate, and the required shear stress is low. These metals, such as aluminum, copper, gold, and silver, have moderate strength and good ductility.

c) The hexagonal close-packed crystal has three slip systems, thus having a low probability of slip. However, more systems become active at elevated temperatures. Metals with hcp structures, such as beryllium, magnesium, and zinc, are generally brittle at room temperature.

Note in Fig. 1.6 that the portions of the single crystal that have slipped have rotated from their original angular position toward the direction of the tensile force. Note also that slip has taken place along certain planes only. With the use of electron microscopy, it has been shown that what appears to be a single slip plane is actually a **slip band**, consisting of a number of slip planes (Fig. 1.7).

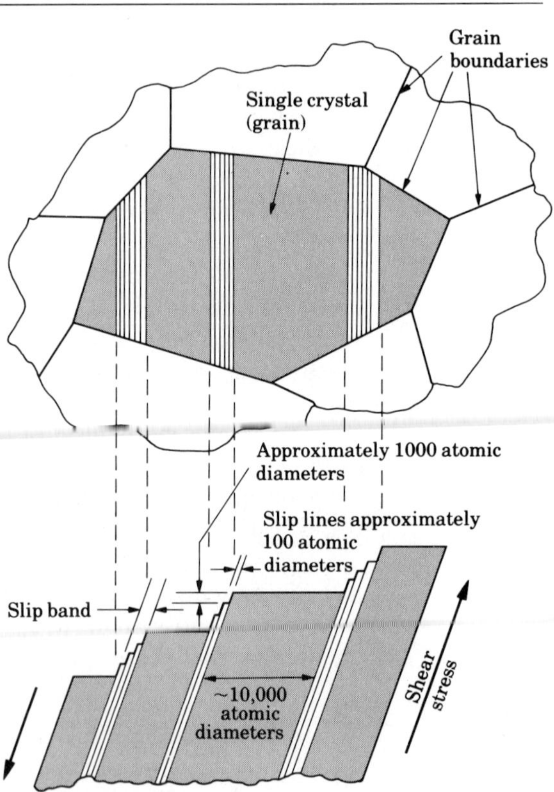

FIGURE 1.7 Schematic illustration of slip lines and slip bands in a single crystal subjected to a shear stress. A slip band consists of a number of slip planes. The crystal at the center of the upper part is an individual grain surrounded by other grains.

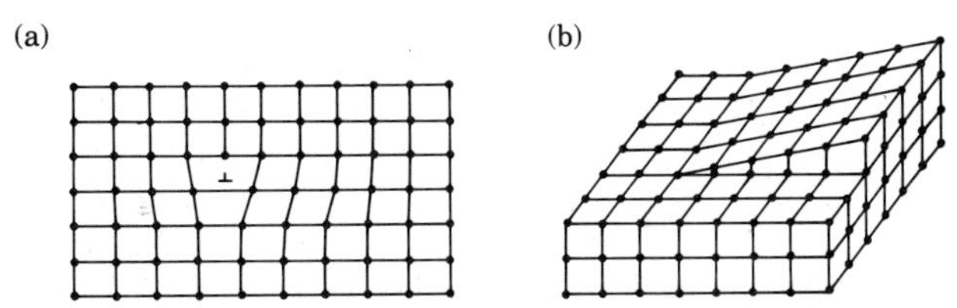

FIGURE 1.8
Types of dislocations in a single crystal: (a) edge dislocation; and (b) screw dislocation.

1.3.1 Imperfections in the crystal structure of metals

The actual strength of metals is approximately one to two orders of magnitude lower than the strength levels obtained from theoretical calculations. This discrepancy has been explained in terms of imperfections in the crystal structure. Unlike the idealized models we have described, actual metal crystals contain a large number of defects and imperfections, which we can categorize as:

a) Line defects, called *dislocations* (Fig. 1.8).
b) Point defects, such as a *vacancy* (missing atom), an *interstitial atom* (extra atom in the lattice), or an *impurity* (foreign atom) that has replaced the atom of the pure metal (Fig. 1.9).
c) Volume or bulk imperfections, such as *voids* or *inclusions* (nonmetallic elements such as oxides, sulfides, and silicates).
d) Planar imperfections, such as *grain boundaries* (Section 1.4).

Dislocations. First observed in the 1930s, **dislocations** are defects in the orderly arrangement of a metal's atomic structure. They are the most significant

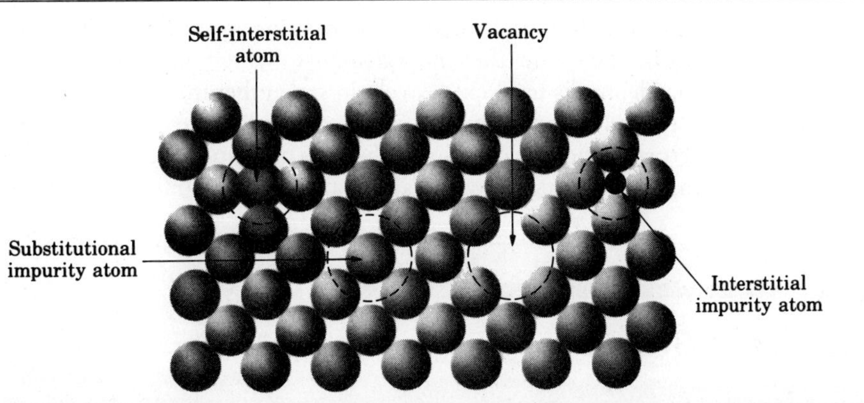

FIGURE 1.9
Defects in a single-crystal lattice. Self-interstitial, vacancy, interstitial, and substitutional. *Source:* After Moffatt et al.

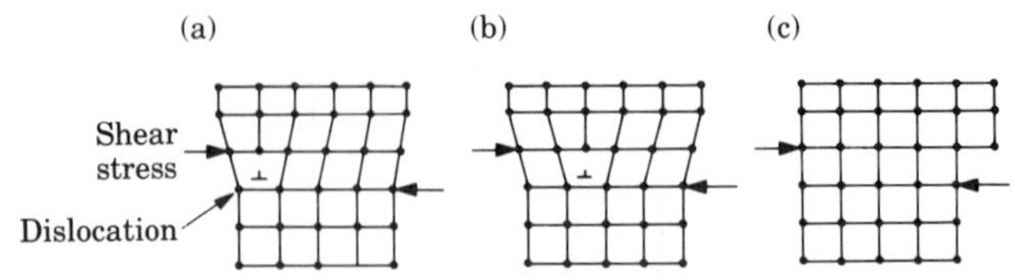

FIGURE 1.10
Movement of an edge dislocation across the crystal lattice under a shear stress. Dislocations help explain why the actual strength of metals is much lower than that predicted by theory.

defects that help explain the discrepancy between the actual and theoretical strength of metals. There are two types of dislocations: *edge* and *screw* (Fig. 1.8).

A slip plane containing a dislocation requires less shear stress to cause slip than a plane in a perfect lattice (Fig. 1.10). One analogy used to describe the movement of an edge dislocation is the earthworm, which moves forward through a hump that starts at the tail and moves toward the head. A second is moving a large carpet by forming a hump at one end and moving it to the other end. The force required to move a carpet in this way is much less than that required to slide the carpet along the floor. Screw dislocations are so named because the atomic planes form a spiral ramp.

1.3.2 Work hardening (strain hardening)

Although the presence of a dislocation lowers the shear stress required to cause slip, dislocations can (1) become entangled and interfere with each other, and (2) be impeded by barriers, such as grain boundaries and impurities and inclusions in the material. Entanglement and impediments increase the shear stress required for slip. This entanglement is like moving two humps at different angles across a carpet, with the two humps interfering with each other's movement, thus making it more difficult to move the carpet.

The increase in shear stress, and hence the increase in the overall strength of the metal, is known as **work hardening** or **strain hardening**. The greater the deformation, the more the entanglements, which increases the metal's strength. Work hardening is used extensively in strengthening metals in metalworking processes at ambient temperature. Typical examples are strengthening wire by reducing its cross-section by drawing it through a die, producing the head on a bolt by forging it, and producing sheet metal for automobile bodies and aircraft fuselages by rolling.

1.4 Grains and Grain Boundaries

Metals commonly used for manufacturing various products are composed of many individual, randomly oriented crystals (**grains**). We are thus dealing with metal structures that are not single crystals but **polycrystals**, that is, many crystals.

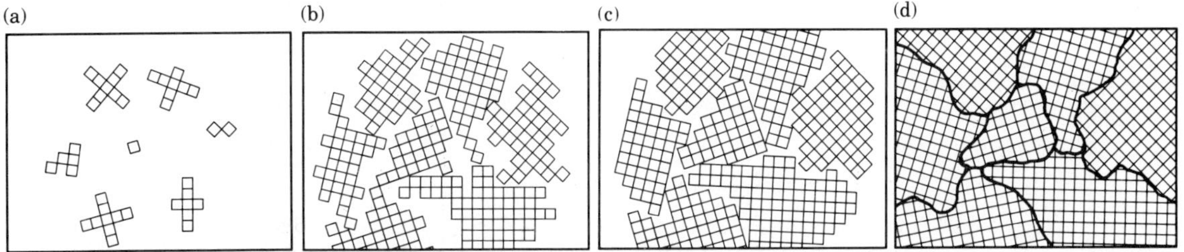

FIGURE 1.11
Schematic illustration of the various stages during solidification of molten metal. Each small square represents a unit cell. (a) Nucleation of crystals at random sites in the molten metal. Note that the crystallographic orientation of each site is different. (b) and (c) Growth of crystals as solidification continues. (d) Solidified metal, showing individual grains and grain boundaries. Note the different angles at which neighboring grains meet each other. *Source:* W. Rosenhain.

When a mass of molten metal begins to solidify, crystals begin to form independently of each other, with random orientations and at various locations within the liquid mass (Fig. 1.11). Each of these crystals grows into a crystalline structure or grain. The number and size of the grains developed in a unit volume of the metal depends on the rate at which *nucleation* (initial stage of formation of crystals) takes place. The number of different sites in which individual crystals begin to form (seven in Fig. 1.11a) and the rate at which these crystals grow are important in the size of grains developed. Generally, rapid cooling produces smaller grains, whereas slow cooling produces larger grains.

If the crystal nucleation rate is high, the number of grains in a unit volume of metal will be greater, and consequently grain size will be small. Conversely, if the rate of growth of the crystals is high, as compared to their nucleation rate, there will be fewer grains per unit volume, and their size will be larger.

Note in Fig. 1.11 how the grains eventually interfere with and impinge upon one another. The surfaces that separate these individual grains are called **grain boundaries.** Each grain consists of either a single crystal, for pure metals, or a polycrystalline aggregate, for alloys. Note that the crystallographic orientation changes abruptly from one grain to the next across the grain boundaries. Recall from Section 1.3 that the behavior of a single crystal or single grain is anisotropic. The ideal behavior of a piece of polycrystalline metal is *isotropic* because the grains have random crystallographic orientations. Thus its properties do not vary with the direction of testing.

1.4.1 Grain size

Grain size significantly influences the mechanical properties of metals. At room temperature, a large grain size is generally associated with low strength, hardness, and ductility. Large grains, particularly in sheet metals, also result in a rough surface appearance after being stretched.

We usually measure grain size by counting the number of grains in a given area or the number of grains that intersect a given length of a line randomly drawn on an enlarged photograph of the grains taken under a microscope on a polished and

TABLE 1.1
GRAIN SIZES

ASTM NO.	GRAINS/mm^2	GRAINS/mm^3
−3	1	0.7
−2	2	2
−1	4	5.6
0	8	16
1	16	45
2	32	128
3	64	360
4	128	1,020
5	256	2,900
6	512	8,200
7	1,024	23,000
8	2,048	65,000
9	4,096	185,000
10	8,200	520,000
11	16,400	1,500,000
12	32,800	4,200,000

etched specimen. Grain size may also be determined by comparing it to a standard chart. The ASTM (The American Society for Testing and Materials) grain size number, n, is related to the number of grains, N, per square inch at a magnification of 100× (equal to 0.0645 mm^2 of actual area) by

$$N = 2^{n-1}. \tag{1.1}$$

Because grains are generally extremely small, many grains occupy a unit volume of metal (Table 1.1). Grain sizes between 5 and 8 are generally considered fine grains. A grain size of 7 is generally acceptable for sheet metals for making car bodies, appliances, and kitchen utensils. Grains can be so large as to be visible by the naked eye, such as those of zinc on the surface of galvanized sheet steels.

1.4.2 Influence of grain boundaries

Grain boundaries have an important influence on the strength and ductility of metals. Because they interfere with the movement of dislocations, grain boundaries also influence strain hardening. These effects depend on temperature, rate of deformation, and the type and amount of impurities present along the grain boundaries. Grain boundaries are more reactive than the grains themselves. This is because the atoms along the grain boundaries are packed less efficiently and are more disordered (see Fig. 1.11) and, hence, have a higher energy than the atoms in the orderly arrangement within the grains.

At elevated temperatures and in materials whose properties depend on the rate of deformation, plastic deformation also takes place by means of *grain boundary sliding*. The **creep** mechanism (elongation under stress over a period of time, usually at elevated temperatures) results from grain-boundary sliding.

1.4.3 Grain boundary embrittlement

When brought into close atomic contact with certain low-melting-point metals, a normally ductile and strong metal can crack under very low stresses. Examples are aluminum wetted with a mercury–zinc amalgam or liquid gallium and copper at elevated temperature wetted with lead or bismuth. These elements weaken the grain boundaries of the metal by **embrittlement.** We use the term *liquid-metal embrittlement* to describe such phenomena because the embrittling element is in a liquid state. However, embrittlement can also occur at temperatures well below the melting point of the embrittling element. This phenomenon is known as *solid-metal embrittlement.*

Hot shortness is caused by local melting of a constituent or an impurity in the grain boundary at a temperature below the melting point of the metal itself. When such a metal is subjected to plastic deformation at elevated temperatures (hot working), the piece of metal crumbles and disintegrates along the grain boundaries. Examples are antimony in copper and leaded steels and brass. To avoid hot shortness, the metal is usually worked at a lower temperature in order to prevent softening and melting along the grain boundaries. Another form of embrittlement is *temper embrittlement* in alloy steels, which is caused by segregation (movement) of impurities to the grain boundaries.

1.5 Plastic Deformation of Polycrystalline Metals

If a piece of polycrystalline metal with uniform equiaxed grains (having equal dimensions in all directions, as shown in the model in Fig. 1.12) is subjected to

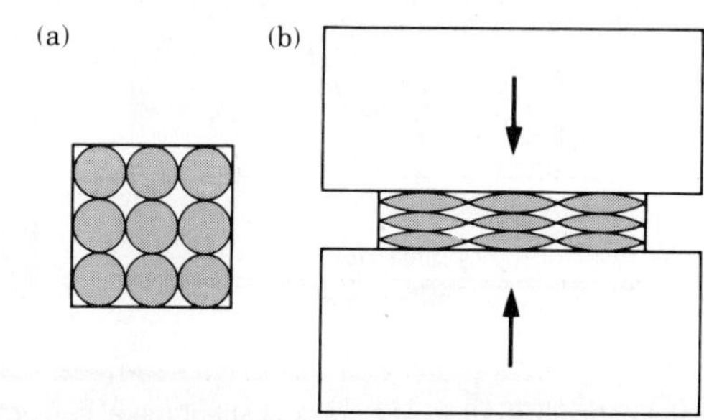

FIGURE 1.12 Plastic deformation of idealized (equiaxed) grains in a specimen subjected to compression, such as is done in rolling or forging of metals: (a) before deformation; and (b) after deformation. Note the alignment of grain boundaries along a horizontal direction, known as preferred orientation.

plastic deformation at room temperature (cold working), the grains become deformed and elongated. The deformation process may be carried out either by compressing the metal, as is done in forging, or by subjecting it to tension, as is done in stretching sheet metal. The deformation within each grain takes place by the mechanisms we described in Section 1.3 for a single crystal.

During plastic deformation, the grain boundaries remain intact and mass continuity is maintained. The deformed metal exhibits greater strength because of the entanglement of dislocations with grain boundaries. The increase in strength depends on the amount of deformation (*strain*) to which the metal is subjected; the greater the deformation, the stronger the metal becomes. Furthermore, the increase in strength is greater for metals with smaller grains because they have a larger grain-boundary surface area per unit volume of metal.

1.5.1 Anisotropy (texture)

Figure 1.12 shows that, as a result of plastic deformation, the grains have elongated in one direction and contracted in the other. Consequently, this piece of metal has become *anisotropic*, and its properties in the vertical direction are different from those in the horizontal direction.

Many products develop anisotropy of mechanical properties after they have been processed by metalworking techniques. The degree of anisotropy depends on how uniformly the metal is deformed. Note from the direction of the crack in Fig. 1.13, for example, that the ductility of the cold-rolled sheet in the vertical (transverse) direction is lower than in its longitudinal direction.

Anisotropy influences both mechanical and physical properties of metals. For example, sheet steel for electrical transformers is rolled in such a way that the resulting deformation imparts anisotropic magnetic properties to the sheet, thus reducing magnetic-hysteresis losses and improving the efficiency of transformers

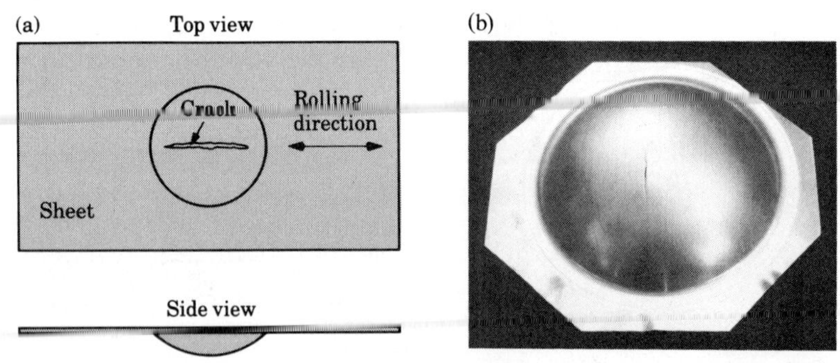

FIGURE 1.13 (a) Schematic illustration of a crack in sheet metal subjected to bulging, such as caused by pushing a steel ball against the sheet. Note the orientation of the crack with respect to the rolling direction of the sheet. This material is anisotropic. (b) Aluminum sheet with a crack (vertical dark line at the center) developed in a bulge test. *Source:* J. S. Kallend.

(see also "Amorphous Alloys," Section 6.14). There are two general types of anisotropy in metals: **preferred orientation** and **mechanical fibering.**

Preferred orientation. Also called *crystallographic anisotropy*, preferred orientation can be best described by reference to Fig. 1.6. When a metal crystal is subjected to tension, the sliding blocks rotate toward the direction of pulling. Thus slip planes and slip bands tend to align themselves with the direction of deformation. Similarly, for a polycrystalline aggregate with grains in various orientations, all slip directions tend to align themselves with the direction of pulling. Conversely, under compression the slip planes tend to align themselves in a direction perpendicular to the direction of compression.

Mechanical fibering. Mechanical fibering results from the alignment of impurities, inclusions (stringers), and voids in the metal during deformation. Note that if the spherical grains in Fig. 1.12 were coated with impurities, these impurities would align themselves generally in a horizontal direction after deformation. Since impurities weaken the grain boundaries, this piece of metal would be weak and less ductile when tested in the vertical direction. An analogy is plywood, which is strong in tension along its planar direction, but peels off easily when tested in tension in its thickness direction.

1.6 Recovery, Recrystallization, and Grain Growth

We have shown that plastic deformation at room temperature results in the deformation of grains and grain boundaries, a general increase in strength, and a decrease in ductility, as well as causing anisotropic behavior. These effects can be reversed and the properties of the metal brought back to their original levels by heating it in a specific temperature range for a period of time. The temperature range and amount of time depend on the material and several other factors. Three events take place consecutively during the heating process:

- *Recovery.* During **recovery**, which occurs at a certain temperature range below the *recrystallization temperature* of the metal, the stresses in the highly deformed regions are relieved. Subgrain boundaries begin to form—called **polygonization**—with no appreciable change in mechanical properties, such as hardness and strength (Fig. 1.14).
- *Recrystallization.* The process in which, at a certain temperature range, new equiaxed and strain-free grains are formed, replacing the older grains, is called **recrystallization.** The temperature for recrystallization ranges approximately between $0.3T_m$ and $0.5T_m$, where T_m is the melting point of the

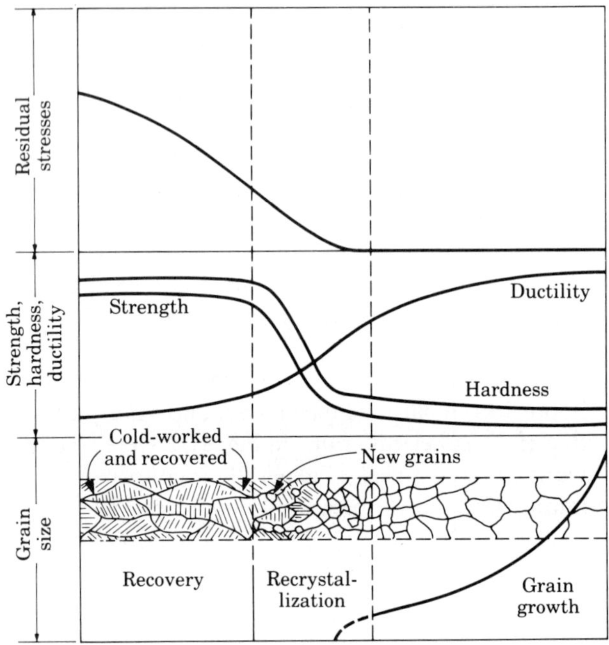

FIGURE 1.14
Schematic illustration of the effects of recovery, recrystallization, and grain growth on mechanical properties and shape and size of grains. Note the formation of small new grains during recrystallization. *Source:* G. Sachs.

metal on the absolute scale (Fig. 1.15). The recrystallization temperature is generally defined as the temperature at which complete recrystallization occurs within approximately one hour. Recrystallization decreases the density of dislocations and lowers the strength and raises the ductility of the metal (Fig. 1.14). Figure 1.15 shows that metals such as lead, tin, cadmium, and zinc recrystallize at about room temperature. Thus when these metals are cold worked, they do not work harden.

Recrystallization depends on the degree of prior cold work (work hardening); the higher the cold work, the lower the temperature required for recrystallization to occur. The reason is that as the amount of cold work increases, the number of dislocations and the amount of energy stored in dislocations (*stored energy*) also increase. This energy supplies the work required for recrystallization. Recrystallization is a function of time because it involves diffusion, that is, movement and exchange of atoms across grain boundaries.

The effects on recrystallization of temperature, time, and reduction in the thickness or height of the workpiece by cold working are as follows:

a) For a constant amount of deformation by cold working, the time required for recrystallization decreases with increasing temperature.

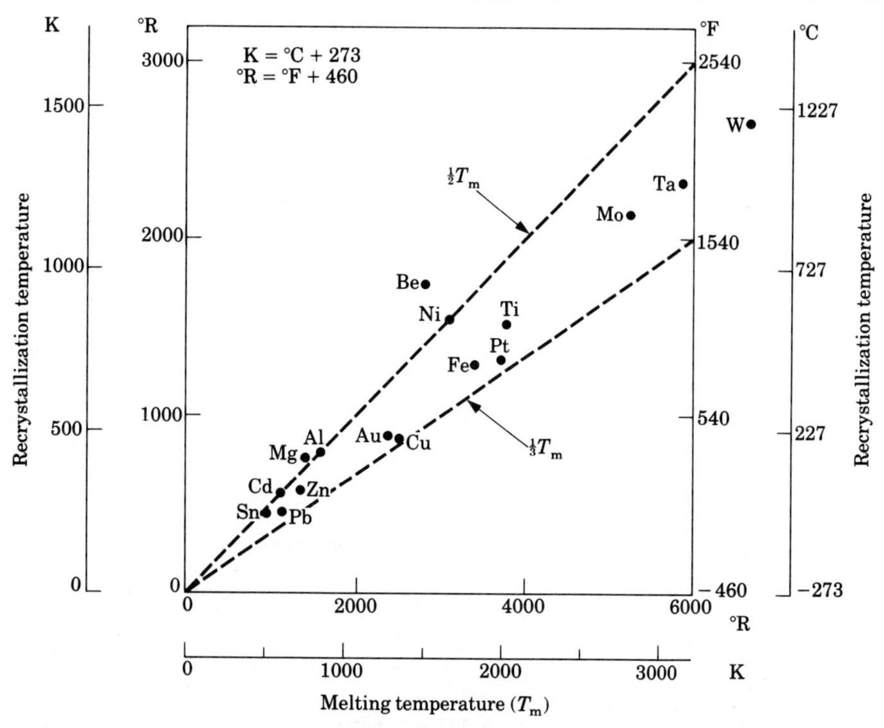

FIGURE 1.15
Recrystallization temperature for various metals. This temperature is approximately between one-half and one-third of the absolute melting point of metals. Note that metals such as lead (Pb) and tin (Sn) recrystallize at room temperature. *Source:* L. H. Van Vlack, *Materials for Engineering.* Addison-Wesley Publishing Company, Inc., Reading, Massachusetts, 1982.

b) The greater the prior cold work done, the lower the temperature required for recrystallization.
c) The higher the amount of deformation, the smaller the grain size becomes during recrystallization. This is a common method of converting a coarse-grained structure to one of fine grain, with improved properties.
d) Anisotropy due to preferred orientation usually persists after recrystallization. To restore isotropy, a temperature higher than that required for recrystallization may be necessary.

- *Grain growth.* If we continue to raise the temperature of the metal, the grains begin to grow and their size may eventually exceed the original grain size. This phenomenon is known as **grain growth,** and it adversely affects mechanical properties slightly (Fig. 1.14). However, large grains produce a rough surface appearance on sheet metals—called **orange peel**—when they are stretched to form a part or when a piece of metal is subjected to compression, such as in forging operations.

1.7

Cold, Warm, and Hot Working

When plastic deformation is carried out at room temperature, it is called **cold working,** and when carried out at above the recrystallization temperature, it is called **hot working.** As the name implies, **warm working** is carried out at higher than room temperature but lower than hot-working temperature. Thus warm working is a compromise between cold and hot working. The temperature range for these three categories of plastic deformation are given in Table 1.2 in terms of a ratio, where T is the working temperature and T_m is the melting point of the metal, both on the absolute scale. Although it is a dimensionless quantity, this ratio is known as the *homologous temperature.* In Part III we will show that there are important technological differences in products that are processed by cold, warm, or hot working.

TABLE 1.2
HOMOLOGOUS TEMPERATURE RANGES FOR VARIOUS PROCESSES

PROCESS	T/T_m
Cold working	<0.3
Warm working	0.3 to 0.5
Hot working	>0.6

SUMMARY

Metals are composed of crystals (grains). There are three basic crystal structures in metals: body-centered cubic, face-centered cubic, and hexagonal close-packed. These crystals are not perfect and contain various defects and imperfections, such as dislocations, vacancies, impurities, inclusions, and grain boundaries. The behavior and properties of metals depend on their crystal structure, grain size, grain boundaries, and the nature and extent of defects.

Plastic deformation takes place by means of a slip mechanism. The stress required to continue deformation increases with deformation and is known as work hardening or strain hardening. It results from the entanglement of dislocations with each other and with impurities and grain boundaries. In addition to increasing strength, plastic deformation at room temperature also causes anisotropy. Both of these effects can be eliminated by recrystallizing the metal at elevated temperatures. Metals may be subjected to cold, warm, or hot working, each of which has its advantages and limitations.

KEY TERMS

Allotropism
Anisotropy
Basal plane
Body-centered cubic
Cold working
Creep
Crystals
Dislocations
Elastic deformation
Embrittlement
Face-centered cubic
Grains
Grain boundaries
Grain growth
Grain size
Hexagonal close-packed
Hot shortness
Hot working
Lattice structure
Mechanical fibering
Orange peel
Plastic deformation
Polycrystals
Polygonization
Polymorphism
Preferred orientation
Recovery
Recrystallization
Shear stress
Slip band
Slip plane
Slip system
Strain hardening
Twinning
Unit cell
Warm working
Work hardening

BIBLIOGRAPHY

Ashby, M.F., and D.R.H. Jones, *Engineering Materials*, Vol. 1, *An Introduction to Their Properties and Applications*, 1980; Vol. 2, *An Introduction to Microstructures, Processing and Design*, 1986. Oxford: Pergamon.

Askeland, D.R., *The Science of Engineering Materials*, 2d ed. Monterey, Calif.: Brooks/Cole, 1989.

Callister, W.D., Jr., *Materials Science and Engineering*, 2d ed. New York: Wiley, 1991.

Dieter, G.E., *Mechanical Metallurgy*, 3d ed. New York: McGraw-Hill, 1986.

Encyclopedia of Materials Science and Engineering. New York: Pergamon, 1986.

Flinn, R.A., and P.K. Trojan, *Engineering Materials and Their Applications*, 4th ed. Boston: Houghton Mifflin, 1990.

Shackelford, J.F., *Introduction to Materials Science for Engineers*, 2d ed. New York: Macmillan, 1988.

Smith, W.F., *Principles of Materials Science and Engineering*, 2d ed. New York: McGraw-Hill, 1990.

Van Vlack, L.H., *Elements of Materials Science and Engineering*, 6th ed. Reading, Mass.: Addison-Wesley, 1989.

REVIEW QUESTIONS

1.1 Explain the difference between a unit cell and a single crystal.

1.2 In tables on crystal structures, iron is listed as having both a bcc and an fcc structure. Why?

1.3 Define anisotropy. What materials can you think of other than metals that exhibit anisotropic behavior?

1.4 What effects does recrystallization have on properties of metals?
1.5 What is strain hardening, and what effects does it have on metals?
1.6 Explain what is meant by structure-sensitive and structure-insensitive properties of metals.
1.7 Make a list of each of the major imperfections in the crystal structure of metals.
1.8 What influence does grain size have on the mechanical properties of metals?
1.9 What is the relationship between the nucleation rate and the number of grains per unit volume of a metal?
1.10 What is a slip system, and what is its significance?
1.11 Explain the difference between recovery and recrystallization.
1.12 What is hot shortness, and what is its significance?
1.13 Explain the differences between cold, warm, and hot working of metals.
1.14 Describe what the orange peel effect is.

QUESTIONS AND PROBLEMS

1.15 Explain your understanding of why we should study the crystal structure of metals.
1.16 What is the significance of some metals undergoing allotropism?
1.17 Describe what happens to the reflectivity of a polished sheet metal as it is stretched.
1.18 Is it possible for two pieces of the same metal to have different recrystallization temperatures? Is it possible for recrystallization to take place in some regions of a part before other regions in the same part? Explain.
1.19 Describe your understanding of why different crystal structures exhibit different strengths and ductilities.
1.20 Plot the data given in Table 1.1 in terms of grains/mm^2 vs. grains/mm^3 and state your observation.
1.21 If the ball of a ball-point pen is 1 mm in diameter and has an ASTM grain size of 10, how many grains are there in the ball?
1.22 Describe the difference between preferred orientation and mechanical fibering.
1.23 A strip of metal is reduced in thickness by cold working from 25 mm to 15 mm. A similar strip is reduced from 25 mm to 10 mm. Which one of these cold-worked strips will recrystallize at a lower temperature? Why?
1.24 A cold-worked piece of metal has been recrystallized. When tested, it is found to be anisotropic. Explain the probable reason.
1.25 Draw some analogies to the phenomenon of hot shortness.
1.26 Explain the advantages and limitations of cold, warm, and hot working, respectively.
1.27 How can you tell the difference between two parts made of the same material, but one was formed by cold working and the other by hot working? Explain the differences you might observe.
1.28 Do you think that it might be important to know whether a raw material for a manufacturing process has anisotropic properties? What about any anisotropy in the finished product? Explain.
1.29 Explain why the strength of a polycrystalline metal at room temperature decreases as its grain size increases.
1.30 Describe the technique by which you would reduce the orange peel effect on the surface of workpieces.

1.31 A paper clip is made of wire that is 4 in. long and $\frac{1}{32}$ in. in diameter. If the ASTM grain size is 7, how many grains are there in the paper clip?

1.32 What is the significance of metals such as lead and tin having recrystallization temperatures at about room temperature?

1.33 The unit cells shown in Figs. 1.2–1.4 can be represented by tennis balls arranged in various configurations in a box. In such an arrangement, *atomic packing factor* (APF) is defined as the ratio of the volume of atoms to the volume of the unit cell. Show that the packing factor for the bcc structure is 0.68 and for the fcc structure 0.74.

1.34 Show that the lattice constant a in Fig. 1.3(a) is related to the atomic radius by $a = 2\sqrt{2}R$, where R is the radius of the atom as depicted by the tennis-ball model.

1.35 Show that for the fcc unit cell, the radius r of the largest hole is given by $r = 0.414R$. Determine the size of the largest hole for the iron atom in the fcc structure.

1.36 Calculate the theoretical (a) shear strength and (b) tensile strength for aluminum, plain-carbon steel, and tungsten. Estimate the ratios of their theoretical to actual strengths.

1.37 A technician determines that the grain size of a certain etched specimen is 6. Upon further checking, he or she finds that the magnification used was 150 ×, instead of 100 × that is required by ASTM standards. Determine the correct grain size.

2

Mechanical Behavior, Testing, and Manufacturing Properties of Materials

2.1 Introduction

In manufacturing operations many parts are formed into various shapes by applying external forces to the workpiece by means of tools and dies. Typical operations are forging of a turbine disk, extruding various parts for an aluminum ladder, or rolling a flat sheet to be processed into a car body. Because deformation in these processes is carried out by mechanical means, an understanding of the behavior of materials in response to applied forces is important. Forming operations may be carried out at room or at elevated temperatures and at slow or high rates of deformation; the forces may be applied in several directions.

The behavior of a manufactured part during its expected service life is an important consideration. The wings of an aircraft, the crankshaft of an automobile engine, and gear teeth in machinery are all subjected to static, as well as fluctuating, forces. If excessive, this condition can lead to cracks and cause total failure of these components by a mechanism called fatigue. Similarly, a turbine disk and the blades in the jet engine of an aircraft are subjected to high stresses and temperature during flight. Over a period of time these components undergo creep, a phenomenon in which the components elongate permanently under applied stresses, which may eventually lead to failure.

In this chapter we consider all aspects of mechanical properties and behavior of materials that are relevant to the design and manufacture of parts. We also describe commonly used tests that are essential in assessing the properties of materials.

2.2 Tension

The **tension** test is the most common test for determining the mechanical properties of materials, such as strength, ductility, and toughness. The test first requires the preparation of a *test specimen*, typically as shown in Fig. 2.1(a). The specimen is prepared according to ASTM specifications in the United States and those of the appropriate organizations in other countries. Although most tension-test specimens are solid and round, some are flat sheet or tubular specimens of various sizes.

Typically, the specimen has an original gage length l_o, generally 50 mm (2 in.), and a cross-sectional area A_o, usually with a diameter of 12.5 mm (0.5 in.). The specimen is mounted between the threaded jaws of a tension testing machine (Fig. 2.1b). These machines are equipped with various controls so that the specimen can be tested at different rates of deformation and temperature.

2.2.1 Stress–strain curves

A typical sequence of the deformation of the tension-test specimen is shown in Fig. 2.2. When the load is first applied, the specimen elongates proportionately with the

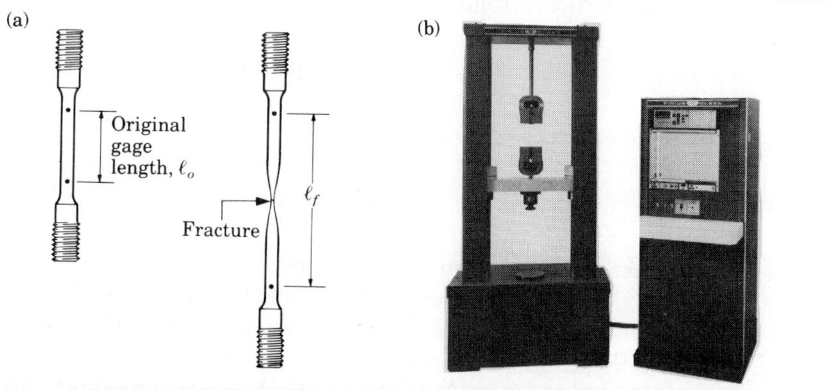

FIGURE 2.1 (a) A standard tensile-test specimen before and after pulling, showing original and final gage lengths. (b) A typical tensile testing machine. The load and elongation of the specimen are recorded on the strip chart on the right. *Source:* Courtesy of Tinius Olsen Testing Machine Co.

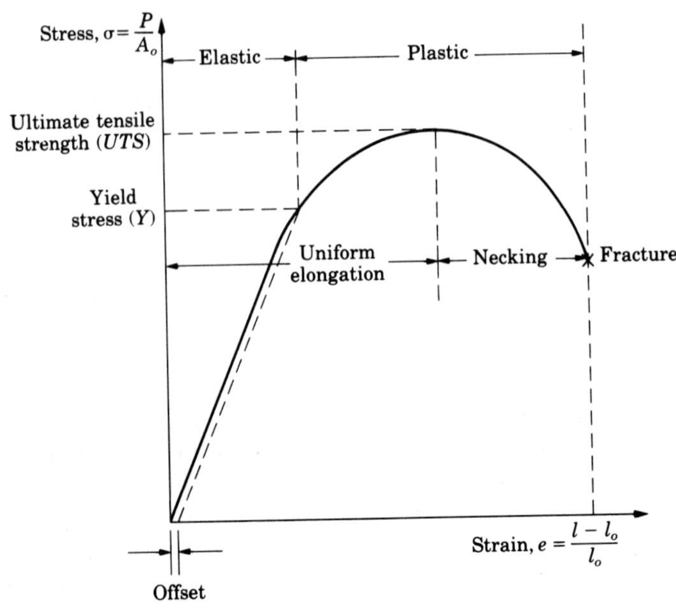

FIGURE 2.2 A typical stress–strain curve obtained from a tension test, showing various features. Tensile test data are important in calculating forces and predicting the behavior of materials in manufacturing processes.

load, which we call *linear elastic behavior*. If the load is removed, the specimen returns to its original length and shape. This process is similar to stretching a rubber band and releasing it.

The **engineering stress**, or *nominal stress*, is defined as the ratio of the applied load P to the original cross-sectional area A_o of the specimen. That is,

$$\text{Engineering stress, } \sigma = \frac{P}{A_o}. \tag{2.1}$$

The **engineering strain** is defined as

$$\text{Engineering strain, } e = \frac{l - l_o}{l_o}, \tag{2.2}$$

where l is the instantaneous length.

As the load is increased, at some level of stress, the specimen begins to undergo permanent (*plastic*) deformation. Beyond this level of stress, the stress and strain are no longer proportional as they were in the elastic region. The stress at which this phenomenon occurs is known as the **yield stress** (Y) of the material. The terms *elastic limit* and *proportional limit* are also used to specify the point where the stress and strain are no longer proportional. The yield stress for various metals is given in Table 2.1.

TABLE 2.1
MECHANICAL PROPERTIES OF VARIOUS MATERIALS AT ROOM TEMPERATURE

METAL	*Y* (MPa)	*UTS* (MPa)	ELONGATION IN 50 mm (%)
Aluminum	35	90	45
Aluminum alloys	35–550	90–600	45–4
Beryllium	185–260	230–350	3.5–1
Columbium (niobium)	205	275	30
Copper	70	220	45
Copper alloys	76–1100	140–1310	65–3
Iron	40–200	185–285	60–3
Steels	205–1725	415–1750	65–2
Lead	7–14	17	50
Lead alloys	14	20–55	50–9
Magnesium	90–105	160–195	15–3
Magnesium alloys	130–305	240–380	21–5
Molybdenum alloys	80–2070	90–2340	40–30
Nickel	58	320	30
Nickel alloys	105–1200	345–1450	60–5
Tantalum alloys	480–1550	550–1550	40–20
Titanium	140–550	275–690	30–17
Titanium alloys	344–1380	415–1450	25–7
Tungsten	550–690	620–760	0
NONMETALLIC			
Ceramics	—	140–2600	0
Glass	—	140	0
Glass, fibers	—	3500–4500	0
Graphite, fibers	—	2100–2500	0
Thermoplastics	—	7–80	5–1000
Thermoplastics, reinforced	—	20–120	1–10
Thermosets	—	35–170	0
Thermosets, reinforced	—	200–520	0

Note: Multiply MPa by 145 to obtain psi.

For soft and ductile materials, we may not be able to easily determine the exact position on the stress–strain curve where yielding occurs because the slope of the straight (elastic) portion of the curve decreases slowly. Therefore we usually define Y as the point on the stress–strain curve that is offset by a strain of 0.002, or 0.2 percent elongation. This simple procedure is shown in Fig. 2.2.

As the specimen continues to elongate further under increasing load beyond Y, its cross-sectional area decreases *permanently* and *uniformly* throughout its gage length. If the specimen is unloaded from a stress level higher than the yield stress Y,

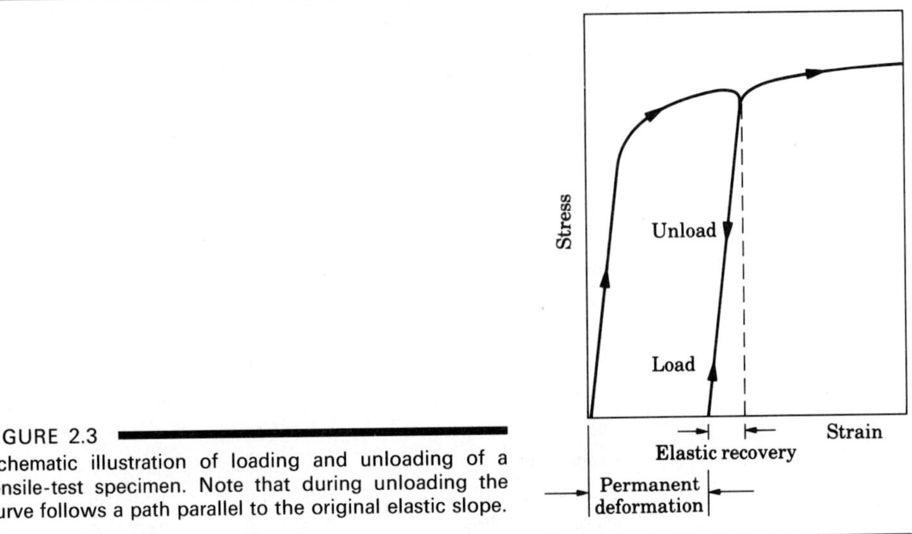

FIGURE 2.3 Schematic illustration of loading and unloading of a tensile-test specimen. Note that during unloading the curve follows a path parallel to the original elastic slope.

the curve follows a straight line downward and parallel to the original slope (Fig. 2.3). As the load—and the engineering stress—is further increased, it eventually reaches a maximum and then begins to decrease (Fig. 2.2). The maximum engineering stress is called the *tensile strength*, or **ultimate tensile strength** (*UTS*), of the material. Values for UTS for some materials are given in Table 2.1.

If the specimen is loaded beyond its ultimate tensile strength, it begins to *neck*, or neck down. The cross-sectional area of the specimen is no longer uniform along the gage length but is smaller in the necked region. As the test progresses, the engineering stress drops further and the specimen finally fractures in the necked region. The engineering stress at fracture is known as *breaking* or *fracture stress.*

The ratio of stress to strain in the elastic region is known as the **modulus of elasticity** (*E*), or Young's modulus (after T. Young, 1773–1829):

$$\text{Modulus of elasticity, } E = \frac{\sigma}{e}. \tag{2.3}$$

This linear relationship is known as *Hooke's law* (after R. Hooke, 1635–1703). The modulus of elasticity is essentially a measure of the *stiffness* of the material. Note in Eq. (2.3) that, because engineering strain is dimensionless, E has the same units as stress. Some typical values of E for various materials are given in Table 2.2. The higher the E value, the higher the load required to stretch it to the same extent and thus the stiffer the material. Compare, for example, the stiffness of a piece of metal with that of a piece of rubber or plastic when you try to stretch them to the same extent.

The elongation of the specimen under tension is accompanied by lateral contraction (Fig. 2.4), as you may observe by stretching a rubber band. The absolute value of the ratio of the lateral strain to longitudinal strain in the specimen

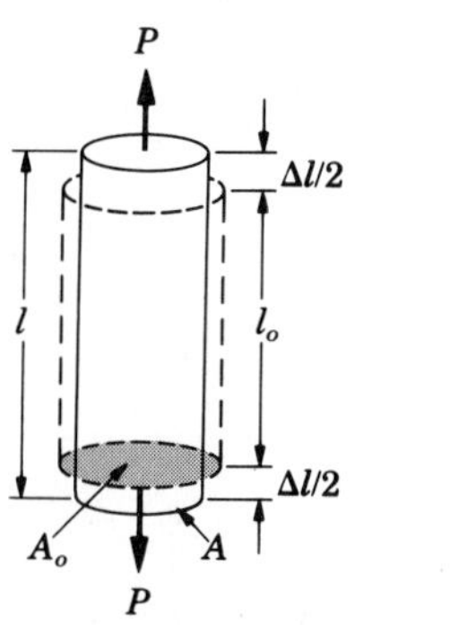

FIGURE 2.4
Contraction of the diameter of a tensile-test specimen as it elongates under a tensile load P.

TABLE 2.2
MODULUS OF ELASTICITY E AND POISSON'S RATIO ν FOR VARIOUS MATERIALS AT ROOM TEMPERATURE

METALS	E (GPa)	ν
Aluminum and its alloys	69–79	0.31–0.34
Cast irons	105–150	0.21–0.30
Copper and its alloys	105–150	0.33–0.35
Lead and its alloys	14	0.43
Magnesium and its alloys	41–45	0.29–0.35
Molybdenum	325	0.32
Nickel and its alloys	180–214	0.31
Steel (plain carbon)	200	0.33
Steel (austenitic stainless)	190–200	0.28
Titanium and its alloys	80–130	0.31–0.34
Tungsten	400	0.27
NONMETALLIC		
Acrylics	1.4–3.4	0.35–0.40
Epoxies	3.5–17	0.34
Nylons	1.4–2.8	0.32–0.40
Rubbers	0.01–0.1	0.5
Plastics, reinforced	2–50	—
Glass and porcelain	70–80	0.24
Diamond	820–1050	—
Graphite	240–390	—

Note: Multiply GPa by 145,000 to obtain psi.

is known as **Poisson's ratio** (after S. D. Poisson, 1781–1840) and is denoted by the letter ν. Values of ν for some materials are given in Table 2.2. Note that rubbers have the highest value; thus they contract more in tension than any of the other materials listed.

2.2.2 Ductility

An important behavior observed during a tension test is **ductility**, that is, the extent of plastic deformation that the material undergoes before fracture. There are two common measures of ductility. The first is the **percent elongation** of the specimen:

$$\text{Elongation} = \frac{l_f - l_o}{l_o} \times 100, \tag{2.4}$$

where l_f and l_o are measured as shown in Fig. 2.1. Note that the percent elongation is based on the original gage length of the specimen.

The other measure is the **reduction of area**:

$$\text{Reduction of area} = \frac{A_o - A_f}{A_o} \times 100, \tag{2.5}$$

where A_o and A_f are the original and final (fracture) cross-sectional areas, respectively, of the test specimen. Thus the ductility of a piece of chalk is zero because it does not stretch at all, whereas a piece of clay stretches considerably before it fails.

2.2.3 True stress and true strain

You have seen that the engineering stress is based on the original cross-sectional area A_o of the specimen. We know, however, that the instantaneous area supporting the load becomes smaller as the specimen elongates, just as the area of a rubber band does. Thus engineering stress does not represent the actual stress to which the specimen is subjected.

True stress is defined as the ratio of load P to the actual (instantaneous), hence true, cross-sectional area A of the specimen. Thus,

$$\text{True stress, } \sigma = \frac{P}{A}. \tag{2.6}$$

As for true strain, we first consider the elongation of the specimen in increments of instantaneous change in length. Then, using calculus, we can show that the **true strain** (*natural* or *logarithmic strain*) is

$$\text{True strain, } \varepsilon = \ln\left(\frac{l}{l_o}\right). \tag{2.7}$$

Note from Eqs. (2.2) and (2.7) that, for small values of strain, the engineering and true strains are approximately equal. However, they diverge rapidly as strain increases. For example, when $e = 0.1$, $\varepsilon = 0.095$, and when $e = 1$, $\varepsilon = 0.69$.

Unlike engineering strains, true strains are consistent with actual physical phenomena in deformation of materials. Let's assume, for example, a hypothetical situation where a specimen 50 mm (2 in.) in height is compressed between flat platens to a final height of zero. In other words, we have deformed the specimen infinitely. According to their definitions, the engineering strain that the specimen undergoes is -1, and the true strain is $-\infty$. We see that true strain describes the extent of deformation correctly, since in this case it is indeed infinite.

2.2.4 Construction of stress–strain curves

The procedure for constructing an engineering stress–strain curve is to take the load-elongation curve (Fig. 2.5a) and divide the load (vertical axis) by the original cross-sectional area A_o, and the elongation (horizontal axis) by the original gage length l_o. Since these two quantities are divided by constants, the engineering stress–strain curve obtained and shown in Fig. 2.5(b) has the same shape as the load-elongation curve. (In this example, $A_o = 0.056\ \text{in}^2$ and $A_f = 0.016\ \text{in}^2$.)

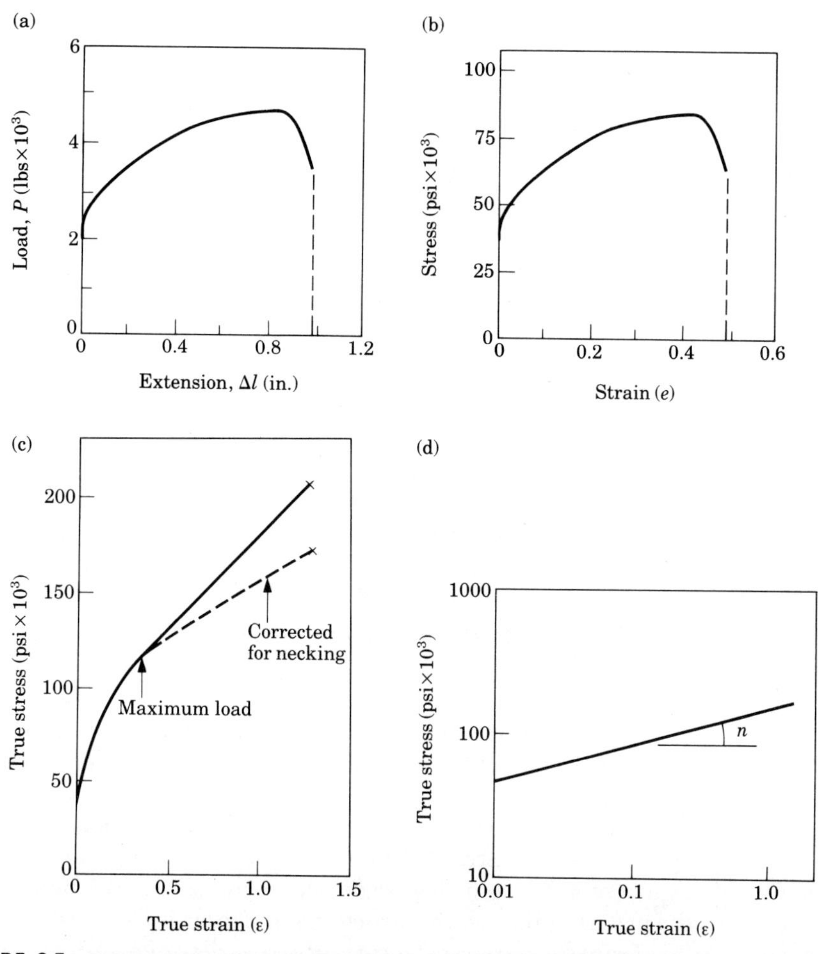

FIGURE 2.5

(a) Load-elongation curve in tension testing. (b) Engineering stress–engineering strain curve obtained from data in Fig. 2.5(a). (c) True stress–true strain curve developed from the data in Fig. 2.5(b). Note that this curve has a positive slope, indicating that the material is becoming stronger as it is strained, which is known as strain hardening or work hardening. (d) True stress–true strain curve plotted on log–log paper and based on the corrected curve in Fig. 2.5(c). The steeper the slope n (strain-hardening exponent), the higher will be the capacity of the material to become stronger and harder as it is worked. The specimen in this figure is stainless steel.

True stress–true strain curves are obtained similarly by dividing the load by the instantaneous cross-sectional area, and the true strain is obtained from Eq. (2.7). The result is shown in Fig. 2.5(c). Note the *correction* to the curve, reflecting the fact that the specimen's necked region is subjected to three-dimensional tensile stresses. This state gives higher stress values than the actual true stress, hence the curve must be corrected downward.

TABLE 2.3
TYPICAL VALUES FOR K AND n AT ROOM TEMPERATURE

	K (MPa)	n
Aluminum		
1100-O	180	0.20
2024-T4	690	0.16
6061-O	205	0.20
6061-T6	410	0.05
7075-O	400	0.17
Brass		
70-30, annealed	900	0.49
85-15, cold-rolled	580	0.34
Cobalt-base alloy, heat-treated	2070	0.50
Copper, annealed	315	0.54
Steel		
Low-C annealed	530	0.26
4135 annealed	1015	0.17
4135 cold-rolled	1100	0.14
4340 annealed	640	0.15
304 stainless, annealed	1275	0.45
410 stainless, annealed	960	0.10

We can represent the true stress–true strain curve in Fig. 2.5(c) by the equation

$$\sigma = K\varepsilon^n, \tag{2.8}$$

where K is known as the **strength coefficient**, and n is the **strain-hardening**, or *work-hardening*, **exponent.** The values for K and n for several metals are given in Table 2.3.

If we plot the corrected curve in Fig. 2.5(c) on a log–log paper, we find that it is approximately a straight line (Fig. 2.5d). The slope of the curve is equal to the exponent n. Thus the higher the slope, the greater the strain-hardening capacity of the material; that is, the stronger and harder it becomes as it is strained.

True stress–true strain curves for a variety of metals are given in Fig. 2.6. (Some differences between Table 2.3 and Fig. 2.6 will exist because of the different sources of data and test conditions.) Note that the elastic regions have been deleted because the slope in this region is very high. Thus the point of intersection of each curve at the vertical axis in this figure is the yield stress Y of the material. The area under the true stress–true strain curve is known as the material's **toughness,** that is, the amount of energy per unit volume that the material dissipates prior to fracture.

2.2.5 Strain at necking in a tension test

As we have noted, the onset of necking of the specimen in a tension test corresponds to the ultimate tensile strength of the material. Note that the slope of the load-elongation curve at this point is zero, and it is here that the specimen begins to neck. The specimen cannot support the load, because the cross-sectional area of the neck is becoming smaller at a rate that is higher than the rate at which the material strain hardens.

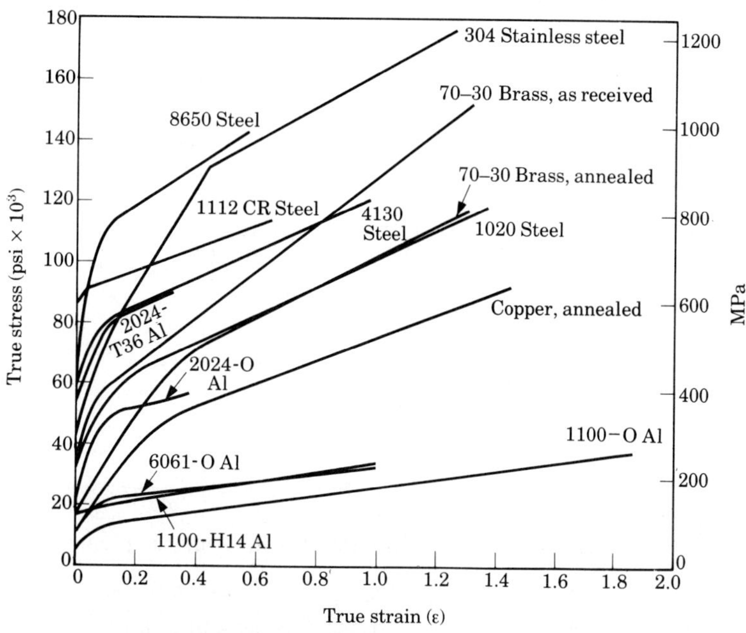

FIGURE 2.6 True stress–true strain curves in tension at room temperature for various metals. The curves start at a finite level of stress because the elastic regions have too steep a slope to be shown in this figure. Thus each curve starts at the yield stress, *Y*, of the material.

The true strain at the onset of necking is numerically equal to the strain-hardening exponent (n) of the material. Thus the higher the value of n, the greater the strain to which a piece of material can be stretched *uniformly* before it begins to neck. This is an important observation, particularly in regard to forming operations that involve stretching of the workpiece material. You can see in Table 2.3 that annealed copper, brass, and stainless steel have high n values, and thus they can be stretched uniformly to a greater extent than the other materials listed.

• Example: Calculation of ultimate tensile strength.

A material has a true stress–true strain curve given by

$$\sigma = 100{,}000\varepsilon^{0.5} \text{ psi}.$$

Calculate the true ultimate tensile strength and the engineering UTS of this material.

SOLUTION. Since the necking strain corresponds to the maximum load and the necking strain for this material is

$$\varepsilon = n = 0.5,$$

we have, as the *true* ultimate tensile strength

$$\begin{aligned}\sigma &= Kn^n \\ &= 100{,}000(0.5)^{0.5} \\ &= 70{,}710 \text{ psi.}\end{aligned}$$

The true area at the onset of necking is obtained from

$$\ln\left(\frac{A_o}{A_{\text{neck}}}\right) = n = 0.5.$$

Thus,

$$A_{\text{neck}} = A_o e^{-0.5}$$

and the maximum load P is

$$\begin{aligned}P &= \sigma A \\ &= \sigma A_o e^{-0.5},\end{aligned}$$

where σ is the true ultimate tensile strength. Hence

$$\begin{aligned}P &= (70{,}710)(0.606)(A_o) \\ &= 42{,}850 A_o \text{ lb.}\end{aligned}$$

Since UTS $= P/A_o$,

$$\text{UTS} = 42{,}850 \text{ psi.}$$

2.2.6 Temperature effects

Increasing temperature generally has the following effects on stress–strain curves (Fig. 2.7): It raises ductility and toughness, and lowers the yield stress and modulus of elasticity (Fig. 2.8). Temperature also affects the strain-hardening exponent of most metals, in that n decreases with increasing temperature. The influence of temperature, however, is best described in conjunction with the rate of deformation.

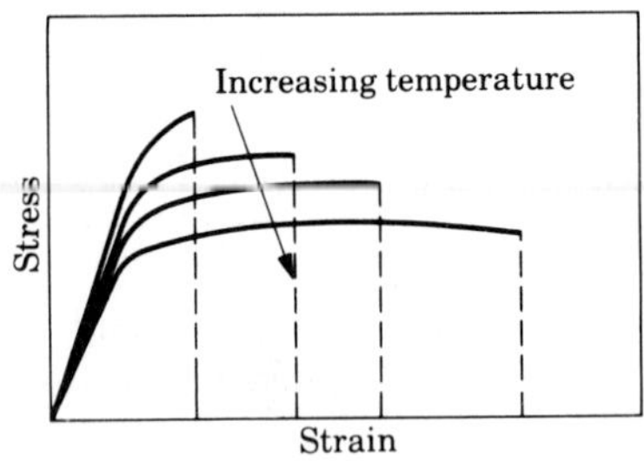

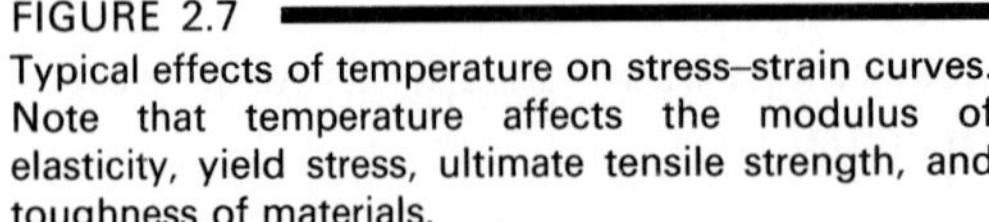

FIGURE 2.7 Typical effects of temperature on stress–strain curves. Note that temperature affects the modulus of elasticity, yield stress, ultimate tensile strength, and toughness of materials.

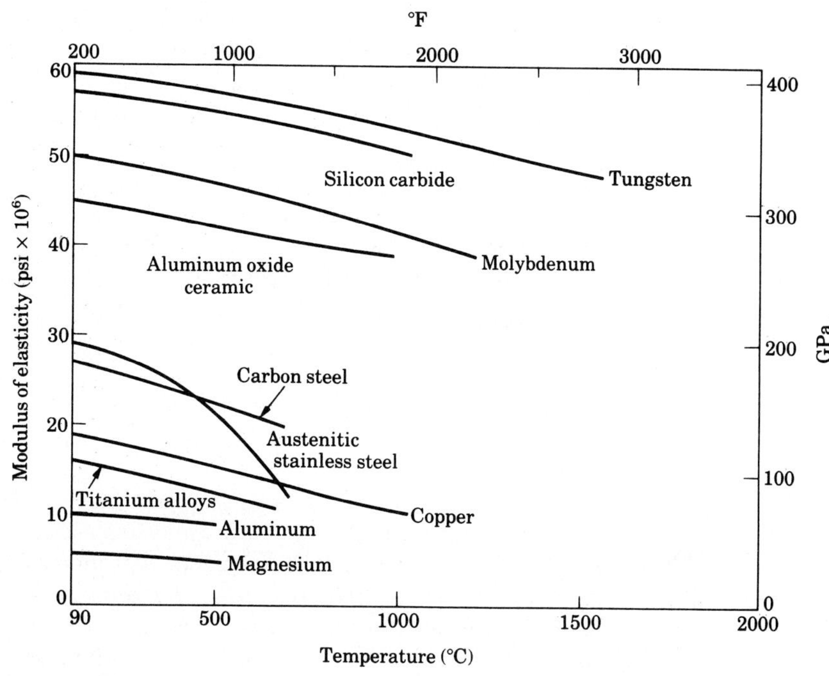

FIGURE 2.8 The effect of temperature on the modulus of elasticity for various materials.

2.2.7 Rate of deformation effects

Just as we can blow up a balloon or stretch a rubber band at different rates, we can shape a piece of material in a manufacturing process at different speeds. Some machines, such as hydraulic presses, form materials at low speeds; others, such as mechanical presses, form at high speeds. Thus the specimen can be strained at different rates to simulate the actual manufacturing processes.

Deformation rate is defined as the speed at which a tension test is being carried out, in units of, say, m/s or ft/min. The **strain rate**, on the other hand, is a function of the specimen length. A short specimen elongates proportionately more during the same time period than does a long specimen. For example, let's take two specimens with 20 mm and 100 mm gage lengths, respectively, and elongate them both by 10 mm within a period of 1 s. The engineering strain in the shorter specimen is $\frac{10}{20} = 0.5$, and in the longer specimen it is $\frac{10}{100} = 0.1$. Thus the strain rates are 0.5/s and 0.1/s, respectively, with the short specimen being subjected to a strain rate five times greater than that for the long specimen, although they are both being stretched at the same deformation rate.

Deformation rates typically employed in various testing and metalworking processes, and the true strains involved, are given in Table 2.4. Note the considerable difference in magnitudes. Because of this wide range, strain rates are usually stated in terms of orders of magnitude, such as 10^2 s, 10^4 s, and so on.

TABLE 2.4
TYPICAL RANGES OF STRAIN AND DEFORMATION RATE IN MANUFACTURING PROCESSES

PROCESS	TRUE STRAIN	DEFORMATION RATE (m/s)
Cold working		
Forging, rolling	0.1–0.5	0.1–100
Wire and tube drawing	0.05–0.5	0.1–100
Explosive forming	0.05–0.2	10–100
Hot working and warm working		
Forging, rolling	0.1–0.5	0.1–30
Extrusion	2–5	0.1–1
Machining	1–10	0.1–100
Sheet-metal forming	0.1–0.5	0.05–2
Superplastic forming	0.2–3	10^{-4}–10^{-2}

The typical effects that temperature and strain rate jointly have on the strength of metals is shown in Fig. 2.9. You can see that increasing the strain rate increases the strength of the material (*strain-rate hardening*). The slope of these curves is called the *strain-rate sensitivity exponent*, m. The value of m is obtained from log–log plots, provided that the vertical and horizontal scales are the same (unlike those in Fig. 2.9). Thus a slope of 45° would indicate a value of $m = 1$. The relationship is given by the equation

$$\sigma = C\dot{\varepsilon}^m, \tag{2.9}$$

where C is the strength coefficient, with the units of stress, and $\dot{\varepsilon}$ is the true strain rate, defined as the true strain that the material undergoes per unit time.

Note in Fig. 2.9 that m increases with increasing temperature. Thus the sensitivity of the strength to strain rate increases with temperature. Note, however, that the slope is relatively flat at room temperature. This condition is generally true for most metals, except those such as lead and tin, that recrystallize at room temperature. Some typical ranges of m for metals are:

Cold working—up to 0.05.

Hot working—0.05 to 0.4.

Superplastic materials—0.3 to 0.85.

The magnitude of the strain-rate sensitivity m significantly influences necking in a tension test. With an increasing m, the material stretches farther before it fails. Thus an increasing m delays necking. Ductility enhancement caused by the high strain-rate sensitivity of materials has been exploited in superplastic forming of sheet metal.

Superplasticity. The term **superplasticity** refers to the capability of some materials to undergo large uniform elongation prior to necking and fracture in tension. The elongation may be on the order of a few hundred percent to as much as

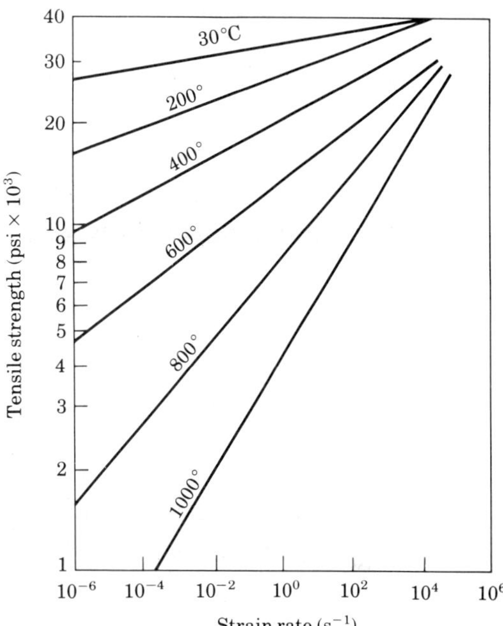

FIGURE 2.9 The effect of strain rate on the ultimate tensile strength for aluminum. Note that as temperature increases, the slopes of the curves increase. Thus strength becomes more and more sensitive to strain rate as temperature increases. *Source:* J. H. Hollomon.

2000 percent. Among metals exhibiting superplastic behavior are very fine-grain alloys of zinc-aluminum and titanium alloys. Other common examples of superplastic behavior are glass and thermoplastics, which when heated and stretched, elongate many times their original length. This is why glass and thermoplastics can be formed successfully into complex shapes, such as beverage bottles and lighted advertising signs. An even more common example of superplastic behavior is bubble gum. Note in Table 2.4 that superplastic behavior is exhibited at very low deformation rates.

2.2.8 Hydrostatic pressure effects

Various tests have been performed under hydrostatic pressure, such as in a small pressurized chamber with pressures up to 3.5 GPa (500 ksi), to determine the effect of hydrostatic pressure on mechanical properties of materials. Test results indicate that increasing the hydrostatic pressure increases the strain at fracture substantially, both for ductile as well as brittle materials. The beneficial effect of hydrostatic pressure has been exploited in metalworking processes, particularly in hydrostatic extrusion. Cupping your hands when making a snowball is a simple demonstration of the effectiveness of hydrostatic pressure.

2.2.9 Radiation effects

In view of the use of many metals and alloys in nuclear applications, studies have also been conducted on the effects of radiation on mechanical properties. Typical changes in the properties of steels and other metals exposed to high-energy radiation are increased yield stress, tensile strength, and hardness, and decreased ductility and toughness. Radiation has similar detrimental effects on the behavior of plastics.

2.3 Compression

Many operations in manufacturing, particularly processes such as forging, rolling, and extrusion, are performed with the workpiece under **compression**. The compression test, in which the specimen is subjected to a compressive load (Fig. 2.10), can give information useful for these processes.

The compression test is usually carried out by compressing a solid cylindrical specimen between two flat platens. However, the deformation of the specimen shown in Fig. 2.10 is ideal. Because of friction between the specimen and the platens, the specimen's cylindrical surface bulges, which we call *barreling* (Fig. 2.11).

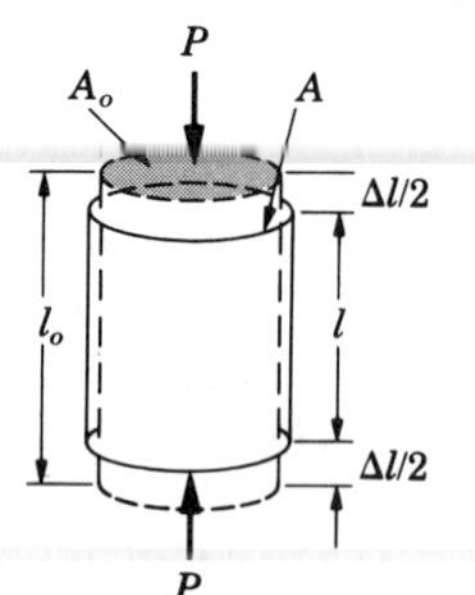

FIGURE 2.10 Idealized compression of a solid cylindrical specimen and the uniform increase in cross-sectional area.

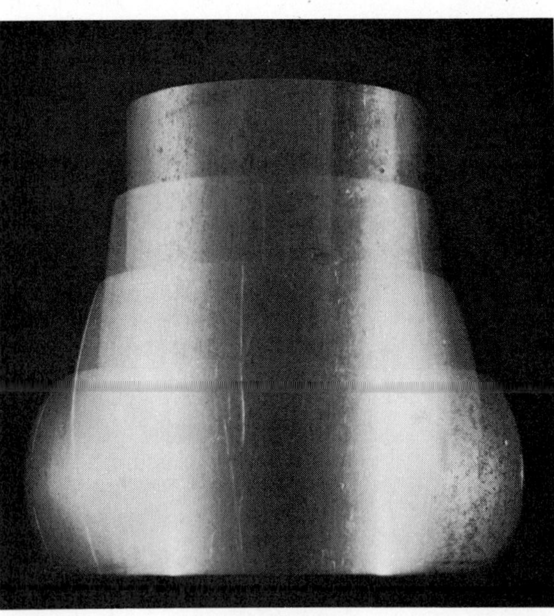

FIGURE 2.11 Barreling of a solid cylindrical aluminum specimen as it is being compressed between flat dies. Note the increase in barreling as the height decreases. *Source:* K. M. Kulkarni and S. Kalpakjian.

Friction prevents the top and bottom surfaces from expanding freely. Since the cross-sectional area of the specimen now changes along its height, being maximum at the center, obtaining stress–strain curves in compression is difficult. Furthermore, since friction dissipates energy, the compressive force is higher than it otherwise would be in order to supply the work required to overcome friction.

With effective lubrication, friction can be minimized and a reasonably constant cross-sectional area can be maintained during the test. When the results of compression tests and tension tests on ductile metals are compared, the true stress–true strain curves for both tests coincide. This, however, is not true for brittle materials, which are generally stronger and more ductile in compression than in tension.

When a metal with a certain tensile yield stress is subjected to tension into the plastic range and then the load is released and applied in compression, the yield stress in compression is lower than that in tension. This phenomenon is known as the **Bauschinger effect** (after J. Bauschinger, who reported it in 1881), and is exhibited by all metals and alloys to varying degrees. Because of the lowered yield stress in the reverse direction of load application, this phenomenon is also called *strain*, or *work*, *softening*.

For brittle materials such as ceramics and glasses, a **disk test** has been developed in which the disk is subjected to compression between two hardened flat platens (Fig. 2.12). When loaded as shown, tensile stresses develop perpendicular to the vertical centerline along the disk, fracture begins, and the disk splits vertically in half.

The tensile stress σ in the disk is uniform along the centerline and can be calculated from the formula

$$\sigma = \frac{2P}{\pi d t}, \tag{2.10}$$

where P is the load at fracture, d is the diameter of the disk, and t is its thickness. In order to avoid premature failure at the contact points, thin strips of soft metal are placed between the disk and the platens. These also protect the platens from being damaged during the test.

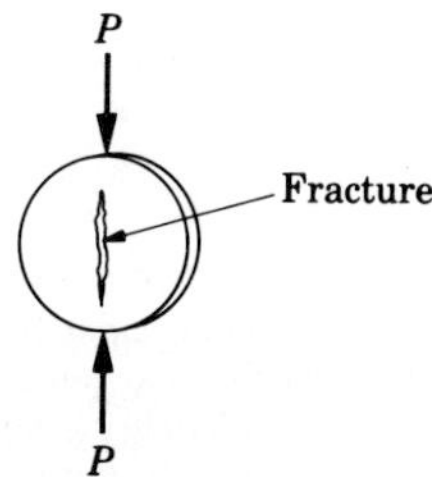

FIGURE 2.12 Disk test on a brittle material, showing the direction of loading and the fracture path. This test is useful for brittle materials, such as ceramics and carbides.

2.4 Torsion

In addition to tension and compression, a workpiece may be subjected to **shear** strains (Fig. 2.13), such as in the punching of holes in sheet metals. The test method generally used for determination of properties of materials in shear is the **torsion test.** In order to obtain an approximately uniform stress and strain distribution along the cross-section, this test is usually performed on a thin tubular specimen.

The torsion specimen usually has a reduced cross-section to confine the deformation, just like the reduced gage-length section of a tensile-test specimen. We calculate the shear stress from the formula

$$\text{Shear stress, } \tau = \frac{T}{2\pi r^2 t}, \tag{2.11}$$

where T is the torque, r is the average radius of the tube, and t is the thickness of the tube.

We calculate the shear strain from the expression

$$\text{Shear strain, } \gamma = \frac{r\phi}{l}, \tag{2.12}$$

where l is the length of tube subjected to torsion and ϕ is the *angle of twist* in radians.

The ratio of the shear stress to shear strain in the elastic range is known as the **shear modulus,** or **modulus of rigidity,** G, a quantity related to the modulus of elasticity, E.

The angle of twist ϕ to fracture in the torsion of solid round bars at elevated temperatures is useful in estimating the forgeability of metals. The greater the number of twists prior to failure, the greater is the forgeability of the metal.

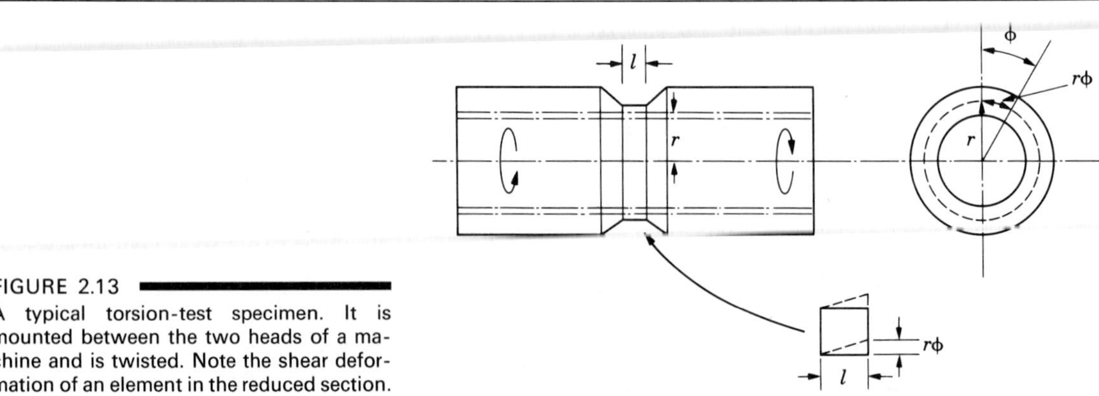

FIGURE 2.13 A typical torsion-test specimen. It is mounted between the two heads of a machine and is twisted. Note the shear deformation of an element in the reduced section.

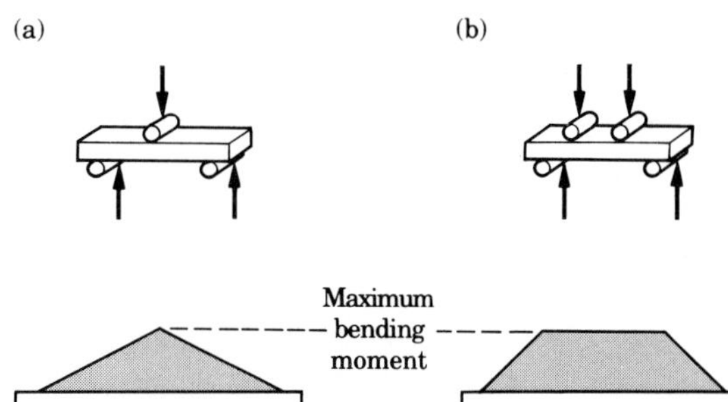

FIGURE 2.14 Two bend-test methods for brittle materials: (a) three-point bending; (b) four-point bending. The shaded areas on the beams represent the bending-moment diagrams, described in texts on mechanics of solids. Note the region of constant maximum bending moment in (b), whereas the maximum bending moment occurs only at the center of the specimen in (a).

2.5 Bending (Flexure)

Preparing specimens from brittle materials, such as ceramics and carbides, is difficult because of the problems involved in shaping and machining them to proper dimensions. Furthermore, because of their sensitivity to surface defects and notches, clamping brittle test specimens for testing is difficult. Improper alignment of the test specimen may result in nonuniform stress distribution along the cross-section of the specimen.

A commonly used test method for brittle materials is the *bend* or *flexure test*, usually involving a specimen with a rectangular cross-section and supported at its ends (Fig. 2.14). The load is applied vertically—either at one or two points; hence these tests are referred to as three-point or four-point bending, respectively. The stresses in these specimens are tensile at their lower surfaces and compressive at their upper surfaces. We calculate these stresses using simple beam equations described in texts on mechanics of solids. The stress at fracture in bending is known as the **modulus of rupture**, or *transverse rupture strength*.

2.6 Hardness

Hardness is a commonly used quantity and gives a general indication of the strength of the material, as well as its resistance to wear and scratching. More specifically, **hardness** is usually defined as *resistance to permanent indentation*. Thus, for example, steel is harder than aluminum, and aluminum is harder than lead. Hardness, however, is not a fundamental property because the resistance to indentation depends on the shape of the indenter and the load applied.

2.6.1 Hardness tests

Several methods have been developed to measure the hardness of materials, using different indenter materials and shapes. Commonly used hardness tests are described below.

Brinell test. Introduced by J. A. Brinell in 1900, this test involves pressing a steel or tungsten carbide ball 10 mm (0.4 in.) in diameter against a surface, with a load of 500 kg, 1500 kg, or 3000 kg (Fig. 2.15). The Brinell hardness number (HB, formerly BHN) is defined as the ratio of the load P to the curved surface area of the indentation.

Depending on the condition of the material, two types of impressions are obtained on the surface after performing a Brinell test, or any of the other hardness tests described below. The impressions in annealed metals generally have a rounded profile, whereas cold-worked metals have a sharp profile (Fig. 2.16). The correct method of measuring the indentation diameter d is shown in the figure.

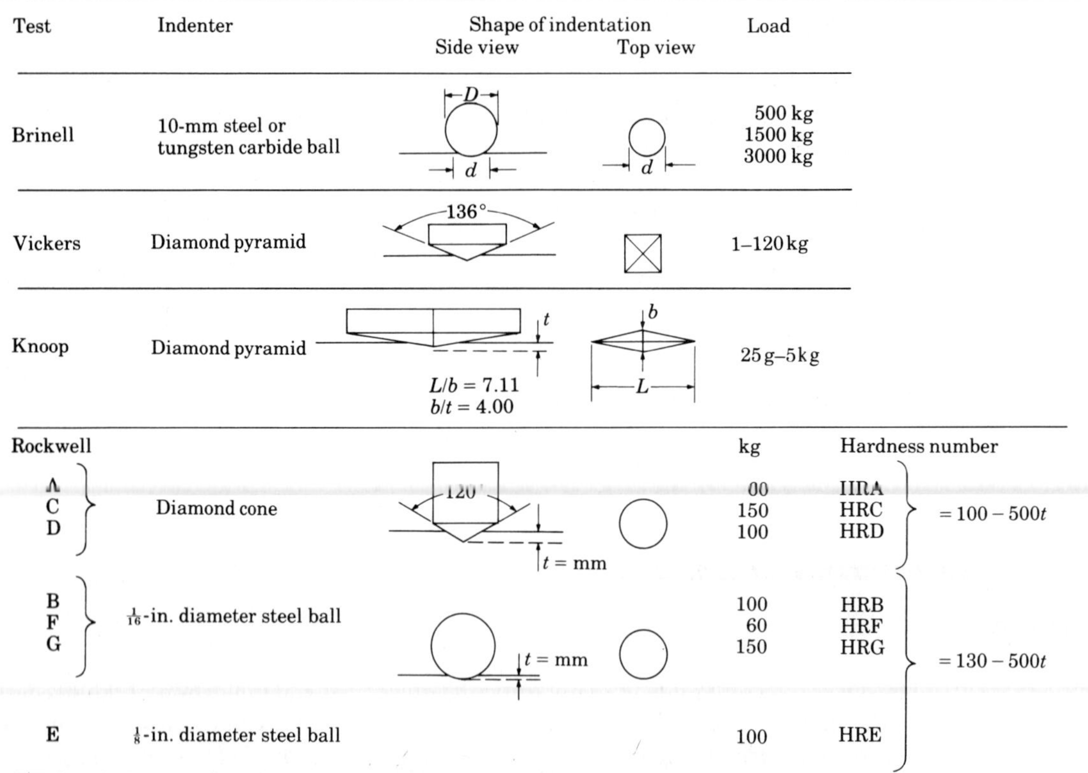

Test	Indenter	Shape of indentation: Side view	Shape of indentation: Top view	Load	Hardness number	
Brinell	10-mm steel or tungsten carbide ball	D, d	d	500 kg 1500 kg 3000 kg		
Vickers	Diamond pyramid	136°		1–120 kg		
Knoop	Diamond pyramid	t $L/b = 7.11$ $b/t = 4.00$	b, L	25 g–5 kg		
Rockwell				kg	Hardness number	
A, C, D	Diamond cone	120° t = mm		60 150 100	HRA HRC HRD	$= 100 - 500t$
B, F, G	$\frac{1}{16}$-in. diameter steel ball	t = mm		100 60 150	HRB HRF HRG	$= 130 - 500t$
E	$\frac{1}{8}$-in. diameter steel ball			100	HRE	

FIGURE 2.15 General characteristics of hardness testing methods. *Source:* H. W. Hayden, et al., *The Structure and Properties of Materials,* Vol. III, John Wiley & Sons, 1965.

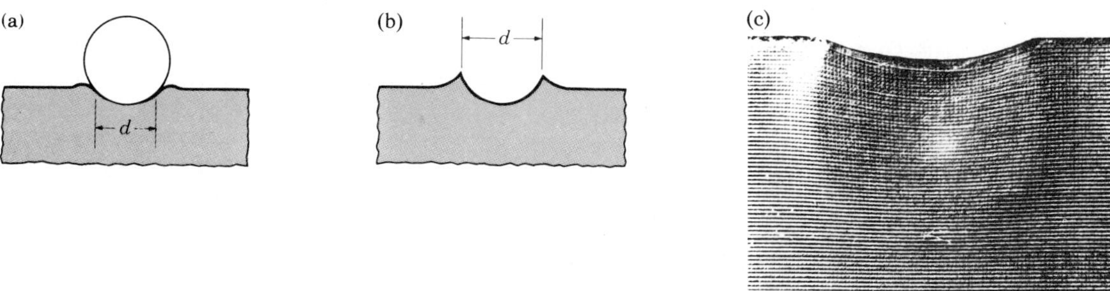

FIGURE 2.16
Indentation geometry in Brinell testing: (a) annealed metal; (b) work-hardened metal; (c) deformation of mild steel under a spherical indenter. Note that the depth of the permanently deformed zone is about one order of magnitude larger than the depth of indentation. For a hardness test to be valid, this zone should be fully developed in the material. *Source*: M. C. Shaw and C. T. Yang.

The indenter, which has a finite elastic modulus, also undergoes elastic deformation under the applied load. Thus hardness measurements may not be as accurate as expected. One method to minimize this effect is to use tungsten carbide balls which, because of their high modulus of elasticity, distort less than steel balls do. Carbide balls are usually recommended for Brinell hardness numbers greater than 500. The harder the material to be tested, the smaller will be the impression, so a 1500 kg or 3000 kg load is usually recommended in order to obtain impressions sufficiently large for accurate measurement.

Rockwell test. Developed by S. P. Rockwell in 1922, this test measures the *depth* of penetration instead of the diameter of the indentation, as the Brinell test does. The indenter is pressed on the surface, first with a minor load and then with a major load. The difference in the depths of penetration is a measure of the hardness of the material (Fig. 2.17a).

(a)

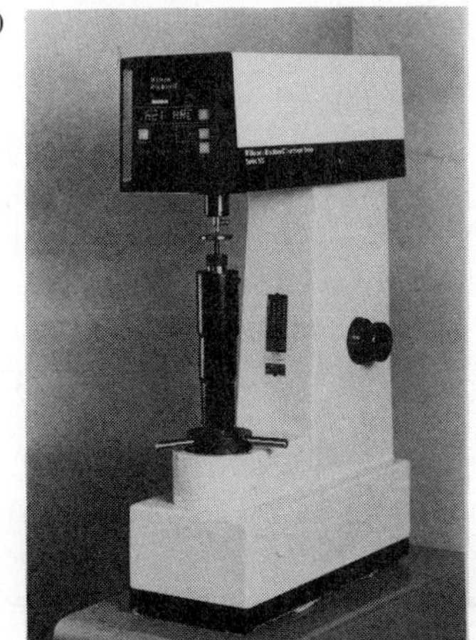

(b)

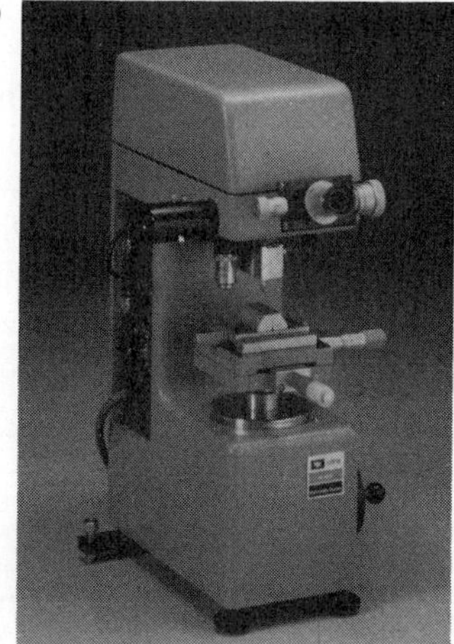

FIGURE 2.17
(a) A Rockwell hardness tester with digital display. This test indicates a hardness of 62.1 HRC. *Source:* Courtesy of Page-Wilson Corporation. (b) A microhardness tester used for Vickers and Knoop tests. *Source:* Courtesy of LECO Corporation.

Some of the more common Rockwell hardness scales and the indenters used are shown in Fig. 2.15. Rockwell superficial hardness tests have also been developed using the same type of indenters but lighter loads.

Vickers test. The Vickers hardness test, developed in 1922 and also known as the *diamond pyramid hardness* test, uses a pyramid shaped diamond indenter (Fig. 2.15) and a load that ranges from 1 kg to 120 kg. The Vickers hardness number is indicated by HV (formerly, DPH). The impressions obtained are typically less than 0.5 mm (0.020 in.) on the diagonal. The Vickers test gives essentially the same hardness number regardless of the load and is suitable for testing materials with a wide range of hardness, including heat-treated steels.

Knoop test. The Knoop test (developed by F. Knoop in 1939) uses a diamond indenter in the shape of an elongated pyramid (Fig. 2.15), with loads ranging generally from 25 g to 5 kg. The Knoop hardness number is indicated by HK (formerly, KHN). Because of the light loads that are applied, it is a **microhardness** test (Fig. 2.17b). Hence it is suitable for very small or thin specimens and for brittle materials, such as carbides, ceramics, and glass. This test is also used for measuring the hardness of the individual grains and components in a metal alloy. The size of the indentation is generally in the range of 0.01 mm to 0.10 mm (0.0004 in. to 0.004 in.), and thus surface preparation is very important. Because the hardness number obtained depends on the applied load, Knoop test results should always cite the load used.

Scleroscope. The *scleroscope* is an instrument in which a diamond-tipped indenter (hammer) enclosed in a glass tube is dropped on the specimen from a certain height. The hardness is related to the rebound of the indenter. The higher the rebound, the harder the material is. The impression made by a scleroscope is very small. The instrument is portable and is simply placed on the surface of the part. Thus it is useful for measuring the hardness of large objects that otherwise would not fit into the limited space of conventional hardness testers.

Mohs hardness. Developed in 1822 by F. Mohs, this test is based on the capability of one material to scratch another. The Mohs hardness is based on a scale of 1 to 10, with 1 for talc and 10 for diamond, the hardest substance known. Thus a material with a higher Mohs hardness number can scratch one with a lower number. Although the Mohs scale is qualitative, it correlates well with Knoop hardness. Soft metals have a Mohs scale of 2 to 3, hardened steels about 6, and aluminum oxide, used as cutting tools and as abrasives in grinding wheels, a scale of 9.

Durometer. The hardness of rubbers, plastics, and similar soft and elastic materials is generally measured with an instrument called a *durometer* (from the Latin *durus*, meaning hard). This is an empirical test in which an indenter is pressed against the surface, with a constant load applied rapidly. The depth of penetration is measured after one second, and the hardness is inversely related to the penetration. There are two different scales for this test. Type A has a blunt indenter and a load of 1 kg, and is used for softer materials. Type D has a sharper indenter and a load of 5 kg, and is used for harder materials. The hardness numbers in these tests range from 0 to 100.

Hot hardness. The hardness of materials at elevated temperatures is important in applications where higher temperatures are involved, such as cutting tools in machining and dies in hot-working and casting operations. Hardness tests can be performed at elevated temperatures with conventional testers with some modifications, such as enclosing the specimen and indenter in a small electric furnace.

2.6.2 Hardness and strength

Since hardness is the resistance to permanent indentation, we can liken it to performing a compression test on a small volume in a block of material (Fig. 2.16c). Studies have shown that, using the same units, the hardness of a cold-worked metal is about three times its yield stress Y, and for annealed metals, it is about five times Y.

A relationship has been established between the ultimate tensile strength (UTS) and Brinell hardness (HB) for steels. In SI units, the relationship is

$$\mathrm{UTS} = 3.5(\mathrm{HB}), \tag{2.13}$$

where UTS is in MPa. In traditional units,

$$\mathrm{UTS} = 500(\mathrm{HB}), \tag{2.14}$$

where UTS is in psi and HB is in kg/mm^2, as measured for a load of 3000 kg.

2.6.3 Hardness testing procedures

For a hardness test to be meaningful and reliable, the zone of deformation under the indenter (see Fig. 2.16c) must be allowed to develop freely. Consequently, the location of the indenter with respect to the edges of the specimen to be tested, and the thickness of the specimen, are important considerations. Generally, the location should be at least two diameters of the indenter from the edge of the specimen, and the thickness of the specimen should be at least 10 times the depth of penetration of the indenter. Successive indentations on the same surface of the workpiece should be far enough apart so as not to influence each other.

Moreover, the indentation should be sufficiently large to give a representative hardness value for the bulk material. If hardness variations are to be detected in a small area, or if the hardness of individual constituents in a matrix or an alloy is to be determined, the indentations should be very small, such as those in Vickers or Knoop tests using light loads.

While surface preparation is not critical for the Brinell test, it is somewhat more important for the Rockwell test and even more important for the other hardness

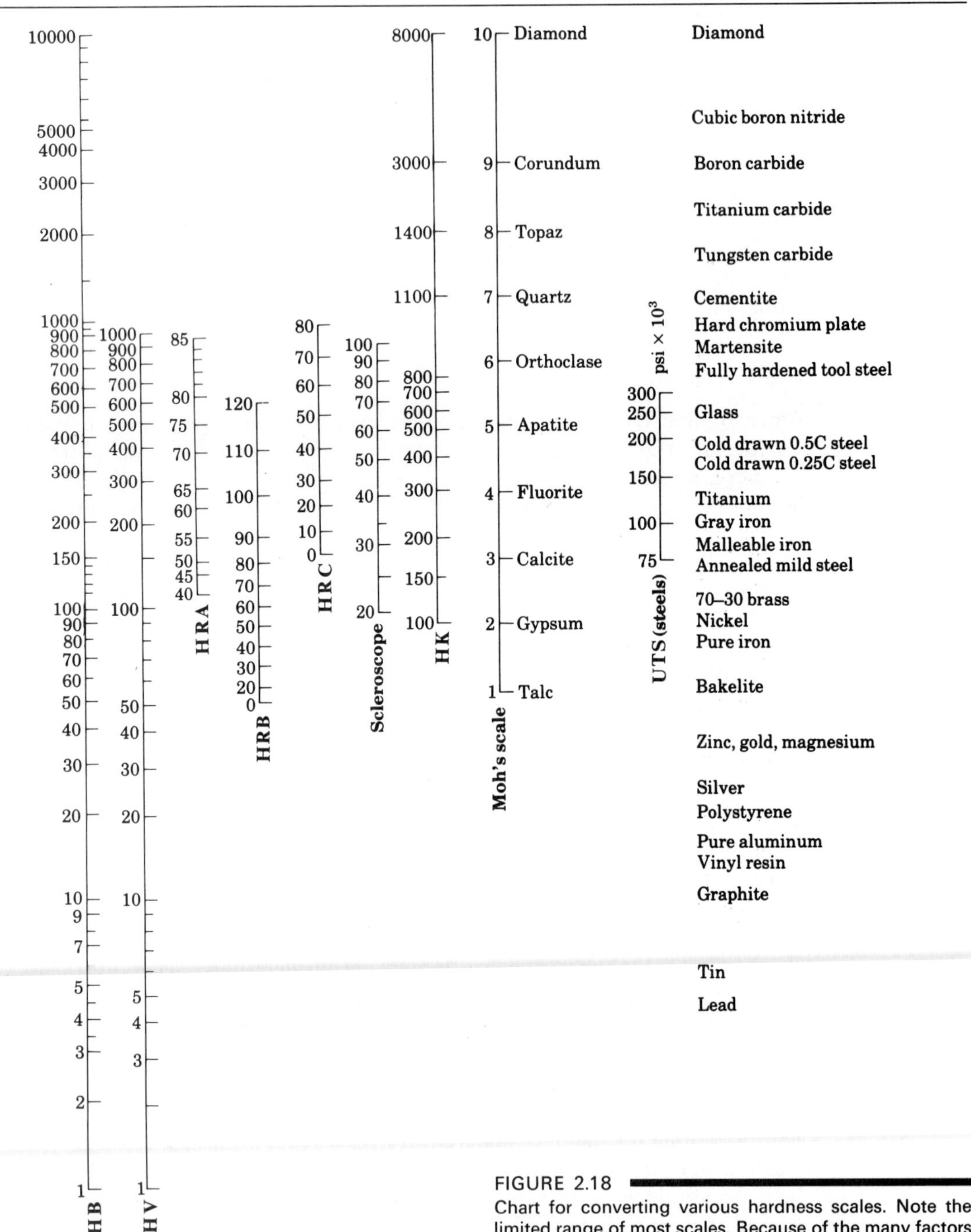

FIGURE 2.18 Chart for converting various hardness scales. Note the limited range of most scales. Because of the many factors involved, hardness conversions are approximate.

tests because of the small size of indentations. Surfaces may have to be polished to allow correct measurement of the impression's dimensions.

We can interrelate and convert the values obtained from different hardness tests to different scales using Fig. 2.18 (see p. 84). Care should be exercised in using these charts because of the many variables in material characteristics and shape of indentation.

2.7 Fatigue

Various structures and components in manufacturing operations, such as tools, dies, gears, cams, shafts, and springs, are subjected to rapidly fluctuating (cyclic or periodic) loads, as well as static loads. Cyclic stresses may be caused by fluctuating mechanical loads, such as gear teeth, or by thermal stresses, such as a cool die coming into repeated contact with hot workpieces. Under these conditions, the part fails at a stress level below which failure would occur under static loading. This phenomenon is known as *fatigue failure*, and is responsible for the majority of failures in mechanical components.

Fatigue test methods involve testing specimens under various states of stress, usually in a combination of tension and compression, or torsion. The test is carried out at various stress amplitudes (S), and the number of cycles (N) to cause total failure of the specimen or part is recorded. Stress amplitude is the maximum stress, in tension and compression, to which the specimen is subjected.

A typical plot, known as *S–N curves*, is shown in Fig. 2.19. These curves are based on complete reversal of the stress, that is, maximum tension, maximum

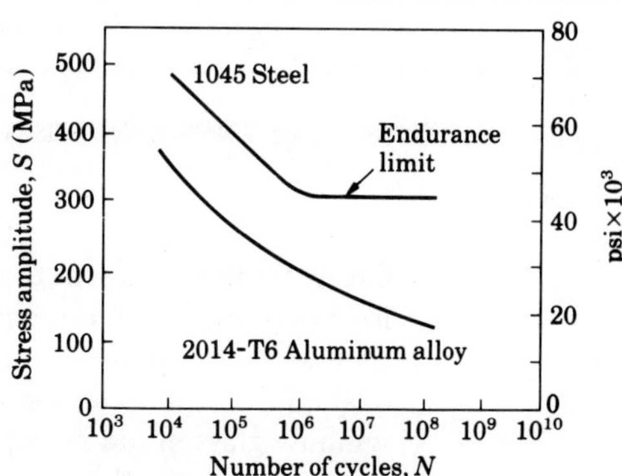

FIGURE 2.19 Typical *S–N* curves for two metals. Note that, unlike steel, aluminum does not have an endurance limit.

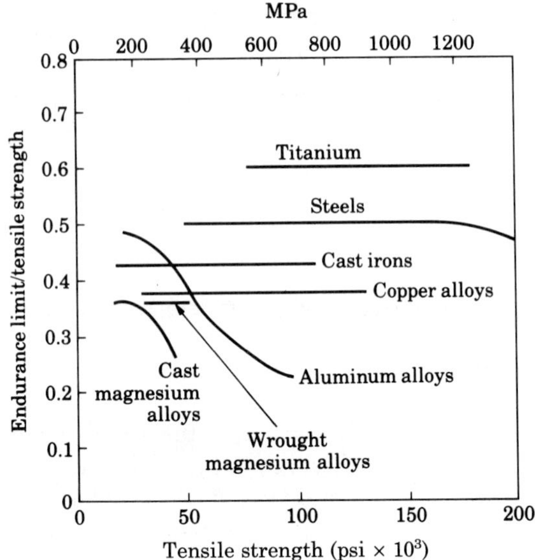

FIGURE 2.20 Ratio of endurance limit to tensile strength for various metals, as a function of tensile strength. Fatigue strength of a metal can thus be estimated from its tensile strength.

compression, maximum tension, and so on, such as that obtained by bending an eraser or piece of wire alternately in one direction, then the other. The test can also be performed on a rotating shaft with a constant downward load. The maximum stress to which the material can be subjected without fatigue failure, regardless of the number of cycles, is known as the *endurance limit* or *fatigue limit.*

The endurance limit for metals is related to their ultimate tensile strength (Fig. 2.20). For steels, the endurance limit is about one-half their tensile strength. Although many metals, especially steels, have a definite endurance limit, aluminum alloys do not have one, and the *S–N* curve continues its downward trend. For metals exhibiting such behavior, the fatigue strength is specified at a certain number of cycles, such as 10^7. In this way the useful service life of the component can be specified.

2.8 Creep

Creep is the permanent elongation of a component under a static load maintained for a period of time. It is a phenomenon of metals and certain nonmetallic materials, such as thermoplastics and rubbers, and can occur at any temperature. Lead, for example, creeps under a constant tensile load at room temperature. The thickness of window glass in old houses has been found to be greater at the bottom than at the top of windows, the glass having undergone creep by its own weight over many years.

For metals and their alloys, creep of any significance occurs at elevated temperatures, beginning at about 200 °C (400 °F) for aluminum alloys, and up to about 1500 °C (2800 °F) for refractory alloys. The mechanism of creep at elevated temperature in metals is generally attributed to grain-boundary sliding. Creep is especially important in high-temperature applications, such as gas-turbine blades and similar components in jet engines and rocket motors. High-pressure steam lines and nuclear-fuel elements also are subject to creep. Creep deformation also can occur in tools and dies that are subjected to high stresses at elevated temperatures during hot-working operations, such as forging and extrusion.

A creep test typically consists of subjecting a specimen to a constant tensile load, hence constant engineering stress, at a certain temperature and measuring the change in length over a period of time. A typical creep curve (Fig. 2.21) usually consists of primary, secondary, and tertiary stages. The specimen eventually fails by necking and fracture, as in a tension test, which is called **rupture** or *creep rupture*. As expected, the creep rate increases with temperature and the applied load.

Design against creep usually involves a knowledge of the secondary (linear) range and its slope, since the creep rate can be determined reliably when the curve has a constant slope. Generally, resistance to creep increases with the melting temperature of a material, which serves as a general guideline for design purposes. Stainless steels, superalloys, and refractory metals and alloys are commonly used in applications where creep resistance is required.

Stress relaxation. Stress relaxation is closely related to creep. In stress relaxation the stresses resulting from external loading of a structural component decrease in magnitude over a period of time, even though the dimensions of the component remain constant. Examples are rivets, bolts, guy wires, and similar parts under tension, compression, or bending. This phenomenon is particularly common and important in thermoplastics.

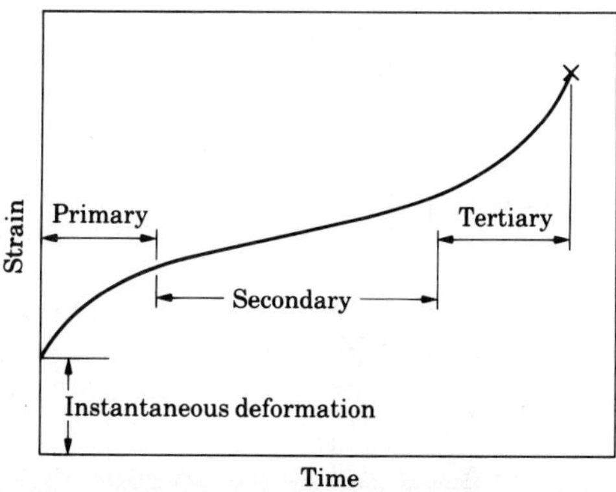

FIGURE 2.21 Schematic illustration of a typical creep curve. The linear segment of the curve (constant slope) is useful in designing components for a specific creep life.

2.9 Impact

In many manufacturing operations, as well as during the service life of components, materials are subjected to **impact** (or *dynamic*) **loading**, as in high-speed metalworking operations such as drop forging. Tests have been developed to determine the impact toughness of metallic and nonmetallic materials under impact loading.

A typical impact test consists of placing a notched specimen in an impact tester and breaking it with a swinging pendulum. In the Charpy test the specimen is supported at both ends (Fig. 2.22a), whereas in the Izod test it is supported at one end like a cantilever beam (Fig. 2.22b). From the amount of swing of the pendulum, the energy dissipated in breaking the specimen is obtained (Fig. 2.22c). This energy is the *impact toughness* of the material.

Impact tests are particularly useful in determining the ductile–brittle transition temperature of materials (see Section 2.10.1). Materials that have impact resistance are generally those that have high strength and high ductility, hence high toughness. Sensitivity to surface defects (*notch sensitivity*) is important, as it lowers impact toughness.

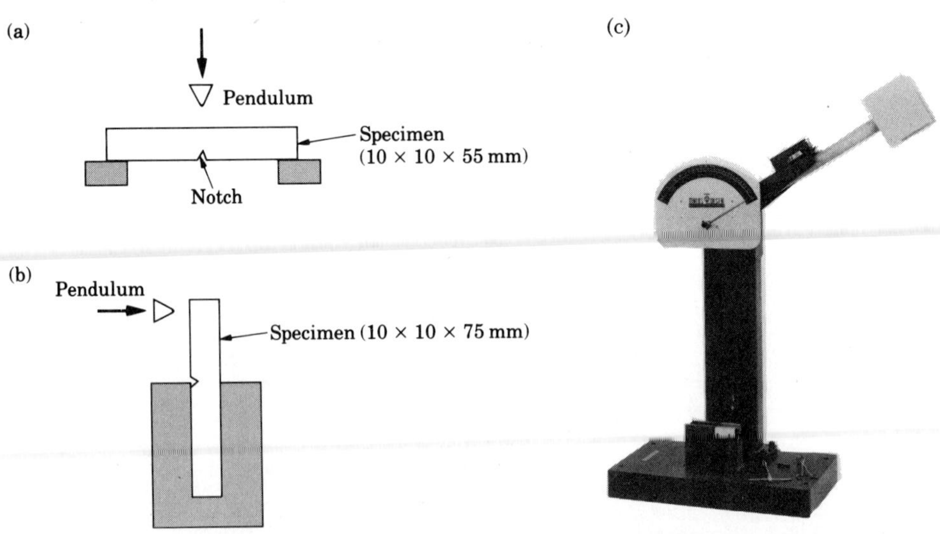

FIGURE 2.22 Impact test specimens: (a) Charpy; (b) Izod; (c) an impact testing machine. The specimen is placed in the fixture at the bottom. *Source:* Courtesy of Tinius Olsen Testing Machine Co.

2.10 Failure and Fracture of Materials in Manufacturing and in Service

Failure is one of the most important aspects of a material's behavior because it directly influences the selection of a material for a certain application, the methods of manufacturing, and the service life of the component. Because of the many factors involved, failure and fracture of materials is a complex area of study. In this section, we consider only those aspects of failure that are of particular significance to selecting and processing materials.

There are two general types of failure: (1) **fracture** and separation of the material, either through internal or external cracking; and (2) **buckling** (Fig. 2.23). Fracture is further classified into two general categories: ductile and brittle (Fig. 2.24).

Although failure of materials is generally regarded as undesirable, certain products are designed in such a way that failure is essential for their function. Typical examples are food and beverage containers with tabs or entire tops, which are removed by tearing the sheet metal along a prescribed path, and screw caps for bottles.

2.10.1 Ductile fracture

Ductile fracture is characterized by plastic deformation, which precedes failure of the part. In a tension test, highly ductile materials such as gold and lead may neck

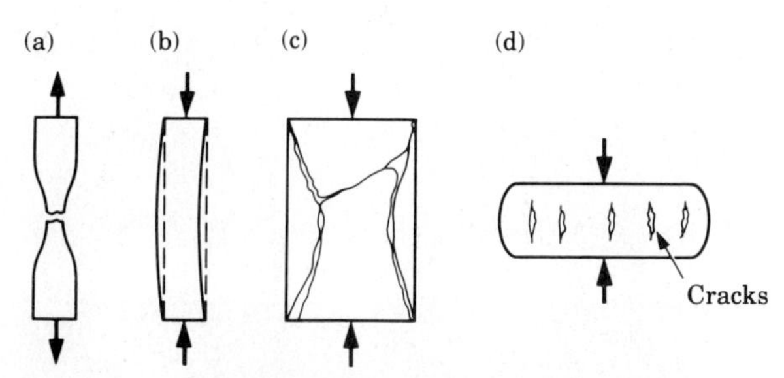

FIGURE 2.23 Schematic illustration of types of failures in materials: (a) necking and fracture of ductile materials; (b) buckling of ductile materials under a compressive load; (c) fracture of brittle materials in compression; (d) cracking on the barreled surface of ductile materials in compression.

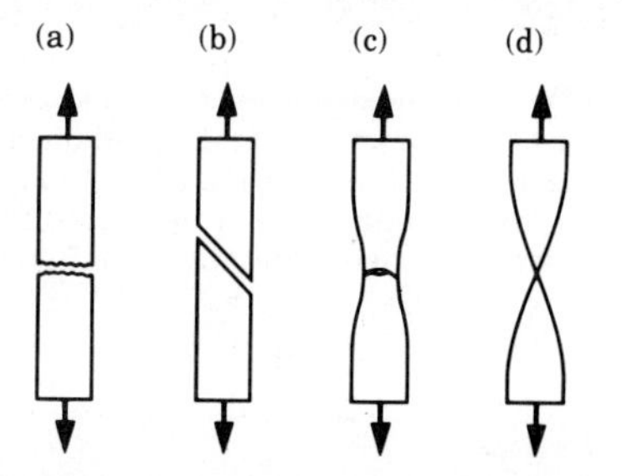

FIGURE 2.24 Schematic illustration of the types of fracture in tension: (a) brittle fracture in polycrystalline metals; (b) shear fracture in ductile single crystals; (c) ductile cup-and-cone fracture in polycrystalline metals; (d) complete ductile fracture in polycrystalline metals, with 100 percent reduction of area.

down to a point and then fail (Fig. 2.24d). Most metals and alloys, however, neck down to a finite area and then fail. Ductile fracture generally takes place along planes on which the *shear stress is a maximum.* In torsion, for example, a ductile metal fractures along a plane perpendicular to the axis of twist, that is, the plane on which the shear stress is a maximum. Fracture in shear is a result of extensive slip along slip planes within the grains.

Upon close examination of the surface of a ductile fracture (Fig. 2.25), you can see a *fibrous* pattern with *dimples*, as if a number of very small tension tests have been carried out over the fracture surface. Failure is initiated with the formation of tiny voids, usually around small inclusions or preexisting *voids*, which then *grow* and *coalesce*, developing cracks which grow in size and lead to fracture.

In a tension-test specimen, fracture begins at the center of the necked region from the growth and coalescence of cavities (Fig. 2.26). The central region becomes one large crack, as shown in the mid-section of the tension-test specimen in Fig. 2.27, and propagates to the periphery of this necked region. Because of its

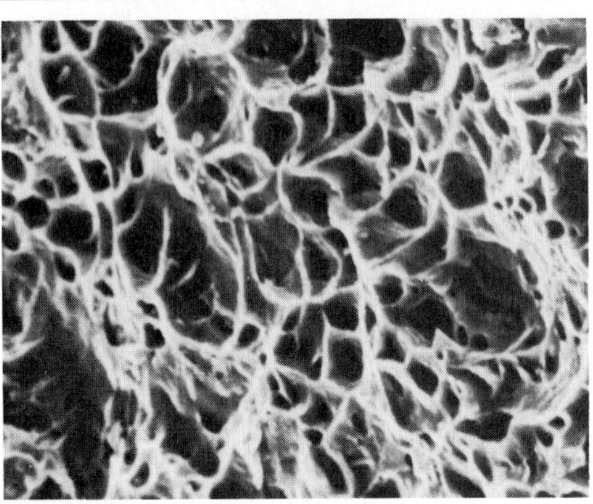

FIGURE 2.25

Surface of ductile fracture in low-carbon steel showing dimples. Fracture is usually initiated at impurities, inclusions, or preexisting voids in the metal. *Source:* K.-H. Habig and D. Klaffke. Photo by BAM Berlin/Germany.

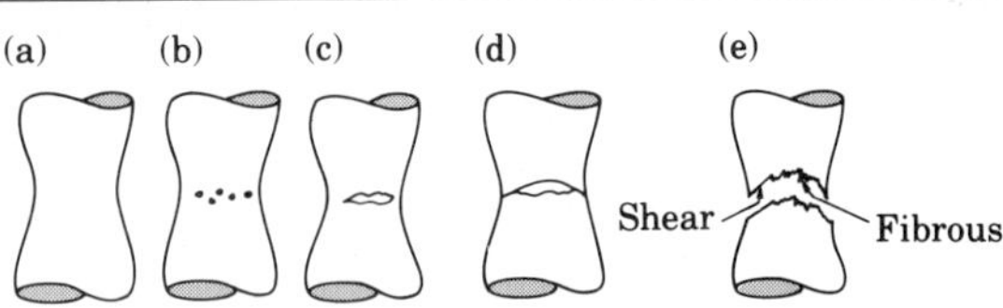

FIGURE 2.26

Sequence of events in necking and fracture of a tensile-test specimen: (a) early stage of necking; (b) small voids begin to form within the necked region; (c) voids coalesce, producing an internal crack; (d) rest of cross-section begins to fail at the periphery by shearing; (e) final fracture surfaces, known as cup- (top fracture surface) and cone- (bottom surface) fracture.

FIGURE 2.27

Typical cup-and-cone fracture in a tensile-test specimen. Because of variations in materials and methods of specimen preparation, not all specimens in tension break symmetrically as in this figure. *Source:* R. E. Reed-Hill, *Physical Metallurgy Principles,* 2d ed. PWS-Kent Publishing Company, 1973.

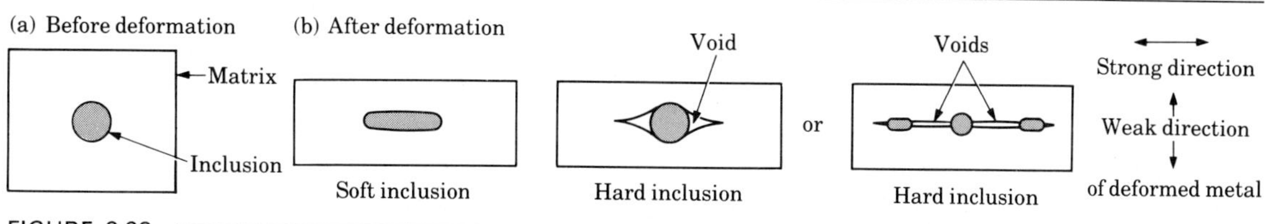

FIGURE 2.28 Schematic illustration of the deformation of soft and hard inclusions and their effect on void formation in plastic deformation. Note that hard inclusions, because they do not comply with the overall deformation of the ductile matrix, can cause voids.

appearance, the fracture surface of a tension-test specimen is called a *cup-and-cone* fracture.

Effects of inclusions. Because they are nucleation sites for voids, **inclusions** have an important influence on ductile fracture and thus on the formability of materials. Inclusions may consist of impurities of various kinds and second-phase particles, such as oxides, carbides, and sulfides. The extent of their influence depends on factors such as their shape, hardness, distribution, and volume fraction. The greater the volume fraction of inclusions, the lower will be the ductility of the material. Voids and porosity developed during processing, such as from casting, reduce the ductility of a material.

Two factors affect void formation. One is the strength of the bond at the interface of an inclusion and the matrix. If the bond is strong, there is less tendency for void formation during plastic deformation. The second factor is the hardness of the inclusion. If it is soft, such as manganese sulfide, it will conform to the overall change in shape of the specimen or workpiece during plastic deformation. If it is hard, such as carbides and oxides, it could lead to void formation (Fig. 2.28). Hard inclusions may also break up into smaller particles during deformation because of their brittle nature. The alignment of inclusions during plastic deformation leads to *mechanical fibering*. Subsequent processing of such a material must therefore involve considerations of the proper direction of working for maximum ductility and strength.

Transition temperature. Many metals undergo a sharp change in ductility and toughness across a narrow temperature range, called the **transition temperature** (Fig. 2.29). This phenomenon occurs in body-centered cubic and some hexagonal close-packed metals; it is rarely exhibited by face-centered cubic metals. The transition temperature depends on factors such as composition, microstructure, grain size, surface finish and shape of the specimen, and rate of deformation. High rates, abrupt changes in shape, and surface notches raise the transition temperature; that is, higher temperatures are required to make materials become less ductile and less tough.

Strain aging. **Strain aging** is a phenomenon in which carbon atoms in steels segregate to dislocations, thereby pinning them and thus increasing the resistance to

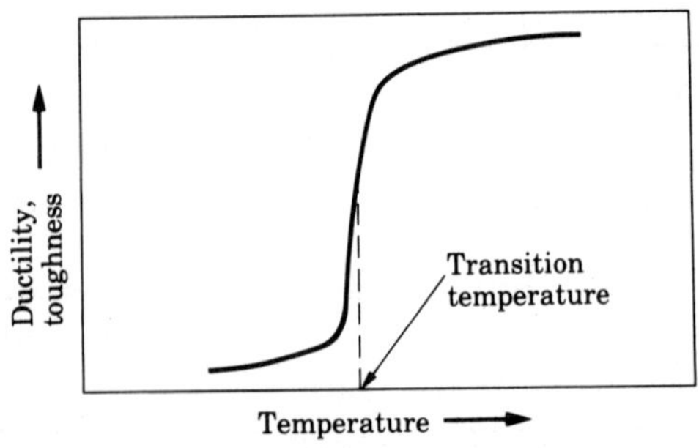

FIGURE 2.29 Schematic illustration of transition temperature. Note the narrow temperature range across which the behavior of the metal undergoes a major transition.

dislocation movement. The result is increased strength and reduced ductility. The effects of strain aging on the shape of the stress–strain curve for low-carbon steel at room temperature are shown in Fig. 2.30. Curve *abc* shows the original curve, with upper and lower yield points typical of these steels. If the tension test is stopped at point *e* and the specimen is unloaded and tested again, curve *dec* is obtained. Note that the upper and lower yield points have disappeared. However, if 4 hours pass before the specimen is strained again, curve *dfg* is obtained. If 126 hours pass, curve

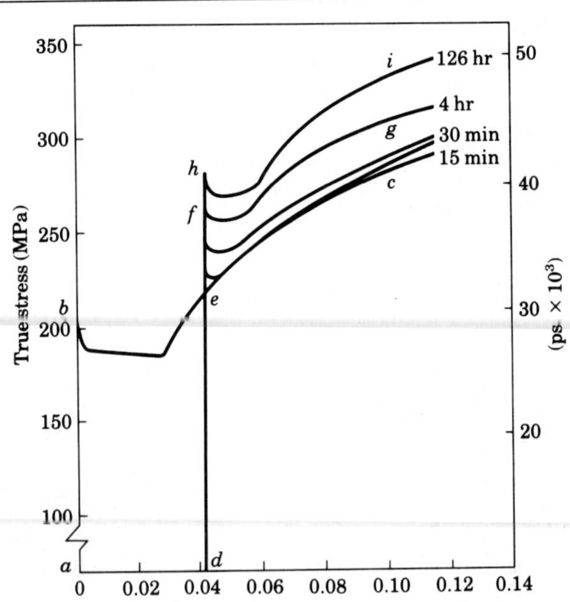

FIGURE 2.30 Strain aging and its effect on the shape of the true stress–true strain curve for 0.03% C rimmed steel at 60 °C (140 °F). *Source:* A. S. Keh and W. C. Leslie.

dhi is obtained. Instead of taking place over several days at room temperature, this same phenomenon can occur in a few hours at a higher temperature, and is called *accelerated strain aging*.

An example of accelerated strain aging in steels is **blue brittleness**, so named because it occurs in the blue-heat range where the steel develops a bluish oxide film. This phenomenon causes a marked decrease in ductility and toughness and an increase in strength of plain carbon and some alloy steels.

2.10.2 Brittle fracture

Brittle fracture occurs with little or no gross plastic deformation preceding the separation of the material into two or more pieces. In tension, fracture takes place along a crystallographic plane, called a *cleavage plane*, on which the normal tensile stress is a maximum. Face-centered cubic metals usually do not fail in brittle fracture, whereas body-centered cubic and some hexagonal close-packed metals fail by cleavage. In general, low temperature and high rates of deformation promote brittle fracture.

In a polycrystalline metal under tension, the fracture surface has a bright granular appearance because of the changes in the direction of the cleavage planes as the crack propagates from one grain to another. An example of the surface of brittle fracture is shown in Fig. 2.31. Brittle fracture of a specimen in compression is more complex and may follow a path that is theoretically 45° to the applied force direction.

Example of fracture along a cleavage plane are the splitting of rock salt or peeling of layers of mica. Tensile stresses normal to the cleavage plane, caused by pulling, initiate and control the propagation of fracture. Another example is the behavior of brittle materials, such as chalk, gray cast iron, and concrete. In tension, they fail in the manner shown in Fig. 2.24(a). In torsion, they fail along a plane at 45° to the axis of twist, that is, along a plane on which the tensile stress is a maximum.

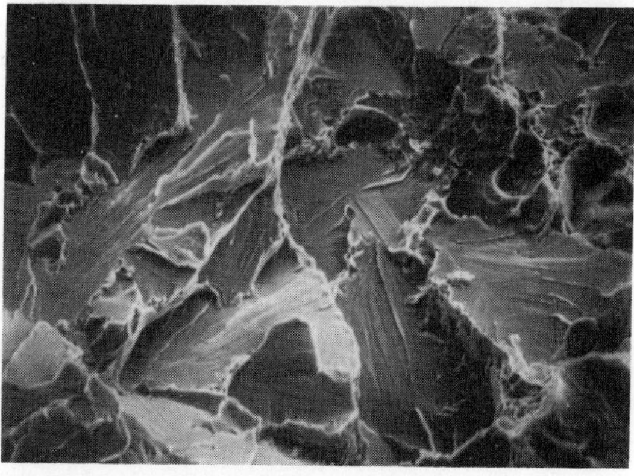

FIGURE 2.31 
Typical surface of steel that has failed in a brittle manner. The fracture path is transgranular (through the grains). Compare with the ductile fracture surface in Fig. 2.25. Magnification: 200 ×. *Source:* Courtesy of B. J. Schulze and S. L. Meiley and Packer Engineering Associates, Inc.

Defects. An important factor in fracture is the presence of defects such as scratches, flaws, and external or internal cracks. Under tension, the tip of the crack is subjected to high tensile stresses, which propagate the crack rapidly because the material has little capacity to dissipate energy.

The presence of defects is essential in explaining why brittle materials are so weak in tension compared to their strength in compression. The difference is on the order of 10 for rocks and similar materials, about 5 for glass, and about 3 for gray cast iron. Under tensile stresses, cracks propagate rapidly, causing what is known as catastrophic failure.

With polycrystalline metals, fracture paths most commonly observed are *transgranular* (transcrystalline or intragranular), meaning that the crack propagates through the grain. *Intergranular* fracture, where the crack propagates along the grain boundaries (Fig. 2.32), generally occurs when the grain boundaries are soft, contain a brittle phase, or have been weakened by liquid- or solid-metal embrittlement.

• Example: Comparison of strengths.

Two pieces of glass rod, one tubular and the other solid and round, are subjected to tension. The rods are of equal length and net cross-sectional area. Which rod is likely to carry the higher tensile load before fracture occurs?

SOLUTION. Glass, like other brittle materials, is very notch sensitive and its strength depends on surface flaws, cracks, and scratches. Hence the tubular

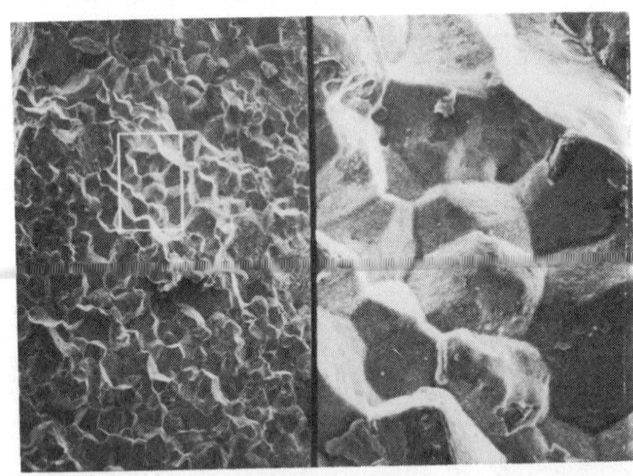

FIGURE 2.32
Intergranular fracture, at two different magnifications. Grains and grain boundaries are clearly visible in this micrograph. The fracture path is along the grain boundaries. Magnification: left, 100×; right, 500×. *Source:* Courtesy of B. J. Schulze and S. L. Meiley and Packer Engineering Associates, Inc.

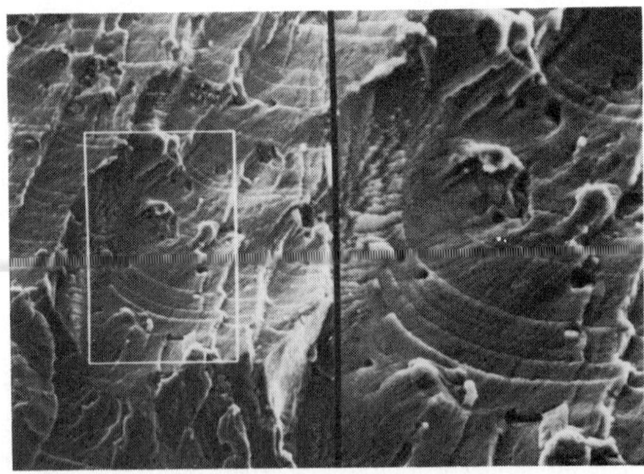

FIGURE 2.33
Typical fatigue fracture surface on metals, showing beach marks. Most components in machines and engines fail by fatigue and not by excessive static loading. Magnification: left, 500×; right, 1000×. *Source:* Courtesy of B. J. Schulze and S. L. Meiley and Packer Engineering Associates, Inc.

specimen, with its larger surface area, has a greater probability of flaws and a greater number of them. Therefore its strength is likely to be lower than that of the solid, round rod. Residual stresses on the surface of glass (due to the nature of manufacturing or preparation of the glass specimens) also play a significant role in the strength of glass.

Fatigue fracture. **Fatigue** fracture is basically of a brittle nature. Minute external or internal cracks develop at flaws or defects in the material, which then propagate and eventually lead to total failure of the part. The fracture surface in fatigue is generally characterized by the term *beach marks*, because of its appearance (Fig. 2.33). With large magnification, such as higher than 1000×, a series of *striations* can be seen on fracture surfaces, each beach mark consisting of a number of striations.

Improving fatigue strength. Fatigue life is greatly influenced by the method of preparation of the surfaces of the part or specimen (Fig. 2.34). The fatigue strength

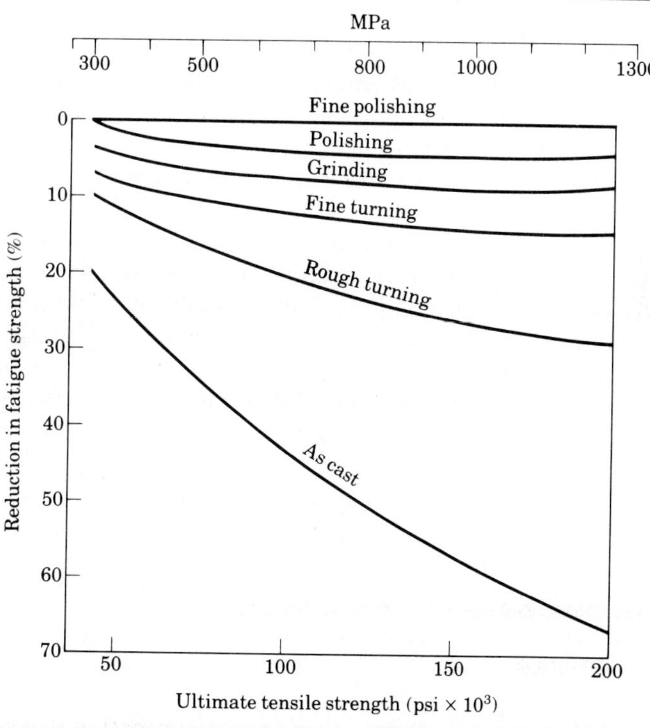

FIGURE 2.34 Reduction in fatigue strength of cast steels subjected to various surface-finishing operations. Note that the reduction is greater as surface roughness and strength of the steel increase. *Source:* M. R. Mitchell.

of manufactured products can be improved generally by the following methods:

1. Inducing compressive residual stresses on surfaces, such as by shot peening or roller burnishing.
2. Surface (case) hardening by various means.
3. Providing fine surface finish, thus reducing effects of notches and other surface imperfections.
4. Selecting appropriate materials and ensuring that they are free from significant amounts of inclusions, voids, and impurities.

Conversely, the following factors and processes can reduce fatigue strength: decarburization, surface pits due to corrosion that act as stress raisers, hydrogen embrittlement, galvanizing, and electroplating.

Stress–corrosion cracking. An otherwise ductile metal can fail in a brittle manner by **stress–corrosion cracking** (*stress cracking* or *season cracking*). Parts free from defects after forming may, either over a period of time or soon after being made into a product, develop cracks. Crack propagation may be intergranular or transgranular.

The susceptibility of metals to stress–corrosion cracking depends mainly on the material, the presence and magnitude of tensile residual stresses, and the environment. Brass and austenitic stainless steels are among metals that are highly susceptible to stress cracking. The environment could be corrosive media such as salt water or other chemicals. The usual procedure to avoid stress–corrosion cracking is to stress relieve the part just after it is formed. Full annealing may also be done, but this treatment reduces the strength of cold-worked parts.

Hydrogen embrittlement. The presence of hydrogen can reduce ductility and cause severe embrittlement in many metals, alloys, and nonmetallic materials and cause premature failure. This phenomenon is known as *hydrogen embrittlement* and is especially severe in high-strength steels. Possible sources of hydrogen are during melting of the metal, during pickling (removing of surface oxides by chemical or electrochemical reaction), through electrolysis in electroplating, from water vapor in the atmosphere, or from moist electrodes and fluxes used during welding.

2.11 Residual Stresses

When workpieces are subjected to deformation that is not uniform throughout the part, they develop **residual stresses**. These are stresses that remain within a part after it has been formed and all external forces have been removed. A typical example is the bending of a piece of metal (Fig. 2.35). The bending moment first

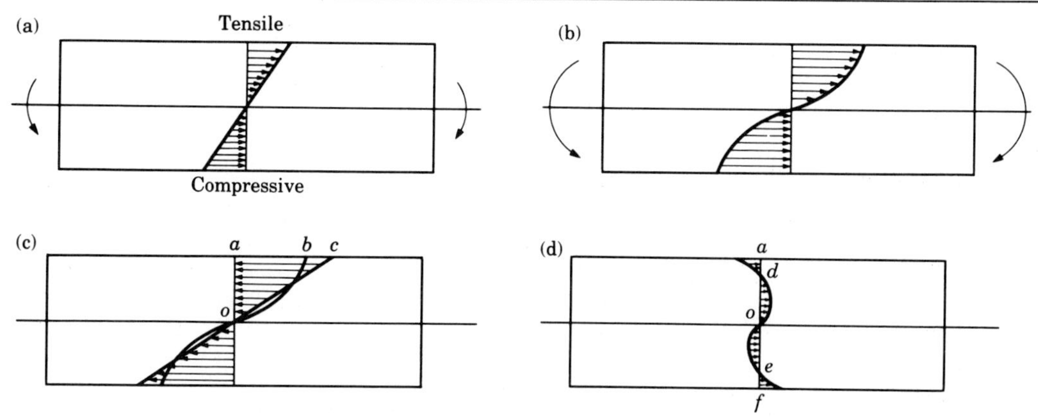

FIGURE 2.35
Residual stresses developed in bending a rectangular beam. Note that the horizontal forces and moments caused by residual stresses in the beam must be internally balanced. Because of nonuniform deformation during metalworking operations, most parts have residual stresses.

produces a linear elastic stress distribution, as in Fig. 2.35(a). As the moment is increased, the outer fibers in the part reach a stress level high enough to cause yielding. For a typical strain-hardening material, the stress distribution shown in Fig. 2.35(b) is eventually obtained, and the part is bent permanently.

Let's now remove the external moment on the part, which is equivalent to applying an equal and opposite moment. Thus the moments of the areas *oab* and *oac* in Fig. 2.35(c) must be equal. Line *oc*, which represents the opposite moment, is linear because all recovery is elastic (see Fig. 2.3). The difference between the two stress distributions gives the residual stress pattern within the part as shown in Fig. 2.35(d). Note the compressive residual stresses in layers *ad* and *oe* and the tensile residual stresses in layers *do* and *ef*.

Since there are no external forces, the internal forces resulting from these residual stresses must be in static equilibrium. Although this example involves only residual stresses in the longitudinal direction of the beam, in most cases in manufacturing operations these stresses are three dimensional.

The equilibrium of residual stresses in Fig. 2.35 may be disturbed by removing a layer of material from the part. The beam will then acquire a new radius of curvature in order to balance the internal forces. Such disturbances of residual stresses lead to *warping* of parts (Fig. 2.36). The equilibrium of residual stresses may also be disturbed by relaxation of these stresses over a period of time, resulting in dimensional changes of the component.

Residual stresses can also be caused by *temperature gradients* within a body, such as during cooling of a casting or forging. The relative expansion and contraction caused by temperature gradients in the material is analogous to nonuniform deformation, as is the case in the bending of a beam.

Tensile residual stresses in the surface of a part are generally considered to be undesirable, since they lower the fatigue life and fracture strength of the part. This

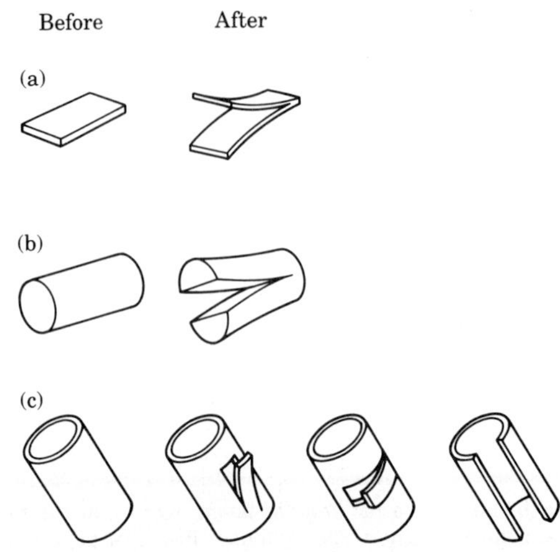

FIGURE 2.36 Distortion of parts with residual stresses after cutting or slitting: (a) rolled sheet or plate, slit partially across the thickness; (b) drawn rod, slit along its diameter and across its length; (c) thin-walled tubing or pipe, slit in different directions.

condition occurs because a surface with tensile residual stresses can sustain lower additional tensile stresses from external forces than can a surface that is free from residual stresses. This condition is particularly true for brittle or less ductile materials where fracture takes place with little or no plastic deformation preceding fracture. Tensile residual stresses can also lead to stress cracking or stress–corrosion cracking of manufactured products over a period of time.

Compressive residual stresses on a surface, on the other hand, are generally desirable. In fact, as we stated earlier, in order to increase the fatigue life of components, compressive residual stresses are imparted to surfaces by techniques such as shot peening and surface rolling.

Residual stresses can be reduced or eliminated either by stress-relief annealing or by further deformation of the part, such as by stretching. Given sufficient time, residual stresses may also diminish at room temperature by relaxation of residual stresses. The time required for relaxation can be greatly reduced by raising the temperature of the workpiece.

2.11.1 Reduction of residual stresses

The mechanism of reduction or elimination of residual stresses by plastic deformation is as follows: Assume that a piece of metal has the residual stresses shown in Fig. 2.37, namely, tensile on the outside and compressive on the inside. The part with these stresses, which are in the elastic range, is in equilibrium. Also assume that the material is elastic–perfectly plastic, as shown in Fig. 2.37(d).

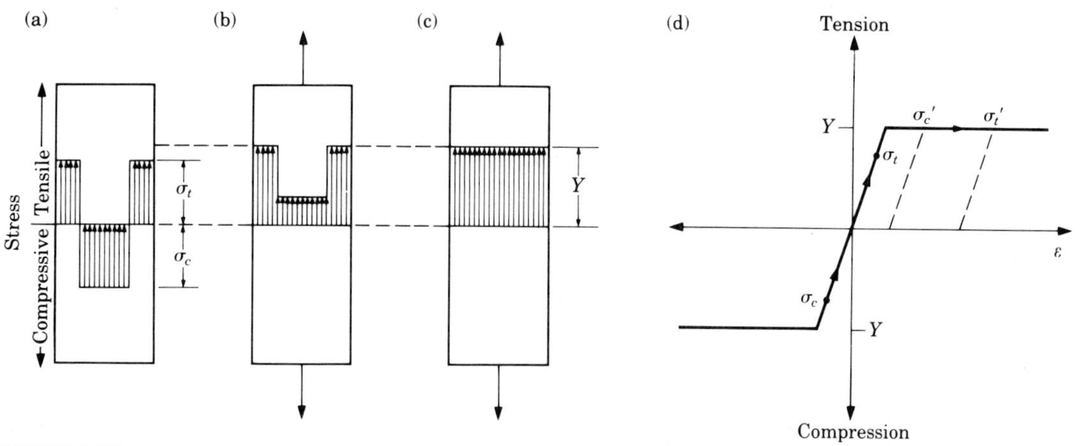

FIGURE 2.37
Elimination of residual stresses by stretching. Residual stresses can also be reduced or eliminated by thermal treatments, such as stress relieving or annealing.

The levels of residual stress are shown on the stress–strain diagram, both being below the yield stress Y. If a uniformly distributed tension is applied to this specimen, points σ_c and σ_t in the diagram move up on the stress–strain curve, as shown by the arrows. The maximum level that these stresses can reach is the tensile yield stress Y. With sufficiently high loading, the stress distribution becomes uniform throughout the part, as shown in Fig. 2.37(c). If the load is then removed, the stresses recover elastically and the part has no residual stresses. Note that very little stretching is required to relieve these residual stresses. The reason is that the elastic portions of the stress–strain curves for metals are very steep; hence the elastic stresses can be raised to the yield stress with very little strain.

The technique for reducing or relieving residual stresses by plastic deformation, such as by stretching as described, requires sufficient straining to establish a uniformly distributed stress in the part. A material such as the elastic, linearly strain-hardening type thus can never reach this condition since the compressive stress $\sigma_{c'}$ will always lag behind $\sigma_{t'}$. If the slope of the stress–strain curve in the plastic region is small, the difference between $\sigma_{c'}$ and $\sigma_{t'}$ will be rather small and little residual stress will be left in the part after unloading.

• Example: Elimination of residual stresses.

Refer to Fig. 2.37 and assume that $\sigma_t = 140$ MPa and $\sigma_c = -140$ MPa. The material is aluminum and the length of the specimen is 0.25 m. Calculate the length to which this specimen should be stretched so that, when unloaded, it will be free from residual stresses. Assume that the yield stress of the material is 150 MPa.

SOLUTION. Stretching should be to the extent that σ_c reaches the yield stress in tension, Y. Thus the total strain should be equal to the sum of the strain required to bring the compressive residual stress to zero and the strain required to bring it to the tensile yield stress. Hence

$$\varepsilon_{\text{total}} = \frac{\sigma_c}{E} + \frac{Y}{E}.$$

For aluminum, let $E = 70$ GPa. Thus,

$$\varepsilon_{\text{total}} = \frac{140}{70 \times 10^3} + \frac{150}{70 \times 10^3} = 0.00414.$$

Hence the stretched length should be

$$\ln\left(\frac{l_f}{0.25}\right) = 0.00414, \quad \text{or} \quad \ell_f = 0.2510 \text{ m}.$$

Since the strains are very small, we can use engineering strains in these calculations. Thus,

$$\frac{\ell_f - 0.25}{0.25} = 0.00414, \quad \text{or} \quad \ell_f = 0.2510 \text{ m}.$$

●

2.11.2 Work, heat, and temperature rise

Almost all the mechanical work of deformation in plastic working is converted into *heat.* This conversion is not 100% because a small portion of this energy is stored within the deformed material as elastic energy. This is known as *stored energy*, and it is generally 5% to 10% of the total energy input. However, it may be as high as 30% in some alloys.

In a simple frictionless process—assuming that work is completely converted into heat—the temperature rise is given by

$$\Delta T = \frac{u}{\rho c}, \tag{2.15}$$

where u is the specific energy (work of deformation per unit volume), ρ is the density, and c is the specific heat of the material. Higher temperatures are associated with large areas under the stress–strain curve and smaller values of specific heat.

The theoretical temperature rise for a true strain of 1 (such as a 27-mm-high specimen compressed down to 10 mm) has been calculated to be as follows: aluminum, 75 °C (165 °F); copper, 140 °C (285 °F); low-carbon steel, 280 °C (535 °F); and titanium 570 °C (1060 °F).

The temperature rise given by Eq. (2.15) is for an ideal situation, where there is no heat loss. In actual operations heat is lost to the environment, to tools and dies, and to any lubricants or coolants used. If the process is performed very rapidly, these losses are relatively small.

Under extreme conditions, an adiabatic process is approached, with very high temperature rise, leading to *incipient melting*. This rise in temperature can be calculated provided the stress–strain curve used is at the appropriate strain rate level. However, if the process is carried out slowly, the actual temperature rise will be a small portion of the calculated value. Properties such as specific heat and thermal conductivity also depend on temperature and they should be taken into account in the calculations.

• Example: Calculation of temperature rise.

Calculate the total work done for the specimen in Fig. 2.5. Calculate the specific energy for an element in the necked area and the theoretical temperature rise.

SOLUTION. The specific energy of an element in the necked area, that is, at fracture, is the area under the true stress–strain curve (corrected) in Fig. 2.5(c). This area is estimated to be 155,000 in. · lb/in^3. The theoretical temperature rise, that is, adiabatic, is obtained from Eq. (2.15), where for stainless steel, $\rho = 0.29$ lb/in^3 and $c = 0.12$ Btu/lb · °F. Thus,

$$\Delta T = \frac{155{,}000}{(0.29)(0.12)(778)(12)} = 477\ ^\circ\text{F} = 247\ ^\circ\text{C}.$$

Note that in these calculations only the work of plastic deformation is considered, because no friction or redundant work is involved in a simple tension test. Also, the actual temperature rise will be lower because of the heat loss from the necked zone to the rest of the specimen and to the environment. •

SUMMARY

The mechanical properties of materials, such as strength and ductility, are important both in manufacturing processes and to the service life of components. They help us estimate forces required in forming processes and predict the behavior of materials in shaping processes. Mechanical properties depend not only on the particular material but also on a number of variables, such as temperature, rate of deformation, surface condition, and environment. Numerous tests have been developed by which these properties can be measured.

An important consideration is fracture of the material during manufacturing and the service life of manufactured components. The two basic types of fracture are ductile and brittle. In addition to the variables just listed, other factors that influence fracture are the presence of flaws, inclusions, and impurities in the material. These can adversely affect strength, ductility, and toughness. The presence of residual stresses in a manufactured component is significant, both in subsequent processing of the part and in its service life. Thus it is important to know or predict the conditions under which a material and component will be processed or used.

SUMMARY TABLE 2.1

RELATIVE MECHANICAL PROPERTIES OF VARIOUS MATERIALS AT ROOM TEMPERATURE, IN DECREASING ORDER. METALS ARE IN THEIR ALLOY FORM.

STRENGTH	HARDNESS	TOUGHNESS	STIFFNESS	STRENGTH/ DENSITY
Glass fibers	Diamond	Ductile metals	Diamond	Reinforced plastics
Graphite fibers	Cubic boron nitride	Reinforced plastics	Carbides	Titanium
Kevlar fibers	Carbides	Thermoplastics	Tungsten	Steel
Carbides	Hardened steels	Wood	Steel	Aluminum
Molybdenum	Titanium	Thermosets	Copper	Magnesium
Steels	Cast irons	Ceramics	Titanium	Beryllium
Tantalum	Copper	Glass	Aluminum	Copper
Titanium	Thermosets		Ceramics	Tantalum
Copper	Magnesium		Reinforced plastics	
Reinforced thermosets	Thermoplastics		Wood	
Reinforced thermoplastics	Tin		Thermosets	
Thermoplastics	Lead		Thermoplastics	
Lead			Rubbers	

TRENDS

- Testing equipment is being designed and manufactured with computer controls for better accuracy and efficiency of operation.
- The influence of flaws, impurities, inclusions, and various other factors continue to be studied in order to improve the quality of manufactured products.
- Reliable test methods for brittle materials such as ceramics and carbides continue to be developed.

KEY TERMS

- Bauschinger effect
- Blue brittleness
- Buckling
- Compression
- Creep
- Deformation rate
- Disk test
- Ductility
- Elongation
- Engineering stress
- Engineering strain
- Fatigue
- Fracture
- Hardness
- Impact loading
- Inclusions
- Microhardness
- Modulus of elasticity
- Modulus of rigidity
- Modulus of rupture
- Poisson's ratio
- Reduction of area
- Residual stresses
- Rupture
- Shear
- Shear modulus
- Strain aging
- Strain-hardening exponent
- Strain rate
- Strength coefficient
- Stress–corrosion cracking
- Superplasticity
- Tension
- Torsion test
- Toughness
- Transition temperature
- True stress
- True strain
- Ultimate tensile strength
- Yield stress

BIBLIOGRAPHY

Ashby, M.F., and D.R.H. Jones, *Engineering Materials*, Vol. 1, *An Introduction to Their Properties and Applications*, 1980; Vol. 2, *An Introduction to Microstructures, Processing and Design*, 1986. New York: Pergamon.

Boyer, H.E. (ed.), *Atlas of Stress–Strain Curves*. Metals Park, Ohio: ASM International, 1986.

Courtney, T.H., *Mechanical Behavior of Materials*. New York: McGraw-Hill, 1990.

Davis, H.E., G.E. Troxell, and G.F.W. Hauck, *The Testing of Engineering Materials*, 4th ed. New York: McGraw-Hill, 1982.

Dieter, G.E. *Mechanical Metallurgy*, 3d ed. New York: McGraw-Hill, 1986.

Encyclopedia of Materials Science and Engineering. New York: Pergamon, 1986.

Hardness Testing. Metals Park, Ohio: ASM International, 1987.

Hsu, T.H., *Stress and Strain Data Handbook*. Houston, Texas: Gulf Publishing Co., 1986.

Metals Handbook, 9th ed., Vol. 8: *Mechanical Testing*. Metals Park, Ohio: American Society for Metals, 1985.

Pohlandt, K., *Material Testing for the Metal Forming Industry*. New York: Springer, 1989.

REVIEW QUESTIONS

2.1 Distinguish between engineering stress and true stress.

2.2 Describe the events that occur when a specimen undergoes a tension test. Sketch a plausible stress–strain curve and identify all significant regions and points between them. Assume that loading continues through fracture.

2.3 What is ductility, and how is it measured?
2.4 What is Poisson's ratio? How is it measured?
2.5 What property of a material does the modulus of elasticity measure?
2.6 In the equation $\sigma = K\varepsilon^n$, which represents the stress–strain curve for a material, what is the significance of the exponent n?
2.7 What is strain-rate sensitivity, and how is it measured?
2.8 What advantages do superplastic materials offer in manufacturing?
2.9 What test can measure the properties of a material undergoing shear strain?
2.10 What testing procedures can be used to measure the properties of brittle materials, such as ceramics and carbides?
2.11 List five tests for measuring hardness, and briefly summarize how each test is carried out.
2.12 What conditions lead to fatigue failure?
2.13 Describe the differences between brittle and ductile fracture.
2.14 Explain the difference between creep and stress relaxation.
2.15 Describe the difference between elastic and plastic behavior.
2.16 Explain what uniform elongation is in tension testing.
2.17 Describe the difference between deformation rate and strain rate. What units does each one have?
2.18 Describe the difficulties involved in making a compression test.
2.19 Define residual stress.
2.20 List the advantages and disadvantages associated with residual stresses.
2.21 Explain the difference between transgranular and intergranular fracture.

QUESTIONS AND PROBLEMS

2.22 Using the same scale for stress, the tensile true stress–true strain curve is higher than the engineering stress–engineering strain curve. Explain whether this condition also holds for a compression test.
2.23 A paper clip is made of wire 1 mm in diameter. If the original material from which the wire is made is a rod 10 mm in diameter, calculate the longitudinal engineering and true strains that the wire has undergone during processing.
2.24 A strip of metal is 15 in. long. It is stretched in two steps, first to a length of 20 in. and then to 25 in. Show that the total true strain is the sum of the true strains in each step; thus the strains can be added. Show that, using engineering strains, the strains cannot be added to obtain the total strain.
2.25 Identify the two materials in Fig. 2.6 that have the lowest and highest uniform elongations, respectively. Calculate these quantities as percentages of the original gage lengths.
2.26 If you remove the layer of material *ad* from the part shown in Fig. 2.35(d), such as by machining or grinding, which way will the specimen curve? (*Hint:* Assume that the part in (d) is composed of four horizontal springs held at the ends. Thus from the top down you have compression, tension, compression, and tension springs.)
2.27 With a simple sketch, explain whether it is necessary to use the offset method to determine the yield stress Y of a material that has been highly cold worked.
2.28 As you know, percent elongation is described in terms of the original gage length, such as 50 mm or 2 in. Explain how percent elongation varies as the gage length of the tensile

specimen increases. (*Hint:* Remember that necking is a local phenomenon, and think of what happens to the elongation as the gage length becomes very small.)

2.29 Which hardness tests and scales would you use for very thin strips of material, such as aluminum foil? Why?

2.30 List and explain the desirable mechanical properties for (a) elevator cable, (b) paper clip, (c) leaf spring for a truck, (d) bracket for a bookshelf, (e) piano wire, (f) wire coat hanger, (g) gas-turbine blade, and (h) staple.

2.31 Explain the significance of the strain-rate sensitivity of materials. Give some examples.

2.32 Have you observed the type of cracks shown in Fig. 2.23(d) when making a hamburger? What can you do to avoid such cracks?

2.33 Which of the two tests, tension or compression, requires the higher capacity testing machine, and why?

2.34 Make a sketch showing the nature and distribution of residual stresses in Fig. 2.36(a) and (b) before they were cut. (*Hint:* Assume that the split parts are free from any stresses. Force these parts back to the shape they had before they were cut.)

2.35 Inspect Table 2.1 and describe your observations as to the difference in the mechanical properties of alloys and the metals on which they are based, such as aluminum alloys versus aluminum.

2.36 List and explain briefly all the conditions that induce brittle fracture in an otherwise ductile metal.

2.37 Explain why the tension test is so commonly used.

2.38 List all the factors you would consider in selecting a hardness test and then interpreting the results from this test.

2.39 Modulus of resilience is defined as the area under the elastic portion of the stress–strain curve, and has the units of stress. If a highly cold-worked piece of steel has a hardness of 300 HB, what is its modulus of resilience?

2.40 You are given the K and n values of two different metals. Is this information sufficient to determine which metal is the tougher? If not, what additional information do you need?

2.41 A cable is made of two strands of different materials, A and B, and cross sections as follows:

For material A: $K = 100{,}000$ psi, $n = 0.5$, $A_o = 0.2\ \text{in}^2$.
For material B: $K = 50{,}000$ psi, $n = 0.5$, $A_o = 0.1\ \text{in}^2$.

Calculate the maximum tensile force that this cable can withstand prior to necking. Let the value of n be numerically equal to the strain at which necking begins.

2.42 Using Fig. 2.6 only, explain why you cannot calculate the percent elongation of the materials listed.

2.43 We have stated that the percent elongation of a tensile specimen (at fracture) decreases as the gage length increases. Explain if it is possible for the elongation to reach zero if the gage length is very high.

2.44 On the basis of the information given in Fig. 2.6, calculate the ultimate tensile strength (engineering) of annealed copper.

2.45 In a disk test performed on a specimen that is 1 in. in diameter and $\frac{1}{8}$ in. thick, the specimen fractures at a stress of 50,000 psi. What was the load on it?

2.46 A piece of steel has a hardness of 300 HB. Calculate the tensile strength in MPa and psi, respectively.

2.47 If you pull and break a tension-test specimen rapidly, where would the temperature be highest, and why?

2.48 Explain why the difference between engineering strain and true strain becomes larger as strain increases. Is this true for both tensile and compressive strains? Explain.

2.49 If a material does not have an endurance limit, such as aluminum, how would you estimate its fatigue life?

2.50 Describe the difference between creep and stress relaxation, giving two engineering examples for each.

2.51 A material has the following properties: UTS = 60,000 psi and $n = 0.4$. Calculate its strength coefficient K.

2.52 A material has a strength coefficient $K = 100{,}000$ psi and $n = 0.2$. Assuming that a tensile-test specimen made from this material begins to neck at a true strain of 0.2, show that the ultimate tensile strength of this material is 59,340 psi.

3

Physical Properties of Materials

3.1 Introduction

In addition to the mechanical properties that you studied in Chapter 2, *physical properties* of materials should also be considered in the selection and processing of materials. Properties of particular interest are density, melting point, specific heat, thermal conductivity and expansion, electrical and magnetic properties, and resistance to oxidation and corrosion.

Why, for example, is electrical wiring generally made of copper? Why are metals such as aluminum, stainless steel, and copper so commonly used in cookware? Why are their handles usually made of wood or plastics, yet other types of handles are made of metal? What kind of material should we choose for the heating elements in toasters? Why are commercial airplanes generally made of aluminum and some titanium, and why are some airplane components being gradually replaced with reinforced plastics?

In this chapter we explain the importance of strength-to-weight and stiffness-to-weight ratios, particularly for aircraft and various aerospace structures. We describe the importance of density. For example, high-speed equipment, such as textile and printing machinery, as well as forming and cutting machines, require lightweight components to reduce inertial forces. We also discuss several other examples of the importance of physical properties. We present each physical property from the viewpoint of material selection and manufacturing and its relevance to the service life of the component.

3.2 Density

Density of a material is its weight per unit volume. If expressed relative to the density of water, it is known as *specific gravity* and thus has no units. The range of densities for a variety of materials at room temperature is given in Table 3.1 (see p. 109). A list of materials in decreasing order of density is given in Table 3.2 (see p. 110).

The density of materials depends on atomic weight, atomic radius, and the packing of the atoms. Alloying elements generally have a minor effect on the density of metals, the effect depending on the density of the alloying elements. Density decreases with increasing temperature because the mass remains constant while the volume increases due to thermal expansion.

The most significant role that density plays is in the **specific strength** (strength-to-weight ratio) and **specific stiffness** (stiffness-to-weight ratio) of materials and structures (Fig. 3.1). Weight saving is particularly important for aircraft and

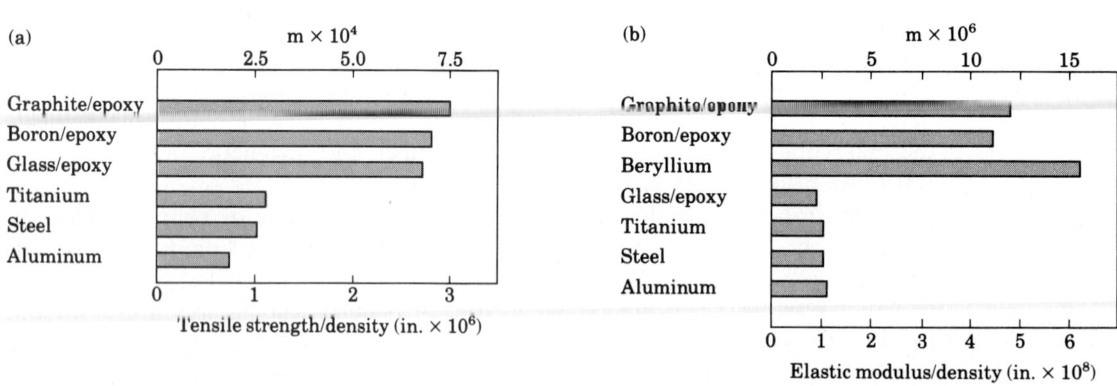

FIGURE 3.1 Strength/density (specific strength) and elastic modulus/density (specific stiffness) ratios for various reinforced plastics and metals at room temperature. (See also Chapter 9.) *Source:* M. J. Salkind.

TABLE 3.1
PHYSICAL PROPERTIES OF SELECTED MATERIALS AT ROOM TEMPERATURE

METAL	DENSITY (kg/m^3)	MELTING POINT (°C)	SPECIFIC HEAT (J/kg K)	THERMAL CONDUCTIVITY (W/m K)
Aluminum	2700	660	900	222
Aluminum alloys	2630–2820	476–654	880–920	121–239
Beryllium	1854	1278	1884	146
Columbium (niobium)	8580	2468	272	52
Copper	8970	1082	385	393
Copper alloys	7470–8940	885–1260	377–435	29–234
Iron	7860	1537	460	74
Steels	6920–9130	1371–1532	448–502	15–52
Lead	11350	327	130	35
Lead alloys	8850–11350	182–326	126–188	24–46
Magnesium	1745	650	1025	154
Magnesium alloys	1770–1780	610–621	1046	75–138
Molybdenum alloys	10210	2610	276	142
Nickel	8910	1453	440	92
Nickel alloys	7750–8850	1110–1454	381–544	12–63
Tantalum alloys	16600	2996	142	54
Titanium	4510	1668	519	17
Titanium alloys	4430–4700	1549–1649	502–544	8–12
Tungsten	19290	3410	138	166
NONMETALLIC				
Ceramics	2300–5500	—	750–950	10–17
Glasses	2400–2700	580–1540	500–850	0.6–1.7
Graphite	1900–2200	—	840	5–10
Plastics	900–2000	110–330	1000–2000	0.1–0.4
Wood	400–700	—	2400–2800	0.1–0.4

aerospace structures, automotive bodies and components, and other products where energy consumption and power limitations are major concerns. Substitution of materials for weight savings and economy is a major factor in advanced equipment and machinery, as well as in consumer products such as automobiles.

Density is an important factor in the selection of materials for high-speed equipment, such as the use of magnesium in printing and textile machinery, many components of which usually operate at very high speeds. To obtain exposure times of 1/4000 s in cameras without sacrificing accuracy, the shutters of some high-quality 35-mm cameras are made of titanium. The resulting light weight of the components in these high-speed operations reduces inertial forces that otherwise

TABLE 3.2
PHYSICAL PROPERTIES OF MATERIALS, IN DESCENDING ORDER

DENSITY	MELTING POINT	SPECIFIC HEAT	THERMAL CONDUCTIVITY	THERMAL EXPANSION	ELECTRICAL CONDUCTIVITY
Platinum	Tungsten	Wood	Silver	Plastics	Silver
Gold	Tantalum	Beryllium	Copper	Lead	Copper
Tungsten	Molybdenum	Porcelain	Gold	Tin	Gold
Tantalum	Columbium	Aluminum	Aluminum	Magnesium	Aluminum
Lead	Titanium	Graphite	Magnesium	Aluminum	Magnesium
Silver	Iron	Glass	Graphite	Copper	Tungsten
Molybdenum	Beryllium	Titanium	Tungsten	Steel	Beryllium
Copper	Copper	Iron	Beryllium	Gold	Steel
Steel	Gold	Copper	Zinc	Ceramics	Tin
Titanium	Silver	Molybdenum	Steel	Glass	Graphite
Aluminum	Aluminum	Tungsten	Tantalum	Tungsten	Ceramics
Beryllium	Magnesium	Lead	Ceramics		Glass
Glass	Lead		Titanium		Plastics
Magnesium	Tin		Glass		Quartz
Plastics	Plastics		Plastics		

could lead to vibrations, inaccuracies, and even part failure over a period of time. Because of their low density, ceramics are being considered for use in components of automated machine tools in high-speed operations.

On the other hand, there are applications where weight is desirable. Examples are counterweights for various mechanisms (using lead and steel) and components for self-winding watches (using high-density materials such as tungsten).

3.3 Melting Point

The **melting point** of a material depends on the energy required to separate its atoms. As Table 3.1 shows, the melting point of an alloy can have a wide range, unlike pure metals which have a definite melting point. The melting points of alloys depend on their particular composition.

The melting point of a material has a number of indirect effects on manufacturing operations. The choice of a material for high-temperature applications is the most obvious consideration, such as in jet engines and furnaces, or where high frictional heat is generated, such as with high sliding speeds of machine components. Sliding down a rope or rubbing your hands together fast is a simple demonstration of how high temperatures can become with increasing sliding speed.

Since the recrystallization temperature of a metal is related to its melting point, operations such as annealing, heat treating, and hot working require a knowledge of the melting points of the materials involved. These considerations, in turn, influence the selection of tool and die materials in manufacturing operations. Another major influence of the melting point is in the selection of the equipment and melting practice in casting operations. The higher the melting point of the material, the more difficult the operation becomes. Also, the selection of die materials, as in die casting, depends on the melting point of the workpiece material. In the electrical-discharge machining process, the melting points of metals are related to the rate of material removal and tool wear.

The temperature range within which a component or structure is designed to function is an important parameter in the selection of materials. Plastics, for example, have the lowest useful temperature range, whereas graphite and refractory-metal alloys have the highest range (Table 3.2).

3.4 Specific Heat

Specific heat is the energy required to raise the temperature of a unit mass of material by one degree. In Table 3.2, materials are listed in decreasing order of specific heat. Alloying elements have a relatively minor effect on the specific heat of metals.

The temperature rise in a workpiece, resulting from forming or machining operations, is a function of the work done and the specific heat of the workpiece material (see Section 2.12). Thus the lower the specific heat, the higher the temperature will rise in the material.

Temperature rise in a workpiece, if excessive, can have detrimental effects on product quality by adversely affecting its surface finish and dimensional accuracy, cause excessive tool and die wear, and result in adverse metallurgical changes in the material. Fortunately, many manufacturing operations are performed at relatively slow speeds with respect to the rate at which energy is imparted. Thus there is sufficient time for the heat generated to be dissipated, and the temperature rise is not significant.

3.5 Thermal Conductivity

Whereas specific heat indicates the temperature rise in the material as a result of energy input, **thermal conductivity** indicates the ease with which heat flows within and through the material.

In Table 3.2, materials are listed in decreasing order of thermal conductivity. Metallically bonded materials (metals) generally have high thermal conductivity, whereas ionically or covalently bonded materials (ceramics and plastics) have poor conductivity. Because of the large difference in their thermal conductivities, alloying elements can have a significant effect on the thermal conductivity of alloys, as you can see in Table 3.1 by comparing the metals with their alloys.

When heat is generated by plastic deformation or friction, the heat should be conducted away at a high enough rate to prevent a severe rise in temperature. The main difficulty experienced in machining titanium, for example, is caused by its very low thermal conductivity. Low thermal conductivity can also result in high thermal gradients and thus cause inhomogeneous deformation in metalworking processes.

3.6 Thermal Expansion

Thermal expansion of materials can have several significant effects. In Table 3.2, materials are listed in decreasing order of thermal expansion. Generally, the coefficient of thermal expansion is inversely proportional to the melting point of the material. Alloying elements have a relatively minor effect on the thermal expansion of metals.

Shrink fits utilize thermal expansion. A part with a hole in it to be installed over a shaft, such as a flange or lever arm, is heated and then slipped over a cool shaft or spindle. When allowed to cool, the part shrinks and the assembly becomes an integral part. Installing steel rims on wooden wheels in horse-drawn carriages is based on this principle.

Other examples where relative expansion or contraction is important are electronic and computer components, glass-to-metal seals, struts on jet engines, and moving parts in machinery that require certain clearances for proper functioning. Improper selection of materials and assembly can cause **thermal stresses**, cracking, warping, or loosening of components in the structure during their service life.

Thermal conductivity, in conjunction with thermal expansion, plays the most significant role in causing thermal stresses, both in manufactured components and in tools and dies. This is particularly important in a forging operation, for example, when hot workpieces are placed over relatively cool dies, making the die surfaces undergo thermal cycling. To reduce thermal stresses, a combination of high thermal conductivity and low thermal expansion is desirable.

Thermal stresses can lead to cracks in ceramic parts and in tools and dies made of relatively brittle materials. *Thermal fatigue* results from thermal cycling and causes a number of surface cracks. *Thermal shock* is the term generally used to describe development of cracks after a single thermal cycle. Thermal stresses may be caused both by temperature gradients and by anisotropy of thermal expansion, which we generally observe in hexagonal close-packed metals and in ceramics.

To alleviate some of the problems with thermal expansion, a family of iron–nickel alloys that have very low thermal-expansion coefficients have been developed, called *low-expansion alloys*. A typical composition is 64 percent iron–36 percent nickel. This alloy is called Invar, because the length of a component made of this alloy is almost invariable with respect to temperature, although over a narrow range. Consequently, these alloys also have good thermal-fatigue resistance.

3.7 Electrical and Magnetic Properties

Electrical conductivity and dielectric properties of materials are of great importance not only in electrical equipment and machinery, but also in manufacturing processes such as magnetic-pulse forming of sheet metals, and electrical-discharge machining and electrochemical grinding of hard and brittle materials.

The **electrical conductivity** of a material can be defined as a measure of the ease with which a material conducts electric current. Materials with high conductivity, such as metals, are generally referred to as *conductors*. The units of electrical conductivity are mho/m or mho/ft, where mho is the inverse of ohm, the unit for electrical resistance. **Electrical resistivity** is the inverse of conductivity. Materials with high resistivity are referred to as **dielectrics** or *insulators*.

In Table 3.2, materials are listed in decreasing order of electrical conductivity. The influence of the type of atomic bonding on the electrical conductivity of materials is the same as that for thermal conductivity. Alloying elements have a major effect on the electrical conductivity of metals: The higher the conductivity of the alloying element, the higher is the conductivity of the alloy.

• Example: Electric wiring in homes.

Because of the increased cost of copper, aluminum electric wiring was widely used in homes built in the United States between 1965 and 1973. It has been alleged that aluminum wiring can be a potential fire hazard because of resistance heating at junctions in switches and outlets. It has been stated that oxidation of the aluminum wire causes loosening of the wire at the terminals, thus increasing its electrical resistance and causing resistance heating. This condition can be detected by noting (a) that the areas around switches and outlets have become warm or hot to the touch, (b) odors of burning plastic insulation, (c) smoking or sparking, (d) flickering of lights, and (e) failing of switches, outlets, and circuits.

Electrical terminals have been redesigned to accommodate aluminum as well as copper wiring, by serrating the surfaces of contact. This modification improves contact (reducing electrical resistance) between the wire surface and the terminals. These devices are marked CO–ALR, meaning copper–aluminum, revised. Labels also state whether a particular switch or outlet is to be used with copper wiring only or may be used for either copper or aluminum wiring.

Dielectric strength of materials is the resistivity to direct electric current. It is defined as the voltage required per unit distance for electrical breakdown, and has the units of V/m or V/ft.

Superconductors. Superconductivity is the phenomenon of almost zero electrical resistivity that occurs in some metals and alloys below a critical temperature. Although the temperatures involved are near absolute zero (0 K, −273 °C, or −460 °F), the highest temperature at which superconductivity has been exhibited to date is at 70 K (−203 °C, −333 °F) with an alloy of lanthanum, strontium, copper, and oxygen. These developments indicate that the efficiency of electrical components such as large high-power magnets, high-voltage power lines, and various other electronic and computer components can be markedly improved.

Semiconductors. The electrical properties of materials such as single-crystal silicon, germanium, and gallium arsenide are extremely sensitive to the presence and type of minute impurities, as well as to temperature. Thus by controlling the concentration and type of impurities (*dopants*), such as phosphorus and boron in silicon, the electrical conductivity can be controlled.

This property is utilized in semiconductor (solid state) devices used extensively in miniaturized electronic circuitry. They are very compact, efficient, and relatively inexpensive, consume little power, and require no warmup time for operation. We discuss the manufacturing of these devices in Chapter 34.

Ferromagnetism. **Ferromagnetism** is large and permanent magnetization due to the alignment of iron, nickel, and cobalt atoms into domains. It is important in applications such as electric motors, generators, transformers, and microwave devices.

Piezoelectric effect. The **piezoelectric effect** (*piezo* from Greek, meaning to press) is exhibited by some materials, such as certain ceramics and quartz crystals, in which there is a reversible interaction between an elastic strain and an electric field. This property is utilized in making transducers, which are devices that convert the strain from an external force to electrical energy. Typical applications are force or pressure transducers, strain gages, sonar detectors, and microphones.

Magnetostriction. The phenomenon of expansion or contraction of a material when subjected to a magnetic field is called **magnetostriction**. Some materials, such as pure nickel and some iron-nickel alloys, exhibit this behavior. Magnetostriction is the principle behind ultrasonic machining equipment.

3.8 Corrosion Resistance

Metals, ceramics, and plastics are all subject to corrosion. We apply the word **corrosion** to deterioration of metals and ceramics, while calling similar phenomena in plastics *degradation.* Corrosion resistance is an important aspect of material selection for applications in the chemical, food, and petroleum industries, as well as in manufacturing operations. In addition to various possible chemical reactions from the elements and compounds present, environmental oxidation and corrosion of components and structures is a major concern, particularly at elevated temperatures and in automobiles and other transportation vehicles.

Corrosion not only leads to deterioration of the surface of components and structures, but also reduces their strength and structural integrity. Replacing corroded parts and preventing corrosion is estimated to cost billions of dollars in each of the industrialized countries. The cost of corrosion to the U.S. economy alone is estimated to be $200 billion per year.

Resistance to corrosion depends on the particular environment, as well as the composition of the material. Corrosive media may be chemicals (acids, alkali, and salts), the environment (oxygen, pollution, and acid rain), and water (fresh or salt). Nonferrous metals, stainless steels, and nonmetallic materials generally have high corrosion resistance. Steels and cast irons generally have poor resistance and must be protected by various coatings and surface treatments that we will study in Chapter 33.

Corrosion can occur over an entire surface, or it can be localized, which we call *pitting*. It can occur along grain boundaries of metals as *intergranular corrosion* and at the interface of bolted or riveted joints as *crevice corrosion.* Two dissimilar metals may form a galvanic cell (after L. Galvani, 1737–1798)—that is, two electrodes in an electrolyte in a corrosive environment (including moisture)—and cause **galvanic corrosion.** Two-phase alloys are more susceptible to galvanic corrosion, because of the two different metals involved, than single-phase alloys or pure metals. Thus heat treatment can have an influence on corrosion resistance.

Moreover, corrosion can act in indirect ways. **Stress-corrosion cracking** is an example of the effect of a corrosive environment on the integrity of a product that, as manufactured, had residual stresses in it. Likewise, metals that are cold worked are likely to contain residual stresses and hence are more susceptible to corrosion, compared to hot-worked or annealed metals.

Tool and die materials also can be susceptible to chemical attack by lubricants and coolants. The chemical reaction alters their surface finish and adversely influences the metalworking operation. An example is carbide tools and dies having cobalt as a binder, in which the cobalt is attacked by elements in the cutting fluid (**selective leaching**). Thus compatibility of the tool, die, and workpiece materials and the metalworking fluid under actual operating conditions is an important consideration in the selection of materials.

Chemical reactions should not be regarded as having adverse effects only. Nontraditional machining processes such as chemical machining and electrochemical machining are indeed based on controlled reactions. These processes remove material by chemical reactions, similar to the etching of metallurgical specimens.

The usefulness of some level of **oxidation** is exhibited by the corrosion resistance of aluminum, titanium, and stainless steel. Aluminum develops a thin (a few atomic layers), strong, and adherent hard oxide film (Al_2O_3) that protects the surface from further environmental corrosion. Titanium develops a film of titanium oxide (TiO_2). A similar phenomenon occurs in stainless steels, which because of the chromium present in the alloy, develop a similar protective film on their surfaces. These processes are known as **passivation.** When the protective film is scratched, thus exposing the metal underneath, a new oxide film forms in due time.

• Example: Selection of materials for coins.

There are five general criteria in the selection of materials for coins.

The first criterion is subjective and involves factors such as the appearance of the coin, its color, weight, and ring (the sound made when striking). Also included in this criterion is the feel of the coin. This term is difficult to describe, as it combines many factors. It is similar to the feel of a fine piece of wood, polished stone, or fine leather.

The second is the life of the coin. This involves resistance to corrosion and wear while the coin is in circulation. These two factors basically determine how long the surface imprint of the coin will be visible and also the ability of the coin to retain its original luster.

The third criterion involves the manufacturing of the coin. This includes factors such as the formability of the candidate coin materials, the life of the dies used in the coining operation, and the capability of the materials and processes to resist counterfeiting.

The fourth criterion concerns the suitability of the coin for use in coin-operated devices, such as vending machines, turnstiles, and pay telephones. These machines are generally equipped with detection devices that test the coins first for proper diameter, thickness, and surface condition and, second, for electrical conductivity and density. The coin is rejected if it fails these tests.

The final criterion relates to costs (raw materials and processing) and sufficient supply of the coin materials.

SUMMARY

The physical and chemical properties of materials have various positive and adverse effects in manufacturing processes and operations and on the service life of components. Many nontraditional machining processes are based on chemical and electrochemical reactions.

Physical and chemical properties and characteristics significantly affect design considerations, service requirements, compatibility with other materials (including tools and workpieces), long-term effects, environmental factors, and economic considerations.

TRENDS

- Greater strength-to-weight ratios of materials, particularly fiber-reinforced systems and with both plastics and metal matrices, are being developed.
- Further progress in developing materials with improved and special physical properties for a variety of electronic and computer applications is being made.
- Because of its major economic impact, corrosion resistance of materials and design of components for corrosion resistance continue to be important areas of research.

KEY TERMS

Corrosion
Density
Dielectrics
Electrical conductivity
Electrical resistivity
Ferromagnetism
Galvanic corrosion
Magnetostriction
Melting point
Oxidation
Passivation
Piezoelectric effect
Selective leaching
Specific heat
Specific stiffness
Specific strength
Stress-corrosion cracking
Thermal conductivity
Thermal expansion
Thermal stresses

BIBLIOGRAPHY

Callister, D.C., Jr., *Materials Science and Engineering*, 2d ed. New York: John Wiley, 1991.

Flinn, R.A., and P.K. Trojan, *Engineering Materials and Their Applications*, 4th ed. Boston: Houghton Mifflin, 1990.

Fontana, M.G., *Corrosion Engineering*, 3d ed. New York: McGraw-Hill, 1986.

Metals Handbook, 9th ed., Vol. 13: *Corrosion*, 1987. Metals Park, Ohio: ASM International.

Schweitzer, P.A., *Corrosion Resistance Tables: Metals, Plastics, Nonmetallics, and Rubbers*. New York: Marcel Dekker, 1986.

———, *Corrosion and Corrosion Protection Handbook*. New York: Marcel Dekker, 1983.

Shackelford, J.F., *Introduction to Materials Science for Engineers*, 2d ed. New York: Macmillan, 1988.

Uhlig, H.H., and R.W. Revie, *Corrosion and Corrosion Control*, 3rd ed. New York: John Wiley, 1985.

Van Vlack, L.H., *Elements of Materials Science and Engineering*, 6th ed. Reading, Mass.: Addison-Wesley, 1989.

REVIEW QUESTIONS

3.1 What is density? What are the different units in which the density of a material is expressed?

3.2 List reasons why density is an important material property in manufacturing.

3.3 Why is the melting point of a material an important factor in manufacturing processes?

3.4 What might be the adverse effects of using a material with a low specific heat in manufacturing?

3.5 What adverse effects can be caused by thermal expansion of materials?

3.6 Identify material classifications based on electrical properties.

3.7 What is the piezoelectric effect?

3.8 What factors lead to the corrosion of a metal?

3.9 Why can thermal expansion of a material lead to problems in manufacturing? Identify a development in materials technology aimed at reducing the adverse effects of thermal expansion.

3.10 What is passivation?

QUESTIONS AND PROBLEMS

3.11 Describe the significance of structures and machine components made of two materials with different coefficients of thermal expansion.

3.12 Explain your understanding of why cracking can occur in parts when suddenly subjected to extremes of temperature, such as a china dinner plate.

3.13 Low-density materials are preferred for high-speed machine components so that inertial forces are reduced. Do you think that these materials lack certain other properties that may be equally desirable in these applications? Explain. Base your answer on the contents of Chapters 2 and 3.

3.14 From your personal knowledge make lists of parts or components of products that are made of the materials listed in each of the categories in Table 3.2. (*Example:* racing bicycles made with magnesium frames.)

3.15 Are any of the properties discussed in this chapter important for a (a) mechanical pencil, (b) cookie sheet for baking, (c) ruler, (d) paper clip, and (e) door hinge? Explain your answers.

3.16 On the basis of your observation of Table 3.1, describe the differences in the physical properties of metals and their alloys.

3.17 Inspect all visible components (such as bumpers, trim, clips, and dashboard) of an automobile a few years old and one that is the latest model. Make a list of the different materials used for the same components. Explain the reasons for these changes.

3.18 From your own experience make a list of parts, components, and products that have corroded and had to be replaced or discarded.

3.19 Make a survey of the corrosion resistance of metals, by referring to books on materials science and corrosion, and name six pairs of metals that will not undergo galvanic corrosion.

3.20 List applications where the following properties would be desirable: (a) high density, (b) low density, (c) high melting point, (d) low melting point, (e) high thermal conductivity, and (f) low thermal conductivity.

3.21 Assume that you are the instructor for the topics covered in this chapter. Prepare three questions that you would ask to check the knowledge and understanding of your students of these topics; then answer these questions yourself. Your questions must be different from those we have asked.

3.22 Describe your observations concerning the data in Table 3.1.

3.23 If we assume that all the work done in plastic deformation is converted into heat, the temperature rise in a workpiece is directly proportional to the work done per unit volume and inversely proportional to the product of the specific heat and the density of the workpiece. Using Fig. 2.6 and letting the areas under the curves be the work done, calculate the temperature rise for (a) 1020 steel, (b) annealed copper, and (c) 1100-O aluminum.

3.24 Calculate the percent reduction in thickness due to corrosion over a period of 5 years for a structural steel panel that was originally 3 mm thick.

3.25 Describe the reasoning behind the changes in the amount and type of materials used for automobiles.

3.26 Why do you think the listing of the materials in the columns in Table 3.2 for thermal and electrical conductivities have a similar order? Do you see similar trends in other columns? Explain.

3.27 You will note in Table 3.1 that the properties of the alloys of metals have a wide range as compared to the properties of the pure metals. What factors are responsible for this?

3.28 Give several applications in which specific strength and specific stiffness, respectively, are important.

3.29 Does corrosion have any beneficial effects in manufacturing? Explain.

3.30 It can be shown that two physical properties that have a major influence on the cracking of workpieces during thermal cycling are thermal conductivity and coefficient of expansion. Explain why. Would the ductility of the material have any influence on the tendency for thermal cracking? Explain.

3.31 The natural frequency of f of a cantilever beam is given by $f = 0.56\sqrt{EIg/wL^4}$, where E is the modulus of elasticity, I is the moment of inertia, g is the gravitational constant, w is the weight of the beam per unit length, and L is the length of the beam. How does the natural frequency of the beam change, if any, as its temperature is increased?

4

Metal Alloys: Their Structure and Strengthening by Heat Treatment

4.1 Introduction

The properties and behavior of metals and alloys during manufacturing and performance during their service life depend on their composition, structure, and processing history and the heat treatment to which they have been subjected. Important properties such as strength, hardness, ductility, and toughness, as well as resistance to wear and scratching, are greatly influenced and modified by alloying elements and by heat-treatment processes (Fig. 4.1). Improvements in nonheat treatable alloy properties are obtained by cold-working operations, such as rolling, forging, and extrusion.

The most common example of property improvement is the heat treatment of steels. Such treatment modifies microstructures, producing a variety of mechanical properties that are important in manufacturing, such as improved formability and machinability. These properties also improve service performance of the metals when used in machine components (such as gears, cams, and shafts), tools, and dies.

FIGURE 4.1 Cross-section of heat-treated (carburized) gear teeth (thin dark outline along gear periphery). Such heat treatments improve the strength, hardness, and wear resistance of metal components. *Source: Advanced Materials and Processes*, March 1990, p. 43. ASM International.

Heat treatment requires an understanding of several concepts, including the fundamentals of the crystal structure of metals and alloys. In this chapter we follow the outline shown in Fig. 4.2, presenting the effects of various alloying elements, the solubility of one element in another, phases, equilibrium phase diagrams, and the influence of composition, temperature, time, and the rates at which these variables are changed during heat treatment. The properties of most nonferrous alloys and

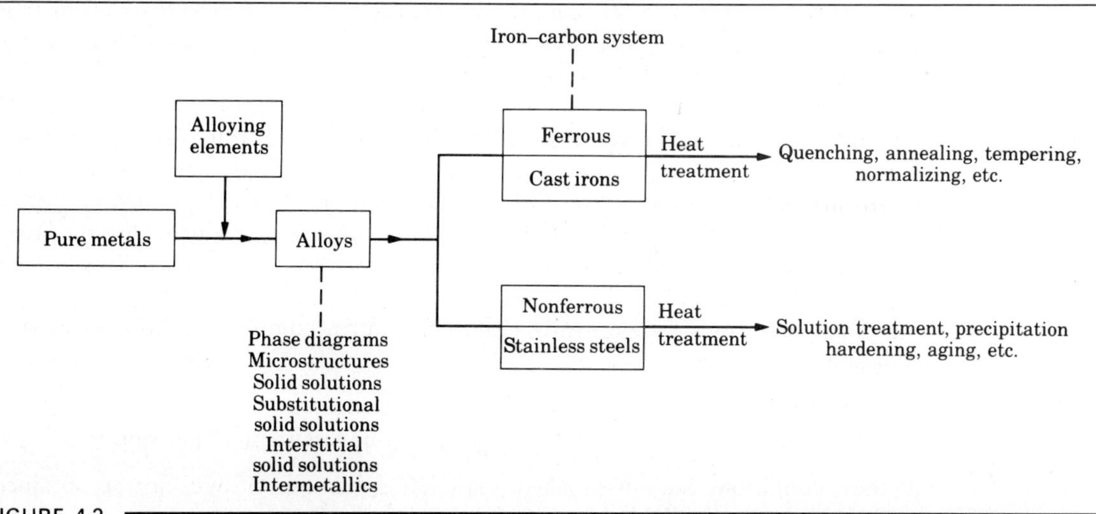

FIGURE 4.2 Outline of topics described in Chapter 4.

stainless steels are also enhanced by heat-treatment techniques, involving mechanisms that are different from those for ferrous alloys. We also discuss methods and techniques of heating, quenching, tempering, and annealing and the characteristics of the equipment involved.

4.2 Structure of Alloys

When describing the basic crystal structure of metals in Chapter 1, we noted that the atoms are all of the *same* type, except for the presence of impurity atoms. These are known as **pure metals**, even though they may not be 100 percent pure. Commercially pure metals are used for various purposes: aluminum for foil, copper for electrical conductors, nickel or chromium for plating, and gold for electrical contacts. Although pure metals have somewhat limited properties, these properties can be enhanced and modified by alloying.

An **alloy** is composed of two or more chemical elements, at least one of which is a metal. The majority of metals used in engineering applications are some form of an alloy. Alloying consists of two basic forms: solid solutions and intermetallic compounds (see also Section 6.14).

4.2.1 Solid solutions

Two terms are essential in describing alloys: **solute** and **solvent.** Solute is the minor element (such as salt or sugar) that is added to the solvent, which is the major element (such as water). In terms of the elements involved in a metal crystal structure, the solute is the element (solute atoms) that is added to the solvent (host atoms). When the particular crystal structure of the solvent is maintained during alloying, the alloy is called a *solid solution.*

Substitutional solid solutions. If the size of the solute atom is similar to that of the solvent atom, the solute atoms can replace solvent atoms and form a *substitutional solid solution* (see Fig. 1.9). An example of this phenomenon is brass, an alloy of zinc and copper, in which zinc (solute atom) is introduced into the lattice of copper (solvent atoms). The properties of brasses can thus be altered over a range by controlling the amount of zinc in copper.

Two conditions are generally required to form complete substitutional solid solutions:

1. The two metals must have similar crystal structures.
2. The difference in their atomic radii should be less than 15 percent.

If these conditions are not satisfied, complete solid solution will not be obtained, and the amount of solid solution formed will be limited.

Interstitial solid solutions. If the size of the solute atom is much smaller than that of the solvent atom, the solute atom occupies an interstitial position and forms an *interstitial solid solution.* The conditions for forming interstitial solutions are:

1. The solvent atom has more than one valence.
2. The atomic radius of the solute atom is less than 59 percent of the solvent atom.

If these conditions are not met, limited or no interstitial solubility may take place.

An important example of interstitial solution is *steel*, which is an alloy of iron and carbon, where carbon atoms are present in an interstitial position between iron atoms. The atomic radius of carbon is 0.71 Å (0.071 nm) and thus is less than 59 percent of the 1.24 Å (0.124 nm) radius of the iron atom. As you will see, we can vary the properties of steel over a wide range by controlling the amount of carbon in iron. This is one reason that, in addition to being inexpensive, steel is such a versatile and useful material with a wide variety of properties and applications.

4.2.2 Intermetallic compounds

Intermetallic compounds are complex structures in which solute atoms are present among solvent atoms in certain proportions. Thus some intermetallic compounds have solid solubility. The type of atomic bonds may range from metallic to ionic. Intermetallic compounds are strong, hard, and brittle.

4.2.3 Two-phase systems

You have seen that a solid solution is one in which two or more elements are soluble in a solid state, forming a single homogeneous solid phase in which the alloying elements are uniformly distributed throughout the solid mass. However, there is a limit to the concentration of solute atoms in a solvent-atom lattice, just as there is a solubility limit to sugar in water. Most alloys consist of two or more solid phases and may be regarded as mechanical mixtures. Thus we call a system with two solid phases a *two-phase system.* Each phase is a homogeneous part of the total mass and has its own characteristics and properties.

Let's consider a mixture of sand and water as an example of a two-phase system. These two different components have their own distinct structure, characteristics, and properties. There is a clear boundary in this mixture between the water (one phase) and sand particles (the second phase). Another example is ice cubes in water. In this case the two phases are a chemical compound of exactly the same chemical elements (hydrogen and oxygen), even though their properties are very different.

A typical example of a two-phase system in metals is lead added to copper in the molten state. After the mixture solidifies, the structure consists of two phases: one having a small amount of lead in solid solution in copper and the other having lead

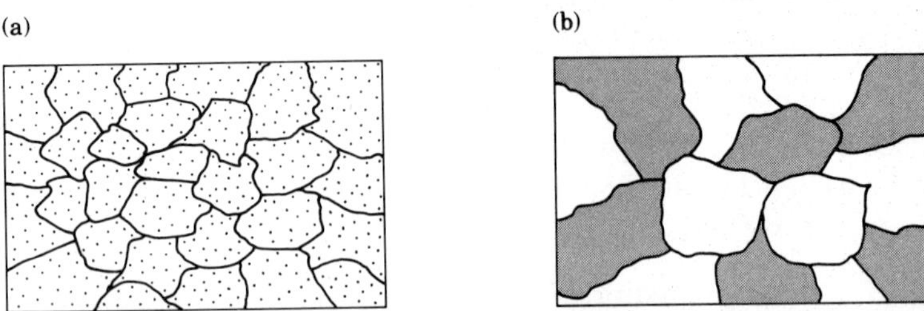

FIGURE 4.3
(a) Schematic illustration of grains, grain boundaries, and particles dispersed throughout the structure of a two-phase system, such as lead–copper alloy. The grains represent lead in solid solution of copper, and the particles are lead as a second phase. (b) Schematic illustration of a two-phase system consisting of two sets of grains: dark and light. Dark and light grains have their own compositions and properties, respectively.

particles, roughly spherical in shape, *dispersed* throughout the structure (Fig. 4.3a). The lead particles are analogous to the sand particles in water that we described. We now find that this copper–lead alloy has properties that are different from those of either copper or lead alone. Lead is also added to steels in this manner to obtain leaded steels, which greatly improves the machinability of steels. Alloying with finely dispersed particles (*second-phase particles*) is an important method of strengthening alloys and controlling their properties. In two-phase alloys the second-phase particles present obstacles to dislocation movement, thus increasing strength.

Another example of a two-phase alloy is the aggregate structure shown in Fig. 4.3(b). In this alloy system there are two sets of grains, each with its own composition and properties. The darker grains may, for example, have a different structure than the lighter grains and be brittle, whereas the lighter grains may be ductile.

Defects may appear during metalworking operations such as forging or extrusion, owing to the lack of ductility of one of the phases in the alloy. In general, two-phase alloys are stronger and less ductile than solid solutions.

4.3 Phase Diagrams

Pure metals have clearly defined melting or freezing points, and solidification takes place at a constant temperature (Fig. 4.4). When the temperature of the molten metal is reduced to the freezing point, the latent heat of solidification is given off while the temperature remains constant. At the end of this thermal cycle, solidification is complete and the solid metal cools to room temperature.

Unlike pure metals, alloys solidify over a *range* of temperatures (Fig. 4.5). Solidification begins when the temperature of the molten metal drops below the

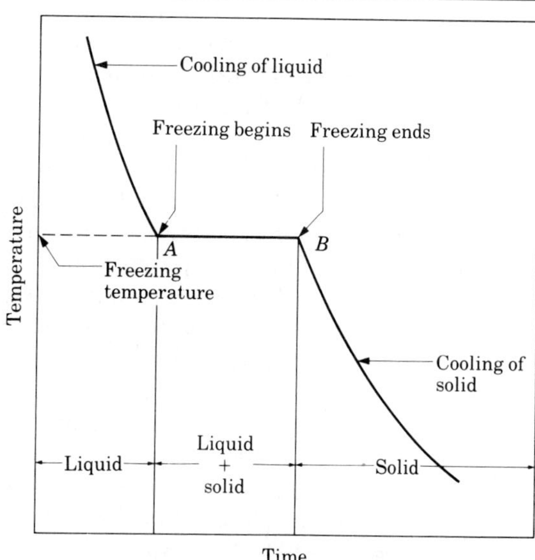

FIGURE 4.4
Cooling curve for the solidification of pure metals. Note that freezing takes place at a constant temperature, during which the latent heat of solidification is given off.

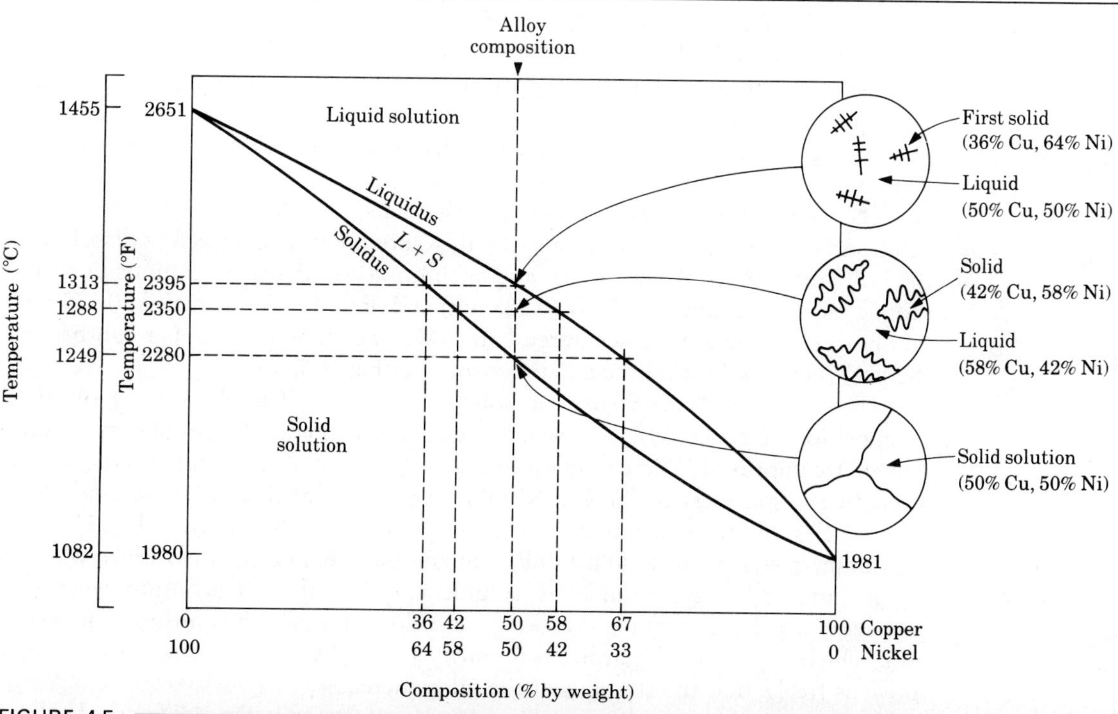

FIGURE 4.5
Phase diagram for nickel–copper alloy system obtained by slow rate of solidification. Note that pure nickel and pure copper each have one freezing or melting temperature. The top circle on the right depicts the nucleation of crystals. The second circle shows the formation of dendrites (see also Section 10.2). The bottom circle shows the solidified alloy with grain boundaries.

liquidus; it is completed when the temperature reaches the *solidus*. Within this temperature range the alloy is in a mushy or pasty state. Its composition and state are described by the particular alloy's phase diagram.

A **phase diagram,** also called an **equilibrium** or a constitutional **diagram,** shows the relationships among temperature, composition, and the phases present in a particular alloy system. *Equilibrium* means that the state of a system remains constant over an indefinite period of time. The word *constitutional* indicates the relationships among the structure, composition, and physical makeup of the alloy.

One example of a phase diagram is shown in Fig. 4.5 for the copper–nickel alloy. It is called a *binary phase diagram* because of the two elements (copper and nickel) in the system. The left boundary of this phase diagram (100% Ni) indicates the melting point of nickel, and the right boundary (100% Cu) the melting point of copper. (All percentages in this discussion are by weight.)

Note that for a composition of, say, 50% Cu–50% Ni, the alloy begins to solidify at a temperature of 1313 °C (2395 °F), and solidification is complete at 1249 °C (2280 °F). Above 1313 °C, a homogeneous liquid of 50% Cu–50% Ni exists. When cooled slowly to 1249 °C, a homogeneous solid solution of 50% Cu–50% Ni results.

However, between the liquidus and solidus curves—at a temperature of 1288 °C (2350 °F)—there is a two-phase region: a *solid phase* composed of 42% Cu–58% Ni, and a *liquid phase* of 58% Cu–42% Ni. To determine the solid composition, go left horizontally to the solidus curve and read down, obtaining 42% Ni. You obtain the liquid composition (58%) similarly by going to the right to the liquidus curve. The procedure for determining the compositions of various phases in phase diagrams is described in detail in texts on materials science and metallurgy.

The completely solidified alloy in the phase diagram shown in Fig. 4.5 is a solid solution because the alloying element (Cu, solute atom) is completely dissolved in the host metal (Ni, solvent atom), and each grain has the same composition. The atomic radius of copper is 1.28 Å (0.128 nm) and that of nickel is 1.25 Å (0.125 nm), and both elements are face-centered cubic. We can show that copper can become the host metal and nickel the solute by rotating Fig. 4.5.

The mechanical properties of solid solutions of Cu–Ni depend on their composition (Fig. 4.6). By increasing the nickel content, the properties of pure copper are improved. There is an optimal percentage of nickel that gives the highest strength and hardness to the Cu–Ni alloy. Figure 4.6 also shows how zinc, as an alloying element in copper, changes the mechanical properties of the alloy. Note the maximum of 40 percent solid solubility for zinc (solute) in copper (solvent), whereas copper and nickel are completely soluble in each other. The improvements in properties are due to pinning (blocking) of dislocations at solute atoms of nickel or zinc, which may also be regarded as impurity atoms. As a result, dislocations cannot move as freely, and the strength of the alloy increases.

Another example of a binary phase diagram is shown in Fig. 4.7 for the lead–tin system. The single phases alpha and beta are solid solutions. Note that the single-phase regions are separated from the liquid phase by two two-phase regions: alpha + liquid and beta + liquid.

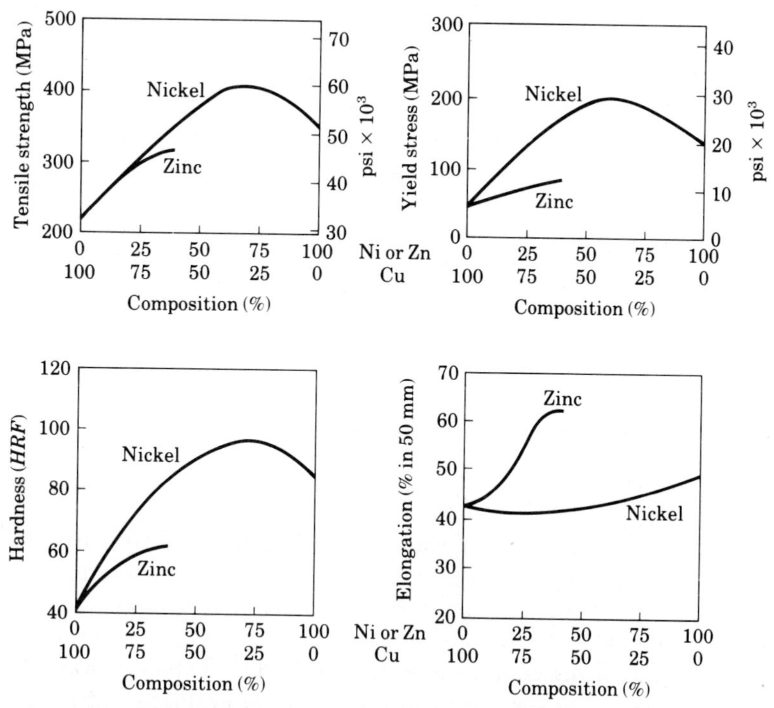

FIGURE 4.6

Mechanical properties of copper–nickel and copper–zinc alloys as a function of their composition. The curves for zinc are short because zinc has a maximum solid solubility of 40 percent in copper. *Source:* L. H. Van Vlack, *Materials for Engineering.* Addison-Wesley Publishing Co., Inc., 1982

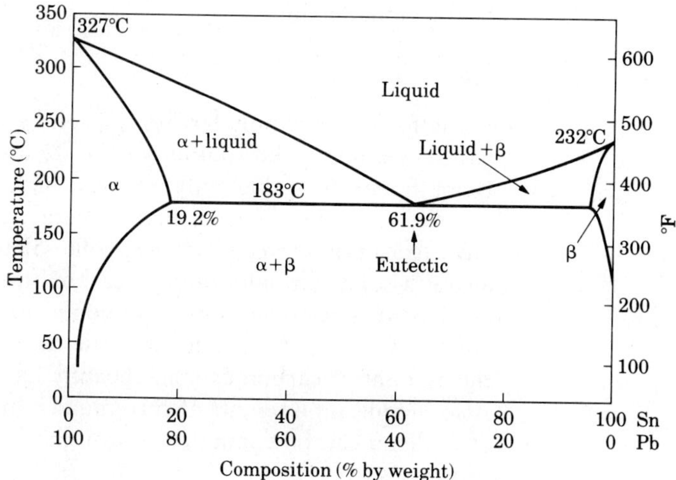

FIGURE 4.7

The lead–tin phase diagram. Note that the composition of the eutectic point for this alloy is 61.9% Sn–28.1% Pb. A composition either lower or higher than this ratio will have a higher liquidus temperature.

Figure 4.7 shows the composition of the alloy (61.9% Sn–38.1% Pb) that has the *lowest* temperature at which the alloy is still completely liquid, namely, 183 °C (361 °F). This point is known as the **eutectic point**. The word *eutectic* is from the Greek *eutektos*, meaning easily melted.

Eutectic points are important in applications such as soldering, where low temperatures may be required to prevent thermal damage to parts during joining. Although there are various types of solders, tin–lead solders are commonly used for general applications, and have a composition ranging from 5% Pb–95% Sn to 70% Pb–30% Sn. Each composition has its own melting point.

4.4 The Iron–Carbon System

As we describe in Chapter 5, steels, ferrous alloys, cast irons, and cast steels are used extensively because of their versatile properties and low cost. Steels and cast irons are represented by the iron–carbon binary system. Commercially pure iron contains up to 0.008% C, steels up to 2.11% C, and cast irons up to 6.67% C, although most cast irons contain less than 4.5% C. In this section we discuss the iron–carbon system so that you can learn how to evaluate and modify the properties of these important materials for specific applications.

The *iron–iron carbide phase diagram* is shown in Fig. 4.8. Although this diagram can be extended to the right—to 100 percent carbon (pure graphite)—the range that is significant to engineering applications is up to 6.67% C where cementite forms.

Pure iron melts at a temperature of 1538 °C (2798 °F), as shown at the left boundary in Fig. 4.8. As iron cools, it first forms delta ferrite, then austenite, and finally alpha ferrite.

4.4.1 Ferrite

Alpha ferrite, or simply **ferrite**, is a solid solution of body-centered cubic iron and has a maximum solid solubility of 0.022% C at a temperature of 727 °C (1341 °F). Delta ferrite is stable only at very high temperatures and is of no practical significance in engineering. Just as there is a solubility limit to salt in water—with any extra amount precipitating as solid salt at the bottom of the container—so there also is a solid solubility limit to carbon in iron.

Ferrite is relatively soft and ductile and is magnetic from room temperature to 768 °C (1414 °F). Although very little carbon can dissolve interstitially in bcc iron, the amount of carbon can significantly affect the mechanical properties of ferrite. Also, significant amounts of chromium, manganese, nickel, molybdenum, tungsten, and silicon can be contained in iron in solid solution, imparting certain desirable properties.

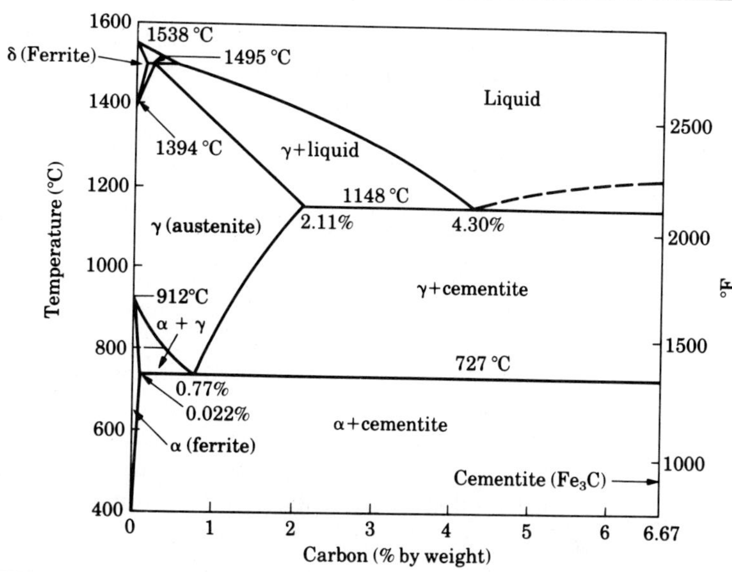

FIGURE 4.8 The iron–iron carbide phase diagram. Because of the importance of steel as an engineering material, this diagram is one of the most important phase diagrams.

4.4.2 Austenite

Between 1394 °C (2541 °F) and 912 °C (1674 °F), iron undergoes a *polymorphic transformation* from the bcc to fcc structure, becoming what is known as gamma iron or, more commonly, **austenite** (after W. R. Austen, 1843–1902). This structure has a solid solubility of up to 2.11% C at 1148 °C (2098 °F). Thus the solid solubility of austenite is about two orders of magnitude higher than that of ferrite, with the carbon occupying interstitial positions (Fig. 4.9).

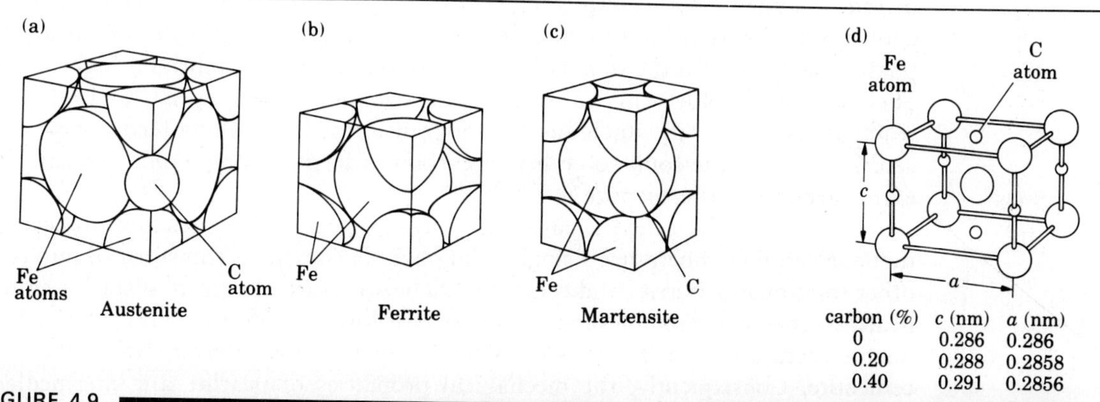

carbon (%)	c (nm)	a (nm)
0	0.286	0.286
0.20	0.288	0.2858
0.40	0.291	0.2856

FIGURE 4.9 The unit cell for (a) austenite, (b) ferrite, and (c) martensite. The effect of percentage of carbon (by weight) on the lattice dimensions for martensite is shown in (d). Note the interstitial position of the carbon atoms (see Fig. 1.9) and the increase in dimension *c* with increasing carbon content. Thus the unit cell of martensite is in the shape of a rectangular prism.

Austenite is an important phase in the heat treatment of steels. It is denser than ferrite and its single-phase fcc structure is ductile at elevated temperatures; thus it possesses good formability. Large amounts of nickel and manganese can also be dissolved in fcc iron to impart various properties. Steel is nonmagnetic in the austenitic form, either at high temperatures or at room temperature for austenitic stainless steels.

4.4.3 Cementite

The right boundary of Fig. 4.8 represents **cementite**, which is 100 percent iron carbide (Fe_3C), with a carbon content of 6.67 percent. Cementite, from the Latin *caementum*, meaning stone chips (as in making mortar), is also called *carbide*. This carbide should not be confused with carbides used for tool and die materials, which are generally tungsten carbide, titanium carbide, or silicon carbide used as dies, cutting tools, and abrasives. Cementite is a very hard and brittle intermetallic compound and has a significant influence on the properties of steels. It can include other alloying elements, such as chromium, molybdenum, and manganese.

4.5 The Iron–Iron Carbide Phase Diagram and Development of Microstructures in Steels

The region of the iron–iron carbide phase diagram that is significant for steels is shown in Fig. 4.10, an enlargement of the lower left portion of Fig. 4.8. Various microstructures can be developed, depending on the carbon content and the method of heat treatment. For example, let's consider iron with a 0.77% C content being cooled very slowly from a temperature of, say, 1100 °C (2000 °F) in the austenite phase. The reason for the *slow* cooling rate is to maintain equilibrium; higher rates of cooling are employed in heat treating, as we explain later. At 727 °C (1341 °F) a reaction takes place in which austenite is transformed into alpha ferrite (bcc) and cementite. Since the solid solubility of carbon in ferrite is only 0.022 percent, the extra carbon forms cementite.

This reaction is called a **eutectoid** (meaning *eutecticlike*) **reaction** indicating that at a certain temperature a single solid phase (austenite) is transformed into two other solid phases (ferrite and cementite). The structure of eutectoid steel is called **pearlite** because it resembles mother of pearl at low magnifications (Fig. 4.11). The microstructure of pearlite consists of alternating layers (*lamellae*) of ferrite and cementite. Consequently, the mechanical properties of pearlite are intermediate between ferrite (soft and ductile) and cementite (hard and brittle).

In iron with less than 0.77% C, the microstructure formed consists of a pearlite phase (ferrite and cementite) and a ferrite phase. The ferrite in the pearlite is called

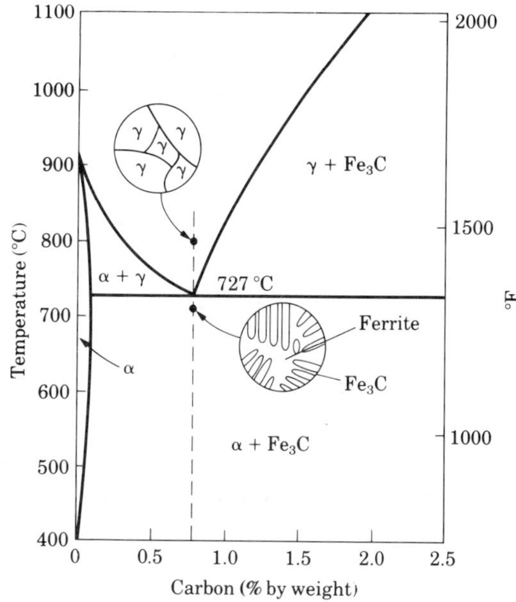

FIGURE 4.10 Schematic illustration of the microstructures for an iron–carbon alloy of eutectoid composition (0.77 percent carbon) above and below the eutectoid temperature of 727 °C (1341 °F).

eutectoid ferrite. The ferrite phase is called *proeutectoid ferrite*, because it forms at a temperature higher than the eutectoid temperature of 727 °C (1341 °F), *pro* meaning before. If the carbon content is greater than 0.77 percent, the austenite transforms into pearlite and cementite. The cementite in the pearlite is called *eutectoid cementite*, and the cementite phase is called *proeutectoid cementite*, because it forms at a temperature higher than the eutectoid temperature.

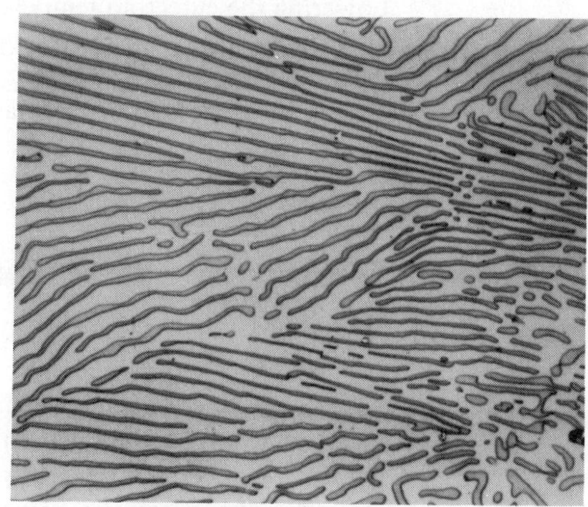

FIGURE 4.11 Microstructure of pearlite in 1080 steel, formed from austenite of eutectoid composition. In this lamellar structure, the lighter regions are ferrite and the darker regions are carbide. Magnification: 2500 ×. *Source:* Courtesy of USX Corporation.

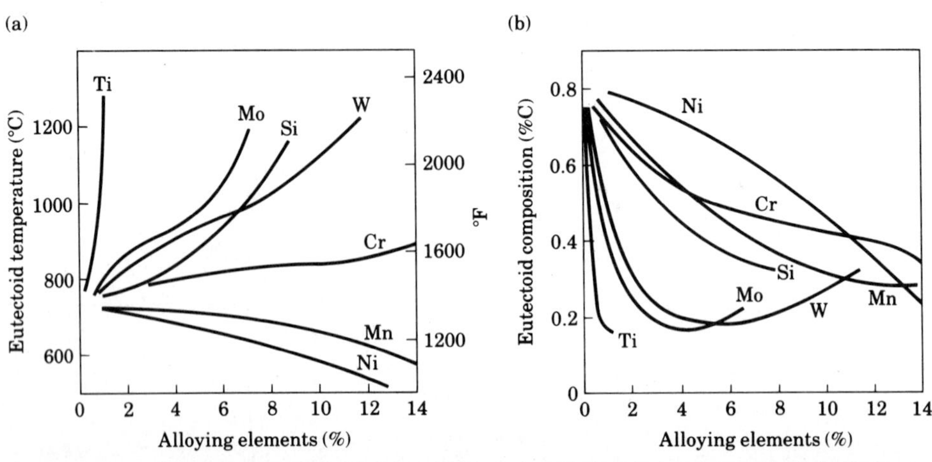

FIGURE 4.12
The effect of alloying elements on the (a) eutectoid temperature and (b) eutectoid composition for steel. *Source:* E. C. Bain.

4.5.1 Effects of alloying elements in iron

Although carbon is the basic element that transforms iron into steel, other elements are also added to impart a variety of desirable properties. The effect of these alloying elements on the iron–iron carbide phase diagram is to shift the eutectoid temperature and eutectoid composition (percentage of carbon in steel at the eutectoid point).

These effects are shown in Figs. 4.12(a) and (b). Note that the eutectoid temperature may be raised or lowered from 727 °C (1341 °F) depending on the particular alloying element. On the other hand, alloying elements always lower the eutectoid composition; that is, the carbon content is lower than 0.77 percent. Lowering the eutectoid temperature means increasing the austenite range. Thus an alloying element such as nickel is known as an *austenite former*. Since nickel has an fcc structure, it is believed that it favors the fcc structure of austenite. Conversely, chromium and molybdenum have the bcc structure, causing these elements to favor the bcc structure of ferrite; these elements are known as *ferrite formers*.

4.6 Cast Irons

The term **cast iron** refers to a family of ferrous alloys composed of iron, carbon (ranging from 2.11 percent to about 4.5 percent), and silicon (up to about 3.5 percent). Cast irons are usually classified according to their solidification morphology, as follows (see also Section 12.3):

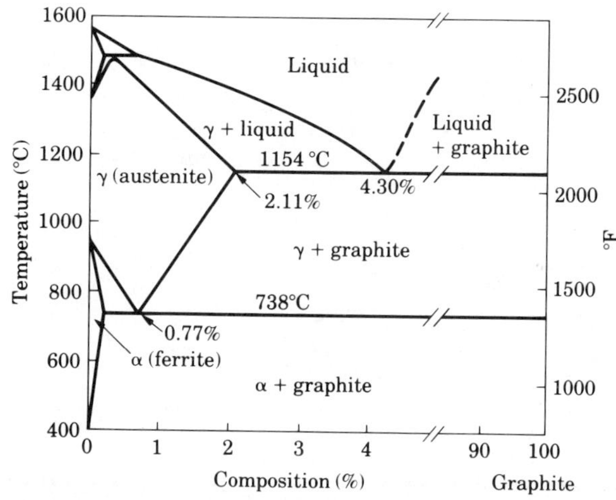

FIGURE 4.13
Phase diagram for the iron–carbon system with graphite, instead of cementite, as the stable phase. Note that this figure is an extended version of Fig. 4.8.

a) Gray cast iron, or gray iron.
b) Nodular cast iron, ductile cast iron, or spheroidal graphite cast iron.
c) White cast iron.
d) Malleable iron.
e) Compacted graphite iron.

Cast irons are also classified by their structure: ferrite, pearlite, quenched and tempered, or austempered.

The equilibrium diagram relevant to cast irons is shown in Fig. 4.13, in which the right boundary is 100% C, that is, pure graphite. The horizontal liquid + graphite line is at 1154 °C (2109 °F). Thus cast irons are completely liquid at temperatures lower than those required for liquid steels. Consequently, cast irons have lower melting temperatures, which is why the *casting process* is so suitable for iron with high carbon content.

Although cementite exists in steels almost indefinitely, it is not completely stable. That is, it is *metastable*, with an extremely low rate of decomposition. However, cementite can be made to decompose into alpha ferrite and graphite. The formation of graphite (*graphitization*) can be controlled, promoted, and accelerated by modifying the composition and the rate of cooling, and by the addition of silicon.

4.6.1 Gray cast iron

In this structure, graphite exists largely in the form of flakes (Fig. 4.14a). It is called *gray cast iron*, or gray iron, because when broken, the fracture path is along the graphite flakes and thus it has a gray sooty appearance. These flakes act as stress

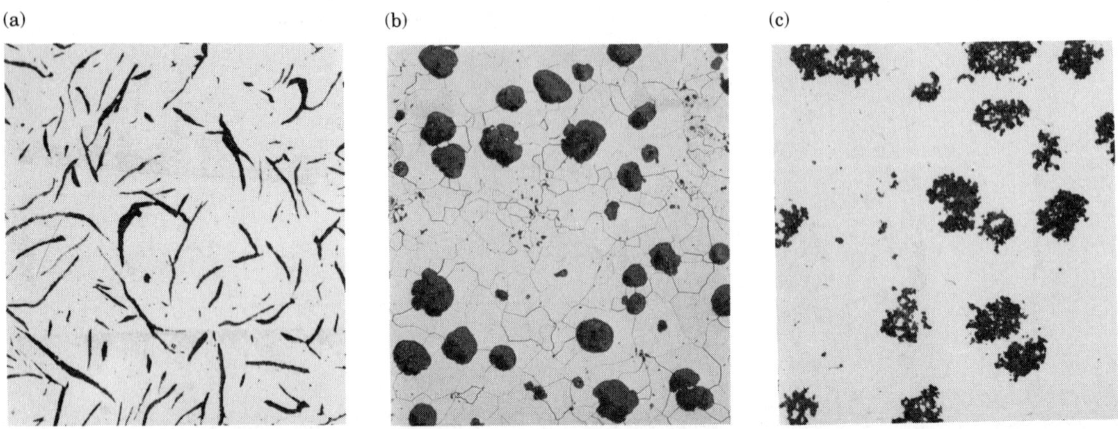

FIGURE 4.14
Microstructure for cast irons. Magnification: 100×. (a) Ferritic gray iron with graphite flakes. (b) Ferritic nodular iron, (ductile iron) with graphite in nodular form. (c) Ferritic malleable iron. This cast iron solidified as white cast iron, with the carbon present as cementite, and was heat treated to graphitize the carbon. *Source:* ASM International.

raisers. Consequently, gray iron has negligible ductility and is weak in tension, although strong in compression, as are other brittle materials. On the other hand, the presence of graphite flakes gives this material the capacity to dampen vibrations caused by internal friction and thus energy dissipation caused by these flakes. This capacity makes gray cast iron a suitable and commonly used material for constructing machine tool bases and structures.

Various types of gray cast iron are called *ferritic*, *pearlitic*, and *martensitic*. Because of the different structures, each has different properties and applications. In ferritic gray iron, also known as fully gray iron, the structure consists of graphite flakes in an alpha ferrite matrix. Pearlitic gray iron has a structure of graphite in a matrix of pearlite. Although still brittle, it is stronger than gray iron. Martensitic gray iron is obtained by austenitizing a pearlitic gray iron, followed by rapid quenching to produce a structure of graphite in a martensite matrix. As a result, this cast iron is very hard.

4.6.2 Nodular iron (ductile iron)

In this structure graphite is in *nodular*, or *spheroid*, form (Fig. 4.14b). This shape permits the material to be somewhat ductile and shock resistant. The shape of graphite flakes is changed into nodules (spheres) by small additions of magnesium and/or cerium to the molten metal prior to pouring. Nodular iron can be made ferritic or pearlitic by heat treatment. It can also be heat treated to obtain a structure of tempered martensite (Section 4.7.6).

4.6.3 White cast iron

This structure is very hard, wear resistant, and brittle because of the presence of large amounts of iron carbide, instead of graphite. White cast iron is obtained either by rapid cooling of gray iron, or by adjusting the composition by keeping the carbon and silicon content low. This cast iron is also called *white iron* because the lack of graphite gives a white crystalline appearance to the fracture surface.

4.6.4 Malleable iron

Malleable iron is obtained by annealing white cast iron in an atmosphere of carbon monoxide and carbon dioxide, between 800 °C and 900 °C (1470 °F and 1650 °F) for up to several hours, depending on the size of the part. During this process the cementite decomposes (*dissociates*) into iron and graphite. The graphite exists as *clusters* (Fig. 4.14c) in a ferrite or pearlite matrix and thus has a structure similar to nodular iron. This structure promotes ductility, strength, and shock resistance; hence the term *malleable* (meaning can be hammered, from the Latin *malleus*).

4.6.5 Compacted-graphite iron

The graphite in this structure is in the form of short, thick, and interconnected flakes with undulating surfaces and rounded extremities. The mechanical and physical properties of this cast iron are intermediate between those of flake graphite and nodular graphite cast irons.

4.7 Heat Treatment of Ferrous Alloys

The various microstructures that we have described thus far can be modified by **heat treatment** techniques, that is, by controlled heating and cooling of the alloys at various rates. These treatments induce **phase transformations** that greatly influence mechanical properties, such as strength, hardness, ductility, toughness, and wear resistance of the alloys.

The effects of thermal treatment depend primarily on the alloy, its composition and microstructure, the degree of prior cold work, and the rates of heating and cooling during heat treatment. The processes of recovery, recrystallization, and grain growth that we described in Section 1.6 are examples of thermal treatment, involving changes in the grain structure of the alloy.

In this section we discuss the microstructural changes in the iron–carbon system. Because of their technological significance, the structures we consider are pearlite, spheroidite, bainite, martensite, and tempered martensite. The heat-treatment processes described are annealing, quenching, and tempering and a commonly used test to determine hardenability of steels.

4.7.1 Pearlite

If the ferrite and cementite lamellae in the pearlite structure of the eutectoid steel shown in Fig. 4.11 are thin and closely packed, the microstructure is called *fine pearlite*. If thick and widely spaced, it is called *coarse pearlite*. The difference between the two depends on the rate of cooling through the eutectoid temperature, a reaction in which austenite is transformed into pearlite. If the rate of cooling is relatively high, as in air, fine pearlite is produced; if slow, as in a furnace, coarse pearlite is produced.

The transformation from austenite to pearlite (and for other structures) is best illustrated by Figs. 4.15(b) and (c). These diagrams are called *isothermal transformation (IT) diagrams*, or *time-temperature-transformation (TTT) diagrams*. They are constructed from the data in Fig. 4.15(a), which shows the percentage of austenite

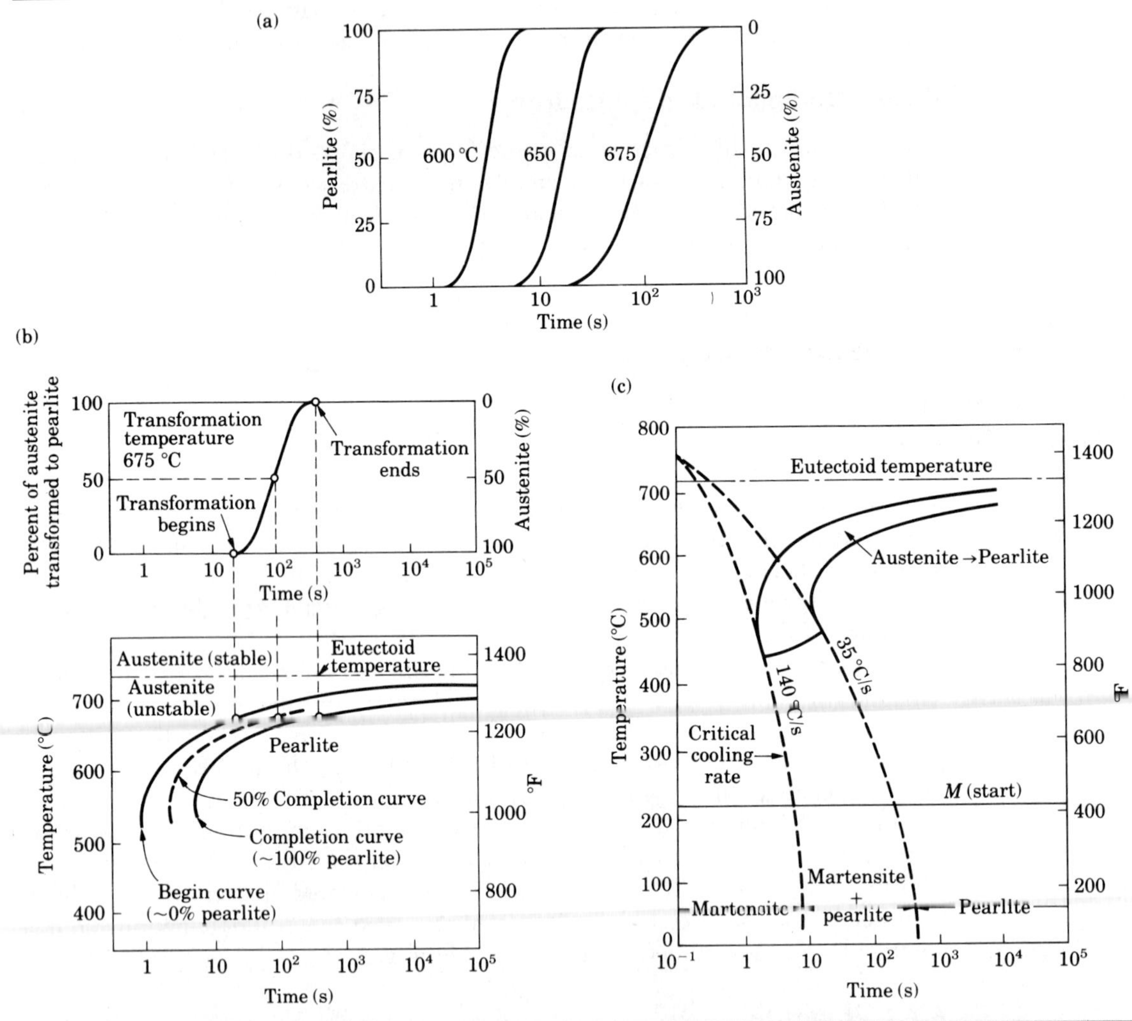

FIGURE 4.15
(a) Austenite to pearlite transformation of iron–carbon alloy as a function of time and temperature. (b) Isothermal transformation diagram obtained from (a) for a transformation temperature of 675 °C (1247 °F). (c) Microstructures obtained for a eutectoid iron–carbon alloy as a function of cooling rate. *Source:* ASM International.

transformed into pearlite as a function of temperature and time. The higher the temperature and/or the longer the time, the greater is the percentage of austenite transformed to pearlite. Note that for each temperature there is a minimum time for the transformation to begin and that some time later all the austenite is transformed to pearlite. You can trace this transformation in Figs. 4.15(b) and (c).

The differences in hardness and toughness of various structures obtained are shown in Fig. 4.16. Fine pearlite is harder and less ductile than coarse pearlite. The effects of various percentages of carbon, cementite, and pearlite on other mechanical properties of steels are shown in Fig. 4.17.

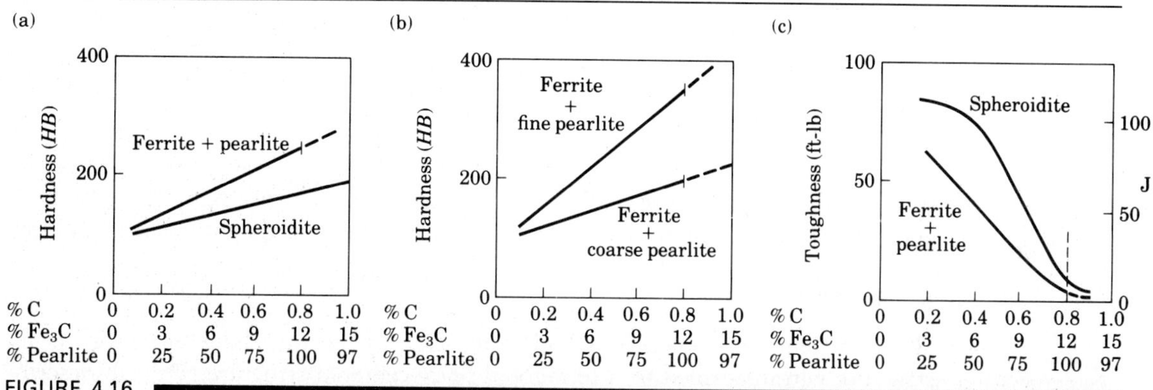

FIGURE 4.16 (a) and (b) Hardness and (c) toughness for annealed plain-carbon steels as a function of carbide shape. Carbides in the pearlite are lamellar. Fine pearlite is obtained by increasing the cooling rate. The spheroidite structure has spherelike carbide particles. Note that the percentage of pearlite begins to decrease after 0.77 percent carbon. *Source:* L. H. Van Vlack, *Materials for Engineering*. Addison-Wesley Publishing Co., Inc., 1982.

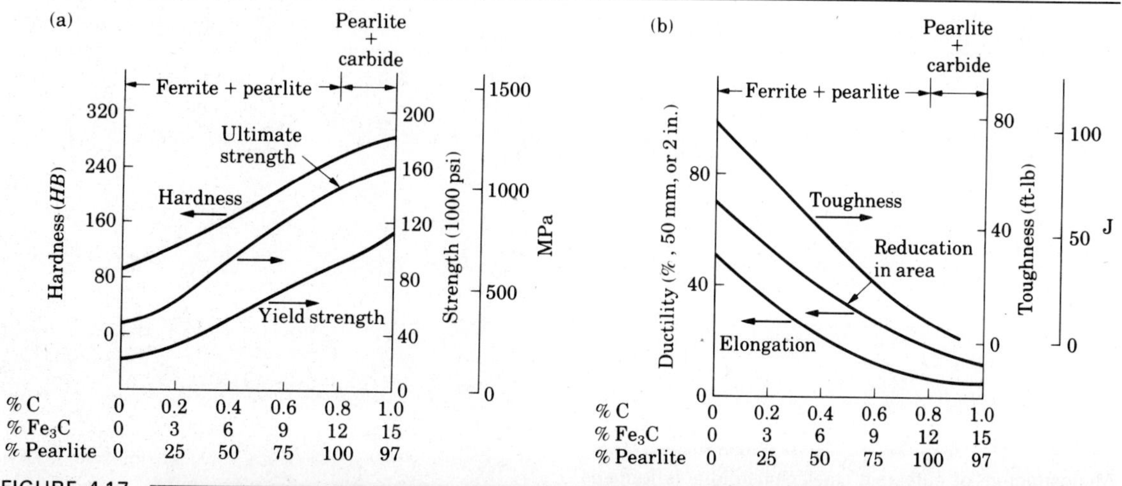

FIGURE 4.17 Mechanical properties of annealed steels as a function of composition and microstructure. Note in (a) the increase in hardness and strength, and in (b) the decrease in ductility and toughness, with increasing amounts of pearlite and iron carbide. *Source:* L. H. Van Vlack, *Materials for Engineering*. Addison-Wesley Publishing Co., Inc., 1982.

4.7.2 Spheroidite

When pearlite is heated to just below the eutectoid temperature and held at that temperature for a period of time, such as for a day at 700 °C (1300 °F), the cementite lamellae transform to *spherical* shapes (Fig. 4.18). Unlike the lamellar shape of cementite, which acts as stress raisers, **spheroidites** (spherical particles) are less conducive to stress concentration because of their rounded shapes. Consequently, this structure has higher toughness and lower hardness than the pearlite structure. In this form it can be cold worked since the ductile ferrite has high toughness, and the spheroidal carbide particles prevent the propagation of cracks within the material.

4.7.3 Bainite

Visible only using electron microscopy, **bainite** is a very fine microstructure, consisting of ferrite and cementite. It can be produced in steels with alloying elements and at cooling rates that are higher than those required for transformation to pearlite. This structure, called *bainitic steels* (after E. C. Bain, 1891–1971), is generally stronger and more ductile than pearlitic steels at the same hardness level.

4.7.4 Martensite

When austenite is cooled at a high rate, such as by quenching in water, its fcc structure is transformed to a tetragonal body-centered structure. We can describe it as a body-centered rectangular prism, which is slightly elongated along one of its principal axes (Fig. 4.9d). This microstructure is called **martensite** (after A. Martens, 1850–1914).

Because this structure does not have as many slip systems as a bcc structure —and the carbon is in interstitial positions—martensite is extremely hard and

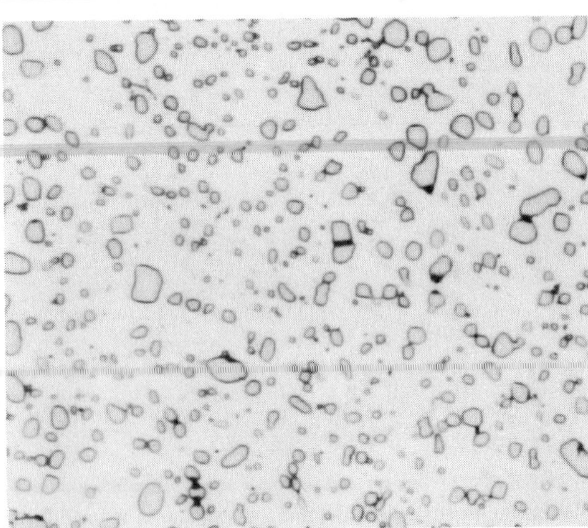

FIGURE 4.18 Microstructure of eutectoid steel. Spheroidite is formed by tempering the steel at 700 °C (1292 °F). Magnification: 1000 ×. *Source:* Courtesy of USX Corporation.

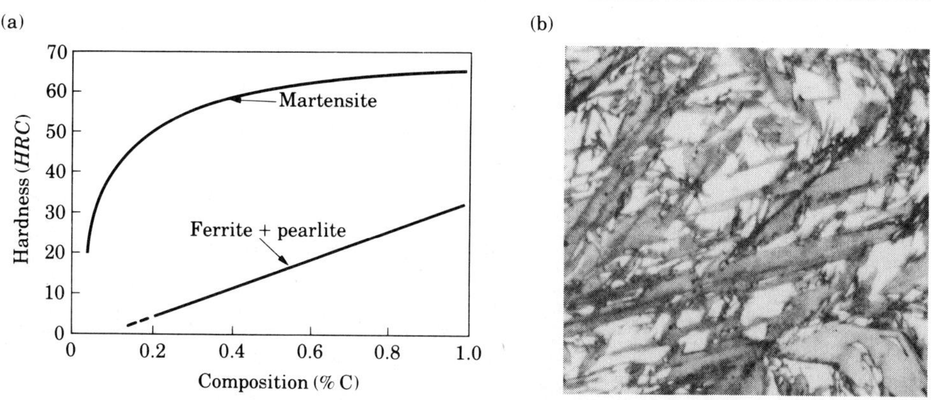

FIGURE 4.19
(a) Hardness of martensite as a function of carbon content. (b) Micrograph of martensite with 0.8 percent carbon. The gray platelike regions are martensite, and have the same composition as the original austenite (white regions). Magnification: 1000 ×. *Source:* Courtesy of USX Corporation.

brittle (Fig. 4.19), lacks toughness, and thus has limited use. Martensite transformation takes place almost instantaneously (Fig. 4.15c) because it does not involve the diffusion process, a time-dependent phenomenon that is the mechanism in other transformations.

Transformations involve volume changes because of the different densities of the various phases present in the structure. For example, when austenite transforms to martensite, its volume increases (its density decreases) by as much as 4 percent. A similar but smaller volume expansion also occurs when austenite transforms to pearlite. These expansions, and the thermal gradients present in a quenched part, cause internal stresses within the body. They may cause parts to crack during heat treatment, such as *quench cracking* of steels caused by rapid cooling during quenching.

4.7.5 Retained austenite

If the temperature to which the alloy is quenched is not sufficiently low, only a portion of the structure is transformed to martensite. The rest is *retained austenite*, which is visible as white areas in the structure along with dark needlelike martensite. Retained austenite can cause dimensional instability and cracking and lowers the hardness and strength of the alloy.

4.7.6 Tempered martensite

In order to improve martensite's properties, it is tempered. **Tempering** is a heating process by which hardness is reduced and toughness is improved. The body-centered tetragonal martensite is heated to an intermediate temperature where it transforms to a two-phase microstructure, consisting of body-centered-cubic alpha

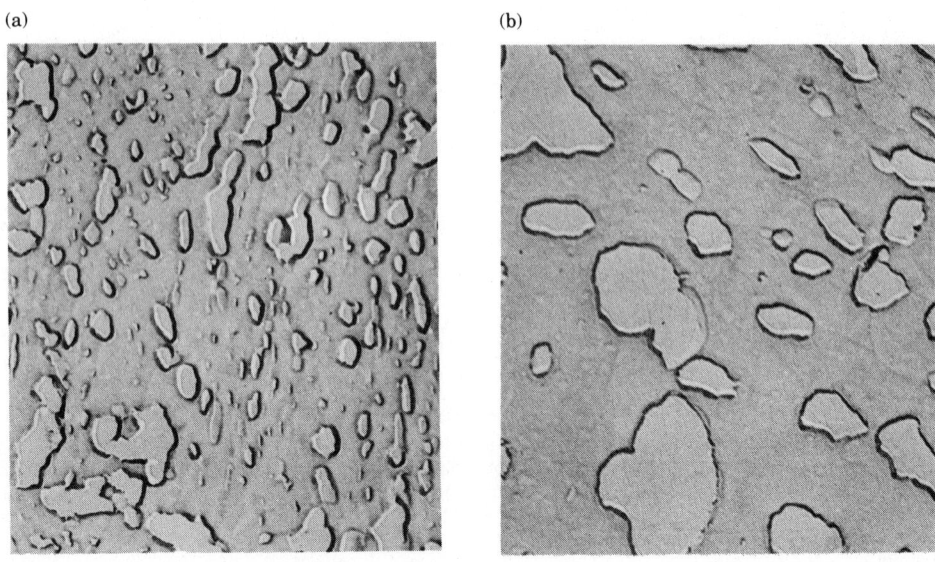

FIGURE 4.20
Growth of carbide particles in tempered martensite: (a) one hour at 590 °C (1094 °F); (b) twelve hours at 675 °C (1247 °F). The two compositions are identical. Magnification: 11,000×. *Source:* Courtesy of ASTM and General Motors Research Laboratories.

ferrite and small particles of cementite (Fig. 4.20a). With increasing tempering time and temperature, the hardness of tempered martensite decreases (Fig. 4.21). The reason is that the cementite particles coalesce and grow, and the distance between the particles in the soft ferrite matrix increases as the less stable, smaller carbide particles dissolve (Fig. 4.20b).

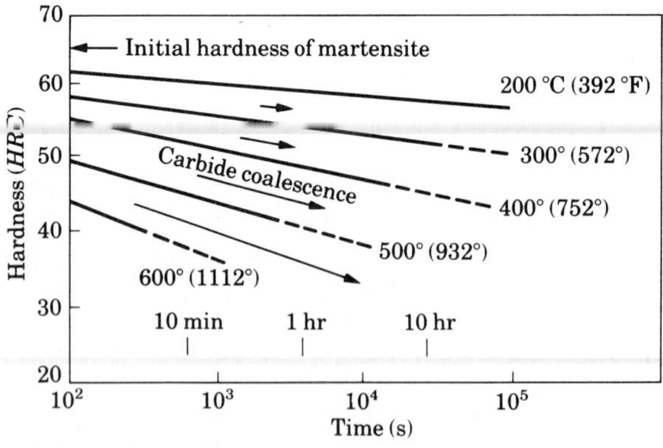

FIGURE 4.21
Hardness of tempered martensite as a function of tempering time for 1080 steel, quenched to 65 HRC. Hardness decreases because the carbide particles coalesce and grow in size, thus increasing the interparticle distance of the softer ferrite.

4.8 Hardenability of Ferrous Alloys

The capability of an alloy to be hardened by heat treatment is called its **hardenability**. It is a measure of the depth of hardness that can be obtained by heating and subsequent quenching. The term hardenability should not be confused with hardness, which is the resistance of a material to indentation or scratching.

From the discussion thus far, you can reason that hardenability of ferrous alloys depends on the carbon content, the grain size of the austenite, the alloying elements present in the material, and the cooling rate. A test—the Jominy test—has been developed in order to determine the hardenability of an alloy.

4.8.1 The end-quench hardenability test

In the commonly used **Jominy test** (after W. E. Jominy, 1893–1976), a round 100 mm (4 in.) long test bar made from the particular alloy is *austenitized*, that is, heated to the proper temperature to form 100 percent austenite. It is then quenched directly at one end (Fig. 4.22a) with a stream of water at 24 °C (75 °F). The cooling

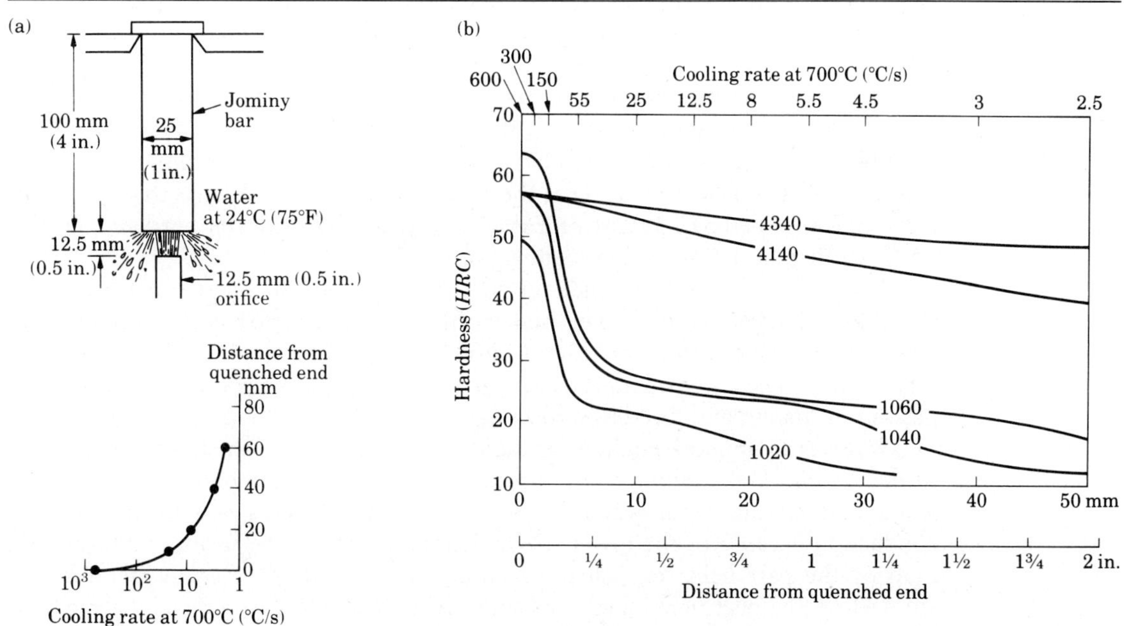

FIGURE 4.22 (a) End-quench test and cooling rate. (b) Hardenability curves for five different steels, as obtained from the end-quench test. Small variations in composition can change the shape of these curves. Each curve is actually a band, and its exact determination is important in heat treatment of metals for better control of properties. *Source:* L. H. Van Vlack, *Materials for Engineering.* Addison-Wesley Publishing Co., Inc., 1982.

rate thus varies throughout the length of the bar, the rate being highest at the lower end which is in contact with the water. The hardness along the length of the bar is then measured at various distances from the quenched end and plotted (Fig. 4.22b).

As expected from our discussion of the effects of cooling rates in Section 4.7, the hardness decreases away from the end of the bar. The greater the depth to which hardness penetrates, the greater is the hardenability of the alloy. Each composition of an alloy has its particular *hardenability band*. Note that the hardness at the quenched end increases with increasing carbon content and that 1040, 4140, and 4340 steels have the same carbon content (0.40 percent) and thus hardness (57 HRC) at the quenched end.

Because small variations in composition and grain size can affect the shape of hardenability curves, each lot of an alloy should be tested individually. The data may be plotted as a band, rather than a single curve. Hardenability curves are necessary in predicting the hardness of heat-treated components, such as gears, cams, and various other components, as a function of their composition.

4.8.2 Quenching media

The fluid used for quenching the heated alloy also has an effect on hardenability. Quenching may be carried out in water, brine (saltwater), oils, molten salts, or air. Caustic solutions, polymer solutions, and gases are also used. Because of the differences in the thermal conductivity, specific heat, and heat of vaporization of these media, the rate of cooling of the alloy (*severity of quench*) will also be different.

In relative terms and in decreasing order, the cooling capacity of several quenching media is: agitated brine 5; still water 1; still oil 0.3; cold gas 0.1; and still air 0.02. Agitation is also a significant factor in the rate of cooling. The more vigorous the agitation, the higher is the rate of cooling. In tool steels the quenching medium is specified by a letter (see Table 5.7), such as W for water hardening, O for oil hardening, and A for air hardening.

The cooling rate also depends on the surface area-to-thickness or surface area-to-volume ratio of the part. The higher this ratio, the higher is the cooling rate. Thus, for example, a thick plate cools more slowly than a thin plate with the same surface area. These considerations are also significant in the cooling of metals and plastics in casting and molding processes.

Water is a common medium for rapid cooling. However, the heated metal may form a *vapor blanket* along its surfaces from water-vapor bubbles that form when water boils at the metal–water interface. This blanket creates a barrier to heat conduction because of the lower thermal conductivity of the vapor. Agitating the fluid or the part helps to reduce or eliminate the blanket. Also, water may be sprayed on the part under high pressure. Brine is an effective quenching medium because salt helps to nucleate bubbles at the interfaces, thus improving agitation. However, brine can corrode the part.

Polymer quenchants have been used for almost thirty years for ferrous as well as nonferrous alloy quenching, and new compositions are being developed regularly.

They have cooling characteristics that are generally between water and petroleum oils. Typical polymer quenchants are polyvinyl alcohol, polyalkaline oxide, polyvinyl pyrrolidone, and polyethyl oxazoline. These quenchants have advantages such as better control of hardness results, elimination of fumes and fire (as when oils are used as a quenchant), and corrosion (as when water is used). The quenching rate can be controlled by varying the concentration of the solutions.

4.8.3 Design considerations for heat treating

In addition to the metallurgical factors described, successful heat treating involves design considerations of avoiding problems such as cracking, warping, and nonuniform properties throughout the heat-treated part. The rate of cooling during quenching may not be uniform, particularly with complex shapes with varying cross-sections and thicknesses, thus producing severe temperature gradients. This leads to variations in contraction, which induces thermal stresses and may cause cracking of the part. Furthermore, nonuniform cooling causes residual stresses in the part, which can lead to stress-corrosion cracking. Thus the method selected and care taken in quenching, as well as the proper choice of quenching media and temperatures, are important considerations.

As a general guideline for part design for heat treating, internal or external sharp corners should be avoided. Otherwise stress concentrations at these corners raise the level of stresses high enough to cause cracking. Parts should have as nearly uniform thicknesses as possible, or the transition between regions of different thicknesses should be smooth. Parts with holes, grooves, keyways, splines, and unsymmetrical shapes may also be difficult to heat treat because they may crack during quenching. Large surfaces with thin cross-sections are likely to warp. Hot forgings and hot steel-mill products may have a *decarburized skin* (loss of carbon, Section 4.10) and thus may not successfully respond to heat treatment.

4.9 Heat Treatment of Nonferrous Alloys and Stainless Steels

Nonferrous alloys and some stainless steels generally cannot be heat treated by the techniques used with ferrous alloys. The reason is that nonferrous alloys do not undergo phase transformations as steels do. The hardening and strengthening mechanisms for these alloys are fundamentally different.

Heat-treatable aluminum alloys, copper alloys, and martensitic and precipitation-hardening stainless steels are hardened and strengthened by a process called **precipitation hardening.** This heat treatment is a technique in which small particles—of a different phase and called *precipitates*—are uniformly dispersed in the matrix of the original phase (Fig. 4.3a). In this process precipitate forms because

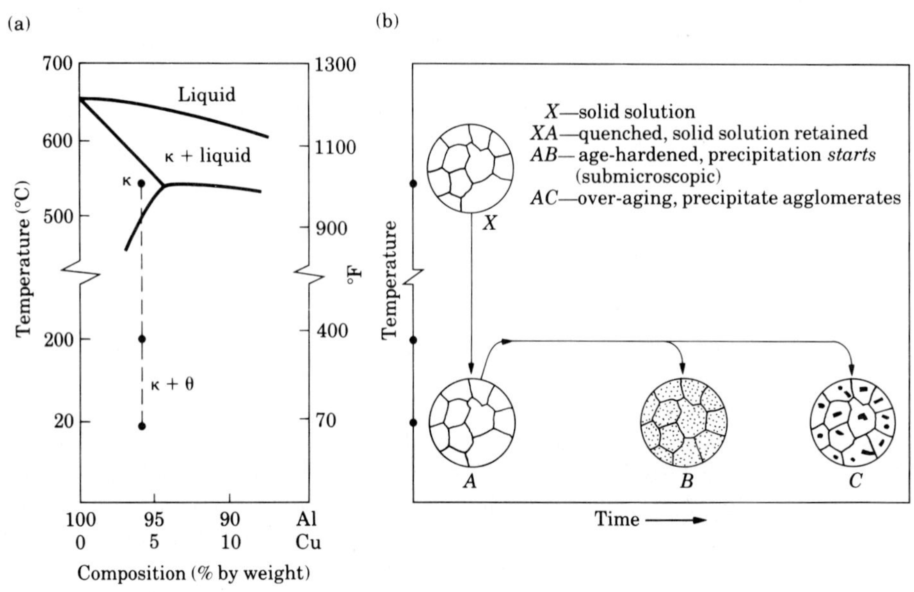

FIGURE 4.23
(a) Phase diagram for the aluminum–copper alloy system. (b) Various microstructures obtained during the age-hardening process. *Source:* L. H. Van Vlack, *Materials for Engineering.* Addison-Wesley Publishing Co., Inc., 1982.

the solid solubility of one element (one component of the alloy) in the other is exceeded.

Three stages are involved in precipitating hardening. We can best describe them by referring to Fig. 4.23, the phase diagram for the aluminum–copper system. For an alloy with the composition of 95.5% Al–4.5% Cu, a single-phase (kappa) substitutional solid-solution of copper (solute) in aluminum (solvent) exists between 500 °C and 570 °C (930 °F and 1060 °F). The kappa phase is aluminum-rich, with an fcc structure, and is ductile. Below the lower temperature, that is, below the lower solubility curve, there are two phases: kappa and theta (a hard intermetallic compound of $CuAl_2$). This alloy can be heat treated and its properties are modified by two different methods: solution treatment and precipitation.

4.9.1 Solution treatment

In **solution treatment** the alloy is heated to within the solid-solution kappa phase, say, 540 °C (1000 °F), and cooled rapidly, such as by quenching in water. The structure obtained soon after quenching (*A* in Fig. 4.23), consists only of the single phase kappa. This alloy has moderate strength and considerable ductility.

4.9.2 Precipitation hardening

The structure obtained in *A* in Fig. 4.23 can be made stronger by precipitation hardening. The alloy is reheated to an intermediate temperature and held there for a period of time, during which precipitation takes place. The copper atoms diffuse to nucleation sites and combine with aluminum atoms, producing the theta phase, which form as submicroscopic precipitates, shown in *B* by the small dots within the grains of the kappa phase. This structure is stronger than that in *A*, although it is less ductile. The increase in strength is attributed to increased resistance to dislocation movement in the region of the precipitates.

Aging. Because the precipitation process is one of time and temperature, it is also called **aging**, and the property improvement is known as **age hardening.** If carried out above room temperature, the process is called *artificial aging*. However, several aluminum alloys harden and become stronger over a period of time at room temperature. This process is known as *natural aging*. Such alloys are first quenched, then if desired, are formed at room temperature and allowed to gain strength and hardness by aging naturally. Natural aging can be slowed down by refrigerating the quenched alloy.

In the precipitation process, if the reheated alloy is held at that temperature for an extended period of time, the precipitates begin to coalesce and grow. They become larger but fewer, as shown by the larger dots in C in Fig. 4.23. This process is called **overaging,** and the alloy is softer and weaker. Thus there is an optimal time–temperature relationship in the aging process in order to obtain desired properties (Fig. 4.24). Obviously, an aged alloy can be used only up to a certain maximum temperature in service; otherwise it will overage and lose its strength and hardness. Although weaker, an overaged part has better dimensional stability.

Maraging. This is a precipitation-hardening treatment for a special group of high-strength iron-base alloys. The word *maraging* is derived from martensite age hardening. In this process one or more intermetallic compounds are precipitated in a matrix of low-carbon martensite. A typical maraging steel may contain 18 percent nickel, in addition to other elements, and aging is done at 480 °C (900 °F). Hardening by maraging does not depend on the cooling rate. Thus full uniform hardness can be obtained throughout large parts with minimal distortion. Typical uses of maraging steels are for dies and tooling for casting, molding, forging, and extrusion.

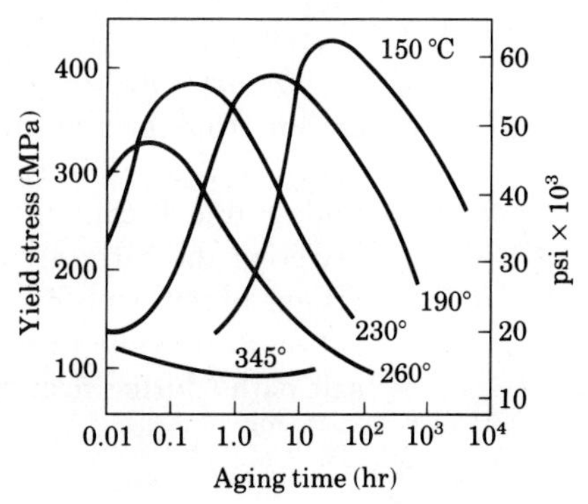

FIGURE 4.24 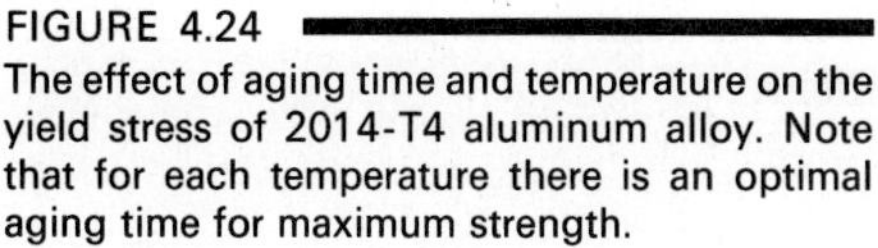
The effect of aging time and temperature on the yield stress of 2014-T4 aluminum alloy. Note that for each temperature there is an optimal aging time for maximum strength.

4.10 Case Hardening

The heat treatment processes we have described so far involve microstructural alterations and property changes in the *bulk* of the material or component by *through hardening*. In many situations, however, alteration of only the *surface* properties of a part—hence the term **case hardening**—is desirable. This method is particularly useful for improving resistance to surface indentation, fatigue, and wear. Typical applications for case hardening are gear teeth, cams, shafts, bearings, fasteners, pins, automotive clutch plates, and tools and dies. Through hardening of these parts would not be desirable, since a hard part lacks the necessary toughness for these applications. A small surface crack can propagate rapidly through the part and cause total failure.

Various surface-hardening processes are available (Table 4.1): *carburizing* (gas, liquid, and pack carburizing), *carbonitriding*, *cyaniding*, *nitriding*, *boronizing*, *flame* and *induction hardening*, and *laser hardening*. Basically, these are heat-treating operations in which the component is heated in an atmosphere containing elements (such as carbon, nitrogen, or boron) that alter the composition, microstructure, and properties of surfaces.

For steels with sufficiently high carbon content, surface hardening takes place without using any of these additional elements. Only the heat-treatment processes described in Section 4.7 are needed to alter the microstructures, usually by flame hardening or induction hardening, as outlined in Table 4.1. Laser beams and electron beams are also used effectively to harden both small and large surfaces. These methods are also used for through hardening of relatively small parts.

Because case hardening is a localized heat treatment, case-hardened parts have a hardness gradient. Typically, the hardness is a maximum at the surface and decreases below the surface, the rate of decrease depending on the composition of the metal and the process variables. Surface-hardening techniques can also be used for tempering, thus modifying the properties of surfaces that have been subjected to heat treatment. Various other processes and techniques for surface hardening, such as shot peening and surface rolling, improve wear resistance and various other characteristics.

Decarburization is the phenomenon in which alloys containing carbon lose carbon from their surfaces as a result of heat treatment or hot working in a medium, usually oxygen, that reacts with the carbon (Fig. 4.25 on p. 148). Decarburization is undesirable because it affects the hardenability of the surfaces of the part by lowering the carbon content. It also adversely affects the hardness, strength, and fatigue life of steels by significantly lowering their endurance limit. Decarburization is best avoided by processing in an inert atmosphere or a vacuum, or using neutral salt baths during heat treatment.

TABLE 4.1
OUTLINE OF HEAT TREATMENT PROCESSES FOR SURFACE HARDENING

PROCESS	METALS HARDENED	ELEMENT ADDED TO SURFACE	PROCEDURE	GENERAL CHARACTERISTICS	TYPICAL APPLICATIONS
Carburizing	Low-carbon steel (0.2% C), alloy steels (0.08–0.2% C)	C	Heat steel at 870–950 °C (1600–1750 °F) in an atmosphere of carbonaceous gases (gas carburizing) or carbon-containing solids (pack carburizing). Then quench.	A hard, high-carbon surface is produced. Hardness 55 to 65 HRC. Case depth <0.5–1.5 mm (<0.020 to 0.060 in.) Some distortion of part during heat treatment.	Gears, cams, shafts, bearings, piston pins, sprockets, clutch plates
Carbonitriding	Low-carbon steel	C and N	Heat steel at 700–800 °C (1300–1600 °F) in an atmosphere of carbonaceous gas and ammonia. Then quench in oil.	Surface hardness 55 to 62 HRC. Case depth 0.07 to 0.5 mm (0.003 to 0.020 in.) Less distortion than in carburizing.	Bolts, nuts, gears
Cyaniding	Low-carbon steel (0.2% C), alloy steels (0.08–0.2% C)	C and N	Heat steel at 760–845 °C (1400–1550 °F) in a molten bath of solutions of cyanide (e.g., 30% sodium cyanide) and other salts.	Surface hardness up to 65 HRC. Case depth 0.025 to 0.25 mm (0.001 to 0.010 in.) Some distortion.	Bolts, nuts, screws, small gears
Nitriding	Steels (1% Al, 1.5% Cr, 0.3% Mo), alloy steels (Cr, Mo), stainless steels, high-speed tool steels	N	Heat steel at 500–600 °C (925–1100 °F) in an atmosphere of ammonia gas or mixtures of molten cyanide salts. No further treatment.	Surface hardness up to 1100 HV. Case depth 0.1 to 0.6 mm (0.005 to 0.030 in.) and 0.02 to 0.07 mm (0.001 to 0.003 in.) for high-speed steel.	Gears, shafts, sprockets, valves, cutters, boring bars, fuel-injection pump parts
Boronizing	Steels	B	Part is heated using boron-containing gas or solid in contact with part.	Extremely hard and wear resistant surface. Case depth 0.025–0.075 mm (0.001–0.003 in.)	Tool and die steels
Flame hardening	Medium-carbon steels, cast irons	None	Surface is heated with an oxyacetylene torch, then quenched with water spray or other quenching methods.	Surface hardness 50 to 60 HRC. Case depth 0.7 to 6 mm (0.030 to 0.25 in.) Little distortion.	Gear and sprocket teeth, axles, crankshaft, piston rod, lathe beds and centers
Induction hardening	Same as above	None	Metal part is placed in copper induction coils and is heated by high-frequency current, then quenched.	Same as above.	Same as above.

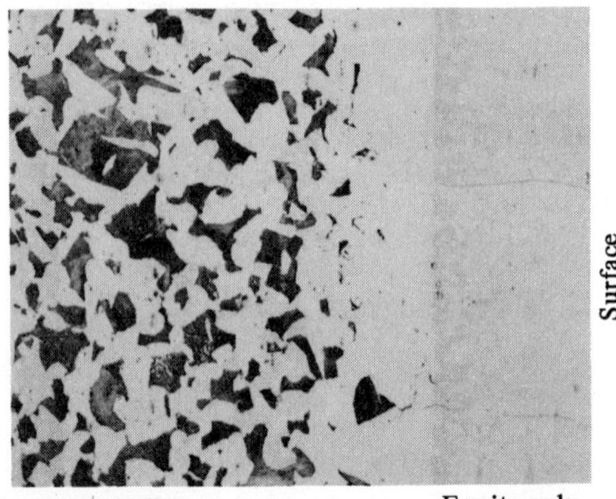

FIGURE 4.25
Decarburization of 1040 steel. As the carbon is preferentially oxidized from the surface, the hardness decreases. Decarburization has adverse effects on hardenability, strength, and fatigue life of steels. Magnification: 100 ×. *Source:* L. H. Van Vlack, *Materials for Engineering.* Addison-Wesley Publishing Co., Inc., 1982.

4.11 Annealing

Annealing is a general term used to describe the restoration of a cold-worked or heat-treated metal or alloy to its original properties, such as to increase ductility, hence formability, and reduce hardness and strength, or to modify the microstructure. Annealing is also used to relieve residual stresses in a manufactured part for improved machinability and dimensional stability. The term annealing also applies to thermal treatment of glasses and similar products, castings, and weldments.

The annealing process consists of (1) heating the workpiece to a specific range of temperature, (2) holding it at that temperature for a period of time (soaking), and (3) cooling it slowly. The process may be carried out in an inert or controlled atmosphere or performed at lower temperatures to prevent or minimize surface oxidation.

Annealing temperatures may be higher than the recrystallization temperature, depending on the degree of cold work. For example, the recrystallization temperature for copper ranges between 200 °C and 300 °C (400 °F and 600 °F), whereas the annealing temperature needed to fully recover the original properties ranges from 260 °C to 650 °C (500 °F to 1200 °F), depending on the degree of prior cold work.

Full annealing is a term applied to annealing ferrous alloys, generally low- and medium-carbon steels. The steel is heated to above A_1 or A_3 (Fig. 4.26), and cooling takes place slowly, such as 10 °C (20 °F) per hour, in a furnace after it is turned off.

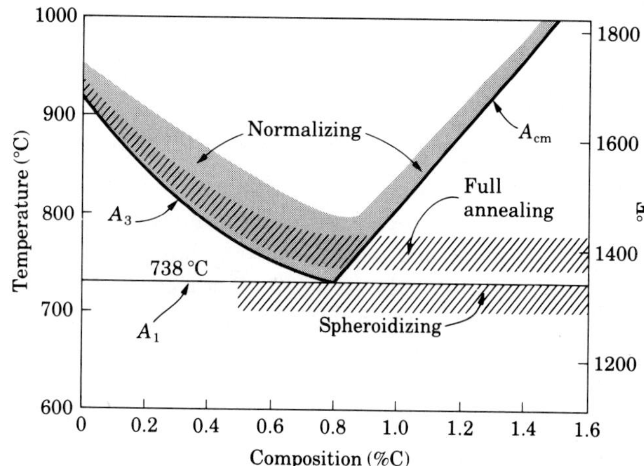

FIGURE 4.26 Heat-treating temperature ranges for plain-carbon steels, as indicated on the iron–iron carbide phase diagram. *Source:* ASM International.

The structure obtained in full annealing is coarse pearlite, which is soft and ductile and has small uniform grains.

To avoid excessive softness in the annealing of steels, the cooling cycle may be done completely in still air. This process is called **normalizing**, in which the part is heated to a temperature above A_3 or A_{cm} to transform the structure to austenite. It results in somewhat higher strength and hardness and lower ductility than in full annealing (Fig. 4.27). The structure obtained is fine pearlite with small uniform grains. Normalizing is generally carried out to refine the grain structure, obtain uniform structure (homogenization), decrease residual stresses, and improve machinability.

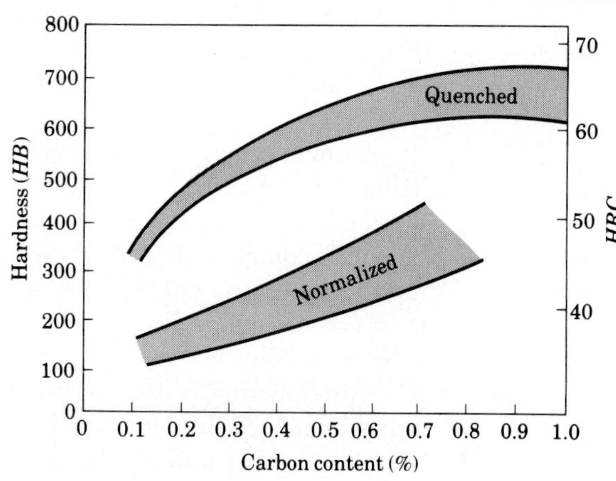

FIGURE 4.27 Hardness of steels in the quenched and normalized conditions, as a function of carbon content.

We described the structure of spheroidizing and the procedure for obtaining it in Section 4.7 and in Figs. 4.18 and 4.26. Spheroidizing annealing improves the cold workability and machinability of steels.

4.11.1 Process annealing

In process annealing (also called *intermediate annealing*, *subcritical annealing*, or *in-process annealing*), the workpiece is annealed to restore its ductility, part or all of which may have been exhausted by work hardening during cold working. In this way, the part can be worked further into the final desired shape. If the temperature is high and/or the time of annealing is long, grain growth may result, with adverse effects on formability of annealed parts (see Section 1.6).

Stress-relief annealing. To reduce or eliminate residual stresses, a workpiece is generally subjected to *stress-relief annealing*, or simply **stress relieving.** The temperature and time required for this process depend on the material and the magnitude of residual stresses present. The residual stresses may have been induced during forming, machining, or other shaping processes, or caused by volume changes during phase transformations. For steels, the part is heated to below A_1, thus avoiding phase transformations. Slow cooling rates, such as in still air, are generally employed. Stress relieving promotes dimensional stability in situations where subsequent relaxing of residual stresses present may cause distortion of the part when it is in service over a period of time. It also reduces the tendency for stress-corrosion cracking.

Tempering. If steels are hardened by heat treatment, tempering, or *drawing* (not to be confused with wire drawing or deep drawing, described in Part III), is used in order to reduce brittleness, increase ductility and toughness, and reduce residual stresses. The term tempering is also used for glasses. In tempering, the steel is heated to a specific temperature, depending on composition, and cooled at a prescribed rate. The results of tempering for an oil-quenched AISI 4340 steel are shown in Fig. 4.28. Alloy steels may undergo *temper embrittlement*, which is caused by the segregation of impurities along the grain boundaries at temperatures between 480 °C and 590 °C (900 °F and 1100 °F).

Austempering. In *austempering*, the heated steel is quenched from the austenitizing temperature rapidly enough to avoid formation of ferrite or pearlite. It is then held at a certain temperature until isothermal transformation from austenite to bainite is complete. It is then cooled to room temperature, usually in still air, at a moderate rate to avoid thermal gradients within the part. The quenching medium most commonly used is molten salt, at temperatures ranging from 160 °C to 750 °C (320 °F to 1380 °F).

Austempering is often substituted for conventional quenching and tempering, either to reduce the tendency for cracking and distortion during quenching or to improve ductility and toughness while maintaining hardness. Because of the shorter

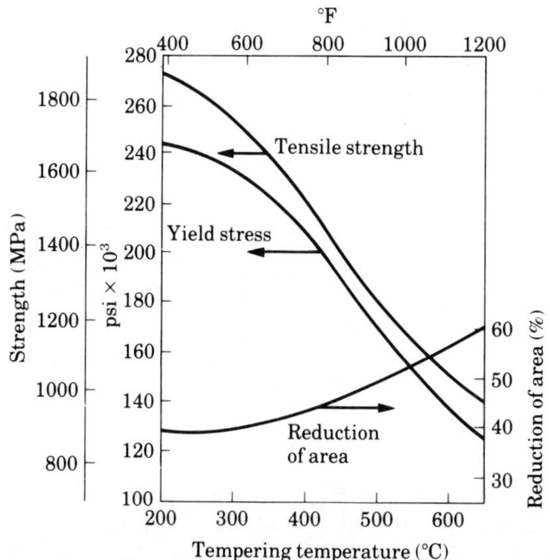

FIGURE 4.28 Mechanical properties of oil-quenched 4340 steel as a function of tempering temperature. *Source:* Courtesy of LTV Steel Company.

cycle time, this process is economical for many applications. In *modified austempering*, a mixed structure of pearlite and bainite is obtained. The best example of this practice is *patenting*, which provides high ductility and moderately high strength, such as the patented wire used in the wire industry.

Martempering (marquenching). In *martempering*, the steel or cast iron is quenched from the austenitizing temperature into a hot-fluid medium, such as hot oil or molten salt. It is held at that temperature until the temperature is uniform throughout the part and then cooled at a moderate rate, such as in air, to avoid temperature gradients within the part. The part is then tempered because the structure thus obtained is primarily untempered martensite and is not suitable for most applications. Martempered steels have less tendency to crack, distort, and develop residual stresses during heat treatment. In modified martempering the quenching temperature is lower, and thus the cooling rate is higher. The process is suitable for steels with lower hardenability.

Ausforming. In *ausforming*, also called *thermomechanical processing*, the steel is formed into desired shapes within controlled ranges of temperature and time to avoid formation of nonmartensitic transformation products. The part is then cooled at various rates to obtain the desired microstructures. Ausformed parts have superior mechanical properties.

4.12 Heat-Treating Furnaces and Equipment

Two basic types of furnaces are used for heat treating: *batch* furnaces and *continuous* furnaces. Because they consume great amounts of energy, their insulation and efficiency are important design considerations, as are their initial cost, manpower needed for operation and maintenance, and safe use. Since uniform temperature and accurate control of temperature–time cycles are important, modern furnaces are equipped with various electronic controls. New developments in furnaces include computer-controlled systems programmed to run through a complete heat-treating cycle, repeatedly and with reproducible accuracy.

Heating-system fuels are usually gas, oil, or electricity (resistance or induction heating). The type of fuel used influences the furnace's atmosphere. Unlike electric heating, gas or oil introduce the products of combustion into the furnace—a disadvantage. However, electrical heating has a slower startup time and is more difficult to adjust and control.

4.12.1 Batch furnaces

In a batch furnace the parts to be heat treated are loaded into and unloaded from the furnace in individual batches. The furnace consists basically of an insulated chamber, a heating system, and an access door or doors. Batch furnaces are of the following types:

- A *box furnace* is a horizontal rectangular chamber, with one or two access doors through which parts are loaded. This type of furnace is commonly used and versatile, simple to construct and use, and available in several sizes. A variation of this type is the *car-bottom furnace*. The parts to be heat treated, usually long or large, are loaded on a flatcar which moves on rails into the furnace.
- A *pit furnace* is a vertical pit below ground level with a lid, into which the parts are lowered. This type of furnace is particularly suitable for long parts, such as rods, shafts, and tubing, since they can be suspended and are less likely to warp during processing than when placed horizontally in a box furnace.
- A *bell furnace* is a round or rectangular box furnace without a bottom. It is lowered over the stacked parts that are to be heat treated. This type of furnace is particularly suitable for coils of wire, rod, or sheet metal.
- In the *elevator furnace*, the parts to be heat treated are loaded on a car platform, rolled into position, and raised into the furnace. This type of

furnace saves space in the plant and is suitable for alloys that have to be quenched rapidly. A quenching tank is placed directly under the furnace.

4.12.2 Continuous furnaces

Parts to be heat-treated continuously move through the furnace on conveyors of various designs, using trays, belts, chains, and other mechanisms. Continuous furnaces are suitable for high production runs and can be designed and programmed so that complete heat-treating cycles can be performed under tight control.

4.12.3 Salt-bath furnaces

Because of their high heating rates and better control of uniform temperature, salt baths are commonly used in various heat-treating operations, particularly for nonferrous strip and wire. Heating rates are high because of the higher thermal conductivity of liquid salts, compared to air or gases. Depending on the electrical conductivity of the salt, heating may be done externally (for nonconducting salts) or by immersed or submerged electrodes using low-voltage alternating current. Direct current cannot be used because it subjects the salt to electrolysis. Salt baths are available for a wide range of temperatures. Lead may be used as the heating medium.

4.12.4 Fluidized beds

Dry, fine, and loose solid particles, usually aluminum oxide, are heated and suspended in a chamber by an upward flow of hot gas at various velocities. The parts to be heat treated are then placed within the floating particles—hence the term *fluidized bed*. Because of its constant agitation, the system is efficient, the temperature distribution is uniform, and the heat-transfer rate is high. These furnaces are used for various batch-type applications.

4.12.5 Induction heating

The part to be heat treated is heated rapidly by the electromagnetic field generated by an induction coil carrying alternating current, which induces eddy currents in the part. The coil, which can be shaped to fit the contour of the part to be heat treated (Fig. 4.29), is made of copper or copper-base alloy, and is usually water cooled. The coil may be designed to also quench the part. Induction heating is desirable for localized heat treating, such as for gear teeth and cams.

4.12.6 Furnace atmospheres

The atmosphere in furnaces can be controlled to avoid oxidation, tarnishing, and decarburization of ferrous alloys heated to elevated temperatures. Oxygen causes oxidation (corrosion, rusting, and scaling). Carbon dioxide has various effects: It may be neutral or decarburizing, depending on its concentration in the furnace

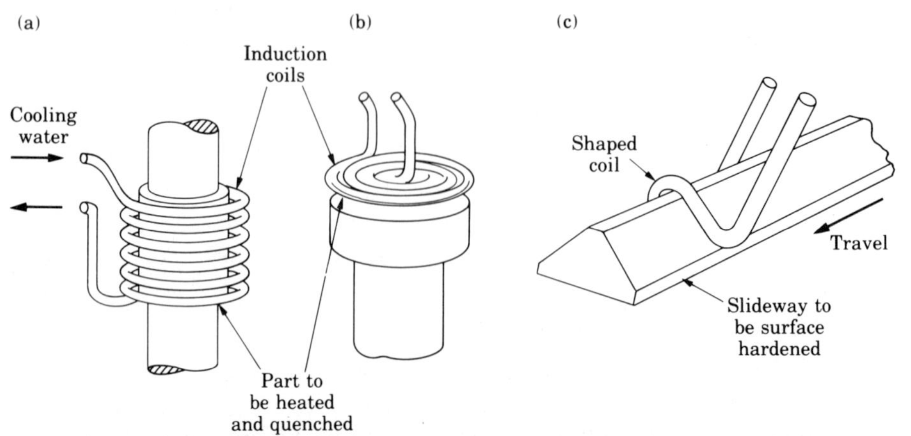

FIGURE 4.29
Types of coils used in induction heating of various surfaces of parts.

atmosphere. Water vapor in the furnace causes oxidation of steels, resulting in a blue color. The term *bluing* is used to describe formation of a thin blue film of oxide on finished parts to improve their appearance and resistance to oxidation. Nitrogen is a common neutral atmosphere, and a vacuum provides a completely neutral atmosphere.

SUMMARY

Commercially pure metals generally do not have sufficient strength for many engineering applications. These metals must be alloyed with various elements, which change their structures and properties. Important concepts in alloying are solubility of the alloying elements in the base metal and the phases that are present at various ranges of temperature and composition.

Phase diagrams show the relationships among temperature, composition, and the phases present in a particular alloy system. Among the binary systems, the most important is the iron–carbon system, which includes steels and cast irons. Various transformations take place in microstructures that have widely varying characteristics and properties, as temperature is decreased at different rates.

Important mechanisms of hardening and strengthening involve thermal treatments, including heat treatment by quenching and precipitation hardening. Also important are the methods and techniques of heat treating, quenching, annealing, control of the furnace atmosphere, and the characteristics of the equipment used, as well as the shape of the parts to be heat treated.

TRENDS

- Studies on better methods of controlling heat-treating furnace atmospheres by interfacing with computers are being made. Improvements are being made in furnace insulation for energy saving.
- Studies are continuing to narrow the hardenability bands of alloys for better property control in heat treating of parts.
- Studies are being conducted of directly quenching hot workpieces after they are hot worked, such as by forging. Hot-forged parts, for example, do not have to be reheated for quenching, improving efficiency and resulting in energy savings.
- Heat-treating facilities are being automated with the aid of computer programs that monitor and control all aspects of their operations.
- Greater emphasis is being placed on vacuum heat-treating furnaces and on induction tempering.
- Electronic sensing systems are being developed to monitor and detect cracks, changes in part shape, and variations in the workpiece material during heat treating.

KEY TERMS

Age hardening
Aging
Alloy
Annealing
Austenite
Bainite
Case hardening
Cast iron
Cementite
Equilibrium diagram
Eutectoid reaction
Eutectic point
Ferrite
Hardenability
Heat treatment
Intermetallic compounds
Jominy test
Martensite
Normalizing
Overaging
Pearlite
Phase diagram
Phase transformations
Precipitation hardening
Pure metals
Solute
Solution treatment
Solvent
Spheroidites
Stress relieving
Tempering

BIBLIOGRAPHY

Boyer, H.E., *Practical Heat Treating*. Metals Park, Ohio: American Society for Metals, 1984.

Brooks, C.R., *Heat Treatment, Structure and Properties of Nonferrous Alloys*. Metals Park, Ohio: American Society for Metals, 1982.

Case Hardening of Steel. Metals Park, Ohio: ASM International, 1987.
Gupton, P.S., (ed.), *The Heat Treating Source Book.* Metals Park, Ohio: American Society for Metals, 1986.
Krauss, G., *Principles of Heat Treatment of Steel.* Metals Park, Ohio: American Society for Metals, 1980.
Metals Handbook, 9th ed., Vol. 4: *Heat Treating.* Metals Park, Ohio: American Society for Metals, 1981.
Semiatin, S.L., and D.E. Stutz, *Induction Heat Treatment of Steel.* Metals Park, Ohio: American Society for Metals, 1985.
Thelning, K.E., *Steel and Its Heat Treatment*, 2d ed., Stoneham, Mass.: Butterworths, 1984.
Unterweiser, P.M., H.E. Boyer, and J.J. Kubbs (eds.), *Heat Treater's Guide.* Metals Park, Ohio: American Society for Metals, 1982.

REVIEW QUESTIONS

4.1 Describe the difference between a solute and a solvent.
4.2 What is a solid solution?
4.3 What are the conditions for obtaining (a) substitutional and (b) interstitial solid solutions?
4.4 What is the purpose of calculating the size of the largest hole in a unit cell?
4.5 What is the difference between a single-phase and a two-phase system?
4.6 Explain what is meant by second-phase particle.
4.7 Describe the features of a phase diagram.
4.8 What do the terms equilibrium and constitutional, as applied to phase diagrams, indicate?
4.9 What is the difference between eutectic and eutectoid?
4.10 Describe the general properties of various types of cast irons.
4.11 What is tempering? Why is it done?
4.12 Explain what is meant by severity of quenching.
4.13 What are precipitates? Why are they significant in precipitation hardening?
4.14 What is the difference between natural and artificial aging?
4.15 Describe the characteristics of ferrite, austenite, and cementite.?
4.16 What is the purpose of annealing?

QUESTIONS AND PROBLEMS

4.17 You may have seen some technical literature on products stating that certain parts in those products are "heat treated." Describe briefly your understanding of this term and why the manufacturer mentions it.
4.18 Describe the engineering significance of the existence of a eutectic point in phase diagrams.
4.19 We stated in this chapter that in parts design sharp corners should be avoided in order to reduce the tendency for cracking during heat treatment. If it is necessary for the part to have sharp corners in order to function properly, and still require heat treatment, what method would you recommend for manufacturing this part?

4.20 Explain the difference between hardness and hardenability.

4.21 Describe your understanding of why the microstructure of metals has been and continues to be studied in such great detail.

4.22 Refer to Table 4.1 and explain why the items listed under typical applications are so suitable for surface hardening.

4.23 Explain the reason for quenching in heat treatment of metals.

4.24 Why is it generally not desirable to use steels in their as-quenched condition?

4.25 You may have noticed that some products, such as kitchen knives, say "high-carbon steel." What is the significance of this statement as far as the customer is concerned?

4.26 Describe the differences between case hardening and through hardening insofar as engineering applications are concerned.

4.27 Describe the characteristics of (a) an alloy, (b) pearlite, (c) austenite, (d) martensite, and (e) cementite.

4.28 Explain why carbon, of all elements, is so effective in imparting strength to iron in the form of steel.

4.29 Explain the usefulness of phase diagrams to someone who is not very knowledgeable about metals.

4.30 How does the shape of graphite in cast iron affect its properties?

4.31 In Section 4.8 we listed several fluids in terms of their cooling capacity in quenching. Which physical properties of these fluids influence their cooling capacity?

4.32 Describe your understanding of precipitation hardening.

4.33 Why is it important to know the characteristics of heat-treating furnaces?

4.34 Using Fig. 4.5, estimate the following quantities for a 20% Cu–80% Ni alloy: (a) liquidus temperature, (b) solidus temperature, (c) the percentage of nickel in the liquid at 1400 °C (2550 °F), (d) the major phase at 1400 °C, and (e) the ratio of solid to liquid at 1400 °C.

4.35 Explain why the percentage of pearlite in the abscissa of Fig. 4.16 goes down after a certain carbon content is reached.

4.36 What is the significance of decarburization? Give some examples.

4.37 Extrapolating the curves in Fig. 4.21, estimate the time that it would take for 1080 steel to soften to 53 HRC at (a) 200 °C and (b) 300 °C.

4.38 A typical steel for tubing is AISI 1040, and for music wire 1085. In view of their applications, explain the reason for the difference in carbon content.

4.39 Figure 4.22 shows hardness distributions in end-quench tests as measured along the *length* of the round bar. Make a simple qualitative sketch showing the hardness distribution *across* the diameter of the bar. Would the shape of the curve depend on carbon content? Explain.

5

Steels: Production, General Properties, and Applications

5.1 Introduction

By virtue of their wide range of mechanical, physical, and chemical properties, *ferrous alloys* are among the most useful of all metals. Ferrous metals and alloys contain iron as their base metal; these metals are carbon and alloy steels, stainless steels, tool and die steels, cast irons, and cast steels. Ferrous alloys are produced as sheet steel for automobiles, appliances, and containers; as plates for ships, boilers, and bridges; as structural members, such as I-beams, bar products for leaf springs, gears, axles, crankshafts, and railroad rails; as stock for tools and dies; as music wire, and as fasteners such as bolts, rivets, and nuts.

A typical U.S. passenger car contains about 800 kg (1750 lb) of steel, accounting for about 55–60 percent of its weight. As an example of their widespread use, ferrous materials comprise 70–85 percent by weight of virtually all structural members and mechanical components. Carbon steels are the least expensive of all metals (Table 5.1; see also Table 6.1), but stainless steels can be costly.

TABLE 5.1
APPROXIMATE COST OF RAW MATERIALS AS A FUNCTION OF THEIR CONDITION, SHAPE, AND SIZE

Carbon-steel plate and sheet	
Hot rolled	60–70
Cold rolled	75–90
Carbon-steel bars	
Hot rolled, round	55–80
Cold finished, round	60–200
Cold finished, square	90–170
Stainless steel sheet	
304	230
316	300–340
410	375
Stainless steel bars	
304 round	310–730
303 square	560–1000
Aluminum plate	
2024 T351	530–590
6061 T651	330–350
7075 T651	560–620
Aluminum sheet	
2024 T3	510–650
3003 H14	275–300
6061 T6	360–400
Aluminum bars	
Round	275–510
Square	575–700
Rectangular	550–1000
Aluminum extrusions	260–310

Note: Prices are in U.S. dollars per 100 kg. Generally, the cost increases with decreasing thickness or cross-section. See also Table 6.1.

The use of iron and steel as structural materials has been one of the most important technological developments. Primitive ferrous tools first appeared in about 4000–3000 B.C. They were made from meteoritic iron obtained from meteorites that had struck the earth. True ironworking began in Asia Minor in about 1100 B.C., and signaled the advent of the iron age. Invention of the blast furnace in about A.D. 1340 made possible the production of large quantities of iron and steel.

In this chapter, we describe the modern, as well as the traditional, methods of steelmaking, as outlined in Fig. 5.1 (see p. 160). The processes used to make the products shown are described in various chapters throughout the remainder of this text. In this chapter we also describe the properties and characteristics of major categories of ferrous alloys, as produced by various methods.

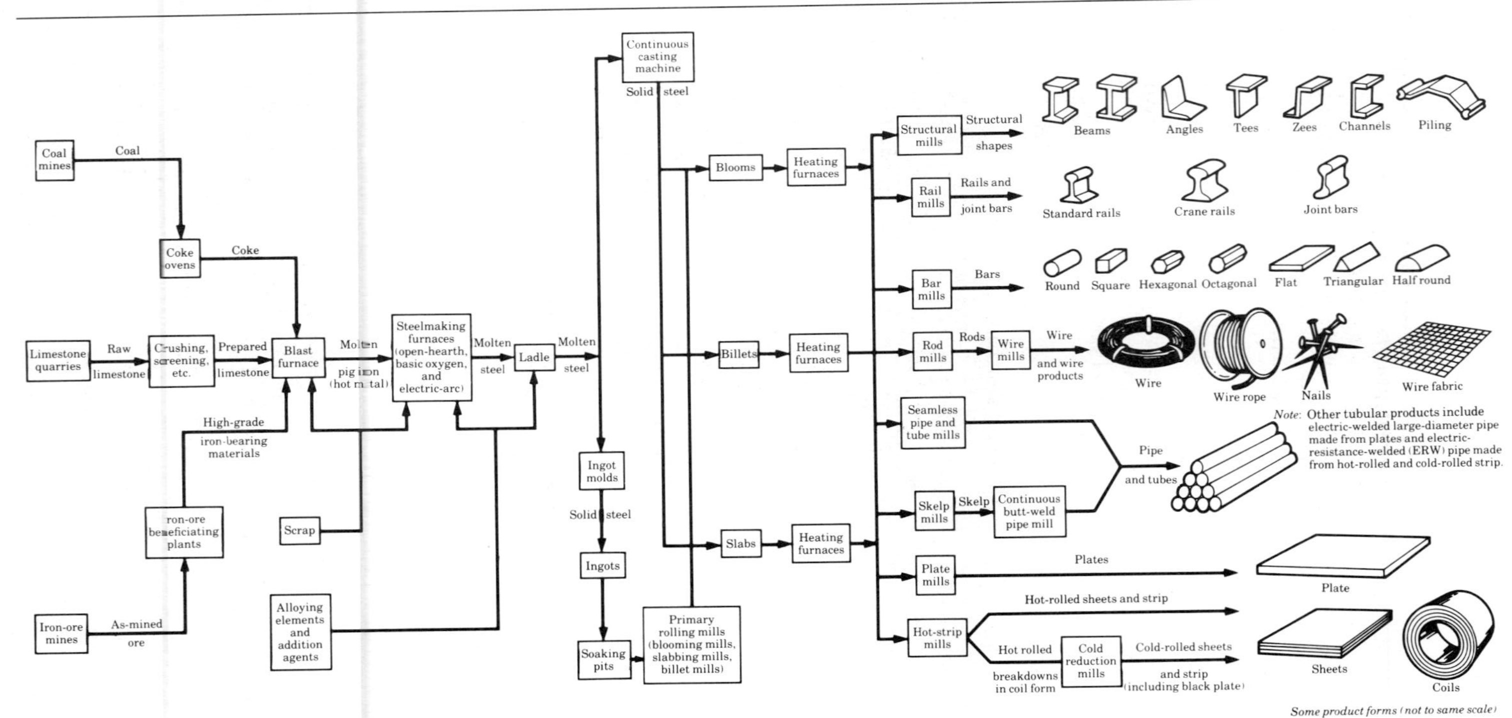

FIGURE 5.1
Chart showing the conversion of raw materials into various shapes of steel, ranging from nails to plates. The processes used in making these products are described in Part III of this text. *Source:* Courtesy of USX Corporation.

5.2 Production of Iron and Steel

Let's begin by reviewing the principles of the iron- and steelmaking processes, beginning with raw materials. This knowledge is essential to an understanding of the quality and characteristics of the steels produced by different methods.

5.2.1 Raw materials

The three basic materials used in iron and steelmaking are *iron ore*, *limestone*, and *coke*. Although it does not occur in a free state in nature, iron is one of the most abundant elements, making up about 5 percent of the earth's crust in the form of various ores. The principal iron ores are *taconite* (a black flintlike rock), *hematite* (an iron oxide mineral), and *limonite* (an iron oxide containing water). After it is mined, the iron ore is crushed into fine particles, the impurities are removed by various means (such as magnetic separation), and it is formed into pellets, balls, or brickets using binders and water. Typically, pellets are about 65 percent pure iron and 25 mm (1 in.) in diameter. The concentrated iron ore is referred to as *beneficiated*, as are other concentrated ores. Some iron-rich ores are used directly without pelletizing.

Coke is obtained from special grades of bituminous coal, which are heated in vertical ovens to temperatures of up to 1150 °C (2100 °F), and cooled with water in quenching towers. Coke has several functions in steelmaking. One is to generate the high level of heat required for chemical reactions to take place in ironmaking. Second, it produces carbon monoxide (a reducing gas) which is then used to reduce iron oxide to iron. Reducing, in this case, means removing oxygen. The chemical by-products of coke are used in making plastics and chemical compounds. Gases evolved during converting coal to coke are used as fuel for plant operations.

The function of limestone (calcium carbonate) is to remove impurities from the molten iron. The limestone reacts chemically with impurities, acting like a *flux* (meaning to flow as a fluid), which causes the impurities to melt at a low temperature. The limestone combines with the impurities and forms a *slag*, which is light, floats over the molten metal, and is subsequently removed. *Dolomite* (an ore of calcium magnesium carbonate) is used as a flux. The slag is later used in making cement, fertilizers, glass, building materials, rock wool insulation, and road ballast.

5.2.2 Ironmaking

The three raw materials are carried to the top of and dumped into the **blast furnace**, called charging the furnace, as shown schematically in Fig. 5.2. The principle of this furnace was developed in Central Europe, and the first furnace built in the United States began operating in 1621. The blast furnace that we are familiar with is

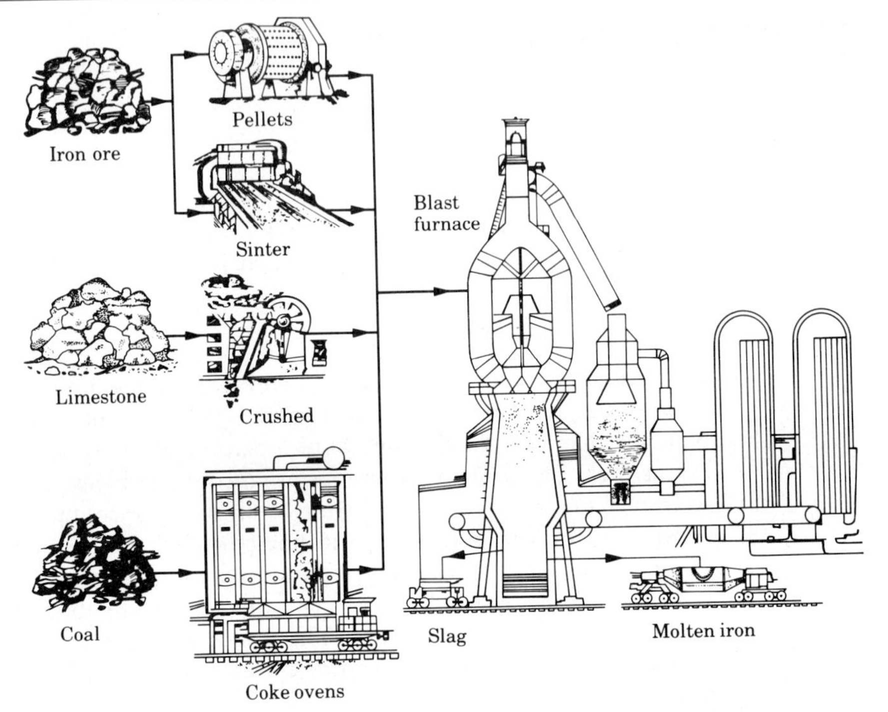

FIGURE 5.2 Schematic illustration of the steelmaking process in a blast furnace. *Source:* Courtesy of American Iron and Steel Institute.

basically a large steel cylinder lined with refractory (heat resistant) brick and has the height of about a ten-story building.

The charge mixture is melted in a reaction at 1650 °C (3000 °F), with air preheated to about 1100 °C (2000 °F) and *blasted* into the furnace (hence the term blast furnace) through nozzles (or *tuyeres*). Although a number of reactions may take place, basically the reaction of iron oxide with carbon produces carbon monoxide, which in turn reacts with the iron oxide, reducing it to iron. Preheating the incoming air is necessary because the burning coke alone does not produce sufficiently high temperatures for the reactions to occur.

The molten metal accumulates at the bottom of the blast furnace, while the impurities float to the top of the metal. At intervals of four to five hours, the molten metal is drawn off, or tapped, into ladle cars, each holding as much as 160 tons of molten iron. The molten metal at this stage is called **pig iron**, or simply *hot metal*. It has a typical composition of 4 percent carbon, 1.5 percent silicon, 1 percent manganese, 0.04 percent sulfur, and 0.4 percent phosphorus, with the rest being pure iron. Use of the word *pig* comes from the early practice of pouring molten iron into small sand molds, arranged like a litter of small pigs around a main channel. The solidified metal is called **pig** and is used in making iron and steels.

5.2.3 Steelmaking

Steel was first produced in China and Japan in about A.D. 600–800. The process is essentially one of refining the pig iron obtained from the blast furnace, by reducing the percentage of manganese, silicon, carbon, and other elements, and controlling its composition by the addition of various elements. The molten metal from the blast furnace is transported into one of three types of furnaces: open hearth, electric, or basic oxygen. The name *open hearth* derives from the shallow hearth shape that is open directly to the flames that melt the metal. Developed in the 1860s, the **open-hearth furnace** industrially is still important but is being replaced by electric furnaces and by the basic-oxygen process. These newer methods are more efficient and produce better quality steels.

Electric furnace. The **electric furnace** was first introduced in the United States in 1906. The source of heat is a continuous electric arc formed between the electrodes and the charged metal (Fig. 5.3). Temperatures as high as 1925 °C (3500 °F) are generated in this type of furnace. There are usually three graphite electrodes, and they can be as large as 75 mm (30 in.) in diameter and 1.5–2.5 m (5–8 ft) in length. Their height in the furnace can be adjusted depending on the amount of metal present and wear of the electrodes.

Steel scrap and a small amount of carbon and limestone are dropped into the electric furnace through the open roof. Electric furnaces can also be charged with 100 percent scrap. The roof is then closed and the electrodes are lowered. Power is turned on, and within a period of about two hours the metal melts. The current is shut off, the electrodes are raised, the furnace is tilted, and the molten metal is poured into a *ladle*, which is a receptacle used for transferring and pouring molten metal. Electric-furnace capacities range from 60–90 tons of steel per day. The quality of steel produced is better than that of open-hearth or basic-oxygen steels.

For smaller quantities, electric furnaces are of the *induction* type. The metal is placed in a *crucible*, which is a large pot made of refractory material, surrounded with a copper coil through which alternating current is passed (Fig. 5.3c). The

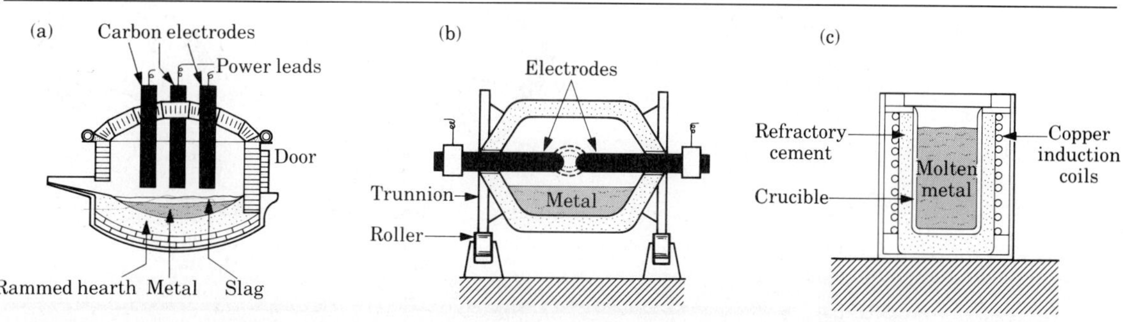

FIGURE 5.3
Schematic illustration of types of furnaces: (a) direct arc, (b) indirect arc, and (c) induction.

induced current in the charge melts the metal. These furnaces are also used for remelting metal for casting.

Basic-oxygen furnace. The **basic-oxygen furnace (BOF)** is the newest and fastest steelmaking process. Typically, 200 tons of molten pig iron and 90 tons of scrap are charged (fed) into a vessel (Fig. 5.4a). Pure oxygen is then blown into the furnace for about 20 minutes under a pressure of about 1250 kPa (180 psi), through a water-cooled *lance*, which is a long tube, as shown in Fig. 5.4(b). Fluxing agents, such as lime, are added through a chute.

The vigorous agitation of the oxygen refines the molten metal by an oxidation process in which iron oxide is first produced. The oxide reacts with the carbon in the molten metal, producing carbon monoxide and carbon dioxide. The lance is retracted and the furnace is tapped by tilting it. Note the opening in Fig. 5.4(c) for the molten metal. The slag is then removed by tilting the furnace in the opposite direction.

The BOF process is capable of refining 250 tons of steel in 35–50 minutes. Most BOF steels, which are of better quality than open-hearth furnace steels and have low

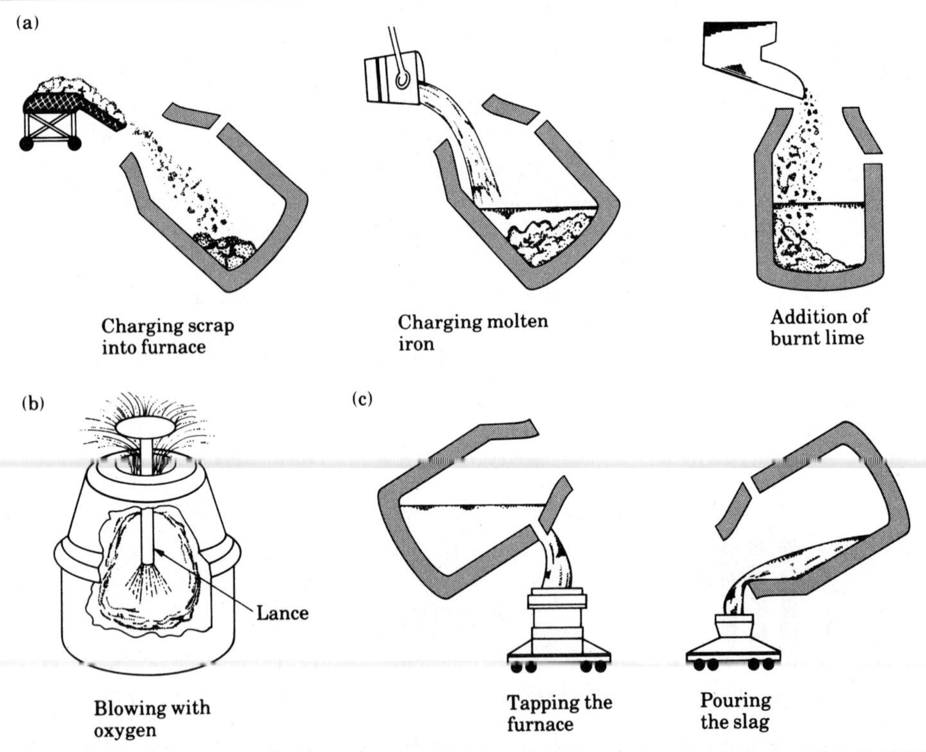

FIGURE 5.4 Schematic illustrations showing (a) charging, (b) melting, and (c) pouring of molten iron in a basic-oxygen process. *Source*: Inland Steel Company.

impurity levels, are processed into plates, sheet, and various structural shapes, such as I-beams and channels.

Vacuum furnace. Steel may also be melted in induction furnaces from which the air has been removed, similar to the one shown in Fig. 5.3(c). Because the process removes gaseous impurities from the molten metal, vacuum melting produces high-quality steels.

5.3 Casting of Ingots

Traditionally, the next step in the steelmaking process is the shaping of the molten steel into a solid form—an **ingot**—for further processing by rolling it into shapes, casting it into semifinished forms, or forging. (The shaping process is being rapidly replaced by **continuous casting**, thus improving efficiency and eliminating the need for ingots.) The molten metal is poured (teemed) from the ladle into ingot molds in which the metal solidifies. Molds are usually made of cupola iron or blast-furnace iron, with 3.5 percent carbon, and are tapered in order to facilitate the removal of the solidified metal. The bottoms of the molds may be closed or open; if open, the molds are placed on a flat surface. The taper may be such that the big end is down.

The cooled ingots are removed (stripped) from the molds and lowered into soaking pits, where they are reheated to a uniform temperature of about 1200 °C (2200 °F) for subsequent processing by rolling. Ingots may be square, rectangular, or round in cross-section, and their weight ranges from a few hundred pounds to 40 tons.

Certain reactions take place during the solidification of an ingot, which in turn have an important influence on the quality of the steel produced. For example, significant amounts of oxygen and other gases can dissolve in the molten metal during steelmaking. However, much of these gases is rejected during solidification of the metal because the solubility limit of gases in the metal decreases sharply as its temperature decreases. The rejected oxygen combines with carbon, forming carbon monoxide, which causes porosity in the solidified ingot; that is, it has small pores in it.

Depending on the amount of gas evolved during solidification, three types of steel ingots can be produced: killed, semi-killed, and rimmed.

Killed steel. **Killed steel** is a fully deoxidized steel; that is, oxygen is removed and thus porosity is eliminated. In the deoxidation process the dissolved oxygen in the molten metal is made to react with elements such as aluminum, silicon, manganese, and vanadium that are added to the melt. These elements have an affinity for oxygen and form metallic oxides. If aluminum is used, the product is called aluminum-killed steel. The term *killed* comes from the fact that the steel lies quietly after being poured into the mold.

The oxide inclusions in the molten bath, if sufficiently large, float out and adhere to, or are dissolved in, the slag. A fully killed steel is thus free of any porosity caused by gases. It is also free of any blowholes (large spherical holes near the surfaces of the ingot). Consequently, the chemical and mechanical properties of killed steels are relatively uniform throughout the ingot. However, because of metal shrinkage during solidification, an ingot of this type develops a *pipe* at the top. It is also called a *shrinkage cavity* and has the appearance of a funnel-like shape. This pipe can comprise a substantial portion of the ingot and has to be cut off and scrapped.

Semi-killed steel. **Semi-killed steel** is a partially deoxidized steel. It contains some porosity, generally in the upper central section of the ingot, but has little or no pipe; thus scrap is reduced. Piping in semi-killed steels is less because it is compensated for by the presence of porosity in that region. Semi-killed steels are economical to produce.

Rimmed steel. In a **rimmed steel**, which generally has a low carbon content (less than 0.15 percent), the evolved gases are only partially killed or controlled by the addition of elements such as aluminum. The gases form blowholes along the outer rim of the ingot—hence the term *rimmed*. Blowholes are generally not objectionable unless they break through the outer skin of the ingot. Rimmed steels have little or no piping, and have a ductile skin with good surface finish. However, blowholes may break through the skin if they are not controlled properly. Furthermore, impurities and inclusions tend to segregate toward the center of the ingot. Products made from this steel may thus be defective and should be inspected.

5.3.1 Refining

The properties and manufacturing characteristics of ferrous alloys are adversely affected by the amount of impurities, inclusions, and other elements present. The removal of impurities is known as **refining**, much of which is done in melting furnaces or ladles, with the addition of various elements. There is an increasing demand for cleaner steels having improved and more uniform properties and consistency of composition. Refining is particularly important in producing high-grade steels and alloys for high-performance and critical applications, such as in aircraft. Moreover, warranty periods on shafts, camshafts, crankshafts for diesel trucks, and similar parts can be increased significantly using higher quality steels.

The trend in steelmaking is for *secondary refining* in ladles (ladle metallurgy) and vacuum chambers. New methods of ladle refining (injection refining) generally consist of melting and processing in a vacuum. Several processes using controlled atmospheres, such as electron-beam melting, vacuum-arc remelting, argon–oxygen decarburization, and vacuum-arc double-electrode remelting, have been developed.

5.4 Continuous Casting

The traditional method of casting ingots is a batch process; that is, each ingot has to be stripped from its mold after solidification and processed individually. Furthermore, piping and microstructural and chemical variations are present throughout the ingot. These problems are alleviated by **continuous casting** processes, which produce higher quality steels at reduced cost.

First conceived in the 1860s, *continuous*, or **strand casting**, was first developed for casting nonferrous metal strip. The process is now used for steel production, with major efficiency and productivity improvements and significant cost reduction. One system for continuous casting is shown schematically in Fig. 5.5. The molten metal in the ladle is cleaned and equalized in temperature by blowing nitrogen gas through it for 5 to 10 minutes. The metal is then poured into a refractory-lined intermediate pouring vessel (*tundish*) where impurities are skimmed off. The tundish holds as much as three tons of metal. The molten metal travels through water-cooled copper molds and begins to solidify as it travels downward along a path supported by rollers (*pinch rolls*).

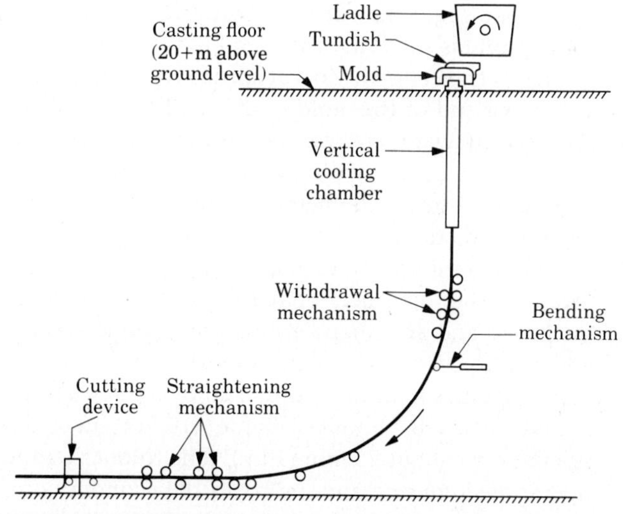

FIGURE 5.5 Schematic illustration of continuous casting of steel. Because of high efficiency and favorable economics, continuous casting is rapidly replacing traditional steelmaking processes, as well as for nonferrous metals. *Source:* USX Corporation.

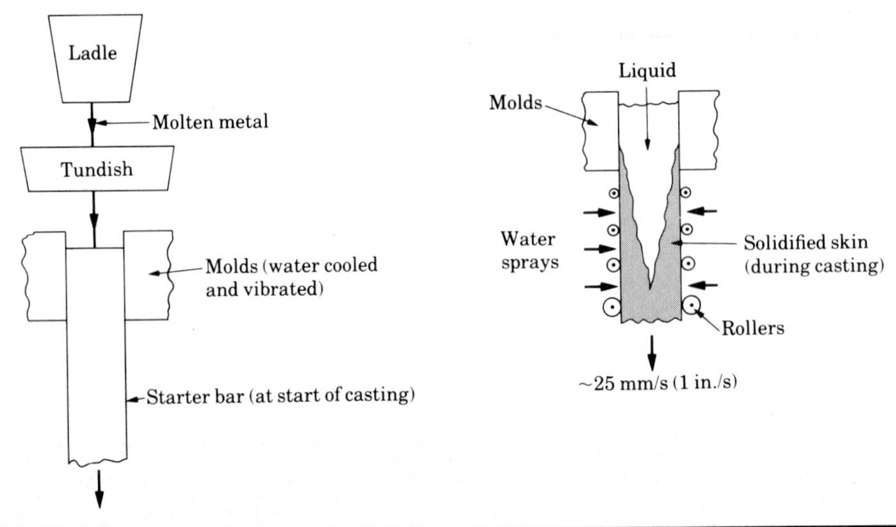

FIGURE 5.6
Method of starting a continuous casting process with a starter bar, and the cross-section of a continuously cast metal as it begins to solidify.

Before starting the casting process, a solid *starter*, or *dummy*, *bar* is inserted into the bottom of the mold. The molten metal is then poured and freezes onto the starter bar (Fig. 5.6). The bar is withdrawn at the same rate the metal is poured. The cooling rate is such that the metal develops a solidified skin (shell) to support itself during its travel downward at speeds typically 25 mm/s (1 in./s). The shell thickness at the exit end of the mold is about 12–18 mm (0.5–0.75 in.). Additional cooling is provided by water sprays along the travel path of the solidifying metal. The molds are generally coated with graphite or similar solid lubricants to reduce friction and adhesion at the mold–metal interfaces. The molds are vibrated to further reduce friction and sticking.

The continuously cast metal may be cut into desired lengths by shearing or torch cutting, or it may be fed directly into a rolling mill, for further reduction in thickness and for shape rolling of products such as channels and I-beams. In addition to costing less, continuously cast metals have more-uniform compositions and properties than those obtained by ingot casting. Although the thickness of steel strand is usually about 250 mm (10 in.), new developments have reduced this thickness to about 25 mm (1 in.). The thinner strand reduces the number of rolling operations required and improves the economy of the overall operation.

After they are hot rolled, steel plates or shapes undergo one or more further processes, such as cleaning and pickling by chemicals to remove surface oxides, cold rolling to improve strength and surface finish, annealing, and coating (galvanizing or aluminizing) for resistance to corrosion.

5.5 Carbon and Alloy Steels

Carbon and alloy steels are among the most commonly used metals and have a wide variety of applications (Table 5.2). The composition and processing of steels are controlled in a manner that makes them suitable for numerous applications. They are available in various basic product shapes: plate, sheet, strip, bar, wire, tube, castings, and forgings.

5.5.1 Effects of various elements in steels

As we have already mentioned, various elements are added to steels to impart properties of hardenability, strength, hardness, toughness, wear resistance, workability, weldability, and machinability. We present these elements in alphabetic order and summarize their beneficial and detrimental effects.

TABLE 5.2
TYPICAL SELECTION OF CARBON AND ALLOY STEELS FOR VARIOUS APPLICATIONS

PRODUCT	STEEL
Aircraft forgings, tubing, fittings	4140, 8740
Automobile bodies	1010
Axles	1040, 4140
Ball bearings and races	52100
Bolts	1035, 4042, 4815
Camshafts	1020, 1040
Chains (transmission)	3135, 3140
Coil springs	4063
Connecting rods	1040, 3141, 4340
Crankshafts (forged)	1045, 1145, 3135, 3140
Differential gears	4023
Gears (car and truck)	4027, 4032
Landing gear	4140, 4340, 8740
Lock washers	1060
Nuts	3130
Railroad rails and wheels	1080
Springs (coil)	1095, 4063, 6150
Springs (leaf)	1085, 4063, 9260, 6150
Tubing	1040
Wire	1045, 1055
Wire (music)	1085

Generally, the higher the percentages of these elements in steels, the higher are the particular properties that they impart. Thus, for example, the higher the carbon content, the higher is the hardenability of the steel and the higher is its strength, hardness, and wear resistance. On the other hand, ductility, weldability, and toughness are reduced with increasing carbon content.

Boron improves hardenability without loss of, or even with some improvement in, machinability and formability.

Calcium deoxidizes steels, improves toughness, and may improve formability and machinability.

Carbon improves hardenability, strength, hardness, and wear resistance; reduces ductility, weldability, and toughness.

Cerium controls the shape of inclusions and improves toughness in high-strength low-alloy steels; deoxidizes steels.

Chromium improves toughness, hardenability, wear and corrosion resistance, and high-temperature strength; increases depth of hardness penetration in heat treatment by promoting carburization.

Cobalt improves strength and hardness at elevated temperatures.

Columbium (niobium) imparts fine grain size, improves strength and impact toughness; lowers transition temperature; may decrease hardenability.

Copper improves resistance to atmospheric corrosion and, to a lesser extent, strength, with little loss in ductility; adversely affects hot-working characteristics and surface quality.

Lead improves machinability; causes liquid metal embrittlement.

Magnesium has the same effects as cerium.

Manganese improves hardenability, strength, abrasion resistance, and machinability; deoxidizes the molten steel and reduces hot shortness; decreases weldability.

Molybdenum improves hardenability, wear resistance, toughness, elevated-temperature strength, creep resistance, and hardness; minimizes temper embrittlement.

Nickel improves strength, toughness, and corrosion resistance; has minor effect on hardenability.

Phosphorus improves strength, hardenability, corrosion resistance, and machinability; severely reduces ductility and toughness.

Selenium improves machinability.

Silicon improves strength, hardness, corrosion resistance, and electrical conductivity; decreases magnetic hysteresis loss, machinability, and cold formability.

Sulfur improves machinability when combined with manganese; lowers impact strength and ductility; impairs surface quality and weldability.

Tantalum has effects similar to those of columbium.

Tellurium improves machinability, formability, and toughness.

Titanium improves hardenability; deoxidizes steels.

Tungsten has the same effects as cobalt.

Vanadium improves strength, toughness, abrasion resistance, and hardness at elevated temperatures; inhibits grain growth during heat treatment.

Zirconium has the same effects as cerium.

5.5.2 Residual elements in steel

During steel production, refining, and processing, some residual elements, or **trace elements**, may still remain. Although we can also consider some of the elements in the preceding list as residuals, the following are generally considered residual elements. Also shown are their effects.

Antimony and *arsenic* cause temper embrittlement.

Hydrogen severely embrittles steels; heating during processing drives out most of the hydrogen.

Nitrogen improves strength, hardness, and machinability; in aluminum-deoxidized steels, it controls the size of inclusions and improves strength and toughness; decreases ductility and toughness.

Oxygen increases strength of rimmed steels slightly; severely reduces toughness.

Tin causes hot shortness and temper embrittlement.

5.5.3 Designations for steels

The American Iron and Steel Institute (AISI) and the Society of Automotive Engineers (SAE) designate carbon and alloy steels by using four digits (Table 5.3). The first two digits indicate the alloying elements and their percentages. The last two digits indicate the carbon content by weight. Thus, for example, 1112 steel is a resulfurized steel with a carbon content of 0.12 percent.

Another numbering system is the American Society for Testing and Materials (ASTM) designations, which incorporate AISI–SAE designations, and include standard specifications for steel products. For ferrous metals, the designation consists of the letter "A" followed by arbitrary numbers (generally three). A more recent numbering system for all metals has been developed cooperatively by several organizations. Known as the *Unified Numbering System* (*UNS*), it consists of a letter indicating the general class of the alloy, followed by five digits, designating its chemical composition.

5.5.4 Carbon steels

Carbon steels are generally classified as:

- *Low-carbon steel*, also called *mild steel*, has less than 0.30 percent carbon. It is generally used for common industrial products, such as bolts, nuts, sheet, plate, tubes, and machine components that do not require high strength.

TABLE 5.3
AISI–SAE DESIGNATIONS FOR STEELS AND THEIR MAJOR ALLOYING ELEMENTS

10xx	...	Plain carbon steels			
11xx	...	Resulphurized carbon steels (free machining)			
12xx	...	Rephosphorized and resulphurized carbon steels (free machining)			
13xx	...	Mn	1.75%		
31xx	...	Ni	1.25	Cr 0.65	
33xx	...	Ni	3.50	Cr 1.55	
40xx	...	Mo	0.20 or 0.25		
41xx	...	Cr	0.50 or 0.95	Mo 0.12 or 0.20	
43xx	...	Ni	1.80	Cr 0.50 or 0.80	Mo 0.25
44xx	...	Mo	0.40		
45xx	...	Mo	0.52		
46xx	...	Ni	1.80	Mo 0.25	
47xx	...	Ni	1.05	Cr 0.45	Mo 0.20 or 0.35
48xx	...	Ni	3.50	Mo 0.25	
50xx	...	Cr	0.25, 0.40, or 0.50		
50xxx	...	C	1.00	Cr 0.50	
51xx	...	Cr	0.80, 0.90, 0.95, or 1.00		
51xxx	...	C	1.00	Cr 1.05	
52xxx	...	C	1.00	Cr 1.45	
61xx	...	Cr	0.60, 0.80, or 0.95		
		V	0.12, 0.10 min., or 0.15 min.		
81xx	...	Ni	0.30	Cr 0.40	Mo 0.12
86xx	...	Ni	0.55	Cr 0.50	Mo 0.20
87xx	...	Ni	0.55	Cr 0.05	Mo 0.25
88xx	...	Ni	0.55	Cr 0.50	Mo 0.35
92xx	...	Mn	0.85	Si 2.00	Cr 0 or 0.35
93xx	...	Ni	3.25	Cr 1.20	Mo 0.12
94xx	...	Ni	0.45	Cr 0.40	Mo 0.12
98xx	...	Ni	1.00	Cr 0.80	Mo 0.25

Note: The last two digits indicate the carbon content. Thus 1020 steel has 0.20% carbon. Leaded steels are identified by the letter "L" inserted between the second and third numerals (10L45).

- *Medium-carbon steel* has 0.30 percent to 0.60 percent carbon. It is generally used in applications requiring higher strength than low-carbon steels, such as machinery, automotive and agricultural equipment parts (gears, axles, connecting rods, crankshafts), railroad equipment, and parts for metalworking machinery.
- *High-carbon steel* has more than 0.60 percent carbon. It is generally used for parts requiring strength, hardness, and wear resistance, such as cutting tools, cable, music wire, springs, and cutlery. After being manufactured into shapes, the parts are usually heat treated and tempered. The higher the

carbon content of the steel, the higher is its hardness, strength, and wear resistance after heat treatment.

Applications and general mechanical properties of carbon and alloy steels are shown in Tables 5.2 (p. 169) and 5.4. The machinability, formability, and weldability of these steels are described in later chapters.

5.5.5 Alloy steels

Steels containing significant amounts of alloying elements are called **alloy steels.** Alloy steels are usually made with more care than are carbon steels. *Structural-grade* alloy steels, as identified by ASTM specifications, are used mainly in the construction and transportation industries because of their high strength. Other alloy steels are used in applications where strength, hardness, creep and fatigue resistance, and toughness are required. These steels may also be heat treated to obtain the desired properties.

5.5.6 High-strength low-alloy steels

In order to improve the strength-to-weight ratio of steels, a number of **high-strength low-alloy** (HSLA) **steels** have been developed. These steels have a low carbon content, usually less than 0.30 percent, and are characterized by a microstructure consisting of fine-grain ferrite and a hard second phase of martensite and austenite. First developed in the 1930s, HSLA steels are usually produced in sheet form by microalloying and controlled hot rolling. Plates, bars, and structural shapes are made from these steels. However, the ductility, formability, and weldability of HSLA steels are generally inferior to conventional low-alloy steels. To improve these properties, dual-phase steels have been developed (see below).

TABLE 5.4

TYPICAL MECHANICAL PROPERTIES OF SELECTED CARBON AND ALLOY STEELS IN THE HOT-ROLLED, NORMALIZED, AND ANNEALED CONDITION

AISI	CONDITION	ULTIMATE TENSILE STRENGTH (MPa)	YIELD STRENGTH (MPa)	ELONGATION (%)	REDUCTION OF AREA (%)	HARDNESS (HB)
1020	As-rolled	448	346	36	59	143
	Normalized	441	330	35	67	131
	Annealed	393	294	36	66	111
1080	As-rolled	1010	586	12	17	293
	Normalized	965	524	11	20	293
	Annealed	615	375	24	45	174
3140	Normalized	891	599	19	57	262
	Annealed	689	422	24	50	197
4340	Normalized	1279	861	12	36	363
	Annealed	744	472	22	49	217
8620	Normalized	632	385	26	59	183
	Annealed	536	357	31	62	149

TABLE 5.5
AISI DESIGNATION FOR HIGH-STRENGTH SHEET STEEL

YIELD STRENGTH psi × 10^3	YIELD STRENGTH MPa	CHEMICAL COMPOSITION	DEOXIDATION PRACTICE
35	240	S = structural quality	F = killed plus sulfide inclusion control
40	275		
45	310		
50	350	X = low alloy	
60	415		K = killed
70	485	W = weathering	
80	550		O = nonkilled
100	690	D = dual phase	
120	830		
140	970		

EXAMPLE

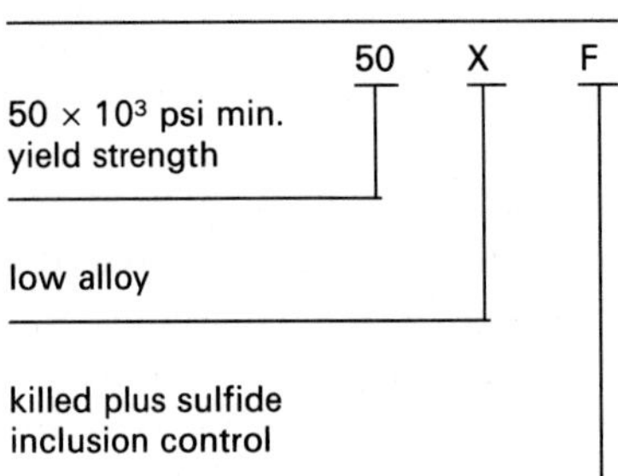

Sheet products of HSLA steels typically are used in certain parts of automobile bodies to reduce weight, and hence fuel consumption, and in transportation equipment, mining, agricultural, and various other industrial applications. Plates are used in ships, bridges, building construction, and shapes such as I-beams, channels, and angles are used in buildings and various other structures.

Designations. Three categories comprise the system of AISI designations for high-strength sheet steel (Table 5.5). Structural quality (S) includes the elements C, Mn, P, and N. Low alloys (X) contain Nb, Cr, Cu, Mo, Ni, Si, Ti, V, and Zr, either singly or in combination. Weathering steels (W) have environmental corrosion resistance that is approximately four times greater than that of conventional low-carbon steels and contain Si, P, Cu, Ni, and Cr in various combinations. Generally, the formability of these sheet steels is graded F (excellent), K (good), and O (fair).

5.5.7 Dual-phase steels

Dual-phase steels, designated with the letter "D" in Table 5.5, are processed specially and have a mixed ferrite and martensite structure. Developed in the late

1960s, these steels have high work-hardening characteristics (high n value in Eq. 2.8), which improve their ductility and formability. The SAE designations for these steels are similar to those given in Table 5.5, with the exception that another letter is added to indicate the carbon content. Thus, 050XF becomes 050XLF, where L indicates the carbon (in this case L meaning low carbon).

5.6 Stainless Steels

Stainless steels are characterized primarily by their corrosion resistance, high strength and ductility, and high chromium content. They are called *stainless* because, in the presence of oxygen (air), they develop a thin, hard adherent film of chromium oxide that protects the metal from corrosion (passivation). This protective film builds up again in the event the surface is scratched. For passivation to occur, the minimum chromium content of the steel should be 10–12 percent by weight.

In addition to chromium, other alloying elements in stainless steels typically are nickel, molybdenum, copper, titanium, silicon, manganese, columbium, aluminum, nitrogen, and sulfur. The higher the carbon content, the lower is the corrosion resistance of stainless steels. The reason is that the carbon combines with chromium in the steel and forms chromium carbide, which lowers the passivity of the steel. The chromium carbides introduce a second phase in the metal, which promotes galvanic corrosion.

Developed in the early 1900s, stainless steels are made by techniques similar to those used in other types of steelmaking, using electric furnaces or the basic-oxygen process. The level of purity is controlled by various refining techniques. Stainless steels are available in a wide variety of shapes. Typical applications are for cutlery, kitchen equipment, health care and surgical equipment, and automotive trim and in the chemical, food processing, and petroleum industries.

Stainless steels are generally divided into five types (Table 5.6).

Austenitic (200 and 300 series). These steels are generally composed of chromium, nickel, and manganese in iron. They are nonmagnetic and have excellent corrosion resistance, but are susceptible to stress-corrosion cracking. Austenitic stainless steels are hardened by cold working. They are the most ductile of all stainless steels, hence can be formed easily. However, with increasing cold work their formability is reduced. These steels are used in a wide variety of applications, such as kitchenware, fittings, welded construction, lightweight transportation equipment, furnace and heat-exchanger parts, and components for severe chemical environments.

Ferritic (400 series). These steels have a high chromium content: up to 27 percent. They are magnetic and have good corrosion resistance, but have lower

TABLE 5.6
ROOM-TEMPERATURE MECHANICAL PROPERTIES AND TYPICAL APPLICATIONS OF ANNEALED STAINLESS STEELS

AISI (UNS)	ULTIMATE TENSILE STRENGTH (MPa)	YIELD STRENGTH (MPa)	ELONGATION (%)	CHARACTERISTICS AND TYPICAL APPLICATIONS
303 (S30300)	550–620	240–260	50–53	Screw machine products, shafts, valves, bolts, bushings, and nuts; aircraft fittings; bolts; nuts; rivets; screws; studs.
304 (S30400)	565–620	240–290	55–60	Chemical and food processing equipment, brewing equipment, cryogenic vessels, gutters, downspouts, and flashings.
316 (S31600)	550–590	210–290	55–60	High corrosion resistance and high creep strength. Chemical and pulp handling equipment, photographic equipment, brandy vats, fertilizer parts, ketchup cooking kettles, and yeast tubs.
410 (S41000)	480–520	240–310	25–35	Machine parts, pump shafts, bolts, bushings, coal chutes, cutlery, finishing, tackle, hardware, jet engine parts, mining machinery, rifle barrels, screws, and valves.
416 (S41600)	480–520	275	20–30	Aircraft fittings, bolts, nuts, fire extinguisher inserts, rivets, and screws.

ductility than austenitic stainless steels. Ferritic stainless steels are hardened by cold working and are not heat treatable. They are generally used for nonstructural applications, such as kitchen equipment and automotive trim.

Martensitic (400 and 500 series). Most martensitic stainless steels do not contain nickel and are hardenable by heat treatment. Their chromium content may be as much as 18 percent. These steels are magnetic and have high strength, hardness and fatigue resistance, and good ductility, but moderate corrosion resistance. Martensitic stainless steels are used for cutlery, surgical tools, instruments, valves, and springs.

Precipitation hardening (PH). These steels contain chromium and nickel, along with copper, aluminum, titanium, or molybdenum. They have good corrosion resistance, ductility, and high strength at elevated temperatures. Their main application is in aircraft and aerospace structural components.

Duplex structure. These steels have a mixture of austenite and ferrite. They have good strength and also have higher resistance to corrosion in most environments and stress-corrosion cracking than the 300 series of austenitic steels. Typical applications are in water-treatment plants and heat-exchanger components.

Example: Use of stainless steels in automobiles.

The types of stainless steels selected by materials engineers for use in automobile parts are usually 301, 409, 430, and 434. Because of its good corrosion resistance and mechanical properties, type 301 is used for wheel covers. Cold working during the forming process increases its yield strength (strain hardening), thus giving the cover a spring action. Type 409 is used extensively for catalytic converters. Type 430 had been used for automotive trim. However, it is not as resistant to de-icing salts, used in winter in colder climates, as type 434, hence its use is limited. In addition to being more corrosion resistant, type 434 closely resembles the color of chromium plating, which makes it an attractive alternative.

5.7 Tool and Die Steels

Tool and die steels are specially alloyed steels (Tables 5.7 and 5.8) designed for high strength, impact toughness, and wear resistance at room and elevated temperatures. They are commonly used in forming and machining of metals (see Parts III and IV).

TABLE 5.7
BASIC TYPES OF TOOL AND DIE STEELS

TYPE	AISI
High speed	M (molybdenum base) T (tungsten base)
Hot work	H1 to H19 (chromium base) H20 to H39 (tungsten base) H40 to H59 (molybdenum base)
Cold work	D (high carbon, high chromium) A (medium alloy, air hardening) O (oil hardening)
Shock resisting	S
Mold steels	P1 to P19 (low carbon) P20 to P39 (others)
Special purpose	L (low alloy) F (carbon–tungsten)
Water hardening	W

TABLE 5.8
PROCESSING AND SERVICE CHARACTERISTICS OF COMMON TOOL AND DIE STEELS

AISI DESIGNATION	RESISTANCE TO DECARBURIZATION	RESISTANCE TO CRACKING	APPROXIMATE HARDNESS (HRC)	MACHINABILITY	TOUGHNESS	RESISTANCE TO SOFTENING	RESISTANCE TO WEAR
M2	Medium	Medium	60–65	Medium	Low	Very high	Very high
T1	High	High	60–65	Medium	Low	Very high	Very high
T5	Low	Medium	60–65	Medium	Low	Highest	Very high
H11, 12, 13	Medium	Highest	38–55	Medium to high	Very high	High	Medium
A2	Medium	Highest	57–62	Medium	Medium	High	High
A9	Medium	Highest	35–56	Medium	High	High	Medium to high
D2	Medium	Highest	54–61	Low	Low	High	High to very high
D3	Medium	High	54–61	Low	Low	High	Very high
H21	Medium	High	36–54	Medium	High	High	Medium to high
H26	Medium	High	43–58	Medium	Medium	Very high	High
P20	High	High	28–37	Medium to high	High	Low	Low to medium
P21	High	Highest	30–40	Medium	Medium	Medium	Medium
W1, W2	Highest	Medium	50–64	Highest	High	Low	Low to medium

Source: Adapted from *Tool Steels*, American Iron and Steel Institute, 1978.

5.7.1 High-speed steels

High-speed steels (*HSS*) are the most highly alloyed tool and die steels and maintain their hardness and strength at elevated operating temperatures. First developed in the early 1900s, there are two basic types of high-speed steels: the molybdenum type (M series) and the tungsten type (T series).

The M-series steels contain up to about 10 percent molybdenum, with chromium, vanadium, tungsten, and cobalt as other alloying elements.

The T-series steels contain 12–18 percent tungsten, with chromium, vanadium, and cobalt as other alloying elements. The M-series steels generally have higher abrasion resistance than the T-series steels, have less distortion in heat treatment, and are less expensive. The M series constitutes about 95 percent of all high-speed steels produced in the United States. High-speed steel tools can be coated with titanium nitride and titanium carbide for better wear resistance.

5.7.2 Hot-work, cold-work, and shock-resisting die steels

Hot-work steels (H series) are designed for use at elevated temperatures and have high toughness and high resistance to wear and cracking. The alloying elements are generally tungsten, molybdenum, chromium, and vanadium.

Cold-work steels (A, D, and O series) are used for cold-working operations. They generally have high resistance to wear and cracking. These steels are available as oil-hardening or air-hardening types.

Shock-resisting steels (S series) are designed for impact toughness and are used in applications such as header dies, punches, and chisels. Other properties of these steels depend on the particular composition.

Various tool and die materials for a variety of manufacturing applications are presented in Table 5.9.

SUMMARY

Their wide range of properties, numerous applications, and low cost have made steels among the most useful of all metallic materials. The major categories of these materials are carbon steels, alloy steels, stainless steels, and tool and die steels. Steelmaking processes have been improved rapidly—notably the continuous casting and secondary refining techniques—resulting in higher quality steels and greater productivity in steelmaking operations.

The mechanical, physical, and chemical properties of steels depend on their method of production, as well as on the type and amount of alloying and trace elements present. These elements, in turn, greatly influence the manufacturing properties of these materials (formability, machinability, and weldability), as well as their performance in service.

TABLE 5.9
TYPICAL TOOL AND DIE MATERIALS FOR METALWORKING

PROCESS	MATERIAL
Die casting	H13, P20
Powder metallurgy	
Punches	A2, S7, D2, D3, M2
Dies	WC, D2, M2
Molds for plastics and rubber	S1, O1, A2, D2, 6F5, 6F6, P6, P20, P21, H13
Hot forging	6F2, 6G, H11, H12
Hot extrusion	H11, H12, H13
Cold heading	W1, W2, M1, M2, D2, WC
Cold extrusion	
Punches	A2, D2, M2, M4
Dies	O1, W1, A2, D2
Coining	52100, W1, O1, A2, D2, D3, D4, H11, H12, H13
Drawing	
Wire	WC, diamond
Shapes	WC, D2, M2
Bar and tubing	WC, W1, D2
Rolls	
Rolling	Cast iron, cast steel, forged steel, WC
Thread rolling	A2, D2, M2
Shear spinning	A2, D2, D3
Sheet metals	
Shearing	
Cold	D2, A2, A9, S2, S5, S7
Hot	H11, H12, H13
Pressworking	Zinc alloys, 4140 steel, cast iron, epoxy composites, A2, D2, O1
Deep drawing	W1, O1, cast iron, A2, D2
Machining	Carbides, high-speed steels, ceramics

Notes: Tool and die materials are usually hardened to 55 to 65 HRC for cold working, and 30 to 55 for hot working.

Tool and die steels contain one or more of the following major alloying elements: chromium, molybdenum, tungsten, and vanadium. For further details see the bibliography at the end of this chapter.

TRENDS

- The highly competitive international market for steel has led to cost control through improved productivity, elimination of obsolete and inefficient equipment and procedures, and reduced labor costs.
- New developments, such as specialty mills and minimills (which use mostly scrap metal), allow production of specialty or one kind of product economically. Another trend is to use iron that is directly reduced (from iron oxide ores), rather than melting it in blast furnaces.

(*continued*)

TRENDS *(continued)*

- Computer controls and methods of process optimization are being implemented to improve efficiency and quality, as well as all other aspects of steelmaking.
- New compositions, treatments, and techniques for refining steels are being developed to improve various properties, such as formability, machinability, weldability, service life, and response to heat treatment.

KEY TERMS

Alloy steels
Basic-oxygen furnace
Blast furnace
Carbon steels
Continuous casting
Dual-phase steels
Electric furnace
High-strength low-alloy steels
Ingot
Killed steel
Open-hearth furnace
Pig
Pig iron
Refining
Rimmed steel
Semi-killed steel
Stainless steels
Strand casting
Trace elements
Tool and die steels

BIBLIOGRAPHY

Aerospace Structural Metals Handbook, 5 vols. Columbus, Ohio: Metals and Ceramics Information Center, Battelle, 1987.

ASM Metals Reference Book, 2d ed. Metals Park, Ohio: American Society for Metals, 1983.

Design Guidelines for the Selection and Use of Stainless Steels. Washington, D. C.: American Iron and Steel Institute, 1977.

Harvey, P.D. (ed.), *Engineering Properties of Steel.* Metals Park, Ohio: American Society for Metals, 1982.

Lankford, W.T., Jr., N.L. Samways, R.F. Craven, and H.E. McGannon, *The Making, Shaping and Treating of Steel*, 10th ed. Pittsburgh: United States Steel Co., 1985.

Lula, R.A., *Stainless Steels.* Metals Park, Ohio: American Society for Metals, 1985.

Material Selector. Annual publication of *Materials Engineering Magazine*, Penton/IPC, Cleveland, Ohio.

Metals Handbook, Desk Edition. Metals Park, Ohio: American Society for Metals, 1985.

Metals Handbook, 10th ed., Vol. 1: *Properties and Selection: Iron, Steels, and High-Performance Alloys.* Materials Park, Ohio: ASM International, 1990.

Roberts, G.A., and R.A. Cary, *Tool Steels*, 4th ed. Metals Park, Ohio: American Society for Metals, 1980.

Tool and Manufacturing Engineers Handbook, 4th ed., Vol. 3: *Materials, Finishing and Coating.* Dearborn, Mich.: Society of Manufacturing Engineers, 1985.

REVIEW QUESTIONS

5.1 What are the major categories of ferrous alloys?
5.2 List the basic raw materials used in making iron and steel and explain their functions.
5.3 List the types of furnaces commonly used in steelmaking and describe their characteristics.
5.4 List and explain the characteristics of three types of steel ingots.
5.5 What does refining mean? How is it done?
5.6 What advantages does continuous casting have over casting ingots?
5.7 Name four alloying elements that have the greatest effect on the properties of steels.
5.8 What are trace elements?
5.9 What are the carbon contents of low-carbon, medium-carbon, and high-carbon steels, respectively?
5.10 How do stainless steels become stainless?
5.11 List the types of stainless steels.
5.12 What are the major alloying elements in tool and die steels and high-speed steels?

QUESTIONS AND PROBLEMS

5.13 Based on the information given in Section 5.5.1, make a table with columns for each improved property, such as hardenability, strength, toughness, and machinability. In each column list the elements that improve that property and identify the element that has the most influence.
5.14 Identify various products that are made of stainless steel and explain why they are made of this material.
5.15 Assume that you are in charge of public relations for a steel-producing company. Outline all the attractive characteristics of steels that you would like your customers to know about.
5.16 Assume that you are in competition with the steel industry and are asked to list all the characteristics of steels that are not attractive. Make a list of these features and explain their engineering relevance.
5.17 As you may know, professional cooks prefer carbon-steel to stainless-steel knives, even though the latter are more popular with consumers. Explain the reasons for this preference.
5.18 Why is the control of ingot structure important?
5.19 Assume that you are in charge of the research department of a large steel-producing company. Make a list of the research projects that you would like to initiate, explaining the reasons for your choices.
5.20 Explain why continuous casting has been such an important technological advancement.
5.21 Certain alloying elements are commonly used in tool and die steels. Explain why these elements are essential in these steels.
5.22 Describe applications in which you would not want to use carbon steels.
5.23 Explain what would happen if the speed of the continuous casting process is (a) higher and (b) lower than that indicated in Fig. 5.6.

5.24 In Table 5.1, why does the cost of mill products of metals increase with decreasing thickness and section size?

5.25 Explain the reasons for the selection of steel compositions for the applications indicated in Table 5.2.

5.26 Prepare a Summary Table for this chapter, similar to those in the preceding chapters.

5.27 Why is sulfide inclusion control identified separately (as F) from killed steel in Table 5.5?

5.28 Describe your observations regarding the information given in Table 5.8.

5.29 How do trace elements adversely affect the ductility of steels?

6

Nonferrous Metals and Alloys: Production, General Properties, and Applications

6.1 Introduction

Nonferrous metals and alloys cover a wide range of materials, from the more common metals, such as aluminum, copper, and magnesium, to high-strength high-temperature alloys, such as tungsten, tantalum, and molybdenum. Although more expensive than ferrous metals (Table 6.1; see also Table 5.1), nonferrous metals and alloys have important applications because of their numerous properties, such as corrosion resistance, high thermal and electrical conductivity, low density, ease of fabrication, and color choices.

A turbofan jet engine for the Boeing 757 aircraft typically contains the following nonferrous metals and alloys: 38 percent titanium, 37 percent nickel, 12 percent chromium, 6 percent cobalt, 5 percent aluminum, 1 percent columbium (niobium), and 0.02 percent tantalum. Without these materials, a jet engine (Fig. 6.1) could not be designed, manufactured, and operated at the energy and efficiency levels required.

TABLE 6.1
APPROXIMATE COST PER UNIT VOLUME FOR WROUGHT METALS AND PLASTICS RELATIVE TO COST OF CARBON STEEL

Gold	60,000	Magnesium alloys	2–4
Silver	600	Aluminum alloys	2–3
Molybdenum alloys	200–250	High-strength low-alloy steels	1.4
Nickel	35	Gray cast iron	1.2
Titanium alloys	20–40	Carbon steel	1
Copper alloys	5–6	Nylons, acetals, and silicon rubber*	1.1–2
Zinc alloys	2.5–3.5	Other plastics and elastomers*	0.2–1
Stainless steels	2–9		

* As molding compounds.
Note: Costs vary significantly with quantity of purchase, supply and demand, size and shape, and various other factors.

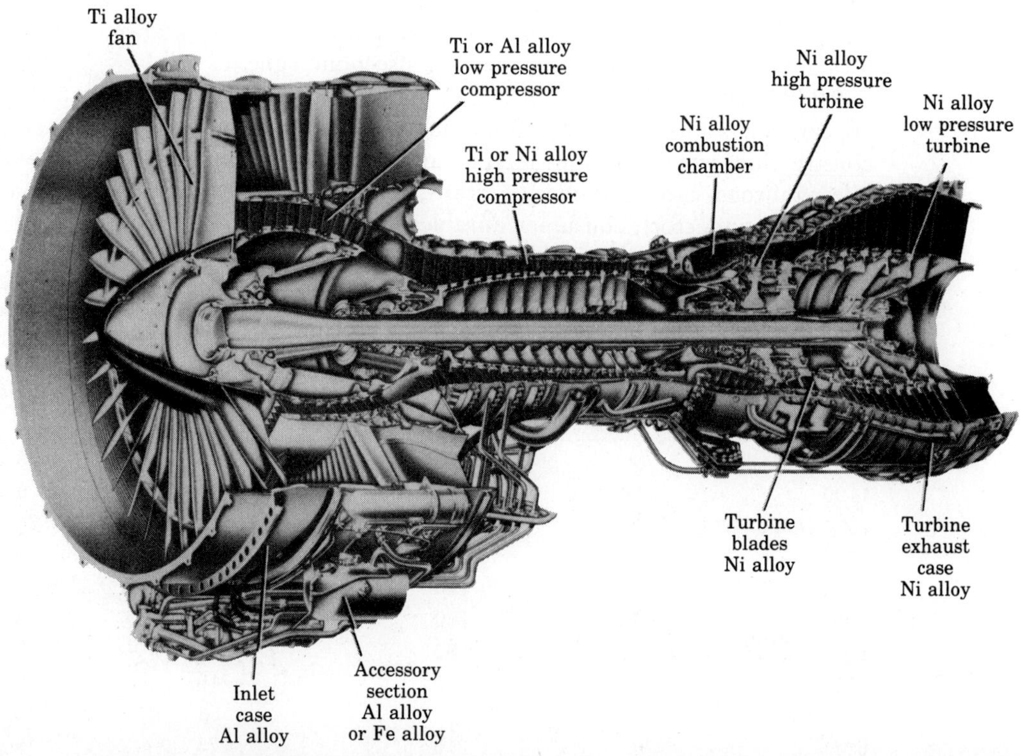

FIGURE 6.1
Cross section of a jet engine (PW2037) showing various components and the alloys used in making them. *Source:* Courtesy of United Aircraft Pratt & Whitney.

Typical examples of the applications of nonferrous metals and alloys are aluminum for cooking utensils and aircraft bodies, copper wire for electricity and copper tubing for water in residences, zinc for carburetors, titanium for jet-engine turbine blades and prosthetic devices (such as artificial limbs), and tantalum for rocket engines.

In this chapter we discuss the general properties, production methods, and important engineering applications of nonferrous metals and alloys. We describe the manufacturing properties of these materials, such as formability, machinability, and weldability, in various chapters throughout this text.

6.2 Aluminum and Aluminum Alloys

The important factors in selecting aluminum (Al) and its alloys are their high strength-to-weight ratio, resistance to corrosion by many chemicals, high thermal and electrical conductivity, nontoxicity, reflectivity, appearance, and ease of formability and machinability; they are also nonmagnetic.

Principal uses of aluminum and its alloys, in decreasing order of consumption, are containers and packaging (aluminum cans and foil), buildings and other types of construction, transportation (aircraft and aerospace applications, buses, automobiles, railroad cars, and marine craft), electrical (economical and nonmagnetic electrical conductor), consumer durables (appliances, cooking utensils, and furniture), and portable tools (Tables 6.2 and 6.3). Nearly all high-voltage transmission

TABLE 6.2
PROPERTIES OF SELECTED ALUMINUM ALLOYS AT ROOM TEMPERATURE

ALLOY (UNS)	TEMPER	ULTIMATE TENSILE STRENGTH (MPa)	YIELD STRENGTH (MPa)	ELONGATION IN 50 mm (%)
1100 (A91100)	O	90	35	35–45
1100	H14	125	120	9–20
2024 (A92024)	O	190	75	20–22
2024	T4	470	325	20–19
3003 (A93003)	O	110	40	30–40
3003	H14	150	145	8–16
5052 (A95052)	O	190	90	25–30
5052	H34	260	215	10–14
6061 (A96061)	O	125	55	25–30
6061	T6	310	275	12–17
7075 (A97075)	O	230	105	17–16
7075	T6	570	500	11

TABLE 6.3
MANUFACTURING PROPERTIES AND TYPICAL APPLICATIONS OF WROUGHT ALUMINUM ALLOYS

	CHARACTERISTICS*			
ALLOY	*CORROSION RESISTANCE*	*MACHINABILITY*	*WELDABILITY*	TYPICAL APPLICATIONS
1100	A	D–C	A	Sheet metal work, spun hollow ware, tin stock
2024	C	C–B	C–B	Truck wheels, screw machine products, aircraft structures
3003	A	D–C	A	Cooking utensils, chemical equipment, pressure vessels, sheet metal work, builders' hardware, storage tanks
5052	A	D–C	A	Sheet metal work, hydraulic tubes, and appliances; bus, truck and marine uses
6061	B	D–C	A	Heavy-duty structures where corrosion resistance is needed, truck and marine structures, railroad cars, furniture, pipelines, bridge railings, hydraulic tubing
7075	C	B–D	D	Aircraft and other structures, keys, hydraulic fittings

* A, excellent; D, poor.

wiring is made of aluminum. Of their structural (load bearing) components, 82 percent of a Boeing 747 aircraft and 79 percent of a Boeing 757 aircraft are made of aluminum.

Aluminum alloys are available as mill products, that is, wrought products made into various shapes by rolling, extrusion, drawing, and forging. Aluminum ingots are available for casting, as are powder metals for powder metallurgy applications. There are two types of wrought alloys of aluminum: (1) alloys that can be hardened by cold working and are not heat treatable, and (2) alloys that are hardenable by heat treatment. Techniques have been developed whereby most aluminum alloys can be machined, formed, and welded with relative ease.

Aluminum was first produced in 1825. It is the most abundant metallic element in the earth's crust, comprising about 8 percent of the crust, and is produced in quantities second only to iron. The principal ore for aluminum is *bauxite*, which is hydrous (containing water) aluminum oxide and includes various other oxides. After the clay and dirt are washed off, the ore is crushed into powder and treated with hot caustic soda (sodium hydroxide) to remove impurities. Alumina (aluminum oxide) is extracted from this solution and dissolved in a molten sodium-fluoride and aluminum-fluoride bath at 940–980 °C (1725–1800 °F).

This mixture is then subjected to direct-current electrolysis. Aluminum metal forms at the cathode (negative pole), while oxygen is released at the anode (positive pole). Commercially pure aluminum is at most 99.5–99.7 percent aluminum. The production process consumes a great deal of electricity, thus contributing significantly to the cost of aluminum.

6.2.1 Designation of aluminum alloys

Wrought aluminum alloys are identified by four digits and a **temper designation,** showing the condition of the material. The major alloying element is identified by the first digit.

1xxx—Commercially pure aluminum: 99.00 percent (minimum) aluminum. Excellent corrosion resistance, high electrical and thermal conductivity, good workability, low strength, not heat treatable.

2xxx—Copper: High strength-to-weight ratio, low resistance to corrosion, heat treatable.

3xxx—Manganese: Good workability, moderate strength, generally not heat treatable.

4xxx—Silicon: Lower melting point, forms dark-gray to charcoal oxide film, generally not heat treatable.

5xxx—Magnesium: Good corrosion resistance and weldability, moderate to high strength, not heat treatable.

6xxx—Magnesium and silicon: Medium strength, good formability, machinability, weldability, and corrosion resistance, heat treatable.

7xxx—Zinc: Moderate to very high strength, heat treatable.

8xxx—Other element.

The second digit in these designations indicates modifications of the alloy. For the 1xxx series, the third and fourth digits indicate the minimum amount of aluminum in the alloy. Thus 1050 indicates a minimum of 99.50 percent aluminum, and 1090 indicates a minimum of 99.90 percent aluminum. In other series, the third and fourth digits identify the different alloys in the group and have no numerical significance.

The designations for *cast* aluminum alloys and ingots also consist of four digits. The first digit indicates the major alloy group, as follows:

1xx.x—Aluminum (99.00 percent minimum).

2xx.x—Aluminum–copper.

3xx.x—Aluminum–silicon, with copper and/or magnesium.

4xx.x—Aluminum–silicon.

5xx.x—Aluminum–magnesium.

6xx.x—Unused series.

7xx.x—Aluminum–zinc.

8xx.x—Aluminum–tin.

In the 1xx.x series, the second and third digits indicate the minimum aluminum content, as in wrought aluminum. For the other series, the second and third digits have no numerical significance. The fourth digit (to the right of the decimal point) indicates product form: 0 denotes discrete castings (such as carburetor or electric-motor housing), and 1 denotes ingot.

Temper designations. The temper designations for both wrought and cast aluminum are:

- F As fabricated (by cold or hot working, or casting).
- O Annealed (from cold-worked or cast state).
- H Strain hardened by cold working (for wrought products only).

  H1—Strain-hardened only, followed by a second digit indicating degree of cold work, ranging from 1 to 9.

  H2—Strain hardened and partially annealed, followed by a second digit indicating degree of strain hardening remaining.

  H3—Strain hardened and stabilized, followed by a second digit indicating degree of strain hardening.
- T Heat treated.

  T1—Cooled from hot working and naturally aged.

  T2—Cooled from hot working, cold worked, and naturally aged.

  T3—Solution treated, cold worked, and naturally aged.

  T4—Solution treated and naturally aged.

  T5—Cooled from hot working and artificially aged.

  T6—Solution treated and artificially aged.

  T7—Solution treated and stabilized.

  T8—Solution treated, cold worked, and artificially aged.

  T9—Solution treated, artificially aged, and cold worked.

  T10—Cooled from hot working, cold worked, and artificially aged.
- W Solution treated only (unstable temper).

We cover cast aluminum alloys in Part II. We have previously described the heat treatment of aluminum alloys in Section 4.9.

6.3 Magnesium and Magnesium Alloys

Magnesium (Mg) is the lightest engineering metal available and also has good vibration-damping characteristics. Its alloys are used in structural and nonstructural applications where weight is of primary importance. Magnesium is also an alloying element in various nonferrous metals.

Typical uses of magnesium alloys are for aircraft and missile components, material-handling equipment, portable power tools (such as drills and sanders), ladders, luggage, bicycles, sporting goods, and general lightweight components. These alloys are available either as castings or wrought products, such as extruded bars and shapes, forgings, and rolled plate and sheet. Magnesium alloys are also used in printing and textile machinery to minimize inertial forces in high-speed components.

Because it is not sufficiently strong in its pure form, magnesium is alloyed with various elements (Table 6.4) to impart certain specific properties, particularly high strength-to-weight ratio. A variety of magnesium alloys have good casting, forming, and machining characteristics. Because they oxidize rapidly—that is, they are *pyrophoric*—a fire hazard exists and precautions must be taken when machining, grinding, or sand casting magnesium alloys. However, products made of magnesium and its alloys are not a fire hazard.

Magnesium is the third most abundant metallic element (2 percent) in the earth's crust, after aluminum and iron. Most magnesium comes from seawater, which contains 0.13 percent magnesium in the form of magnesium chloride. First produced in 1808, magnesium metal can be obtained electrolytically and by thermal reduction. In the first method, seawater is mixed with lime (calcium hydroxide) in settling tanks. Magnesium hydroxide precipitates to the bottom, and is filtered and mixed with hydrochloric acid. This solution is subjected to electrolysis (as is done with aluminum), producing magnesium metal, which is then cast into ingots for further processing into various shapes.

In the thermal reduction method, mineral rock containing magnesium (dolomite, magnesite, and other rocks) is broken down with reducing agents, such as powdered ferrosilicon (an alloy of iron and silicon), by heating the mixture in a vacuum chamber. As a result of this reaction, vapors of magnesium form and condense into magnesium crystals. These crystals are then melted, refined, and poured into ingots to be processed further into various shapes.

TABLE 6.4
PROPERTIES AND TYPICAL FORMS OF WROUGHT MAGNESIUM ALLOYS

	COMPOSITION (%)					ULTIMATE TENSILE STRENGTH	YIELD STRENGTH	ELONGATION IN 50 mm	
ALLOY	Al	Zn	Mn	Zr	CONDITION	(MPa)	(MPa)	(%)	TYPICAL FORMS
AZ31B	3.0	1.0	0.2		F	260	200	15	Extrusions
					H24	290	220	15	Sheet and plates
AZ80A	8.5	0.5	0.2		T5	380	275	7	Extrusions and forgings
HK31A			3Th	0.7	H24	255	200	8	Sheet and plates
ZK60A		5.7		0.55	T5	365	300	11	Extrusions and forgings

6.3.1 Designation of magnesium alloys

Magnesium alloys are designated as follows:

1. One or two prefix letters, indicating the principal alloying elements from

   A Aluminum
   E Rare earth
   H Thorium
   K Zirconium
   M Manganese
   P Lead
   Q Silver
   S Silicon
   T Tin
   Z Zinc

2. Two or three numerals, indicating the percentage of the principal alloying elements, rounded off to the nearest decimal.
3. A letter of the alphabet, except the letters I and O, indicating the standardized alloy, with minor variations in composition.
4. The temper of the material, indicated by the same symbols used for aluminum alloys.

For example, the alloy AZ91C-T6 indicates that:

- The principal alloying elements are aluminum (A, 9 percent, rounded off) and zinc (Z, 1 percent).
- The letter C, the third letter of the alphabet, indicates that this is the third alloy standardized, after A and B, which were the first and second alloys, respectively, that were standardized.
- T6 indicates that this alloy has been solution treated and artificially aged.

6.4 Copper and Copper Alloys

First produced in about 4000 B.C., copper (Cu) and its alloys have properties somewhat similar to those of aluminum alloys. In addition, they are among the best conductors of electricity and heat and have good corrosion resistance. Because of these properties, copper and its alloys are among the most important metals. They can be processed easily by various forming, machining, casting and joining techniques.

TABLE 6.5
PROPERTIES AND TYPICAL APPLICATIONS OF WROUGHT COPPER AND BRASSES

TYPE AND UNS NUMBER	NOMINAL COMPOSITION (%)	ULTIMATE TENSILE STRENGTH (MPa)	YIELD STRENGTH (MPa)	ELONGATION IN 50 mm (%)	TYPICAL APPLICATIONS
Electrolytic tough pitch copper (C11000)	99.90 Cu, 0.04 O	220–450	70–365	55–4	Downspouts, gutters, roofing, gaskets, auto radiators, busbars, nails, printing rolls, rivets.
Red brass, 85% (C23000)	85.0 Cu, 15.0 Zn	270–725	70–435	55–3	Weather-stripping, conduit, sockets, fasteners, fire extinguishers, condenser and heat exchanger tubing.
Cartridge brass, 70% (C26000)	70.0 Cu, 30.0 Zn	300–900	75–450	66–3	Radiator cores and tanks, flashlight shells, lamp fixtures, fasteners, locks, hinges, ammunition components, plumbing accessories.
Free-cutting brass (C36000)	61.5 Cu, 3.0 Pb, 35.5 Zn	340–470	125–310	53–18	Gears, pinions, automatic high-speed screw machine parts.
Naval brass (C46400 to C46700)	60.0 Cu, 39.25 Zn, 0.75 Sn	380–610	170–455	50–17	Aircraft turnbuckle barrels, balls, bolts, marine hardware, propeller shafts, rivets, valve stems, condenser plates.

TABLE 6.6
PROPERTIES AND TYPICAL APPLICATIONS OF WROUGHT BRONZES

TYPE AND UNS NUMBER	NOMINAL COMPOSITION (%)	ULTIMATE TENSILE STRENGTH (MPa)	YIELD STRENGTH (MPa)	ELONGATION IN 50 mm (%)	TYPICAL APPLICATIONS
Architectural bronze (C38500)	57.0 Cu, 3.0 Pb, 40.0 Zn	415 (As extruded)	140	30	Architectural extrusions, store fronts, thresholds, trim, butts, hinges.
Phosphor bronze, 5% A (C51000)	95.0 Cu, 5.0 Sn, trace P	325–960	130–550	64–2	Bellows, clutch disks, cotter pins, diaphragms, fasteners, wire brushes, chemical hardware, textile machinery.
Free-cutting phosphor bronze (C54400)	88.0 Cu, 4.0 Pb, 4.0 Zn, 4.0 Sn	300–520	130–435	50–15	Bearings, bushings, gears, pinions, shafts, thrust washers, valve parts.
Low silicon bronze, B (C65100)	98.5 Cu, 1.5 Si	275–655	100–475	55–11	Hydraulic pressure lines, bolts, marine hardware, electrical conduits, heat exchanger tubing.
Nickel silver, 65–10 (C74500)	65.0 Cu, 25.0 Zn, 10.0 Ni	340–900	125–525	50–1	Rivets, screws, slide fasteners, hollow ware, nameplates.

Copper alloys often are attractive for applications where combined properties, such as electrical, mechanical, corrosion resistance, nonmagnetic, thermal conductivity, and wear resistance, are required. Applications include electrical and electronic components, springs, cartridges for small arms, plumbing, heat exchangers, and marine hardware, as well as consumer goods, such as cooking utensils, jewelry and other decorative objects.

Copper alloys can acquire a wide variety of properties by the addition of alloying elements and by heat treatment to improve their manufacturing characteristics. The most common copper alloys are brasses and bronzes. **Brass**, which is an alloy of copper and zinc, is one of the earliest alloys developed and has numerous applications, including decorative objects (Table 6.5). **Bronze** is an alloy of copper and tin (Table 6.6). There are also other bronzes, such as aluminum bronze, which is an alloy of copper and aluminum, and tin bronzes. Beryllium–copper, or *beryllium bronze*, and *phosphor bronze* have good strength and hardness for applications such as springs and bearings. Other major copper alloys are copper nickels and nickel silvers.

Copper is found in several types of ores, the most common being sulfide ores. The ores are generally low grade (although some contain up to 15 percent copper) and are usually obtained from open-pit mines. The ore is first crushed and formed into a slurry (a watery mixture with insoluble solid particles). The slurry is ground into fine particles in ball mills (rotating cylinders with metal balls inside to crush the ore). Chemicals and oil are then added and the mixture is agitated. The mineral particles form froth, which is scraped and dried.

The dry copper concentrate, containing as much as one-third copper, is traditionally **smelted** (melted and fused) and refined. This process is known as *pyrometallurgy*, because heat is used to refine the metal. For applications such as electrical conductors, the copper is further refined electrolytically to a purity of at least 99.95 percent (oxygen-free electrolytic copper). A more recent technique for processing copper is *hydrometallurgy*, a process involving chemical and electrolytic reactions.

6.4.1 Designation of copper alloys

In addition to being identified by their composition, copper and copper alloys are known by various names (Tables 6.5 and 6.6). The temper designations, such as 1/2 hard, extra hard, extra spring, and so on, are based on percentage reduction by cold working, such as by rolling or drawing.

• Example: Selection of metals for cookware.

We know that cooking pots and pans should, among other considerations, conduct heat evenly and rapidly. Several materials meet this requirement, but each has its advantages and limitations. Copper is an excellent material for this application, as verified by the fact that cooks have always preferred copper utensils. However, copper is more expensive than other suitable materials.

Aluminum also is a good conductor of heat. Note, for example, the layer of aluminum on the outside bottom of some newer models of stainless-steel cooking pots. Aluminum, however, is attacked by acids and develops small pores on the cooking surfaces where dirt and bacteria can gather, thus affecting the flavor of the food. One solution is to coat the aluminum pot to prevent corrosion. The resistance of coatings to food products and damage, such as by scratching, is then a concern, as are potentially adverse effects of the coating on health.

Carbon steel is another candidate material for this application, and it is inexpensive. However, it tarnishes and corrodes. Stainless steel has good corrosion resistance and is easy to clean because of its smooth surface. Also, it has good forming characteristics. However, stainless steel does not conduct heat as well as the other materials.

One solution is three-ply construction. A core of carbon steel is sandwiched between two layers of stainless steel. The outside may be buffed to a fine finish. The inner cooking surface could have a coarser finish, so that scratches made during cooking and cleaning are not readily visible.

●

6.5 Nickel and Nickel Alloys

Nickel (Ni), a silver-white metal discovered in 1751, is a major alloying element that imparts strength, toughness, and corrosion resistance. It is used extensively in stainless steels and nickel-base alloys, also called **superalloys**. These alloys are used for high-temperature applications, such as jet engine components, rockets, and nuclear power plants, as well as in food-handling and chemical processing equipment, coins, and marine applications. Because nickel is magnetic, nickel alloys are also used in electromagnetic applications, such as solenoids. The principal use of nickel as a metal is in electroplating for appearance and for improving corrosion and wear resistance.

Nickel alloys have high strength and corrosion resistance at elevated temperatures. Alloying elements in nickel are chromium, cobalt, and molybdenum. The behavior of nickel alloys in machining, forming, casting, and welding can be modified by various other alloying elements.

A variety of nickel alloys having a range of strength at different temperatures have been developed (Table 6.7). *Monel* is a nickel–copper alloy, and *Inconel* is a nickel–chromium alloy with a tensile strength of up to 1400 MPa (200 ksi). A nickel–molybdenum–chromium alloy (*Hastelloy*) has good corrosion resistance and high strength at elevated temperatures. *Nichrome*, an alloy of nickel, chromium, and iron, has high oxidation and electrical resistance, and is used for electrical heating elements. *Invar*, an alloy of iron and nickel, has relatively low sensitivity to temperature (see Section 3.6).

TABLE 6.7
PROPERTIES AND TYPICAL APPLICATIONS OF NICKEL ALLOYS (ALL ARE TRADE NAMES)

ALLOY [CONDITION]	PRINCIPAL ALLOYING ELEMENTS (%)	ULTIMATE TENSILE STRENGTH (MPa)	YIELD STRENGTH (MPa)	ELONGATION IN 50 mm (%)	TYPICAL APPLICATIONS
Nickel 200 [annealed]	None	380–550	100–275	60–40	Chemical and food processing industry, aerospace equipment, electronic parts.
Duranickel 301 [age hardened]	4.4 Al, 0.6 Ti	1300	900	28	Springs, plastics extrusion equipment, molds for glass, diaphragms.
Monel R-405 [hot-rolled]	30 Cu	525	230	35	Screw-machine products, water meter parts.
Monel K-500 [age hardened]	29 Cu, 3 Al	1050	750	30	Pump shafts, valve stems, springs.
Inconel 600 [annealed]	15 Cr, 8 Fe	640	210	48	Gas turbine parts, heat-treating equipment, electronic parts, nuclear reactors.
Hastelloy C-4 [solution-treated and quenched]	16 Cr, 15 Mo	785	400	54	High temperature stability, resistance to stress-corrosion cracking.

The main sources of nickel are sulfide and oxide ores, all of which have low concentrations of nickel. Nickel metal is produced by sedimentary and thermal processes, followed by electrolysis, yielding 99.95 percent pure nickel. Although it is also present in the ocean in significant amounts, undersea mining of nickel is not yet economical.

6.6 Superalloys

Superalloys are important in high-temperature applications, hence they are also known as *heat-resistant*, or *high-temperature*, *alloys*. Major applications of superalloys are in jet engines and gas turbines, with other applications in reciprocating engines, rocket-engines, tools and dies for hot working of metals, and in the nuclear, chemical, and petrochemical industries.

These alloys are referred to as *iron-base*, *cobalt-base*, or *nickel-base superalloys*. They contain nickel, chromium, cobalt, and molybdenum as major alloying elements. Other alloying elements are aluminum, tungsten, and titanium. Superalloys are generally identified by trade names or by special numbering systems and are available in a variety of shapes. Most superalloys have a maximum service

TABLE 6.8
PROPERTIES AND TYPICAL APPLICATIONS OF SELECTED NICKEL-BASE SUPERALLOYS AT 870 °C (1600 °F) (ALL ARE TRADE NAMES)

ALLOY	CONDITION	ULTIMATE TENSILE STRENGTH (MPa)	YIELD STRENGTH (MPa)	ELONGATION IN 50 mm (%)	TYPICAL APPLICATIONS
Astroloy	Wrought	770	690	25	Forgings for high temperature.
Hastelloy X	Wrought	255	180	50	Jet engine sheet parts.
IN-100	Cast	885	695	6	Jet engine blades and wheels.
IN-102	Wrought	215	200	110	Superheater and jet engine parts.
Inconel 625	Wrought	285	275	125	Aircraft engines and structures, chemical processing equipment.
Inconel 718	Wrought	340	330	88	Jet engine and rocket parts.
MAR-M 200	Cast	840	760	4	Jet engine blades.
MAR-M 432	Cast	730	605	8	Integrally cast turbine wheels.
René 41	Wrought	620	550	19	Jet engine parts.
Udimet 700	Wrought	690	635	27	Jet engine parts.
Waspaloy	Wrought	525	515	35	Jet engine parts.

temperature of about 1000 °C (1800 °F) for structural applications. The temperatures can be as high as 1200 °C (2200 °F) for nonload-bearing components. Superalloys generally have good resistance to corrosion, mechanical and thermal fatigue, mechanical and thermal shock, creep, and erosion at elevated temperatures.

Iron-base superalloys generally contain 32–67 percent iron, 15–22 percent chromium, and 9–38 percent nickel. Common alloys in this group are the *Incoloy* series.

Cobalt-base superalloys generally contain 35–65 percent cobalt, 19–30 percent chromium, and up to 35 percent nickel. Cobalt (Co) is a white-colored metal that resembles nickel. These superalloys are not as strong as nickel-base superalloys, but they retain their strength at higher temperatures.

Nickel-base superalloys are the most common of the superalloys and are available in a wide variety of compositions (Table 6.8). The range of nickel is from 38–76 percent. They also contain up to 27 percent chromium and 20 percent cobalt. Common alloys in this group are the *Hastelloy*, *Inconel*, *Nimonic*, *René*, *Udimet*, *Astroloy*, and *Waspaloy* series.

6.7 Titanium and Titanium Alloys

Titanium (Ti), named after the giant Greek god Titan, was discovered in 1791 but was not commercially produced until the 1950s. Although expensive, the high strength-to-weight ratio of titanium and its corrosion resistance at room and elevated temperatures make it attractive for applications such as aircraft, jet engine (see Fig. 6.1), racing-car, chemical, petrochemical, and marine components, submarine hulls, and biomaterials, such as prosthetic devices (Table 6.9; see p. 198). Titanium alloys have been developed for service at 550 °C (1000 °F) for long periods of time and up to 750 °C (1400 °F) for shorter periods.

Unalloyed titanium, known as commercially pure titanium, has excellent corrosion resistance for applications where strength considerations are secondary. Aluminum, vanadium, molybdenum, manganese, and other alloying elements are added to titanium alloys to impart properties such as improved workability, strength, and hardenability.

The properties and manufacturing characteristics of titanium alloys are extremely sensitive to small variations in both alloying and residual elements. Thus control of composition and processing are important, including prevention of surface contamination by hydrogen, oxygen, or nitrogen during processing. These elements cause embrittlement of titanium, resulting in reduced toughness and ductility.

The body-centered cubic structure of titanium (beta-titanium, above 880 °C, 1600 °F) is ductile, whereas its hexagonal close-packed structure (alpha-titanium) is somewhat brittle and is very sensitive to stress corrosion. A variety of other structures (alpha, near alpha, alpha–beta, and beta) can be obtained by alloying and heat treating, such that the properties can be optimized for specific applications.

New developments include the so-called *titanium aluminide intermetallics*, TiAl and Ti_3Al. They have higher stiffness and lower density and can withstand temperatures higher than conventional titanium alloys.

Ores containing titanium are first reduced to titanium carbide in an arc furnace, then converted to titanium chloride in a chlorine atmosphere. This compound is reduced further by distillation and leaching (dissolving), which forms sponge titanium. The sponge is then pressed into billets, melted, and poured into ingots to be processed later into various shapes. The complexity of these operations adds considerably to the cost of titanium.

• Example: Cracking of marble columns in Greece.

Marble columns in the Parthenon and other ancient monuments on the Acropolis in Athens were reinforced by embedding iron bars into the columns in order to keep them from deteriorating further. This was done in the early 1900s, using an estimated 10 tons of iron. Over a period of about 60 years, however, the iron bars corroded. The resulting expansion, due to the oxide layer on the bars, caused the columns to crack, imparting serious damage to the monuments. The rusted iron bars are now being removed and replaced with bars made of titanium, which does not corrode and is lightweight, hence easier to handle during restoration of the monuments. •

TABLE 6.9
PROPERTIES AND TYPICAL APPLICATIONS OF WROUGHT TITANIUM ALLOYS

			ROOM TEMPERATURE				VARIOUS TEMPERATURES					
NOMINAL COMPOSITION (%)	UNS	CONDITION	*ULTIMATE TENSILE STRENGTH* (MPa)	*YIELD STRENGTH* (MPa)	*ELONGATION* (%)	*REDUCTION OF AREA* (%)	*TEMP.* (°C)	*ULTIMATE TENSILE STRENGTH* (MPa)	*YIELD STRENGTH* (MPa)	*ELONGATION* (%)	*REDUCTION OF AREA* (%)	TYPICAL APPLICATIONS
99.5 Ti	R50250	Annealed	330	240	30	55	300	150	95	32	80	Airframes; chemical, desalination, and marine parts; plate type heat exchangers.
5 Al, 2.5 Sn	R54520	Annealed	860	810	16	40	300	565	450	18	45	Aircraft engine compressor blades and ducting; steam turbine blades.
6 Al, 4 V	R56400	Annealed	1000	925	14	30	300 425 550	725 670 530	650 570 430	14 18 35	35 40 50	Rocket motor cases; blades and disks for aircraft turbines and compressors; structural forgings and fasteners.
		Solution + age	1175	1100	10	20	300	980	900	10 12 22	28 35 45	
13 V, 11 Cr, 3 Al	R58010	Solution + age	1275	1210	8	—	425	1100	830	12	—	High strength fasteners; aerospace components; honeycomb panels.

6.8 Refractory Metals and Alloys

Refractory metals are molybdenum, columbium, tungsten, and tantalum. They are called *refractory* because of their high melting point. Although refractory metal elements were discovered about 200 years ago—and have been used as important alloying elements in steels and superalloys—their use as engineering metals and alloys did not begin until about the 1940s.

More than most other metals and alloys, these metals maintain their strength at elevated temperatures. Thus they are of great importance and use in rocket engines, gas turbines, and various other aerospace applications, in the electronics, nuclear power, and chemical industries, and as tool and die materials. The temperature range for some of these applications is on the order of 1100 °C to 2200 °C (2000 °F to 4000 °F), where strength and oxidation are of major concern.

6.8.1 Molybdenum

Molybdenum (Mo), a silvery white metal, was discovered in the 18th century. It has a high melting point, a high modulus of elasticity, good resistance to thermal shock, and good electrical and thermal conductivity. Typical applications of molybdenum are in solid-propellent rockets, jet engines, honeycomb structures, electronic components, heating elements, and dies for die casting.

Principal alloying elements for molybdenum are titanium and zirconium. Molybdenum is used in greater amounts than any other refractory metal. It is an important alloying element in casting and wrought alloys, such as steels and heat-resistant alloys, and imparts strength, toughness, and corrosion resistance. A major disadvantage of molybdenum alloys is their low resistance to oxidation at temperatures above 500 °C (950 °F), thus necessitating the use of protective coatings.

The main source for molybdenum is the mineral molybdenite (molybdenum disulfide). The ore is processed, concentrated, and reduced by reaction with oxygen and then hydrogen. Powder metallurgy techniques are also used to produce ingots for further processing into various shapes.

6.8.2 Columbium (niobium)

Columbium (after the mineral columbite and Nb, for niobium, after Niobe, the mythical Greek king Tantalus's daughter) possesses good ductility and formability and has greater oxidation resistance than other refractory metals. With various alloying elements, columbium alloys can be produced with moderate strength and good fabrication characteristics. These alloys are used in rockets and missiles and nuclear, chemical, and superconductor applications.

Columbium, first identified in 1801, is also an alloying element in various alloys and superalloys. It is processed from ores by reduction and refinement and from powder by melting and shaping into ingots.

6.8.3 Tungsten

Tungsten (W, from wolframite) was first identified in 1781 and is the most plentiful of all refractory metals. In Swedish, *tung* means heavy and *sten* means stone. Tungsten has the highest melting point of any metal (3410 °C, 6170 °F), and thus it is characterized by high strength at elevated temperatures. On the other hand, it has high density, brittleness at low temperatures, and poor resistance to oxidation.

Tungsten and its alloys are used for applications involving temperatures above 1650 °C (3000 °F), such as nozzle throat liners in missiles and in the hottest parts of jet and rocket engines, circuit breakers, welding electrodes, and spark-plug electrodes. The filament wire in incandescent light bulbs is made of pure tungsten, using powder metallurgy and wire drawing techniques. Because of its high density, tungsten is used as balancing weights and counterbalances in mechanical systems, including self-winding watches. Tungsten is an important element in tool and die steels, imparting strength and hardness at elevated temperatures. Tungsten carbide, with cobalt as a binder for the carbide particles, is one of the most important tool and die materials.

Tungsten is processed from ore concentrates by chemical decomposition and is then reduced. It is further processed by powder metallurgy techniques in a hydrogen atmosphere.

6.8.4 Tantalum

Identified in 1802, tantalum (Ta, after the mythical Greek king Tantalus) is characterized by a high melting point (3000 °C, 5425 °F), good ductility, and resistance to corrosion. However, it has high density and poor resistance to chemicals at temperatures above 150 °C (300 °F). Tantalum is also used as an alloying element.

Tantalum is used extensively in electrolytic capacitors and various components in the electrical, electronic, and chemical industries, as well as for thermal applications, such as in furnaces and acid-resistant heat exchangers. A variety of tantalum-base alloys is available in many forms for use in missiles and aircraft. Tantalum is processed by techniques similar to those used for processing columbium.

6.9 Beryllium

Steel gray in color, beryllium (Be) has a high strength-to-weight ratio. Unalloyed beryllium is used in nuclear and x-ray applications, because of its low neutron absorption characteristics, and in rocket nozzles, space and missile structures, aircraft disc brakes, and precision instruments and mirrors. Beryllium is also an alloying element, and its alloys of copper and nickel are used in applications such as springs (beryllium–copper), electrical contacts, and nonsparking tools for use in explosive environments, such as mines and metal powder production. Beryllium and its oxide are toxic and should not be inhaled.

6.10 Zirconium

Zirconium (Zr) is silvery in appearance, has good strength and ductility at elevated temperatures, and has good corrosion resistance because of an adherent oxide film. The element is used in electronic components and nuclear power reactor applications because of its low neutron absorption.

6.11 Low-Melting Alloys

The major metals in this category are lead, zinc, and tin.

6.11.1 Lead

Lead (Pb, after plumbium, the root for the word plumber) has properties of high density, resistance to corrosion (by virtue of the stable lead oxide layer that forms and protects the surface), softness, low strength, ductility, and good workability. Alloying with various elements, such as antimony and tin, enhances lead's properties, making it suitable for piping, collapsible tubing, bearing alloys, cable sheathing, roofing, and lead–acid storage batteries.

Lead is also used for damping sound and vibrations, radiation shielding against x-rays, printing (type metals), ammunition, and weights and in the chemical and paint industries. The oldest lead artifacts were made in about 3000 B.C. Lead pipes made by the Romans and installed in the Roman baths in Bath, England, two millenia ago are still in use. Lead is also an alloying element in solders, steels, and copper alloys and promotes corrosion resistance and machinability. Because of its toxicity, environmental contamination by lead is a significant concern.

The important mineral for lead is galena (PbS). It is mined, smelted, and refined by chemical treatments.

6.11.2 Zinc

Industrially, zinc (Zn), bluish-white in color, is the fourth most utilized metal, after iron, aluminum, and copper. Although known for many centuries, zinc was not studied and developed until the 18th century. It has two major uses: one is for galvanizing iron and steel sheet and wire, and the other is as an alloy base for casting. In *galvanizing*, zinc serves as the anode and protects the steel (cathode) from corrosive attack should the coating be scratched or punctured. Zinc is also used as an alloying element. Brass, for example, is an alloy of copper and zinc.

The major use of zinc is structural, but pure zinc is rarely used for this purpose. Major alloying elements in zinc are aluminum, copper, and magnesium. They impart strength and provide dimensional control during casting of the metal. Zinc-base alloys are used extensively in die casting for making products such as carburetors, fuel pumps, and grills for automobiles, components for household appliances (such as vacuum cleaners, washing machines, and kitchen equipment), various other machine parts, and photoengraving plates.

Another use for zinc is in superplastic alloys, which have good formability characteristics by virtue of their capacity to undergo large deformation without failure. Very fine-grained 78% Zn–22% Al sheet is a common example of a superplastic zinc alloy that can be formed by methods used for forming plastics or metals.

A number of minerals containing zinc are found in nature, although the principal source is the zinc sulfide, called zincblende. The ore is first roasted in air and converted to zinc oxide. It is then reduced to zinc either electrolytically, using sulfuric acid, or by heating in a furnace with coal, whereby the molten zinc is separated.

• Example: Failure of hard disk drive mechanism.

The drive mechanism for hard disk drives manufactured by a company malfunctioned. Investigations indicated that the clamping screw for one of the mechanisms did not apply sufficient torque to the stepper-motor drive shaft to which it was attached. Increasing the torque on the clamping screw did not alleviate the problem. After several tests it was determined that the material chosen for the mechanism was not suitable for the intended purpose. The mechanism was made of zinc die casting because, as we describe in Section 11.13, screw threads can be formed readily in zinc castings. However, the creep characteristics of zinc are such that the stresses developed during tightening of the clamping screw relax significantly over a few days. Consequently, the torque on the shaft is reduced, the mechanism slips around the shaft, and the disk drive malfunctions. It is therefore essential that stress relaxation (see Section 2.8) of zinc be considered in its use for such applications. •

6.11.3 Tin

Although used in small amounts, tin (Sn, after stannum) is an important metal. The most extensive use of tin, a silvery white, lustrous metal, is as a protective coating on steel sheet (tin plate), which is used in making containers—*tin cans*—for food and various other products. The low shear strength of the tin coatings on steel sheet improves its performance in deep drawing and general pressworking operations. However, unlike galvanized steels, if this coating is punctured, the steel corrodes because it is anodic to tin (cathode).

Unalloyed tin is used in applications such as lining material for water distillation plants, and as a molten layer of metal over which plate glass is made. Tin-base alloys (also called *white metals*) generally contain copper, antimony, and lead. The alloying elements impart hardness, strength, and corrosion resistance. Because of their low friction coefficients, which result from low shear strength and low adhesion, tin alloys are used as journal bearing materials. These alloys are known as **babbitts** (after I. Babbitt, 1799–1862) and contain tin, copper, and antimony. **Pewter** is an alloy of tin, copper, and antimony. It was developed in the 15th century and is used for tableware, hollowware, and decorative artifacts. Organ pipes are made of tin alloys.

Tin is an alloying element for type metals, dental alloys, and bronze (copper–tin alloy), titanium, and zirconium alloys. Tin–lead alloys are common *soldering* materials, with a wide range of compositions and melting points.

The most important tin mineral is cassiterite (tin oxide), which is low grade. The ore is mined, concentrated by various techniques, smelted, refined, and cast into ingots for further processing.

6.12 Precious Metals

Gold, silver, and platinum are the most important **precious** (that is, costly) **metals** and are also called *noble metals*. Gold (Au, after aurum) is soft and ductile, and has good corrosion resistance at any temperature. Typical applications include jewelry, coinage, reflectors, gold leaf for decorative purposes, dental work, and electroplating, as well as important applications as electrical contacts and terminals.

Silver (Ag, after argentum) is a ductile metal and has the highest electrical and thermal conductivity of any metal. It does, however, develop an oxide film that affects its surface properties and appearance. Typical applications for silver include tableware, jewelry, coinage, electroplating, and photographic film, as well as electrical contacts, solders, bearings, and food and chemical equipment. Sterling silver is an alloy of silver and 7.5 percent copper.

Platinum (Pt) is a soft, ductile, grayish-white metal that has good corrosion resistance, even at elevated temperatures. Platinum alloys are used as electrical contacts, spark-plug electrodes, catalysts for automobile pollution-control devices, filaments, nozzles, dies for extruding glass fibers, and thermocouples, and in the electrochemical industry. Other applications include jewelry and dental work.

6.13 Shape-Memory Alloys

Shape-memory alloys, after being plastically deformed at room temperature into various shapes, return to their original shapes upon heating. For example, a piece of straight wire made of this material can be wound into a helical spring. When heated with a match, the spring uncoils and returns to the original straight shape. A typical shape-memory alloy is 55% Ni–45% Ti. Other alloys under development are iron-, silicon-, and manganese-based. Shape-memory alloys, which have general properties such as good ductility, corrosion resistance, and electrical conductivity, can be used to generate motion and/or force in temperature-sensitive actuators, couplings, and fasteners. One potential application for these alloys is in structures, such as an antenna, for space travel. They would be folded at room temperature into a small volume to save space and for ease of transportation. After reaching their destination, the structures would be heated by some suitable means and return to their original shapes.

6.14 Amorphous Alloys

A class of metal alloys which, unlike metals, do not have a long-range crystalline structure are called **amorphous alloys**. They have no grain boundaries, and the atoms are randomly and tightly packed. Because their structure resembles that of glasses, these alloys are also called **metallic glasses**. Amorphous alloys typically contain iron, nickel, and chromium, alloyed with carbon, phosphorus, boron, aluminum, and silicon. They are available in the form of wire, ribbon, strip, and powder.

These alloys exhibit excellent corrosion resistance, good ductility, high strength, and very low loss from magnetic hysteresis. The latter property is utilized in making magnetic steel cores for transformers, generators, motors, lamp ballasts, magnetic amplifiers, and linear accelerators, with greatly improved efficiency.

The amorphous structure was first obtained in the late 1960s by extremely rapid cooling of the molten alloy. One method is called *splat cooling*, in which the alloy is propelled at very high speed against a rotating metal surface. Since the rate of cooling is on the order of 10^6 to 10^8 K/s, the molten alloy does not have sufficient time to crystallize. If an amorphous alloy is raised in temperature and then cooled, it develops a crystalline structure.

SUMMARY

Various nonferrous metals and alloys are available. They have a wide range of properties, such as strength, toughness, hardness, ductility, and resistance to oxidation, in addition to physical and chemical properties. Among their attractive properties are high strength-to-weight ratio and resistance to high temperature and corrosion.

The selection of a nonferrous material for a particular application requires careful consideration of many factors. Among these are design and service requirements, long-term effects, chemical affinity to other materials, environmental attack, and cost.

SUMMARY TABLE 6.1

OUTLINE OF TOPICS DESCRIBED IN CHAPTER 6

TOPIC	CHARACTERISTICS
Nonferrous alloys	More expensive than steels and plastics; wide range of mechanical, physical, and electrical properties; good corrosion resistance; high-temperature applications.
Aluminum	High strength-to-weight ratio; high thermal and electrical conductivity; good corrosion resistance; good manufacturing properties.
Magnesium	Lightest metal; good strength-to-weight ratio.
Copper	High electrical and thermal conductivity; good corrosion resistance; good manufacturing properties.
Superalloys	Good strength and resistance to corrosion at elevated temperatures; can be iron-, cobalt-, and nickel-base.
Titanium	Highest strength-to-weight ratio of all metals; good strength and corrosion resistance at high temperatures.
Refractory metals	Molybdenum, columbium (niobium), tungsten, and tantalum; high strength at elevated temperatures.
Precious metals	Gold, silver, and platinum; generally good corrosion resistance.

TRENDS

- Aluminum is competing strongly to become an important structural metal to reduce the weight and improve corrosion resistance of automobiles. Aluminum–lithium alloys are being developed, particularly for aircraft components, for increased stiffness and reduced density.
- The purity and corrosion resistance of magnesium alloys are being improved, particularly for automotive applications.
- High-purity titanium is being developed for electronic and aerospace applications.
- Techniques for refining superalloys are being developed to improve their mechanical and physical properties and corrosion resistance. The aim is to produce cleaner metals and alloys, using various melting techniques.
- Single crystal nickel-base alloys are another group of materials with important high-temperature applications.
- Superalloys that are intermetallic compounds of nickel, chromium, molybdenum, and aluminum are being developed for high-temperature applications in critical gas-turbine components.

KEY TERMS

Amorphous alloys	Metallic glasses	Shape-memory alloys
Babbitts	Pewter	Smelting
Brass	Precious metals	Superalloys
Bronze	Refractory metals	Temper designation

BIBLIOGRAPHY

Aerospace Structural Metals Handbook, 5 vols. Columbus, Ohio: Metals and Ceramics Information Center, Battelle, 1987.

Aluminum Standards and Data, revised periodically. Washington, D.C.: The Aluminum Association.

Boyer, H.E. (ed.), *Selection of Materials for Component Design: Source Book*. Metals Park, Ohio: American Society for Metals, 1986.

Donachie, M.J. (ed.), *Superalloys—Source Book*. Metals Park, Ohio: American Society for Metals, 1983.

———, *Titanium and Titanium Alloys—Source Book*. Metals Park, Ohio: American Society for Metals, 1982.

Easterling, K., *Tomorrow's Materials*, 2d ed. London: The Institute of Metals, 1990.
Hatch, J.E. (ed.), *Aluminum: Properties and Physical Metallurgy*. Metals Park, Ohio: American Society for Metals, 1984.
Luborsky, F.E., *Amorphous Metallic Alloys*. Woburn, Mass.: Butterworth, 1983.
Material Selector, annual publication of *Materials Engineering Magazine*, Penton/IPC, Cleveland, Ohio.
Metals Handbook, 9th ed., Vol. 2: *Properties and Selection: Nonferrous Alloys and Pure Metals*. Metals Park, Ohio: American Society for Metals, 1979.
Sedlacek, V., *Nonferrous Metals and Alloys*. New York: Elsevier, 1986.
Source Book on Copper and Copper Alloys. Metals Park, Ohio: American Society for Metals, 1979.

REVIEW QUESTIONS

6.1 List the major manufacturing properties of aluminum. Given the abundance of aluminum in the earth's crust, explain why it is more expensive than steel.

6.2 Summarize the designation system for aluminum and aluminum alloys.

6.3 Why is magnesium often used as a structural material in power hand tools? Why are alloys used instead of pure magnesium?

6.4 What are the major uses of copper? What are the alloying elements in brass and bronze, respectively?

6.5 What are superalloys? Why are they so named?

6.6 What properties of titanium make it attractive for use in race-car and jet-engine components? Why is titanium not widely used for engine components in passenger cars?

6.7 What are the individual properties of each of the major refractory metals that define their most useful applications?

6.8 Summarize the major uses of zinc.

6.9 What are metallic glasses? Why is the word glass used in defining these materials?

6.10 What is the composition of (a) babbitts, (b) pewter, and (c) sterling silver?

6.11 Which of the materials described in this chapter has the highest (a) density, (b) electrical conductivity, (c) thermal conductivity, (d) strength, and (e) cost?

QUESTIONS AND PROBLEMS

6.12 From your own experience and observations, list as many applications as you can for the following metals and their alloys: (a) aluminum, (b) copper, (c) lead, (d) gold, and (e) silver. (Since we have not yet covered plating processes, your answers should pertain to solid metals only.)

6.13 Explain why cooking utensils are generally made of stainless steels, aluminum, or copper.

6.14 Describe the advantages of making products with multilayer materials, such as the use of aluminum attached externally to the bottom of stainless steel pots.

6.15 Name products that would not have been developed to their advanced stages, as we find them today, if alloys with high strength and corrosion and creep resistance at elevated temperatures had not been developed.

6.16 List and explain the differences in the requirements for material characteristics for use in household versus industrial products.

6.17 As you know, many ferrous and nonferrous metals may not be easily identified unless you have some prior knowledge and experience with their use and application. For instance, by looking at them, how can you tell whether a piece of metal is silver or stainless steel, titanium or zinc, brass or bronze, or one kind of brass or another? With the information given thus far in this text, and with the help of other sources in the library or by consulting those with more experience, make a list of the metals and alloys described in Chapters 5 and 6 and state how best they can be identified by their color and appearance.

6.18 Assume that you are a technical sales manager of a company that produces nonferrous metals. Choose any one of the metals and alloys described in this chapter and prepare a brochure, including some illustrations, for use as sales literature by your staff in their contact with potential customers.

6.19 Why were gold, copper, and bronze the metals used earliest by man?

6.20 Based on the tables in this chapter, make a table in which the left column lists applications and the right column lists metals suitable for those applications.

6.21 Inspect several metal products and components and make an educated guess as to what materials they are made from. Give reasons for your guess. If you list two or more possibilities, explain your reasoning.

6.22 Would it be advantageous to plot the data in Table 6.1 in terms of cost per unit weight, rather than cost per unit volume? Explain.

6.23 Inspect Table 6.2 and comment on which of the two hardening processes (heat treating and work hardening) is more effective in improving the strength of aluminum alloys.

6.24 Why is there a range for the strengths for copper in Table 6.5? For brasses and bronzes?

6.25 Other than mechanical strength, what other factors should be considered in selecting metals and alloys for high temperature applications?

6.26 Explain why you would want to know the ductility of metals and alloys before selecting them.

6.27 What kind of material is ZK60A and what is its composition?

6.28 Explain the techniques you would use to strengthen aluminum alloys.

6.29 Give applications for (a) shape-memory alloys, (b) amorphous metals, (c) precious metals, and (d) low melting alloys.

6.30 Assume that the price of copper increases rapidly for geopolitical reasons. Name two metals with similar mechanical and physical properties that can be substituted for copper.

7

Polymers: Structure, General Properties, and Applications

7.1 Introduction

Hardly a product on the market today does not have some component made of **polymers** (*plastics*). These consumer and industrial products include food and beverage containers, packaging, signs, housewares, textiles, safety shields, toys, appliances, lenses, gears, electronic and electrical products and automobile bodies and components. Because of their many unique and diverse properties, plastics have increasingly replaced metallic components in applications such as automobiles, civilian and military aircraft, sporting goods, and office equipment. These substitutions reflect the advantages of plastics in terms of high strength-to-weight ratio, design possibilities, wide choice of colors and transparencies, ease of manufacturing, and relatively low cost (see Table 6.1).

Compared to metals, plastics are generally characterized by low density, low strength and stiffness (Table 7.1), low electrical and thermal conductivity, good resistance to chemicals, and high coefficient of thermal expansion. However, the useful temperature range for most plastics is generally low—up to about 300 °C (600 °F)—and they are not as dimensionally stable in service, over a period of time, as metals are.

TABLE 7.1
RANGE OF MECHANICAL PROPERTIES FOR VARIOUS ENGINEERING PLASTICS AT ROOM TEMPERATURE

MATERIAL	*UTS* (MPa)	*E* (GPa)	ELONGATION (%)	POISSON'S RATIO (ν)
ABS	28–55	1.4–2.8	75–5	—
ABS, reinforced	100	7.5	—	0.35
Acetal	55–70	1.4–3.5	75–25	—
Acetal, reinforced	135	10	—	0.35–0.40
Acrylic	40–75	1.4–3.5	50–5	—
Cellulosic	10–48	0.4–1.4	100–5	—
Epoxy	35–140	3.5–17	10–1	—
Epoxy, reinforced	70–1400	21–52	4–2	—
Fluorocarbon	7–48	0.7–2	300–100	0.46–0.48
Nylon	55–83	1.4–2.8	200–60	0.32–0.40
Nylon, reinforced	70–210	2–10	10–1	—
Phenolic	28–70	2.8–21	2–0	—
Polycarbonate	55–70	2.5–3	125–10	0.38
Polycarbonate, reinforced	110	6	6–4	—
Polyester	55	2	300–5	0.38
Polyester, reinforced	110–160	8.3–12	3–1	—
Polyethylene	7–40	0.1–1.4	1000–15	0.46
Polypropylene	20–35	0.7–1.2	500–10	—
Polypropylene, reinforced	40–100	3.5–6	4–2	—
Polystyrene	14–83	1.4–4	60–1	0.35
Polyvinyl chloride	7–55	0.014–4	450–40	—

The word **plastics**, first used around 1909, is from the Greek word *plastikos*, meaning it can be molded and shaped. Plastics can be machined, cast, formed, and joined into many shapes with relative ease. Little or no additional surface-finishing operations are required, which is an important advantage over metals. Plastics are commercially available as sheet, plate, film, rods, and tubing of various cross-sections.

The earliest polymers were made of *natural organic materials* from animal and vegetable products, cellulose being the most common example. With various chemical reactions, cellulose is modified into cellulose acetate, used in making photographic films (celluloid), sheets for packaging, textile fibers, as well as cellulose nitrate for plastics, explosives, rayon (a cellulose textile fiber), and varnishes. The earliest *synthetic* (manmade) polymer was a phenol-formaldehyde, a thermoset developed in 1906 and called Bakelite (a trade name, after L. H. Baekeland, 1863–1944). The word *polymer* was first used in 1866.

The development of modern plastics technology began in the 1920s with raw materials extracted from coal and petroleum products. Ethylene was the first example of the building block for polyethylene. Ethylene is the product of the

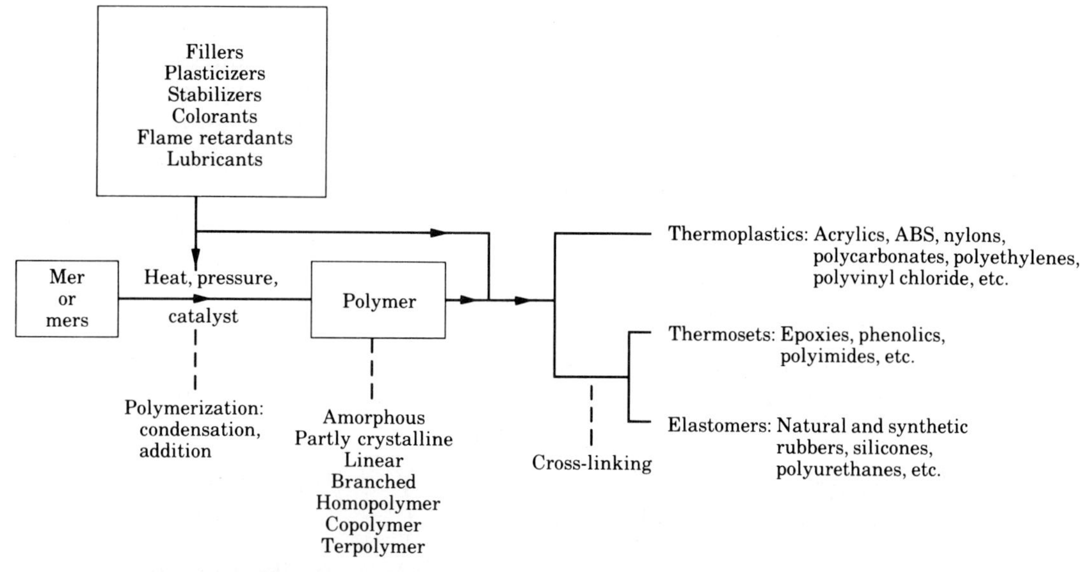

FIGURE 7.1
Outline of topics described in Chapter 7.

reaction between acetylene and hydrogen, and acetylene is the product of the reaction between coke and methane. These materials are known as *synthetic organic polymers*. Although in polyethylene only carbon and hydrogen atoms are involved, other compounds can be obtained with chlorine, fluorine, sulfur, silicon, nitrogen, and oxygen. As a result, an extremely wide range of polymers with an equally wide range of properties have been developed.

In this chapter, we describe the structure of plastics to enable you to understand and evaluate their properties and behavior in manufacturing processes, as well as during their service life under various physical and environmental conditions. An outline of the topics presented is given in Fig. 7.1. We also discuss the properties and engineering applications of plastics, as well as rubbers and elastomers, in this chapter. Reinforced plastics are described in Chapter 9 and processing methods for plastics and reinforced plastics in Chapter 18.

7.2 The Structure of Polymers

Plastics are composed of polymer molecules and various additives. Polymers are *long-chain molecules*, also called *macromolecules* or *giant molecules*, which are formed by polymerization, that is, by linking and cross-linking of different monomers. A **monomer** is the basic building block of polymers. The word **mer**, from the Greek *meros*, meaning part, indicates the smallest repetitive unit, similar to the term unit cell used in connection with crystal structures of metals.

The term polymer means many mers or units, generally repeated hundreds or thousands of times in a chainlike structure. Most monomers are organic materials in which carbon atoms are joined in *covalent* bonds (electron sharing) with other atoms, such as hydrogen, oxygen, nitrogen, fluorine, chlorine, silicon, and sulfur. A simple monomer is the ethylene molecule, which consists of carbon and hydrogen atoms (Fig. 7.2a).

7.2.1 Polymerization

Molecules can be linked in repeating units to make longer and larger molecules by **polymerization** processes, which are chemical reactions. In these reactions, the

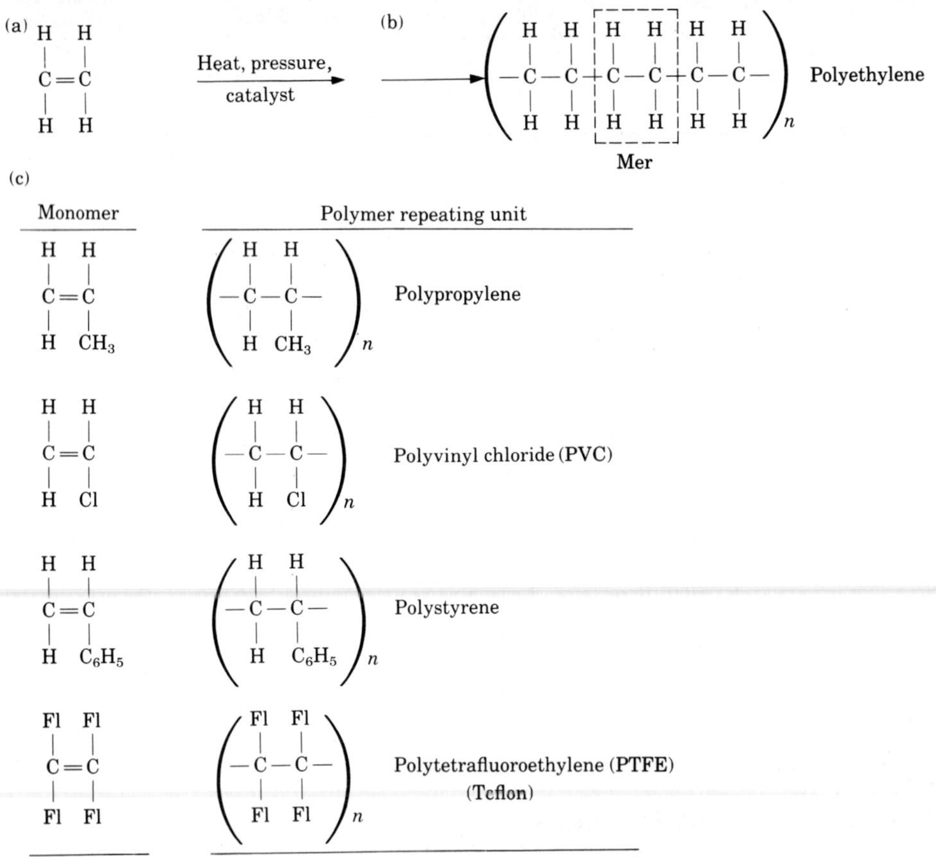

FIGURE 7.2 Basic structure of polymer molecules: (a) ethylene molecule; (b) polyethylene, a linear chain of many ethylene molecules; (c) molecular structure of various polymers. These are examples of the basic building blocks for plastics.

double bonds between the carbon atoms open, and the molecules arrange themselves end to end linearly. Thus ethylene molecules, for example, link to become the polymer known as polyethylene (Fig. 7.2b). Polymerization processes are complex and we can describe them only briefly here. Although there are many variations, two basic polymerization processes are condensation and addition polymerization.

In *condensation polymerization*, the bonds between the molecules are formed by the application of heat and pressure, with the aid of a catalyst, or initiator. The reaction byproducts, such as water, are condensed; hence the term condensation. This process is also known as *step-growth* or *step-reaction* polymerization because the polymer molecule grows step by step.

In *addition polymerization*, also known as *chain-growth* or *chain-reaction* polymerization, bonding takes place without reaction byproducts. It is called chain-reaction because of the high rate at which long molecules form simultaneously, usually within a few seconds. This rate is much higher than that for condensation polymerization.

Molecular weight. The sum of the molecular weights of the mers in the polymer chain is the **molecular weight** of the polymer. These chains are of different lengths and their arrangement is *amorphous*, that is, without any long-range order. The amorphous arrangement of polymer molecules is often described as a bowl of spaghetti, or worms in a bucket, all intertwined with each other. Because the lengths of the various chains differ, we determine and express the *average* molecular weight of a polymer on a statistical basis by averaging. For a polymer to have sufficient strength and toughness, the chains must be at least a certain length.

Degree of polymerization. The **degree of polymerization** is defined as the ratio of the molecular weight of the polymer to the molecular weight of the mer, or the number of mers per average molecule. This number can range from a few hundred to millions. The higher the number, the higher is the polymer's resistance to flow, a behavior similar to that obtained by increasing fluid viscosity. This behavior, in turn, affects the processing and shaping of the polymer into useful products.

Bonding. Within each molecular chain are covalent bonds, also known as **primary bonds** because of their strength. The bonds between different chains, and between the overlapping portions of the same chain, are known as **secondary bonds**, such as van der Waals bonds, hydrogen bonds, and ionic bonds. Secondary bonds are much weaker than the covalent bonds within the chain, the difference in strength being from one to two orders of magnitude.

Linear polymers. The chainlike polymers shown in Fig. 7.2 are called *linear polymers* because of their linear structure (Fig. 7.3a). A linear molecule is not straight in shape. In addition to those shown, other linear polymers are polyamides (nylon 6,6) and polyvinyl fluoride. Generally, a polymer consists of more than one type of structure. Thus a linear polymer may contain some branched and cross-linked chains.

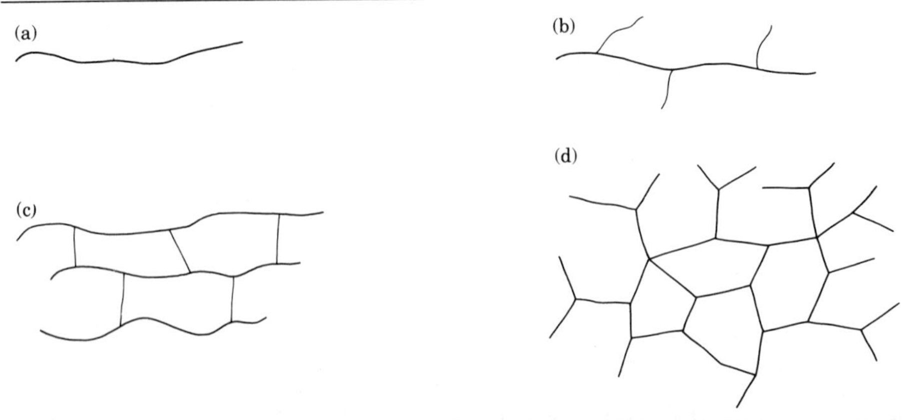

FIGURE 7.3
Schematic illustration of polymer chains. (a) Linear structure. Thermoplastics such as acrylics, nylons, polyethylene, and polyvinyl chloride have linear structures. (b) Branched structure, such as in polyethylene. (c) Cross-linked structure. Many rubbers or elastomers have this structure. Vulcanization of rubber produces this structure. (d) Network structure, which is basically highly cross-linked. Examples are thermosetting plastics, such as epoxies and phenolics.

Copolymers and terpolymers. If the repeating units in a polymer chain are all of the same type, we call the molecule a *homopolymer*. However, as with solid-solution metal alloys, two or three different types of monomers can be combined to impart certain special properties and characteristics to the polymer, such as improving both the strength and toughness as well as the formability of the polymer. *Copolymers* contain two types of polymers, such as styrene-butadiene, used widely for automobile tires. *Terpolymers* contain three types, such as ABS (acrylonitrile-butadiene-styrene) used for helmets, telephones, and refrigerator liners.

Branched polymers. The properties of a polymer depend not only on the type of monomers, but also on their arrangement in the molecular structure. In **branched polymers** (Fig. 7.3b), side-branch chains are attached to the main chain during the synthesis of the polymer. Branching interferes with the relative movement of the molecular chains. It affects the resistance to deformation of the polymer by making it stronger. The density of branched polymers is lower than that of linear-chain polymers. The behavior of branched polymers can be compared to that of linear-chain polymers by making an analogy with a pile of tree branches (branched polymers) and a bundle of straight logs (linear). You will note that it is more difficult to move a branch within the pile of branches than to move a log in its bundle. The three-dimensional entanglements of branches make movements more difficult, a phenomenon akin to increased strength.

Cross-linked polymers. Generally three-dimensional in structure, **cross-linked polymers** have adjacent chains linked by covalent bonds (Fig. 7.3c). Polymers with

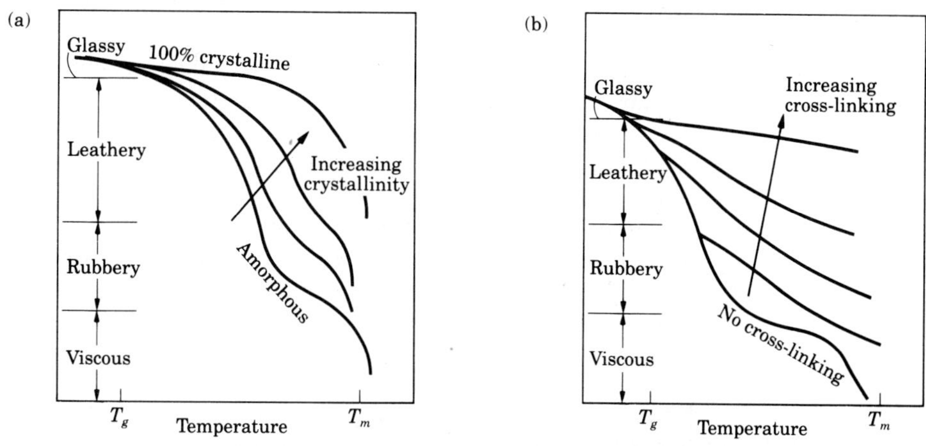

FIGURE 7.4
Behavior of polymers as a function of temperature and (a) degree of crystallinity and (b) cross-linking.

cross-linked chain structure are called **thermosets**, or *thermosetting* plastics, such as epoxies, phenolics, and silicones. Cross-linking has a major influence on the properties of polymers (generally imparting hardness, strength, stiffness, brittleness, and better dimensional stability; see Fig. 7.4), as well as in the vulcanization of rubber. **Network polymers** consist of spatial (three-dimensional) networks of three active covalent bonds (Fig. 7.3d). A highly cross-linked polymer is also considered a network polymer. Thermoplastic polymers that have already been formed or shaped can be cross linked to obtain greater strength by subjecting them to high-energy radiation, such as ultraviolet light, x-rays, and electron beams. However, excessive radiation can cause degradation of the polymer.

Example: Degree of polymerization in polyvinyl chloride (PVC).

Determine the molecular weight of a polyvinyl chloride mer. If a PVC polymer has an average molecular weight of 50,000, what is the degree of polymerization?

SOLUTION. From Fig. 7.2(c), we note that each PVC mer has 3 hydrogen atoms, 2 carbon atoms, and 1 chlorine atom. Since the atomic number of each element is 1, 12, and 35.5, respectively, we have the weight of a PVC mer as $(3)(1) + (2)(12) + (1)(35.5) = 62.5$. Hence the degree of polymerization is $50{,}000/62.5 = 800$.

7.2.2 Crystallinity

In the preceding section, we described the arrangement of the long-chain molecules as amorphous and randomly intertwined. However, it is possible to impart some crystallinity to polymers and thereby modify their characteristics. This may be done either during the synthesis of the polymer or by deformation during its subsequent processing.

The crystalline regions in polymers are called **crystallites** (Fig. 7.5). These crystals are formed when the long molecules fold over themselves in an orderly manner, similar to folding a fire hose in a cabinet or facial tissues in a box. Thus we can regard a partially crystalline polymer as a two-phase material, one phase being crystalline and the other amorphous.

By controlling the rate of solidification during cooling and chain configuration, it is possible to impart different degrees of crystallinity to polymers, although never completely 100 percent. Crystallinity ranges from an almost complete crystal, up to about 95 percent by volume in the case of polyethylene, to slightly crystallized but mostly amorphous polymers.

Effects of crystallinity. The mechanical and physical properties of polymers are greatly influenced by the *degree of crystallinity*. As it increases, polymers become stiffer, harder, less ductile, more dense, less rubbery, and more resistant to solvents and heat (Fig. 7.4). The increase in density with increasing crystallinity is caused by crystallization shrinkage and a more efficient packing of the molecules in the crystal lattice. For example, the highly crystalline form of polyethylene, known as high-density polyethylene (HDPE), has a specific gravity of 0.97 and is stronger, stiffer, tougher, and less ductile than low-density polyethylene (LDPE), which is about 60 percent crystalline and has a specific gravity of 0.915.

Optical properties are also affected by the degree of crystallinity. The reflection of light from the boundaries between the crystalline and amorphous regions in the polymer causes opaqueness. Furthermore, because the index of refraction is proportional to density, the greater the density difference between the amorphous

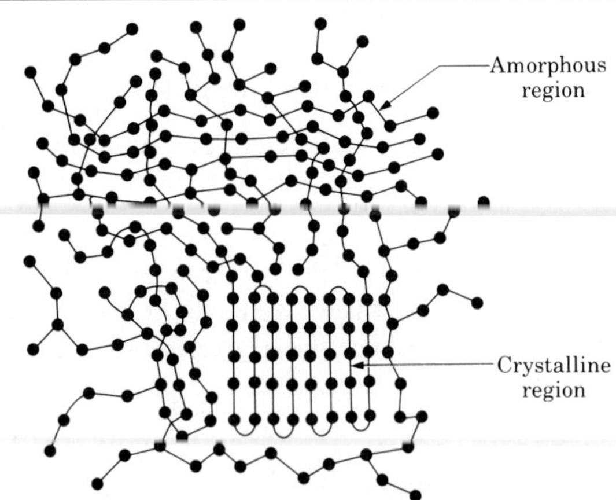

FIGURE 7.5 Amorphous and crystalline regions in a polymer. The crystalline region (crystallite) has an orderly arrangement of molecules. The higher the crystallinity, the harder, stiffer, and less ductile is the polymer.

and crystalline phases, the greater is the opaqueness of the polymer. Polymers that are completely amorphous can be transparent, such as polycarbonate and acrylics.

7.2.3 Glass-transition temperature

Amorphous polymers do not have a specific melting point, but they undergo a distinct change in their mechanical behavior across a narrow range of temperature. At low temperatures they are hard, rigid, brittle, and glassy and at high temperatures are rubbery or leathery. The temperature at which this transition occurs is called the **glass-transition temperature,** T_g, and is also called the *glass point* or *glass temperature.* The term glass is included in this definition because glasses, which are amorphous solids, behave in the same manner (as you can see by holding a glass rod over a flame and observing its behavior).

Although most amorphous polymers exhibit this behavior, there are some exceptions, such as polycarbonate, which is not rigid or brittle below its glass-transition temperature. Polycarbonate is tough at ambient temperature and is thus used for safety helmets and shields.

To determine T_g, we measure the specific volume of the polymer and plot it against temperature to find the sharp change in the slope of the curve (Fig. 7.6). However, in the case of highly cross-linked polymers, the slope of the curve changes gradually near T_g, making it difficult to determine T_g for these polymers. The

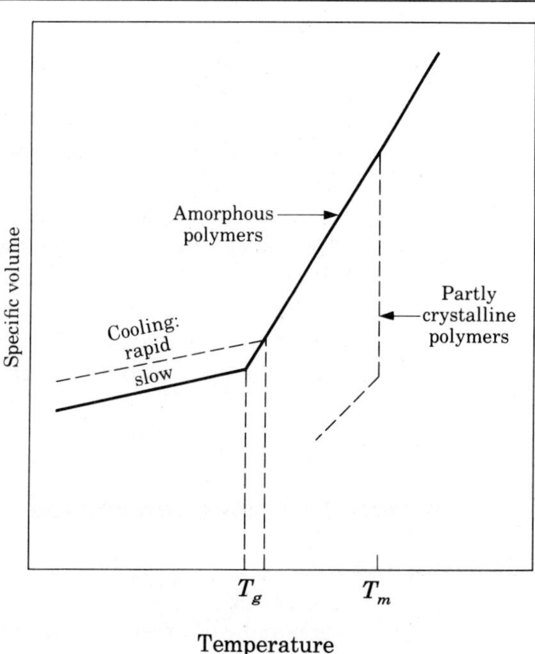

FIGURE 7.6 Specific volume of polymers as a function of temperature. Amorphous polymers, such as acrylic and polycarbonate, have a glass-transition temperature, T_g, but do not have a specific melting point, T_m. Partly crystalline polymers, such as polyethylene and nylons, contract sharply at their melting points during cooling.

TABLE 7.2
GLASS-TRANSITION AND MELTING TEMPERATURES OF SOME POLYMERS

MATERIAL	T_g(°C)	T_m(°C)
Nylon 6,6	57	265
Polycarbonate	150	265
Polyester	73	265
Polyethylene		
High density	−90	137
Low density	−110	115
Polymethylmethacrylate	105	—
Polypropylene	−14	176
Polystyrene	100	239
Polytetrafluoroethylene	−90	327
Polyvinyl chloride	87	212
Rubber	−73	—

glass-transition temperature varies with different polymers (Table 7.2). For example, room temperature is above T_g for some polymers and below it for others. Unlike amorphous polymers, partly crystalline polymers have a distinct melting point, T_m (Fig. 7.6; see also Table 7.2). Because of the structural changes occurring, the specific volume of the polymer drops suddenly as its temperature is reduced.

7.2.4 Blends

To improve the brittle behavior of amorphous polymers below their glass-transition temperature, we can blend them (mix them with another polymer), usually with small quantities of an **elastomer**. These tiny particles are dispersed throughout the amorphous polymer, enhancing its toughness and impact strength by improving its resistance to crack propagation. Such polymers are known as *rubber modified*. More recent trends in blending involve several components, or **polyblends,** that utilize the favorable properties of different polymers. Some advances have been made in *miscible blends* (mixing without separation of two phases), a process similar to alloying of metals, enabling polymer blends to become more ductile. Polymer blends account for about 20 percent of all polymer production.

7.3 Thermoplastics

We noted earlier that in the amorphous structure of a polymer, the bonds between adjacent long-chain molecules (secondary bonds) are much weaker than the covalent bonds (primary bonds) within each molecule. Hence, it is the strength of the secondary bonds that determines the overall strength of the polymer. Linear and branched polymers have weak secondary bonds.

If we now raise the temperature of this polymer above its glass-transition temperature, we find that it becomes softer and easier to form or mold into shape. The mobility of the long molecules (thermal vibrations) increases at T_g and above. If we then cool the polymer, it returns to its original hardness and strength. In other words, the process is reversible. Polymers that exhibit this behavior are known as **thermoplastics.** Typical examples are acrylics, cellulosics, nylons, polyethylenes, and polyvinyl chloride.

In addition to the structure and composition of thermoplastics, their behavior depends on a number of variables. Among the most important are temperature and rate of deformation. Below the glass-transition temperature, the polymer is glassy (brittle) and behaves like an elastic solid; that is, the relationship between stress and strain is linear. For example, below its T_g, polymethylmethacrylate (PMMA) is glassy, whereas polycarbonate is not glassy below its T_g.

The glassy behavior can be represented by a spring whose stiffness is equivalent to the modulus of elasticity of the polymer. When the applied stress is increased, the polymer fractures, just as a piece of glass does at ambient temperature. The mechanical properties of several polymers listed in Table 7.1 indicate that thermoplastics are about two orders of magnitude less stiff than metals. Their ultimate tensile strength is about one order of magnitude lower than that of metals (see Tables 2.1 and 2.2).

Typical stress–strain curves for some thermoplastics and thermosets at room temperature are shown in Fig. 7.7. Note that these plastics exhibit different behaviors, which we may describe as rigid, soft, brittle, flexible, and so on. Plastics undergo fatigue and creep phenomena, just as metals do.

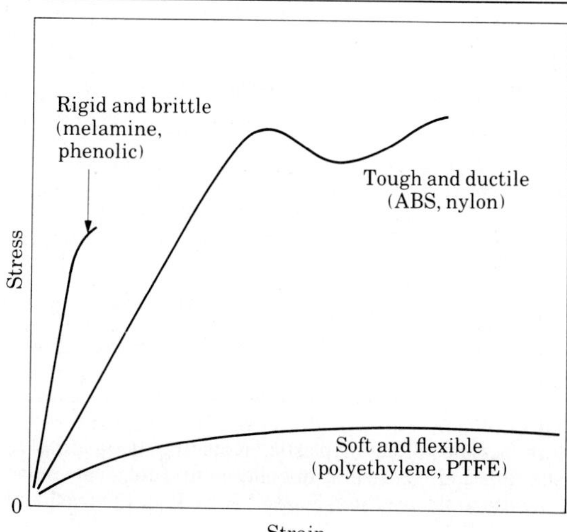

FIGURE 7.7 General terminology describing the behavior of three types of plastics. PTFE (polytetrafluoroethylene) is Teflon, a trade name. *Source:* R. L. E. Brown.

Effects of temperature. If we raise the temperature of a thermoplastic polymer above its T_g, it becomes a viscous fluid, its viscosity decreasing with increasing temperature. The response of a thermoplastic can be likened to ice cream. It can be softened, molded into shapes, refrozen, softened, and remolded a number of times. In practice, however, repeated heating and cooling causes *degradation*, or *thermal aging*, of thermoplastics.

The typical effects of temperature on the strength and elastic modulus of thermoplastics is similar to those for metals. Thus with increasing temperature, the strength and modulus of elasticity decrease, and toughness increases (Fig. 7.8). The effect of temperature on impact strength is shown in Fig. 7.9. Note the large difference in the impact behavior of various polymers.

• Example: Lowering the viscosity of a polymer.

On the basis of experimental observations that, at the glass-transition temperature T_g, polymers have a viscosity η of about 10^{12} Pa · s, an empirical relationship between viscosity and temperature has been developed for linear thermoplastics:

$$\log \eta = 12 - \frac{17.5\,\Delta T}{52 + \Delta T}, \tag{7.1}$$

where $\Delta T = T - T_g$, in K or °C. Thus we can estimate the viscosity of the polymer

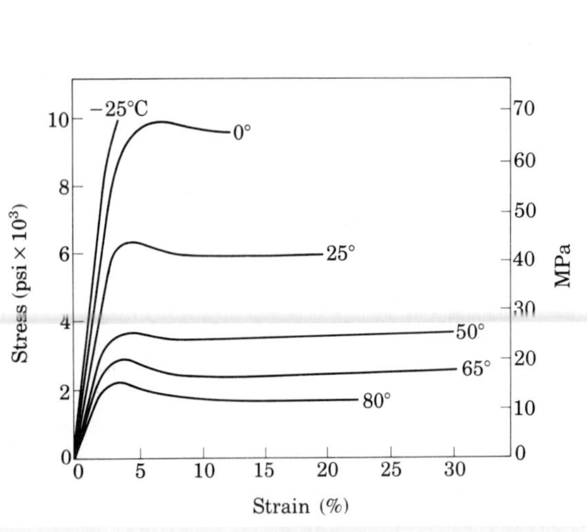

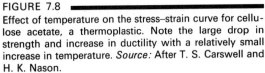

FIGURE 7.8 Effect of temperature on the stress–strain curve for cellulose acetate, a thermoplastic. Note the large drop in strength and increase in ductility with a relatively small increase in temperature. *Source:* After T. S. Carswell and H. K. Nason.

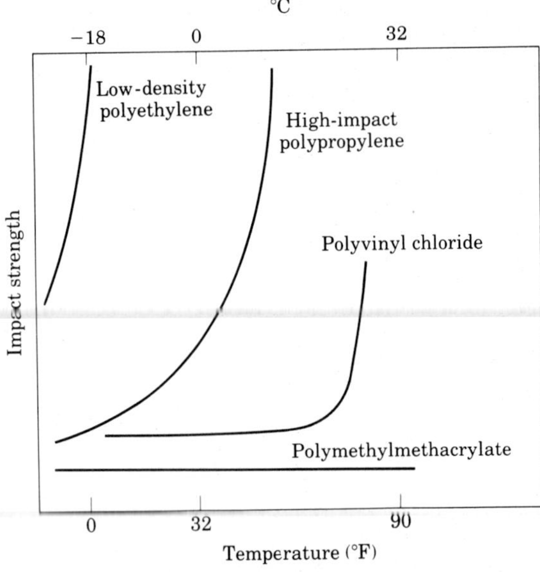

FIGURE 7.9 Effect of temperature on the impact strength of various plastics. Small changes in temperature can have a significant effect on impact stength. *Source:* P. C. Powell.

at any temperature. In processing a batch of polycarbonate at 170 °C to make a part, we find that its viscosity is twice that desired. We want to determine the temperature at which this polymer should be processed.

SOLUTION. From Table 7.2, we find that T_g for polycarbonate is 150 °C. We find its viscosity at 170 °C by using Eq. (7.1):

$$\log \eta = 12 - \frac{17.5(20)}{52 + 20} = 7.14.$$

Hence $\eta = 13.8$ MPa · s. Because this magnitude is twice what we want, the new viscosity should be 6.9 MPa · s. Therefore,

$$\log (6.9 \times 10^6) = 12 - \frac{17.5(\Delta T)}{52 + \Delta T},$$

and $\Delta T = 21.7$, or the new temperature is $150 + 21.7 = 171.7$ °C. Note that we rounded the numbers for temperature and that viscosity is very sensitive to temperature.

Effect of rate of deformation. The behavior of thermoplastics is similar to the strain-rate sensitivity of metals, indicated by the strain-rate sensitivity exponent m in Eq. (2.9). Thermoplastics have high m values, indicating that they can undergo large uniform deformations in tension before fracture (Fig. 7.10; see p. 222). Note how—unlike in ordinary metals—the necked region elongates considerably. You can easily demonstrate this phenomenon by stretching a piece of the plastic holder for 6-pack beverage cans. Observe that the sequence of necking and stretching behavior is as shown in Fig. 7.10(a). This characteristic, which is the same as in superplastic metals, enables the forming of thermoplastics into complex shapes such as bottles for soft drinks, cookie and meat trays, and lighted signs.

Orientation. When thermoplastics are deformed, say by stretching, the long-chain molecules align in the general direction of elongation. This process is called **orientation** and, just as in metals, the polymer becomes anisotropic. The specimen becomes stronger and stiffer in the elongated (stretched) direction than in its transverse direction. This is an important technique to enhance the strength and toughness of polymers. However, orientation weakens the polymer in the transverse direction.

Crazing. Some thermoplastics, such as polystyrene and polymethylmethacrylate, develop localized, wedge-shaped, narrow regions of highly deformed material when subjected to tensile stresses or to bending. This phenomenon is called **crazing.** Although they may appear to be like cracks, crazes are spongy material, typically containing about 50 percent voids. With increasing tensile load on the specimen, these voids coalesce and eventually lead to fracture of the polymer. Crazing has been observed both in transparent glassy polymers and in other types.

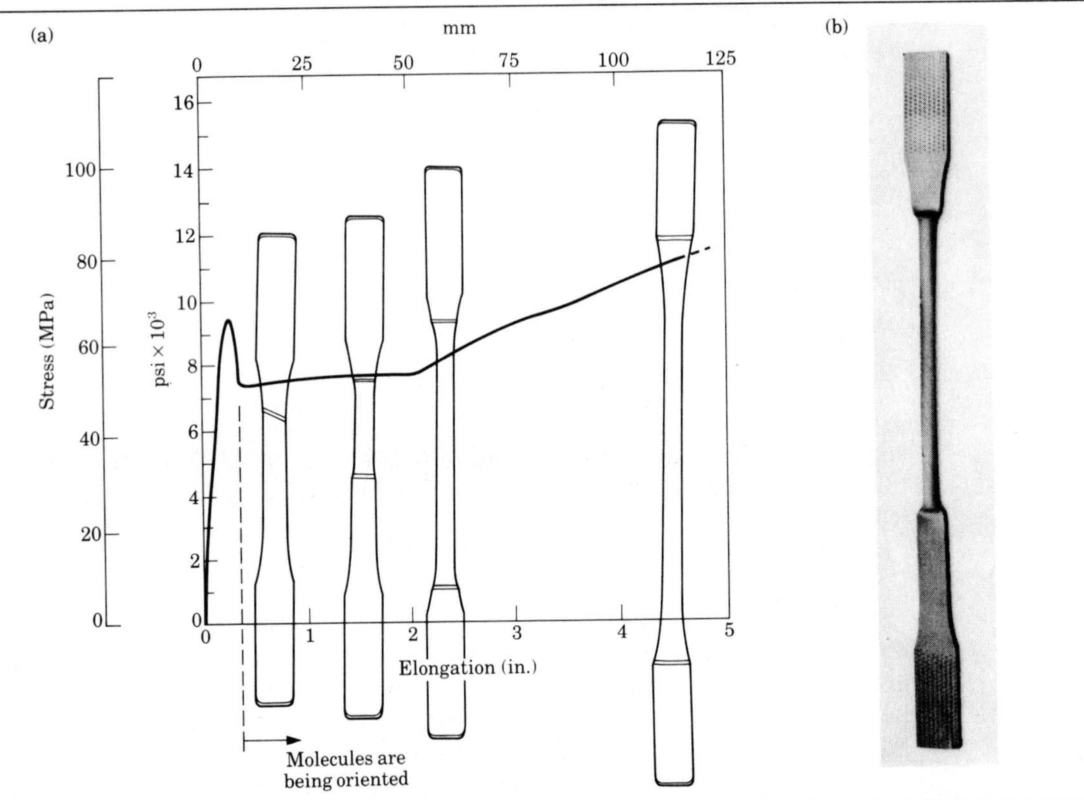

FIGURE 7.10
(a) Load-elongation curve for polycarbonate, a thermoplastic. *Source:* R. P. Kambour and R. E. Robertson. (b) High-density polyethylene tensile-test specimen, showing uniform elongation (the long, narrow region in the specimen).

The environment and the presence of solvents, lubricants, and water vapor enhance the formation of crazes (environmental stress cracking and solvent crazing). Residual stresses in the material contribute to crazing and cracking of the polymer. Radiation and visible light can adversely affect the strength of polymers.

A related phenomenon is **stress whitening.** When subjected to tensile stresses, such as by folding or bending, the plastic becomes lighter in color. This phenomenon is usually attributed to the formation of microvoids in the material. As a result, the material becomes less translucent (transmits less light), or more opaque. You can easily demonstrate this result by bending plastic components commonly found in household products and toys.

Water absorption. An important limitation of some polymers, such as nylons, is their ability to absorb water. Water acts as a plasticizing agent; that is, it makes the polymer more plastic. Thus in a sense, it lubricates the chains in the amorphous region. Typically, with increasing moisture absorption, the glass-transition temperature, the yield stress, and the elastic modulus of the polymer are lowered severely.

Dimensional changes also occur because of water absorption, such as in a humid environment.

Thermal and electrical properties. Compared to metals, plastics are generally characterized by low thermal and electrical conductivity, low specific gravity (ranging from 0.90 to 2.2), and relatively high coefficient of thermal expansion, which is about an order of magnitude higher.

Most polymers are electrical insulators and hence are used for insulators and as packaging material for electronic components. However, the electrical conductivity of some polymers can be increased by doping (introducing certain impurities in the polymer, such as metal powder, salts, and iodides). One of the earliest conducting polymers developed was polyacetylene. The electrical conductivity of polymers increases with moisture absorption.

7.4 Thermosets

When the long-chain molecules in a polymer are cross-linked in a three-dimensional arrangement, the structure in effect becomes one giant molecule with strong covalent bonds. As we previously stated, these polymers are called thermosetting plastics, or thermosets, because during polymerization the network is completed and the shape of the part is permanently set. This *curing* reaction, unlike that in thermoplastics, is irreversible. We can liken the response of a thermosetting plastic to temperature to baking a cake or boiling an egg. Once the cake is baked and cooled, or the egg boiled and cooled, reheating it will not change its shape.

Some thermosets, such as epoxy, polyester, and urethane, cure at room temperature. Although curing takes place at ambient temperature, the heat of the reaction cures the plastic. Thermosetting polymers do not have a sharply defined glass-transition temperature. The polymerization process for thermosets generally takes place in two stages. The first is at the chemical plant where the molecules are partially polymerized into linear chains. The second stage is at the parts-producing plant where cross-linking is completed under heat and pressure during the molding and shaping of the part.

Because of the nature of their bonds, the strength and hardness of thermosets, unlike thermoplastics, are not affected by temperature or rate of deformation. A typical thermoset is phenolic, which is a product of the reaction between phenol and formaldehyde. Common products made from this polymer are the handles and knobs on cooking pots and pans and components of light switches and outlets. Thermosetting plastics generally possess better mechanical, thermal, and chemical properties, electrical resistance, and dimensional stability than do thermoplastics. However, if the temperature is increased sufficiently, the thermosetting polymer begins to burn up, degrade, and char.

7.5 Additives

In order to impart certain specific properties, polymers are usually compounded with **additives**. These additives modify and improve certain characteristics of polymers, such as their stiffness, strength, color, weatherability, flammability, arc resistance for electrical applications, and ease of subsequent processing.

Fillers used are generally wood flour (fine sawdust), silica flour (fine silica powder), clay, powdered mica, and short fibers of cellulose, glass, and asbestos. Because of their low cost, fillers are important in reducing the overall cost of polymers. Depending on their type, fillers improve the strength, hardness, toughness, abrasion resistance, dimensional stability, and/or stiffness of plastics. These properties are greatest at various percentages of different types of polymer–filler combinations. As in reinforced plastics, a filler's effectiveness depends on the nature of the bond between the filler material and the polymer chains.

Plasticisers are added to some polymers to impart flexibility and softness by lowering their glass-transition temperature. Plasticisers are low molecular weight solvents with high boiling points (nonvolatile). They reduce the strength of the secondary bonds between the long-chain molecules, thus making the polymer soft and flexible. The most common use of plasticizers is in polyvinyl chloride (PVC), which remains flexible during its many uses. Other applications of plasticizers are in thin sheet, film, tubing, shower curtains, and clothing materials.

Most polymers are adversely affected by ultraviolet radiation (sunlight) and oxygen, which weaken and break the primary bonds, resulting in the scission (splitting) of the long-chain molecules. The polymer then degrades and becomes brittle and stiff. On the other hand, degradation may be beneficial, as in the disposal of plastic objects by subjecting them to environmental attack.

A typical example of protection against ultraviolet radiation is the compounding of rubber with *carbon black* (soot). The carbon black absorbs a high percentage of the ultraviolet radiation. Protection against degradation by oxidation, particularly at elevated temperatures, is done by adding antioxidants to the polymer. Various coatings are another means of protection of polymers.

The wide variety of colors available in plastics is obtained by adding *colorants*. They are either organic (dyes) or inorganic (pigments) materials. The selection of a colorant depends on service temperature and exposure to light. Pigments, which are dispersed particles, generally have greater resistance than dyes to temperature and light.

If the temperature is sufficiently high, most polymers will ignite. The flammability (ability to support combustion) of polymers varies considerably, depending on their composition (such as the chlorine and fluorine content). Polymethylmethacrylate, for example, continues to burn when ignited, whereas polycarbonate self-extinguishes. The flammability of polymers can be reduced either by making them from less flammable raw materials, or by the addition of **flame retardants**, such as compounds of chlorine, bromine, and phosphorus.

Lubricants may be added to polymers to reduce friction during their subsequent processing into useful products and to prevent parts from sticking to the molds. Lubrication is also important in preventing thin polymer films from sticking to each other.

7.6 General Properties and Applications of Thermoplastics

In this section we outline the general characteristics and typical applications of major thermoplastics, particularly as they relate to manufacturing plastic products and their service life. General recommendations for various plastics applications are given in Table 7.3.

Acetals (from acetic and alcohol) have good strength, stiffness, and resistance to creep, abrasion, moisture, heat, and chemicals. Typical applications are mechanical parts and components where high performance is required over a long period: bearings, cams, gears, bushings, rollers, impellers, wear surfaces, pipes, valves, shower heads, and housings. Common trade name: Delrin.

TABLE 7.3
GENERAL RECOMMENDATIONS FOR PLASTIC PRODUCTS

DESIGN REQUIREMENT	APPLICATIONS	PLASTICS
Mechanical strength	Gears, cams, rollers, valves, fan blades, impellers, pistons	Acetal, nylon, phenolic, polycarbonate
Functional and decorative	Handles, knobs, camera and battery cases, trim moldings, pipe fittings	ABS, acrylic, cellulosic, phenolic, polyethylene, polypropylene, polystyrene, polyvinyl chloride
Housings and hollow shapes	Power tools, pumps, housings, sport helmets, telephone cases	ABS, cellulosic, phenolic, polycarbonate, polyethylene, polypropylene, polystyrene
Functional and transparent	Lenses, goggles, safety glazing, signs, food-processing equipment, laboratory hardware	Acrylic, polycarbonate, polystyrene, polysulfone
Wear resistance	Gears, wear strips and liners, bearings, bushings, roller-skate wheels	Acetal, nylon, phenolic, polyimide, polyurethane, ultrahigh molecular weight polyethylene

Acrylics (polymethylmethacrylate, PMMA) possess moderate strength, good optical properties, and weather resistance. They are transparent but can be made opaque, are generally resistant to chemicals, and have good electrical resistance. Typical applications are lenses, lighted signs, displays, window glazing, skylights, bubble tops, automotive lenses, windshields, lighting fixtures, and furniture. Common trade names: Plexiglas, Lucite.

Acrylonitrile-butadiene-styrene (ABS) is dimensionally stable and rigid and has good impact, abrasion, and chemical resistance, strength and toughness, low-temperature properties, and electrical resistance. Typical applications are pipes, fittings, chrome-plated plumbing supplies, helmets, tool handles, automotive components, boat hulls, telephones, luggage, housing, appliances, refrigerator liners, and decorative panels.

Cellulosics have a wide range of mechanical properties, depending on composition. They can be made rigid, strong, and tough. However, they weather poorly and are affected by heat and chemicals. Typical applications are tool handles, pens, knobs, frames for eyeglasses, safety goggles, machine guards, helmets, tubing and pipes, lighting fixtures, rigid containers, steering wheels, packaging film, signs, billiard balls, toys, and decorative parts.

Fluorocarbons possess good resistance to temperature, chemicals, weather, and electricity. They also have unique nonadhesive properties and low friction. Typical applications are linings for chemical process equipment, nonstick coatings for cookware, electrical insulation for high-temperature wire and cable, gaskets, low friction surfaces, bearings, and seals. Common trade name: Teflon.

Polyamides (from the words poly, amine, and carboxyl acid) are available in two main types: nylons and aramids. *Nylons* (a coined word) have good mechanical properties and abrasion resistance. They are self-lubricating and resistant to most chemicals. All nylons are hygroscopic (absorb water). Moisture absorption reduces mechanical properties and increases part dimensions. Typical applications are gears, bearings, bushings, rollers, fasteners, zippers, electrical parts, combs, tubing, wear-resistant surfaces, guides, and surgical equipment. *Aramids* (aromatic polyamides) have very high tensile strength and stiffness. Typical applications are fibers for reinforced plastics (composite materials), bulletproof vests, cables, and radial tires. Common trade name: Kevlar.

Polycarbonates are versatile and have good mechanical and electrical properties. They also have high impact resistance, and can be made resistant to chemicals. Typical applications are safety helmets, optical lenses, bullet-resistant window glazing, signs, bottles, food-processing equipment, windshields, load-bearing electrical components, electrical insulators, medical apparatus, business machine components, guards for machinery, and parts requiring dimensional stability. Common trade name: Lexan.

Polyesters (thermoplastic; see also Section 7.7) have good mechanical, electrical, and chemical properties, good abrasion resistance, and low friction. Typical applications are gears, cams, rollers, load-bearing members, pumps, and electromechanical components. Common trade names: Dacron, Mylar, Kodel.

Polyethylenes possess good electrical and chemical properties. Their mechanical properties depend on composition and structure. Three major classes are low density (LDPE), high density (HDPE), and ultrahigh molecular weight (UHMWPE). Typical applications for LDPE are housewares, bottles, garbage cans, ducts, bumpers, luggage, toys, tubing, bottles, and packaging material; for HDPE, machinery parts, belts and straps, wear-resistant surfaces, sleds, canoes, and camper tops; and for UHMWPE, parts requiring high-impact toughness and abrasive wear resistance.

Polyimides have the structure of a thermoplastic but the nonmelting characteristic of a thermoset. See Section 7.7.

Polypropylenes have good mechanical, electrical, and chemical properties and good resistance to tearing. Typical applications are automotive trim and components, medical devices, appliance parts, wire insulation, TV cabinets, pipes, fittings, drinking cups, dairy-product and juice containers, luggage, ropes, and weather stripping.

Polystyrenes have properties that depend on composition. They are inexpensive, have generally average properties, and are somewhat brittle. Typical applications are disposable containers, packaging, trays for meats, cookies, and candy, foam insulation, appliances, automotive and radio/TV components, housewares, and toys and furniture parts (as a wood substitute).

Polysulfones have excellent resistance to heat, water, and steam, are highly resistant to some chemicals, but are attacked by organic solvents. Typical applications are steam irons, coffeemakers, hotwater containers, medical equipment that requires sterilization, power-tool and appliance housings, aircraft cabin interiors, and electrical insulators.

Polyvinyl chloride (PVC) has a wide range of properties, is inexpensive and water resistant, and can be made rigid or flexible. It is not suitable for applications requiring strength and heat resistance. Rigid PVC is tough and hard. Typical applications are rigid PVC for the construction industry, such as pipes and conduits, automobile windshields, and signs; flexible PVC for wire and cable coatings, low-pressure flexible tubing and hose, footware, imitation leather, upholstery, records, gaskets, seals, trim, film, sheet, and coatings. Common trade names: Saran, Tygon.

Vinyls. See polyvinyl chloride.

7.7 General Properties and Applications of Thermosetting Plastics

In this section we outline the general characteristics and typical applications of major thermosetting plastics.

Alkyds (from alkyl, meaning alcohol, and acid) possess good electrical insulating properties, impact resistance, and dimensional stability and have low water absorption. Typical applications are electrical and electronic components.

Aminos (urea and melamine) have properties that depend on composition. Generally, aminos are hard and rigid and are resistant to abrasion, creep, and electrical arcing. Typical applications are small appliance housings, countertops, appliance housings, handles, and distributor caps. Urea is used for electrical and electronic components, melamine for dinnerware.

Epoxies have excellent mechanical and electrical properties, dimensional stability, strong adhesive properties, and good resistance to heat and chemicals. Typical applications are electrical components requiring mechanical strength and high insulation, tools and dies, and adhesives. Fiber-reinforced epoxies have excellent mechanical properties and are used in pressure vessels, rocket motor casings, tanks, and similar structural components (see Chapter 9).

Phenolics, although brittle, are rigid, dimensionally stable, and have high resistance to heat, water, electricity, and chemicals. Typical applications are knobs, handles, laminated panels, telephones, bond material to hold abrasive grains together in grinding wheels, and electrical components, such as wiring devices, connectors, and insulators.

Polyesters (thermosetting; see also Section 7.6) have good mechanical, chemical, and electrical properties. Polyesters are generally reinforced with glass or other fibers. Typical applications are boats, luggage, chairs, automotive bodies, swimming pools, material for impregating cloth, paper, and decorations. They also are available as casting resins.

Polyimides possess good mechanical, physical, and electrical properties at elevated temperatures. They also have creep resistance and low friction and wear characteristics. Polyimides have the nonmelting characteristics of a thermoset but the structure of a thermoplastic. Typical applications are pump components (bearings, seals, valve seats, retainer rings, and piston rings), electrical connectors for high-temperature use, aerospace parts, high-strength impact-resistant structures, sports equipment, and safety vests.

Silicones have properties that depend on composition. Generally, they possess excellent electrical properties over a wide range of humidity and temperature, weather well, and resist chemicals and heat (see also Section 7.8). Typical applications are electrical components requiring strength at elevated temperatures, oven gaskets, heat seals, and waterproof materials.

• Example: Materials for refrigerator door liner.

In the selection of candidate materials for a door liner for refrigerators—where eggs, butter, salad dressings, and small bottles are stored—the following factors should be considered:

Mechanical requirements: Strength, toughness (to withstand impact, door slamming, racking), stiffness, resilience, and resistance to scratching and wear at operating temperatures.

Physical requirements: Dimensional stability and electrical insulation.

Chemical requirements: Staining, odor, chemical reactions with food and beverages, and resistance to cleaning fluids.

Appearance: Color, stability of color, surface finish, texture, and feel.

Manufacturing properties: Methods of manufacturing and assembly, effects of processing on material properties and behavior over a period of time, compatibility with other components in the door, and cost of materials and manufacturing.

An extensive study, considering all the factors involved, identified two candidate materials for door liners: ABS (acrylonitrile-butadiene-styrene) and HIPS (high-impact polystyrene). One aspect of the study involved the effect of vegetable oils, such as salad dressing stored in the door shelf, on the strength of these plastics. Experiments showed that the presence of vegetable oils significantly reduced the load-bearing capacity of HIPS. It was found that HIPS becomes brittle in the presence of oils (stress cracking), whereas ABS is not affected to any significant extent.

7.8 Elastomers (Rubbers)

Elastomers comprise a large family of amorphous polymers having a low glass-transition temperature. They have the characteristic ability to undergo large elastic deformations without rupture; they are soft and have a low elastic modulus. The term elastomer is derived from the words *elastic* and *mer*. These polymers are highly kinked (tightly twisted or curled). They stretch but then return to their original shape after the load is removed. Cross-linking also takes place, the best example being the elevated temperature **vulcanization** of rubber with sulfur, discovered by Charles Goodyear in 1839 and named for Vulcan, the Roman god of fire. Once the elastomer is cross-linked, it cannot be reshaped. For example, an automobile tire, which is one giant molecule, cannot be reshaped.

The terms rubber and elastomer are often used interchangeably. Generally, an **elastomer** is defined as being capable of recovering substantially in shape and size after the load has been removed. **Rubber** is defined as being capable of recovering from large deformations quickly.

The hardness of elastomers, which is measured with a durometer, increases with increasing cross-linking of the molecular chains. A variety of additives can be blended with elastomers to impart specific properties, as with plastics. Elastomers have a wide range of applications, such as high-friction and nonskid surfaces, protection against corrosion and abrasion, electrical insulation, and shock and vibration insulation. Examples include tires, hoses, weather stripping, footwear, linings, gaskets, seals, printing rolls, and flooring.

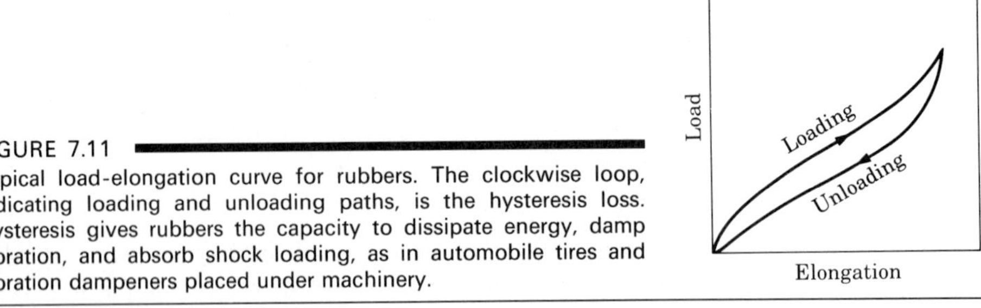

FIGURE 7.11 Typical load-elongation curve for rubbers. The clockwise loop, indicating loading and unloading paths, is the hysteresis loss. Hysteresis gives rubbers the capacity to dissipate energy, damp vibration, and absorb shock loading, as in automobile tires and vibration dampeners placed under machinery.

A characteristic of elastomers is their hysteresis loss in stretching or compression (Fig. 7.11). The clockwise loop indicates energy loss, whereby mechanical energy is converted into heat. This property is desirable for absorbing vibrational energy (damping) and sound deadening.

Natural rubber. The base for natural rubber is **latex,** a milklike sap obtained from the inner bark of a tropical tree. It has good resistance to abrasion and fatigue and high frictional properties, but it has low resistance to oil, heat, ozone, and sunlight. Typical applications are tires, seals, shoe heels, coupling, and engine mounts.

Synthetic rubbers. Further developed natural rubbers are the synthetic rubbers. Examples are synthetic natural rubber, butyl, styrene butadiene, polybutadiene, and ethylene propylene. Compared to natural rubbers, they have improved resistance to heat, gasoline, and chemicals and higher useful temperature range. Examples of synthetic rubbers that are resistant to oil are neoprene, nitrile, urethane, and silicone. Typical applications of synthetic rubbers are tires, shock absorbers, seals, and belts.

Silicones (see also Section 7.7) have the highest useful temperature range, up to 315 °C (600 °F), but their other properties—such as strength and resistance to wear and oils—are generally inferior to other elastomers. Typical applications are seals, gaskets, thermal insulation, high-temperature electrical switches, and electronic apparatus.

Polyurethane has very good overall properties of high strength, stiffness, and hardness and exceptional resistance to abrasion, cutting, and tearing. Typical applications are seals, gaskets, cushioning, diaphragms for rubber forming of sheet metals, and auto body parts.

SUMMARY

Plastics are an important class of materials because they possess a very wide range of mechanical, physical, and chemical properties. Compared to metals, plastics are generally characterized by lower density, strength, elastic modulus, and thermal and electrical conductivity and a higher coefficient of thermal expansion.

Plastics are composed of polymer molecules and various additives. The smallest repetitive unit in a polymer chain is called a mer. Monomers are linked by polymerization processes to form larger molecules. Two major classes of polymers are thermoplastics and thermosets. Thermoplastics become soft and are easy to form at elevated temperatures; they return to their original properties when cooled. Thermosets, which are obtained by cross-linking polymer chains, do not become soft to any significant extent with increasing temperature. Polymer structures can be modified by various means to impart a wide range of properties to plastics.

Elastomers have the characteristic ability to undergo large elastic deformations and return to their original shapes when unloaded. Consequently, they have important applications as tires, seals, footware, hose, belts, and shock absorbers.

TRENDS

- Developments are taking place to produce ultrahigh-purity polymers, polymer blends, high-strength fibers, optical fibers, and multilayer films for optical applications, lenses, and recording media.
- Recyclability of plastics is—and will continue to be—an important topic in terms of environmental protection. Biodegradable plastics are also under development.
- Improvements are being made in water absorption, flammability, and degradation of plastics, as well as in their physical properties.

KEY TERMS

Additives
Branched polymers
Catalyst
Crazing
Cross-linked polymers
Crystallites
Degree of polymerization
Elastomer
Fillers
Flame retardants
Glass-transition temperature
Linear polymer
Latex
Mer
Molecular weight
Monomer
Network polymers
Orientation
Plasticizers
Plastics
Polyblends
Polymers
Polymerization
Primary bonds
Rubber
Secondary bonds
Stress whitening
Thermoplastics
Thermosets
Vulcanization

BIBLIOGRAPHY

Ash, M., and I. Ash, *Encyclopedia of Plastics, Polymers and Resins*, 3 vols. New York: Chemical Publishing Co., 1980–81.

Chanda, M., and S.K. Roy, *Plastics Technology Handbook*. New York: Marcel Dekker, 1987.

Cheremisinoff, N.P., *Product Design and Testing of Polymeric Materials*. New York: Marcel Dekker, 1990.

Easterling, K., *Tomorrow's Materials*, 2d ed. London: The Institute of Metals, 1990.

Engineered Materials Handbook, Vol. 2: *Engineering Plastics*. Metals Park, Ohio: ASM International, 1988.

Engineering Plastics and Composites. Metals Park, Ohio: ASM International, 1990.

Frados, J. (ed.), *Plastics Engineering Handbook*, 4th ed. New York: Van Nostrand Reinhold, 1976.

Goodman, S.H., *Handbook of Thermoset Plastics*. Park Ridge, N.J.: Noyes Publications, 1986.

Hall, C., *Polymer Materials*. New York: Macmillan, 1981.

Kaufman, H.S., *Introduction to Polymer Science and Technology*. New York: Wiley, 1986.

Kroschwitz, J.I. (ed.), *Concise Encyclopedia of Polymer Science and Engineering*. New York: Wiley, 1990.

MacDermott, C.P., *Selecting Thermoplastics for Engineering Applications*. New York: Marcel Dekker, 1984.

Margolis, J.M., *Engineering Thermoplastics: Properties and Applications*. New York: Marcel Dekker, 1985.

McCrum, N.G., C.P. Buckley, and C.B. Bucknall, *Principles of Polymer Engineering*. Oxford: Oxford University Press, 1988.

Moore, G.R., and D.E. Kline, *Properties and Processing of Polymers for Engineers*. Englewood Cliffs, N.J.: Prentice-Hall, 1984.

Rosen, S.L., *Fundamental Principles of Polymeric Materials*, 2d ed. New York: Wiley, 1982.

Rubin, I.I. (ed.), *Handbook of Plastic Materials and Technology*. New York: Wiley, 1990.

Rudin, A., *The Elements of Polymer Science and Engineering*. New York: Academic Press, 1982.

Seymour, R.B., *Engineering Polymer Sourcebook*. New York: McGraw-Hill, 1990.

——, *Polymers for Engineering Applications*. Metals Park, Ohio: ASM International, 1987.

REVIEW QUESTIONS

7.1 Summarize the most important mechanical and physical properties of plastics.

7.2 What are the major differences between the properties of plastics and metals?

7.3 What are (a) polymerization and (b) degree of polymerization? What properties are influenced by the degree of polymerization?

7.4 What is the difference between linear, branched, and cross-linked polymers?

7.5 Why would we want to synthesize a polymer with a high degree of crystallinity?

7.6 What is glass-transition temperature?

7.7 Compare and contrast thermoplastics and thermosets.

7.8 What is crazing?
7.9 What are the water absorption characteristics of polymers? How do they relate to properties important for application?
7.10 List and describe five polymer additives.
7.11 What are polyblends?
7.12 What are the properties of elastomers?

QUESTIONS AND PROBLEMS

7.13 Inspect various plastic components in your automobile and state whether you think they are made of thermoplastic or thermosetting plastic materials.
7.14 What design considerations are involved in replacing a metal container for a beverage with one made of plastic?
7.15 Assume that you are manufacturing a product in which all the gears are made of metal. A salesman visits you and asks you to consider replacing some of these metal gears with plastic ones. Make a list of the questions that you would raise before making a decision.
7.16 Explain the significance of orientation in polymers.
7.17 Calculate the areas under the stress–strain curve (toughness) for the material in Fig. 7.8 and plot them as a function of temperature, and describe your observations.
7.18 Give applications for which flammability of plastics would be a major concern.
7.19 Explain why the plastics industry has grown phenomenally over the past few decades.
7.20 What is the significance of the glass-transition temperature?
7.21 In Sections 7.6 and 7.7, we listed several plastics and their applications. Rearrange this information by making a table of products (gears, helmets, luggage, electrical parts, etc.) and the type of plastics that can be used to make these products.
7.22 What properties do elastomers have that thermoplastics in general do not have?
7.23 Review the tables on mechanical properties of thermoplastics in this chapter and metals in Chapters 5 and 6. What generalizations can you state after comparing these different groups of materials?
7.24 Do you think that the substitution of plastics for metals in products traditionally made of metal is viewed negatively by the public at large? If so, why?
7.25 Make a list of products or parts that are not currently made of plastics and explain the possible reasons why they are not.
7.26 Review Table 7.1 and name three plastics that have high strengths and three that have low strengths.
7.27 Name three plastics that are suitable for use at elevated temperatures.
7.28 Review the three curves in Fig. 7.7 and name applications for each type of material. Explain your choices.
7.29 Repeat Question 7.28 for the curves in Fig. 7.9.
7.30 Is it possible for a material to have a hysteresis behavior that is the opposite of that shown in Fig. 7.11, whereby the arrows are counterclockwise? Explain.
7.31 Observe the behavior of the specimen shown in Fig. 7.10, and state whether the material has a high or low m value. Explain why.
7.32 Add more to the applications column in Table 7.3.
7.33 Discuss the significance of the glass-transition temperature, T_g, in engineering applications.

7.34 Why does cross-linking improve the strength of polymers?

7.35 Describe the differences between and significance of covalent, ionic, hydrogen, and van der Waals bonds. Identify the primary and secondary bonds.

7.36 Describe the methods by which optical properties of polymers can be altered.

7.37 Can polymers be made to conduct electricity? How?

7.38 Explain the reasons why elastomers were developed. Are there any substitutes for elastomers? Explain.

8

Ceramics, Graphite, and Diamond: Structure, General Properties, and Applications

8.1 Introduction

None of the materials we have discussed thus far is suitable for certain engineering applications. Which material, for example, would you select for an electrical insulator to be used at high temperatures? For floor tiles to resist spills, scuffing, and abrasion? For a transparent baking dish? For a small ball that is light, rigid, hard, and resists high temperatures? How do we protect the surfaces of the space shuttle orbiter, made of aluminum, when its skin temperature reaches 1450 °C (2650 °F) as it takes off and reenters the atmosphere?

When trying to answer these questions, we soon realize that we are looking for materials with properties such as high-temperature strength, hardness, inertness to chemicals, food, and environment, resistance to wear, and low electrical and thermal conductivity. Ceramics generally have such desirable properties.

Ceramics are compounds of metallic and nonmetallic elements. The term **ceramics** refers both to the material and to the ceramic product itself, in Greek the word *keramos* meaning potter's clay and *keramikos* meaning clay products.

TABLE 8.1
CATEGORIES AND USES OF CERAMICS

	TRADITIONAL CERAMICS
Abrasive products	Abrasive wheels, emery cloth and sand paper, nozzles for sandblasting, ball milling
Clay products	Brick, pottery, sewer pipe
Construction	Brick, concrete, tile, plaster, glass
Glass	Bottles, laboratory ware, glazing
Refractories	Brick, crucibles, molds, cement
Whitewares	Dishes, tiles, plumbing, enamels
	ENGINEERING CERAMICS
Automotive and aerospace	Turbine components, heat shields and exchangers, reentry components, seals
Electronics	Semiconductors, insulators, transducers, lasers, dielectrics, heating elements
High temperature	Refractories, brazing fixtures, kilns
Manufacturing	Cutting tools, wear and corrosion resistant components, glass ceramics, magnets, fiber optics
Medical	Laboratory ware, controls, prosthetics, dental

Because of the large number of possible combinations of elements, a great variety of ceramics is now available for widely different consumer and industrial applications (Table 8.1).

The earliest use of ceramics was in pottery and bricks, dating back to before 4000 B.C. Ceramics have been used for many years in automotive spark plugs as an electrical insulator and for high-temperature strength. They are becoming increasingly important in heat engines and various other applications (Table 8.1), as well as tool and die materials. More recent applications of ceramics are in automotive components, such as exhaust-port liners, coated pistons, and cylinder liners, with the desirable properties of strength and corrosion resistance at high operating temperatures.

We divide ceramics into two general categories: *traditional* and *industrial* ceramics, also called engineering, high-tech, or fine ceramics. In this chapter, we present the general characteristics and applications of ceramics, glasses, and glass ceramics that are of importance in engineering applications and manufacturing. Because of their unique characteristics, we also discuss the properties and uses of two forms of carbon, namely, graphite and diamond. (We describe the manufacturing of ceramic and glass components, and various shaping and finishing operations, in Chapter 19. We discuss composites, which are an important group of materials composed of ceramics, metals, and polymers, in Chapter 9.)

8.2 The Structure of Ceramics

The structure of ceramic crystals is among the most complex of all materials, containing various elements of different sizes. The bonding between these atoms is generally covalent (electron sharing, hence strong bonds) and ionic (primary bonding between oppositely charged ions, thus strong bonds). These bonds are much stronger than metallic bonds. Consequently, the properties of ceramics are significantly higher than those for metals, particularly their hardness and thermal and electrical resistance.

Ceramics are available as a single crystal or in polycrystalline form, consisting of many grains. Grain size has a major influence on the strength and properties of ceramics. The finer the grain size, the higher are the strength and toughness—hence the term *fine ceramics*.

8.2.1 Raw materials

Among the oldest raw materials for ceramics is **clay**, a fine-grained sheetlike structure, the most common example being *kaolinite* (from Kao-ling, a hill in China). It is a white clay, consisting of silicate of aluminum with alternating weakly bonded layers of silicon and aluminum ions (Fig. 8.1). When added to kaolinite, water attaches itself to the layers (adsorption), makes them slippery, and gives wet clay its well-known softness and plastic properties (*hydroplasticity*) that make it formable.

Other major raw materials for ceramics that are found in nature are **flint** (rock of very fine grained silica, SiO_2) and **feldspar** (a group of crystalline minerals consisting of aluminum silicates, potassium, calcium, or sodium). In their natural state, these raw materials generally contain impurities of various kinds, which have to be removed prior to further processing of the materials into useful products with reliable performance. Highly refined raw materials produce ceramics with improved properties.

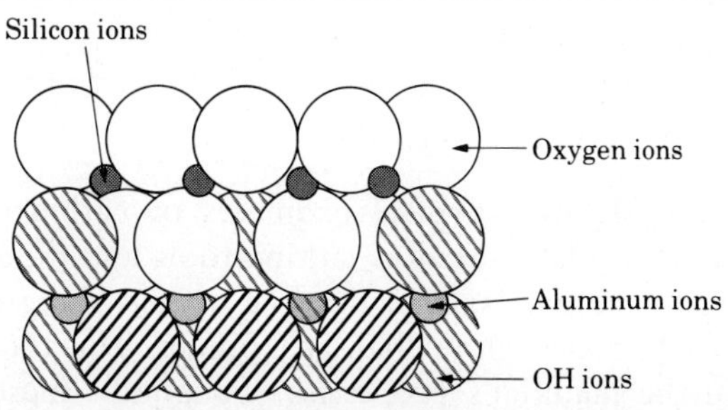

FIGURE 8.1
The crystal structure of kaolinite, commonly known as clay.

8.2.2 Oxide ceramics

Alumina. Also called *corundum* or *emery*, **alumina** (aluminum oxide, Al_2O_3) is the most widely used *oxide ceramic*, either in pure form or as a raw material to be mixed with other oxides. It has high hardness and moderate strength. Although alumina exists in nature, it contains unknown amounts of impurities and possesses nonuniform properties. As a result, its behavior is unreliable. Aluminum oxide, as well as silicon carbide and many other ceramics, are now almost totally manufactured synthetically so that we can control their quality.

First made in 1893, synthetic aluminum oxide is obtained by the fusion of molten bauxite (an aluminum oxide ore that is the principal source of aluminum), iron filings, and coke in electric furnaces. It is then crushed and graded by size by passing the particles through standard screens. Parts made of aluminum oxide are cold pressed and sintered (*white ceramics*). Their properties are improved by minor additions of other ceramics, such as titanium oxide and titanium carbide. Structures containing various alumina and other oxides are known as *mullite* and *spinel* and are used as refractory materials for high-temperature applications. The mechanical and physical properties of alumina are particularly suitable for applications such as electrical and thermal insulation and as cutting tools and abrasives.

Zirconia and partially stabilized zirconia. **Zirconia** (zirconium oxide, ZrO_2, white in color) has good toughness, resistance to thermal shock, wear, and corrosion, low thermal conductivity, and low friction coefficient. A more recent development is **partially stabilized zirconia (PSZ)**, which has high strength and toughness and better reliability in performance than zirconia. It is obtained by doping the zirconia with oxides of calcium, yttrium, or magnesium. This process forms a material with fine particles of tetragonal zirconia in a cubic lattice.

Another important characteristic of PSZ is the fact that its coefficient of thermal expansion is only about 20 percent lower than that of cast iron, and its thermal conductivity is about one-third that of other ceramics. Consequently, it is very suitable for heat-engine components, such as cylinder liners and valve bushings, to keep the cast-iron engine assembly intact. New developments to further improve the properties of PSZ include *transformation-toughened zirconia* (*TTZ*), which has higher toughness because of dispersed tough phases in the ceramic matrix.

8.2.3 Other ceramics

Carbides. Typical examples of **carbides** are those of tungsten (WC) and titanium (TiC), used as cutting tools and die materials, and silicon carbide (SiC), used as abrasives, as in grinding wheels. *Tungsten carbide* consists of tungsten–carbide particles with cobalt as a binder. The amount of binder has a major influence on the material's properties. Toughness increases with cobalt content, whereas hardness, strength, and wear resistance decrease. *Titanium carbide* has nickel and molybdenum as the binder and is not as tough as tungsten carbide.

Silicon carbide (SiC) has good wear, thermal shock, and corrosion resistance. It has a low friction coefficient and retains strength at elevated temperatures. It is suitable for high-temperature components in heat engines and is also used as an abrasive. First produced in 1891, synthetic silicon carbide is made from silica sand, coke, and small amounts of sodium chloride and sawdust. The process is similar to making synthetic aluminum oxide.

Nitrides. Another important class of ceramics is the **nitrides**, particularly cubic boron nitride (CBN), titanium nitride (TiN), and silicon nitride (Si_3N_4).

Cubic boron nitride, the second hardest known substance, after diamond, has special applications, such as abrasives in grinding wheels and as cutting tools. It does not exist in nature and was first made synthetically in the 1970s, with techniques similar to those used in making synthetic diamond.

Titanium nitride is used widely as coatings on cutting tools. It improves tool life by virtue of its low frictional characteristics.

Silicon nitride has high resistance to creep at elevated temperatures, low thermal expansion, high thermal conductivity, and hence resists thermal shock. It is suitable for high-temperature structural applications, such as in automotive engine and gas-turbine components.

Sialon. **Sialon** consists of silicon nitride, with various additions of aluminum oxide, yttrium oxide, and titanium carbide. The word sialon is derived from silicon, aluminum, oxygen, and nitrogen. It has higher strength and thermal-shock resistance than silicon nitride and thus far is used primarily as a cutting-tool material.

Cermets. **Cermets** are combinations of ceramics bonded with a metallic phase. Introduced in the 1960s, they combine the high-temperature oxidation resistance of ceramics and the toughness, thermal-shock resistance, and ductility of metals. An application of cermets is cutting tools, a typical composition being 70 percent aluminum oxide and 30 percent titanium carbide. Other cermets contain various oxides, carbides, and nitrides. They have been developed for high-temperature applications such as nozzles for jet engines and aircraft brakes. Cermets can be regarded as composite materials and can be used in various combinations of ceramics and metals bonded by powder-metallurgy techniques.

8.2.4 Silica

Abundant in nature, **silica** is a polymorphic material; that is, it can have different crystal structures. The cubic structure is found in refractory bricks used for high-temperature furnace applications. Most glasses contain more than 50 percent silica. The most common form of silica is *quartz*, which is a hard, abrasive hexagonal crystal. It is used extensively as oscillating crystals of fixed frequency in communications applications, since it exhibits the piezoelectric effect.

Silicates are products of the reaction of silica with oxides of aluminum, magnesium, calcium, potassium, sodium, and iron. Examples are clay, asbestos, mica, and silicate glasses. *Lithium aluminum silicate* has very low thermal expansion and thermal conductivity and good thermal-shock resistance. However, it has very low strength and fatigue life. Thus it is suitable only for nonstructural applications, such as catalytic converters, regenerators, and heat-exchanger components.

8.3 General Properties and Applications of Ceramics

Compared to metals, ceramics have the following relative characteristics: brittle, high strength and hardness at elevated temperatures, high elastic modulus, low toughness, low density, low thermal expansion, and low thermal and electrical conductivity. However, because of the wide variety of ceramic material composition and grain size, the mechanical and physical properties of ceramics vary significantly. For example, the electrical conductivity of ceramics can be modified from poor to good, which is the principle behind semiconductors.

Because of their sensitivity to flaws, defects, and cracks (surface or internal), the presence of different types and levels of impurities, and different methods of manufacturing, ceramics can have a wide range of properties. Although we stated the individual characteristics of ceramics in Section 8.2, we still need to describe their general mechanical and physical properties.

8.3.1 Mechanical properties

The mechanical properties of several engineering ceramics are presented in Table 8.2. Note that their strength in tension (transverse rupture strength) is approximately one order of magnitude lower than their compressive strength. The reason is their sensitivity to cracks, impurities, and porosity. Such defects lead to the initiation and propagation of cracks under tensile stresses, severely reducing tensile strength. Thus reproducibility and reliability (acceptable performance over a specified period of time) is an important aspect in the service life of ceramic components. Tensile strength of polycrystalline ceramic parts increases with decreasing grain size.

Tensile strength is empirically related to porosity as follows:

$$\mathrm{UTS} \simeq \mathrm{UTS}_0 e^{-nP}, \tag{8.1}$$

where P is the volume fraction of pores in the solid, UTS_0 is the tensile strength at zero porosity, and the exponent n ranges between 4 and 7.

The modulus of elasticity is likewise affected by porosity, as given by

$$E \simeq E_0(1 - 1.9P + 0.9P^2), \tag{8.2}$$

where E_0 is the modulus at zero porosity. Equation (8.1) is valid up to 50 percent porosity. Common earthenware has a porosity ranging between 10 percent and 15 percent, whereas hard porcelain is about 3 percent.

TABLE 8.2
PROPERTIES OF VARIOUS CERAMICS AT ROOM TEMPERATURE

MATERIAL	SYMBOL	TRANSVERSE RUPTURE STRENGTH (MPa)	COMPRESSIVE STRENGTH (MPa)	ELASTIC MODULUS (GPa)	HARDNESS (HK)	POISSON'S RATIO (ν)	DENSITY (kg/m^3)
Aluminum oxide	Al_2O_3	140–240	1000–2900	310–410	2000–3000	0.26	4000–4500
Cubic boron nitride	CBN	725	7000	850	4000–5000	—	3480
Diamond	—	1400	7000	830–1000	7000–8000	—	3500
Silica, fused	SiO_2	—	1300	70	550	0.25	—
Silicon carbide	SiC	100–750	700–3500	240–480	2100–3000	0.14	3100
Silicon nitride	Si_3N_4	480–600	—	300–310	2000–2500	0.24	3300
Titanium carbide	TiC	1400–1900	3100–3850	310–410	1800–3200	—	5500–5800
Tungsten carbide	WC	1030–2600	4100–5900	520–700	1800–2400	—	10,000–15,000
Partially stabilized zirconia	PSZ	620	—	200	1100	0.30	5800

Note: These properties vary widely depending on the condition of the material.

Although there are exceptions and unlike most metals and thermoplastics, ceramics generally lack impact toughness and thermal-shock resistance because of their inherent lack of ductility. Once initiated, a crack propagates rapidly. In addition to undergoing fatigue failure under cyclic loading, ceramics (and particularly glasses) exhibit a phenomenon called **static fatigue**. When subjected to a static tensile load over a period of time, these materials may suddenly fail. This phenomenon occurs in environments where water vapor is present. Static fatigue, which does not occur in a vacuum or dry air, has been attributed to a mechanism similar to stress-corrosion cracking of metals.

Ceramic components that are to be subjected to tensile stresses may be prestressed, much like prestressed concrete. Prestressing shaped ceramic components subjects them to compressive stresses. Methods used include (a) heat treatment and chemical tempering, (b) laser treatment of surfaces, (c) coating with ceramics with different thermal expansion coefficients, and (d) surface-finishing operations, such as grinding, in which compressive residual stresses are induced on the surfaces.

Significant advances are being made in improving the toughness and other properties of ceramics. Among these are proper selection and processing of raw materials, control of purity and structure, and use of reinforcements and with particular emphasis during design on advanced methods of stress analysis in ceramic components.

8.3.2 Physical properties

Most ceramics have relatively low specific gravity, ranging from about 3 to 5.8 for oxide ceramics, compared to 7.86 for iron. They have very high melting or decomposition temperature. Thermal conductivity of ceramics varies by as much as three orders of magnitude, depending on their composition, whereas metals vary by one order. Thermal conductivity of ceramics, as well as other materials, decreases with increasing temperature and **porosity** because air is a poor thermal conductor.

The thermal conductivity k is related to porosity by

$$k = k_0(1 - P), \tag{8.3}$$

where k_0 is the thermal conductivity at zero porosity.

The thermal expansion characteristics of ceramics are shown in Fig. 8.2. Thermal expansion and thermal conductivity induce thermal stresses that can lead to thermal shock or thermal fatigue. The tendency for *thermal cracking* (called *spalling* when a piece or a layer from the surface breaks off) is lower with low

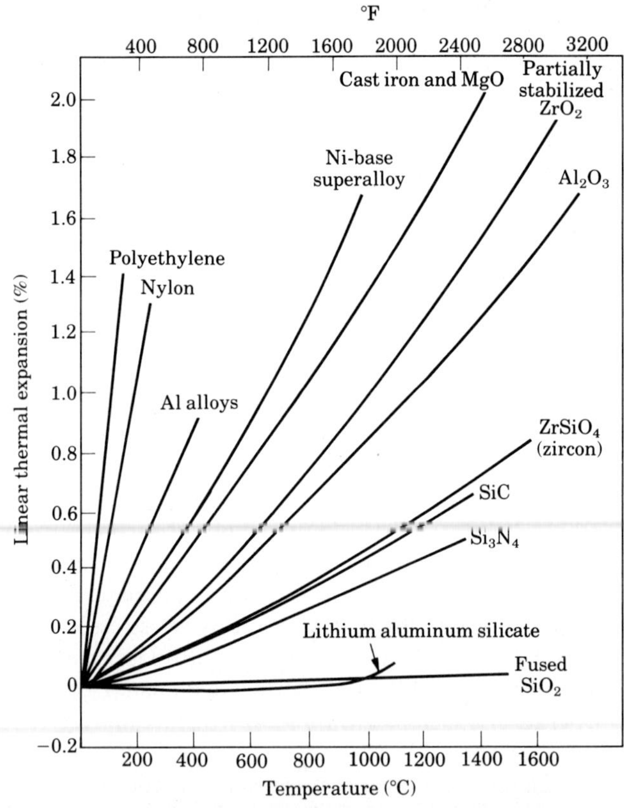

FIGURE 8.2 Effect of temperature on thermal expansion for several ceramics, metals, and plastics. Note that the expansion for cast iron and for partially stabilized zirconia (PSZ) are within about 20 percent. This makes the two materials compatible for internal combustion engine applications.

thermal expansion and high thermal conductivity. For example, fused silica has high thermal shock resistance because of its virtually zero thermal expansion.

A familiar example that illustrates the importance of low thermal expansion is the heat-resistant ceramics for cookware and stove tops. They can sustain high thermal gradients, from hot to cold, and vice versa. Moreover, the relative thermal expansion of ceramics and metals is an important reason for the use of ceramic components in heat engines. The fact that the thermal conductivity of partially stabilized zirconia components is close to that of the cast iron in engine blocks is an additional advantage in the use of PSZ in heat engines.

An additional characteristic is the *anisotropy of thermal expansion* exhibited by oxide ceramics, whereby thermal expansion varies in different directions of the ceramic, by as much as 50 percent for quartz. This behavior causes thermal stresses that can lead to cracking of the ceramic component.

The optical properties of ceramics can be controlled by various formulations and control of structure, imparting different degrees of transparency and colors. Single-crystal sapphire, for example, is completely transparent, zirconia is white, and fine-grained polycrystalline aluminum oxide is a translucent gray. Porosity influences the optical properties of ceramics, much like trapped air in ice cubes, which makes the ice less transparent and gives it a white appearance.

Although basically they are resistors, it has been shown that ceramics can be made electrically conducting by alloying them with certain elements, thus making the ceramic act like a semiconductor or even a superconductor.

Example: Effect of porosity on properties.

If a fully dense ceramic has the properties of $UTS_0 = 100$ MPa, $E_0 = 400$ GPa, and $k_0 = 0.5$ W/m · K, what are these properties at 10% porosity? Let $n = 5$ and $P = 0.1$.

SOLUTION. Using Eqs. (8.1)–(8.3), we have

$$UTS = 100e^{-(5)(0.1)} = 61 \text{ MPa},$$

$$E = 400[1 - (1.9)(0.1) + (0.9)(0.1)^2] = 328 \text{ GPa},$$

and

$$k = 0.5(1 - 0.1) = 0.45 \text{ W/m} \cdot \text{K}.$$

8.3.3 Applications

As we observe in Tables 8.1 and 8.3, ceramics have numerous consumer and industrial applications. Several types of ceramics are used in the electrical and electronics industry because of their high electrical resistivity, dielectric strength (voltage required for electrical breakdown per unit thickness), and magnetic properties suitable for applications such as magnets for speakers. An example is

TABLE 8.3
TYPICAL CERAMICS FOR SPECIFIC FUNCTIONS AND APPLICATIONS

FUNCTION	APPLICATION	MATERIAL(S)
Mechanical	Various	Silicon carbide, silicon nitride, transformation-toughened alumina and zirconia, silicon-carbide whisker reinforced alumina, and silicon-carbide fiber reinforced glass ceramic
Tribological	Various	Monolithic alumina, silicon carbide
	Coatings	Carbide, nitride, oxide, and diamond
Electromechanical	Transducers and micropositioners	Pb-Zr-Ti-O
Electronic	Electronic substrates	Alumina
	Capacitors	Barium titanate
	Piezoelectric filters	Pb-Zr-Ti-O
	Radio-frequency absorbers	Mg-Al-Si-O
	Solid electrolytes	Na-Al-O
	Over-voltage protectors	ZnO
Magnetic	Soft magnets	Zn-Mn-Fe-O
	Hard magnets	Sr-Fe-O
Optical	Lasers	Alumina-Cr
	Light guides	Silica
	Infrared detectors	Ba-Na-Nb-O
	Memory systems	Pb-La-Zr-Ti-O
Thermal/chemical	Hot-gas filters	Microporous alumina
	Oxygen sensor	Zirconia
	Refractories	Fibrous silica/alumina
	Thermal-barrier coatings	Zirconia

Source: Advanced Materials and Processes, July 1989.

porcelain, which is a white ceramic composed of kaolin, quartz, and feldspar. Certain ceramics also have good piezoelectric properties.

The capability of ceramics to maintain their strength and stiffness at elevated temperatures (Figs. 8.3 and 8.4) makes them very attractive for high-temperature applications. Their high resistance to wear makes them suitable for applications such as cylinder liners, bushings, seals, and bearings. The higher operating temperatures made possible by the use of ceramic components means more efficient fuel burning and reduced emissions. Currently, internal combustion engines are only about 30 percent efficient, but with the use of ceramic components the operating performance can be improved by at least 30 percent.

Ceramics being used successfully, especially in gasoline and diesel engine components and as rotors (Fig. 8.5), are silicon nitride, silicon carbide, and partially stabilized zirconia. Coating metal with ceramics is another application, which may be done to reduce wear, prevent corrosion, and provide a thermal barrier. The tiles on the space shuttle, for example, are made of silica fibers with an open cellular structure that consists of 5 percent silica. The rest of the tile structure is air, thus making the tile not only very lightweight but also an excellent heat barrier. The tiles (34,000 on each shuttle) are bonded to the aluminum structure of the space shuttle with several layers of silicon-based adhesives. The skin temperature on the shuttle reaches 1400 °C (2550 °F) due to frictional heat with the atmosphere.

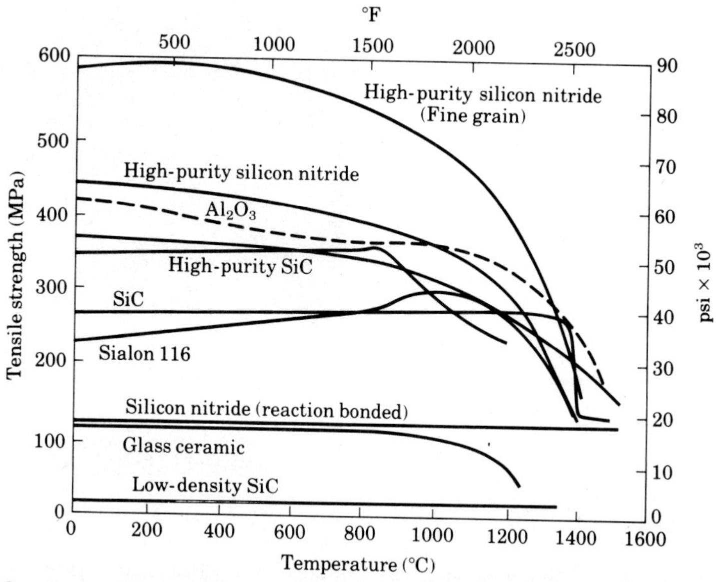

FIGURE 8.3 Effect of temperature on the strength of various engineering ceramics. Note that much of the strength is maintained at high temperatures.

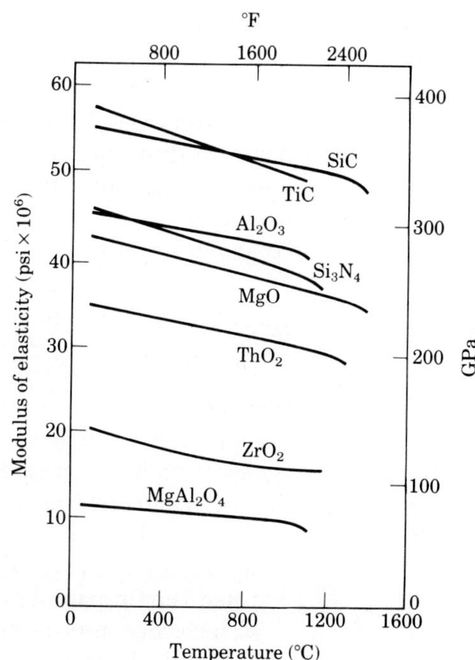

FIGURE 8.4 Effect of temperature on the modulus of elasticity for several ceramics. *Source:* D. W. Richerson, *Modern Ceramic Engineering.* New York: Marcel Dekker, Inc., 1982.

FIGURE 8.5
Ceramic stator for a small gas turbine. This part, made in one piece from silicon carbide, is 130 mm (5 in.) in diameter. *Source:* Courtesy of Norton Company—Advanced Ceramics.

Other attractive properties of ceramics are their low density and high elastic modulus. Thus engine weight can be reduced and, in other applications, the inertial forces generated by moving parts are lower. Ceramic turbochargers, for example, are about 40 percent lighter than conventional ones. High-speed components for machine tools are also candidates for ceramics. Furthermore, the higher elastic modulus of ceramics makes them attractive for improving the stiffness, while reducing the weight, of machines. Ceramics are also used as ball bearings and rollers. Because of their strength and inertness, ceramics are used as biomaterials (*bioceramics*) to replace joints in the human body, as prosthetic devices, and for dental work. Furthermore, ceramic implants can be made porous, thus enabling bones to grow into the porous structure (as is also done with porous titanium implants), developing a strong bond with high structural integrity between them. Commonly used bioceramics are aluminum oxide, silicon nitride, and various compounds of silica.

• Example: Selection of materials for the outer surfaces of the space shuttle orbiter.

During ascent and reentry, the space shuttle orbiter is subjected to surface temperatures as high as 1400 °C (2500 °F). To protect the surfaces of the craft against such high temperatures, two design approaches were considered. One was the use of ablative materials. Ablation is the process of extracting heat from a surface by allowing the surface to erode by melting and evaporation at high temperatures. However, this means that material is lost from the surface of the craft and that certain parts have to be replaced after each space mission.

The second approach, which was the one selected, was to use materials that can withstand high temperatures, so that parts do not have to be replaced after each mission. Because most materials undergo severe oxidation at these temperatures, carbon-reinforced composite materials were selected, with a silicon-carbide base coating. Silicon carbide is a hard refractory ceramic, with a decomposition temperature of 2300–2500 °C (4200–4500 °F) and is therefore suitable for this application.

Each orbiter has more than 34,000 ceramic tiles, each with a special shape, bonded into individual nesting places on its aluminum skin. The bonding material consists of several layers of silicon-base adhesive.

8.4 Glasses

Glass is an amorphous solid with the structure of a liquid. In other words, it has been *supercooled*, that is, cooled at a rate too high for crystals to form. Generally, we define glass as an inorganic product of fusion that has cooled to a rigid condition without crystallizing. Glass has no distinct melting or freezing point; thus its behavior is similar to amorphous polymers (Section 7.2).

Glass beads were produced in about 2000 B.C., followed by glass blowing in about 200 B.C. Silica was used for all glass products until the late 1600s. Rapid developments in glasses began in the early 1900s. Presently there are some 750 different types of commercially available glasses. The uses of glass range from window glass, bottles, and cookware to glasses with special mechanical, electrical, high-temperature, chemical, corrosion, and optical characteristics. Special glasses are used in fiber optics for communication by light with little loss in signal power and in glass fibers with very high strength for reinforced plastics (Section 9.2).

All glasses contain at least 50 percent silica, which is known as a **glass former**. The composition and properties of glasses, except strength, can be modified greatly by the addition of oxides of aluminum, sodium, calcium, barium, boron, magnesium, titanium, lithium, lead, and potassium. Depending on their function, these oxides are known as *intermediates*, or *modifiers*. Glasses are generally resistant to chemical attack, and are ranked by their resistance to acid, alkali, or water corrosion.

8.4.1 Types of glasses

Almost all commercial glasses are categorized by type (Table 8.4):

1. Soda–lime glass (the most common).
2. Lead–alkali glass.
3. Borosilicate glass.
4. Aluminosilicate glass.
5. 96 percent silica glass.
6. Fused silica.

TABLE 8.4
VARIOUS PROPERTIES OF GLASSES

	SODA–LIME GLASS	LEAD GLASS	BOROSILICATE GLASS	96 PERCENT SILICA	FUSED SILICA
Density	High	Highest	Medium	Low	Lowest
Strength	Low	Low	Moderate	High	Highest
Resistance to thermal shock	Low	Low	Good	Better	Best
Electrical resistivity	Moderate	Best	Good	Good	Good
Hot workability	Good	Best	Fair	Poor	Poorest
Heat treatability	Good	Good	Poor	None	None
Chemical resistance	Poor	Fair	Good	Better	Best
Impact-abrasion resistance	Fair	Poor	Good	Good	Best
Ultraviolet-light transmission	Poor	Poor	Fair	Good	Good
Relative cost	Lowest	Low	Medium	High	Highest

Glasses are also classified as colored, opaque (white and translucent), multiform (variety of shapes), optical, photochromatic (darkens when exposed to light, as in sunglasses), photosensitive (changing from clear to opal), fibrous (drawn into long fibers, as in fiberglass), and foam or cellular glass (containing bubbles, thus a good thermal insulator).

Glasses are referred to as hard and soft, usually in the sense of a thermal property rather than mechanical, as in hardness. Thus a soft glass softens at a lower temperature than does a hard glass. Soda–lime and lead–alkali glasses are considered soft and the rest as hard.

8.4.2 Mechanical properties

For all practical purposes, we regard the behavior of glass, as for most ceramics, as perfectly elastic and brittle. The modulus of elasticity range for most commercial glasses is 55–90 GPa (8–13 million psi), and their Poisson's ratio 0.16–0.28. Hardness of glasses, as a measure of resistance to scratching, ranges from 5 to 7 on the Mohs scale, equivalent to a range of approximately 350–500 HK.

Glass in bulk form has a strength of less than 140 MPa (20 ksi). The relatively low strength of bulk glass is attributed to the presence of small flaws and microcracks on its surface, some or all of which may be introduced during normal handling of the glass by inadvertent abrading. These defects reduce the strength of glass by two to three orders of magnitude, compared to its ideal (defect free) strength. Glasses can be strengthened by thermal or chemical treatments to obtain high strength and toughness.

The strength of glass can theoretically reach as high as 35 GPa (5 million psi). When molten glass is freshly drawn into fibers (fiberglass), its tensile strength ranges from 0.2 GPa to 7 GPa (30 ksi to 1000 ksi), with an average value of about 2 GPa (300 ksi). Thus glass fibers are stronger than steel and are used to reinforce plastics in applications such as boats, automobile bodies, furniture, and sports equipment. The strength of glass is usually measured by bending it. The surface of the glass is first thoroughly abraded (roughened) to ensure that the test gives a reliable strength level in actual service under adverse conditions. The phenomenon of static fatigue observed in ceramics is also exhibited by glasses. If a glass item must withstand a load for 1000 hours or longer, the maximum stress that can be applied to it is approximately one-third the maximum stress that the same item can withstand during the first second of loading.

8.4.3 Physical properties

Glasses have low thermal conductivity and high electrical resistivity and dielectric strength. Their thermal expansion coefficient is lower than those for metals and plastics, and may even approach zero. Titanium silicate glass (a clear synthetic high-silica glass), for example, has a near-zero coefficient of expansion. Fused silica, a clear synthetic amorphous silicon dioxide of very high purity, also has a near-zero coefficient of expansion (see Fig. 8.2). Optical properties of glasses, such as reflection, absorption, transmission, and refraction, can be modified by varying their composition and treatment.

8.5 Glass Ceramics

Although glasses are amorphous, **glass ceramics** (such as Pyroceram, a trade name) have a high crystalline component to their microstructure. Glass ceramics contain large proportions of several oxides, and thus their properties are a combination of those for glass and ceramics. Most glass ceramics are stronger than glass. These products are first shaped and then heat treated, with **devitrification** (recrystallization) of the glass occurring. Unlike most glasses, which are clear, glass ceramics are generally white or gray in color.

The hardness of glass ceramics ranges approximately from 520 HK to 650 HK. They have a near-zero coefficient of thermal expansion; hence they have good thermal shock resistance and are strong because of the absence of porosity usually found in conventional ceramics. The properties of glass ceramics can be improved by modifying their composition and by heat-treatment techniques. First developed in 1957, glass ceramics are suitable for cookware, heat exchangers for gas-turbine engines, radomes (housings for radar antenna), and electrical and electronics applications.

8.6 Graphite

Graphite is a crystalline form of carbon having a *layered structure* of basal planes or sheets of close-packed carbon atoms. Consequently, graphite is weak when sheared along the layers. This characteristic, in turn, gives graphite its low frictional properties as a solid lubricant. However, frictional properties are low only in an environment of air or moisture; graphite is abrasive and a poor lubricant in vacuum.

Although brittle, graphite has high electrical and thermal conductivity and resistance to thermal shock and high temperature (although it begins to oxidize at 500 °C (930 °F)). It is therefore an important material for applications such as electrodes, heating elements, brushes for motors, high-temperature fixtures and furnace parts, mold materials such as crucibles for melting and casting of metals, and seals (because of low friction and wear). Unlike other materials, the strength and stiffness of graphite increase with temperature.

An important use of graphite is as fibers in composite materials and reinforced plastics. A characteristic of graphite is its resistance to chemicals; hence it is used as filters for corrosive fluids. Also, its low thermal neutron absorption cross-section and high scattering cross-section make graphite suitable for nuclear applications. Ordinary pencil "lead" is a graphite and clay mixture.

Graphite is generally graded in decreasing order of grain size: industrial, fine grain, and micrograin. As in ceramics, the mechanical properties of graphite improve with decreasing grain size. Micrograin graphite can be impregnated with copper and is used as electrodes for electrical discharge machining and for furnace fixtures. Amorphous graphite is known as lampblack (black soot) and is used as a pigment. Graphite is usually processed by molding or forming, oven baking, and then machining to the final shape. It is available commercially in square, rectangular, or round shapes of various sizes.

8.7 Diamond

The second principal form of carbon is **diamond**, which has a covalently bonded structure. It is the hardest substance known (7000–8000 HK). This characteristic makes diamond an important cutting-tool material, as a single crystal or in polycrystalline form, as an abrasive in grinding wheels for grinding hard materials, and for dressing of grinding wheels (sharpening of abrasive grains). Diamond is also used as a die material for drawing thin wire of less than 0.06 mm (0.0025 in.) in diameter. Diamond is brittle and it begins to decompose in air at about 700 °C (1300 °F). In nonoxidizing environments, it resists higher temperatures.

Synthetic, or **industrial diamond**, was first made in 1955 and is used extensively for industrial applications. One method of manufacturing it is to subject graphite to a hydrostatic pressure of 14 GPa (2 million psi) and a temperature of 3000 °C (5400 °F). Synthetic diamond is identical to natural diamond, and has superior properties because of lack of impurities. It is available in various sizes and shapes, the most common abrasive grain size being 0.01 mm (0.004 in.) in diameter.

Gem-quality synthetic diamond is now being made with electrical conductivity 50 times higher than that for natural diamond and 10 times more resistant to laser damage. Potential applications are for heat sinks for computers, telecommunications, integrated-circuit industries, and windows for high-power lasers.

SUMMARY

Several nonmetallic materials are of great importance in engineering applications and manufacturing processes. Ceramics, which are compounds of metallic and nonmetallic materials, are generally characterized by high hardness and compressive strength, high-temperature resistance, and chemical inertness. These properties make ceramics particularly attractive for heat-engine components, cutting tools, and components requiring wear and corrosion resistance.

Glasses are supercooled liquids; that is, the rate of cooling is so high that they do not have time to solidify into a crystalline structure. They do not have a clearly defined melting point. Glasses are available in a wide variety of compositions and mechanical, physical, and optical properties. Their strength can be improved by thermal or chemical treatments. Glass ceramics are predominantly crystalline in structure and have better properties than glasses.

Graphite has high-temperature and electrical applications and is also used as fibers to reinforce plastics. Diamond, the hardest substance known, is used as cutting tools, dies for wire drawing, and abrasives for grinding wheels.

SUMMARY TABLE 8.1

TYPES AND GENERAL CHARACTERISTICS OF CERAMICS

TYPE	GENERAL CHARACTERISTICS
Oxide ceramics	
Alumina	High hardness, moderate strength; most widely used ceramic; cutting tools, abrasives, electrical and thermal insulation.
Zirconia	High strength and toughness; thermal expansion close to cast iron; suitable for heat engine components.
Carbides	
Tungsten carbide	Hardness, strength, and wear resistance depend on cobalt binder content; commonly used for dies and cutting tools.
Titanium carbide	Not as tough as tungsten carbide; has nickel and molybdenum as the binder; used as cutting tools.
Silicon carbide	High-temperature strength and wear resistance; used for heat engines and as abrasives.

(continued)

SUMMARY TABLE 8.1 (*continued*)

TYPES AND GENERAL CHARACTERISTICS OF CERAMICS

TYPE	GENERAL CHARACTERISTICS
Nitrides	
Cubic boron nitride	Second hardest substance known, after diamond; used as abrasives and cutting tools.
Titanium nitride	Used as coatings because of low frictional characteristics; gold in color.
Silicon nitride	High resistance to creep and thermal shock; used in heat engines.
Sialon	Consists of silicon nitrides and other oxides and carbides; used as cutting tools.
Cermets	Consist of oxides, carbides, and nitrides; high temperature applications.
Silica	High temperature resistance; quartz exhibits piezoelectric effect; silicates, containing various oxides, are used in high-temperature nonstructural applications.
Glasses	Contain at least 50 percent silica; amorphous structures; several types available with a range of mechanical and physical properties.
Glass ceramics	Have a high crystalline component to their structure; good thermal-shock resistance and strong.
Graphite	Crystalline form of carbon; high electrical and thermal conductivity, good thermal-shock resistance.
Diamond	Hardest substance known; available as single crystal or polycrystalline form; used as cutting tools and abrasives and as dies for fine wire drawing.

TRENDS

- Major efforts are under way to improve the strength, toughness, fatigue, and resistance to corrosion, wear, and thermal shock of ceramics, as well as the reproducibility of their properties and reliability in service.
- Standardized and nondestructive test methods and techniques are being developed for inspection and detection of flaws in ceramic components and assessment impact and fatigue damage to components.
- New methods, such as laser treatments and ion implantation, are being developed to modify ceramic surfaces and their properties.
- The use of ceramics is being extended to numerous structural (load bearing) and prosthetic applications.
- Mass-production techniques are being developed to reduce the costs of ceramic components.
- Further developments are taking place in the use of fiber optics and glass ceramics.

KEY TERMS

Alumina
Carbides
Ceramics
Cermets
Clay
Devitrification
Diamond
Feldspar
Flint
Glass
Glass ceramics
Glass former
Graphite
Industrial diamond
Nitrides
Partially stabilized zirconia
Porcelain
Porosity
Sialon
Silica
Static fatigue
Zirconia

BIBLIOGRAPHY

Doremus, R.H., *Glass Science*. New York: Wiley, 1973.

Encyclopedia of Glass, Ceramics, Clay and Cement. New York: Wiley, 1984.

Kingery, W.D., H.K. Bowen, and D.R. Uhlmann, *Introduction to Ceramics*, 2d ed. New York: Wiley, 1976.

Kirchner, H.P., *Strengthening of Ceramics: Treatments, Tests, and Design Applications*. New York: Marcel Dekker, 1979.

McColm, I.J., *Ceramic Science for Materials Technologists*. Glasgow: Leonard Hill (Chapmann and Hall), 1983.

McLellen, G.W., and E.B. Shand, *Glass Engineering Handbook*. New York: McGraw-Hill, 1984.

Norton, F.H., *Elements of Ceramics*, 2d ed. Reading, Mass.: Addison-Wesley, 1974.

Richerson, D.W., *Modern Ceramic Engineering*. New York: Marcel Dekker, 1982.

Samsonov, C.V., and J.M. Vinitsku, *Handbook of Refractory Compounds*. New York: Plenum, 1980.

Schwartz, M.M. (ed.), *Engineering Applications of Ceramic Materials, Source Book*. Metals Park, Ohio: American Society for Metals, 1985.

REVIEW QUESTIONS

8.1 Compare the major differences between the properties of ceramics and those of metals and plastics.

8.2 List the major types of ceramics that are useful in engineering applications.

8.3 What do the following materials consist of? (a) Carbides. (b) Cermets. (c) Sialon.

8.4 List the major limitations of ceramics.

8.5 What is porcelain?

8.6 What is glass? Why is it called a supercooled material?

8.7 What is devitrification?
8.8 List the major types of glasses and their applications.
8.9 What is static fatigue?
8.10 Describe the major uses of graphite.

QUESTIONS AND PROBLEMS

8.11 Explain why ceramics are weaker in tension than in compression.
8.12 Make a list of all the ceramics parts that you can find around your house and in your car. Explain why those parts are made of ceramics.
8.13 Assume that you are in technical sales and are fully familiar with all the advantages and limitations of ceramics. Which of the traditional markets using nonceramic materials do you think ceramics can penetrate? What would you say to your potential customers during your sales visits? What kind of questions do you think they will ask?
8.14 What are the advantages of cermets? Suggest additional applications to those given in the text.
8.15 Explain why the electrical and thermal conductivity of ceramics decreases with increasing porosity.
8.16 Describe applications where a ceramic material with near-zero coefficient of thermal expansion would be desirable.
8.17 Write a brief technical brochure on glass, including its advantages and limitations, suitable for high school students whose professional interests are not in technical areas.
8.18 We have shown that the modulus of elasticity of ceramics is maintained at elevated temperatures. What engineering applications could benefit from this characteristic?
8.19 Explain why the mechanical property data in Table 8.2 has such a broad range. What is its significance in engineering practice?
8.20 What reasons can you think of that encouraged the development of synthetic diamond?
8.21 List the factors that you would take into account when replacing a metal component with a ceramic component.
8.22 Explain why the mechanical properties of ceramics are generally higher than those for metals.
8.23 How are ceramics made tougher?
8.24 Describe applications in which static fatigue can be important.
8.25 How does porosity affect mechanical properties of ceramics, and why?
8.26 What properties are important in making heat-resistant ceramics for oven tops?
8.27 Describe the differences between the properties of glasses and ceramics.
8.28 We have stated that a large variety of glasses are available. Why is this so?
8.29 What is the difference between the structures of graphite and diamond? Is it important? Explain.
8.30 Assume that you are in charge of the research and development facilities of a company that manufactures ceramic products. Make a list of topics that you would want your staff to investigate. Give your reasons for choosing those topics.
8.31 Plot the UTS, E, and k values for ceramics as a function of porosity P, and describe and explain the trends that you observe in their behavior.

9

Composite Materials: Structure, General Properties, and Applications

9.1 Introduction

Among the most important developments in materials in recent years are **composite materials**. We define these materials as a combination of two or more chemically distinct and insoluble phases whose properties and structural performance are superior to those of the constituents acting independently. We showed that plastics, for example, possess mechanical properties (particularly strength, stiffness, and creep resistance) that are generally inferior to those for metals and alloys. These properties can be improved by embedding reinforcements of various types, such as glass or graphite fibers, to produce **reinforced plastics** (Fig. 9.1). As shown in Table 7.1, reinforcements improve the strength, stiffness, and creep resistance of plastics—and their strength-to-weight and stiffness-to-weight ratios.

Composite materials have found increasingly wider applications in aircraft (Fig. 9.2), automobiles, boats, ladders, and sporting goods. Metals and ceramics also can be embedded with fibers or particles to improve their properties.

The oldest example of composites is the addition of straw to clay for making

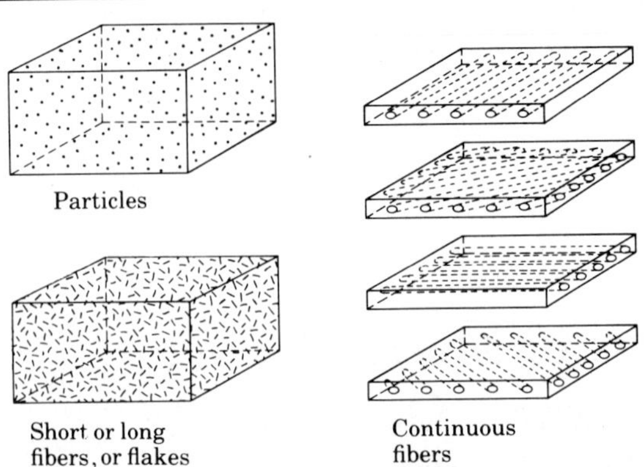

FIGURE 9.1
Schematic illustration of several methods of reinforcing plastics. The plastic is the matrix in these structures. The four layers in the illustration of continuous fibers are assembled into a laminate, similar to plywood.

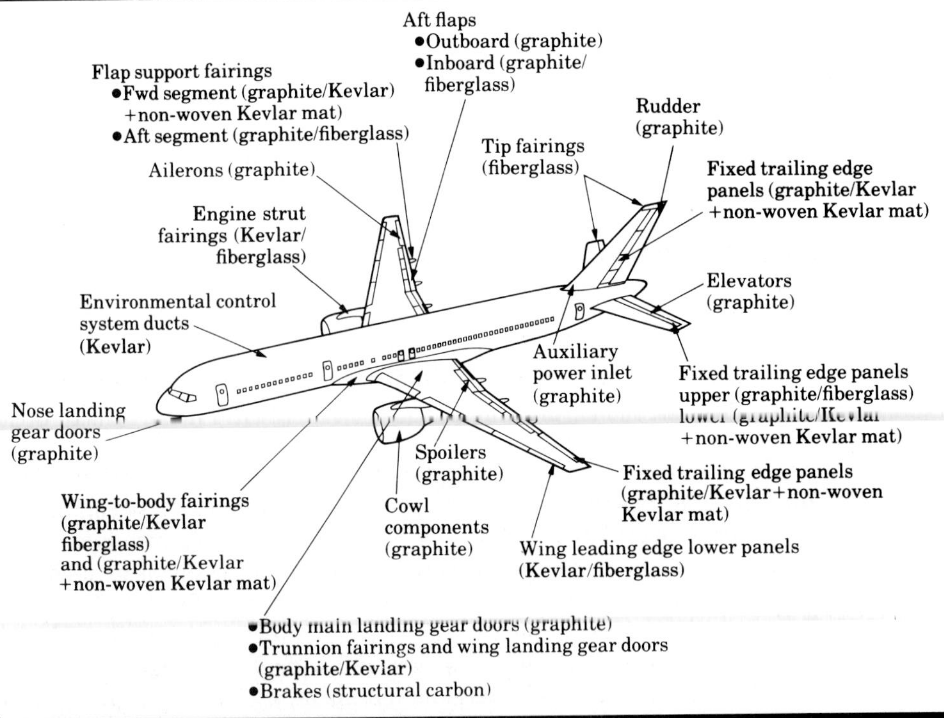

FIGURE 9.2
Application of advanced composite materials in Boeing 757-200 commercial aircraft. *Source:* Boeing Commercial Airplane Company.

mud huts and bricks for structural use, dating back to 4000 B.C. In that application the straws are the reinforcing fibers, and the clay is the matrix. Another example of a composite material is the reinforcing of masonry and concrete with iron rods, begun in the 1800s. In fact, concrete itself is a composite material, consisting of cement, sand, and gravel. In reinforced concrete, steel rods impart the necessary tensile strength to the composite, since concrete is brittle and generally has little or no useful tensile strength.

In this chapter, we describe the structure of composite materials, the type of reinforcing fibers used and their characteristics, and various applications of these materials. We discuss the processing and shaping of composite materials in Chapter 18.

9.2 Structure of Reinforced Plastics

Reinforced plastics consist of **fibers** (the discontinuous or dispersed phase) in a plastic **matrix** (the continuous phase). Commonly used fibers are glass, graphite, aramids, and boron. These fibers are strong and stiff (Table 9.1) and have high specific strength (strength-to-weight ratio) and specific modulus (stiffness-to-weight ratio), as shown in Fig. 9.3. However, they are generally brittle and abrasive and lack toughness. Thus fibers, by themselves, have little structural value. The plastic matrix is less strong and less stiff but tougher than the fibers. Thus reinforced plastics combine the advantages of each of the two constituents. When more than one type of fiber is used in a reinforced plastic, the composite is called a *hybrid*, which generally has better properties yet.

TABLE 9.1
PROPERTIES OF REINFORCING FIBERS

TYPE	TENSILE STRENGTH (MPa)	ELASTIC MODULUS (GPa)	DENSITY (kg/m^3)	RELATIVE COST
Boron	3500	380	2600	Highest
Carbon				
High strength	3000	275	1900	Low
High modulus	2000	415	1900	Low
Glass				
E type	3500	73	2480	Lowest
S type	4600	85	2540	Lowest
Kevlar				
29	2800	62	1440	High
49	2800	117	1440	High

Note: These properties vary significantly depending on the material and method of preparation.

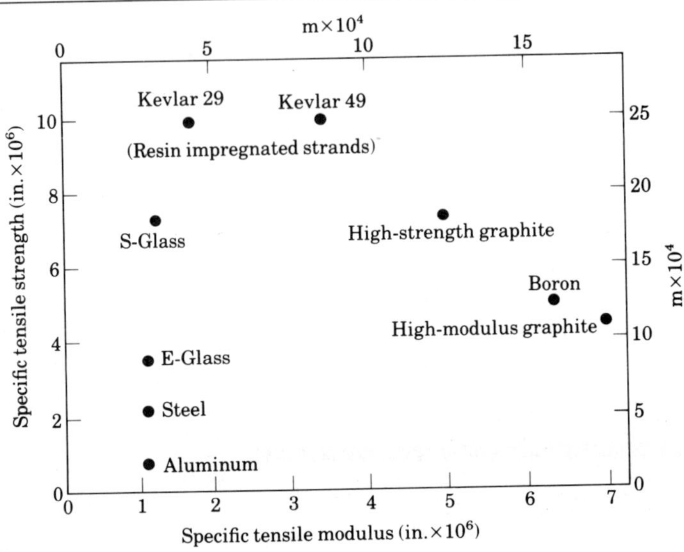

FIGURE 9.3 Specific tensile strength (tensile strength-to-density ratio) and specific tensile modulus (modulus of elasticity-to-density ratio) for various fibers used in reinforced plastics. Note the wide range of specific strengths and stiffnesses available.

In addition to high specific strength and specific modulus, reinforced plastic structures have improved fatigue resistance, greater toughness, and higher creep resistance than unreinforced plastics. These structures are relatively easy to design, fabricate, and repair.

The percentage of fibers (by volume) in reinforced plastics usually ranges between 10 percent and 60 percent. Practically, the percentage of fiber in a matrix is limited by the average distance between adjacent fibers or particles. The highest practical fiber content is 65 percent; higher percentages generally result in diminished structural properties.

9.2.1 Reinforcing fibers

Glass. Glass fibers are the most widely used and least expensive of all fibers. The composite material is called *glass-fiber reinforced plastic* (GFRP) and may contain between 30 percent and 60 percent glass fibers by volume. Glass fibers are made by drawing molten glass through small openings in a platinum die. There are two principal types of glass fibers: (1) the E type, a borosilicate glass, which is used most; and (2) the S type, a magnesia-alumina-silicate glass, which has higher strength and stiffness and is more expensive.

Graphite. Graphite fibers (Fig. 9.4a), although more expensive than glass fibers, have a combination of low density, high strength, and high stiffness. The product is called *carbon-fiber reinforced plastic* (CFRP). All graphite fibers are made by pyrolysis of organic **precursors**, commonly polyacrylonitrile (PAN)

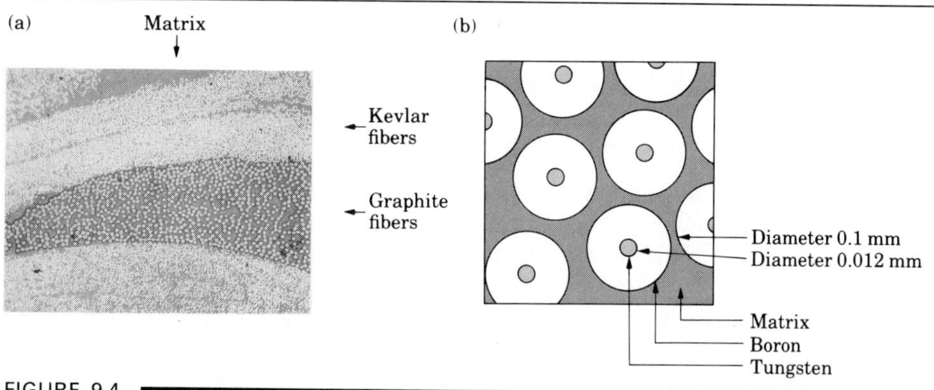

FIGURE 9.4
(a) Cross-section of a tennis racket, showing graphite and aramid (Kevlar) reinforcing fibers. *Source:* J. Dvorak, Mercury Marine Corporation, and F. Garrett, Wilson Sporting Goods Co. (b) Cross-section of boron-fiber reinforced composite material.

because of its lower cost. Rayon and pitch (the residue from catalytic crackers in petroleum refining) can also be used as precursors. **Pyrolysis** is the term for inducing chemical changes by heat, such as burning a length of yarn, which becomes carbon and black in color. The temperatures for carbonizing range up to about 3000 °C (5400 °F).

The difference between carbon and graphite, although the words are often used interchangeably, depends on the temperature of pyrolysis and the purity of the material. Carbon fibers are generally 93–95 percent carbon, and graphite fibers are usually more than 99 percent carbon.

Aramids. Marketed under the trade name Kevlar, **aramids** are the toughest fibers available and have the highest specific strength of any fiber (Fig. 9.3). They can undergo some plastic deformation before fracture and thus have higher toughness than brittle fibers. However, aramids absorb moisture, which reduces their properties and complicates their application, as hygrothermal stresses must be considered.

Boron. Boron fibers consist of boron deposited (by chemical vapor-deposition techniques) on tungsten fibers (Fig. 9.4b), although boron can also be deposited on carbon fibers. These fibers have favorable properties, such as high strength and stiffness in tension and compression and resistance to high temperatures. However, because of the use of tungsten, they have high density and are expensive, thus increasing the cost and weight of the reinforced plastic component.

Other fibers. Other fibers that are being used are nylon, silicon carbide, silicon nitride, aluminum oxide, sapphire, steel, tungsten, molybdenum, boron carbide, boron nitride, and tantalum carbide. **Whiskers** are also used as reinforcing fibers. They are tiny needlelike single crystals that grow to 1 μm to 10 μm (40 μin. to 400 μin.) in diameter and have aspect ratios (length to diameter) ranging from 100 to 15,000. Because of their small size, either they are free of imperfections or the imperfections they contain do not significantly affect their strength, which approaches the theoretical strength of the material.

9.2.2 Fiber size and length

The mean diameter of fibers used in reinforced plastics is usually less than 0.01 mm (0.0004 in.). The fibers are very strong and rigid in tension. The reason is that the molecules in the fibers are oriented in the longitudinal direction, and their cross-sections are so small that the probability is low that any defects exist in the fiber. Glass fibers, for example, can have tensile strengths as high as 4600 MPa (650 ksi), whereas the strength of glass in bulk form is much lower. Thus glass fibers are stronger than steel.

Fibers are classified as short or long, both also called *chopped fibers*. Short fibers generally have an aspect ratio between 20 and 60 and long fibers between 200 and 500. In addition to the discrete fibers that we have described, reinforcements in composites may be in the form of continuous *roving* (slightly twisted strand of fibers), *yarn* (twisted strand), *woven* fabric (similar to cloth), and *mats* of various combinations. Reinforcing elements may also be in the form of particles and flakes.

9.2.3 Matrix materials

The matrix in reinforced plastics has three functions:

1. Support and transfer the stresses to the fibers, which carry most of the load.
2. Protect the fibers against physical damage and the environment.
3. Prevent propagation of cracks in the composite by virtue of the ductility and toughness of the plastic matrix.

Matrix materials are usually epoxy, polyester, phenolic, fluorocarbon, polyethersulfone, and silicon. The most commonly used are epoxies (80 percent of all reinforced plastics) and polyesters, which are less expensive than epoxies. Polyimides, which resist exposure to temperatures in excess of 300 °C (575 °F), are being developed for use with graphite fibers. Some thermoplastics, such as polyetheretherketone, are also being developed as matrix materials. They generally have higher toughness than thermosets, but their resistance to temperature is lower, being limited to 100–200 °C (200–400 °F).

9.3 Properties of Reinforced Plastics

The properties of reinforced plastics depend on the kind, shape, and orientation of the reinforcing material, the length of the fibers, and the volume fraction (percentage) of the reinforcing material. Short fibers are less effective than long fibers (Fig.

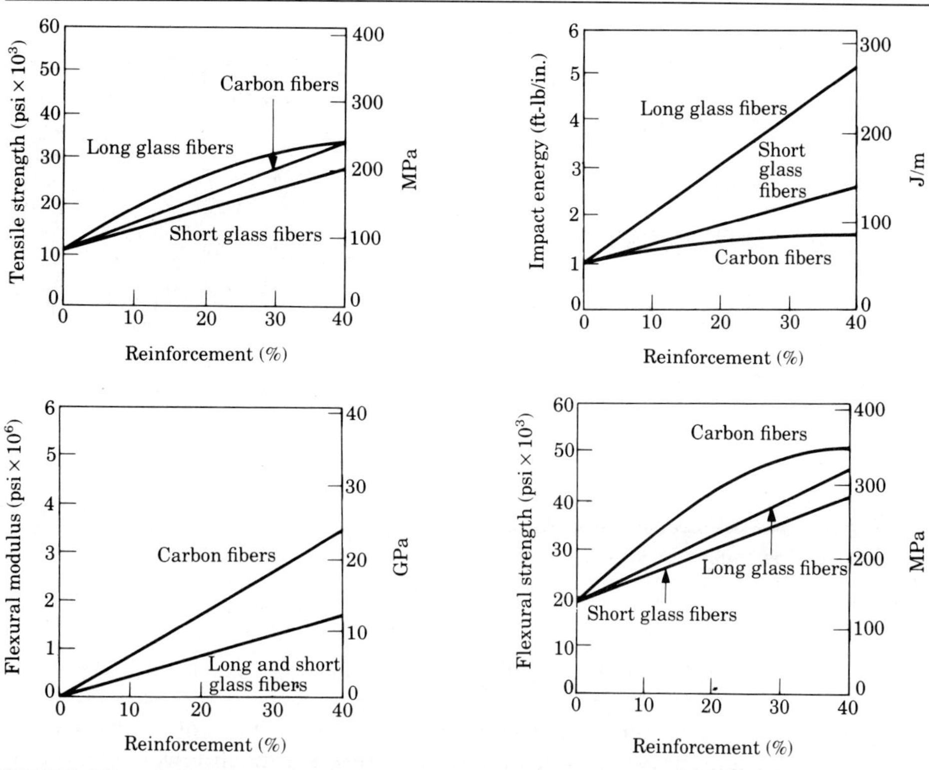

FIGURE 9.5 Effect of the amount of reinforcing fibers and fiber length on the mechanical properties of reinforced nylon. Note the significant improvement with increasing percentage of fiber reinforcement. *Source:* Courtesy of Wilson Fiberfill International.

9.5), and their properties are strongly influenced by time and temperature. Long fibers transmit the load through the matrix better and thus are commonly used in critical applications, particularly at elevated temperatures. Fiber reinforcement also affects the physical and other properties of composites (Fig. 9.6).

A critical factor in reinforced plastics is the strength of the bond between the fiber and the polymer matrix, since the load is transmitted through the fiber–matrix interface. Weak bonding causes **fiber pullout** and **delamination** of the structure, particularly under adverse environmental conditions. Poor bonding in composites is analogous to a brick structure with poor bonding between the bricks and the mortar.

Bonding can be improved by special surface treatments for better adhesion at the interface, such as coatings and the use of coupling agents. Glass fibers, for example, are treated with a chemical called silane (SiH_4) for improved wetting and bonding between the fiber and the matrix. You can appreciate the importance of proper bonding by looking at Figs. 9.7(a) and (b), which show the fracture surfaces of reinforced plastics.

Generally, the greatest stiffness and strength in reinforced plastics is obtained when the fibers are aligned in the direction of the tension force. This composite, of

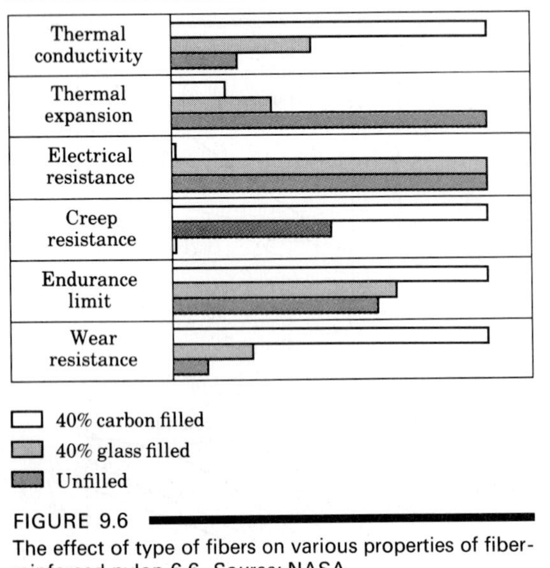

FIGURE 9.6
The effect of type of fibers on various properties of fiber-reinforced nylon 6,6. *Source:* NASA.

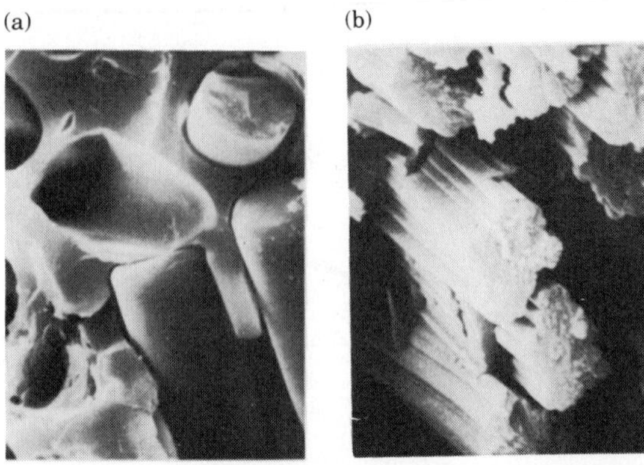

FIGURE 9.7
(a) Fracture surface of glass-fiber reinforced epoxy composite. The fibers are 10 μm (400 μin.) in diameter and have random orientation. (b) Fracture surface of a graphite-fiber reinforced epoxy composite. The fibers, 9–11 μm in diameter, are in bundles and are all aligned in the same direction. *Source:* L. J. Broutman.

course, is highly anisotropic (Fig. 9.8). As a result, other properties of the composite, such as stiffness, creep resistance, thermal and electrical conductivity, and thermal expansion, are also anisotropic. The transverse properties of such a unidirectionally reinforced structure are much lower than the longitudinal. Note, for example, how easily you can split a fiber-reinforced packaging tape, yet how strong it is when you pull on it (tension).

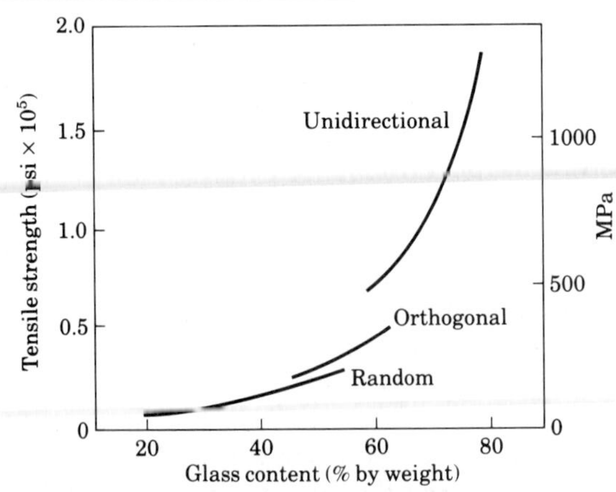

FIGURE 9.8
The tensile strength of glass-reinforced polyester as a function of fiber content and fiber direction in the matrix. *Source:* R. M. Ogorkiewicz, *The Engineering Properties of Plastics.* Oxford: Oxford University Press, 1977.

For a specific service condition, we can give a reinforced plastic part an optimal configuration. For example, if the reinforced plastic part is to be subjected to forces in different directions (such as thin-walled, pressurized vessels), the fibers are crisscrossed in the matrix. Reinforced plastics may also be made with various other materials and shapes of the polymer matrix in order to impart specific properties, such as permeability (ability to diffuse through) and dimensional stability, as well as making processing easier and reducing costs.

• Example: Strength and elastic modulus of reinforced plastics.

The strength of a reinforced plastic with longitudinal fibers can be determined in terms of the strength of the fibers and matrix, respectively, and the volume fraction of fibers in the composite. In the following equations, c refers to the composite, f to the fiber, and m to the matrix. The total strength P_c of the composite is

$$P_c = P_f + P_m, \tag{9.1}$$

which we can write as

$$\sigma_c A_c = \sigma_f A_f + \sigma_m A_m. \tag{9.2}$$

However, we know that $A_c = A_f + A_m$. Let's now denote x as the area fraction of the fibers in the composite. (Note that x also represents the volume fraction, because the fibers are uniformly longitudinal in the matrix.) We may now rewrite Eq. (9.2) as follows:

$$\sigma_c = x\sigma_f + (1 - x)\sigma_m. \tag{9.3}$$

We can now calculate the fraction of the total load carried by the fibers. First, we note that, in the composite under a tension load, the strains sustained by the fibers and the matrix are the same, and then recall from Section 2.2 that

$$e = \frac{\sigma}{E} = \frac{P}{AE}.$$

Consequently,

$$\frac{P_f}{P_m} = \frac{A_f E_f}{A_m E_m}. \tag{9.4}$$

As we know the relevant quantities for a specific situation, and using Eq. (9.1), we can determine the fraction P_f/P_c. Then, using the foregoing relationships, we can also calculate the elastic modulus E_c of the composite by replacing σ in Eq. (9.3) with E. Thus,

$$E_c = xE_f + (1 - x)E_m. \tag{9.5}$$

Let us assume that a graphite–epoxy reinforced plastic with longitudinal fibers contains 20 percent graphite fibers with a strength of 2500 MPa and an elastic modulus of 300 GPa. The strength of the epoxy matrix is 120 MPa, with an elastic modulus of 100 GPa. Calculate the elastic modulus of the composite and fraction of the load supported by the fibers.

SOLUTION. The data given are $x = 0.2$, $E_f = 300$ GPa, $E_m = 100$ GPa, $\sigma_f = 2500$ MPa, and $\sigma_m = 120$ MPa. Using Eq. (9.5), we find that

$$E_c = 0.2(300) + (1 - 0.2)100 = 60 + 80 = 140 \text{ GPa}.$$

We obtain the load fraction P_f/P_m from Eq. (9.4):

$$\frac{P_f}{P_m} = \frac{0.2(300)}{0.8(100)} = \frac{60}{80} = 0.75.$$

Since

$$P_c = P_f + P_m \quad \text{and} \quad P_m = \frac{P_f}{0.75},$$

we find that

$$P_c = P_f + \frac{P_f}{0.75} = 2.33P_f, \quad \text{or} \quad P_f = 0.43P_c.$$

Thus the fibers support 43 percent of the load, even though they occupy only 20 percent of the cross-sectional area (hence volume) of the composite. ●

9.4 Applications

The first application of reinforced plastics (in 1907) was for an acid-resistant tank, made of a phenolic resin with asbestos fibers. Formica, commonly used for counter tops, was developed in the 1920s. Epoxies were first used as a matrix in the 1930s and beginning in the 1940s, boats were made with fiberglass, and reinforced plastics were used for aircraft, electrical equipment, and sporting goods. Major developments in composites began in the 1970s, and these materials are now called *advanced composites.*

Reinforced plastics are typically used in military and commercial aircraft and rocket components, helicopter blades, helmets, automotive bodies, leaf springs, drive shafts, pipes, ladders, pressure vessels, sporting goods, boat hulls, and various other structures. Applications of reinforced plastics include components in DC-10, L-1011, and Boeing 727, 757, and 767 commercial aircraft. By virtue of the resulting weight savings (Table 9.2), reinforced plastics have reduced fuel consumption by about 2 percent. Substituting aluminum in large commercial aircraft with graphite–epoxy reinforced plastics could reduce both weight and production costs by 30 percent, with improved fatigue and corrosion resistance. The structure of the Lear

TABLE 9.2
APPROXIMATE WEIGHT SAVINGS IN THE SUBSTITUTION OF COMMERCIAL AIRCRAFT COMPONENTS WITH GRAPHITE/EPOXY REINFORCED PLASTICS

COMPONENT	WEIGHT SAVING (%)
McDonnell–Douglas DC-10 vertical tail	20
Lockheed L-1011 aileron	26
Boeing 727 elevator	26
McDonnell–Douglas DC-10 rudder	27
Boeing 737 horizontal tail	27
Lockheed L-1011 vertical tail	28

Fan 2100 passenger aircraft is almost totally made of graphite–epoxy reinforced plastic. Nearly 90 percent of the structure of the lightweight Voyager craft, which circled the earth without refueling, was made of carbon-reinforced plastic. Boron-fiber reinforced composites are used in military aircraft, golf club shafts, tennis rackets, fishing rods, and surfboards (Fig. 9.9).

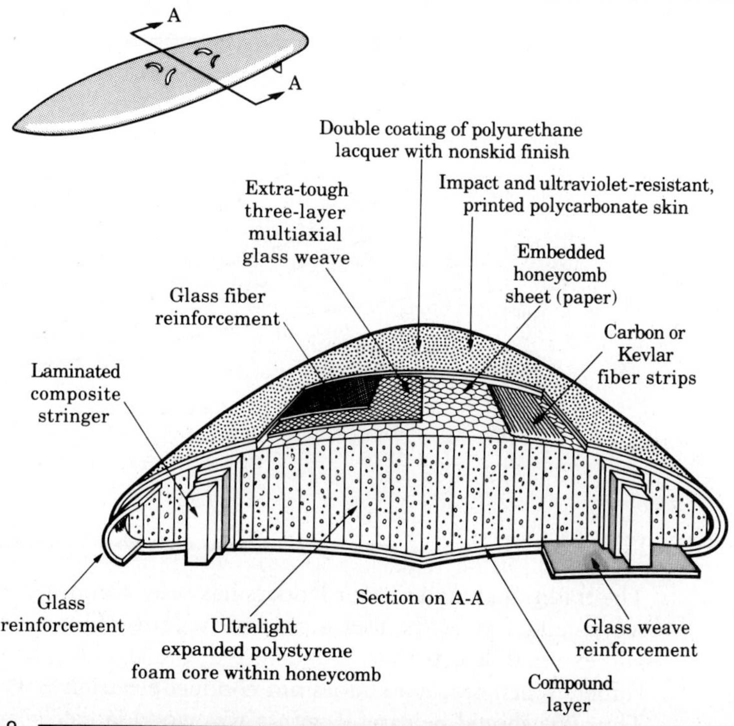

FIGURE 9.9
Cross-section of composite surfboard, an example of advanced materials construction. *Source:* Ken Easterling, *Tomorrow's Materials*, 2d ed., p. 133. Institute of Metals, 1990.

Careful inspection and testing of reinforced plastics is essential in critical applications, in order to ensure that good bonding between the reinforcing fiber and the matrix has been obtained throughout. In some instances, the cost of inspection can be as high as one quarter of the total cost of the composite product.

• Example: Composite blades for helicopters.

Metal blades on helicopters are now being replaced with blades made of composite materials, principally S-glass fibers in an epoxy matrix. These blades have high stiffness, strength, resilience, and temperature and fatigue resistance (see the accompanying figure). They also have high impact strength and consequently, as compared to metal blades, composites can withstand a major ballistic impact without catastrophic failure and the helicopter can return safely to its base. Furthermore, repairs of damaged blades can be made on the field whereas failure of metal blades requires a more extensive maintenance. Note in the figure that S glass has a higher fatigue life than E glass (see also Table 9.1). One million cycles in the figure represents the equivalent of 20- to 30-year blade life on a helicopter. It has also been shown that glass-reinforced composite blades have a better cost/performance basis, and are superior to aramid- or carbon-reinforced composites.

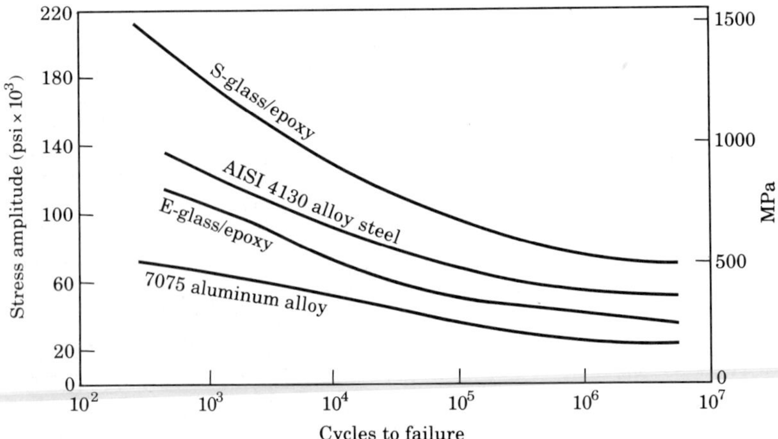

• Example: Fiberglass ladders.

The traditional material for ladders has been wood. Partly because of its weight, a wood ladder gives the user a sense of security. However, wood can have internal and external defects that, if undetected, can significantly reduce the strength of the ladder. When dry, wood does not conduct electricity; when wet, however, it does. Thus one should be careful with a wet wood ladder near electrical wiring.

Aluminum ladders are lightweight and can be designed for high strength and stiffness. They are longer lasting than wood ladders and require little maintenance.

They do, however, conduct electricity.

Composite materials are now used in making ladders for various uses. Fiberglass is the most common reinforcing fiber, with epoxies and polyesters as the matrix. Reinforced plastics are used for the rails of ladders, while the rungs are made of aluminum. These ladders have advantages similar to aluminum, and since they do not conduct electricity they are preferred by electricians. Fiberglass-reinforced ladders have a rugged feel, which gives the user a sense of security, and can be made in various colors. However, fiberglass ladders can absorb moisture and undergo surface weathering when exposed to outdoor environments, especially in warm and moist climates. There is a loss of color and gloss, and the surface appearance changes by fiber prominence due to erosion of the matrix resin and fiber blooming (exposed fibers). Fiberglass ladders can be coated with polyurethane for surface protection. Proper maintenance and periodic inspection of fiberglass ladders are thus important.

9.5 Metal-Matrix and Ceramic-Matrix Composites

New developments in composite materials are continually taking place, with a wide range and form of polymeric, metallic, and ceramic materials being used both as fibers and as matrix materials. Research and development activities in this area are concerned with improving strength, toughness, stiffness, resistance to high-temperature, and reliability in service.

9.5.1 Metal-matrix composites

The advantage of a metal matrix over a polymer matrix is its higher resistance to elevated temperatures and higher ductility and toughness. The limitations are higher density and greater difficulty in processing components. Matrix materials in these composites are usually aluminum, aluminum–lithium, magnesium, and titanium, although other metals are also being investigated. Fiber materials are graphite, aluminum oxide, silicon carbide, and boron, with beryllium and tungsten as other possibilities.

Because of their high specific stiffness, light weight, and high thermal conductivity, boron fibers in an aluminum matrix have been used for structural tubular supports in the space shuttle orbiter. Other applications are bicycle frames and sporting goods. Studies are in progress of techniques for optimal bonding of fibers to the metal matrix. Typical applications for metal-matrix composites are given in Table 9.3. See also Fig. 9.10, which shows trends in the use of these composites for jet engines.

TABLE 9.3
METAL-MATRIX COMPOSITE MATERIALS AND APPLICATIONS

FIBER	MATRIX	APPLICATIONS
Graphite	Aluminum	Satellite, missile, and helicopter structures
	Magnesium	Space and satellite structures
	Lead	Storage-battery plates
	Copper	Electrical contacts and bearings
Boron	Aluminum	Compressor blades and structural supports
	Magnesium	Antenna structures
	Titanium	Jet-engine fan blades
Alumina	Aluminum	Superconductor restraints in fission power reactors
	Lead	Storage-battery plates
	Magnesium	Helicopter transmission structures
Silicon carbide	Aluminum, titanium	High-temperature structures
	Superalloy (cobalt-base)	High-temperature engine components
Molybdenum, tungsten	Superalloy	High-temperature engine components

9.5.2 Ceramic-matrix composites

Ceramic-matrix composites are another important development in engineered materials because of their resistance to high temperatures and corrosive environments. As we described in Chapter 8, ceramics are strong and stiff, resist high temperatures, but generally lack toughness. New matrix materials that retain their strength to 1700 °C (3000 °F) are silicon carbide, silicon nitride, aluminum oxide, and mullite (a compound of aluminum, silicon, and oxygen). Also under develop-

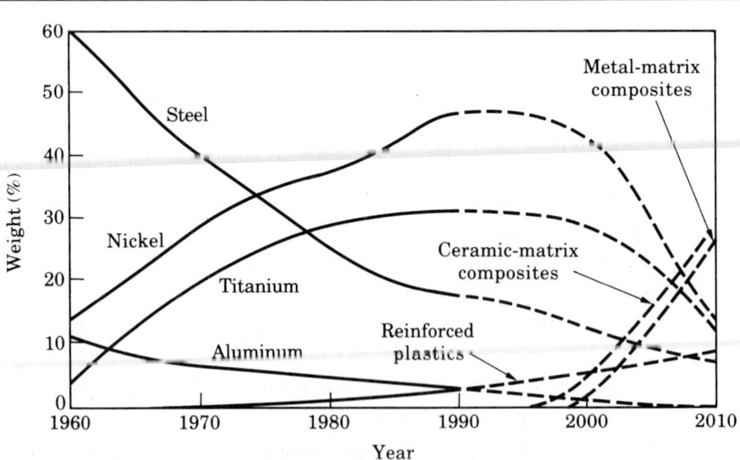

FIGURE 9.10
Trends in jet engine material usage. *Source:* Courtesy of Rolls-Royce plc.

ment are carbon–carbon matrix composites that retain much of their strength up to 2500 °C (4500 °F), although they lack oxidation resistance at high temperatures.

Present applications for ceramic-matrix composites are in jet and automotive engines, deep-sea mining equipment, pressure vessels, and various structural components.

9.6 Other Composites

Composites may also consist of coatings of various kinds on base metals or substrates. Examples are plating of aluminum and other metals over plastics for decorative purposes and enamels, dating to before 1000 B.C., or similar vitreous (glasslike) coatings on metal surfaces for various functional or ornamental purposes.

Composites are made into cutting tools and dies, such as cemented carbides, usually tungsten carbide and titanium carbide, with cobalt and nickel, respectively, as a binder. Other composites are grinding wheels made of aluminum oxide, silicon carbide, diamond, or cubic boron nitride abrasive particles, held together with various organic, inorganic, or metallic binders. Another category of composites is cermets.

A composite developed relatively recently is granite particles in an epoxy matrix. It has high strength, good vibration damping capacity (better than gray cast iron), and good frictional characteristics. It is used as machine-tool beds for some precision grinders (Section 24.4).

9.7 Honeycomb Structures

The **honeycomb structure** (Fig. 9.11) and similar sandwich, or laminate, structures are another form of composites having high specific strength and specific stiffness. The structure consists basically of a core of honeycomb or other corrugated shapes bonded to two thin outer skins. You will readily recognize that the simplest example of such a structure is corrugated cardboard, which has a high stiffness-to-weight ratio and is used extensively in packaging for consumer and industrial goods.

Honeycomb structures are most commonly made of 3000-series aluminum but are also made of titanium, stainless steels, and nickel alloys. New developments include the use of reinforced plastics, such as aramid–epoxy. Bonding between the core and the skins is accomplished either with adhesives or by brazing.

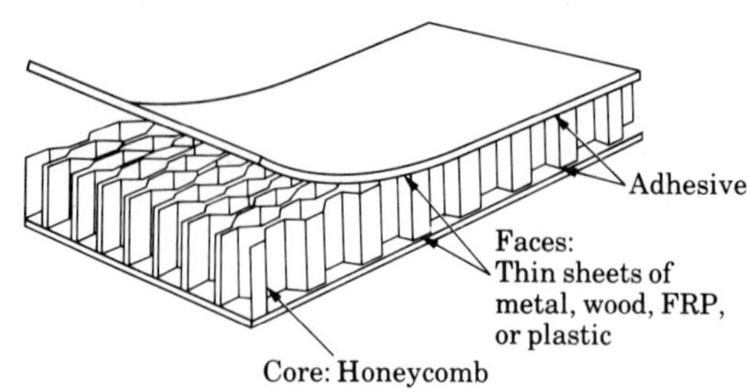

FIGURE 9.11
Schematic illustration of a honeycomb structure. *Source:* Courtesy of American Cyanamid Company.

Because of their light weight and high resistance to bending forces, honeycomb structures are used for aircraft and aerospace components (Fig. 9.12), as well as in buildings and transportation equipment. Figure 9.12 also illustrates various possible designs for reducing the weight of aircraft, while maintaining strength and stiffness of the aircraft body. The core in honeycomb structures can be filled with fiberglass, or similar batting materials, which serve as sound- and vibration-absorbing media, thus reducing engine noise levels in the fuselage.

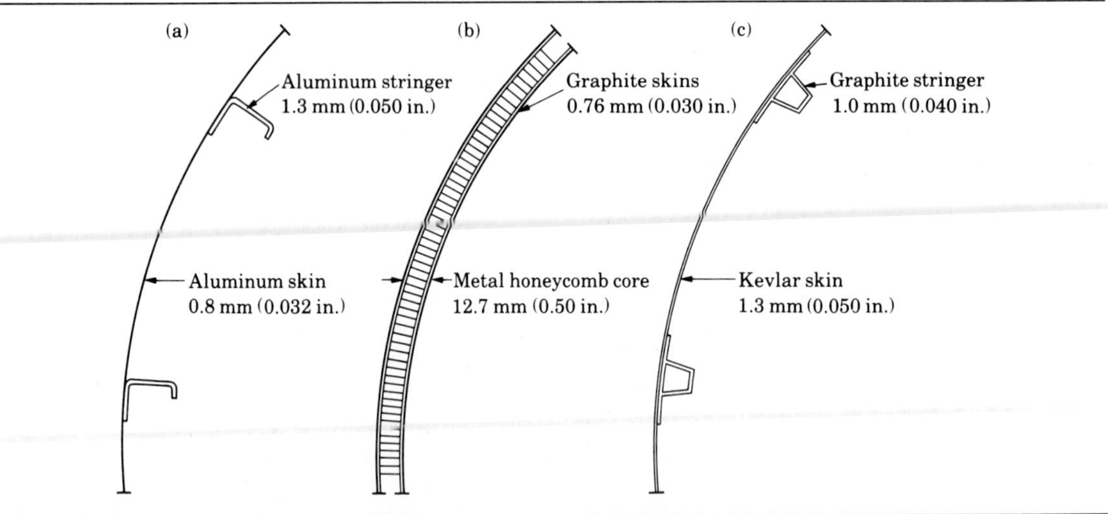

FIGURE 9.12
Three types of structures for aircraft fuselage: (a) conventional metal structure; (b) honeycomb structure; and (c) reinforced plastic structure. *Source:* Sikorsky Aircraft, Division of United Technologies.

SUMMARY

Composite materials and reinforced plastics are an important class of materials that have superior mechanical properties and are lightweight. The reinforcing fibers are usually glass, graphite, aramids, and boron; epoxies commonly serve as a matrix material. Reinforced plastics are being developed rapidly, with a wide variety of present and future applications in aircraft, transportation, structural components, containers, and sporting goods.

New developments concern metal-matrix and ceramic-matrix composites and honeycomb structures. These materials have several important aircraft and aerospace applications, as well as high-temperature or lightweight industrial applications.

SUMMARY TABLE 9.1

TYPES AND GENERAL CHARACTERISTICS OF COMPOSITE MATERIALS

MATERIAL	CHARACTERISTICS
Fibers	
Glass	High strength, low stiffness, high density; lowest cost; E (borosilicate) and S (magnesia-alumina-silicate) types commonly used.
Graphite	Available as high-modulus or high-strength; low cost; less dense than glass.
Boron	High strength and stiffness; highest density; highest cost; has tungsten filament at its center.
Aramids (Kevlar)	Highest strength-to-weight ratio of all fibers; high cost.
Other fibers	Nylon, silicon carbide, silicon nitride, aluminum oxide, boron carbide, boron nitride, tantalum carbide, steel, tungsten, molybdenum.
Matrix materials	
Thermosets	Epoxy and polyester, with the former most commonly used; others are phenolics, fluorocarbons, polyethersulfone, silicon, and polyimides.
Thermoplastics	Polyetheretherketone; tougher than thermosets but lower resistance to temperature.
Metals	Aluminum, aluminum-lithium, magnesium, and titanium; fibers are graphite, aluminum oxide, silicon carbide, and boron.
Ceramics	Silicon carbide, silicon nitride, aluminum oxide, and mullite; fibers are various ceramics.

TRENDS

- Improvements are being made in methods of fabrication, quality assurance, reproducibility, reliability, and predictability of the behavior of composite materials during their service life.
- Developments are taking place in techniques for three-dimensional reinforcement of plastics and improvements in their resistance to compression and buckling.
- An important area of study is reduction in the costs of raw materials and fabrication of composite materials.
- Extended applications of composite materials for aircraft include floor beams, decks, and engine components, and for automobiles include drive shafts and body and chassis components.
- Major developments in metal-matrix and ceramic-matrix composites are anticipated for applications requiring high strength and stiffness, and low density, particularly for aerospace and automotive use.
- Metal-matrix composites are being developed for superconductivity applications.
- Ceramic-matrix cutting tools are being developed, made of silicon-carbide reinforced alumina, with greatly improved tool life. Carbon-whisker reinforced ceramic-matrix materials are being developed for valves and bushing because of their high hardness and favorable tribological properties.

KEY TERMS

Aramids	Fibers	Precursor
Ceramic-matrix	Honeycomb structure	Pyrolysis
Composite materials	Matrix	Reinforced plastics
Delamination	Metal-matrix	Whiskers
Fiber pullout		

BIBLIOGRAPHY

Clegg, D.W., *Mechanical Properties of Reinforced Thermoplastics*. New York: Elsevier, 1985.

Delmonte, J., *Technology of Carbon and Graphite Fiber Composites*. New York: Van Nostrand Reinhold, 1981.

Easterling, K., *Tomorrow's Materials*, 2d ed. London: The Institute of Metals, 1990.
Engineered Materials Handbook, Vol. 1: *Composites*. Metals Park, Ohio: ASM International, 1987.
Engineering Plastics and Composites. Metals Park, Ohio: ASM International, 1990.
Grayson, M. (ed.), *Encyclopedia of Composite Materials and Components*. New York: Wiley, 1983.
Hull, D., *An Introduction to Composite Materials*. Cambridge, England: Cambridge University Press, 1981.
Kaelble, D.H., *Computer-Aided Design of Polymers and Composites*. New York: Marcel Dekker, 1985.
Kelly, A. (ed.), *Concise Encyclopedia of Composite Materials*. New York: Pergamon, 1989.
Kowata, K., and T. Akasaka (eds.), *Composite Materials: Mechanical Properties and Fabrication*. London: Applied Science Publishers, 1982.
Lubin, G. (ed.), *Handbook of Composites*. New York: Van Nostrand Reinhold, 1982.
Schwartz, M., *Composite Materials Handbook*. New York: McGraw-Hill, 1984.
Sheldon, R.P., *Composite Polymeric Materials*. London: Applied Science Publishers, 1982.
Shook, G. (ed.), *Reinforced Plastics for Commercial Composites: Source Book*. Metals Park, Ohio: American Society for Metals, 1986.
Weeton, J.W. (ed.), *Engineers' Guide to Composite Materials*. Metals Park, Ohio: American Society for Metals, 1986.

REVIEW QUESTIONS

9.1 Distinguish between composites and alloys.

9.2 Describe the functions of the matrix and the reinforcing fibers. What fundamental differences are there in the characteristics of the two materials?

9.3 What major reinforcing fibers are used to make composites? Which type of fiber is the strongest? Which type is the weakest?

9.4 What is the range of length and diameter of reinforcing fibers?

9.5 List the important factors that determine the properties of reinforced plastics.

9.6 Compare the advantages and disadvantages of metal-matrix composites, reinforced plastics, and ceramic-matrix composites.

9.7 What are the basic functions and components of a honeycomb structure?

QUESTIONS AND PROBLEMS

9.8 Calculate the average increase in the properties of plastics in Table 7.1 as a result of their reinforcement.

9.9 How do you think the use of straw in clay came about in making brick for dwellings?

9.10 What products have you personally seen that are made of reinforced plastics? How can you tell that they are reinforced?

9.11 Identify metals and alloys that have strengths comparable to those of reinforced plastics.

9.12 You studied the many advantages of composite materials in this chapter. What limitations or disadvantages do these materials have? What suggestions would you make to overcome these limitations?

9.13 In Fig. 9.5 we note the influence of the length of the fiber on the properties of reinforced plastics. Why does length have an influence?

9.14 What applications for composite material can you think of other than those listed in Section 9.4? Why do you think your applications are suitable for these materials?

9.15 What factors contribute to the cost of reinforcing fibers?

9.16 Give examples of composite materials other than those stated in this chapter.

9.17 A hybrid composite is defined as one containing two or more different types of reinforcing fibers. What advantages would such a composite have over others?

9.18 Using the information in this chapter, develop special designs and shapes for new applications of composite materials.

9.19 Would a composite material with a strong and stiff matrix and soft and flexible reinforcement have any practical uses? Explain.

9.20 Make a list of products for which the use of composite materials could be advantageous because of their anisotropic properties.

9.21 Inspect Fig. 9.2 and explain what other components of an aircraft, including parts in the cabin, could be made of composites.

9.22 Name applications in which both specific strength and specific stiffness (Fig. 9.3) are important.

9.23 Explain why the behavior of the materials shown in Fig. 9.6 is as shown.

9.24 Suggest lightweight structures for aircraft fuselage other than those shown in Fig. 9.12.

9.25 Why are fibers capable of supporting a major portion of the load in composite materials?

9.26 What advantages do metal-matrix composites have over reinforced plastics?

9.27 Give reasons for the development of ceramic-matrix composites. Name some possible applications.

9.28 Do you think honeycomb structures could be used in passenger cars? If so, where? Explain.

9.29 Other than those described in this chapter, what materials can you think of that can be regarded as composite materials?

9.30 What applications for composite materials can you think of in which high thermal conductivity would be desirable?

9.31 Calculate the average increase in the properties of plastics in Table 7.1 as a result of their reinforcement and describe your observations.

9.32 In the example on p. 263, what would be the percentage of the load supported by the fibers if their strength is 1250 MPa and the matrix strength is 240 MPa?

PART II

Metal-Casting Processes and Equipment

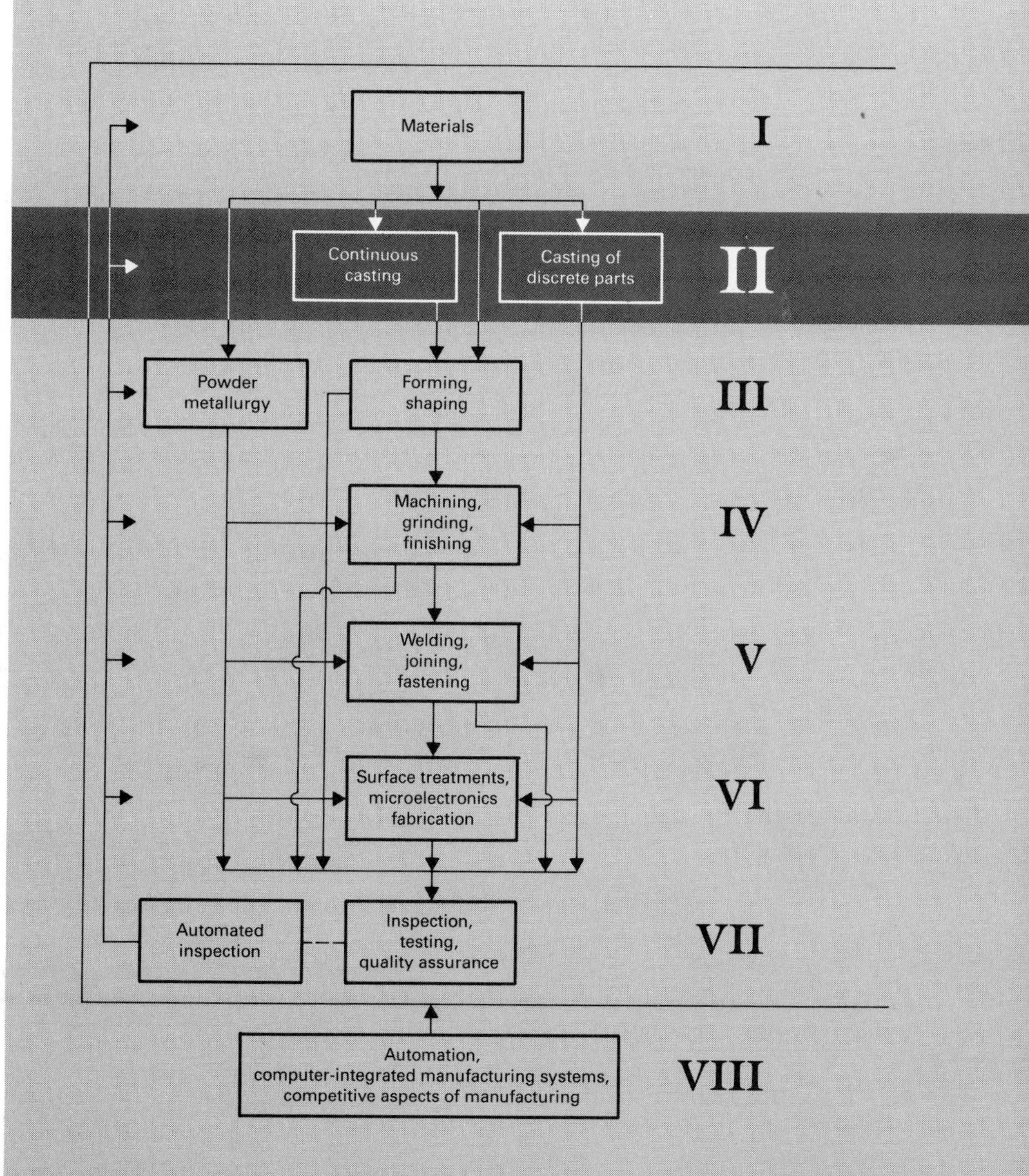

• INTRODUCTION •

We can use several different methods to shape materials into useful products. Making parts by casting molten metal into a mold and letting it solidify is a logical choice. Indeed, casting is among the oldest methods of manufacturing and was first used in about 4000 B.C. to make ornaments, copper arrowheads, and various other objects.

In the broad sense, casting is a process that is applied not only to metals but also to nonmetallic materials. Plastics, ceramics, and glasses are also cast into shapes. Although there are certain similarities in the processes of casting different materials, we defer our discussion of casting nonmetallic materials to Chapters 18 and 19.

Basically, metal-casting processes involve the introduction of molten metal into a mold cavity where, upon solidification, the metal takes the shape of the cavity. The casting process is thus capable of producing intricate shapes in a single piece, including those with internal cavities. Very large or hollow parts can be produced economically by casting techniques. Typical cast products are engine blocks, crankshafts, pistons, valves, railroad wheels, and ornamental artifacts.

Figure II.1 shows cast components in a typical automobile, a product that we used in the Introduction to Part I to illustrate the selection and use of materials. We

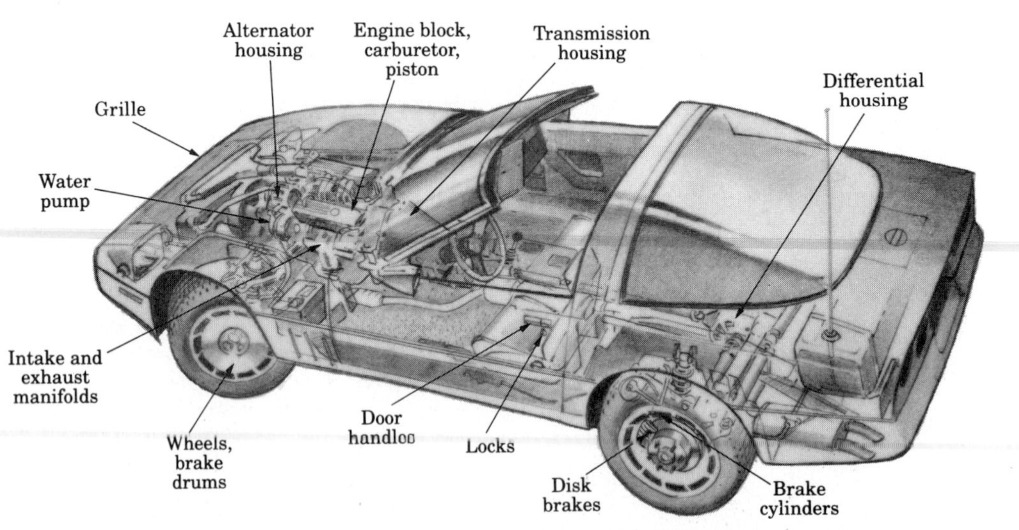

FIGURE II.1
Cast parts in a typical automobile.

use the automobile theme throughout this text, in order to give you a comprehensive view of manufacturing engineering and technology.

Although casting processes allow a great deal of versatility in part size and shape, they most often are selected over other manufacturing methods for the following reasons:

- To produce complex shapes with internal cavities or hollow sections.
- To produce very large parts.
- To utilize workpiece materials that are difficult to process by other means.
- Economic considerations.

Almost all metals can be cast in (or nearly in) the final shape desired, often with only minor finishing required. With appropriate control of material and process parameters and subsequent heat treatment, parts can be cast with uniform properties throughout. With modern processing techniques and control of chemical composition, mechanical properties of castings can equal those for forged or welded components. Furthermore, the anisotropic mechanical properties of castings can be a desirable characteristic.

Several casting processes have been developed to date (Fig. II.2). As in all manufacturing, each process has its own characteristics, applications, advantages, and limitations. Regardless of the method of casting used, we must understand and control certain fundamental aspects of the process in order to economically produce castings of good quality, dimensional accuracy, and surface finish. These aspects are solidification of metals, heat transfer, and flow of the molten metal into the mold cavities. They are influenced by mold material, casting design, and various other material and process variables. In this Part, we present these fundamental aspects of casting and casting processes, design, and characteristics of metals cast.

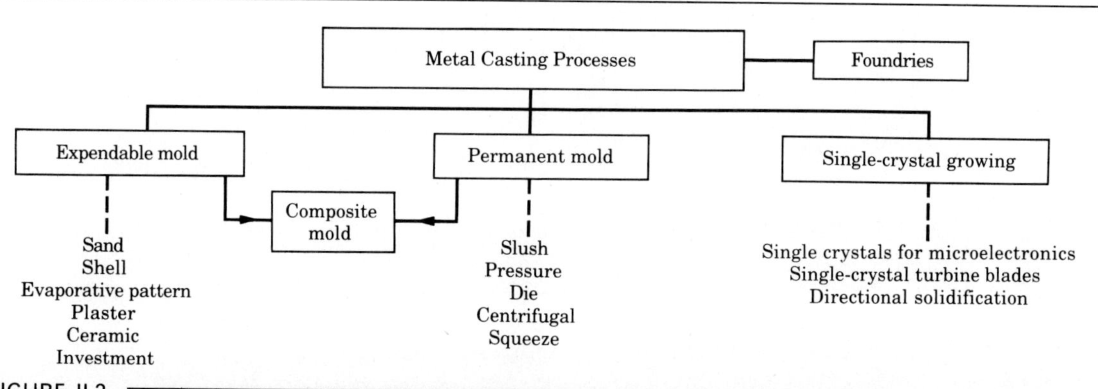

FIGURE II.2
An outline of metal casting processes described in Part II.

The knowledge gained in this Part will enable you to better evaluate the technical and economic feasibility of using casting to manufacture certain components. After studying other manufacturing processes in the remainder of this book, you will be able to decide whether a particular component should be produced by casting or by forming, machining, welding, or by a combination of manufacturing processes.

10

Fundamentals of Metal Casting

10.1 Introduction

The **casting** process basically involves pouring molten metal into a mold patterned after the part to be manufactured, allowing it to cool, and removing the metal from the mold. As with all other manufacturing processes, certain fundamental relationships are essential to the production of good quality and economical castings. Knowledge of these relationships helps us establish proper techniques for mold design and casting practice. Our objective is to produce castings that are free from defects and that meet requirements for strength, dimensional accuracy, and surface finish.

The important factors in casting operations are:

- The flow of the molten metal into the mold cavity.
- Solidification of the metal from its molten state.
- Heat transfer during solidification and cooling of the metal in the mold.
- Influence of the type of mold material.

Note the similarities between metal casting and pouring cake mix into a mold (pan) and baking it. We first select the kind and size of mold to use, control the composition of the mix, carefully pour the cake mix into the mold, set the proper baking temperature, set the timer for the proper baking time, and leave the baked cake in the mold a certain amount of time before we remove it.

Although the methods and procedures involved are similar, we leave the discussion of casting plastics and ceramics to Chapters 18 and 19, respectively.

10.2 Solidification of Metals

After molten metal is poured into a **mold**, a series of events takes place during **solidification** of the casting and its cooling to ambient temperature. These events greatly influence the size, shape, and uniformity of the grains formed throughout the casting, which in turn influence its overall properties. The significant factors affecting these events are the type of metal, thermal properties of both the metal and the mold, the geometric relationship between volume and surface area of the casting, and the shape of the mold.

10.2.1 Pure metals

Because a pure metal has a clearly defined melting or freezing point, it solidifies at a *constant* temperature. Pure aluminum, for example, solidifies at 660 °C (1220 °F), iron at 1537 °C (2798 °F), and tungsten at 3410 °C (6170 °F). (See Fig. 4.4.)

When the temperature of the molten metal is reduced to its freezing point, its temperature remains constant while the latent heat of fusion is given off. The solidification front (solid–liquid interface) moves through the molten metal, solidifying from the mold walls in toward the center. Once solidification has taken place at any point, cooling resumes. The solidified metal, which we now call the **casting**, is then taken out of the mold and begins to cool to ambient temperature.

The grain structure of a pure metal cast in a square mold is shown in Fig. 10.1(a). At the mold walls, the metal cools rapidly since the walls are at ambient temperature. Rapid cooling produces a solidified **skin**, or *shell*, of fine equiaxed grains. The grains grow in the direction opposite to the heat transfer out through the mold. Those grains that have favorable orientation will grow preferentially and are called **columnar grains** (Fig. 10.2). As the driving force of the heat transfer is reduced away from the mold walls, the grains become equiaxed and coarse. Those grains that have substantially different orientations are blocked from further growth. Such grain development is known as **homogeneous nucleation,** meaning that the grains (crystals) grow upon themselves, starting at the mold wall.

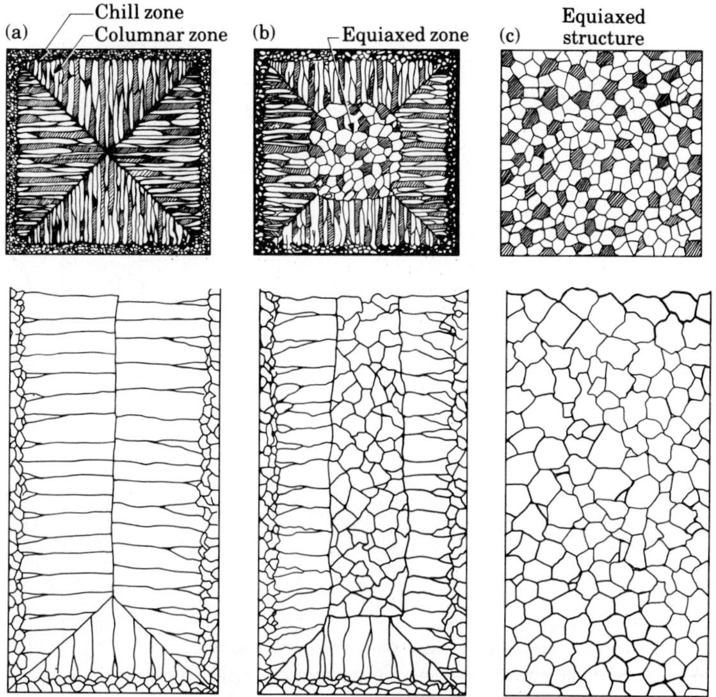

FIGURE 10.1

Schematic illustration of three cast structures of metals solidified in a square mold: (a) pure metals; (b) solid-solution alloys; and (c) structure obtained by using nucleating agents. *Source:* G. W. Form, J. F. Wallace, J. L. Walker, and A. Cibula.

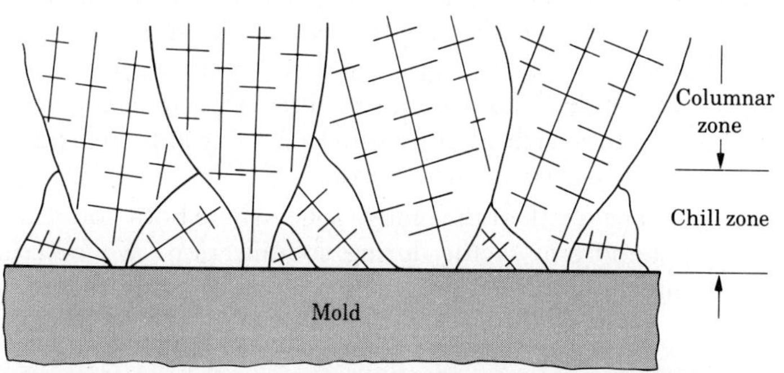

FIGURE 10.2

Development of a preferred texture at a cool mold wall. Note that only favorably oriented grains grow away from the surface of the mold.

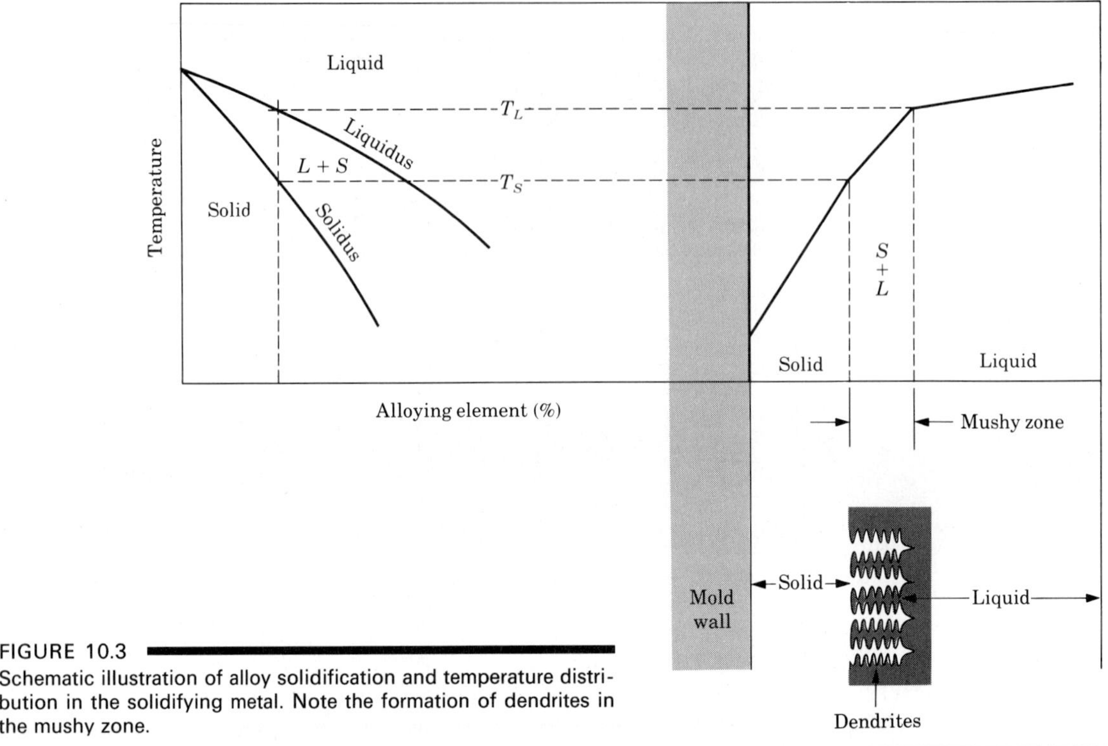

FIGURE 10.3 Schematic illustration of alloy solidification and temperature distribution in the solidifying metal. Note the formation of dendrites in the mushy zone.

10.2.2 Alloys

Solidification in alloys begins when the temperature drops below the liquidus, T_L, and is complete when it reaches the solidus, T_S (Fig. 10.3). Within this temperature range, the alloy is in a mushy or pasty state with **columnar dendrites** (from the Greek *dendron* meaning akin to, and *drys* meaning tree). Note the presence of liquid metal between the dendrite arms. Dendrites have three-dimensional arms and branches (secondary arms) and they eventually interlock, as shown in Fig. 10.4. The study of dendritic structures, although complex, is important because of detrimental factors such as compositional variations, segregation, and microporosity that generally exist within a casting.

The width of the mushy zone, where both liquid and solid phases are present, is an important factor during solidification. We describe this zone in terms of a temperature difference, known as the **freezing range**, as follows:

$$\text{Freezing range} = T_L - T_S. \tag{10.1}$$

Figure 10.3 shows that pure metals have a freezing range that approaches zero and that the solidification front moves as a plane front, without forming a mushy zone. Eutectics solidify in a similar manner with an approximately plane front. The type of solidification structure developed depends on the composition of the

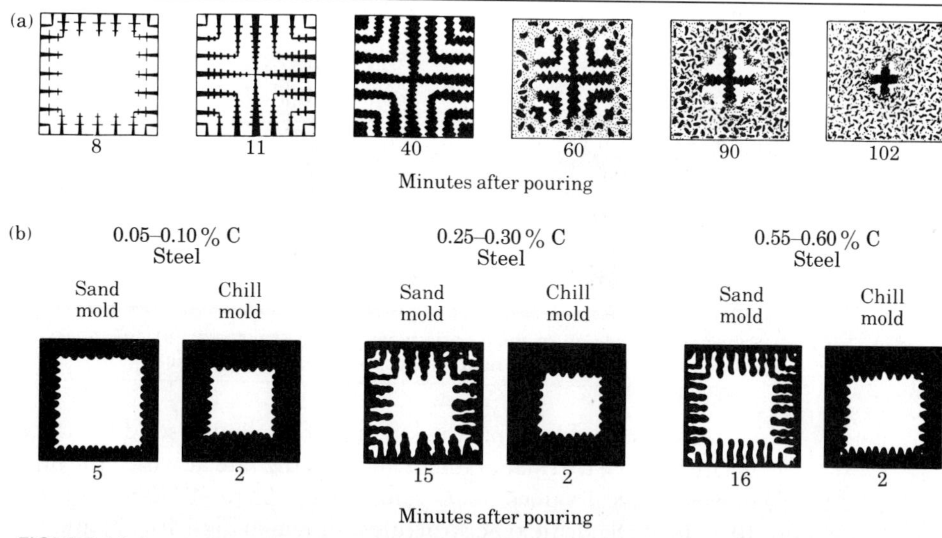

FIGURE 10.4

(a) Solidification patterns for gray cast iron in a 180-mm (7-in.) square casting. Note that after 11 min of cooling, dendrites reach each other, but the casting is still mushy throughout. It takes about two hours for this casting to solidify completely. (b) Solidification of carbon steels in sand and chill (metal) molds. Note the difference in solidification patterns as the carbon content increases. *Source:* H. F. Bishop and W. S. Pellini.

eutectic. For alloys with a nearly symmetrical phase diagram, the structure is generally lamellar with two or more solid phases present, depending on the alloy system. When the volume fraction of the minor phase of the alloy is less than about 25 percent, the structure generally becomes fibrous. These conditions are particularly important for cast irons.

For alloys, although it is not precise, a short freezing range generally involves a temperature difference of less than 50 °C (90 °F), and a long freezing range greater than 110 °C (200 °F). Ferrous castings generally have narrow mushy zones, whereas aluminum and magnesium alloys have wide mushy zones. Consequently, these alloys are in a mushy state throughout most of the solidification process.

Effects of cooling rates. Slow cooling rates—on the order of 10^2 K/s—or long local solidification times result in coarse dendritic structures with large spacing between the dendrite arms. For faster cooling rates—on the order of 10^4 K/s—or short local solidification times, the structure becomes finer with smaller dendrite arm spacing. For still faster cooling rates—on the order of 10^6 to 10^8 K/s—the structures developed are amorphous, as described in Section 6.14. The structures developed and the resulting grain size, in turn, influence the properties of the casting. As grain size decreases, (a) the strength and ductility of the cast alloy increase, (b) microporosity (interdendritic shrinkage voids) in the casting decreases, and (c) the tendency for the casting to crack (*hot tearing*) during solidification decreases. Lack of uniformity in grain size and distribution results in castings with anisotropic properties.

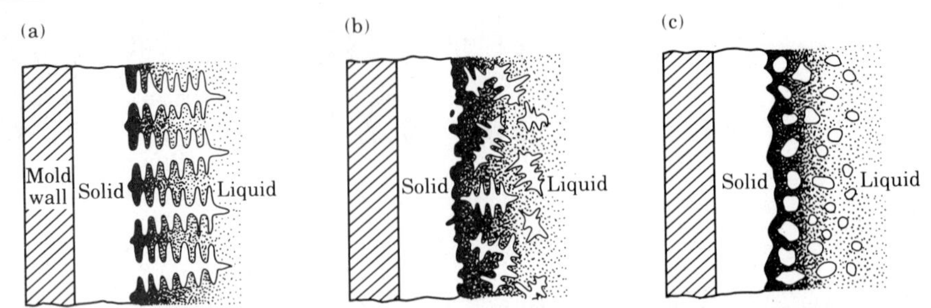

FIGURE 10.5 Schematic illustration of three basic types of cast structures: (a) columnar dendritic; (b) equiaxed dendritic; and (c) equiaxed nondendritic. *Source:* After D. Apelian.

A criterion describing the kinetics of the liquid–solid interface is the ratio G/R, where G is the thermal gradient and R is the rate at which the liquid–solid interface moves. Typical values for G range from 10^2 to 10^3 K/m and for R from 10^{-3} to 10^{-4} m/s. Dendritic type structures, shown in Figs. 10.5(a) and (b), typically have a G/R ratio in the range of 10^5 to 10^7, whereas ratios of 10^{10} to 10^{12} produce a plane front nondendritic liquid–solid interface (Fig. 10.6.)

10.2.3 Structure–property relationships

Because all castings are expected to possess certain properties to meet design and service requirements, the relationships between these properties and the structures developed during solidification are important aspects of casting. In this section, we describe these relationships in terms of dendrite morphology and the concentration of alloying elements in various regions.

The compositions of dendrites and the liquid metal are given by the phase diagram of the particular alloy. When the alloy is cooled very slowly, each dendrite develops a uniform composition. Under normal cooling encountered in practice, however, *cored dendrites* are formed and have a surface composition different from that at their centers (concentration gradient). The surface has a higher concentration of alloying elements than at the core of the dendrite owing to solute rejection from the core toward the surface during solidification of the dendrite, which is called **microsegregation.** The darker shading in the interdendritic liquid near the

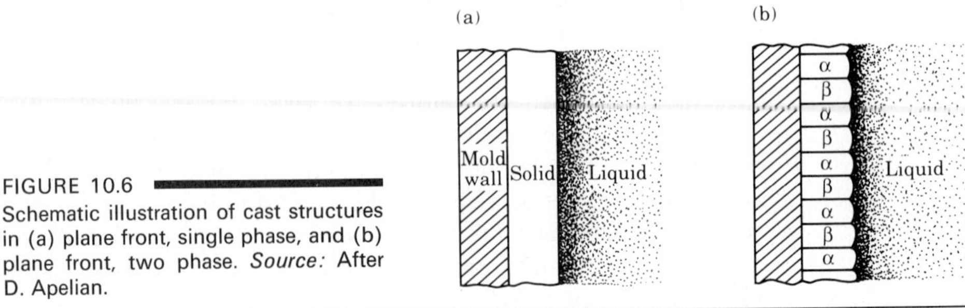

FIGURE 10.6 Schematic illustration of cast structures in (a) plane front, single phase, and (b) plane front, two phase. *Source:* After D. Apelian.

dendrite roots in Fig. 10.5 indicates that these regions have a higher solute concentration. Thus microsegregation in these regions is much more pronounced than in others.

In contrast to microsegregation, **macrosegregation** involves differences in composition throughout the casting itself. In situations where the solidifying front moves away from the surface of a casting as a plane front (Fig. 10.6), lower melting point constituents in the solidifying alloy are driven toward the center (*normal segregation*). Consequently, such a casting has a higher concentration of alloying elements at its center than at its surfaces. In dendritic structures such as those for solid-solution alloys (Fig. 10.1b), the opposite occurs: The center of the casting has a lower concentration of alloying elements (*inverse segregation*). The reason is that liquid metal (having a higher concentration of alloying elements) enters the cavities developed from solidification shrinkage in the dendrite arms, which have solidified sooner. Another form of segregation is the result of gravity (*gravity segregation*), whereby higher density inclusions or compounds sink, and lighter elements (such as antimony in an antimony–lead alloy) float to the surface.

A typical cast structure of a solid-solution alloy with an inner zone of equiaxed grains is shown in Fig. 10.1(b). This inner zone can be extended throughout the casting, as shown in Fig. 10.1(c), by adding an **inoculant** (nucleating agent) to the alloy. The inoculant induces nucleation of grains throughout the liquid metal (**heterogeneous nucleation**).

Because of the presence of thermal gradients in a solidifying mass of liquid metal and because of gravity (hence density differences), convection has a strong influence on the structures developed. Convection promotes the formation of an outer chill zone, refines grain size, and accelerates the transition from columnar to equiaxed grains. The structure shown in Fig. 10.5(b) can also be obtained by increasing convection within the liquid metal, whereby dendrite arms separate (*dendrite multiplication*). Conversely, reducing or eliminating convection results in coarser and longer columnar dendritic grains. The dendrite arms are not particularly strong, and they can be broken up by agitation or mechanical vibration in the early stages of solidification (*rheocasting*). This results in finer grain size, with equiaxed nondendritic grains distributed more uniformly throughout the casting (Fig. 10.5c). Convection can be enhanced by the use of mechanical or electromagnetic methods. Experiments are now being conducted during space flights concerning the effects of gravity on the microstructure of castings.

10.3 Fluid Flow and Heat Transfer

10.3.1 Fluid flow

To emphasize the importance of fluid flow, let's briefly describe the basic casting system, as shown in Fig. 10.7. The molten metal is poured through a **pouring basin**

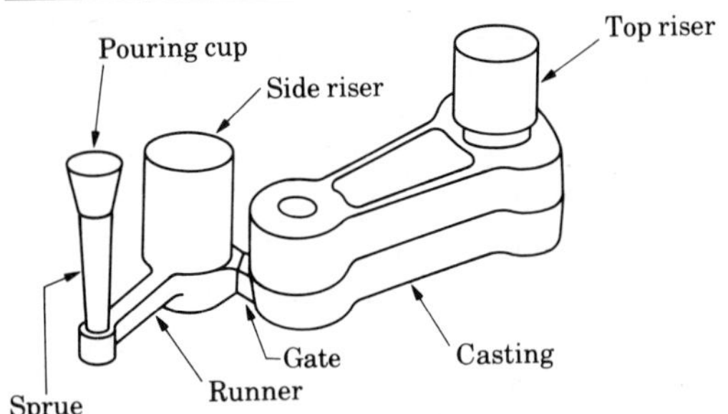

FIGURE 10.7 Schematic illustration of a typical riser-gated casting. Risers serve as reservoirs, supplying molten metal to the casting as it shrinks during solidification. *Source:* American Foundrymen's Society.

or *cup*. It then flows through the **sprue, runners,** and **gates** into the mold cavity. **Risers** serve as reservoirs of molten metal to supply the metal necessary to prevent shrinkage. Although such a *gating system* appears to be relatively simple, successful casting requires careful design and control of the solidification process to ensure adequate fluid flow in the system. For example, one of the most important functions of the gating system is to trap contaminants (such as oxides and other inclusions) in the molten metal by having the contaminants adhere to the walls of the gating system, thereby preventing their reaching the actual mold cavity. Furthermore, the proper design of the gating system avoids or minimizes problems such as premature cooling, turbulence, and gas entrapment. Even before it reaches the mold cavity, the molten metal must be handled carefully to avoid forming oxides on molten metal surfaces from exposure to the environment or introduction of impurities into the molten metal.

Two basic principles of fluid flow are relevant to gating design: Bernoulli's theorem and the law of mass continuity.

Bernoulli's theorem. **Bernoulli's theorem** is based on the principle of conservation of energy and relates pressure, velocity, elevation of the fluid at any location in the system, and the frictional losses in a system that is full of liquid:

$$h + \frac{p}{\rho g} + \frac{v^2}{2g} = \text{constant}, \tag{10.2}$$

where h is the elevation above a certain reference plane, p is the pressure at that elevation, v is the velocity of the liquid at that elevation, ρ is the density of the fluid (assuming that it is incompressible), and g is the gravitational constant. Conservation of energy requires that, at a particular location in the system, the following relationship be satisfied:

$$h_1 + \frac{p_1}{\rho g} + \frac{v_1^2}{2g} = h_2 + \frac{p_2}{\rho g} + \frac{v_2^2}{2g} + f, \tag{10.3}$$

where the subscripts 1 and 2 represent two different elevations, respectively, and f represents the frictional loss in the liquid as it travels downward through the system. The frictional loss includes such factors as energy loss at the liquid–mold wall interfaces and turbulence in the liquid.

Continuity. The continuity law states that, for incompressible liquids and in a system with impermeable walls, the rate of flow is constant:

$$Q = A_1 v_1 = A_2 v_2, \tag{10.4}$$

where Q is the rate of flow, such as m^3/s, A is the cross-sectional area of the liquid stream, and v is the velocity of the liquid in that cross-sectional location. The subscripts 1 and 2 pertain to two different locations in the system. Thus the flow rate must be maintained anywhere in the system. The permeability of the walls of the system is important, because otherwise some liquid will permeate through (for example, in sand molds) and the flow rate will decrease as the liquid moves through the system.

Application. An application of the two principles stated is the traditional tapered design of sprues (Fig. 10.7). We can determine the shape of the sprue by using Eqs. (10.3) and (10.4). Assuming that the pressure at the top of the sprue is equal to the pressure at the bottom and that there are no frictional losses, at any point in the sprue relationship between height and cross-sectional area is given by the parabolic relationship:

$$\frac{A_1}{A_2} = \sqrt{\frac{h_2}{h_1}}, \tag{10.5}$$

where, for example, the subscript 1 denotes the top of the sprue and 2 the bottom. Moving downward from the top, the cross-sectional area of the sprue must decrease. Recall that in a free-falling liquid, such as water from a faucet, the cross-sectional area of the stream decreases as it gains velocity downward. If we design a sprue with a constant cross-sectional area and pour the molten metal, regions may develop where the liquid loses contact with the sprue walls. As a result, **aspiration** may take place whereby air will be sucked in or entrapped in the liquid. On the other hand, in many systems tapered sprues are now replaced by straight-sided sprues with a *choking* mechanism at the bottom, consisting of either a choke core or a runner choke (see Fig. 11.3).

Flow characteristics. An important consideration in fluid flow in gating systems is the presence of **turbulence**, as opposed to *laminar flow* of fluids. We use the *Reynolds number* to quantify this aspect of fluid flow. The Reynold's number, Re, represents the ratio of the inertia to the viscous forces in fluid flow, and is defined as

$$\text{Re} = \frac{vD\rho}{\eta}, \tag{10.6}$$

where v is the velocity of the liquid, D is the diameter of the channel, and ρ and η are the density and viscosity, respectively, of the liquid. The higher this number, the greater is the tendency for turbulent flow. In ordinary gating systems, Re ranges from 2000 to 20,000. An Re value of up to 2000 represents laminar flow; between

2000 and 20,000 it is a mixture of laminar and turbulent flow and is generally regarded as harmless in gating systems. However, Re values in excess of 20,000 represent severe turbulence, resulting in air entrainment and *dross* formation (the scum that forms on the surface of molten metal) from the reaction of the liquid metal with air and other gases. Techniques for minimizing turbulence generally involve avoidance of sudden changes in flow direction and in the geometry of channel cross-sections in the design of the gating system.

The elimination of dross or slag is another important consideration in fluid flow. It can be achieved by skimming, properly designing pouring basins and runner systems, or using filters. Filters are usually made of ceramics, mica, or fiberglass and their proper location and placement are important for effective filtering of dross and slag. Filters also can eliminate turbulent flow in the runner system.

10.3.2 Fluidity of molten metal

A term commonly used to describe the capability of the molten metal to fill mold cavities is **fluidity**. This term consists of two basic factors: (1) characteristics of the molten metal and (2) casting parameters. The following characteristics of molten metal influence fluidity.

a) *Viscosity.* As viscosity and its sensitivity to temperature (*viscosity index*) increase, fluidity decreases.
b) *Surface tension.* A high surface tension of the liquid metal reduces fluidity. Oxide films developed on the surface of the molten metal thus have a significant adverse effect on fluidity. For example, the oxide film on the surface of pure molten aluminum triples the surface tension.
c) *Inclusions.* As insoluble particles, inclusions can have a significant adverse effect on fluidity. This effect can be verified by observing the viscosity of a liquid such as oil with and without sand particles in it; the latter has higher viscosity.
d) *Solidification pattern of the alloy.* The manner in which solidification occurs, as described in Section 10.2, can influence fluidity. Moreover, fluidity is inversely proportional to the freezing range (see Eq. 10.1). Thus the shorter the range (as in pure metals and eutectics), the higher the fluidity becomes. Conversely, alloys with long freezing ranges (such as solid-solution alloys) have lower fluidity.

The following casting parameters influence fluidity and also influence the fluid flow and thermal characteristics of the system.

a) *Mold design.* The design and dimensions of components such as the sprue, runners, and risers all influence fluidity.
b) *Mold material and its surface characteristics.* The higher the thermal conductivity of the mold and the rougher its surfaces, the lower the fluidity of the molten metal becomes. Heating the mold improves fluidity, even though it slows down solidification of the metal and the casting develops coarse grains; hence it has less strength.

c) *Degree of superheat.* Defined as the increment of temperature above the melting point of an alloy, superheat improves fluidity by delaying solidification.
d) *Rate of pouring.* The slower the rate of pouring the molten metal into the mold, the lower the fluidity becomes because of the faster rate of cooling.
e) *Heat transfer.* This factor directly affects the viscosity of the liquid metal.

Although the interrelationships are complex, we use the term *castability* generally to describe the ease with which a metal can be cast to obtain a part with good quality. Obviously, this term includes not only fluidity but casting practices as well.

Tests for fluidity. Although none is accepted universally, several tests have been developed to quantify fluidity. One such test is shown in Fig. 10.8, where the molten metal is made to flow along a channel at room temperature. The distance of metal flow before it solidifies and stops is a measure of its fluidity. Obviously this length is a function of the thermal properties of the metal and the mold, as well as the design of the channel. Such tests are useful and simulate casting situations to a reasonable degree.

10.3.3 Heat transfer

An important consideration in casting is the heat transfer during the complete cycle from pouring to solidification and cooling to room temperature. Heat flow at different locations in the system is a complex phenomenon; it depends on many factors relating to the casting material and the mold and process parameters. For instance, in casting thin sections the metal flow rates must be high enough to avoid premature chilling and solidification. But, the flow rate must not be so fast as to cause excessive turbulence, with its detrimental effects on the casting process.

A typical temperature distribution in the mold–liquid metal interface is shown in Fig. 10.9. Heat from the liquid metal is given off through the mold wall and the

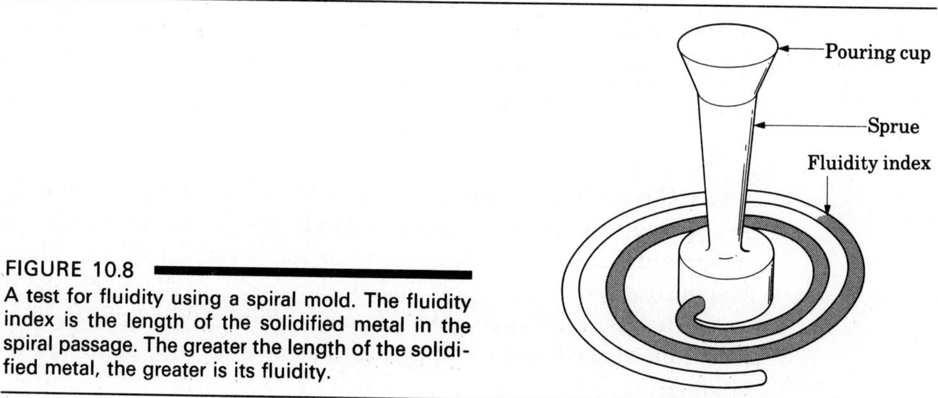

FIGURE 10.8 A test for fluidity using a spiral mold. The fluidity index is the length of the solidified metal in the spiral passage. The greater the length of the solidified metal, the greater is its fluidity.

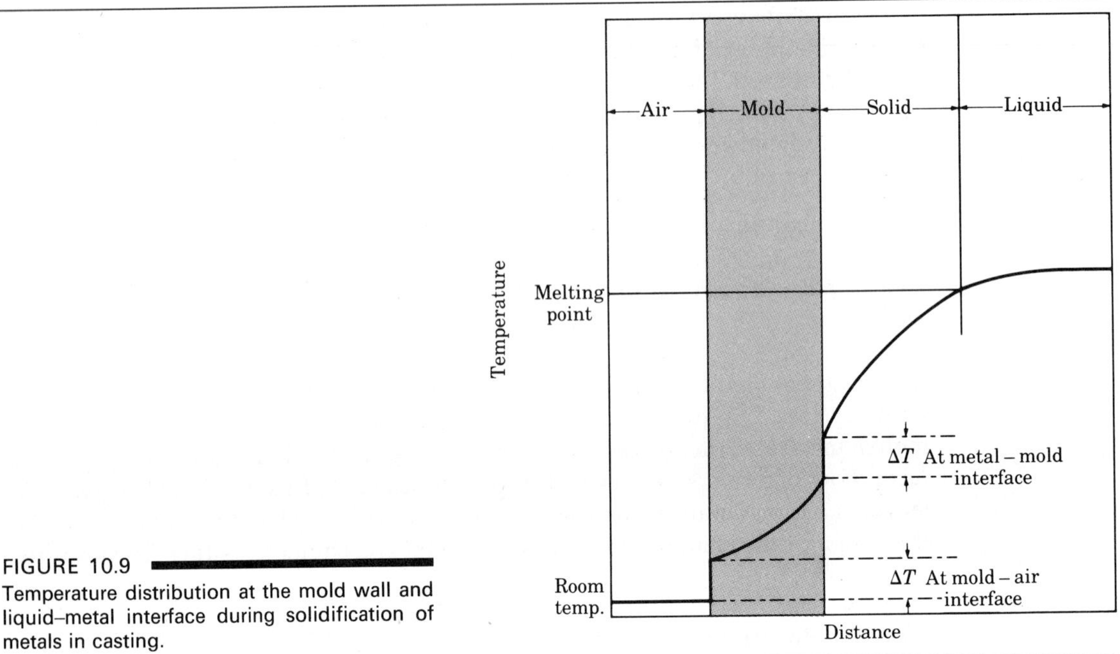

FIGURE 10.9 Temperature distribution at the mold wall and liquid–metal interface during solidification of metals in casting.

surrounding air. The temperature drop at the air–mold and mold–metal interfaces is caused by the presence of boundary layers and imperfect contact at these interfaces. The shape of the curve depends on the thermal properties of the molten metal and the mold.

10.3.4 Solidification time

During the early stages of solidification, a thin solidified skin begins to form at the cool mold walls and, as time passes, the skin thickens. With flat mold walls, this thickness is proportional to the square root of time. Thus doubling the time will make the skin $\sqrt{2} = 1.41$ times, or 41 percent, thicker.

The *solidification time* is a function of the volume of a casting and its surface area (*Chvorinov's rule*):

$$\text{Solidification time} = C\left(\frac{\text{volume}}{\text{surface area}}\right)^2, \qquad (10.7)$$

where C is a constant that reflects mold material, metal properties (including latent heat), and temperature. Thus a large sphere solidifies and cools to ambient temperature at a much lower rate than does a smaller sphere. The reason is that the volume of a sphere is proportional to the cube of its diameter, and the surface area is proportional to the square of its diameter. Similarly, we can show that molten metal in a cube-shaped mold will solidify faster than in a spherical mold of the same volume.

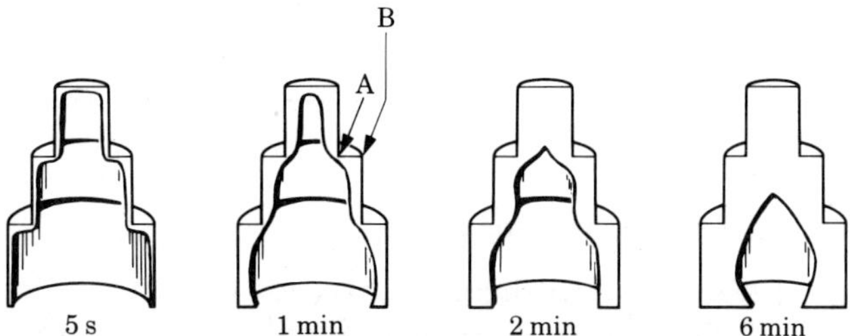

FIGURE 10.10
Solidified skin on a steel casting. The remaining molten metal is poured out at the times indicated in the figure. Hollow ornamental and decorative objects are made by a process called slush casting, which is based on this principle. *Source:* H. F. Taylor, J. Wulff, and M. C. Flemings.

The effects of mold geometry and elapsed time on skin thickness and shape are shown in Fig. 10.10. As illustrated, the unsolidified molten metal has been poured from the mold at different time intervals, ranging from 5 s to 6 min. Note that the skin thickness increases with elapsed time but that the skin is thinner at internal angles (location A in the figure) than at external angles (location B). This latter condition is caused by slower cooling at internal angles than at external angles.

• Example: Solidification times for various shapes.

Three pieces being cast have the same volume but different shapes. One is a sphere, one a cube, and the other a cylinder with a height equal to its diameter. Which piece will solidify the fastest and which one the slowest?

SOLUTION. The volume is unity, so we have from Eq. (10.7):

$$\text{Solidification time} \propto \frac{1}{(\text{surface area})^2}.$$

The respective surface areas are

$$\text{Sphere:} \quad V = \left(\frac{4}{3}\right)\pi r^3, \qquad r = \left(\frac{3}{4\pi}\right)^{1/3}, \qquad \text{and} \quad A = 4\pi r^2 = 4\pi\left(\frac{3}{4\pi}\right)^{2/3} = 4.84;$$

$$\text{Cube:} \quad V = a^3, \qquad a = 1, \quad \text{and} \quad A = 6a^2 = 6;$$

$$\text{Cylinder:} \; V = \pi r^2 h = 2\pi r^3, \qquad r = \left(\frac{1}{2\pi}\right)^{1/3}, \quad \text{and}$$

$$A = 2\pi r^2 + 2\pi r h = 6\pi r^2 = 6\pi\left(\frac{1}{2\pi}\right)^{2/3} = 5.54.$$

Thus the respective solidification times t are

$$t_{\text{sphere}} = 0.043C, \qquad t_{\text{cube}} = 0.028C, \quad \text{and} \quad t_{\text{cylinder}} = 0.033C.$$

Hence the cube-shaped casting will solidify the fastest and the sphere-shaped casting will solidify the slowest.

10.3.5 Shrinkage

Because of their thermal expansion characteristics, metals shrink (contract) during solidification and cooling. **Shrinkage**, which causes dimensional changes—and, sometimes, cracking—is the result of:

1. contraction of the molten metal as it cools prior to its solidification;
2. contraction of the metal during phase change from liquid to solid (latent heat of fusion); and
3. contraction of the solidified metal (the casting) as its temperature drops to ambient temperature.

The largest amount of shrinkage occurs during cooling of the casting. The amount of contraction for various metals during solidification is shown in Table 10.1. Note that gray cast iron expands. The reason is that graphite has a relatively high specific volume, and when it precipitates as graphite flakes during solidification, it causes a net expansion of the metal.

10.3.6 Defects

As we will see in this section as well as other sections throughout Parts II–VI, various defects can result in manufacturing processes, depending on factors such as materials, part design, and processing techniques. While some defects affect only the appearance of parts, others can have major adverse effects on the structural integrity of parts made.

Several defects can develop in castings. Because different names have been used to describe the same defect, the International Committee of Foundry Technical

TABLE 10.1
SOLIDIFICATION CONTRACTION FOR VARIOUS CAST METALS

METAL OR ALLOY	VOLUMETRIC SOLIDIFICATION CONTRACTION (%)	METAL OR ALLOY	VOLUMETRIC SOLIDIFICATION CONTRACTION (%)
Aluminum	6.6	70% Cu–30% Zn	4.5
Al–4.5% Cu	6.3	90% Cu–10% Al	4
Al–12% Si	3.8	Gray iron	Expansion to 2.5
Carbon steel	2.5–3	Magnesium	4.2
1% carbon steel	4	White iron	4–5.5
Copper	4.9	Zinc	6.5

Source: After R. A. Flinn.

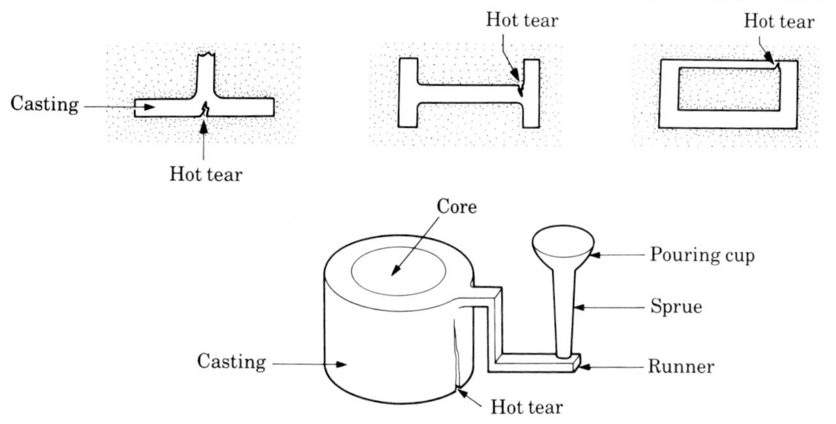

FIGURE 10.11
Examples of hot tears in castings. These defects occur because the casting cannot shrink freely during cooling, owing to constraints in various portions of the molds and cores. Exothermic (heat producing) compounds may be used (exothermic padding) to control cooling at critical sections to avoid hot tearing.

Associations has developed standardized nomenclature consisting of seven basic categories of casting defects (Figs. 10.11 and 10.12):

A. *Metallic projections*, consisting of fins, flash, or massive projections such as swells and rough surfaces.
B. *Cavities*, consisting of rounded or rough internal or exposed cavities, including blowholes, pinholes, and shrinkage cavities (see porosity below).
C. *Discontinuities*, such as cracks, cold or hot tearing, and cold shuts. If the solidifying metal is constrained from shrinking freely, cracking and tearing can occur. Although many factors are involved in tearing, coarse grain size and the presence of low-melting segregates along the grain boundaries (intergranular) increase the tendency for hot tearing. Incomplete castings result from the molten metal being at too low a temperature or pouring the metal too slowly. Cold shut is an interface in a casting that lacks complete fusion because of the meeting of two streams of liquid metal from different gates.
D. *Defective surface*, such as surface folds, laps, scars, adhering sand layers, and oxide scale.
E. *Incomplete casting*, such as misruns (due to premature solidification), insufficient volume of metal poured, and runout (due to loss of metal from mold after pouring).
F. *Incorrect dimensions or shape*, owing to factors such as improper shrinkage allowance, pattern mounting error, irregular contraction, deformed pattern, or warped casting.

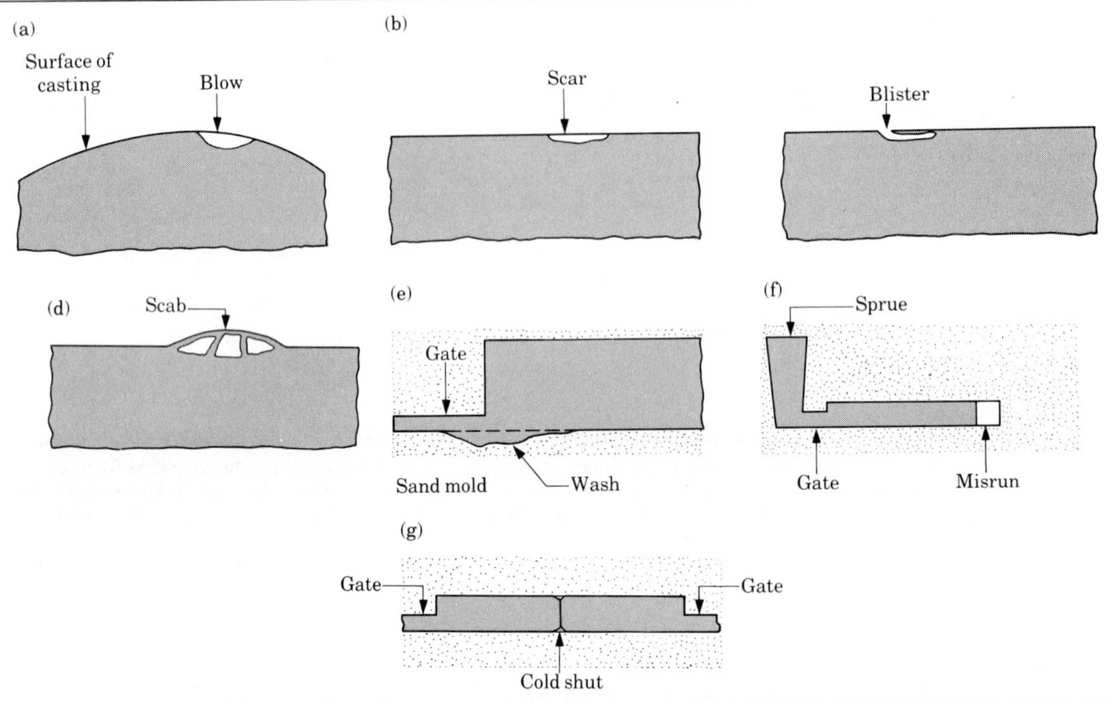

FIGURE 10.12 Examples of common defects in castings. These defects can be minimized or eliminated by proper design and preparation of molds and control of pouring procedures. *Source:* J. Datsko.

G. *Inclusions*, which form during melting, solidification, and molding. Generally nonmetallic, they are regarded as harmful because they act like stress raisers and reduce the strength of the casting. They can be filtered out during processing of the molten metal. Inclusions may form during melting because of reaction of the molten metal with the environment (usually oxygen) or the crucible material. Chemical reactions among components in the molten metal may produce inclusions; slags and other foreign material entrapped in the molten metal also become inclusions. Reactions between the metal and the mold material may produce inclusions. Spalling of the mold and core surfaces also produces inclusions, indicating the importance of the quality and maintenance of molds.

Porosity. Porosity in a casting may be caused by shrinkage or gases, or both. Porosity is detrimental to the ductility of a casting (Fig. 10.13) and its surface finish, making it permeable and thus affecting pressure tightness of a cast pressure vessel.

Porous regions can develop in castings because of *shrinkage* of the solidified metal. Thin sections in a casting solidify sooner than thicker regions. As a result, molten metal cannot be fed into the thicker regions that have not yet solidified.

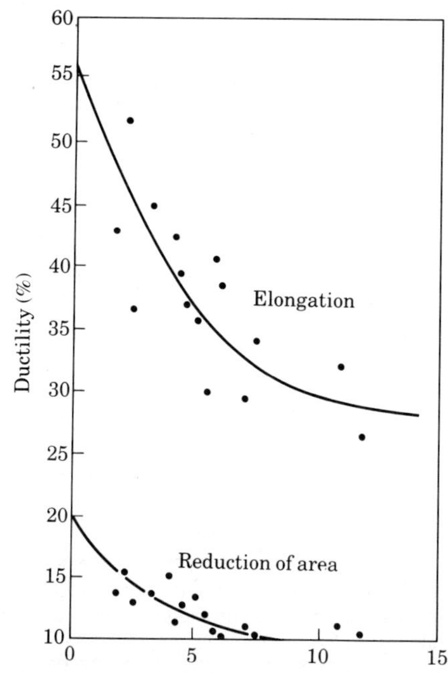

FIGURE 10.13
The effect of microporosity on the ductility of quenched and tempered 1% Cr–0.25% Mo cast steel. *Source:* T. Grousatier.

Because of contraction, as the surfaces of the thicker region begin to solidify, porous regions develop at their centers. Microporosity can also develop when the liquid metal solidifies and shrinks between dendrites and between dendrite branches.

Porosity caused by shrinkage can be reduced or eliminated by various means. Basically, adequate liquid metal should be provided to avoid cavities caused by shrinkage. Internal or external **chills**, used in sand casting (Fig. 10.14), also are an effective means of reducing shrinkage porosity. The function of chills is to increase the rate of solidification in critical regions. Internal chills are usually made of the same material as the castings. However, there may be problems with proper fusion of the internal chills with the casting. Foundries try to avoid use of internal chills for this reason. External chills may be made of the same material or may be iron, copper, or graphite. With alloys, porosity can be reduced or eliminated by making the temperature gradient steep, using, for example, mold materials that have high thermal conductivity. Subjecting the casting to hot isostatic pressing is another method of reducing porosity.

Liquid metals have much greater solubility for *gases* than do solid metals (Fig. 10.15). When a metal begins to solidify, the dissolved gases are expelled from the solution. Gases may also result from reactions of the molten metal with the mold materials. Gases either accumulate in regions of existing porosity, such as in interdendritic areas, or they cause microporosity in the casting, particularly in cast

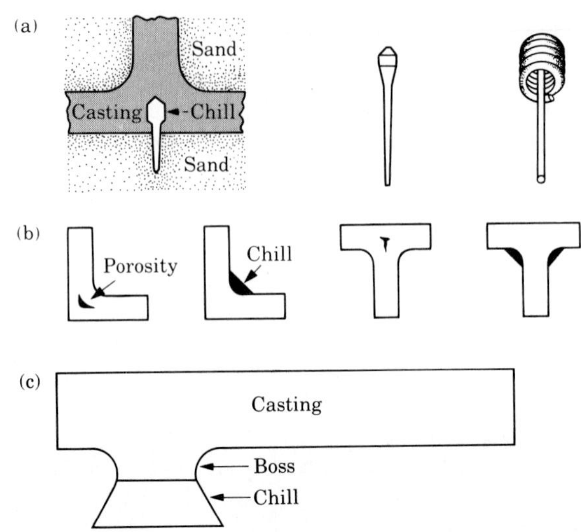

FIGURE 10.14 Various types of (a) internal and (b) external chills (dark areas at corners), used in castings to eliminate porosity caused by shrinkage. Chills are placed in regions where there is a larger volume of metal, as shown in (c).

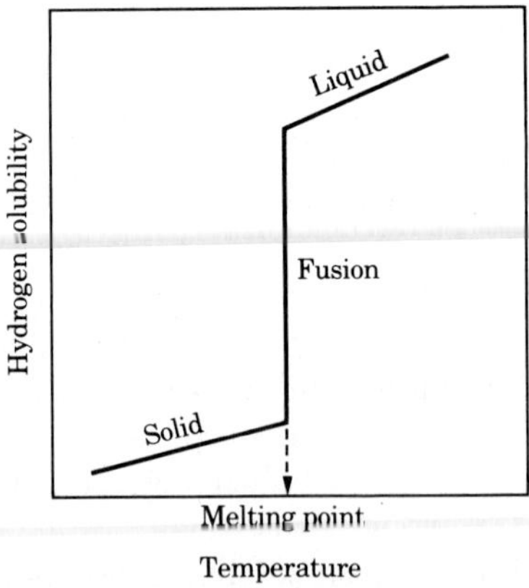

FIGURE 10.15 Solubility of hydrogen in aluminum. Note the sharp decrease in solubility as the molten metal begins to solidify.

iron, aluminum, and copper. Dissolved gases may be removed from the molten metal by flushing or purging with an inert gas or by melting and pouring the metal in a vacuum. If the dissolved gas is oxygen, the molten metal can be *deoxidized.* Steel is usually deoxidized with aluminum, silicon, copper-base alloys with phosphorus copper, titanium, and zirconium-bearing materials. High-quality steels are being produced by argon–oxygen decarburization (AOD).

Whether microporosity is a result of shrinkage or is caused by gases may be difficult to determine. If the porosity is spherical and has smooth walls, much like the shiny surfaces of holes in Swiss cheese, it is generally from gases. If the walls are rough and angular, porosity is likely from shrinkage between dendrites. Gross porosity is from shrinkage and is usually called *shrinkage cavities.*

10.4 Melting Practice

Melting practice is an important aspect of casting operations, because it has a direct bearing on the quality of castings. Furnaces are charged with melting stock consisting of metal, alloying elements, and various other materials such as flux and slag-forming constituents. **Fluxes** are inorganic compounds, such as limestone and dolomite, and may include secondary fluxes, such as sodium carbonate and calcium fluoride (for cast iron) and borax–silica mixtures (for copper alloys). These compounds refine the molten metal by removing dissolved gases and various impurities. However, fluxes can have a detrimental effect on the lining of some furnaces, particularly induction furnaces.

To protect the surface of the molten metal against atmospheric reaction and contamination—and to refine the melt—the pour must be insulated against heat loss. Insulation is usually provided by covering the surface or mixing the melt with compounds that form a *slag*. In casting steels, the composition of the slag includes CaO, SiO_2, MnO, and FeO. A small quantity of liquid metal is usually tapped and its composition analyzed. Necessary additions or inoculations are then made prior to pouring the metal into the molds.

The metal charge may be composed of commercially pure *primary* metals, which are remelted scrap. Clean scrapped castings, gates, and risers may also be included in the charge. If the melting points of the alloying elements are sufficiently low, pure alloying elements are added to obtain the desired composition in the melt. If the melting points are too high, the alloying elements do not mix readily with the low-melting-point metals. In this case, **master alloys**, or *hardeners*, are often used. They usually consist of lower melting-point alloys with higher concentrations of one or two of the needed alloying elements. Differences in specific gravity of master alloys should not be great enough to cause segregation in the melt.

10.4.1 Furnaces

Furnace selection requires careful consideration of several factors that can significantly influence the quality of castings, as well as the economics of casting operations. A variety of furnaces meet requirements for melting and casting metals and alloys in foundries, so proper selection of a furnace generally depends on:

a) economic considerations, such as initial cost and operating and maintenance costs;
b) the composition and melting point of the alloy to be cast and ease of controlling metal chemistry;
c) control of the atmosphere to avoid contamination of the metal;
d) capacity and the rate of melting required;
e) environmental considerations, such as air pollution and noise;

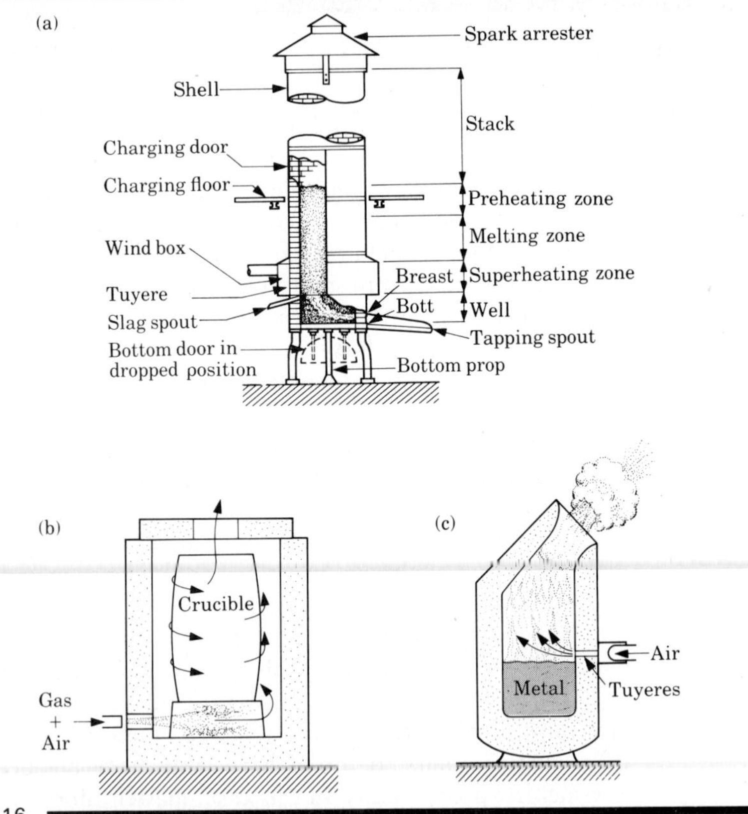

FIGURE 10.16 Types of melting furnaces used in foundries: (a) cupola; (b) crucible; and (c) converter. The selection of a furnace for a particular application depends on many technical and economic factors. *Source:* R. A. Flinn, *Fundamentals of Casting*. Addison-Wesley Publishing Co., Inc., Reading, Massachusetts, 1983.

f) power supply and its availability and cost of fuels;
g) ease of superheating the metal; and
h) type of charge material that can be used.

The most commonly used furnaces in foundries today are cupolas, electric-arc, induction, and gas-fired crucibles. *Cupolas* are basically refractory-lined vertical steel vessels that are charged with alternating layers of metal, coke, and flux (Fig. 10.16a). Although they require major investments, cupolas produce large amounts of molten metal (as much as 40 tons per hour). They are counterflow furnaces in which gases move up and the metal drops to the bottom as melting takes place. They operate continuously and have high melting rates. However, they are being replaced by *induction* furnaces (see Fig. 5.3c) to form composition-controlled smaller melts for casting shapes, especially in smaller foundries. *Gas-fired* furnaces (Fig. 10.16b) are very common for nonferrous metals, particularly aluminum. *Converters* (Fig. 10.16c) need hot metal and are generally used for refining or treating molten metal.

SUMMARY

Casting is a solidification process in which molten metal is poured into a mold and allowed to cool to ambient temperature. Fluid flow and heat transfer are important considerations, as is the design of the mold and gating system to ensure proper flow of the metal into the mold cavities.

Because metals contract during solidification and cooling, cavities can form in the casting. Porosity caused by gases evolving during solidification is a significant problem, particularly because of its adverse effect on the mechanical properties of castings. The grain structure of castings can be controlled by various means to obtain the desired properties. The choice of furnaces is important and involves economic considerations. Melting practices are also important, including proper melting of the materials, preparation for alloying, removal of slag and dross, and pouring the molten metal into the molds. Various defects can develop in castings from lack of control of material and process variables.

TRENDS

- Argon–oxygen decarburization and deoxidation, as well as electroslag remelting, ladle metallurgy, plasma refining, and calcium wire injection for producing high-quality steels, are now being practiced by modern foundries.
- Electromagnetic stirring of the molten metal in the mold and vibrating the molds to obtain smaller and more uniform grain size during solidification, counter-gravity pouring, and shrouding the pouring stream are being investigated.

(*continued*)

TRENDS *(continued)*

- Computer-aided design and manufacturing techniques are being used to predict solidification patterns, prevent casting defects, and calculate weights, volumes, and dimensions for proper mold design and economic production.
- Improvements in the efficiency of furnaces, molten metal quality, and purifying and filtering techniques are being made.

KEY TERMS

Aspiration
Bernoulli's theorem
Casting
Chills
Columnar dendrites
Columnar grains
Continuity defects
Dendrites
Fluidity
Fluxes
Freezing range
Heterogeneous nucleation
Homogeneous nucleation
Inoculant
Macrosegregation
Master alloys
Microsegregation
Mold
Mushy zone
Porosity
Pouring basin
Risers
Runners
Segregation
Shrinkage
Skin
Solidification
Sprue
Turbulence

BIBLIOGRAPHY

Flemings, M.C., *Solidification Processing*. New York: McGraw-Hill, 1974.

Flinn, R.A., *Fundamentals of Metal Casting*. Reading, Mass.: Addison-Wesley, 1963.

Heine, R.W., C.R. Loper, Jr., and C. Rosenthal, *Principles of Metal Casting*, 2d ed. New York: McGraw-Hill, 1967.

Metals Handbook, 9th ed., Vol. 15: *Casting*. Metals Park, Ohio: ASM International, 1988.

Mikelonis, P.J. (ed.), *Foundry Technology: A Source Book*. Metals Park, Ohio: American Society for Metals, 1982.

Minkoff, I., *Solidification and Cast Structure*. New York: Wiley, 1986.

Szekely, J., *Fluid Flow Phenomena in Metals Processing*. New York: Academic Press, 1979.

REVIEW QUESTIONS

10.1 Why is casting an important manufacturing process?

10.2 What is the difference between the solidification of pure metals and metal alloys?

10.3 What are dendrites?

10.4 State the difference between short and long freezing ranges. How is range determined?

10.5 Describe the parameters on which solidification time depends.
10.6 Define shrinkage and porosity. How can you tell whether cavities in a casting are due to porosity or to shrinkage?
10.7 What is the function of chills?
10.8 How are dissolved gases removed from molten metal?
10.9 Describe the features of a gating system.
10.10 How is fluidity defined? Why is it important?
10.11 Name the factors involved in the selection of furnaces.
10.12 What are master alloys?
10.13 Explain the reasons for hot tearing in castings.
10.14 Name various defects in castings.
10.15 Why is it important to remove dross or slag during the pouring of molten metal into the mold? What methods are used to remove them?
10.16 What are the effects of mold materials on fluid flow and heat transfer?
10.17 Describe the stages involved in the contraction of metals during casting.

QUESTIONS AND PROBLEMS

10.18 Explain the reasons why heat transfer and fluid flow are so important in metal casting.
10.19 We know that pouring metal at a high rate into a mold has certain disadvantages. Are there any disadvantages to pouring it very slowly?
10.20 Describe the events depicted in Fig. 10.4.
10.21 Would you be concerned about the fact that parts of internal chills are left within the casting? What materials do you think chills should be made of, and why?
10.22 Can you think of fluidity tests other than that shown in Fig. 10.8? Explain your tests.
10.23 What practical illustrations can you offer to indicate the relationship of solidification time to volume and surface area?
10.24 Do you think early formation of dendrites in a mold can impede the free flow of molten metal into the mold? Give an illustration.
10.25 Explain why you may want to subject a casting to various heat treatments.
10.26 Casting is one of the earliest manufacturing processes, dating back to about 4000 B.C. Why do you think this is so?
10.27 Why does porosity have detrimental effects on the mechanical properties of castings? Would physical properties such as thermal and electrical conductivity also be affected by porosity? Explain.
10.28 Assume that the Summary for this chapter is missing. Write a two-page summary.
10.29 Explain the use of risers. Why can blind risers be smaller than open-top risers?
10.30 A spoked handwheel is to be cast in gray iron. In order to prevent hot tearing of the spokes, would you insulate the spokes or chill them? Explain.
10.31 Which of the following considerations are important for a riser to function properly? (a) Have a surface area larger than the part being cast. (b) Be kept open to atmospheric pressure. (c) Solidify first. Why?
10.32 A round casting is 0.1 m in diameter and 0.5 m in length. Another casting of the same metal is elliptical in cross-section, with a major-to-minor axis ratio of 2, and has the same length and cross-sectional area as the round casting. Both pieces are cast under

the same conditions. What is the difference in the solidification times of the two castings?

10.33 Explain why the constant C in Eq. (10.7) depends on mold material, metal properties, and temperature.

10.34 Are external chills as effective as internal chills? Explain.

10.35 Explain why gray cast iron undergoes expansion, rather than contraction, during solidification, as shown in Table 10.1.

10.36 Referring to Fig. 10.10, explain why internal corners (as A) develop a thinner skin than external corners (as B) during solidification.

10.37 Note the shape of the two risers in Fig. 10.7, and discuss your observations with respect to Eq. (10.7).

10.38 Is there any difference in the tendency for shrinkage void formation for metals with short freezing and long freezing ranges, respectively? Explain.

10.39 What is the influence of the cross-sectional area of the spiral channel in Fig. 10.8 on fluidity test results? What is the effect of sprue height? If this test is run with the test setup heated to elevated temperatures, would the test results be useful? Explain.

10.40 In Section 10.4, we have outlined the factors involved in furnace selection for melting metals and alloys. Explain why the type of furnace selected depends on these factors.

10.41 Make a list of safety considerations and precautions that should be taken concerning all aspects of melting and casting of metals, including the equipment involved.

10.42 Sketch a graph of specific volume versus temperature for a metal that shrinks as it cools from the liquid state to room temperature. On the graph, mark the area where shrinkage is compensated for by risers.

10.43 The pulley shown in the figure below is an aluminum sand casting. The sections of both the rim and the spokes are also shown. Residual stresses have been found to exist in this part. Describe why these stresses exist and determine whether the stresses tend to compress the spokes or pull them apart.

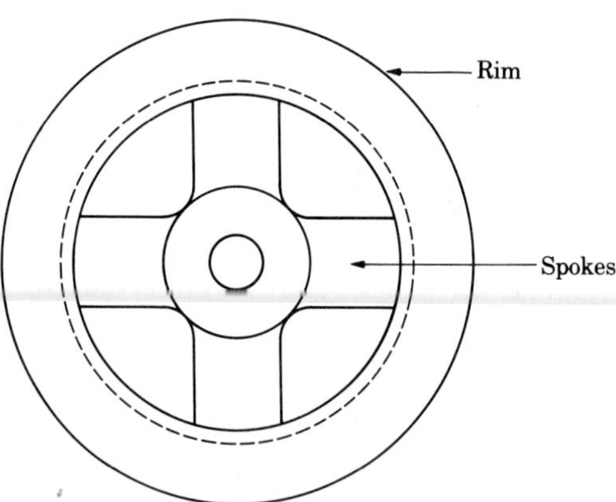

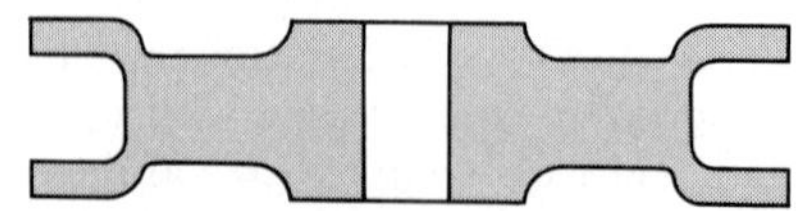

10.44 The optimum shape of a riser is spherical, to ensure that it cools more slowly than the casting it feeds. Spherically shaped risers, however, are difficult to cast. (a) Sketch the shape of a blind riser that is easy to mold, but also has the smallest possible surface area-to-volume ratio. (b) Compare the solidification time of the riser in part (a) to that of a riser shaped like a right circular cylinder. Assume that the volume of each riser is the same, and that for each the height is equal to the diameter. (See the example in Section 10.3.4.)

10.45 A 50-mm (2-in.) thick square plate and a right circular cylinder with a radius of 100 mm (4 in.) and height of 50 mm each have the same volume. If each is to be cast using a cylindrical riser, will each part require the same size riser to ensure proper feeding? Explain.

11

Metal-Casting Processes

11.1 Introduction

In Chapter 10, we presented the fundamentals that underlie all casting processes. In this chapter, we move to consideration of the major metal-casting processes and their principles, advantages, and limitations. Many parts and components are made by casting, including carburetors, frying pans, engine blocks, crankshafts, railroad-car wheels, plumbing fixtures, power tools, gun barrels, and machine bases. Various casting processes have been developed over a long period of time, each with its own characteristics and applications, to meet specific engineering and service requirements. In fact, the first castings were made during the period of 4000–3000 B.C., using stone and metal molds for casting copper.

Two trends currently are having a large impact on the casting industry. The first is continuing *mechanization* and *automation* of the casting process, which has led to significant changes in the use of equipment and labor. Advanced machinery and automated process-control systems have replaced traditional methods of casting. Moreover, casting processes that especially lend themselves to advances in technology are developing significant economic advantages over other processes.

The second major trend affecting the casting industry is the increasing demand for *high-quality* castings with *close tolerances*. This demand is spurring the further development of casting processes that produce high-quality castings (Table 11.1; see also Fig. 13.11). We emphasize the significance of these trends as we discuss the major casting processes.

This chapter is organized around the major classifications of casting practices (see Fig. II.2 in the Introduction to Part II). These classifications are related to mold materials, molding processes, and methods of feeding the mold with the molten metal. The two major categories are expendable-mold and permanent-mold casting.

Expendable molds are made of sand, plaster, ceramics, and similar materials, which are generally mixed with various binders, or bonding agents. As we described in Chapter 8, these materials are refractories, that is, they have the capability to withstand the high temperatures of molten metals. After the casting has solidified, the molds in these processes are broken up to remove the casting.

Permanent molds, as the name implies, are used repeatedly and are designed in such a way that the casting can be easily removed and the mold used for the next casting. These molds are made of metals that maintain their strength at high temperatures and thus can be used repeatedly. Because metal molds are better heat conductors than expendable molds, the solidifying casting is subjected to a higher rate of cooling, which in turn affects the microstructure and grain size within the casting.

Composite molds are made of two or more different materials, such as sand, graphite, and metal, combining the advantages of each material. They are used in various casting processes to improve mold strength, cooling rates, and overall economics of the process.

We consider the major expendable-mold processes in Sections 11.2–11.9, then continue with permanent-mold processes in Sections 11.10–11.15. Because of their unique characteristics and applications, particularly in manufacturing microelectronic devices, we also describe basic crystal-growing techniques. We conclude the chapter with an overview of modern foundries.

11.2 Sand Casting

The traditional method of casting metals is in sand molds and has been used for millenia. Simply stated, **sand casting** consists of placing a pattern (having the shape of the desired casting) in sand to make an imprint, incorporating a gating system, filling the resulting cavity with molten metal, allowing the metal to cool until it

TABLE 11.1
GENERAL CHARACTERISTICS OF CASTING PROCESSES

PROCESS	TYPICAL MATERIALS CAST	WEIGHT (kg)		TYPICAL SURFACE FINISH (μm, R_a)	POROSITY*	SHAPE COMPLEXITY*	DIMENSIONAL ACCURACY*	SECTION THICKNESS (mm)	
		MINIMUM	*MAXIMUM*					*MINIMUM*	*MAXIMUM*
Sand	All	0.05	No limit	5–25	4	1–2	3	3	No limit
Shell	All	0.05	100+	1–3	4	2–3	2	2	—
Evaporative pattern	All	0.05	No limit	5–20	4	1	2	2	No limit
Plaster	Nonferrous (Al, Mg, Zn, Cu)	0.05	50+	1–2	3	1–2	2	1	—
Investment	All (High-melting pt.)	0.005	100+	1–3	3	1	1	1	75
Permanent mold	All	0.5	300	2–3	2–3	3–4	1	2	50
Die	Nonferrous (Al, Mg, Zn, Cu)	<0.05	50	1–2	1–2	3–4	1	0.5	12
Centrifugal	All	—	5000+	2–10	1–2	3–4	3	2	100

* Relative rating: 1 best, 5 worst.

Note: These ratings are only general; significant variations can occur, depending on the methods used.

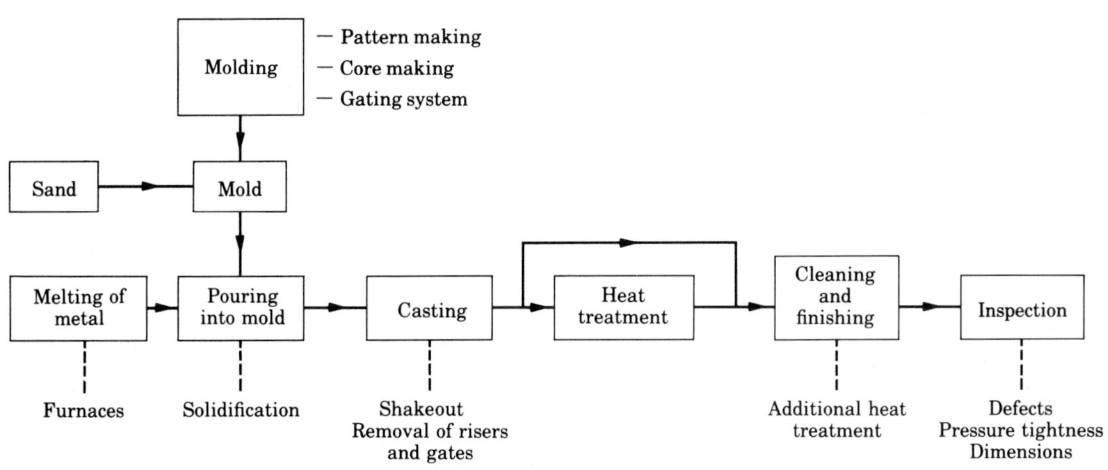

FIGURE 11.1
Outline of production steps in a typical sand-casting operation.

solidifies, breaking away the sand mold, and removing the casting (Fig. 11.1). Although the origins of sand casting date to ancient times, it is still the most prevalent form of casting. In the United States alone, about 15 million tons of metal are cast by this method each year. Typical parts made by sand casting are machine-tool bases, engine blocks, cylinder heads, and pump housings (Fig. 11.2).

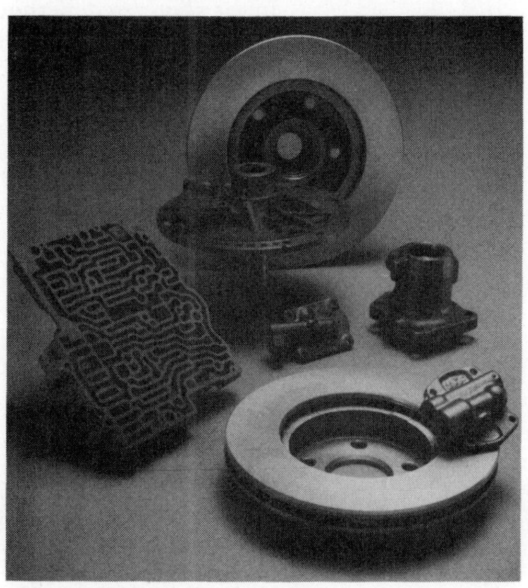

FIGURE 11.2
Typical gray iron castings used in automobiles, including transmission valve body (left); hub rotor with disk-brake cylinder (front). *Source:* Courtesy of Central Foundry Division of General Motors Corporation.

11.2.1 Sands

Most sand casting operations use silica sand (SiO_2). Sand is the product of the disintegration of rocks over extremely long periods of time. It is inexpensive and is suitable as mold material because of its resistance to high temperatures. There are two general types of sand: *naturally bonded* (*bank* sands) and *synthetic* (*lake* sands). Because its composition can be controlled more accurately, synthetic sand is preferred by most foundries. Several factors are important in the selection of sand for molds. Sand having fine, round grains can be closely packed and forms a smooth mold surface. Good permeability of molds and cores allows gases and steam evolved during casting to escape easily. The mold should have good collapsibility (the casting shrinks while cooling) to avoid defects in the casting, such as hot tearing and cracking. The selection of sand involves certain tradeoffs with respect to properties. For example, fine-grained sand enhances mold strength, but the fine grains also lower mold permeability. Sand is typically conditioned before use. *Mulling* machines are used to uniformly mull (mix thoroughly) sand with additives. Clay is used as a cohesive agent to bond sand particles, giving the sand strength.

11.2.2 Types of sand molds

Sand molds are characterized by the types of sand that comprise them and by the methods used to produce them. There are three basic types of sand molds: green-sand, cold-box, and no-bake molds.

The most common mold material is **green molding sand**. The term *green* refers to the fact that the sand in the mold is moist or damp while the metal is being poured into it. Green molding sand is a mixture of sand, clay, and water. Green-sand molding is the least expensive method of making molds.

In the *skin-dried* method, the mold surfaces are dried, either by storing the mold in air or drying it with torches. Skin-dried molds are generally used for large castings because of their higher strength. Sand molds are also oven dried (*baked*) prior to receiving the molten metal. They are stronger than green-sand molds and impart better dimensional accuracy and surface finish to the casting. However, distortion of the mold is greater, the castings are more susceptible to hot tearing because of the lower collapsibility of the mold, and the production rate is slower because of the drying time required.

In the *cold-box mold* process, various organic and inorganic **binders** are blended into the sand to bond the grains chemically for greater strength. These molds are dimensionally more accurate than green-sand molds but are more expensive.

In the *no-bake mold* process, a synthetic liquid resin is mixed with the sand, and the mixture hardens at room temperature. Because bonding of the mold in this and the cold-box process takes place without heat, they are called *cold-setting processes*.

Major components of sand molds, as shown in Fig. 11.3, are:

- The mold itself, which is supported by a *flask*. Two-piece molds consist of a *cope* on top and a *drag* on the bottom. The seam between them is the

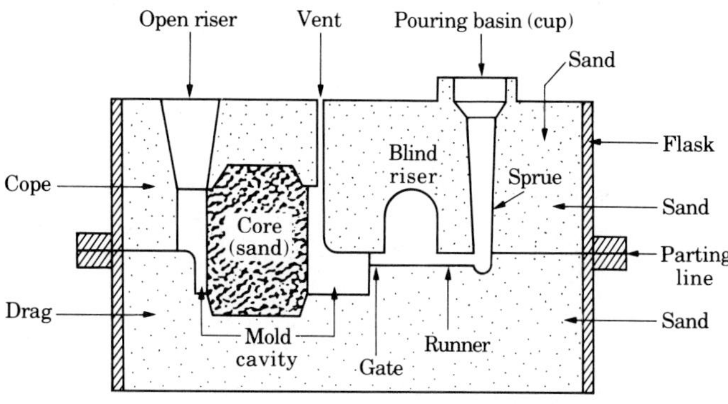

FIGURE 11.3
Schematic illustration of a sand mold showing various features.

parting line. When more than two pieces are used, the additional parts are called *cheeks*.

- A *pouring basin* or *pouring cup*, into which the molten metal is poured.
- A *sprue*, through which the molten metal flows downward.
- The *runner system*, which has channels that carry the molten metal from the sprue to the mold cavity. *Gates* are the inlets into the mold cavity.
- *Risers*, which supply additional metal to the casting as it shrinks during solidification. Figure 11.3 shows two different types of risers: a *blind riser* and an *open riser*.
- *Cores*, which are inserts made from sand. They are placed in the mold to form hollow regions or otherwise define the interior surface of the casting. Cores are also used on the outside of the casting to form features such as lettering on the side of a casting or deep external pockets.
- *Vents*, which are placed in molds to carry off gases produced when the molten metal comes into contact with the sand in the molds and core. They also exhaust air from the mold cavity as the molten metal flows into the mold.

11.2.3 Patterns

Patterns are used to mold the sand mixture into the shape of the casting. They may be made of wood, plastic, or metal. The selection of a pattern material depends on the size and shape of the casting, the dimensional accuracy and the quantity of castings required, and the molding process to be used (Table 11.2). Because patterns are used repeatedly to make molds, the strength and durability of the material selected for patterns must reflect the number of castings that the mold will produce. They may be made of a combination of materials to reduce wear in critical regions. Patterns are usually coated with a **parting agent** to facilitate their removal from the molds.

TABLE 11.2
CHARACTERISTICS OF PATTERN MATERIALS

	RATING[a]				
CHARACTERISTIC	*WOOD*	*ALUMINUM*	*STEEL*	*PLASTIC*	*CAST IRON*
Machinability	E	G	F	G	G
Wear resistance	P	G	E	F	E
Strength	F	G	E	G	G
Weight[b]	E	G	P	G	P
Repairability	E	P	G	F	G
Resistance to:					
Corrosion[c]	E	E	P	E	P
Swelling[c]	P	E	E	E	E

Source: D. C. Ekey and W. P. Winter, *Introduction to Foundry Technology*. New York: McGraw-Hill, 1958.
[a] E, excellent; G, good; F, fair; P, poor.
[b] As a factor in operator fatigue.
[c] By water.

Patterns can be designed with a variety of features to fit application and economic requirements. *One-piece patterns* are generally used for simpler shapes and low-quantity production. They are generally made of wood and are inexpensive. *Split patterns* are two-piece patterns made so that each part forms a portion of the cavity for the casting. In this way castings having complicated shapes can be produced. *Match-plate patterns* are a popular type of mounted pattern in which two-piece patterns are constructed by securing each half of one or more split patterns to the opposite sides of a single plate (Fig. 11.4). In such constructions, the gating system can be mounted on the drag side of the pattern. This type of pattern is used most often in conjunction with molding machines and large production runs to produce smaller castings.

Pattern design is a crucial aspect of the total casting operation. The design should provide for metal shrinkage, ease of removal from the sand mold by means of a taper or draft (Fig. 11.5), and proper metal flow in the mold cavity. We discuss these topics in greater detail in Chapter 12.

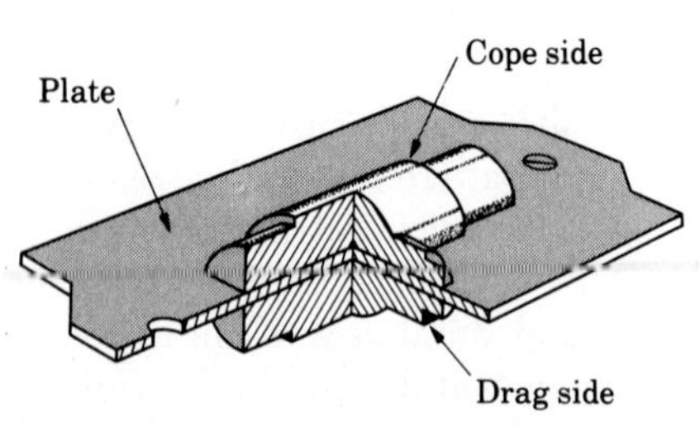

FIGURE 11.4
A typical metal match-plate pattern used in sand casting.

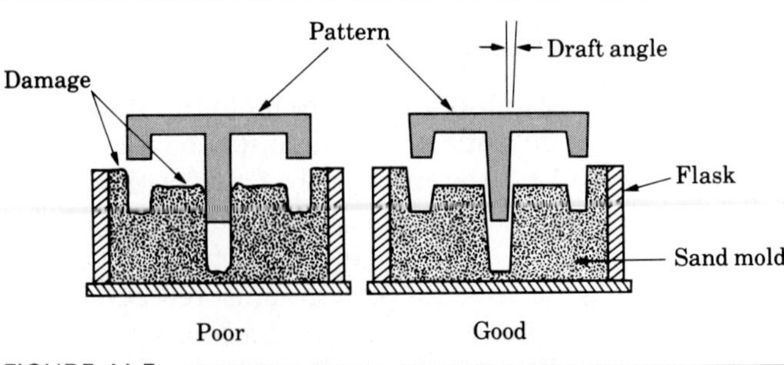

FIGURE 11.5
Taper on patterns for ease of removal from the sand mold.

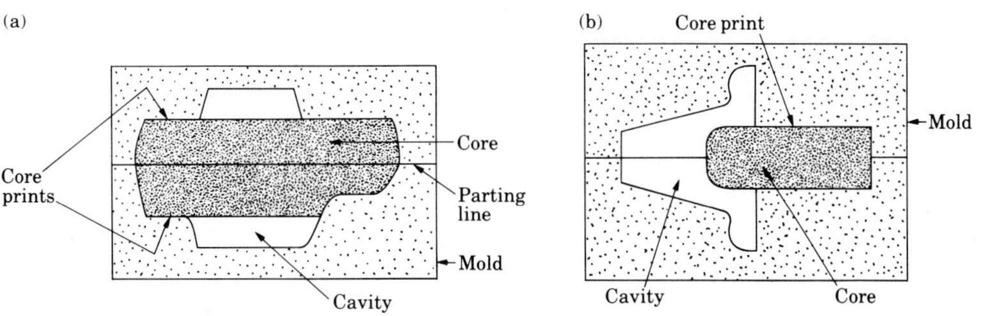

FIGURE 11.6
Examples of sand cores supported by core prints.

11.2.4 Cores

For castings with internal cavities or passages, such as in an automotive engine block or a valve body, cores are utilized. Cores are placed in the mold cavity before casting to form the interior surfaces of the casting and are removed from the finished part during shakeout and further processing. Like molds, cores must possess strength, permeability, ability to withstand heat, and collapsibility. Therefore cores are made of sand aggregates.

The core is anchored by *core prints*. These are recesses that are added to the pattern to support the core and to provide vents for the escape of gases (Fig. 11.6). A common problem with cores is that for certain casting requirements, as in the case where a recess is required, they may lack sufficient structural support in the cavity. To keep the core from shifting, metal supports, known as **chaplets**, may be used to anchor the core in place.

Cores are generally made in a manner similar to that used in making molds, and the majority are made with shell, no-bake, or cold-box processes. Cores are formed in *core boxes*, which are used much like patterns are used to form sand molds. The sand can be packed into the boxes with sweeps or blown into the box by compressed air from *core blowers*. Core blowers have the advantages of producing uniform cores and operating at a very high production rate.

11.2.5 Sand-molding machines

The oldest known method of molding, which is still used for simple castings, is to compact the sand by hand hammering or ramming it around the pattern. For most operations, however, the sand mixture is compacted around the pattern by molding machines (Fig. 11.7). These machines eliminate arduous labor, offer higher quality casting by improving the application and distribution of forces, manipulate the mold in a carefully controlled fashion, and increase the rate of production.

Mechanization of the molding process can be further assisted by *jolting* the assembly. The flask, molding sand, and pattern are placed on a pattern plate

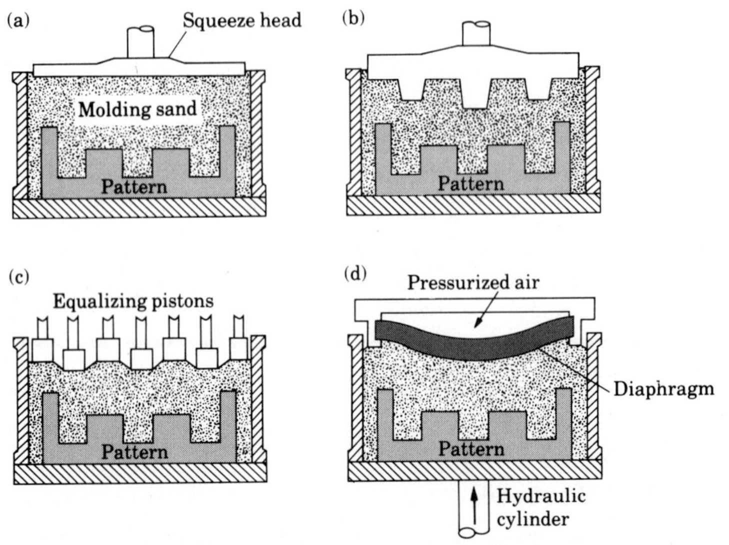

FIGURE 11.7 Various designs of squeeze heads for mold making: (a) conventional flat head; (b) profile head; (c) equalizing squeeze pistons; and (d) flexible diaphragm. *Source:* © Institute of British Foundrymen. Used with permission.

mounted on an anvil, and jolted upward by air pressure at rapid intervals (Fig. 11.8). The inertial forces compact the sand around the pattern. Jolting produces the highest compaction at the horizontal parting line, whereas in squeezing, compaction is highest at the squeezing head (Fig. 11.7). Thus more uniform compaction can be obtained by combining them (Fig. 11.8b).

In *vertical flaskless molding*, the halves of the pattern form a vertical chamber wall against which sand is blown and compacted (Fig. 11.9). Then the mold halves are packed horizontally, with the parting line oriented vertically and moved along a pouring conveyor. This operation is simple and eliminates the need to handle flasks, making potential production rates very high, particularly when other aspects of the operation, such as coring and pouring, are automated.

Sandslingers fill the flask uniformly with sand under a stream of high pressure. They are used to fill large flasks and are typically operated by machine. An impeller in the machine throws sand from its blades or cups at such high speeds that the machine not only places the sand but also rams it appropriately.

In *impact molding*, the sand is compacted by controlled explosion or instantaneous release of compressed gases. This method produces molds with uniform strength and good permeability.

In *vacuum molding*, also known as the "*V*" *process*, the pattern is covered tightly by a thin sheet of plastic. A flask is placed over the coated pattern and is filled with dry binderless sand. A second sheet of plastic is placed on top of the sand, and a vacuum action hardens the sand so that the pattern can be withdrawn. Both halves

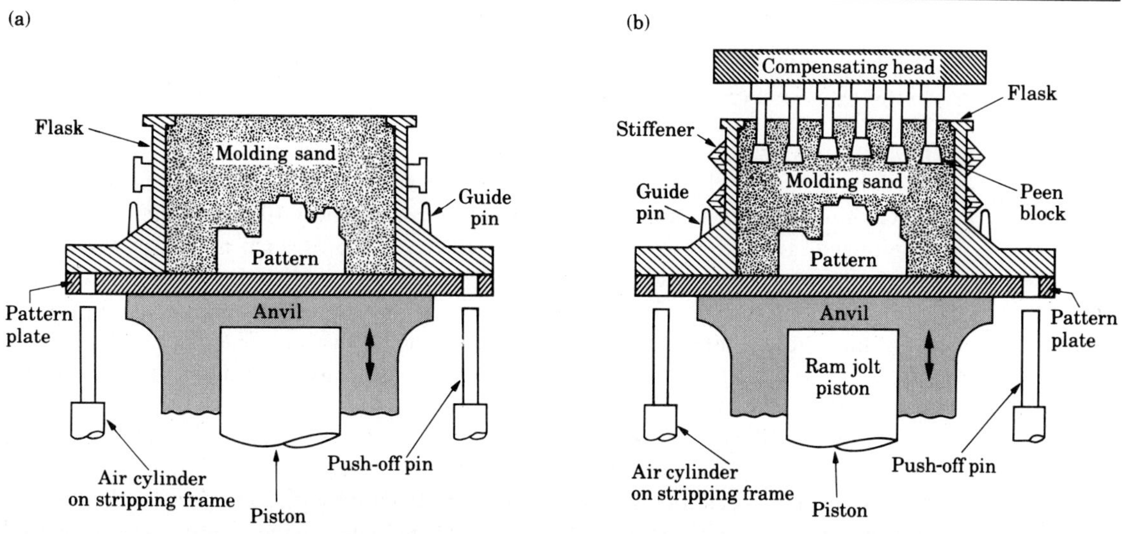

FIGURE 11.8
(a) Schematic illustration of a jolt-type mold-making machine. (b) Schematic illustration of a mold-making machine which combines jolting and squeezing.

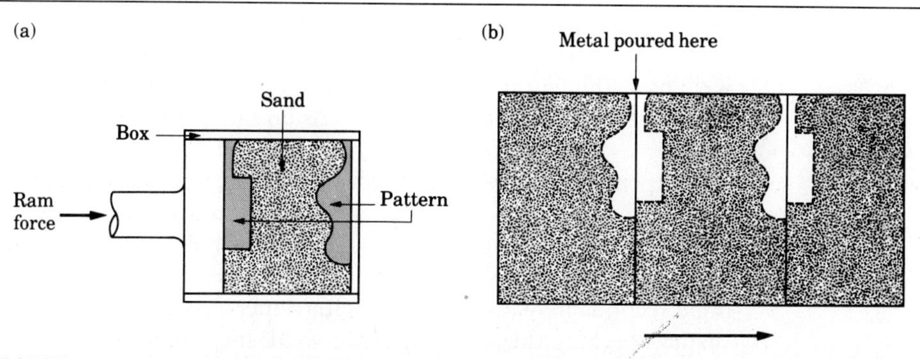

FIGURE 11.9
Vertical flaskless molding. (a) Sand is squeezed between two halves of the pattern. (b) Assembled molds pass along an assembly line for pouring.

of the mold are made this way and then assembled. During pouring, the mold remains under a vacuum but the casting cavity does not. When the metal has solidified, the vacuum is turned off and the sand falls away, releasing the casting. Vacuum molding produces castings having very good detail and accuracy. It is especially well suited for large, relatively flat castings.

11.2.6 The sand-casting operation

After the mold has been shaped and the cores have been placed in position, the two halves (cope and drag) are closed, clamped, and weighted down. They are weighted to prevent the separation of the mold sections under the pressure exerted when the molten metal is poured into the mold cavity.

The design of the gating system is important for proper delivery of the molten metal into the mold cavity. As we described in Section 10.3, turbulence must be minimized, air and gases must be allowed to escape by such means as vents, and proper temperature gradients must be established and maintained to minimize shrinkage and porosity. The design of risers is also important in order to supply the necessary molten metal during solidification of the casting. The pouring basin may also serve as a riser. A complete sequence of operations in sand casting is shown in Fig. 11.10.

After solidification, the casting is shaken out of its mold, and the sand and oxide layers adhering to the casting are removed by vibration (using a shaker) or by sand blasting. Ferrous castings are also cleaned by blasting with steel shot (shot blasting) or grit. The risers and gates are cut off by oxyfuel-gas cutting, sawing, shearing, and abrasive wheels, or they are trimmed in dies. Gates and risers on steel castings are also removed with air carbon-arc or powder-injection torches. Castings may be cleaned by electrochemical means or by pickling with chemicals to remove surface oxides.

The surface of castings is important in subsequent machining operations, because machinability can be adversely affected if the castings are not cleaned properly and sand particles remain on the surface. If regions of the casting have not formed properly or have formed incompletely, the defects may be repaired by welding by filling them with weld metal. Sand-mold castings generally have rough, grainy surfaces, although that depends on the quality of the mold and the materials used.

Depending on the metal used, the casting may subsequently be heat treated to improve certain properties needed for its intended service use; these processes are particularly important for steel castings. Finishing operations may involve straightening or forging with dies to obtain final dimensions, and machining. Minor surface imperfections may also be filled with a metal-filled epoxy, especially for cast-iron castings because they are difficult to weld. Inspection is an important final step and is carried out to ensure that the casting meets all design and quality-control requirements.

Almost all commercially used metals can be sand cast. The surface finish obtained is largely a function of the materials used in making the mold. Dimensional accuracy is not as good as that of other casting processes. However, intricate shapes can be cast by this process, such as cast-iron engine blocks and very large propellers for ocean liners and impellers (Fig. 11.11). Sand casting can be economical for relatively small production runs, and equipment costs are generally low. The characteristics of sand casting and other casting processes are given in Table 11.1 (see also Chapter 12).

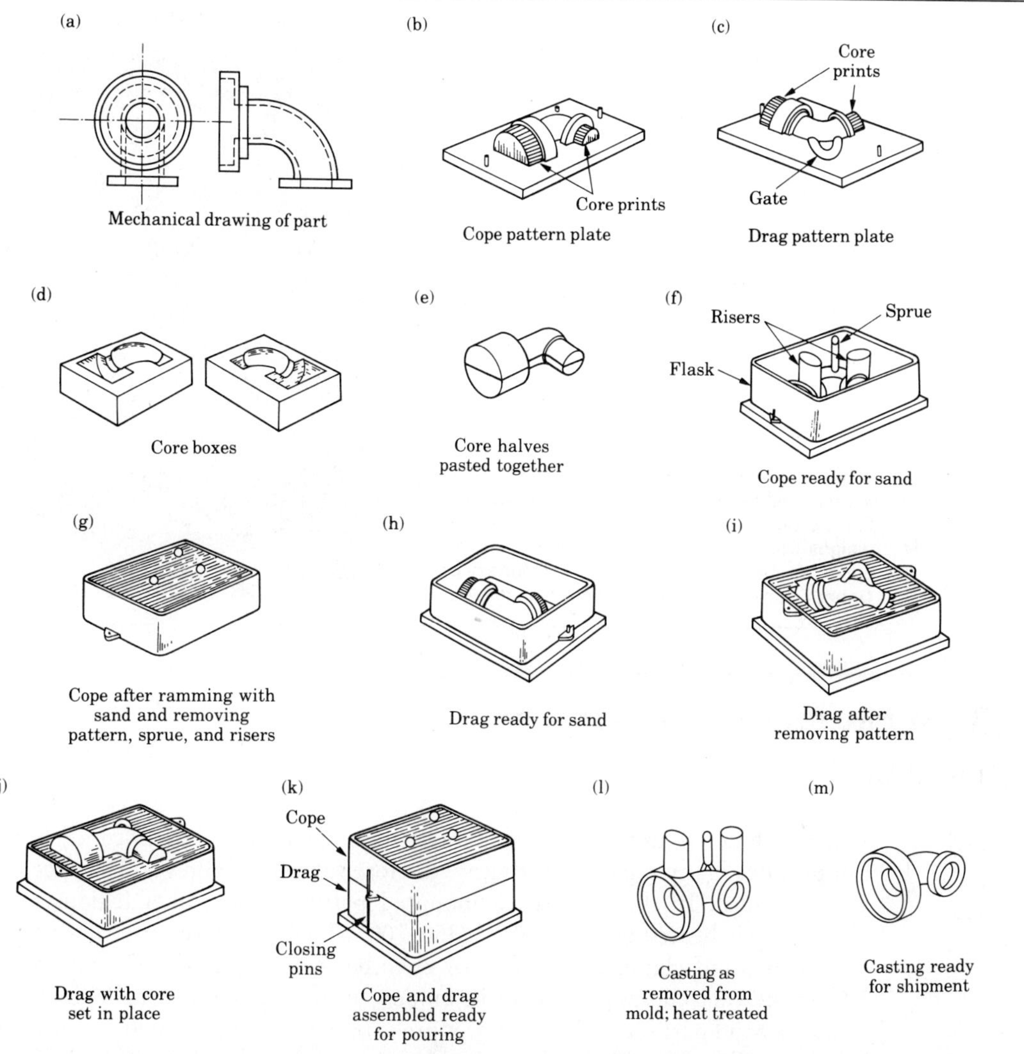

FIGURE 11.10

Schematic illustration of the sequence of operations for sand casting. *Source:* Steel Founders' Society of America. (a) A mechanical drawing of the part is used to generate a design for the pattern. Considerations such as part shrinkage and draft must be built into the drawing. (b–c) Patterns have been mounted on plates equipped with pins for alignment. Note the presence of core prints designed to hold the core in place. (d–e) Core boxes produce core halves, which are pasted together. The cores will be used to produce the hollow area of the part shown in (a). (f) The cope half of the mold is assembled by taking the cope pattern plate, securing it to the flask through aligning pins, and attaching inserts to form the sprue and risers. (g) The flask is rammed with sand and the plate and inserts are removed. (h) The drag half is produced in a similar manner, with the pattern inserted. A bottom board is placed below the drag and aligned with pins. (i) The pattern, flask, and bottom board are inverted, and the pattern is withdrawn, leaving the appropriate imprint. (j) The core is set in place within the drag cavity. (k) The mold is closed by placing the cope on top of the drag and securing the assembly with pins. The flasks are then subjected to pressure to counteract buoyant forces in the liquid, which might lift the cope. (l) After the metal solidifies, the casting is removed from the mold. (m) The sprue and risers are cut off and recycled, and the casting is cleaned, inspected, and heat treated (when necessary).

FIGURE 11.11
A cast-steel runner (buckets and shroud) for a Francis-type hydraulic turbine. This casting weighs 50,000 kg (110,000 lb) and has a diameter of 4.6 m (180 in.). *Source:* Courtesy of Georg Fischer AG, Schaffhausen, Switzerland.

11.3 Shell Molding

Shell molding was first developed in the 1940s and has grown significantly because it can produce many types of castings with close tolerances and good surface finishes at a low cost. In this process, a mounted pattern, made of a ferrous metal or aluminum, is heated to 175–370 °C (350–700 °F), coated with a parting agent such as silicone, and clamped to a box or chamber containing a fine sand containing 2.5–4.0 percent thermosetting resin binder, such as phenol-formaldehyde, which coats the sand particles. The sand mixture is blown over the heated pattern, coating it evenly. The assembly is then placed in an oven for a short period of time to complete the curing of the resin. The oven in most shell-molding machines is a metal box with gas-fired burners that swings over the shell mold to cure it. The shell hardens around the pattern and is removed from the pattern using built-in ejector pins. Two half-shells are made in this manner and are bonded or clamped together in preparation for pouring.

The thickness of the shell can be accurately determined by controlling the time that the pattern is in contact with the mold. In this way, the shell can be formed with the required strength and rigidity to hold the weight of the molten liquid. The shells are light and thin, usually 5–10 mm (0.2–0.4 in.), and consequently their thermal characteristics are different from those for thicker molds. Shell sand has a much lower permeability than sand used for green-sand molding, because a much smaller grain-size sand is used for shell molding. The decomposition of the shell sand binder also produces a high volume of gas. Unless the molds are properly vented, trapped air and gas can cause serious problems in shell molding of ferrous castings.

Shell molds are generally poured with the parting line horizontal and may also be supported by sand. The walls of the mold are relatively smooth, offering low resistance to flow of the molten metal and producing castings with sharper corners, thinner sections, and smaller projections than are possible in green-sand molds. With the use of multiple gating systems, several castings can be made in a single mold. Nearly any metal suited for sand casting may be cast by the shell-molding process.

Shell molding may be more economical than other casting processes, depending on various production factors. The cost of the resin binders is offset somewhat by the fact that only 5 percent as much sand is used, compared to sand casting. The relatively high cost of metal patterns becomes a smaller factor as the size of production runs increases. The high quality of the finished casting can significantly reduce cleaning, machining, and other finishing costs. Complex shapes can be produced with less labor, and the process can be automated fairly easily. Shell-molding applications include small mechanical parts requiring high precision, such as gear housings, cylinder heads, and connecting rods. Shell molding is also widely used in producing high-precision molding cores.

11.3.1 Composite molds

Composite molds are made of two or more different materials and are used in shell molding and other casting processes. They are generally employed in casting complex shapes, such as impellers for turbines. Examples of composite molds are shown in Figs. 11.12 and 11.13. Molding materials commonly used are shells (made

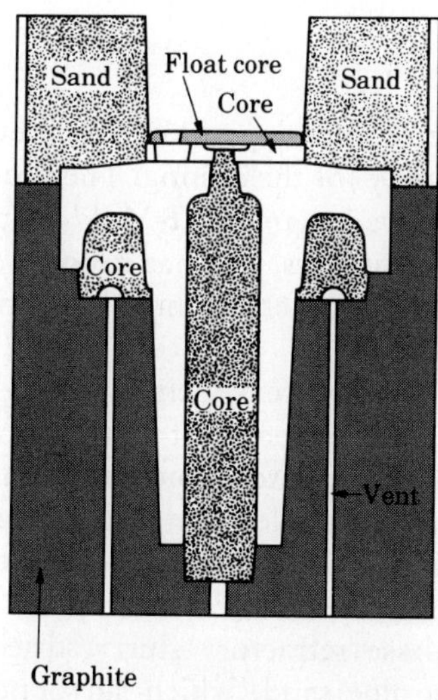

FIGURE 11.12 Schematic illustration of a semipermanent composite mold. *Source: Steel Castings Handbook,* 5th ed. Steel Founders' Society of America, 1980.

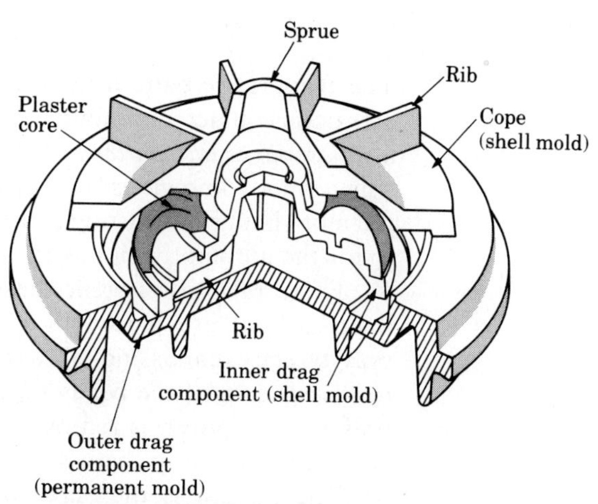

FIGURE 11.13 A composite mold used in casting an aluminum-alloy torque converter. This part was previously cast in an all-plaster mold. *Source: Metals Handbook,* vol. 5, 8th ed.

as previously described), plaster, sand with binder, metal, and graphite. Composite molds may also include cores and chills to control the rate of solidification in critical areas of castings. Composite molds increase the strength of the mold, improve the dimensional accuracy and surface finish of castings, and may help reduce overall costs and processing time.

11.4 Sodium Silicate Process

The mold material in the **sodium silicate process** is a mixture of sand and 1.5–6 percent sodium silicate (waterglass) as the binder for sand. This mixture is packed around the pattern and hardened by blowing carbon dioxide (CO_2) gas through it. This process, also known as *silicate-bonded sand* or the *carbon-dioxide* process, was first used in the 1950s and has been developed further, such as by using various other chemicals for binders. Cores made by this process reduce the tendency to tear because of their compliance at elevated temperatures.

11.5 Evaporative Pattern Casting (Lost Foam)

The **evaporative pattern casting process** uses a polystyrene pattern, which evaporates upon contact with molten metal to form a cavity for the casting. The process is also known as *lost-pattern* casting, and under the trade name *Full-Mold* process. It was formerly known as the expanded polystyrene process, and has become one of the more important casting processes for ferrous and nonferrous metals, particularly for the automotive industry.

First, raw polystyrene beads, containing 5–8 percent pentane (a volatile hydrocarbon) are placed in a preheated die, usually made of aluminum. The polystyrene expands and takes the shape of the die cavity. Additional heat is applied to fuse and bond the beads together. The die is then cooled and opened, and the polystyrene pattern is removed. Complex patterns may also be made by bonding various individual pattern sections by hot-melt adhesive.

The pattern is then coated with a water-base refractory slurry, dried, and placed in a flask. The flask is filled with loose fine sand, which surrounds and supports the pattern. The sand may be dried or mixed with bonding agents to give it additional strength. The sand is periodically compacted by various means (see Figs.

11.7 and 11.8). Then, without removing the polystyrene pattern, the molten metal is poured into the mold. This action immediately vaporizes the pattern (an ablation process) and fills the mold cavity, completely replacing the space previously occupied by the polystyrene pattern. The heat degrades (depolymerizes) the polystyrene and the degradation products are vented into the surrounding sand.

The flow velocity in the mold depends on the rate of degradation of the polymer. Studies have shown that the flow of the molten metal is basically laminar, with Reynold numbers in the range of 400 to 3000 (Section 10.3.1). The velocity of the molten metal at the metal-polymer pattern front is estimated to be in the range of 0.1–1 m/s. The velocity can be controlled by producing patterns with cavities or hollow sections; thus the velocity will increase as the molten metal crosses these hollow regions, similar to pouring in an empty cavity as is done in sand casting. Furthermore, because the polymer requires considerable energy to degrade, large thermal gradients are present at the metal-polymer interface; in other words, the molten metal cools faster than it would if it were poured into a cavity. Consequently, fluidity (see Section 10.3.2) is less than in sand casting. This has important effects on the microstructure throughout the casting and also leads to directional solidification of the metal.

The evaporative pattern process has a number of advantages over other casting methods:

a) It is a relatively simple process because there are no parting lines, cores, or riser systems; hence it has design flexibility.
b) Inexpensive flasks are sufficient for the process.
c) Polystyrene is inexpensive and can be easily processed into patterns having very complex shapes, various sizes, and fine surface detail.
d) The casting requires minimum finishing and cleaning operations.
e) The process is economical for long production runs. A major factor is the cost to produce the die for expanding the polystyrene beads to make the pattern.
f) The process can be automated.

Typical applications for this process are cylinder heads, crankshafts, brake components and manifolds for automobiles, and machine bases. The aluminum engine blocks and other components of the General Motors Saturn automobile are made by this process. Recent developments include the use of polymethylmethacrylate (PMMA) and polyalkylene carbonate as pattern materials for ferrous castings. In a modification of the evaporative pattern process, a polystyrene pattern is surrounded by a ceramic shell (Replicast process). The pattern is burned out prior to pouring the molten metal into the mold.

New developments in evaporative pattern casting include production of metal-matrix composites (Section 9.5). During the process of molding the polymer pattern, it is embedded throughout with fibers or particles, which then become an integral part of the casting. Further studies include modification and grain refinement of the casting by the use of grain refiners and modifier master alloys (Section 10.4) within the pattern while it is being molded.

11.6 Plaster-Mold Casting

In the **plaster-mold casting process**, the mold is made of plaster of paris (gypsum, or calcium sulfate), with the addition of talc and silica flour to improve strength and control the time required for the plaster to set. These components are mixed with water, and the resulting slurry is poured over the pattern. After the plaster sets, usually within 15 minutes, the pattern is removed and the mold is dried at 120–260 °C (250–500 °F) to remove the moisture. Higher drying temperatures may be used depending on the type of plaster. The mold halves are then assembled to form the mold cavity and preheated to about 120 °C (250 °F). The molten metal is then poured into the mold.

Because plaster molds have very low permeability, gases evolved during solidification of the metal cannot escape. Consequently, the molten metal is poured either in a vacuum or under pressure. Plaster-mold permeability can be increased substantially by the *Antioch* process: The molds are dehydrated in an autoclave (pressurized oven) for 6–12 hours, then rehydrated in air for 14 hours. Another method of increasing permeability is to use foamed plaster, containing trapped air bubbles.

Patterns for plaster molding are generally made of aluminum alloys, thermosetting plastics, brass, or zinc alloys. Wood patterns are not suitable for making a large number of molds, because the patterns are repeatedly subjected to the water-based plaster slurry. Since there is a limit to the maximum temperature that the plaster mold can withstand, generally about 1200 °C (2200 °F), plaster-mold casting is used only for aluminum, magnesium, zinc, and some copper-base alloys. The castings have fine details with good surface finish. Because plaster molds have lower thermal conductivity than others, the castings cool slowly, and more uniform grain structure is obtained with less warpage. Wall thickness of parts can be 1–2.5 mm (0.04–0.1 in.).

This process and the ceramic-mold and investment casting processes (described below) are known as **precision casting** because of the high dimensional accuracy and good surface finish obtained. Typical parts made are lock components, gears, valves, fittings, tooling, and ornaments. Castings usually weigh less than 10 kg (22 lb) and are typically in the range of 125–250 g (1/4–1/2 lb), although parts as light as 1 g (0.035 oz) have been made.

11.7 Ceramic-Mold Casting

The **ceramic-mold casting process** is similar to the plaster-mold process, with the exception that it uses refractory mold materials suitable for high-temperature

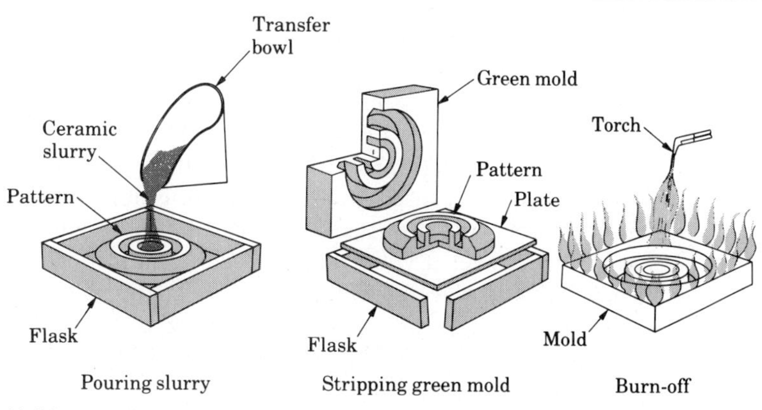

FIGURE 11.14
Sequence of operations in making a ceramic mold. *Source: Metals Handbook,* vol. 5, 8th ed.

applications. The process is also called *cope-and-drag investment casting*. The slurry is a mixture of fine-grained zircon ($ZrSiO_4$), aluminum oxide, and fused silica, which are mixed with bonding agents and poured over the pattern (Fig. 11.14), which has been placed in a flask.

The pattern may be made of wood or metal. After setting, the molds (ceramic

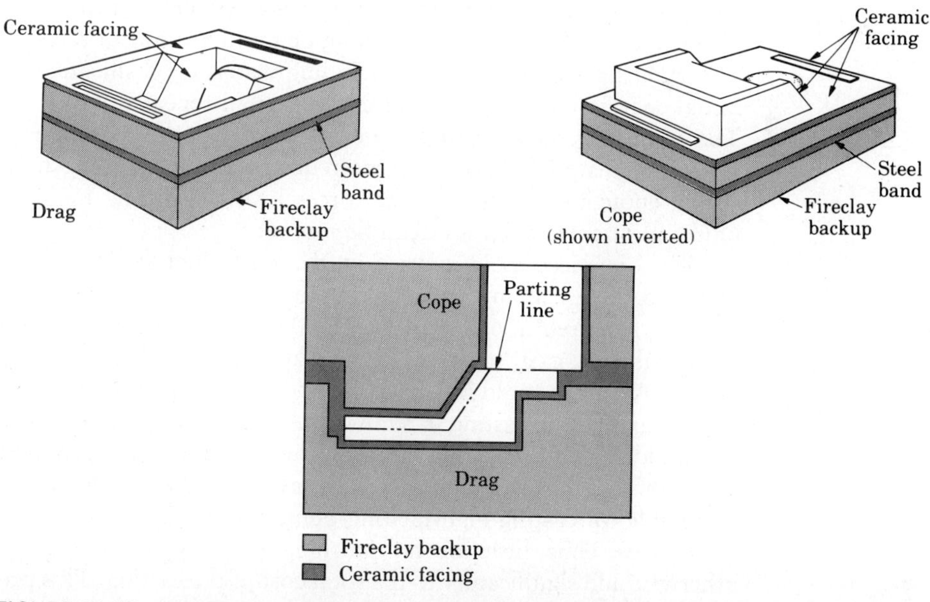

FIGURE 11.15
A typical ceramic mold (Shaw process) for casting steel dies used in hot forging. *Source: Metals Handbook,* vol. 5, 8th ed.

facings) are removed, dried, burned off to remove volatile matter, and baked. The molds are clamped firmly and used as all-ceramic molds. In the *Shaw* process, the ceramic facings are backed by fireclay (clay used in making firebricks that resist high temperatures) to give strength to the mold. The facings are then assembled into a complete mold, ready to be poured (Fig. 11.15).

The high-temperature resistance of the refractory molding materials allows these molds to be used in casting ferrous and other high-temperature alloys, stainless steels, and tool steels. The castings have good dimensional accuracy and surface finish over a wide range of sizes and intricate shapes, but the process is somewhat expensive. Typical parts made are impellers, cutters for machining, dies for metalworking, and molds for making plastic or rubber components. Parts weighing as much as 700 kg (1500 lb) have been cast by this process.

11.8 Investment Casting

The **investment-casting process**, also called the *lost-wax* process, was first used during the period 4000–3000 B.C. The pattern is made of wax or a plastic such as polystyrene. The sequences involved in investment casting are shown in Fig. 11.16. The pattern is made by injecting molten wax or plastic into a metal die in the shape of the pattern. The pattern is then dipped into a slurry of refractory material, such as very fine silica and binders, including water, ethyl silicate, and acids. After this initial coating has dried, the pattern is coated repeatedly to increase its thickness. The term *investment* comes from the fact that the pattern is invested with the refractory material. Wax patterns require careful handling because they are not strong enough to withstand the forces involved during mold making. However, unlike plastic patterns, wax can be reused.

The one-piece mold is dried in air and heated to a temperature of 90–175 °C (200–375 °F)—in an inverted position to melt out the wax—for about 12 hours. The mold is then fired to 650–1050 °C (1200–1900 °F) for about 4 hours, depending on the metal to be cast, to drive off the water of crystallization (chemically combined water). After the mold has been poured and the metal has solidified, the mold is broken up and the casting is removed. A number of patterns can be joined to make one mold, called a *tree* (Fig. 11.16c), thus increasing the production rate.

Although the labor and materials involved make the lost-wax process costly, it is suitable for casting high-melting-point alloys with good surface finish and close tolerances. Thus little or no finishing operations are required, which would otherwise add significantly to the total cost of the casting. This process is capable of producing intricate shapes, with parts weighing from 1 g to 35 kg (0.035 oz to 75 lb), from a wide variety of ferrous and nonferrous metals and alloys. Typical parts made are components for office equipment and mechanical components such as gears, cams, valves, and ratchets.

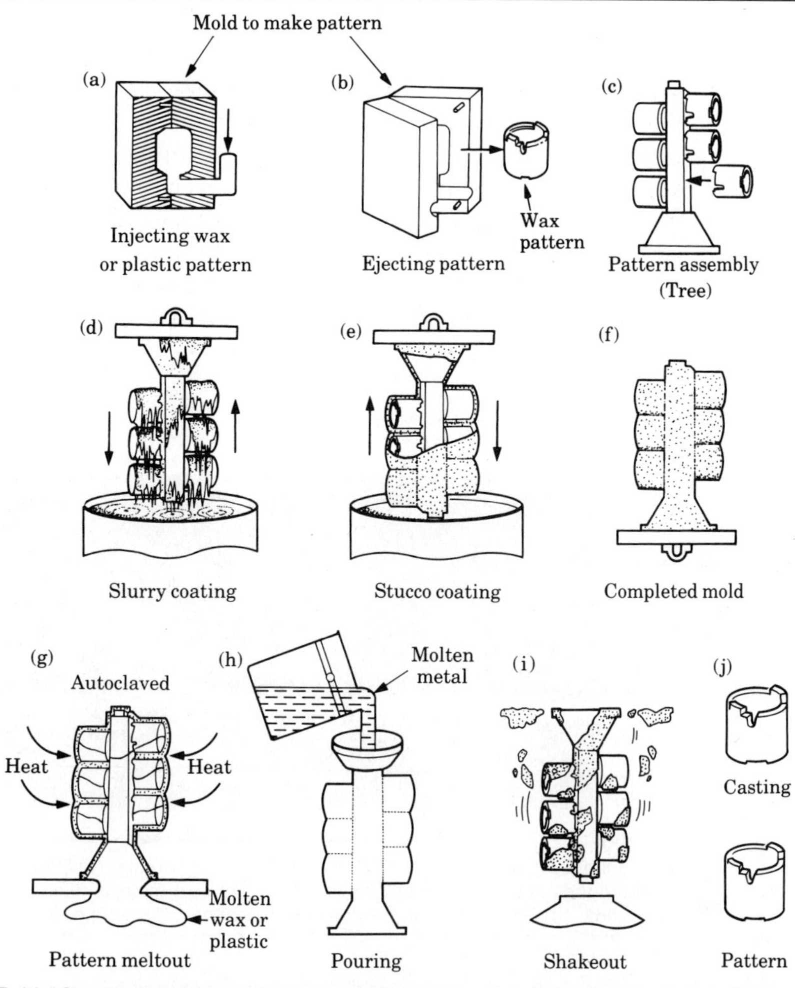

FIGURE 11.16 Schematic illustration of investment casting (lost-wax process). Castings by this method can be made with very fine detail and from a variety of metals. *Source:* Steel Founders' Society of America.

• Example: Eliminating porosity.

In investment casting of an aluminum-alloy valve body, porosity developed at the core–casting interface. The mold was originally heated to 200 °C (400 °F), which was too high for the metal around the core to solidify at a sufficiently high rate. Thus the casting began to solidify from the outside wall toward the core, and the gas (hydrogen) expelled during freezing of the metal accumulated at the area near the core–metal interface, thus producing porosity. By lowering the mold temperature to around 90 °C (200 °F), the metal around the core solidified at a high enough rate to prevent expulsion of gases around the core area, thus eliminating porosity. •

11.8.1 Ceramic-shell investment casting

A variation of the investment-casting process is ceramic-shell casting. It uses the same type of wax or plastic pattern, which is dipped first in ethyl silicate gel and then into a fluidized bed of fine-grained fused silica or zircon flour. The pattern is then dipped into coarser grain silica to build up additional coatings and thickness to withstand the thermal shock of pouring. The rest of the procedure is similar to investment casting. This process is economical and is used extensively for precision casting of steels and high-temperature alloys.

The sequence of operations involved in making a turbine disk by this method is shown in Fig. 11.17. Another complex part—an exhaust duct of a gas turbine —made by this process is shown in Fig. 11.18. If ceramic cores are used in the

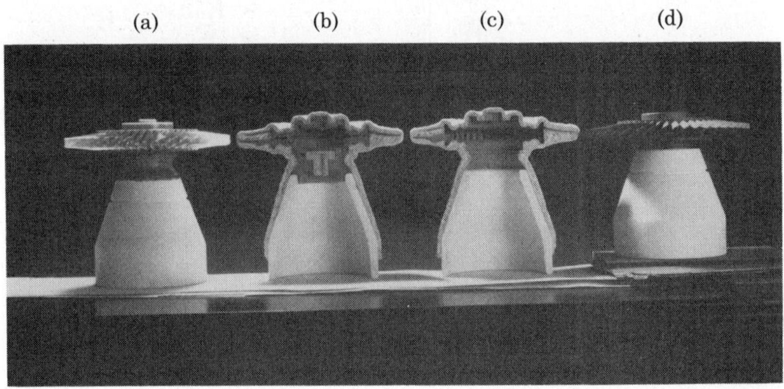

FIGURE 11.17
Investment casting of an integrally cast rotor for a gas turbine. (a) Wax pattern assembly. (b) Ceramic shell around wax pattern. (c) Wax is melted out and mold is filled, under a vacuum, with molten superalloy. (d) The cast rotor, produced to net or near-net shape. *Source:* Howmet Corporation.

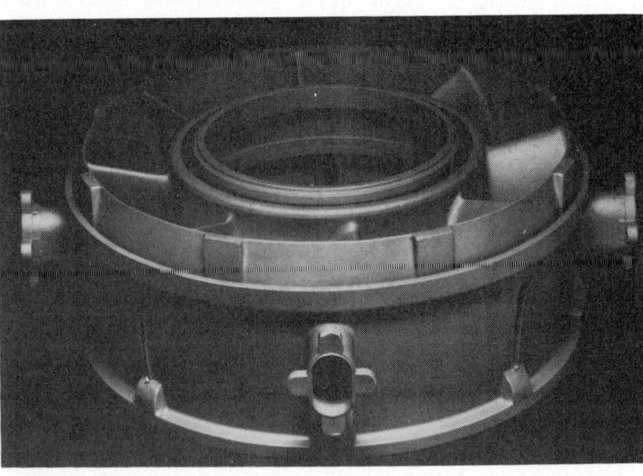

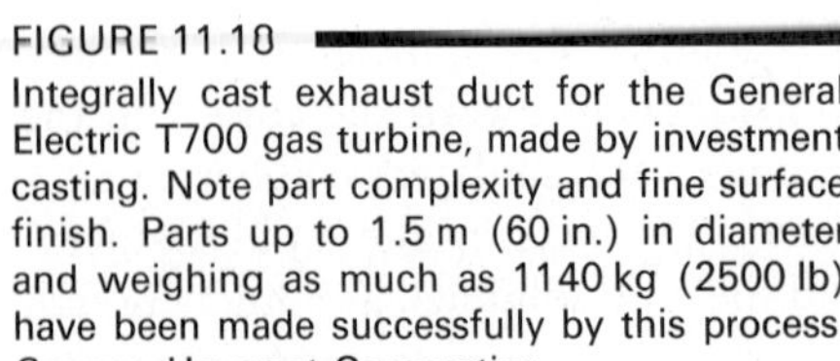

FIGURE 11.18
Integrally cast exhaust duct for the General Electric T700 gas turbine, made by investment casting. Note part complexity and fine surface finish. Parts up to 1.5 m (60 in.) in diameter and weighing as much as 1140 kg (2500 lb) have been made successfully by this process. *Source:* Howmet Corporation.

casting, they are removed by leaching with caustic solutions under high pressure and temperature. The molten metal may be poured in a vacuum to extract evolved gases and reduce oxidation, thus improving the quality of the casting. To further reduce microporosity, the castings made by this and other processes are subjected to hot isostatic pressing (see Section 17.3). Aluminum castings, for example, are subjected to a gas pressure up to 100 MPa (15 ksi) at 500 °C (900 °F).

Example: Investment-cast superalloy components for gas turbines.

Since the 1960s, investment-cast superalloys have been replacing wrought counterparts in high-performance gas turbines. Much development has been taking place in producing cleaner superalloys (nickel-base and cobalt-base), and improvements in melting and casting techniques, such as by vacuum-induction melting, using microprocessor controls. Impurity and inclusion levels have continually been reduced, thus improving the strength and ductility of these components. Such control is essential because these parts operate at a temperature only about 50 °C (90 °F) below the solidus.

The microstructure of an integrally investment-cast, gas-turbine rotor is shown in the upper portion of the accompanying photograph. Note the fine, uniform, equiaxed grain size throughout the cross section. Recent techniques to obtain this result include the use of a nucleant addition to the molten metal, as well as close control of its superheat, pouring techniques, and control of cooling rate of the casting. In contrast, the lower portion of the photograph shows the same type of rotor cast conventionally; note the coarse grain structure. This rotor will have inferior properties compared to the fine-grained rotor. Due to developments in these processes, the proportion of the weight of cast parts in aircraft engines has increased from 20 percent to about 45 percent. (*Source of photograph*: *Advanced Materials and Processes*, October 1990, p. 25. ASM International.)

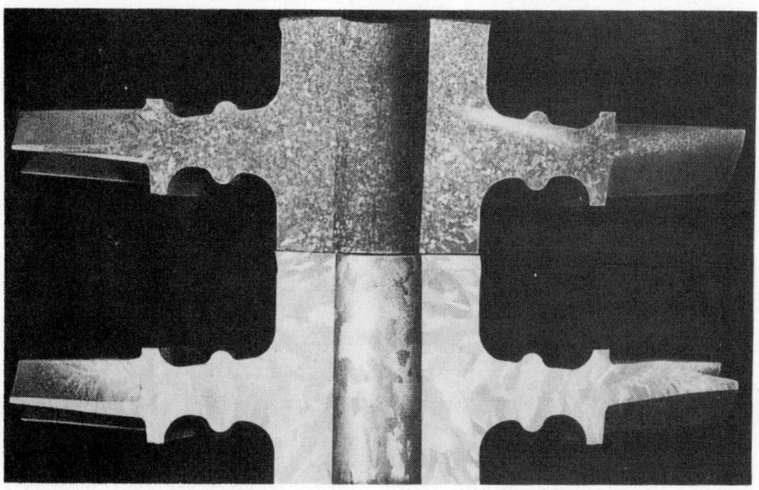

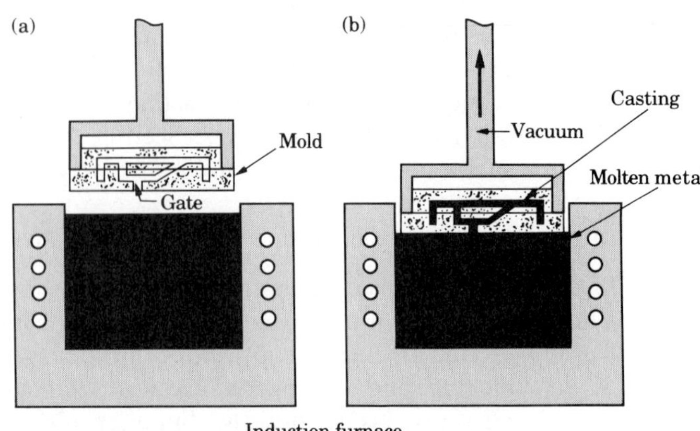

FIGURE 11.19 Schematic illustration of the vacuum-casting process. Note that the mold has a bottom gate. (a) Before and (b) after immersion of the mold into the molten metal. *Source:* From Dr. Robert Blackburn, "Vacuum Casting Goes Commercial," *Advanced Materials and Processes*, February 1990, p. 18. ASM International.

11.9 Vacuum Casting

A schematic illustration of the **vacuum-casting process** (not to be confused with the vacuum-molding process described in Section 11.2.5) is shown in Fig. 11.19. A mixture of fine sand and urethane is molded over metal dies and cured with amine vapor. Then the mold is held with a robot arm and partially immersed into molten metal held in an induction furnace. The vacuum reduces the air pressure inside the mold to about two thirds of atmospheric pressure, drawing the molten metal into the mold cavities through a gate in the bottom of the mold. The molten metal in the furnace is at a temperature usually 55 °C (100 °F) above the liquidus temperature; consequently, it begins to solidify within a fraction of a second. After the mold is filled, it is withdrawn from the molten metal.

This process is an alternative to investment, shell-mold, and green-sand casting, and is particularly suitable for thin-wall (1.75 mm; 3/32 in.) complex shapes with uniform properties. Carbon and low- and high-alloy steel and stainless steel parts, weighing as much as 70 kg (155 lb), have been vacuum cast by this method. The process can be automated and production costs are similar to those for green-sand casting.

11.10 Permanent-Mold Casting

In the **permanent-mold casting process**, also called *hard mold* casting, two halves of a mold are made from materials such as cast iron, steel, bronze, graphite, or refractory metal alloys. The mold cavity and gating system are machined into the mold and thus become an integral part of it. To produce castings with internal cavities, cores made of metal or sand aggregate are placed in the mold prior to casting. Typical core materials are oil-bonded or resin-bonded sand, plaster, graphite, gray iron, low-carbon steel, and hot-work die steel. Gray iron is the most commonly used, particularly for large molds for aluminum and magnesium castings. Inserts are also used for various parts of the mold.

In order to increase the life of permanent molds, the surfaces of the mold cavity are usually coated with a refractory slurry, such as sodium silicate and clay, or sprayed with graphite every few castings. These coatings also serve as parting agents and as thermal barriers, controlling the rate of cooling of the casting. Mechanical ejectors, such as pins located in various parts of the mold, may be needed for removal of complex castings. Ejectors usually leave small round impressions on castings.

The molds are clamped together by mechanical means and heated to about 150–200 °C (300–400 °F) to facilitate metal flow and reduce thermal damage to the dies. The molten metal is then poured through the gating system. After solidification, the molds are opened and the casting is removed. Special means employed to cool the mold include water or the use of fins, similar to those found on motorcycle or lawnmower engines that cool the engine block.

Although the permanent-mold casting operation can be performed manually, the process can be automated for large production runs. This process is used mostly for aluminum, magnesium, and copper alloys and gray iron because of their generally lower melting points. Steels can also be cast using graphite or heat-resistant metal molds.

This process produces castings with good surface finish, close tolerances, uniform and good mechanical properties, and at high production rates. Typical parts made by permanent-mold casting are automobile pistons (Fig. 11.20), cylinder

FIGURE 11.20 An automotive piston made by permanent-mold casting. *Source:* Courtesy of Zollner Corporation.

heads, connecting rods, gear blanks for appliances, and kitchenware. Parts that can be made economically generally weigh less than 25 kg (55 lb), although special castings weighing a few hundred kilograms have been made by this process.

Although equipment costs can be high because of die costs, the process can be mechanized, thus keeping labor costs low. Permanent-mold casting is not economical for small production runs. Furthermore, because of the difficulty in removing the casting from the mold, intricate shapes cannot be cast by this process. However, easily collapsed sand cores can be used and removed from castings to leave intricate internal cavities. The process is then called *semipermanent-mold* casting.

11.11 Slush Casting

We noted in Fig. 10.10 that a solidified skin first develops in a casting and that this skin becomes thicker with time. Hollow castings with thin walls can be made by permanent-mold casting using this principle. This process is called **slush casting.** The molten metal is poured into the metal mold, and after the desired thickness of solidified skin is obtained, the mold is inverted or slung, and the remaining liquid metal is poured out. The mold halves are then opened and the casting is removed. The process is suitable for small production runs and is generally used for making ornamental and decorative objects and toys from low-melting-point metals, such as zinc, tin, and lead alloys.

11.12 Pressure Casting

In the two permanent-mold processes that we have just described, the molten metal flows into the mold cavity by gravity. In the **pressure-casting process,** also called *pressure pouring* or *low-pressure casting* (Fig. 11.21a), the molten metal is forced upward by gas pressure into a graphite or metal mold. The pressure is maintained until the metal has completely solidified in the mold. The molten metal may also be forced upward by a vacuum, which also removes dissolved gases and gives the casting lower porosity.

Pressure casting is generally used for high-quality castings. An example of this process is steel railroad-car wheels. These wheels may also be cast in sand molds or semipermanent molds made of graphite and sand (Fig. 11.21b).

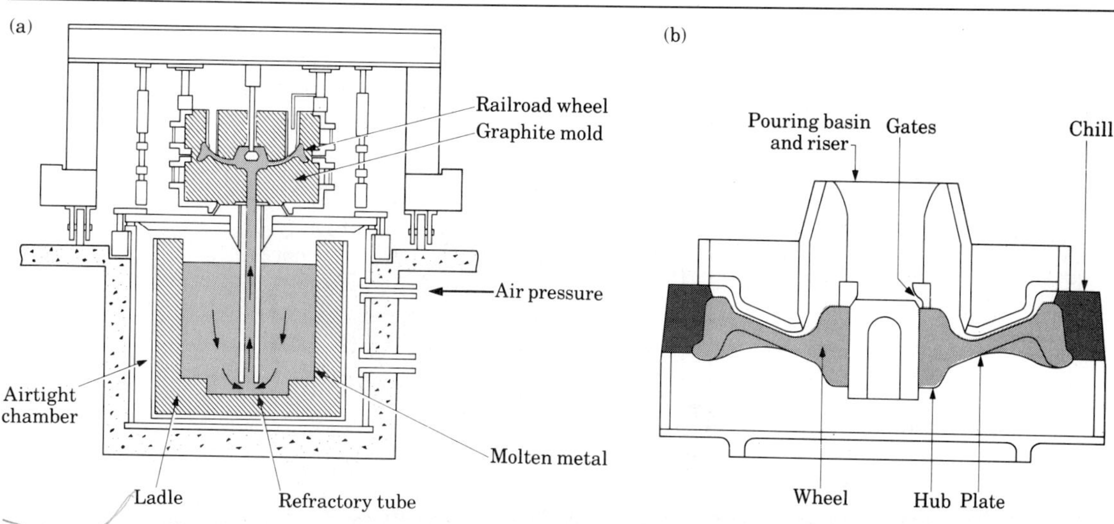

FIGURE 11.21 (a) The bottom-pressure casting process utilizes graphite molds for the production of steel railroad wheels. *Source:* The Griffin Wheel Division of Amsted Industries Incorporated. (b) Gravity-pouring method of casting a railroad wheel. Note that the pouring basin also serves as a riser. Railroad wheels can also be manufactured by forging.

11.13 Die Casting

The **die-casting process**, developed in the early 1900s, is a further example of permanent-mold casting. The molten metal is forced into the die cavity at pressures ranging from 0.7–700 MPa (0.1–100 ksi). The European term *pressure die casting*, or simply die casting, that we describe in this section, is not to be confused with the term *pressure casting* that we described in Section 11.12. Typical parts made by die casting are carburetors, motors, business-machine and appliance components, hand tools, and toys. The weight of most castings ranges from less than 90 g (3 oz) to about 25 kg (55 lb). There are two basic types of die-casting machines: hot-chamber and cold-chamber.

11.13.1 Hot-chamber process

The *hot-chamber process* (Fig. 11.22) involves the use of a piston, which traps a certain volume of molten metal and forces it into the die cavity through a gooseneck and nozzle. The pressures range up to 35 MPa (5000 psi), with an average of about 15 MPa (2000 psi). The metal is held under pressure until it solidifies in the die. To improve die life and to aid in rapid metal cooling—thus reducing cycle time—dies are usually cooled by circulating water or oil through various passageways in the die block. Cycle times usually range up to 900 shots (individual injections) per hour for

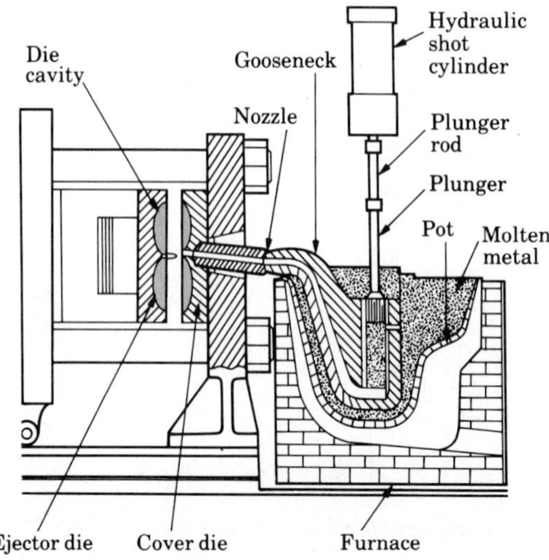

FIGURE 11.22
Sequence of steps in die casting of a part in the hot-chamber process. *Source:* Courtesy of *Foundry Management and Technology*.

zinc, although very small components such as zipper teeth can be cast at 18,000 shots per hour. Low-melting-point alloys such as zinc, tin, and lead are commonly cast by this process.

11.13.2 Cold-chamber process

In the *cold-chamber process* (Fig. 11.23), molten metal is poured into the injection cylinder (*shot chamber*) with a ladle. The shot chamber is not heated—hence the term *cold* chamber. The metal is forced into the die cavity at pressures usually ranging from 20 MPa to 70 MPa (3 ksi to 10 ksi), although they may be as high as 150 MPa (20 ksi). The machines may be horizontal (Fig. 11.24) or vertical, in which the shot chamber is vertical and the machine is similar to a vertical press.

High-melting-point alloys of aluminum, magnesium, and copper are normally cast by this method, although other metals (including ferrous metals) can also be cast in this manner. Molten-metal temperatures start at about 600 °C (1150 °F) for aluminum and magnesium alloys and increase considerably for copper-base and iron-base alloys.

11.13.3 Process capabilities and machine selection

Because of the high pressures involved, the dies have a tendency to part unless clamped together tightly. Die-casting machines are rated according to the clamping force that can be exerted to keep the dies closed. The capacities of commercially available machines range from about 25 tons to 3000 tons. Other factors involved in the selection of die-casting machines are die size, piston stroke, shot pressure, and cost.

Die-casting dies (Fig. 11.25) may be made single cavity, multiple cavity (several

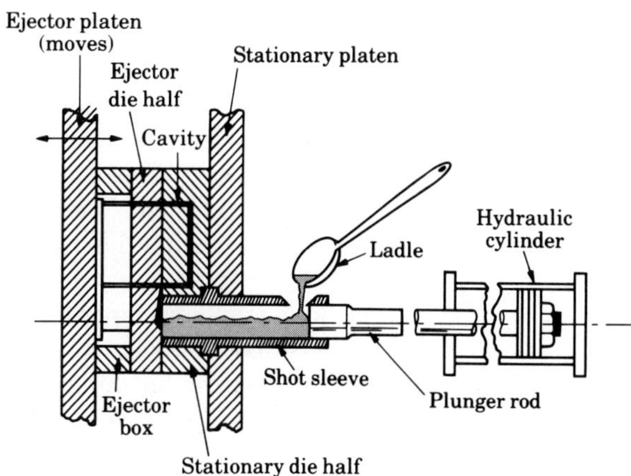

FIGURE 11.23 Sequence of operations in die casting of a part in the cold-chamber process. *Source:* Courtesy of *Foundry Management and Technology.*

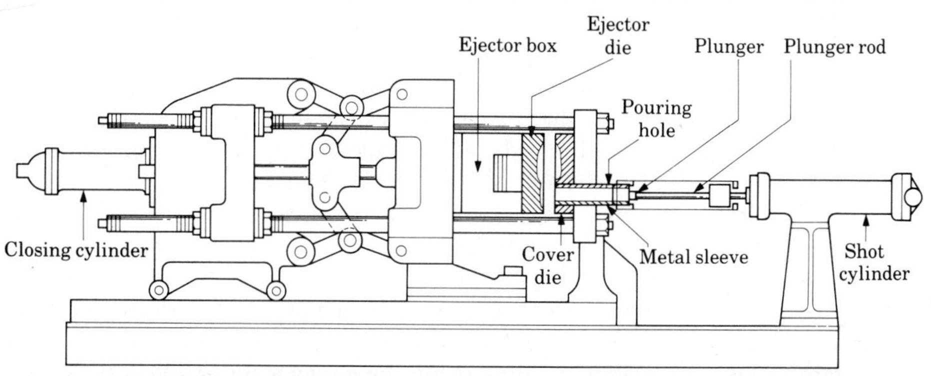

FIGURE 11.24 Schematic illustration of a die-casting machine. These machines are large compared to the size of the casting because large forces are required to keep the two halves of the dies closed. Otherwise, the pressure of the molten metal in the die cavities may force the dies apart.

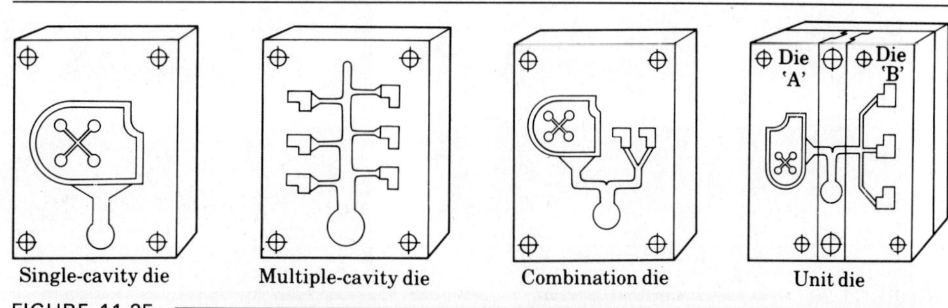

FIGURE 11.25 Various types of cavities in die-casting dies. *Source:* Courtesy of American Die Casting Institute.

identical cavities), combination cavity (several different cavities), or unit dies (simple small dies that can be combined in two or more units in a master holding die). Typically, the ratio of die weight to part weight is 1000 to 1. Thus the die for a casting weighing 2 kg will weigh about 2000 kg. Dies are usually made of hot-work die steels or mold steels. Die wear increases with the temperature of the molten metal. *Heat checking* of dies (surface cracking from repeated heating and cooling of the die) can be a problem. When die materials are selected and maintained properly, dies may last more than half a million shots before any significant die wear takes place.

Die design includes taper (draft) to allow the removal of the casting. The sprues and runners may be removed either manually or by using trim dies in a press. The entire die casting and finishing process can be highly automated. Lubricants (parting agents) are usually applied as thin coatings on die surfaces. Alloys, except magnesium alloys, generally require lubricants. These are usually water-base lubricants with graphite or other compounds in suspension. Because of the high cooling capacity of water, these fluids are also effective in keeping die temperatures low.

Die casting has the capability for high production rates with good strength, high-quality parts with complex shapes, and good dimensional accuracy and surface details, thus requiring little or no subsequent machining or finishing operations. Because of the high pressures involved, wall thicknesses as small as 0.5 mm (0.02 in.) are produced and are smaller than those obtained by other casting methods. Components such as pins, shafts, and fasteners can be cast integrally (*insert molding*). This process is similar to putting wooden sticks (the pin) in popsicles (the casting). Ejector marks remain, as do small amounts of flash (thin material squeezed out between the dies) at the die parting line.

Typical small and large parts made by die casting are shown in Figs. 11.26(a) and (b), respectively. Note the intricate shape and fine surface detail. In the fabrication of certain parts, die casting can compete favorably with other manufac-

(a)

(b)

FIGURE 11.26

(a) Die-cast body for a 35-mm camera, made of copper–aluminum alloy. Note the fine details. *Source:* Nikon, Inc. (b) Lawnmower housing made by die casting. This part is light and thin, and has fine detail. *Source:* J. D. Hanawalt and Dow Chemical Co.

TABLE 11.3
PROPERTIES AND TYPICAL APPLICATIONS OF COMMON DIE CASTING ALLOYS

ALLOY	ULTIMATE TENSILE STRENGTH (MPa)	YIELD STRENGTH (MPa)	ELONGATION IN 50 mm (%)	APPLICATIONS
Aluminum 380 (3.5 Cu–8.5 Si)	320	160	2.5	Appliances, automotive components, electrical motor frames and housings
13 (12 Si)	300	150	2.5	Complex shapes with thin walls; parts requiring strength at elevated temperatures
Brass 858 (60 Cu)	380	200	15	Plumbing fixtures, lock hardware, bushings, ornamental castings
Magnesium AZ91B (9 Al–0.7 Zn)	230	160	3	Power tools, automotive parts, sporting goods
Zinc No. 3 (4 Al)	280	—	10	Automotive parts, office equipment, household utensils, building hardware, toys
5 (4 Al–1 Cu)	320	—	7	Appliances, automotive parts, building hardware, business equipment

Source: Data from American Die Casting Institute.

turing methods, such as sheet-metal stamping and forging, or other casting processes. In addition, because the molten metal chills rapidly at the die walls, the casting has a fine-grain, hard skin with higher strength. Consequently, the strength-to-weight ratio of die-cast parts increases with decreasing wall thickness. With good surface finish and dimensional accuracy, die casting can produce bearing surfaces that are normally machined.

Equipment costs, particularly the cost of dies, are somewhat high, but labor costs are generally low because the process is semi- or fully automated. Die casting is economical for large production runs. The properties and typical applications of common die-casting alloys are given in Table 11.3.

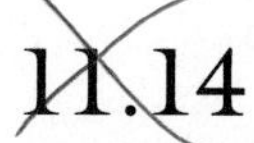

11.14 Centrifugal Casting

As its name implies, the **centrifugal-casting process** utilizes the inertial forces caused by rotation to distribute the molten metal into the mold cavities. This method was first suggested in the early 1800s. There are three types of centrifugal casting: true centrifugal casting, semicentrifugal casting, and centrifuging.

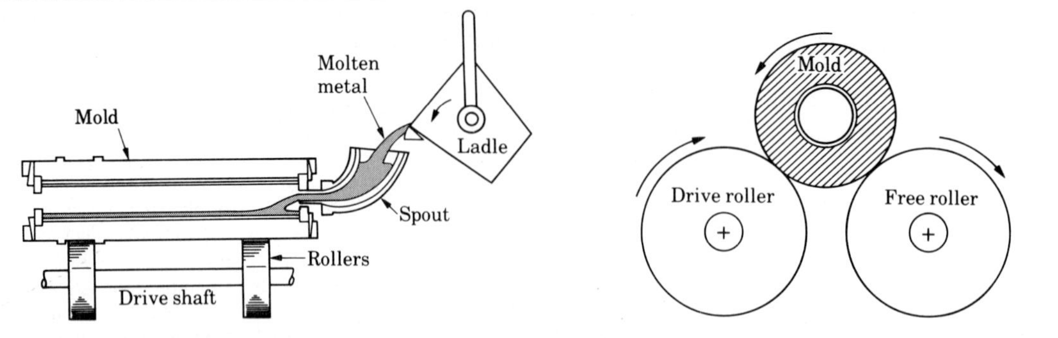

FIGURE 11.27 Schematic illustration of the centrifugal casting process. Pipes, cylinder liners, and similarly shaped parts can be cast by this process.

11.14.1 True centrifugal casting

In *true centrifugal casting*, hollow cylindrical parts, such as pipes, gun barrels, and streetlamp posts, are produced by the technique shown in Fig. 11.27, in which molten metal is poured into a rotating mold. The axis of rotation is usually horizontal but can be vertical for short workpieces. Molds are made of steel, iron, or graphite and may be coated with a refractory lining to increase mold life. The mold surfaces can be shaped so that pipes with various outer shapes, including square or polygonal, can be cast. The inner surface of the casting remains cylindrical because the molten metal is uniformly distributed by centrifugal forces. However, because of density differences, lighter elements such as dross, impurities, and pieces of the refractory lining tend to collect on the inner surface of the casting.

Cylindrical parts ranging from 13 mm (0.5 in.) to 3 m (10 ft) in diameter and 16 m (50 ft) long can be cast centrifugally, with wall thicknesses ranging from 6 mm to 125 mm (0.25 in. to 5 in.). The pressure generated by the centrifugal force is high, as much as 150 g's, and are necessary for casting thick-walled parts. Castings of good quality, dimensional accuracy, and external surface detail are obtained by this process. In addition to pipes, typical parts made are bushings, engine-cylinder liners, and bearing rings with or without flanges.

11.14.2 Semicentrifugal casting

An example of *semicentrifugal casting* is shown in Fig. 11.28(a). This method is used to cast parts with rotational symmetry, such as a wheel with spokes.

11.14.3 Centrifuging

In *centrifuging*, also called *centrifuge casting*, mold cavities of any shape are placed at a certain distance from the axis of rotation. The molten metal is poured from the center and is forced into the mold by centrifugal forces (Fig. 11.28b). The properties of castings vary by distance from the axis of rotation.

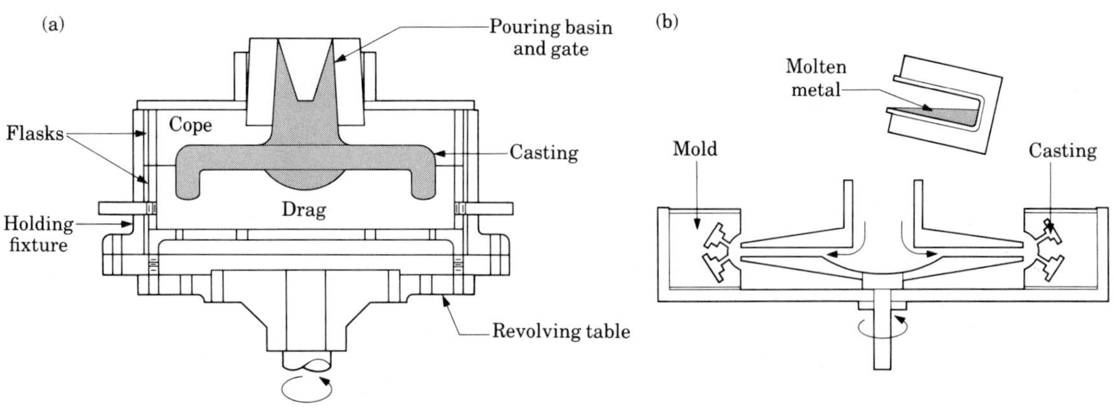

FIGURE 11.28
(a) Schematic illustration of the semicentrifugal casting process. Wheels with spokes can be cast by this process. (b) Schematic illustration of casting by centrifuging. The molds are placed at the periphery of the machine, and the molten metal is forced into the molds by centrifugal forces.

11.15 Squeeze Casting

The **squeeze-casting process**, developed in the 1960s, involves solidification of the molten metal under high pressure. Thus it is a combination of casting and forging (Fig. 11.29). The machinery includes a die, punch, and ejector pin. The pressure applied by the punch keeps the entrapped gases in solution, and the contact under high pressure at the die–metal interface promotes rapid heat transfer, resulting in a fine microstructure with good mechanical properties. The application of pressure also overcomes feeding problems that can arise when casting metals with long freezing range.

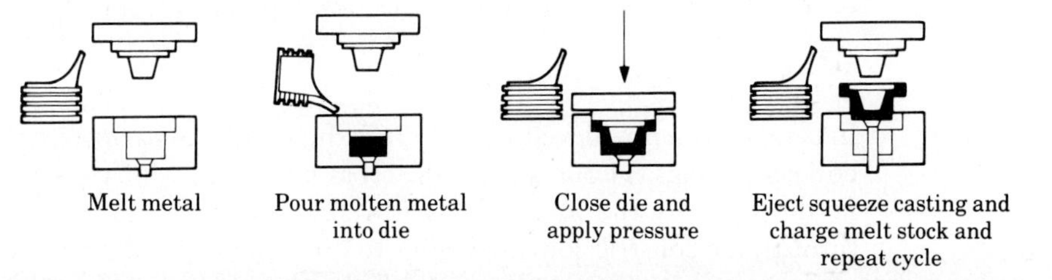

FIGURE 11.29
Sequence of operations in the squeeze-casting process. This process combines the advantages of casting and forging.

Parts can be made to near-net shape, with complex shapes and fine surface detail, from both nonferrous and ferrous alloys. Typical products made are automotive wheels and mortar bodies (a short-barreled cannon). The pressures required in squeeze casting are lower than those for hot or cold forging.

11.16 Casting Techniques for Single-Crystal Components

In Chapter 1 we discussed the advantages and limitations of single-crystal and polycrystalline structures in metals. Let's now consider techniques for producing single-crystal (*monocrystal*) components. We illustrate these techniques by describing the casting of gas turbine blades, which are generally made of nickel-base superalloys. The procedures involved can also be used for other alloys and components.

11.16.1 Conventional casting of turbine blades

The *conventional casting process* uses a ceramic mold. The molten metal is poured into the mold and begins to solidify at the ceramic walls. The grain structure developed is polycrystalline and is similar to that shown in Fig. 10.1(c). The presence of grain boundaries makes this structure susceptible to creep and cracking along those boundaries under the centrifugal forces at elevated temperatures.

11.16.2 Directionally solidified blades

In the *directional solidification process* (Fig. 11.30a), first developed in 1960, the ceramic mold is preheated by radiant heating. The mold is supported by a water-cooled chill plate. After the metal is poured into the mold, the assembly is lowered slowly. Crystals begin to grow at the chill-plate surface and upward, like the columnar grains shown in Fig. 10.2. The blade is thus directionally solidified, with longitudinal but no transverse grain boundaries. Consequently, the blade is stronger in the direction of centrifugal forces developed in the gas turbine.

11.16.3 Single-crystal blades

In *crystal growing*, developed in 1967, the mold has a constriction in the shape of a corkscrew (Figs. 11.30b and c), the cross-section of which allows only one crystal to fit through. As the assembly is lowered slowly, a single crystal grows upward through the constriction and begins to grow in the mold. Strict control of the rate of movement is necessary. The solidified mass in the mold is a single-crystal blade. Although more expensive than other blades, the lack of grain boundaries makes these blades resistant to creep and thermal shock. Thus they have a longer and more reliable service life.

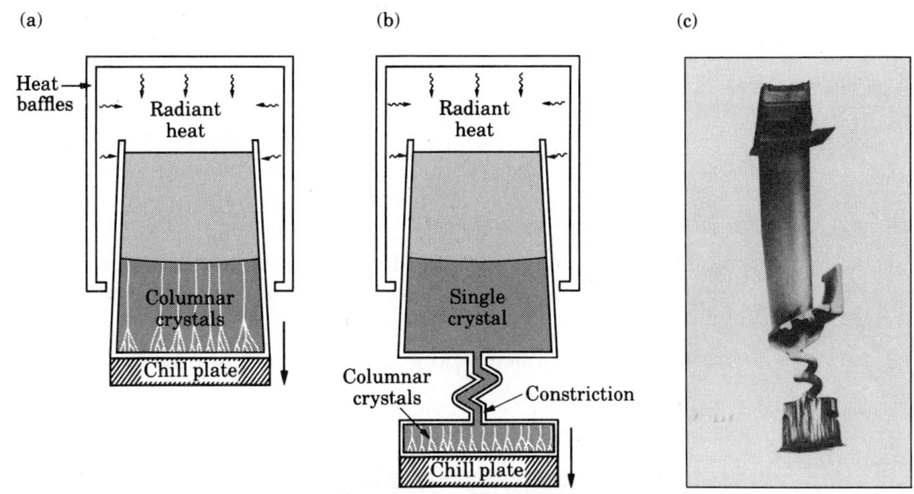

FIGURE 11.30 Methods of casting turbine blades: (a) directional solidification; (b) method to produce a single-crystal blade; and (c) a single-crystal blade with the constriction portion still attached. *Source:* (a) and (b) B. H. Kear, *Scientific American*, October 1986; (c) *Advanced Materials and Processes*, October 1990, p. 29, ASM International.

11.16.4 Single-crystal growing

With the advent of the semiconductor industry, single-crystal growing has become a major activity in the manufacture of microelectronic devices (see also Chapter 34). There are basically two methods of crystal growing.

In the *crystal pulling* method, known as the *Czochralski* process (Fig. 11.31a), a

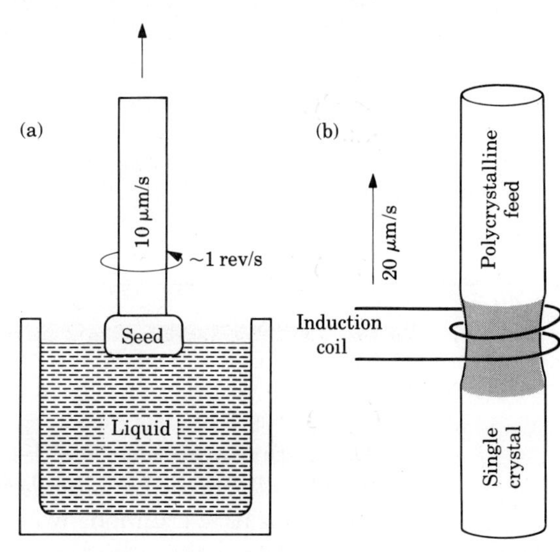

FIGURE 11.31 Two methods of crystal growing: (a) crystal pulling (Czochralski process) and (b) floating-zone method. Crystal growing is especially important in the semiconductor industry. *Source:* L. H. Van Vlack, *Materials for Engineering.* Addison-Wesley Publishing Co., Inc., 1982.

seed crystal is dipped into the molten metal and then pulled out slowly , at a rate of about 10 μm/s, while being rotated. The liquid metal begins to solidify on the seed, and the crystal structure of the seed is continued throughout. Dopants (alloying elements) may be added to the liquid metal to impart special electrical properties. Single crystals of silicon, germanium, and various other elements are grown by this process. Single-crystal ingots typically 50–150 mm (2–6 in.) in diameter and over 1 m (40 in.) in length have been produced by this technique.

The second technique for crystal growing is the *floating-zone* method (Fig. 11.31b). Starting with a rod of polycrystalline silicon resting on a single crystal, an induction coil heats these two pieces while moving slowly upward. The single crystal grows upward while maintaining its orientation. Thin wafers are then cut from the rod, cleaned, and polished for use in microelectronic device fabrication.

11.17 Inspection of Castings

Several methods are available for inspection of castings to determine quality and the presence of any defects. Castings can be inspected visually or optically for surface defects. Subsurface and internal defects are investigated using various *nondestructive* techniques that we describe in Section 36.2. In *destructive* testing, test specimens are removed from various sections of a casting and tested for strength, ductility, and other mechanical properties, and to determine the presence and location of any defects.

Pressure tightness of cast components (valves, pumps, pipes) is usually determined by sealing the openings in the casting and pressurizing it with water, oil, or air. The casting is then inspected for leaks while the pressure is maintained. Because air is compressible, its use is dangerous in such tests because of the possibility of a sudden explosion if there is a major flaw in the casting.

Unacceptable or defective castings are remelted for reprocessing. Because of the major economic impact, the types of defects present in castings and their causes must be investigated. Control of all stages during casting, from mold preparation to the removal of castings from molds or dies, is important in maintaining good quality.

11.18 Foundries

The casting operations we described in this chapter are usually carried out in *foundries* (from the Latin *fundere*, meaning melting and pouring). Although casting operations have traditionally involved much manual labor, modern foundries have automated and computer-integrated facilities for all aspects of their operations. They produce a wide variety and sizes of castings at high production rates, at low cost, and with good quality control.

As outlined in Fig. 11.1, foundry operations initially involve two separate activities. The first is pattern and mold making, which are now beginning to utilize computer-aided design and manufacturing techniques, minimizing trial and error and improving efficiency. A variety of automated machinery is used to minimize labor costs, which can be significant in the production of castings. The second activity is melting the metals, controlling their composition and impurities, and pouring them into appropriate molds.

The rest of the operations, such as pouring into molds carried along conveyors, shakeout, cleaning, heat treatment, and inspection, are also automated. Automation minimizes labor, reduces the possibility of human error, increases the production rate, and attains higher quality levels. These topics are described in Part VIII.

The level of automation in foundries is an important economic consideration, particularly in view of the fact that many foundries are small businesses. The degree of automation depends on the type of products made. A die-casting facility or a foundry making parts for the automotive industry may involve production runs in the hundreds of thousands; thus a high level of automation is desirable. Such facilities can afford automation. On the other hand, a jobbing foundry producing short production runs may not be as automated.

Furthermore, foundries in general still tend to be somewhat dirty, hot, and labor-intensive operations. It is difficult to find qualified personnel to work in such an environment. Consequently, automation has become increasingly necessary to compensate for the decline in worker competency.

SUMMARY

In this chapter we described expendable-mold and permanent-mold casting processes. The most common processes in the first group are sand, shell, plaster, and ceramic mold and investment casting. Compared to permanent-mold casting, castings made by these processes usually involve relatively low mold and equipment costs. However, they generally tend to produce castings having high porosity and low dimensional accuracy at a low production rate.

The molds used in permanent-mold casting are made of metal or graphite and are used repeatedly to produce many parts. Because metals are good heat conductors but do not allow gases to escape, permanent molds perform in fundamentally different ways from sand and other aggregate mold materials.

Processes that use permanent molds are pressure, slush, die, centrifugal, and squeeze casting. Die and equipment costs are relatively high, but the processes are economical for large production runs. Scrap loss is low and dimensional accuracy is relatively high, with good surface details.

SUMMARY TABLE 11.1

SUMMARY OF CASTING PROCESSES, THEIR ADVANTAGES AND LIMITATIONS

PROCESS	ADVANTAGES	LIMITATIONS
Sand	Almost any metal is cast; no limit to size, shape or weight; low tooling cost.	Some finishing required; somewhat coarse finish; wide tolerances.
Shell mold	Good dimensional accuracy and surface finish; high production rate.	Part size limited; expensive patterns and equipment required.
Evaporative pattern	Most metals cast with no limit to size; complex shapes.	Patterns have low strength and can be costly for low quantities.
Plaster mold	Intricate shapes; good dimensional accuracy and finish; low porosity.	Limited to nonferrous metals; limited size and volume of production; mold making time relatively long.
Ceramic mold	Intricate shapes; close tolerance parts; good surface finish.	Limited size.
Investment	Intricate shapes; excellent surface finish and accuracy; almost any metal cast.	Part size limited; expensive patterns, molds, and labor.
Permanent mold	Good surface finish and dimensional accuracy; low porosity; high production rate.	High mold cost; limited shape and intricacy; not suitable for high-melting-point metals.
Die	Excellent dimensional accuracy and surface finish; high production rate.	Die cost is high; part size limited; usually limited to nonferrous metals; long lead time.
Centrifugal	Large cylindrical parts with good quality; high production rate.	Equipment is expensive; part shape limited.

TRENDS

- Computer-aided design and manufacturing of castings, molds and dies, and gating and runner systems are now being implemented at a rapid rate.
- Automation of molding and casting processes to reduce costs, using computer controls, sensors, and industrial robots, is an important area of continuing development and implementation.
- Research is in progress on automated inspection of castings, such as with machine vision and the use of fiber optics to inspect internal surfaces of castings that cannot be viewed directly.
- Better control of casting quality, dimensional accuracy, and surface finish continues to be of major interest.

(*continued*)

TRENDS (*continued*)

- Improvements in melting and remelting techniques and refining the molten metal prior to pouring are in progress, including the use of exothermic materials to prevent choking of runners.
- Investment casting continues to be an efficient metal-shaping technology, with larger and more complex structural castings being made, particularly for high-temperature aerospace applications.
- Other areas under investigation include evaporative pattern and vacuum casting and hot isostatic pressing of castings for improved quality and reliability.

KEY TERMS

Binders
Centrifugal-casting process
Ceramic-mold casting process
Chaplets
Composite molds
Cores
Die-casting process
Evaporative pattern casting process
Expendable molds
Green molding sand
Investment-casting process
Parting agent
Patterns
Permanent-mold casting process
Permanent molds
Plaster-mold casting process
Precision casting
Pressure-casting process
Sand casting
Shell molding
Slush casting
Sodium silicate process
Squeeze-casting process
Vacuum-casting process

BIBLIOGRAPHY

Allsop, D.F., and D. Kennedy, *Pressure Die Casting—Part II: The Technology of the Casting and the Die*. Oxford: Pergamon, 1983.

An Introduction to Die Casting. Des Plaines, Ill.: American Die Casting Institute, 1981.

Bradley, E. F., *High-Performance Castings: A Technical Guide*. Columbus, Ohio: Edison Welding Institute, 1989.

Flinn, R.A., *Fundamentals of Metal Casting*. Reading, Mass.: Addison-Wesley, 1963.

Heine, R.W., C.R. Loper, Jr., and C. Rosenthal, *Principles of Metal Casting*, 2d ed. New York: McGraw-Hill, 1967.

Investment Casting Handbook. Chicago, Ill.: Investment Casting Institute, 1979.

Kaye, A., and A.C. Street, *Die Casting Metallurgy*. London: Butterworths, 1982.

The Metallurgy of Die Castings. River Grove, Ill.: Society of Die Casting Engineers, 1986.

Metals Handbook, 9th ed., Vol. 15: *Casting*. Metals Park, Ohio: ASM International, 1988.

Mikelonis, P.J. (ed.), *Foundry Technology: A Source Book*. Metals Park, Ohio: American Society for Metals, 1982.

Romanoff, R., *Centrifugal Casting*, 2d ed. Blue Ridge Summit, Penn.: TAB Books, 1981.

Street, A.C., *The Diecasting Book*, 2d ed. Surrey, England: Portcullis Press, 1986.

Upton, B., *Pressure Die Casting—Part I: Metals-Machines-Furnaces*. Oxford: Pergamon, 1982.

Wieser, P.F. (ed.), *Steel Castings Handbook*, 5th ed. Des Plaines, Ill.: Steel Founders' Society of America, 1980.

REVIEW QUESTIONS

11.1 Describe the differences between expendable and permanent molds.
11.2 Name the important factors in selecting sand for molds.
11.3 What are the major types of sand molds? What are their characteristics?
11.4 List the considerations for selecting pattern materials.
11.5 What is the function of a core? What are core prints?
11.6 Name and describe the characteristics of the types of sand-molding machines.
11.7 What is the difference between sand and shell molding?
11.8 What are composite molds? Why are they used?
11.9 Describe the features of plaster-mold casting.
11.10 Why is the investment-casting process capable of producing fine surface detail on castings?
11.11 Name the type of materials used for permanent-mold casting processes.
11.12 What are the advantages of pressure casting?
11.13 List the advantages and limitations of die casting.
11.14 What are parting agents?
11.15 Describe the methods used for producing single-crystal parts.

QUESTIONS AND PROBLEMS

11.16 Explain why a casting may have a slightly different shape than the pattern used to make the mold.
11.17 What are the reasons for the large variety of casting processes that have been developed over the years? Explain your answer with specific examples.
11.18 If you need only five units of a casting, which process(es) would you use? Why?
11.19 Describe the advantages and limitations of hot-chamber and cold-chamber die casting processes, respectively.
11.20 Explain why processes such as sand, shell, plaster, and investment casting can produce parts with greater shape complexity than others, such as permanent-mold, die, and centrifugal casting.
11.21 Why is it that die casting can produce the smallest parts?
11.22 What differences, if any, would you expect in the properties of castings made by permanent mold versus sand casting methods?
11.23 Would you recommend preheating the molds in permanent-mold casting? Also, would you remove the casting soon after it has solidified? Explain your reasons.
11.24 Describe the advantages of composite molds. Where would you use them?

11.25 Referring to Fig. 11.3, do you think it is necessary to weigh down or clamp the two halves of the mold? Explain your reasons. Do you think the kind of metal cast, such as gray cast iron versus aluminum, should make a difference on the clamping force?

11.26 Give a step-by-step procedure for the (a) investment-casting process and (b) die-casting process.

11.27 Explain why squeeze casting produces parts with better mechanical properties, dimensional accuracy, and surface finish than expendable-mold processes.

11.28 In a sand-casting operation, what factors determine the time at which you would remove the casting from the mold?

11.29 Which of the casting processes would be suitable for making small toys? Why?

11.30 What would you do to improve the surface finish in expendable-mold casting processes?

11.31 How would you attach the individual wax patterns on a "tree" in investment casting?

11.32 Describe the measures that you would take to reduce core shifting in sand casting.

11.33 You have seen that even though die casting produces thin parts, there is a limit to the thickness. Why can't even thinner parts be made by this process?

11.34 What is the function of a blind riser?

11.35 Describe the procedures that would be involved in making a bronze statue. Which casting processes would be suitable? Why?

11.36 Write a brief report on the permeability of molds and the techniques that are used to determine permeability.

11.37 Describe the characteristics of chaplet materials. Should they melt while molten metal is being poured and solidified in the mold? Explain.

11.38 Estimate the clamping force for a die casting machine in which the casting is rectangular with projected dimensions of 100 mm × 200 mm (4 in. × 8 in.). Would your answer depend on whether or not it is a hot-chamber or cold-chamber process? Explain.

11.39 Make a list of the mold and die materials used in the casting processes described in this chapter. Under each type of material, list the casting processes that are used, and explain why these processes are suitable for that particular mold or die material.

11.40 How are hollow parts with various cavities made by die casting? Are cores used? If so, how? Explain.

11.41 Explain why the strength-to-weight ratio of die-cast parts increases with decreasing wall thickness.

11.42 The blank for the spool shown in the figure below is to be sand cast out of A-319, an

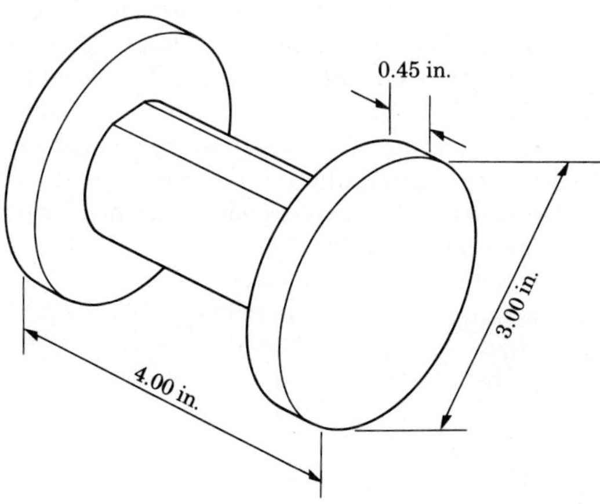

aluminum casting alloy. Make a sketch of the wooden pattern for this part. Be sure to include all necessary allowances for shrinkage and machining.

11.43 Repeat Problem 11.42, but assume that the aluminum spool is to be cast using evaporative pattern casting. Explain the important differences between the two patterns.

11.44 In sand casting, it is important that the cope mold half be held down with enough force to keep it from floating when the molten metal is poured in. For the casting shown in the figure below, calculate the minimum amount of weight necessary to keep the cope from floating up as the molten metal is poured in. (*Hint*: The buoyancy force exerted by the molten metal on the cope is dependent on the effective height of the metal head above the cope.)

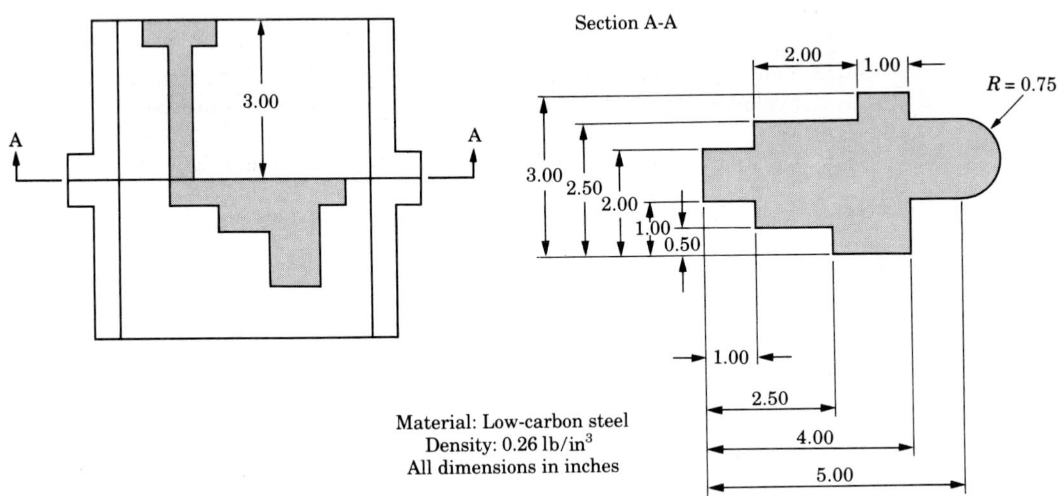

11.45 In shell molding, the curing process is very critical to the quality of the finished mold. In this part of the process, the shell mold assembly and cores are placed in an oven for a short period of time to complete the curing of the resin binder. List probable causes of unevenly cured cores or of uneven core thicknesses.

11.46 Sketch an automated casting line (using conveyors, sensors, machinery, robots, etc.) that could automatically perform the evaporative pattern casting process.

11.47 Besides being an excellent way to produce hollow cylindrical parts, centrifugal casting provides a way of centrifuging lightweight, nonmetallic contaminants (slag) to the inside diameter of the part. Here the contaminants may be easily machined away after casting, rather than dispersed throughout the part. (a) Explain how this happens. (b) List several variables in the process that govern the effectiveness of this separation process, and (c) explain why if the metal is poured too cold, the separation effect may not occur.

11.48 Describe a test that could be performed to test the permeability of sand for sand casting.

12

Casting Design; Casting Alloys; Economics of Casting

12.1 Introduction

In the preceding two chapters, we saw that successful casting practice requires careful control of a large number of variables. These variables pertain to the particular characteristics of the metals and alloys cast, method of casting, mold and die materials, mold design, and various process parameters. The flow of the molten metal in the mold cavities, gating systems, the rate of cooling, and the gases evolved all influence the quality of a casting.

In this chapter, we discuss general design considerations and guidelines for metal casting. As we have pointed out, poor casting practices and lack of control of process variables can lead to defective castings. We present suggestions for avoiding defects. We also describe the alloys that are commonly cast, together with their characteristics and typical applications.

As we have stated before, the economics of manufacturing processes are just as important as the technical considerations that we have been describing in detail. In this chapter, we outline the basic economic factors relevant to casting operations and make some economic comparisons between different casting processes.

12.2 Design Considerations

As in all engineering practice and manufacturing operations, certain guidelines and **design principles** pertaining to casting have been developed over many years. Although these principles were established primarily through practical experience, analytical methods and computer-aided design and manufacturing techniques are now coming into wider use, improving productivity and the quality of castings. Moreover, careful design can result in significant cost savings.

12.2.1 Designing for expendable-mold casting

The following guidelines generally apply to all types of castings. The most significant design considerations are identified and addressed.

Corners, angles, and section thickness. Sharp corners, angles, and fillets should be avoided (Fig. 12.1), as they may cause cracking and tearing during solidification of the metal. Fillet radii should be selected to reduce stress concentrations and to ensure proper liquid-metal flow during the pouring process. Fillet radii usually range from 3 mm to 25 mm (1/8 in. to 1 in.), although smaller radii may be permissible in small castings and in limited applications. On the other hand, if the fillet radii are too large, the volume of the material in those regions is also large and, consequently, the rate of cooling is less.

Section changes in castings should smoothly blend into each other. The location of the largest circle that can be inscribed in a particular region is critical so far as shrinkage cavities are concerned (Figs. 12.2a and b). Because the cooling rate in regions with the larger circles is less, they are called **hot spots**. These regions could develop *shrinkage cavities* and *porosity* (Figs. 12.2c and d). Cavities at hot spots can be eliminated with small cores. Although they produce cored holes in the casting

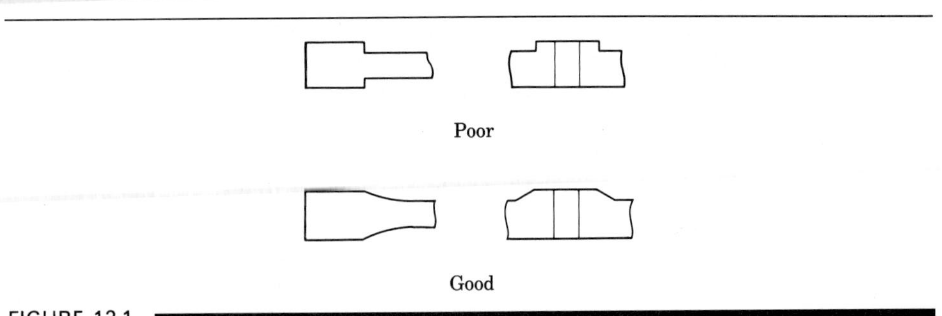

FIGURE 12.1
Suggested design modifications to avoid defects in castings. Note that sharp corners are avoided to reduce stress concentrations.

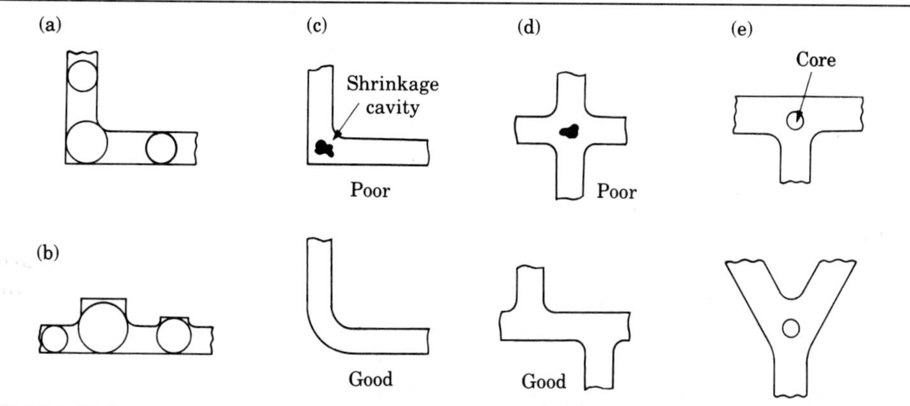

FIGURE 12.2
Examples of designs showing the importance of maintaining uniform cross-sections in castings to avoid hot spots and shrinkage cavities.

(Fig. 12.2e), these holes do not affect the strength of the casting significantly. Other examples of design principles that can be used to avoid shrinkage cavities are shown in Fig. 12.3. Although they increase the cost of production, metal paddings in the mold can eliminate or minimize hot spots. These paddings act as external chills, such as that shown for casting of a hollow cylindrical part with internal ribs in Fig. 12.4. From these illustrations you can see the importance of maintaining, insofar as

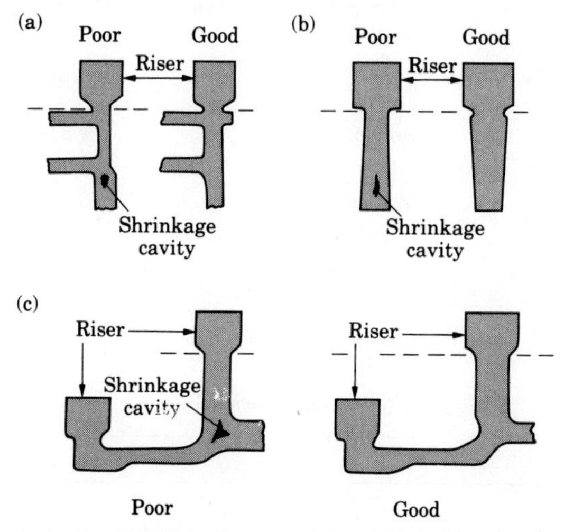

FIGURE 12.3
Examples of design modifications to avoid shrinkage cavities in castings. *Source: Steel Castings Handbook,* 5th ed. Steel Founders' Society of America, 1980. Used with permission.

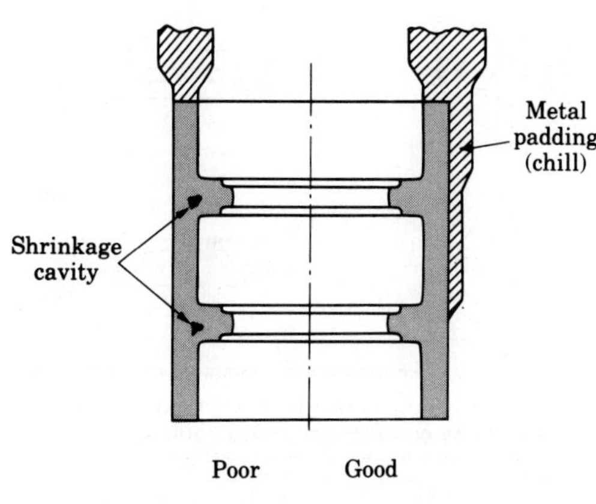

FIGURE 12.4
The use of metal padding (chills) to increase the rate of cooling in thick regions in a casting to avoid shrinkage cavities. *Source: Steel Castings Handbook,* 5th ed. Steel Founders' Society of America, 1980. Used with permission.

possible, uniform cross-sections and wall thicknesses throughout the casting to avoid shrinkage cavities.

Flat areas. Large flat areas (plain surfaces) should be avoided, as they may warp because of temperature gradients during cooling or develop poor surface finish because of uneven flow of metal during pouring. Flat surfaces can be broken up with ribs and serrations.

Shrinkage. Allowances for shrinkage during solidification should be provided for, so as to avoid cracking of the casting. Figure 12.5(a) depicts a wheel with spokes. If the spokes are curved, the tensile stress in them resulting from contraction during solidification—and hence the tendency for cracking—is reduced. Another example is shown in Fig. 12.5(b), in which the original design has been altered slightly. In castings with intersecting ribs (Fig. 12.6), the tensile stresses can be reduced by staggering the ribs, as shown in Fig. 12.2(d), or by changing the intersection geometry from an X configuration to a Y configuration.

Pattern dimensions should also provide for shrinkage of the metal during solidification and cooling. Allowances for shrinkage, also known as **patternmaker's shrinkage allowances**, usually range from about 10 mm/m to 20 mm/m (1/8 in./ft to 1/4 in./ft). Table 12.1 gives the normal shrinkage allowance for metals commonly sand cast.

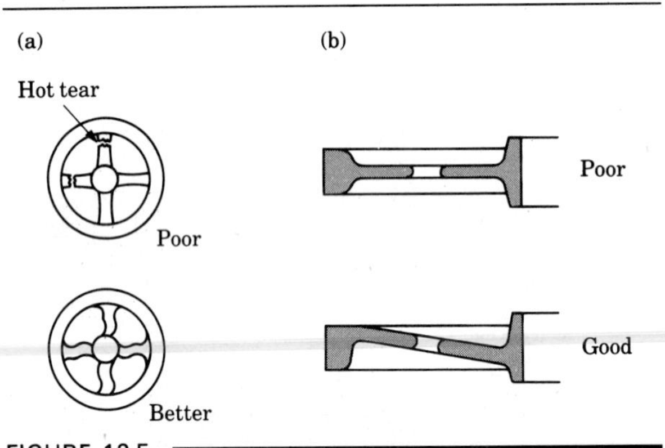

FIGURE 12.5 Two examples of poor and good casting design practice to avoid tears caused by contraction during cooling.

TABLE 12.1 NORMAL SHRINKAGE ALLOWANCE FOR SOME METALS CAST IN SAND MOLDS

METAL	PERCENT
Gray cast iron	0.83–1.3
White cast iron	2.1
Malleable cast iron	0.78–1.0
Aluminum alloys	1.3
Magnesium alloys	1.3
Yellow brass	1.3–1.6
Phosphor bronze	1.0–1.6
Aluminum bronze	2.1
High-manganese steel	2.6

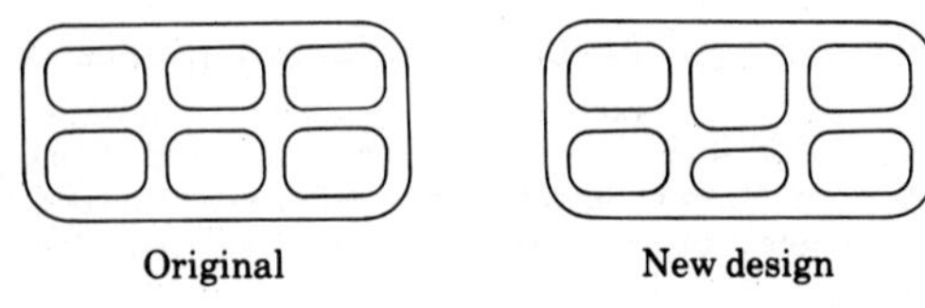

FIGURE 12.6 Modification of a design to avoid shrinkage cavities in castings. Note the staggering of intersecting regions in the improved design (see also Fig. 12.2d).

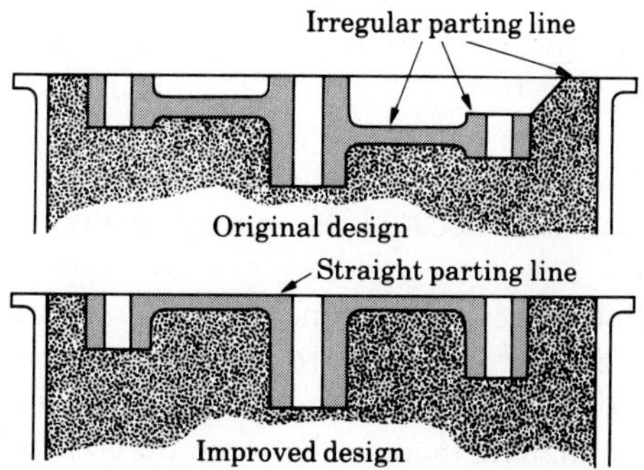

FIGURE 12.7
Redesign of a casting by making the parting line straight to avoid defects. *Source: Steel Casting Handbook,* 5th ed. Steel Founders' Society of America, 1980. Used with permission.

Parting line. Recall that the parting line is the line, or plane, separating the upper (cope) and lower (drag) halves of molds (see Figs. 11.3 and 12.7). In general, it is desirable for the parting line to be along a flat plane, rather than contoured. Whenever possible, the parting line should be at the corners or edges of castings, rather than on flat surfaces in the middle of the casting. In this way, the *flash* at the parting line (material squeezing out between the two halves of the mold) will not be as visible. The location of the parting line is important because it influences mold design, ease of molding, number and shape of cores, method of support, and the gating system. Three examples of casting design modifications are shown in Fig. 12.8.

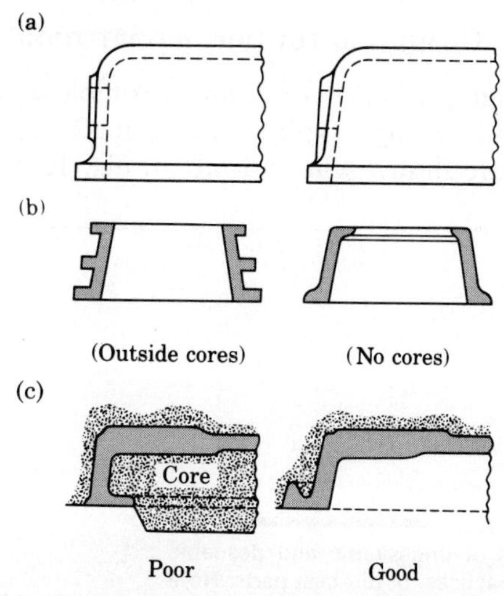

FIGURE 12.8
Examples of casting design modifications. *Source: Steel Castings Handbook,* 5th ed. Steel Founders' Society of America, 1980. Used with permission.

Draft. As we saw in Fig. 11.5, a small *draft* (taper) is provided in sand-mold patterns to enable removal of the pattern without damaging the mold. Typical drafts range from 5 mm/m to 15 mm/m (1/16 in./ft to 3/16 in./ft). Depending on the quality of the pattern, draft angles usually range from 0.5° to 2°. The angles on inside surfaces are typically twice this range. They have to be higher than those for outer surfaces because the casting shrinks inward toward the core.

Tolerances. Tolerances—the permissible variations in the dimensions of a part—depend on the particular casting process, size of the casting, and type of pattern used. Tolerances are smallest within one part of the mold and, because they are cumulative, increase between different parts of the mold. Tolerances should be as wide as possible, within the limits of good part performance; otherwise the cost of the casting increases. In commercial practice, tolerances usually are in the range of ±0.8 mm (1/32 in.) for small castings and increase with the size of castings, say, to 6 mm (1/4 in.) for large castings.

Machining allowance. Because most expendable-mold castings require some additional finishing operations, such as machining, allowances should be made in casting design for these operations. **Machining allowances**, which are included in pattern dimensions, depend on the type of casting and increase with the size and section thickness of castings. Allowances usually range from about 2 mm to 5 mm (0.1 in. to 0.2 in.) for small castings, to more than 25 mm (1 in.) for large castings.

Residual stresses. The different cooling rates within the body of a casting cause residual stresses. Stress relieving may thus be necessary to avoid distortions in critical applications.

12.2.2 Designing for permanent-mold casting

The design principles for permanent-mold casting are similar to those for expendable-mold casting. Typical design guidelines and examples for permanent-mold casting are shown schematically in Fig. 12.9 for die casting. Note that the cross-

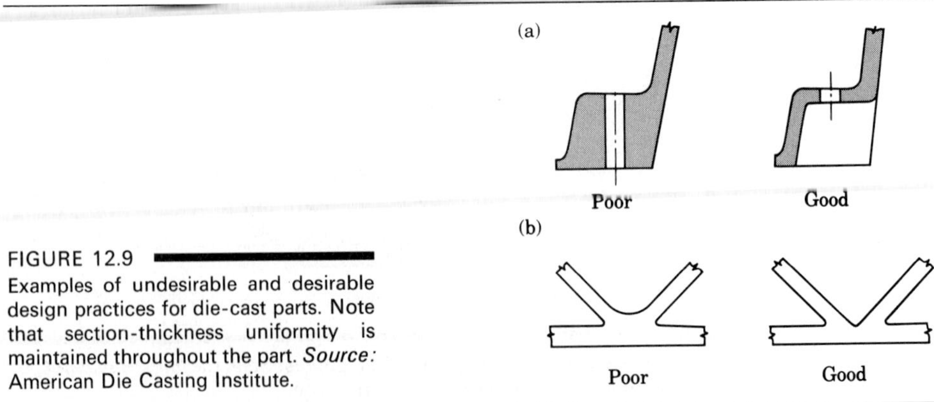

FIGURE 12.9 Examples of undesirable and desirable design practices for die-cast parts. Note that section-thickness uniformity is maintained throughout the part. *Source:* American Die Casting Institute.

sections have been reduced in order to decrease the time for solidification and save material.

Special considerations are involved in designing and tooling for die casting, where sharp edges rather than smooth transitions at the intersection of two members may be desirable for longer mold life. Furthermore, designs may be modified to eliminate the draft for better dimensional accuracy.

12.3 Casting Alloys

We classify **casting alloys** as ferrous and nonferrous (see Chapters 5 and 6). Parts made of aluminum-base and magnesium-base alloys are known as *light-metal* castings. We describe the basic properties and characteristics of various casting alloys in this section. We summarize the mechanical properties of cast metals in Fig. 12.10 and Tables 12.2–12.5 (see pp. 353 and 354).

12.3.1 Nonferrous casting alloys

Aluminum-base alloys. Alloys with an aluminum base have a wide range of mechanical properties, mainly because of various hardening mechanisms and heat treatments that can be used with them (see Section 4.9). Their fluidity depends on oxides and alloying elements in the metal. These alloys have high electrical conductivity and generally good atmospheric corrosion resistance. However, their resistance to some acids and all alkalis is poor and care must be taken to prevent galvanic corrosion. They are nontoxic and lightweight and have good machinability. However, except for alloys with silicon, they generally have low resistance to wear and abrasion. Aluminum-base alloys have many applications, including architectural and decorative use. Engine blocks of some automobiles are made of aluminum-alloy castings.

Magnesium-base alloys. The lowest density of all commercial casting alloys are those in the magnesium-base group. They have good corrosion resistance and moderate strength, depending on the particular heat treatment used.

Copper-base alloys. Although somewhat expensive, copper-base alloys have the advantages of good electrical and thermal conductivity, corrosion resistance, nontoxicity, and wear properties suitable for bearing materials. Mechanical properties and fluidity are influenced by the alloying elements.

Zinc-base alloys. A low-melting-point alloy group, zinc-base alloys have good fluidity and sufficient strength for structural applications. These alloys are commonly used in die casting.

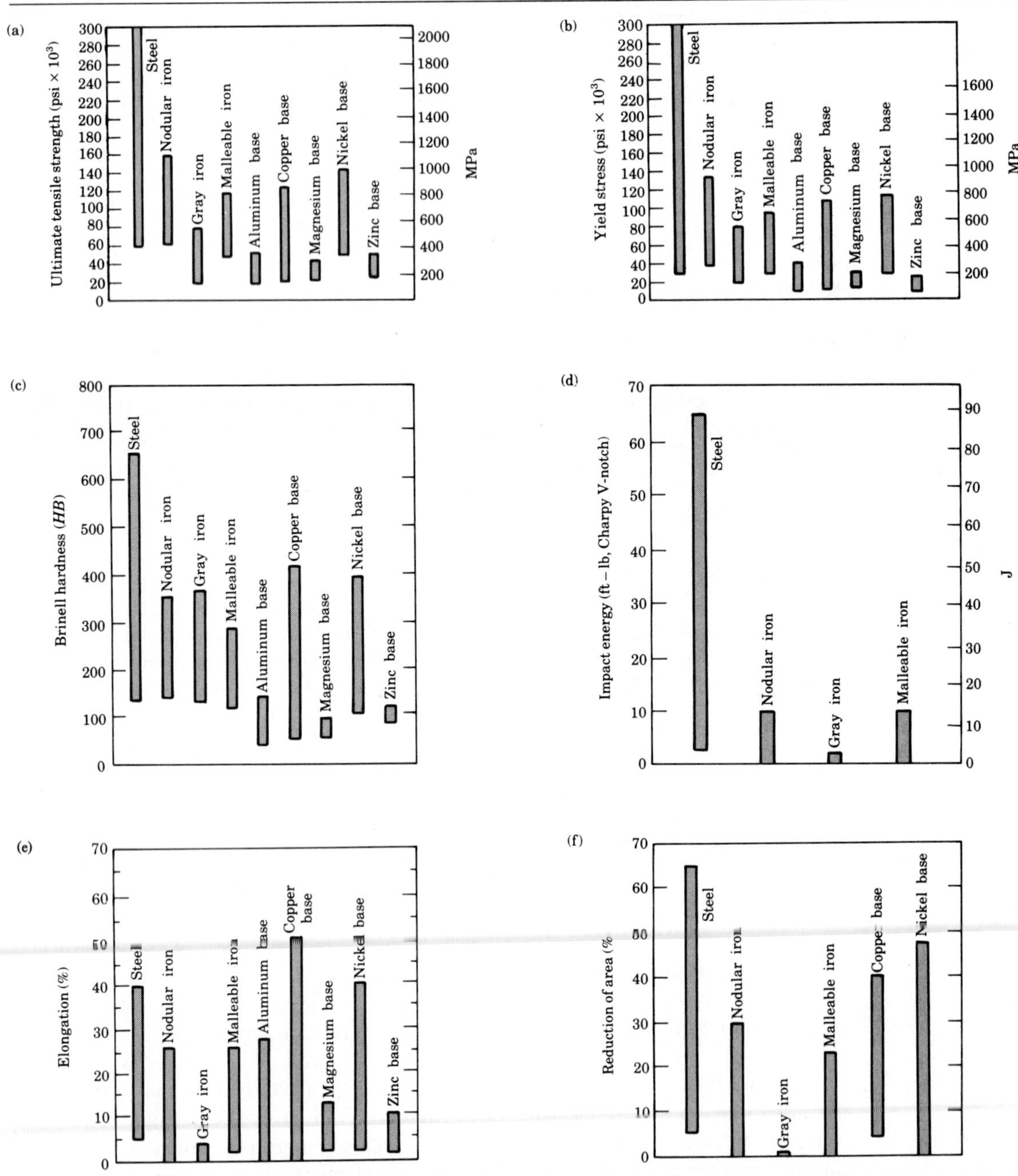

FIGURE 12.10 Mechanical properties for various groups of cast alloys. Note that gray iron has very little ductility and toughness, compared to most other cast alloys, some of which undergo considerable elongation and reduction of area in tension. Note also that even within the same group, the properties of cast alloys vary over a wide range, particularly for cast steels. *Source:* Steel Founders' Society of America.

TABLE 12.2
TYPICAL APPLICATIONS FOR CASTINGS AND CASTING CHARACTERISTICS

TYPE OF ALLOY	APPLICATION	CASTABILITY*	WELDABILITY*	MACHINABILITY*
Aluminum	Pistons, clutch housings, exhaust manifolds	E	F	G–E
Copper	Pumps, valves, gear blanks, marine propellers	F–G	F	F–G
Gray iron	Engine blocks, gears, brake disks and drums, machine bases	E	D	G
Magnesium	Crankcase, transmission housings	G–E	G	E
Malleable iron	Farm and construction machinery, heavy-duty bearings, railroad rolling stock	G	D	G
Nickel	Gas turbine blades, pump and valve components for chemical plants	F	F	F
Nodular iron	Crankshafts, heavy-duty gears	G	D	G
Steel (carbon and low alloy)	Die blocks, heavy-duty gear blanks, aircraft undercarriage members, railroad wheels	F	E	F
Steel (high alloy)	Gas turbine housings, pump and valve components, rock crusher jaws	F	E	F
White iron	Mill liners, shot blasting nozzles, railroad brake shoes, crushers and pulverizers	G	VP	VP
Zinc	Door handles, radiator grills, carburetor bodies	E	D	E

* E, excellent; G, good; F, fair; VP, very poor; D, difficult.

TABLE 12.3
PROPERTIES AND TYPICAL APPLICATIONS OF CAST IRONS

CAST IRON	TYPE	ULTIMATE TENSILE STRENGTH (MPa)	YIELD STRENGTH (MPa)	ELONGATION IN 50 mm (%)	TYPICAL APPLICATIONS
Gray	Ferritic	170	140	0.4	Pipe, sanitary ware
	Pearlitic	275	240	0.4	Engine blocks, machine tools
	Martensitic	550	550	0	Wearing surfaces
Nodular (Ductile)	Ferritic	415	275	18	Pipe, general service
	Pearlitic	550	380	6	Crankshafts, highly stressed parts
	Tempered martensite	825	620	2	High-strength machine parts, wear resistance
Malleable	Ferritic	365	240	18	Hardware, pipe fittings, general engineering service
	Pearlitic	450	310	10	Railroad equipment, couplings
	Tempered martensite	700	550	2	Railroad equipment, gears, connecting rods
White	Pearlitic	275	275	0	Wear-resistance, mill rolls

TABLE 12.4
MECHANICAL PROPERTIES OF GRAY CAST IRONS

ASTM CLASS	ULTIMATE TENSILE STRENGTH (MPa)	COMPRESSIVE STRENGTH (MPa)	ELASTIC MODULUS (GPa)	HARDNESS (HB)
20	152	572	66 to 97	156
25	179	669	79 to 102	174
30	214	752	90 to 113	210
35	252	855	100 to 119	212
40	293	965	110 to 138	235
50	362	1130	130 to 157	262
60	431	1293	141 to 162	302

TABLE 12.5
PROPERTIES AND TYPICAL APPLICATIONS OF CAST NONFERROUS ALLOYS

ALLOYS (*UNS*)	CONDITION	ULTIMATE TENSILE STRENGTH (MPa)	YIELD STRENGTH (MPa)	ELONGATION IN 50 mm (%)	TYPICAL APPLICATIONS
ALUMINUM ALLOYS					
195 (AO1950)	Heat treated	220–280	110–220	8.5–2	Sand castings
319 (AO3190)	Heat treated	185–250	125–180	2–1.5	Sand castings
356 (AO3560)	Heat treated	260	185	5	Permanent mold castings
COPPER ALLOYS					
Red brass (C83600)	Annealed	235	115	25	Pipe fittings, gears
Yellow brass (C86400)	Annealed	275	95	25	Hardware, ornamental
Manganese bronze (C86100)	Annealed	480	195	30	Propeller hubs, blades
Leaded tin bronze (C92500)	Annealed	260	105	35	Gears, bearings, valves
Gun metal (C90500)	Annealed	275	105	30	Pump parts, fittings
Nickel silver (C97600)	Annealed	275	175	15	Marine parts, valves
MAGNESIUM ALLOYS					
AZ91A	F	230	150	3	Die castings
AZ63A	T4	275	95	12	Sand and permanent mold castings
AZ91C	T6	275	130	5	High strength
EZ33A	T5	160	110	3	Elevated temperature
HK31A	T6	210	105	8	Elevated temperature
QE22A	T6	275	205	4	Highest strength

High-temperature alloys. High-temperature alloys have a wide range of properties and typically require temperatures of up to 1650 °C (3000 °F) for casting titanium and superalloys—and higher for refractory alloys. Special techniques are used to cast these alloys into parts for jet and rocket engine components. Some of these alloys are more suitable and economical for casting than for shaping by other manufacturing methods, such as forging.

12.3.2 Ferrous casting alloys

Cast irons. Cast irons represent the largest amount of all metals cast. They generally possess several desirable properties, such as wear resistance, hardness, and good machinability. These alloys can easily be cast into intricate shapes.

The term **cast iron** refers to a family of alloys. As we described in Section 4.6, they are classified as gray cast iron (gray iron), nodular (ductile or spheroidal) iron, white cast iron, malleable iron, and compacted graphite iron. We discuss the characteristics of each of these cast irons in this section. We show their general properties and typical applications in Tables 12.3 and 12.4.

Gray cast iron. Castings of gray cast iron have relatively few shrinkage cavities and little porosity. Recall that various forms of gray cast iron are termed ferritic, pearlitic, and martensitic. Because of differences in their structures, each type has different properties. Typical uses of gray cast iron are for engine blocks, machine bases, electric-motor housings, pipes, and wear surfaces for machines. Gray cast irons are specified by a two-digit ASTM designation. Class 20, for example, specifies that the material must have a minimum tensile strength of 20 ksi (140 MPa). We show the mechanical properties for several classes of gray cast iron in Table 12.4.

Nodular (ductile) iron. Typically used for machine parts, pipe, and crankshafts (Fig. 12.11), nodular cast irons are specified by a set of two-digit numbers. Thus, class or grade 80-55-06, for example, indicates that the material has a minimum tensile strength of 80 ksi (550 MPa), a minimum yield strength of 55 ksi (380 MPa), and 6 percent elongation in 2 in. (50 mm).

White cast iron. Because of its extreme hardness and wear resistance, white cast iron is used mainly for liners for machinery to process abrasive materials, rolls for rolling mills, and railroad-car brake shoes.

Malleable iron. The principal use of malleable iron is for railroad equipment and various types of hardware. Malleable irons are specified by a five-digit designation. Thus 35018, for example, indicates that the yield strength of the material is 35 ksi (240 MPa), and its elongation is 18 percent in 2 in.

Compacted graphite iron. First produced commercially in 1976, compacted graphite iron has properties that fall between those of gray and nodular irons. Its machinability is better than nodular iron. Typical applications are automotive engine blocks and heads.

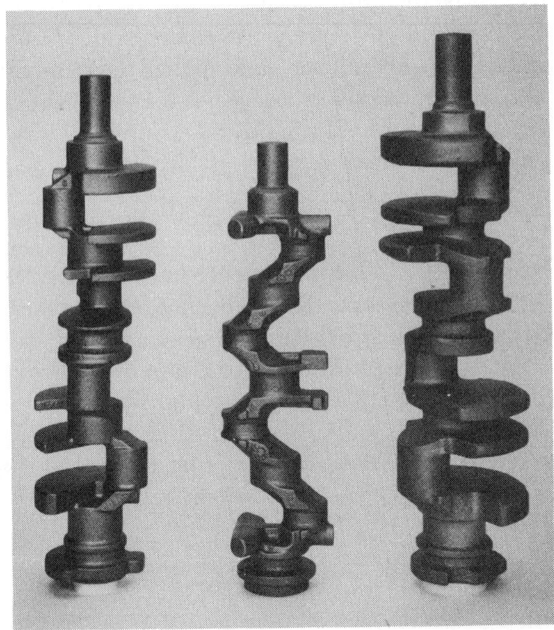

FIGURE 12.11
Nodular cast iron crankshafts, made by shell-molding or full-mold processes. These parts can also be made by sand casting. First put into service in the 1950s, most cars and trucks are now equipped with cast crankshafts. *Source:* Courtesy of Central Foundry Division, General Motors Corporation.

Cast steels. Because of the high temperatures required to melt cast steels, up to about 1650 °C (3000 °F), their casting requires considerable knowledge and experience. The high temperatures involved present difficulties in the selection of mold materials—particularly in view of the high reactivity of steels with oxygen—in melting and pouring the metal. Steel castings possess properties that are more uniform (isotropic) than those made by mechanical working processes, which we discuss in Part III. Cast steels can be welded; however, welding alters the cast microstructure in the heat-affected zone (see Section 29.2), influencing the strength, ductility, and toughness of the base metal. Subsequent heat treatment must be performed to restore the mechanical properties of the casting. Cast weldments have gained importance where complex configurations, or the size of the casting, may prevent casting the part economically in one place.

Cast stainless steels. Casting of stainless steels involves considerations similar to those for steels in general. Stainless steels generally have a long freezing range and high melting temperatures. They develop various structures, depending on their composition and the process parameters. Cast stainless steels are available in various compositions and can be heat treated and welded. These cast products have high heat and corrosion resistance. Nickel-base casting alloys are used for severely corrosive environments and very high temperature service.

12.4 Economics of Casting

When looking at various casting processes, we noted that some require more labor than others, some require expensive dies and machinery, and some take a great deal of time to complete. These important characteristics are outlined in Table 12.6. Each of the individual factors listed affects to varying degrees the overall cost of a casting operation.

As we describe in greater detail in Chapter 40, the cost of a product involves the costs of materials, labor, tooling, and equipment. Preparations for casting a product include making molds and dies that require raw materials, time, and effort, which we can translate into costs. As you can see in Table 12.6, relatively little cost is involved in molds for sand casting. On the other hand, die-casting dies require expensive materials and a great deal of machining and preparation. In addition to molds and dies, facilities are required for melting and pouring the molten metal into the molds or dies. These facilities include furnaces and related machinery; their costs depend on the level of automation desired. Finally, costs are involved in heat treating, cleaning, and inspecting castings. Heat treating is an important part of the production of many alloy groups, especially ferrous castings, and is necessary to produce improved mechanical properties. However, heat treating also introduces another set of production problems, such as scale formation and warpage, and can be a significant part of the production costs.

The amount of labor required for these operations can vary considerably, depending on the particular process and level of automation. Investment casting, for example, requires a great deal of labor because of the large number of steps involved in this operation. On the other hand, operations such as highly automated die casting can maintain high production rates with little labor required.

TABLE 12.6
GENERAL COST CHARACTERISTICS OF CASTING PROCESSES

PROCESS	COST*			PRODUCTION RATE (Pc/hr)
	DIE	*EQUIPMENT*	*LABOR*	
Sand	L	L	L–M	<20
Shell	L–M	M–H	L–M	<50
Plaster	L–M	M	M–H	<10
Investment	M–H	L–M	H	<1000
Permanent mold	M	M	L–M	<60
Die	H	H	L–M	<200
Centrifugal	M	H	L–M	<50

* L, low; M, medium; H, high.

We should note, however, that the cost of equipment per casting (**unit cost**) will decrease as the number of parts cast increases. Thus sustained high production rates can justify the high cost of dies and machinery. However, if demand is relatively small, the cost per casting increases rapidly. It then becomes more economical to manufacture the parts by sand casting—or by other manufacturing processes.

The two processes (sand and die casting) we compared produce castings with significantly different dimensional and surface-finish characteristics (see Table 11.1). Thus not all manufacturing decisions are based purely on economic considerations. In fact, as you will see in Parts III and IV, parts can usually be made by more than one or two processes. The final decision rests on both economic and technical considerations. We describe competitive aspects of manufacturing processes in Chapter 40.

SUMMARY

General principles have been established to aid designers in producing castings that are free from defects and meet tolerances and service requirements. These principles concern shape of casting and various techniques to minimize hot spots that could lead to shrinkage cavities. Because of the large number of variables involved, close control of all parameters is essential, particularly those related to the nature of liquid metal flow into the molds and dies and the rate of cooling in different regions of the mold or die.

Several nonferrous and ferrous casting alloys are available. They have a wide range of properties, casting characteristics, and applications. Because many castings are designed and produced to be assembled with other mechanical components and structures, various other considerations, such as weldability, machinability, and surface conditions, are also important.

Within the limits of good performance, the economics of casting is just as important as technical considerations. Factors affecting the overall cost are the cost of materials, molds, dies, equipment, and labor, each of which varies with the particular casting process. An important parameter is the cost per casting which, for large production runs, can justify large expenditures for automated machinery.

TRENDS

- Methods of casting high-strength gray iron, high-temperature ductile iron, and austempered ductile iron are being developed.
- Developments of high-strength low-alloys through microalloying, and carbon steels without heat treating, are being pursued.
- Cast stainless steels with improved corrosion resistance and higher strength have been introduced in the last few years and are being used in many applications.

(continued)

TRENDS *(continued)*

- Nickel-base alloys for casting to match the corrosion resistance of wrought alloys have recently been introduced.
- Computer-aided design and manufacturing of molds and dies are important areas being studied and implemented.
- Modern casting plants are being built with state-of-the-art facilities, including computer control of all aspects of casting and inspection, in order to produce quality castings economically.

KEY TERMS

Cast iron
Casting alloys
Design principles
Hot spots
Machining allowance
Patternmaker's shrinkage allowance
Unit cost

BIBLIOGRAPHY

Analysis of Casting Defects. Des Plaines, Ill.: American Foundrymen's Society, 1974.

Angus, H.T., *Cast Iron: Physical and Engineering Properties.* New York: Butterworths, 1976.

Casting Defects Handbook. Des Plaines, Ill.: American Foundrymen's Society, 1972.

Kaye, A., and A.C. Street, *Die Casting Metallurgy.* Woburn, Mass.: Butterworths, 1982.

Metals Handbook, 9th ed., *Vol. 15: Casting.* Metals Park, Ohio: ASM International, 1988.

Mikelonis, P.J. (ed.), *Foundry Technology: A Source Book.* Metals Park, Ohio: American Society for Metals, 1982.

Rowley, M.T. (ed.), *International Atlas of Casting Defects.* Des Plaines, Ill.: American Foundrymen's Society, 1974.

Walton, C.F., and T.J. Opar (eds.), *Iron Castings Handbook*, 3d ed. Des Plaines, Ill.: Iron Castings Society, 1981.

Wieser, P.F. (ed.), *Steel Castings Handbook*, 5th ed. Des Plaines, Ill.: Steel Founders' Society of America, 1980.

REVIEW QUESTIONS

12.1 List the general design considerations in casting.

12.2 What are hot spots?

12.3 What is shrinkage allowance? Machining allowance?

12.4 Why are drafts necessary in some molds?
12.5 What are light-metal castings?
12.6 Name the types of cast irons and list their major characteristics.
12.7 Why are steels more difficult to cast than cast irons?
12.8 Name the important factors involved in the economics of casting operations.
12.9 Describe your observations concerning Figs. 12.2, 12.3, and 12.4.

QUESTIONS AND PROBLEMS

12.10 Describe the procedure you would follow to determine whether a defect in a casting is a shrinkage cavity or porosity caused by gases.
12.11 Explain how you would go about avoiding hot tearing.
12.12 If you need only a few castings of the same design, which three processes would be the most expensive per piece?
12.13 Do you generally agree with the cost ratings in Table 12.6? If so, why?
12.14 Explain how ribs and serrations are helpful in casting flat surfaces that otherwise may warp. Give an illustration.
12.15 In view of the material covered in preceding chapters and based on the information given in Fig. 12.10, what general statements would you make about the mechanical properties of castings?
12.16 Describe the nature of the design changes made in Fig. 12.8. What general principles do you observe?
12.17 In Fig. 12.10 we note that the ductility of some cast alloys is nil. Do you think this should be a significant concern in engineering applications of castings?
12.18 Assume that the Introduction to this chapter is missing. Write a brief introduction to highlight the importance of the topics covered in this chapter.
12.19 Do you think there will be fewer or more defects in a casting made by gravity pouring or made by pouring under pressure?
12.20 Why are allowances provided for in making patterns? What do they depend on?
12.21 Explain the difference in importance in draft in green-sand casting versus permanent-mold casting.
12.22 What type of cast iron would be suitable for a stationary heavy machine base? Why?
12.23 Explain the advantages and limitations of sharp and rounded fillets, respectively, in casting design.
12.24 Referring to Tables 10.1 and 12.1, do you think that there is a contradiction regarding the behavior of gray cast iron? Explain.
12.25 Explain why the elastic modulus E of gray cast iron varies so much, as shown in Table 12.4.
12.26 When designing patterns for casting, patternmakers use special rulers that automatically incorporate solid shrinkage allowances into their designs. Therefore, a 12-in. patternmaker's ruler is longer than a foot. How long is a patternmaker's ruler designed for the making of patterns for (a) aluminum castings? (b) high-manganese steel?

12.27 Porosity developed in the boss of the casting is shown in the figure below. Show how simply repositioning the parting line of this casting can eliminate this problem.

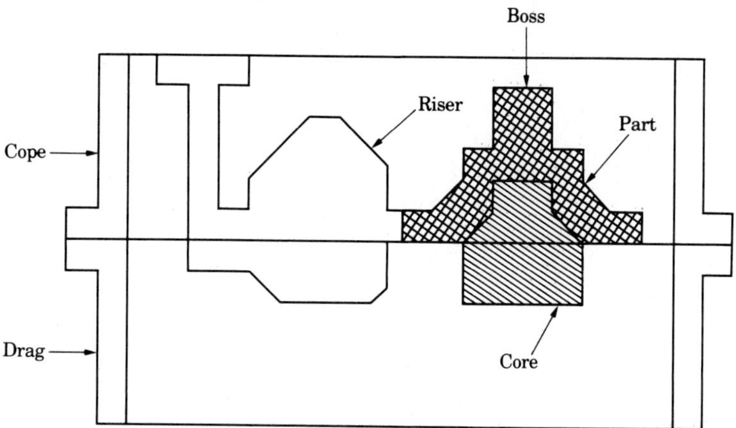

12.28 For the wheel shown in the figure below, show how (a) riser placement, (b) core placement, (c) padding, and (d) chills may be used to help feed and eliminate porosity in the isolated hub boss.

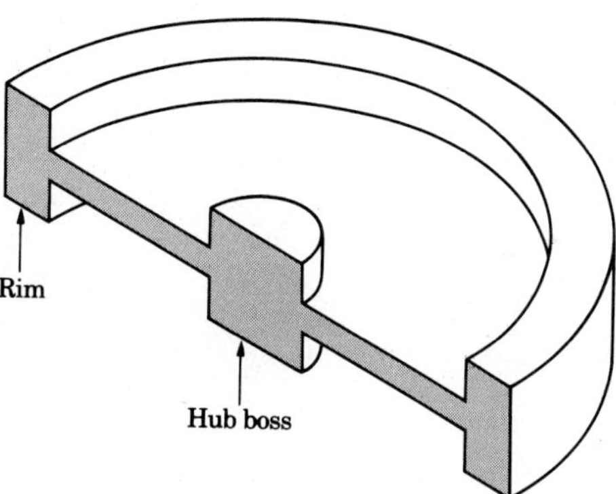

PART III

Forming and Shaping Processes and Equipment

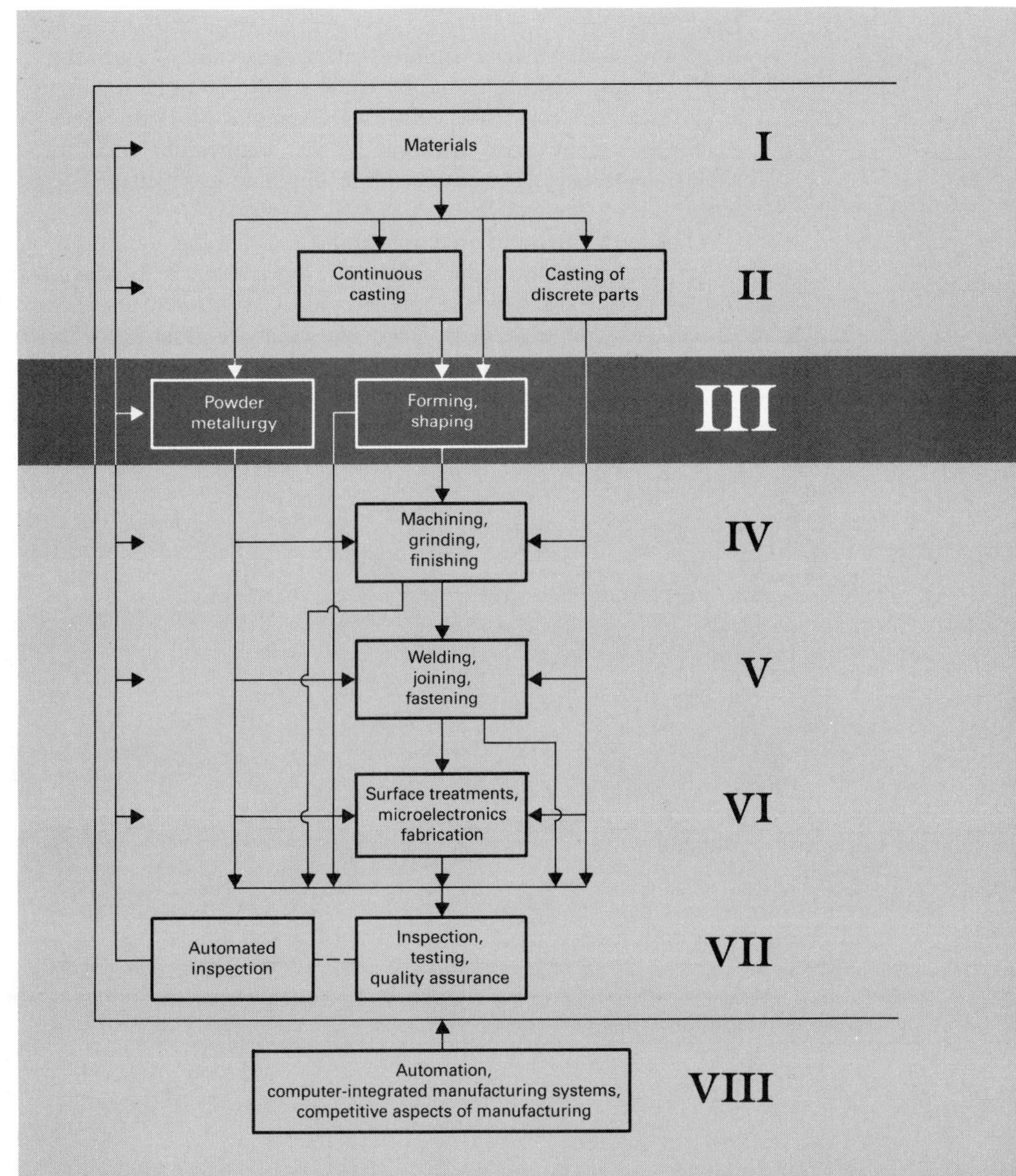

• INTRODUCTION •

We generally tend to take for granted many of the products that we use or come across every day—and the materials and components from which they are made. When we look at these products carefully, we soon realize that a wide variety of materials and processes have been used to make them (Fig. III.1). We also note that some products consist of a few parts (stapler, pipe wrench, light fixtures), and others consist of thousands of parts (automobiles, computers, airplanes). Some products are thin (aluminum foil, plastic film, electrical-resistance wire for toasters), whereas others are thick (ship hulls, boiler plates, machine bases).

Some products have simple shapes with smooth curvatures (bicycle handles, ball bearings, cooking pots), but others have complex configurations and detailed surface features (coins, silverware, engine blocks). Some products are used in critical applications (turbine blades, connecting rods for engines, elevator cables), whereas others are routine applications (watering cans, spoons, paper clips).

The initial material used in forming and shaping metals is usually molten metal, which is cast into individual ingots or, more recently, continuously cast into slabs, rod, or pipe, as discussed in Parts I and II. Cast structures are converted to wrought structures by the deformation processes that we describe in Part III. In addition to cast and wrought structures, the raw material for making products may consist of

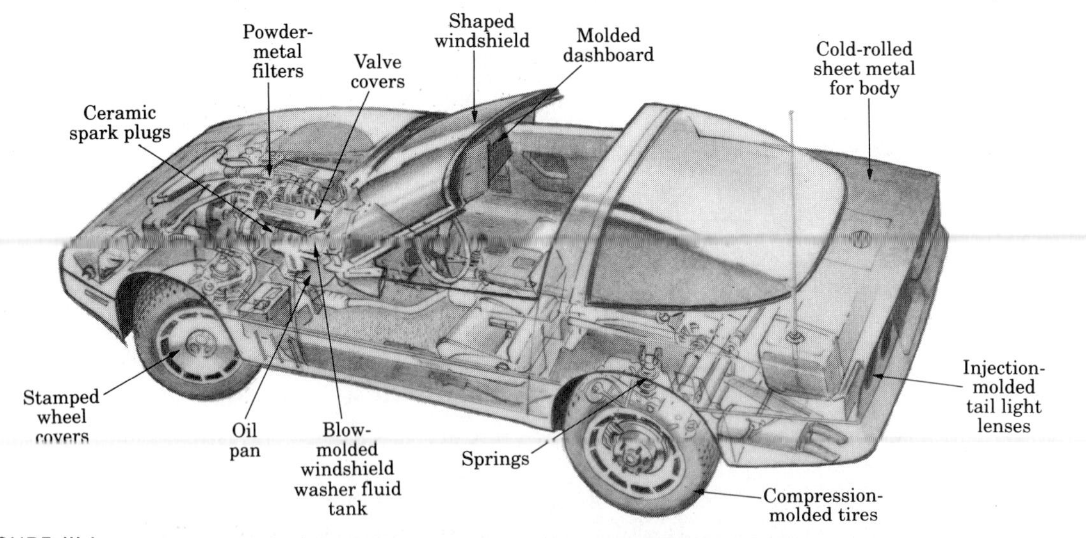

FIGURE III.1
Formed and shaped parts in a typical automobile.

metal powders. For plastics, the starting material is usually pellets, flakes, or powder, and for ceramics it is clays and oxides, obtained from ores or produced synthetically.

Note that we use the words *forming* and *shaping* together in the part title. Although the distinction between the two is not rigid, by forming we mean changing the shape of an existing solid body. The body, which we call the workpiece, stock, or blank throughout the rest of this text, may be in the shape of a plate, sheet, bar, rod, wire, or tubing of various cross-sections. An ordinary wire coat hanger, for example, is made by forming a straight piece of wire by bending and twisting it into the shape of a hanger. A metal body for an automobile is made by forming sheet metal in dies, starting with flat sheet. The initial material for a typical translucent advertising sign is a flat piece of plastic, and is formed by heating and forcing it under a vacuum or air pressure into a mold. These are called **forming processes**.

Shaping processes usually involve molding and casting. The resulting product is usually at or near the final desired shape, and may require little or no further finishing operations. A plastic coat hanger, for example, is made by forcing molten plastic into a mold in the shape of the hanger. Telephone receivers, keys for keyboards, and beverage bottles are made by shaping molten plastic in a mold. The white ceramic insulator for an automotive spark plug is made by shaping clay in a mold, then drying and firing it in a furnace.

Some of these manufacturing operations produce long *continuous* products, such as plates, sheet, tubing, and bars with various cross-sections. Rolling, extrusion, and drawing processes are capable of making such products of metal, as well as nonmetallic materials, including reinforced plastics, which are then cut into the desired lengths. On the other hand, processes such as forging, powder metallurgy, and most forming and shaping processes for nonmetallic materials produce *discrete* products, such as turbine disks, gears, and containers.

In this Part, we describe the important factors involved in each forming and shaping process and how material and process variables influence the quality of a product (Table III.1). We explain why some materials can be processed by only certain manufacturing methods and not by others, and why parts with particular shapes can be processed by only certain techniques and not by others. The characteristics of the machinery and equipment used in these processes also significantly affect the quality of the product, its rate of production, and the economics of the manufacturing operation.

We describe rolling of flat and shaped products (sheet metal for car bodies and appliances, railroad rails), forging of discrete parts (connecting rods, gears, turbine blades), extrusion of long pieces with various cross-sections (tubing, door frames, railings), and drawing of rod, wire, and tube (electrical wiring, nails, shopping carts). These processes are called **bulk-deformation processes** because the workpieces and products have relatively high ratios of volume to surface area or volume to thickness. We then describe sheet-forming processes, powder metallurgy techniques, and processing of polymers, composites, and ceramics and glasses.

TABLE III.1

GENERAL CHARACTERISTICS OF FORMING AND SHAPING PROCESSES

PROCESS	CHARACTERISTICS
Rolling	
Flat	Production of flat plate, sheet, and foil in long lengths, at high speeds, and with good surface finish, especially in cold rolling; requires high capital investment; low to moderate labor cost.
Shape	Production of various structural shapes, such as I-beams, at high speeds; includes thread rolling; requires shaped rolls and expensive equipment; low to moderate labor cost; moderate operator skill.
Forging	Production of discrete parts with a set of dies; some finishing operations usually required; similar parts can be made by casting and powder-metallurgy techniques; usually performed at elevated temperatures; die and equipment costs are high; moderate to high labor cost; moderate to high operator skill.
Extrusion	Production of long lengths of solid or hollow products with constant cross-section; usually performed at elevated temperatures; product is then cut into desired lengths; can be competitive with roll forming; cold extrusion has similarities to forging and is used to make discrete products; moderate to high die and equipment cost; low to moderate labor cost; low to moderate operator skill.
Drawing	Production of long rod and wire, with round or various cross-sections; smaller cross-sections than extrusions; good surface finish; low to moderate die, equipment, and labor costs; low to moderate operator skill.
Sheet-metal forming	Production of a wide variety of shapes with thin walls and simple or complex geometries; generally low to moderate die, equipment, and labor costs; low to moderate operator skill.
Powder metallurgy	Production of simple or complex shapes by compacting and sintering metal powders; can be competitive with casting, forging, and machining processes; moderate die and equipment cost; low labor cost and skill.
Processing of plastics and composite materials	Production of a variety of continuous or discrete products by extrusion, molding, casting, and fabricating processes; can be competitive with sheet metal parts; moderate die and equipment costs; high operator skill in processing of composite materials.
Forming and shaping of ceramics	Production of discrete ceramic products by a variety of shaping, drying, and firing processes; low to moderate die and equipment cost; moderate to high operator skill.

The knowledge gained in this Part will enable you to assess the relative merits and limitations of various forming and shaping processes. Decisions have to be made about whether a particular component should be manufactured of metal, plastic, ceramic, or composite materials and whether it should be made by forging, casting, powder metallurgy, sheet forming, or a combination of processes. The examples and summary tables in each chapter emphasize the importance of properly selecting materials and processes for manufacturing quality products that meet service requirements and are also economical to produce.

13

Rolling

13.1 Introduction

Rolling is the process of reducing the thickness or changing the cross-section of a long workpiece by compressive forces applied through a set of **rolls** (Fig. 13.1), similar to rolling dough with a rolling pin to reduce its thickness. Rolling, which accounts for about 90 percent of all metals produced by metalworking processes, was first developed in the late 1500s. The basic operation is flat rolling, or simply rolling, where the rolled products are flat plate and sheet.

Plates, which are generally regarded as having a thickness greater than 6 mm (1/4 in.), are used for structural applications such as ship hulls, boilers, bridges, girders, machine structures, and nuclear vessels. Plates can be as much as 0.3 m (12 in.) thick for the supports for large boilers, 150 mm (6 in.) for reactor vessels, and 100–125 mm (4–5 in.) for battleships and tanks.

Sheets are generally less than 6 mm thick and are used for automobile bodies, appliances, containers for food and beverages, and kitchen and office equipment.

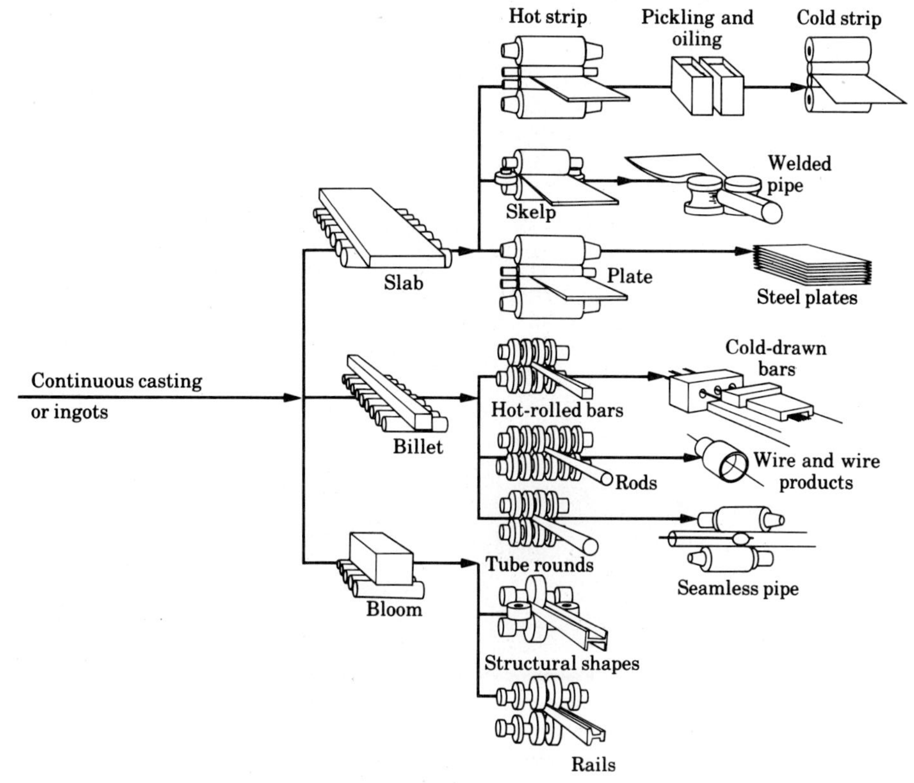

FIGURE 13.1 Schematic outline of various flat- and shape-rolling processes. *Source:* American Iron and Steel Institute.

Commercial aircraft fuselages are usually made of about 1 mm (0.040 in.) thick aluminum-alloy sheet, while beverage cans are made of 0.15 mm (0.006 in.) aluminum sheet. Aluminum **foil** used to wrap candy and cigarettes has a thickness of 0.008 mm (0.0003 in.). Sheets are provided as flat pieces or as strip in coils (Fig. 13.2) to manufacturing facilities for further processing into products.

In this chapter we discuss the basic process of flat rolling and related rolling operations and the important factors involved in rolling practices. Traditionally, the initial material form for rolling is an ingot. However, as we stated in Section 5.4, this practice is now being rapidly replaced by continuous casting and rolling, with much higher efficiency and lower cost. Rolling is first carried out at elevated temperatures (hot rolling), wherein the coarse-grained, brittle, and porous structure of the ingot or continuously cast metal is broken down into a **wrought structure**, with finer grain size.

FIGURE 13.2
Coiled rolls of sheet steel produced in a rolling mill. Most of these coils weigh about 7000 kg (15,000 lb), although some weigh as much as 30 tons. *Source:* Inland Steel Company.

13.2 Flat Rolling

A schematic illustration of the *flat rolling* process is shown in Fig. 13.3(a). A strip of thickness h_0 enters the *roll gap* and is reduced to h_f by a pair of rotating rolls, each roll being powered through its own shaft by electric motors. The surface speed of the roll is V_r. The velocity of the strip increases from its initial value of V_0 as it moves through the roll gap, just as fluid flows faster as it moves through a converging channel. The velocity of the strip is highest at the exit of the roll gap and is denoted as V_f.

Since the surface speed of the roll is constant, there is relative sliding between the roll and the strip along the arc of contact in the roll gap L. At one point along the contact length, the velocity of the strip is the same as that of the roll. This is called the *neutral*, or *no-slip*, *point*. To the left of this point, the roll moves faster than the strip, and to the right of this point, the strip moves faster than the roll. Hence the frictional forces, which oppose motion, act on the strip as shown in Fig. 13.3(b).

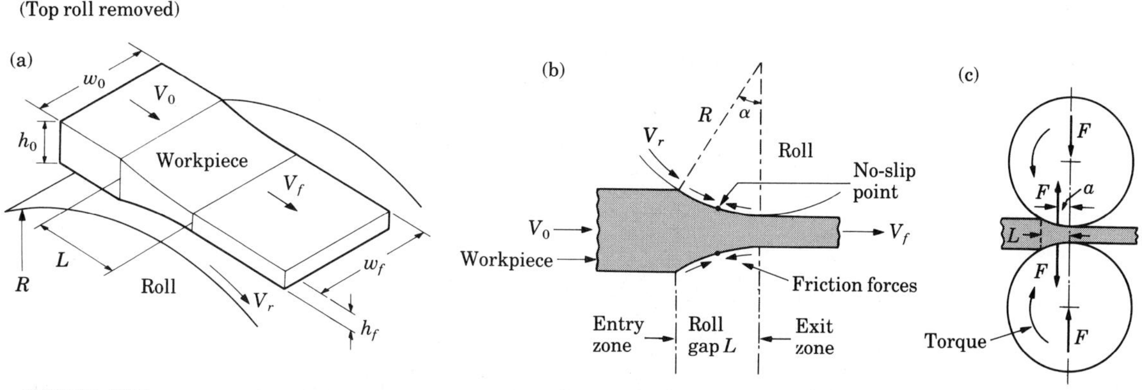

FIGURE 13.3
(a) Schematic illustration of the flat-rolling process. (b) Friction forces acting on strip surfaces. (c) Roll force F and torque acting on the rolls.

13.2.1 Frictional forces

The rolls pull the material into the roll gap through a net frictional force on the material. You can see that this net frictional force must be to the right in Fig. 13.3(b), and consequently the frictional force on the left of the neutral point must be higher than the force on the right.

As you can see, friction is needed to roll materials. However, energy is dissipated in overcoming friction, so increasing friction means increasing forces and power consumption. Furthermore, high friction could damage the surface of the rolled product. We therefore have to make a compromise, obtaining low coefficients of friction with effective lubricants.

The maximum possible **draft**, defined as the difference between the initial and final thicknesses $(h_0 - h_f)$, is a function of the coefficient of friction, μ, and the roll radius, R:

$$h_0 - h_f = \mu^2 R. \tag{13.1}$$

Thus the higher the friction and the larger the roll radius, the greater the maximum draft and reduction in thickness become. This situation is similar to the use of large tires and rough treads on farm tractors and off-road earth-moving equipment, permitting the equipment to travel over rough terrain without skidding.

13.2.2 Roll force and power requirement

Because the rolls apply pressure on the material to reduce its thickness, we need a normal force perpendicular to the arc of contact (Fig. 13.3c). For low coefficients of friction, we can estimate the roll force F in flat rolling from the formula

$$F = LwY_{\text{avg}}, \tag{13.2}$$

where L is the roll-strip contact length, w is the width of the strip, and Y_{avg} is the average true stress of the strip in the roll gap. Because the strip thickness is very small compared to the roll diameter, we assume that roll forces act perpendicular to the plane of the sheet.

We can estimate the power required per roll by assuming that the force F acts in the middle of the arc of contact. Hence $a = L/2$ in Fig. 13.3(c). Thus *torque per roll* is the product of F and a. Consequently, the *power per roll* is

$$\text{Power} = \frac{2\pi FLN}{60{,}000}\ \text{kW}, \tag{13.3}$$

where F is in newtons, L is in meters, and N is the rpm of the roll. We can also express the power as

$$\text{Power} = \frac{2\pi FLN}{33{,}000}\ \text{hp}, \tag{13.4}$$

where F is in lb and L is in ft. We apply these formulas in the following example.

• Example: Calculation of roll force and torque.

An annealed copper strip, 9 in. (228 mm) wide and 1.00 in. (25 mm) thick, is rolled to a thickness of 0.80 in. (20 mm). The roll radius is 12 in. (300 mm) and rotates at 100 rpm. Calculate the roll force and the power in this operation.

SOLUTION. The roll force is determined from Eq. (13.2), in which L is the roll-strip contact length. It can be shown from simple geometry that this length is approximately given by

$$L = \sqrt{R(h_0 - h_f)} = \sqrt{12(1.00 - 0.80)} = 1.55\ \text{in.} = 0.13\ \text{ft.}$$

We now must determine Y_{avg} for annealed copper. We note that the absolute value of the true strain that the strip undergoes in this operation is

$$\varepsilon = \ln (1.00/0.80) = 0.223.$$

Referring to Fig. 2.6, we note that annealed copper has a true stress of about 12,000 psi in the unstrained condition; and at a true strain of 0.223, the true stress is 40,000 psi. Thus the average stress Y_{avg} is about 26,000 psi. We can now determine the roll force F as follows:

$$F = LwY_{avg} = (1.55)(9)(26{,}000) = 363{,}000\ \text{lb} = 1.6\ \text{MN}.$$

The power per roll is calculated from Eq. (13.4), where $N = 100$ and L is 0.13 ft. Thus,

$$\begin{aligned}\text{Power} &= 2\pi FLN/33{,}000 = 2\pi(363{,}000)(0.13)(100)/33{,}000\\ &= 898\ \text{hp} = 670\ \text{kW}.\end{aligned}$$

Thus the total power in rolling is 1796 hp = 1340 kW.

•

Making exact calculations of forces and power requirements in rolling is difficult because of the problems involved in determining the exact contact geometry and estimating the coefficient of friction and the strength of the material in the roll gap accurately, particularly for hot rolling, and the sensitivity of the material to deformation at elevated temperatures.

Roll forces can cause deflection and roll flattening, which in turn, adversely affect the rolling operation. Roll forces can be reduced by (a) reducing friction, (b) using smaller diameter rolls to reduce the contact area, (c) taking smaller reductions per pass to reduce the contact area, and (d) rolling at elevated temperature to reduce the strength of the material. Another effective method of reducing roll forces is to apply longitudinal *tensions* to the strip during rolling. As a result, the material requires less compressive stress to deform plastically. Because they require high roll forces, tensions are particularly important in rolling high-strength metals. Tensions can be applied to the strip either at the entry zone (*back tension*) or at the exit (*front tension*), or both.

Roll deflections. Just as a straight beam deflects under a transverse load, roll forces tend to bend the rolls elastically during rolling (Fig. 13.4a). As a result, the rolled strip tends to be thicker at its center than at its edges, giving it a **crown**. The usual method of avoiding this problem is to grind the rolls so that their diameter at the center is slightly larger than at their edges, giving them **camber**. Thus when the roll bends, its contact along the width of the strip is straight.

For rolling sheet metals, the radius of the camber is generally 0.25 mm (0.01 in.) greater than at the edges of the roll. When properly designed, cambered rolls produce flat strips (Fig. 13.4b). However, a particular camber is correct only for a certain load and strip width. To reduce the effects of deflection, the rolls can be

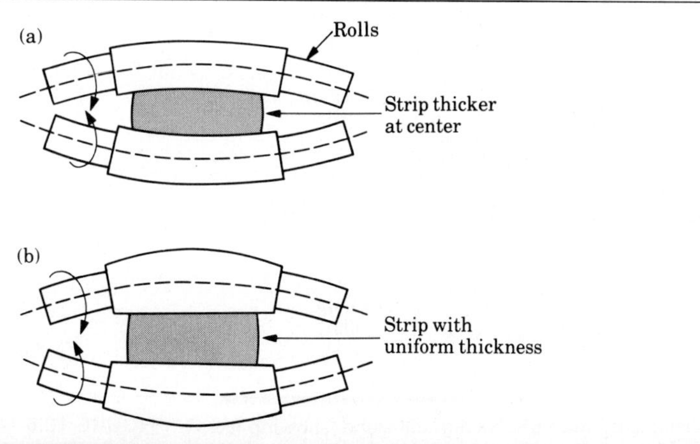

FIGURE 13.4
(a) Bending of straight cylindrical rolls caused by the roll force. (b) Bending of rolls ground with camber, producing a strip with uniform thickness.

subjected to bending, by applying moments at their bearings (similar to bending a wooden stick at its ends), thus simulating camber.

Because of the heat generated by plastic deformation during rolling, rolls can become slightly barrel shaped (thermal camber). With stiff rolls, this condition produces strips that are thinner at the center. In fact, camber can be controlled by varying the location of the coolant on the rolls during hot rolling. Roll forces also tend to *flatten* the rolls elastically, much like the flattening of automobile tires. Flattening of the rolls means a larger roll radius and hence a larger contact area for the same draft. Thus the roll force, in turn, increases with increased flattening.

The *roll stand*, including the housing, chocks, and bearings (Fig. 13.5), may stretch under the roll forces to such an extent that the roll gap can open up significantly. Consequently, the rolls have to be set closer to compensate for this deflection and obtain the desired final thickness.

Spreading. In rolling plates and sheet with high width-to-thickness ratios (see Fig. 13.3a), the width of the material remains essentially constant during rolling. However, with smaller ratios, such as with a square cross-section, the width increases considerably in the roll gap, just as you would observe when rolling dough. The increase in width is called *spreading* (Fig. 13.6). Therefore, in calculating the roll force, we take the width w in Eq. (13.2) as an average width. Spreading can be prevented by the use of vertical rolls in contact with the edges of the rolled product. Because bars have a tendency to spread, they are rolled in *edger mills* in this manner.

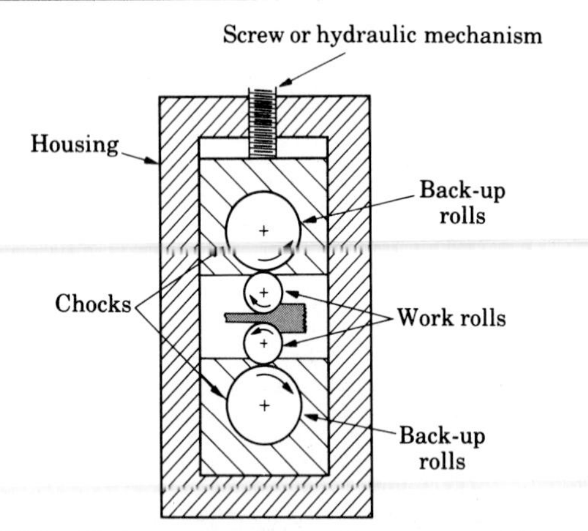

FIGURE 13.5 Schematic illustration of a rolling-mill stand, showing its various features. The stiffnesses of the housing, rolls, and roll bearings are important in controlling and maintaining the thickness of the rolled strip.

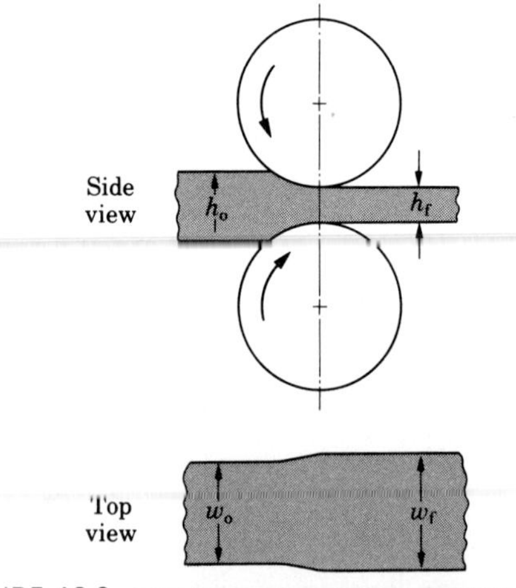

FIGURE 13.6 Increase in the width (spreading) of a strip in flat rolling (see also Fig. 13.3a). Spreading can be similarly observed when dough is rolled with a rolling pin.

13.3 Flat-Rolling Practice

The initial breaking down of an ingot (or continuous casting) by hot rolling converts the coarse-grained, brittle, and porous cast structure to a wrought structure (Fig. 13.7). This structure has finer grains and enhanced ductility resulting from breaking up brittle grain boundaries and closing up internal defects, such as porosity. As we described in Section 5.4, traditional methods of rolling ingots are now being rapidly replaced by continuous casting.

Temperature ranges for hot rolling are typically about 450 °C (850 °F) for aluminum alloys, up to 1250 °C (2300 °F) for alloy steels, and up to 1650 °C (3000 °F) for refractory alloys (see also Table 14.1). The width of rolled products may range up to 5 m (200 in.) and be as thin as 0.0025 mm (0.0001 in.). Rolling speeds may range up to 25 m/s (5000 ft/min) for cold rolling—higher with highly automated and computer-controlled facilities.

The product of the first hot-rolling operation is called a **bloom** or **slab** (see Fig. 13.1). A bloom usually has a square cross-section, at least 150 mm (6 in.) on the side; a slab is usually rectangular in cross-section. Blooms are processed further by *shape rolling* into structural shapes, such as I-beams and railroad rails. Slabs are rolled into plates and sheet. **Billets** are usually square, with a cross-sectional area smaller than blooms, and are rolled into various shapes, such as round rods and bars, using shaped rolls. Hot-rolled round rods are used as the starting material for rod and wire drawing and are called *wire rods*.

In hot rolling blooms, billets, and slabs, the surface of the material is usually *conditioned* (prepared for a subsequent operation) prior to rolling. Conditioning is done by various means, such as with a torch (*scarfing*) to remove heavy scale, or by rough grinding to smoothen surfaces. Prior to cold rolling, the scale developed

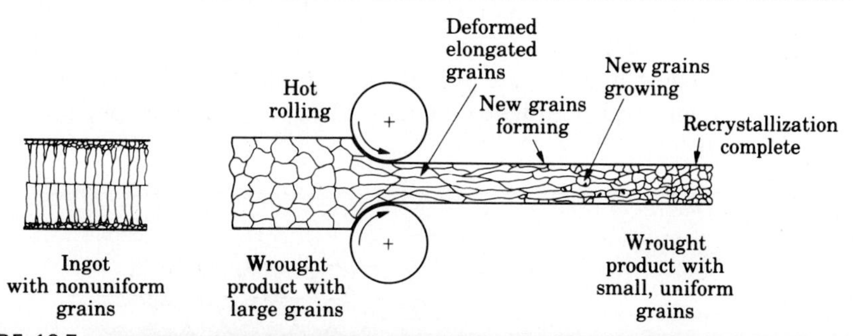

FIGURE 13.7 Changes in the grain structure of cast or large-grain wrought metals during hot rolling. Hot rolling is an effective way to reduce grain size in metals for improved strength and ductility. Cast structures of ingots or continuous castings are converted to a wrought structure by hot working.

during hot rolling or other defects may be removed by pickling with acids or by mechanical means, such as blasting with water, or by grinding.

Pack rolling is a flat-rolling operation in which two or more layers of metal are rolled together, thus improving productivity. Aluminum foil, for example, is pack rolled in two layers. You probably have noticed that one side of aluminum foil is matte and the other side shiny. The foil-to-foil side has a matte and satiny finish, and the foil-to-roll side is shiny and bright because it has been in contact with the polished roll.

Mild steel, when stretched during sheet-forming operations, undergoes *yield-point elongation.* This phenomenon causes surface irregularities called *stretcher strains* or *Lueder's bands* (see also Section 16.3). To avoid this situation, the sheet metal is subjected to a light pass of 0.5–1.5 percent reduction, known as **temper rolling.**

A rolled sheet may not be sufficiently flat as it leaves the roll gap because of variations in the material or in the processing parameters during rolling. To improve flatness, the strip is then passed through a series of *leveling rolls.* Each roll is usually driven separately with individual electric motors. The strip is flexed in opposite directions as it passes through the sets of rollers. Several different roller arrangements are used, one of which is shown in Fig. 13.8.

13.3.1 Defects in rolled plates and sheet

Defects may be present on the surfaces of rolled plates and sheets, or they may be internal (**structural defects**). Defects are undesirable not only because of surface appearance, but because they may adversely affect the strength, formability, and other manufacturing characteristics of plates and sheets. A number of surface defects, such as scale, rust, scratches, gouges, pits, and cracks, have been identified for sheet metals. These defects may be caused by inclusions and impurities in the original material or various other conditions related to material preparation and the rolling operation.

Wavy edges on sheets (Fig. 13.9a) are the result of roll bending, whereby the strip is thinner along its edges than at its center (see Fig. 13.4a). Since the edges elongate more than the center, they buckle because they are restrained from expanding freely in the longitudinal (rolling) direction. The cracks shown in Figs. 13.9(b) and (c) are usually the result of poor material ductility at the rolling temperature. **Alligatoring** (Fig. 13.9d) is a complex phenomenon and may be caused

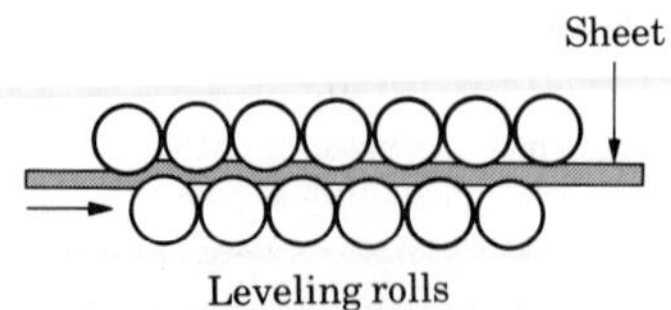

FIGURE 13.8
A method of roller leveling to flatten rolled sheets.

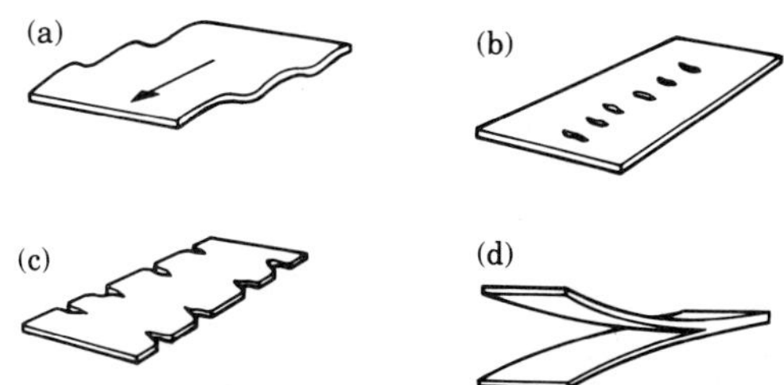

FIGURE 13.9
Schematic illustration of typical defects in flat rolling: (a) wavy edges; (b) zipper cracks in the center of the strip; (c) edge cracks; and (d) alligatoring.

by nonuniform deformation during rolling or by the presence of defects in the original cast billet.

13.3.2 Other characteristics

Residual stresses. Because of nonuniform deformation of the material in the roll gap, residual stresses can develop in rolled plates and sheets. Small-diameter rolls or small reductions per pass tend to deform the metal plastically at its surfaces (Fig. 13.10a), producing compressive residual stresses on the surfaces and tensile stresses in the middle. On the other hand, large-diameter rolls and high reductions tend to deform the bulk more than the surfaces (Fig. 13.10b). The reason is the frictional constraint at the surfaces along the arc of contact between the roll and the

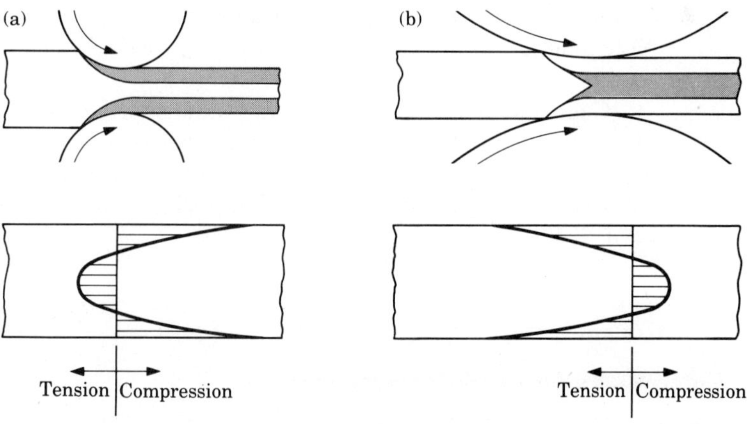

FIGURE 13.10
(a) Residual stresses developed in rolling with small rolls or at small reductions in thickness per pass. (b) Residual stresses developed in rolling with large rolls or at high reductions per pass. Note the reversal of the residual stress patterns.

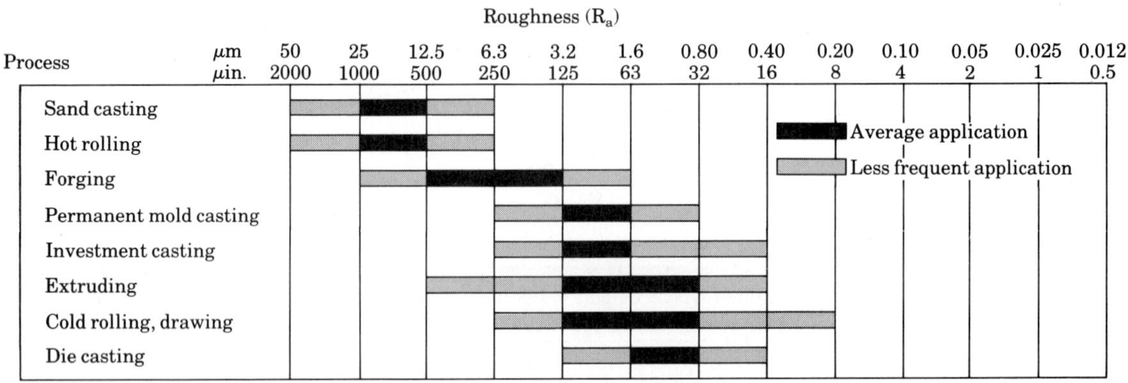

FIGURE 13.11 Surface roughness in various metalworking and casting processes. Note the large difference in roughness between hot and cold rolling.

strip. This situation produces residual stresses that are opposite of those in the case of small-diameter rolls.

Tolerances. Thickness tolerances for cold-rolled sheet usually range from ±0.1 mm to 0.35 mm (0.004 in. to 0.014 in.). Tolerances are much greater for hot-rolled plates. Flatness tolerances are usually within ±15 mm/m (3/16 in./ft) for cold rolling and ±55 mm/m (5/8 in./ft) for hot rolling.

Surface roughness. The ranges of surface roughness in cold and hot rolling are given in Fig. 13.11, which also includes ranges for some of the other manufacturing processes. Note that cold rolling can produce very fine surface finish, indicating that products made of cold-rolled sheet may not require additional finishing operations. Note that hot rolling and sand casting produce the same range of surface roughness.

Gage numbers. The thickness of a sheet is usually identified by a **gage number**: the smaller the number the thicker the sheet. Several numbering systems are used, depending on the type of sheet metal being classified. Rolled sheets of copper and brass are also identified by thickness changes during rolling, such as 1/4 hard, 1/2 hard, and so on.

13.4 Rolling Mills

The design, construction, and operation of **rolling mills** require a major investment. Highly automated mills produce close-tolerance, high-quality plates and sheet at

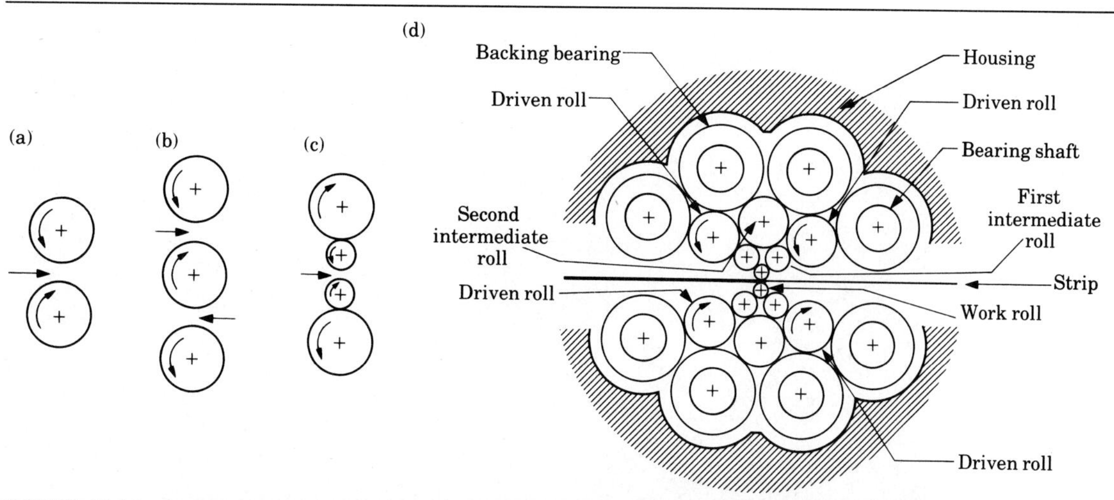

FIGURE 13.12
Schematic illustration of various roll arrangements: (a) two-high; (b) three-high; (c) four-high; and (d) cluster (Sendzimir) mill.

high production rates and low cost per unit weight, particularly when integrated with continuous casting. Many different types of rolling mills and equipment are built and use diverse roll arrangements. The major types of roll arrangement are shown in Fig. 13.12. The equipment for hot and cold rolling is essentially the same, although there are differences in the roll materials, lubricants, cooling systems, and process parameters.

Two-high or *three-high* rolling mills are used for hot rolling in initial breakdown passes (primary roughing) on cast ingots or in continuous casting, with roll diameters ranging from 0.6 m to 1.4 m (24 in. to 55 in.). In the three-high, or *reversing* mill, the direction of material movement is reversed after each pass. The plate being rolled is repeatedly raised to the upper roll gap, rolled, and then lowered to the lower roll gap by elevators and various manipulators.

Four-high mills (Fig. 13.12c) and *cluster* (*Sendzimir*) mills (Fig. 13.12d) are based on the principle that small-diameter rolls lower roll forces and power requirements and reduce spreading. Moreover, when worn or broken, small rolls can be replaced at less cost than can large ones. However, small rolls deflect more under roll forces and have to be supported by other rolls, as is done in four-high and cluster mills. The Sendzimir mill is particularly suitable for cold rolling thin strips of high-strength metals.

In *tandem rolling* the strip is rolled continuously through a number of *stands* to smaller gages with each pass. Each stand consists of a set of rolls with its own housing and controls. A group of stands is called a *train*. The control of the gage and the speed at which the sheet travels through each roll gap is critical. Electronic and computer controls, along with extensive hydraulic controls (particularly in precision rolling), are used in tandem rolling operations (Fig. 13.13).

FIGURE 13.13
View of the control pulpit and roll stands in a 2-m (80-in.) wide hot-strip mill. The hot strip (light areas between the stands) is being tandem rolled. *Source:* Inland Steel Company.

Example: A tandem rolling operation.

A typical tandem sheet-rolling operation is shown in the accompanying figure, indicating the thickness and the speed of the sheet after each reduction in the stands. The 2.25-mm (0.088-in.) sheet is supplied from a pay-off reel. The surface speed of the sheet after the first reduction (stand 5) is 4.1 m/s (820 ft/min). Four additional reductions are taken through the rest of the stands. The final thickness of the sheet is 0.26 mm (0.010 in.) and is taken up by the take-up reel at a speed of 30 m/s (6000 ft/min). The total reduction taken is $(2.25 - 0.26)/2.25 = 0.88$, or 88%.

Tensions may be applied to the sheet entering stand 5 by braking the pay-off reel by some suitable means (back tension), and by increasing the rotational speed of the take-up reel (front tension).

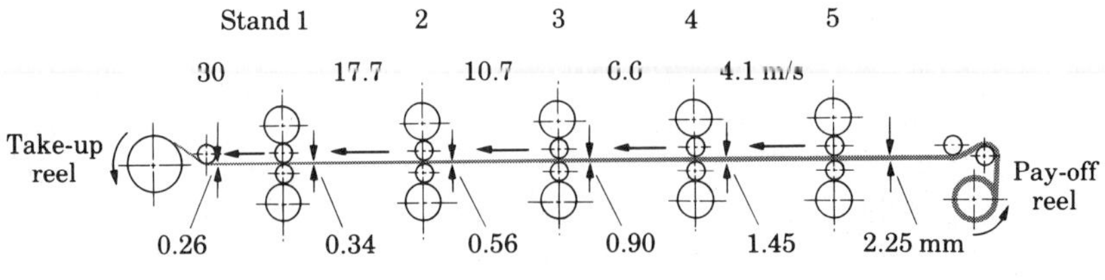

13.4.1 Rolls

The main requirements for roll materials are strength and resistance to wear. Three common roll materials are cast iron, cast steel, and forged steel. Forged-steel rolls, although more costly, have greater strength, stiffness, and toughness than cast rolls. Rolls for cold rolling are ground to a fine finish and, for special applications, are polished. You may have noticed that the bottom surface of aluminum beverage cans appears to have longitudinal scratches on it. This surface is a replica of the ground-roll surface, and in this way you can determine the rolling direction. Rolls made for cold rolling should not be used for hot rolling, because they may crack from thermal cycling (heat checking) and spalling (cracking or flaking of surface layers). Note that the elastic modulus of the rolls influences roll deflection and flattening.

13.4.2 Lubricants

Hot rolling of ferrous alloys is usually carried out without lubricants, although graphite may be used (Table 13.1; see p. 382). Water-base solutions are used to cool the rolls and to break up the scale on the rolled material. Nonferrous alloys are hot rolled with a variety of compounded oils, emulsions, and fatty acids. Cold rolling is carried out with low-viscosity lubricants, such as mineral oils, emulsions, paraffin, and fatty oils (see also Chapter 32).

13.5 Shape-Rolling Operations

In addition to flat rolling, various shapes can also be produced by rolling. Straight and long structural shapes, such as solid bars with various cross-sections, channels, I-beams, and railroad rails, are rolled by passing the stock through a set of specially designed rolls (Fig. 13.14; see also Fig. 13.1). Because the material's cross-section is to be reduced nonuniformly, designing a series of rolls (*roll-pass design*) requires considerable experience in order to avoid external and internal defects, hold tolerances, and reduce roll wear.

13.5.1 Ring rolling

In the *ring-rolling* process, a thick ring is expanded into a larger diameter ring with a reduced cross-section. The ring is placed between two rolls, one of which is driven (Fig. 13.15a), and its thickness is reduced by bringing the rolls closer together as they rotate. Since the volume of the workpiece remains constant during deformation, the reduction in thickness is compensated by an increase in the ring's diameter. Various shapes can be ring rolled using shaped rolls (Fig. 13.15b). A typical procedure for

TABLE 13.1
LUBRICANTS COMMONLY USED IN METALWORKING OPERATIONS (SEE ALSO CHAPTER 32)

MATERIAL AND ALLOY	TEMPERATURE	ROLLING	FORGING	EXTRUSION	ROD AND WIRE DRAWING	SHEET METALWORKING
Aluminum	Cold	FA + MO, MO	FA + MO, S	D, G, MS, L, S	MO, E, Wa, FA + MO	FO + MO, E, L, Wa, S
	Hot	FA + CO, FA + E	MO + G, MS	D, G, PI	—	G
Beryllium	Hot	G	MO + G, J	MS, G, J	G	—
Copper	Cold	E, MO	S, E, T	S, T, L, Wa, G, MS	FO + E + S, MO, Wa	FO + E, FO + MO, S + Wa
	Hot	D, E	G	D, MO + G	—	G
Lead	Cold	FA + MO, EM	FO + MO	D, S, T, E, FO	FO	FO + MO
	Hot	—	—	D	—	—
Magnesium	Cold	D, FA + MO	FA + MO, CC + S	D, T	—	FO + MO
	Hot	MO + FA + EM, D, G	W + G, MO + G	D, PI	—	G + MO, S + Wa, T + G
Nickel	Cold	MO + CL	—	CC + S	CC + S, MO + CL	E, MO + EP, CL, CC + S
	Hot	W, E	MO + G, W + G, GI	GI, J	—	G, MS
Refractory	Hot	G + MS	GI, G, MS	J + GI	G + MS, PI	MS, G
Steels (carbon and low alloy)	Cold	E, CO, MO	EP + MO, CC + S	CC + S, T, W + MS	S, CC + S	E, S, Wa, FO + MO, PI, CSN
	Hot	W, MO, CO, G + E	MO + G, Sa, GI	GI, G	—	G
Steels (stainless)	Cold	CL + EM, CL + MO	CL + MO, CC + S	CC + S, CP, MO	CC + S, CL + MO	FO + MO, Wa, PI, EP + MO, S
	Hot	D, W, E	MO + G, GI	GI	—	G
Titanium	Cold	MO, CC + FO, G	S, MO	CC + G, CC + S	CC + PI	CL, S, PI, Wa
	Hot	PI, CC, G, MS	W + G, GI, MS	J + GI, GI	—	G, MS

CC Conversion coatings
CL Chlorinated paraffin
CO Compounded oil
CP Copper plate
CSN Chemicals and synthetics
D Dry
E Emulsion
EP Extreme pressure
FA Fatty acid
FO Fatty oil
G Graphite
GI Glass
J Canned or jacketed
L Lanolin
MO Mineral oil
MS Molybdenum disulfide
PI Polymer
S Soap
Sa Salts
T Tallow
W Water
Wa Wax

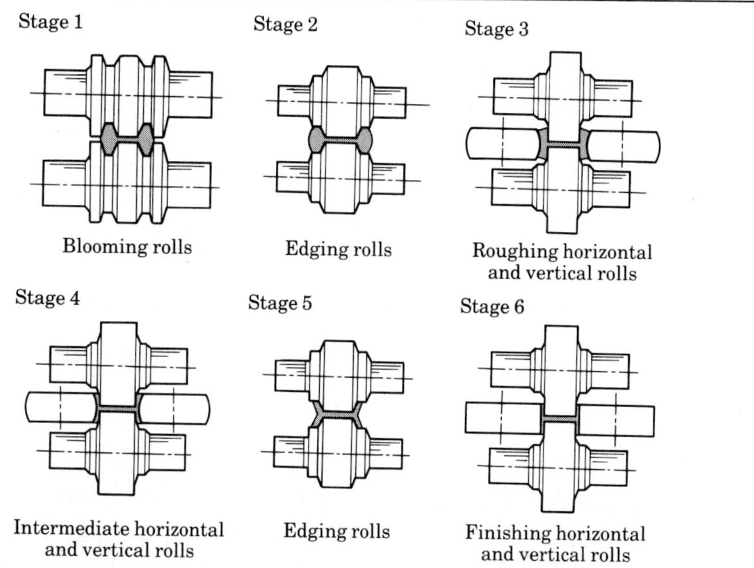

FIGURE 13.14 Stages in shape rolling of an H-section part. Various other structural sections, such as channels and I-beams, are also rolled by this process.

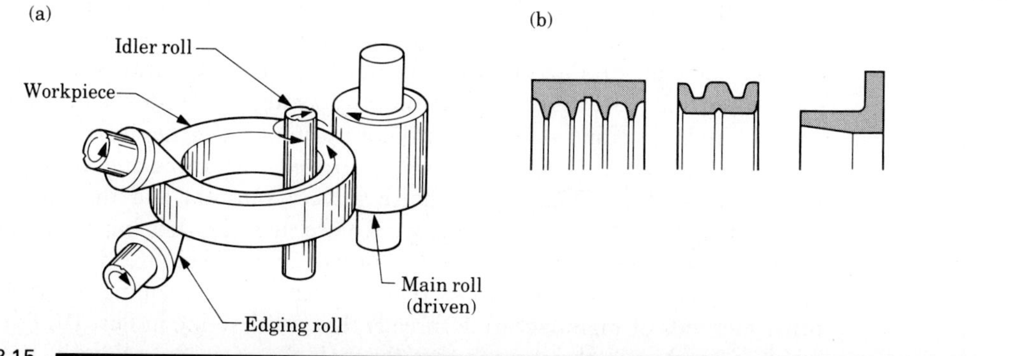

FIGURE 13.15 (a) Schematic illustration of a ring-rolling operation. Thickness reduction results in an increase in the part diameter. (b) Examples of cross-sections that can be formed by ring rolling.

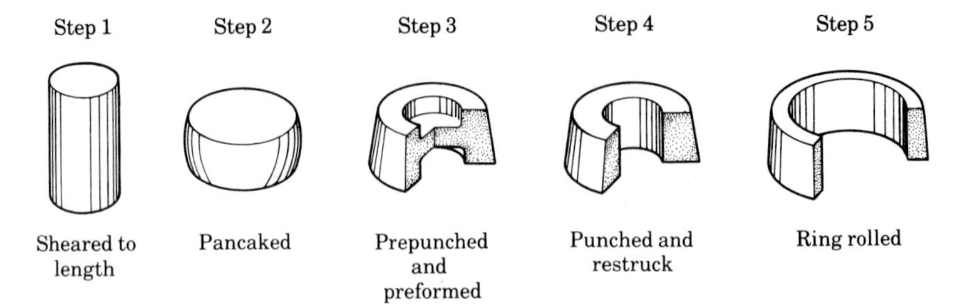

FIGURE 13.16 Stages in the production of a tapered ring, using a combination of forming processes. The sequence shown in steps 2–4 is a common method of producing holes in relatively small parts.

producing a seamless ring for a tapered roller-bearing race is shown in Fig. 13.16. Typical applications of ring rolling are large rings for rockets and turbines, gearwheel rims, ball-bearing and roller-bearing races, flanges, and reinforcing rings for pipes.

The ring-rolling process can be carried out at room or elevated temperature, depending on the size, strength, and ductility of the workpiece material. The advantages of this process, compared to other manufacturing processes capable of making the same part, are short production times, material savings, close tolerances, and favorable grain flow in the product.

13.5.2 Thread rolling

The *thread-rolling* process is a cold-forming process by which straight or tapered threads are formed on round rods, by passing them between dies. Threads are formed on the rod or wire with each stroke of a pair of flat reciprocating dies (Fig. 13.17a). Typical products are screws, bolts, and similar threaded parts. The process is capable of generating similar shapes on other surfaces, such as grooves and various gear forms, and can be used in the production of almost all threaded fasteners at high production rates. In another method, threads are formed with rotary dies (Fig. 13.17b) at production rates as high as 80 pieces per second.

The thread-rolling process has the advantages of generating threads without any loss of material (scrap) and with improved tolerances and strength resulting from cold working. The surface finish is very smooth, and the process induces compressive residual stresses on the workpiece surfaces, thus improving fatigue life. Because of volume constancy during deformation, a rolled thread has a larger major diameter than a machined thread (Fig. 13.18a). Thread rolling is superior to the other methods of manufacturing threads that we describe in Part IV. For example, whereas machining the threads cuts through the grain-flow lines of the material, rolled threads have a grain-flow pattern that improves the strength of the thread (Fig. 13.18b). Internal thread rolling can be carried out with a fluteless *forming tap*.

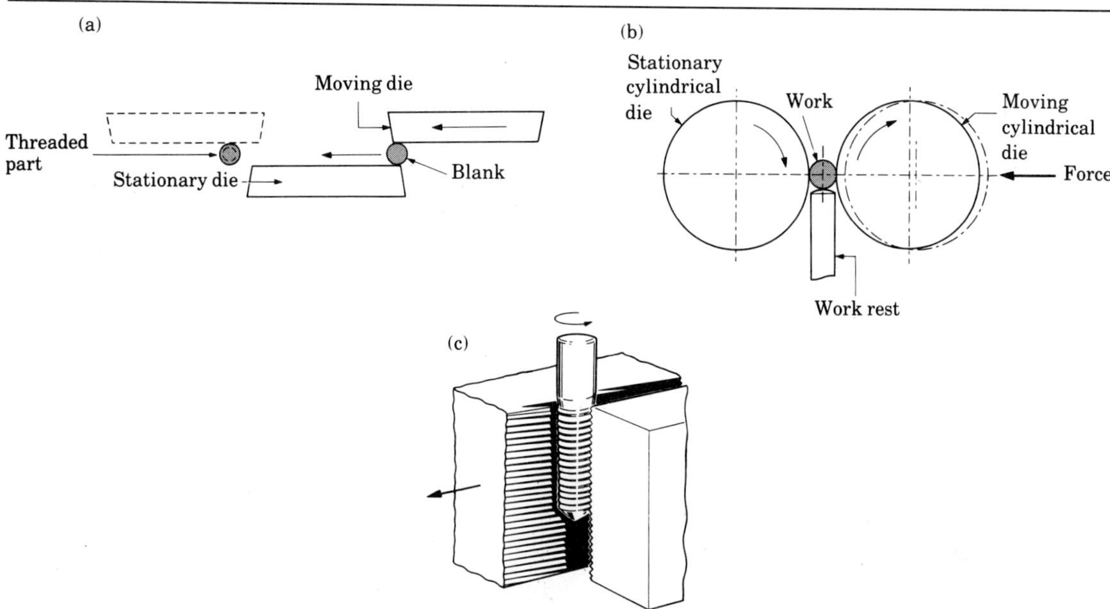

FIGURE 13.17 Thread-rolling processes: (a) reciprocating flat dies; and (b) two-roller dies. Threaded fasteners, such as bolts, are made economically by these processes at high rates of production.

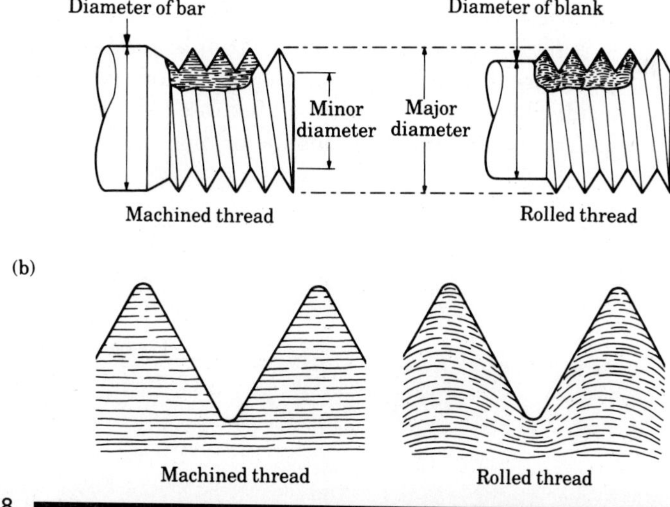

FIGURE 13.18 (a) Differences in the diameters of machined and rolled threads. (b) Grain flow in machined and rolled threads. Unlike machining, which cuts through the grains of the metal, rolled threads have improved strength because of cold working and favorable grain flow.

This operation is similar to external thread rolling and produces accurate internal threads having good strength.

Spur and helical gears can be produced by cold-rolling processes similar to thread rolling. The process may be carried out on solid cylindrical blanks or on precut gears. Cold rolling of gears has extensive applications in automatic transmissions and power tools.

Lubrication is important for good surface finish and to minimize defects. Usually made of hardened steel, dies are expensive to make because of their complex shape and usually cannot be reground after they become worn. However, with proper die materials and preparation, die life may range up to millions of pieces.

13.6 Production of Seamless Tubing and Pipe

Rotary tube piercing is a hot-working process for making long, thick-walled *seamless* tubing and pipe. It is based on the principle that when a round bar is subjected to radial compressive forces (Fig. 13.19a), tensile stresses develop at the center of the bar; then when subjected to cyclic compressive stresses (Fig. 13.19b), a cavity begins to form at the center of the bar. (You may experiment with a short piece of round eraser by rolling it on a flat surface as shown in Fig. 13.19b.)

Rotary tube piercing (the *Mannesmann* process) is carried out using an arrangement of rotating rolls (Fig. 13.19c). The axes of the rolls are skewed in order to pull the round bar through the rolls by the axial component of the rotary motion. An internal mandrel assists the operation by expanding the hole and sizing the inside diameter of the tube. Because of the severe deformation that the bar undergoes, the material must be of high quality and free from defects.

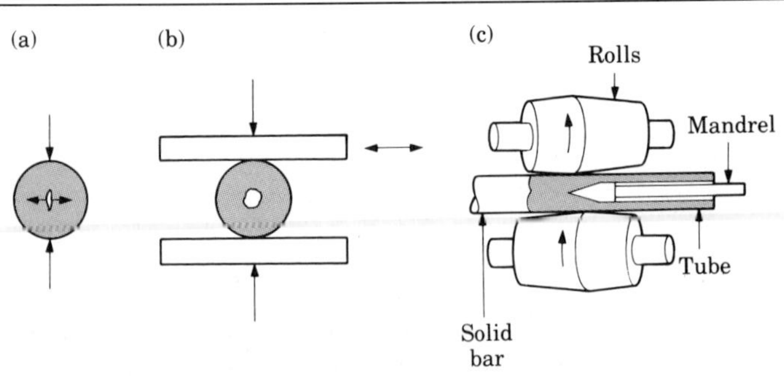

FIGURE 13.19 Cavity formation in a solid round bar and its utilization in the rotary tube piercing process for making seamless pipe and tubing (Mannesmann mill developed in the 1880s).

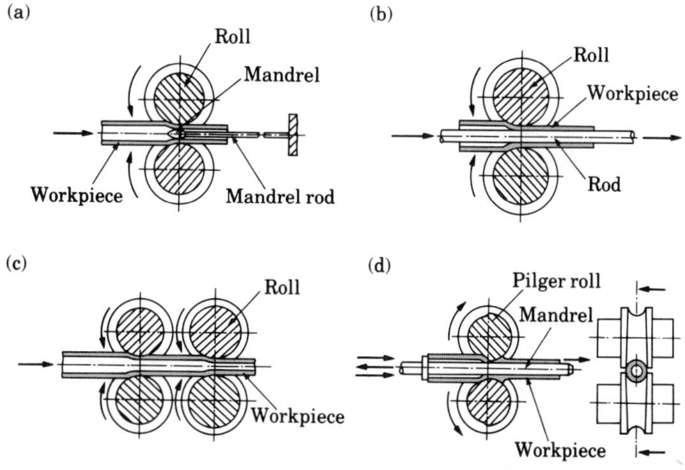

FIGURE 13.20
Schematic illustration of various tube rolling processes: (a) fixed mandrel; (b) moving mandrel; (c) rolling without mandrel; and (d) Pilger rolling over a mandrel and a pair of shaped rolls. Tube diameters and thicknesses can also be changed by other processes, such as drawing, extrusion, and spinning.

The diameter of tubes and pipes can be reduced by *tube rolling* (Fig. 13.20). These processes allow the production of tubes and pipe with different diameters and thicknesses from the same initial tube stock. In Chapter 15, we discuss other processes for tube manufacturing. Some of those operations can be carried out either with or without an internal mandrel. In the *pilger mill* (Fig. 13.20d), the tube and mandrel undergo a reciprocating motion. The rolls are specially shaped and are rotated continuously. During the gap cycle on the roll, the tube is advanced and rotated, starting another cycle of tube reduction.

13.7 Continuous Casting and Rolling

In Section 5.4, we discussed the advantages of continuous casting. This operation, also called *strand casting*, is highly automated, which reduces the product's cost significantly. Continuous casting is an important development and most companies are rapidly converting their facilities to this type of operation.

Competition in the steel industry has led to other developments having major potential economic significance, particularly for minimills. Some of these techniques are shown in Fig. 13.21. These processes involve *direct casting* of sheet and plate steel, in thicknesses ranging from less than 1 mm (0.04 in.) to 50 mm (2 in.). Because the number of operations involved can make conventional production of tubes and pipes costly, a method of spray casting is also being developed (Fig. 13.22).

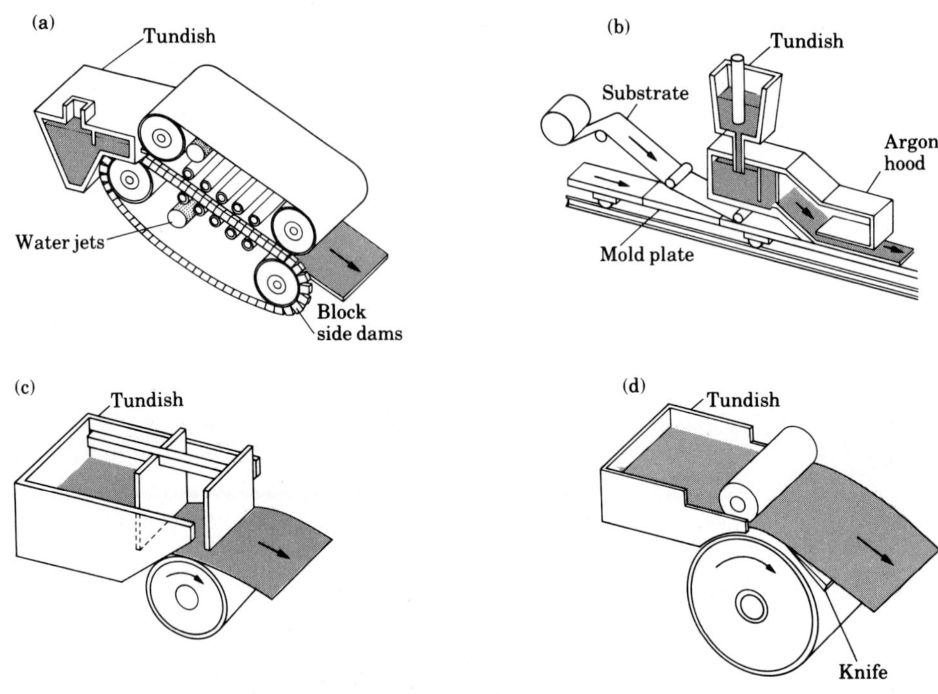

FIGURE 13.21
Examples of direct casting techniques under development for economical production of plate and sheet. *Source:* J. Szekely, *Scientific American,* July 1987.

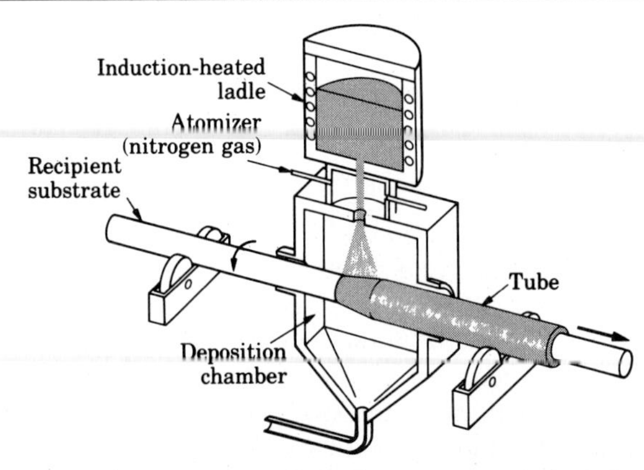

FIGURE 13.22
Spray casting (Osprey process) in which molten metal is sprayed over a rotating mandrel to produce seamless tubing and pipe. *Source:* J. Szekely, *Scientific American,* July 1987.

13.7.1 Minimills

The **minimill** is a relatively new and important concept. In a minimill, scrap metal is melted in electric-arc furnaces, cast continuously, and rolled directly into a specific line of products. Each minimill produces essentially one kind of rolled product (rod, bar, or structural sections, such as angle iron) from basically one type of metal or alloy. The scrap metal, which is obtained locally to reduce transportation costs, is usually old machinery, cars, and farm equipment.

Minimills presently have the economic advantages of low-investment optimal operations for each type of metal and product line, and low labor and energy costs. The products are usually oriented to markets in the mill's particular geographic area. One minimill in the northeast United States, for instance, supplies 95 percent of the domestic steel rod (for reinforced concrete) to that region.

SUMMARY

Rolling is one of the most important processes in the primary working of metals. Rolled plates, sheets, pipe and tubing, and foil are used in a wide variety of products, ranging from beverage cans and packaging to car bodies, boilers, and ship hulls. In addition to flat rolling, shape rolling is used to make products with various cross-sections, such as bars and rods and structural shapes for buildings and transportation equipment. An important development is continuous casting of billets, which are rolled directly into semifinished products.

As in all metalworking processes, rolling involves a number of process and material variables that should be controlled in order to roll products having proper quality, properties, surface finish, and dimensional accuracy. These variables include rolling temperature and speed, lubrication, and the condition of the rolls and the characteristics of the equipment.

TRENDS

- Continuous casting and rolling of both ferrous and nonferrous metals are being implemented at a rapid rate.
- Further developments are taking place in minimills to roll specialized products efficiently and economically.
- Computer controls are being implemented in all aspects of rolling and subsequent processing of plates, sheets, and shapes.

KEY TERMS

Alligatoring	Foil	Sheet
Billet	Gage number	Slab
Bloom	Minimill	Structural defects
Bulk-deformation processes	Plate	Temper rolling
Camber	Rolling	Wrought structure
Crown	Rolling mill	
Draft	Roll	

BIBLIOGRAPHY

Blazynski, T.Z., *Plasticity and Modern Metal-forming Technology*. New York: Elsevier, 1989.

Lange, K. (ed.), *Handbook of Metal Forming*. New York: McGraw-Hill, 1985.

Larke, E.C., *The Rolling of Strip, Sheet, and Plate*, 2d ed. London: Chapman and Hall, 1963.

Roberts, W.L., *Cold Rolling of Steel*. New York: Marcel Dekker, 1978.

———, *Hot Rolling of Steel*. New York: Marcel Dekker, 1983.

Starling, C.W., *The Theory and Practice of Flat Rolling*. London: The University of London Press, 1962.

Underwood, L.R., *The Rolling of Metals*, Vol. 1. New York: Wiley, 1950.

Wusatowski, Z., *Fundamentals of Rolling*. New York: Pergamon, 1969.

REVIEW QUESTIONS

13.1 What is the difference between a rolled plate and sheet?

13.2 Define (a) roll gap, (b) neutral point, and (c) draft.

13.3 What factors contribute to spreading in flat rolling?

13.4 Explain the types of deflection that rolls undergo.

13.5 Describe the difference between bloom, slab, and billet.

13.6 Why is roller leveling necessary?

13.7 List the defects commonly observed in flat rolling.

13.8 Explain the features of different types of rolling mills.

13.9 What is the advantage of tandem rolling?

13.10 Make a list of parts made by shape rolling.

13.11 How are seamless tubes produced?

13.12 Explain the features and advantages of continuous casting.

QUESTIONS AND PROBLEMS

13.13 Rolling reduces the thickness of plates and sheets. However, it is possible to reduce the thickness by simply stretching the material. Would this be a feasible process? Explain.

13.14 What are the advantages and limitations of using small-diameter rolls?

13.15 Explain how a cast structure is converted into a wrought structure by hot rolling.

13.16 It is said that necessity is the mother of invention. Explain why the rolling process was invented and developed.

13.17 In Section 13.2, we described three factors that influence spreading in flat rolling: workpiece width-to-thickness ratio, friction, and ratio of roll radius to strip thickness. Explain how these factors affect spreading.

13.18 Explain how the residual stress patterns shown in Fig. 13.10 become reversed when the roll radius or reduction per pass is changed.

13.19 Explain whether it would be practical to apply the roller-leveling technique shown in Fig. 13.8 to plates.

13.20 Describe the comparative advantages and limitations of hot and cold rolling.

13.21 How does applying tensions affect flat-rolling practice?

13.22 Using simple geometric relationships and the inclined-plane principle for friction, prove Eq. (13.1).

13.23 Sketch cross-sections that can be made by shape rolling.

13.24 Describe the factors that influence the roll force F in Fig. 13.3(c).

13.25 Describe the advantages and limitations of the new techniques shown in Figs. 13.21 and 13.22.

13.26 Show that the maximum angle α, known as the *angle of acceptance* (see Fig. 13.3b), at which a plate can be pulled into the roll gap is equal to $\tan^{-1} \mu$, where μ is the coefficient of friction.

13.27 Explain how you would go about applying front and back tensions to sheet metals during rolling.

13.28 In Section 13.2.2, we noted that rolls tend to flatten under roll forces. Describe the methods by which flattening can be reduced. Which property of the roll material can be increased to reduce flattening?

13.29 Spreading in flat rolling increases with (a) decreasing width-to-thickness ratio of the entering material, (b) decreasing friction, and (c) decreasing ratio of roll radius-to-strip thickness. Explain why this is so.

13.30 Explain the technical and economic reasons for taking larger rather than smaller reductions per pass in flat rolling.

13.31 In Fig. 13.11, we note that surface roughness in hot rolling is much higher than in cold rolling. Explain why.

13.32 Calculate the roll force and torque for AISI 1020 carbon steel strip, 400 mm wide and 10 mm thick, rolled to a thickness of 7 mm. The roll radius is 200 mm, and it rotates at 200 rpm.

13.33 Flat rolling can be carried out by front tension only, using idling rolls (*Steckel rolling*). Thus the torque on the rolls is zero. Where then is the energy coming from to supply the work of deformation in rolling?

13.34 What is the consequence of applying too high a back tension in rolling?

13.35 In Fig. 13.12(d), we note that the driven rolls (powered rolls) are the third set from the work roll. Why isn't power supplied through the work roll? Is it possible?

13.36 Describe the importance of control of roll speeds, roll gaps, temperature, etc., in a tandem rolling operation shown in the example in Section 13.4.

14

Forging

14.1 Introduction

Forging is the name for processes whereby the workpiece is shaped by compressive forces applied through various dies and tools. It is one of the oldest metalworking operations, dating back at least to 4000 B.C. and, perhaps, as far back as 8000 B.C. Forging was first used to make jewelry, coins, and various implements by hammering metal with tools made of stone. Simple forging can be done with a heavy hand hammer and an anvil, as traditionally done by blacksmiths. Most forgings, however, require a set of dies and equipment such as a press or a forging hammer.

Unlike rolling operations, which generally produce continuous plates, sheets, strip, or various structural cross-sections, forging operations produce discrete parts. Metal flow and grain structure can be controlled, so forged parts have good strength

(a)

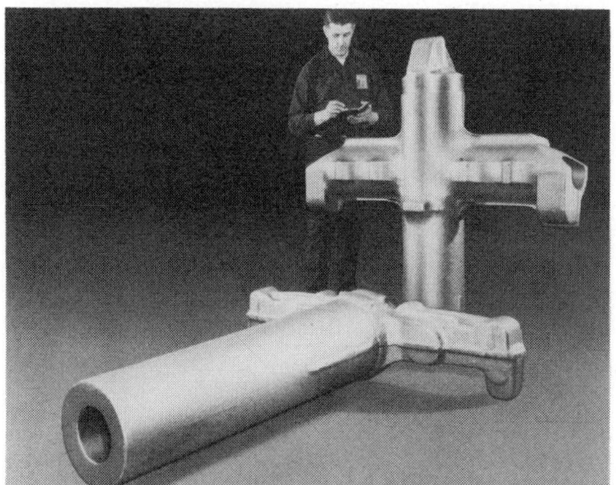

(b)

FIGURE 14.1
(a) Landing-gear components for the C5A and C5B aircraft, made by forging. *Source:* Wyman-Gordon Company. (b) A hot-forged rotor for a steam turbine weighing 300,000 kg (700,000 lb). *Source:* Courtesy of General Electric Company.

and toughness. Thus they can be used reliably for highly stressed and critical applications, such as landing gear for aircraft (Fig. 14.1a) and jet-engine shafts and disks. Typical forged products are bolts and rivets, connecting rods, shafts for turbines (Fig. 14.1b), gears, hand tools, and structural components for machinery, railroads, and a variety of other transportation equipment.

Forging may be done at room temperature (*cold forging*) or at elevated temperatures (*warm* or *hot forging*, depending on the temperature). Because of the higher strength of the material, cold forging requires greater forces, and the workpiece materials must have sufficient ductility at room temperature. Cold-forged parts have good surface finish and dimensional accuracy. Hot forging requires smaller forces, but dimensional accuracy and surface finish are not as good. Forgings generally require additional finishing operations, such as heat treating to modify properties and machining for accurate finished dimensions.

A component that may be forged successfully may also be manufactured economically by other methods, such as casting, machining, or powder metallurgy. However, as you might expect, each process will produce a part having different characteristics and limitations, particularly with regard to strength, toughness, dimensional accuracy, surface finish, and internal or external defects.

We begin this chapter by discussing the principles of forging, following the outline in Fig. 14.2. We continue by describing various forging techniques and the characteristics and capabilities of the equipment used. We also compare the economics of forging with that of the other processes covered thus far.

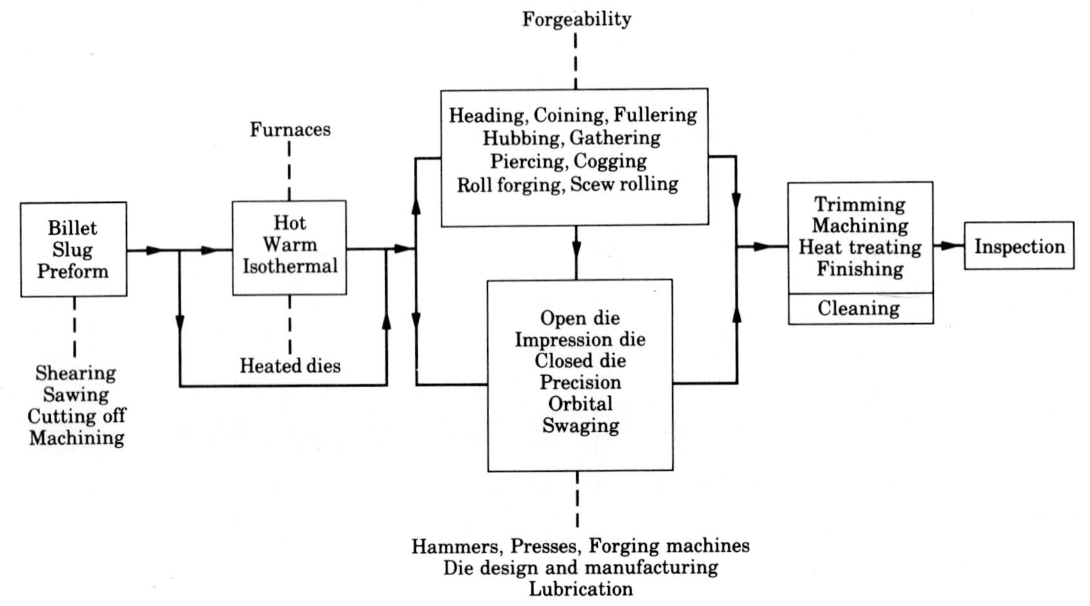

FIGURE 14.2
Outline of forging and related operations.

14.2 Open-Die Forging

Open-die forging is the simplest forging process. A solid workpiece is placed between two flat dies and reduced in height by compressing it (Fig. 14.3a). This process is also called *upsetting* or *flat-die* forging. The die surfaces in open-die forging may have simple cavities to produce relatively simple forgings. The deformation of the workpiece under *ideal* conditions is shown in Fig. 14.3(b). Since a constant volume has to be maintained, any reduction in height increases the diameter of the part.

Note that in Fig. 14.3(b) the workpiece is deformed *uniformly*. In actual operations, the part develops a barrel shape (Fig. 14.3c), also known as pancaking. **Barreling** is caused primarily by frictional forces at the die–workpiece interfaces that oppose the outward flow of the materials at these interfaces. Thus barreling can be minimized if an effective lubricant is used. Barreling can also occur in upsetting hot workpieces between cold dies. The material at and near the interfaces cools rapidly, while the rest of the workpiece is relatively hot. Thus the material at the ends of the workpiece has greater resistance to deformation than at its center.

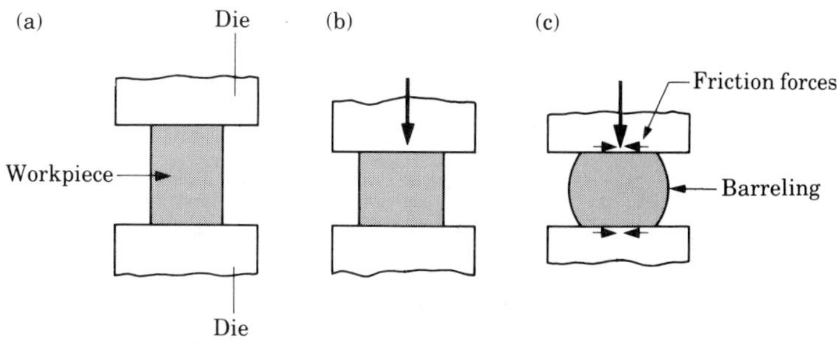

FIGURE 14.3 (a) Solid cylindrical billet upset between two flat dies. (b) Uniform deformation of the billet without friction. (c) Deformation with friction. Note barreling of the billet caused by friction forces at the billet–die interfaces.

Consequently, the central portion of the workpiece deforms to a greater extent than its ends. Barreling from thermal effects can be reduced or eliminated if heated dies or a thermal barrier, such as glass cloth at the die–workpiece interfaces, are used.

Simple forgings can be made by the open-die process. The large rotor for a steam turbine shown in Fig. 14.1(b), for example, is made from a long cast ingot that is hot forged as shown in Fig. 14.4. The ingot, which may be square in cross-section, rests lengthwise on a flat die and is reduced in diameter a little at a time. The workpiece is rotated intermittently, with the use of large mechanical manipulators, after each step of deformation. This process is known as *breaking down* the cast ingot. It changes the microstructure of the workpiece from a cast to a wrought structure (see Fig. 13.7). Ring-shaped parts may also be reduced in thickness in this manner with the use of an internal mandrel. Although most open-die forgings generally weigh 15–500 kg (30–1000 lb), forgings as large as 300 tons have been made. Sizes may range from very small to 23 m (75 ft) long shafts in the case for ship propellers.

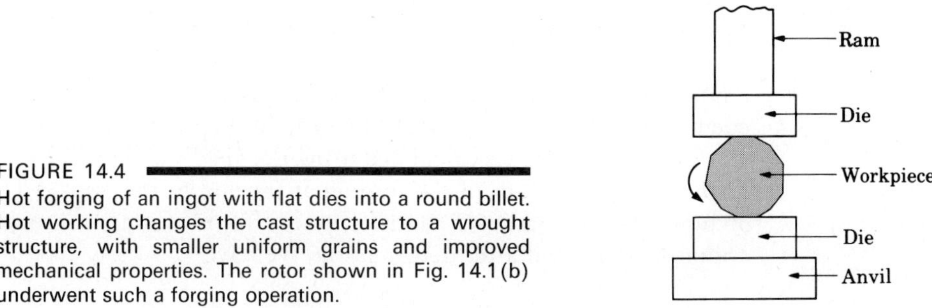

FIGURE 14.4 Hot forging of an ingot with flat dies into a round billet. Hot working changes the cast structure to a wrought structure, with smaller uniform grains and improved mechanical properties. The rotor shown in Fig. 14.1(b) underwent such a forging operation.

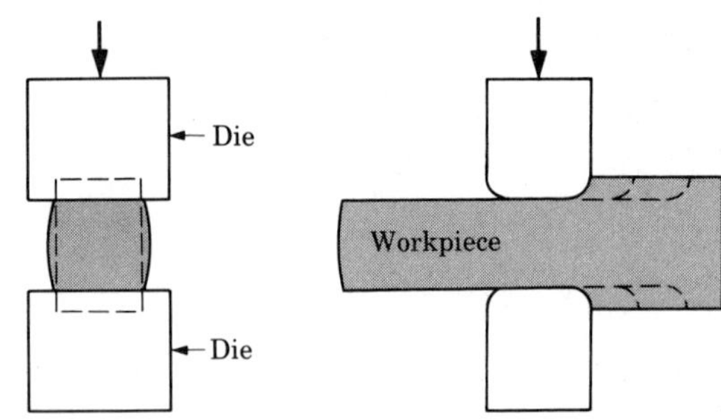

FIGURE 14.5
Two views of a cogging operation on a rectangular bar. Blacksmiths use this process to reduce the thickness of bars by hammering the part on an anvil. Note the barreling of the workpiece.

Cogging (also called *drawing out*) is basically an open-die forging operation in which the thickness of a bar is reduced by successive forging steps at certain intervals (Fig. 14.5). Because the contact area per stroke is small, a long section of a bar can be reduced in thickness without requiring large forces or machinery. You may have seen blacksmiths performing such an operation with a hammer and anvil on hot pieces of metal. Iron fences of various design are often made by this process.

14.2.1 Forging force

We can estimate the force F required for an open-die forging operation on a solid cylindrical piece from the following formula:

$$F = Y_f \pi r^2 \left(1 + \frac{2\mu r}{3h}\right), \tag{14.1}$$

where Y_f is the *flow stress* of the material (see the example below), μ is the coefficient of friction, and r and h are the radius and height of the workpiece, respectively. Derivation of this formula and others for various forging processes are given in the Bibliography at the end of the chapter.

• Example: Calculation of forging force.

A solid cylindrical workpiece made of 304 stainless steel is 150 mm (6 in.) in diameter and 100 mm (4 in.) high. It is forged by open-die forging at room temperature with flat dies to a 50-percent reduction in height. Assuming that the coefficient of friction is 0.2, calculate the forging force at the end of the stroke.

The forging force is calculated using Eq. (14.1), in which the dimensions pertain to the final dimensions of the forging at the end of the stroke. Thus, $h = 100/2 = 50$ mm, and the radius r is determined from volume constancy. Equating the volumes before and after deformation, we have

$$(\pi)(75)^2(100) = (\pi)(r)^2(50).$$

Therefore, $r = 106$ mm (4.17 in.). The quantity Y_f in Eq. (14.1) is the flow stress of the material, which is the stress required to continue plastic deformation of the workpiece at a particular true strain. The *absolute* value of the true strain that the workpiece has undergone in this operation is

$$\varepsilon = \ln (100/50) = 0.69.$$

Referring to Fig. 2.6, we see that the flow stress for 304 stainless steel at a true strain of 0.69 is about 1000 MPa (140 ksi). We now calculate the forging force, noting that for this problem the units in Eq. (14.1) must be in N and m. Thus,

$$F = (1000)(10^6)(\pi)(0.106)^2\left[1 + \frac{(2)(0.2)(0.106)}{(3)(0.050)}\right]$$

$$= 4.5 \times 10^7 \text{ N} = 45 \text{ MN} = 10^7 \text{ lb} = 5000 \text{ tons.}$$

See Section 14.9 for forging machine capacities.

14.3 Impression-Die and Closed-Die Forging

In **impression-die forging**, the workpiece acquires the shape of the die cavities (impressions), while being forged between two shaped dies (Fig. 14.6). During forging, some of the material flows outward and forms a **flash**. The flash has a significant role in the flow of material in impression-die forging. The thin flash cools rapidly and, with its frictional résistance, subjects the material in the die cavity to high pressures, thus encouraging the filling of the die cavity.

In a typical operation the blank is first *cropped*, or *sheared*, from a bar stock (Fig. 14.7) and placed on the lower die. The blank's shape gradually changes, as shown for the forging of a connecting rod in Fig. 14.8(a). Performing processes, such as fullering and edging (Figs. 14.8b and c), distribute the material in various regions of the blank, much as in shaping dough to make pastry. In *fullering*, material is distributed away from an area; in *edging*, it is gathered into a localized area. The

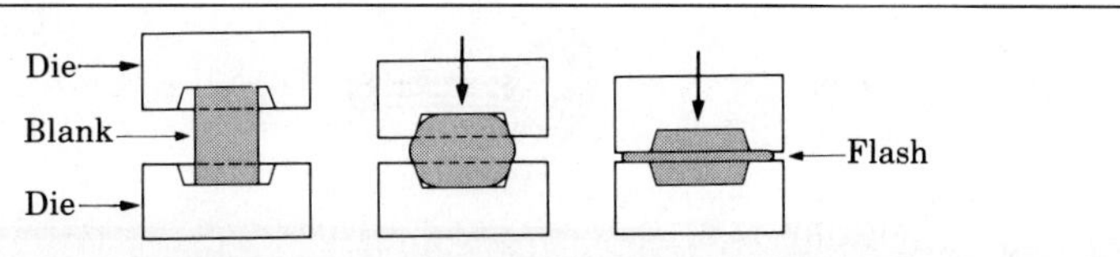

FIGURE 14.6 Stages in impression-die forging of a solid round billet. Note the formation of flash, which is excess metal that is subsequently trimmed off (see Fig. 14.9).

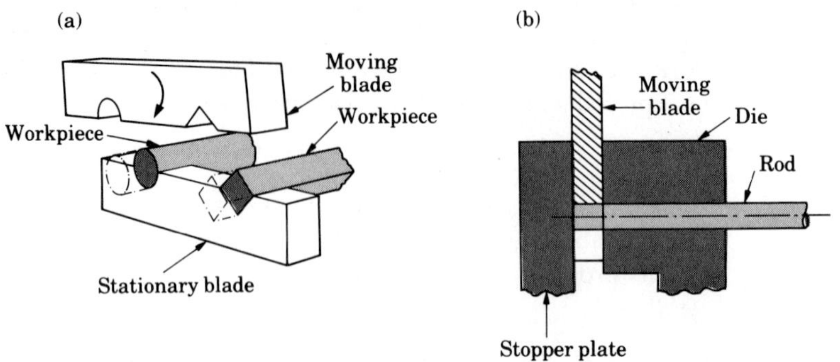

FIGURE 14.7

(a) Cropping (shearing) of a long bar to prepare blanks for forging. (b) Cropping with a stopper plate. The part cropped in this setup produces more square edges because it is constrained laterally by the stopper plate. (See also Figs. 16.2–16.7.)

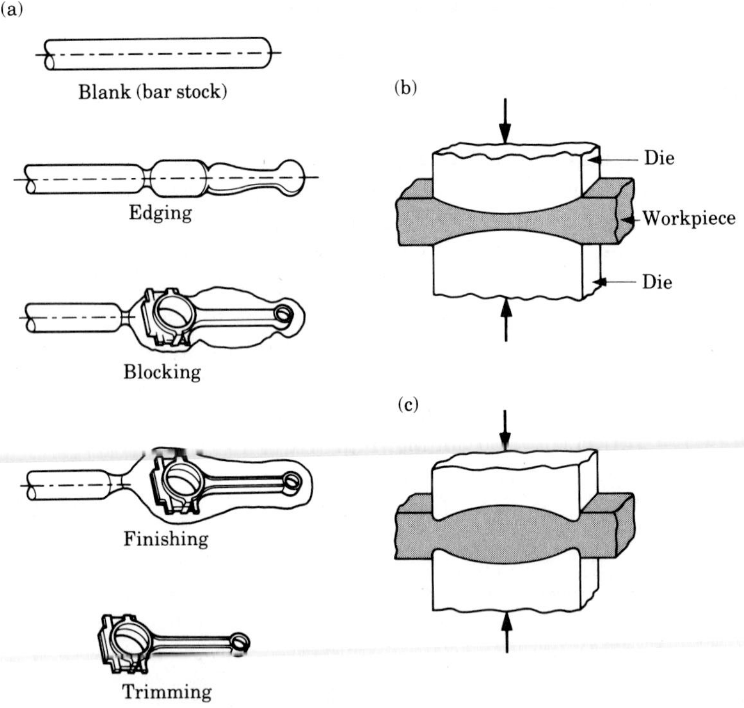

FIGURE 14.8

(a) Stages in forging a connecting rod for an internal combustion engine. Note the amount of flash required to ensure proper filling of the die cavities. (b) Fullering and (c) edging processes to distribute the material for preshaping the blank for forging.

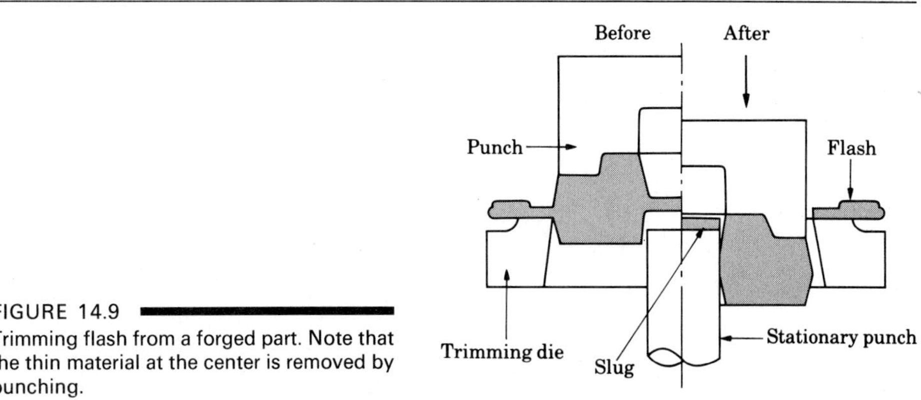

FIGURE 14.9 Trimming flash from a forged part. Note that the thin material at the center is removed by punching.

part is then formed into the rough shape of a connecting rod by a process called *blocking*, using blocker dies. The final operation is finishing the forging in impression dies that give the forging its final shape. The flash is removed by a trimming operation (Fig. 14.9).

The examples shown in Figs. 14.6 and 14.8(a) are also referred to as *closed-die* forgings. However, in true closed-die, or *flashless*, forging, flash does not form, and the workpiece completely fills the die cavity. Accurate control of the volume of material and proper die design are essential in order to obtain a closed-die forging of desired dimensions and tolerances. Undersize blanks prevent the complete filling of the die cavity. Conversely, oversize blanks generate excessive pressures and may cause dies to fail prematurely or jam.

14.3.1 Precision forging

For economic reasons the trend in forging operations today is toward greater precision, thus requiring fewer additional finishing operations. Operations in which the part formed is close to the final dimensions of the desired component are known as *near-net-shape* or *net-shape forging*. Any excess material on the forged part is usually small and is removed, generally by trimming or grinding.

In **precision forging**, special dies allow parts to be machined to greater accuracies than in impression-die forging. The process requires higher capacity equipment because of the greater forces required to obtain fine details on the part. Aluminum and magnesium alloys are particularly suitable for precision forging because of the low forging loads and temperatures required. Also, little die wear takes place, and the surface finish is good. Steels and other alloys are more difficult to precision forge.

Precision forging requires special dies, precise control of material volume and shape, and proper positioning in the die cavity. However, less material is wasted and much less machining is required, since the part is closer to the final desired shape. Thus the choice between conventional forging and precision forging requires an economic analysis.

14.3.2 Coining

Coining essentially is a closed-die forging process typically used in minting coins, medallions, and jewelry (Figs. 14.10a and b). The slug is coined in a completely closed die cavity. The pressures required can be as high as 5–6 times the strength of the material in order to produce fine details. Note, for example, the detail on newly minted coins. Several coining operations may be required in order to produce fine details on some parts. Lubricants cannot be tolerated in coining, since they can be entrapped in the die cavities and, being incompressible, prevent the full reproduction of die-surface details.

The coining process is also used with forgings and other products to improve surface finish and impart the desired dimensional accuracy. This process, called *sizing*, involves high pressures with little change in part shape taking place during sizing. *Marking* of parts with letters and numbers can be done rapidly by a process similar to coining (Fig. 14.11).

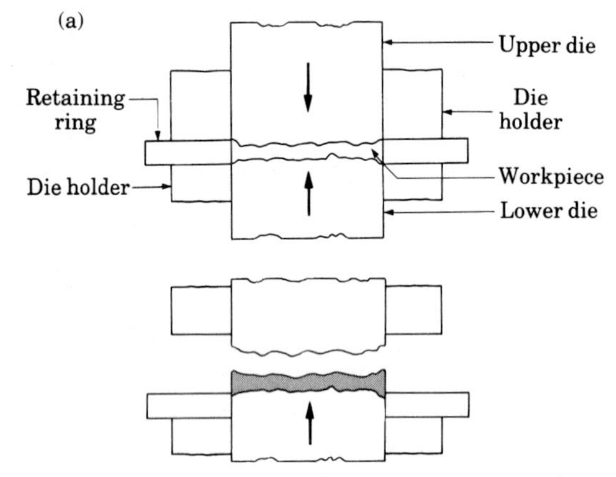

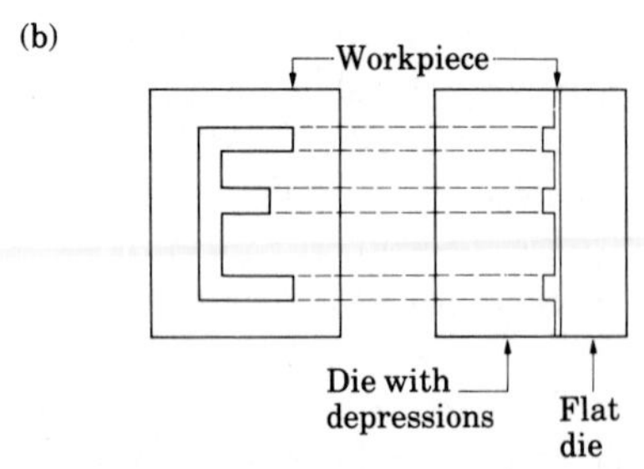

FIGURE 14.10
(a) Schematic illustration of the coining process. The earliest coins were made by open-die forging and lacked sharp details. (b) An example of a coining operation.

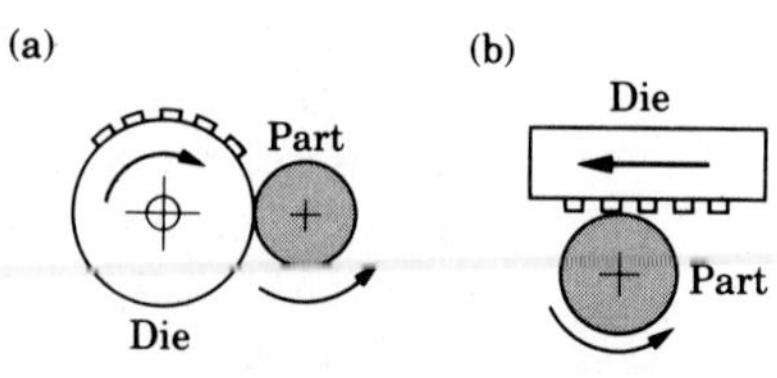

FIGURE 14.11
(a) Rotary and (b) roll-on marking of round parts by coining. *Source:* Pannier Corp.

TABLE 14.1
RANGE OF k VALUES FOR EQ. (14.2)

Simple shapes, without flash	3–5
Simple shapes, with flash	5–8
Complex shapes, with flash	8–12

14.3.3 Forging force

We can estimate the forging force F required to carry out an impression-die forging operation from the formula

$$F = kY_f A, \tag{14.2}$$

where k is a multiplying factor to be obtained from Table 14.1, Y_f is the flow stress of the material at the forging temperature, and A is the projected area of the forging, including the flash. In hot-forging operations, the actual forging pressure for most metals ranges from 550 MPa to 1000 MPa (80 ksi to 140 ksi).

14.4 Related Forging Operations

A number of other forging operations are carried out in order to impart the desired shape and features to forged products. We describe these operations in this section.

14.4.1 Heading

Heading is essentially an upsetting operation, usually performed at the end of a round rod or wire in order to produce a larger cross-section. Typical examples are the heads of bolts, screws, rivets, nails, and other fasteners (Fig. 14.12). Heading processes can be carried out cold, warm, or hot. They are performed on machines called *headers*, which are usually highly automated, with production rates of hundreds of pieces per minute for small parts. These machines tend to be noisy; thus soundproof enclosures or the use of ear protectors may be required. Heading operations can be combined with cold-extrusion processes (see Section 15.5) to make various parts.

An important aspect of heading is the tendency for the bar to buckle if its unsupported length-to-diameter ratio is too high. This ratio is usually limited to less than 3 : 1 but can be higher, depending on die geometry. For example, greater ratios can be accommodated if the diameter of the die cavity is not more than 1.5 times the diameter of the bar.

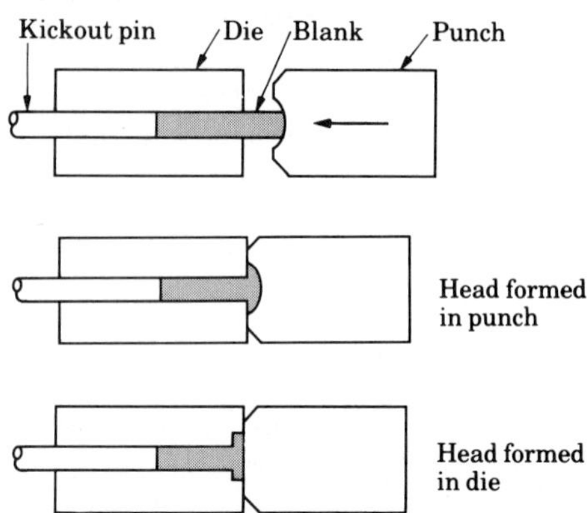

FIGURE 14.12
Forming heads on fasteners such as bolts and rivets. These processes are called heading.

• Example: Manufacturing of a bolt by heading operations.

The starting material for the steel bolt shown in the accompanying figure is a round rod 147 mm (5.8 in.) long and 38 mm (1.5 in.) in diameter, sheared from a long drawn rod. The first operation consists of preforming by gathering material at one end of the rod to prepare it for heading. The second operation produces a round head, while reducing the long section to 34 mm (1.34 in.). The last operation produces a hexagonal head. All operations are performed at room temperature, thus cold working the material, improving its mechanical properties, and producing a good surface finish.

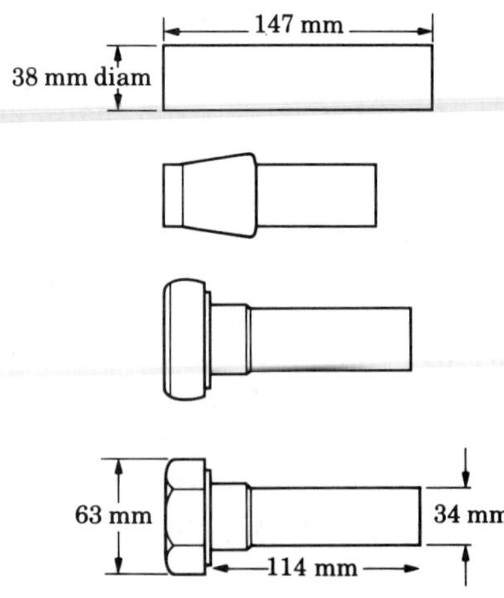

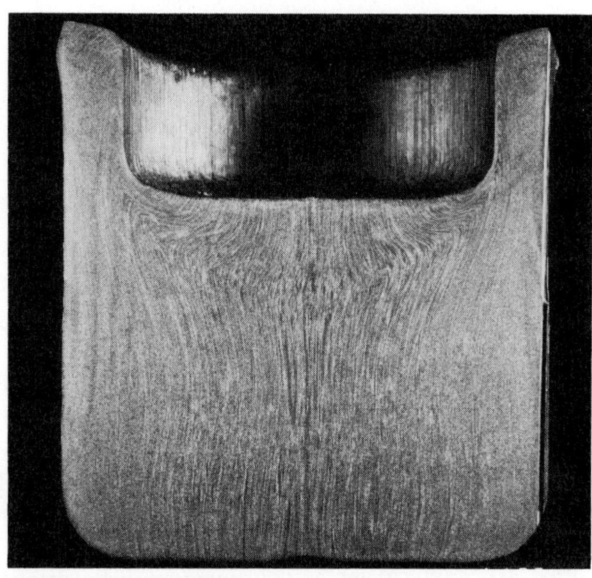

FIGURE 14.13
A pierced round billet, showing grain flow pattern.
Source: Courtesy of Ladish Co., Inc.

14.4.2 Piercing

Piercing is a process of indenting—but not breaking through—the surface of a workpiece with a punch in order to produce a cavity or an impression (Fig. 14.13). The workpiece may be confined in a die cavity, or it may be unconstrained. Piercing may be followed by punching to produce a hole in the part (see Fig. 13.16). Piercing is also performed to produce hollow regions in forgings (Fig. 14.14; see p. 404), using side-acting auxiliary equipment.

The *piercing force* depends on the punch's cross-sectional area and tip geometry, the strength of the material, and friction. The pressure may range from three to five times the strength of the material, or the same level of stress required to make an indentation in hardness testing.

14.4.3 Other operations

Hubbing consists of pressing a hardened punch with a particular tip geometry into the surface of a block of metal (Fig. 14.14b). The cavity produced is then used as a die for forming operations, such as for making tableware. The die cavity is usually shallow, but for deeper cavities some material may be removed from the surface by machining prior to hubbing.

In **roll forging**, the cross-section of a bar is reduced or shaped by passing it through a pair of rolls with shaped grooves (Fig. 14.15; see p. 404). Roll forging is used to produce tapered shafts and leaf springs, table knives, and hand tools. It may also be used as a preliminary forming operation, followed by other forging processes.

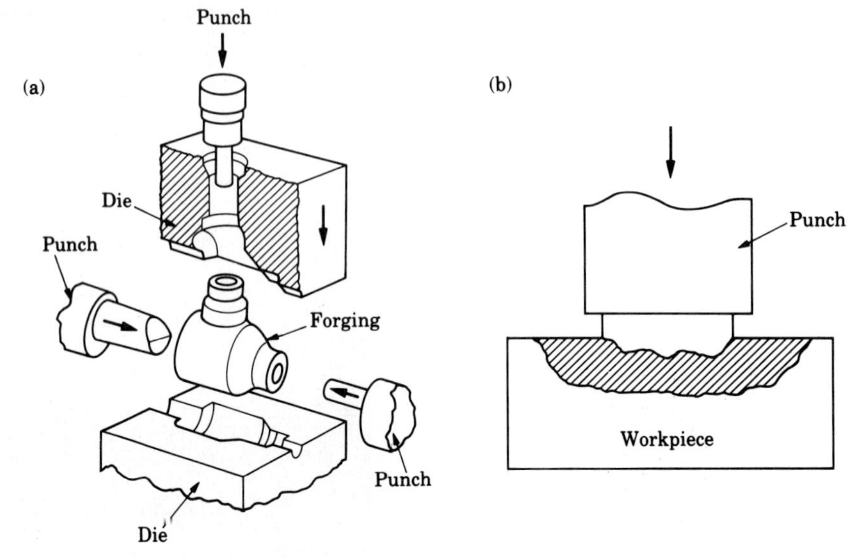

FIGURE 14.14
(a) Forging a complex shape, including piercing, using side punches. (b) Hubbing to produce a cavity. The workpiece is then heat treated, and the cavity serves as a female die for shallow forming operations.

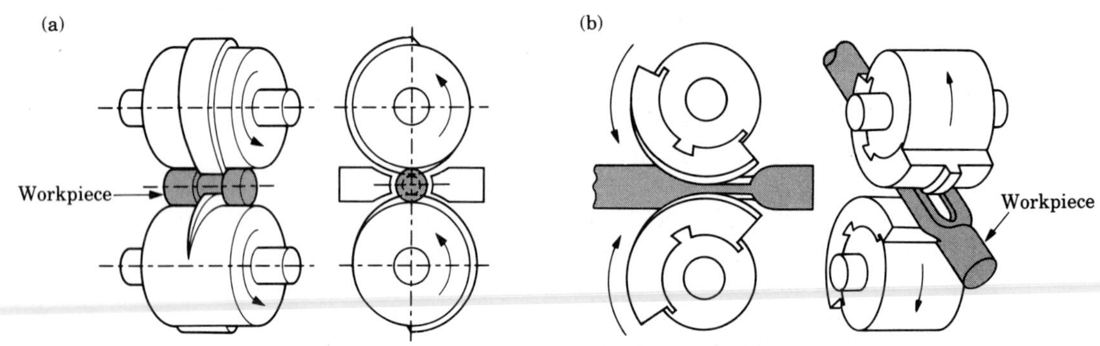

FIGURE 14.15
Two examples of the roll-forging operation, also known as cross-rolling. Tapered leaf springs and knives can be made by this process. *Source:* (a) J. Holub; (b) reprinted with permission of General Motors Corporation.

A process similar to roll forging is *skew rolling*, typically used for making ball bearings (Fig. 14.16a). Round wire or rod stock is fed into the roll gap, and roughly spherical blanks are formed continuously by the rotating rolls. Another method of forming spherical blanks for ball bearings is by shearing pieces from a round bar and upsetting them between two dies with hemispherical cavities (Fig. 14.16b) in ball headers. The balls are then ground and polished in special machinery.

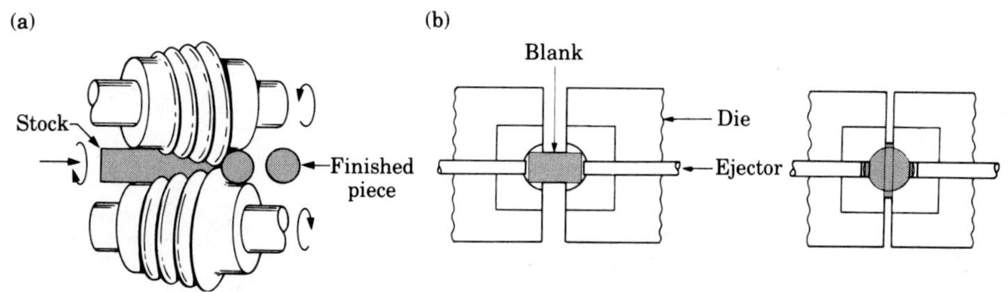

FIGURE 14.16 (a) Production of steel balls by the skew-rolling process. (b) Production of steel balls by upsetting a cylindrical blank. Note the formation of flash. The balls made by these processes are subsequently ground and polished for use in ball bearings.

Orbital forging is a process in which the upper die, by adjustable mechanical means, moves along an orbital path (Fig. 14.17a) and forms the part *incrementally*. The operation is similar to the action of a mortar and pestle. Typical components forged by this process are disk-shaped parts (Fig. 14.17b) and conical parts, such as bevel gears. The forging force is relatively small because, at any instant, the die contact is concentrated on a small area of the workpiece. The operation is relatively quiet, and parts can be formed within 10–20 cycles of the orbiting die.

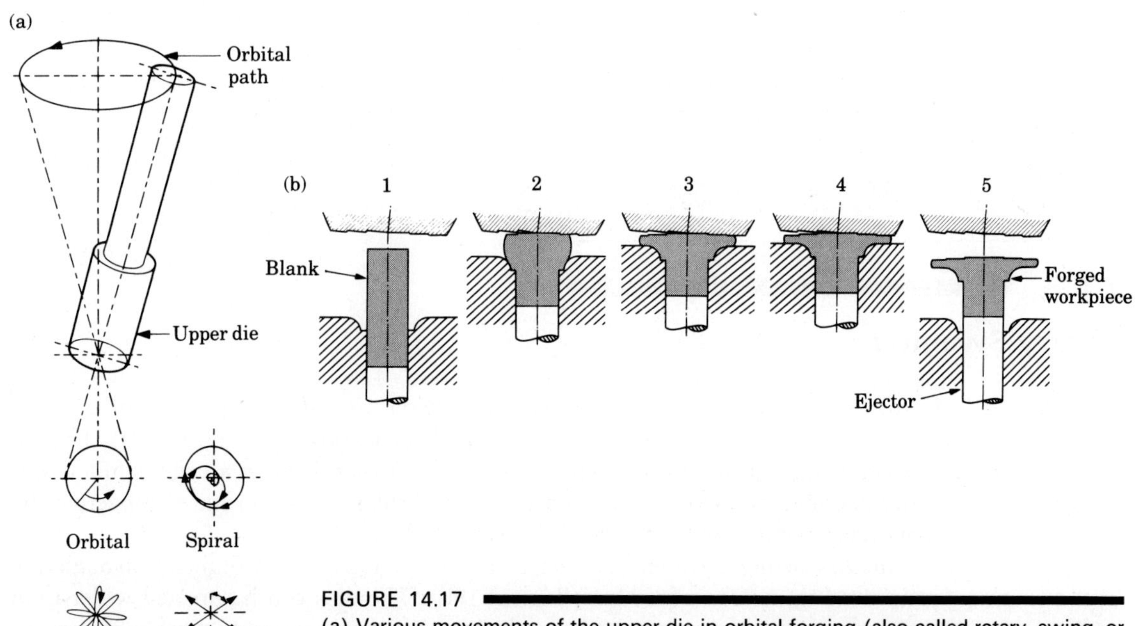

FIGURE 14.17 (a) Various movements of the upper die in orbital forging (also called rotary, swing, or rocking-die forging), similar to the action of a mortar and pestle. (b) An example of orbital forging. Bevel gears, wheels, and rings for bearings can be made by this process.

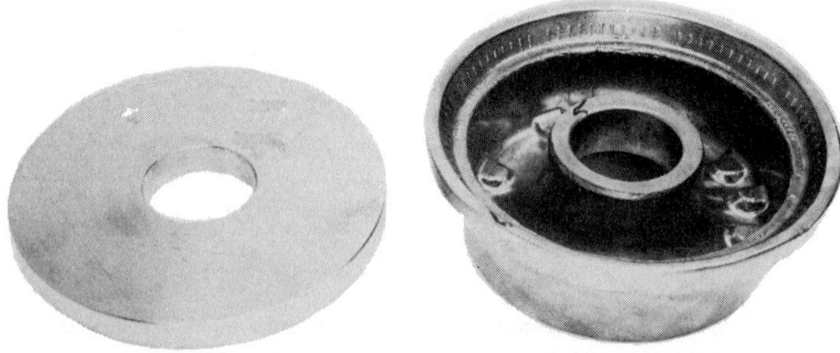

FIGURE 14.18
Isothermal forging of a nose wheel for an aircraft. The blank on the left, made of Ti-6Al-6V-2Sn titanium alloy, was formed at 900 °C (1650 °F) into a wheel with a single stroke in a hydraulic press, with a force of 8 MN (900 tons). *Source:* IIT Research Institute.

In **isothermal forging**—also known as *hot-die forging*—the dies are heated to the same temperature as that of the hot workpiece. Because the workpiece remains hot during forging, its low strength and ductility are maintained; thus the forging load is low and material flow within the die cavity is improved. Complex parts (Fig. 14.18) with good dimensional accuracy can be forged to near-net shape in one stroke in a hydraulic press.

The dies are generally made of nickel or molybdenum alloys. Isothermal forging is expensive and production is slow. However, it can be economical for intricate forgings made of materials such as titanium and superalloys, provided that the quantity required is sufficiently high to justify the die costs.

14.5 Rotary Swaging

In *rotary swaging*, also known simply as **swaging** (or *radial forging*), a solid rod or tube is subjected to radial impact forces by reciprocating dies (Fig. 14.19a). The die movements generally are obtained through a set of rollers in a cage, similar to a roller bearing. The workpiece is stationary and the set of dies rotate, impacting the workpiece at rates as high as 20 strokes per second.

In *die-closing swaging machines*, die movements are obtained through the reciprocating motion of wedges (Fig. 14.19b). The dies can be opened wider than rotary swagers, thus accommodating large-diameter or variable-diameter parts. In another type of machine the dies do not rotate but move radially in and out. Typical products are screwdriver blades and soldering-iron tips (Fig. 14.19c).

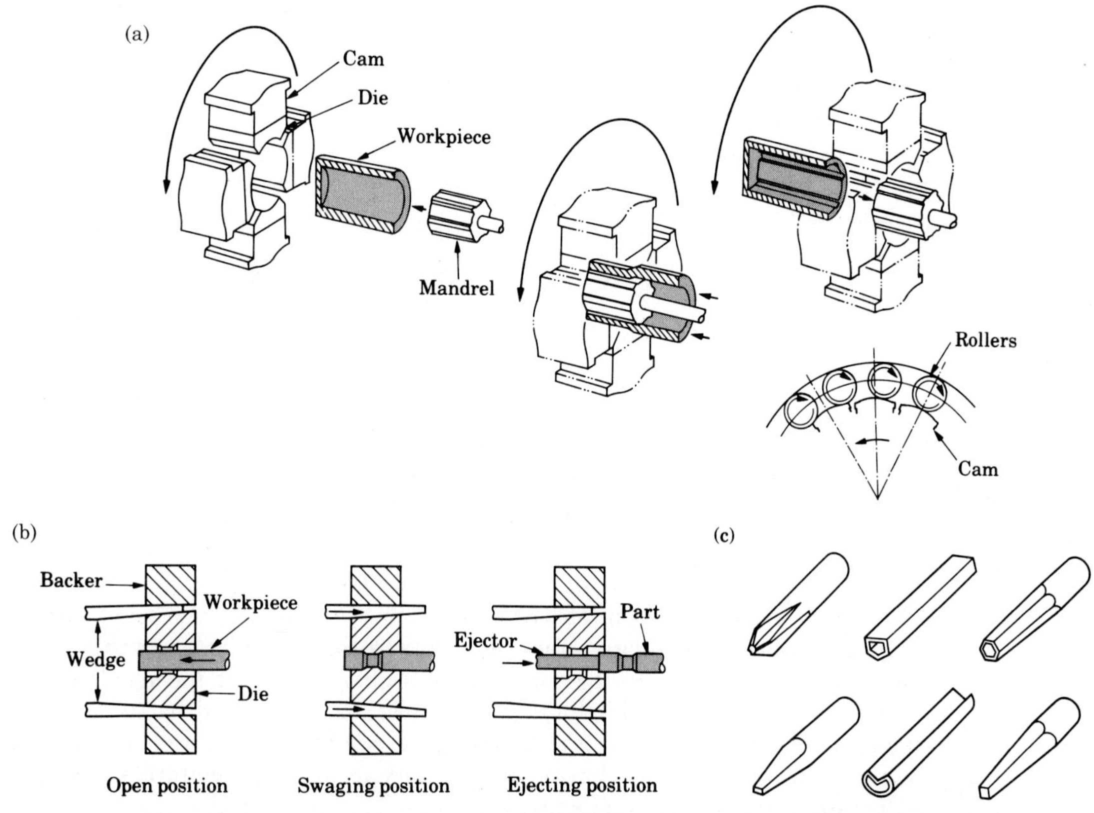

FIGURE 14.19 (a) Schematic illustration of the rotary-swaging process. Note the formation of internal profiles on the tubular workpiece. (b) A die-closing type swaging machine, showing forming of a stepped shaft. (c) Typical parts made by swaging.

In *tube swaging*, the internal diameter and/or the thickness of the tube can be controlled with or without the use of internal mandrels (Figs. 14.20a and b). Mandrels can also be made with longitudinal grooves, thus allowing internally shaped tubes to be swaged (Fig. 14.21). The rifling in gun barrels, for example, is made by swaging a tube over a mandrel with spiral grooves. For small-diameter tubing, high-strength wire is used as a mandrel. Special machinery has been built to swage gun barrels and other parts with starting diameters as large as 350 mm (14 in.).

The swaging process can be used to assemble fittings over cables and wire, in which case the fitting is swaged directly on the cable. This process is also used for operations such as *pointing* (tapering the tip of a cylindrical part) and *sizing* (finalizing the dimensions of a part).

Swaging is usually limited to a maximum stock diameter of about 150 mm (6 in.), although parts as small as 0.5 mm (0.02 in.) have been swaged. Tolerances

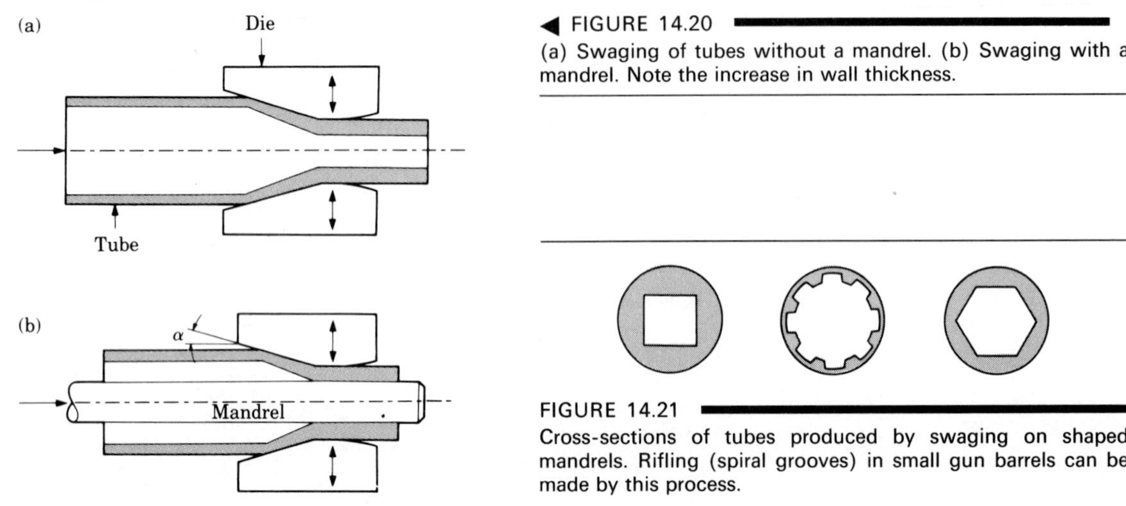

◀ FIGURE 14.20
(a) Swaging of tubes without a mandrel. (b) Swaging with a mandrel. Note the increase in wall thickness.

FIGURE 14.21
Cross-sections of tubes produced by swaging on shaped mandrels. Rifling (spiral grooves) in small gun barrels can be made by this process.

usually range from ±0.05 mm to ±0.5 mm (0.002 in. to 0.02 in.). Swaging is suitable for medium to high rates of production, with rates as high as 50 parts per minute possible, depending on the complexity of the part. It is a versatile process and is limited in length only by the length of the bar supporting the mandrel, if one is needed. As in other cold-working processes, parts produced by swaging have improved mechanical properties. Lubricants are used for improved surface finish and die life. For workpieces having low ductility at room temperature, swaging can be performed at elevated temperatures.

14.6 Forging-Die Design

The design of forging dies requires a knowledge of the strength and ductility of the workpiece material, its sensitivity to deformation rate and temperature, frictional characteristics, and the shape and complexity of the workpiece. Die distortion under high forging loads can be an important consideration, particularly if close tolerances are required.

The most important rule in die design is that the part or workpiece material flow in the direction of least resistance. Thus the material (*intermediate shape*) is commonly distributed so as to properly fill the die cavities. An example of the intermediate shapes for a connecting rod are shown in Fig. 14.8(a); another example is shown in Fig. 14.22. You can appreciate the importance of preforming by noting how a piece of dough is preshaped to make a pie crust or how ground meat is preshaped to make a hamburger.

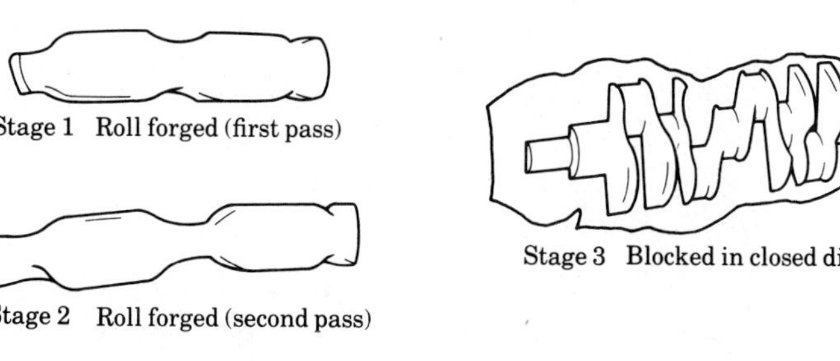

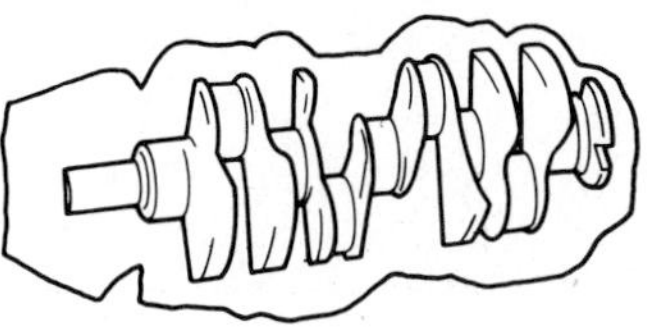

Stage 4 Finish forged in closed dies

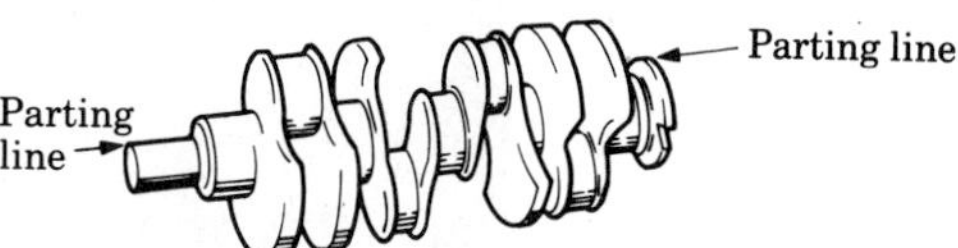

Stage 5 Trimmed (before twisting)

FIGURE 14.22
Stages in forging a crankshaft. Note how the material in the blank is distributed prior to being blocked (see also Fig. 14.8a).

Example: Forging railroad-car wheels.

Railroad-car wheels can be cast, as shown in Fig. 11.21(b), or they can be forged. The sequence of operations involved in making a railroad-car wheel by forging is shown in the accompanying figure. Starting with a round block of material, an upsetting operation preforms the blank with a flat top die and a lower die having a shallow impression. The second forging operation begins to shape the wheel. The center hole in the wheel is produced by a punching operation, which produces a slug. The thin sections (web) of the wheel are produced by a rolling operation while the wheel is being rotated. Note the increase in the diameter of the wheel from the material displaced during rolling. In the final operation, the wheel is given its conical shape by pressing it. All these operations are carried out at elevated temperature to enhance the ductility of the material and reduce forging forces.

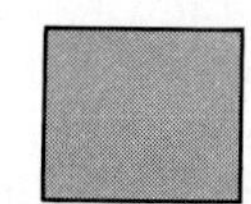

Step 1 Wheel block before forging

Step 2 Wheel blank after first forging

Step 3 Wheel blank after second forging

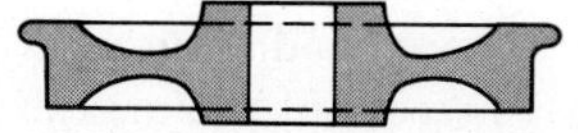

Step 4 Wheel blank after being punched

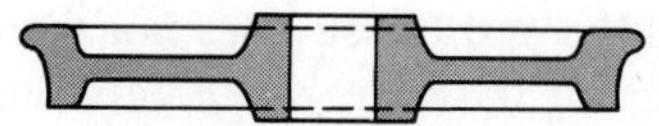

Step 5 Wheel after rolling

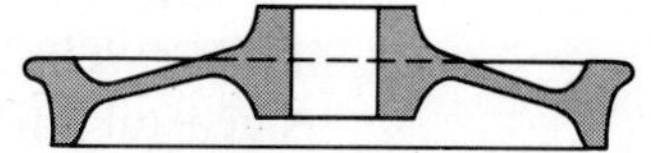

Step 6 Wheel after coning

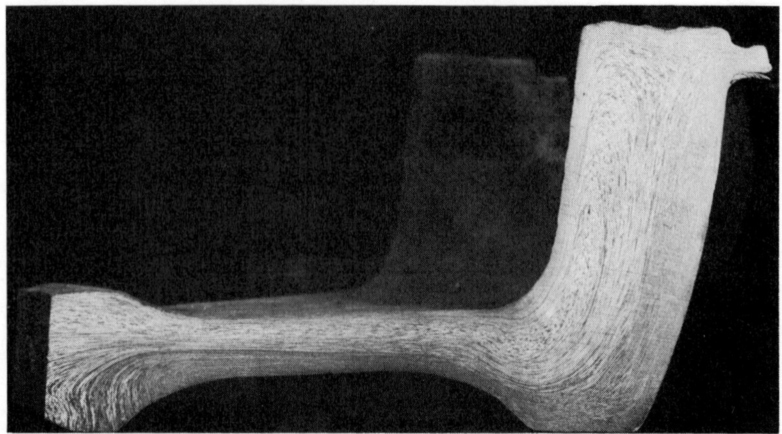

FIGURE 14.23
Grain flow in a forged carbon-steel gear blank. The blank, only the right half of which is shown, is 0.45 m (18 in.) in diameter. *Source:* Courtesy of Ladish Co., Inc.

In *preshaping* the material should not flow easily into the flash, grain flow pattern should be favorable, and excessive sliding at the workpiece–die interfaces should be minimized in order to reduce wear. Selection of shapes requires considerable experience and involves calculations of cross-sectional areas at each location in the forging. Computer-aided design techniques have been developed to expedite these calculations, as well as to predict the material-flow pattern in the die cavity and the formation of defects.

As you can see from the grain flow in the steel gear blank in Fig. 14.23, the material undergoes different degrees of deformation in various regions in the die cavity. Consequently, mechanical properties depend on the particular location in the forging. Test samples taken from a round forging are shown in Fig. 14.24, showing the direction of the specimens (tangential, radial, and axial) and their mechanical properties after forging.

The terminology for forging dies is shown in Fig. 14.25, and the significance of various parameters are described in the following paragraphs. Some of these considerations are similar to those for casting that we presented in Chapter 12.

For most forgings the *parting line* is usually at the largest cross-section of the part. For simple symmetrical shapes the parting line is usually a straight line at the center of the forging, but for more complex shapes the line may not lie in a single plane. The dies are then designed in such a way that they lock during engagement in order to avoid side thrust, thus balancing forces and maintaining die alignment.

After sufficiently constraining lateral flow, the flash material is allowed to flow into a *gutter*. Thus the extra flash does not increase the forging load unnecessarily. A general guideline for flash clearance between dies is 3 percent of the maximum thickness of the forging. The length of the *land* is usually two to five times the flash thickness. Several gutter designs are available.

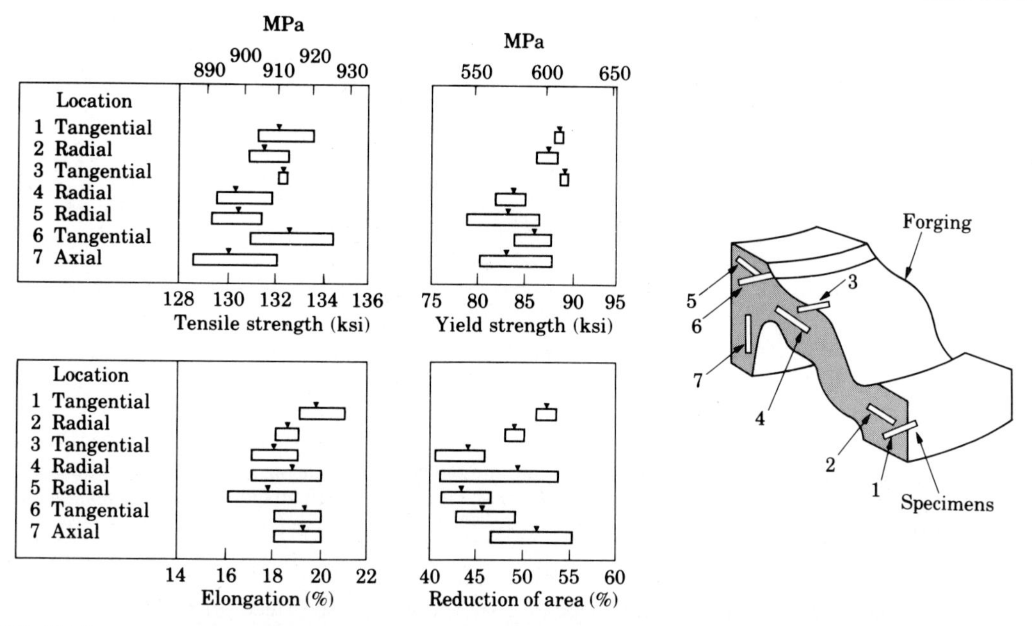

FIGURE 14.24

Variation in mechanical properties of a forged wheel, as determined from several tensile-test specimens removed from the wheel. Arrows on the charts indicate average values. Material: AISI 1050 steel, quenched and tempered. Diameter of part: 0.6 m (24 in.). Weight of part: 70 kg (150 lb). *Source: Steel Casting Handbook,* 5th ed. Steel Founders' Society of America, 1980. Used with permission.

Draft angles are necessary in almost all forging dies in order to facilitate the removal of the part from the die. Upon cooling, the forging shrinks both radially and longitudinally, so internal draft angles are made larger than external ones. Internal angles are about 7° to 10° and external angles about 3° to 5°.

Selection of the proper radii for corners and fillets is important in order to ensure smooth flow of the metal in the die cavity and to improve die life. Small radii

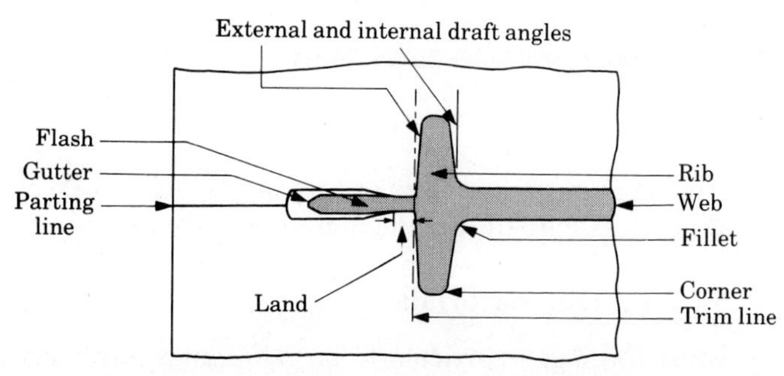

FIGURE 14.25

Standard terminology for various features of a typical impression-forging die.

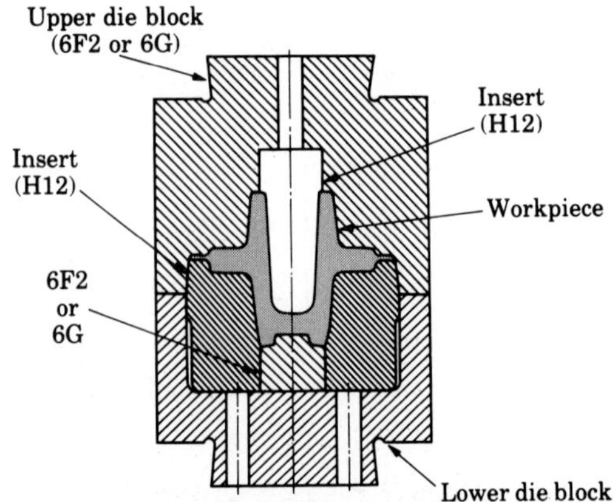

FIGURE 14.26
Die inserts used in dies for forging an automotive axle housing. (See Tables 5.7 to 5.9 for die materials.) *Source: Metals Handbook, Desk Edition.* ASM International, Metals Park, Ohio, 1985. Used with permission.

are generally undesirable because of their adverse effect on metal flow and their tendency to wear rapidly because of stress concentration and thermal cycling. Small fillet radii can cause fatigue cracking of the dies. Thus, as a general rule, radii should be as large as permissible by the design of the forging.

Instead of being made in one piece, dies may be assembled with *die inserts* (Fig. 14.26), which reduces the cost of making the dies. The inserts can be made of stronger and harder materials and can be changed easily in case of wear or failure in a particular section of the die.

As with patterns in casting, *allowances* are provided in forging-die design, since machining of the forging may be necessary to obtain final desired dimensions and tolerances. Machining allowance should be provided at flanges, holes, and mating surfaces.

14.7 Die Materials and Lubrication

Among important aspects of forging, as well as other manufacturing processes that we discuss Part III, are die materials and lubrication.

14.7.1 Die materials

Most forgings—particularly large forgings—are carried out at elevated temperatures. General requirements for die materials therefore are (a) strength and toughness at elevated temperatures, (b) hardenability and ability to harden uniformly, (c) mechanical and thermal shock resistance, and (d) wear resistance, particularly abrasive wear resistance, because of the presence of scale in hot forging.

Selecting proper die materials depends on factors such as die size, composition and properties of the workpiece, complexity of shape, forging temperature, type of forging operation, cost of die materials, and the number of forgings required. Heat transfer from the hot workpiece to the dies and subsequent distortion of the dies are also important factors. Common die materials are tool and die steels containing chromium, nickel, molybdenum, and vanadium (see Tables 5.7 and 5.8). Dies are made from die blocks, which themselves are forged from castings and are machined and finished to the desired shape and surface finish.

Die failures usually result from improper heat treatment, improper design, overheating and heat checking (thermal cracks caused by temperature cycling), excessive wear, and overloading. To reduce heat checking and fracture, dies should be preheated to temperatures of about 150–250 °C (300–500 °F).

14.7.2 Lubrication

Lubricants greatly influence friction and wear and, consequently, affect the forces required (see Eq. 14.1) and the flow of the metal in die cavities. They can also act as a thermal barrier between the hot workpiece and the relatively cool dies, thus slowing the rate of cooling of the workpiece and improving metal flow. Another important role of the lubricant is to serve as a parting agent, that is, to prevent the forging from sticking to the dies and to help in its release from the die.

A wide variety of lubricants can be used in forging (see Table 13.1). For hot forging, graphite, molybdenum disulfide, and (sometimes) glass are often used. For cold forging, mineral oils and soaps are common lubricants, preceded by conversion coating of the blanks. In hot forging, the lubricant is usually applied to the dies and in cold forging to the workpiece. The method of application and uniformity of lubricant thickness are important to product quality. (See also Chapter 32.)

14.8 Forgeability

Forgeability is generally defined as the capability of a material to undergo deformation without cracking. A number of tests have been developed to quantify forgeability, although none is accepted universally. A commonly used test is to *upset* a solid cylindrical specimen and observe any cracking on the barreled surfaces (see Fig. 2.23d). The greater the deformation prior to cracking, the greater is the forgeability of the metal. Upsetting tests can be performed at various temperatures and deformation rates. If notch sensitivity of the material is high, surface defects will affect the results by causing premature cracking. A typical surface defect is a seam, which may be a longitudinal scratch, string of inclusions, or folds introduced during prior working of the material.

TABLE 14.2
CLASSIFICATION OF METALS IN ORDER OF INCREASING FORGING DIFFICULTY

METAL OR ALLOY	APPROXIMATE RANGE OF FORGING TEMPERATURE (°C)
Aluminum alloys	400–550
Magnesium alloys	250–350
Copper alloys	600–900
Carbon and low-alloy steels	850–1150
Martensitic stainless steels	1100–1250
Austenitic stainless steels	1100–1250
Titanium alloys	700–950
Iron-base superalloys	1050–1180
Cobalt-base superalloys	1180–1250
Tantalum alloys	1050–1350
Molybdenum alloys	1150–1350
Nickel-base superalloys	1050–1200
Tungsten alloys	1200–1300

In the *hot-twist test*, a round specimen is twisted continuously in the same direction until it fails. The test is performed on a number of specimens at various temperatures, and the number of turns that each specimen undergoes before failure is observed. The optimum forging temperature is then determined. The hot-twist test is particularly useful for steels.

The forgeabilities of several metals and alloys are given in Table 14.2, in decreasing order of forgeability. These ratings should be regarded only as general guidelines. They are based on considerations of ductility and strength, forging temperature required, frictional behavior, and the quality of the forgings produced. Because of differences in ductility at different temperatures, two-phase alloys (such as titanium) are more difficult to forge than single-phase alloys and require careful selection and control of forging temperature.

Typical forging temperature ranges for various metals and alloys also are included in Table 14.2. Note that higher forging temperature does not necessarily indicate greater difficulty in forging that material. For warm forging, temperatures range from 200 °C to 300 °C (400 °F to 600 °F) for aluminum alloys and 550 °C to 750 °C (1000 °F to 1400 °F) for steels.

14.8.1 Forging defects

In addition to surface cracking during forging, other defects can develop as a result of the material flow pattern in the die. Excess material in the web may buckle during forging and develop laps (Fig. 14.27a). But if the web is thick, the excess material flows past the already forged portions of the forging and develops internal cracks

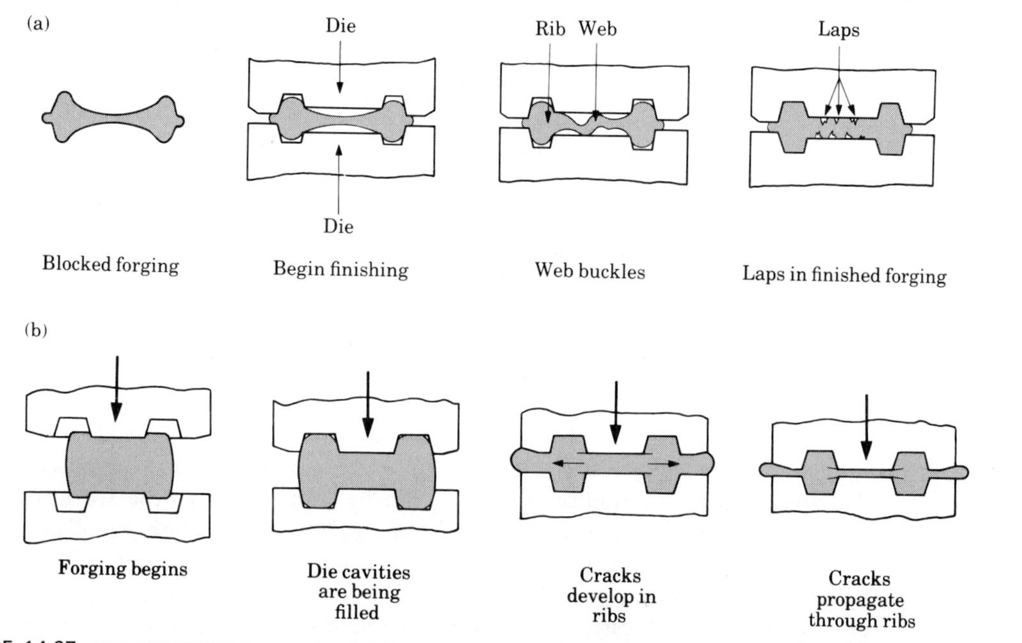

FIGURE 14.27 (a) Laps formed by web buckling during forging. Web thickness should be increased to avoid this problem. (b) Internal defects caused by oversized billet. Die cavities are filled prematurely, and the material at the center flows past the filled regions as the dies close.

(Fig. 14.27b). The various radii in the forging die cavity can significantly influence the formation of such defects. Internal defects may also develop from nonuniform deformation of the material in the die cavity, temperature variations throughout the workpiece during forging, and microstructural changes caused by phase transformations.

Forging defects can cause fatigue failures and lead to other problems such as corrosion and wear during the service life of the component. The importance of inspecting forgings before they are put into service, particularly in critical applications, is obvious. We describe inspection techniques in Chapter 36.

14.9 Forging Machines

A variety of forging machines are in use, with a range of capacities, speeds, and speed-stroke characteristics (Table 14.3). We generally classify these machines as **presses** or **hammers**.

TABLE 14.3
SPEED RANGE OF FORGING EQUIPMENT

EQUIPMENT	m/s
Hydraulic press	0.06–0.30
Mechanical press	0.06–1.5
Screw press	0.6–1.2
Gravity drop hammer	3.6–4.8
Power drop hammer	3.0–9.0
Counterblow hammer	4.5–9.0

14.9.1 Presses

Hydraulic presses. *Hydraulic presses* operate at constant speeds and are *load limited*, or load restricted. In other words, a press stops if the load required exceeds its capacity. Large amounts of energy can be transmitted to the workpiece by a constant load throughout the stroke, the speed of which can be controlled. Because hydraulic-press forging takes longer than other types of forging machines, the workpiece may cool rapidly unless heated dies are used. Compared to mechanical presses, hydraulic presses are slower, involve higher initial cost, but require less maintenance.

A hydraulic press typically consists of a frame with two or four columns, pistons, cylinders (Fig. 14.28a), rams, and hydraulic pumps driven by electric

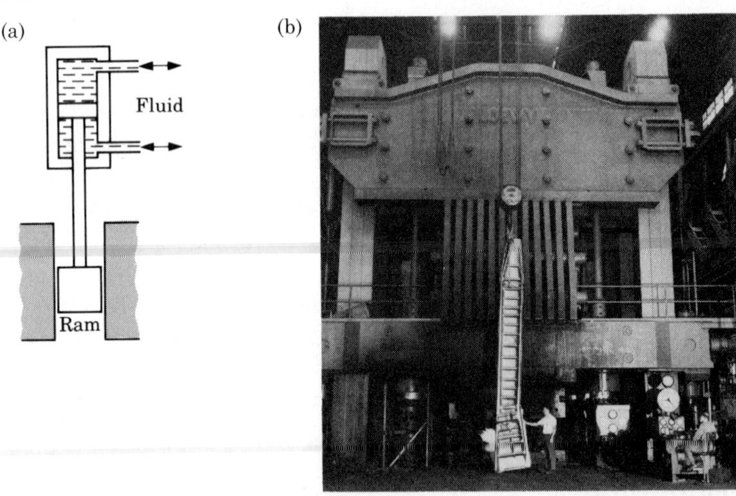

FIGURE 14.28
(a) Schematic illustration of the principle of a hydraulic press. (b) A general view of a 445-MN (50,000-ton) hydraulic press. The part shown is a Ti-6Al-4V titanium-alloy forging for the main landing-gear support beam for the Boeing 747 aircraft. Weight of part: 1350 kg (3000 lb). Length of part: 6.7 m (22 ft). *Source:* Courtesy of Wyman-Gordon Company.

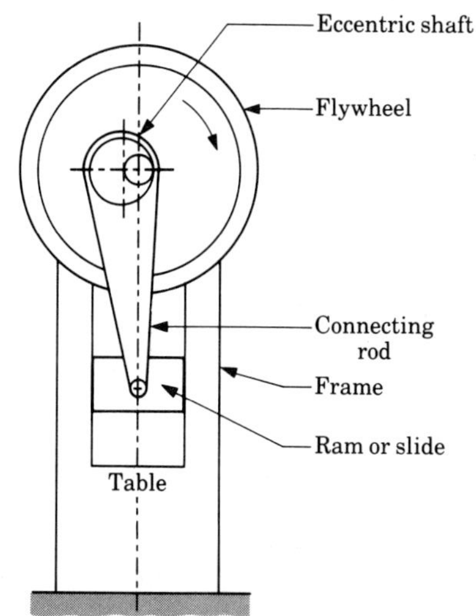

FIGURE 14.29 Schematic illustration of a mechanical press with an eccentric drive. The eccentric shaft can be replaced by a crankshaft to give the up-and-down motion to the ram.

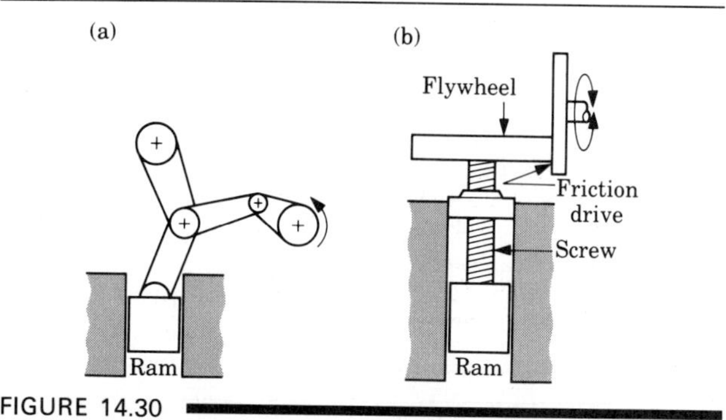

FIGURE 14.30 (a) Principle of a knucklejoint press and (b) a screw press.

motors. The ram speed can be varied during the stroke. Press capacities range up to 125 MN (14,000 tons) for open-die forging and 670 MN (75,000 tons) for closed-die forging. The main landing-gear support beam for the Boeing 747 aircraft is forged in a 450-MN (50,000-ton) hydraulic press, shown in Fig. 14.28(b) with the part in the forefront. This part is made of titanium alloy and weighs approximately 1350 kg (1.5 tons).

Mechanical presses. *Mechanical presses* are basically either the crank or eccentric type (Fig. 14.29), with speeds that vary from a maximum at the center of the stroke to zero at the bottom of the stroke. These presses are thus *stroke limited.* The energy in a mechanical press is generated by a large flywheel powered by an electric motor. A clutch engages the flywheel to an eccentric shaft. A connecting rod translates the rotary motion to a reciprocating linear motion. A *knuckle-joint* mechanical press is shown in Fig. 14.30(a). Because of the linkage design, very high forces can be applied in this type of press (see also Fig. 11.24).

The force available in a mechanical press depends on the stroke position and becomes extremely high at the bottom dead center. Thus proper setup is essential to avoid breaking the dies or equipment components. Press capacities generally range from 2.7 MN (300 tons) to 107 MN (12,000 tons). Mechanical presses have high production rates, are easier to automate, and require less operator skill than other types of forging machines.

Screw presses. *Screw presses* (Fig. 14.30b) derive their energy from a flywheel; hence they are *energy limited.* The forging load is transmitted through a vertical screw. The ram comes to a stop when the flywheel energy is dissipated. If the dies do not close at the end of the cycle, the operation is repeated until the forging is

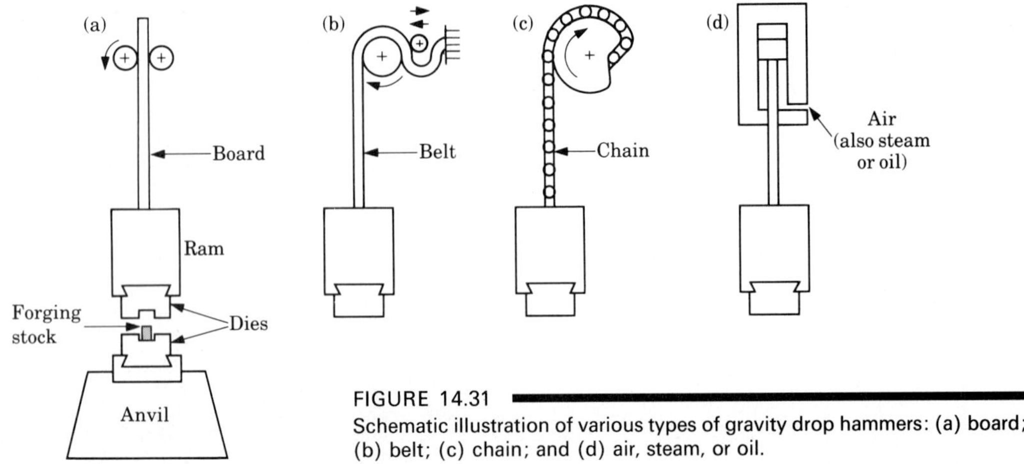

FIGURE 14.31 Schematic illustration of various types of gravity drop hammers: (a) board; (b) belt; (c) chain; and (d) air, steam, or oil.

completed. Capacities range from 1.4 MN to 280 MN (160 tons to 31,500 tons). Screw presses are used for various open-die and closed-die forging operations, and are particularly suitable for small production quantities and precision parts, such as turbine blades.

14.9.2 Hammers

Hammers derive their energy from the potential energy of the ram, which is converted to kinetic energy; thus they are *energy limited.* Unlike hydraulic presses, they operate at high speeds, and the low forming times minimize cooling of the hot forging. Low cooling rates allow the forging of complex shapes, particularly those with thin and deep recesses. To complete the forging, several successive blows are usually made in the same die. Hammers are the most versatile and least expensive type of forging equipment (Fig. 14.31).

Gravity drop hammers. In the *gravity drop hammer* (hence the term *drop forging*), the energy is derived from the free-falling ram. The available energy of the hammer is the product of the ram's weight and the height of its drop. Ram weights range from 180 kg to 4500 kg (400 lb to 10,000 lb), with energy capacities ranging up to 120 kJ (90,000 ft-lb).

Power drop hammers. In the *power drop hammer*, the ram's downstroke is accelerated by steam, air, or hydraulic pressure at about 700 kPa (100 psi). Ram weights range from 225 kg to as much as 22,500 kg (500 lb to 50,000 lb), with energy capacities ranging up to 1150 kJ (850,000 ft-lb).

Counterblow hammers. A *counterblow hammer* has two rams that simultaneously approach each other, either horizontally or vertically, to forge the part. As in open-die forging operations, the part may be rotated between blows for proper shaping of the workpiece during forging. Counterblow hammers operate at high speeds and transmit less vibration to their bases. Capacities range up to 1200 kJ (900,000 ft-lb).

High-energy-rate machines. In a high-energy-rate machine, the ram is accelerated by inert gas at high pressure, and the part is forged in one blow at very high speeds. Although there are several types for these machines, various problems associated with their operation and maintenance, die breakage, and safety considerations limit their actual use in forging plants.

14.9.3 Selection of forging machines

Important to the selection of forging machines are force or energy requirements, size, shape and complexity of the forging, strength of the workpiece material, and sensitivity of the material to rate of deformation. Other factors include production rate, accuracy, maintenance, operating skills required, noise level, and cost. In general, presses are usually preferred for use with aluminum, magnesium, beryllium, bronze, and brass. Hammers are usually preferred for use with steels, titanium, copper, and refractory metal alloys. A forging may be made on two or more pieces of equipment, that is, first on a hammer followed by hydraulic or mechanical presses.

14.10 Forging Practice and Process Capabilities

A typical forging operation involves the following sequence of steps:

1. Prepare slug, billet, or preform by shearing, sawing, or cutting off, either cold or hot. If necessary, clean surfaces by means such as shot blasting.
2. For hot forging, heat workpiece in a suitable furnace and, if necessary, descale after heating with a wire brush, water jet, or steam or by scraping. Descaling may also be done during the initial stages of forging when the scale, which is usually brittle, falls off during deformation of the part.
3. For hot forging, preheat and lubricate the dies. For cold forging, lubricate the blank.
4. Forge in appropriate dies and in the proper sequence. Then if necessary, remove excess material such as flash by trimming, machining, or grinding.
5. Clean the forging, check dimensions, and if necessary, machine to final dimensions and tolerances.

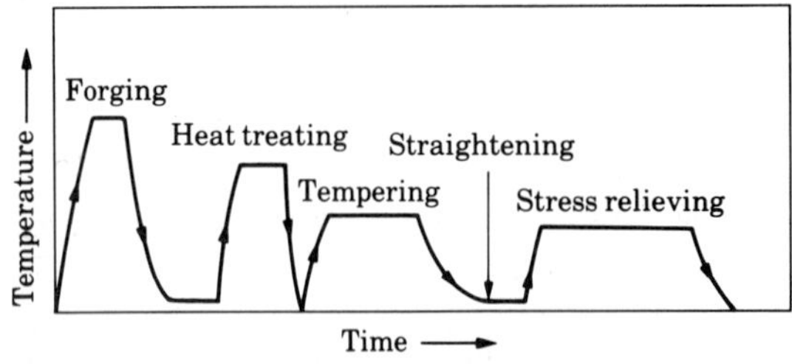

FIGURE 14.32
Temperature–time cycles in hot forging and subsequent heat treatment and finishing operations. In a recent development, hot forgings are heat-treated directly, without cooling and reheating.

6. Perform additional operations, such as straightening and heat treating, for improved mechanical properties (Fig. 14.32). Perform any finishing operations required.
7. Inspect forging for external and internal defects (see Sections 36.2 and 36.3).

The quality, tolerances, and surface finish of a forging depend on how well these operations are performed and controlled. Generally, tolerances range between ± 0.5 percent and ± 1 percent of the dimensions of the forging. In good practice, tolerances for hot forging of steel are usually less than ± 6 mm (1/4 in.), and can be as low as ± 0.25 mm (0.01 in.) in precision forging. Other factors that contribute to dimensional accuracy are draft angles, radii, die wear, die closure, and mismatching of dies. Surface finish of the forging depends on the effectiveness of the lubricant, preparation of the blank, and die surface finish and die wear.

14.10.1 Automation in forging

Many forging machines and facilities have been automated and operations are now computer controlled. Blanks and forgings are handled, including their loading into and unloading from furnaces, by robots and other automatic handling equipment. Mechanical manipulators are used to move and position billets in the dies. Lubrication and other operations, such as trimming, heat treating, and transporting billets, have been automated.

Production rates have been increased through better control of all aspects of forging operations. Automation has been particularly effective in producing high-quantity parts, such as gears, axles, nuts, bolts, and bearing races.

Plant layout for forging depends on factors such as the size of forging and the equipment involved. A typical layout for high-production hot forging of relatively small pieces from bar stock is shown in Fig. 14.33. New developments in forging include elimination of intermediate steps by direct melting and forging into net or near-net shapes (Fig. 14.34). Such improvements can significantly reduce the cost of forgings by saving labor, equipment, and material.

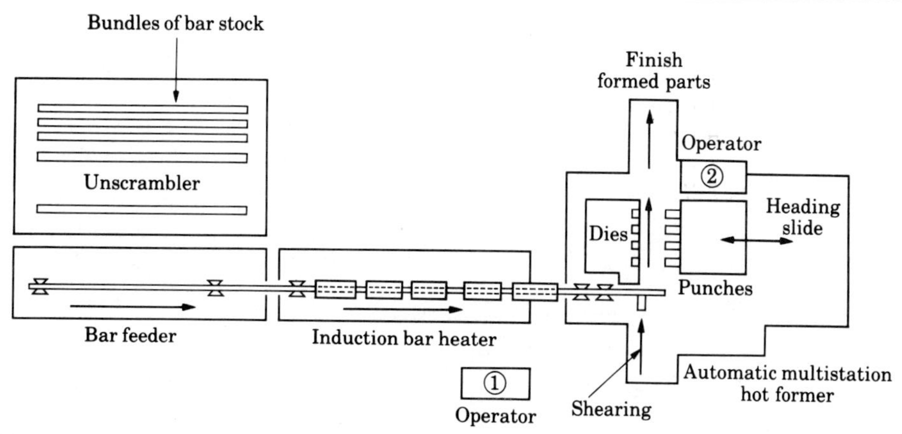

FIGURE 14.33 Schematic illustration of a forging facility for hot heading of bars. These facilities are automated for economical production of parts at high production rates. *Source:* Girard Associates, Inc.

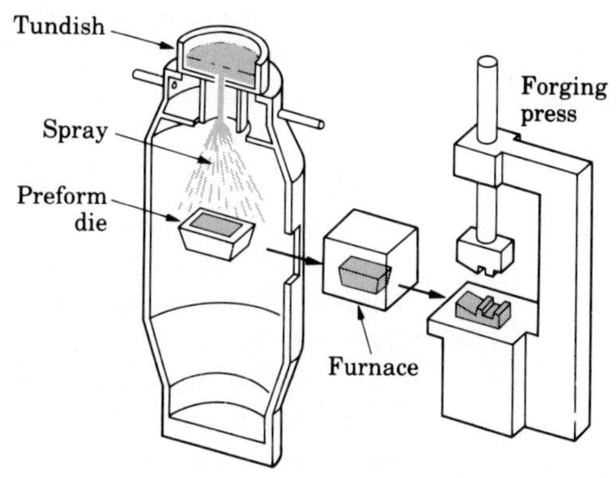

FIGURE 14.34 Schematic illustration of a net or near-net shape casting and forging operation (see also squeeze casting, Fig. 11.29). *Source:* J. Szekely, *Scientific American,* July 1987.

14.11 Economics of Forging

A number of factors are involved in the cost of forgings. We show their relationships in Fig. 14.35. Tool and die costs range from moderate to high, depending on the complexity of the forging. However, as you will also see for other manufacturing operations, this cost is spread over the number of parts forged with that particular die set. So, even though the cost of material per piece is constant, setup and tooling costs per piece decrease as the number of pieces forged increases.

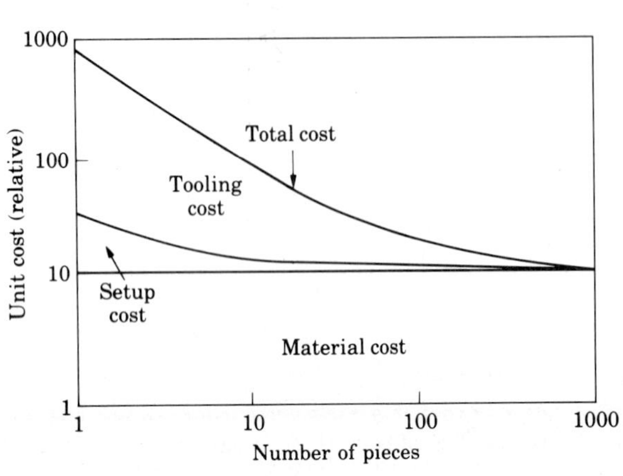

FIGURE 14.35 Typical unit cost (cost per piece) in forging. Note how setup and tooling costs per piece decrease with the number of pieces forged in the same die.

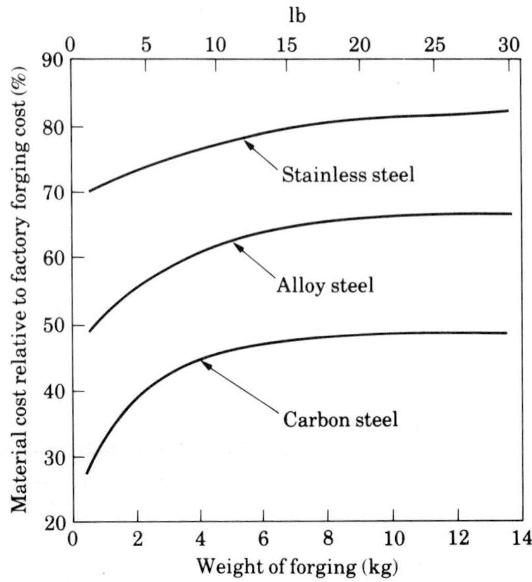

FIGURE 14.36 Material cost for three types of steel as a percentage of the total cost of forging each material.

The ratio of cost of material to the total cost of forging the part increases with the weight of forgings. We show these ratios in Fig. 14.36 for three different steels. The three curves are not related; they show only the percentages for each workpiece material. The more expensive the material, the higher the cost of the material relative to the total cost, hence the higher the curve. Because dies must be made and forging operations must be performed regardless of the size of forging, the cost of dies and the forging operation relative to material cost is high for small parts, and, conversely, material costs are relatively low. As part size increases, the share of material cost in the total cost increases but at a lower rate. This is because the incremental increase in die cost is relatively small, the machinery and operations involved are essentially the same regardless of part size, and the amount of labor involved per piece is not that much higher. Furthermore, because the total cost involved in a forging operation is not influenced to any major extent by the type of materials shown in Fig. 14.36, the three curves have similar shapes.

Labor costs in forging are generally moderate and can be reduced significantly with automated and computer-controlled operations. Die design and manufacturing are now being performed by computer-aided design and manufacturing techniques, with significant savings in time and effort.

The cost of forging a part compared to making it by various casting techniques, powder metallurgy, machining, or other methods is an important consideration in a competitive marketplace. For example, all other factors being the same—and

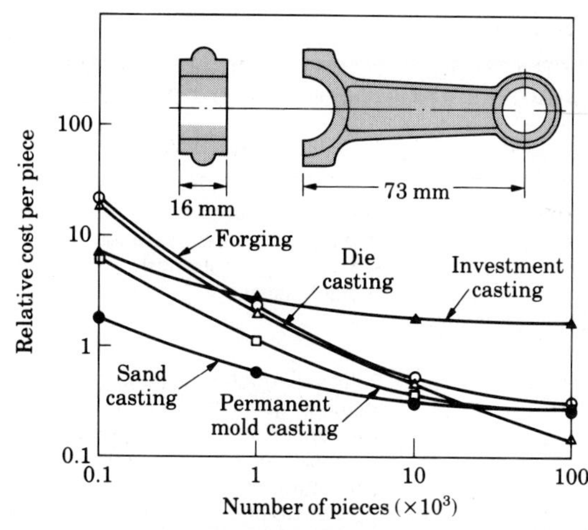

FIGURE 14.37
Relative unit costs of a small connecting rod made by various forging and casting processes. Note that for large quantities forging is more economical. Sand casting is the most economical process for quantities less than about 20,000 pieces.

depending on the number of pieces required—manufacturing a certain part by, say, expendable-mold casting may well be more economical than by forging (Fig. 14.37). This casting method does not require expensive molds and tooling, whereas forging requires expensive dies. We discuss competitive aspects of manufacturing in greater detail in Chapter 40.

SUMMARY

Forging is an ancient and important metalworking process. It is capable of producing a wide variety of parts with favorable characteristics of strength, toughness, dimensional accuracy, and reliability in service. Material behavior during deformation, as well as friction, heat transfer, and material-flow characteristics in a die cavity, are important considerations. Also important is the proper selection of die materials, lubricants, temperatures, speeds, and equipment. Defects can develop if the process is not controlled properly.

A variety of forging machines are available, each with its own characteristics and capabilities. Forging processes have been highly automated with industrial robots and computer controls. Computer-aided design and manufacturing techniques are now being used extensively in die design and manufacturing, as well as in preform design and predicting material flow and possibility of defects during forging.

SUMMARY TABLE 14.1

CHARACTERISTICS OF FORGING PROCESSES

PROCESS	ADVANTAGES	LIMITATIONS
Open die	Simple, inexpensive dies; useful for small quantities; wide range of sizes available; good strength characteristics	Limited to simple shapes; difficult to hold close tolerances; machining to final shape necessary; low production rate; relatively poor utilization of material; high degree of skill required
Closed die	Relatively good utilization of material; generally better properties than open-die forgings; good dimensional accuracy; high production rates; good reproducibility	High die cost for small quantities; machining often necessary
Blocker type	Low die costs; high production rates	Machining to final shape necessary; thick webs and large fillets necessary
Conventional type	Requires much less machining than blocker type; high production rates; good utilization of material	Somewhat higher die cost than blocker type
Precision type	Close tolerances; machining often unnecessary; very good material utilization; very thin webs and flanges possible	Requires high forces, intricate dies, and provision for removing forging from dies

TRENDS

- Computer-aided design and manufacturing are being implemented increasingly in all aspects of forging design and manufacturing. Techniques being used include modeling of the deformation of the workpiece during forging in the dies, die design, preform design, calculation of forces, and prediction of forging defects and die failure.
- Further developments in automation and use of industrial robots for material handling and processing are taking place rapidly.
- To improve operational efficiency and save energy costs, heat treatment of forgings immediately following a hot-forging operation is being implemented.
- Improvements in heating furnace design, computer control, and overall efficiency of hot-forging operations are taking place.

KEY TERMS

Barreling
Coining
Flash
Forging
Hammers
Heading
Hubbing
Impression-die forging
Isothermal forging
Open-die forging
Orbital forging
Piercing
Precision forging
Presses
Roll forging
Swaging

BIBLIOGRAPHY

Altan, T., S.I. Oh, and H.C. Gegel, *Metal Forming—Fundamentals and Applications.* Metals Park, Ohio: American Society for Metals, 1983.

———, *Forging: Equipment, Materials and Practices.* Columbus, Ohio: Battelle Memorial Institute, 1973.

Avitzur, B., and C.J. van Tyne (eds.), *Production to Near Net Shape: Source Book.* Metals Park, Ohio: American Society for Metals, 1983.

Blazynski, T.Z., *Plasticity and Modern Metal-forming Technology.* New York: Elsevier, 1989.

Byrer, T.G. (ed.), *Forging Handbook.* Metals Park, Ohio: American Society for Metals, 1985.

Lange, K. (ed.), *Handbook of Metal Forming.* New York: McGraw-Hill, 1985.

Metals Handbook, 9th ed., *Vol. 14: Forming and Forging.* Metals Park, Ohio: ASM International, 1988.

Open Die Forging Manual, 3d ed. Cleveland: Forging Industry Association, 1982.

Sabroff, A.M., F.W. Boulger, and H.J. Henning, *Forging Materials and Practices.* New York: Reinhold, 1968.

Thomas, A., *DFRA Forging Handbook: Die Design.* Sheffield, England: Drop Forging Research Association, 1980.

REVIEW QUESTIONS

14.1 What is the difference between cold, warm, and hot forging?

14.2 What is upsetting?

14.3 Explain the difference between open-die and impression-die forging.

14.4 What does breaking down a cast ingot mean?

14.5 Explain the difference between fullering, edging, and blocking.

14.6 What is involved in precision forging?

14.7 Name the operations related to forging.

14.8 Describe orbital forging and explain how it differs from conventional forging operations.

14.9 What type of parts is rotary swaging capable of producing?

14.10 Explain the features of a typical forging die.

14.11 Why is intermediate shape important in forging operations?

14.12 How is forgeability defined?

14.13 Explain what is meant by load limited, energy limited, and stroke limited as these terms pertain to forging machines.

14.14 Describe the sequence of operations in forging a part.

QUESTIONS AND PROBLEMS

14.15 Explain the function of flash in impression-die forging.

14.16 How can you tell whether a certain part is forged or cast? Explain the features that you would investigate.

14.17 Why is the control of the volume of the blank important in closed-die forging?

14.18 What are the advantages and limitations of a cogging operation?

14.19 Describe your observations concerning Fig. 14.22.

14.20 What are the advantages and limitations of using die inserts?

14.21 Explain why inner draft angles are larger than outer angles. Is this also true for permanent-mold casting?

14.22 How would you make the sequence of operations shown in Fig. 14.32 more energy efficient?

14.23 Explain why there are so many different types of forging machines.

14.24 In Part II we noted that very large parts can be made by casting. Is this also true for forging? Explain the technical and economic aspects of such operations.

14.25 Devise an experimental method whereby you can measure the force required for forging only the flash in impression-die forging.

14.26 Assume that you represent the forging industry and are facing a representative of the casting industry. What would you tell that person about the merits of forging processes? How would you prepare yourself to face questions about any limitations of forging over casting?

14.27 Calculate the forging force for a solid cylindrical workpiece, made of annealed copper, that is 4 in. high and 4 in. in diameter and is reduced in height by 30 percent. Let the coefficient of friction be 0.15.

14.28 Using Eq. (14.2), estimate the forging force for the workpiece in Problem 14.27, assuming that it is a complex forging and that the projected area of the flash is 20 percent greater than the projected area of the forged workpiece.

14.29 Determine the temperature rise in the specimen in Problem 14.27, assuming that the process is adiabatic and the temperature is uniform throughout the specimen.

14.30 Take two solid cylindrical specimens of equal diameter but different heights and compress them (frictionless) to the same percent reduction in height. Show that the final diameters will be the same.

14.31 The accompanying illustration shows a round impression-die forging made from a round blank as shown on the left. As we described in this chapter, such parts are made in a sequence of forging operations. Suggest a sequence of intermediate forging steps to make this part and sketch the shape of the dies needed.

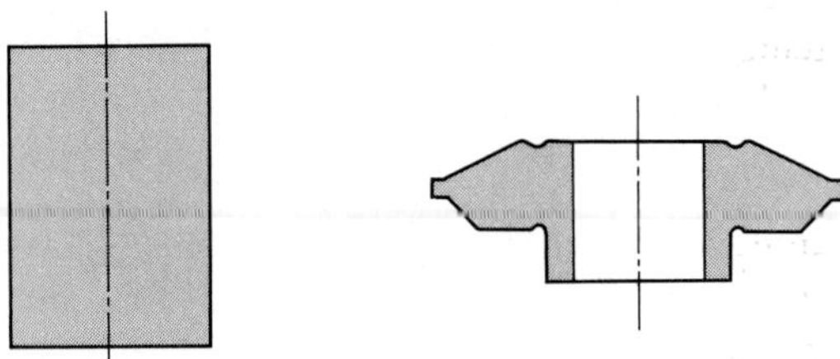

14.32 How would you use the type of information given in Fig. 14.24? Give specific examples.

15

Extrusion and Drawing

15.1 Introduction

In the **extrusion** process, material is forced through a die (Fig. 15.1), which is similar to squeezing toothpaste from a tube. Almost any cross-section may be produced by extrusion (Fig. 15.2). Since the die geometry remains the same throughout the operation, extruded products have a constant cross-section. Depending on the ductility of the material, extrusion may be carried out at room or elevated temperatures. Because a chamber is involved, each billet is extruded individually, and thus extrusion is a batch or semicontinuous process, producing essentially semifinished parts.

Typical products made by extrusion are door and window frames, railings for sliding doors, tubing having various cross-sections, and structural and architectural shapes. Extruded products can be cut into desired lengths, which then become discrete parts, such as door handles, brackets, and gears (Fig. 15.2). Commonly extruded materials are aluminum, copper, steel, magnesium, and lead (lead pipes

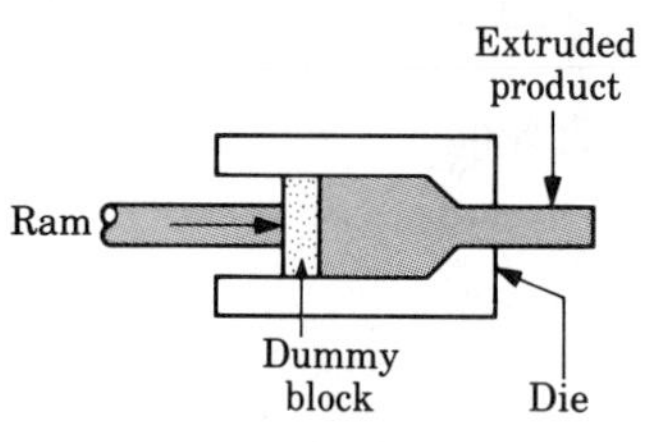

FIGURE 15.1 Schematic illustration of the direct extrusion process.

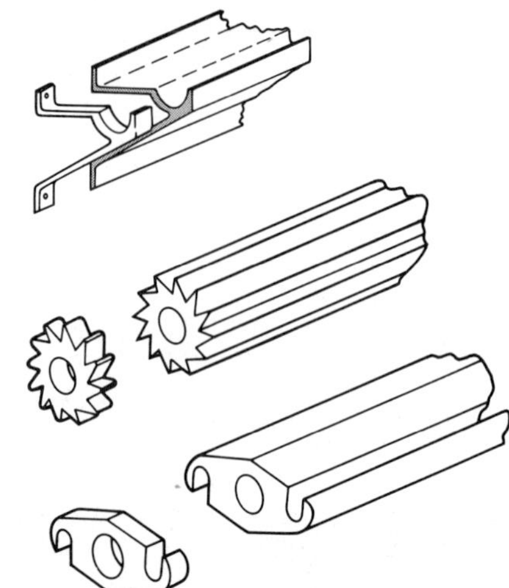

FIGURE 15.2 Extrusions and examples of products made by sectioning off extrusions. *Source:* Kaiser Aluminum.

were made by extrusion in the eighteenth century). Other metals and alloys can be extruded with various levels of difficulty. We describe extrusion of plastics in Section 18.2.

Extrusion may be done by various means and is often combined with forging operations, in which case it is generally known as *cold extrusion.* It has numerous important applications, including components for automobiles, bicycles, and motorcycles, fasteners, heavy machinery, and transportation equipment.

Drawing is an operation in which the cross-section of solid rod, wire, or tubing is reduced or changed in shape by *pulling* it through a die. Drawn rods are used for shafts, spindles, and small pistons and as the raw material for fasteners such as rivets, bolts, and screws. In addition to round rods, various profiles are also drawn. Developed between A.D. 1000 and A.D. 1500, drawing is similar to extrusion. However, in drawing, the material is subjected to a tensile force, whereas in extrusion the billet is under compression. The term *drawing* is also used to refer to making cup-shaped parts by sheet forming operations (Section 16.9).

The distinction between the terms **rod** and **wire** is somewhat arbitrary, rod being relatively larger in cross-section than wire. In industry, wire is generally defined as a rod that has been drawn through a die at least once. Wire drawing involves smaller diameters than rod drawing, with sizes down to 0.01 mm (0.0005 in.) for magnet wire and even smaller for use in very low current fuses. Wire and wire products cover a wide range of applications, such as electrical and electronic wiring, cables, screens, tension-loaded structural members, welding electrodes, springs, paper clips, spokes for bicycle wheels, and string musical instruments.

15.2 The Extrusion Process

In the basic extrusion process, called *direct*, or forward, *extrusion*, a round billet is placed in a *chamber* (or container) and forced through a die opening by a hydraulically driven ram (Fig. 15.1). The die opening may be round, or it may have various other shapes. Other types of extrusion are indirect, hydrostatic, and impact extrusion. In *indirect extrusion* (reverse, inverted, or backward extrusion), the die moves toward the billet (Fig. 15.3a). In *hydrostatic extrusion* (Fig. 15.3b), the billet is smaller in diameter than the chamber, which is filled with a fluid, and the pressure is transmitted to the billet by a ram. Unlike in direct extrusion, there is no friction to overcome along the container walls. Another type of extrusion is *lateral*, or side, *extrusion* (Fig. 15.3c).

As you can see in Fig. 15.4, the geometric variables in extrusion are the die angle, α, and the ratio of the cross-sectional area of the billet to that of the extruded product, A_0/A_f, called the **extrusion ratio.** A parameter describing the shape of the extruded product is the **circumscribing-circle diameter** (CCD), which is the

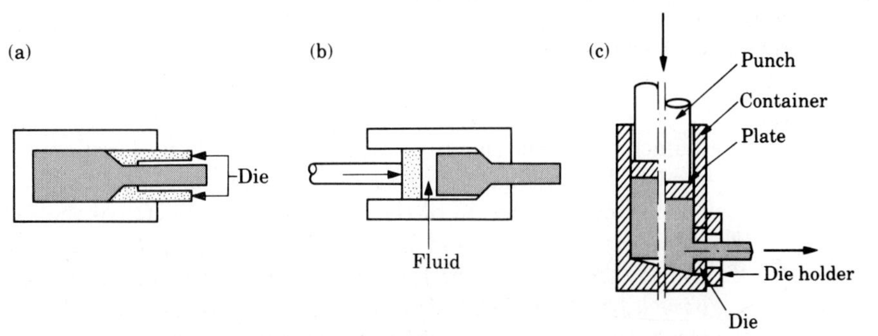

FIGURE 15.3 Types of extrusion: (a) indirect; (b) hydrostatic; and (c) lateral.

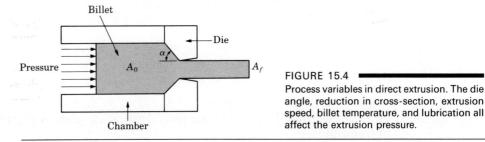

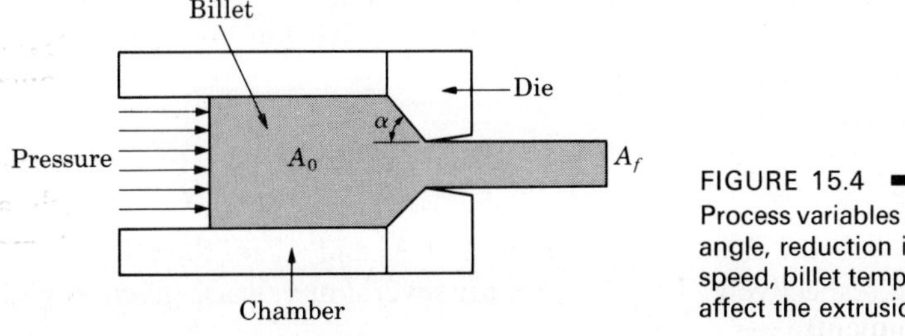

FIGURE 15.4 Process variables in direct extrusion. The die angle, reduction in cross-section, extrusion speed, billet temperature, and lubrication all affect the extrusion pressure.

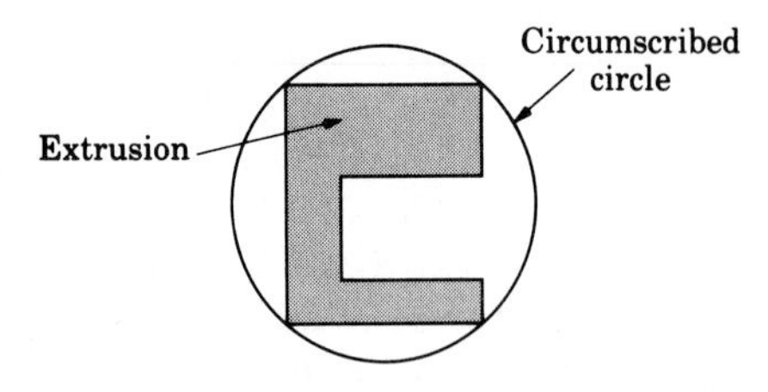

FIGURE 15.5
Method of determining the circumscribing-circle diameter (CCD) of an extruded cross-section.

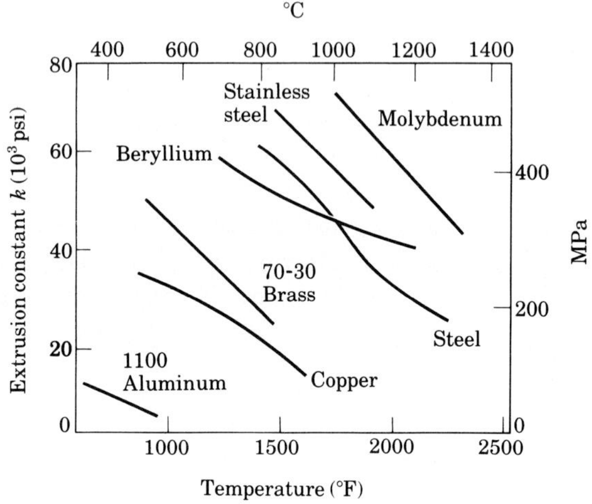

FIGURE 15.6
Extrusion constant k for various metals at different temperatures. *Source:* P. Loewenstein.

diameter of the circle into which the extruded cross-section will fit (Fig. 15.5). Thus the CCD for a square cross-section is its diagonal dimension. The complexity of an extrusion is a function of the ratio of the perimeter of the extruded product to its cross-sectional area, and is known as the *shape factor*. You can see that a solid round extrusion has the lowest shape factor, whereas the parts shown in Fig. 15.2 have high shape factors. Other extrusion-process variables are the temperature of the billet, the speed at which the ram travels, and the type of lubricant used.

15.2.1 Extrusion force

The force required for extrusion depends on the strength of the billet material, the extrusion ratio, friction in the chamber and die, and process variables such as the temperature of the billet and speed of extrusion. We can estimate the *extrusion force F* from the formula

$$F = A_0 k \ln\left(\frac{A_0}{A_f}\right), \tag{15.1}$$

where k is the **extrusion constant**, and A_0 and A_f are the billet and extruded product areas, respectively. The k values for several metals are given in Fig. 15.6 for a range of temperatures.

Example: Calculation of force in hot extrusion.

A round billet made of 70-30 brass is extruded at a temperature of 1250 °F (675 °C). The billet diameter is 5 in. (125 mm), and the diameter of the extrusion is 2 in. (50 mm). Calculate the extrusion force required.

The extrusion force is calculated using Eq. (15.1), in which the extrusion constant k is obtained from Fig. 15.6. For this material, we find that $k = 35{,}000$ psi (250 MPa) at the extrusion temperature. We now have the data to calculate the force. Thus,

$$F = \pi(2.5)^2(35{,}000)\ \ln\left[\pi(2.5)^2/\pi(1.0)^2\right]$$

$$= 1.26 \times 10^6 \text{ lb} = 630 \text{ tons} = 5.5 \text{ MN}.$$

See Section 15.9 for capacities of extrusion presses.

15.2.2 Metal flow in extrusion

The metal flow pattern in extrusion, as in other forming processes, is important because of its influence on the quality and mechanical properties of the final product. The material flows longitudinally much like fluid flow in a channel, so extruded products have an elongated grain structure (preferred orientation). As we describe in Section 15.8, improper metal flow during extrusion can produce various defects.

A common technique for investigating the flow pattern is to section the round billet in half lengthwise and mark one face with a square grid pattern. The two halves are placed in the chamber together and extruded. They are then taken apart and studied. Figure 15.7 shows typical flow patterns obtained by this technique in direct extrusion with *square* dies (90° die angle). The conditions under which these

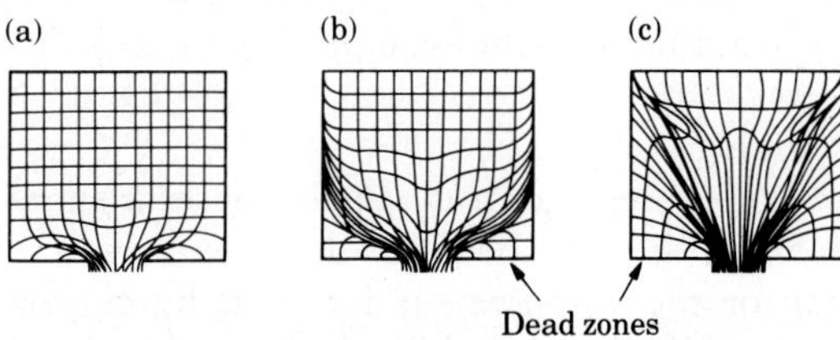

FIGURE 15.7 Types of metal flow in extruding with square dies. (a) Flow pattern obtained with low friction, or in indirect extrusion. (b) Pattern obtained with high friction in the billet–chamber interfaces. (c) Pattern obtained with high friction and/or cooling of the outer regions of the billet in the chamber. This type, observed with metals whose strength increases rapidly with decreasing temperature, leads to a defect known as pipe, or extrusion defect (see Fig. 15.16).

different flow patterns occur are described in the figure caption. Note the **dead-metal zones** in Figs. 15.7(b) and (c), where the metal at the corners is essentially stationary. This situation is similar to stagnation of fluid flow in channels that have sharp angles and turns.

15.3 Extrusion Practice

Because they have sufficient ductility, aluminum, copper, magnesium, and their alloys and steels and stainless steels are extruded with relative ease into numerous shapes. Other metals such as titanium and refractory metals can be extruded but with some difficulty and considerable die wear. Although it is a batch or semicontinuous process, extrusion can be economical for large as well as short production runs. Tool costs are generally low, particularly for producing simple solid cross-sections.

Extrusion ratios usually range from about 10 : 1 to 100 : 1. They may be higher for special applications (400 : 1) or lower for less ductile materials, although they are usually at least 4 : 1 to work the material through the bulk of the workpiece. The circumscribed-circle diameter for aluminum ranges from 6 mm to 1 m (0.25 in. to 40 in.), although most are within 0.25 m (10 in.). Because of the higher forces required, the maximum CCD for steel is usually limited to 0.15 m (6 in.).

Ram speeds may range up to 0.5 m/s (100 ft/min). Generally, slower speeds are preferred for aluminum, magnesium, and copper and higher speeds for steels, titanium, and refractory alloys. Extruded products are usually less than 7.5 m (25 ft) long because of the difficulty in handling greater lengths but have been as long as 30 m (100 ft).

Most extruded products, particularly those with small cross-sections, require straightening and twisting. This is done by stretching the extruded product, usually in a hydraulic stretcher equipped with jaws. Tolerances in extrusion are usually in the range of ± 0.25–2.5 mm (± 0.01–0.1 in.) and increase with increasing cross-section.

The presence of a die angle causes a small portion of the end of the billet to remain in the chamber after the operation has been completed. This piece—called scrap or the *butt end*—is removed by cutting off the extrusion at the die exit. Another billet or a graphite block may be placed in the chamber to extrude the piece remaining from the previous extrusion.

Coaxial extrusion (or *cladding*) is also possible. In this operation, coaxial billets are extruded together when the strength and ductility of the two metals are compatible. An example is copper clad with silver. *Stepped extrusions* are produced by extruding the billet partially in one die, then in one or more larger dies. (See also Cold Extrusion, Section 15.5.)

TABLE 15.1
EXTRUSION TEMPERATURE RANGES FOR VARIOUS METALS

	°C
Lead	200–250
Aluminum and its alloys	375–475
Copper and its alloys	650–975
Steels	875–1300
Refractory alloys	975–2200

15.4 Hot Extrusion

Extrusion is carried out at elevated temperatures for metals and alloys that do not have sufficient ductility at room temperature—or in order to reduce forces required (Table 15.1). As in all other hot working operations, hot extrusion has special requirements because of the high operating temperatures. Die wear can be excessive. Cooling of the hot billet in the cool container can be a problem, resulting in highly nonuniform deformation (Fig. 15.7c). To reduce cooling of the billet and to prolong die life, extrusion dies may be preheated, as is done with forging dies.

Because the billet is hot, it develops an oxide film unless heated in an inert-atmosphere furnace. This film can affect the flow pattern of the material because of its frictional characteristics. It also results in an extruded product that may be unacceptable when good surface finish is important. Lateral extrusion (Fig. 15.3c) is used for sheathing of wire and coating of electric wire with plastic (see Fig. 18.3a). In order to avoid formation of oxide films on the extruded product, the dummy block placed ahead of the ram (Fig. 15.1) is made a little smaller in diameter than the container. After extrusion, a thin cylindrical shell (*skull*), consisting mainly of the oxidized layer, is left in the container and the extruded product is thus free of oxides. This shell is later removed from the chamber.

15.4.1 Die design and materials

Die design (Fig. 15.8) requires considerable experience. Square dies (**shear dies**) are used in extrusion of nonferrous metals, especially aluminum. Square dies develop dead-metal zones, which in turn, form a die geometry (see Fig. 15.7) along which the material flows in the deformation zone. The dead-metal zones produce extrusions with bright finishes.

Tubing is extruded from a solid or hollow billet to wall thicknesses as small as 1 mm (0.040 in.). For solid billets, the ram is fitted with a mandrel that pierces a hole in the billet. Billets with a previously pierced hole may also be extruded in this

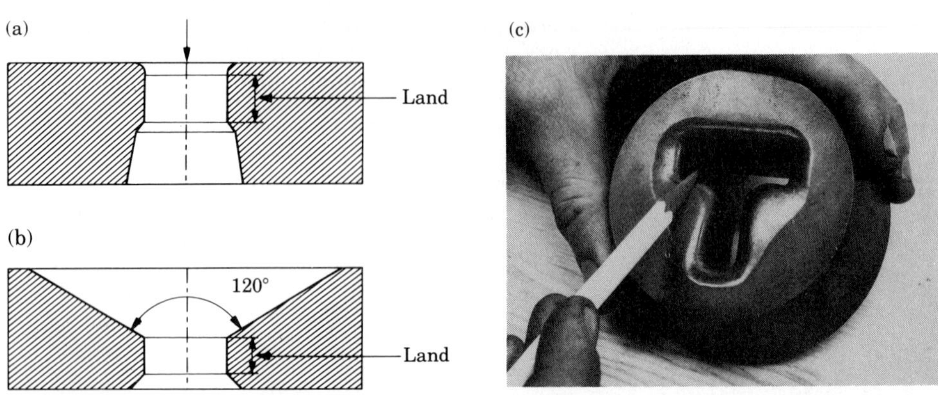

FIGURE 15.8
Typical extrusion-die configurations: (a) die for nonferrous metals; (b) die for ferrous metals; and (c) die for T-shaped extrusion, made of hot-work die steel and used with molten glass as a lubricant. *Source* for (c): Courtesy of LTV Steel Company.

way. Because of friction and severity of deformation, thin-walled extrusions are more difficult to produce than thick-walled extrusions. Wall thickness is usually limited to 1 mm (0.040 in.) for aluminum, 3 mm (0.125 in.) for carbon steels, and 5 mm (0.20 in.) for stainless steels.

Hollow cross-sections (Fig. 15.9a) can be extruded by welding-chamber methods and the use of various special dies known as **spider dies**, **porthole dies**, and **bridge dies** (Fig. 15.9b). The metal divides and flows around the supports for the internal mandrel into strands. These strands are then rewelded under the high pressures in the welding chamber before exiting through the die. This condition is much like air flowing around a moving car, and rejoining downstream. The welding-chamber process is suitable only for aluminum and some of its alloys because of their capacity for developing a strong weld under pressure. Lubricants cannot be used because they prevent rewelding of the metal in the die.

Guidelines for proper die design in extrusion are illustrated in Fig. 15.10. Note the importance of symmetry of cross-section and the avoidance of sharp corners and extreme changes in dimension within the cross-section.

Die materials for hot extrusion are usually hot-work die steels. Coatings such as zirconia may also be applied to the dies to extend die life. Partially stabilized zirconia dies are also being used for hot extrusion.

15.4.2 Lubrication

Lubrication is important in hot extrusion (see Table 13.1 and Chapter 32). Glass is an excellent lubricant for steels, stainless steels, and high-temperature metals and alloys. In a process developed in the 1940s and known as the **Séjournet process**, a circular glass pad is placed at the die entrance. This pad acts as a reservoir of molten glass and supplies it as a lubricant as extrusion progresses. Before the billet is placed

FIGURE 15.9

(a) An extruded 6063-T6 aluminum ladder lock for aluminum extension ladders. This part is 8 mm (5/16 in.) thick and is sawed from the extrusion (see Fig. 15.2). (b) Components of various dies for extruding intricate hollow shapes. *Source* for (b): K. Laue and H. Stenger, *Extrusion—Processes, Machinery, Tooling.* American Society for Metals, Metals Park, Ohio, 1981. Used with permission.

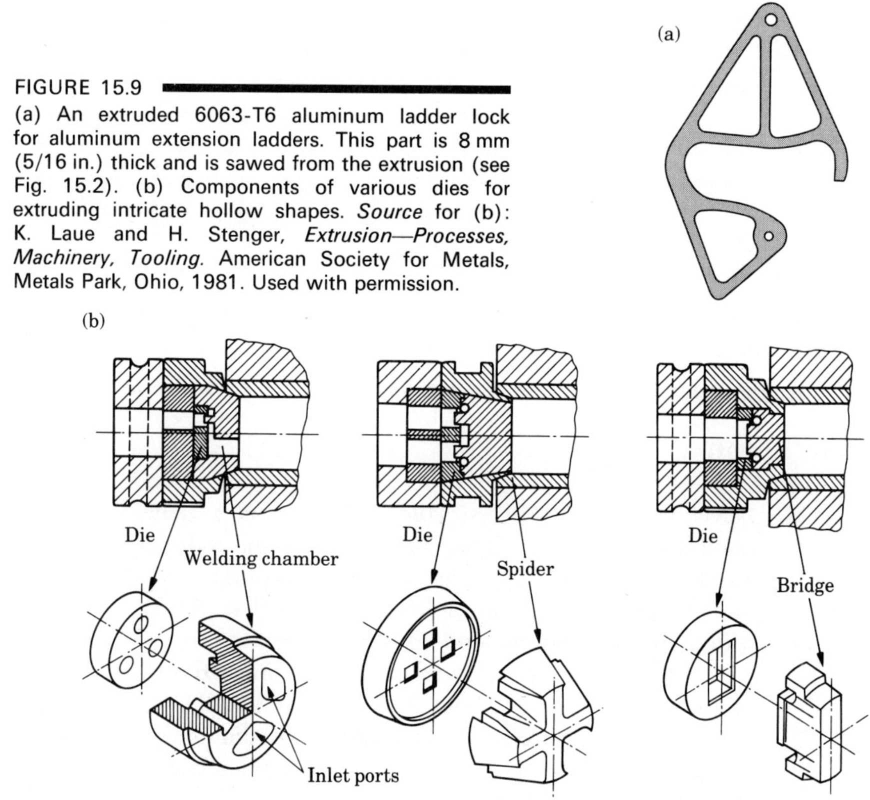

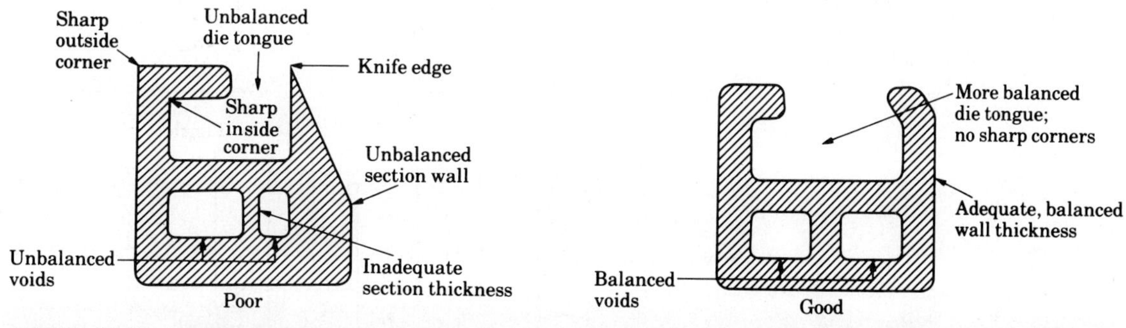

FIGURE 15.10

Poor and good examples of cross-sections to be extruded. Note the importance of eliminating sharp corners and keeping section thicknesses uniform. *Source:* J. G. Bralla (ed.), *Handbook of Product Design for Manufacturing.* New York: McGraw-Hill Publishing Company, 1986. Used with permission.

in the chamber, its cylindrical surface is coated with a layer of powdered glass to reduce friction at the billet–chamber interface.

For metals that have a tendency to stick to the container and the die, the billet can be enclosed in a thin-walled container made of a softer, lower strength metal, such as copper or mild steel. This procedure is called **jacketing** (or *canning*). In addition to acting as a low-friction interface, this jacket prevents contamination of the billet by the environment or, if the billet material is toxic or radioactive, from contaminating the environment. This technique can also be used for extruding metal powders (see Section 17.3).

15.5 Cold Extrusion

Developed in the 1940s, *cold extrusion* is a general term often denoting a combination of operations, such as direct and indirect extrusion and forging (Fig. 15.11). Cold extrusion has gained wide acceptance in industry, particularly for tools and components for automobiles, motorcycles, bicycles, appliances, and transportation and farm equipment.

This process uses slugs cut from cold-finished or hot-rolled bar, wire, or plate. Slugs that are less than about 40 mm (1.5 in.) in diameter are sheared and their ends are squared by grinding or upsetting. Larger diameter slugs are machined from bars in specific lengths. Although most cold-extruded parts weigh less, parts weighing as much as 45 kg (100 lb) and having lengths of up to 2 m (80 in.) have been made. Powder-metal slugs (preforms) are also cold extruded.

Cold extrusion has the following advantages over hot extrusion:

- Improved mechanical properties resulting from work-hardening, provided that the heat generated by deformation and friction does not recrystallize the extruded metal.

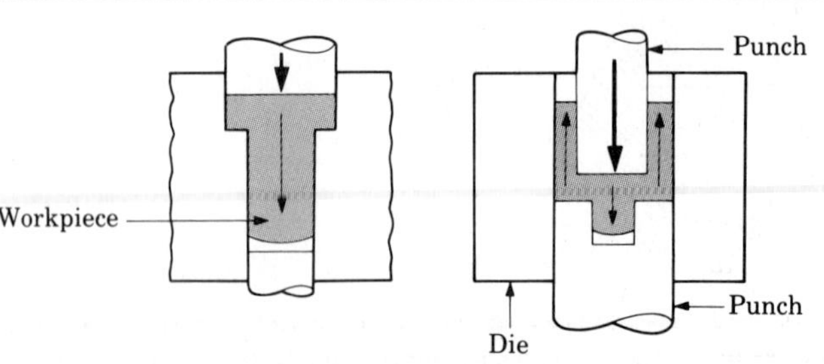

FIGURE 15.11
Two examples of cold extrusion. Thin arrows indicate the direction of metal flow during extrusion.

- Good control of tolerances, thus requiring little subsequent machining or finishing operations.
- Improved surface finish, due partly to lack of an oxide film, provided that lubrication is effective.
- Production rates and costs competitive with those of other methods of producing the same part. Some machines are capable of producing more than 2000 parts per hour.

However, the magnitude of the stresses on the tooling in cold extrusion is very high, especially with steel workpieces, being on the order of the hardness of the workpiece material. The punch hardness usually ranges between 60 and 65 HRC and 58 and 62 HRC for the die. Punches are a critical component, as they must have not only sufficient strength but also toughness and wear and fatigue resistance.

The design of tooling and selection of appropriate tool and die materials is crucial to the success in cold extrusion (see Table 5.9). Also important is the control of workpiece material with regard to its quality, accuracy of slug dimensions, and surface condition. Lubrication is critical, especially with steels, because of the possibility of sticking (*seizure*) between the workpiece and the tooling if the lubrication breaks down. The most effective means of lubrication is application of phosphate *conversion coatings* on the workpiece, followed by a coating of soap or wax (Section 32.12).

• Example: Cold-extruded part.

The following is a typical cold-extruded product, similar to the metal component of an automotive spark plug.

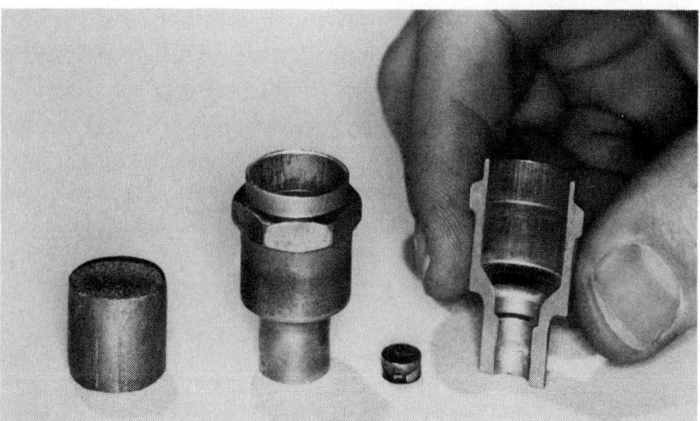

First a slug is sheared off the end of a round rod (left). The slug is then cold extruded (middle) in an operation similar to those shown in Fig. 15.11 but with a blind hole. The material at the bottom of the blind hole is then punched out, producing the small slug shown. Note the diameters of the slug and the hole at the bottom of the sectioned part, respectively. Studying material flow during deformation is important. Defects can be spotted, and improvements in punch and die design can be identified. The part is usually sectioned in the mid plane, then polished and etched to show the grain flow, as follows. *Source for illustrations*: National Machinery Company.

Example: Cold-extruded bevel-gear shaft.

The sequence of operations involved in making a bevel-gear shaft by cold extrusion is shown in the accompanying figure. After shearing or cutting off an appropriate length of round stock, the billet is partially extruded to form the small lower end of the shaft. The upper end is then upset to preform it for the next operation, which is piercing the part from the top, whereby the part is backward extruded and a cavity is produced. The final operation is upsetting of the upper section, making it larger in diameter and forming the bevel section. Gear teeth can be produced either by additional forming operations or by machining.

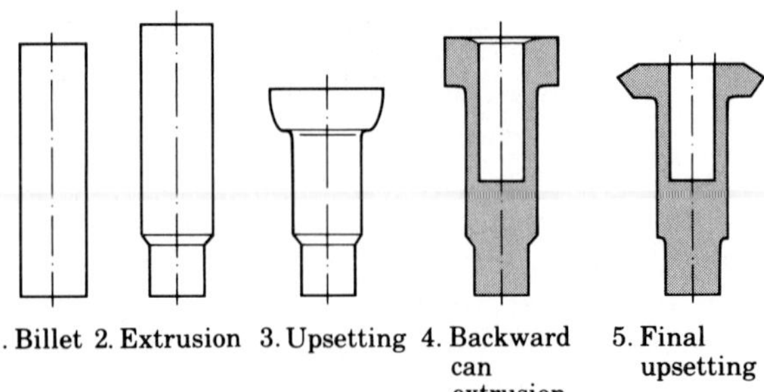

15.6 Impact Extrusion

Impact extrusion is similar to indirect extrusion and is often included in the cold-extrusion category. The punch descends rapidly on the blank (slug), which is extruded backward (Fig. 15.12). The thickness of the tubular extruded section is a function of the clearance between the punch and the die cavity. Typical products made by impact extrusion are shown in Fig. 15.13. The diameter of the parts made can approach 150 mm (6 in.). Another example is the production of collapsible tubes, such as for toothpaste (Fig. 15.14). Most nonferrous metals can be impact extruded using vertical presses at production rates as high as two parts per second.

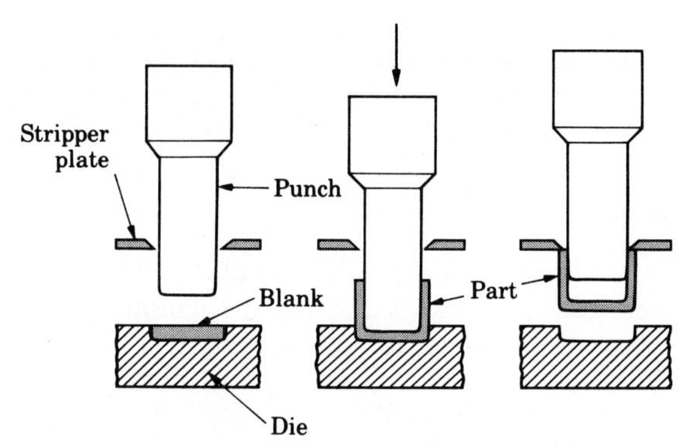

FIGURE 15.12 Schematic illustration of the impact-extrusion process. The extruded parts are stripped using a stripper plate, as they tend to stick to the punch.

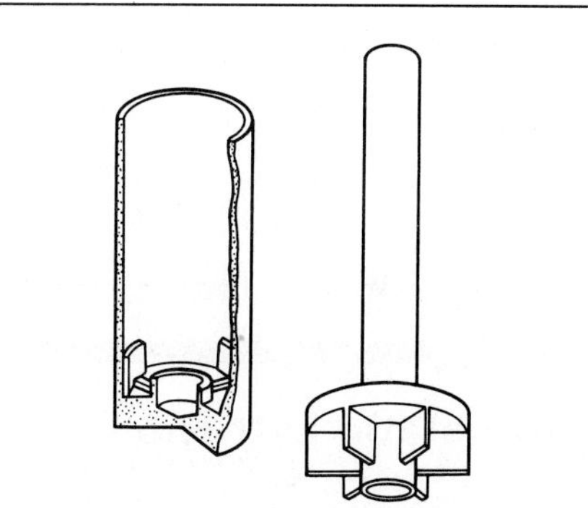

FIGURE 15.13 Two examples of products made by impact extrusion. These parts may also be made by casting, forging, and machining, depending on the dimensions and materials involved and the properties desired. Economic considerations are also important in final process selection.

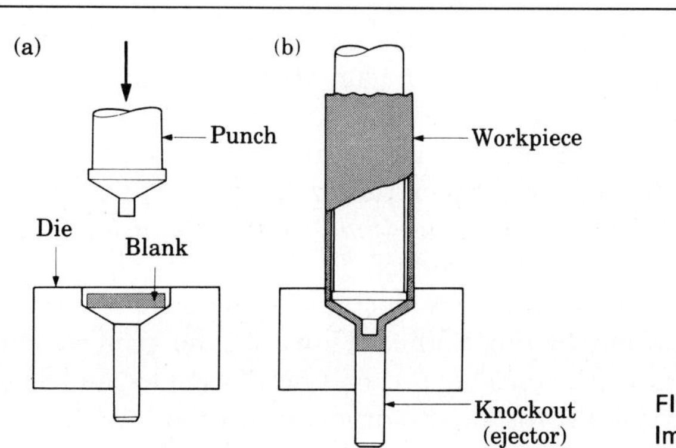

FIGURE 15.14 Impact extrusion of a collapsible tube (Hooker process).

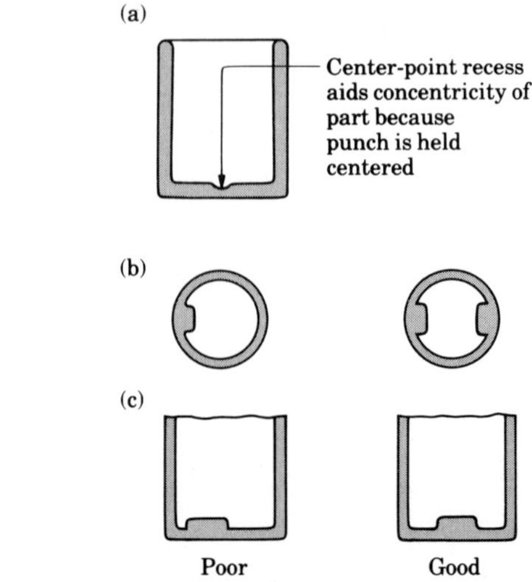

FIGURE 15.15
Design guidelines for impact extrusion. Note the importance of symmetry to keep material flow uniform. *Source:* J. L. Everhart, *Impact and Cold Extrusion of Metals.* Reprinted by permission of Chemical Publishing Co., Inc., 1964.

The impact-extrusion process can produce thin-walled tubular sections, with ratios of thickness to diameter as small as 0.005. Consequently, the symmetry of the part and concentricity of the punch and the blank are important (Fig. 15.15).

15.7 Hydrostatic Extrusion

In *hydrostatic extrusion* the pressure required for extrusion is supplied through a fluid medium surrounding the billet (Fig. 15.3b). Consequently, there is no container-wall friction. Pressures are usually about 1400 MPa (200 ksi). The high pressure in the chamber transmits some of the fluid to the die surfaces, thus significantly reducing friction and forces. Hydrostatic extrusion, which was developed in the early 1950s, has been improved by extruding the part into a second pressurized chamber, which is under lower pressure (*fluid-to-fluid extrusion*). This operation reduces the defects in the extruded product.

Because the hydrostatic pressure increases the ductility of the material, brittle materials can be extruded successfully by this method. However, the main reasons for this success appear to be low friction and use of low die angles and high extrusion ratios. Most commercial hydrostatic-extrusion operations use ductile materials. However, a variety of metals and polymers, solid shapes, tubes and other hollow shapes, as well as honeycomb and clad profiles, have been extruded successfully.

Hydrostatic extrusion is usually carried out at room temperature, typically using vegetable oils as the fluid, particularly castor oil because it is a good lubricant and its viscosity is not influenced significantly by pressure. For elevated-temperature extrusion, waxes, polymers, and glass are used as the fluid. These materials also serve as thermal insulators and help maintain the billet temperature during extrusion.

In spite of the success obtained, hydrostatic extrusion has had limited industrial applications, largely because of the somewhat complex nature of tooling, the experience required with high pressures and design of specialized equipment, and the long cycle times required.

15.8 Extrusion Defects

Depending on material condition and process variables, several types of defects can develop in extruded products, which can significantly affect their strength and product quality. There are three principal extrusion defects: surface cracking, pipe, and internal cracking. Some defects are visible to the naked eye; others can be detected only by the techniques we describe in Chapter 36.

15.8.1 Surface cracking

If extrusion temperature, friction, or speed is too high, surface temperatures rise significantly, which may cause surface cracking and tearing (**fir-tree cracking** or **speed cracking**). These cracks are intergranular (along the grain boundaries) and are usually caused by *hot shortness.* These defects occur especially in aluminum, magnesium, and zinc alloys, although they may also occur in high-temperature alloys. This situation can be avoided by lowering the billet temperature and extrusion speed.

Surface cracking may also occur at lower temperatures. These cracks have been attributed to periodic sticking of the extruded product along the die land. When the product being extruded sticks to the die land, the extrusion pressure increases rapidly. Shortly thereafter the product moves forward again and the pressure is released. The cycle is then repeated continuously, producing periodic circumferential cracks on the surface. Because of its similarity to the surface of a bamboo stem, it is known as **bamboo defect**.

15.8.2 Pipe

The type of metal-flow pattern shown in Fig. 15.7(c) tends to draw surface oxides and impurities toward the center of the billet, much like a funnel (Fig. 15.16). This

FIGURE 15.16 Tailpipe (extrusion defect) in an extruded refractory-metal alloy, showing the tail end of the product. The type of metal flow shown in Fig. 15.7(c) usually causes this type of defect. *Source:* V. DePierre.

defect is known as **pipe defect** (also *tailpipe* or *fishtailing*). As much as one third of the length of the extruded product may contain this type of defect and have to be cut off as scrap. Piping can be minimized by modifying the flow pattern to a more uniform one, such as by controlling friction and minimizing temperature gradients. Another method is to machine the billet's surface prior to extrusion so that scale and surface impurities are removed.

15.8.3 Internal cracking

The center of the extruded product can develop cracks, which are variously called **center cracking**, *center-burst*, *arrowhead fracture*, or *chevron cracking* (Fig. 15.17a). These cracks are attributed to a state of hydrostatic tensile stress at the centerline in the deformation zone in the die (Fig. 15.17b), a situation similar to the necked region in a tensile-test specimen (see Fig. 2.26). The tendency for center

(a)

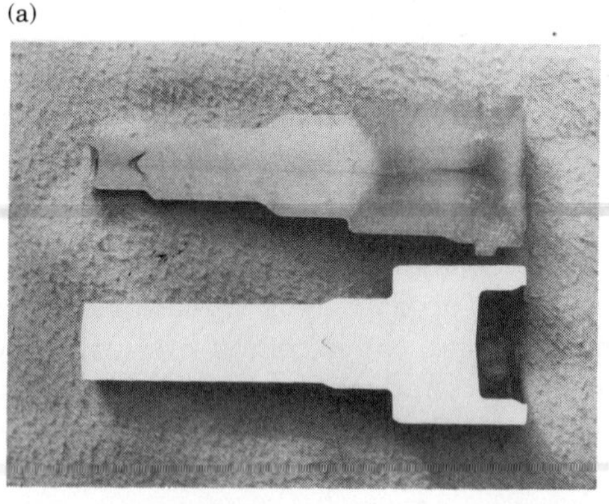

(b)

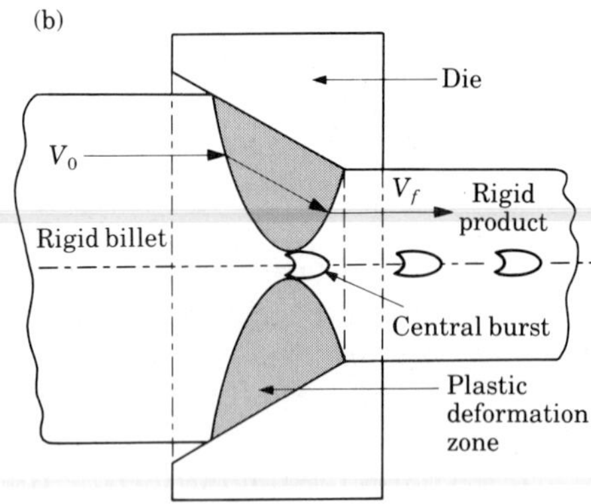

FIGURE 15.17 (a) Chevron cracking (central burst) in extruded round steel bars. Unless inspected, such internal defects may remain undetected and possibly cause failure of the part in service. These defects can also develop in drawing rod, wire, and tubes. (b) Schematic illustration of rigid and plastic zones in extrusion. The tendency for chevron cracking increases if the two plastic zones do not meet. Note that the plastic zone can be made larger by decreasing the die angle and/or increasing the reduction in cross-section. *Source:* B. Avitzur.

cracking increases with increasing die angle and amount of impurities and decreases with increasing extrusion ratio and friction. These cracks have also been observed in tube extrusion and in spinning of tubes (Section 16.11), appearing on the inside surfaces of tubes for the same reasons.

15.9 Extrusion Equipment

The basic equipment for extrusion is a hydraulic press (Fig. 15.18). These presses are suitable for extrusion because the stroke and speed of the operation can be controlled. They are capable of applying a constant force over a long stroke; thus long billets can be used and the production rate increased. Hydraulic presses with a ram-force capacity as high as 120 MN (14,000 tons) have been built and are used for hot extrusion of large billets.

Vertical hydraulic presses are generally used for cold extrusion. They generally have less capacity than those used for hot extrusion, but take up less floor space. In addition to presses, crankjoint and knucklejoint mechanical presses are also used for cold extrusion and impact extrusion to mass produce small components. Multistage operations, where the cross-sectional area is reduced in a number of operations, are also carried out on specially designed presses.

FIGURE 15.18
General view of a 9-MN (1000-ton) hydraulic-extrusion press. *Source:* Courtesy of Jones & Laughlin Steel Corporation.

15.10 The Drawing Process

In drawing, the cross-section typically of a round rod or wire is reduced or changed by pulling it through a die (Fig. 15.19). The major variables in drawing are similar to those in extrusion, that is, reduction in cross-sectional area, die angle, friction along the die–workpiece interfaces, and speed. The die angle influences the drawing force and the quality of the drawn product. We can show that for a certain reduction in diameter and frictional condition, there is one die angle at which the drawing force is a minimum. However, this does not mean that the process should be carried out at this optimum angle because, as you will see, there are other product-quality considerations.

Because more work has to be done to overcome friction and to reduce the diameter, the force increases with increasing friction and reduction of cross-sectional area. However, there has to be a limit to the magnitude of the drawing force as reduction increases, because when the tensile stress due to the drawing force reaches the yield stress of the drawn metal, it will simply yield. In that case, the product will undergo further deformation after it leaves the die, which is not acceptable. Ideally, the maximum reduction in cross-sectional area per pass is 63 percent. Thus, for example, a 10-mm diameter rod can at most be reduced to a diameter of 6.1 mm in one pass.

15.10.1 Drawing of other shapes

As in extrusion, various solid cross-sections can be produced by drawing through dies with various profiles (Fig. 15.20). The initial cross-section is usually round or square. Proper die design and the selection of reduction sequence per pass require considerable experience to ensure proper material flow in the die in order to reduce defects and improve surface quality.

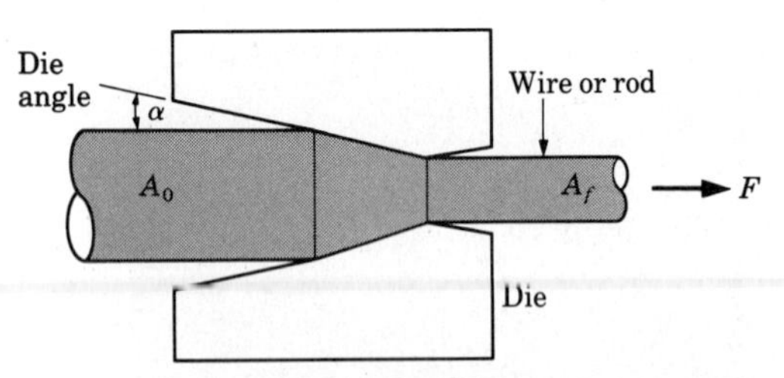

FIGURE 15.19 Wire-drawing process variables. The die angle, reduction in cross-sectional area per pass, speed of drawing, temperature, and lubrication all affect the drawing force F. (Compare with Fig. 15.4.)

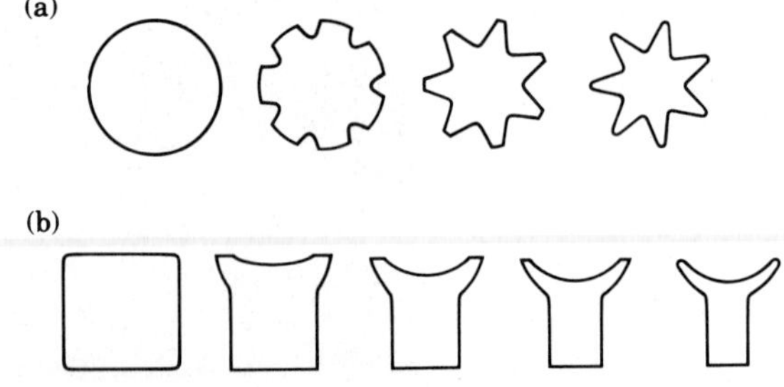

FIGURE 15.20 Examples of solid cross-sections that can be produced by drawing in several stages.

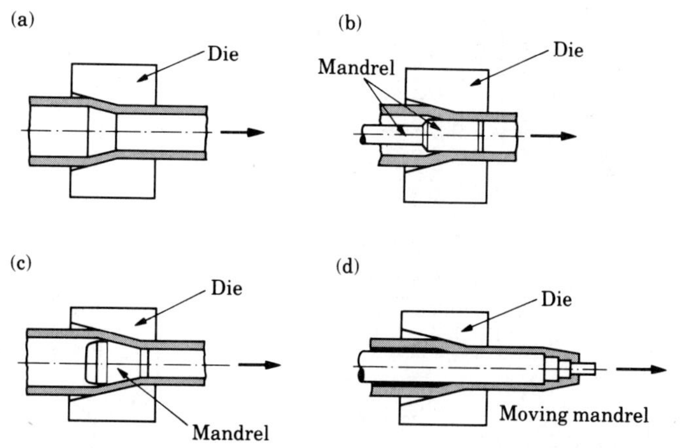

FIGURE 15.21
Examples of tube-drawing operations, with or without an internal mandrel. Note that a variety of diameters and wall thicknesses can be produced from the same initial tube stock.

The wall thickness, diameter, or shape of tubes produced by extrusion or other processes can be reduced by tube drawing processes (Fig. 15.21). Tubes as large as 0.3 m (12 in.) in diameter can be drawn by these techniques. Mandrels of various profiles are available for these operations.

In drawing flat strips, the dies are wedge shaped. This process, although not of major industrial significance, is the fundamental deformation process in *ironing*, which is used extensively in making aluminum beverage cans (see Section 16.9).

15.11 Drawing Practice

As in all metalworking processes, successful drawing operations require careful selection of process parameters and consideration of many factors. The procedure and sequence used to make wire are shown schematically in Fig. 15.22.

Drawing speeds depend on the material and cross-sectional area. They may range from 1 m/s to 2.5 m/s (200 ft/min to 500 ft/min) for heavy sections to as much as 50 m/s (10,000 ft/min) for very fine wire, such as that used for electromagnets. Because it does not have sufficient time to dissipate, temperature can rise substantially at high drawing speeds and can have detrimental effects on product quality.

Reductions in cross-sectional area per pass range from near zero to about 45 percent. Usually, the smaller the cross-section is to begin with, the smaller will be the reduction per pass. Fine wires are usually drawn at 15–25 percent reduction per pass, and larger sizes at 20–45 percent. Reductions of more than 45 percent may

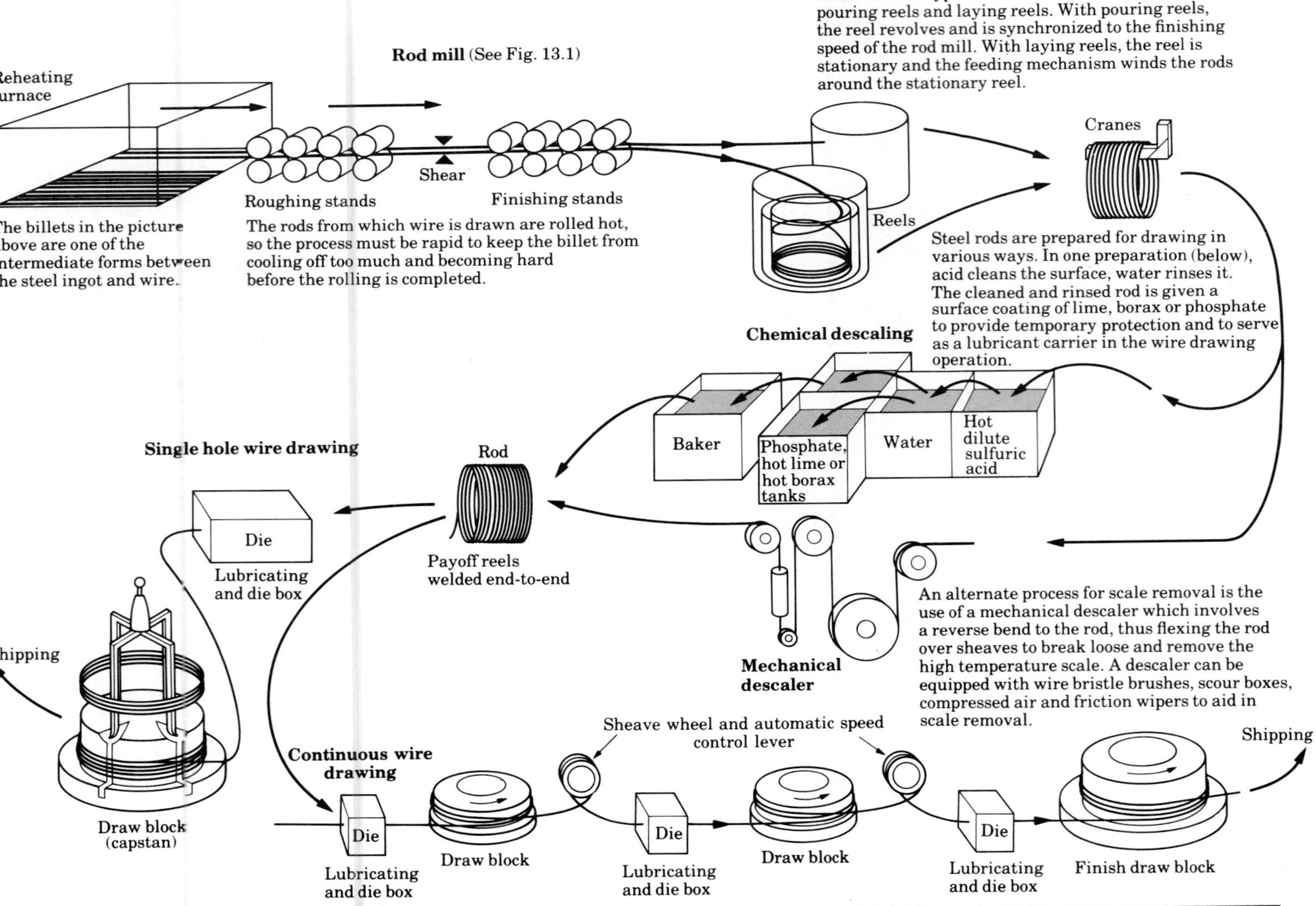
Rod mill (See Fig. 13.1)
Reheating furnace
Roughing stands
Shear
Finishing stands
The billets in the picture above are one of the intermediate forms between the steel ingot and wire.
The rods from which wire is drawn are rolled hot, so the process must be rapid to keep the billet from cooling off too much and becoming hard before the rolling is completed.
There are two types of reels for receiving finished rods: pouring reels and laying reels. With pouring reels, the reel revolves and is synchronized to the finishing speed of the rod mill. With laying reels, the reel is stationary and the feeding mechanism winds the rods around the stationary reel.
Reels
Cranes
Steel rods are prepared for drawing in various ways. In one preparation (below), acid cleans the surface, water rinses it. The cleaned and rinsed rod is given a surface coating of lime, borax or phosphate to provide temporary protection and to serve as a lubricant carrier in the wire drawing operation.
Chemical descaling
Baker
Phosphate, hot lime or hot borax tanks
Water
Hot dilute sulfuric acid
Single hole wire drawing
Rod
Payoff reels welded end-to-end
Die
Lubricating and die box
Mechanical descaler
An alternate process for scale removal is the use of a mechanical descaler which involves a reverse bend to the rod, thus flexing the rod over sheaves to break loose and remove the high temperature scale. A descaler can be equipped with wire bristle brushes, scour boxes, compressed air and friction wipers to aid in scale removal.
Shipping
Draw block (capstan)
Continuous wire drawing
Sheave wheel and automatic speed control lever
Die
Lubricating and die box
Draw block
Die
Lubricating and die box
Draw block
Die
Lubricating and die box
Finish draw block
Shipping

FIGURE 15.22
Schematic diagram of the procedures involved in drawing steel wire from straight round rods. *Source:* American Iron and Steel Institute.

result in lubrication breakdown and surface-finish deterioration. Drawing large solid or hollow sections can be done at elevated temperatures.

A light reduction—called a **sizing pass**—may also be taken on rods to improve surface finish and dimensional accuracy. However, because they basically deform the surface layers, light reductions usually produce highly nonuniform deformation of the material and its microstructure. Consequently, the properties of the material vary with location in the cross-section.

Because of work hardening, intermediate annealing between passes may be necessary to maintain sufficient ductility during cold drawing. Drawn copper and brass wires are designated by their temper, such as 1/4 hard, 1/2 hard, etc.

High-carbon steel wires for springs and musical instruments are made by heat treating, or **patenting** the drawn wire, whereby the microstructure obtained is fine pearlite. These wires have ultimate tensile strengths as high as 5 GPa (700 ksi) and tensile reduction of area of about 20 percent.

15.11.1 Die design

The characteristic features of a typical die design for drawing are shown in Fig. 15.23. Die angles usually range from 6° to 15°. Note, however, that there are two angles (entering and approach) in a typical die. This shape is arrived at through experience. The purpose of the land is to set the final diameter of the product, called *sizing*. Also, when the worn die is reground, the land maintains the exit dimension of the die opening.

A set of dies is required for profile drawing for various stages of deformation (see Fig. 15.20). Designing these dies requires considerable experience in order to produce defect-free products. The dies may be made in one piece or, depending on the complexity of the profile, with several segments held together in a ring. Computer-aided design techniques are being implemented to design dies that smooth material flow through the dies and minimize defects.

A set of idling rolls is also used in drawing rods or bars of various shapes (Fig. 15.24). This arrangement (*Turk's head*) is more versatile than that of ordinary

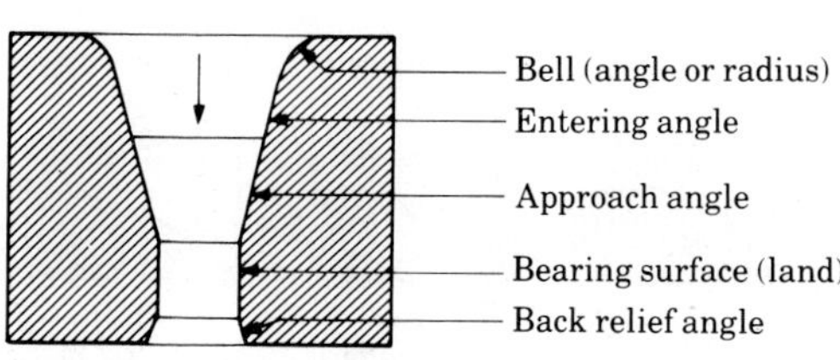

FIGURE 15.23 Terminology for a typical die for drawing round rod or wire.

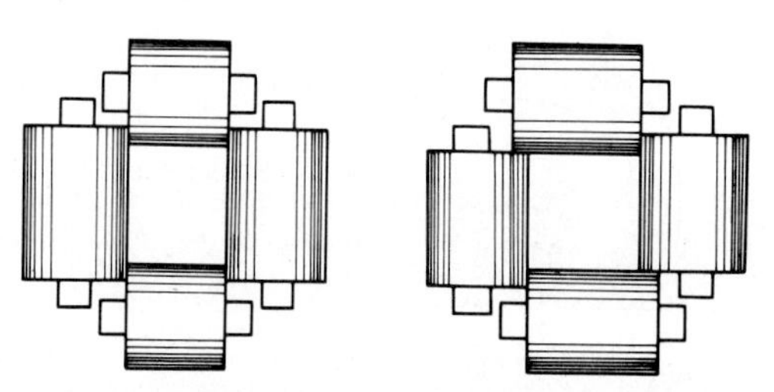

FIGURE 15.24 Arrangement of rolls in a Turk's head. Note that bars with different cross-sections can be produced by repositioning the rolls.

drawing dies, since the rolls can be adjusted to different positions for different products. The rolls can be shaped for drawing various profiles.

15.11.2 Die materials

Die materials for drawing are usually tool steels and carbides—and diamond for fine wire (Table 5.9). For improved wear resistance, steel dies may be chromium plated, and carbide dies may be coated with titanium nitride. Mandrels for tube drawing are usually made of hardened tool steels or carbides.

Diamond dies are used for drawing fine wire with diameters ranging from 2 μm to 1.5 mm (0.0001 in. to 0.06 in.). They may be made from a single-crystal diamond or in polycrystalline form, with diamond particles in a metal matrix (compacts). Because of their cost and lack of tensile strength and toughness, carbide and diamond dies are used as inserts or nibs, which are supported in a steel casing (Fig. 15.25). For hot drawing, cast-steel dies are the best because of their high resistance to wear. The wear of drawing dies after drawing certain lengths of wire is shown schematically in Fig. 15.26. Note the increase in exit diameter, which increases the diameter of the drawn wire. Such a worn die is reground for a larger diameter.

15.11.3 Lubrication

Proper lubrication is essential in rod, tube, and wire drawing in order to improve die life, reduce drawing forces and temperature, and improve the surface finish of the drawn product. In tube drawing, lubrication is critical because of the difficulty in maintaining a sufficiently thick lubricant film at the mandrel–tube interface. The following are the basic types of lubrication used:

- *Wet drawing*, in which the dies and the rod are completely immersed in the lubricant, typically oils and emulsions containing fatty or chlorinated additives and various chemical compounds (see Table 13.1).
- *Dry drawing*, in which the surface of the rod to be drawn is coated with a lubricant such as soap by passing it through a box filled with the lubricant (*stuffing box*).

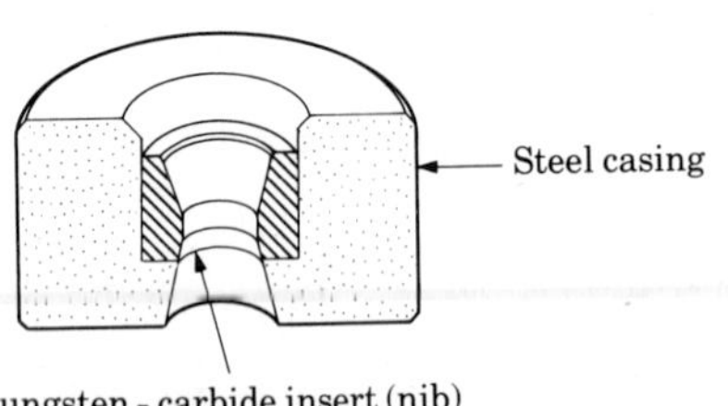

FIGURE 15.25
Tungsten-carbide die insert in a steel casing. Diamond dies, used in drawing thin wire, are encased in a similar manner.

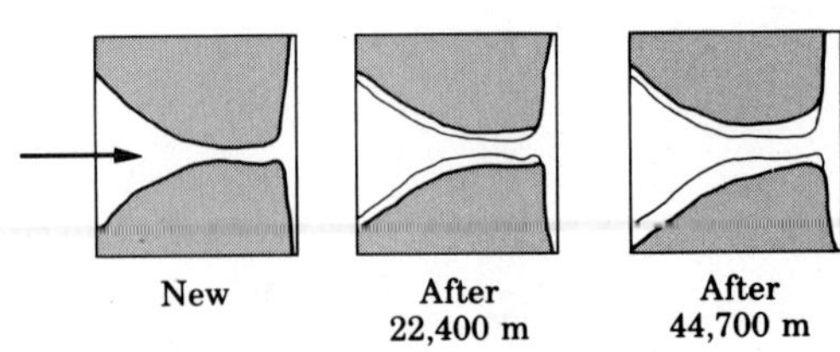

FIGURE 15.26
Development of wear pattern in a draw die, after drawing wire to the lengths indicated. *Source:* J. G. Wistreich.

- *Coating* the rod or wire with a soft metal that acts as a solid lubricant. Copper or tin, for example, is chemically deposited on the surface of the metal.
- *Ultrasonic vibration* of the dies and mandrels, which has been utilized successfully to reduce friction in drawing solid or hollow sections. When done properly, vibrations reduce forces, improve surface finish and die life, and allow larger reductions per pass without failure.

15.12 Defects and Residual Stresses

Defects in drawn rod and wire are similar to those observed in extrusion, especially center cracking (see Fig. 15.17). An additional type of defect in drawing is **seams**, which are longitudinal scratches or folds in the material. Seams may open up during subsequent forming operations, such as upsetting, heading, thread rolling, or bending of the rod or wire, and cause serious quality-control problems in production. Various other surface defects, such as scratches and die marks, can also result from improper selection of process parameters, poor lubrication, or die condition.

Because they undergo nonuniform deformation, residual stresses are usually present in cold-drawn products. For light reductions of a few percent, the longitudinal surface residual stresses are compressive, while the middle is in tension, thus improving fatigue life. Heavier reductions induce tensile surface stresses, while the middle is in compression. Residual stresses can be significant in stress-corrosion cracking of the part over a period of time. Moreover, they cause the component to warp if a layer of material is subsequently removed, such as by slitting, machining, or grinding (see Fig. 2.36).

Rods and tubes that may not be sufficiently straight, or are supplied as coil, can be straightened by passing them through an arrangement of rolls placed at different axes (Fig. 15.27). The rolls subject the product to a series of bending and unbending operations, a process similar to roller leveling (see Fig. 13.8).

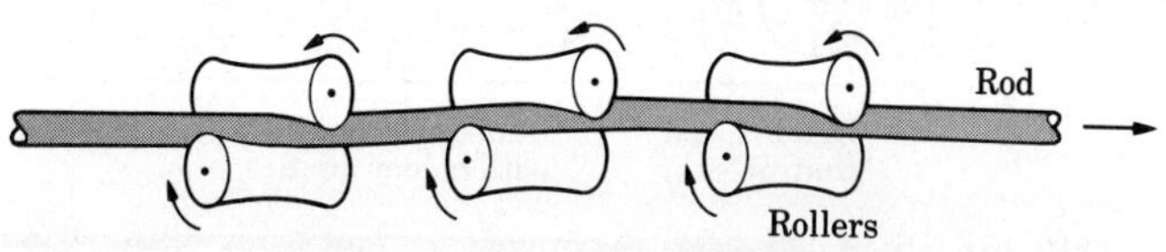

FIGURE 15.27
Schematic illustration of roll straightening drawn round rods (see also Fig. 13.8).

15.13

Drawing Equipment

Although there are several designs, the equipment for drawing is basically of two types: draw bench and bull block. A **draw bench** contains a single die and is similar to a long horizontal tension-testing machine (Fig. 15.28). The pulling force is

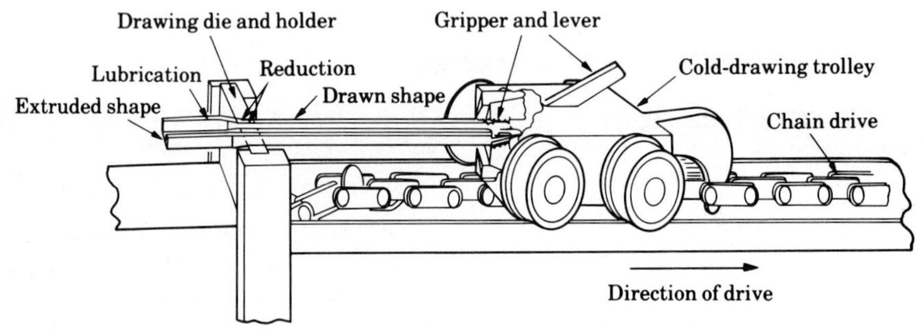

FIGURE 15.28
Cold drawing an extruded channel on a draw bench to reduce its cross-section. Individual lengths of straight rods or cross-sections are drawn by this method. *Source:* Courtesy of The Babcock and Wilcox Company, Tubular Products Division.

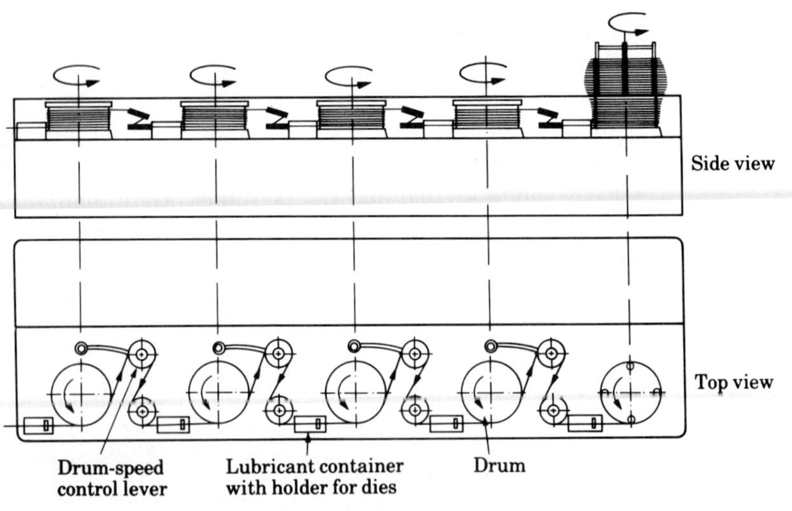

FIGURE 15.29
Two views of a multistage wire-drawing machine, used typically in making copper wire for electrical wiring. *Source:* H. Auerswald.

supplied by a chain drive or is activated hydraulically. Draw benches are used for single-length drawing of straight rods and tubes with diameters larger than 20 mm (0.75 in.). Lengths may be up to 30 m (100 ft). Machine capacities reach 1.3 MN (300 klb) of pulling force, with a speed range of 10 m/s to 100 m/s (20 ft/min to 200 ft/min).

Very long rod and wire (many kilometers) and smaller cross-sections, usually less than 13 mm (0.5 in.), are drawn by a rotating *drum* (**bull block, capstan,** Fig. 15.29). The tension in this setup provides the force required for drawing the wire, usually through multiple dies.

SUMMARY

The extrusion process is capable of producing lengths of solid and hollow sections with constant cross-sectional area. Important factors in successful extrusion are die design, extrusion ratio, lubrication, billet temperature, and extrusion speed. Cold extrusion, which is a combination of various extrusion and forging operations, is capable of producing parts economically and with good mechanical properties.

Rod, wire, and tube have numerous important applications. These products are made basically by a drawing process in which the material is pulled through one or more dies. Although the cross-sections of most drawn products are round, rectangular and other shapes can be drawn. Drawing tubular products usually requires internal mandrels. Proper die design and selection of materials and lubricants are essential to obtaining a product with good quality and surface finish, dimensional accuracy, and strength.

TRENDS

- Computer-aided die design and manufacturing are being implemented to improve material flow in extrusion and drawing and to reduce defects.
- Ceramic dies are being used in high-temperature extrusion of small cross-sections.
- Improvements are taking place in die materials and coatings to extend die life.

KEY TERMS

Bamboo defect	Extrusion	Rod
Bridge die	Extrusion constant	Seam
Bull block	Extrusion defect	Séjournet process
Capstan	Extrusion ratio	Shear die
Center cracking	Fir-tree cracking	Sizing pass
Circumscribing-circle diameter	Jacketing	Speed cracking
Dead-metal zone	Patenting	Spider die
Draw bench	Pipe defect	Wire
Drawing	Porthole die	

BIBLIOGRAPHY

Extrusion

Alexander, J.M., and B. Lengyel, *Hydrostatic Extrusion*. London: Mills and Boon, 1971.

Blazynski, T.Z., *Plasticity and Modern Metal-forming Technology*. New York: Elsevier, 1989.

Everhart, J.E., *Impact and Cold Extrusion of Metals*. New York: Chemical Publishing Company, 1964.

Inoue, N., and M. Nishihara (eds.), *Hydrostatic Extrusion: Theory and Applications*. New York: Elsevier, 1985.

Lange, K. (ed.), *Handbook of Metal Forming*. New York: McGraw-Hill, 1985.

Laue, K., and H. Stenger, *Extrusion—Processes, Machinery, Tooling*. Metals Park, Ohio: American Society for Metals, 1981.

Metals Handbook, 9th ed., Vol. 14: *Forming and Forging*. Metals Park, Ohio: ASM International, 1988.

Michaeli, W., *Extrusion Dies*. New York: Macmillan, 1984.

Pearson, C.E., and R.N. Parkins, *The Extrusion of Metals*, 2d ed. New York: Wiley, 1961.

Source Book on Cold Forming. Metals Park, Ohio: American Society for Metals, 1975.

Drawing

Bernhoeft, C.P., *The Fundamentals of Wire Drawing*. London: The Wire Industry Ltd., 1962.

Developments in the Drawing of Metals, Book no. 301. London: The Metals Society, 1983.

Lange, K. (ed.), *Handbook of Metal Forming*. New York: McGraw-Hill, 1985.

Nonferrous Wire Handbook, 2 vols. Branford, Conn.: The Wire Association International, Inc., 1977 and 1981.

Steel Wire Handbook, vol. 1, 1968; vol. 2, 1969; vol. 3, 1972; vol. 4, 1980. Guilford, Conn.: Wire Association International.

REVIEW QUESTIONS

15.1 How does extrusion vary from rolling and forging?
15.2 What is the difference between extrusion and drawing?
15.3 Describe the difference between rod and wire.
15.4 Define extrusion ratio.
15.5 Define circumscribing-circle diameter (CCD).
15.6 What is shape factor? Why is it important?
15.7 Describe the types of metal flow that occur in extrusion. Why are they important?
15.8 What is a dead-metal zone?
15.9 Define (a) cladding, (b) dummy block, (c) shear dies, (d) skull, and (e) canning.
15.10 What is the principle of welding-chamber dies?
15.11 Why is glass a good lubricant in hot extrusion?
15.12 Explain why cold extrusion has become an important manufacturing process.
15.13 What types of defects may occur in extrusion and drawing?
15.14 Name the important process variables in drawing rod and wire.
15.15 Describe the difference between wet and dry drawing.

QUESTIONS AND PROBLEMS

15.16 Describe the advantages and limitations of direct, indirect, and hydrostatic extrusion.
15.17 Explain why extrusion is a batch or semicontinuous process. Do you think it can be made into a continuous process? Explain.
15.18 Explain the different ways that changing the die angle affects the extrusion process.
15.19 Assume that you are the technical director of an association of extruders and extruding machines. Prepare a leaflet for potential customers stating all the advantages of extrusion.
15.20 Since glass is a good lubricant in hot extrusion, would you use glass for impression-die forging also? Explain.
15.21 Extrusion ratio, speed, and temperature all affect the extrusion force. Explain why they do.
15.22 How would you go about avoiding centerburst defects in extrusion? Explain why your methods would be effective.
15.23 Assume that you are reducing the diameter of two rods, one by simple tension and the other by indirect extrusion. Which method will require more force? Why?
15.24 How would you go about making an extrusion that has increasingly larger cross-section along its length? Is it possible? Would your process be economical and suitable for high production rates?
15.25 Describe your observations concerning Fig. 15.10.
15.26 What do you think will happen if you skip some of the steps in making the parts shown in Fig. 15.20?
15.27 Table 15.1 gives temperature ranges for extruding various metals. Describe the consequence of extruding at a temperature (a) below and (b) above these ranges.
15.28 List the products that you could make based on the examples shown in Fig. 15.2.
15.29 Will the force in direct extrusion vary as the billet gets shorter and shorter? If so, why?

15.30 Calculate the force required in extruding copper at 800 °C if the billet diameter is 100 mm and the extruded part is 25 mm in diameter.

15.31 Assume that the Summary to this chapter is missing. Write a one-page summary to highlight the wire-drawing process.

15.32 What changes would you expect in the strength, hardness, and ductility of metals after being drawn through dies?

15.33 Assuming an ideal drawing process, what is the smallest diameter to which a 60-mm diameter rod can be drawn?

15.34 If you include friction in Problem 15.33, would the final diameter be different? Explain.

15.35 Calculate the extrusion force for a round billet 100 mm in diameter, made of beryllium and extruded at 1000 °C to a diameter of 20 mm.

15.36 Show that for a perfectly plastic material with a yield stress Y and under frictionless conditions, the pressure p in direct extrusion is

$$p = Y \ln\left(\frac{A_o}{A_f}\right).$$

15.37 Show that for the conditions stated in Problem 15.36, the drawing stress σ_d in wire drawing is

$$\sigma_d = Y \ln\left(\frac{A_o}{A_f}\right).$$

15.38 Plot the equations given in Problems 15.36 and 15.37 as a function of the percent reduction of area. Describe your observations.

16

Sheet-Metal Forming

16.1 Introduction

Products made by sheet-metal forming processes are all around us. They include metal desks, file cabinets, appliances, car bodies, aircraft fuselages, and beverage cans. Sheet forming dates back to 5000 B.C., when household utensils and jewelry were made by hammering and stamping gold, silver, and copper. Compared to casting and forging, sheet-metal parts offer the advantages of light weight and versatile shape. Because of low cost and generally good strength and formability characteristics, low-carbon steel is the most commonly used sheet metal.

In this chapter, we first describe the methods by which blanks are cut from large rolled sheets, which are processed further into desired shapes. We also discuss the characteristic features of sheet metals and their formability. Finally, we explain the major processes of sheet forming and the equipment used to make sheet-metal products, as outlined in Fig. 16.1.

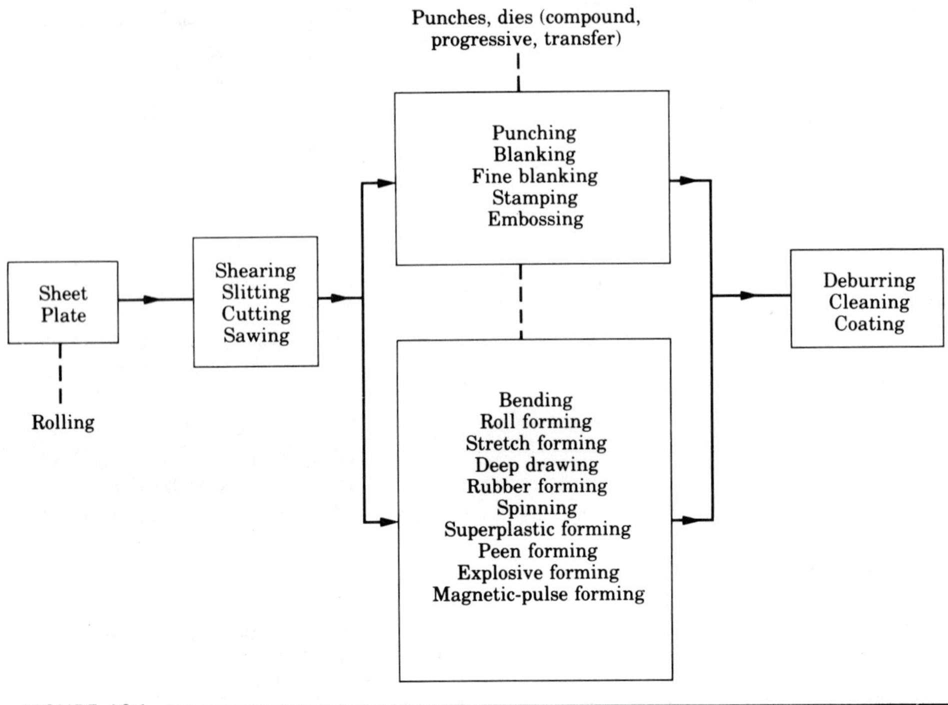

FIGURE 16.1
Outline of sheet-metal forming processes.

16.2 Shearing

Before a sheet-metal part is made, a *blank* of suitable dimensions is first removed from a large sheet, usually from a coil by **shearing**. That is, the sheet is cut by subjecting it to shear stresses typically between a punch and a die (Fig. 16.2a). The features of sheared edges of the sheet and the slug are shown in Figs. 16.2(b) and (c), respectively. Note that the edges are neither smooth nor perpendicular to the plane of the sheet. Shearing usually starts with the formation of cracks on both the top and bottom edges of the workpiece (A and B, and C and D in Fig. 16.2a). These cracks eventually meet each other and separation occurs. The rough **fracture surfaces** are due to these cracks. The smooth and shiny **burnished surfaces** on the hole and the slug are from the contact and rubbing of the sheared edge against the walls of the punch and die, respectively.

The major processing parameters in shearing are the shape and materials for the punch and die, the speed of punching, lubrication, and the **clearance** c between the

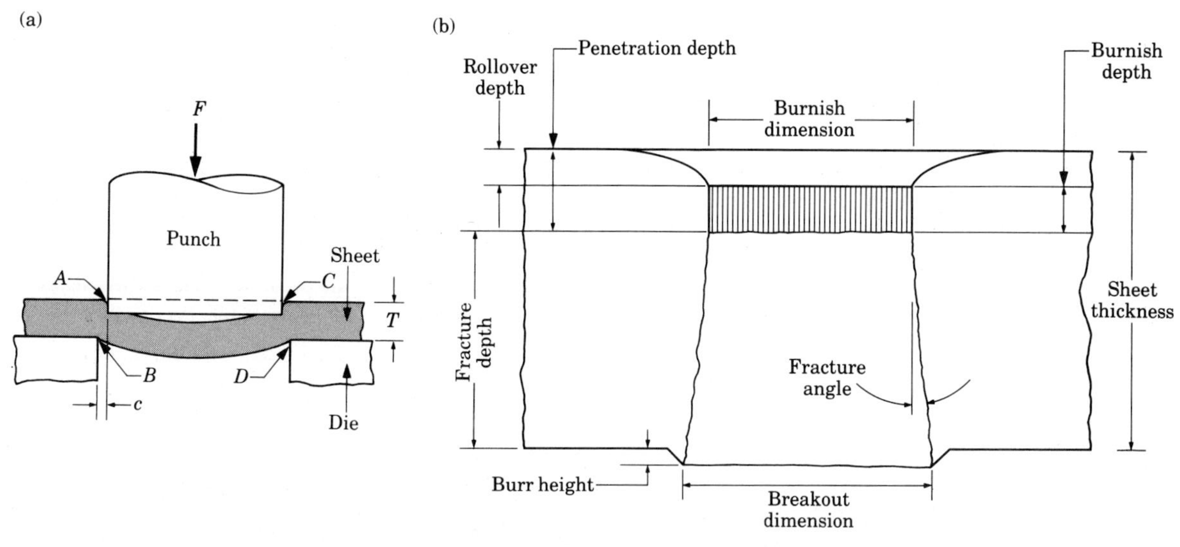

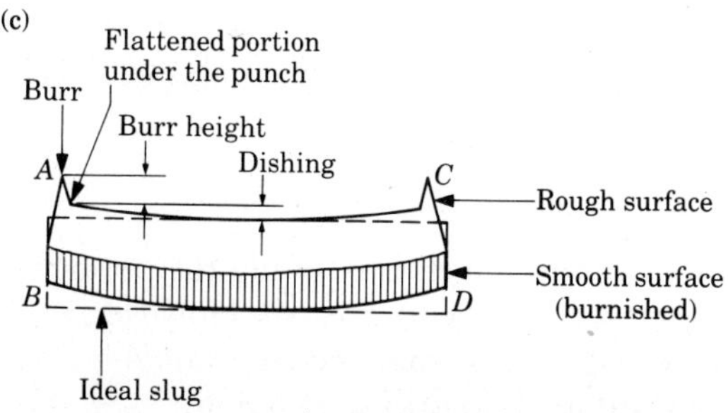

FIGURE 16.2
(a) Schematic illustration of shearing with a punch and die, indicating some of the process variables. Characteristic features of (b) a punched hole and (c) the slug.

punch and the die. The clearance is a major factor in determining the shape and quality of the sheared edge. As clearance increases, the sheared edge becomes rougher and the zone of deformation (Fig. 16.3) becomes larger. The sheet tends to get pulled into the clearance zone, and the edges of the sheared zone become rougher. Unless such edges are acceptable as produced, secondary operations may be necessary to make them smoother—thus adding to cost.

The ratio of the burnished to rough areas on the sheared edge increases with increasing ductility of the sheet metal, and it decreases with increasing sheet thickness and clearance. The width of the deformation zone in Fig. 16.3 depends on punch speed. With increasing speed, the heat generated by plastic deformation is

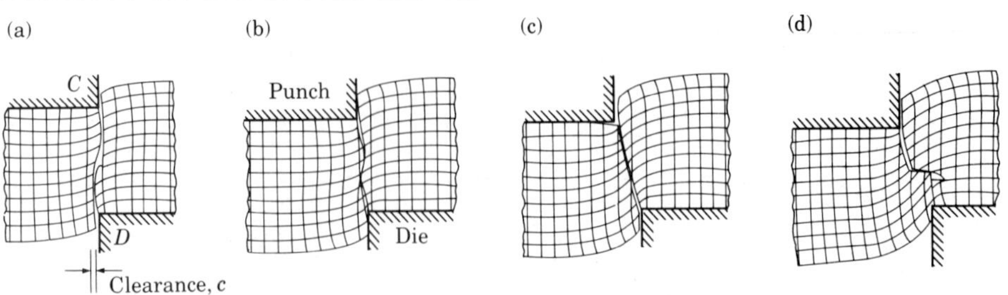

FIGURE 16.3
Effect of increasing the clearance c on the deformation zone in shearing. Note that the quality of the sheared edge deteriorates as clearance increases.

increasingly confined to a smaller zone and, consequently, the sheared surface is narrower and smoother.

Note also the formation of a **burr**, a thin edge or ridge, in Figs. 16.2(b) and (c). Burr height increases with increasing clearance and increasing ductility of the sheet metal. Dull tool edges also contribute greatly to burr formation. The height, shape, and size of the burr can significantly affect subsequent forming operations. We describe several deburring operations in Section 25.10.

16.2.1 Punch force

The force required to punch is basically the product of the shear strength of the sheet metal and the area being sheared. However, friction between the punch and the workpiece can increase this force substantially. We can estimate the **maximum punch force** F from the following equation:

$$F = 0.7TL(\text{UTS}), \quad \textbf{(16.1)}$$

where T is sheet thickness, L is the total length sheared (perimeter of hole), and UTS is the ultimate tensile strength of the material.

• Example: Calculation of punch force.

Estimate the force required in punching a 1-in. (25-mm) diameter hole through a $\frac{1}{8}$-in. (3.2-mm) thick annealed titanium-alloy Ti-6Al-4V sheet at room temperature.

SOLUTION. The force is estimated from Eq. (16.1), where the UTS for this alloy is found from Table 6.9 to be 1000 MPa, or 140,000 psi. Thus,

$$F = 0.7(\tfrac{1}{8})(\pi)(1)(140{,}000) = 38{,}500 \text{ lb}$$

$$= 19.25 \text{ tons} = 0.17 \text{ MN}.$$

•

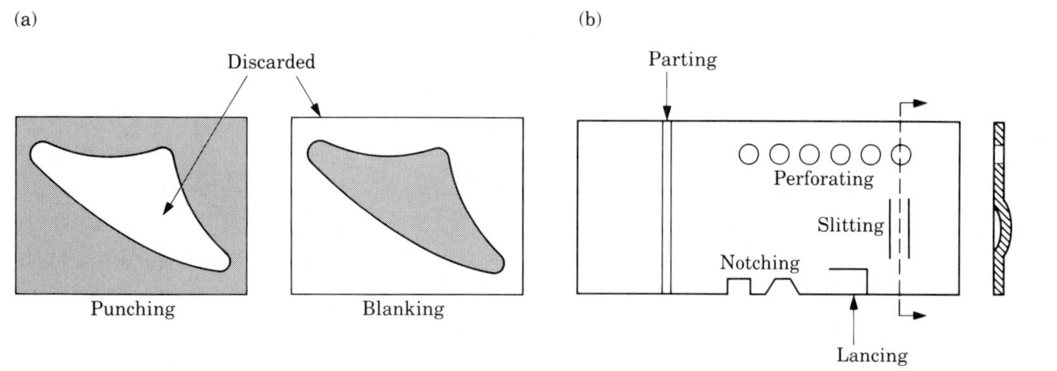

FIGURE 16.4
(a) Punching (piercing) and blanking. (b) Examples of various shearing operations on sheet metal.

16.2.2 Shearing operations

Various operations based on the shearing process are performed. We first define two terms. In **punching**, the sheared slug is discarded (Fig. 16.4a). In **blanking**, the slug is the part and the rest is scrap.

Die cutting. Die cutting consists of the following operations (Fig. 16.4b): (a) *perforating*, or punching a number of holes in a sheet; (b) *parting*, or shearing the sheet into two or more pieces; (c) *notching*, or removing pieces or various shapes from the edges; and (d) *lancing*, or leaving a tab without removing any material. Parts produced by these processes have various uses, particularly in assembly with other components.

Fine blanking. Very smooth and square edges can be produced by *fine blanking* (Fig. 16.5a). One basic die design is shown in Fig. 16.5(b). A V-shaped stinger, or impingement, locks the sheet metal tightly in place and prevents the type of distortion of the material shown in Fig. 16.3. The fine-blanking process involves clearances on the order of 1 percent of the sheet thickness, which may range from 0.13 mm to 13 mm (0.005 in. to 0.5 in.). The operation is usually carried out on triple-action hydraulic presses where the movements of the punch, pressure pad, and die are controlled individually. Fine blanking usually involves parts having holes, which are punched simultaneously with blanking.

Slitting. Shearing operations can be carried out with a pair of circular blades (*slitting*), similar to those in a can opener (Fig. 16.6). The blades follow either a straight line or a circular or curved path. A slit edge normally has a burr, which may be removed by rolling. There are two types of slitting equipment. In the *driven* type, the blades are powered. In the *pull-through* type, the strip is pulled through idling blades. Slitting operations, if not performed properly, may cause various distortions of the sheared part.

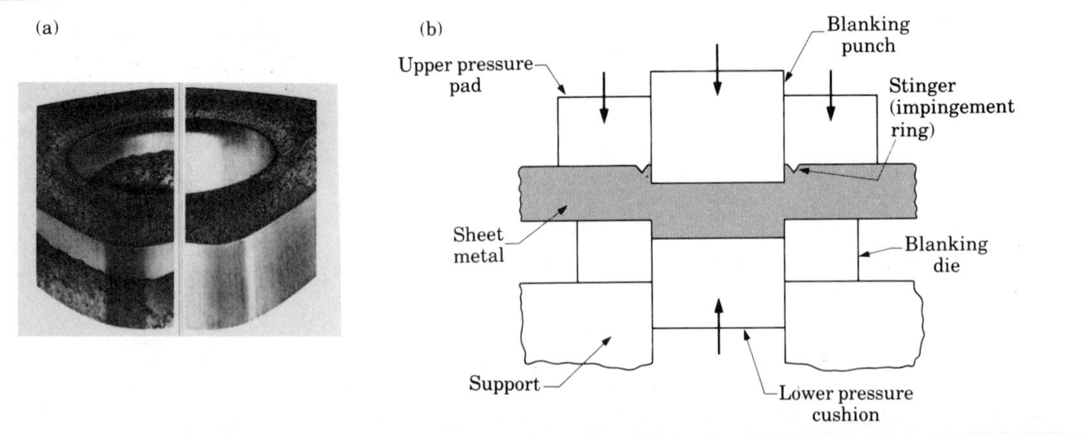

FIGURE 16.5
(a) Comparison of sheared edges by conventional (left) and fine-blanking (right) techniques. (b) Schematic illustration of setup for fine blanking. *Source:* Feintool U.S. Operations.

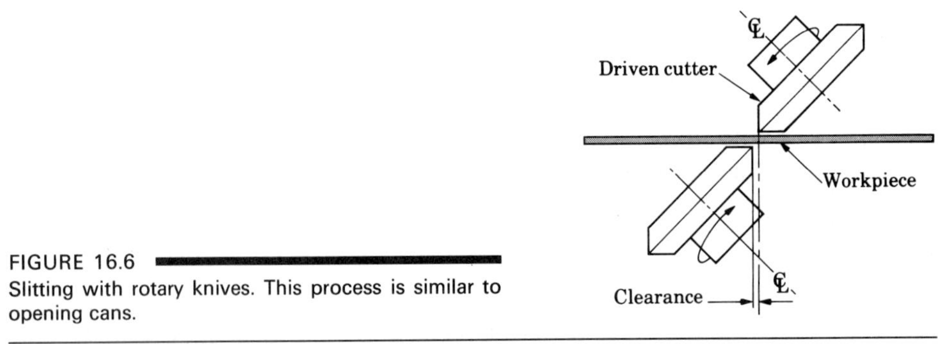

FIGURE 16.6
Slitting with rotary knives. This process is similar to opening cans.

Steel rules. Soft metals, paper, leather, and rubber can be blanked with *steel-rule dies*. Such a die consists of a thin strip of hardened steel bent to the shape to be sheared (similar to a cookie cutter) and held on its edge on a flat wooden base. The die is pressed against the sheet, which rests on a flat surface, and cuts the sheet to the shape of the steel rule.

Nibbling. In *nibbling*, a machine called a nibbler moves a straight punch up and down rapidly into a die. The sheet metal is fed through the gap, and a number of overlapping holes are made. This operation is similar to making a large elongated hole by successively punching holes with a paper punch. Sheets can be cut along any desired path by manual control. The process is economical for small production runs since no special dies are required.

Scrap in shearing. The amount of scrap produced, or *trim loss*, in shearing operations can be significant. On large stampings it can be as high as 30 percent. A significant factor in manufacturing cost, scrap can be reduced significantly by proper arrangement of the shapes on the sheet to be cut. Computer-aided design techniques have been developed to minimize scrap in shearing operations.

16.2.3 Shearing dies

Because the formability of the sheared part can be influenced by the quality of its sheared edges, clearance control is important. In practice, clearances usually range between 2 percent and 8 percent of the sheet's thickness. The thicker the sheet is, the larger the clearance (as much as 10 percent). However, the smaller the clearance, the better is the quality of the edge. In a process called *shaving* (Fig. 16.7), the extra material from a rough sheared edge is trimmed by cutting.

Punch and die shapes. Note in Fig. 16.2(a) that the surfaces of the punch and die are flat. Thus the punch force builds up rapidly during shearing because the entire thickness is sheared at the same time. The area being sheared at any moment can be controlled by beveling the punch and die surfaces (Fig. 16.8). The geometry is similar to that of a paper punch, which you can see by looking closely at the tip.

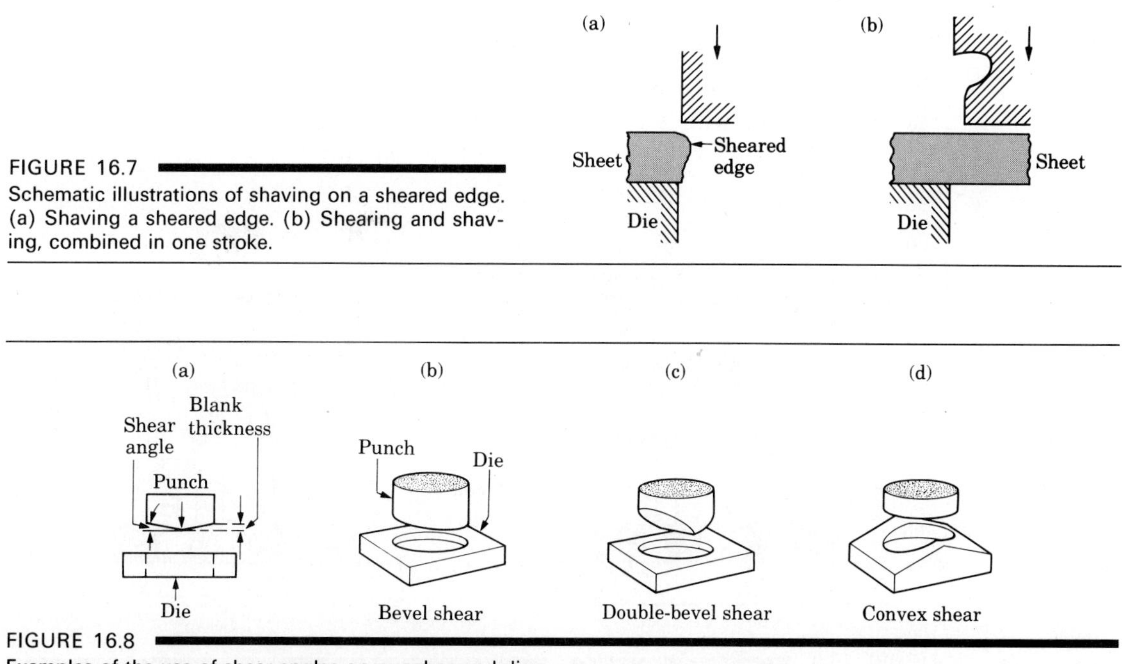

FIGURE 16.7 Schematic illustrations of shaving on a sheared edge. (a) Shaving a sheared edge. (b) Shearing and shaving, combined in one stroke.

FIGURE 16.8 Examples of the use of shear angles on punches and dies.

This geometry is particularly suitable for shearing thick blanks because it reduces the force at the beginning of the stroke. It also reduces the operation's noise level.

Compound dies. Several operations on the same strip may be performed in one stroke with a **compound die** in one station (Fig. 16.9). These operations are usually limited to relatively simple shearing because they are somewhat slow, and the dies are more expensive than those for individual shearing operations.

Progressive dies. Parts requiring multiple operations, such as punching, blanking, and notching, are made at high production rates in **progressive dies**. The sheet metal is fed through a coil strip, and a different operation is performed at the same station with each stroke of a series of punches (Fig. 16.10a). An example of a part made in progressive dies is shown in Fig. 16.10(b).

Transfer dies. In a **transfer die** setup, the sheet metal undergoes different operations at different stations, which are arranged along a straight line or a circular path. After each operation, the part is transferred to the next station for additional operations.

Tool and die materials. Tool and die materials for shearing are generally tool steels and, for high production rates, carbides (see Table 5.9). Lubrication is important for reducing tool and die wear and improving edge quality.

16.2.4 Other methods of cutting sheet metal

There are several other methods for cutting sheet, particularly plates. The sheet or plate may be cut with a *band saw*, a chip removal process that we describe in

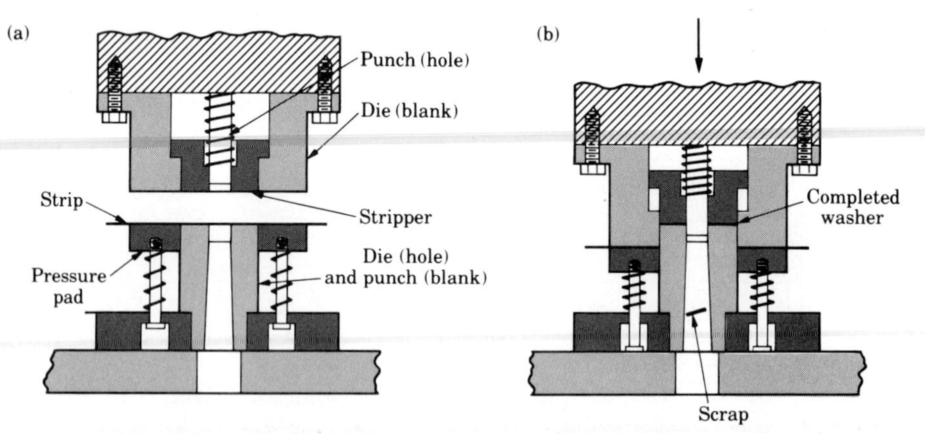

FIGURE 16.9 Schematic illustrations of (a) before and (b) after blanking a common washer in a compound die. Note the separate movements of the die (for blanking) and punch (for punching the hole in the washer).

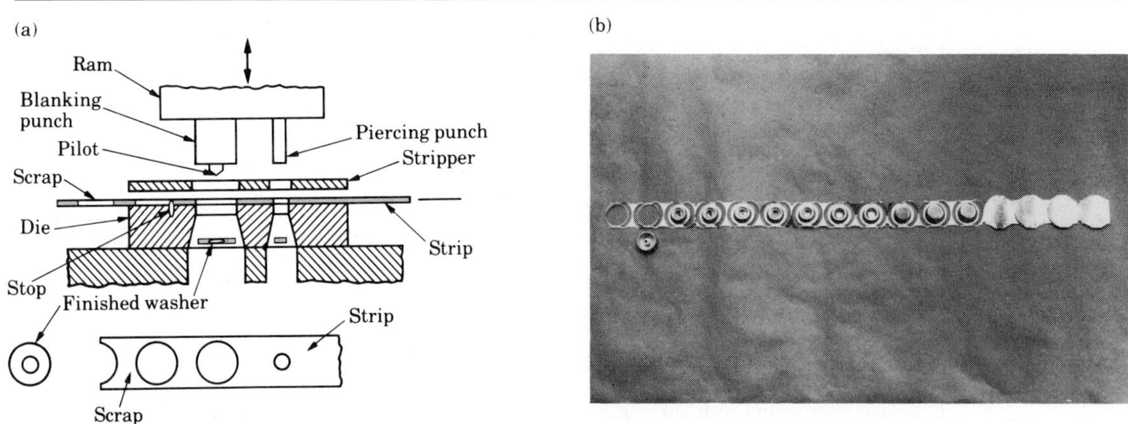

FIGURE 16.10
(a) Schematic illustration of making a washer in a progressive die. (b) Forming of the top piece of an aerosol spray can in a progressive die. Note that the part is attached to the strip until the last operation is completed.

Section 23.7. *Flame cutting* is another common method, particularly for thick steel plates (see Section 27.7), and is widely used in shipbuilding and heavy structural components. *Laser cutting* has also become an important process and is used with computer-controlled equipment to cut a variety of shapes consistently. In *friction sawing*, a disk or blade rubs against the sheet or plate at high surface speeds (see Section 23.7.2).

16.3 Sheet-Metal Characteristics

After a blank is removed from a larger sheet, it is formed into various shapes. Basically, all sheet-forming processes involve various dies and tooling to stretch and bend the sheet. Before considering these processes, however, we need to describe certain characteristics of sheet metals because of their effect on the overall operation.

16.3.1 Elongation

Although sheet-forming operations rarely involve simple uniaxial stretching, as in a tension test, our earlier observations about tensile testing can be useful to understanding the behavior of sheet metals. Recall from Section 2.2 that a specimen subjected to tension first undergoes uniform elongation, and when the load reaches the ultimate tensile strength, the specimen begins to neck.

Since the material is being stretched in sheet forming, high uniform elongation is desirable for good formability. The true strain at which necking begins is numerically equal to the *strain-hardening exponent* (n) in Eq. (2.8). Thus a large n value indicates large uniform elongation (see Table 2.3). Necking may be localized or it may be diffuse, depending on the *strain-rate sensitivity* (m) of the material, as in Eq. (2.9). The higher the value of m, the more diffuse the neck becomes, which is desirable in sheet-forming operations.

In addition to uniform elongation and necking, the *total elongation* of the specimen, such as in 50 mm (2 in.), is also a significant factor in the formability of sheet metals. Obviously, the total elongation of the material increases with increasing values of both n and m.

16.3.2 Yield-point elongation

Low-carbon steels exhibit a behavior called *yield-point elongation*, with upper and lower yield points (Fig. 16.11a). This behavior indicates that after the material yields, the sheet stretches farther in certain regions without any increase in the lower yield point, while other regions in the sheet have not yet yielded.

This behavior produces **Lueder's bands** (*stretcher strain marks* or *worms*) on the sheet (Fig. 16.11b). These marks are elongated depressions on the surface of the sheet. They may be objectionable in the final product because of coarse surface appearance and difficulties in subsequent coating and painting operations. You can find stretcher strain marks on the bottom of cans used for common household products (Fig. 16.11c).

The usual method of avoiding these marks is to eliminate or to reduce yield-point elongation by reducing the thickness of the sheet 0.5–1.5 percent by cold rolling (*temper*, or *skin*, *rolling*). However, because of strain aging (see Fig. 2.30), the yield-point elongation reappears after a few days at room temperature—or after

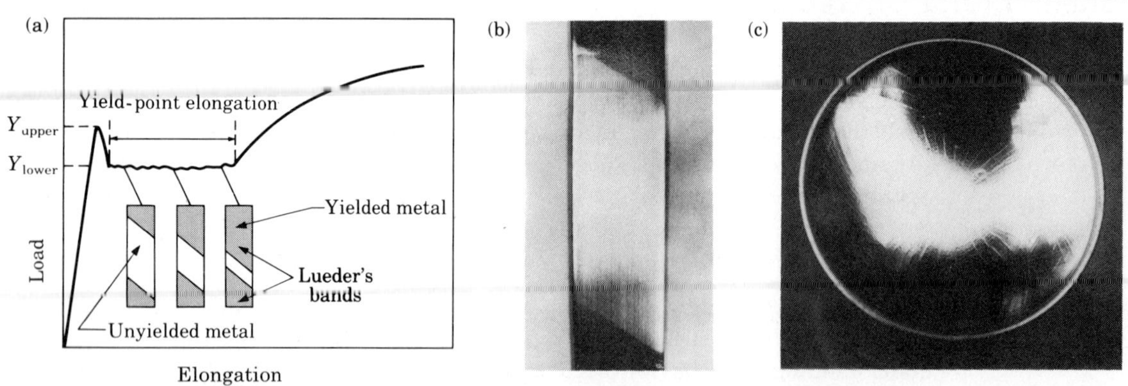

FIGURE 16.11 (a) Yield-point elongation in a sheet-metal specimen. (b) Lueder's bands in a low-carbon steel sheet. *Source:* Courtesy of Caterpillar Inc. (c) Stretcher strains at the bottom of a steel can for household products.

a few hours at higher temperatures. Thus the material should be formed within a certain time limit, which depends on the type of steel, in order to avoid the reappearance of stretcher strains.

16.3.3 Anisotropy

An important factor that influences sheet-metal forming is *anisotropy* (directionality) of the sheet. Recall that anisotropy is acquired during the thermo-mechanical processing of the sheet and (from Section 1.5.1) that there are two types of anisotropy: *crystallographic anisotropy* (because of preferred orientation of the grains) and *mechanical fibering* (because of the alignment of impurities, inclusions, and voids throughout the thickness of the sheet). We discuss this subject further in Section 16.9.

16.3.4 Grain size

Sheet-metal grain size is important for two reasons. As we described in Section 1.4, grain size affects mechanical properties, and it influences the surface appearance of the formed part (*orange peel*). The coarser the grain, the rougher is the surface appearance. An ASTM grain size of No. 7 or finer is preferred for general sheet-metal forming operations.

16.4 Test Methods for Formability of Sheet Metals

You undoubtedly are familiar with the cooking utensils shown in Fig. 16.12. The initial forms of these utensils are flat sheets, usually aluminum or stainless steel. These sheet metals are formed by various processes. Before deciding on a process, however, we have to determine whether a particular sheet metal can be formed into desired shapes without failure. Note that these utensils have undergone various deformations without tearing or wrinkling.

The sheet-metal **formability** is of great technological and economic interest. We generally define it as the ability of the metal to undergo the desired shape change without failure, such as by necking or tearing. In this section we describe the methods that are generally, but not always, used in manufacturing industries to predict formability.

16.4.1 Cupping tests

Because sheet forming is basically a process of stretching, the earliest tests developed to predict formability were cupping tests (Fig. 16.13). The sheet-metal specimen is clamped between two circular flat dies, and a steel ball or round punch

FIGURE 16.12
Kitchen utensils made of sheet metal. *Source:* Courtesy of the Vollrath Company.

FIGURE 16.13
A cupping test (Erichsen test) to determine formability of sheet metals. The greater the distance *d* before failure of the sheet, the greater its formability.

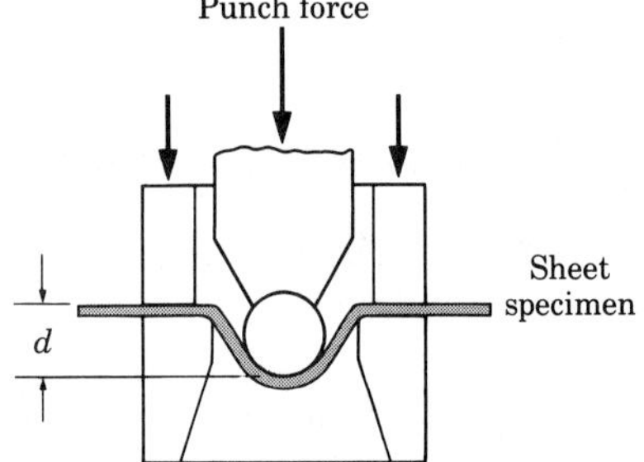

is pushed hydraulically into the sheet metal until a crack begins to appear on the stretched specimen. The greater the value of *d*, the greater is the formability of the sheet. Although such tests are easy to perform and are approximate indicators of formability, they do not simulate the exact conditions of actual operations.

16.4.2 Forming-limit diagrams

An important development in testing the formability of sheet metals is the **forming-limit diagram** (FLD). The sheet is marked with a grid pattern of circles, typically 2.5–5 mm (0.1–0.2 in.) in diameter, using electrochemical or photoprinting techniques. The blank is then stretched over a punch, and the deformation of the circles is observed and measured in regions where failure (necking and tearing) has occurred. For improved accuracy of measurement, the circles are made as small as practicable.

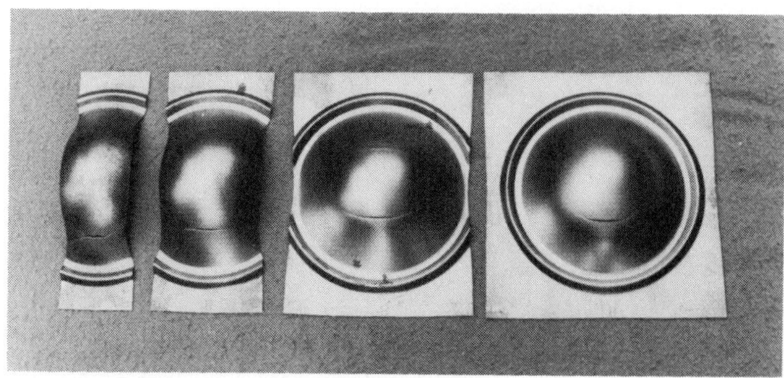

FIGURE 16.14
Bulge-test results on steel sheets of various widths. The specimen farthest left was subjected basically to simple tension. The specimen farthest right was subjected to equal biaxial stretching. *Source:* Inland Steel Company.

In order to develop unequal stretching, as in actual sheet-forming operations, the specimens are cut to varying widths (Fig. 16.14). Note that a square specimen (farthest right in the figure) produces equal biaxial stretching (such as achieved in blowing up a spherical balloon), whereas a narrow specimen (farthest left in the figure) approaches a state of simple uniaxial stretching. After a series of such tests is performed on a particular sheet metal, a forming-limit diagram showing the boundaries between failure and safe regions is constructed (Fig. 16.15).

In order to plot the forming-limit diagram, the major and minor engineering strains, as measured from the deformation of the original circles, are obtained as follows. Note in Fig. 16.15(a) that the original circle has deformed into an ellipse. The major axis of the ellipse represents the major direction and magnitude of stretching. The major strain is the engineering strain in this direction, and is always *positive*, because of sheet-metal stretching. The minor axis of the ellipse represents the magnitude of the stretching or shrinking in the transverse direction.

Note that the minor strain can be either *negative* or *positive*. If, for example, we place a circle in the center of a tensile-test specimen and stretch it, the specimen becomes narrower as it is stretched (Poisson effect), and the minor strain is negative. You can easily demonstrate this behavior by stretching a rubber band. On the other hand, if we place a circle on a spherical rubber balloon and inflate it, the minor strain is positive and is equal in magnitude to the major strain.

By comparing the surface areas of the original circle and the deformed circle on the formed sheet, we can also determine whether the thickness of the sheet has changed. If the area of the deformed circle is larger than the original circle, then because the volume remains constant in plastic deformation, we know that the sheet has become thinner. You can observe this phenomenon by blowing up a balloon and noting how thin it gets as it is stretched.

The data obtained from different locations in each of the samples shown in Fig. 16.14 are plotted in the form shown in Fig. 16.15(b). The curves represent the boundaries between failure and safe zones. Thus if a circle underwent major and

(a)

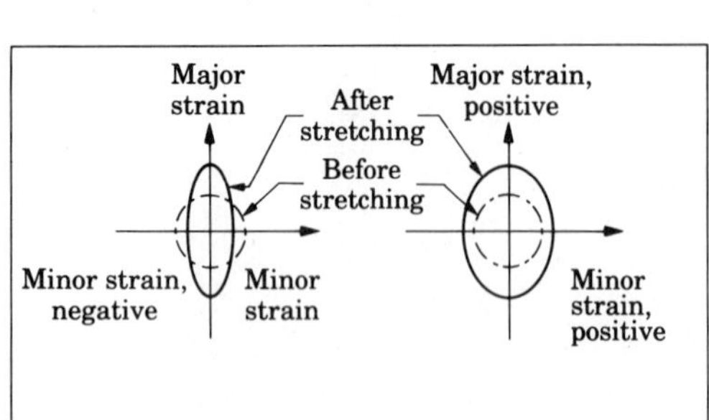

(b)

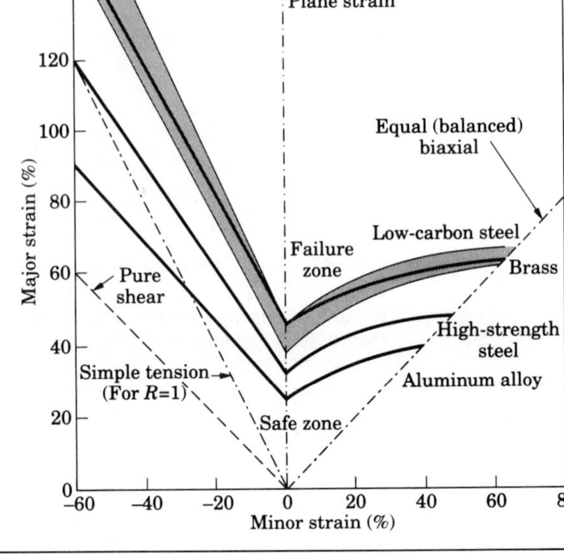

FIGURE 16.15 (a) Strains in deformed circular grid patterns. (b) Forming-limit diagram (FLD) for various sheet metals. Whereas the major strain is always positive (stretching), the minor strain may be either positive or negative. *Source:* S. S. Hecker and A. K. Ghosh.

minor strains of plus and minus 40 percent, respectively, there was no tear in that region of the specimen. On the other hand, if the strains in an aluminum-alloy specimen were plus 80 percent and minus 40 percent, respectively, there was a tear in that region of the specimen. An example of a formed sheet-metal part with a grid pattern is shown in Fig. 16.16. Note the tear and the deformation of the circular patterns in the vicinity of the tear region.

Figure 16.15(b) shows that different materials have different forming-limit diagrams and that the higher the curve, the better is the formability of the material. It is important to note that a compressive minor strain of, say, 20 percent is associated with a higher major strain than a tensile (positive) minor strain of the same magnitude. In other words, it is desirable for the minor strain to be negative (shrinking in the minor direction). Special tooling has been designed for forming sheet metals to take advantage of the beneficial effect of negative minor strains on formability.

The effect of sheet-metal thickness on forming-limit diagrams is to raise the curves in Fig. 16.15(b). Thus the thicker the sheet, the higher its formability curve is and the more formable it is. On the other hand, in actual forming operations, a thick blank may not bend as easily around small radii without cracking.

Friction and lubrication at the punch–sheet metal interface also are important factors in the test results. With well-lubricated interfaces, the strains are more uniformly distributed over the punch. Surface scratches, deep gouges, and blemishes can reduce formability and cause premature tearing and failure. Although forming-

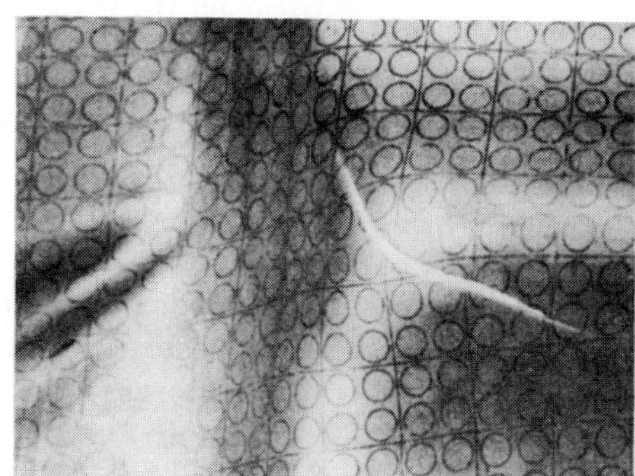

FIGURE 16.16 Deformation of grid pattern and tearing of sheet metal during forming. The major and minor axes of the circles are used to determine the coordinates on the forming-limit diagram in Fig. 16.15(b). *Source:* S. P. Keeler.

limit diagrams are an important development, in-house tests simulating a manufacturer's actual product often must be devised for quicker determination of formability.

16.5 Bending Sheet and Plate

Bending is one of the most common forming operations. You merely have to look at the components in an automobile or appliance—or at a paper clip or file cabinet—to appreciate how many parts are shaped by bending. Bending is used not only to form parts, such as flanges, seams, and corrugations, but also to impart stiffness to the part by increasing its moment of inertia.

The terminology used in bending is shown in Fig. 16.17. In bending, the outer fibers of the material are in tension and the inner fibers are in compression. Because of the Poisson's ratio, the width of the part (dimension L) in the outer region is smaller and in the inner region it is larger than the original width (Fig. 16.18). You may easily observe this phenomenon by bending a rectangular rubber eraser.

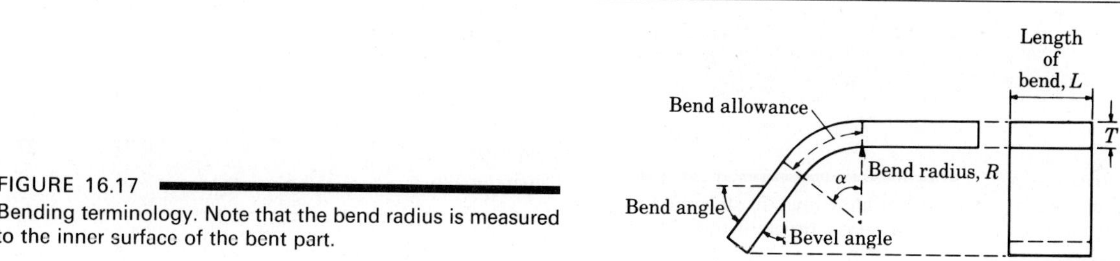

FIGURE 16.17 Bending terminology. Note that the bend radius is measured to the inner surface of the bent part.

16.5.1 Minimum bend radius

The engineering strain on a sheet during bending is

$$e = \frac{1}{(2R/T) + 1}, \tag{16.2}$$

where R is the bend radius and T is the sheet thickness. As R/T decreases (bend radius becomes smaller), the tensile strain at the outer fiber increases, and the material eventually cracks (Fig. 16.18).

The radius at which a crack appears on the outer surface of the bend is referred to as the **minimum bend radius** for the material. It is usually expressed in terms of its thickness, such as $2T$, $3T$, $4T$, and so on. Thus a $3T$ bend radius indicates that the smallest radius to which the sheet can be bent, without cracking, is three times its thickness. The minimum bend radius for various materials is given in Table 16.1.

There is an inverse relationship between bendability and the tensile reduction of area of the material (Fig. 16.19). The minimum bend radius is, approximately,

$$R = T\left(\frac{50}{r} - 1\right), \tag{16.3}$$

where r is the tensile reduction of area of the sheet metal. Note that at $r = 50$, the minimum bend radius is zero; that is, the sheet can be folded over itself, much like folding a piece of paper. To increase the bendability of metals, we may increase their tensile reduction of area either by heating or by bending in a high-pressure environment to improve ductility. As with all other formability tests, bendability tests may not always apply to a specific situation in a manufacturing plant. Thus in-house tests simulating actual conditions may have to be developed.

FIGURE 16.18 Outer-surface cracking of an aluminum strip bent to an angle of 90°. Note the narrowing of the top surface because of the Poisson effect.

TABLE 16.1 MINIMUM BEND RADIUS FOR VARIOUS MATERIALS AT ROOM TEMPERATURE

MATERIAL	CONDITION SOFT	HARD
Aluminum alloys	0	6T
Beryllium copper	0	4T
Brass, low-leaded	0	2T
Magnesium	5T	13T
Steels		
Austenitic stainless	0.5T	6T
Low-carbon, low-alloy, and HSLA	0.5T	4T
Titanium	0.7T	3T
Titanium alloys	2.6T	4T

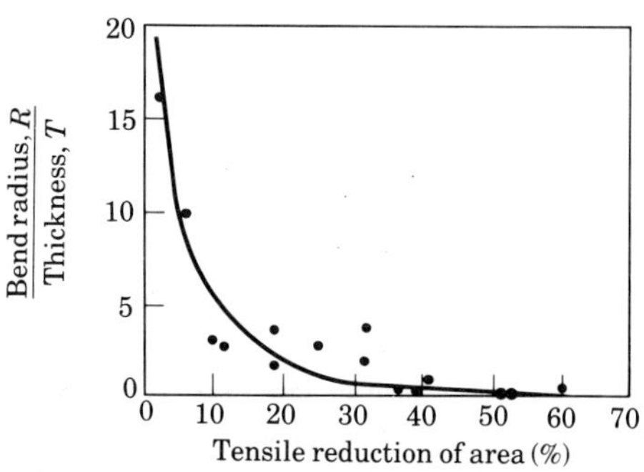

FIGURE 16.19
Relationship between R/T ratio and tensile reduction of area for sheet metals. Note that sheet metal with a 50 percent reduction of area can be bent over itself, like folding a piece of paper, without cracking.

Bendability depends on the edge condition of the sheet. Since rough edges are points of stress concentration, bendability decreases as edge roughness increases. Another significant factor in edge cracking is the amount and shape of inclusions in the sheet metal and the amount of cold working that the edges undergo during shearing. Inclusions in the form of stringers are more detrimental than globular shaped inclusions because of their pointed shape. The removal of the cold-worked regions, say, by machining, or annealing the part to improve ductility, greatly improves the resistance to edge cracking.

Anisotropy of the sheet is an important factor in bendability. As shown in Fig. 1.13, cold rolling produces anisotropy by preferred orientation and the alignment of impurities, inclusions, and voids. Prior to bending such a sheet, caution should be exercised in cutting it in the proper direction from a rolled sheet, although this may not always be possible. Whether a sheet is anisotropic can be determined by observing the direction of cracking in the cupping test (Fig. 16.13). If the crack is as shown in Fig. 1.13, the sheet is anisotropic; if it is circular, the sheet is isotropic.

16.5.2 Springback

Since all materials have a finite modulus of elasticity, plastic deformation is followed by some elastic recovery when the load is removed. In bending, this recovery is called **springback**, which you can easily observe by bending a piece of sheet metal or wire. As we note in Fig. 16.20, the final bend angle after springback is smaller and the final bend radius is larger than before. Springback occurs not only in flat sheets or plate, but also in rod, wire, and bar with any cross-section.

We can calculate springback approximately, in terms of the radii R_i and R_f (Fig. 16.20), as

$$\frac{R_i}{R_f} = 4\left(\frac{R_i Y}{ET}\right)^3 - 3\left(\frac{R_i Y}{ET}\right) + 1. \tag{16.4}$$

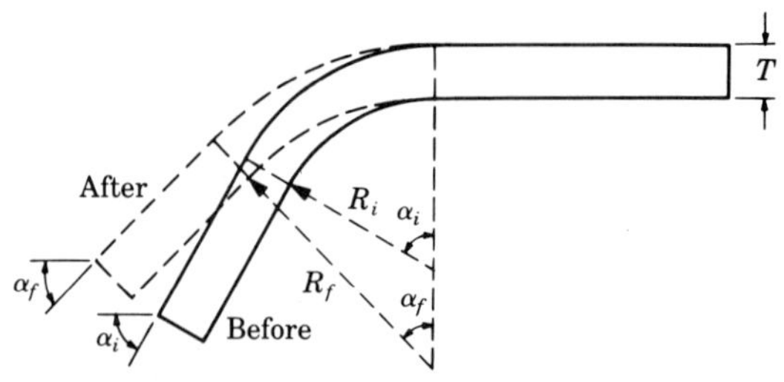

FIGURE 16.20 Springback in bending. In springback, the part tends to recover elastically after bending, and its bend radius becomes larger.

Note that springback increases as the R/T ratio and yield stress Y of the material increase and the elastic modulus E decreases.

In V-die bending, it is possible for the material to exhibit *negative*, as well as positive, springback. This condition is caused by the nature of deformation as the punch completes the bending operation. Negative springback does not occur in air bending (free bending) because of the lack of constraints in a V-die.

16.5.3 Compensation for springback

In forming practice, springback is usually compensated for by *overbending* the part (Figs. 16.21a and b). Several trials may be necessary to obtain the desired results. Another method is to coin the bend area by subjecting it to high localized compressive stresses between the tip of the punch and the die surface (Figs. 16.21c and d)—known as *bottoming* the punch. Another method is *stretch bending*, in which the part is subjected to tension while being bent. Because springback decreases as yield stress decreases, all other parameters being the same, bending may also be carried out at elevated temperatures to reduce springback.

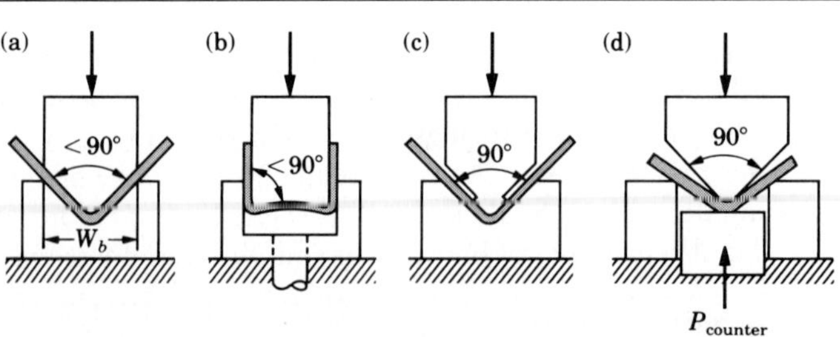

FIGURE 16.21 Methods of reducing or eliminating springback in bending operations. *Source:* V. Cupka, T. Nakagawa, and H. Tyamoto.

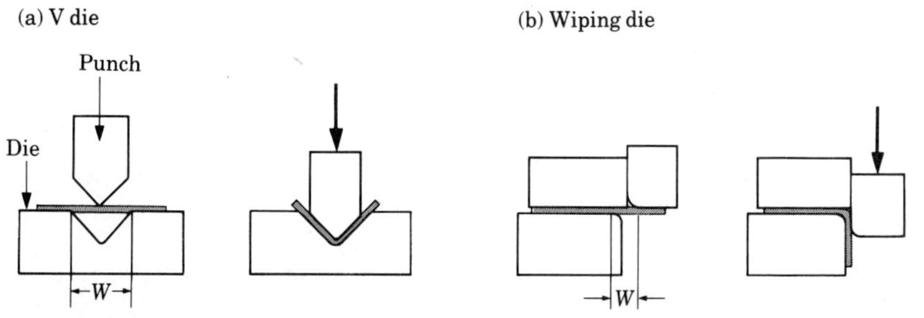

FIGURE 16.22
Common die-bending operations, showing the die-opening dimension W used in calculating bending forces (see Eq. 16.5).

16.5.4 Bending forces

We can calculate the **bending force** F approximately as

$$F = \frac{LT^2(\text{UTS})}{W}, \qquad (16.5)$$

where L is the length (see Fig. 16.17), T is the thickness, and W is the die opening (Fig. 16.22). The bending force varies with punch travel. In air bending, the force increases from zero to a maximum and then decreases as the bend is completed. In V-die bending, the force increases sharply as the punch bottoms.

16.6 Common Bending Operations

In this section we describe common bending operations. Some of these processes are performed on discrete sheet-metal parts; others are done continuously, as in roll forming of coiled sheet stock.

16.6.1 Press brake forming

Sheet metal or plate can be bent easily with simple fixtures using a press. Long (7 m—20 ft—or more) and relatively narrow pieces are usually bent in a *press brake* (Fig. 16.23). This machine utilizes long dies in a mechanical or hydraulic press and is suitable for small production runs. The tooling is simple and adaptable to a wide variety of shapes. Die materials may range from hardwood, for low-strength materials and small production runs, to carbides. For most applications, carbon-steel or gray-iron dies are generally used.

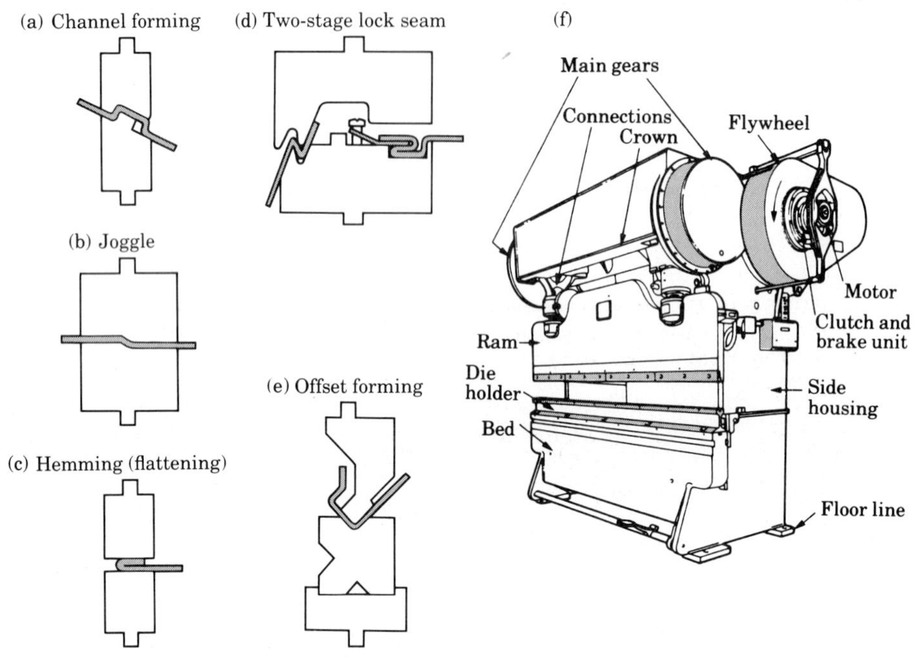

FIGURE 16.23
(a–e) Schematic illustrations of various bending operations in a press brake. (f) Schematic illustration of a press brake. *Source:* Verson Allsteel Company.

16.6.2 Other bending operations

Sheet metal may be bent by a variety of processes (Fig. 16.24). Plates are bent by the *roll-bending* process shown in Fig. 16.24(b). By adjusting the distance between the three rolls, various curvatures can be obtained.

Beading. In *beading*, the edge of the sheet metal is bent into the cavity of a die (Figs. 16.25a and b). The bead gives stiffness to the part by increasing the moment of

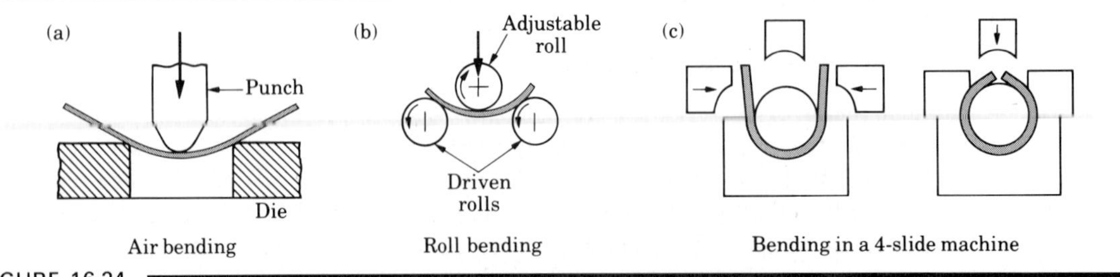

FIGURE 16.24
Examples of various bending operations.

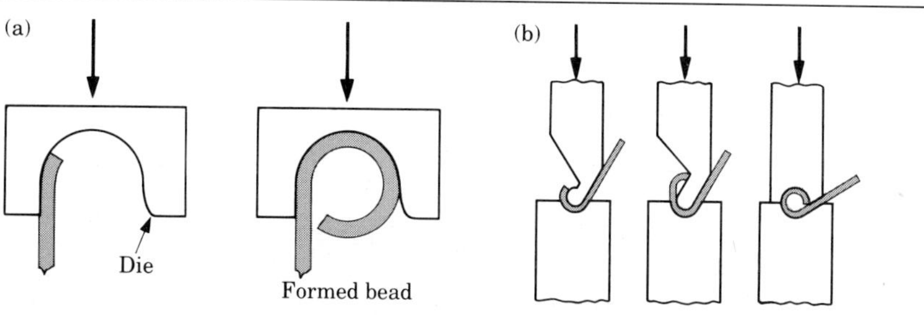

FIGURE 16.25
(a) Bead forming with a single die. (b) Bead forming with two dies in a press brake.

inertia of the edges. Also, it improves the appearance of the part and eliminates exposed sharp edges.

Flanging. Flanging is a process of bending the edges of sheet metals. In *shrink flanging* (Fig. 16.26a), the flange is subjected to compressive hoop stresses which, if excessive, cause the flange edges to wrinkle. The wrinkling tendency increases with decreasing radius of curvature of the flange. In *stretch flanging*, the flange edges are subjected to tensile stresses and, if excessive, can lead to cracking at the edges.

In the *dimpling* operation (Fig. 16.26b), a hole is first punched and then expanded into a flange, or a shaped punch pierces the sheet metal and expands the hole. Flanges may be produced by *piercing* with a shaped punch (Fig. 16.26c). The ends of tubes are flanged by a similar process (Fig. 16.26d). When the angle of bend is less than 90°, as in fittings with conical ends, the process is called *flaring*.

The condition of the edges is important in these operations. Stretching the material causes high tensile stresses at the edges, which could lead to cracking and tearing of the flange. As the ratio of flange to hole diameters increases, the strains increase proportionately. The rougher the edge, the greater will be the tendency for cracking. Sheared or punched edges may be shaved with a sharp tool (see Fig. 16.7) to improve the surface finish of the edge and reduce the tendency for cracking.

Hemming. In the *hemming* process (also called *flattening*), the edge of the sheet is folded over itself (Fig. 16.23c). Hemming increases the stiffness of the part, improves its appearance, and eliminates sharp edges. *Seaming* involves joining two edges of sheet metal by hemming (Fig. 16.23d). Double seams are made by a similar process, using specially shaped rollers, for watertight and airtight joints, such as in food and beverage containers.

Roll forming. For bending continuous lengths of sheet metal and for large production runs, *roll forming* is used (contour roll forming or cold roll forming). The metal strip is bent in stages by passing it through a series of rolls (Fig. 16.27).

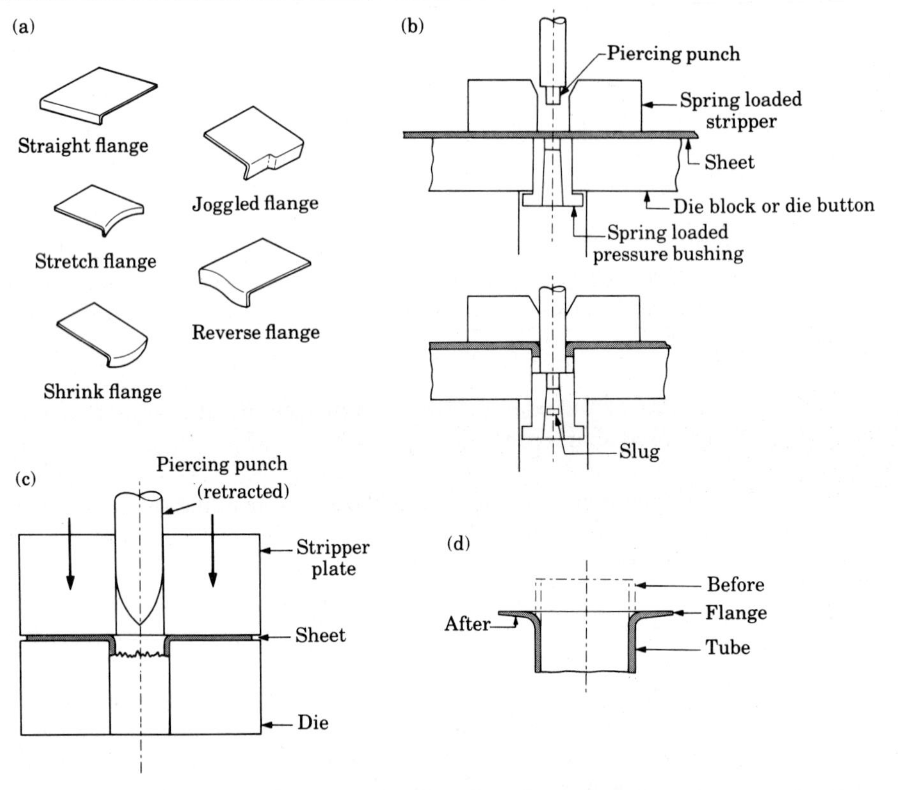

FIGURE 16.26

Various flanging operations. (a) Flanges on flat sheet. (b) Dimpling. (c) Piercing sheet metal to form a flange. In this operation a hole does not have to be prepunched before the punch descends. Note, however, the rough edges along the circumference of the flange. (d) Flanging of a tube. Note thinning of the edges of the flange.

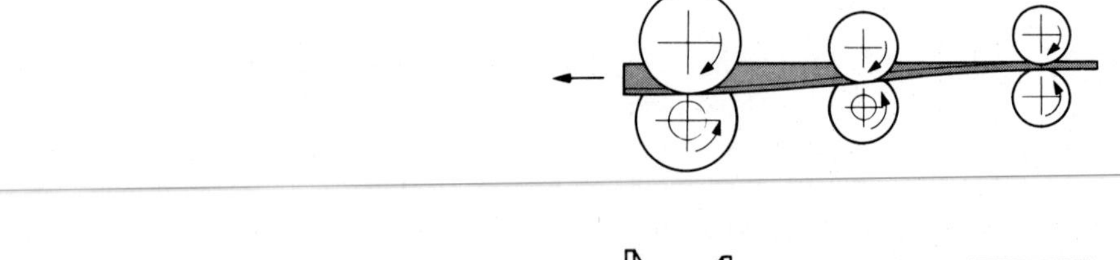

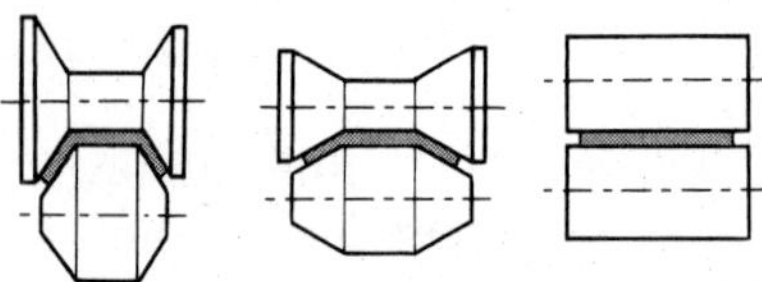

FIGURE 16.27

The roll-forming process.

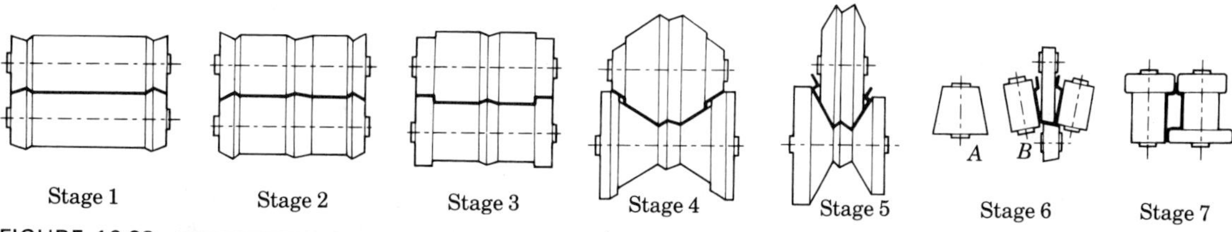

FIGURE 16.28
Stages in roll forming of sheet-metal door frame. In stage 6, the rolls may be shaped as in *A* or *B*. *Source:* G. Oehler.

Typical products are channels, gutters, siding, panels, frames (Fig. 16.28), and pipes and tubing with lock seams. The length of the part is limited only by the amount of material supplied from the coiled stock. The parts are usually sheared and stacked continuously. The sheet thickness typically ranges from about 0.125 mm to 20 mm (0.005 in. to 0.75 in.). Forming speeds are generally below 1.5 m/s (300 ft/min), although they can be much higher for special applications.

Proper design and sequencing of the rolls, which usually are mechanically driven, requires considerable experience. Tolerances, springback, and tearing and buckling of the strip have to be considered. The rolls are generally made of carbon steel or gray iron and may be chromium plated—for better surface finish of the product and wear resistance of the rolls. Lubricants may be used to improve roll life and surface finish and to cool the rolls and the workpiece.

16.7 Tube Bending and Forming

Bending and forming tubes and other hollow sections require special tooling to avoid buckling and folding. The oldest and simplest method of bending a tube or pipe is to pack the inside with loose particles, commonly sand, and bend the part in a suitable fixture. This technique prevents the tube from buckling. After the tube has been bent, the sand is shaken out. Tubes can also be plugged with various flexible internal mandrels, as shown in Fig. 16.29, which also illustrates various bending techniques and fixtures for tubes and sections. A relatively thick tube having a large bend radius can be bent without filling it with particulates or using plugs.

The basic forming process of *bulging* involves placing a tubular, conical, or curvilinear part in a split-female die and expanding it with, say, a polyurethane plug (Fig. 16.30a). The punch is then retracted, the plug returns to its original shape, and the part is removed by opening the dies. Typical products are coffee or water pitchers, barrels, and beads on drums. For parts with complex shapes, the plug,

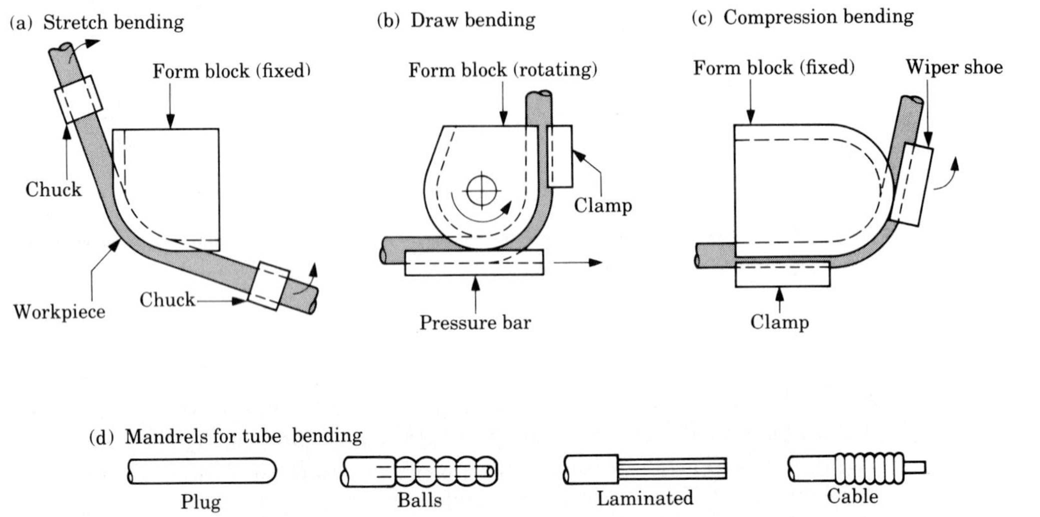

FIGURE 16.29
Methods of bending tubes. Internal mandrels, or filling tubes with particulate materials such as sand, are often necessary to prevent collapsing of the tubes during bending. Solid rods and structural shapes are also bent by these techniques.

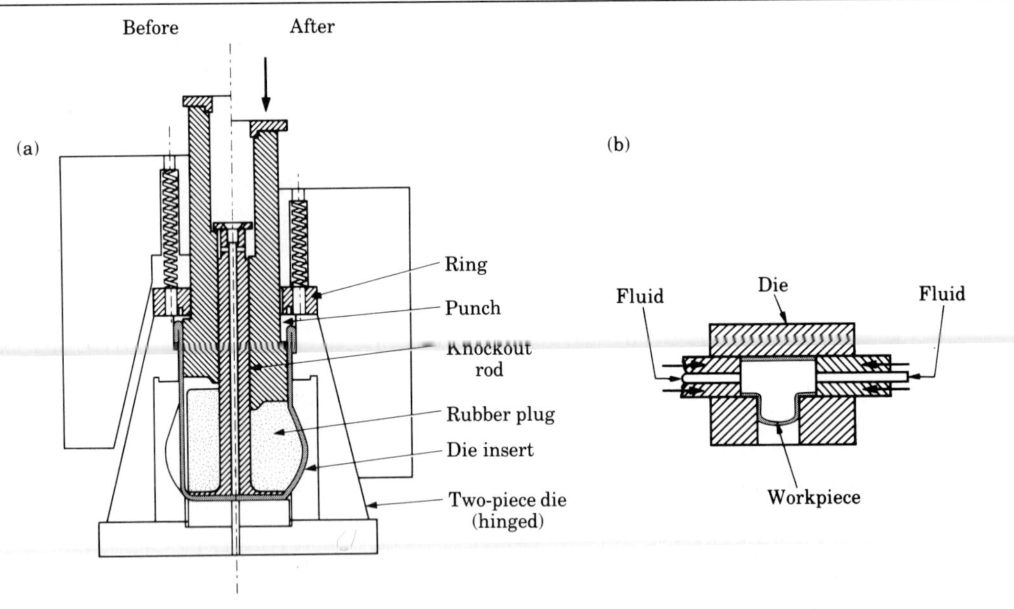

FIGURE 16.30
(a) Bulging of a tubular part with a flexible plug. Water pitchers can be made by this method. (b) Production of fittings for plumbing by expanding tubular blanks with internal pressure. The bottom of the piece is then punched out to produce a "T." *Source:* J. A. Schey, *Introduction to Manufacturing Processes,* 2d ed. New York: McGraw-Hill Publishing Company, 1987.

instead of being cylindrical, may be shaped in order to apply greater pressure at critical points. The major advantage of using polyurethane plugs is that they are very resistant to abrasion, wear, and lubricants, and they do not damage the surface finish of the part being formed.

Hydraulic pressure may be used in bulging operations, although they require sealing and need hydraulic controls (Fig. 16.30b). *Segmented dies* that are expanded and retracted mechanically may also be used. These dies are relatively inexpensive and can be used for large production runs.

16.8 Stretch Forming

In **stretch forming**, the sheet metal is clamped around its edges and stretched over a die or form block, which moves upward, downward, or sideways, depending on the particular machine (Fig. 16.31). Stretch forming is used primarily to make aircraft wing-skin panels, automobile door panels, and window frames. Although this process is generally used for low-volume production, it is versatile and economical. Aluminum skins for the Boeing 767 and 757 aircraft are made by stretch forming, with a tensile force of 9 MN (2 million lb). The rectangular sheets are 12 m × 2.5 m × 6.4 mm (40 ft × 8.3 ft × 0.25 in.).

In most operations, the blank is a rectangular sheet, clamped along its narrower edges and stretched lengthwise, thus allowing the material to shrink in width. Controlling the amount of stretching is important to avoid tearing. Stretch forming cannot produce parts with sharp contours or re-entrant corners (depressions on the surface of the die).

Dies for stretch forming are generally made of zinc alloys, steel, plastics, or wood. Most applications require little or no lubrication. Various accessory

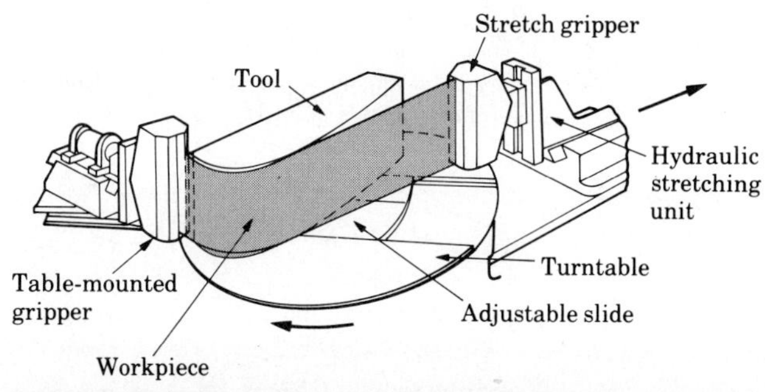

FIGURE 16.31
Schematic illustration of a stretch-forming process. Aluminum skins for aircraft can be made by this process. *Source:* Cyril Bath Co.

equipment can be used in conjunction with stretch forming, including additional forming with both male and female dies while the part is in tension.

• Example: Manufacturing bellows.

Bellows are manufactured by a bulging process, as shown in the accompanying figure. After the tube is bulged at several equidistant locations, it is compressed axially to collapse the bulged regions, thus forming bellows. The tube material must be able to undergo the large strains involved during the collapsing process.

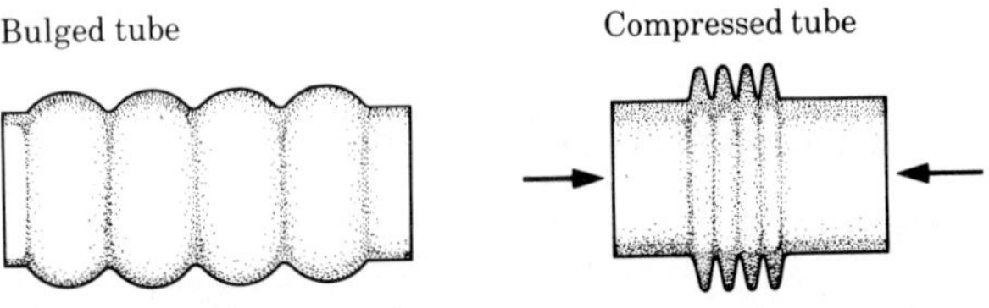

• Example: Manufacturing a rear-axle housing.

A rear-axle housing from an AISI 1035 steel tubular blank is manufactured as follows:

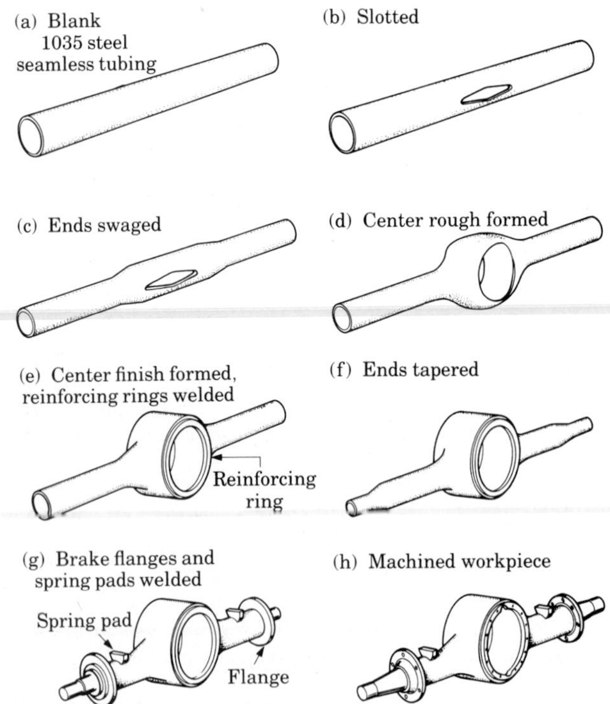

First, a slot is made in the center of the tube. The ends of the tube are then swaged, or tapered, by forcing a shaped die into the end of the tube, as shown in the following drawing. The center section is heated and expanded by inserting special tooling through the slot. Various other processes are then used to complete the part, as indicated in the first figure.

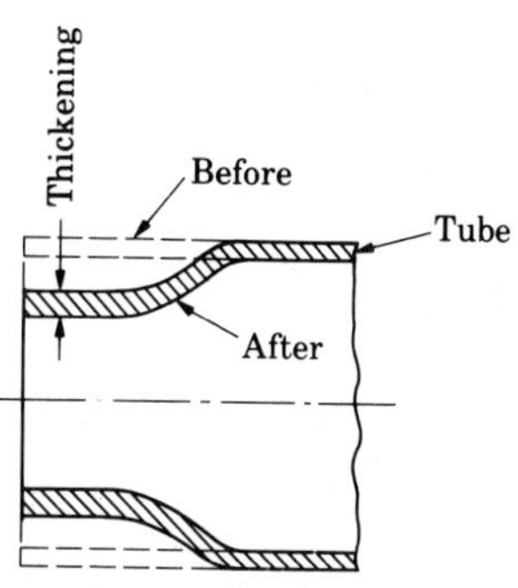

16.9 Deep Drawing

Undoubtedly, you have observed that many parts, components, and products made of sheet metal are cylindrically or box shaped: pots and pans, containers for food and beverages, kitchen sinks, and automotive fuel tanks. Such parts are made by a process in which a punch forces a flat sheet-metal blank into a die cavity (Fig. 16.32a). Although the process is generally called **deep drawing**, meaning deep parts, it is also used to make parts that are shallow or have moderate depth.

16.9.1 The process

Because of its relative simplicity, let's first consider the deep drawing of a cylindrical cup (Fig. 16.32b). A round sheet-metal blank is placed over a circular die opening, and is held in place with a *blankholder*, or hold-down ring. The punch travels downward and forces the blank into the die cavity, forming a cup. The important variables in deep drawing are the properties of the sheet metal, the ratio of blank diameter (D_0) to punch diameter (D_p), the clearance (c) between punch and die, the punch radius (R_p) and die-corner radius (R_d), the blankholder force, friction, and lubrication.

During the drawing operation, the sheet is subjected to the stresses shown in Fig. 16.33. On element A in the blank, the radial tensile stress is caused by the blank being pulled into the cavity. The compressive stress perpendicular to element A is caused by blankholder pressure. Pulling the blank into the die cavity induces

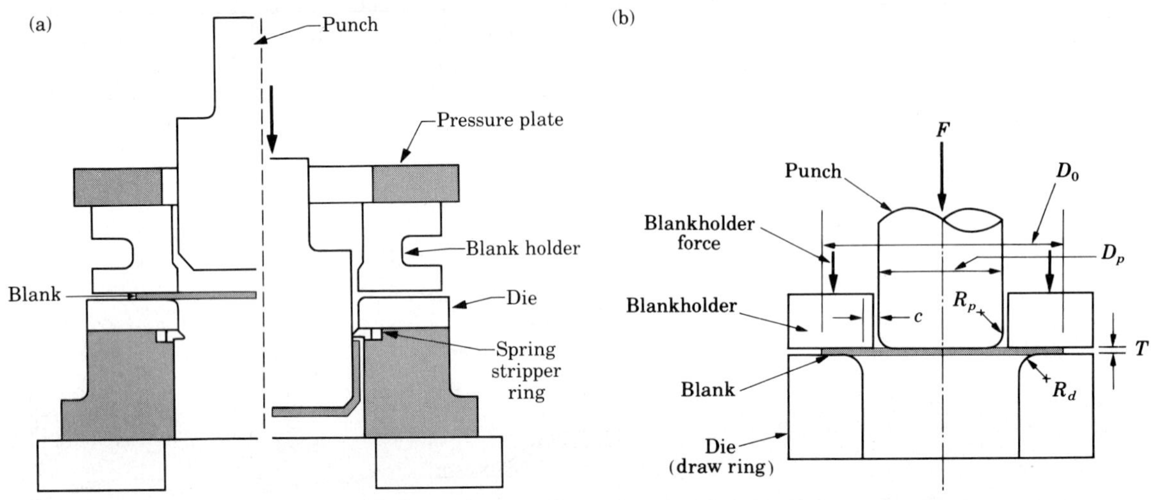

FIGURE 16.32
(a) Schematic illustration of a deep-drawing process using a circular sheet-metal blank. The stripper ring facilitates the removal of the formed cup from the punch. (b) Process variables in deep drawing. Except for the punch force, all parameters indicated are independent variables.

compressive circumferential (hoop) stresses on element *A*, which tend to cause the flange to **wrinkle** during drawing. You can demonstrate wrinkling by trying to force a circular piece of paper into a round cavity such as a drinking glass. Wrinkling can be reduced or eliminated if a blankholder is used under a certain force.

The cup wall, which is already formed, is subjected principally to a longitudinal tensile stress, as shown in element *B*. The punch transmits the force *F* (Fig. 16.32b) through the walls of the cup and to the flange being drawn into the die cavity. Under

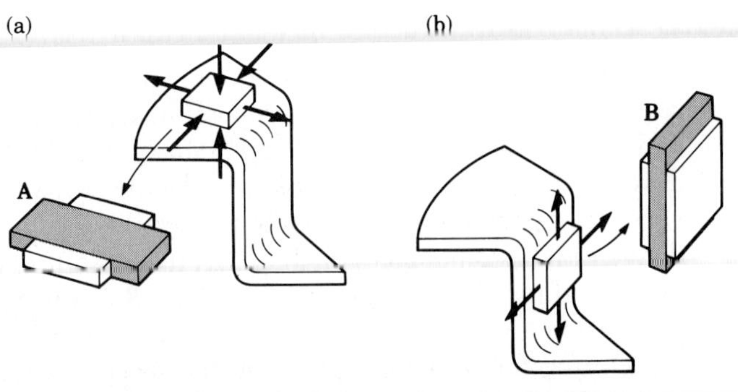

FIGURE 16.33
Deformation of and stresses acting on elements in (a) the flange and (b) the cup wall during deep drawing.

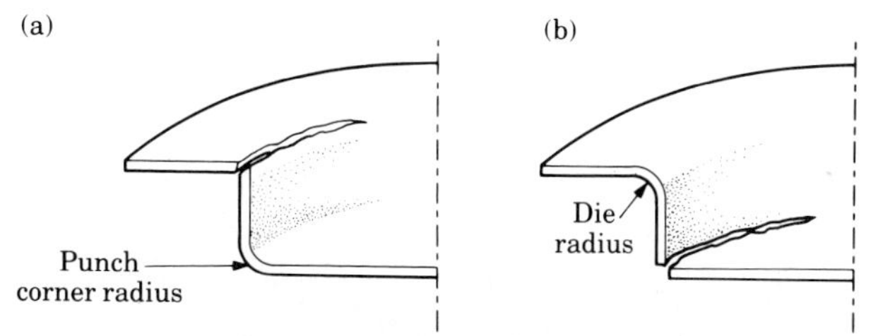

FIGURE 16.34
(a) Fracture of a cup in deep drawing caused by too small a die radius. (b) Fracture caused by too small a punch corner radius. Fracture also results from high longitudinal tensile stresses in the cup due to high ratios of blank diameter to punch diameter.

this stress state, element *B* tends to elongate in the longitudinal direction. Elongation causes the cup wall to thin, which if excessive, causes tearing (Fig. 16.34). The tensile hoop stress on element *B* indicates that the cup may be tight on the punch because of its contraction under tensile stresses in the cup wall (Poisson's effect). Because of the many variables involved, the punch force *F* is difficult to calculate. It increases with increasing strength, diameter, and thickness of the blank.

16.9.2 Deep drawability

In a deep-drawing operation, failure generally results from thinning of the cup wall under high longitudinal tensile stresses. If you follow the movement of the material into the die cavity, you can see that the sheet metal must be capable of undergoing a reduction in width by being reduced in diameter. Yet, it should resist thinning under the longitudinal tensile stresses in the cup wall. Generally, we express *deep drawability* by the **limiting drawing ratio** (LDR):

$$\text{LDR} = \frac{\text{Maximum blank diameter}}{\text{Punch diameter}} = \frac{D_0}{D_p}. \tag{16.6}$$

How can we determine whether we can successfully deep draw a particular sheet metal into a round cup-shaped part? Much effort has gone into identifying the mechanical properties that influence drawability. It has been found that the most important property is normal anisotropy.

Normal anisotropy. **Normal anisotropy** (R), or *plastic anisotropy*, of a sheet metal is defined in terms of the true strains that the specimen undergoes in tension (Fig. 16.35). Thus

$$\text{Normal anisotropy, } R = \frac{\text{Width strain}}{\text{Thickness strain}} = \frac{\varepsilon_w}{\varepsilon_t}. \tag{16.7}$$

In order to obtain the value of R, a specimen is prepared and subjected to an elongation of 15–20 percent, and the true strains are calculated in the manner

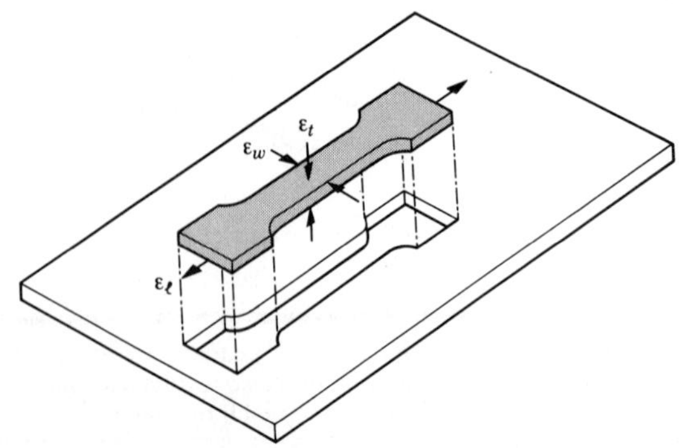

FIGURE 16.35
Strains on a tensile-test specimen removed from a piece of sheet metal. These strains are used in determining the normal and planar anisotropy of the sheet metal.

discussed in Section 2.2. Because cold-rolled sheets generally have anisotropy in their planar direction, the R value of a specimen cut from a rolled sheet will depend on its orientation with respect to the rolling direction of the sheet. In this case, an average value (R_{avg}) is calculated (see Question 16.41). Some typical R_{avg} values are given in Table 16.2.

The direct relationship between R_{avg} and the limiting drawing ratio, as determined experimentally, is shown in Fig. 16.36. No other mechanical property of sheet metal shows as consistent a relationship to LDR as R_{avg} does. Thus by using tension-test results and determining the normal anisotropy of the sheet metal, we can determine the limiting drawing ratio of materials.

TABLE 16.2
TYPICAL RANGE OF AVERAGE NORMAL ANISOTROPY (R_{avg}) FOR VARIOUS SHEET METALS

Zinc	0.2
Hot-rolled steel	0.8–1.0
Cold-rolled rimmed steel	1.0–1.35
Cold-rolled aluminum-killed steel	1.35–1.8
Aluminum	0.6–0.8
Copper and brass	0.8–1.0
Titanium	4–6

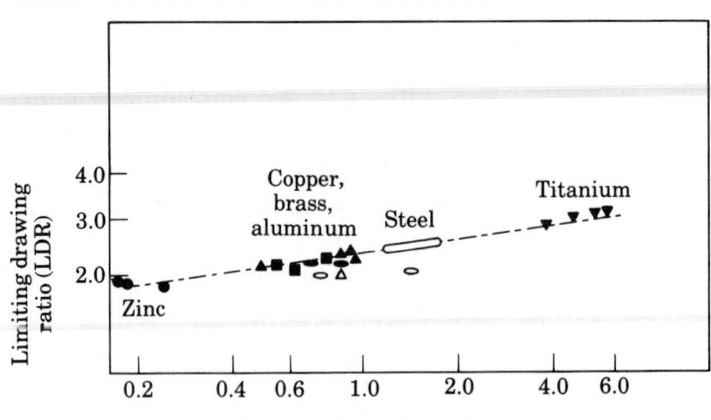

FIGURE 16.36
Relationship between average normal anisotropy and the limiting drawing ratio for various sheet metals. *Source*: M. Atkinson.

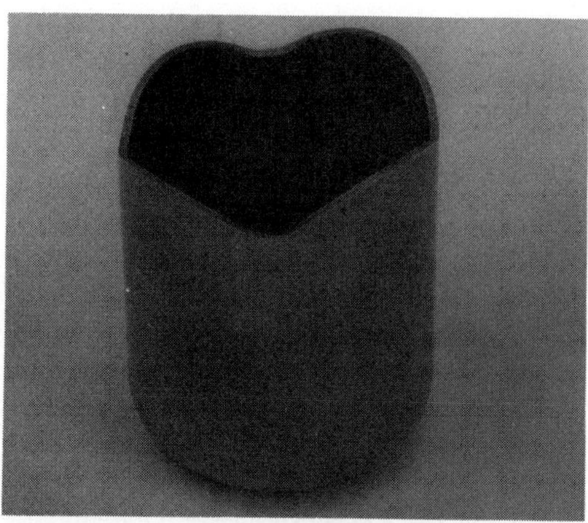

FIGURE 16.37
Earing in a drawn steel cup, caused by the planar anisotropy of the sheet metal.

Earing and planar anisotropy. In drawing, the edges of cups may become wavy, which is called *earing* (Fig. 16.37). This condition is caused by the **planar anisotropy** of the sheet, indicated by ΔR. This value is defined in terms of directional R values (see Question 16.42). The height of the ears increases as ΔR increases. When $\Delta R = 0$, no ears form. Ears are objectionable on drawn cups because they have to be trimmed off, resulting in scrap. The number of ears produced may be two, four, or six.

As you can see, deep drawability is enhanced by a high R_{avg} value and a low ΔR value. Generally, however, sheet metals with high R_{avg} values also have high ΔR values. Sheet-metal textures are being developed to improve drawability by controlling the type of alloying elements in the material and process parameters during rolling of the sheet.

16.9.3 Deep-drawing practice

Certain guidelines have been established for successful deep drawing practice. The blankholder pressure is generally chosen as 0.7–1.0 percent of the sum of the yield and ultimate tensile strengths of the sheet metal. Too high a blankholder force increases the punch force and causes the cup wall to tear. On the other hand, wrinkling will occur if the blankholder force is too low.

Clearances are usually 7–14 percent greater than sheet thickness. If the clearance is too small, the blank may simply be pierced or sheared by the punch. The corner radii of the punch and the die are important. If they are too small, they can cause fracture at the corners (see Fig. 16.34a); if they are too large, the cup wall may wrinkle (*puckering*).

Drawbeads are often necessary to control the flow of the blank into the die cavity (Fig. 16.38a). Beads restrict the flow of the sheet metal by bending and unbending it during drawing. They are especially necessary in drawing box-shaped

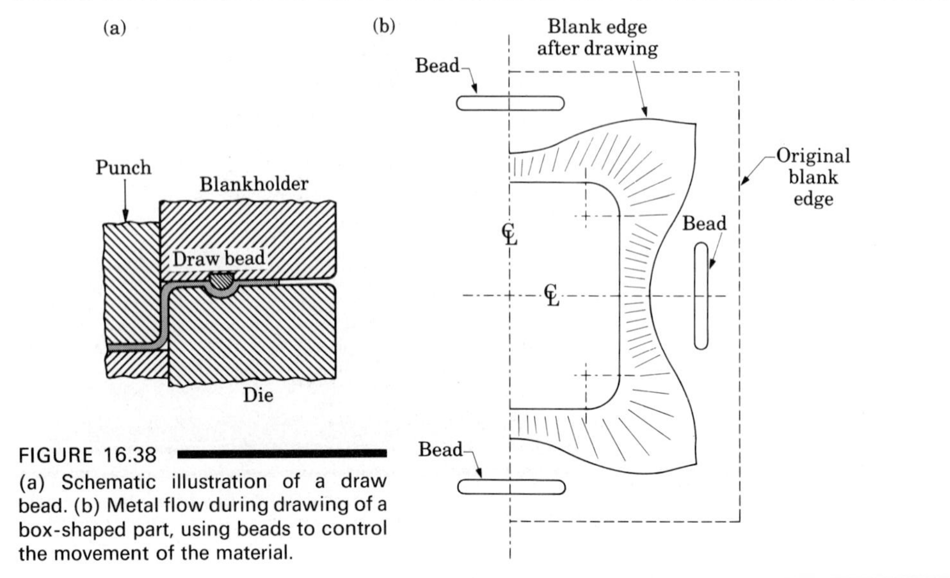

FIGURE 16.38 (a) Schematic illustration of a draw bead. (b) Metal flow during drawing of a box-shaped part, using beads to control the movement of the material.

and nonsymmetric parts (Fig. 16.38b). Drawbeads also help reduce the required blankholder forces, because the beaded sheet has a higher stiffness and hence less tendency to wrinkle. Drawbead diameters may range from 13 mm to 20 mm (0.50 in. to 0.75 in.) for large stampings such as automotive panels.

Ironing. Note in Fig. 16.32(b) that if the clearance is large, the drawn cup will have thicker walls at its rim than at its base. The reason is that the rim consists of material from the outer diameter of the blank, which was reduced in diameter more than the rest of the cup wall. As a result, the cup will have nonuniform wall thickness. **Ironing** is a process in which the wall thickness of a drawn cup is reduced uniformly, either by controlling the clearance or by a separate operation (Fig. 16.39). Aluminum beverage cans, for example, undergo two or three ironing operations in which the drawn cup is pushed through a set of ironing rings.

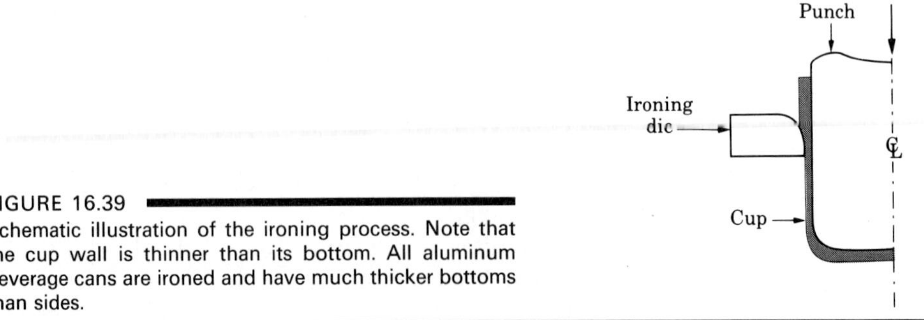

FIGURE 16.39 Schematic illustration of the ironing process. Note that the cup wall is thinner than its bottom. All aluminum beverage cans are ironed and have much thicker bottoms than sides.

Redrawing. Containers or shells that are too difficult to draw in one operation are generally *redrawn* (Fig. 16.40). In *reverse redrawing*, the cup is subjected to bending in the direction opposite to its original bending configuration. Because of volume constancy, the cup becomes longer as it is redrawn to smaller diameters. Aluminum beverage cans generally are redrawn once before being ironed.

Drawing without blankholder. Deep drawing may also be carried out without a blankholder, provided that the sheet metal is thick enough to prevent wrinkling. The dies are specially contoured for this operation (Fig. 16.41).

Various other drawing operations. Many types of parts are made with shallow or moderate depths. Some of these parts involve drawing, stretching, or a combination of these operations. Parts may be embossed with male and female dies or by other means. **Embossing** is an operation consisting of shallow draws, made with matching dies, on a sheet (Fig. 16.42). The process is used principally for decorative purposes and for stiffening flat panels.

Lubrication. In deep drawing, lubrication lowers forces, increases drawability, and reduces defects in parts and wear of the tooling. In general, lubrication of the punch should be held to a minimum, as friction between the punch and the cup

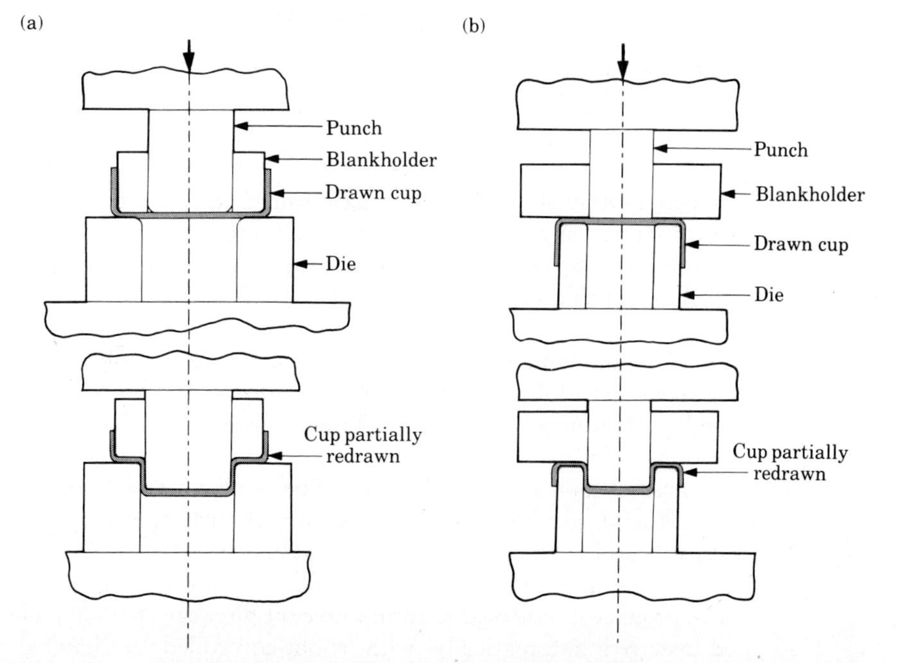

FIGURE 16.40 Reducing the diameter of drawn cups by (a) conventional redrawing and (b) reverse redrawing.

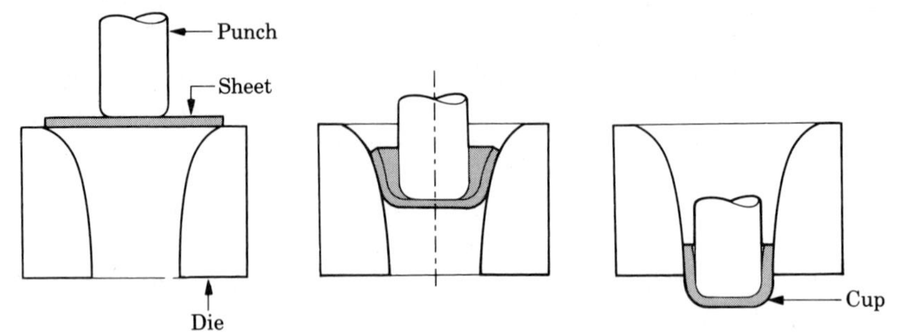

FIGURE 16.41 Drawing without a blankholder, using a die with a tractrix profile. The sheet metal must be relatively thick for these processes to be successful; otherwise the flange will wrinkle (see Question 16.43).

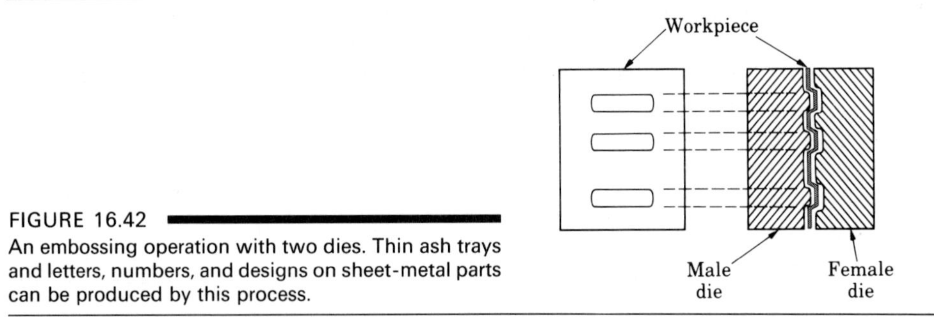

FIGURE 16.42 An embossing operation with two dies. Thin ash trays and letters, numbers, and designs on sheet-metal parts can be produced by this process.

improves drawability by reducing tensile stresses in the cup. For general applications, commonly used lubricants are mineral oils, soap solutions, and heavy-duty emulsions. For more difficult applications, coatings, wax, and solid lubricants are used (see Table 13.1).

Tooling and equipment. The most common tool and die materials for deep drawing are tool steels and cast irons, although other materials, such as carbides and plastics, may also be used. The equipment for deep drawing is usually a double-action hydraulic press or mechanical press, the latter being generally favored because of its higher punch speed. The double-action hydraulic press controls the punch and blankholder independently. Punch speeds generally range between 0.1 m/s and 0.3 m/s (20 ft/min and 60 ft/min).

Modern production facilities are highly automated. For example, a single plant can produce up to 100,000 automotive oil-filter cans per day. The blanks are fed and transferred automatically with robot-controlled mechanical fingers. Lubricant spraying is synchronized with press stroke. Parts are transferred with magnetic or vacuum devices. Inspection systems monitor the entire operation (see Section 36.4).

16.10 Rubber Forming

In the processes described in the preceding sections, we noted that the dies are made of solid materials. However, in **rubber forming** one of the dies in a set can be made of flexible material, such as a rubber or polyurethane membrane. Polyurethanes are used widely because of their resistance to abrasion, resistance to cutting by burrs or sharp edges of the sheet metal, and long fatigue life.

In bending and embossing sheet metal, as shown in Fig. 16.43, the female die has been replaced with a rubber pad. Note that the outer surface of the sheet is protected from damage or scratches because it is not in contact with a hard metal surface during forming. Pressures are usually on the order of 10 MPa (1500 psi).

In the *hydroform* or *fluid-forming* process (Fig. 16.44), the pressure over the rubber membrane is controlled throughout the forming cycle, with maximum pressures of up to 100 MPa (15,000 psi). This procedure allows close control of the part during forming to prevent wrinkling or tearing. Deeper draws are obtained than in conventional deep drawing because the pressure around the rubber membrane forces the cup against the punch. The friction at the punch–cup interface reduces the longitudinal tensile stresses in the cup and thus delays fracture. The control of frictional conditions in rubber forming as well as other sheet-forming operations can be a critical factor in making parts successfully. Selection of proper lubricants and application methods is important.

When selected properly, rubber forming processes have the advantages of low tooling cost, flexibility and ease of operation, low die wear, no damage to the surface of the sheet, and capability to form complex shapes. Parts can also be formed with laminated sheets of various nonmetallic materials or coatings.

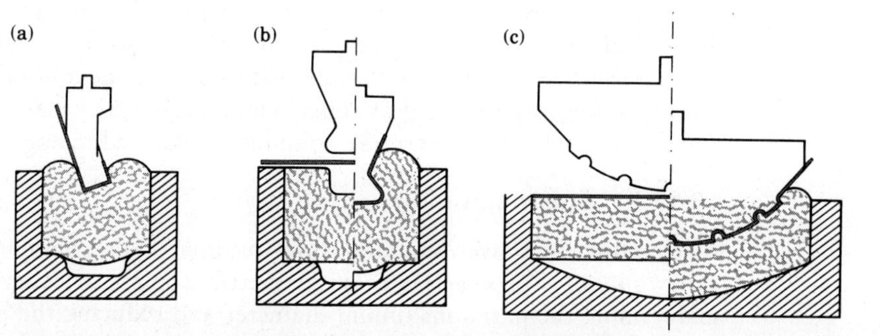

FIGURE 16.43 Examples of bending and embossing sheet metal with a metal punch and flexible pad serving as the female die. *Source*: Polyurethane Products Corporation.

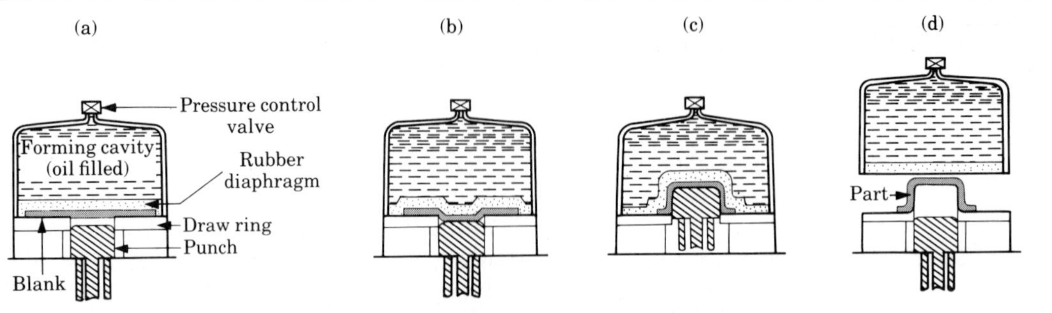

FIGURE 16.44
The hydroform, or fluid forming, process. Note that unlike the ordinary deep-drawing process, the dome pressure forces the cup walls against the punch. The cup travels with the punch and thus deep drawability is improved.

16.11 Spinning

Spinning is an old process and involves the forming of axisymmetric parts over a mandrel with tools or rollers. This process is somewhat similar to forming clay on a potter's wheel. There are three basic types of spinning processes: conventional (or manual), shear, and tube spinning. The equipment used in these processes is similar to a lathe, but with special features.

16.11.1 Conventional spinning

In *conventional spinning*, a circular blank of flat or preformed sheet metal is held against a mandrel and rotated while a rigid tool deforms and shapes the material over the mandrel (Fig. 16.45a). The tools may be activated manually or by a computer-controlled hydraulic mechanism. The process involves a sequence of passes and requires considerable skill. Conventional spinning is particularly suitable for conical and curvilinear shapes, which otherwise would be difficult or uneconomical to produce (Fig. 16.45b). Part diameters may range up to 6 m (20 ft). Although most spinning is performed at room temperature, thick parts or metals with low ductility or high strength require spinning at elevated temperatures.

16.11.2 Shear spinning

Also known as power spinning, flow turning, hydrospinning, and spin forging, *shear spinning* produces an axisymmetric conical or curvilinear shape while maintaining the part's maximum diameter and reducing the part's thickness (Fig. 16.46). Although a single roller can be used, two rollers are preferable in order to balance the forces acting on the mandrel. Typical parts are rocket motor casings and missile nose cones.

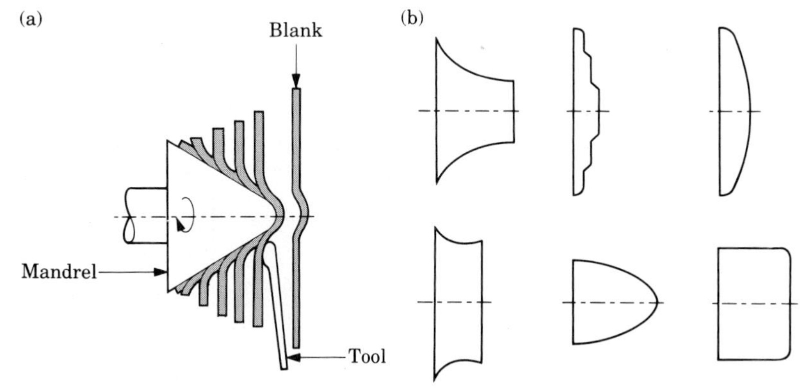

FIGURE 16.45
(a) Schematic illustration of the conventional spinning process. (b) Types of parts conventionally spun. All parts are axisymmetric.

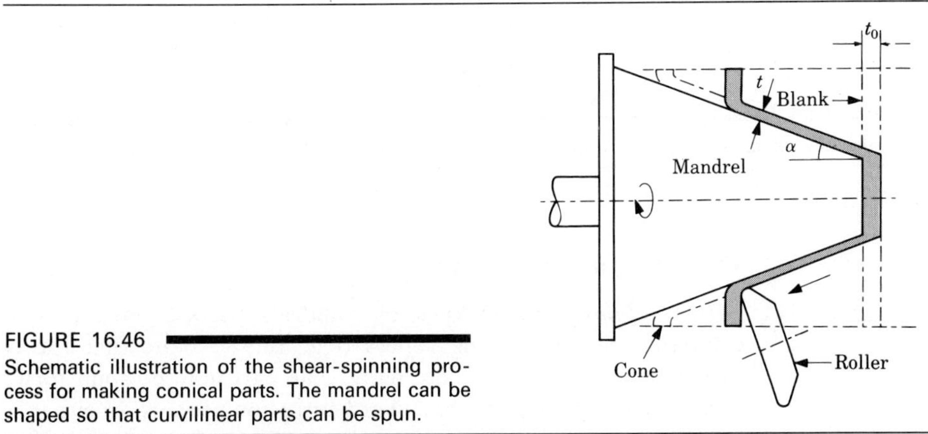

FIGURE 16.46
Schematic illustration of the shear-spinning process for making conical parts. The mandrel can be shaped so that curvilinear parts can be spun.

Parts up to about 3 m (10 ft) in diameter can be formed by shear spinning. The operation wastes little material and can be completed in a relatively short time. Various shapes can be spun with relatively simple tooling, which is generally made of tool steel. Because of the large deformations involved, this process generates considerable heat, thus necessitating the use of water-base coolants during spinning.

The **spinnability** of metal is defined as the maximum reduction in thickness to which a part can be subjected by spinning without fracture. Spinnability is related to the tensile reduction of area of the material, just as it is in bending (see Fig. 16.19). If a metal has a tensile reduction of area of about 50 percent or more, its thickness can be reduced in one spinning pass by as much as 80 percent. Materials with low ductility are spun at elevated temperatures.

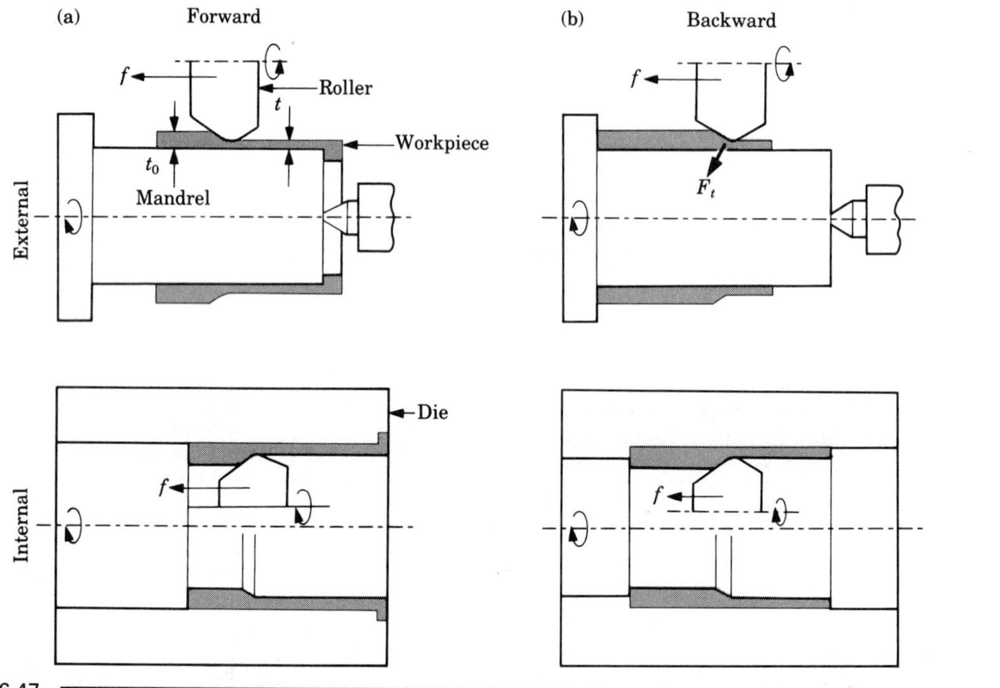

FIGURE 16.47
Examples of external and internal and forward and backward tube-spinning processes.

16.11.3 Tube spinning

In *tube spinning*, the thickness of cylindrical parts is reduced by spinning them on a cylindrical mandrel using rollers (Fig. 16.47). Note that the operation may be carried out externally or internally. The part may be spun forward or backward, similar to a drawing or a backward extrusion process. In either case, the reduction in wall thickness results in a longer tube. The maximum reduction per pass in tube spinning is related to the tensile reduction of area of the material, similar to shear spinning. Tube spinning can be used to make pressure vessels, automotive components, and rocket and missile parts. Some jet-engine parts are made by this process.

• Example: Manufacturing a jet-engine compressor shaft.

The Olympus jet-engine compressor shaft for the supersonic Concorde aircraft is made by a combination of forging and spinning operations, as shown in the accompanying figure. First, a preform is made by a series of forging operations on a superalloy blank. It is then machined to fit over the mandrel of a horizontal spinning machine. The first forming operation consists of tube spinning (pass 1) the small end of the forged blank with a pair of rollers. In pass 2 the annular section is formed into a conical shape by shear spinning it over the conical section of the mandrel. The last three passes consist of gradually reducing the thickness of the large end by tube spinning. Note the profile produced on the outer diameter of the shaft. The formed part is then machined to obtain desired tolerances and geometric features.

A number of alternative methods had been considered for manufacturing this and similar axisymmetric parts for this jet engine. However, the most economical method was that shown in the figure. Forging the part to closer dimensions to the final shape, thus reducing the number of spinning passes, was not economical. The cost of forging such parts increased significantly as their length-to-diameter ratio increased, particularly for thin-walled components.

Several other parts of this jet engine also are made by spinning them on a two-roller, 100-hp (75-kW) spinning machine. The operations are performed at room temperature, using a water-based lubricant that acts mainly as a coolant to carry away the substantial heat generated during the spinning process. The parts are made of superalloys.

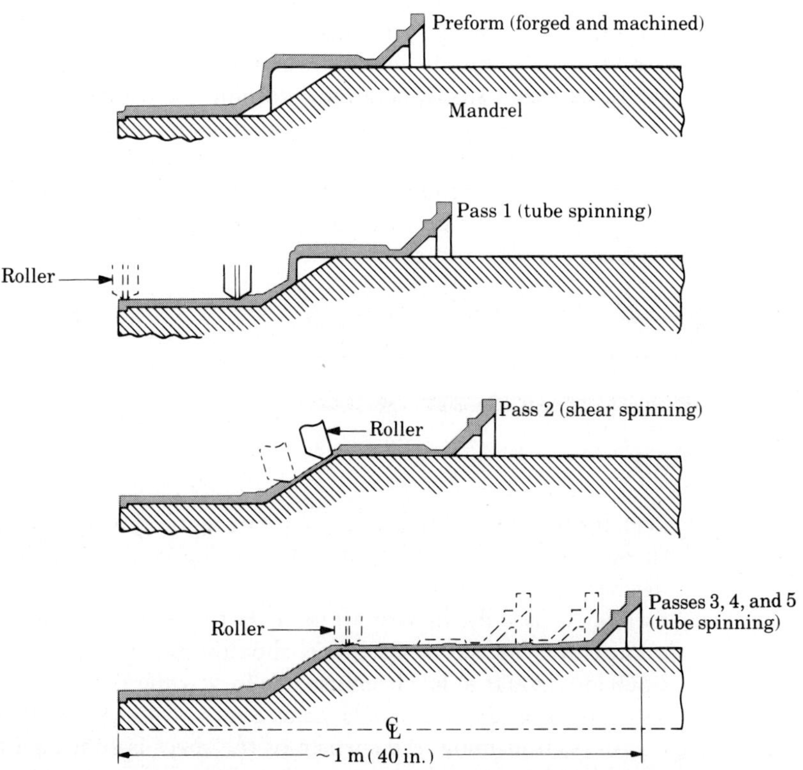

16.12

Superplastic Forming

In Section 2.2, we described that superplastic behavior has been observed in certain very fine-grained alloys (less than 10–15 μm) where elongations up to 2000 percent have been obtained at certain temperatures and low strain rates. Thus these materials have excellent formability. Sheet metals suitable for **superplastic forming** are Ti-6Al-4V, 7475 aluminum (modified 7075), IN-100 and INCO 718 nickel alloys, Zn-22Al, and iron-base, high-carbon alloys.

Complex shapes with fine detail and close tolerances can be formed in one piece, thus eliminating secondary operations. The relatively low strength of superplastic alloys at forming temperatures means that low-strength tooling can be used, thus reducing die costs. Furthermore, little or no residual stresses are present in formed parts. On the other hand, the sheet metal must not be superplastic at service temperatures. Furthermore, because of the extreme strain-rate sensitivity of the superplastic metal (high m values), it must be formed at sufficiently low rates, typically at strain rates of 10^{-4}–10^{-2}/s. Forming times range from a few seconds to several hours. Thus cycle times are much longer than in conventional forming processes. Commonly used die materials in superplastic forming are low-alloy steels, cast tool steels, ceramics, graphite, and plaster of paris, depending on the strength and temperature of the superplastic alloy to be formed.

Superplastic alloys can be formed by methods used to form plastics, such as thermoforming, vacuum forming, and blow molding, which we describe in Chapter 18. They can also be formed by bulk-deformation processes, such as closed-die forging, coining, and extrusion.

16.13

Peen Forming

Peen forming is used to produce curvatures on thin sheet metals by *shot peening* one surface of the sheet (Fig. 16.48). Peening is done with cast-iron or steel shot, discharged either from a rotating wheel or with an air blast from a nozzle. Peen forming is used by the aircraft industry to generate smooth and complex curvatures on aircraft wing skins. Cast-steel shot about 2.5 mm (0.1 in.) in diameter at speeds of 60 m/s (200 ft/s) has been used to form wing panels 25 m (80 ft) long. For heavy sections, shot diameters as large as 6 mm (1/4 in.) may be used.

In peen forming, the surface of the sheet is subjected to compressive stresses, which tend to expand the surface layer. Since the material below the peened surface remains rigid, the surface expansion causes the sheet to develop a curvature. The

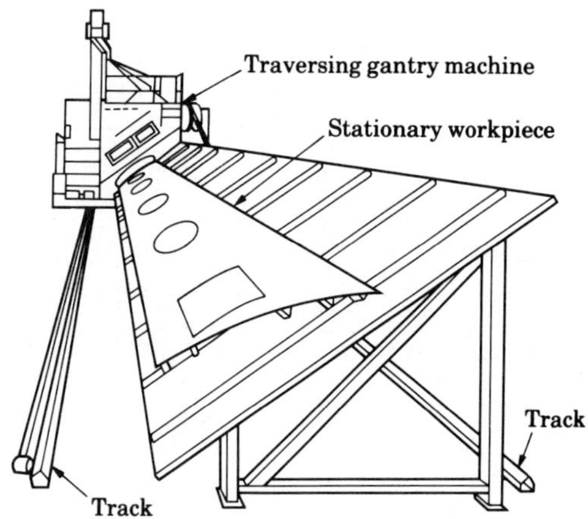

FIGURE 16.48
Peen-forming machine to form a large sheet-metal part, such as an aircraft-skin panel. The sheet is stationary and the machine traverses it. *Source:* Metal Improvement Company.

process also induces compressive surface residual stresses, with improved fatigue strength of the sheet. The peen forming process is also used for straightening twisted or bent parts. Out-of-round rings, for example, can be straightened by this method.

16.14 Explosive Forming

We know that explosives are used for many destructive purposes, such as in demolition work and warfare. However, by controlling their quantity, we can use explosives as a source of energy. This energy was first utilized to form metals in the early 1900s. Typically, in **explosive forming** the sheet-metal blank is clamped over a die and the entire assembly is lowered into a tank filled with water (Fig. 16.49a). The air in the die cavity is evacuated, an explosive charge is placed at a certain height, and the charge is detonated. The rapid conversion of the explosive charge into gas generates a shock wave. The pressure of this wave is sufficient to form sheet metals. A variety of shapes can be formed, provided the material is ductile at high rates of deformation.

The process is versatile, with no limit to the size of the workpiece, and is particularly suitable for low-quantity production runs of large parts, as in aerospace applications. Steel plates 25 mm (1 in.) thick and 3.6 m (12 ft) in diameter have been formed by this method. Tubes having walls as thick as 25 mm (1 in.) have been bulged by explosive forming techniques. The mechanical properties of parts made

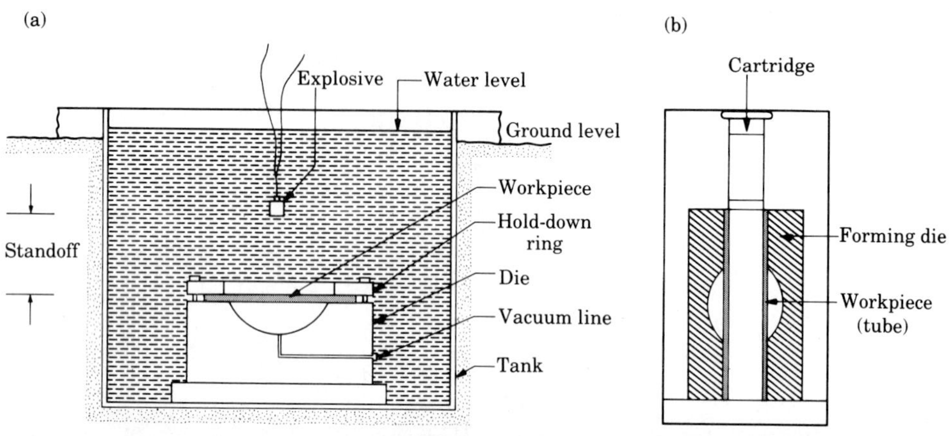

FIGURE 16.49
(a) Schematic illustration of the explosive forming process. (b) Illustration of the confined method of explosive bulging of tubes.

by this process are basically the same as those made by conventional forming methods. Depending on the number of parts to be produced, dies may be made of aluminum alloys, steel, ductile iron, zinc alloys, reinforced concrete, wood, plastics, or composite materials.

Another explosive forming method is shown in Fig. 16.49(b), in which only a *cartridge* (canned explosive) is used as the source of energy. This process can also be used for bulging and expanding thin-walled tubes.

16.15 Magnetic-Pulse Forming

In **magnetic-pulse forming**, the energy stored in a capacitor bank is discharged rapidly through a magnetic coil. In a typical example, a ring-shaped coil is placed over a tubular workpiece that is to be formed. The tube is then collapsed over another solid piece by magnetic forces, making the assembly an integral part (Fig. 16.50).

The magnetic field produced by the coil crosses the metal tube (a conductor), generating *eddy currents* in the tube. This current, in turn, produces its own *magnetic field* (Fig. 16.50a). The forces produced by the two magnetic fields oppose each other; thus there is a repelling force between the coil and the tube. The forces generated collapse the tube over the inner piece. The higher the electrical conductivity of the workpiece, the higher are the magnetic forces. It is not necessary for the workpiece material to have magnetic properties.

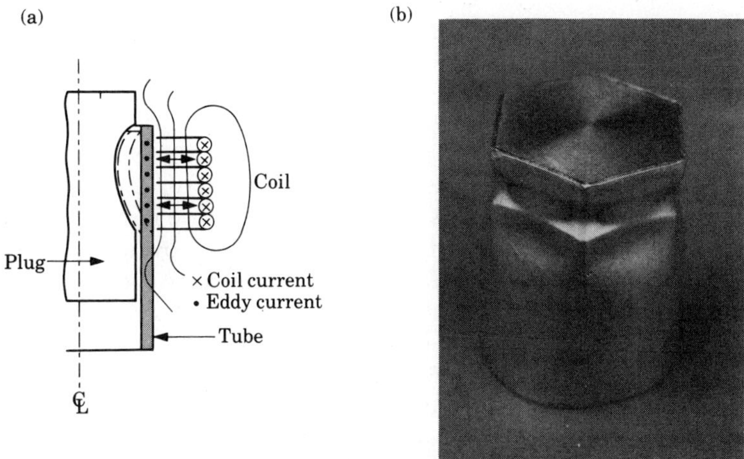

FIGURE 16.50

(a) Schematic illustration of the magnetic pulse forming process for forming a tube over a plug. (b) Aluminum tube collapsed over a hexagonal plug by the magnetic pulse forming process.

Magnetic-pulse forming is used to collapse thin-wall tubes over rods, cables, and plugs, and in bulging and flaring operations. Flat magnetic coils are also available for embossing and shallow drawing operations on sheet metals.

16.16 Other Sheet-Forming Methods

Although not commonly used, there are several other types of sheet-forming processes. In *electrohydraulic forming* (underwater spark or electric-discharge forming), the source of energy is a spark from electrodes connected with a thin wire. The rapid discharge of the energy in a capacitor bank through the electrodes generates a shock wave, much like explosives do, and forms the part. This process has been used in making relatively small parts, at energy levels lower than in explosive forming.

Gas mixtures in a closed container have been utilized as an energy source. When ignited, the pressures generated are sufficient to form parts. The principle is similar to the generation of pressure in an internal combustion engine.

Liquefied gases, such as liquid nitrogen, may be used to develop pressures high enough to form sheet metals. When allowed to reach room temperature in a closed container, liquefied nitrogen becomes gaseous and expands, developing the necessary pressure to form the part.

16.17

Dent Resistance of Sheet-Metal Parts

You have seen dents on cars, appliances, office furniture, and kitchen utensils. How can we design parts and select appropriate materials to minimize dents? Dents are usually caused by dynamic forces from moving objects that hit the sheet metal. In typical automotive panels, for example, velocities at impact range up to 45 m/s (150 ft/sec). Thus it is the dynamic yield stress (yield stress under high rates of deformation) rather than the static yield stress that is the significant strength parameter. Dynamic forces tend to cause localized dents, whereas static forces tend to spread the dented area. Try to dent a piece of flat sheet metal, first by pushing a hammer against it, and then by striking it with the hammer. Note how localized the dent is in the latter case.

Dent resistance of sheet-metal parts has been found to increase as the sheet's yield stress and thickness increase. Dent resistance decreases as the sheet's elastic modulus and overall panel stiffness increase. Thus panels rigidly held at their edges have lower dent resistance.

16.18

Equipment for Sheet-Metal Forming

We have described a number of sheet-forming operations, each requiring its own set of tooling and equipment. For most pressworking operations, the basic equipment consists of mechanical, hydraulic, pneumatic, or pneumatic-hydraulic presses. We described basic features and characteristics of such presses in Section 14.9. Typical designs for press frames are shown in Fig. 16.51 (see also Figs. 14.28–14.30). Proper design, construction, and stiffness of such equipment is essential to efficient operation, high production rate, good dimensional control, and high product quality.

The traditional *C-frame* structure (Fig. 16.51a) has been widely used because of easy tool and workpiece accessibility. However, it is not as stiff as the *box-type* (O-type) pillar and double-column frame structures (Fig. 16.51b). On the other hand, advances in automation and use of industrial robots and computer controls have made accessibility less important.

Press selection for sheet-metal forming operations depends on several factors:

a) Type of forming operation, and dies and tooling required.
b) Size and shape of workpiece.
c) Length of stroke of the slide or slides, strokes per minute, speed, and shut height (distance from the top of the bed to the bottom of the slide, with the stroke down).

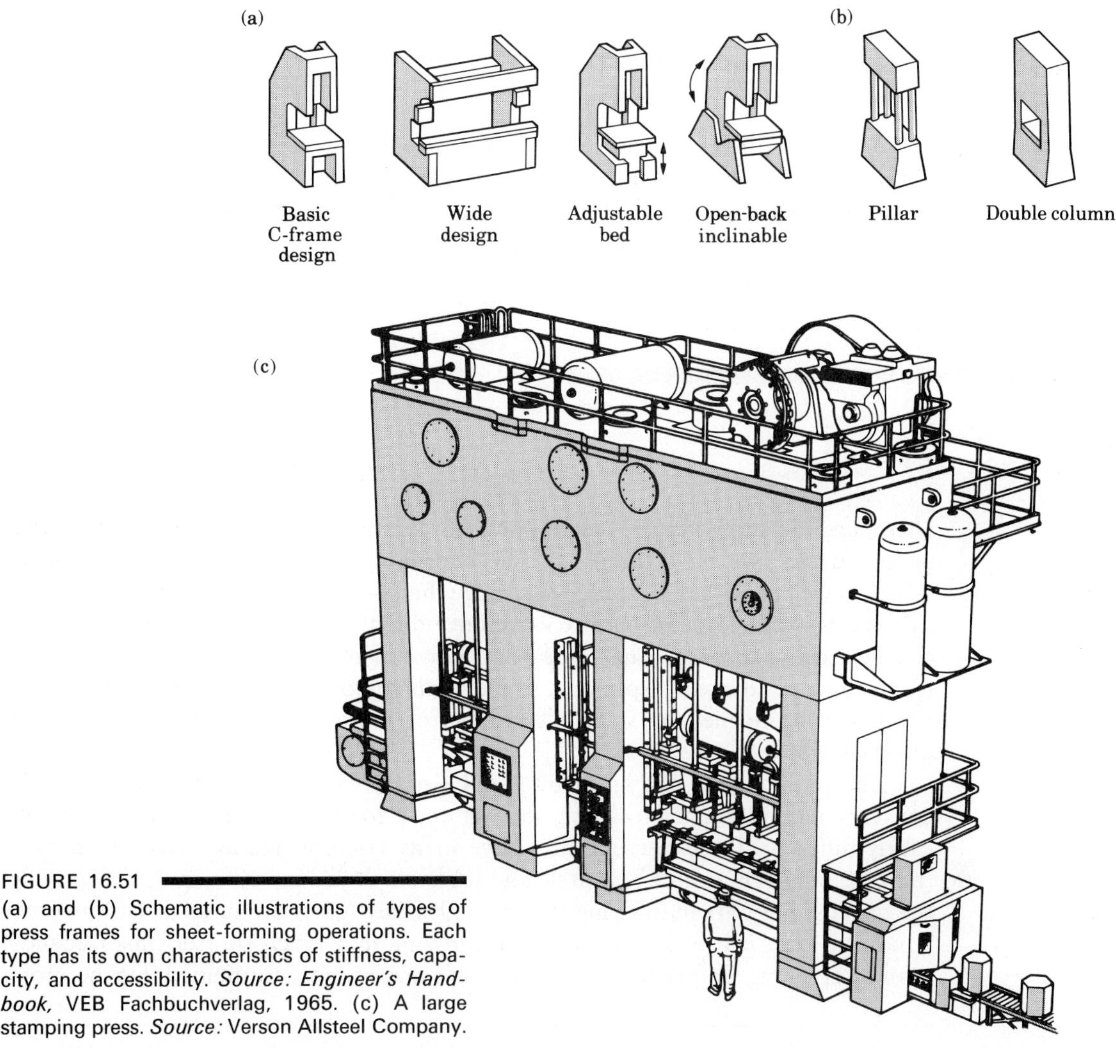

FIGURE 16.51 (a) and (b) Schematic illustrations of types of press frames for sheet-forming operations. Each type has its own characteristics of stiffness, capacity, and accessibility. *Source: Engineer's Handbook,* VEB Fachbuchverlag, 1965. (c) A large stamping press. *Source:* Verson Allsteel Company.

d) Number of slides. Single-action presses have one reciprocating slide. Double-action presses have two slides, reciprocating in the same direction, and are used typically for deep drawing, one slide for the punch and the other for the blankholder. Triple-action presses have three slides, and are used typically for reverse redrawing and other forming operations.
e) Maximum force required (press capacity, tonnage rating).
f) Type of controls.
g) Die changing features. Because the time required for changing dies in presses can be significant (as much as a few hours), thus affecting productivity, rapid die-changing systems have been developed. Die setups, called *single-minute exchange of dies* (SMED), can now be changed in less than ten minutes,

using automated hydraulic or pneumatic systems. These techniques are particularly important in automated and computer-integrated manufacturing systems (Chapters 38 and 39).

h) Safety features (see Section 37.3).

Because a press is a major capital investment, its use for a variety of parts and applications should be investigated. Thus versatility and multiple use is an important factor in its selection, particularly for product modifications and new products to respond to changes in market demand.

16.19 Economics of Sheet-Metal Forming

Sheet-metal forming involves economic considerations similar to those for the other processes that we have discussed. Sheet-forming operations compete with each other, as well as with other processes, more than other processes do. We have noted that sheet-forming operations are versatile and that a number of different processes can be used to produce the same part. For example, a cup-shaped part can be formed by deep drawing, spinning, rubber forming, or explosive forming. Similarly, a cup can be formed by impact extrusion or casting, or by fabricating it from different pieces.

The part shown in Fig. 16.52 can be made either by deep drawing or by conventional spinning. However, the die costs for the two processes are significantly different. Deep-drawing dies have many components and cost much more than the relatively simple mandrels and tools employed in spinning. Consequently, the die cost per part in drawing will be high if few parts are needed. On the other hand, this part can be formed by deep drawing in a much shorter time (seconds) than by spinning (minutes), even if the latter operation is automated. Furthermore,

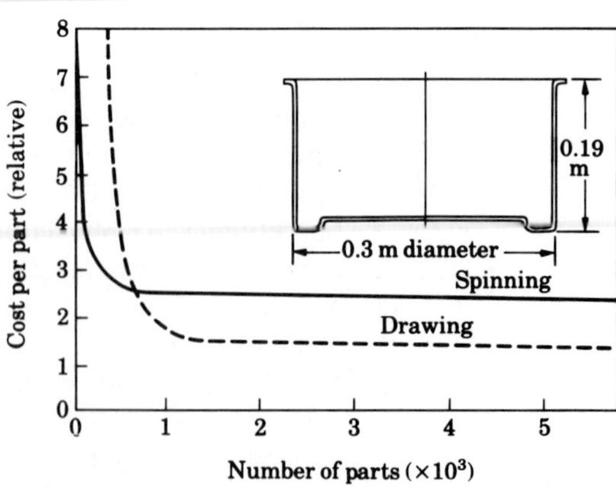

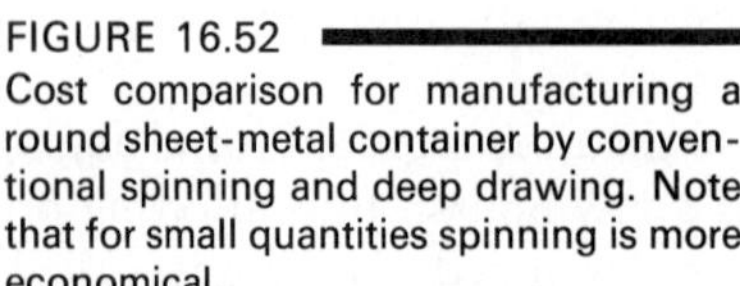
FIGURE 16.52 Cost comparison for manufacturing a round sheet-metal container by conventional spinning and deep drawing. Note that for small quantities spinning is more economical.

spinning requires more skilled labor. Considering all these factors, we find that the breakeven point is at about 700 parts. Thus deep drawing is more economical for quantities greater than that. We describe further details of economics of manufacturing in Chapter 40.

SUMMARY

Sheet-metal forming processes are among the most versatile of all forming operations and generally are used with workpieces having high ratios of surface area to thickness. Unlike bulk deformation processes, such as forging and extrusion, the cross-sectional area of the material in sheet forming is generally prevented from being reduced, avoiding necking and tearing. The important material parameters are the capacity of the sheet metal to stretch uniformly and its resistance to thinning. Various tests have been developed to predict the formability of sheet metals.

Because of the thin sheet used, springback, buckling, and wrinkling are significant problems in sheet forming. These problems can be avoided by proper tool and die design and by minimizing the unsupported length of the material during processing. The forces and energy required in sheet forming can be transmitted to the workpiece through solid tools and dies, flexible rubber or polyurethane members, and electrical, chemical, magnetic, and gaseous means.

SUMMARY TABLE 16.1

CHARACTERISTICS OF METALS IMPORTANT IN SHEET FORMING

CHARACTERISTIC	IMPORTANCE
Elongation	Determines the capability of the sheet metal to stretch without necking and failure; high strain-hardening exponent (n) and strain-rate sensitivity exponent (m) desirable.
Yield-point elongation	Observed with mild-steel sheets; also called Lueder's bands and stretcher strains; causes flamelike depressions on the sheet surfaces; can be eliminated by temper rolling, but sheet must be formed within a certain time after rolling.
Anisotropy (planar)	Exhibits different behavior in different planar directions; present in cold-rolled sheets because of preferred orientation or mechanical fibering; causes earing in drawing; can be reduced or eliminated by annealing but at lowered strength.
Anisotropy (normal)	Determines thinning behavior of sheet metals during stretching; important in deep-drawing operations.
Grain size	Determines surface roughness on stretched sheet metal; the coarser the grain, the rougher the appearance (orange peel).
Residual stresses	Caused by nonuniform deformation during forming; causes part distortion when sectioned and can lead to stress-corrosion cracking; reduced or eliminated by stress relieving.
Springback	Caused by elastic recovery of the plastically-deformed sheet after unloading; causes distortion of part and loss of dimensional accuracy; can be controlled by techniques such as overbending and bottoming of the punch.

(*continued*)

SUMMARY TABLE 16.1 (*continued*)

CHARACTERISTICS OF METALS IMPORTANT IN SHEET FORMING

CHARACTERISTIC	IMPORTANCE
Wrinkling	Caused by compressive stresses in the plane of the sheet; can be objectionable or can be useful in imparting stiffness to parts; can be controlled by proper tool and die design.
Quality of sheared edges	Depends on process used; edges can be rough, not square, and contain cracks, residual stresses, and a work-hardened layer, which are all detrimental to the formability of the sheet; quality can be improved by control of clearance, tool and die design, fine blanking, shaving, and lubrication.
Surface condition of sheet	Depends on rolling practice; important in sheet forming as it can cause tearing and poor surface quality; see, also, Section 13.3.

SUMMARY TABLE 16.2

CHARACTERISTICS OF SHEET-METAL FORMING PROCESSES

PROCESS	CHARACTERISTICS
Roll forming	Long parts with constant complex cross-sections; good surface finish; high production rates; high tooling costs.
Stretch forming	Large parts with shallow contours; suitable for low-quantity production; high labor costs; tooling and equipment costs depend on part size.
Drawing	Shallow or deep parts with relatively simple shapes; high production rates; high tooling and equipment costs.
Stampings	Includes a variety of operations, such as punching, blanking, embossing, bending, flanging, and coining; simple or complex shapes formed at high production rates; tooling and equipment costs can be high, but labor cost is low.
Rubber forming	Drawing and embossing of simple or complex shapes; sheet surface protected by rubber membranes; flexibility of operation; low tooling costs.
Spinning	Small or large axisymmetric parts; good surface finish; low tooling costs, but labor costs can be high unless operations are automated.
Superplastic forming	Complex shapes, fine detail and close tolerances; forming times are long, hence production rates are low; parts not suitable for high-temperature use.
Peen forming	Shallow contours on large sheets; flexibility of operation; equipment costs can be high; process is also used for straightening parts.
Explosive forming	Very large sheets with relatively complex shapes, although usually axisymmetric; low tooling costs, but high labor cost; suitable for low-quantity production; long cycle times.
Magnetic-pulse forming	Shallow forming, bulging, and embossing operations on relatively low strength sheets; most suitable for tubular shapes; high production rates; requires special tooling.

TRENDS

- Work is continuing on techniques to determine formability of sheet metals under various conditions.
- Various metal textures are being developed and controlled to improve formability.
- Computer-aided die design and manufacturing are being implemented to reduce costs and improve productivity.
- Automated systems for changing press dies quickly are being implemented.

KEY TERMS

Bending	Forming-limit diagram	Progressive dies
Bending force	Fracture surface	Punching
Blanking	Ironing	Rubber forming
Burnished surface	Lueder's bands	Shearing
Burr	Limiting drawing ratio	Spinnability
Clearance	Magnetic-pulse forming	Spinning
Compound dies	Maximum punch force	Springback
Deep drawing	Minimum bend radius	Stretch forming
Embossing	Normal anisotropy	Superplastic forming
Explosive forming	Peen forming	Transfer dies
Formability	Planar anisotropy	Wrinkling

BIBLIOGRAPHY

Blazynski, T.Z., *Plasticity and Modern Metal-forming Technology*. New York: Elsevier, 1989.

Bowman, H.B., *Handbook of Precision Sheet, Strip and Foil*. Metals Park, Ohio: American Society for Metals, 1980.

Eary, D.F., and E.A. Reed, *Techniques of Pressworking Sheet Metal*, 2d ed. Englewood Cliffs, N.J.: Prentice-Hall, 1974.

Hoffmann, E.G., *Fundamentals of Tool Design*, 2d ed. Dearborn, Mich.: Society of Manufacturing Engineers, 1984.

Metals Handbook, 8th ed., Vol. 4: *Forming*. Metals Park, Ohio: American Society for Metals, 1969.

Metals Handbook, 9th ed., Vol. 14: *Forming and Forging*. Metals Park, Ohio: ASM International, 1988.

Morgan, E., *Tinplate and Modern Canmaking Technology*. Oxford, England: Pergamon, 1985.

Pacquin, J.R., and R.E. Crowley, *Die Design Fundamentals*. New York: Industrial Press, 1987.

Sachs, G., *Principles and Methods of Sheet Metal Fabricating*, 2d ed. New York: Reinhold, 1966.

Smith, D., *Die Design Handbook*, 3d ed. Dearborn, Mich: Society of Manufacturing Engineers, 1990.

Source Book on Forming of Steel Sheet. Metals Park, Ohio: American Society for Metals, 1976.

Strasser, F., *Metal Stamping Plant Productivity Handbook*. New York: Industrial Press, 1983.

Tool and Manufacturing Engineers Handbook, 4th ed., Vol. 2: *Forming*. Dearborn, Mich.: Society of Manufacturing Engineers, 1984.

REVIEW QUESTIONS

16.1 How does sheet-metal forming differ from rolling, forging, and extrusion?

16.2 Name the important process variables in shearing.

16.3 What causes burrs? How can they be reduced or eliminated?

16.4 Explain the difference between punching and blanking.

16.5 List the various operations performed by die cutting. What types of applications do these processes have in manufacturing?

16.6 What is a steel-rule die?

16.7 Explain what nibbling is. What are its advantages?

16.8 Describe the difference between compound, progressive, and transfer dies.

16.9 Name the various methods by which sheet-metal blanks can be cut from a large sheet.

16.10 List the characteristics of sheet metals that are important in sheet-forming operations. Explain why they are important.

16.11 What is the significance of anisotropy?

16.12 Describe the features of forming-limit diagrams (FLD).

16.13 How is the minimum bend radius for a material defined in practice?

16.14 List the properties of materials that influence springback. Explain why they do.

16.15 Make a list of common bending operations, giving one application for each.

16.16 Why do tubes buckle when bent?

16.17 Outline the features of a deep-drawing operation.

16.18 Explain why normal anisotropy is important in determining the deep drawability of a material.

16.19 Describe earing and why it occurs.

16.20 What are the advantages of rubber forming?

16.21 Explain the difference between conventional and shear spinning.

16.22 What is superplastic forming of metals?

16.23 Explain the features of unconventional forming methods for sheet metal.

QUESTIONS AND PROBLEMS

16.24 Outline the differences that you have observed between products made of sheet metals and those made by casting and forging.

16.25 Make a survey of some of the products in your home that are made of sheet metal and discuss the process or processes by which you think they were made.

16.26 Describe the cutting process of a pair of scissors.

16.27 Identify the material and process variables that influence the punch force in shearing and explain how each of these affects this force.

16.28 Explain why springback in bending depends on yield stress, elastic modulus, sheet thickness, and bend radius.

16.29 What precautions would you take to reduce burrs in shearing?

16.30 What is the significance of the size of the circles in the grid pattern shown in Fig. 16.16?

16.31 Explain why cupping tests may not predict the formability of sheet metals in actual forming processes.

16.32 In the text we stated that the thicker the sheet metal, the higher the curves in Fig. 16.15(b) become. Why do you think this is so?

16.33 Inspect sheet-metal food or coffee cans and describe how the bulges could be produced on a high-volume production setup.

16.34 Identify the factors that influence the deep-drawing force F in Fig. 16.32 and explain why they do.

16.35 Inspect the earing shown in Fig. 16.37 and identify the direction in which the blank was cut from a cold-rolled sheet.

16.36 Why are the beads in Fig. 16.38(b) placed in those particular locations?

16.37 Describe the factors that influence the size and length of beads in drawing operations.

16.38 Many missile components are made by spinning. What other processes could you use if spinning processes were not available?

16.39 Duplicate the peen forming process by hammering an aluminum sheet with a ball-peen hammer. Describe your observations about the curvature produced.

16.40 Describe the features of the different types of presses shown in Fig. 16.51. What are typical applications for each one?

16.41 The average R value for a sheet metal with planar anisotropy is expressed as

$$R_{avg} = \frac{R_0 + 2R_{45} + R_{90}}{4},$$

where the angles are relative to the rolling direction of the sheet. Calculate the average value of R for steel where the R values for the 0°, 45°, and 90° directions are 0.9, 1.3, and 1.9, respectively. What is the limiting drawing ratio for this material?

16.42 The planar anisotropy ΔR for sheet metal is defined as

$$\Delta R = \frac{R_0 - 2R_{45} + R_{90}}{2}.$$

Calculate this value for the case in Question 16.41. Will any ears form when this material is deep drawn? Explain.

16.43 A general rule for dimensional relationships for successful drawing without a blankholder is given by

$$(D_0 - D_p) < 5T.$$

Explain what happens if this limit is exceeded.

16.44 Estimate the limiting drawing ratio (LDR) for the materials listed in Table 16.2.

16.45 Prove Eq. (16.2).

16.46 Regarding Eq. (16.2), it has been stated that actual values of strain e at the outer fibers (in tension) in bending are higher than those for the inner fibers (in compression), the reason being that the neutral axis shifts during bending. With an appropriate sketch, explain this phenomenon.

16.47 Describe the (a) similarities and (b) dissimilarities between the bulk deformation processes described in Chapters 13–15 and the sheet-forming processes covered in this chapter.

16.48 Inspect a common paper punch and comment on the shape of the punch end, compared to those shown in Fig. 16.8.

16.49 It has been shown that the peak pressure p in explosive forming, using TNT as the explosive, can be approximated by the formula

$$p = 21{,}600(W^{1/3}/R)^{1.15},$$

where p is in psi, W is the weight of the explosive in lb, and R is the standoff (distance of the explosive from the workpiece surface) in ft. What should be the weight of the explosive in order to generate a pressure of 10,000 psi at a standoff distance of 5 ft?

16.50 Using the formula in Problem 16.49, plot the pressure as a function of weight W and R, respectively. Describe your observations.

17

Powder Metallurgy

17.1 Introduction

In the manufacturing processes we described in the preceding chapters, the raw materials used are either in a molten state or in solid form. In this chapter we describe how metal parts can be made by compacting metal powders in suitable dies and sintering them (heating without melting). This process is called **powder metallurgy (P/M)**. One of its first uses was in the early 1900s to make the tungsten filaments for incandescent light bulbs. The availability of a wide range of powder compositions, the capability to produce parts to net dimensions (net-shape forming), and the economics of the overall operation make this process attractive for many applications.

Typical products made by powder metallurgy techniques are gears, cams, bushings, cutting tools, porous products such as filters and oil-impregnated bearings, and automotive components, such as piston rings, valve guides, connecting rods, and hydraulic pistons (Fig. 17.1 and Table 17.1). Advances in technology now permit structural parts of aircraft—such as landing gear, engine-mount

TABLE 17.1
TYPICAL APPLICATIONS FOR METAL POWDERS

APPLICATION	METALS	USES
Abrasives	Fe, Sn, Zn	Cleaning, abrasive wheels
Aerospace	Al, Be, Nb	Jet engines, heat shields
Automotive	Cu, Fe, W	Valve inserts, bushings, gears
Electrical/electronic	Ag, Au, Mo	Contacts, diode heat sinks
Heat treating	Mo, Pt, W	Furnace elements, thermocouples
Joining	Cu, Fe, Sn	Solders, electrodes
Lubrication	Cu, Fe, Zn	Greases, abradable seals
Magnetic	Co, Fe, Ni	Relays, magnets
Manufacturing	Cu, Mn, W	Dies, tools, bearings
Medical/dental	Ag, Au, W	Implants, amalgams
Metallurgical	Al, Ce, Si	Metal recovery, alloying
Nuclear	Be, Ni, W	Shielding, filters, reflectors
Office equipment	Al, Fe, Ti	Electrostatic copiers, cams
Plastics	Al, Fe, Mg	Tools, dies, fillers, cements

Source: R. M. German.

supports, engine disks, impellers, and engine nacelle frames—to be made by P/M.

Pure metals, alloys, or mixtures of metallic and nonmetallic materials can be used in powder metallurgy. The most commonly used metals are iron, copper, aluminum, tin, nickel, titanium, and refractory metals. For parts made of brass,

(a)

(b)

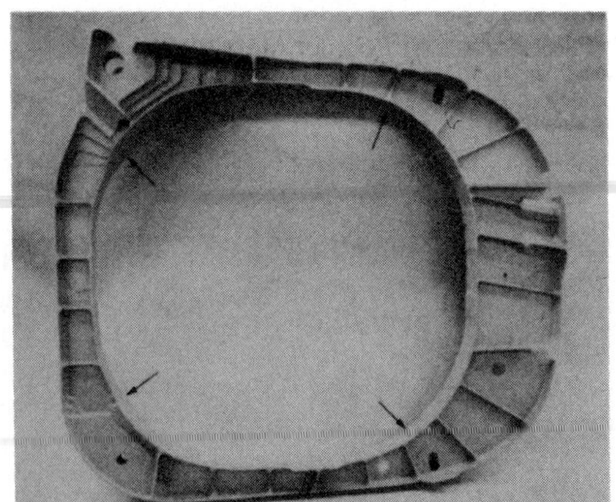

FIGURE 17.1
Examples of parts made by powder-metallurgy processes. (a) Left, steel chain-drive sprocket; right, race and cam for automatic transmission. *Source:* Metal Powder Industries Federation. (b) Titanium-alloy nacelle frame for F-14 fighter aircraft made of four pieces and welded (at arrows) by the electron-beam process. *Source:* Colt-Crucible.

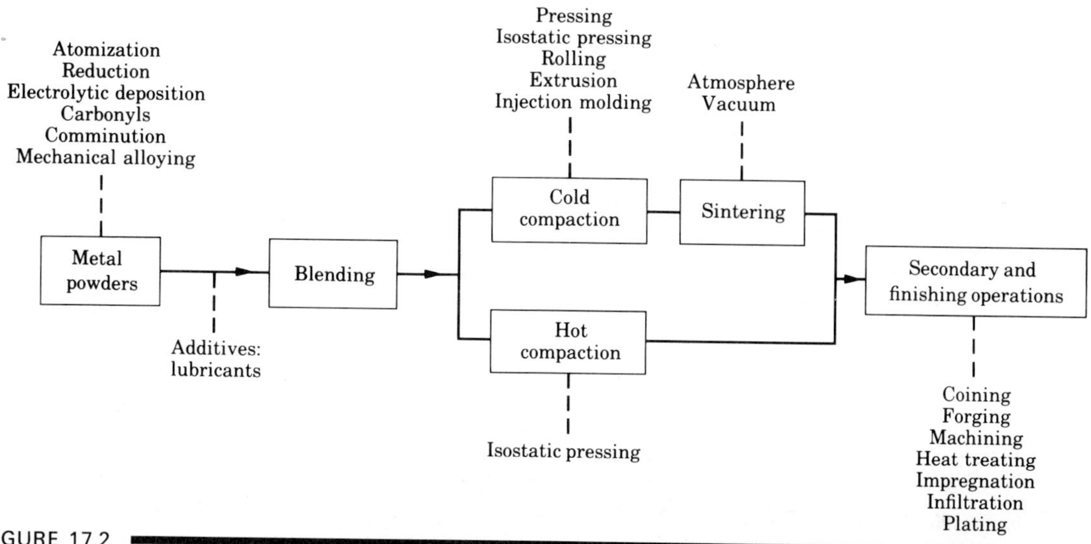

FIGURE 17.2
Outline of processes and operations involved in making powder-metallurgy parts.

bronze, steels, and stainless steels, prealloyed powders are used, where each powder particle itself is an alloy.

Powder metallurgy has become competitive with processes such as casting, forging, and machining, particularly for relatively complex parts made of high-strength and hard alloys. Parts made by this process have good dimensional accuracy, and their sizes range from tiny balls for ball-point pens to parts weighing about 50 kg (100 lb), although most parts weigh less than 2.5 kg (5 lb). Figure 17.2 shows the steps involved in the P/M process.

17.2 Production of Metal Powders

Basically, the powder metallurgy process consists of the following operations:

a) Powder production
b) Blending
c) Compaction
d) Sintering
e) Finishing operations.

For improved quality and dimensional accuracy, or for special applications, additional processing such as coining, sizing, forging, machining, infiltration, and resintering may be carried out.

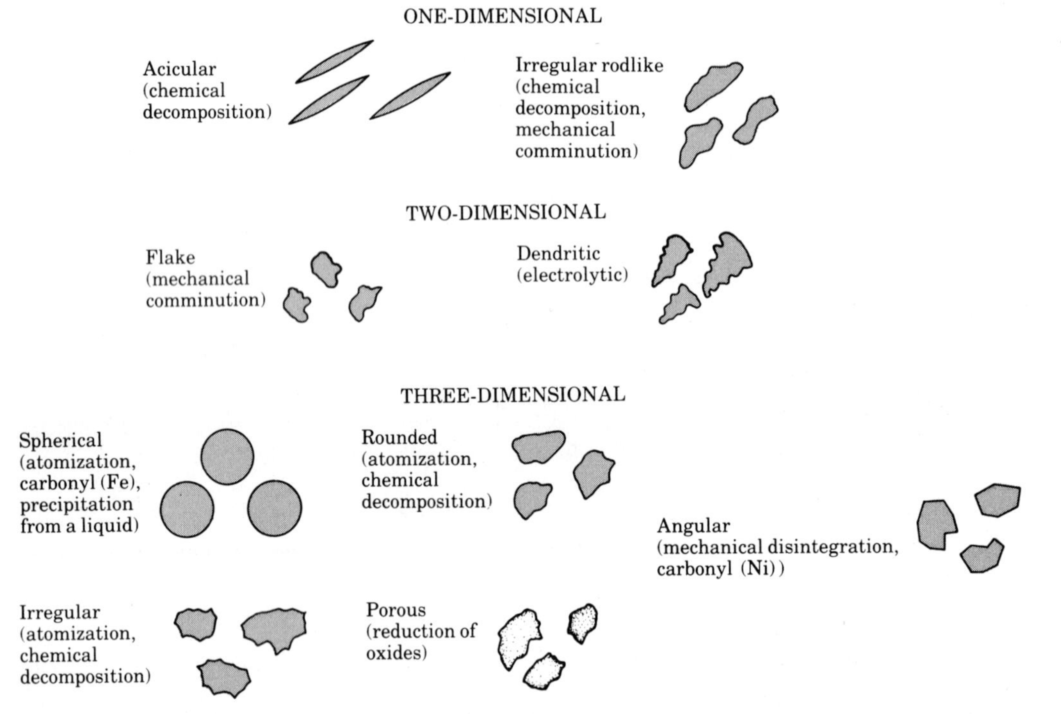

FIGURE 17.3
Particle shapes in metal powders and the processes by which they are produced. Iron powders are produced by many of these processes. *Source:* After P. K. Johnson.

17.2.1 Methods of powder production

There are several methods of producing metal powder. Metal sources are generally bulk metals and alloys, ores, salts, and other compounds. Most metal powders can be produced by more than one method, the choice depending on the requirements of the end product. Particle sizes range from 0.1 to 1000 μm (4 μin. to 0.04 in.).

The shape, size distribution, porosity, chemical purity, and bulk and surface characteristics of the particles depend on the particular process used (Fig. 17.3). These characteristics are important because they significantly affect permeability and flow characteristics during compaction and in subsequent sintering operations.

Atomization. Atomization produces a liquid-metal stream by injecting molten metal through a small orifice (Fig. 17.4a). The stream is broken up by jets of inert gas, air, or water. The size of particles formed depends on the temperature of the metal, the rate of flow, nozzle size, and jet characteristics. In one variation of this method, a consumable electrode is rotated rapidly in a helium-filled chamber (Fig. 17.4b). The centrifugal force breaks up the molten tip of the electrode, producing metal particles.

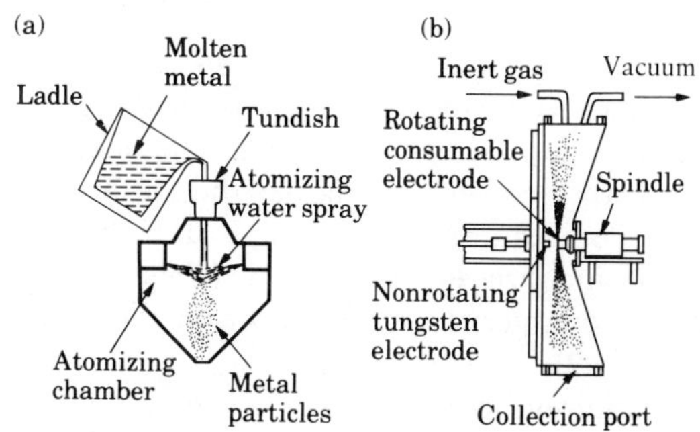

FIGURE 17.4 Methods of metal powder production by atomization: (a) melt atomization; and (b) atomization with a rotating consumable electrode.

Reduction. **Reduction** of metal oxides (removal of oxygen) uses gases, such as hydrogen and carbon monoxide, as reducing agents. Thus very fine metallic oxides are reduced to the metallic state. The powders produced by this method are spongy and porous and have uniformly sized spherical or angular shapes.

Electrolytic deposition. **Electrolytic deposition** utilizes either aqueous solutions or fused salts. The powders produced are among the purest.

Carbonyls. Metal **carbonyls**, such as iron carbonyl, $Fe(CO)_5$, and nickel carbonyl, $Ni(CO)_4$, are formed by letting iron or nickel react with carbon monoxide. The reaction products are then decomposed to iron and nickel, producing small, dense, and uniform spherical particles of high purity.

Comminution. Mechanical **comminution** (*pulverization*) involves crushing, milling in a *ball mill*, or grinding brittle or less ductile metals into small particles. With brittle materials, the powder particles have angular shapes, whereas with ductile metals they are flaky and are not particularly suitable for powder metallurgy applications.

Mechanical alloying. In **mechanical alloying,** developed in the 1960s, powders of two or more pure metals are mixed in a ball mill. Under the impact of the hard balls, the powders fracture and weld together by diffusion, forming alloy powders.

Other methods. Other less commonly used methods are *precipitation* from a chemical solution, production of fine metal chips by *machining*, and *vapor condensation*. New developments include techniques based on high-temperature extractive metallurgical processes. Metal powders are being produced using high-temperature processing techniques based on the reaction of volatile halides with liquid metals and on controlled reduction and reduction/carburization of solid oxides.

17.2.2 Particle size, distribution, and shape

Particle *size* is measured usually by screening, that is, passing the metal powder through screens (sieves) of various mesh sizes. Screen analysis is achieved using a vertical stack of screens with increasing mesh size as the powder flows downward through the screens. The larger the mesh size, the smaller is the opening in the screen. For example, a mesh size of 30 has an opening of 600 μm, size 100 has 150 μm, and size 400 has 38 μm. (This method is similar to numbering abrasive grains; the larger the number, the smaller is the size of the abrasive particle; see Section 25.2).

In addition to screen analysis, several other methods are also used for particle size analysis:

a) *Sedimentation*, which involves measuring the rate at which particles settle in a fluid.
b) *Microscopic* analysis, including the use of transmission and scanning electron microscopy.
c) *Light scattering* from a laser that illuminates a sample consisting of particles suspended in a liquid medium. The particles cause the light to be scattered, which is then focused on a detector that digitizes the signals and computes the particle size distribution.
d) *Optical* means such as particles blocking a beam of light, which is then sensed by a photocell.
e) Suspending particles in a liquid and detecting particle size and distribution by *electrical sensors.*

The *size distribution* of particles is an important consideration since it affects the processing characteristics of the powder. The distribution of particle size is

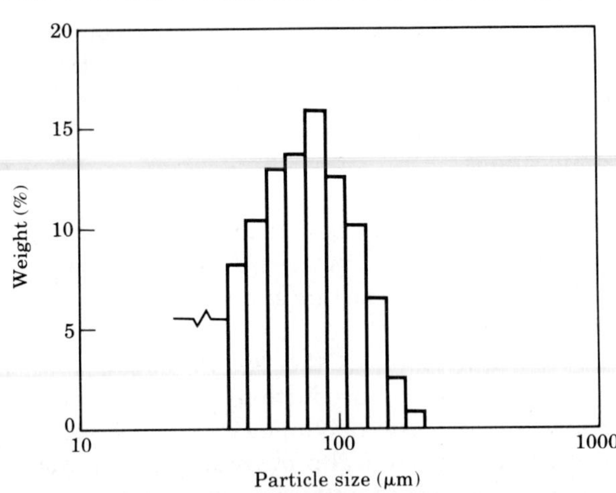

FIGURE 17.5
A plot of the weight of particles as a function of particle size. The most populous size is termed the *mode*. In this case, it is between 75 μm and 90 μm. *Source:* Reprinted with permission from Randall M. German, *Powder Metallurgy Science.* Princeton NJ: Metal Powder Industries Federation, 1984.

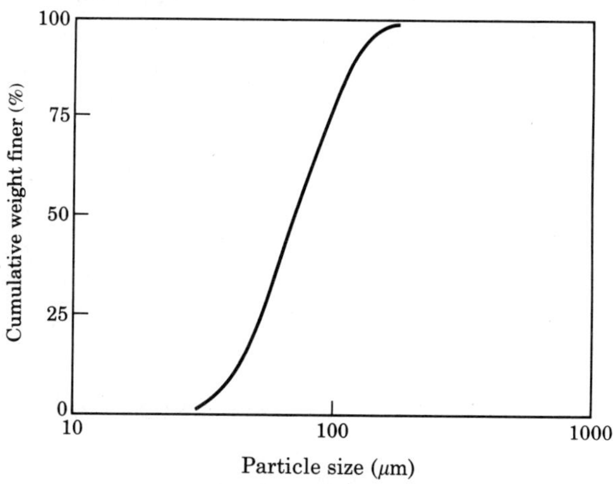

FIGURE 17.6
Cumulative particle size distribution as a function of weight. *Source:* Reprinted with permission from Randall M. German, *Powder Metallurgy Science.* Princeton NJ: Metal Powder Industries Federation, 1984.

given in terms of a *frequency distribution* plot, as shown in Fig. 17.5 (see Section 36.7 for details). In this plot, note that the highest percentage of the particles (by weight) have a size in the range of 75 μm to 90 μm. This maximum is called the *mode size.* The particle size data are also plotted in the form of a *cumulative distribution* (Fig. 17.6). Thus, for example, about 75 percent of the weight of the particles have a size finer than 100 μm. Note that, as expected, as the cumulative weight reaches 100 percent, almost all particles have a size below about 200 μm.

The *shape* of particles has a major influence on their processing characteristics. The shape is usually described in terms of aspect ratio or shape index. *Aspect ratio* is the ratio of the largest dimension to the smallest dimension of the particle. This ratio ranges from unity for a spherical particle, to about 10 for flakelike or needlelike particles. *Shape index* or *factor* (SF) is a measure of the surface area to the volume of the particle with reference to a spherical particle of equivalent diameter. Thus the shape factor for a flake is higher than that for a sphere.

• Example: Particle shape-factor determination.

Determine the shape factor for (a) a spherical particle, (b) a cubic particle, and (c) a cylinder with a length-to-diameter ratio of 2. Let the smallest dimension be unity.

SOLUTION.

a) The surface area A of a sphere of diameter D is $A = \pi D^2$ and its volume is $V = \pi D^3/6$, so the shape factor will be

$$\text{SF} = A/V = 6/D = 6.$$

Note that $D = (6V/\pi)^{1/3}$.

b) The surface area of a cube with lateral dimensions of unity is 6 and its volume is 1. Thus, $A/V = 6$. The equivalent diameter for a sphere is

$$D = (6V/\pi)^{1/3} = (6/\pi)^{1/3} = 1.24.$$

Hence,

$$\text{SF} = (1.24)(6) = 7.44.$$

c) The surface area of this cylinder is

$$A = 2\pi D^2/4 + 2(\pi D) = 2.5\pi.$$

Its volume is $2\pi D^2/4 = \pi/2$. Thus, $A/V = 5$. The equivalent diameter for a sphere is

$$D = [(6\pi/2)(1/\pi)]^{1/3} = 1.44.$$

Hence,

$$\text{SF} = (1.44)(6) = 7.21.$$

17.2.3 Blending metal powders

Blending (mixing) powders is the second step in powder metallurgy processing and is carried out for the following purposes:

- Because the powders made by various processes may have different sizes and shapes, they must be mixed to obtain uniformity. The ideal mix is one in which all the particles of each material are distributed uniformly.
- Powders of different metallic and other materials may be mixed in order to impart special physical and mechanical properties and characteristics to the P/M product.
- Lubricants may be mixed with the powders to improve their flow characteristics. The results are reduced friction between the metal particles, improved flow of the powder metals into the dies, and longer die life. Lubricants typically are stearic acid or zinc stearate, in proportions of 0.25–5 percent by weight.

Powder mixing must be carried out under controlled conditions to avoid contamination or deterioration. Deterioration is caused by excessive mixing, which may alter the shape of the particles and work-harden them, thus making the subsequent compacting operation more difficult. Powders can be mixed in air or in inert atmospheres (to avoid oxidation) and in liquids, which act as lubricants and make the mix more uniform. Several types of blending equipment are available (Fig. 17.7). These operations are being increasingly controlled by microprocessors to improve and maintain quality.

Hazards. Because of their high surface area-to-volume ratio, metal powders are explosive, particularly aluminum, magnesium, titanium, zirconium, and thor-

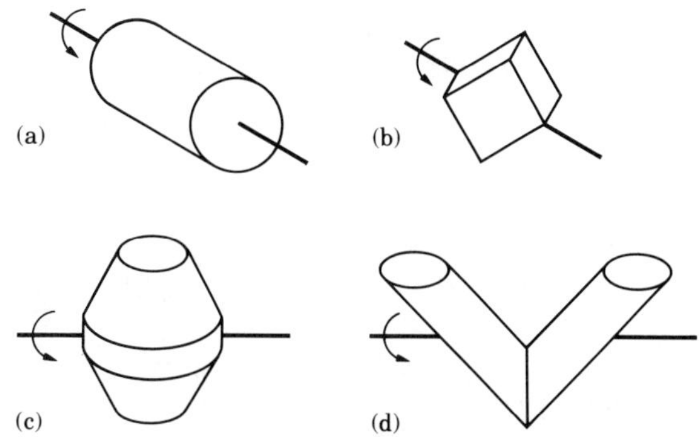

FIGURE 17.7
Some common equipment geometries for mixing or blending powders: (a) cylindrical, (b) rotating cube, (c) double cone, and (d) twin shell. *Source:* Reprinted with permission from Randall M. German, *Powder Metallurgy Science.* Princeton NJ: Metal Powder Industries Federation, 1984.

ium. Great care must be exercised both during blending and during storage and handling. Precautions include grounding equipment and avoiding sparks (by using nonsparking tools and avoiding friction as a source of heat), dust clouds, open flames, and chemical reactions.

17.3 Compaction of Metal Powders

Compaction is the step in which the blended powders are pressed into shapes in dies (Figs. 17.8a and b), using presses that are either hydraulically or mechanically activated. The purposes of compaction are to obtain the required shape, density, and particle-to-particle contact and to make the part strong enough to be processed further. The pressed powder is known as a *green compact.* The powder must flow easily to properly feed into the die cavity. Pressing is generally carried out at room temperature, although it can be done at elevated temperatures.

The *density* of the green compact depends on the pressure applied (Fig. 17.9a). As the compacting pressure is increased, the density approaches that of the theoretical density of the metal in its bulk form. Another important factor is the size distribution of the particles. If all the particles are the same size, there will always be some porosity when they are packed together—theoretically, at least 24 percent by volume. Imagine, for example, a box filled with tennis balls; there are always open spaces between the balls. However, introducing smaller particles will fill the spaces between the larger particles and thus result in a higher density of the compact.

The higher the density, the higher will be the strength and elastic modulus of the part (Fig. 17.9b). The reason is that the higher the density, the higher will be the

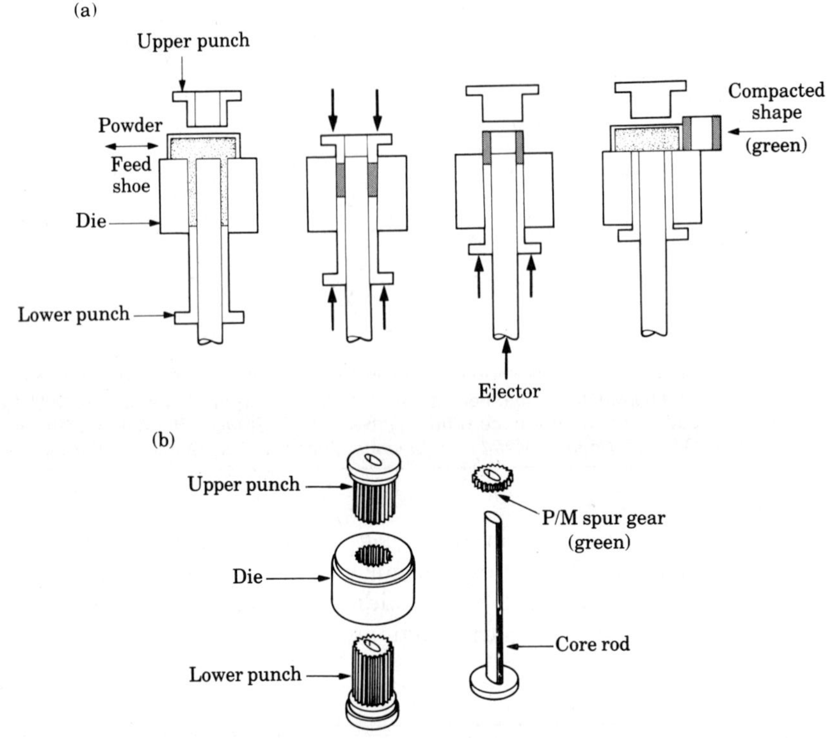

FIGURE 17.8
(a) Compaction of metal powder to form a bushing. The pressed powder part is called green compact. (b) Typical tool and die set for compacting a spur gear. *Source:* Metal Powder Industries Federation.

amount of solid metal in the same volume—hence the greater its resistance to external forces. Because of friction between the metal particles in the powder, and the friction between the punches and the die walls, the density can vary considerably within the part. This variation can be minimized by proper punch and die design and friction control. For example, it may be necessary to use multiple punches, with separate movements in order to ensure that the density is more uniform throughout the part (Fig. 17.10). Recall our similar discussion regarding compaction of sand in mold making (see Fig. 11.7).

17.3.1 Equipment

The pressure required for pressing metal powders ranges from 70 MPa (10 ksi) for aluminum to 800 MPa (120 ksi) for high-density iron parts (Table 17.2; see p. 518). The compacting pressure required depends on the characteristics and shape of the particles, method of blending, and lubrication.

Press capacities are on the order of 1.8–2.7 MN (200–300 tons), although presses with much higher capacities are used for special applications. Most

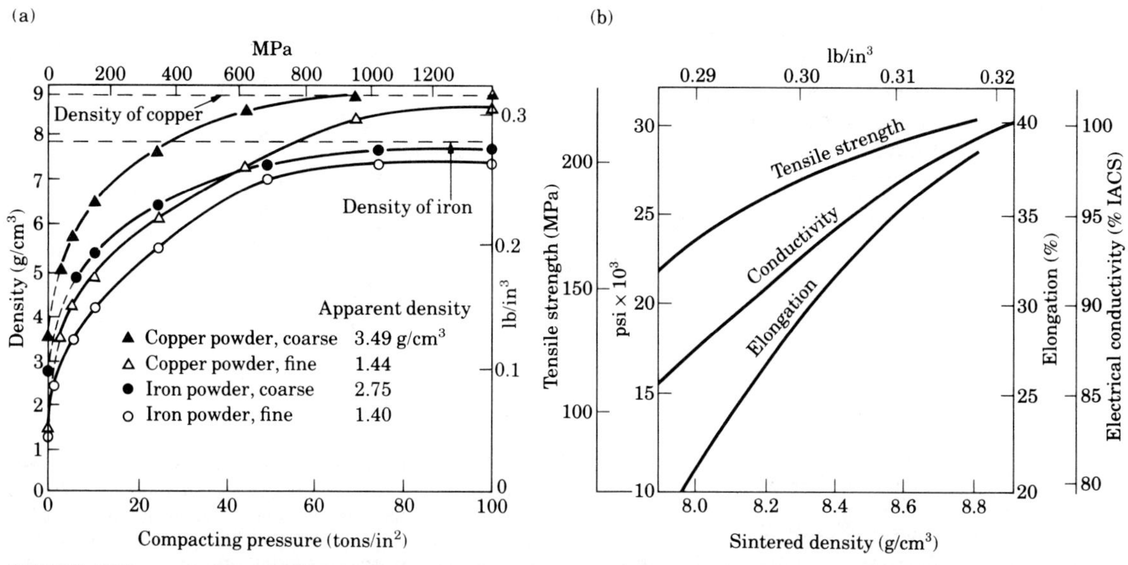

FIGURE 17.9 (a) Density of copper and iron powder compacts as a function of compacting pressure. Density greatly influences the mechanical and physical properties of P/M parts. *Source:* F. V. Lenel, *Powder Metallurgy: Principles and Applications.* Princeton, N.J.: Metal Powder Industries Federation, 1980. (b) Effect of density on tensile strength, elongation, and electrical conductivity of copper powder. IACS means International Annealed Copper Standard for electrical conductivity. *Source:* After J. L. Everhart.

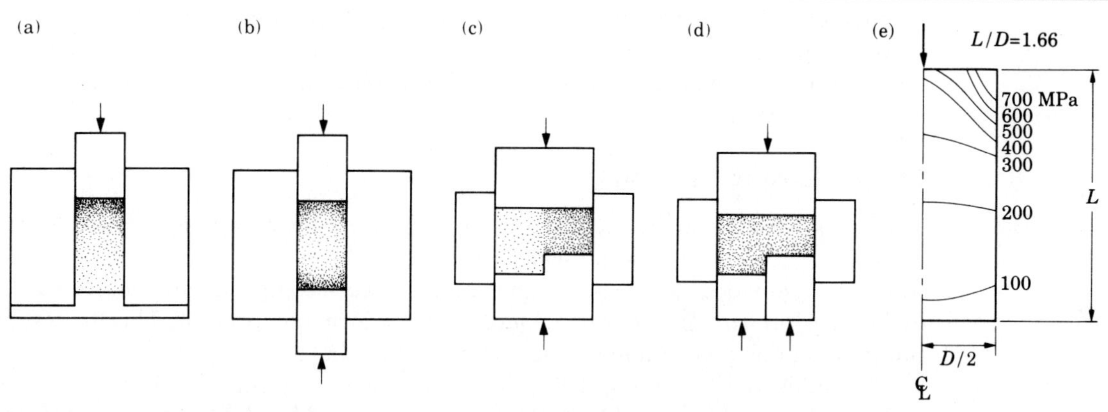

FIGURE 17.10 Density variation in compacting metal powders in different dies: (a) and (c) single-action press; (b) and (d) double-action press. Note in (d) the greater uniformity of density in pressing with two punches with separate movements compared with (c). There are situations in which density variation, hence property variation, within a part may be desirable. (e) Pressure contours in compacted copper powder in a single-action press. *Source:* P. Duwez and L. Zwell.

TABLE 17.2
COMPACTING PRESSURES FOR VARIOUS METAL POWDERS

METAL	PRESSURE (MPa)
Aluminum	70–275
Brass	400–700
Bronze	200–275
Iron	350–800
Tantalum	70–140
Tungsten	70–140
OTHER MATERIALS	
Aluminum oxide	110–140
Carbon	140–165
Cemented carbides	140–400
Ferrites	110–165

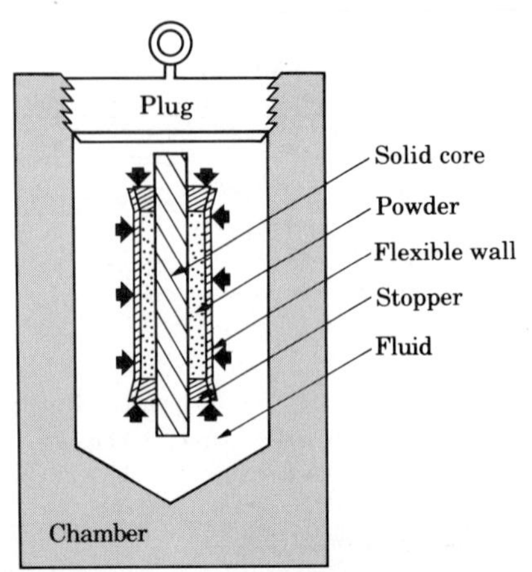

FIGURE 17.11 Schematic diagram of cold isostatic pressing as applied to forming a tube. The powder is enclosed in a flexible container around a solid core rod. Pressure is applied isostatically to the assembly inside a high pressure chamber. *Source:* Reprinted with permission from Randall M. German, *Powder Metallurgy Science*. Princeton, NJ: Metal Powder Industries Federation, 1984.

applications require less than 100 tons. For small tonnage, crank or eccentric type mechanical presses are used; for higher capacities, toggle or knucklejoint presses are employed (see Fig. 14.30). Hydraulic presses can be used with capacities as high as 45 MN (5000 tons) for large parts.

The selection of the press depends on part size and configuration, density requirements, and production rate. An important consideration with regard to pressing speed is the entrapment of air in the die cavity. The presence of air will prevent proper compaction and the higher the speed, the greater is the tendency for the press to trap air.

17.3.2 Isostatic pressing

Compaction can also be carried out or improved by a number of additional processes, such as isostatic pressing, rolling, and forging. Because the density of compacted powders can vary significantly, green compacts are subjected to *hydrostatic pressure* in order to achieve more uniform compaction. This process is similar to cupping your hands when making snow balls.

In **cold isostatic pressing (CIP)**, the metal powder is placed in a flexible rubber mold made of neoprene rubber, urethane, polyvinyl chloride, or other elastomers (Fig. 17.11). The assembly is then pressurized hydrostatically in a chamber, usually with water. The most common pressure is 400 MPa (60 ksi), although pressures of up to 1000 MPa (150 ksi) have been used. The applications of CIP and other compacting methods, in terms of size and complexity of part, are shown in Fig. 17.12.

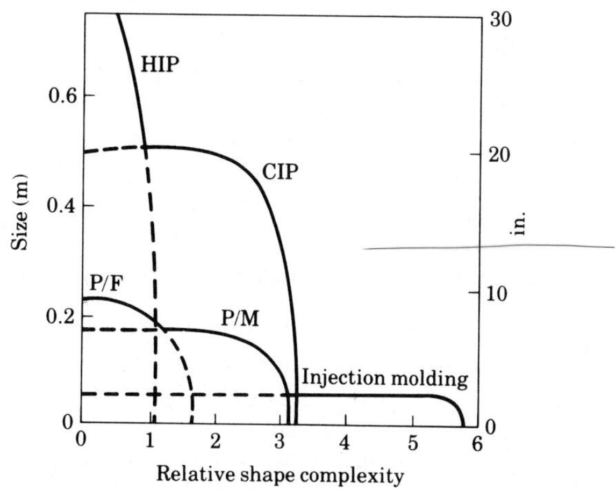

FIGURE 17.12 Capabilities of part size and shape complexity according to various P/M operations. P/F means powder forging. *Source:* Metal Powder Industries Federation.

In **hot isostatic pressing (HIP)**, the container is usually made of a high-melting-point sheet metal, and the pressurizing medium is inert gas or vitreous (glasslike) fluid (Fig. 17.13). Common conditions for HIP are 100 MPa (15 ksi) at 1100 °C (2000 °F), although the trend is toward higher pressures and temperatures. The main advantage of HIP is its ability to produce compacts with essentially 100 percent density, good metallurgical bond among the particles, and good mechanical properties.

The HIP process is relatively expensive and is used mainly in making superalloy components for the aerospace industry. It is routinely used as a final densification step for tungsten-carbide cutting tools and P/M tool steels. The process is also used

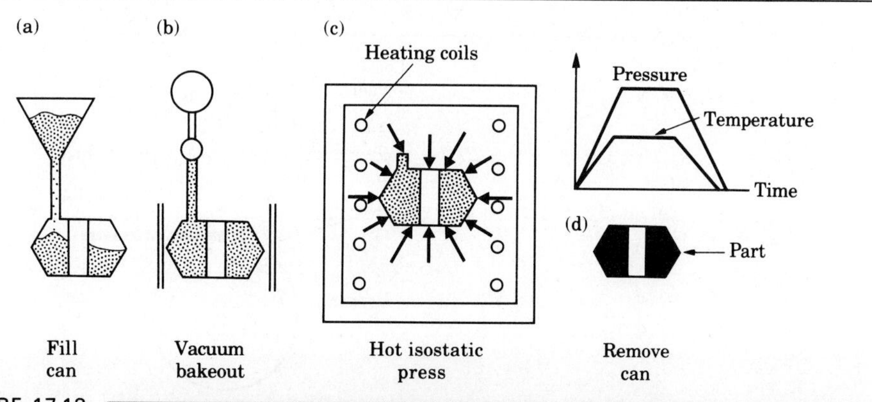

FIGURE 17.13 Schematic illustration of hot isostatic pressing. The pressure and temperature variations versus time are shown in the diagram. *Source:* Reprinted with permission from Randall M. German, *Powder Metallurgy Science*. Princeton, NJ: Metal Powder Industries Federation, 1984.

to close internal porosity and improve properties in superalloy and titanium-alloy castings for the aerospace industry.

The main advantage of isostatic pressing is that, because of uniformity of pressure from all directions and the absence of die-wall friction, it produces compacts of practically uniform grain structure and density, irrespective of shape. Parts with high length-to-diameter ratios have been produced, with very uniform density, strength, and toughness and good surface details.

17.3.3 Other compacting and shaping processes

Rolling. In *powder rolling*, also called *roll compaction*, the powder is fed to the roll gap in a two-high rolling mill (see Fig. 13.12), and is compacted into a continuous strip at speeds of up to 0.5 m/s (100 ft/min). The process can be carried out at room or elevated temperatures. Sheet metal for electrical and electronic components and for coins can be made by powder rolling. An example is shown in Fig. 17.14.

Extrusion. Powders can be compacted by *extrusion*; the powder is encased in a metal container and extruded. After sintering, preformed P/M parts may be reheated and forged in a closed die to their final shape. Superalloy powders, for example, are hot extruded for improved properties.

Injection molding. In *injection molding*, very fine metal powders are blended with a polymer. The mixture then undergoes a process similar to die casting (see also injection-molding of plastics, in Section 18.3). The molded greens are placed in a low-temperature oven to burn off the plastic and then sintered in a furnace. The major advantage of injection molding over conventional compaction is that relatively complex shapes, with wall thicknesses as small as 5 mm (0.2 in.), can be molded and removed easily from the dies. Typical parts made are components for watches, guns, and automobiles, and surgical knives.

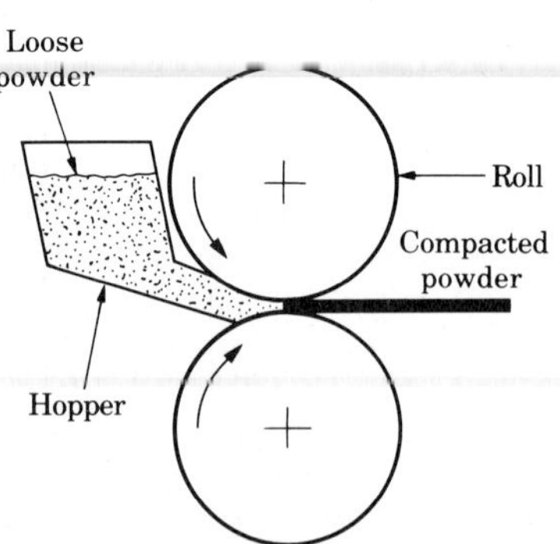

FIGURE 17.14
An example of powder rolling. *Source: Metals Handbook*, 9th ed., Vol. 7, American Society for Metals.

Pressureless compaction. In *pressureless compaction*, the die is filled with metal powder by gravity, and the powder is sintered directly in the die. Because of the resulting low density, pressureless compaction is used principally for porous parts, such as filters.

Ceramic molds. Molds for shaping metal powders are made by the technique used in investment casting. After the ceramic mold is made, it is filled with metal powder and placed in a steel container. The space between the mold and the container is filled with particulate material. The container is then evacuated, sealed, and subjected to hot isostatic pressing. Titanium-alloy compressor rotors for missile engines have been made by this process.

17.3.4 Punch and die materials

The selection of punch and die materials for P/M depends on the abrasiveness of the powder metal and the number of parts to be made. Most common die materials are air- or oil-hardening tool steels, such as D2 or D3, with a hardness range of 60–64 HRC. Because of their greater hardness and wear resistance, tungsten-carbide dies are used for more severe applications. Punches are generally made of similar materials.

Close control of die and punch dimensions is essential for proper compaction and die life. Too large a clearance between the punch and the die will allow the metal powder to enter the gap, interfere with the operation, and result in eccentric parts. Diametral clearances are generally less than 25 μm (0.001 in.). Die and punch surfaces must be lapped or polished—and in the direction of tool movements for improved die life and overall performance.

17.4 Sintering

Sintering is the process whereby compressed metal powder is heated in a controlled-atmosphere furnace to a temperature below its melting point, but sufficiently high to allow bonding (fusion) of the individual particles. Prior to sintering, the compact is brittle and its strength, known as *green strength*, is low. The nature and strength of the bond between the particles, and hence of the sintered compact, depend on the mechanisms of diffusion, plastic flow, evaporation of volatile materials in the compact, recrystallization, grain growth, and pore shrinkage.

The principal governing variables in sintering are temperature, time, and the atmosphere in the sintering furnace. Sintering temperatures (Table 17.3) are generally within 70–90 percent of the melting point of the metal or alloy. Sintering times range from a minimum of about 10 minutes for iron and copper alloys to as much as 8 hours for tungsten and tantalum. Continuous sintering furnaces are used for most production today. These furnaces have three chambers: (1) a burn-off

TABLE 17.3
SINTERING TEMPERATURE AND TIME FOR VARIOUS METALS

MATERIAL	TEMPERATURE (°C)	TIME (MIN)
Copper, brass, and bronze	760–900	10–45
Iron and iron-graphite	1000–1150	8–45
Nickel	1000–1150	30–45
Stainless steels	1100–1290	30–60
Alnico alloys (for permanent magnets)	1200–1300	120–150
Ferrites	1200–1500	10–600
Tungsten carbide	1430–1500	20–30
Molybdenum	2050	120
Tungsten	2350	480
Tantalum	2400	480

chamber to volatilize the lubricants in the green compact in order to improve bond strength and prevent cracking; (2) a high-temperature chamber for sintering; and (3) a cooling chamber.

Proper control of the furnace atmosphere is essential for successful sintering and to obtain optimum properties. An oxygen-free atmosphere is essential to control the carburization and decarburization of iron and iron-base compacts and to prevent oxidation of powders. A vacuum is generally used for sintering refractory metal alloys and stainless steels. The gases most commonly used for sintering a variety of other metals are hydrogen, dissociated or burned ammonia, partially combusted hydrocarbon gases, and nitrogen.

Sintering mechanisms are complex and depend on the composition of metal particles as well as processing parameters (Fig. 17.15). As temperature increases, two adjacent particles begin to form a bond by diffusion (*solid-state bonding*). As a result, the strength, density, ductility, and thermal and electrical conductivities of the compact increase (Fig. 17.16). At the same time, however, the compact shrinks; hence allowances should be made for shrinkage, as in casting.

If two adjacent particles are of different metals, *alloying* can take place at the interface of the two particles. One of the particles may be a lower melting-point metal than the other. In that case, one particle may melt and, because of surface tension, surround the particle that has not melted (*liquid-phase sintering*). An example is cobalt in tungsten-carbide tools and dies. Stronger and denser parts can be obtained in this way.

Depending on temperature, time, and processing history, different structures and porosities can be obtained in a sintered compact. However, porosity cannot be completely eliminated because voids remain after compaction and gases evolve during sintering. Porosities can consist of either a network of interconnected pores

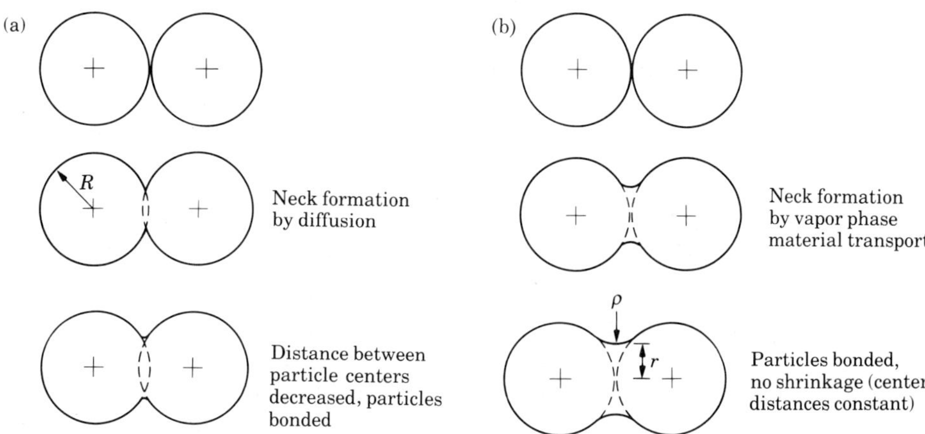

FIGURE 17.15 Schematic illustration of two mechanisms for sintering metal powders: (a) solid-state material transport; and (b) liquid-phase material transport. R = particle radius, r = neck radius, and ρ = neck profile radius.

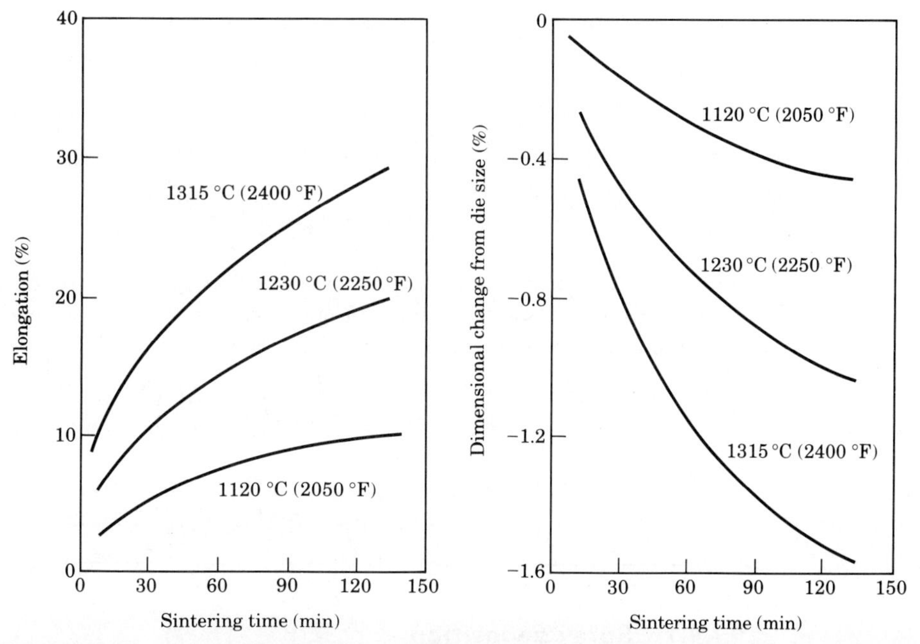

FIGURE 17.16 Effect of sintering temperature and time on elongation and dimensional change during sintering of type 316L stainless steel. *Source:* ASM.

or closed holes. Their presence is an important consideration in making P/M filters and bearings.

Another method, which is still at an experimental stage, is *spark sintering*. In this process, loose metal powders are placed in a graphite mold, heated by an electric current, subjected to a high-energy discharge, and compacted, all in one step. The rapid discharge strips any oxide coating, such as found on aluminum, or contaminants from the surfaces of the particles, and thus encourages good bonding during compaction at elevated temperatures.

Typical mechanical properties and applications for several sintered P/M alloys are given in Table 17.4. Note the effect of heat treating on the properties of aluminum. To evaluate the differences between the properties of P/M, wrought, and cast metals and alloys, compare this table with tables in Parts I and II, such as Tables 5.6, 6.2, 6.5, and Fig. 12.10. The effects of various processes on the mechanical properties of a titanium alloy are shown in Table 17.5 (p. 526). Note that hot isostatically pressed (HIP) titanium has properties that are similar to those for cast and forged titanium. It should be remembered, however, that forged components are likely to require additional machining processes (unless precision forged to net shape), which the P/M component may not. Thus powder metallurgy can be a competitive alternative.

• Example: Shrinkage in sintering.

In solid-state bonding during sintering of a powder-metal green compact, the linear shrinkage is 4%. If the desired sintered density is 95% of the theoretical density of the metal, what should be the density of the green compact?

SOLUTION. We define linear shrinkage as $\Delta L/L_o$, where L_o is the original length. We can then express the volume shrinkage during sintering as

$$V_{\text{sint}} = V_{\text{green}}\left(1 - \frac{\Delta L}{L_o}\right)^3. \tag{17.1}$$

The volume of the green compact must be larger than that for the sintered part. However, the mass does not change during sintering, so we can rewrite this expression in terms of the density ρ as

$$\rho_{\text{green}} = \rho_{\text{sint}}\left(1 - \frac{\Delta L}{L_o}\right)^3. \tag{17.2}$$

Thus,

$$\rho_{\text{green}} = 0.95(1 - 0.04)^3 = 0.84, \quad \text{or} \quad 84\%.$$

•

17.4.1 Spray deposition

Spray deposition is a shape generation process, an example of which is shown in Fig. 13.22. The basic components of a spray deposition process for metal powders

TABLE 17.4
PROPERTIES AND TYPICAL APPLICATIONS OF POWDER METALLURGY PARTS

MATERIAL	MPIF TYPE	CONDITION	DENSITY (kg/m^3)	ULTIMATE TENSILE STRENGTH (MPa)	HARDNESS	ELASTIC MODULUS (GPa)	ELONGATION IN 25 mm (%)	TYPICAL APPLICATIONS
P/M brass (leaded) 77.0–80.0 Cu, 1.0–2.0 Pb, 0.3 Fe max, 0.1 Sn max, bal Zn	T	Sintered	7400	165	55 HRH	83	13	Mechanical components resistant to atmospheric corrosion.
	U	Sintered	7800	195	68 HRH	90	19	Ordnance components, builders' hardware, lock parts, housings, nuts, gears.
	W	Sintered	8200	220	75 HRH	95	23	
601AB (Alcoa) 0.25 Cu, 0.6 Si, 1.0 Mg, 1.5 lubricant, bal Al		Sintered	2550	145	65–70 HRH	—	6	Similar to wrought 6061; strength, ductility, corrosion resistance.
		Heat treated	2550	240	80–85 HRE	—	2	
P/M iron 0.3 C max	N	Sintered	5800	110	10 HRH	72	2	Structural (lightly loaded gears); magnetic (motor pole pieces); self-lubricating bearings; structural, wear resisting (small levers and cams) as carbonitrided.
	P	Sintered	6200	130	70 HRH	90	2.5	
	R	Sintered	6600	165	80 HRH	110	5	
	S	Sintered	7000	205	15 HRB	130	9	
	T	Sintered	7400	275	30 HRB	160	15	
P/M austenitic stainless steels								Type 303, mechanical components; type 410, structural, corrosion resisting components.
303	P	Sintered	6200	240	—	—	1	
303	R	Sintered	6600	360	—	—	2	
410	N	Sintered	5800	290	—	—	<1	
410	P	Sintered	6200	380	—	—	<1	

Source: Metal Powder Industries Federation

TABLE 17.5
MECHANICAL PROPERTY COMPARISON FOR Ti-6Al-4V

PROCESS (*)	DENSITY (%)	YIELD STRENGTH (MPa)	ULTIMATE STRENGTH (MPa)	ELONGATION (%)	REDUCTION OF AREA (%)
Cast	100	840	930	7	15
Cast and forged	100	875	965	14	40
Blended elemental (P + S)	98	786	875	8	14
Blended elemental (HIP)	> 99	805	875	9	17
Prealloyed (HIP)	100	880	975	14	26

(*) P + S = pressed and sintered, HIP = hot isostatically pressed.
Source: R. M. German.

are an atomizer, a spray chamber with inert atmosphere, and a mold for producing preforms. Although there are several variations, the best known is the *Osprey* process. After the metal is atomized, it is deposited on a cooled preform mold, usually made of copper or ceramic, where it solidifies. The metal particles weld together, developing a density that is normally above 99 percent of the solid metal density. The mold may be of various shapes, such as billets, tubes, disks, and cylinders. Spray-deposited preforms may be subjected to additional shaping and consolidation processes, such as forging, rolling, and extrusion. The grain size is fine and the mechanical properties are comparable to those for wrought products of the same alloy.

17.5

Secondary and Finishing Operations

In order to further improve the properties of sintered P/M products or to give them special characteristics, several additional operations may be carried out after sintering. *Coining* and *sizing* are additional compacting operations, performed under high pressure in presses. The purposes of these operations are to impart dimensional accuracy to the sintered part and to improve its strength and surface finish by additional densification.

An important recent development is the use of *preformed* and sintered alloy powder compacts, which are subsequently cold or hot forged to the desired final shapes. These products have good surface finish and tolerances, with uniform fine grain size. The superior properties obtained make this technology particularly suitable for applications such as highly stressed automotive and jet-engine components.

The inherent porosity of powder metallurgy components can be utilized by *impregnating* them with a fluid. A typical application is to impregnate the sintered part with oil, which is usually done by immersing the part in heated oil. Bearings and bushings that are internally lubricated, with up to 30 percent oil by volume, are made by this method. Internally lubricated components have a continuous supply of lubricant during their service lives. Universal joints are now being made with grease-impregnated P/M techniques, no longer requiring grease fittings.

Infiltration is a process whereby a slug of lower melting-point metal is placed against the sintered part, and the assembly is heated to a temperature sufficient to melt the slug. The molten metal infiltrates the pores by capillary action, resulting in a relatively pore-free part with good density and strength. The most common application is the infiltration of iron-base compacts with copper. The advantages are that hardness and tensile strength are improved and the pores are filled, thus preventing moisture penetration, which could cause corrosion. Infiltration may also be done with lead whereby, because of the low shear strength of lead, the infiltrated part has lower frictional characteristics than the uninfiltrated one.

Powder-metal parts may be subjected to other finishing operations, including:

a) Heat treating, for improved hardness and strength.
b) Machining by milling, drilling, and tapping to produce threaded holes.
c) Grinding, for improved dimensional accuracy and surface finish.
d) Plating, for wear resistance, improved appearance, and corrosion resistance.

Example: Powder metallurgy gears for a garden tractor.

Components such as gears, bushings, and some structural parts of garden tractors have been made by P/M techniques, replacing the traditional casting or forging methods. Gears have been manufactured competitively using high-quality powders with high compressibility and requiring low compacting pressures. These parts range from medium to high density and are suitable for severe applications with high loads and for high-wear surfaces.

In one application, a reduction gear for a garden tractor was made from iron powder and infiltrated with copper. Although its strength was acceptable, the wear rate under high loads was very high. This resulted in loss of tooth profile, side loading on the bearings, and a high noise level. To improve wear resistance and strength, a new powder was selected containing 2.0 percent nickel, 0.5 percent graphite, 0.5 percent molybdenum, and the balance atomized iron powder. Because of the size of the part, the pressing loads were very high. The part was redesigned and the tooling was made with three punches to compact the part with three different densities.

By this method, high density was obtained in those sections of the part requiring high strength and wear resistance. The high density also permitted carburization for hardness improvement. The presintered part was re-pressed to a density of 7.3–7.5 g/cm^3 in the tooth area for improved strength, while the hub of the gear remained at its pressed density of 6.4–6.6 g/cm^3. The entire part required a compacting load of 4 MN (450 tons). The weight of the gear was 1 kg (2.25 lb).

Example: Production of tungsten carbide for tools and dies.

Tungsten carbide is an important tool and die material mostly because of its hardness, strength, and wear resistance over a wide range of temperatures. Powder metallurgy techniques are used in making these carbides. First, powders of tungsten and carbon are blended together in a ball mill or rotating mixer. The mixture (basically 94 percent tungsten and 6 percent carbon, by weight) is heated to approximately 1500 °C (2800 °F) in a vacuum induction furnace. As a result, the tungsten is carburized, forming tungsten carbide in fine powder form. A binding agent (usually cobalt) is then added to the tungsten carbide (with an organic fluid such as hexane), and the mixture is ball milled to produce a uniform and homogenous mix, a process that can take several hours and even days.

The mixture is then dried and consolidated, usually by cold compaction, at pressures in the range of 200 MPa (30,000 psi). It is then sintered in a hydrogen-atmosphere or vacuum furnace at a temperature of 1350 °C to 1600 °C (2500 °F to 2900 °F), depending on its composition. At this temperature, the cobalt is in a liquid phase and acts as a binder for the carbide particles. (Powders may also be hot pressed at the sintering temperature, using graphite dies.) During sintering, the tungsten carbide undergoes a linear shrinkage of about 16 percent, corresponding to a volume shrinkage of about 40 percent (see also the example in Section 19.2.5). Thus control of size and shape is important to produce tools with accurate dimensions. Carbides are also made with a combination of other carbides such as titanium carbide and tantalum carbide, using carbide mixtures made by the method described.

17.6 Design Considerations

Because of the unique properties of metal powders, their flow characteristics in the die, and the brittleness of green compacts, there are certain design principles that should be followed (Fig. 17.17):

- The shape of the compact must be kept as simple and uniform as possible. Sharp changes in contour, thin sections, variations in thickness, and high length-to-diameter ratios should be avoided.
- Provision must be made for ejection of the green compact from the die without damaging the compact. Thus holes or recesses should be parallel to the axis of punch travel. Chamfers should also be provided.
- As with most other processes, P/M parts should be made with the widest tolerances, consistent with their intended applications, in order to increase tool and die life and reduce production costs.

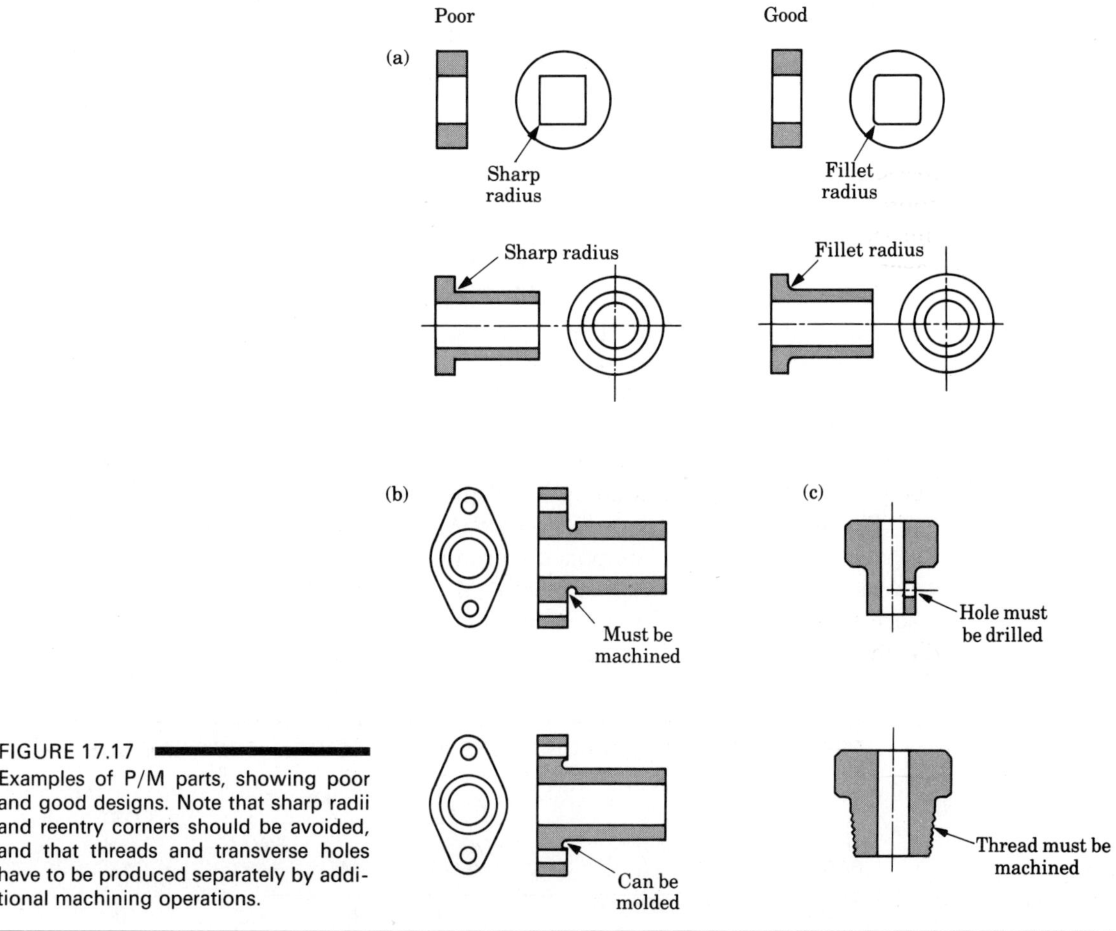

FIGURE 17.17 Examples of P/M parts, showing poor and good designs. Note that sharp radii and reentry corners should be avoided, and that threads and transverse holes have to be produced separately by additional machining operations.

Tolerances of sintered P/M parts are usually on the order of $\pm$0.05–0.1 mm ($\pm$0.002–0.004 in.). Tolerances improve significantly with additional operations such as sizing, machining, and grinding.

17.7 Selective Laser Sintering

As we described in the General Introduction, there is a growing trend in making prototypes of parts and components (*desktop manufacturing*) rapidly and economically, without the need for complex and expensive equipment. There are several such processes, one of which is *selective laser sintering* of metal powders. Because of the somewhat common nature of these rapid-prototyping processes, this technique is described in Section 39.6 with others.

17.8 Economics of Powder Metallurgy

Because P/M can produce parts at or near net shape, thus eliminating many secondary manufacturing and assembly operations, it has become increasingly competitive with casting, forging, and machining. However, because of the high initial cost of punches, dies, and equipment for P/M processing, production volume must be high enough to warrant this expenditure. Although there are exceptions, the process is generally economical for quantities above 10,000 pieces.

The near net-shape capability of P/M reduces or eliminates scrap. Weight comparisons of aircraft components produced by forging and P/M processes are shown in Table 17.6. Note that these P/M parts are subjected to material-removal processes; thus the final parts weigh less than those made by either of the two processes.

TABLE 17.6
FORGED AND P/M TITANIUM PARTS AND POTENTIAL COST SAVINGS

PART	WEIGHT (kg)			POTENTIAL COST SAVING (%)
	FORGED BILLET	*P/M*	*FINAL PART*	
F-14 Fuselage brace	2.8	1.1	0.8	50
F-18 Engine mount support	7.7	2.5	0.5	20
F-18 Arrestor hook support fitting	79.4	25.0	12.9	25
F-14 Nacelle frame	143	82	24.2	50

• Example: A powder-metallurgy application.

The preload arm for a hard disk drive is shown in the accompanying figure. The function of this part is to support a rail that guides the tape head. This part was formerly made of an aluminum casting, but it did not have sufficient strength. Making it from wrought aluminum for improved strength would require extensive machining. Other manufacturing methods were considered, but the required strength and the complex shape were not economically attainable for the medium- to high-volume production needed for this part.

A new part was made of 304 stainless steel using powder-metallurgy techniques. The tensile strength was 350 MPa (50 ksi), with a yield strength of 275 MPa (40 ksi)

and an elongation of 5 percent. The secondary operations required were drilling five side holes and tapping the center hole. The tapped hole and the two larger ones next to it were produced as part of the powder compacting operation. The P/M process was capable of producing this part with acceptable dimensional accuracy and strength. *Source*: CEROMET, Inc.

SUMMARY

The powder-metallurgy process consists of preparing metal powders, compacting them into shapes, and sintering them to impart strength, hardness, and toughness. Although size and weight are limited, the process is capable of producing relatively complex parts economically, in net-shape form to close tolerances, from a wide variety of metal and alloy powders. Control of powder shape and quality, process variables, and sintering atmospheres is an important consideration in product quality. Parts may be subjected to additional metalworking, machining, and finishing operations to impart certain geometric features and to improve properties and dimensional accuracy.

Density and mechanical and physical properties can be controlled by tooling design and compacting pressure. Some critical parts, such as jet-engine components, are now being made by P/M techniques. By controlling porosity, products such as filters and oil-impregnated bearings can be made. The P/M process is suitable for medium- to high-volume production runs, and has competitive advantages over other methods of production, such as casting, forging, and machining.

SUMMARY TABLE 17.1

CHARACTERISTICS OF POWDER-METALLURGY PROCESSING

ADVANTAGES

- Availability of a wide range of compositions to obtain special mechanical and physical properties, such as stiffness, damping characteristics, hardness, density, toughness, and electrical and magnetic properties. Some of the highly alloyed new superalloys can be manufactured into parts only by P/M processing.
- A technique for making parts from high-melting-point refractory metals, which would be difficult or uneconomical to make by other methods.
- High production rates on relatively complex parts, with automated equipment requiring little labor.
- Good dimensional control and, in many instances, elimination of machining and finishing operations, thus eliminating scrap and waste and saving energy.
- Capability for impregnation and infiltration for special applications.

LIMITATIONS

- Size and complexity of shape of parts and press capacity.
- High cost of powder metals compared to other raw materials.
- High cost of tooling and equipment for small production runs.
- Mechanical properties, such as strength and ductility, that are generally lower than those obtained by forging. However, the properties of full-density P/M parts made by HIP or additional forging can be better than those made by other processes.

TRENDS

- The trend in P/M applications is toward making larger parts and increased use of hot isostatic pressing.
- New alloys are being developed for P/M parts. These alloys cannot easily be processed into shapes by other processes.
- Powders are being made by mechanical alloying and rapid-solidification techniques to impart desirable properties. Machinable prealloyed powders are being developed.
- Studies are being conducted on powder consolidation by shock waves.
- Although traditionally P/M products are aimed mostly (70 percent) at the automotive industry, major efforts are being expended to produce parts for the aerospace and other industries.
- Selective laser sintering of metal powders is being used and further developed for rapid prototyping of parts and components.

KEY TERMS

Atomization	Comminution	Mechanical alloying
Blending	Compaction	Powder metallurgy
Carbonyls	Electrolytic deposition	Reduction
Cold isostatic pressing	Hot isostatic pressing	Sintering

BIBLIOGRAPHY

Bradbury, S. (ed.), *Powder Metallurgy Equipment Manual.* Princeton, N.J.: Powder Metallurgy Equipment Association, 1986.

———, *Source Book on Powder Metallurgy.* Metals Park, Ohio: American Society for Metals, 1979.

German, R.M., *Powder Injection Molding.* Princeton, N.J.: Metal Powder Industries Federation, 1990.

——, *Powder Metallurgy Science.* Princeton, N.J.: Metal Powder Industries Federation, 1984.

Gessinger, G.H., *Powder Metallurgy of Superalloys.* London: Butterworths, 1984.

Hanes, E.D., D.A. Seifert, and C.R. Watts, *Hot Isostatic Processing.* New York: Springer-Verlag, 1980.

Hausner, H.H., and M.K. Mal, *Handbook of Powder Metallurgy.* New York: Chemical Publishing Company, 1982.

James, P.J. (ed.), *Isostatic Pressing Technology.* London: Applied Science Publishers, 1983.

Klar, E. (ed.), *Powder Metallurgy: Applications, Advantages, and Limitations.* Metals Park, Ohio: American Society for Metals, 1983.

Kuhn, H.A., and B.L. Ferguson, *Powder Forging.* Princeton, N.J.: Metal Powder Industries Federation, 1990.

Lenel, F.V., *Powder Metallurgy: Principles and Applications.* New York: American Powder Metallurgy Institute, 1980.

Metals Handbook, 9th ed., Vol. 7: *Powder Metallurgy.* Metals Park, Ohio: American Society for Metals, 1984.

Powder Metallurgy Design Guidebook. New York: American Powder Metallurgy Institute, revised periodically.

REVIEW QUESTIONS

17.1 Describe briefly the production steps involved in making powder-metallurgy parts.

17.2 Name the various methods of powder production, and explain the types of powders produced.

17.3 Explain why metal powders are blended.

17.4 What is meant by "green"? Is green strength important? Explain.

17.5 Name the methods used in metal powder compaction. Why is there a density variation in compacting powders?
17.6 What is the magnitude of forces involved in compaction?
17.7 Describe the hazards involved in P/M processing.
17.8 Give the reasons why injection molding of powders is becoming an important process.
17.9 What requirements should punches and dies have in P/M processing?
17.10 Describe what happens during sintering.
17.11 Why are secondary and finishing operations performed on P/M parts?
17.12 Explain the difference between impregnation and infiltration. Give some examples for each.
17.13 What is mechanical alloying? What are its advantages over conventional alloying of metals?
17.14 Why are protective atmospheres necessary in sintering? What would be the effects on the properties of P/M parts if such atmospheres were not used?

QUESTIONS AND PROBLEMS

17.15 Describe the design considerations in making powder-metallurgy parts. How different are these compared to casting and forging of metals?
17.16 Why do mechanical and physical properties depend on the density of P/M parts?
17.17 What are the effects of different shapes and sizes of metal particles in P/M processing?
17.18 Describe the relative advantages and limitations of cold and hot isostatic pressing.
17.19 What are the advantages of injection molding of powders compared to other shaping processes?
17.20 Are the requirements for punch and die materials in powder metallurgy different from those for forging and extrusion? Explain.
17.21 Explain why powder metallurgy has become highly competitive with casting, forging, and machining processes.
17.22 Are there applications in which you would not use a P/M product? Explain.
17.23 Describe in detail other methods of manufacturing the parts shown in Fig. 17.1.
17.24 Explain the reasons for the shapes of the curves and their relative position shown in Fig. 17.9.
17.25 Should green compacts be brought up to the sintering temperature slowly or rapidly? Explain the advantages and limitations of each.
17.26 Because they undergo special processing, metal powders are more expensive than the same metals in bulk form. How is the additional cost justified in powder-metallurgy parts?
17.27 Explain the effects of using fine powders and coarse powders, respectively, in making P/M parts.
17.28 What type of press is required to compact parts by the set of punches shown in Fig. 17.10(d)?
17.29 Estimate the maximum tonnage required to compact a 2-in. diameter slug made of brass. Would the height of the slug make any difference in your answer? Explain.
17.30 Referring to Fig. 17.9, what should be the volume of loose, fine iron powder in order to make a solid cylindrical compact 20 mm in diameter and 10 mm high?

17.31 Determine the shape factors for (a) a cylinder with dimensional ratios of 1:1:1 and (b) a flake with ratios of 1:10:10.

17.32 In Fig. 17.10(e), we note that the pressure is not uniform across the diameter of the compact at a particular distance from the punch. What is the reason for it?

17.33 Estimate the number of particles in a 500-g sample of iron powder, if the particle size is 100 μm.

17.34 Assume that the surface of a copper particle is covered with an oxide layer 0.1 μm in thickness. What is the volume occupied by this layer if the copper particle itself is 50 μm in diameter?

18

Forming and Shaping Plastics and Composite Materials

18.1 Introduction

The processing of plastics involves operations similar to those used to form and shape metals, as we described in the preceding chapters. Plastics can be molded, cast, and formed, as well as machined and joined, into many shapes with relative ease and with little or no additional operations required. As we showed in Chapter 7, plastics melt or cure at relatively low temperatures and hence, unlike metals, are easy to handle and require less energy to process.

Plastics are usually shipped to manufacturing plants as pellets or powders and are melted just before the shaping process. Plastics are also available as sheet, plate, rod, and tubing, which may be formed into a variety of products. Liquid plastics are used especially in making reinforced plastic parts.

In this chapter, we follow the outline shown in Fig. 18.1, explaining the basic processes of forming and shaping plastics, including reinforced plastics. Some of the same processes are used for shaping elastomers.

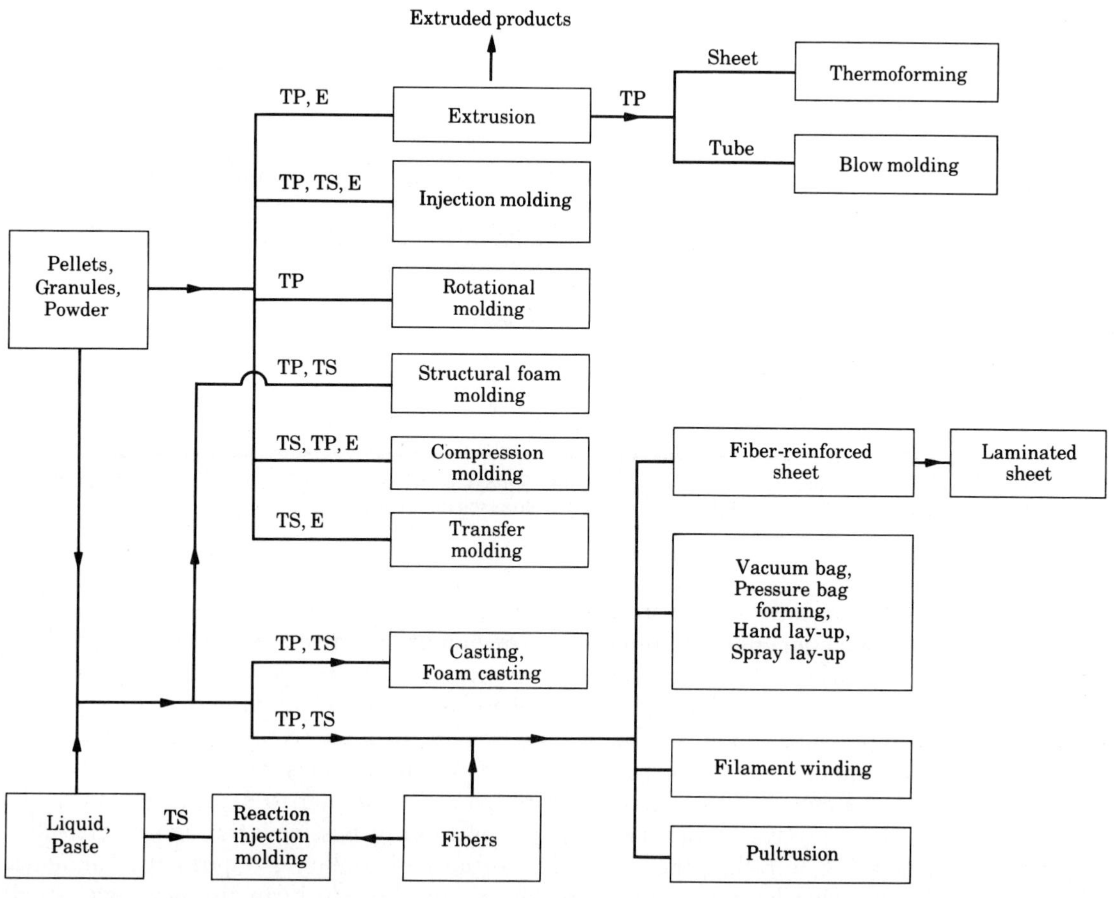

FIGURE 18.1
Outline of forming and shaping processes for plastics, elastomers, and composite materials. (TP, Thermoplastic; TS, Thermoset; E, Elastomer.)

18.2 Extrusion

In extrusion, raw materials in the form of thermoplastic pellets, granules, or powder are placed into a hopper and fed into the extruder barrel (Fig. 18.2). The barrel is equipped with a *screw* that blends and conveys the pellets down the barrel. The internal friction from the mechanical action of the screw, along with heaters around the extruder's barrel, heats the pellets and liquefies them. The screw action also builds up pressure in the barrel.

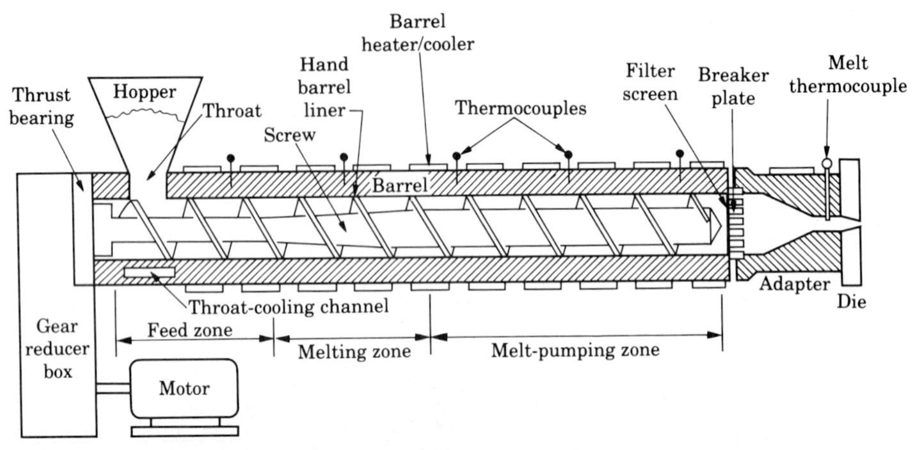

FIGURE 18.2
Schematic illustration of a typical extruder. *Source: Encyclopedia of Polymer Science and Engineering,* 2d ed., Vol. 7, p. 92. New York: Wiley-Interscience, 1985.

Screws have three distinct sections: (1) a feed section that conveys the material from the hopper area into the central region of the barrel; (2) a melt, or transition, section where the heat generated from shearing of the plastic causes melting to begin; and (3) a pumping section where additional shearing and melting occurs, with pressure buildup at the die. The lengths of these sections can be changed to accommodate the melting characteristics of different plastics.

The molten plastic is forced through a die—similar to extruding metals. The extruded product is then cooled, either by air or by passing it through a water-filled channel. Controlling the rate and uniformity of cooling is important to minimize product shrinkage and distortion. The extruded product can also be drawn (sized) by a puller after it has cooled. The extruded product is then coiled or cut into desired lengths. Complex shapes with constant cross-section can be extruded with relatively inexpensive tooling. This process is also used to extrude elastomers.

Because this operation is continuous, long products with cross-sections of solid rods, channels, pipe, window frames, and architectural components, as well as sheet, are extruded through dies of various geometries. Plastic-coated wire, cable, or strips for electrical applications are also extruded and coated by this process (Fig. 18.3). The wire is fed into the die opening at a controlled rate with the extruded plastic.

Pellets, which are used for other plastics-processing methods described in this chapter, are also made by extrusion. Here the extruded product is a small-diameter rod and is chopped into short lengths, or **pellets**, as it is extruded. With some modifications, extruders can also be used as simple melters for other shaping processes, such as injection molding and blow molding.

Process parameters such as extruder screw speed, barrel-wall temperatures, die design, and cooling and drawing speeds should be controlled carefully in order to

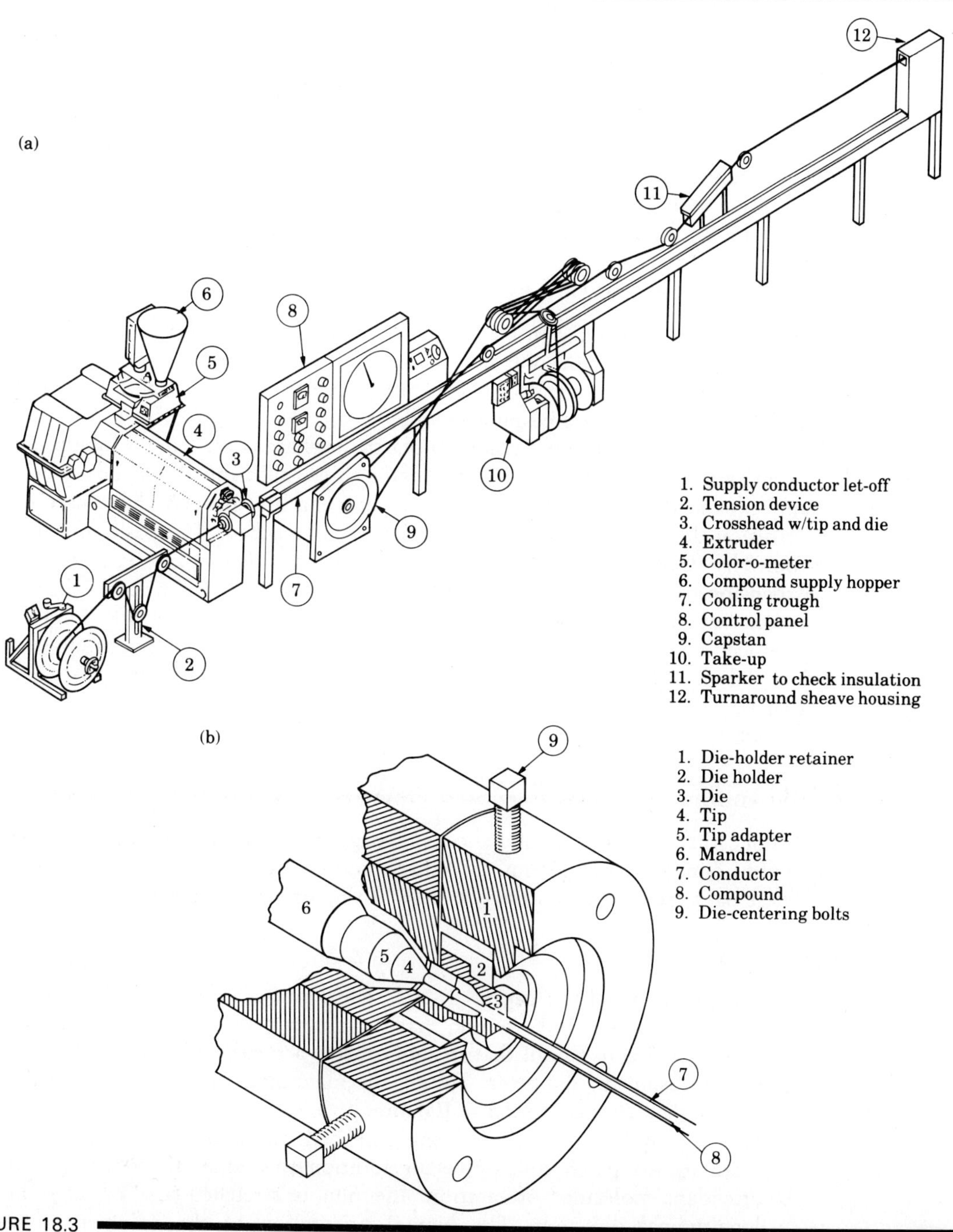

FIGURE 18.3

(a) Coating of electrical wire with plastic in an extruder. (b) Cross-section of extrusion die where wire is continuously coated with plastic. Other shapes can also be coated by this operation. *Source*: Belden Corporation.

extrude products having uniform dimensional accuracy. To filter out unmelted or congealed resin, a metal screen is usually placed just before the die and is replaced periodically.

Extruders are generally rated by the diameter of the barrel and by the length-to-diameter (L/D) ratio of the barrel. Typical commercial units are 25–200 mm (1–8 in.) in diameter, with L/D ratios of 5 to 30. Production-size extrusion equipment costs $30,000–$80,000, with an additional $30,000 cost for equipment for downstream cooling and winding of the extruded product. Thus large production runs are generally required to justify such an expenditure.

18.2.1 Sheet and film extrusion

Polymer sheet and film can be produced using a flat extrusion die, such as that shown in Fig. 18.4. The polymer is extruded by forcing it through a specially designed die, and the extruded sheet is taken up first on water-cooled rolls and then by a pair of rubber-covered pull-off rolls.

Thin polymer films and common plastic bags are made from a tube produced by an extruder (Fig. 18.5). In this process, a thin-walled tube is extruded vertically and expanded into a balloon shape by blowing air through the center of the extrusion die until the desired film thickness is reached. The balloon is usually cooled by air from a cooling ring around it, which can also act as a barrier to further expansion of the balloon. Blown film is sold as wrapping film (after slitting the cooled bubble) or as bags (where the bubble is pinched and cut off). Film is also produced by shaving solid round billets of plastics, especially polytetrafluoroethylene (PTFE), by skiving, that is, shaving with specially designed knives (see Section 20.4.1).

• Example: Blown film.

Assume that a typical plastic shopping bag, made by blown film, has a lateral (width) dimension of 400 mm. (a) What should be the extrusion die diameter? (b) These bags are relatively strong, so how is this strength achieved?

SOLUTION.

a) The perimeter of the bag is (2)(400) = 800 mm. As the original cross-section of the film was round, we find that the blown diameter should be $\pi D = 800$, or $D = 255$ mm. Recall that in this process a tube is expanded 1.5–2.5 times the extrusion die diameter. Taking the maximum value of 2.5, we calculate the die diameter as $255/2.5 = 100$ mm.

b) Note in Fig. 18.3 that after being extruded, the bubble is being pulled upward by the pinch rolls. Thus in addition to diametral stretching and the attendant molecular orientation, the film is stretched and oriented in the longitudinal direction. The biaxial orientation of the polymer molecules significantly improves the strength and toughness of the blown film. •

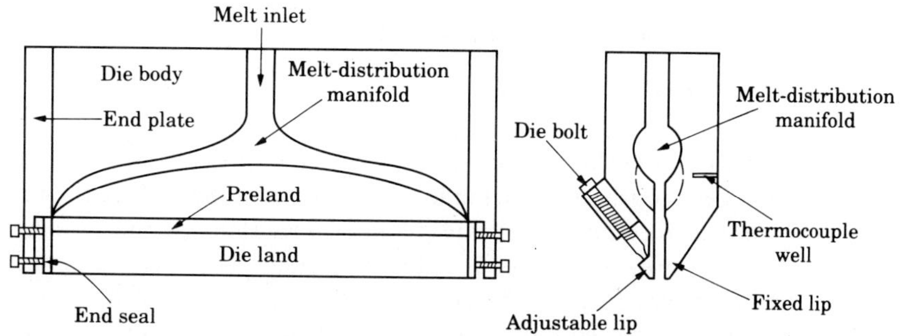

FIGURE 18.4

Die geometry (coat-hanger die) for extruding sheet. *Source: Encyclopedia of Polymer Science and Engineering*, 2d ed., Vol. 7, p. 93. New York: Wiley-Interscience, 1985.

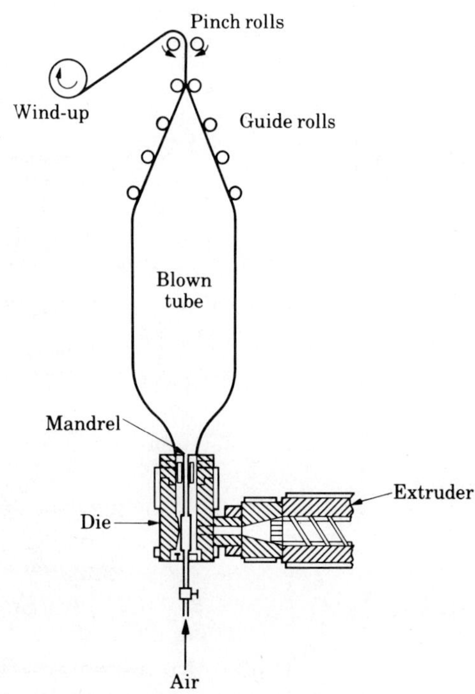

FIGURE 18.5

Schematic illustration of production of thin film and plastic bags from tube produced by an extruder and then blown by air. *Source:* D. C. Miles and J. H. Briston, *Polymer Technology*, 1979. Reproduced by permission of Chemical Publishing Co., Inc.

18.3 Injection Molding

Injection molding is essentially the same process as hot-chamber die casting. The pellets or granules are fed into a heated cylinder, where they are melted. The melt is then forced into a split-die chamber (Fig. 18.6a), either by a hydraulic plunger or by the rotating screw system of an extruder. Newer equipment is of the *reciprocating screw* type (Fig. 18.6b). As the pressure builds up at the mold entrance, the rotating screw begins to move backward under pressure to a predetermined distance, thus controlling the volume of material to be injected. The screw stops rotating and is pushed forward hydraulically, forcing the molten plastic into the mold cavity. Injection-molding pressures usually range from 70 MPa to 200 MPa (10,000 psi to 30,000 psi), which is on the same order of magnitude as in cold-chamber die casting.

Typical injection-molded products are cups, containers, housings, tool handles, knobs, electrical and communication components (such as telephone receivers), toys, and plumbing fittings. Although for thermoplastics the molds are relatively cool, thermosets are molded in heated molds where *polymerization* and *cross-linking* take place. In either case, after the part is sufficiently cooled (for thermoplastics) or set or cured (for thermosets), the molds are opened and the part is ejected. The molds are then closed and the process is repeated automatically. Elastomers also are injection molded by these processes.

Because the material is molten when injected into the mold, complex shapes and good dimensional accuracy can be achieved, as in die casting. Molds with moving

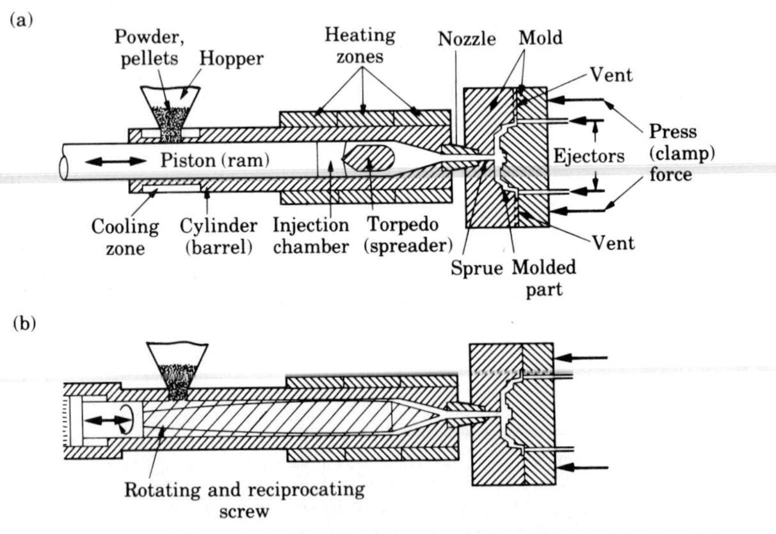

FIGURE 18.6
Injection molding with (a) plunger and (b) reciprocating rotating screw. Telephone receivers, plumbing fittings, tool handles, and housings are made by injection molding.

and unscrewing mandrels are not unusual, and allow the molding of parts with multiple cavities and internal and external threads.

Injection molds have several components, depending on part design, such as runners (as in metal-casting dies), cores, cavities, cooling channels, inserts, knockout pins, and ejectors (Fig. 18.7). There are three basic types of molds: (a) *cold-*

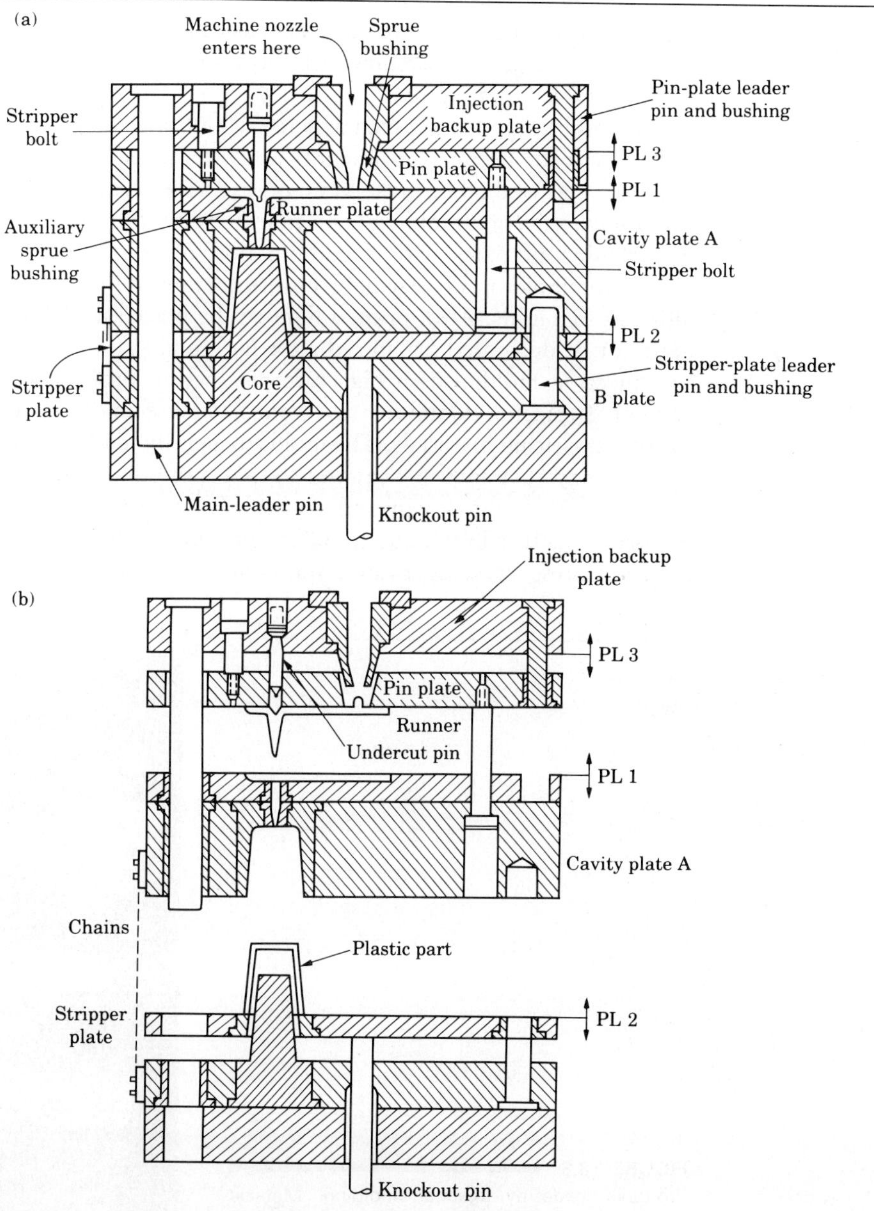

FIGURE 18.7 A three-plate injection mold in (a) closed position and (b) open position. *Source: Encyclopedia of Polymer Science and Engineering,* 2d ed. New York: Wiley-Interscience, 1985.

runner two-plate mold, which is the basic and simplest mold design, (b) *cold-runner three-plate* mold, in which the runner system is separated from the part when the mold opens, and (c) *hot-runner* mold, also called *runnerless* mold, in which the molten plastic is kept hot in a heated runner plate. In cold-runner molds, the solidified plastic in the channels that connect the mold cavity to the end of the barrel must be removed, usually by trimming. This scrap can be chopped and recycled. In hot-runner molds, which are more expensive, there are no gates, runners, or sprues attached to the part. Cycle times are shorter because only the injection-molded part must be cooled and ejected.

Metallic components, such as screws, pins, and strips, can also be placed in the mold cavity to become an integral part of the injection-molded product (*insert molding*, Fig. 18.8). The most common examples are electrical components. *Multi-component* injection molding (also called *co-injection* or *sandwich* molding) allows the forming of parts with a combination of colors and shapes. Examples are multicolor molding of rear-light covers for automobiles and ball joints made of different materials. Also, printed film can be placed in the mold cavity; thus parts need not be decorated or labeled after molding.

Injection molding is a high-rate production process, with good dimensional control. Typical cycle times range from 5 to 60 seconds, but can be several minutes for thermosetting materials. The molds, generally made of tool steels or beryllium–copper, may have multiple cavities so that more than one part can be made in one cycle. Proper mold design and control of material flow in the die cavities are important factors in the quality of the product. Other factors affecting quality are injection pressure, temperature, and condition of the resin. Modern machines are

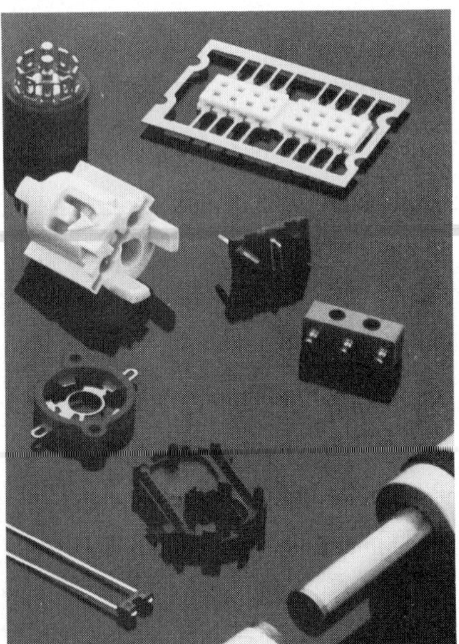

FIGURE 18.8
Products made by injection molding. Metallic components are embedded in these parts during molding. *Source:* Plainfield Molding Inc.

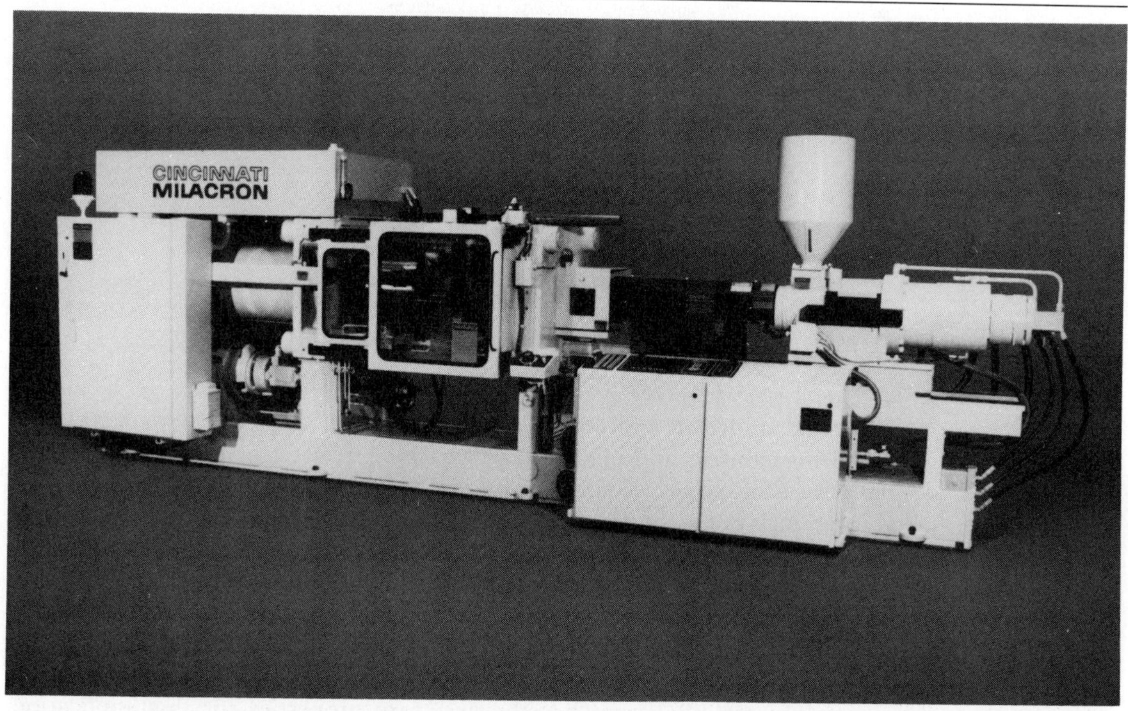

FIGURE 18.9
A 2.2-MN (250-ton) injection-molding machine. The tonnage is the force applied to keep the dies closed during injection of molten plastic into the mold cavities. *Source:* Courtesy of Cincinnati Milacron, Plastics Machinery Division.

equipped with microprocessors and microcomputers in a control panel and monitor all aspects of the operation.

Injection-molding machines are usually horizontal (Fig. 18.9) and the clamping force on the dies is supplied generally by hydraulic means, although electrical types are now available. Electrically driven models weigh less and are more quiet than hydraulic machines. Vertical machines are used for making small, close-tolerance parts, and for insert molding. Injection-molding machines are rated according to the capacity of the mold and the clamping force. Although in most machines this force generally ranges from 0.9 MN to 2.2 MN (100 tons to 250 tons), the largest machine in operation has a capacity of 45 MN (5000 tons), and can produce parts weighing 25 kg (55 lb). However, parts typically weigh 100–600 g (3–20 oz). Because of the high cost of dies, ranging from $20,000 to $200,000, high-volume production is required to justify such an expenditure.

• Example: Coin delivery cup.

One type of coin delivery cup for a vending machine was made of an aluminum casting. This is the cup into which change is returned. This casting had to be trimmed of flash, then hand buffed to remove burrs and provide a smooth surface.

Holes were then drilled and tapped for four machine screws. The part was then degreased for painting and loaded into a masking device to allow painting in the desired areas using a textured paint. The cup support was a flange made from sheet metal in a punch press, which was then degreased, chrome plated, and mounted on the cup.

Close examination of this part revealed that it could be made by injection molding using polyester, a thermoplastic. The production of this part by injection molding eliminated all the operations required in the previous method. The part was molded in one piece with the desired surface texture and in the desired color (black in this case). Polyester has the necessary properties for this application, namely, mechanical strength, low friction, and good abrasion resistance. Injection molding of this part is 35 percent more energy efficient than metal casting. Elimination of handling, plating, and other operations can reduce costs by as much as 75 percent. *Source for photograph*: Courtesy of Plaspros, Inc.

• Example: Injection molding of parts.

A 250-ton injection molding machine is to be used to make 4.5-in. diameter spur gears, 0.5 in. thick. The gears have fine teeth profile. How many gears can be injection molded in one set of molds? Does the thickness of the gears influence the answer?

SOLUTION. Because of the fine detail involved, the pressures required in the mold cavity will probably be on the order of 100 MPa (15 ksi). The cross-sectional (projected) area of the gear is $\pi(4.5)^2/4 = 15.9$ in^2. If we assume that the parting plane of the two halves of the mold is in the middle of the gear, the force required is $(15.9)(15{,}000) = 238{,}500$ lb. The capacity of the machine is 200 tons, so we have $(250)(2000) = 500{,}000$ lb of clamping force available. Therefore, the mold can accommodate two cavities, thus producing two gears per cycle. Because it does not influence the cross-sectional area of the gear, the thickness of the gear does not directly influence the pressures involved, and the answer is the same. •

18.3.1 Reaction-injection molding

In the **reaction-injection molding** (RIM) process, a mixture of two or more reactive fluids is forced into the mold cavity (Fig. 18.10). Chemical reactions take place rapidly in the mold and the polymer solidifies, producing a thermoset part. Major applications are automotive bumpers and fenders, thermal insulation for refrigerators and freezers, and stiffeners for structural components. Various reinforcing fibers, such as glass or graphite, may also be used to improve the product's strength and stiffness. Mold costs are low.

18.3.2 Structural foam molding

The *structural foam molding* process is used to make plastic products that have a solid skin and a cellular inner structure. Typical products are furniture components, TV cabinets, business-machine housings, and storage-battery cases. Although there are several foam molding processes, they are basically similar to injection molding or extrusion. Both thermoplastics and thermosets can be used for foam molding, but thermosets are in the liquid processing form, similar to polymers for reaction-injection molding.

In *injection foam molding*, thermoplastics are mixed with a blowing agent (usually an inert gas such as nitrogen), which expands the material. The core of the part is cellular and the skin is rigid. The thickness of the skin can be as much as 2 mm (0.08 in.), and part densities are as low as 40 percent of the density of the solid plastic. Thus parts have a high stiffness-to-weight ratio and can weigh as much as 55 kg (120 lb).

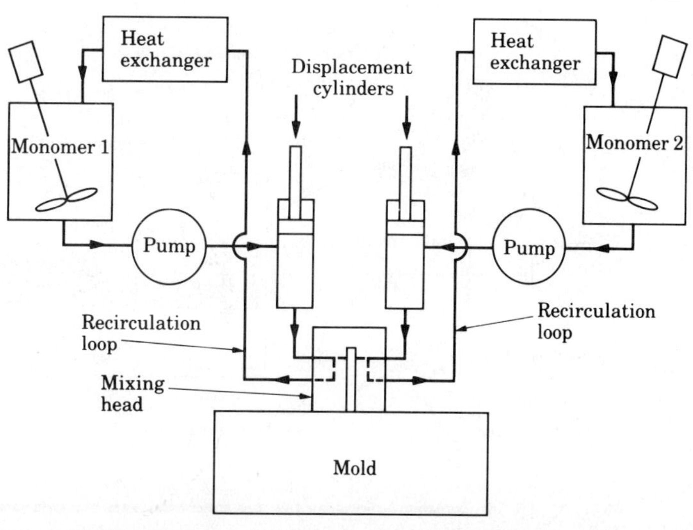

FIGURE 18.10 Schematic illustration of the reaction-injection molding process. *Source:* Modern Plastics Encyclopedia.

18.4 Blow Molding

Blow molding is a modified extrusion and injection molding process. In *extrusion blow molding*, a tube is extruded (usually turned so that it is vertical), clamped into a mold with a cavity much larger than the tube diameter, and then blown outward to fill the mold cavity (Fig. 18.11a). Blowing is usually done with an air blast, at a pressure of 350–700 kPa (50–100 psi). In some operations the extrusion is continuous, and the molds move with the tubing. The molds close around the tubing, close off both ends (thereby breaking the tube into sections), and then move away as air is

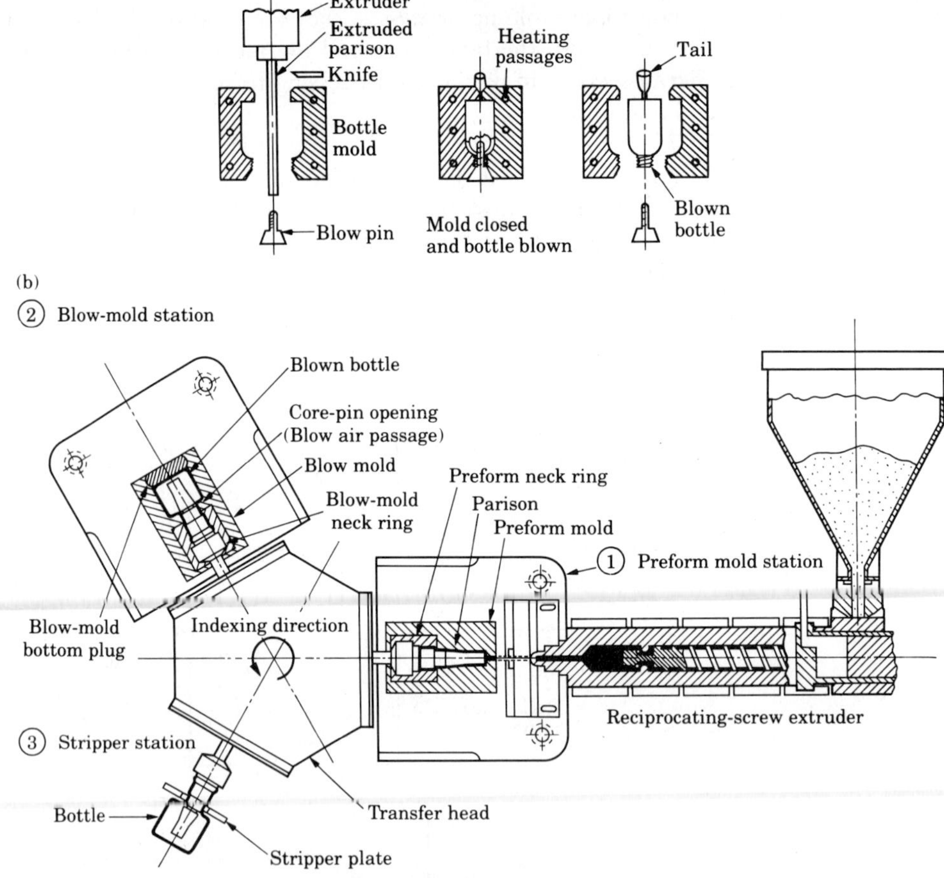

FIGURE 18.11
Schematic illustrations of (a) the blow-molding process for making plastic beverage bottles and (b) a three-station injection blow-molding machine. *Source: Encyclopedia of Polymer Science and Engineering*, 2d ed., Vol. 2, p. 450. New York: Wiley-Interscience, 1985.

injected into the tubular piece. The part is then cooled and ejected. Corrugated pipe and tubing is made by continuous blow molding, whereby the pipe or tube is extruded horizontally and blown into moving molds.

In *injection blow molding*, a short tubular piece (**parison**) is first injection molded (Fig. 18.11b). The dies then open and the parison is transferred to a blow-molding die. Hot air is injected into the parison, which then expands and fills the mold cavity. Typical products are plastic beverage bottles and hollow containers.

Multilayer blow molding involves the use of coextruded tubes or parisons, thus allowing the use of multilayer structures. Typical examples of multilayer structures are plastic packaging for food and beverage with characteristics such as odor and permeation barrier, taste and aroma protection, scuff resistance, printing capability, and ability to be filled with hot fluids. Other applications are in cosmetics and pharmaceutical industries.

18.5 Rotational Molding

Most thermoplastics and some thermosets can be formed into large hollow parts by **rotational molding.** The thin-walled metal mold is made of two pieces (split female mold) and is designed to be rotated about two perpendicular axes (Fig. 18.12). A

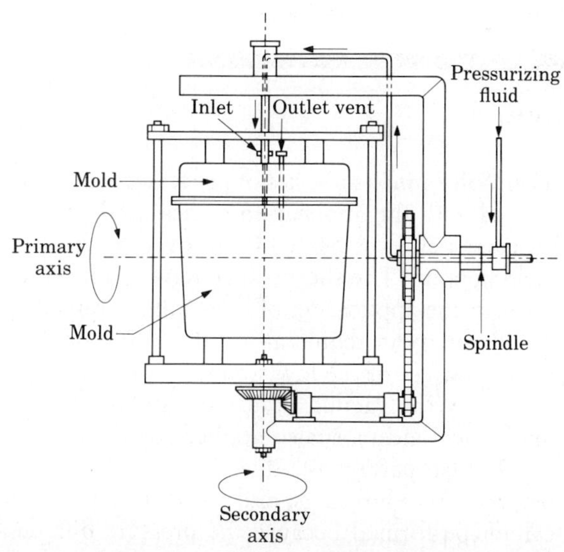

FIGURE 18.12 The rotational molding (rotomolding or rotocasting) process. Trash cans, buckets, and plastic footballs can be made by this process.

premeasured quantity of powdered plastic material is placed inside a warm mold. The powder is obtained from a polymerization process that precipitates a powder from a liquid. The mold is then heated, usually in a large oven, while it is rotated about the two axes. This action tumbles the powder against the mold where heating fuses the powder without melting it. In some parts, a chemical cross-linking agent is added to the powder, and cross-linking occurs after the part is formed in the mold by continued heating. Typical parts made by rotational molding are tanks of various sizes, trash cans, boat hulls, buckets, housings, toys, carrying cases, and footballs. Various metallic or plastic inserts may also be molded into the parts made by this process.

Liquid polymers, called **plastisols** (vinyl plastisols being the most common), can also be used in a process called *slush molding*. The mold is heated and rotated simultaneously. The particles of plastic material are forced against the inside walls of the heated mold by the tumbling action. Upon contact, the material melts and coats the walls of the mold. The part is cooled while still rotating and is then removed by opening the mold.

Rotational molding can produce parts with complex hollow shapes, with wall thicknesses as small as 0.4 mm (0.016 in.). Parts as large as 1.8 m × 1.8 m × 3.6 m (6 ft × 6 ft × 12 ft) have been formed. The outer surface finish of the part is a replica of the surface finish of the mold walls. Cycle times are longer than in other processes, but equipment costs are low. Quality control considerations usually involve proper weight of powder placed in the mold, proper rotation of the mold, and the temperature–time relationship during the oven cycle.

18.6 Thermoforming

Thermoforming is a series of processes for forming thermoplastic sheet or film over a mold with the application of heat and pressure differentials (Fig. 18.13). In this process, a sheet is heated in an oven to the sag (softening) point—but not to the melting point. The sheet is then removed from the oven and placed over a mold and through the application of a vacuum is pulled against the mold. Since the mold is usually at room temperature, the shape of the plastic is set upon contacting the mold. Because of the low strength of the materials formed, the pressure differential caused by the vacuum is usually sufficient for forming, although air pressure or mechanical means is also applied for some parts.

Typical parts made this way are advertising signs, refrigerator liners, packaging, appliance housings, and panels for shower stalls. Parts with openings or holes cannot be formed because the pressure differential cannot be maintained during forming. Because thermoforming is a drawing and stretching operation, much like sheet-metal forming, the material should exhibit high uniform elongation—otherwise, it will neck and fail. Thermoplastics have a high capacity for uniform

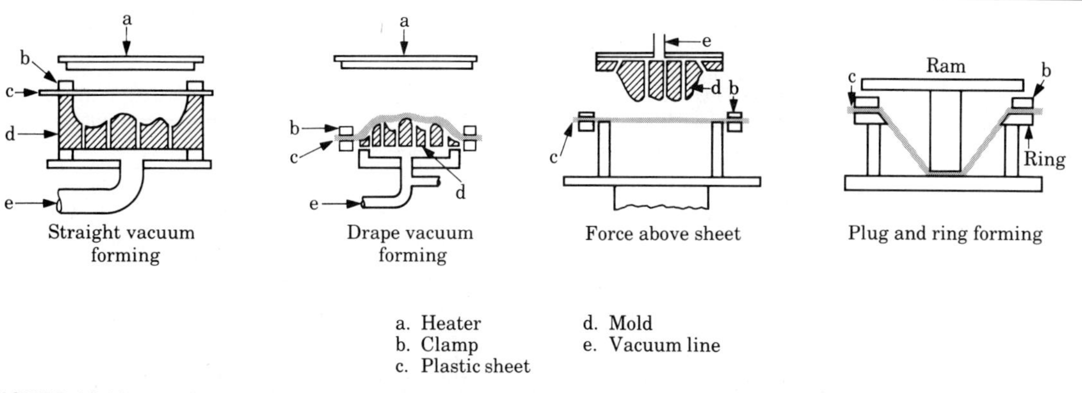

FIGURE 18.13
Various thermoforming processes for thermoplastic sheet. These processes are commonly used in making advertising signs, cookie and candy trays, panels for shower stalls, and packaging.

elongation by virtue of their high strain-rate sensitivity exponent, m. The sheets used in thermoforming are made by sheet extrusion.

Molds for thermoforming are usually made of aluminum since high strength is not a requirement. The holes in the molds are usually less than 0.5 mm (0.02 in.) in order not to leave any marks on the formed sheets. Tooling is inexpensive. Quality considerations include tears, nonuniform wall thickness, improperly filled molds, and poor part definition (surface details).

18.7 Compression Molding

In **compression molding**, a preshaped charge of material, a premeasured volume of powder, or a viscous mixture of liquid resin and filler material, is placed directly in a heated mold cavity. Forming is done under pressure with a plug or the upper half of the die (Fig. 18.14). Compression molding results in flash formation; it is removed by trimming or some other means. Typical parts made are dishes, handles, container caps, fittings, electrical and electronic components, washing-machine agitators, and housings. Fiber-reinforced parts with long chopped fibers are exclusively formed by this process.

Compression molding is used mainly with thermosetting plastics, with the original material in a partially polymerized state. Cross-linking is completed in the heated die, with curing times ranging from 0.5 min to 5 min, depending on the material and part geometry and its thickness. The thicker the material, the longer it will take to cure. Elastomers also are shaped by compression molding.

Because of their relative simplicity, die costs in compression molding are generally lower than in injection molding. Three types of compression molds are

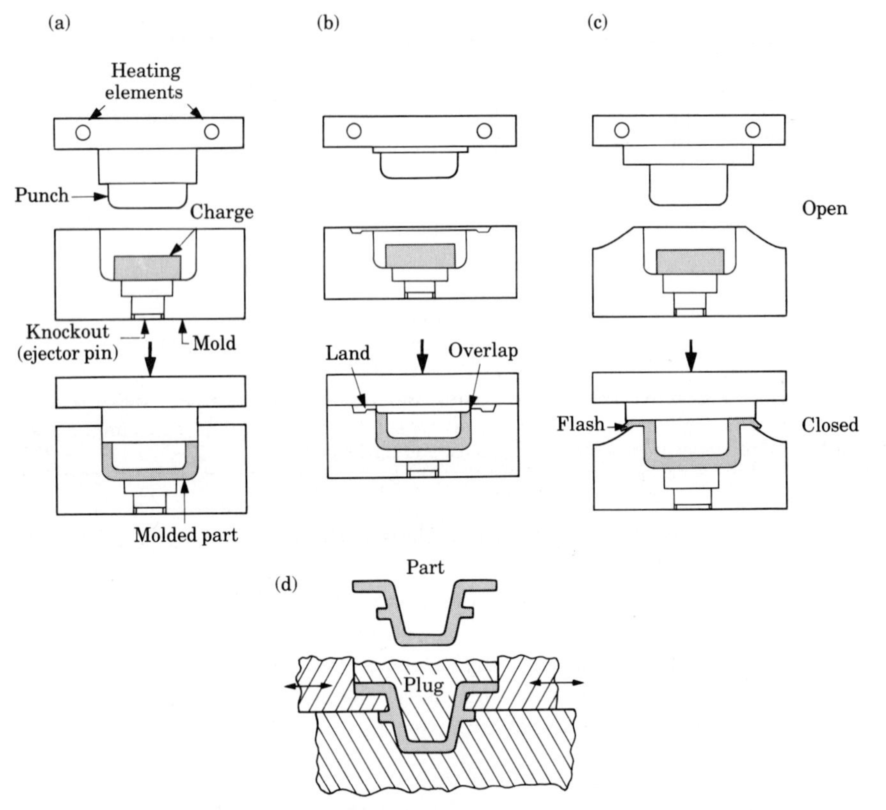

FIGURE 18.14 Types of compression molding, a process similar to forging: (a) positive, (b) semipositive, and (c) flash. The flash in part (c) has to be trimmed off. (d) Die design for making a compression-molded part with undercuts. Such designs are also used in other molding and shaping operations.

available: (1) *flash-type* for shallow or flat parts, (2) *positive* for high density, and (3) *semipositive* for quality production. Undercuts in parts are not recommended; however, dies can be designed to open sideways (Fig. 18.14d) to allow removal of the part. In general, the part complexity is less than with injection molding and dimensional control is better.

18.8 Transfer Molding

Transfer molding represents a further development of compression molding. The uncured thermosetting material is placed in a heated transfer pot or chamber (Fig. 18.15). After the material is heated, it is injected into heated, closed molds.

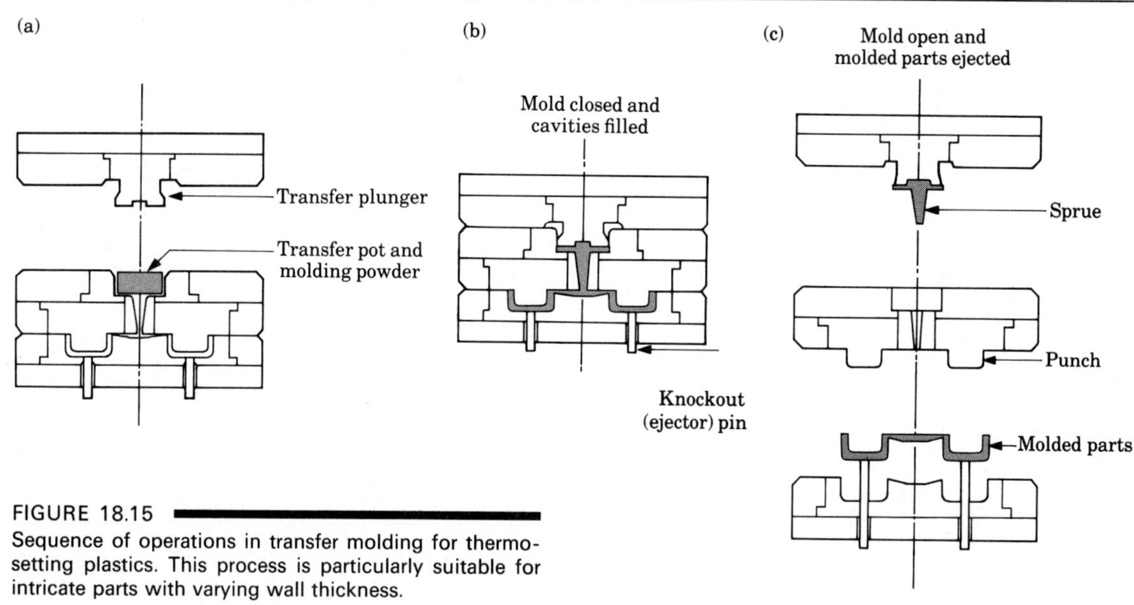

FIGURE 18.15 Sequence of operations in transfer molding for thermosetting plastics. This process is particularly suitable for intricate parts with varying wall thickness.

Depending on the type of machine used, a ram, plunger, or rotating screw-feeder forces the material to flow through the narrow channels into the mold cavity. This flow generates considerable heat, which raises the temperature of the material and homogenizes it. Curing takes place by cross-linking. Because the resin is molten as it enters the molds, the complexity of the part and dimensional control approach those for injection molding.

Typical parts made by transfer molding are electrical and electronic components and rubber and silicone parts. The process is particularly suitable for intricate shapes having varying wall thicknesses. Molds tend to be more expensive than those for compression molding, and material is wasted in the channels of the mold during filling.

18.9 Casting

Some thermoplastics, such as nylons and acrylics, and thermosetting plastics, such as epoxies, phenolics, polyurethanes, and polyester, can be cast in rigid or flexible molds into a variety of shapes (Fig. 18.16a). Typical parts cast are large gears, bearings, wheels, thick sheets, and components requiring resistance to abrasive wear.

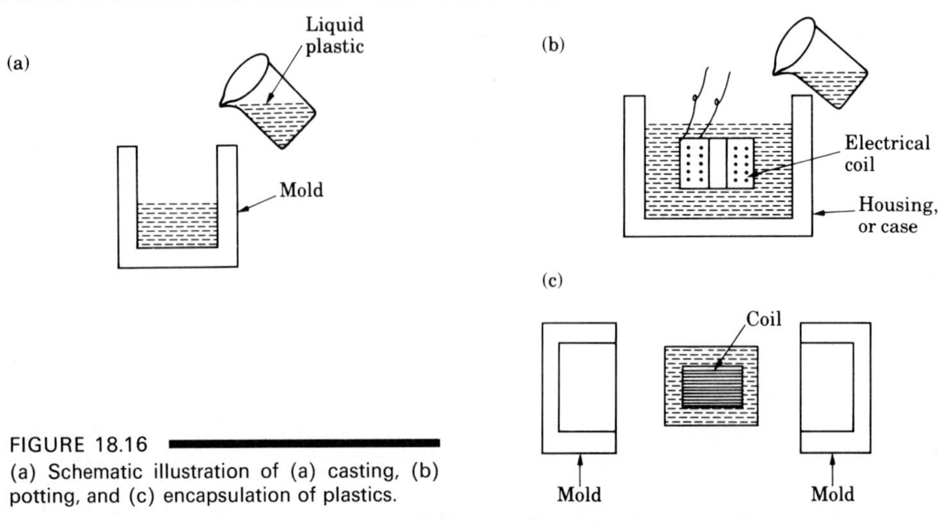

FIGURE 18.16
(a) Schematic illustration of (a) casting, (b) potting, and (c) encapsulation of plastics.

In casting thermoplastics, a mixture of monomer, catalyst, and various additives is heated and poured into the mold. The part is formed after polymerization takes place at ambient pressure. Intricate shapes can be formed with flexible molds, which are then peeled off. Centrifugal casting is also used with plastics, including reinforced plastics with short fibers. Thermosets are cast in a similar manner. Typical parts produced are similar to those made by thermoplastic castings. Degassing may be necessary for product integrity.

18.9.1 Potting and encapsulation

A variation of casting that is important to the electrical and electronics industry is potting and encapsulation. This involves casting the plastic around an electrical component, thus embedding it in the plastic. **Potting** (Fig. 18.16b) is done in a housing or case, which is an integral part of the product. In **encapsulation** (Fig. 18.16c), the component is covered with a layer of the solidified plastic. In both applications the plastic serves as a dielectric (nonconductor). Structural members, such as hooks and studs, may also be partly encapsulated.

18.9.2 Foam molding/casting

Products such as styrofoam cups and food containers, insulating blocks, and shaped packaging materials (such as for cameras, appliances, and electronics), are made by **foam molding.** The material is made of expandable polystyrene in which polystyrene (obtained by polymerization of styrene monomer) *beads*, containing a blowing agent, are placed in a mold and exposed to heat, usually by steam. As a result, the beads expand to as much as 50 times their original size and take the shape of the mold. The amount of expansion can be controlled through temperature and time. A common method of molding is using pre-expanded beads, in which beads are expanded by steam (although hot air, hot water, or an oven can also be used) in an open-top chamber. They are then placed in a storage bin and allowed to stabilize for a period of 3–12 hr. The beads can then be molded into shapes as described above.

Polystyrene beads are available in three sizes: small for cups, medium for molded shapes, and large for molding of insulating blocks (which can then be cut to size). Thus the bead size chosen depends on the minimum wall thickness of the product. Beads can also be colored prior to expansion, or are available as integrally colored.

Polyurethane foam processing, for products such as cushions and insulating blocks, involves several processes but basically they consist of mixing two or more chemical components. The reaction forms a cellular structure, which solidifies in a mold. Various low-pressure and high-pressure machines, with computer controls for proper mixing, are available.

18.10 Cold Forming and Solid-Phase Forming

Processes that have been used in cold working of metals (Chapters 13–16) can also be used to form many thermoplastics at room temperature, such as rolling, deep drawing, extrusion, closed-die forging, coining, and rubber forming. Typical materials formed are polypropylene, polycarbonate, ABS, and rigid PVC. The important considerations are that (a) the material be sufficiently ductile at room temperature (hence polystyrenes, acrylics, and thermosets cannot be formed); and (b) the material's deformation must be nonrecoverable (to minimize springback and creep).

The advantages of cold forming of plastics over other methods of shaping are:

- Strength, toughness, and uniform elongation are increased.
- Plastics with high molecular weight can be used to make parts with superior properties.
- Forming speeds are not affected by part thickness since there is no heating or cooling involved. Typical cycle times are shorter than molding processes.

Solid-phase forming is carried out at a temperature about 10–20 °C (20–40 °F) below the melt temperature of the plastic, if a crystalline polymer, and formed while still in a solid state. The advantages over cold forming are that forming forces and springback are lower. These processes are not as widely used as hot-processing methods and are generally restricted to special applications.

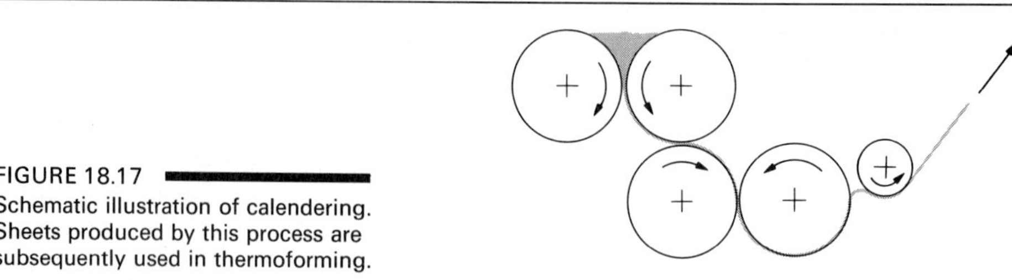

FIGURE 18.17 Schematic illustration of calendering. Sheets produced by this process are subsequently used in thermoforming.

18.11 Processing Elastomers

Elastomers can be formed by a variety of processes used for shaping thermoplastics. Thermoplastic elastomers are commonly shaped by extrusion and injection molding, extrusion being the most economical and the fastest process. In terms of its processing characteristics, a thermoplastic elastomer is a polymer; in terms of its function and performance, it is a rubber. These polymers can also be formed by blow molding and thermoforming. Thermoplastic polyurethane can be shaped by all conventional methods. It can also be blended with thermoplastic rubbers, polyvinyl chloride compounds, ABS, and nylon. Dryness of the materials is important. For extrusion, the temperatures are in the range of 170 °C to 230 °C (340 °F to 440 °F), and for molding up to 60 °C (140 °F). Typical extruded products are tubing, hoses, molding, and inner tubes. Injection-molded products cover a broad range of applications such as components for automobiles and appliances (see Section 7.8).

Rubber and some thermoplastic sheets are formed by the **calendering** process (Fig. 18.17), wherein a warm mass of the compound is fed through a series of rolls (*masticated*) and is then stripped off in the form of a sheet. The rubber may also be formed over both surfaces of a fabric liner. Discrete rubber products, such as gloves, are made by dipping a form repeatedly into a liquid compound that adheres to the form. It is then vulcanized, usually in steam, and stripped from the form.

18.12 Processing Composite Materials

As we described in Chapter 9, reinforced plastics, also called *composites*, are among the most important materials and can be engineered to meet specific design requirements such as high strength-to-weight and stiffness-to-weight ratios and creep resistance. Because of their unique structure, reinforced plastics require special methods to shape them into useful products (Fig. 18.18).

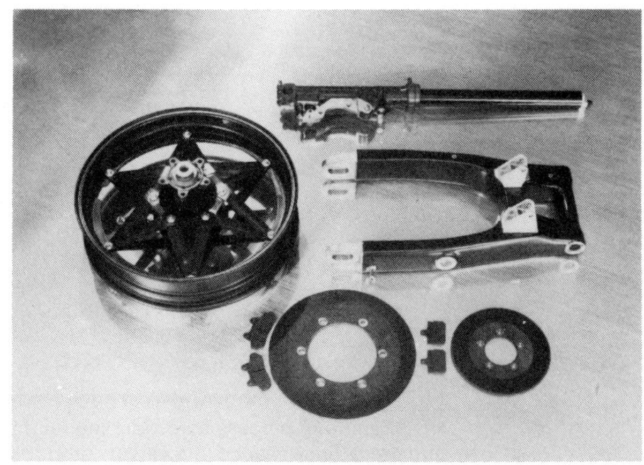

FIGURE 18.18 Reinforced plastic components for a Honda motorcycle. Parts shown are front and rear forks, rear swingarm, wheel, and brake disks.

Reinforced plastics can usually be fabricated by the methods described in this chapter, with some provision for the presence of more than one type of material in the composite. The reinforcement may be chopped fibers, woven fabric or mat, roving (slightly twisted fiber), or continuous lengths of fiber. In order to obtain good bonding between the reinforcing fibers and the polymer matrix, it is necessary to impregnate and coat the reinforcement with the polymer. Short fibers are commonly added to thermoplastics for injection molding, milled fibers can be used for reaction-injection molding, and longer chopped fibers are primarily used in compression molding of reinforced plastics.

When the impregnation is done as a separate step, the resulting partially dry sheets are called **prepregs**, **bulk-molding compound** (BMC), or **sheet-molding compound** (SMC), depending on their form. They should be stored at a sufficiently low temperature to delay curing. Alternatively, the resin and the fibers can be mixed together at the time they are placed in the mold. More recently, a **thick molding compound** (**TMC**, a trademark) has been developed that combines the characteristics of BMC (lower cost) and SMC (higher strength). Thick molding compounds are usually injection molded using chopped fibers of various lengths. One application is for electrical components because of the high dielectric strength of TMC.

Commercial reinforced plastics are available as sheet-molding compounds. Continuous strands of reinforcing fiber are chopped into short fibers (Fig. 18.19) and deposited over a layer of resin paste, usually a polyester mixture, carried on a polymer film such as polyethylene. A second layer of resin paste is deposited on top, and the sheet is pressed between rollers. The product is gathered into rolls, or placed into containers in layers, and stored until it undergoes a maturation period, reaching the desired molding viscosity. The maturing process is under controlled temperature and humidity, and usually takes one day. The matured SMC, which has a leatherlike feel and is tack-free, has a shelf life of about 30 days and must be processed within this period.

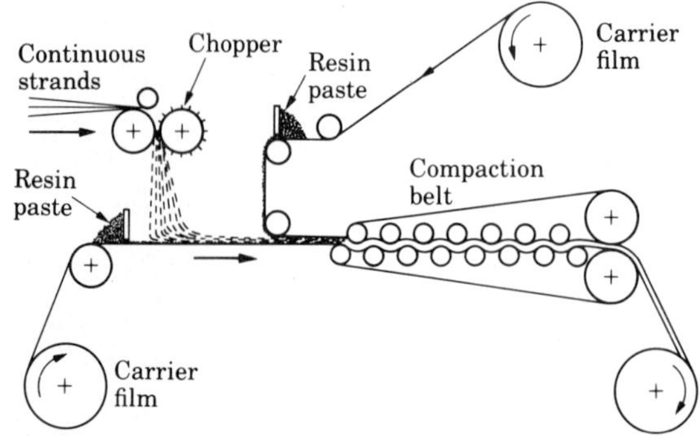

FIGURE 18.19 Manufacturing process for producing reinforced plastic sheets. The sheet is still viscous at this stage and can later be shaped into various products. *Source:* T.-W. Chou, R. L. McCullough, and R. B. Pipes.

A typical procedure for making reinforced plastic prepregs is shown in Fig. 18.20(a). The continuous fibers are aligned and subjected to surface treatment to enhance adhesion to the polymer matrix. They are then coated by dipping them in a resin bath and made into a sheet or tape (Fig. 18.20b). Individual pieces of the sheet are then assembled into laminated structures (Fig. 18.21a), such as the horizontal stabilizer for the F-14 fighter aircraft (Figs. 18.21b and c). Special computer-controlled tape-laying machines have been developed for this purpose (Fig. 18.22). Typical products are flat or corrugated architectural paneling, panels for construction and electric insulation, and structural components of aircraft.

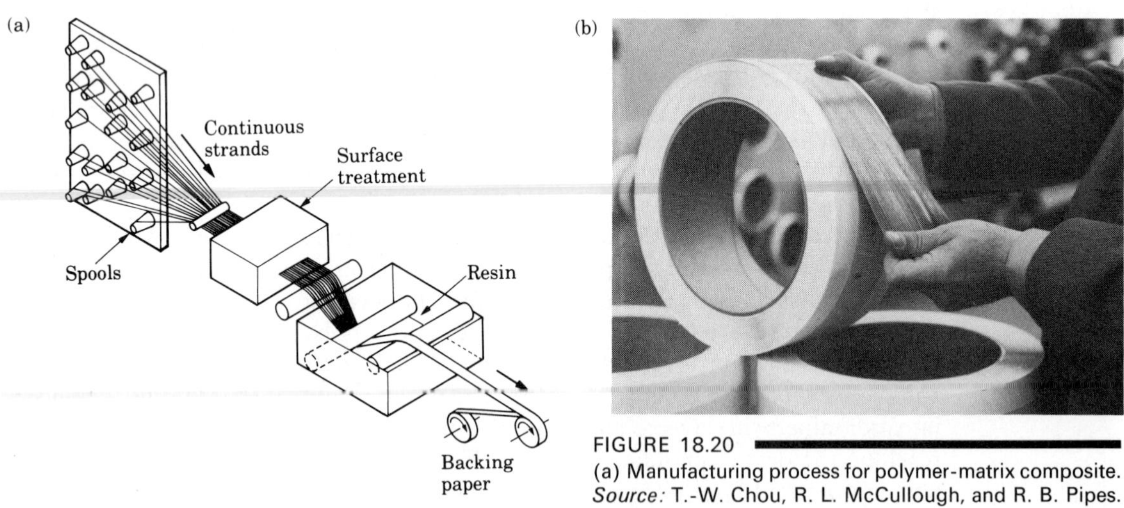

FIGURE 18.20 (a) Manufacturing process for polymer-matrix composite. *Source:* T.-W. Chou, R. L. McCullough, and R. B. Pipes. (b) Boron-epoxy prepreg tape. *Source:* Avco Specialty Materials/Textron.

(a)

(b)

(c)

FIGURE 18.21

(a) Single-ply layup of boron-epoxy tape for the horizontal stabilizer for F-14 fighter aircraft. (b) Horizontal stabilizer with boron-epoxy wing skin. (c) F-14 fighter aircraft, showing two horizontal stabilizers. *Source:* Grumman Aircraft Corporation.

FIGURE 18.22

A 10-axis computer-numerical controlled tape laying system. This machine is capable of laying up 75 mm and 150 mm (3 in. and 6 in.) wide tapes, on contours of up to $\pm 30°$ and at speeds up to 0.5 m/s (1.7 ft/s). *Source:* Courtesy of The Ingersoll Milling Machine Company.

• Example: Tennis rackets made of composite materials.

In order to impart certain desirable characteristics, such as light weight and stiffness, composite-material tennis rackets are being manufactured with graphite, fiberglass, boron, ceramic (silicon carbide), and Kevlar as the reinforcing fibers. The accompanying figure shows the cross-sections of five types of rackets and their composition. All rackets have a foam core; some have unidirectional and others have braided reinforcement. We can assess the contribution of various types of fibers by referring back to Table 9.1, where we note that boron fibers have the highest stiffness, followed by graphite (carbon), glass, and Kevlar. The racket with the lowest stiffness has 80 percent fiberglass, whereas the stiffest has 95 percent graphite and 5 percent boron. Thus it has the highest percentage of inexpensive reinforcing fiber and the smallest percentage of the most expensive fiber. The stiffest racket weighs 360 g (12.70 oz) and the least stiff 354 g (12.50 oz).

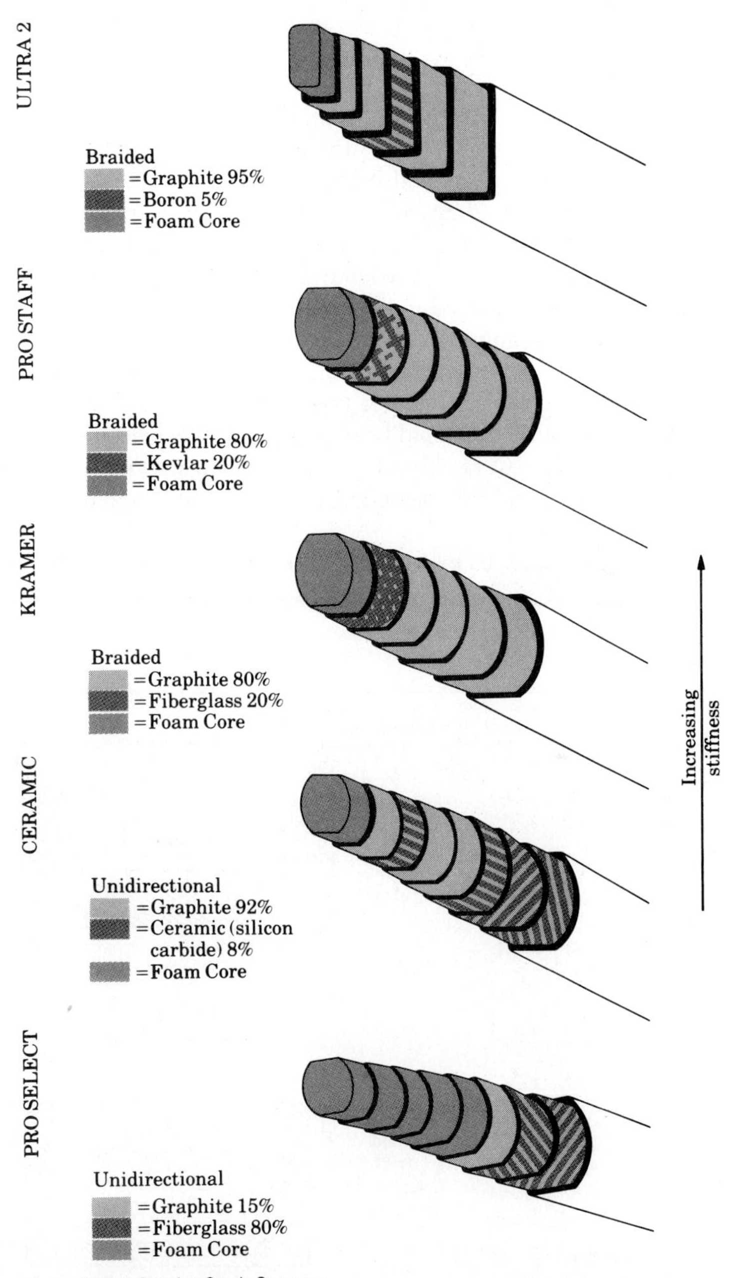

Source: Wilson Sporting Goods Company.

18.12.1 Molding

In compression molding, the material is placed between two molds and pressure is applied. Depending on the material, the molds may be either at room temperature or heated to accelerate hardening. The material may be in bulk form (bulk-molding compound), which is a viscous, sticky mixture of polymers, fibers, and additives. It is generally shaped into a log which is cut into the desired mass. Fiber lengths generally range from 3 mm to 50 mm (0.125 in. to 2 in.), although longer fibers (75 mm; 3 in.) may also be used.

Sheet-molding compounds can also be used in molding. Sheet-molding compound is similar to BMC, except that the resin–fiber mixture is laid between plastic sheets to make a sandwich that can be easily handled. The sheets are removed when the SMC is placed in the mold.

In **vacuum-bag molding** (Fig. 18.23), prepregs are laid in a mold to form the desired shape. In this case, the pressure required to form the shape and good bonding is obtained by covering the lay-up with a plastic bag and creating a vacuum. If additional heat and pressure are desired, the entire assembly is put in an autoclave. Care should be exercised to maintain fiber orientation, if specific fiber orientations are desired. In chopped-fibers materials, no specific orientation is intended.

In order to prevent the resin from sticking to the vacuum bag and to facilitate removal of excess resin, several sheets of various materials (*release cloth*, *bleeder*

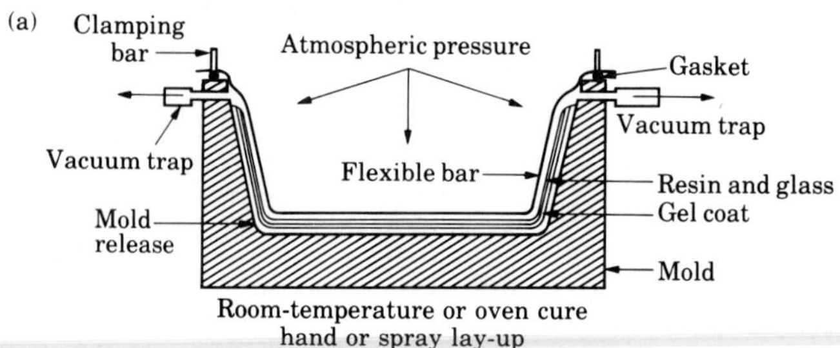

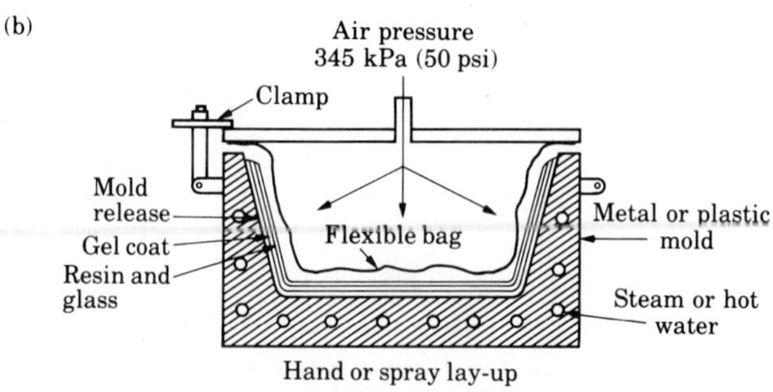

FIGURE 18.23
(a) Vacuum-bag forming. (b) Pressure-bag forming. *Source:* T. H. Meister.

cloth) are placed on top of the prepreg sheets. The molds can be made of metal, usually aluminum, but more often are made from the same resin (with reinforcement) as the material to be cured. This eliminates any problem with differential thermal expansion between the mold and the part.

Contact molding processes use a single male or female mold (Fig. 18.24) made of materials such as reinforced plastics, wood, or plaster. Contact molding is used in making products with high surface area-to-thickness ratios, such as swimming pools, boats, tub and shower units, and housings. This is a "wet" method, in which the reinforcement is impregnated with the resin at the time of molding. The simplest method is called *hand lay-up*. The materials are placed and formed in the mold by hand (Fig. 18.24a), and the squeezing action expels any trapped air and compacts the part.

Molding may also be done by spraying (*spray-up*; Fig. 18.24b). Although spraying can be automated, these processes are relatively slow and labor costs are high. However, they are simple and the tooling is inexpensive. Only the mold-side surface of the part is smooth and the choice of materials is limited. Many types of boats are made by this process.

Resin transfer molding is based on transfer molding (Section 18.8) whereby a resin, mixed with a catalyst, is forced by a piston-type positive displacement pump into the mold cavity filled with fiber reinforcement. The process is a viable alternative to hand lay-up, spray-up, and compression molding for intermediate volume production.

18.12.2 Filament winding, pultrusion, and pulforming

Filament winding is a process whereby the resin and fibers are combined at the time of curing. Axisymmetric parts, such as pipes and storage tanks, are produced on a rotating mandrel. The reinforcing filament, tape, or roving is wrapped continuously

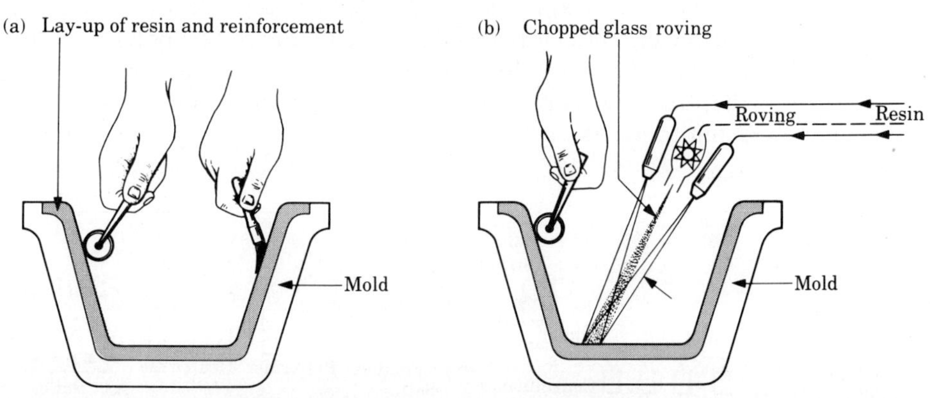

FIGURE 18.24 Manual methods of processing reinforced plastics: (a) hand lay-up and (b) spray-up. These methods are also called *open-mold* processing.

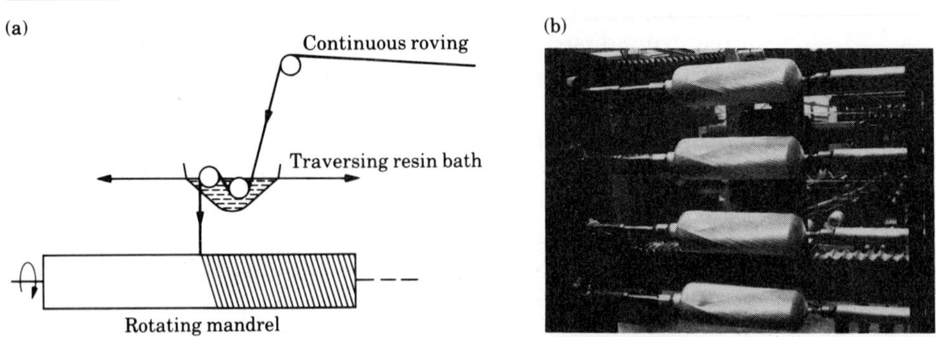

FIGURE 18.25
(a) Schematic illustration of the filament-winding process. (b) Fiberglass being wound over aluminum liners for slide-raft inflation vessels for the Boeing 767 aircraft. *Source:* Brunswick Corporation, Defense Division.

around the form. The reinforcements are impregnated by passing them through a polymer bath (Fig. 18.25a). The process can be modified by wrapping the mandrel with prepreg material.

The products made by filament winding are very strong because of their highly reinforced structure. Filament winding has also been used for strengthening cylindrical or spherical pressure vessels (Fig. 18.25b) made of materials such as aluminum and titanium. The presence of a metal inner lining makes the part impermeable. Filament winding can be used directly over solid-rocket propellant forms.

Pultrusion. Long shapes with various constant profiles, such as rods, profiles, or tubing (similar to extruded metal products), are made by the **pultrusion** process. Typical products are golf clubs, drive shafts, and structural members such as ladders, walkways, and handrails. In this process the continuous reinforcement (roving or fabric) is pulled through a thermosetting polymer bath, and then through a long heated steel die (Fig. 18.26). The product is cured during its travel through the die. The most common material used in pultrusion is polyester with glass reinforcements.

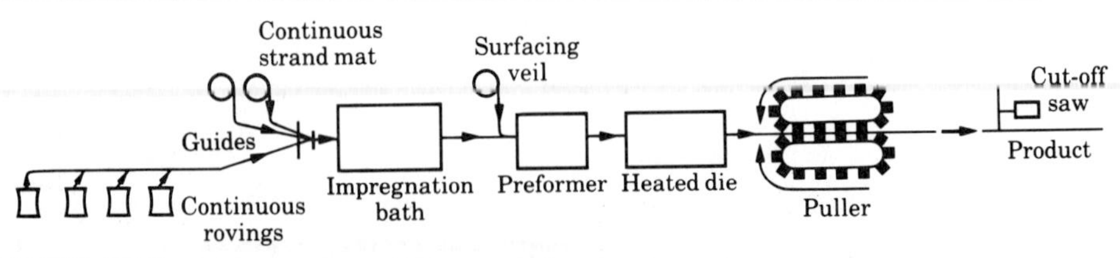

FIGURE 18.26
Schematic illustration of the pultrusion process.

Pulforming. Continuously reinforced products other than constant cross-sectional profiles are made by **pulforming.** After being pulled through the polymer bath, the composite is clamped between the two halves of a die and cured into a finished product. The dies recirculate and shape the products successively. Common examples are glass-fiber reinforced hammer handles and curved automotive leaf springs.

18.12.3 Quality considerations

The major quality considerations for the processes described in this section involve internal voids and gaps between successive layers of material. Microcracks also may develop due to improper curing or during transportation and handling. These defects can be detected using ultrasonic scanning and other techniques (see Section 36.2).

18.13 Manufacturing Honeycomb Materials

There are two principal methods of manufacturing honeycomb materials. In the *expansion* process (Fig. 18.27a)—the most common method—sheets are cut from a coil and an adhesive is applied at intervals (node lines). The sheets are stacked and

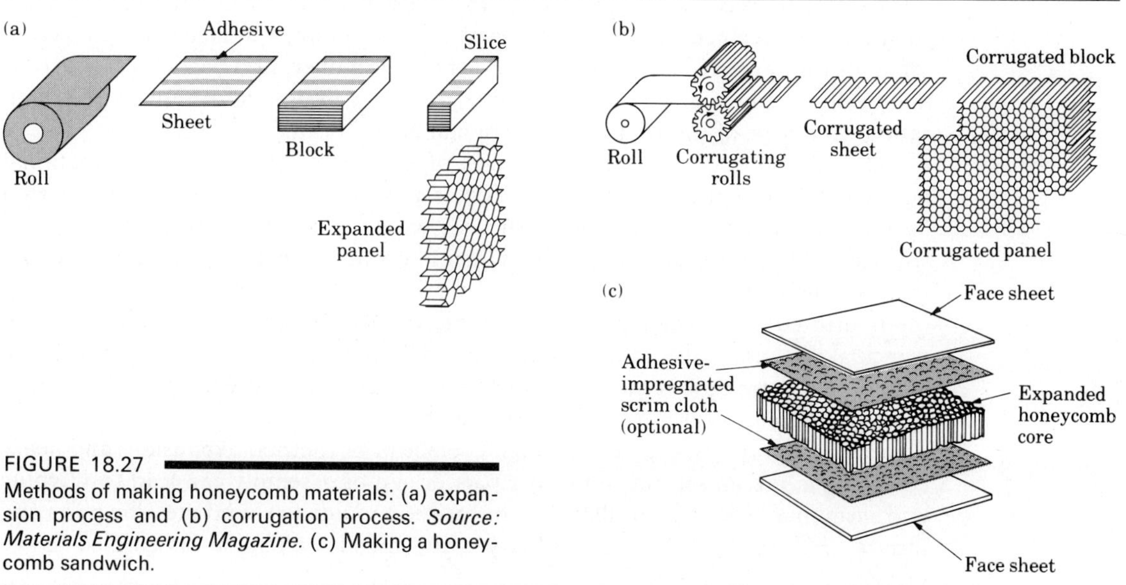

FIGURE 18.27 Methods of making honeycomb materials: (a) expansion process and (b) corrugation process. *Source: Materials Engineering Magazine.* (c) Making a honeycomb sandwich.

cured in an oven, whereby strong bonds develop at the adhesive joints. The block is then cut into slices of desired dimension and stretched to produce a honeycomb structure. This procedure is similar to expanding folded paper structures into the shape of decorative objects.

In the *corrugation* process (Fig. 18.27b), the sheet passes through a pair of specially designed rolls, which produce corrugated sheets that are then cut into desired lengths. Again, adhesive is applied to the node lines, and the block is cured. Note that no expansion process is involved. The honeycomb material is then made into a sandwich structure (Fig. 18.27c). Face sheets are subsequently joined with adhesives to the top and bottom surfaces.

18.14 Design Considerations

Design considerations for forming and shaping plastics are somewhat similar to those for processing of metals. However, the mechanical and physical properties of plastics should be carefully considered during design and material and process selection.

Selection of an appropriate material from an extensive list requires consideration of service requirements and possible long-range effects on properties and behavior, such as dimensional stability and wear. Compared to metals, plastics have lower strength and stiffness, although the strength-to-weight and stiffness-to-weight ratio for reinforced plastics is higher than for many metals. Thus section sizes should be selected accordingly, with a view to maintaining a sufficiently high section modulus for improved stiffness. Reinforcement with fibers and particles can also be highly effective in achieving this objective, as can be designing sections with a high ratio of moment of inertia to cross-sectional area.

One of the major design advantages of reinforced plastics is the directional nature of the strength of the material. Forces applied to the material are transferred by the resin matrix to the fibers, which as we described in Chapter 9, are much stronger and stiffer than the matrix. When fibers are all oriented in one direction, the resulting material is exceptionally strong in the fiber direction. This property is often utilized in designing reinforced plastic structures. For strength in two principal directions, the unidirectional materials are often laid at different angles to each other. If strength in the third (thickness) direction is desired, a different type of material is used to form a sandwich structure.

Physical properties, especially high coefficient of thermal expansion, and hence contraction, are important. Improper part design or assembly can lead to warping and shrinking (Fig. 18.28a). Plastics can easily be molded around metallic parts and inserts. However, their compatibility with metals when so assembled is an important consideration.

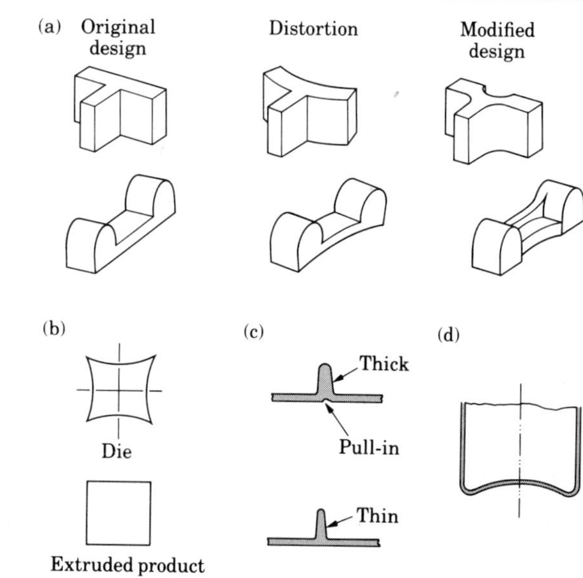

FIGURE 18.28 Examples of design modifications to eliminate or minimize distortion of plastic parts. (a) Suggested design changes to minimize distortion. *Source:* F. Strasser. (b) Die design (exaggerated) for extrusion of square sections. Without this design, product cross-sections swell because of the recovery of the material, which is known as die swell. (c) Design change in a rib to minimize pull-in caused by shrinkage during cooling. (d) Stiffening of bottom of thin plastic containers by doming, which is similar to the process used to make the bottoms of aluminum beverage cans.

The overall part geometry often determines the particular forming or molding process. Table 18.1 is a guide to making this selection. Even after a particular process is selected, the design of the part and die should be such that it will not cause problems concerning shape generation (Fig. 18.28b), dimensional control, and surface finish. As in casting metals and alloys, material flow in the mold cavities should be controlled properly. The effects of molecular orientation during processing should also be considered, especially in processes such as extrusion, thermoforming, and blow molding.

Large variations in section sizes (Fig. 18.28c) and abrupt changes in geometry should be avoided for better product quality and increased mold life. Furthermore, contraction in large cross-sections tends to cause porosity in plastic parts. Conversely, because of a lack of stiffness, removing thin sections from molds after shaping may be difficult. The low elastic modulus of plastics further requires that shapes be selected properly for improved stiffness of the component (Fig. 18.28d), particularly when saving materials is important. These considerations are similar to those in designing metal castings and forgings.

The properties of the final product depend on the original material and its processing history. Cold working of polymers improves their strength and toughness. On the other hand, because of the nonuniformity of deformation (even in simple rolling), residual stresses develop in polymers, as they do in metals. Residual stresses can also be generated by thermal cycling of the part. Also, as in metals, the magnitude and direction of residual stresses, however produced, are important factors. These stresses can relax over a period of time and cause distortion of the part during its service life.

TABLE 18.1
CHARACTERISTICS OF VARIOUS MOLDING AND FORMING PROCESSES FOR PLASTICS

	SHAPE LIMITATIONS	INTRICATE, COMPLICATED SHAPES	CONTROLLED WALL THICKNESS	OPEN, HOLLOW SHAPES	ENCLOSED, HOLLOW SHAPES	LARGE ENCLOSED VOLUME	VERY SMALL ITEMS	PLAN AREA, 1 m^2 (10 ft^2)	FACTOR LIMITING MAXIMUM SIZE	INSERTS	MOLDED-IN HOLES	THREADS
NONREINFORCED MATERIALS												
Compression molding	Moldable	Yes	Yes	Yes					Press	Yes	Yes	Yes
Transfer molding	Moldable	Yes	Yes	Yes			Yes		Press	Yes	Yes	Yes
Injection molding	Moldable	Yes	Yes	Yes			Yes		Press	Yes	Yes	Yes
Extrusion	Constant cross-section	Yes	Yes						Die	Yes		
Rotational molding	Hollow			Yes	Yes	Yes		Yes	Available machine	Yes	Yes	Yes
Blow molding	Hollow, Thin-wall			Yes	Yes	Yes		Yes	Mold			Yes
Thermoforming	Thin-wall			Yes				Yes	Available machine			
Casting	Moldable	Yes	Yes						Mold	Yes	Yes	
Forging	Moldable	Yes	Yes						Die			
Foam molding	Moldable	Yes	Yes	Yes				Yes	Press		Yes	
FIBER-REINFORCED MATERIALS												
Injection molding	Moldable	Yes	Yes	Yes					Press	Yes	Yes	Yes
Hand lay-up and spray-up	Large, thin-wall	Yes	Yes	Yes		Yes, by joining		Yes	Mold, or transport of parts	Yes	Yes	
Compression-type molding	Moldable	Yes	Yes	Yes					Press	Yes	Yes	Yes
Preform molding	Moldable	Yes	Yes	Yes				Yes	Press			
Cold-press molding	Moldable	Yes	Yes	Yes				Yes	Press			
Filament winding	Surface of revolution		Yes						Available machine			
Pultrusion	Constant cross-section	Yes	Yes						Die			

Source: After R. L. E. Brown, *Design and Manufacture of Plastic Parts.* Copyright © 1980 by John Wiley & Sons, Inc. Reprinted by permission of John Wiley & Sons, Inc.

18.15 The Economics of Forming and Shaping Plastics

We have described a number of processes used to form and shape plastics and composite materials. As in all other processes, design and manufacturing decisions are ultimately based on performance and cost, including the costs of equipment, tooling, and production. The final selection of a process depends greatly on production volume. High equipment and tooling costs can be acceptable if the production run is large, as is the case in casting and forging.

Various types of equipment are used in plastics forming and shaping processes. The most expensive are injection-molding machines, with cost directly proportional to clamping force. A machine with a 2000-kN (225-ton) clamping force costs about \$100,000, and one with a 20,000-kN (2250-ton) clamping force costs about \$450,000.

The optimum number of cavities in the die for making the product in one cycle is an important consideration, as in die casting. For small parts, a number of cavities can be made in a die, with runners to each cavity. If the part is large, then only one cavity may be accommodated. As the number of cavities increases, so does the cost of the die. Larger dies may be considered for larger numbers of cavities, increasing die cost even further. On the other hand, more parts will be produced per machine cycle, thus increasing the production rate. Hence a detailed analysis has to be made to determine the optimum number of cavities, die size, and machine capacity. Similar considerations apply to other plastics processing methods. Tables 18.2 and

TABLE 18.2
COMPARATIVE COSTS AND PRODUCTION VOLUMES FOR PROCESSING OF PLASTICS

	EQUIPMENT CAPITAL COST	PRODUCTION RATE	TOOLING COST	TYPICAL PRODUCTION VOLUME, NUMBER OF PARTS (10, 10^2, 10^3, 10^4, 10^5, 10^6, 10^7)
Machining	Medium	Medium	Low	10–10^2
Compression molding	High	Medium	High	10^3–10^6
Transfer molding	High	Medium	High	10^3–10^6
Injection molding	High	High	High	10^4–10^7
Extrusion	Medium	High	Low	
Rotational molding	Low	Low	Low	10^2–10^3
Blow molding	Medium	Medium	Medium	10^3–10^7
Thermoforming	Low	Low	Low	10^2–10^3
Casting	Low	Very low	Low	10–10^2
Forging	High	Low	Medium	10–10^2
Foam molding	High	Medium	Medium	10^4–10^7

Source: After R. L. E. Brown, *Design and Manufacture of Plastic Parts.* Copyright © 1980 by John Wiley & Sons, Inc. Reprinted by permission of John Wiley & Sons, Inc.

TABLE 18.3
ECONOMIC PRODUCTION QUANTITIES FOR VARIOUS MOLDING METHODS*

MOLDING METHOD	RELATIVE INVESTMENT REQUIRED		RELATIVE PRODUCTION RATE	ECONOMIC PRODUCTION QUANTITY
	EQUIPMENT	*TOOLING*		
Hand lay-up	VL	L	L	VL
Spray-up	L	L	L	L
Casting	M	L	L	L
Vacuum-bag molding	M	L	VL	VL
Compression-molded BMC	H	VH	H	H
SMC and preform	H	VH	H	H
Pressure-bag molding	H	H	L	L
Centrifugal casting	H	H	M	M
Filament winding	H	H	L	L
Pultrusion	H	H	H	H
Rotational molding	H	H	L	M
Injection molding	VH	VH	VH	VH

* VL, very low; L, low; M, medium; H, high; VH, very high.
Source: After J. G. Bralla (ed.), *Handbook of Product Design for Manufacturing*. New York: McGraw-Hill, 1986.

18.3 are general guides to economical processing of plastics and composite materials. Note, for example, the high capital costs for molding plastics and the wide range of production rates. For composite materials, equipment and tooling costs for most molding operations are generally high. Production rates and economic production quantities vary widely.

• Example: Polymer automotive body panels shaped by various processes.

The trend in the use of polymeric materials for automobile panels has been increasing at an accelerated rate. Among presently U.S. manufactured automobiles using various polymeric body panels are minivans such as Pontiac Trans Sport, Chevrolet Lumina, and Oldsmobile Silhouette. Upcoming models are the General Motors Saturn, Chevrolet Camaro, and Pontiac Firebird cars. Typical parts are vertical panels such as fenders and quarter panels and front and rear fascias, as well as flat horizontal components such as hoods. The three commonly used and competing processing methods are (a) injection-molded thermoplastics and elastomers, (b) reaction-injection molded polyurea/polyurethanes, and (c) compression-molded sheet-molding compound (SMC); also, some resin-transfer-molded polyester and vinylester.

Saturn has *injection-molded* body panels and other large exterior components. Front fenders and rear quarter panels are made of polyphenylene-ether/nylon, door outer panels are made of polycarbonate/ABS, and fascias are made of thermoplastic polyolefin. These materials are selected for design flexibility, impact strength and toughness, corrosion resistance, high durability, and low mass. Vertical panels and fascias are made in multicavity molds on 5000-ton injection-molding machines, and are assembled mechanically to a steel frame. The 1991 Buick Park Avenue has copolyester elastomer for front and rear fascias, chosen for its moldability, impact resistance (in 5-mph crash tests), paintability, surface appearance, and overall cost effectiveness.

Large exterior body parts are also made of *reaction-injection molded* (RIM) polyurethane, although polyureas are becoming important for body panels and bumpers. Most thermoset fascias are made of reinforced RIM polyurethane, such as those for General Motors All-Purpose Vehicle (APV); however, preference is now being given to new polyureas because of their higher thermal stability, low-temperature toughness, and lower cycle times possible in molding them.

Large horizontal exterior body panels, such as hoods, roofs, and rear decks, are made of reinforced polyester or vinylester, using *compression-molded* sheet molding compounds. However, lower-volume parts are now being made by resin transfer molding (RTM), largely because of the relatively low tooling costs.

SUMMARY

Plastics can be formed and shaped by a variety of processes, such as extrusion, molding, casting, and thermoforming. The starting material is usually in the form of pellets and powders. Thermosets are generally molded and cast, and thermoplastics are formed by these processes as well as by thermoforming and techniques used for metalworking. The high strain-rate sensitivity of thermoplastics allows extensive stretching in forming operations; thus complex and deep shapes can be produced. The design of plastic parts should include considerations of their low strength and stiffness, as well as physical properties such as high thermal expansion and low resistance to temperature.

Reinforced plastics are shaped into important structural components using liquid plastics, prepregs, and bulk- and sheet-molding compounds. Fabricating techniques include various molding methods, filament winding, and pultrusion. Important factors in fabricating reinforced-plastic components are the type and orientation of the fibers, and the strength of the bond between fibers and matrix and between different layers of materials. Inspection techniques are available to check the integrity of these products.

SUMMARY TABLE 18.1

CHARACTERISTICS OF FORMING AND SHAPING PROCESSES FOR PLASTICS AND COMPOSITE MATERIALS

PROCESS	CHARACTERISTICS
Extrusion	Long, uniform, solid or hollow complex cross-sections; high production rates; low tooling costs; wide tolerances.
Injection molding	Complex shapes of various sizes, eliminating assembly; high production rates; costly tooling; good dimensional accuracy.
Structural foam molding	Large parts with high stiffness-to-weight ratio; less expensive tooling than in injection molding; low production rates.
Blow molding	Hollow thin-walled parts of various sizes; high production rates and low cost for making containers.
Rotational molding	Large hollow shapes of relatively simple shape; low tooling cost; low production rates.
Thermoforming	Shallow or relatively deep cavities; low tooling costs; medium production rates.
Compression molding	Parts similar to impression-die forging; relatively inexpensive tooling; medium production rates.
Transfer molding	More complex parts than compression molding and higher production rates; some scrap loss; medium tooling cost.
Casting	Simple or intricate shapes made with flexible molds; low production rates.
Processing of composite materials	Long cycle times; tolerances and tooling cost depend on process.

TRENDS

- Reinforced plastic parts are being developed at a rapid pace, replacing components requiring high stiffness and strength-to-weight ratios.
- Computer-aided mold design is being implemented, particularly for injection molding of complex shapes. Various analytical and computer techniques are being developed to study material flow in molds.
- Computer-controlled machinery is being developed to improve the quality and production rate of reinforced plastic structures.
- Multiple and coextrusion processes are being developed for multilayer sheet and films with special properties.

KEY TERMS

Blow molding	Injection molding	Reaction injection molding
Bulk-molding compound	Parison	Resin transfer molding
Calendering	Pellets	Rotational molding
Compression molding	Plastisols	Sheet-molding compound
Contact molding	Potting	Thermoforming
Encapsulation	Prepregs	Thick molding compound
Filament winding	Pulforming	Transfer molding
Foam molding	Pultrusion	Vacuum-bag molding

BIBLIOGRAPHY

Ash, M., and I. Ash, *Encyclopedia of Plastics, Polymers and Resins*, 3 vols. New York: Chemical Publishing Co., 1980–1981.

Astarita, G., and L. Nicolais, *Polymer Processing and Properties*. New York: Plenum, 1985.

Beck, R.D., *Plastic Product Design*, 2d ed. New York: Van Nostrand, 1980.

Benjamin, B.S., *Structural Design with Plastics*, 2d ed. New York: Van Nostrand Reinhold, 1982.

Brown, R.L.E., *Design and Manufacture of Plastic Parts*. New York: Wiley, 1980.

Chanda, M., and S.K. Roy, *Plastics Technology Handbook*. New York: Marcel Dekker, 1987.

Cheremisinoff, N.P., *Product Design and Testing of Polymeric Materials*. New York: Marcel Dekker, 1990.

Dym, J.B., *Product Design with Plastics: A Practical Manual*. New York: Industrial Press, 1982.

Encyclopedia of Polymer Science and Engineering, 2d ed. New York: Wiley-Interscience, 1985.

Engineering Materials Handbook, Vol. 1: *Composites*. Metals Park, Ohio: ASM International, 1987.

Florian, J., *Practical Thermoforming: Principles and Applications*. New York: Marcel Dekker, 1987.

Frados, J. (ed.), *Plastics Engineering Handbook*, 4th ed. New York: Van Nostrand Reinhold, 1976.

Grayson, M. (ed.), *Encyclopedia of Composite Materials and Components*. New York: Wiley, 1983.

Kelly, A. (ed.), *Concise Encyclopedia of Composite Materials*. New York: Pergamon, 1989.

Levy, S., *Plastics Extrusion Technology Handbook*. New York: Industrial Press, 1981.

Levy, S., and J.H. DuBois, *Plastics Product Design Engineering Handbook*, 2d ed. New York: Van Nostrand Reinhold, 1985.

Macosko, C.W., *Fundamentals of Reaction-Injection Molding*. New York: Oxford University Publishers, 1989.

Meyer, R.W., *Handbook of Pultrusion Technology*. New York: Methuen, 1985.

Miller, E. (ed.), *Plastics Products Design Handbook. Part A: Materials and Components*, 1981; *Part B: Processes and Design for Processes*. 1983. New York: Marcel Dekker.

Modern Plastics Encyclopedia. New York: McGraw-Hill, annual.

Morena, J.J., *Advanced Composite Mold Making*. New York: Van Nostrand Reinhold, 1988.

Rosato, D.V., and D.V. Rosato, *Injection Molding Handbook*. New York: Van Nostrand Reinhold, 1986.

Schwartz, M. (ed.), *Composite Materials Handbook*. New York: McGraw-Hill, 1984.

———, *Fabrication of Composite Materials: Source Book*. Metals Park, Ohio: American Society for Metals, 1985.

Strong, A.B., *Fundamentals of Composites Manufacturing: Materials, Methods, and Applications*. Dearborn, Mich.: Society of Manufacturing Engineers, 1989.

Sweeney, F.M., *Reaction Injection Molding Machinery and Processes*. New York: Marcel Dekker, 1987.

Tadmor, Z., and C.G. Gogos, *Principles of Polymer Processing*. New York: Wiley, 1979.

Wendle, B.C., *Structural Foam*. New York: Marcel Dekker, 1985.

REVIEW QUESTIONS

18.1 What are the forms of raw materials for processing plastics into products?

18.2 Describe the features of an extruder.

18.3 Why is injection molding capable of producing parts with complex shapes and fine detail?

18.4 How are injection-molding machines rated?

18.5 Describe the blow molding process.

18.6 What is a (a) parison, (b) plastisol, (c) prepreg?

18.7 How is thin plastic film produced?

18.8 List several products that can be made by thermoforming.

18.9 What similarities are there between compression molding and closed-die forging?

18.10 Explain the difference between potting and encapsulation.

18.11 Describe the advantages of cold forming of plastics over other processing methods.

18.12 Name the major methods used in processing reinforced plastics.

18.13 What are the characteristics of filament-wound products?

18.14 Describe the methods used to make honeycomb structures.

18.15 List the major design considerations in forming and shaping reinforced plastics.

QUESTIONS AND PROBLEMS

18.16 Describe the advantages of applying traditional metalworking techniques to forming plastics.

18.17 Explain the reasons why some forming processes are more suitable for certain plastics than for others.

18.18 Inspect various plastic components in your car and identify the processes that could be used in making them.

18.19 What are the design considerations in replacing a metal beverage container with one made of plastic?

18.20 List all possible applications for filament-wound plastics.

18.21 Explain the difference between extrusion and pultrusion.

18.22 Would you use thermosetting plastics for injection molding? Explain.

18.23 By inspecting plastic containers, such as for baby powder, you can see that the lettering on them is raised rather than sunk. Can you offer an explanation as to why they are molded in that way?

18.24 Describe the differences among three types of compression molding.

18.25 Outline the precautions that you would take in shaping reinforced plastics.

18.26 What are the factors that contribute to the cost of each forming and shaping process discussed in this chapter?

18.27 What is the purpose of the parison in blow molding?

18.28 An injection-molded nylon gear is found to contain small pores. It is recommended that the material be dried before molding it. Explain why drying will solve this problem.

18.29 Explain why operations such as blow molding and film-bag making are done vertically.

18.30 Describe your observations concerning the structure of tennis rackets illustrated in Section 18.12.

18.31 Describe the operation of the mold shown in Fig. 18.7.

18.32 Give examples of several parts suitable for insert molding. How would you manufacture these parts if insert molding were not available?

18.33 Comment on the principle of operation of the tape laying machine shown in Fig. 18.22.

18.34 Describe the sequence of manufacturing and assembly operations that would be involved in making the tennis rackets illustrated in Section 18.12.

18.35 Give other examples of design modifications in addition to those shown in Fig. 18.28.

18.36 Comment on the typical production volumes given in Table 18.2. Why is there such a wide range?

19

Forming and Shaping Ceramics and Glass

19.1 Introduction

We described the many desirable properties and characteristics of ceramics for engineering applications in Chapter 8. In this chapter we describe several techniques for processing ceramics into useful products. Generally, the procedure involves the following steps: crushing or grinding the raw materials into very fine particles, mixing them with additives to impart certain desirable characteristics, and shaping, drying, and firing the material (Fig. 19.1).

Glass is processed by melting the glass and forming it in molds and various devices, or by blowing. Shapes produced include flat sheet and plate, rods, tubing, discrete products such as bottles and headlights, and glass fibers. The strength of glass can be improved by thermal and chemical treatments, which induce compressive surface residual stresses, or by laminating with a thin sheet of tough plastic.

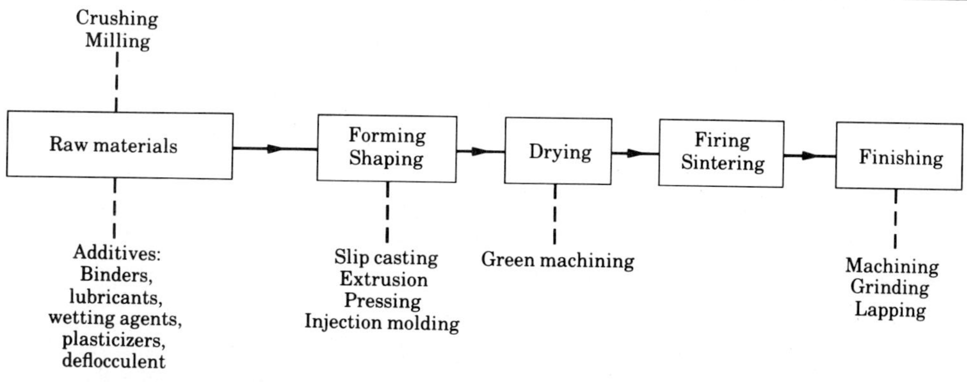

FIGURE 19.1
Processing steps involved in making ceramic parts.

19.2 Shaping Ceramics

The first step in processing ceramics is *crushing* (also called comminution or milling) of the raw materials (see Section 8.2.1). Crushing is generally done in a ball mill (Fig. 19.2), either dry or wet. Wet crushing is more effective because it keeps the particles together and prevents the suspension of fine particles in air. The ground particles are then mixed with *additives*, the functions of which are one or more of the following:

a) Binder for the ceramic particles.
b) Lubricant for mold release and to reduce internal friction between particles during molding.
c) Wetting agent to improve mixing.
d) Plasticizer to make the mix more plastic and formable.

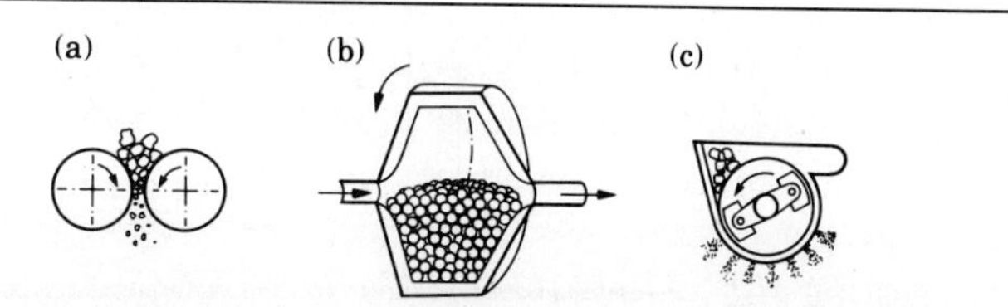

FIGURE 19.2
Methods of crushing ceramics to obtain very fine particles: (a) roll crushing, (b) ball mill, and (c) hammer milling.

e) Deflocculent to make the ceramic–water suspension. Deflocculation changes the electrical charges on the particles of clay so that they repel instead of attracting each other. Water is added to make the mixture more pourable and less viscous. Typical deflocculants are Na_2CO_3 and Na_2SiO_3 in amounts of less than 1 percent.
f) Various agents to control foaming and sintering.

The three basic shaping processes for ceramics are casting, plastic forming, and pressing.

19.2.1 Casting

The most common casting process is **slip casting**, also called *drain casting* (Fig. 19.3). A **slip** is a suspension of ceramic particles in a liquid, generally water. In this process, the slip is poured into a porous mold made of plaster of paris. The slip must have sufficient fluidity and low viscosity to flow easily into the mold, much like the fluidity of molten metals we described in Section 10.3.

After the mold has absorbed some of the water from the outer layers of the suspension, it is inverted and the remaining suspension is poured out (for making hollow objects, as in slush casting of metals described in Section 11.11). The top of the part is then trimmed, the mold is opened, and the part is removed.

Large and complex parts, such as plumbing ware, art objects, and dinnerware, can be made by slip casting. Although dimensional control is limited and the production rate is low, mold and equipment costs are low. In some applications, components of the product (such as handles for cups and pitchers) are made separately and then joined, using the slip as an adhesive.

Thin sheets of ceramics, less than 1.5-mm (0.06-in.) thick, can be made by a casting technique called the **doctor-blade process**. The slip is cast over a moving plastic belt and its thickness is controlled by a blade. Other processes include *rolling* the slip between pairs of rolls and casting the slip over a paper tape, which is then burned off during firing.

For solid ceramic parts, the slip is supplied continuously into the mold to replenish the absorbed water; the suspension is not drained from the mold. At this

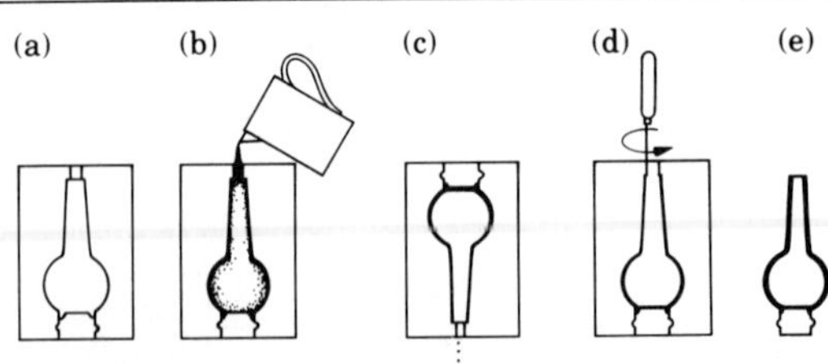

FIGURE 19.3 Sequence of operations in slip casting a ceramic part. After the slip has been poured, the part is dried and fired in an oven to give it strength and hardness. *Source:* F. H. Norton, *Elements of Ceramics.* Copyright © 1974, Addison-Wesley Publishing Company, Inc.

stage the part is a soft solid or semirigid. The higher the concentration of solids in the slip, the less water has to be removed. The part, called *green*, as in powder metallurgy, is then fired.

19.2.2 Plastic forming

Plastic forming (also called *soft*, *wet*, or *hydroplastic forming*) can be done by various methods, such as extrusion, injection molding, or molding and jiggering (as done on a potter's wheel). Plastic forming tends to orient the layered structure of clays along the direction of material flow. This leads to anisotropic behavior of the material, both in subsequent processing and in the final properties of the ceramic product.

In *extrusion*, the clay mixture, containing 20–30 percent water, is forced through a die opening by screw-type equipment. The cross-section of the extruded product is constant, and there are limitations to wall thickness for hollow extrusions. Tooling costs are low and production rates are high. The extruded products may be subjected to additional shaping operations.

19.2.3 Pressing

Dry pressing. Similar to powder-metal compaction, *dry pressing* is used for relatively simple shapes. Typical parts are whiteware, refractories, and abrasive products. The process has the same high production rates and close control of tolerances as in P/M. The moisture content of the mixture is generally below 4 percent but may be as high as 12 percent. Organic and inorganic binders, such as stearic acid, wax, starch, and polyvinyl alcohol, are usually added to the mixture and also act as lubricants. The pressure for pressing is between 35 MPa and 200 MPa (5 ksi and 30 ksi). Modern presses used for dry pressing are highly automated. Dies, usually made of carbides or hardened steel, must have high wear resistance to withstand the abrasive ceramic particles and can be expensive.

Density can vary significantly in dry-pressed ceramics because of friction between particles and at the mold walls, as in P/M compaction. Density variations cause warping during firing. Warping is particularly severe for parts having high length-to-diameter ratios; the recommended maximum ratio is 2:1. Several methods may be used to minimize density variations. Design of tooling is important. Vibratory pressing and impact forming are used, particularly for nuclear-reactor fuel elements. Isostatic pressing also reduces density variations.

Wet pressing. In *wet pressing* the part is formed in a mold while under high pressure in a hydraulic or mechanical press. This process is generally used to make intricate shapes. Moisture content usually ranges from 10 percent to 15 percent. Production rates are high but part size is limited, dimensional control is difficult because of shrinkage during drying, and tooling costs can be high.

Isostatic pressing. Used extensively in powder metallurgy, as you have seen, *isostatic pressing* is also used for ceramics in order to obtain uniform density distribution throughout the part. Automotive spark-plug insulators are made by this method. Silicon-nitride vanes for high-temperature use are made by hot isostatic pressing (Fig. 19.4).

Jiggering. A combination of processes is used to make ceramic plates. Clay slugs are first extruded, then formed into a *bat* over a plaster mold, and finally jiggered on a rotating mold (Fig. 19.5). **Jiggering** is a motion in which the clay bat is formed with templates or rollers. The part is then dried and fired. The process is limited to axisymmetric parts and has limited dimensional accuracy, but the operation can be automated.

FIGURE 19.4 Silicon-nitride vane for a gas-turbine engine made by hot isostatic pressing. *Source:* Battelle Columbus Laboratories.

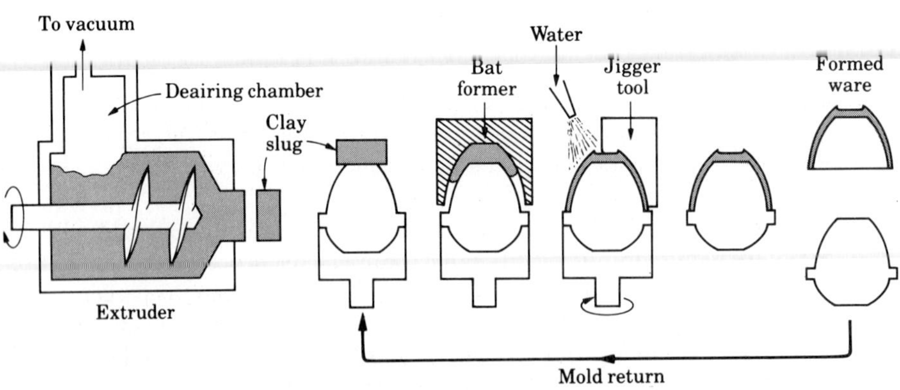

FIGURE 19.5 Extruding and jiggering operations. *Source:* R. F. Stoops.

Injection molding. We have previously described the advantages of *injection molding* of plastics and powder metals. This process is now being used extensively for precision forming of ceramics for high-technology applications, as in rocket engine components. The raw material is mixed with a binder, such as a thermoplastic polymer (polypropylene, low-density polyethylene, ethylene vinyl acetate, or wax). The binder is usually removed by pyrolysis, and the part is sintered by firing. This process can produce thin sections, typically less than 10–15 mm (0.4–0.6 in.), using most engineering ceramics such as alumina, zirconia, silicon nitride, and silicon carbide. Thicker sections require careful control of the materials used and processing parameters to avoid internal voids and cracks, such as those due to shrinkage.

Hot pressing. In **hot pressing**, also called *pressure sintering*, pressure and temperature are applied simultaneously. This method reduces porosity, making the part denser and stronger. Hot isostatic pressing may also be used in this operation, particularly to improve the quality of high-technology ceramics. Because of the presence of both pressure and temperature, die life in hot pressing can be short. Protective atmospheres are usually employed, and graphite is a commonly used punch and die material.

19.2.4 Drying and firing

After the ceramic has been shaped by any of the methods described, the next step is to dry and fire the part to give it the proper strength. *Drying* is a critical stage because of the tendency for the part to warp or crack from variations in moisture content and thickness within the part and the complexity of its shape. Control of atmospheric humidity and temperature is important in order to reduce warping and cracking. Loss of moisture results in shrinkage of the part by as much as 15–20 percent of the original moist size (Fig. 19.6). In a humid environment, the evaporation rate is low, and consequently the moisture gradient across the thickness of the part is lower than that in a dry environment. The low moisture gradient, in turn, prevents a large, uneven gradient in shrinkage from the surface to the interior during drying.

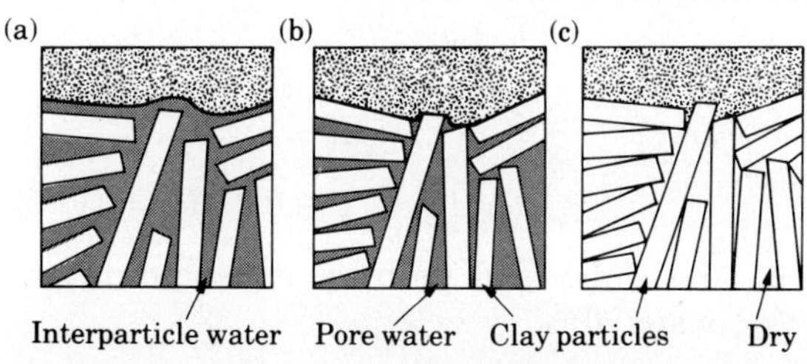

FIGURE 19.6
Shrinkage of wet clay caused by removal of water during drying. Shrinkage may be as much as 20 percent by volume. *Source:* F. H. Norton, *Elements of Ceramics.* Copyright © 1974, Addison-Wesley Publishing Company, Inc.

Firing (sintering) involves heating the part to an elevated temperature in a controlled environment, similar to sintering in powder metallurgy. Some shrinkage occurs during firing. Firing gives the ceramic part its strength and hardness. The improvement in properties results from (a) development of a strong bond between the complex oxide particles in the ceramic and (b) reduced porosity.

19.2.5 Finishing operations

After firing, additional operations may be performed to give the part its final shape, remove surface flaws, and improve surface finish and tolerances. The processes used can be grinding, lapping, and ultrasonic, chemical, and electrical-discharge machining. The choice of the process is important in view of the brittle nature of most ceramics and the additional costs involved in these processes. The effect of the finishing operation on properties of the product must also be considered. Because of notch sensitivity, the finer the finish, the higher the part's strength will be. To improve appearance and strength, and to make them impermeable, ceramic products are often coated with a *glaze* material, which forms a glassy coating after firing (see Section 33.12).

• Example: Dimensional changes during shaping of ceramic components.

A solid cylindrical ceramic part is to be made whose final length must be $L = 20$ mm. It has been established that for this material, linear shrinkages during drying and firing are 7 percent and 6 percent, respectively, based on the dried dimension L_d. Calculate (a) the initial length L_o of the part and (b) the dried porosity P_d if the porosity of the fired part, P_f, is 3 percent.

SOLUTION.

a) On the basis of the information given and remembering that firing is preceded by drying, we can write

$$(L_d - L)/L_d = 0.06$$

or

$$L = (1 - 0.06)L_d;$$

hence,

$$L_d = 20/0.94 = 21.28 \text{ mm}$$

and

$$L_o = (1 + 0.07)L_d = (1.07)(21.28) = 22.77 \text{ mm}.$$

b) Since the final porosity is 3 percent, the actual volume V_a of the ceramic material is

$$V = (1 - 0.03)V_f = 0.97V_f,$$

where V_f is the fired volume of the part. Since the linear shrinkage during firing is 6 percent, we can determine the dried volume V_d of the part as

$$V_d = \frac{V_f}{(1 - 0.06)^3} = 1.2V_f.$$

Hence

$$\frac{V_a}{V_d} = \frac{0.97}{1.2} = 0.81 = 81 \text{ percent.}$$

Therefore, the porosity P_d of the dried part is 19 percent.

19.3 Forming and Shaping Glass

Glass products can generally be categorized as:

- Flat sheet or plate, ranging in thickness from about 0.8 mm to 10 mm (0.03 in. to 0.4 in.), such as window glass, glass doors, and table tops.
- Rods and tubing used for chemicals, neon lights, and decorative artifacts.
- Discrete products, such as bottles, vases, headlights, and television tubes.
- Glass fibers to reinforce composite materials and for fiber optics.

All glass forming and shaping processes begin with molten glass, which has the appearance of red-hot viscous syrup, supplied from a melting furnace or tank.

19.3.1 Flat sheet and plate

Flat sheet glass can be made by drawing or rolling from the molten state, or by a floating method, all of which are continuous processes. Figure 19.7(a) shows the

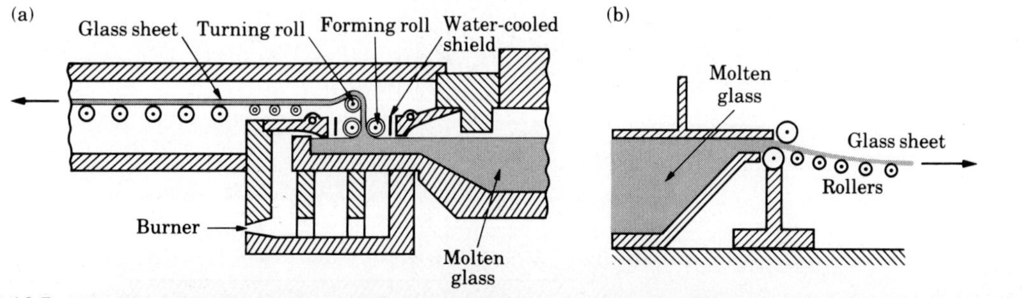

FIGURE 19.7 (a) Continuous process for drawing sheet glass from a molten bath. *Source:* W. D. Kingery, *Introduction to Ceramics.* New York: John Wiley & Sons, Inc., 1976. (b) Rolling glass to produce flat sheet.

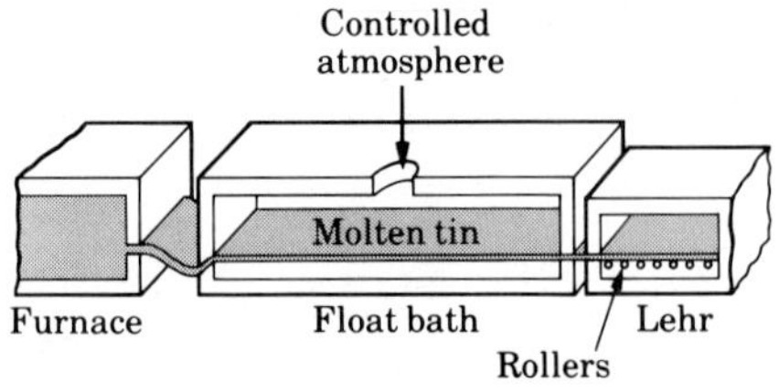

FIGURE 19.8
The float method of forming sheet glass. *Source:* Corning Glass Works.

drawing process for making flat sheet or plate by a machine in which the molten glass passes through a pair of rolls, similar to an old-fashioned clothes wringer. The solidifying glass is squeezed between these rolls, forming a sheet, which is then moved forward over a set of smaller rolls. In the *rolling* process (Fig. 19.7b), the molten glass is squeezed between rollers, forming a sheet. The surfaces of the glass can be embossed with a pattern by shaping the roller surfaces accordingly. Glass sheet produced by drawing and rolling has a rough surface appearance. In making plate glass, both surfaces have to be ground parallel and polished.

In the *float* method of production (Fig. 19.8), molten glass from the furnace is fed into a bath in which the glass, under controlled atmosphere, floats on a bath of molten tin. The glass then moves over rollers into another chamber (called a *lehr*) and solidifies. *Float glass* has a smooth (fire-polished) surface and needs no further grinding or polishing.

19.3.2 Tubing and rods

Glass tubing is manufactured by the process shown in Fig. 19.9. Molten glass is wrapped around a rotating hollow cylindrical or cone-shaped mandrel and is drawn out by a set of rolls. Air is blown through the mandrel to keep the glass tube from

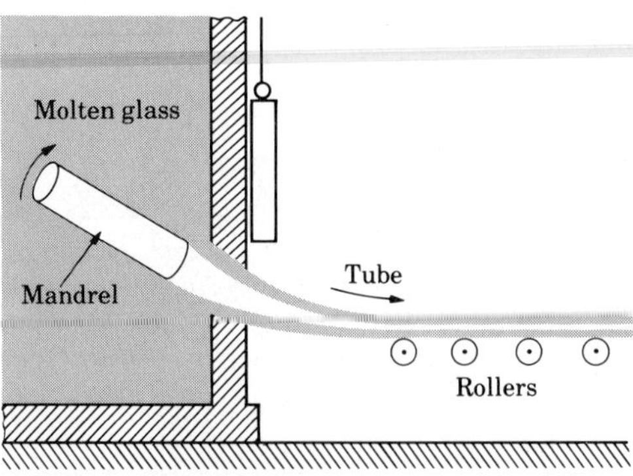

FIGURE 19.9
Manufacturing process for glass tubing. Air is blown through the mandrel to keep the tube from collapsing. *Source:* Corning Glass Works.

collapsing. These machines may be horizontal, vertical, or slanted downward. Glass rods are made in a similar manner, but air is not blown through the mandrel; thus the drawn product becomes a solid rod.

19.3.3 Glass fibers

Continuous fibers are drawn through multiple (200–400) orifices in heated platinum plates, at speeds as high as 500 m/s (1700 ft/s). Fibers as small as 2 μm (80 μin.) in diameter can be produced by this method. In order to protect their surfaces, the fibers are subsequently coated with chemicals. Short glass fibers, used as thermal insulating material (*glass wool*) or for acoustic insulation, are made by a *centrifugal spraying process* in which molten glass is fed into a rotating head.

19.3.4 Manufacturing of discrete glass products

Several processes are used in making discrete glass objects. These processes include blowing, pressing, centrifugal casting, and sagging.

Blowing. The **blowing** process is used to make hollow thin-walled glass items, such as bottles and flasks, and is similar to blow molding of thermoplastics. The steps involved in the production of an ordinary glass bottle by the blowing process are shown in Fig. 19.10. Blown air expands a hollow gob of heated glass against the walls of the mold. The molds are usually coated with a parting agent, such as oil or emulsion, to prevent the part from sticking to the mold.

The surface finish of products made by the blowing process is acceptable for most applications. Although it is difficult to control the wall thickness of the product, the process is used for high rates of production. Light bulbs are made in automatic blowing machines, at a rate of over 1000 bulbs per minute.

Pressing. In **pressing**, a gob of molten glass is placed in a mold and is pressed into shape with the use of a plunger. The mold may be made in one piece (Fig. 19.11), or it may be a split mold (Fig. 19.12). After pressing, the solidifying glass acquires the shape of the mold-plunger cavity. Because of the confined environment, the product has greater dimensional accuracy than can be obtained with blowing. However, pressing cannot be used on thin-walled items, or for parts such as bottles from which the plunger cannot be retracted.

Centrifugal casting. Also known as *spinning* in the glass industry (Fig. 19.13), the **centrifugal casting** process is similar to that for metals. The centrifugal force pushes the molten glass against the mold wall where it solidifies. Typical products are TV picture tubes and missile nose cones.

Sagging. Shallow dish-shaped or lightly embossed glass parts can be made by the **sagging** process. A sheet of glass is placed over the mold and is heated. The glass sags by its own weight and takes the shape of the mold. The process is similar to thermoforming with thermoplastics but without pressure or a vacuum. Typical applications are dishes, sunglass lenses, mirrors for telescopes, and lighting panels.

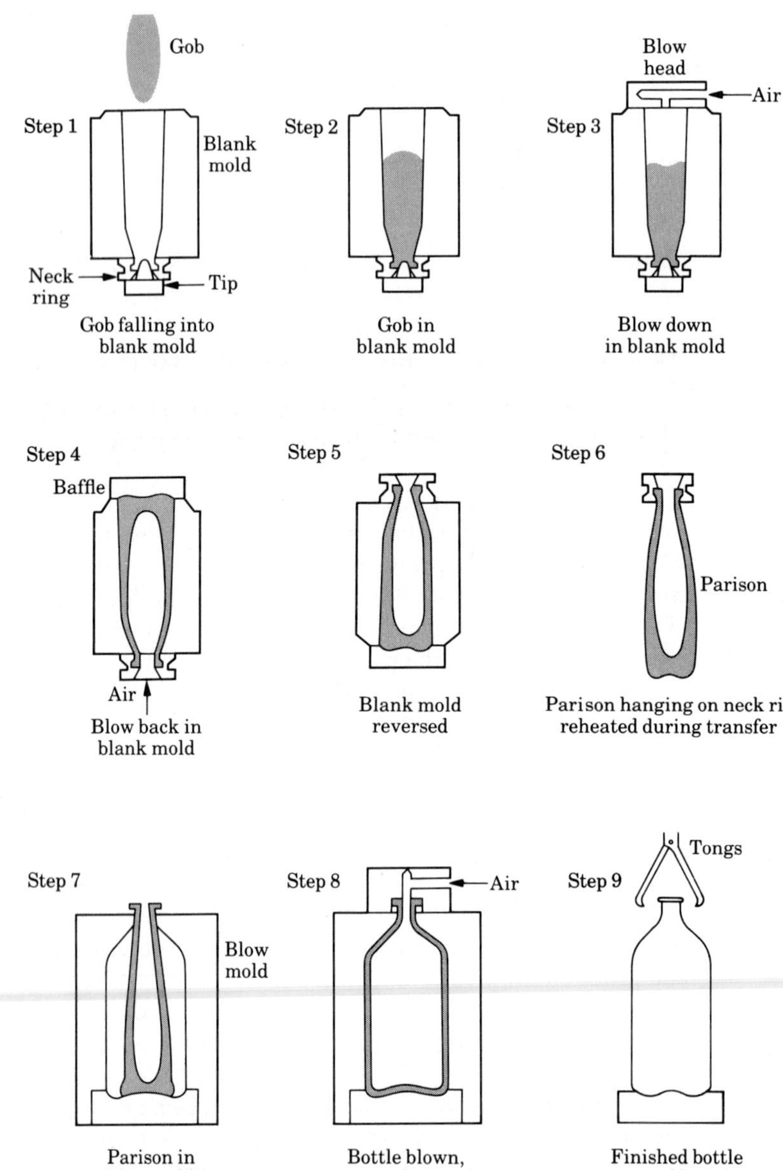

FIGURE 19.10
Stages in manufacturing an ordinary glass bottle. *Source:* F. H. Norton, *Elements of Ceramics.* Copyright © 1974, Addison-Wesley Publishing Company, Inc.

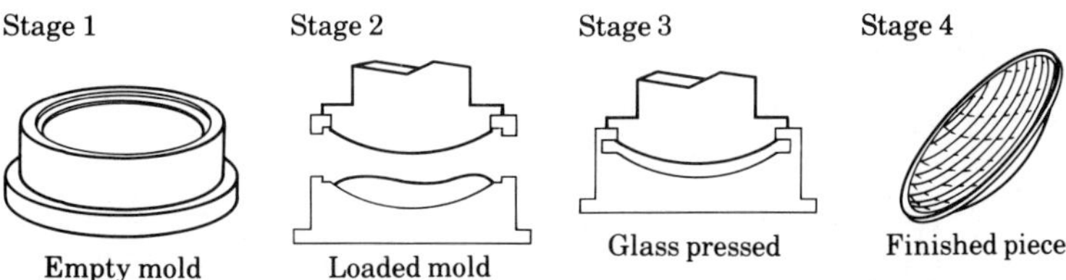

FIGURE 19.11
Manufacturing a glass item by pressing in a mold. *Source:* Corning Glass Works.

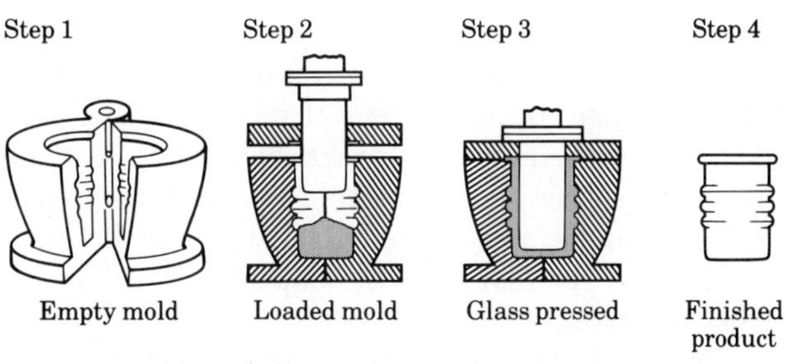

FIGURE 19.12
Pressing glass in a split mold. *Source:* E. B. Shand, *Glass Engineering Handbook.* New York: McGraw-Hill, 1958.

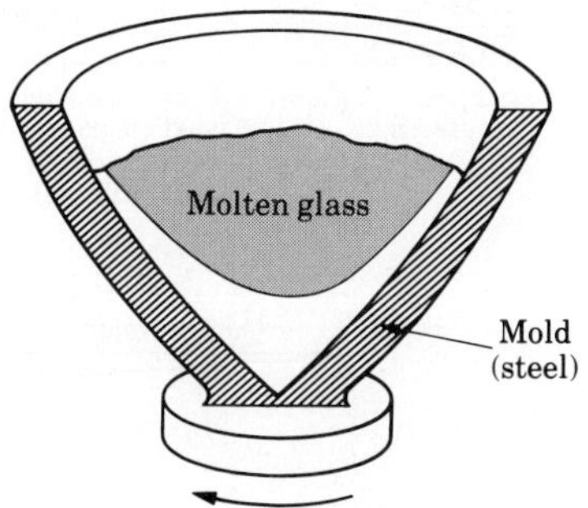

FIGURE 19.13
Centrifugal casting of glass. Television-tube funnels are made by this process. *Source:* Corning Glass Works.

19.4 Techniques for Treating Glass

As produced by the methods we described, glass can be strengthened by thermal tempering, chemical tempering, and laminating. Glass products may also undergo annealing and other finishing operations.

19.4.1 Strengthening

In **thermal tempering** (also called *physical tempering* or *chill tempering*), the surfaces of the hot glass are cooled rapidly (Fig. 19.14). As a result, the surfaces shrink, and tensile stresses develop on the surfaces. As the bulk of the glass begins to cool, it contracts. The solidified surfaces are forced to contract, thus developing residual compressive surface stresses and interior tensile stresses. Compressive surface stresses improve the strength of the glass, as they do in other materials. Note that the higher the coefficient of thermal expansion of the glass and the lower its thermal conductivity, the higher the level of residual stresses developed and hence the stronger the glass becomes. Thermal tempering takes a relatively short time (minutes) and can be applied to most glasses. Because of the large amount of energy stored from residual stresses, **tempered glass** shatters into a large number of pieces when broken.

In **chemical tempering**, the glass is heated in a bath of molten KNO_3, K_2SO_4, or $NaNO_3$, depending on the type of glass. Ion exchange takes place, with larger atoms replacing the smaller atoms on the surface of the glass. As a result, residual compressive stresses are developed on the surface. This condition is similar to that

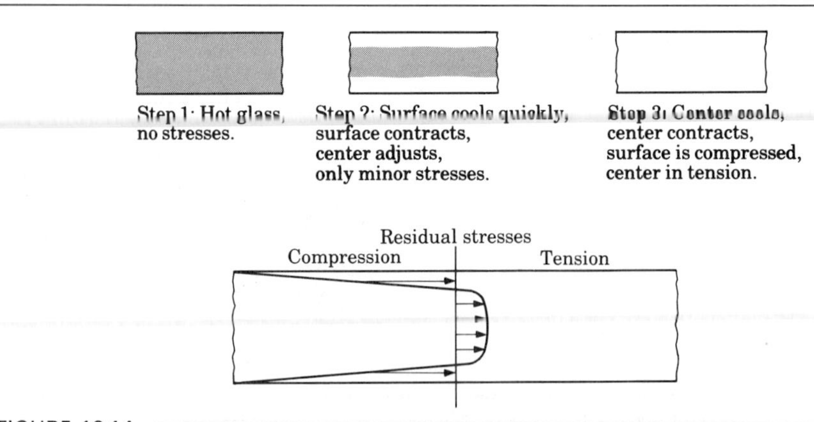
Step 1: Hot glass, no stresses.
Step 2: Surface cools quickly, surface contracts, center adjusts, only minor stresses.
Step 3: Center cools, center contracts, surface is compressed, center in tension.
Residual stresses
Compression
Tension

FIGURE 19.14 Residual stresses in tempered glass plate and stages involved in inducing compressive surface residual stresses for improved strength.

created by forcing a wedge between two bricks in a wall. The time required for chemical tempering is longer (about one hour) than for thermal tempering. Chemical tempering may be performed at various temperatures. At low temperatures, part distortion is minimal and complex shapes can be treated. At elevated temperatures, there may be some distortion of the part, but the product can be used at higher temperatures without loss of strength.

Another strengthening method is to make **laminated glass**, which is two pieces of flat glass with a thin sheet of tough plastic between them. The process is also called *laminate strengthening*. When laminated glass is broken, its pieces are held together by the plastic sheet. You probably have seen this phenomenon in a shattered automobile windshield.

19.4.2 Finishing operations

As in metal products, residual stresses can develop in glass products if they are not cooled slowly enough. In order to ensure that the product is free from these stresses, it is *annealed* by a process similar to stress-relief annealing of metals. The glass is heated to a certain temperature and cooled gradually. Depending on the size, thickness, and type of glass, annealing times may range from a few minutes, to as long as 10 months, as in the case of a 600-mm (24-in.) mirror for a telescope.

In addition to annealing, glass products may be subjected to further operations, such as cutting, drilling, grinding, and polishing. Sharp edges and corners can be smoothened by grinding, as in glass tops for desks and shelves, or by holding a torch against the edges (*fire polishing*), which rounds them by localized softening and surface tension.

19.5 Design Considerations

Ceramic and glass products require careful selection of composition, processing methods, finishing operations, and methods of assembly into other components. Limitations such as general lack of tensile strength, sensitivity to defects, and low impact toughness are important. These limitations have to be balanced against desirable characteristics, such as their hardness, scratch resistance, compressive strength at room and elevated temperatures, and diverse physical properties.

Control of processing parameters and the quality and level of impurities in the raw materials used are important. As in all design decisions, there are priorities, limitations, and various factors that should be considered, including the number of parts needed and the costs of tooling, equipment, and labor.

Dimensional changes and warping and cracking possibilities during processing are significant factors in selecting methods for shaping these materials. When a ceramic or glass component is part of a larger assembly, compatibility with other components is another important consideration. Particularly important are thermal expansion (as in seals) and the type of loading. The potential consequences of part failure are always a significant factor in designing ceramic products.

SUMMARY

Ceramic products are shaped by various casting, plastic forming, or pressing techniques. They are then dried and fired to impart strength and hardness. Because of their inherent brittleness, ceramics are processed with due consideration of distortion and cracking. Control of raw-material quality and processing parameters are important factors.

Glasses are available in a wide variety of forms, compositions, and mechanical, physical, and optical properties. Glass products are made by a variety of shaping processes, which are similar to those used for plastics and ceramics. The strength of glass can be improved by thermal and chemical treatments.

SUMMARY TABLE 19.1

CHARACTERISTICS OF CERAMICS PROCESSING

PROCESS	ADVANTAGES	LIMITATIONS
Slip casting	Large parts, complex shapes; low equipment cost.	Low production rate; limited dimensional accuracy.
Extrusion	Hollow shapes and small diameters; high production rate.	Parts have constant cross-section; limited thickness.
Dry pressing	Close tolerances; high production rate with automation.	Density variation in parts with high length-to-diameter ratios; dies require high abrasive-wear resistance; equipment can be costly.
Wet pressing	Complex shapes; high production rate.	Part size limited; limited dimensional accuracy; tooling costs can be high.
Hot pressing	Strong, high-density parts.	Protective atmospheres required; die life can be short.
Isostatic pressing	Uniform density distribution.	Equipment can be costly.
Jiggering	High production rate with automation; low tooling cost.	Limited to axisymmetric parts; limited dimensional accuracy.
Injection molding	Complex shapes; high production rate.	Tooling can be costly.

TRENDS

- Shaping processes for ceramics are being controlled more precisely to minimize defects, improve dimensional accuracy, and impart higher strength and reliability of the product.
- Machining, grinding, and finishing operations for ceramics are being improved for better precision and enhanced surface properties.
- Surfaces of ceramic parts are being subjected to laser treatments for improved properties and friction and wear characteristics.
- Two-phase ceramics are being developed with superplastic behavior characteristics, which allow them to be shaped into products, as in superplastic metals.

KEY TERMS

Blowing	Hot pressing	Sagging
Centrifugal casting	Jiggering	Slip
Chemical tempering	Laminated glass	Slip casting
Doctor-blade process	Plastic forming	Tempered glass
Float method	Pressing	Thermal tempering

BIBLIOGRAPHY

Ford, R.W., *Ceramics Drying*. Elmsford, New York: Pergamon, 1986.

Hlavac, J., *The Technology of Glass and Ceramics*. Amsterdam: Elsevier, 1983.

Kingery, W.D., H.K. Bowen, and D.R. Uhlmann, *Introduction to Ceramics*, 2d ed. New York: Wiley, 1976.

Morena, J.J., *Advanced Composite Mold Making*. New York: Van Nostrand Reinhold, 1988.

Norton, F.H., *Elements of Ceramics*, 2d ed. Reading, Mass.: Addison-Wesley, 1974.

Richerson, D.W., *Modern Ceramic Engineering*. New York: Marcel Dekker, 1982.

Scholes, S.R., and C.H. Greene, *Modern Glass Practice*, 7th ed. Boston: Cahners Books, 1975.

Schwartz, M.M. (ed.), *Engineering Applications of Ceramic Materials—Source Book*. Metals Park, Ohio: American Society for Metals, 1985.

Wang, F.F.Y. (ed.), *Ceramic Fabrication Processes*. New York: Academic Press, 1984.

REVIEW QUESTIONS

19.1 What are the steps involved in processing ceramics?
19.2 List and describe the functions of additives.
19.3 Describe the slip-casting process.
19.4 What is the doctor-blade process?
19.5 Explain the relative advantages of dry, wet, and isostatic pressing.
19.6 What is jiggering? What shapes does it produce?
19.7 Name the factors that are important in drying ceramic products.
19.8 What types of finishing operations are used on ceramics?
19.9 List the categories by which glass products are generally classified.
19.10 Describe briefly the methods by which flat sheet glass is made.
19.11 How are glass tubing and glass rods produced?
19.12 Explain the glass-blowing process.
19.13 What is the difference between physical and chemical tempering of glass?
19.14 What are the advantages of laminated glass?

QUESTIONS AND PROBLEMS

19.15 What should be the requirements for the metal balls in a ball mill?
19.16 What is the reason for shrinkage during drying of ceramic products?
19.17 How different, if any, are the design considerations for ceramics from other materials?
19.18 Which property of glasses allows them to be expanded and shaped into bottles by blowing, as shown in Fig. 19.10?
19.19 Explain why ceramic parts may distort or warp during drying. What precautions would you take to avoid this situation?
19.20 What properties should plastic sheet have when used in laminated glass? Why?
19.21 We have stated in Section 19.4.1 that the higher the coefficient of thermal expansion of the glass and the lower its thermal conductivity, the higher the level of residual stresses developed. Explain why.
19.22 In the example in Section 19.2, calculate (a) the porosity of the dried part if the porosity of the fired part is to be 9 percent and (b) the initial length L_o of the part if the linear shrinkages during drying and firing are 8 and 7 percent, respectively.

PART IV

Material-Removal Processes and Machines

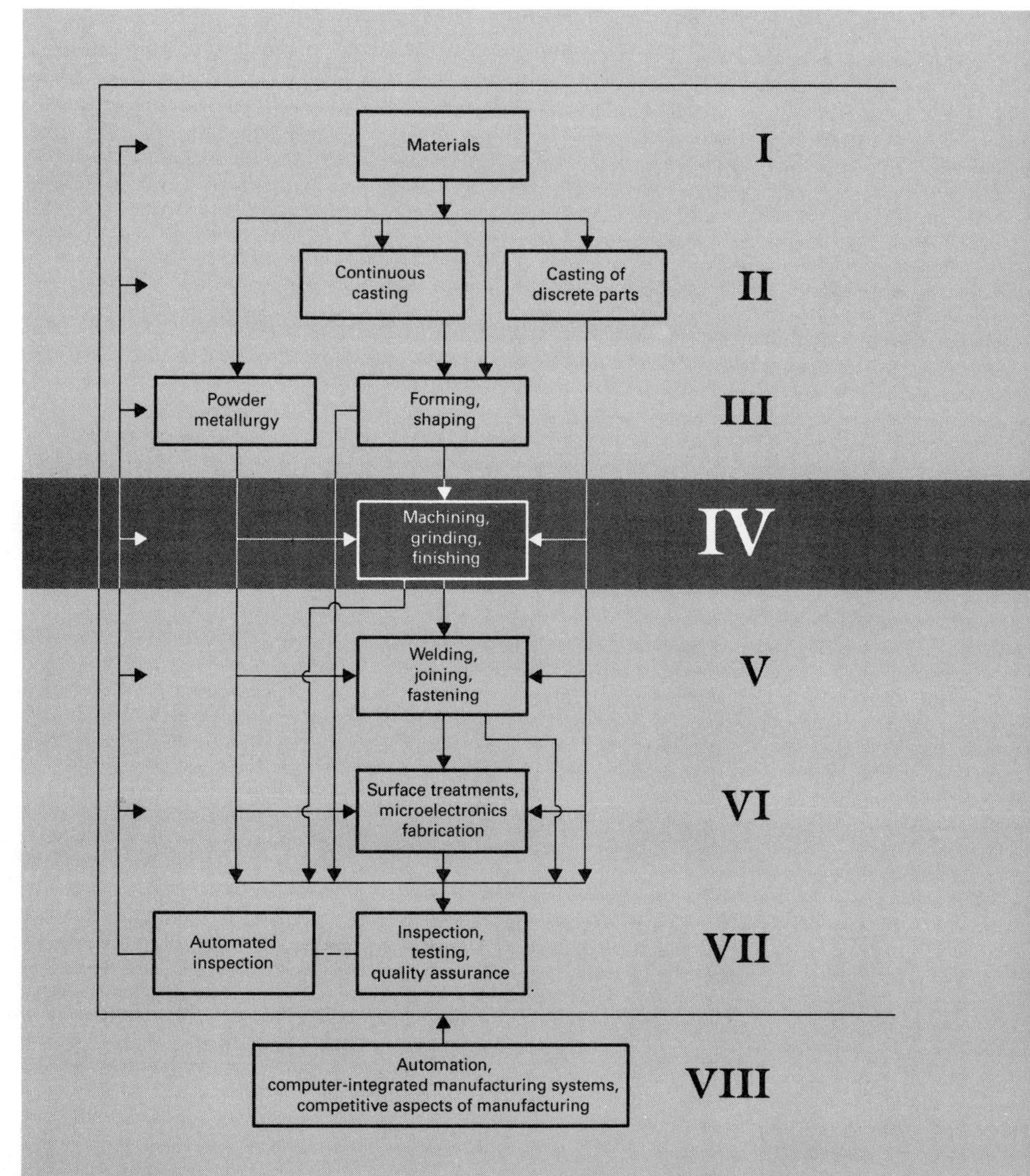

• INTRODUCTION •

Parts manufactured by casting, forming, and shaping processes, which we described in Parts II and III, often require further operations before the product is ready for use. Moreover, in many engineering applications, parts must be interchangeable to function properly and reliably during their expected service lives, as in the case of automobile parts (Fig. IV.1). Thus we have to obtain certain dimensional accuracy. Note, for example, the tolerances specified on the part shown in Fig. IV.2, and the presence of a threaded end. A brief review will show you that none of the processes described thus far can produce a part with such accuracy. Parts must also be produced economically. Thus critical choices have to be made about the extent of shaping and forming versus the extent of machining to be done on a workpiece to produce an acceptable part.

Although *machining* is the broad term used to describe *removal* of material from a workpiece, it covers several processes, which we usually divide into the following categories:

- Cutting, generally involving single-point or multipoint cutting tools, each with a clearly defined geometry.
- Abrasive processes, such as grinding.
- Nontraditional machining processes, utilizing electrical, chemical, and optical sources of energy.

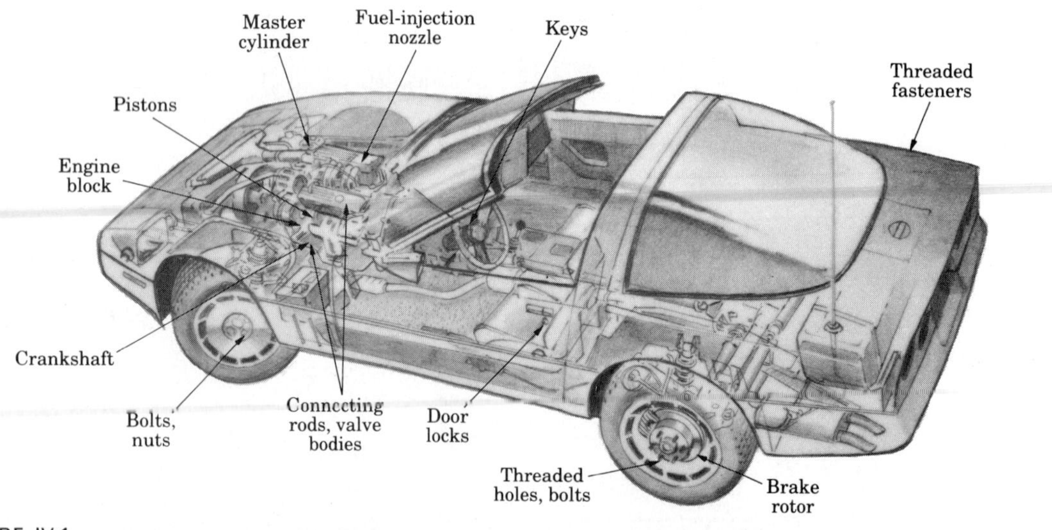

FIGURE IV.1
Typical machined parts on an automobile.

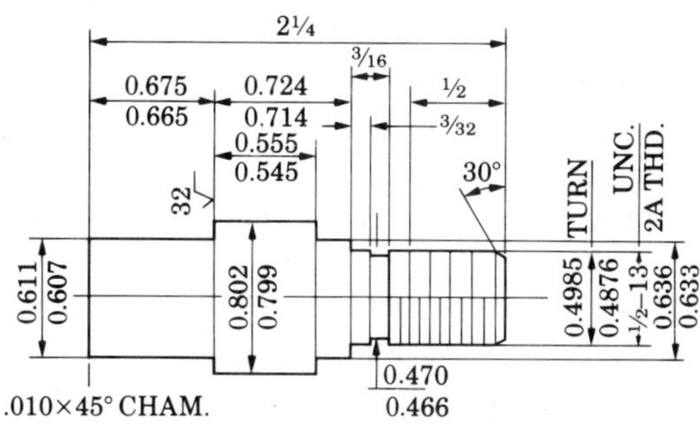

FIGURE IV.2
A machined part showing various dimensions and tolerances. All dimensions are in inches.

We can summarize why material-removal processes are desirable or even necessary in manufacturing operations as follows:

- Closer dimensional accuracy may be required than is available from casting, forming, or shaping processes alone. In the forged crankshaft shown in Fig. IV.3, for example, the bearing surfaces and the holes cannot be produced with good dimensional accuracy and surface finish by forming and shaping processes alone.
- Parts may have external and internal profiles, as well as sharp corners and flatness, that cannot be produced by forming and shaping processes.
- Some parts are heat treated for improved hardness and wear resistance. Since heat-treated parts may undergo distortion and surface discoloration, they generally require additional finishing operations, such as grinding, to obtain the desired final dimensions and surface finish.

FIGURE IV.3
A forged crankshaft before and after machining of bearing surfaces. *Source:* Courtesy of Wyman-Gordon Company.

- Special surface characteristics or texture that cannot be produced by other means may be required on all or part of the surfaces of the product. Copper mirrors with very high reflectivity, for example, are made by machining with a diamond cutting tool.
- Machining the part may be more economical than manufacturing it by other processes, particularly if the number of parts desired is relatively small.

Against these advantages, material-removal processes have certain limitations:

- Removal processes inevitably waste material and generally require more energy, capital, and labor than forming and shaping operations. Thus they should be avoided whenever possible.
- Removing a volume of material from a workpiece generally takes longer than to shape it by other processes.
- Unless carried out properly, material-removal processes can have adverse effects on the surface quality and properties of the product.

In spite of these limitations, material-removal processes and machines are indispensable to manufacturing technology. Ever since lathes were introduced in the 1700s, these processes have developed continuously. We now have available a variety of computer controlled machines, as well as new techniques using electrical, chemical, and optical energy sources.

Following the outline shown in Fig. IV.4, we first describe the basic mechanics of chip formation in cutting processes. Among the most important aspects of cutting operations is the type of tools used. Improper tool selection can have a major economic impact on the operation. Tool materials have been developed to meet the challenges of machining new materials with high strength and toughness, including composite materials.

A variety of shapes can be produced by machining. We describe all the major cutting processes and their capabilities, typical applications, and limitations. The machines on which material-removal operations are performed are generally called *machine tools*. Their construction and characteristics influence greatly these operations and product quality. It is important to view machining, as well as all manufacturing operations, as a *system* consisting of the workpiece, the tool, and the machine. Cutting operations cannot be carried out efficiently and economically without a knowledge of the interactions among these elements.

We also identify important machine-tool characteristics, including their structure and stiffness and new developments in machine-tool design and the materials used in their construction. Among these new developments are *machining centers*, which are versatile machine tools controlled by computers and capable of performing a variety of machining operations efficiently. We then describe the processes in which removal of material is carried out by *abrasive* processes. The most common example is a grinding wheel in which the abrasive particles are held together with a bond. Other examples of abrasive operations are sanding with coated abrasives (sandpaper, emery paper), and honing, lapping, buffing, polishing, shot-blasting, and ultrasonic machining.

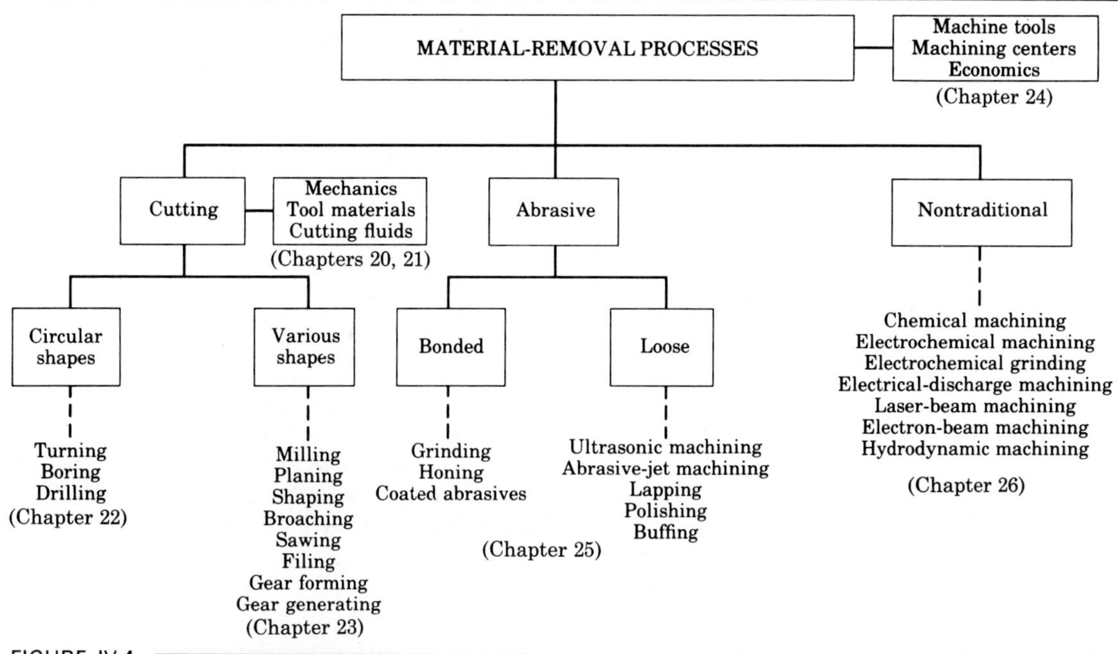

FIGURE IV.4
Outline of processes described in Part IV.

For technical and economic reasons, certain parts cannot be manufactured satisfactorily either by cutting or by abrasive processes. Since the 1940s, important developments have taken place in electrical, chemical, and optical means of material removal. As a result, chemical, electrochemical, electrical-discharge, laser-beam, and electron-beam machining have become major manufacturing processes.

The knowledge gained in this Part will make you appreciate one of the oldest and most important groups of manufacturing processes. It will also enable you to assess the capabilities and limitations of material-removal processes and equipment, their proper selection for maximum productivity and low product cost, and how these processes fit into the broader scheme of manufacturing processes.

20

Fundamentals of Cutting

20.1 Introduction

Cutting processes, such as turning on a lathe, drilling, milling, or thread cutting remove material from the surface of the workpiece by producing **chips**. The basic mechanics of chip formation is essentially the same for all these operations, which we represent by the two-dimensional model in Fig. 20.1. In this model a tool moves along the workpiece at a certain velocity V and a depth of cut t_0. A chip is produced ahead of the tool by shearing the material continuously along the shear plane.

The major *independent variables* (those that we can change directly) in the cutting process are:

- Tool material and its condition.
- Tool shape, surface finish, and sharpness.
- Workpiece material, condition, and temperature.
- Cutting conditions, such as speed and depth of cut.

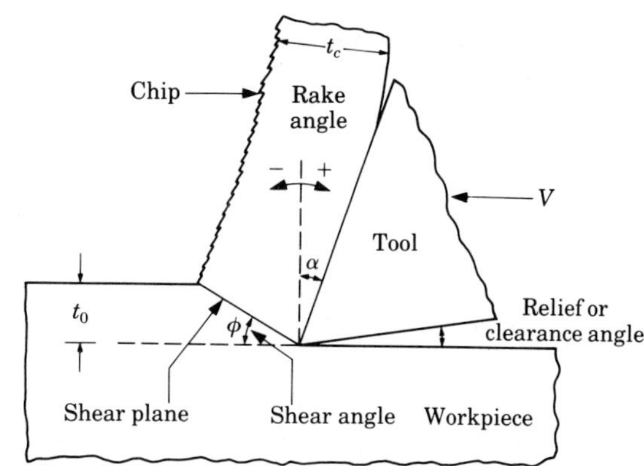

FIGURE 20.1
Schematic illustration of a two-dimensional cutting process (also called orthogonal cutting).

- Use of a cutting fluid.
- The characteristics of the machine tool, such as its stiffness and damping.

Dependent variables are those that are influenced by changes in the independent variables and are:

- Type of chip produced.
- Force and energy dissipated in the cutting process.
- Temperature rise in the workpiece, the chip, and the tool.
- Wear and failure of the tool.
- The surface finish produced on the workpiece after machining.

Ask yourself the following questions. If, for example, the surface finish of the workpiece being cut is poor and unacceptable, which of the independent variables do you change first? The angle of the tool? If so, do you increase it or decrease it? If the tool wears and becomes dull rapidly, do you change the cutting speed, the depth of cut, or the tool material? If the cutting tool begins to vibrate, what should be done to eliminate this vibration?

In this chapter we will describe the mechanics of chip formation in order to establish the effects of various parameters on the overall cutting process. With this knowledge you will be able to plan efficient and economical machining operations and be able to select the proper equipment and tooling. You will also be able to assess the machinability of materials, as we did for castability, forgeability, and formability.

The specific topics covered are the mechanics of producing chips, chip types, force and power requirements, temperature rise caused by the cutting action, tool wear and failure, surface finish, and machinability.

20.2 Mechanics of Chip Formation

Although almost all cutting processes are three-dimensional in nature, as you will see in Chapters 22 and 23, the model shown in Fig. 20.1 is very useful in studying the basic mechanics of cutting. In this model, known as *orthogonal cutting*, the tool has a **rake angle** of α (positive, as shown in the figure) and a **relief**, or **clearance, angle**. Note that the sum of the rake, relief, and included angles of the tool is 90°.

Microscopic examinations have revealed that chips are produced by the shearing process shown in Fig. 20.2(a), and that shearing takes place along a *shear plane* making an angle ϕ, called the **shear angle**, with the surface of the workpiece. Below the shear plane the workpiece is undeformed, and above it is the chip, already formed and climbing up the face of the tool as cutting progresses. Because of the relative velocity, there is friction between the chip and the rake face of the tool. Note also that this shearing process is like cards in a deck sliding against each other.

You can see that the thickness of the chip, t_c, can be determined by knowing t_o, α, and ϕ. The ratio of t_o to t_c is known as the **cutting ratio**, r, which we can express as

$$r = \frac{t_o}{t_c} = \frac{V_c}{V} = \frac{\sin \phi}{\cos (\phi - \alpha)}. \tag{20.1}$$

The chip thickness is always greater than the depth of cut; hence the value of r is less than unity. The reciprocal of r is known as the *chip compression ratio* and is a measure of how thick the chip has become compared to the depth of cut. Thus the chip compression ratio is always greater than unity.

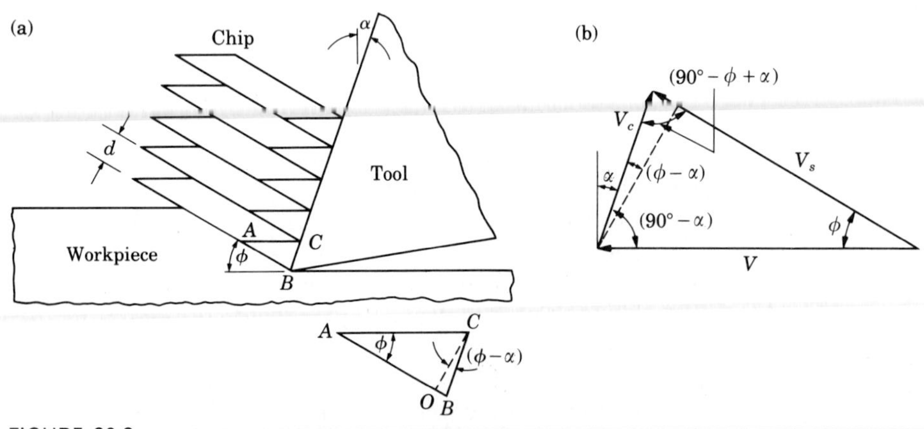

FIGURE 20.2
(a) Schematic illustration of the basic mechanism of chip formation in cutting. (b) Velocity diagram in the cutting zone.

On the basis of Fig. 20.2(a), we can express the **shear strain** γ (see also Fig. 2.13) that the material undergoes as

$$\gamma = \frac{AB}{OC} = \frac{AO}{OC} + \frac{OB}{OC},$$

or

$$\gamma = \cot \phi + \tan (\phi - \alpha). \tag{20.2}$$

Thus large shear strains are associated with low shear angles and low or negative rake angles. Shear strains of 5 or larger have been observed in actual cutting operations. Compared to forming and shaping processes, therefore, the material undergoes greater deformation during cutting. Also, deformation in cutting takes place within a very narrow deformation zone; that is, the dimension $d = OC$ in Fig. 20.2(a) is very small. Hence the rate at which deformation takes place is high.

The shear angle has great significance in the mechanics of cutting operations. It influences chip thickness, forces and power requirements, and temperature (because of the work of deformation). Consequently, much attention has been focused on determining the relationship of the shear angle to material properties and process variables. One of the earliest analyses was based on the assumption that the shear angle adjusts itself to minimize the cutting force, or that the shear plane is a plane of maximum shear stress. As a result, the following expression was obtained:

$$\phi = 45^\circ + \frac{\alpha}{2} - \frac{\beta}{2}, \tag{20.3}$$

where β is the **friction angle** (see Fig. 20.12) and is defined as

$$\mu = \tan \beta, \tag{20.4}$$

where μ is the coefficient of friction at the tool–chip interface.

Equation (20.3) indicates that as the rake angle decreases and/or as the friction at the tool–chip interface increases, the shear angle decreases and thus the chip is thicker. Thicker chips mean more energy dissipation because the shear strain is higher (see Eq. 20.2). Because work done during cutting is converted into heat, temperature rise is also higher. We describe the effects of these phenomena throughout the rest of this chapter.

From Fig. 20.1 we note that since chip thickness t_c is greater than the depth of cut t_o, the velocity of the chip, V_c, has to be lower than the cutting speed, V. Since mass continuity has to be maintained, we have

$$Vt_o = V_c t_c \qquad \text{or} \qquad V_c = Vr.$$

Hence

$$V_c = V \frac{\sin \phi}{\cos (\phi - \alpha)}. \tag{20.5}$$

We can also construct a velocity diagram (Fig. 20.2b) and from trigonometric relationships obtain the following relationships:

$$\frac{V}{\cos(\phi - \alpha)} = \frac{V_s}{\cos\alpha} = \frac{V_c}{\sin\phi}, \tag{20.6}$$

where V_s is the velocity at which shearing takes place in the shear plane. Note also that $t_o/t_c = V_c/V$.

20.3 Types of Chips Produced in Cutting

When we observe actual chip formation under different metal-cutting conditions, we find significant deviations from the ideal model shown in Figs. 20.1 and 20.2(a). Some types of metal chips commonly observed in practice are shown schematically in Fig. 20.3, with micrographs in Fig. 20.4 (p. 604). Because the type of chips produced significantly influences the surface finish produced and the overall cutting operation, we will discuss the type of chips in the following order:

1. Continuous
2. Built-up edge
3. Serrated
4. Discontinuous

Let's first note that a chip has two surfaces: one that is in contact with the tool face (rake face), and the other from the original surface of the workpiece. The tool side of the chip surface is shiny, or *burnished* (Fig. 20.5 on p. 604), which is caused by rubbing of the chip as it climbs up the tool face. The other surface of the chip does not come into contact with any solid body. This surface has a jagged, steplike appearance (Figs. 20.2a and 20.4a), which is caused by the shearing mechanism of chip formation shown in Fig. 20.2(a).

20.3.1 Continuous chips

Continuous chips are usually formed at high cutting speeds and/or high rake angles (Figs. 20.3a and 20.4a). The deformation of the material takes place along a narrow shear zone, called the **primary shear zone**. Continuous chips may develop a **secondary shear zone** at the tool–chip interface (Figs. 20.3b and 20.4b), caused by friction. The secondary zone becomes deeper as tool–chip friction increases.

In continuous chips, deformation may also take place along a wide primary shear zone with curved boundaries (Fig. 20.3c). Note that the lower boundary is below the machined surface, which subjects the machined surface to distortion, as depicted by the distorted vertical lines. This situation occurs particularly in machining soft metals at low speeds and low rake angles. It can produce poor

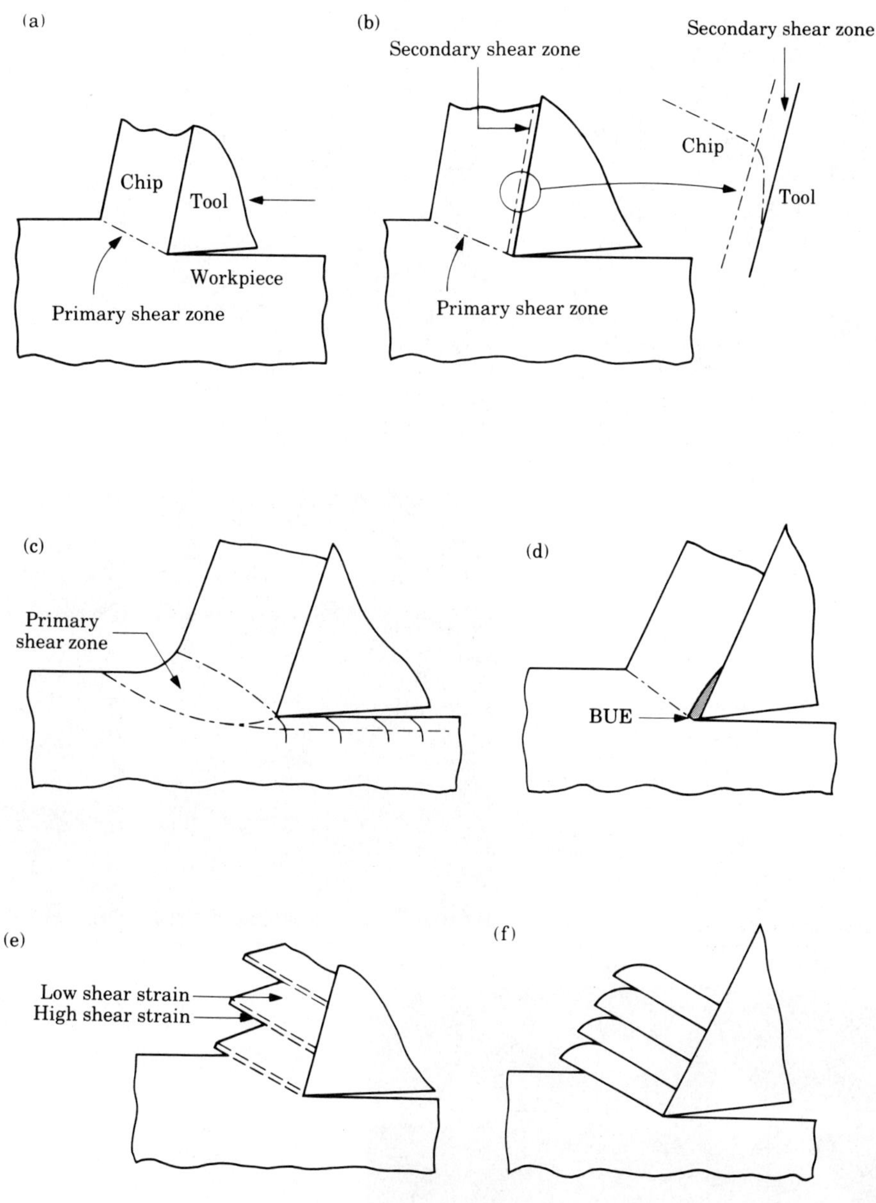

FIGURE 20.3

Basic types of chips produced in metal cutting: (a) continuous chip with narrow, straight primary shear zone; (b) secondary shear zone at the chip–tool interface; (c) continuous chip with large primary shear zone; (d) continuous chip with built-up edge; (e) segmented or nonhomogeneous chip; and (f) discontinuous chip. *Source:* After M. C. Shaw.

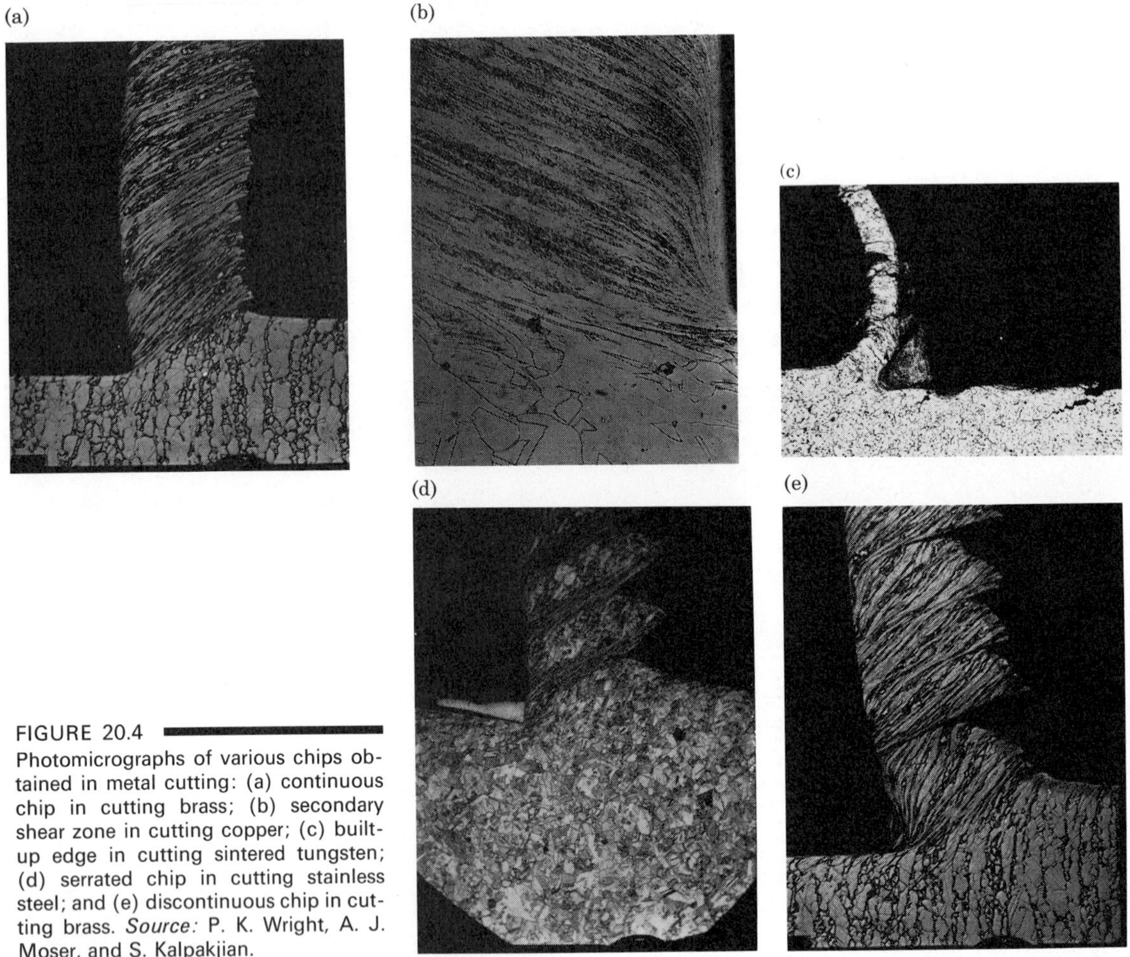

(a) (b) (c) (d) (e)

FIGURE 20.4

Photomicrographs of various chips obtained in metal cutting: (a) continuous chip in cutting brass; (b) secondary shear zone in cutting copper; (c) built-up edge in cutting sintered tungsten; (d) serrated chip in cutting stainless steel; and (e) discontinuous chip in cutting brass. *Source:* P. K. Wright, A. J. Moser, and S. Kalpakjian.

FIGURE 20.5

Shiny (burnished) surface on the tool-side of a continuous chip produced in turning.

surface finish and induce residual surface stresses, which may be detrimental to the properties of the machined part.

Although they generally produce good surface finish, continuous chips are not always desirable, particularly in automated machine tools. They tend to get tangled around the tool holder, and the operation has to be stopped to clear away the chips. This problem can be alleviated with chip breakers (see Section 20.3.7).

20.3.2 Built-up edge chips

A **built-up edge (BUE)** may form at the tip of the tool during cutting (Figs. 20.3d and 20.4c). This edge consists of layers of material from the workpiece that are gradually deposited on the tool (hence the term *built-up*). As it becomes larger, the BUE becomes unstable and eventually breaks up. Part of the BUE material is carried away by the tool side of the chip; the rest is deposited randomly on the workpiece surface. The process of BUE formation and destruction is repeated continuously during the cutting operation.

Built-up edge formation is similar to what happens when you walk through mud. It sticks to your boot soles, changing the shape of the bottom of the boot as well as the surface texture of the ground. The chunk of mud grows in size, and when it eventually falls off, the process is repeated.

The built-up edge is commonly observed in practice. It is one of the factors that most adversely affects surface finish in cutting, as you can see in Figs. 20.4(c) and 20.6. A built-up edge, in effect, changes the geometry of cutting. Note, for example, the large tip radius of the BUE and the rough surface finish produced. Because of work hardening and deposition of successive layers of material, BUE hardness increases significantly (Fig. 20.6a). Although BUE is generally undesirable, a thin, stable BUE is usually regarded as desirable because it protects the tool's surface.

As cutting speed increases, the size of the BUE decreases—or it doesn't form at all. The tendency for BUE to form is reduced by decreasing the depth of cut, increasing the rake angle, and using a sharp tool and an effective cutting fluid. A cold-worked metal also has less tendency to form BUE.

20.3.3 Serrated chips

Serrated chips (also called *segmented* or *nonhomogeneous* chips) are semicontinuous chips, with zones of low and high shear strain (Figs. 20.3e and 20.4d). Metals with low thermal conductivity and strength that decreases sharply with temperature, such as titanium, exhibit this behavior. The chips have a sawtoothlike appearance.

20.3.4 Discontinuous chips

Discontinuous chips consist of segments that may be firmly or loosely attached to each other (Figs. 20.3f and 20.4e). Discontinuous chips usually form under the following conditions: (a) brittle workpiece materials, because they do not have the capacity to undergo the high shear strains developed in cutting; (b) materials that

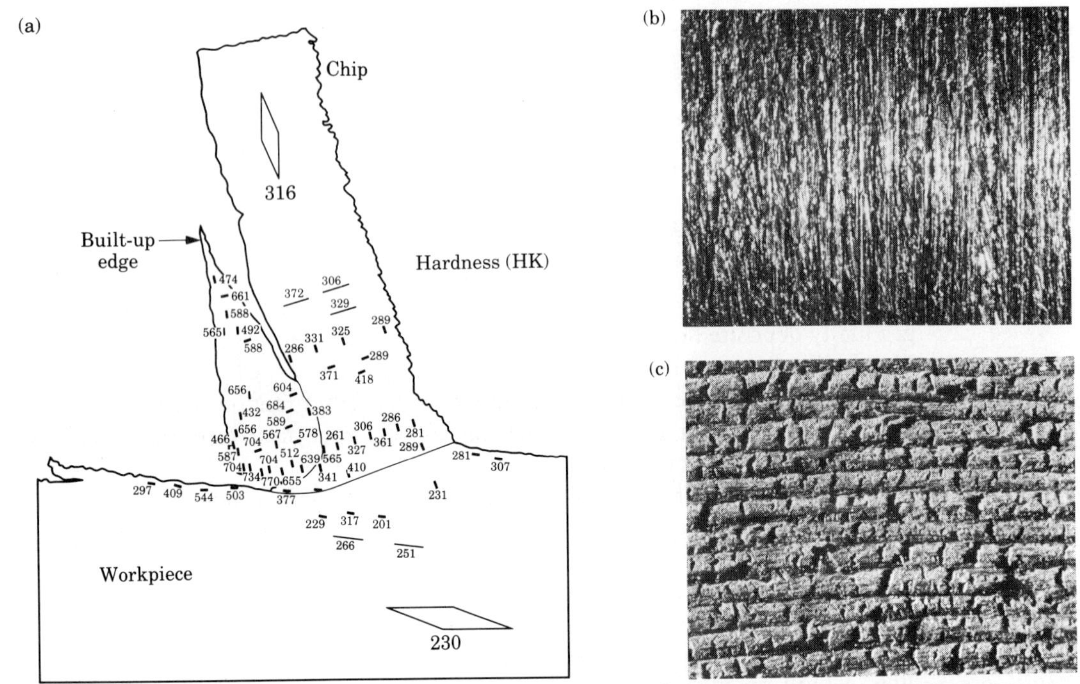

FIGURE 20.6
(a) Hardness distribution in the cutting zone for 3115 steel. Note that some regions in the built-up edge are as much as three times harder than the bulk metal. (b) Surface finish in turning 5130 steel with a built-up edge. (c) Surface finish on 1018 steel in face milling. Magnifications: 15 ×. *Source:* Courtesy of Metcut Research Associates, Inc.

contain hard inclusions and impurities; (c) very low or very high cutting speeds; (d) large depths of cut and low rake angles; (e) low stiffness of the machine tool; and (f) lack of an effective cutting fluid.

Because of the discontinuous nature of chip formation, forces continually vary during cutting. Consequently, the stiffness of the cutting-tool holder and the machine tool is important in cutting with discontinuous-chip as well as serrated-chip formation. If not stiff enough the machine tool may begin to vibrate and chatter. This, in turn, adversely affects the surface finish and dimensional accuracy of the machined component and may damage or cause excessive wear of the cutting tool. This result is another example of why it is important to understand the mechanics of cutting processes.

20.3.5 Chip formation in nonmetallic materials

Many of the discussions thus far for metals are also generally applicable to nonmetallic materials. A variety of chips are obtained in cutting thermoplastics,

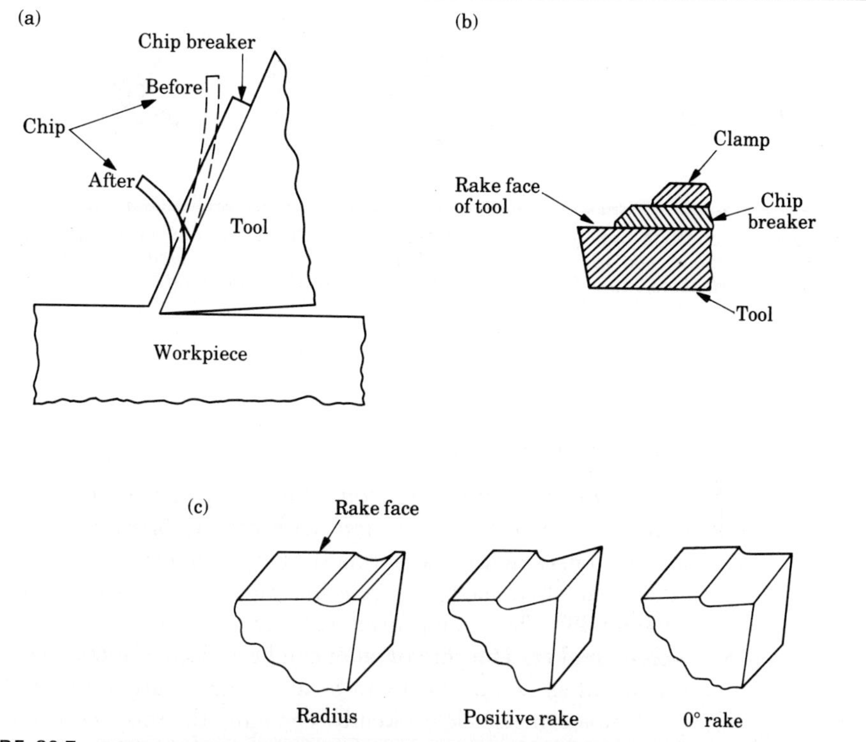

FIGURE 20.7
(a) Schematic illustration of the action of a chip breaker. Note that the chip breaker decreases the radius of curvature of the chip. (b) Chip breaker clamped on the rake face of a cutting tool. (c) Grooves in cutting tools acting as chip breakers. See also Fig. 21.3.

depending on the type of polymer and process parameters such as depth of cut, tool geometry, and cutting speed. Because they are brittle, thermosetting plastics and ceramics generally produce discontinuous chips.

20.3.6 Chip curl

Chip curl (Fig. 20.7a) is common to all cutting operations with metals, as well as with nonmetallic materials, such as plastics and wood. The reasons for chip curl are still not clearly understood. Among the possible factors contributing to it are the distribution of stresses in the primary and secondary shear zones, thermal effects, and the work-hardening characteristics of the workpiece material.

Process variables also affect chip curl. Generally, the radius of curvature decreases (the chip becomes curlier) with decreasing depth of cut, increasing rake angle, and decreasing friction at the tool–chip interface. The use of cutting fluids and various additives in the workpiece material also influence chip curl.

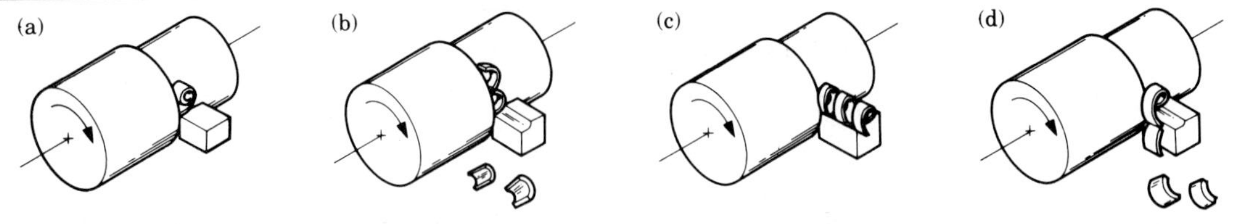

FIGURE 20.8
Various chips produced in turning: (a) tightly curled chip; (b) chip hits workpiece and breaks; (c) continuous chip moving away from workpiece; and (d) chip hits tool shank and breaks off. *Source:* G. Boothroyd, *Fundamentals of Metal Machining and Machine Tools.* Copyright © 1975; McGraw-Hill Publishing Company. Used with permission.

20.3.7 Chip breakers

As we stated earlier, long, continuous chips are undesirable because they tend to become entangled and interfere with cutting operations and can become a safety hazard. This situation is especially troublesome in high-speed automated machinery and in untended machining cells in computer-integrated manufacturing systems (see Chapter 39). The usual procedure to avoid it is to break the chip intermittently with a **chip breaker**. The chip breaker can be a piece of metal clamped to the rake face of the tool (Fig. 20.7b), or it can be an integral part of the tool (Fig. 20.7c).

Chips can also be broken by changing the tool geometry, thus controlling chip flow, as in the turning operations shown in Fig. 20.8. Various cutting tools with chip-breaker features are available. In interrupted cutting operations, such as milling, chip breakers are generally not necessary, since the chips already have finite lengths resulting from the intermittent nature of the operation.

20.4 Mechanics of Oblique Cutting

Thus far we have described the cutting process two-dimensionally. However, the majority of cutting operations involve tool shapes that are three-dimensional (*oblique*). The basic difference between two-dimensional and oblique cutting is shown in Fig. 20.9(a). As you have seen, in orthogonal cutting the tool edge is perpendicular to the movement of the tool, and the chip slides directly up the face of the tool. In oblique cutting, the cutting edge is at an angle, i, the **inclination angle** (Fig. 20.9b). Note the lateral direction of chip movement in oblique cutting. This situation is similar to an angled snow-plow blade, which throws the snow sideways.

Note that the chip in Fig. 20.9(a) flows up the rake face of the tool at angle α_c (*chip flow angle*), measured in the plane of the tool face. Angle α_n is known as the

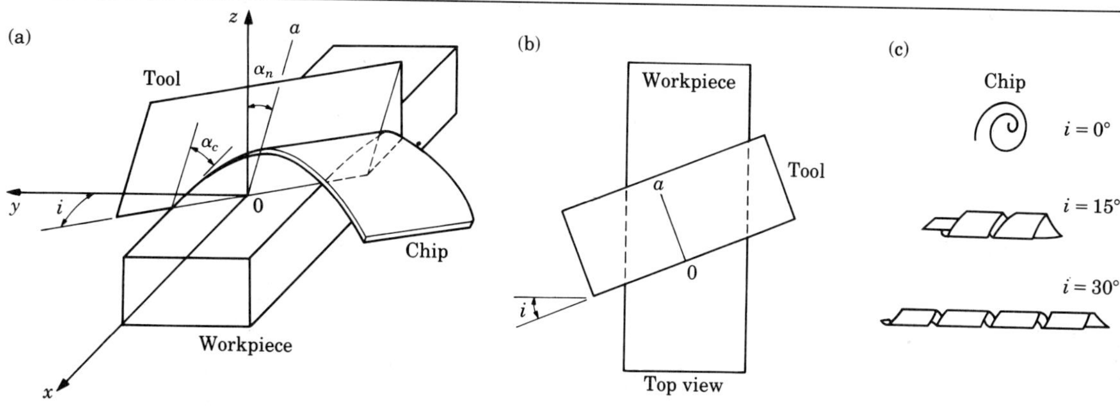

FIGURE 20.9
(a) Schematic illustration of cutting with an oblique tool. (b) Top view showing the inclination angle *i*. (c) Types of chip produced with different inclination angles.

normal rake angle, which is a basic geometric property of the tool. This is the angle between the normal o_z to the workpiece surface and the line o_a on the tool face.

The workpiece material approaches the tool at a velocity V, and leaves the surface (as a chip) with a velocity V_c. We calculate the effective rake angle α_e in the plane of these two velocities. Assuming that the chip flow angle α_c is equal to the inclination angle i (which is experimentally found to be approximately correct), the effective rake angle α_e is

$$\alpha_e = \sin^{-1}(\sin^2 i + \cos^2 i \sin \alpha_n). \tag{20.7}$$

Since we can measure both i and α_n directly, we can calculate the effective rake angle. As i increases, the effective rake angle increases, and the chip becomes thinner and longer. The effect of the inclination angle on chip shape is shown in Fig. 20.9(c).

A typical single-point turning tool used on a lathe is shown in Fig. 20.10. Note the various angles involved, each of which has to be selected properly for efficient cutting. We discuss various three-dimensional cutting tools in greater detail in Chapters 22 and 23. These include tools for drilling, tapping, milling, planing, shaping, broaching, sawing, and filing.

20.4.1 Shaving and skiving

Thin layers of material can be removed from straight or curved surfaces by a process similar to the use of a plane to shave wood. *Shaving* is particularly useful in improving the surface finish and dimensional accuracy of sheared parts and punched slugs (Fig. 20.11; see also Fig. 16.7). Another application of shaving is in finishing gears with a cutter that has the shape of the gear (see Section 23.9). Parts that are long or have a combination of shapes are shaved by *skiving*. The skiving tool moves tangentially across the length of the workpiece.

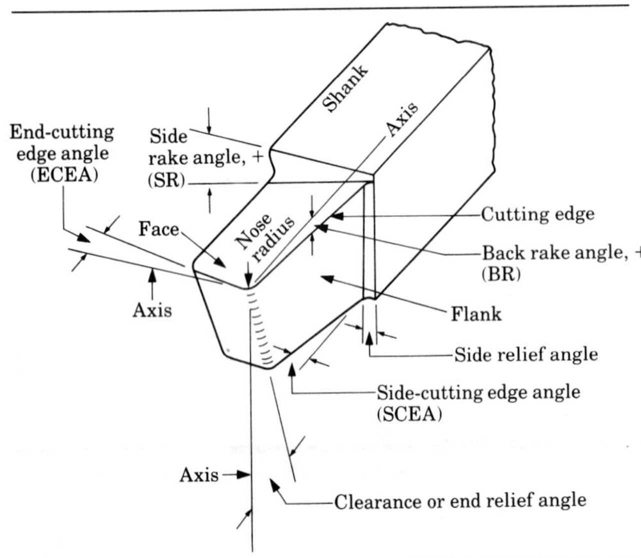

FIGURE 20.10
Terminology for a right-hand cutting tool for turning.

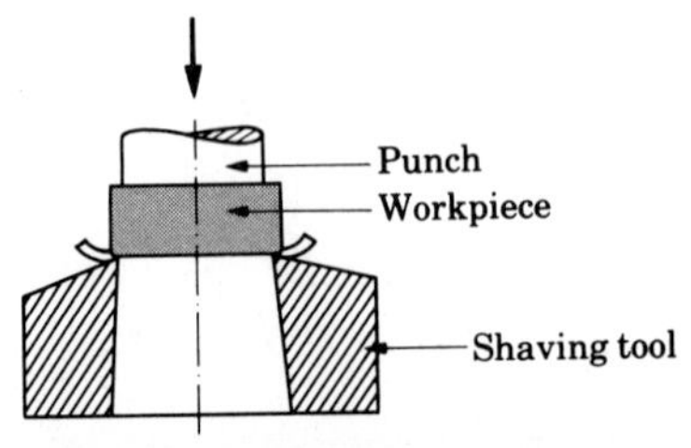

FIGURE 20.11
Schematic illustration of the shaving process for improving the accuracy of a slug. Note that the tool has a small positive rake angle (see also Fig. 16.7).

20.5 Cutting Forces, Stresses, and Power

Knowledge of the forces and power involved in cutting operations is important for the following reasons:

- Power requirements have to be determined so that a motor of suitable capacity can be installed in the machine tool.
- Data on forces are necessary for the proper design of machine tools for cutting operations that avoid excessive distortion of the machine elements and maintain desired tolerances for the machined part.
- Whether the workpiece can withstand the cutting forces without excessive distortion has to be determined in advance.

The forces acting on the tool in orthogonal cutting are shown in Fig. 20.12. The **cutting force,** F_c, acts in the direction of the cutting speed V and supplies the energy required for cutting. The **thrust force,** F_t, acts in the direction normal to the cutting velocity, that is, perpendicular to the workpiece. These two forces produce the *resultant force*, R.

Note that the resultant force can be resolved into two components on the tool face: a *friction force*, F, along the tool–chip interface, and a *normal force*, N,

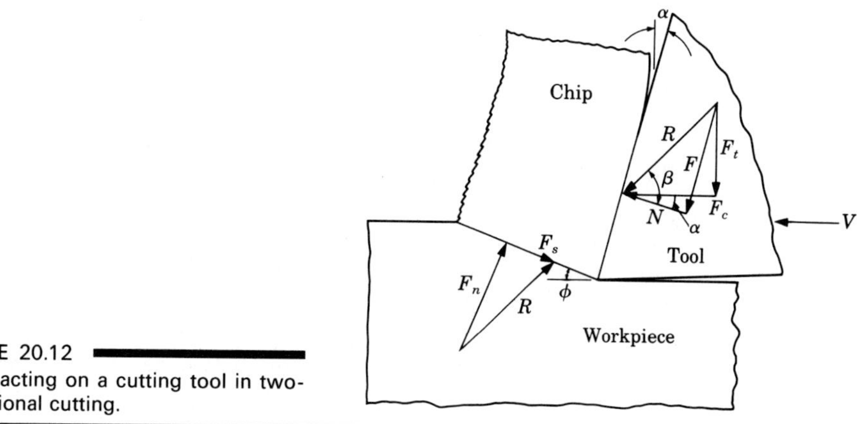

FIGURE 20.12 Forces acting on a cutting tool in two-dimensional cutting.

perpendicular to it. From Fig. 20.12, we can show the force F to be

$$F = R \sin \beta, \tag{20.8}$$

and the force N as

$$N = R \cos \beta. \tag{20.9}$$

Note that the resultant force is balanced by an equal and opposite force along the shear plane and is resolved into a *shear force*, F_s, and a *normal force*, F_n. It can be shown that these forces can be expressed as follows:

$$F_s = F_c \cos \phi - F_t \sin \phi \tag{20.10}$$

and

$$F_n = F_c \sin \phi + F_t \cos \phi. \tag{20.11}$$

Because we can calculate the area of the shear plane by knowing the shear angle and the depth of cut, we can determine the shear and normal stresses in the shear plane.

The ratio of F to N is the *coefficient of friction*, μ, at the tool–chip interface, and the angle β is the friction angle (see also Eq. 20.4). We can express μ as

$$\mu = \frac{F}{N} = \frac{F_t + F_c \tan \alpha}{F_c - F_t \tan \alpha}. \tag{20.12}$$

The coefficient of friction in metal cutting generally ranges from about 0.5 to 2.0, thus indicating that the chip encounters considerable frictional resistance while climbing up the face of the tool.

Although the magnitude of forces in actual cutting operations is generally on the order of a few hundred newtons, the local stresses in the cutting zone and the pressures on the tool are very high because the contact areas are very small. The chip–tool contact length (Fig. 20.1), for example, is typically on the order of 1 mm

(0.04 in.). Thus the tool is subjected to very high stresses, which lead to wear and sometimes chipping and fracture of the tool.

20.5.1 Thrust force

A knowledge of the thrust force in cutting is important because the tool holder and the machine tool must be sufficiently stiff to minimize deflections caused by this force. For example, if the thrust force is too high or if the machine tool is not sufficiently stiff, the tool will be pushed away from the surface being machined. This movement will, in turn, reduce the depth of cut, causing lack of dimensional accuracy in the machined part.

Refer again to Fig. 20.12 and note that the thrust force acts downward. As the rake angle increases, and/or friction at the rake face decreases, this force can act upward. You can visualize this situation by observing that when $\mu = 0$ and $\beta = 0$, the resultant force R coincides with the normal force N. In this case, R will have a thrust-force component that is upward. Also note that when $\alpha = 0$ and $\beta = 0$, the thrust force is zero.

Example: Direction of thrust force.

We note from Fig. 20.12 that

$$F_t = R \sin (\beta - \alpha) \qquad \text{or} \qquad F_t = F_c \tan (\beta - \alpha).$$

Because the magnitude of F_c is always positive (say, as shown in Fig. 20.12), the sign of F_t can be either positive or negative, depending on the relative magnitudes of β and α. When $\beta > \alpha$, the sign of F_t is positive (downward) and when $\beta < \alpha$, it is negative (upward). It is thus possible to have an upward thrust force at high rake angles and/or with low friction at the tool–chip interface.

20.5.2 Measuring forces

We can measure forces in cutting operations by using suitable dynamometers (with resistance-wire strain gages) or force transducers (such as piezoelectric crystals), mounted on the machine tool. We can also calculate forces from the power consumption during cutting, such as with a power monitor, provided that we can determine the efficiency of the machine tool.

20.5.3 Power

Power is the product of force and velocity. Referring to Fig. 20.12, we can see that the power input in cutting is

$$\text{Power} = F_c V. \tag{20.13}$$

This power is dissipated mainly in the shear zone (because of the energy required to shear the material) and on the rake face (because of tool–chip interface friction).

Referring to Figs. 20.2(b) and 20.12, you can see that the power dissipated in the shear plane is

$$\text{Power for shearing} = F_s V_s. \tag{20.14}$$

If we let w be the width of cut, the *specific energy for shearing*, u_s, is

$$u_s = \frac{F_s V_s}{w t_o V}. \tag{20.15}$$

Similarly, the power dissipated in friction is

$$\text{Power for friction} = F V_c, \tag{20.16}$$

and the *specific energy for friction*, u_f, is

$$u_f = \frac{F V_c}{w t_o V} = \frac{F r}{w t_o}. \tag{20.17}$$

The *total specific energy*, u_t, is thus

$$u_t = u_s + u_f. \tag{20.18}$$

Because of the many factors involved, the reliable prediction of cutting forces and power is still based largely on experimental data, such as those shown in Table 20.1. The wide range of values shown can be attributed to differences in strength within each material group and various other factors, such as friction and processing variables. The sharpness of the tool tip also influences forces and power. The duller the tool, the higher are the forces and power required.

TABLE 20.1
APPROXIMATE ENERGY REQUIREMENTS IN VARIOUS CUTTING OPERATIONS (at drive motor, corrected for 80% efficiency)

MATERIAL	SPECIFIC ENERGY ($W \cdot s/mm^3$)*
Aluminum alloys	0.4–1.1
Cast irons	1.6–5.5
Copper alloys	1.4–3.3
High-temperature alloys	3.3–8.5
Magnesium alloys	0.4–0.6
Nickel alloys	4.9–6.8
Refractory alloys	3.8–9.6
Stainless steels	3.0–5.2
Steels	2.7–9.3
Titanium alloys	3.0–4.1

* Divide by 2.73 to obtain $hp \cdot min/in^3$.

• Example: Relative energies in cutting.

An orthogonal cutting process is being carried out, where $t_o = 0.005$ in., $V = 400$ ft/min, $\alpha = 10°$, and the width of cut = 0.25 in. It is observed that $t_c = 0.009$ in., $F_c = 125$ lb, and $F_t = 50$ lb. Calculate the percentage of the total energy that goes into overcoming friction at the tool–chip interface.

SOLUTION. We can express the percentage as

$$\frac{\text{Friction energy}}{\text{Total energy}} = \frac{FV_c}{F_c V} = \frac{Fr}{F_c},$$

where

$$r = \frac{t_o}{t_c} = \frac{5}{9} = 0.555,$$

$$F = R \sin \beta,$$

$$F_c = R \cos (\beta - \alpha),$$

and

$$R = \sqrt{F_t^2 + F_c^2} = \sqrt{50^2 + 125^2} = 135 \text{ lb}.$$

Thus

$$125 = 135 \cos (\beta - 10),$$

from which

$$\beta = 32°$$

and

$$F = 135 \sin 32° = 71.5 \text{ lb}.$$

Hence

$$\text{Percentage} = \frac{(71.5)(0.555)}{125} = 0.32 \quad \text{or} \quad 32\,\%.$$

20.6 Temperature in Cutting

As in all metalworking operations, the energy dissipated in cutting operations is converted into heat, which in turn, raises the temperature in the cutting zone.

Knowledge of the temperature rise in cutting is important because it

- adversely affects the strength, hardness, and wear resistance of the cutting tool;
- causes dimensional changes in the part being machined, making control of dimensional accuracy difficult; and
- can induce thermal damage to the machined surface, adversely affecting its properties.

Because of the work done in shearing and in overcoming friction on the rake face of the tool, the main sources of heat generation are the primary shear zone and the tool–chip interface. Additionally, if the tool is dull or worn, heat is also generated by the tool tip rubbing against the machined surface.

Temperature increases with the strength of the workpiece material, cutting speed, and depth of cut; it decreases with increasing specific heat and thermal conductivity of the workpiece material. The *mean temperature* in turning on a lathe (Fig. 20.13) is found to be proportional to cutting speed and feed as follows:

$$\text{Mean temperature} \propto V^a f^b, \tag{20.19}$$

where a and b are constants, V is the cutting speed, and f is the feed of the tool, that is, how far the tool travels per revolution of the workpiece. Approximate values for the constants a and b are:

Tool material	a	b
Carbide	0.2	0.125
High-speed steel	0.5	0.375

A typical temperature distribution in the cutting zone is shown in Fig. 20.14. Note the severe temperature gradients and that the maximum temperature is about halfway up the face of the tool. The particular temperature pattern depends on factors such as specific heat and thermal conductivity of the tool and workpiece material, cutting speed, depth of cut, and the type of cutting fluid used, if any.

The temperatures developed in a turning operation on steel are shown in Fig. 20.15. Note that temperature increases with cutting speed and that the highest temperature is almost 1100 °C (2000 °F). The presence of such high temperatures can be verified by observing the dark bluish color of chips (caused by oxidation) produced at high cutting speeds. Chips can in fact become red hot, thus creating a safety hazard to the operator.

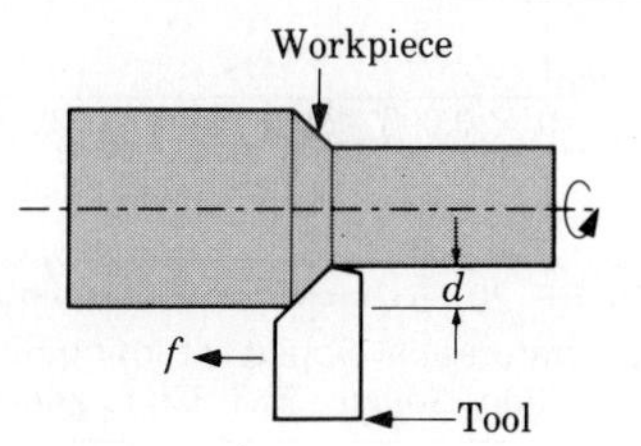

FIGURE 20.13 Terminology for a turning operation on a lathe. Feed, *f*, is the distance the tool travels per revolution of the workpiece. Note that feed in turning is equivalent to the depth of cut, t_o, in orthogonal cutting, and depth of cut, *d*, in turning is equivalent to the width of cut in orthogonal cutting.

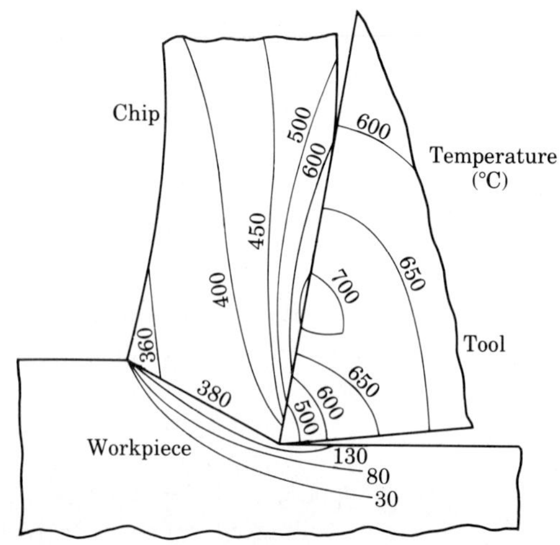

FIGURE 20.14 Typical temperature distribution in the cutting zone. Note the steep temperature gradients within the tool and the chip. *Source:* G. Vieregge.

From Eq. (20.19) and the preceding values for a, you can see that cutting speed greatly influences temperature. As speed increases, the time for heat dissipation decreases—and thus temperature rises. You can easily demonstrate this effect by rubbing your hands together faster and faster. The chip is a good heat sink because it carries away much of the heat generated (Fig. 20.16). As cutting speed increases, a

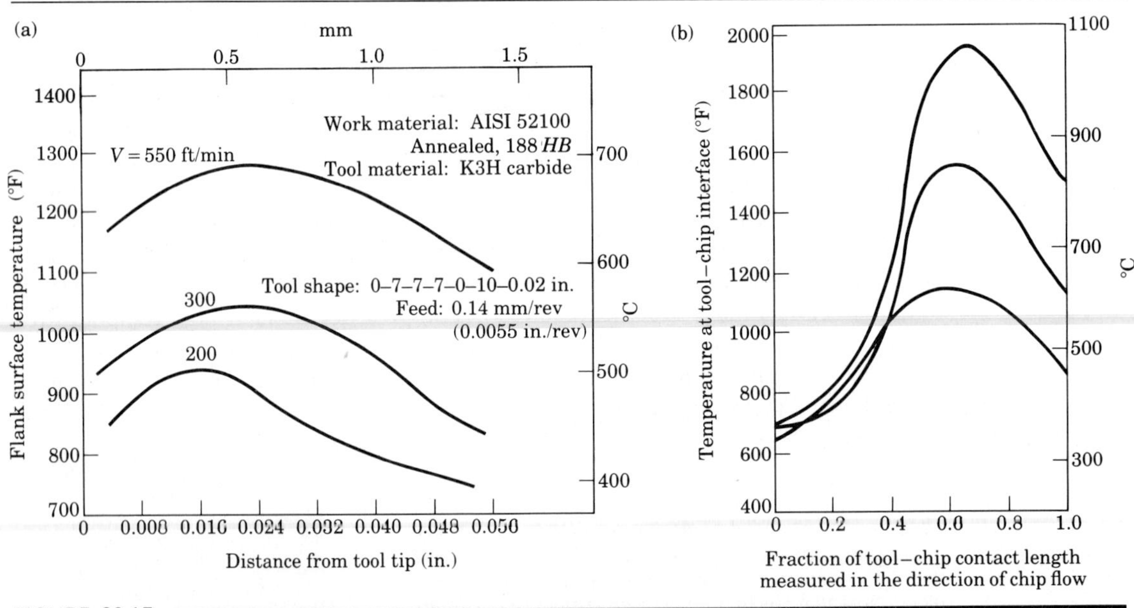

FIGURE 20.15 Temperatures developed in turning 52100 steel: (a) flank temperature distribution; and (b) tool–chip interface temperature distribution. *Source:* B. T. Chao and K. J. Trigger.

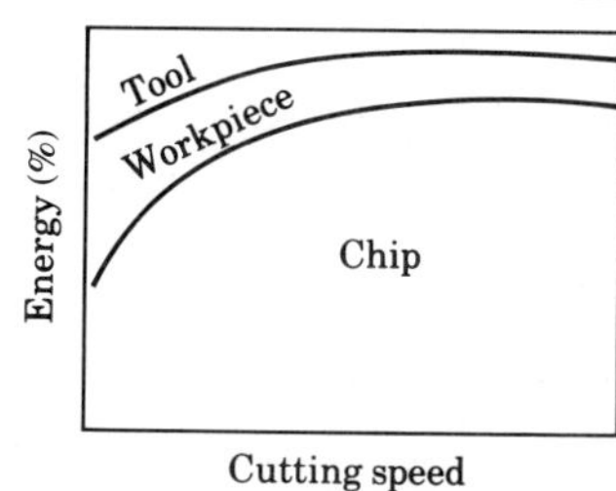

FIGURE 20.16
Percentage of the heat generated going into the workpiece, tool, and chip, as a function of cutting speed. Note that the chip carries away most of the heat.

larger proportion of the heat generated is carried away by the chip, and little heat goes into the workpiece.

20.6.1 Techniques for measuring temperature

Temperatures and their distribution in the cutting zone may be determined from thermocouples embedded in the tool and/or the workpiece. This technique has been used successfully, although it involves considerable effort. A simpler technique for determining the average temperature is by the thermal emf (electromotive force) at the tool–chip interface, which acts as a hot junction between two different (tool and chip) materials. The infrared radiation from the cutting zone may also be monitored with a radiation pyrometer. However, this technique indicates only surface temperatures, and the accuracy of the results depends on the emissivity of the surfaces, which is difficult to determine accurately.

20.7 Tool Wear and Failure

In the preceding sections, we have shown that cutting tools are subjected to high localized stresses, high temperatures, sliding of the chip along the rake face, and sliding of the tool flank along the freshly cut surface. These conditions induce tool *wear*, which in turn, adversely affects the quality of the machined surface, its dimensional accuracy, and consequently the economics of cutting operations.

Tool wear is generally a gradual process, much like the wear of the tip of an ordinary pencil. The rate of wear depends on tool and workpiece materials, tool shape, cutting fluids, process parameters (such as cutting speed, feed, and depth of cut), and machine-tool characteristics. There are two basic regions of wear in a cutting tool: flank wear and crater wear.

20.7.1 Flank wear

Flank wear occurs on the relief face of the tool (Fig. 20.17a; see also Fig. 20.1 and the side relief angle in Fig. 20.10) and is generally attributed to (a) rubbing of the tool along the machined surface and (b) high temperatures developed. In a classic study by F. W. Taylor on machining steels (published in 1907), the following approximate relationship was established:

$$VT^n = C, \tag{20.20}$$

where V is the cutting speed, T is the time (in minutes) that it takes to develop a certain flank wear land (VB in Fig. 20.17c), n is an exponent that depends on tool and workpiece materials and cutting conditions, and C is a constant. Thus each combination of workpiece and tool materials and each cutting condition has its own n and C values, both of which are determined experimentally. The range of n values observed in practice is given in Table 20.2.

Although cutting speed has been found to be the most significant process variable in tool life, depth of cut and feed rate are also important. Thus Eq. (20.20) can be modified as follows:

$$VT^n d^x f^y = C, \tag{20.21}$$

where d is the depth of cut and f is the feed rate (in mm/rev or in./rev) in turning.

The exponents x and y must be determined experimentally for each cutting condition. For example, taking $n = 0.15$, $x = 0.15$, and $y = 0.6$ as typical values encountered in practice, we see that cutting speed, feed rate, and depth of cut are of decreasing order of importance.

Equation (20.21) can be rewritten as

$$T = C^{1/n} V^{-1/n} d^{-x/n} f^{-y/n} \tag{20.22}$$

or

$$T \simeq C^7 V^{-7} d^{-1} f^{-4}. \tag{20.23}$$

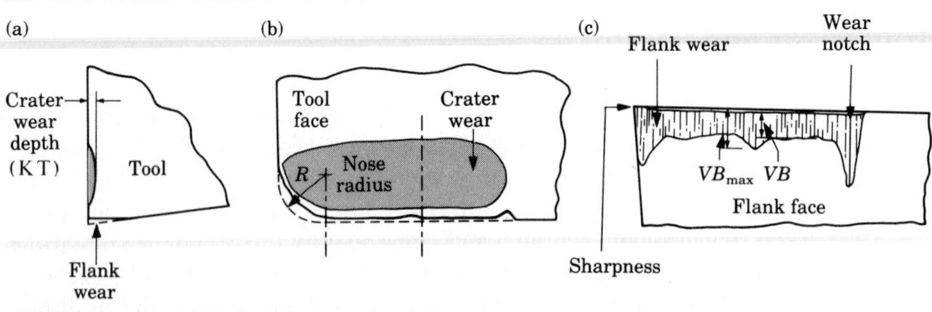

FIGURE 20.17
(a) Flank and crater wear in a cutting tool. Tool moves to the left. (b) View of the rake face of a turning tool, showing nose radius R and crater wear pattern on the rake face of the tool. (c) View of the flank face of a turning tool, showing the average flank wear land VB and the wear notch (depth-of-cut line).

For a constant tool life, the following observations can be made from Eq. (20.23).

a) If the feed rate or the depth of cut is increased, the cutting speed must be decreased, and vice versa.
b) Depending on the exponents, a reduction in speed can then result in an increase in the volume of the material removed because of the increased feed rate and/or depth of cut.

Tool-life curves. **Tool-life curves** are plots of experimental data obtained in cutting tests (Fig. 20.18). Note the rapid decrease in tool life as cutting speed increases and the strong influence of the condition of the workpiece material on tool life. Also note the large difference in tool life for different workpiece microstructures. Heat treatment is important largely because of increasing workpiece hardness. For example, ferrite has a hardness of about 100 HB, pearlite 200 HB, and martensite 300–500 HB. Impurities and hard constituents in the material are also important considerations because they reduce tool life by their abrasive action.

We usually plot tool-life curves on log–log paper, from which we can easily determine the exponent n (Fig. 20.19). These curves are usually linear over a certain range of cutting speeds but are rarely so over a wide range. Moreover, the exponent n can indeed become negative at low cutting speeds. Thus tool-life curves may actually reach a maximum and then curve downward. Therefore, caution should be exercised when using tool-life equations beyond the range of cutting speeds for which they are applicable.

Because temperature affects the physical and mechanical properties of materials, we can expect that temperature strongly influences wear. Investigations have indeed confirmed that as temperature increases, flank wear increases rapidly.

TABLE 20.2
RANGE OF n VALUES FOR EQ. (20.18)

High-speed steels	0.08–0.2
Cast alloys	0.1–0.15
Carbides	0.2–0.5
Ceramics	0.5–0.7

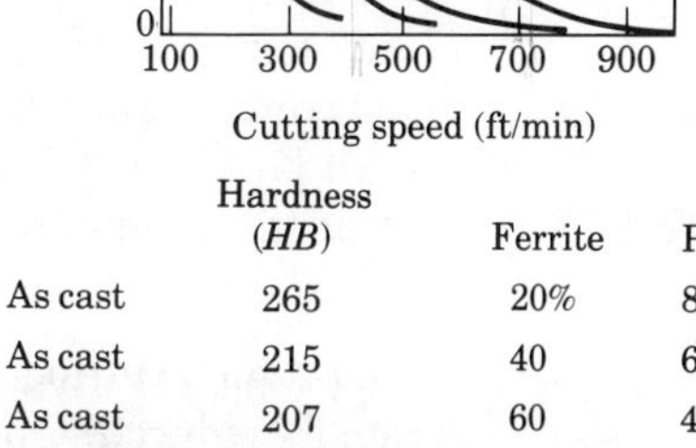

	Hardness (HB)	Ferrite	Pearlite
a. As cast	265	20%	80%
b. As cast	215	40	60
c. As cast	207	60	40
d. Annealed	183	97	3
e. Annealed	170	100	—

FIGURE 20.18 ▶ Effect of microstructure and hardness on tool life in turning ductile cast iron. Note the rapid decrease in tool life as cutting speed increases.

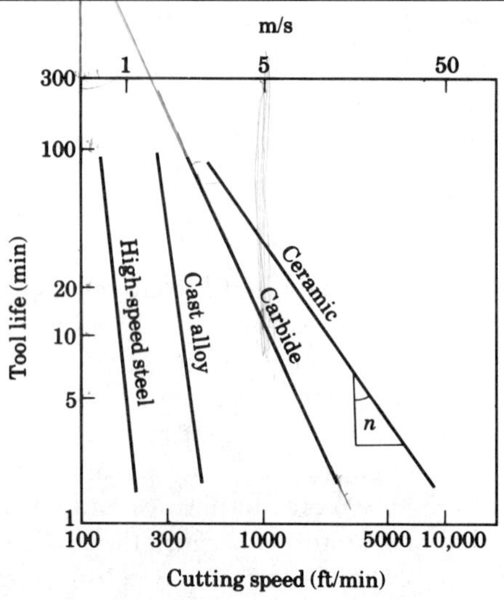

FIGURE 20.19
Tool-life curves for a variety of cutting-tool materials. The negative inverse of the slope of these curves is the exponent *n* in tool-life equations.

Although cutting speed is the most significant process variable in tool life, feed *f* and depth of cut *d* also are important factors. As *f* and *d* increase, tool life decreases, with feed having a greater effect than depth of cut.

Allowable wear land. When do we decide to sharpen a knife or a pair of scissors? We sharpen them when the quality of the cut begins to deteriorate or the forces required increase too much. Similarly, cutting tools should be resharpened or replaced when the surface finish of the machined workpiece begins to deteriorate and temperature rises excessively. The **allowable wear land** (*VB* in Fig. 20.17c) for various conditions is given in Table 20.3.

For improved dimensional accuracy and surface finish, the allowable wear land may be made smaller than the values given in Table 20.3. The recommended cutting speed for a high-speed-steel tool is generally the one that gives a tool life of 60–120 min and for carbide tools 30–60 min. Cutting speeds selected can vary significantly from these values depending on the particular workpiece and the operation, as well as high productivity considerations using modern computer-controlled machine tools.

Optimum cutting speed. You have seen that as cutting speed increases, tool life is rapidly reduced. On the other hand, if cutting speeds are low, tool life is long but the rate at which material is removed is also low. Thus there is an optimum cutting speed, as we will see in Section 24.5.

TABLE 20.3
ALLOWABLE AVERAGE WEAR LAND (*VB*) FOR CUTTING TOOLS IN VARIOUS OPERATIONS

	ALLOWABLE WEAR LAND (mm)	
OPERATION	*HIGH-SPEED STEELS*	*CARBIDES*
Turning	1.5	0.4
Face milling	1.5	0.4
End milling	0.3	0.3
Drilling	0.4	0.4
Reaming	0.15	0.15

• Example: Effect of cutting speed on material removal.

We can appreciate the effect of cutting speed on the volume of metal removed between tool resharpenings or replacements by analyzing Fig. 20.18. Assume that we are machining the material in the "a" condition, that is, as cast with a hardness of 265 HB. If our cutting speed is 1 m/s (200 ft/min), we note that tool life is about 40 min. Hence the tool travels a distance of (1 m/s)(60 s/min)(40 min) = 2400 m before it is resharpened or replaced. If we change the cutting speed to 2 m/s, tool life is about 5 min. Hence the tool travels (2)(60)(5) = 600 m.

Since the volume of material removed is directly proportional to the distance the tool has traveled, we see that by decreasing the cutting speed, we can remove more material between tool changes. Note, however, that the lower the cutting speed, the longer is the time required to machine a part. These variables have important economic impact, and we will discuss them further in Section 24.5. •

20.7.2 Crater wear

Crater wear occurs on the rake face of the tool (Figs. 20.17a and b and 20.20) and changes the chip–tool interface geometry, thus affecting the cutting process. The most significant factors influencing crater wear are temperature at the tool–chip interface and the chemical affinity between the tool and workpiece materials. Additionally, the factors influencing flank wear also influence crater wear.

Crater wear has been described in terms of a *diffusion* mechanism, that is, the movement of atoms across the tool–chip interface. Since diffusion rate increases with increasing temperature, crater wear increases as temperature increases. Note in Fig. 20.21 how sharply crater wear increases within a narrow temperature range.

When we compare Figs. 20.14 and 20.17(a), we can see that the location of maximum depth of crater wear coincides with the location of the maximum temperature at the tool–chip interface. An actual cross-section of the tool–chip interface in cutting steel at high speed is shown in Fig. 20.22. Compare the location

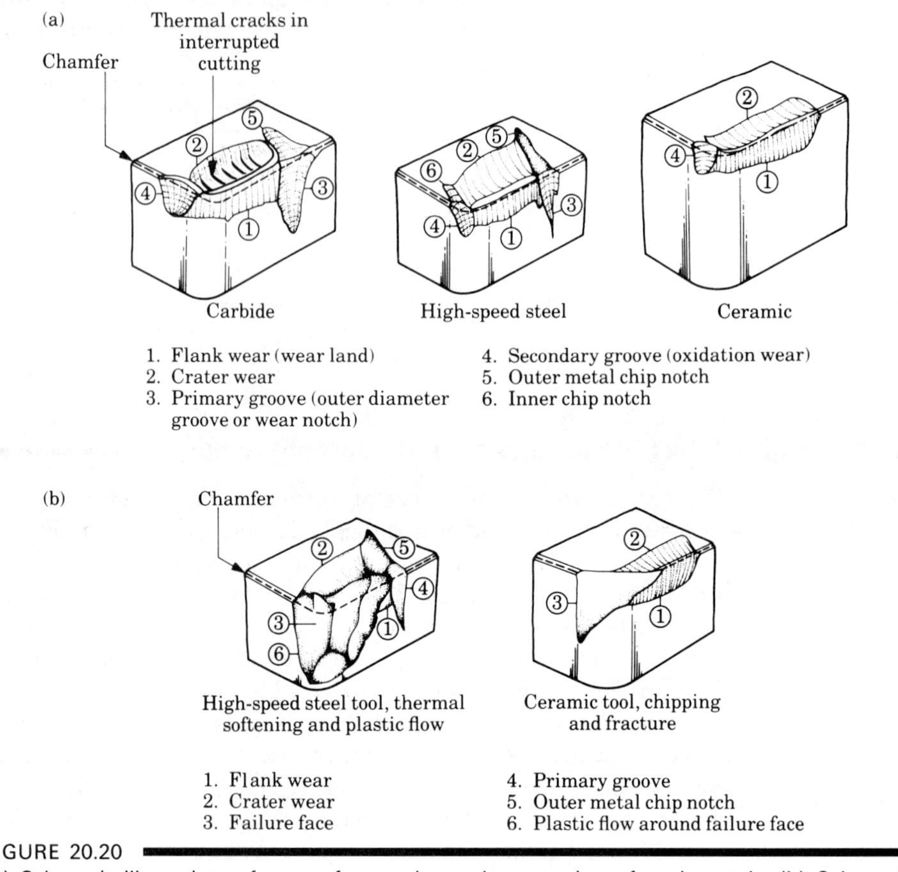

FIGURE 20.20
(a) Schematic illustrations of types of wear observed on a variety of cutting tools. (b) Schematic illustrations of catastrophic tool failures. *Source:* V. C. Venkatesh.

of the crater-wear pattern and the discoloration pattern of the tool by high temperatures and observe their similarity.

20.7.3 Chipping

Chipping is the term used to describe the breaking away of a small piece from the cutting edge of the tool, a phenomenon similar to breaking the tip of a sharp pencil. The chipped pieces from the cutting tool may be very small (*micro-* or *macrochipping*), or they may be in relatively large fragments (*gross chipping* or *fracture*). Unlike wear, which is a gradual process, chipping results in a sudden loss of tool material and shape, and has a major detrimental effect on surface finish.

Two main causes of chipping are *mechanical shock* (impact by interrupted cutting, as in milling or turning a splined shaft), and *thermal fatigue* (cyclic variations in temperature of the tool in interrupted cutting). Chipping may occur in

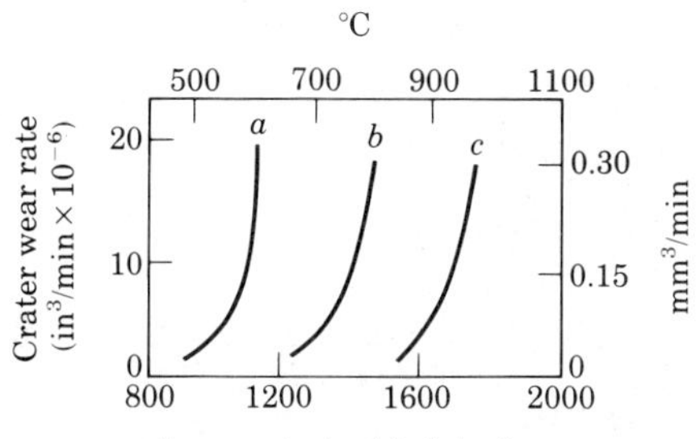

FIGURE 20.21 Relationship between crater-wear rate and average tool–chip interface temperature: (a) High-speed steel; (b) C-1 carbide; and (c) C-5 carbide. Note how rapidly crater-wear rate increases as the temperature increases. *Source:* B. T. Chao and K. J. Trigger.

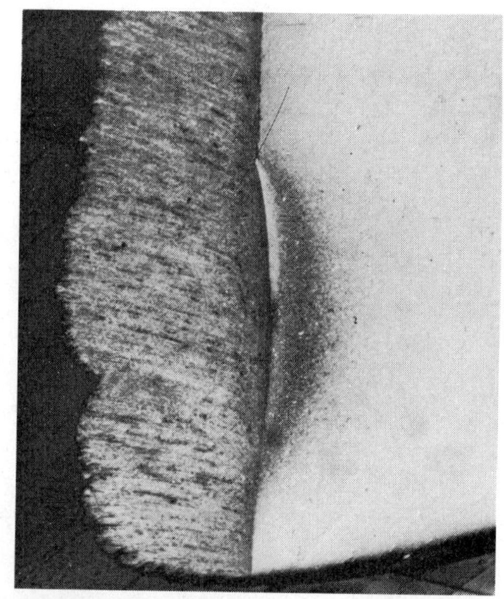

FIGURE 20.22 Tool (right) and chip (left) interface in cutting plain-carbon steel. The discoloration on the tool indicates high temperatures. Note that the crater-wear pattern coincides with the discoloration pattern. (See also Fig. 20.14.) *Source:* P. K. Wright.

a region in the tool where a small crack or defect already exists. Thermal cracks are usually perpendicular to the cutting edge of the tool (Fig. 20.20a).

High positive rake angles can contribute to chipping, because of the small included angle of the tool tip. It is possible for the crater-wear region to progress toward the tool tip, thus weakening it and causing chipping. Chipping or fracture can be reduced by selecting tool materials with high impact and thermal shock resistance, which we describe in Chapter 21.

20.7.4 General observations on tool wear

Because of the many factors involved, including the characteristics of the machine tool, the wear behavior of cutting tools varies significantly. In addition to the wear processes we have already described, other phenomena also occur in tool wear (Fig. 20.20). For example, as a result of the high temperatures generated during cutting, tools could soften and undergo plastic deformation because their yield strength decreases. This type of deformation generally occurs in machining high-strength metals and alloys. Thus tools must maintain their strength and hardness at the elevated temperatures encountered in cutting.

The various regimes of wear for a carbide tool in cutting austenitic stainless steel are shown in Fig. 20.23. A built-up edge forms at low cutting speeds. As speed increases the BUE decreases, but the tool may experience crater wear and plastic deformation because of higher temperatures.

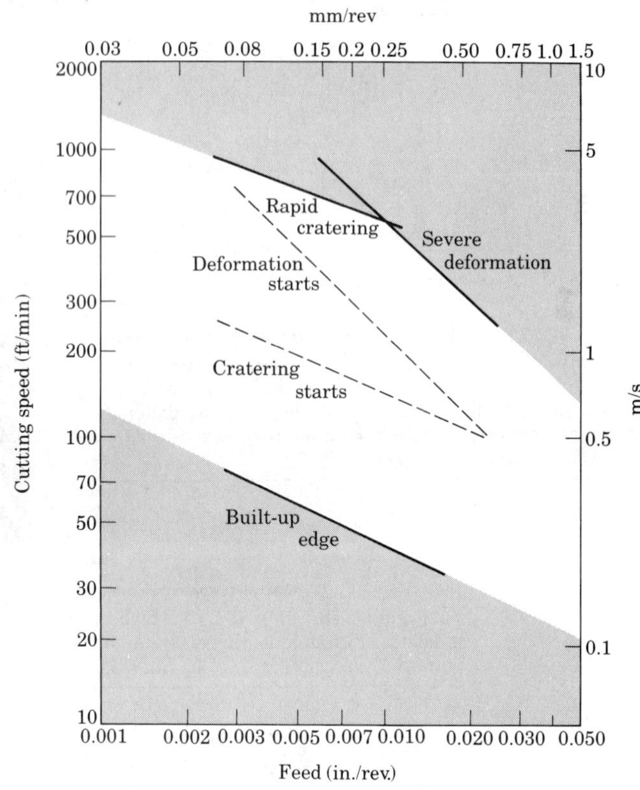

FIGURE 20.23 Trends in wear, built-up edge, and plastic deformation of a carbide tool as a function of cutting speed and feed in turning. *Source:* P. A. Dearnley and E. M. Trent.

The wear groove or notch on cutting tools (Fig. 20.20) has been attributed to the fact that this region is the boundary where the chip is no longer in contact with the tool (see Fig. 20.13). This boundary, or **depth-of-cut line**, oscillates because of inherent variations in the cutting operation and accelerates the wear process. Furthermore, this region is in contact with the machined surface from the previous cut. Since a machined surface may develop a thin work-hardened layer, this contact could contribute to the formation of the wear groove.

Because they are hard and abrasive, scale and oxide layers on a workpiece surface increase wear (see Table 31.1). In such cases, the depth of cut should be greater than the thickness of the oxide film or the work-hardened layer. Thus the value of d in Fig. 20.13 should be greater than the thickness of the scale on the workpiece. In other words, light cuts should not be taken on rusted workpieces.

20.7.5 Techniques for measuring and monitoring wear

Tool-wear measuring techniques fall into two categories: direct and indirect. The *direct* method involves optical measurement of wear, such as by periodically observing changes in the tool profile. This is the most common and reliable technique and is done using a microscope (toolmakers' microscope). This procedure, however, requires that the cutting operation be stopped. Another direct method involves observing the tool-side face of the chip (see Fig. 20.5) for the presence and amount of crater wear particles using special instrumentation.

Indirect methods of measuring wear involve the correlation of wear with process variables such as forces, power, temperature rise, surface finish, and vibrations. One recent development is the *acoustic emission technique*, which utilizes a piezoelectric transducer attached to a tool holder. The transducer picks up signals that are acoustic emissions resulting from the stress waves generated during cutting. By analyzing the signals, we can monitor tool wear and chipping. We can also record and monitor forces and vibrations during cutting, and changes in their patterns as the cut progresses, for indications of tool wear and fracture.

Because direct observation methods interrupt the steady-state nature of the cut, they influence the economics of machining operations in manufacturing plants. Implementing on-line monitoring of the rate of tool wear, that is, while the cutting operation is taking place, is therefore more desirable, particularly for computer-controlled machine tools. The indirect methods stated above, including monitoring of power, is being used for this purpose. There are, however, difficulties involved concerning reliability and calibration in the use of these techniques. Continued progress is being made in refining measurement techniques. Some instrumentation for tool-condition monitoring is now commercially available.

20.8 Surface Finish and Integrity

Surface finish influences not only the dimensional accuracy of machined parts, but also their properties. Whereas **surface finish** describes the geometric features of surfaces, **surface integrity** pertains to properties such as fatigue life and corrosion resistance, which are influenced strongly by the type of surface produced. Factors influencing surface integrity are temperatures generated during processing, residual stresses, metallurgical (phase) transformations, and surface plastic deformation, tearing, and cracking.

The built-up edge, with its significant effect on tool profile, has the greatest influence on surface roughness. Figure 20.24 shows surfaces obtained in two different cutting operations. Note the considerable damage to the surfaces from BUE. Ceramic and diamond tools generally produce better surface finish than other tools, largely because of their much lower tendency to form BUE.

A tool that is not sharp has a large tip radius along its edges (see Fig. 20.17c), just like a dull pencil or knife does. If this radius, not to be confused with radius R in Fig. 20.17(b), is large in relation to the depth of cut, the tool will rub over the machined surface (Fig. 20.25). This rubbing generates heat and induces surface

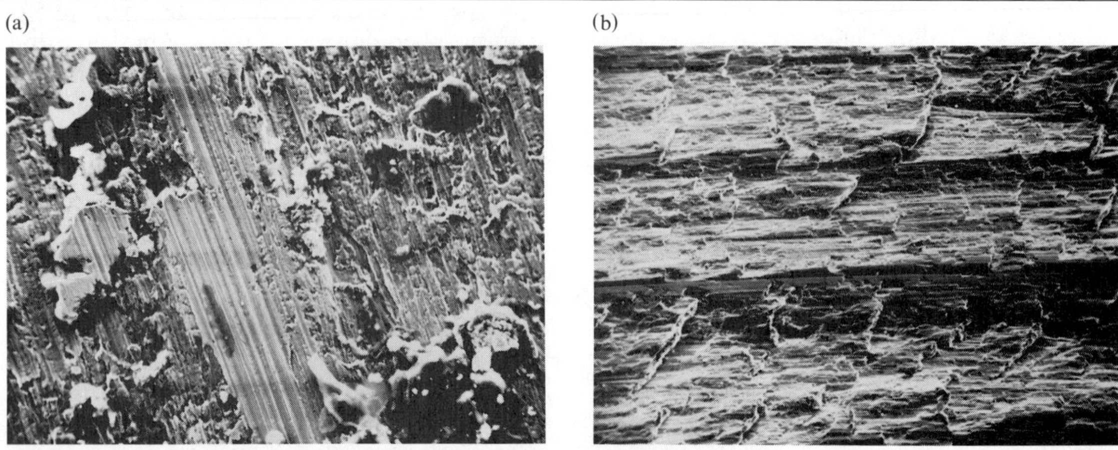

FIGURE 20.24
Surfaces produced on steel by cutting, as observed with a scanning electron microscope: (a) turned surface and (b) surface produced by shaping. *Source:* J. T. Black and S. Ramalingam.

residual stresses, which in turn, may cause surface damage such as tearing and cracking.

In turning, as in other cutting operations, the tool leaves a spiral profile—**feed marks**—on the machined surface as it moves across the workpiece (Figs. 20.26 and 20.27). You can see that the higher the feed f and the smaller the radius R, the more prominent these marks will be. Although not significant in rough machining operations, these marks are important in finish machining. We discuss feed marks for individual machining processes in Chapters 22–26.

We describe vibration and chatter in some detail in Section 24.3. For now, we should recognize that if the tool vibrates or chatters during cutting, it will adversely

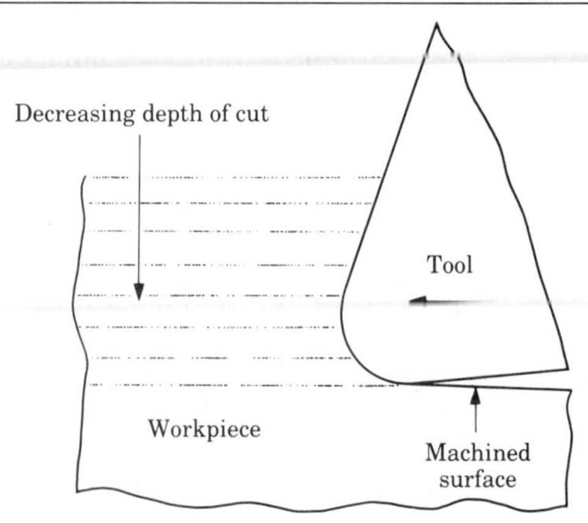

FIGURE 20.25
Schematic illustration of a dull tool in orthogonal cutting. Note that at small depths, the rake angle can effectively become negative. The tool may simply ride over the workpiece surface, burnishing it.

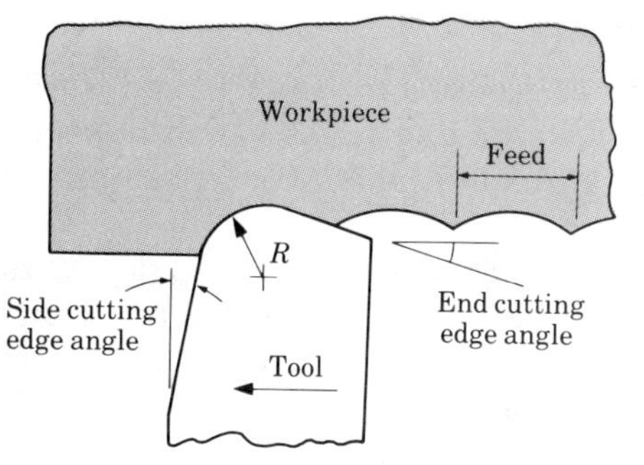

FIGURE 20.26
Schematic illustration of feed marks in turning.

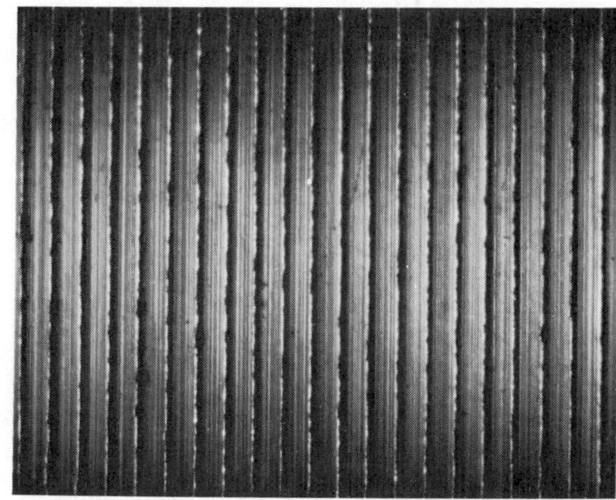

FIGURE 20.27
Feed marks in turning tungsten with a ceramic tool. *Source:* A. J. Moser and S. Kalpakjian.

affect surface finish. The reason is that a vibrating tool changes the dimensions of the cut periodically. Excessive chatter can also cause chipping and premature failure of the more brittle cutting tools such as ceramics and diamond.

20.9 Machinability

We usually define **machinability** of a material in terms of three factors: (1) surface finish and integrity of the machined part, (2) tool life obtained, and (3) force and power requirements. Thus good machinability indicates good surface finish and integrity, long tool life, and low force and power requirements. An additional parameter is chip curl. As we stated earlier, long, thin curled chips, if not broken up, can severely interfere with the cutting operation by becoming entangled in the cutting zone. Thus the type of chip a material produces is also a factor in its machinability.

Because of the complex nature of cutting operations, establishing relationships to define quantitatively the machinability of a material is difficult. In manufacturing plants, tool life and surface roughness are generally considered to be the most important factors in machinability. Although not used much any more, approximate *machinability ratings* are available.

• Example: Machinability ratings.

Machinability ratings are based on a tool life of $T = 60$ min. The standard is AISI 1112 steel, which is given a rating of 100. Thus for a tool life of 60 min, this steel should be machined at a cutting speed of 100 ft/min (0.5 m/s). Higher speeds will reduce tool life, and lower speeds will increase it.

For example, 3140 steel has a machinability rating of 55. This means that when it is machined at a cutting speed of 55 ft/min (0.275 m/s), tool life will be 60 min. Nickel has a rating of 200, indicating that it should be machined at 200 ft/min (1 m/s) to obtain a tool life of 60 min. Machinability ratings for various materials are as follows: free-cutting brass 300, 2011 wrought aluminum 200, pearlitic gray iron 70, Inconel 30, and precipitation-hardening 17-7 steel 20.

As the hardness of the material increases, its machinability rating decreases proportionately. However, such ratings are only approximate and should be used with caution.

•

20.9.1 Machinability of steels

Because steels are among the most important engineering materials, their machinability has been studied extensively. The machinability of steels has been improved mainly by adding lead and sulfur to obtain so-called *free-machining steels.*

Leaded steels. Lead is added to the molten steel and takes the form of dispersed fine lead particles. Lead is insoluble in iron, copper, and aluminum and their alloys. Thus during cutting, the lead particles are sheared and smeared over the tool–chip interface. Because of its low shear strength, the lead acts as a solid lubricant.

This behavior has been verified by the presence of high concentrations of lead on the tool-side face of chips in machining leaded steels. In addition to this effect, lead probably lowers the shear stress in the primary shear zone, thus reducing cutting forces and power consumption. Lead may be used in either nonsulfurized or resulfurized steels. If the presence of lead is objectionable, it may be replaced by bismuth. Leaded steels are identified by the letter *L* between the second and third numerals; thus, for example, 10L45. However, in stainless steels, similar use of the letter *L* in their identification means low carbon, which improves their corrosion resistance.

Resulfurized steels. Sulfur in steels forms manganese sulfide inclusions (second-phase particles), which act as stress raisers in the primary shear zone. As a result, the chips produced are small and break up easily, thus improving machinability. The shape, orientation, distribution, and concentration of these inclusions significantly influence machinability. Elements such as tellurium and selenium (both chemically similar to sulfur) in resulfurized steels act as inclusion modifiers.

Calcium-deoxidized steels. An important development is calcium-deoxidized steels (see Chapter 5) in which oxide flakes of calcium aluminosilicate (CaO, SiO_2, and Al_2O_3) are formed. These flakes, in turn, reduce the strength of the secondary shear zone, thus decreasing tool–chip interface friction and wear, hence temperature. Consequently, these steels produce less crater wear, especially at high cutting speeds.

Effects of other elements in steels on machinability. The presence of aluminum and silicon in steels is always harmful because they combine with oxygen and form aluminum oxide and silicates. These compounds are hard and abrasive, thus increasing tool wear and reducing machinability.

Carbon, manganese, and phosphorus have various effects on the machinability of steels, depending on their composition. As the carbon content increases, machinability decreases. However, plain low-carbon steels (less than 0.15% C) can produce poor surface finish by forming a built-up edge. Cast steels are more abrasive, although their machinability is similar to wrought steels. Tool and die steels are very difficult to machine and usually require annealing prior to machining. Machinability of most steels is generally improved by cold working, which reduces the tendency for built-up edge formation.

Other alloying elements, such as nickel, chromium, molybdenum, and vanadium, which improve the properties of steels, generally reduce machinability. The effect of boron is negligible. The role of gaseous elements such as oxygen, hydrogen, and nitrogen has not been clearly established. Any effect that they may have would depend on the presence and quantity of other alloying elements.

In selecting various elements to improve machinability, we should consider the possible detrimental effects of these elements on the properties and strength of the machined part in service. At elevated temperatures, for example, lead causes embrittlement of steels (hot shortness), although at room temperature it has no effect on mechanical properties. Sulfur can severely reduce hot workability of steels, because of the presence of iron sulfide, unless sufficient manganese is present to prevent the formation of iron sulfide. At room temperature, the mechanical properties of resulfurized steels depend on the orientation of the deformed manganese sulfide inclusions (anisotropy).

Stainless steels. Austenitic (300 series) steels are generally difficult to machine. Chatter could be a problem, thus requiring machine tools with high stiffness. However, ferritic stainless steels (also 300 series) have good machinability. Martensitic (400 series) steels are abrasive and tend to form built-up edge, and require tool materials with high hot hardness and crater-wear resistance. Precipitation-hardening stainless steels are strong and abrasive, requiring hard and abrasion-res[illegible] tool materials.

20.9.2 Machinability of various other metals

Aluminum is generally easy to machine. Howeve[illegible] built-up edge. High cutting speeds, high r[illegible] recommended. Wrought alloys with high silicon content and cast aluminum alloys may be abrasive and hence require harder tool materials. Dimensional control may be a problem in machining aluminum since it has a low elastic modulus and a relatively high thermal coefficient of expansion.

Beryllium is similar to cast irons but is more abrasive and toxic. Hence it requires machining in a controlled environment.

Gray cast irons are generally machinable but are abrasive. Free carbides in castings reduce their machinability and cause tool chipping or fracture, thus requiring tools with high toughness. Nodular and malleable irons are machinable with hard tool materials.

Cobalt-base alloys are abrasive and highly work hardening. They require sharp and abrasion-resistant tool materials and low feeds and speeds.

Wrought copper can be difficult to machine because of built-up edge formation, although cast copper alloys are easy to machine. Brasses are easy to machine, especially with the addition of lead (leaded free-machining brass). Bronzes are more difficult to machine than brass.

Magnesium is very easy to machine, has good surface finish, and prolongs tool life. However, care should be exercised because of its high rate of oxidation and the danger of fire.

Molybdenum is ductile and work hardening. Hence it can produce poor surface finish, thus requiring sharp tools.

Nickel-base alloys are work hardening, abrasive, and strong at high temperatures. Their machinability is similar to that of stainless steels.

Tantalum is very work hardening, ductile, and soft. Hence it produces a poor surface finish; tool wear is high.

The poor thermal conductivity of *titanium* (lowest of all engineering metals) and its alloys causes significant temperature rise and built-up edge. Thus it can be difficult to machine.

Because *tungsten* is brittle, strong, and very abrasive, its machinability is low. Machinability improves greatly at elevated temperatures.

Zirconium has good machinability. However, it requires a coolant-type cutting fluid because of the danger of explosion and fire.

20.9.3 Machinability of various materials

Graphite is abrasive. It requires hard, abrasion-resistant, sharp tools. Thermoplastics generally have low thermal conductivity, low elastic modulus, and low softening temperature. Consequently, machining them requires tools with positive rake to reduce cutting forces, large relief angles, small depths of cut and feed, relatively high speeds, and proper support of the workpiece. Tools should be sharp. External cooling of the cutting zone may be necessary to keep the chips from becoming "gummy" and sticking to the tools. Cooling can usually be done with a jet of air, vapor mist, or water-soluble oils. Residual stresses may develop during machining. To relieve residual stresses, machined parts can be annealed at temperatures ranging from 80 °C to 160 °C (175 °F to 315 °F) for a period of time and then cooled slowly and uniformly to room temperature.

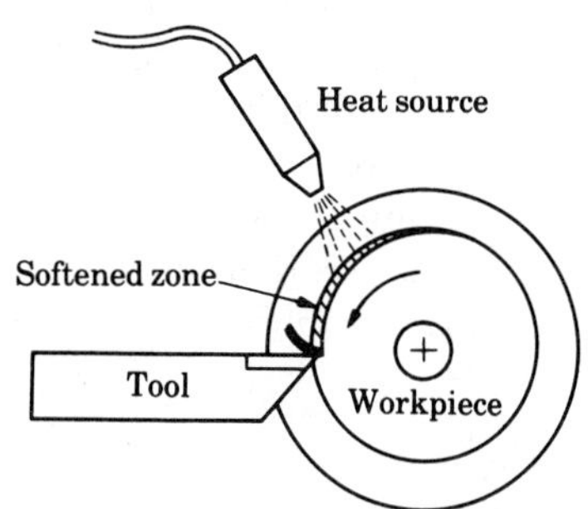

FIGURE 20.28
Schematic illustration of thermally assisted machining on a lathe. The heat source may be a torch or high-energy beams such as lasers.

Thermosetting plastics are brittle and sensitive to thermal gradients during cutting. Their machinability is generally similar to that of thermoplastics.

Because of the fibers present, reinforced plastics are very abrasive and are difficult to machine. Fiber tearing and pulling is a significant problem. Furthermore, composites require careful removal of machining debris to avoid contact with and inhaling of fibers.

20.9.4 Thermally assisted machining

Metals and alloys that are difficult to machine at room temperature can be machined more easily at elevated temperatures, thus lowering cutting forces and increasing tool life. In **thermally assisted machining** (*hot machining*), the source of heat is a torch, high-energy beam (such as laser or electron beam), or plasma arc, focused to an area just ahead of the cutting tool (Fig. 20.28). Most applications in hot machining are in turning. Heating and maintaining a uniform temperature distribution within the workpiece may be difficult to control. Except in isolated cases, thermally assisted machining offers no significant advantage over machining at room temperature with the use of appropriate cutting tools and fluids.

SUMMARY

Cutting processes are among the most important of manufacturing operations. They are often necessary in order to impart the desired surface finish and dimensional accuracy to components, particularly those with complex shapes that cannot be produced economically or properly by other techniques.

A large number of variables have significant influence on the mechanics of chip formation in cutting operations. Commonly observed chip types are continuous, built-up edge, discontinuous, and segmented. Among important process variables are tool shape and material, cutting conditions such as speed, feed, and depth of cut, use of cutting fluids, and the characteristics of the machine tool, as well as the characteristics of the workpiece material. Parameters influenced by these variables are forces and power consumption, tool wear, surface finish and integrity, temperature, and dimensional accuracy of the workpiece. Machinability of materials depends not only on their intrinsic properties, but also on proper selection and control of process variables.

SUMMARY TABLE 20.1

FACTORS INFLUENCING CUTTING PROCESSES

PARAMETER	INFLUENCE AND INTERRELATIONSHIP
Cutting speed, depth of cut, feed, cutting fluids	Forces, power, temperature rise, tool life, type of chip, surface finish.
Tool angles	As above; influence on chip flow direction; resistance to tool chipping.
Continuous chip	Good surface finish; steady cutting forces; undesirable in automated machinery.
Built-up edge chip	Poor surface finish; thin stable edge can protect tool surfaces.
Discontinuous chip	Desirable for ease of chip disposal; fluctuating cutting forces; can affect surface finish and cause vibration and chatter.
Temperature rise	Influences tool life, particularly crater wear, and dimensional accuracy of workpiece; may cause thermal damage to workpiece surface.
Tool wear	Influences surface finish, dimensional accuracy, temperature rise, forces and power.
Machinability	Related to tool life, surface finish, forces and power.

TRENDS

- Studies of cutting processes are continuing, particularly for new metallic and nonmetallic materials, as well as engineered materials, to find better ways of machining.
- Because of their importance in automated manufacturing and in planning tool changes, reliable tool-life-testing techniques and accurate prediction of tool life continue to be investigated.
- On-line tool-wear sensing techniques and devices for computer-controlled machine tools are being developed.
- Control of chip flow and removal has become a significant problem, particularly in high-production machining.

KEY TERMS

Built-up edge
Chip breaker
Chip curl
Chip
Clearance angle
Continuous chip
Crater wear
Cutting force
Cutting ratio
Depth-of-cut line
Discontinuous chip
Feed marks
Flank wear
Friction angle
Inclination angle
Machinability
Power
Primary shear zone
Rake angle
Relief angle
Secondary shear zone
Serrated chip
Shear angle
Shear strain
Surface finish
Surface integrity
Thermally assisted machining
Thrust force
Tool-life curves
Wear land

BIBLIOGRAPHY

Armarego, E.J.A., and R.H. Brown, *The Machining of Metals.* Englewood Cliffs, N.J.: Prentice-Hall, 1969.

Boothroyd, G., *Fundamentals of Metal Machining and Machine Tools.* Washington, D. C.: Scripta/McGraw-Hill, 1975.

——, and W.A. Knight, *Fundamentals of Machining and Machine Tools*, 2d ed. New York: Marcel Dekker, 1989.

Machinability Testing and Utilization of Machining Data. Metals Park, Ohio: American Society for Metals, 1979.

Metals Handbook, 8th ed., *Vol. 3: Machining.* Metals Park, Ohio: American Society for Metals, 1967.

Mills, B., and A.H. Redford, *Machinability of Engineering Materials.* London: Applied Science Publishers, 1983.

Oxley, P.L.B., *Mechanics of Machining—An Analytical Approach to Assessing Machinability.* New York: Wiley, 1989.

Shaw, M.C., *Metal Cutting Principles.* New York: Oxford, 1984.

Tool and Manufacturing Engineers Handbook, 4th ed., *Vol. 1: Machining.* Dearborn, Mich.: Society of Manufacturing Engineers, 1983.

Trent, E.M., *Metal Cutting*, 2d ed. London: Butterworths, 1984.

Venkatesh, V.C., and H. Chandrasekaran, *Experimental Techniques in Metal Cutting*, rev. ed. New Delhi: Prentice-Hall, 1987.

REVIEW QUESTIONS

see notes

20.1 List the (a) independent variables and (b) dependent variables in cutting. Why are they so named?

20.2 Differentiate between positive and negative rake angles.

20.3 What variables does shear strain depend on? Why?

20.4 Explain the difference between discontinuous chips and segmented chips.

20.5 Why are continuous chips not always desirable?

20.6 Is there any advantage in having a built-up edge? Explain.

20.7 Name the factors that contribute to the formation of discontinuous chips.

20.8 What is the function of chip breakers?

20.9 Why is a built-up edge hard?

20.10 Identify all the forces in a cutting operation. Which force contributes to the power required?

20.11 Why does temperature depend on cutting speed and feed?

20.12 Identify all the angles on a single-point cutting tool.

20.13 Explain the features of different kinds of tool wear.

20.14 Why is it not always advisable to increase the cutting speed in order to increase production rate?

20.15 What are the consequences of tools chipping?

20.16 Name the techniques used for measuring tool wear. Describe their advantages and limitations.

20.17 Are the locations of maximum temperature and crater wear related? If so, why?

20.18 List the factors that contribute to poor surface finish in cutting.

20.19 What effects does a dull tool have in cutting?

20.20 Explain the term machinability and what it involves. Why has titanium poor machinability?

QUESTIONS AND PROBLEMS

20.21 Describe briefly your understanding of the need to study the mechanics of cutting processes in detail.

20.22 Show that for the same shear angle, there are two rake angles that give the same cutting ratio.

20.23 Do you agree with the following statement? If the cutting speed, the shear angle, and the rake angle are known, the chip velocity up the face of the tool can be calculated. Explain.

20.24 Explain why studying the types of chips produced is important in understanding cutting operations.

20.25 The shear strains that a material undergoes in metal cutting without failure are higher than those calculated from material properties and behavior that you studied in Chapter 2. Offer an explanation.

20.26 Why do you think the maximum temperature in orthogonal cutting is located at about the middle of the tool–chip interface? Remember that the two sources of heat are shearing in the primary shear plane and friction at the tool–chip interface.

20.27 What are the effects of lowering the friction at the tool–chip interface, say with a lubricant, on the mechanics of cutting operations?

20.28 Tool life can be almost infinite at low cutting speeds. Would you recommend that all machining be done at low speeds? Explain any limitations on doing so.

20.29 What are the consequences of allowing temperatures to rise to high levels in cutting?

20.30 The cutting force increases with depth of cut and decreasing rake angle. Why?

20.31 How would you expect the cutting force to vary in the case of serrated chip formation?

20.32 Wood is a highly anisotropic material. Explain the effects of cutting wood at different angles to the grain direction on chip formation.

20.33 Describe the advantages of oblique cutting.

20.34 Using the Taylor equation for tool wear, let $n = 0.5$ and $C = 400$. What is the percent increase in tool life if the cutting speed is reduced by (a) 20% and (b) 50%?

20.35 What are the effects of performing a cutting operation with a dull tool?

20.36 To what factors do you attribute the difference in the specific energies involved in machining the materials shown in Table 20.1? Why is there a range of energies for each group of material?

20.37 Assume that in orthogonal cutting the rake angle is 10° and the coefficient of friction is 0.5. Using Eq. (20.3), determine the percentage increase in chip thickness when the friction is doubled.

20.38 Derive Eq. (20.12).

20.39 Taking carbide as an example and using Eq. (20.19), determine how much the feed should be reduced in order to keep the mean temperature constant when the cutting speed is doubled.

20.40 Explain why it is possible to remove more material between tool resharpenings by lowering the cutting speed.

20.41 With appropriate diagrams show how the use of a cutting fluid can change the magnitude of the thrust force F_t.

20.42 An 8-in. diameter stainless steel bar is being turned on a lathe at 500 rpm and a depth of cut $d = 0.1$ in. If the power of the motor is 2 hp, what is the maximum feed (in./rev) that you can have before the motor stalls? (See Fig. 20.13.)

20.43 Using trigonometric relationships, obtain an expression for the ratio of the shear energy to frictional energy in orthogonal cutting, in terms of the angles α, β, and ϕ only.

20.44 An orthogonal cutting operation is being carried out under the following conditions: $t_o = 0.1$ mm, $t_c = 0.2$ mm, width of cut = 5 mm, $V = 2$ m/s, rake angle = 10°, $F_c = 500$ N, and $F_t = 200$ N. Calculate the percentage of the total energy that is dissipated in the shear plane.

20.45 Determine the C and n values for the four tool materials shown in Fig. 20.19.

20.46 Noting that the dimension d in Fig. 20.2(a) is very small, being on the order of less than 0.01 mm, explain why the shear strain rate $\dot{\gamma}$ in metal cutting is so high. (See also Table 2.4.)

20.47 Explain the significance of Eq. (20.7).

20.48 The data in the table below are obtained in orthogonal cutting of AISI 4130 steel using a high-speed-steel tool, at a cutting speed $V = 90$ ft/min, depth of cut $t_o = 0.0025$ in., and width of cut = 0.475 in. Calculate the missing quantities in the table. Describe your observations concerning variations of forces and energies with increasing rake angle.

α	ϕ	γ	μ	β	F_c, lb	F_t, lb	u_t, $\frac{\text{in.-lb}}{\text{in}^3}$	u_s	u_f	$\frac{u_f}{u_t}$, %
25°	20.9°		1.46		380	224				
35	31.6		1.53		254	102				
40	35.7		1.54		232	71				
45	41.9		1.83		232	68				

20.49 We note in Eq. (20.19) that cutting speed V has a greater influence on mean temperature than the feed f. Why?

20.50 Comment on the location of various phenomena indicated in the plots in Fig. 20.23.

20.51 In Fig. 20.23, we note that all curves have a downward trend as the feed increases. Why?

20.52 Describe the problems involved, if any, in utilizing the thermally assisted machining process (Fig. 20.28).

20.53 Explain the uses of the equations developed in Sections 20.2 and 20.5 in actual production.

20.54 Would you have anticipated the general temperature distribution in metal cutting, as shown in Fig. 20.14? Explain.

20.55 Design an experiment whereby orthogonal cutting can be simulated in a turning operation on a lathe.

20.56 Describe the consequences of exceeding the allowable wear land (Table 20.3) for cutting tool materials.

21

Cutting-Tool Materials and Cutting Fluids

21.1 Introduction

Cutting-tool materials and their proper selection are among the most important factors in machining operations, as are mold and die materials for forming and shaping processes. We noted in Chapter 20 that the tool is subjected to high temperatures, contact stresses, and rubbing on the workpiece surface, as well as by the chip climbing up the rake face of the tool. Consequently, a cutting tool must have certain characteristics in order to produce good-quality and economical parts. These characteristics are:

- *Hardness*, particularly at elevated temperatures (hot hardness), so that the hardness and strength of the tool are maintained at the temperatures encountered in cutting operations (Fig. 21.1).
- *Toughness*, so that impact forces on the tool in interrupted cutting operations, such as milling or turning a splined shaft, do not chip or fracture the tool.

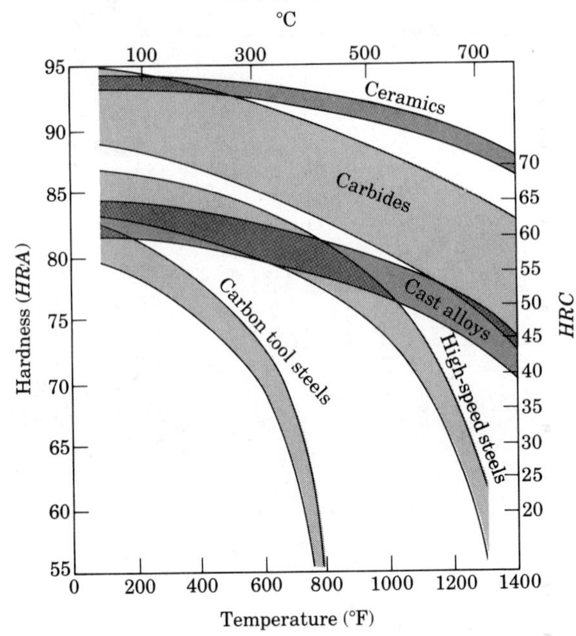

FIGURE 21.1 Hardness of various cutting-tool materials as a function of temperature (hot hardness). The wide range in each group of materials results from the variety of tool compositions and treatments available for that group.

- *Wear resistance*, so that an acceptable tool life is obtained before the tool is resharpened or replaced.
- *Chemical stability* or *inertness* with respect to the workpiece material, so that any adverse reactions contributing to tool wear are avoided.

Various cutting-tool materials having a wide range of properties are available (Tables 21.1 and 21.2 on pp. 639 and 640). Tool materials are usually divided into the following general categories, which are listed in the approximate chronological order in which they were developed:

1. Carbon and medium-alloy steels
2. High-speed steels
3. Cast-cobalt alloys
4. Carbides
5. Coated tools
6. Ceramics
7. Cubic boron nitride
8. Silicon nitride
9. Diamond

TABLE 21.1
TYPICAL PROPERTIES OF TOOL MATERIALS

PROPERTY	HIGH-SPEED STEELS	CAST ALLOYS	CARBIDES		CERAMICS	CUBIC BORON NITRIDE	DIAMOND*
			WC	TiC			
Hardness	83–86 HRA	82–84 HRA 46–62 HRC	90–95 HRA 1800–2400 HK	91–93 HRA 1800–3200 HK	91–95 HRA 2000–3000 HK	4000–5000 HK	7000–8000 HK
Compressive strength							
MPa	4100–4500	1500–2300	4100–5850	3100–3850	2750–4500	6900	6900
psi × 10^3	600–650	220–335	600–850	450–560	400–650	1000	1000
Transverse rupture strength							
MPa	2400–4800	1380–2050	1050–2600	1380–1900	345–950	700	1350
psi × 10^3	350–700	200–300	150–375	200–275	50–135	105	200
Impact strength							
J	1.35–8	0.34–1.25	0.34–1.35	0.79–1.24	<0.1	<0.5	<0.2
in.-lb	12–70	3–11	3–12	7–11	<1	<5	<2
Modulus of elasticity							
GPa	200	—	520–690	310–450	310–410	850	820–1050
psi × 10^6	30	—	75–100	45–65	45–60	125	120–150
Density							
kg/m^3	8600	8000–8700	10,000–15,000	5500–5800	4000–4500	3500	3500
lb/in^3	0.31	0.29–0.31	0.36–0.54	0.2–0.22	0.14–0.16	0.13	0.13
Volume of hard phase, %	7–15	10–20	70–90	—	100	95	95
Melting or decomposition temperature							
°C	1300	—	1400	1400	2000	1300	700
°F	2370	—	2550	2550	3600	2400	1300
Thermal conductivity, W/m. K	30–50	—	42–125	17	29	13	500–2000
Coefficient of thermal expansion, $\times 10^{-6}$/°C	12	—	4–6.5	7.5–9	6–8.5	4.8	1.5–4.8

* Single crystal. The values for polycrystalline diamond are generally lower, except impact strength, which is higher.

TABLE 21.2

GENERAL CHARACTERISTICS OF CUTTING-TOOL MATERIALS. THESE TOOL MATERIALS HAVE A WIDE RANGE OF COMPOSITIONS AND PROPERTIES. THUS OVERLAPPING CHARACTERISTICS EXIST IN MANY CATEGORIES OF TOOL MATERIALS.

	CARBON AND LOW- TO MEDIUM-ALLOY STEELS	HIGH-SPEED STEELS	CAST-COBALT ALLOYS	CEMENTED CARBIDES	COATED CARBIDES	CERAMICS	POLYCRYSTALLINE CUBIC BORON NITRIDE	DIAMOND
Hot hardness			Increasing →					
Toughness			← Increasing					
Impact strength			← Increasing					
Wear resistance			Increasing →					
Chipping resistance			← Increasing					
Cutting speed			Increasing →					
Depth of cut	Light to medium	Light to heavy	Light to heavy	Light to heavy	Light to heavy	Light to heavy	Light to heavy	Very light for single-crystal diamond
Finish obtainable	Rough	Rough	Rough	Good	Good	Very good	Very good	Excellent
Method of processing	Wrought	Wrought, cast, HIP* sintering	Cast and HIP sintering	Cold pressing and sintering	CVD†	Cold pressing and sintering or HIP sintering	High-pressure, high-termperature sintering	High-pressure, high-temperature sintering
Fabrication	Machining and grinding	Machining and grinding	Grinding	Grinding		Grinding	Grinding and polishing	Grinding and polishing
Thermal-shock resistance				← Increasing				
Tool material cost			Increasing →					

* Hot-isostatic pressing.

† Chemical-vapor deposition.

Source: Komanduri, R., *Kirk–Othmer Encyclopedia of Chemical Technology,* 3d ed. New York: Wiley, 1978.

Note that—as we described in Parts II and III—many of these materials are also used for dies and molds. In this chapter, we present the characteristics, applications, and limitations of these tool materials in machining operations. We discuss characteristics such as hot hardness, toughness, impact strength, wear resistance, thermal shock resistance, and costs, as well as the range of cutting speeds and depth of cut for optimum performance. We show that proper selection of tools is a critical factor in the quality of surfaces produced and the economics of machining (see Section 24.5).

21.2 Carbon and Medium-Alloy Steels

Carbon steels are the oldest of tool materials and have been used widely for drills, taps, broaches, and reamers since the 1880s. Low-alloy and medium-alloy steels were developed later for similar applications but with longer tool life. Although inexpensive and easily shaped and sharpened, these steels do not have sufficient hot hardness and wear resistance for cutting at high speeds where, as you have seen, the temperature rises significantly. Note in Fig. 21.1, for example, how rapidly the hardness of carbon steels decreases as the temperature increases. Consequently, the use of these steels is limited to low-speed cutting operations.

21.3 High-Speed Steels

High-speed steel (HSS) tools are so named because they were developed to cut at high speeds. First produced in the early 1900s, **high-speed steels** are the most highly alloyed of the tool steels. They can be hardened to various depths, have good wear resistance, and are relatively inexpensive. Because of their high toughness and resistance to fracture, high-speed steels are especially suitable for high positive-rake-angle tools (small included angle) and for machine tools with low stiffness that are subject to vibration and chatter.

There are two basic types of high-speed steels: *molybdenum* (M series) and *tungsten* (T series). The M series contains up to about 10 percent molybdenum, with chromium, vanadium, tungsten, and cobalt as alloying elements. The T series contains 12–18 percent tungsten, with chromium, vanadium, and cobalt as alloying elements. The M series generally has higher abrasion resistance than the T series, undergoes less distortion during heat treating, and is less expensive. Consequently,

95 percent of all high-speed steel tools produced in the United States are made of M-series steels.

High-speed steel tools are available in wrought, cast, and sintered (powder-metallurgy) forms. They can be coated for improved performance (see Section 21.6). High-speed steels account for the largest tonnage of tool materials used today, followed by various die steels and carbides. They are used in a wide variety of cutting operations requiring complex tool shapes such as drills, reamers, taps, and gear cutters (Table 21.3).

21.4 Cast-Cobalt Alloys

Introduced in 1915, **cast-cobalt alloys** have the following ranges of composition: 38–53 percent cobalt, 30–33 percent chromium, and 10–20 percent tungsten. Because of their high hardness, typically 58–64 HRC, they have good wear resistance and maintain their hardness at elevated temperatures.

Commonly known as *Stellite* tools, these alloys are cast and ground into relatively simple tool shapes. However, they are not as tough as high-speed steels and are sensitive to impact forces. Consequently, they are less suitable than high-speed steels for interrupted cutting operations. These tools are now used only for special applications that involve deep, continuous roughing operations at relatively high feeds and speeds—as much as twice the rates possible with high-speed steels.

21.5 Carbides

The three groups of tool materials we have just described (alloy steels, high-speed steels, and cast alloys) have the necessary toughness, impact strength, and thermal shock resistance but have important limitations, such as strength and hardness, particularly hot hardness. Consequently, they cannot be used as effectively where high cutting speeds, hence high temperatures, are involved and thus tool life can be short.

To meet the challenge of higher speeds for higher production rates, **carbides** (also known as *cemented* or *sintered carbides*) were introduced in the 1930s. Because of their high hardness over a wide range of temperatures (see Fig. 21.1), high elastic modulus and thermal conductivity, and low thermal expansion, carbides are among the most important tool and die materials. The two basic

TABLE 21.3
OPERATING CHARACTERISTICS OF CUTTING-TOOL MATERIALS

TOOL MATERIALS	MACHINING OPERATION AND (CUTTING-SPEED RANGE)	MODES OF TOOL WEAR OR FAILURE	LIMITATIONS
Carbon steels	Tapping, drilling, reaming (low speed)	Buildup, plastic deformation, abrasive wear, microchipping	Low hot hardness, limited hardenability, and limited wear resistance
Low/medium alloy steels	Tapping, drilling, reaming (low speed)	Buildup, plastic deformation, abrasive wear, microchipping	Low hot hardness, limited hardenability, and limited wear resistance
High-speed steels	Turning, drilling, milling, broaching (medium speed)	Flank wear, crater wear	Low hot hardness, limited hardenability, and limited wear resistance
Cemented carbides	Turning, drilling, milling, broaching (medium speed)	Flank wear, crater wear	Cannot use at low speed because of cold welding of chips and microchipping
Coated carbides	Turning (medium to high speed)	Flank wear, crater wear	Cannot use at low speed because of cold welding of chips and microchipping
Ceramics	Turning (high speed to very high speed)	Depth-of-cut line notching, microchipping, gross fracture	Low strength, low thermomechanical fatigue strength
Cubic boron nitride	Turning, milling (medium to high speed)	Depth-of-cut line notching, chipping, oxidation, graphitization	Low strength, low chemical stability at higher temperature
Diamond	Turning, milling (high to very high speed)	Chipping, oxidation, graphitization	Low strength, low chemical stability at higher temperature

Source: Komanduri, R., *Kirk–Othmer Encyclopedia of Chemical Technology,* 3d ed. New York: Wiley, 1978.

TABLE 21.4
CLASSIFICATION OF TUNGSTEN CARBIDES ACCORDING TO MACHINING APPLICATION

CLASSIFICATION NUMBER	MATERIALS TO BE MACHINED	MACHINING OPERATION	TYPE OF CARBIDE	CHARACTERISTICS OF		TYPICAL PROPERTIES	
				CUT	*CARBIDE*	*HARDNESS (HRA)*	*TRANSVERSE RUPTURE STRENGTH (MPa)*
C-1	Cast iron, nonferrous metals, and nonmetallic materials requiring abrasion resistance	Roughing cuts	Wear-resistant grades; generally straight WC–Co with varying grain sizes	Increasing cutting speed	Increasing hardness and wear resistance	89.0	2,400
C-2		General purpose				92.0	1,725
C-3		Finishing				92.5	1,400
C-4		Precision boring and fine finishing		Increasing feed rate	Increasing strength and binder content	93.5	1,200
C-5	Steels and steel alloys requiring crater and deformation resistance	Roughing cuts	Crater-resistant grades; various WC–Co compositions with TiC and/or TaC alloys	Increasing cutting speed	Increasing hardness and wear resistance	91.0	2,070
C-6		General purpose				92.0	1,725
C-7		Finishing				93.0	1,380
C-8		Precision boring and fine finishing		Increasing feed rate	Increasing strength and binder content	94.0	1,035

groups of carbides used for machining operations are tungsten carbide and titanium carbide.

21.5.1 Tungsten carbide

Tungsten carbide (WC) is generally used for cutting nonferrous abrasive materials and cast irons. It is a composite material, consisting of tungsten-carbide particles bonded together in a cobalt matrix, hence the term *cemented* carbides. These tools are manufactured by powder-metallurgy techniques, in which WC powders are crushed together with cobalt in a ball mill, with the cobalt coating the WC particles. These particles, which are 1–5 μm (40–200 μin.) in size, are then pressed and sintered into the desired insert shapes (see Chapter 17).

The amount of cobalt significantly affects the properties of carbide tools. As the cobalt content increases, the strength, hardness, and wear resistance of WC decrease, while its toughness increases because of the higher toughness of cobalt. To improve hot hardness and crater-wear resistance, WC may be compounded with carbides of titanium and tantalum. These carbides can then be used for machining steels.

Table 21.4 gives typical applications for tungsten-carbide cutting tools according to the C-system used in the United States. The ISO (International Standards Organization) system of classification uses the symbols P, M, and K to identify various carbides (Table 21.5).

21.5.2 Titanium carbide

Titanium carbide (TiC) has higher wear resistance than tungsten-carbide but is not as tough. With a nickel–molybdenum alloy as the matrix, TiC is suitable for machining hard materials, mainly steels and cast irons, and for cutting at speeds higher than those for tungsten carbide (Fig. 21.2).

TABLE 21.5
ISO CLASSIFICATION OF CARBIDE CUTTING TOOLS ACCORDING TO USE

SYMBOL	WORKPIECE MATERIAL	COLOR	DESIGNATION IN INCREASING ORDER OF WEAR RESISTANCE AND DECREASING ORDER OF TOUGHNESS IN EACH CATEGORY
P	Ferrous metals with long chips	Blue	P01, P10, P20, P30, P40, P50
M	Ferrous metals with long or short chips; nonferrous metals	Yellow	M10, M20, M30, M40
K	Ferrous metals with short chips; nonferrous metals; nonmetallic materials	Red	K01, K10, K20, K30, K40

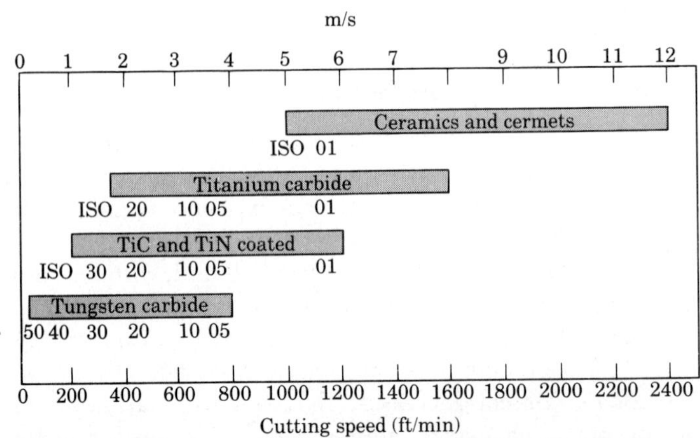

FIGURE 21.2
Approximate cutting-speed ranges for optimum use of various cutting-tool materials. Increasing the cutting speed has a significant economic impact in machining operations. The various grades of tools are indicated by the ISO numbers.

21.5.3 Inserts

The more traditional cutting tools are made of carbon steels and high-speed steels. These tools are formed in one piece and ground to various shapes (see Fig. 20.10). Other such tools include drills and milling cutters. After the cutting edge wears, the tool has to be removed from its holder and reground. Although a supply of sharp or resharpened tools is usually available from tool rooms, tool-changing operations are not efficient. The need for a more effective method has led to the development of **inserts**, which are individual cutting tools with a number of cutting points (Fig. 21.3).

FIGURE 21.3
Typical carbide inserts with various shapes and chip-breaker features. The holes in the inserts are standardized for interchangeability. *Source:* Courtesy of Kyocera Engineered Ceramics, Inc., and *Manufacturing Engineering Magazine,* Society of Manufacturing Engineers.

Inserts are usually clamped on the tool *shank* with various locking mechanisms (Figs. 21.4a–c), or they may be *brazed* to the tool shank (Fig. 21.4d). However, because of the difference in thermal expansion between the insert and the tool-shank materials, brazing must be done carefully to avoid cracking or warping. Clamping is the preferred method because each insert has a number of cutting edges, and after one edge is worn, it is *indexed* (rotated in its holder) for another cutting edge.

Carbide inserts are available in a variety of shapes, such as square, triangle, diamond, and round, with or without *chip-breaker* features for chip-flow control. The strength of the cutting edge of an insert depends on its shape. The smaller the angle (Fig. 21.5), the less strength the edge has. In order to further improve edge strength and prevent chipping, all insert edges are usually honed, chamfered, or produced with a negative land (Fig. 21.6). Most inserts are honed to a radius of about 0.025 mm (0.001 in.).

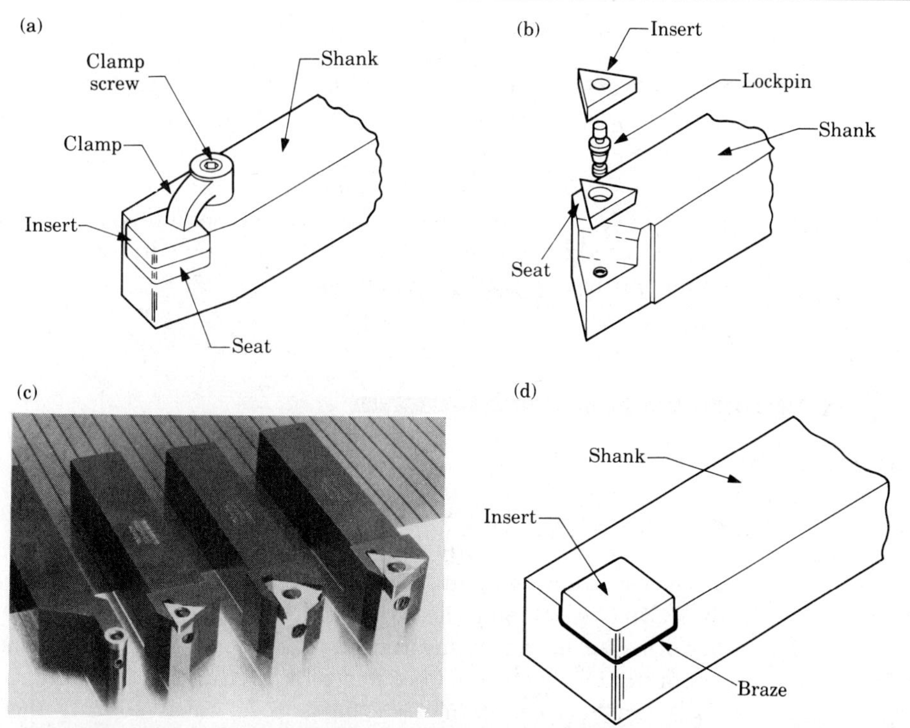

FIGURE 21.4 Methods of attaching inserts to tool shank: (a) clamping; and (b) wing lockpins. (c) Examples of inserts attached to tool shank with threadless lockpins, which are secured with side screws. *Source:* Courtesy of GTE Valenite. (d) Brazed insert on a tool shank.

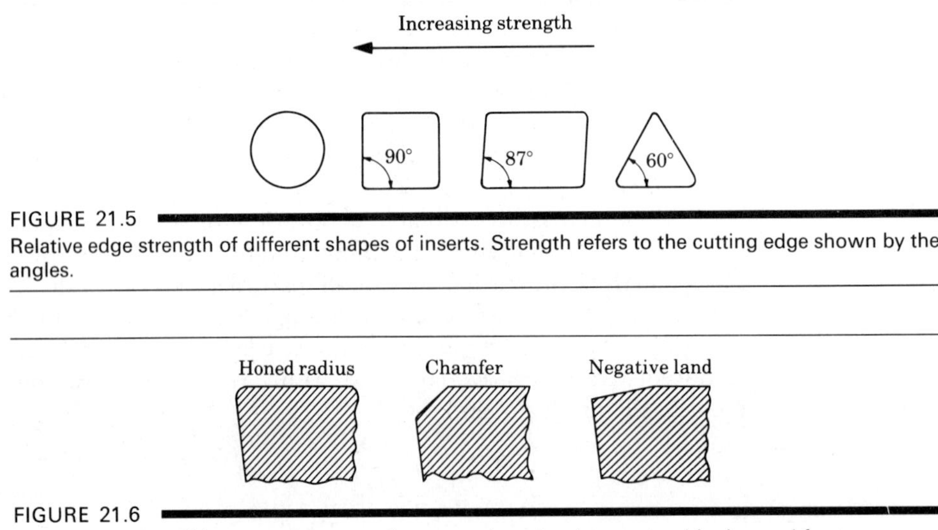

FIGURE 21.5
Relative edge strength of different shapes of inserts. Strength refers to the cutting edge shown by the angles.

FIGURE 21.6
Edge preparation of inserts to improve edge strength and resistance to chipping and fracture.

Stiffness of the machine tool is of major importance in using carbide tools (see Chapter 24). Light feeds, low speeds, and chatter are detrimental because they tend to damage the tool's cutting edge. Light feeds, for example, concentrate the forces and temperature closer to the edges of the tools—hence the tendency for the edges to chip off. Low cutting speeds tend to encourage cold welding of the chip to the tool. Cutting fluids are generally not needed, but if used to minimize the heating and cooling of the tool in interrupted cutting operations, they should be applied continuously and in large quantities.

21.6 Coated Tools

As we described in Part I, new alloys and engineered materials have been developed continuously, particularly since the 1960s. These materials have high strength but are generally abrasive and are highly reactive chemically with tool materials. The difficulty of machining these materials efficiently—and the need for improving the performance in machining the more common engineering materials—has led to important developments in **coated tools**. Because of their unique properties, coated tools can be used at high cutting speeds, thus reducing the time required for machining operations and costs. Figure 21.7 shows that cutting time has been reduced by more than 100 times since 1900. Since the late 1950s alone, coated tools have reduced cutting time some 4 times. Coated tools are now being used for cutting operations, with tool life as much as 10 times that of uncoated tools.

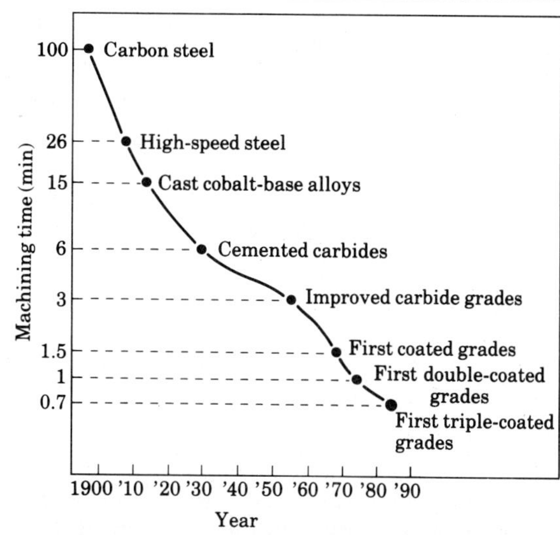

FIGURE 21.7 Relative time required to machine with various cutting-tool materials, indicating the year the tool materials were introduced. *Source:* Sandvik Coromant.

21.6.1 Coating materials

Coating materials commonly used are titanium nitride, titanium carbide, and ceramics. Other materials, such as hafnium nitride, are being investigated. Coatings on tools, generally 5–10 μm (200–400 μin.) in thickness, are applied by various techniques (see Chapter 33). Honing is an important procedure used to maintain the strength of the coating along the edges of the tool. Otherwise the coating may chip off at sharp edges.

Coatings for cutting tools, as well as for dies, should have the following general characteristics:

a) High hardness at elevated temperatures.
b) Chemical stability and inertness to the workpiece material.
c) Low thermal conductivity.
d) Good bonding to the substrate to prevent flaking or spalling.
e) Little or no porosity.

The effectiveness of coatings, in turn, are enhanced by hardness, toughness, and high thermal conductivity of the substrate.

Titanium nitride. Titanium-nitride (TiN) coatings have low coefficient of friction, high hardness, resistance to high temperature, and good adhesion to the substrate. Consequently, they greatly improve the life of high-speed-steel tools, as well as the lives of carbide tools, drills, and cutters. Titanium-nitride coated tools, which are gold in color, perform well at higher cutting speeds and feeds.

Flank wear is significantly lower than for uncoated tools (Fig. 21.8), and flank surfaces can be reground after use, since regrinding does not remove the coating on the rake face of the tool. However, coated tools do not perform as well at low cutting speeds because the coating can be worn off by chip adhesion. Hence the use of appropriate cutting fluids to discourage adhesion is important.

Titanium carbide. Titanium-carbide (TiC) coatings on tungsten-carbide inserts have high flank-wear resistance in machining abrasive materials.

Ceramics. Their resistance to high temperature, chemical inertness, low thermal conductivity, and resistance to flank and crater wear make ceramics suitable coatings for tools. The most commonly used ceramic coating is aluminum oxide (Al_2O_3). However, because they are very stable (not chemically reactive), oxide coatings generally bond weakly with the substrate.

Multiple coatings. The desirable properties of the coatings just described can be combined and thus optimized in multiple coatings. Tungsten-carbide tools are now available with two or three layers of such coatings and are particularly effective in machining cast irons and steels.

The first layer over the substrate is TiC, followed by Al_2O_3, and then TiN. The first layer should bond well with the substrate; the outer layer should resist wear and have low thermal conductivity; the intermediate layer should bond well and be compatible with both layers. Typical applications of multiple-coated tools are:

a) High-speed, continuous cutting: TiC/Al_2O_3.
b) Heavy-duty, continuous cutting: TiC/Al_2O_3/TiN.
c) Light, interrupted cutting: TiC/TiC + TiN/TiN.

A new development in coatings is *alternating multiple layers*, with layers that are thinner than in ordinary multiple-layer tools (Fig. 21.9). The reason for using thinner coatings is that coating hardness increases with decreasing grain size, a

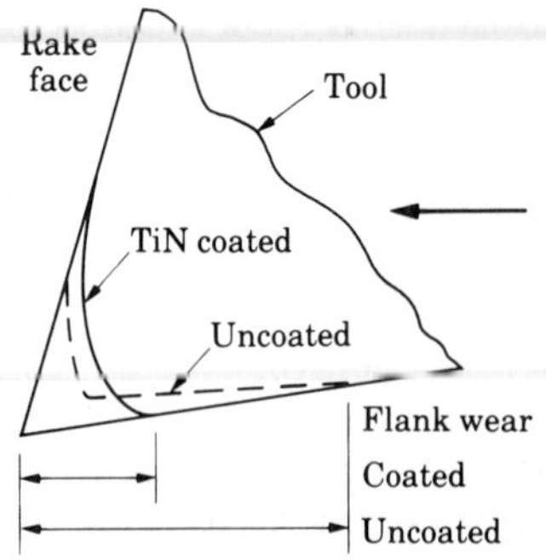

FIGURE 21.8
Wear patterns on high-speed-steel uncoated and titanium-nitride coated tools. Note that flank wear is lower for the coated tool.

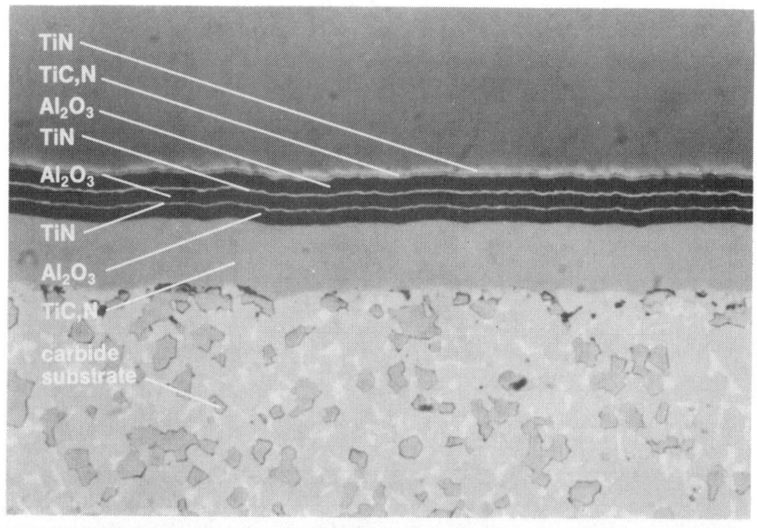

FIGURE 21.9
Multilayer coatings on a tungsten-carbide substrate. Three alternating layers of aluminum oxide are separated by very thin layers of titanium nitride. Inserts with as many as thirteen layers of coatings have been made. *Source:* Courtesy of Kennametal Inc., and *Manufacturing Engineering Magazine,* Society of Manufacturing Engineers.

phenomenon that is similar to the increase in strength of metals with decreasing grain size. Thus thinner layers are harder than thicker layers. The thickness of the layers shown in Fig. 21.9 are in the range of 1 μm to 6 μm.

Diamond coatings. We describe the properties and applications of diamond and diamond coatings in Sections 8.7 and 33.13, respectively, and we discuss their use as cutting tools in Section 21.10. A recent development concerns the use of diamond powder as coatings for cutting tools, particularly tungsten carbide and silicon nitride inserts. Increases in tool life of as much as tenfold have been reported, although problems exist regarding adherence of the diamond film to the substrate and the difference in thermal expansion between diamond and substrate materials.

21.7 Ceramics

These tool materials, introduced in the early 1950s, consist primarily of fine-grained, high-purity aluminum oxide. They are cold pressed under high pressure, sintered at high temperature, and called *white*, or *cold-pressed*, ceramics. Additions of titanium carbide and zirconium oxide help improve properties such as toughness and thermal-shock resistance.

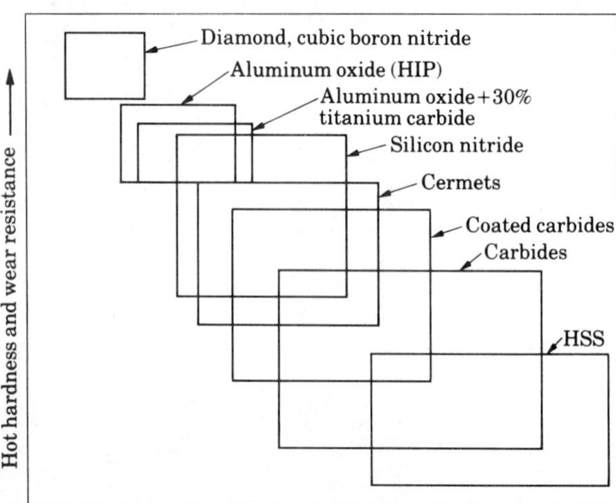

FIGURE 21.10 Ranges of properties for various groups of tool materials (see also various tables in this chapter).

Ceramic tools have very high abrasion resistance and hot hardness (Fig. 21.10). Chemically, they are more stable than high-speed steels and carbides. Thus they have less tendency to adhere to metals during cutting and hence less tendency to form a built-up edge. Consequently, good surface finish is obtained with ceramic tools in cutting cast irons and steels. However, ceramics lack toughness, resulting in premature tool failure by chipping or catastrophic failure (see Fig. 20.20).

Ceramic inserts are available in shapes similar to carbide inserts. They are effective in very high speed, uninterrupted cutting operations, such as finishing or semifinishing by turning. To reduce thermal shock, cutting should be performed either dry or with a copious amount of steady stream of cutting fluid. Improper or intermittent applications of fluid can cause thermal shock in the ceramic tool.

Ceramic tool shape and setup are important. Negative rake angles, hence large included angles, are generally preferred in order to avoid chipping. Tool failure can be reduced by increasing the stiffness and damping capacity of machine tools and mountings, thus reducing vibration and chatter.

Black, or *hot-pressed*, ceramics (carboxides) were introduced in the 1960s. They typically contain 70 percent aluminum oxide and 30 percent titanium carbide, and are also called *cermets* (from ceramic and metal). Other cermets contain molybdenum carbide, niobium carbide, and tantalum carbide (see Section 8.2.3). Their brittleness and high cost have been a problem when cermets are used as cutting tools. However, further refinements have resulted in improved strength, toughness, and reliability.

• Example: Tool materials and their *n* value in the Taylor tool-life equation.

In Table 20.2, we note that the *n* values for the cutting tools listed range from 0.08 to 0.7. To what factors can this difference be attributed?

SOLUTION. From the tool-life equation $T \propto 1/V^{1/n}$, we note that n is a measure of the sensitivity of the tool life to cutting speed. The smaller the value of n, the higher the sensitivity, and hence the shorter the tool life or the higher the wear rate. As we describe in Chapter 32, wear is related to hardness; the higher the hardness, the lower the wear. Thus, as a first approximation, tool life will increase with tool hardness. Furthermore, because temperature affects hardness, we note in Fig. 21.1 that, in decreasing order of hardness, we have ceramics, carbides, and high-speed steels.

We next consider the chemical stability or inertness of the tool materials. While ceramics have high and carbides have good chemical stability, high-speed steels can react with the workpiece material at elevated temperatures. Also, from Eq. (20.19) we note that temperature rise is higher with high-speed steels than with carbides, due partly to the higher friction and lower thermal conductivity of steels compared to carbides (see Table 21.1). Combining these observations, we can now present a general and relative rating of three tool materials in terms of their characteristics, as follows:

Material	Hot hardness	Inertness	Temperature rise
High-speed steels	Low	Low	High
Carbides	High	High	Low
Ceramics	Highest	Highest	Low

An inspection of this table indicates that the value of n for high-speed steels should be the lowest, and for ceramics the highest.

21.8 Cubic Boron Nitride

Next to diamond, **cubic boron nitride** (CBN) is the hardest material presently available. The CBN cutting tool was introduced in 1962. It is made by bonding a 0.5–1-mm (0.02–0.04-in.) layer of polycrystalline cubic boron nitride to a carbide substrate by sintering under pressure (Fig. 21.11). While the carbide provides shock resistance, the CBN layer provides very high wear resistance and cutting-edge strength (Fig. 21.12). Cubic boron nitride tools are also made in small sizes without a substrate. Because CBN tools are brittle, stiffness of the machine tool is important.

At elevated temperatures, CBN is chemically inert to iron and nickel and its resistance to oxidation is high. It is therefore particularly suitable for cutting hardened ferrous and high-temperature alloys. Cubic boron nitride is also used as an abrasive (see Chapter 25).

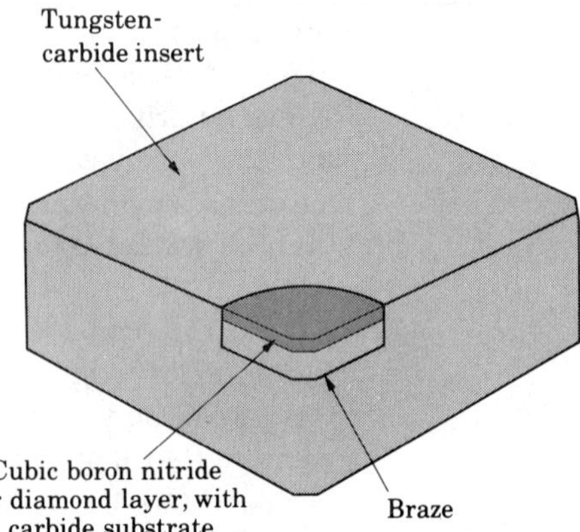

FIGURE 21.11
Construction of polycrystalline cubic boron nitride or diamond layer on a tungsten-carbide insert.

21.9 Silicon-Nitride Base Tools

Developed in the 1970s, silicon-nitride (SiN) base tool materials consist of silicon nitride with various additions of aluminum oxide, yttrium oxide, and titanium carbide. These tools have high toughness, hot hardness, and good thermal-shock resistance.

An example of an SiN-base material is **sialon**, so called after the elements silicon, aluminum, oxygen, and nitrogen in its composition. It has higher thermal-shock resistance than silicon nitride and is recommended for machining cast irons and nickel-base superalloys at intermediate cutting speeds. Because of chemical affinity, SiN-base tools are not suitable for machining steels.

FIGURE 21.12
Inserts with cubic boron nitride tips (top row) and solid CBN inserts (bottom row). *Source:* Courtesy of GTE Valenite.

New developments include composite ceramic tool materials. Consisting of silicon nitride reinforced with silicon-carbide whiskers, these materials have high fracture toughness and resistance to thermal shock.

21.10 Diamond

The hardest substance of all known materials is diamond. It has low friction, high wear resistance, and the ability to maintain a sharp cutting edge. It is used when good surface finish and dimensional accuracy are required, particularly with soft nonferrous alloys and abrasive nonmetallic materials. *Single-crystal* diamonds of various carats are used for special applications, such as machining copper-front surface mirrors.

Because diamond is brittle, tool shape is important. Low rake angles (large included angles) are normally used to provide a strong cutting edge. Special attention should be given to proper mounting and crystal orientation in order to obtain optimum tool use. Diamond wear may occur by microchipping, caused by thermal stresses and oxidation, and transformation to carbon, caused by the heat generated during cutting.

Single-crystal diamond tools have been largely replaced by **polycrystalline-diamond tools** (*compacts*), which are also used as wire-drawing dies for fine wire. These materials consist of very small synthetic crystals, fused by a high-pressure, high-temperature process to a thickness of about 0.5–1 mm (0.02–0.04 in.) and bonded to a carbide substrate, similar to CBN tools (see Fig. 21.11). The random orientation of the diamond crystals prevents the propagation of cracks though the structure, thus improving its toughness.

Diamond tools can be used satisfactorily at almost any speed, but are suitable mostly for light, uninterrupted finishing cuts. In order to minimize tool fracture, the diamond must be resharpened as soon as it becomes dull. Because of its strong chemical affinity, diamond is not recommended for machining plain-carbon steels and titanium, nickel, and cobalt-base alloys. Diamond is also used as an abrasive in grinding and polishing operations (see Chapter 25) and as coatings (see Sections 21.6 and 33.13).

21.11 Cutting-Tool Reconditioning

When tools, particularly high-speed steels, become worn, they are *reconditioned* (resharpened) for further use. They are usually ground on tool and cutter grinders in toolrooms having special fixtures. The reconditioning may be carried out either by hand, which requires considerable operator skill or on computer-controlled tool and cutter grinders. Other methods (see Chapter 26) may also be used to recondition tools and cutters. Reconditioning may also involve recoating used tools with titanium nitride.

Consistency and precision in reconditioning is important. Resharpened tools should be inspected for their shape and surface finish. As we described earlier, inserts are usually discarded after use. However, whether tools should be reconditioned or discarded depends on the relative costs involved. Skilled labor is costly, as are computer-controlled grinders. Thus an additional consideration is the possible recycling of tool materials, since many contain expensive materials of strategic importance, such as tungsten and cobalt.

21.12 Cutting Fluids

Also called lubricants and coolants, **cutting fluids** are used extensively in machining operations to:

- Reduce friction and wear, thus improving tool life and surface finish.
- Reduce forces and energy consumption.
- Cool the cutting zone, thus reducing workpiece temperature and distortion.
- Wash away the chips.
- Protect the newly machined surfaces from environmental corrosion.

A cutting fluid can interchangeably be a coolant and a lubricant. Its effectiveness in cutting operations depends on a number of factors, such as the method of application, temperature, cutting speed, and type of machining operation. As we have shown, temperature increases as cutting speed increases. Thus cooling of the cutting zone is of major importance at high cutting speeds. On the other hand, if the speed is low, such as in broaching or tapping, lubrication—not cooling—is the important factor. Lubrication reduces the tendency for built-up edge formation and thus improves surface finish.

The relative severity of various machining operations is shown qualitatively in Fig. 21.13, which also includes the relative cutting speeds employed. Severity is defined as the magnitude of temperatures and forces encountered, the tendency for built-up edge formation, and the ease with which chips are disposed of from the cutting zone. Note how important cutting-fluid effectiveness is as severity increases. Recommendations for cutting fluids for specific machining operation applications are presented in Chapters 22 and 23.

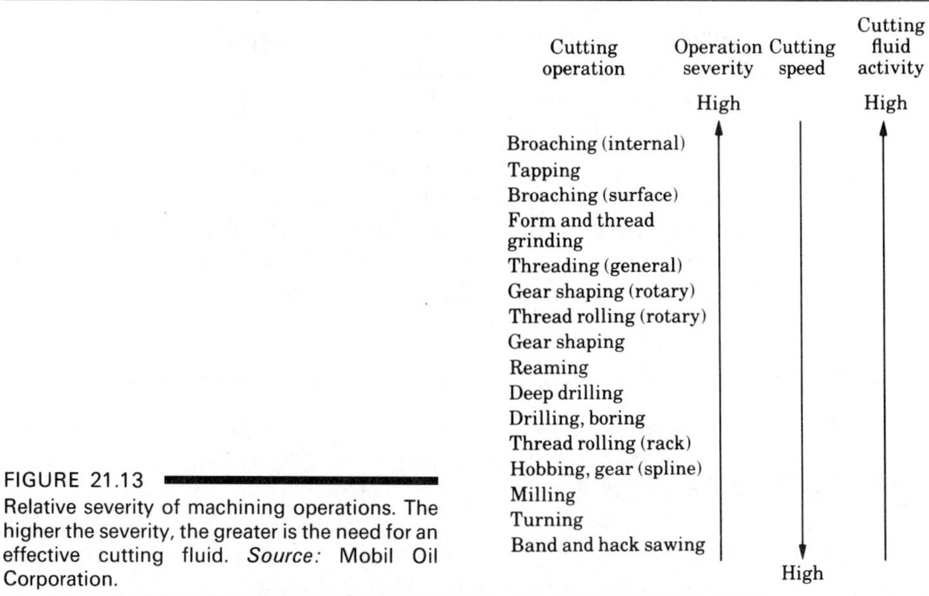

FIGURE 21.13 Relative severity of machining operations. The higher the severity, the greater is the need for an effective cutting fluid. *Source:* Mobil Oil Corporation.

21.12.1 Cutting-fluid action

Although we describe the basic lubrication mechanisms in metalworking operations in greater detail in Section 32.9, we need to briefly discuss here the mechanisms by which cutting fluids influence machining operations. In view of the high contact pressures and relative sliding at the tool–chip interface, how does a cutting fluid penetrate this interface to influence the cutting process?

It appears that the fluid is drawn into the tool–chip interface by the *capillary action* of the interlocking network of surface asperities (Fig. 21.14). Studies have

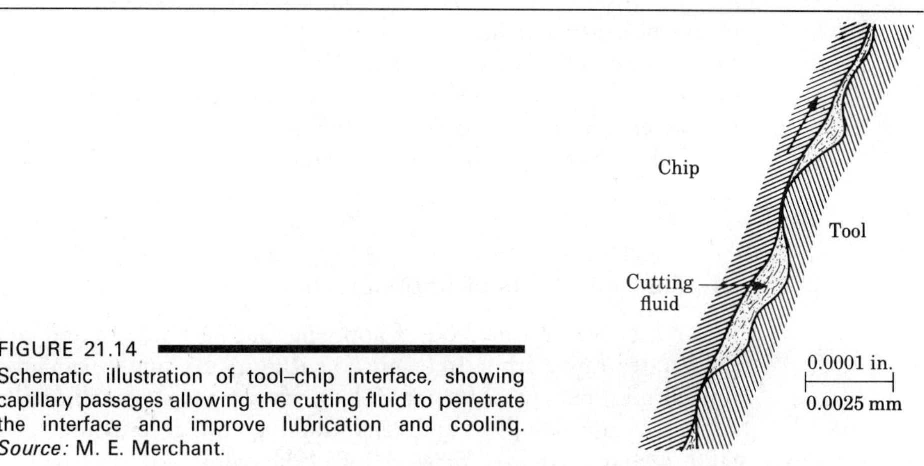

FIGURE 21.14 Schematic illustration of tool–chip interface, showing capillary passages allowing the cutting fluid to penetrate the interface and improve lubrication and cooling. *Source:* M. E. Merchant.

shown that the cutting fluid gains access to the interface by seeping from the sides of the chip. Because of the small size of this capillary network, the cutting fluid should have a small molecular size and proper wetting (surface tension) characteristics.

There are situations, however, in which the use of cutting fluids can be detrimental. In interrupted cutting operations, such as milling, the cooling action of the cutting fluid extends the range of temperature amplitude to which the cutter teeth are subjected. This condition can lead to thermal cracks because of thermal cycling of the tool. Cutting fluids may also cause the chip to become curlier, thus concentrating the stresses near the tool tip. These stresses, in turn, concentrate the heat closer to the tip and reduce tool life.

• Example: Effect of cutting fluids on machining.

A machining operation is being carried out with an effective cutting fluid. Explain the changes in the mechanics of the cutting operation and total energy consumption if the fluid is shut off.

SOLUTION. An effective cutting fluid is a good lubricant. Thus when the fluid is shut off, the friction at the tool–chip interface will increase. The following chain of events then takes place:

a) Fluid is shut off.
b) Friction at the tool–chip interface increases.
c) The shear angle decreases.
d) The shear strain increases.
e) The chip is thicker.
f) A built-up edge is likely to form.

As a consequence:

a) The shear energy in the primary zone increases.
b) The friction energy in the secondary zone increases.
c) Hence the total energy increases.
d) Surface finish is likely to deteriorate.
e) The temperature in the cutting zone increases; hence tool wear increases.
f) Tolerances may be difficult to maintain because of the increased temperature and expansion of the workpiece during machining.

•

21.12.2 Methods of application

The most common method of applying cutting fluid is **flood cooling** (Fig. 21.15). Flow rates range from 10 L/min (3 gal/min) for single-point tools, to 225 L/min (60 gal/min) per cutter for multiple-tooth cutters, such as in milling. In operations such as gun drilling and end milling, fluid pressures of 700–14,000 kPa (100–2000 psi) are used to wash away the chips.

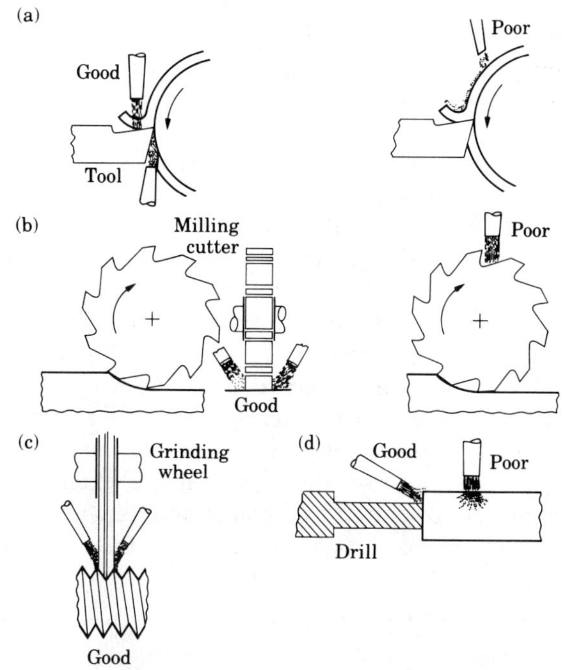

FIGURE 21.15 Schematic illustration of proper methods of applying cutting fluids in various machining operations: (a) turning, (b) milling, (c) thread grinding, and (d) drilling.

Mist cooling is another method of applying cutting fluids and is used particularly with water-base fluids. Although it requires venting (to prevent inhaling of fluid particles by the machine operator and others nearby) and has limited cooling capacity, mist cooling supplies fluid to otherwise inaccessible areas and provides better visibility of the workpiece being machined. It is particularly effective in grinding operations and at air pressures of 70–600 kPa (10–80 psi).

With increasing speed and power of machine tools, heat generation in machining operations has become a significant factor. New developments include the use of high-pressure refrigerated coolant systems to improve the rate of heat removal from the cutting zone.

21.12.3 Effects of cutting fluids

The selection of a cutting fluid should also include other considerations. These are effects on workpiece material and machine tools and biological and environmental effects.

Effects on workpiece material. When selecting a cutting fluid, we should consider whether the machined component will be subjected to environmental attack and high service stresses, thus possibly leading to stress-corrosion cracking.

This consideration is particularly important for cutting fluids having sulfur and chlorine additives. Additional considerations are staining of the workpiece by cutting fluids, especially on copper and aluminum. Machined parts should be cleaned and washed in order to remove any cutting-fluid residue.

Effects on machine tools. Just as a cutting fluid may adversely affect workpiece material, it can similarly affect the machine tool and its various components, such as bearings and slideways. The choice of fluid must therefore include consideration of its compatibility with the machine-member materials.

Biological and environmental effects. Because the machine-tool operator is usually in close proximity to cutting fluids, the effects of operator contact with fluids should be of primary concern. Fumes, smoke, and odors from cutting fluids can cause severe skin reactions, as well as respiratory problems. Considerable progress has been made in ensuring the safe use of cutting fluids in manufacturing plants. Additionally, the effect on the external environment, particularly with regard to degradation of the fluid and its ultimate disposal, is important. Disposal practices must comply with federal, state, and local laws and regulations.

SUMMARY

A variety of cutting-tool materials have been developed over the past century for specific applications in machining operations. These materials have a wide range of mechanical and physical properties, such as hot hardness, toughness, chemical stability, and resistance to chipping and wear. Various coatings have been developed, resulting in major improvements in tool life. The selection of appropriate tool materials depends not only on the material to be machined, but also on process parameters and the characteristics of the machine tool.

Cutting fluids are an important factor in machining operations. Generally, slower operations with high tool pressures require a fluid with good lubricating characteristics. In high-speed operations with significant temperature rise, fluids with cooling capacity are preferred. Selection should include consideration of the various adverse effects of cutting fluids on products, machinery, personnel, and the environment.

SUMMARY TABLE 21.1

OPERATING CHARACTERISTICS OF CUTTING TOOL MATERIALS. (Komanduri, R., *Kirk–Othmer Encyclopedia of Chemical Technology*, 3d ed. New York: Wiley, 1978.)

TOOL MATERIALS	MACHINING OPERATION AND (CUTTING SPEED RANGE)	MODES OF TOOL WEAR OR FAILURE	LIMITATIONS
High-speed steels	Turning, drilling, milling, broaching (medium speed)	Flank wear, crater wear	Low hot hardness, limited hardenability, and limited wear resistance
Cemented carbides	Turning, drilling, milling, broaching (medium speed)	Flank wear, crater wear	Cannot use at low speed due to cold welding of chips and microchipping
Coated carbides	Turning (medium to high speed)	Flank wear, crater wear	Cannot use at low speed due to cold welding of chips and microchipping
Ceramics	Turning (high speed to very high speed)	Depth-of-cut line notching, microchipping, gross fracture	Low strength, low thermomechanical fatigue strength
Cubic boron nitride	Turning, milling (medium to high speed)	Depth-of-cut line notching, chipping, oxidation, graphitization	Low strength, low-chemical stability at higher temperature
Diamond	Turning, milling (high to very high speed)	Chipping, oxidation, graphitization	Low strength, low chemical stability at higher temperature

TRENDS

- Continual progress is being made in developing new cutting-tool materials with improved properties, particularly for machining high-strength, high-temperature materials and composites containing abrasive fibers and particles. Cutting-edge strength is also an important area of study.
- Laser treatment of tools is being studied as a way to improve their hardness and wear resistance. New tool materials include hafnium nitride, borides, and composite ceramics.
- The purity and porosity of tool materials, particularly in ceramics, is an important area of study, which is aimed at improving toughness, reliability, and consistency of performance.

(*continued*)

TRENDS *(continued)*

- Attempts are being made to standardize cutting-tool materials, testing for properties, and their applications.
- Various coatings, applied either in single or multiple layers, are being developed to improve tool performance under various conditions, particularly at high cutting speeds. Adherence of coatings to the substrate is being improved.
- Cutting fluids are continually being developed, not only for better performance, but also with improved biological and ecological characteristics.

KEY TERMS

Carbides
Cast-cobalt alloys
Coated tools
Cubic boron nitride
Cutting fluids
Flood cooling
High-speed steels
Inserts
Mist cooling
Polycrystalline-diamond tools
Sialon

BIBLIOGRAPHY

Boothroyd, G., and W.A. Knight, *Fundamentals of Machining and Machine Tools*, 2d ed. New York: Marcel Dekker, 1989.

Cutting Tool Materials. Metals Park, Ohio: American Society for Metals, 1981.

Kalpakjian, S. (ed.), *New Developments in Tool Materials and Applications*. Chicago: Illinois Institute of Technology, 1977.

Komanduri, R. (ed.), *Advances in Hard Material Tool Technology*. Pittsburgh: Carnegie Press, 1976.

Komanduri, R., "Tool Materials," In *Kirk–Othmer Encyclopedia of Chemical Technology*, 3d ed., v. 23. New York: Wiley, 1978.

Machinery's Handbook. New York: Industrial Press, revised periodically.

Machining Data Handbook, 3d ed., 2 vols. Cincinnati: Machinability Data Center, 1980.

Metals Handbook, 8th ed., *Vol. 3: Machining*. Metals Park, Ohio: American Society for Metals, 1967.

Metals Handbook, 9th ed., *Vol. 3: Properties and Selection: Stainless Steels, Tool Materials and Special Purpose Metals*. Metals Park, Ohio: American Society for Metals, 1980.

Nachtman, E.S., and S. Kalpakjian, *Lubricants and Lubrication in Metalworking Operations*. New York.: Marcel Dekker, 1985.

Roberts, G.A., and R.A. Cary, *Tool Steels*, 4th ed. Metals Park, Ohio: American Society for Metals, 1980.

Schey, J.A., *Tribology in Metalworking—Friction, Wear and Lubrication*. Metals Park, Ohio: American Society for Metals, 1983.

Trent, E.M., *Metal Cutting*, 2d ed. Stoneham, Mass.: Butterworths, 1984.
Tool and Manufacturing Engineers Handbook, 4th ed., *Vol. 1: Machining*. Dearborn, Mich.: Society of Manufacturing Engineers, 1983.

REVIEW QUESTIONS

21.1 What are the major requirements for cutting-tool materials?
21.2 What differences are there between carbon-steel and high-speed steel tools?
21.3 List the major elements in cast-cobalt tools.
21.4 State the composition of carbide tools. What are the major characteristics of these tools?
21.5 Why were cutting-tool inserts developed?
21.6 Why are tools coated? What are the common coating materials?
21.7 Explain the applications and limitations of ceramic tools.
21.8 Describe the differences between the properties and characteristics of cubic boron nitride and diamond. What is the difference between a single-crystal and a polycrystalline tool?
21.9 What is the composition of sialon?
21.10 How are cutting tools reconditioned? List the considerations involved in whether a tool should be reconditioned or discarded.
21.11 List the functions of cutting fluids.
21.12 What are the two major types of cutting fluids? What are their major applications?
21.13 Explain how cutting fluids penetrate the cutting zone.
21.14 List the methods by which cutting fluids are applied in actual machining operations.
21.15 Can cutting fluids have any adverse effects? If so, what are they?

QUESTIONS AND PROBLEMS

21.16 Explain why there are so many different types of cutting-tool materials.
21.17 Which tool materials would be suitable for interrupted cutting operations? Why?
21.18 How would you go about measuring the hot hardness of cutting tools?
21.19 Describe the reasons for coating tools with multiple layers of different materials.
21.20 Explain the advantages and limitations of inserts.
21.21 Make a list of alloying elements used in high-speed steels. Explain why they are so effective in cutting tools.
21.22 Assume that you are in charge of a laboratory for developing new or improved cutting fluids. On the basis of the discussions presented in this and the previous chapter, suggest a list of topics for your staff to work on.
21.23 Repeat Question 21.22, but for cutting tools.
21.24 What are the purposes of chamfers on inserts?
21.25 What is the economic impact of the trend shown in Fig. 21.7?
21.26 Why does temperature have such an important effect on the life of cutting tools?
21.27 Ceramic and cermet cutting tools have certain advantages over carbide tools. Why then are they not completely replacing carbide tools?
21.28 What precautions would you take in cutting with brittle tool materials?

21.29 Decribe the trends you see in Table 21.2.

21.30 Why are chemical stability and inertness important in cutting tools?

21.31 Explain the effects of increasing the cobalt content in tungsten carbides on their mechanical properties.

21.32 Why do cutting fluids have different effects at different cutting speeds?

21.33 How would you go about measuring the effectiveness of cutting fluids?

21.34 Which of the two cutting-tool materials, diamond or cubic boron nitride, is more suitable for cutting steels? Why?

21.35 Titanium-nitride coatings on tools reduce the coefficient of friction at the tool–chip interface. What is the significance of this?

21.36 Describe the properties that the substrate for multiple-coated tools should have.

21.37 Make a listing of the tool materials described in this text in terms of decreasing stiffness. What is the significance of such a list in selecting cutting tools?

21.38 Describe the manufacturing processes involved in making the cutting tools described in this chapter.

21.39 Describe the conditions that are critical in utilizing the capabilities of diamond and cubic boron nitride cutting tools.

21.40 What would be the advantages of coating high-speed steel tools?

21.41 Explain the limits of application when comparing tungsten-carbide and titanium-carbide cutting tools.

21.42 Negative rake angles are generally preferred for ceramic, diamond, and cubic boron nitride tools. Why?

21.43 Do you think that there is a relationship between the cost of a cutting tool and its hot hardness? Explain.

21.44 Make a recommendation for the U.S. grade of a carbide tool for a finishing pass in turning a steel shaft at a 0.050-in. depth of cut.

21.45 Same as Problem 21.44 but for a turning operation on high-strength aluminum in which the same tool is to be used for both roughing and finishing passes.

22

Cutting Processes for Producing Round Shapes

22.1 Introduction

In this chapter we describe processes that produce parts that are basically round in shape. Typical products made include parts as small as miniature screws for eyeglass-frame hinges and as large as shafts, pistons, cylinders, gun barrels, and turbines for hydroelectric power plants. These processes are usually performed by turning the workpiece on a lathe. *Turning* means that the part is rotating while it is being machined. The starting material is usually a workpiece that has been made by other processes, such as casting, shaping, forging, extrusion, and drawing. Turning processes are versatile and capable of producing a wide variety of shapes, as we outline in Fig. 22.1:

- **Turning** straight, conical, curved, or grooved workpieces (Figs. 22.1a–d), such as shafts, spindles, pins, handles, and various machine components.

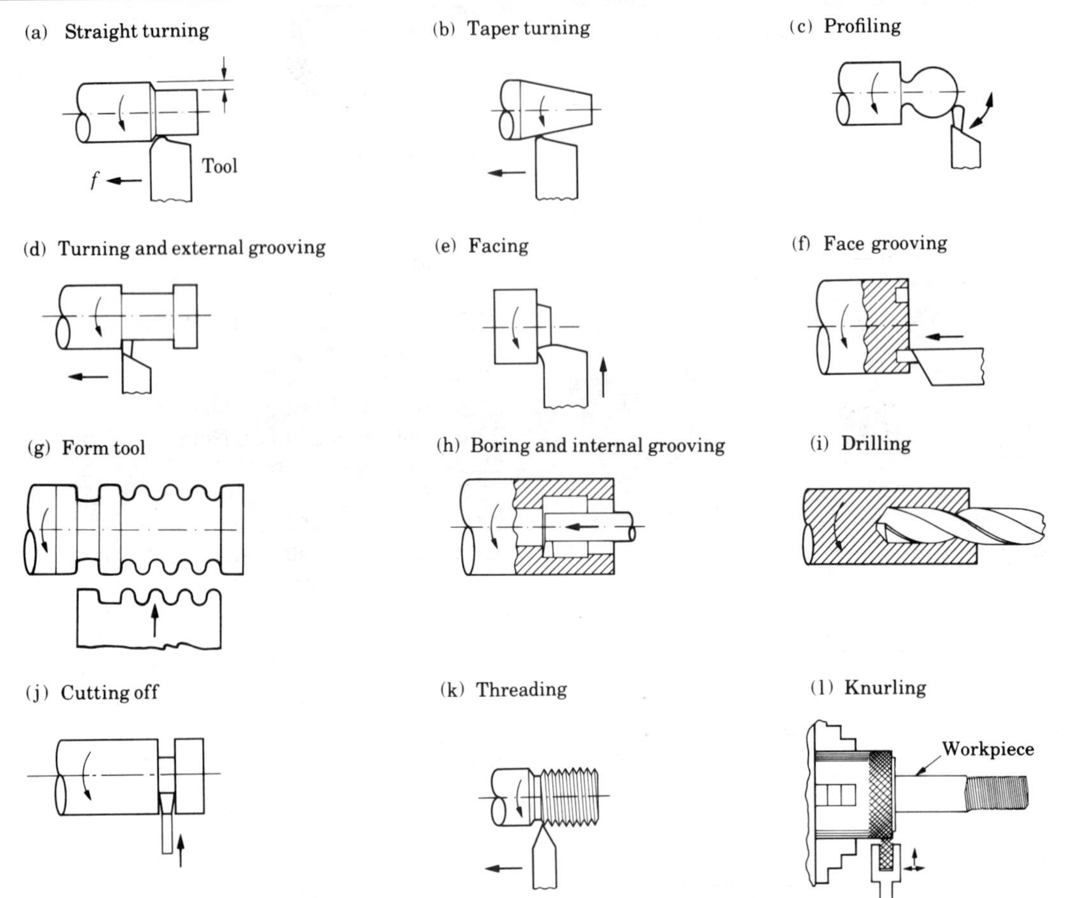

FIGURE 22.1
Various cutting operations that can be performed on a lathe.

- **Facing,** to produce a flat surface at the end of the part (Fig. 22.1e), such as parts that are attached to other components, or to produce grooves for O-ring seats (Fig. 22.1f).
- Producing various shapes by **form tools** (Fig. 22.1g), such as for functional purposes or for appearance.
- **Boring,** to enlarge a hole made by a previous process or in a tubular work-piece or to produce internal grooves (Fig. 22.1h).
- **Drilling,** to produce a hole (Fig. 22.1i), which may be followed by boring to improve its accuracy and surface finish.
- **Parting,** also called *cutting off*, to cut a piece from the end of a part, as in

making slugs or blanks for additional processing into discrete products (Fig. 22.1j).

- **Threading**, to produce external and internal threads in workpieces (Fig. 22.1k).
- **Knurling**, to produce a regularly shaped roughness on cylindrical surfaces, as in making knobs (Fig. 22.1ℓ).

These operations may be performed at various rotational speeds of the workpiece, depths of cut, d, and feed, f, depending on the workpiece and tool materials, the surface finish and dimensional accuracy required, and the capacity of the machine tool. Cutting speeds are usually in the range of 0.15–4 m/s (30–800 ft/min). *Roughing cuts*, which are performed for large-scale material removal, usually involve depths of cut greater than 0.5 mm (0.02 in.) and feeds on the order of 0.2–2 mm/rev (0.008–0.08 in./rev). *Finishing cuts* usually involve lower depths of cut and feed and higher cutting speeds.

In this chapter we describe processes, machine tools, cutting tools, process parameters, and process capabilities required to produce parts with round shapes. We begin our discussion with lathes, which are among the most important and versatile machine tools.

22.2 Lathes

Lathes are generally considered to be the oldest machine tools. Although woodworking lathes were first developed during the period 1000–1 B.C., metalworking lathes with lead screws were not built until the late 1700s. The most common lathe, shown schematically in Fig. 22.2, was originally called an *engine lathe* because it was powered with overhead pulleys and belts from nearby engines. Although simple and versatile, an engine lathe requires a skilled machinist because all controls are manipulated by hand. Consequently, it is inefficient for repetitive operations and for large production runs. However, various types of automation can be added to improve efficiency, as we describe in the rest of this section.

22.2.1 Lathe components

Lathes are equipped with a variety of components and accessories. The basic components of a common lathe are described in the following paragraphs.

Bed. The bed supports all the other major components of the lathe. Beds have a large mass and are built rigidly, usually from gray or nodular cast iron. The top portion of the bed has two *ways*, with various cross-sections, that are hardened and machined accurately for wear resistance and dimensional accuracy during use.

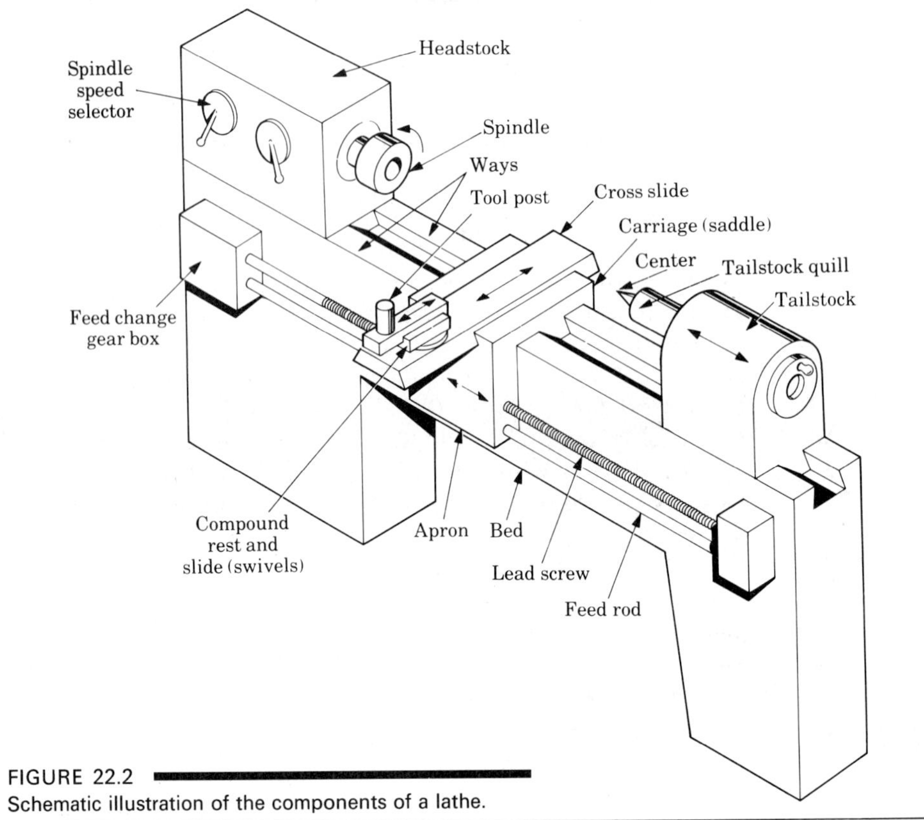

FIGURE 22.2
Schematic illustration of the components of a lathe.

Carriage. The carriage or carriage assembly (Fig. 22.3) slides along the ways and consists of an assembly of the cross-slide, tool post, and apron. The cutting tool is mounted on the *tool post*, usually with a *compound rest* that swivels for tool positioning and adjustment. The cross-slide moves radially in and out, thus controlling the radial position of the cutting tool, as in facing operations. The *apron* is equipped with mechanisms for both manual and mechanized movement of the carriage and the cross-slide, by means of the lead screw.

Headstock. The headstock is fixed to the bed and is equipped with motors, pulleys, and V-belts that supply power to the *spindle* at various rotational speeds. The speeds can be set through manually controlled selectors. Most headstocks are equipped with a set of gears, and some have various drives to provide a continuously variable speed range to the spindle. Headstocks have a hollow spindle to which workholding devices, such as *chucks* and *collets*, are attached, and long bars can be fed through for various turning operations.

Tailstock. The tailstock, which can slide along the ways and be clamped at

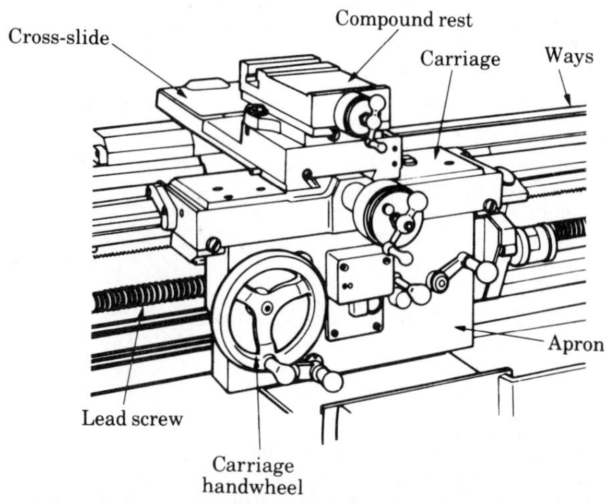

FIGURE 22.3
A lathe carriage showing various components. *Source:* LeBlond Makino Machine Tool Company.

any position, supports the other end of the workpiece. It is equipped with a center that may be fixed (*dead center*), or it may be free to rotate with the workpiece (*live center*). Drills and reamers can be mounted on the tailstock *quill* (a hollow cylindrical part with a tapered hole) to produce axial holes in the workpiece.

Feed rod and lead screw. The feed rod is powered by a set of gears from the headstock. It rotates during operation of the lathe and provides movement to the carriage and the cross-slide by means of gears, a friction clutch, and a keyway along the length of the rod. The lead screw is used for cutting threads accurately (see Section 22.4). Closing a split nut around the lead screw engages it with the carriage.

22.2.2 Lathe specifications

A lathe is usually specified by its *swing*, that is, the maximum diameter of the workpiece that can be machined (Table 22.1), by the maximum distance between the headstock and tailstock centers, and by the length of the bed. Thus, for example, a lathe may have the following size: 360 mm (14 in.) (swing) by 760 mm (30 in.) (between centers) by 1830 mm (6 ft) (length of bed). Lathes are available in a variety of styles, types of construction, stiffnesses, and power.

Bench lathes are placed on a workbench; they have fractional horsepower, are usually operated by hand feed, and are used for precision machining small workpieces. *Toolroom lathes* have high precision, thus enabling the machining of parts to close tolerances. *Engine lathes* are available in a wide range of sizes and are used for a variety of turning operations. In *gap lathes* a section of the bed in front of the headstock can be removed to accommodate larger diameter workpieces. *Special-purpose lathes* are used for applications such as railroad wheels, gun barrels, and rolling-mill rolls, with workpiece sizes as large as 1.7 m in diameter by 8 m long

TABLE 22.1
TYPICAL CAPACITIES AND MAXIMUM WORKPIECE DIMENSIONS FOR MACHINE TOOLS

MACHINE TOOL	MAXIMUM DIMENSION (m)	POWER (kW)	MAXIMUM rpm
Lathes (swing/length)			
Bench	0.3/1	<1	3000
Engine	3/5	70	4000
Turret	0.5/1.5	60	3000
Automatic screw	0.1/0.3	20	10,000
Boring machines (work diameter/length)			
Vertical spindle	4/3	200	300
Horizontal spindle	1.5/2	70	1000
Drilling machines			
Bench and column (drill diameter)	0.1	10	12,000
Radial (column to spindle distance)	3	—	—
Numerical control (table travel)	4	—	—

Note: Larger capacities are available for special applications.

(66 in. × 25 ft), and capacities of 300 kW (400 hp). Maximum spindle speeds are usually 2000 rpm, although they may range from 4000 rpm to 10,000 rpm for special applications, but may be only about 200 rpm for large lathes. Cost of engine lathes ranges from about $2000 for bench types to about $100,000 for larger units.

22.2.3 Workholding devices

Workholding devices are important in machine tools. In a lathe, one end of the workpiece is clamped to the spindle by a chuck, collet, face plate, or mandrel—or between centers.

A *chuck* is usually equipped with three or four *jaws*. Three-jaw chucks generally have a geared-scroll design that makes the jaws self-centering (Fig. 22.4) and hence are used for round workpieces, such as bar stock, pipes, and tubing. Workpieces can be centered within 0.025 mm (0.001 in.). Four-jaw chucks (independent chucks) have jaws that can be moved and adjusted independently of each other and thus can be used for square or rectangular, as well as odd-shaped, workpieces. They are more ruggedly constructed than three-jaw chucks and hence are used for heavy workpieces.

The jaws in both types of chuck can be reversed to permit clamping of the workpieces on either outside surfaces (as shown in Fig. 22.4) or on inside surfaces of hollow workpieces, such as pipes and tubing. Chucks are available in various designs and sizes. Their selection depends on the type and speed of operation, workpiece size, production and accuracy requirements, and the jaw forces required. The magnitude of jaw forces is important to ensure that the part does not slip in the chuck during machining. High spindle speeds can reduce jaw forces significantly because of centrifugal forces. Chucks are actuated manually with a special key or

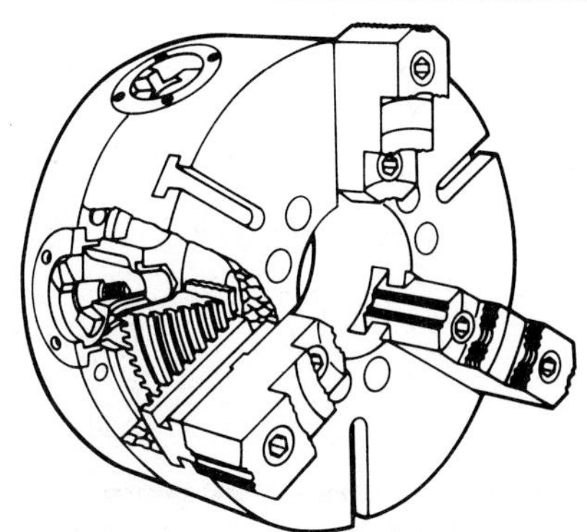

FIGURE 22.4
A typical three-jaw self-centering chuck of gear-scrolled design. *Source:* Cushman Industries, Inc.

are power actuated. Because it takes longer to operate them, manually actuated chucks are generally used for toolroom and limited production runs.

To meet the increasing demands for stiffness, precision, versatility, power, and high cutting speeds in modern machine tools, major advances have been made in the design of workholding devices. *Power chucks*, actuated pneumatically or hydraulically, are now used in automated equipment for high production rates. Also available are several types of power chucks with lever or wedge type mechanisms to actuate the jaws. These chucks have jaw movements that are limited to about 13 mm (0.5 in.).

A *collet* is basically a longitudinally split, tapered bushing. The workpiece is placed inside the collet, and the collet is pulled (draw-in collet; Fig. 22.5a) or pushed (push-out collet; Fig. 22.5b) into the spindle by mechanical means. The tapered surfaces shrink the segments of the collet radially, tightening the workpiece. Collets are used for round workpieces and are available in a range of internal diameters. Because the radial movement of the collet segments is small, workpieces should generally be within 0.125 mm (0.005 in.) of the nominal size of the collet.

Face plates are used for clamping irregularly shaped workpieces. The plates are round and have several slots and holes through which the workpiece is bolted or clamped. *Mandrels* (Fig. 22.6) are placed inside hollow or tubular workpieces and are used to hold workpieces that require machining on both ends or their cylindrical surfaces.

22.2.4 Accessories and attachments

Several devices are available as accessories and attachments for lathes. Among these devices are (1) carriage and cross-slide stops with various designs to stop the carriage at a predetermined distance along the bed; (2) devices for turning parts

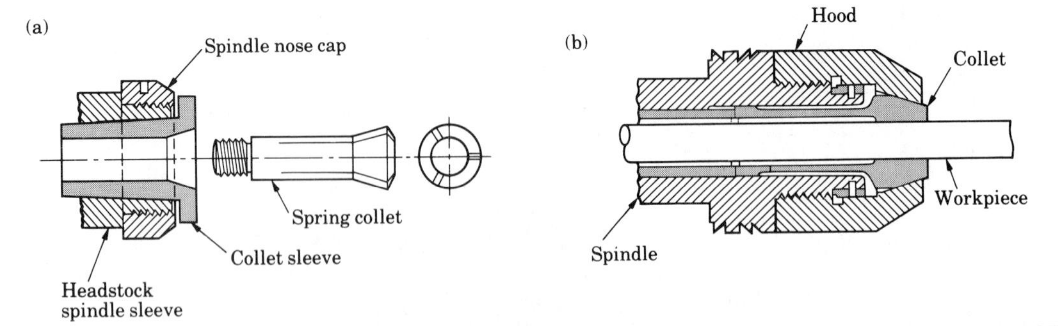

FIGURE 22.5
(a) Schematic illustrations of a draw-in type collet. The round workpiece is placed in the collet hole, and the conical surfaces of the collet are forced inward by pulling it with a draw bar into the sleeve. (b) A push-out type collet.

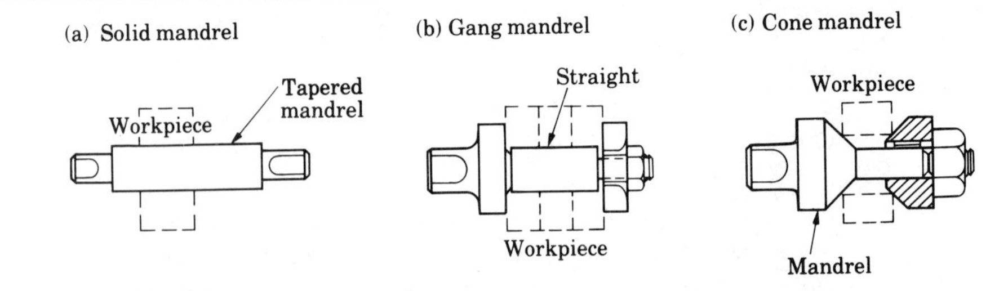

FIGURE 22.6
Various types of mandrels to hold workpieces for turning.

with various tapers or radii; (3) milling, sawing, gear-cutting, and grinding attachments; and (4) various attachments for boring, drilling, and thread cutting.

22.2.5 Lathe operations

In a typical turning operation, the workpiece is clamped by any one of the workholding devices that we have described. Long and slender parts are supported by a *steady rest* or *follow rest* placed on the bed; otherwise the part will deflect under the cutting forces. These rests are equipped with three adjustable fingers, which support the workpiece while allowing it to rotate freely. Steady rests are clamped to the ways, whereas follow rests travel with the carriage. The cutting tool, attached to the tool post and driven by the lead screw, removes material by traveling along the bed. A right-hand tool travels toward the headstock and a left-hand tool toward the tailstock. Workpiece facing is done by moving the tool radially, with the cross-slide, and clamping the carriage for better dimensional accuracy.

Form tools are used to produce various shapes on round workpieces by turning (see Fig. 22.1g). The tool moves radially inward to machine the part. Machining by form cutting is not suitable for deep and narrow grooves or sharp corners. To avoid vibration, the formed length should not be greater than about 2.5 times the minimum diameter of the part.

The boring operation on a lathe is similar to turning. Boring is performed on hollow workpieces or in a hole made previously by drilling or other means. The workpiece is held in a chuck or some other suitable workholding device. We describe boring large workpieces in Section 22.5.

Drilling can be performed on a lathe by mounting the drill in a drill chuck into the tailstock quill (a tubular shaft). The workpiece is placed in a workholder on the headstock, and the quill is advanced by rotating the hand wheel. Holes drilled in this manner may not be concentric because of the drill's radial drifting. The concentricity of the hole is improved by boring the drilled hole. Drilled holes may be reamed on lathes in a manner similar to drilling.

The tools for parting, grooving, thread cutting, and various other operations are specially shaped for the particular purpose or are available as inserts. Knurling is performed on a lathe with hardened rolls (see Fig. 22.1ℓ). The surface of the rolls is a replica of the profile to be generated. The rolls are pressed radially against the rotating workpiece while the tool moves axially along the part.

22.2.6 Tracer lathes

Tracer lathes are machine tools with attachments that are capable of turning parts with various contours (Fig. 22.7). Also called *duplicating lathes* or *contouring lathes*, the cutting tool follows a path that duplicates the contour of a template, similar to a pencil following the shape of a plastic template used in engineering

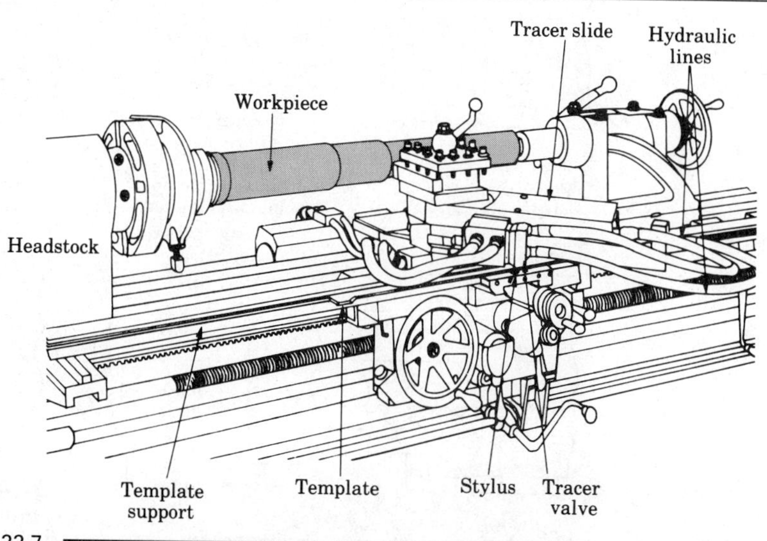

FIGURE 22.7
Schematic illustration of a hydraulically operated tracer lathe.

drawing. A tracer finger follows the template and, through a hydraulic or electrical system, guides the cutting tool along the workpiece without operator interference. Operations performed on a tracer lathe can also be done on numerical-control lathes (see Section 38.3).

22.2.7 Automatic lathes

Lathes have been increasingly automated over the years. Manual machine controls have been replaced by various mechanisms that enable cutting operations to follow a certain prescribed sequence. In a fully automatic machine, parts are fed and removed automatically, whereas in semiautomatic machines these functions are performed by the operator. *Automatic lathes*, which are usually vertical and do not have tailstocks, are also called chucking machines, or chuckers. They are used for machining individual pieces of regular or irregular shapes.

• Example: Machining outer bearing races on a computer numerical control automatic lathe.

Outer bearing races (see the accompanying figure) are machined on an automatic lathe equipped with two-axis computer numerical control (CNC) attachments. The starting material is hot-rolled 52100 steel tube 91 mm (3.592 in.) OD and 75.5 mm

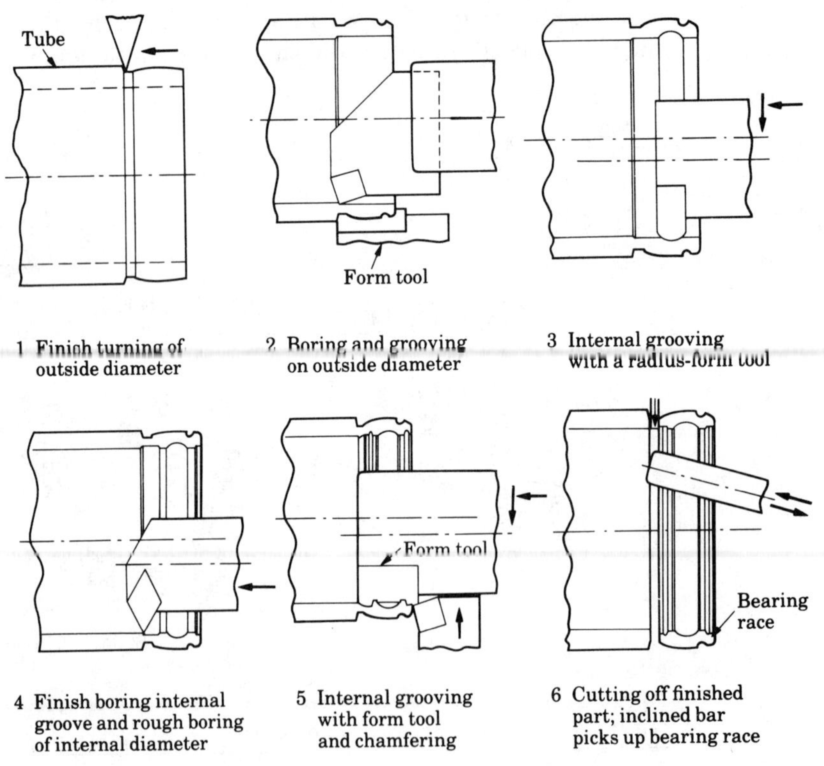

1 Finish turning of outside diameter

2 Boring and grooving on outside diameter

3 Internal grooving with a radius-form tool

4 Finish boring internal groove and rough boring of internal diameter

5 Internal grooving with form tool and chamfering

6 Cutting off finished part; inclined bar picks up bearing race

(2.976 in.) ID. The cutting speed is 1.6 m/s (313 ft/min) for all operations. All tools are carbide, including the cutoff tool (last operation) which is 3.18 mm ($\frac{1}{8}$ in.) instead of 4.76 mm ($\frac{3}{16}$ in.) for the high-speed steel cutoff tool that was formerly used. The material saved by this change is significant because the width of the race is small. The CNC automatic lathe was able to machine these races at high speeds and with repeatable tolerances of ± 0.025 mm (0.001 in.). *Source:* McGill Manufacturing Company.

22.2.8 Turret lathes

Turret lathes are capable of performing multiple cutting operations on the same workpiece, such as turning, boring, drilling, thread cutting, and facing (Fig. 22.8a). Several cutting tools—usually as many as six—are mounted on the hexagonal *main turret* (Fig. 22.8b), which is rotated for each specific cutting operation. Also, the lathe usually has a *square turret* on the cross-slide, with as many as four cutting tools mounted on it. The workpiece, generally a long round rod, is advanced a preset distance through the chuck. After the part is machined, it is cut off by a tool mounted on the square turret, which moves radially into the workpiece. The rod is then advanced the same preset distance into the work area, and the next part is machined.

Turret lathes are versatile, and operations may be carried out either by hand, using the turnstile (capstan wheel), or automatically. Once set up properly by a setup person, these machines do not require skilled operators. Turret lathe sizes are specified by the swing and the maximum diameter of the bar that can fit through the hole in the spindle. The turret lathe shown in Fig. 22.8(a) is known as a *ram-type* turret lathe in which the ram slides in a separate base on the saddle (Fig. 22.9a). The short stroke of the turret slide limits this machine to relatively short workpieces and light cuts, in both small and medium quantity production.

In another style, called the *saddle type*, the main turret is installed directly on the saddle (Fig. 22.9b), which slides directly on the bed. Hence the length of the stroke is limited only by the length of the bed. This type of lathe is more heavily constructed and is used to machine large workpieces. Because of the large mass of the components, saddle-type lathe operations are slower than ram-type lathe operations. Vertical turret lathes are also available that are more suitable for short and heavy workpieces, with diameters as large as 1.2 m (48 in.). Turret lathes cost $100,000 and up, depending on their size and level of automation.

22.2.9 Computer-controlled lathes

In the most advanced lathes, movement and control of the machine and its components are actuated by computer numerical controls (CNC). The features of such a lathe are shown in Fig. 22.10. These lathes are usually equipped with one or more turrets (Figs. 22.11a and b). Each turret is equipped with a variety of tools and performs several operations on different surfaces of the workpiece. These machines

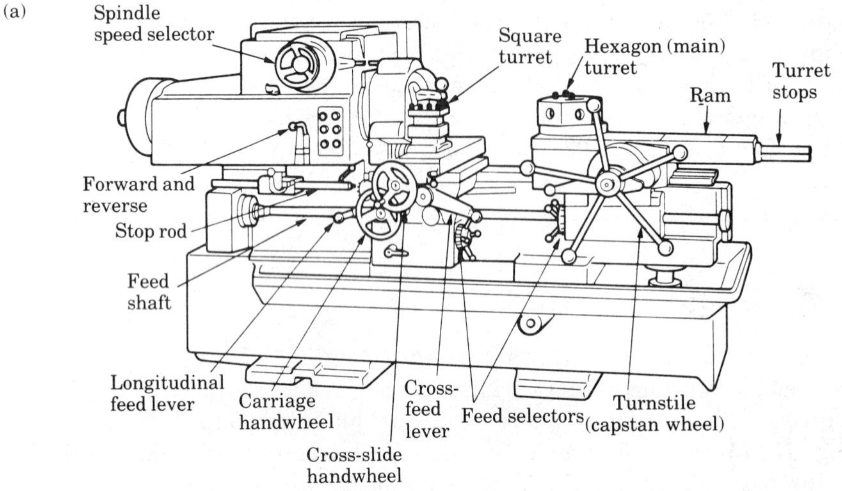

(a)
Spindle speed selector
Square turret
Hexagon (main) turret
Turret stops
Ram
Forward and reverse
Stop rod
Feed shaft
Longitudinal feed lever
Carriage handwheel
Cross-feed lever
Feed selectors
Turnstile (capstan wheel)
Cross-slide handwheel

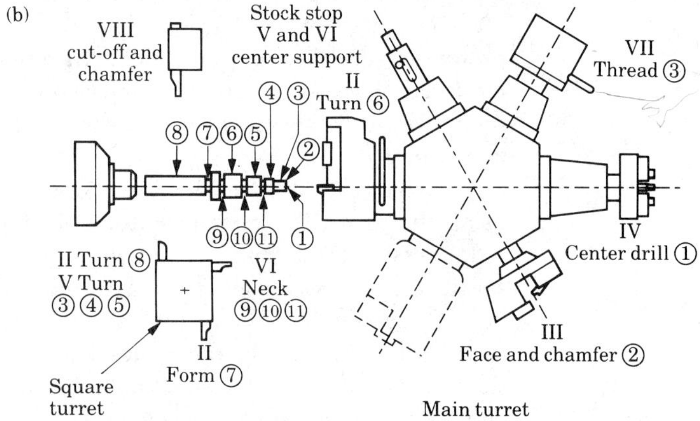

(b)
VIII cut-off and chamfer
Stock stop V and VI center support
VII Thread ③
II Turn ⑥
④ ③
⑧ ⑦ ⑥ ⑤
②
⑨ ⑩ ⑪ ①
IV Center drill ①
II Turn ⑧
V Turn ③ ④ ⑤
VI Neck ⑨ ⑩ ⑪
II Form ⑦
III Face and chamfer ②
Square turret
Main turret

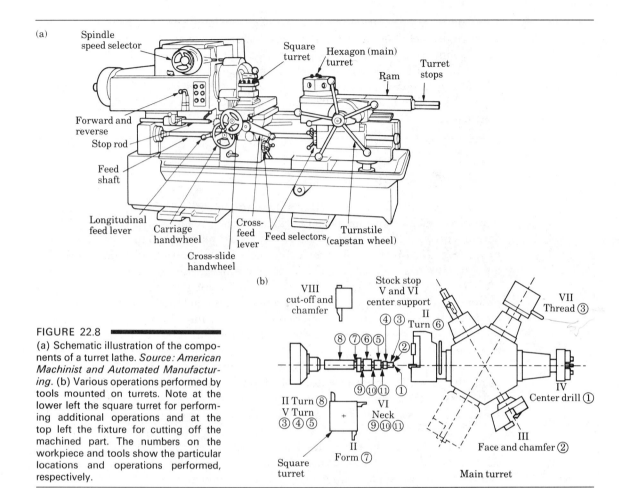

FIGURE 22.8 (a) Schematic illustration of the components of a turret lathe. *Source: American Machinist and Automated Manufacturing.* (b) Various operations performed by tools mounted on turrets. Note at the lower left the square turret for performing additional operations and at the top left the fixture for cutting off the machined part. The numbers on the workpiece and tools show the particular locations and operations performed, respectively.

(a)
(b)
Hexagon turret
Ram
Saddle
Capstan wheel
Turret
Saddle

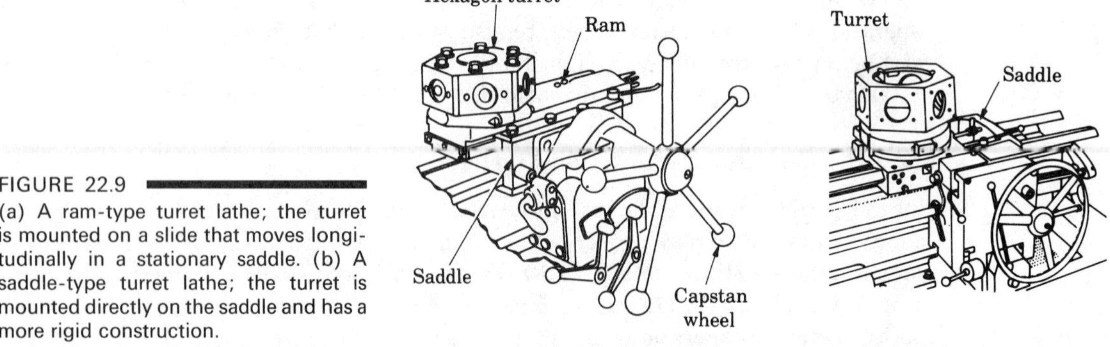

FIGURE 22.9 (a) A ram-type turret lathe; the turret is mounted on a slide that moves longitudinally in a stationary saddle. (b) A saddle-type turret lathe; the turret is mounted directly on the saddle and has a more rigid construction.

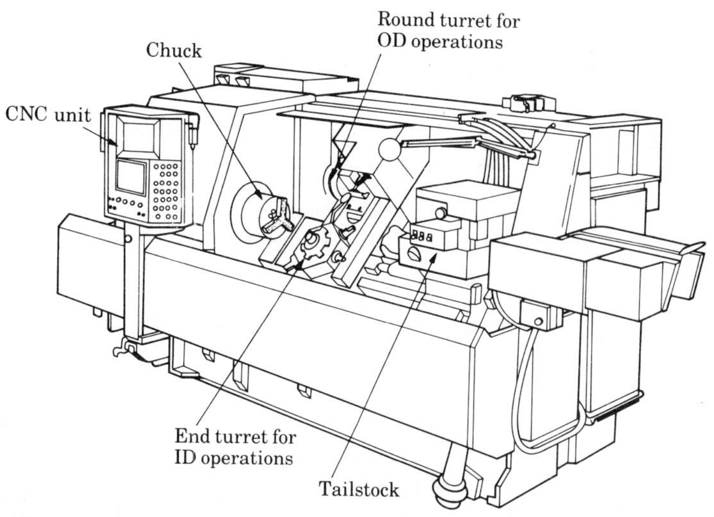

FIGURE 22.10
A computer numerical control lathe. Note the two turrets on this machine. *Source:* Jones & Lamson, Textron, Inc.

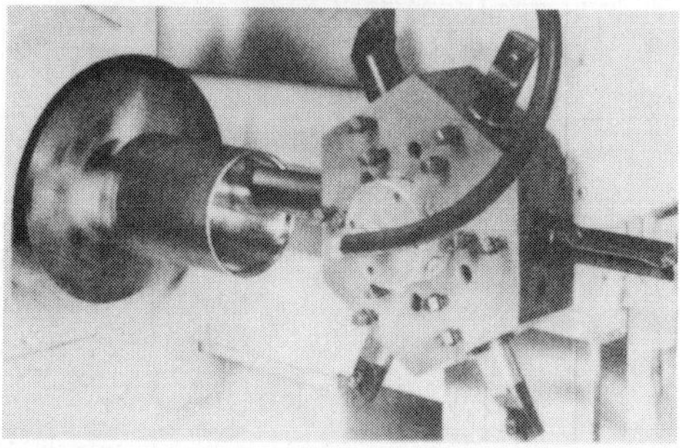
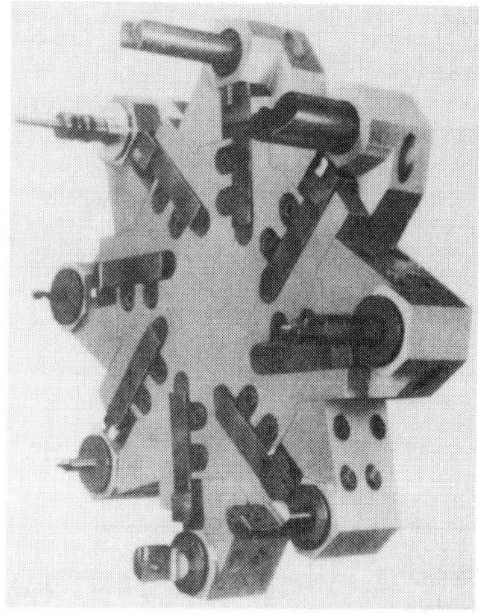

FIGURE 22.11
(a) Turret with six different tools for inside-diameter and outside-diameter cutting and threading operations. (b) Turret with eight different cutting tools. *Source:* Monarch Machine Tool Company.

are highly automated, the operations are repetitive and maintain the desired accuracy, and less-skilled labor is required. We present the details of computer controls in Chapters 38 and 39 (see also Section 24.2).

• **Example: Typical parts made on computer numerical control turning machine tools.**

The capabilities of CNC turning machine tools are illustrated in the accompanying figure. Material and number of cutting tools used and machining times are indicated for each part. *Source:* Monarch Machine Tool Company.

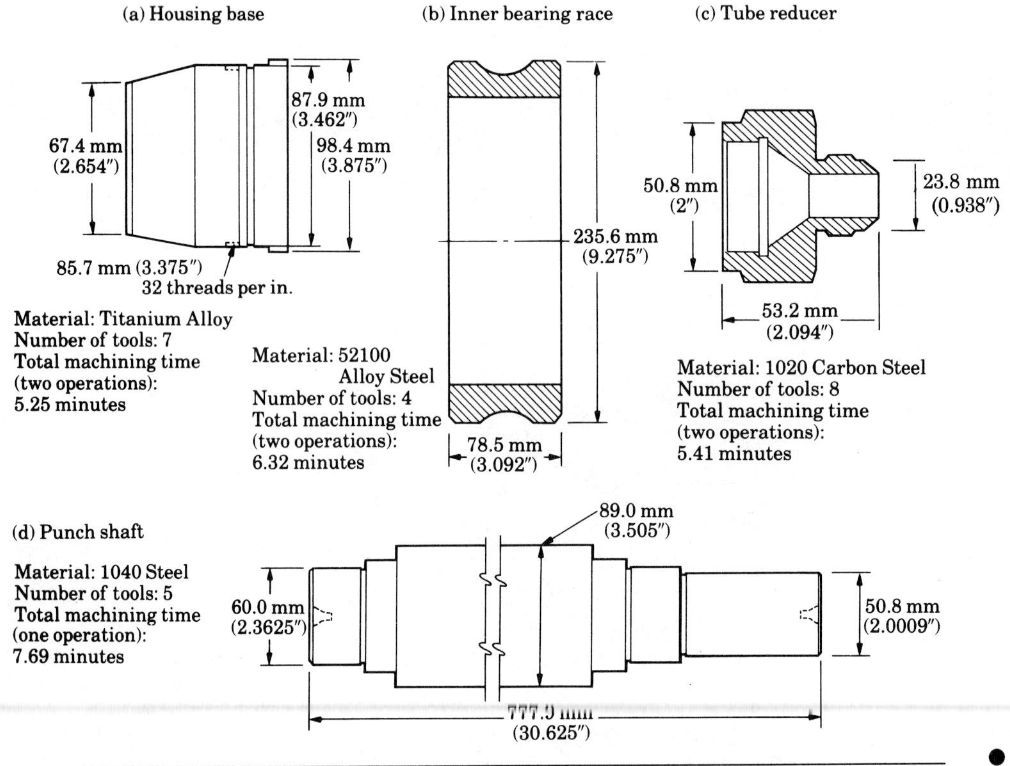

22.3 Turning Parameters and Process Capabilities

The majority of turning operations involve simple single-point cutting tools. We showed the geometry of a typical right-hand cutting tool for turning in Fig. 20.10. Such tools are described by a standardized nomenclature (Fig. 22.12). Each group of

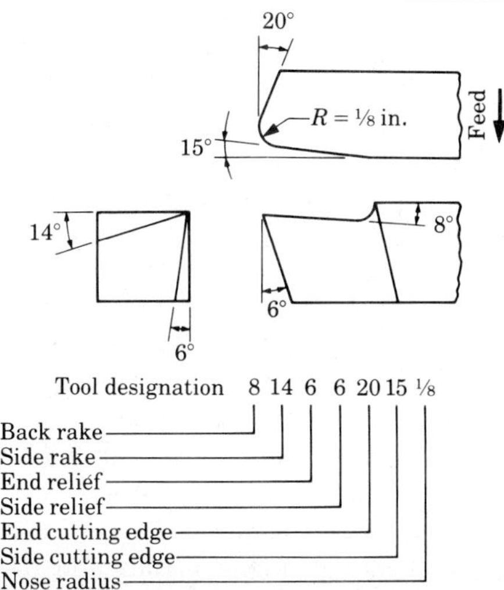

FIGURE 22.12
Designation and symbols for a right-hand cutting tool.

tool and workpiece materials has an optimum set of tool angles, which was developed largely through experience (Table 22.2).

22.3.1 Turning parameters

Rake angles are important in controlling the direction of chip flow and to the strength of the tool tip. As we described in Chapter 20, positive angles improve the

TABLE 22.2
GENERAL RECOMMENDATIONS FOR TURNING TOOLS

	HIGH-SPEED STEEL					CARBIDE (INSERTS)				
MATERIAL	*BACK RAKE*	*SIDE RAKE*	*END RELIEF*	*SIDE RELIEF*	*SIDE AND END CUTTING EDGE*	*BACK RAKE*	*SIDE RAKE*	*END RELIEF*	*SIDE RELIEF*	*SIDE AND END CUTTING EDGE*
Aluminum and magnesium alloys	20	15	12	10	5	0	5	5	5	15
Copper alloys	5	10	8	8	5	0	5	5	5	15
Steels	10	12	5	5	15	−5	−5	5	5	15
Stainless steels	5	8–10	5	5	15	−5–0	−5–5	5	5	15
High-temperature alloys	0	10	5	5	15	5	0	5	5	45
Refractory alloys	0	20	5	5	5	0	0	5	5	15
Titanium alloys	0	5	5	5	15	−5	−5	5	5	5
Cast irons	5	10	5	5	15	−5	−5	5	5	15
Thermoplastics	0	0	20–30	15–20	10	0	0	20–30	15–20	10
Thermosets	0	0	20–30	15–20	10	0	15	5	5	15

cutting operation by reducing forces and temperatures. However, positive angles produce a small included angle of the tool tip, which, depending on the toughness of the tool material, may cause premature tool failure. Side rake angle is more important than back rake angle, although the latter usually controls the direction of chip flow.

Relief angles control interference and rubbing at the tool-workpiece interface. If the relief angle is too large, the tool may chip off; if too small, flank wear may be excessive. *Cutting edge angles* affect chip formation, tool strength, and cutting forces to various degrees. *Nose radius* affects surface finish and tool-tip strength. The sharper the radius, the rougher will be the surface finish of the workpiece and the lower will be the strength of the tool. However, large nose radii can lead to tool chatter (see Section 24.3). Recommendations for cutting-tool selection are given in Table 22.3. The most commonly used tool materials are M2 and M3 tool steels and C-6 and C-7 carbides.

Conventional cutting-speed ranges for a variety of materials are presented in Table 22.4. Although many metals and nonmetallic materials can be machined without a cutting fluid, in many cases the application of fluids can improve the operation (see also Section 21.12). Recommendations for cutting fluids are given in Table 22.5.

The **material removal rate** (MRR) is the volume of material removed per unit time. For turning operations, we can calculate it by referring to Fig. 22.13 (p. 682) and noting that $D_{avg} = (D_o + D_f)/2$. Thus,

$$MRR = (\pi)(D_{avg})(d)(f)(N), \tag{22.1}$$

where f is the feed in mm/rev or in./rev, and N is the workpiece rotational speed per unit time. For cases when $D_o \gg d$, we can replace D_{avg} by D_o.

TABLE 22.3
COMMON TOOL MATERIALS FOR CUTTING OPERATIONS

WORKPIECE	TURNING	DRILLING, REAMING, TAPPING	MILLING	BROACHING, GEAR CUTTING
Aluminum and copper alloys	HSS, WC, CA, CC, D	HSS, WC	HSS, WC, CA	HSS
Carbon steels and cast irons	HSS, WC, CA, CC, C	HSS, WC	HSS, WC, CA	HSS, WC, CA
Alloys steels and alloy cast irons	WC, CA, CC, C, CBN	HSS, WC	WC, CA	HSS, WC
High-temperature alloys and titanium	HSS, WC, CA, CC, C	HSS, WC	WC, CA	HSS, WC
Plastics and composites	WC, CA, CBN, D	WC	WC	HSS

Note: HSS, high-speed steel; WC, tungsten carbide; CA, cast alloy; CC, coated carbide; C, ceramic and cermet; CBN, cubic boron nitride; and D, diamond.

TABLE 22.4
RECOMMENDATIONS FOR CUTTING SPEEDS IN TURNING

WORKPIECE MATERIAL	CUTTING SPEED (m/s) *HSS*	*WC*
Aluminum alloys	3–4	5–7
Magnesium alloys	4	10
Copper alloys	0.5–2	1–5
Steels	0.5–1	1–3
Stainless steels	0.15–0.5	1–2
High-temperature alloys	0.05–0.1	0.15–0.3
Titanium alloys	0.15–1	0.5–2
Cast irons	0.15–0.5	0.5–2
Thermoplastics	1.5–2	2–3
Thermosets	1–2	1–4

Note: (a) Depth of cut is usually 4 mm for rough turning and 0.7 mm for finish turning.
(b) Feeds for rough turning range from 0.2 mm/rev for materials with high hardness, to 2 mm/rev for lower hardness. Finishing cuts require lower feeds.
(c) Cutting speeds are for uncoated tools. Speeds for coated tools are from 25–75 percent higher.
(d) Cutting speeds for ceramic tools can be 2–3 times higher than the values indicated.
(e) Cutting speed for diamond tools is usually 4–15 m/s, depth of cut 0.05–0.2 mm, and feed 0.02–0.05 mm/rev.
(f) As hardness increases, cutting speed, feed, and depth of cut should be decreased.
(g) Speeds for free-machining metals are higher than those indicated.
(h) Speeds for other cutting processes are generally lower by as much as 75 percent.

TABLE 22.5
GENERAL RECOMMENDATIONS FOR CUTTING FLUIDS FOR MACHINING

MATERIAL	TYPE OF FLUID
Aluminum	D, MO, E, MO + FO, CSN
Beryllium	MO, E, CSN
Copper	D, E, CSN, MO + FO
Lead	D
Magnesium	D, MO, MO + FO
Nickel	MO, E, CSN
Refractory	MO, E, EP
Steels (carbon and low alloy)	D, MO, E, CSN, EP
Steels (stainless)	D, MO, E, CSN
Titanium	CSN, EP, MO
Zinc	D, MO, E, CSN
Zirconium	D, E, CSN

Note: CSN, chemicals and synthetics; D, dry; E, emulsion; EP, extreme pressure; FO, fatty oil; and MO, mineral oil.

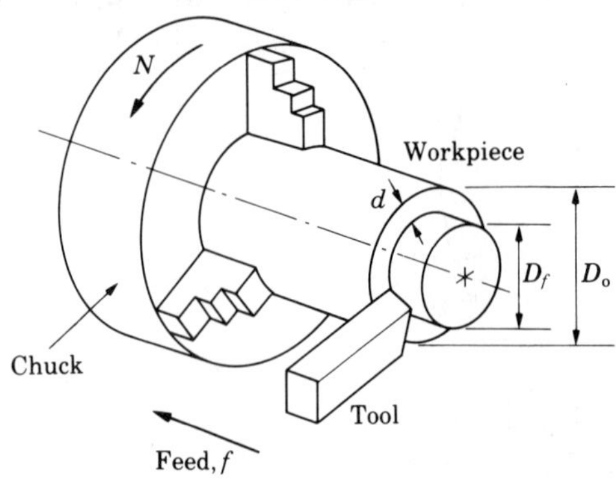

FIGURE 22.13
Schematic illustration of a turning operation showing depth of cut, *d*, and feed, *f*. Cutting speed is the surface speed of the workpiece.

Similarly, the cutting time, t, for a workpiece of length L is

$$t = \frac{L}{fN}. \qquad (22.2)$$

This time does not include the time required for tool approach and retraction. Because the time spent in noncutting cycles of a machining operation is nonproductive and adds to the overall economics, the time involved in approaching and retracting tools to and from the workpiece is an important consideration. Machine tools are now being designed and built to minimize this time—one method is first by rapid movements of the tools, then a slower movement as the tool engages the workpiece. These equations are also applicable to boring operations, where D_o is the bore diameter.

• Example: Material removal rate and cutting force in turning.

A 6-in. long, $\frac{1}{2}$-in. diameter 304 stainless steel rod is being reduced in diameter to 0.480 in. by turning on a lathe. The spindle rotates at $N = 400$ rpm, and the tool is traveling at an axial speed of 8 in./min. Calculate the cutting speed, material removal rate, time to cut, power dissipated, and cutting force.

SOLUTION. The cutting speed is the tangential speed of the workpiece. The maximum cutting speed is at the outer diameter D_o and is obtained from the expression

$$V = \pi D_o N.$$

Thus,

$$V = (\pi)(0.500)(400) = 628 \text{ in./min} = 52 \text{ ft/min}.$$

The cutting speed at the machined diameter is

$$V = (\pi)(0.480)(400) = 603 \text{ in./min} = 50 \text{ ft/min}.$$

From the information given, we note that the depth of cut is

$$d = (0.500 - 0.480)/2 = 0.010 \text{ in.},$$

and the feed is

$$f = 8/400 = 0.02 \text{ in./rev}.$$

Thus according to Eq. (22.1), the material removal rate is

$$\text{MRR} = (\pi)(0.490)(0.010)(0.02)(400) = 0.123 \text{ in}^3/\text{min}.$$

The actual time to cut, according to Eq. (22.2), is

$$t = 6/(0.02)(400) = 0.75 \text{ min}.$$

We can calculate the power dissipated by referring to Table 20.1 and taking an average value for stainless steel as $4 \text{ W} \cdot \text{s/mm}^3 = 4/2.73 = 1.47 \text{ hp} \cdot \text{min/in}^3$. Thus power dissipated is

$$\text{Power} = (1.47)(0.123) = 0.181 \text{ hp},$$

and since 1 hp = 396,000 in.-lb/min, the power dissipated is 71,700 in.-lb/min. The cutting force F_c is the tangential force exerted by the tool. Since power is the product of torque T and rotational speed in radians per unit time, we have

$$T = (71{,}700)/(400)(2\pi) = 29 \text{ in.-lb}.$$

Since $T = (F_c)(D_{avg}/2)$, we have

$$F_c = 29/0.245 = 118 \text{ lb} = 5.17 \text{ kN}.$$

22.3.2 Turning process capabilities

Turning process production rates are shown in Table 22.6. The surface finish and dimensional accuracy obtained in turning and related operations (Figs. 22.14 and 22.15 on p. 686; see also Summary Table 23.1) depends on factors such as the characteristics and condition of the machine tool, stiffness, vibration and chatter, process parameters, tool shape and wear, cutting fluids, machinability of the workpiece material, and operator skill. As a result, a wide range of surface finishes is obtained, as shown in Fig. 22.14 (see also Fig. 26.4).

22.3.3 Design considerations for turning operations

Certain considerations are important in designing parts to be manufactured economically by turning operations. Because cutting takes considerable time, wastes material, and is not as economical as forming or shaping parts to final

TABLE 22.6
TYPICAL PRODUCTION RATES FOR VARIOUS CUTTING OPERATIONS

OPERATION	RATE
Turning	
Engine lathe	Very low to low
Tracer lathe	Low to medium
Turret lathe	Low to medium
Computer-control lathe	Low to medium
Single-spindle chuckers	Medium to high
Multiple-spindle chuckers	High to very high
Boring	Very low
Drilling	Low to medium
Milling	Low to medium
Planing	Very low
Shaping	Very low
Gear cutting	Low to medium
Broaching	Medium to high
Sawing	Very low to low

Note: Production rates indicated are relative: *Very low* is about one or more parts per hour; *medium* is approximately 100 parts per hour; *very high* is 1000 or more parts per hour.

dimensions, machining should be avoided as much as possible. When turning operations are necessary, the following general design guidelines should be used:

a) Parts should be designed so that they can be fixtured and held in work-holding devices with relative ease. Thin, slender workpieces are difficult to support properly to withstand clamping and cutting forces.
b) Dimensional accuracy and surface finish specified should be as wide as permissible for the part to function properly.
c) Sharp corners, tapers, and major dimensional variations in the part should be avoided.
d) Blanks to be machined should be as close to final dimensions as possible, so as to reduce production cycle time.
e) Parts should be designed so that cutting tools can travel across the workpiece without obstruction.
f) Design features should be such that standard, commercially available cutting tools and inserts can be used.
g) Materials should, as much as possible, be selected for their machinability (see Section 20.9).

22.3.4 Guidelines for turning operations

A general guide to the probable causes of problems in turning operations is given in Table 22.7. Recall that in Chapters 20 and 21 we described the factors influencing

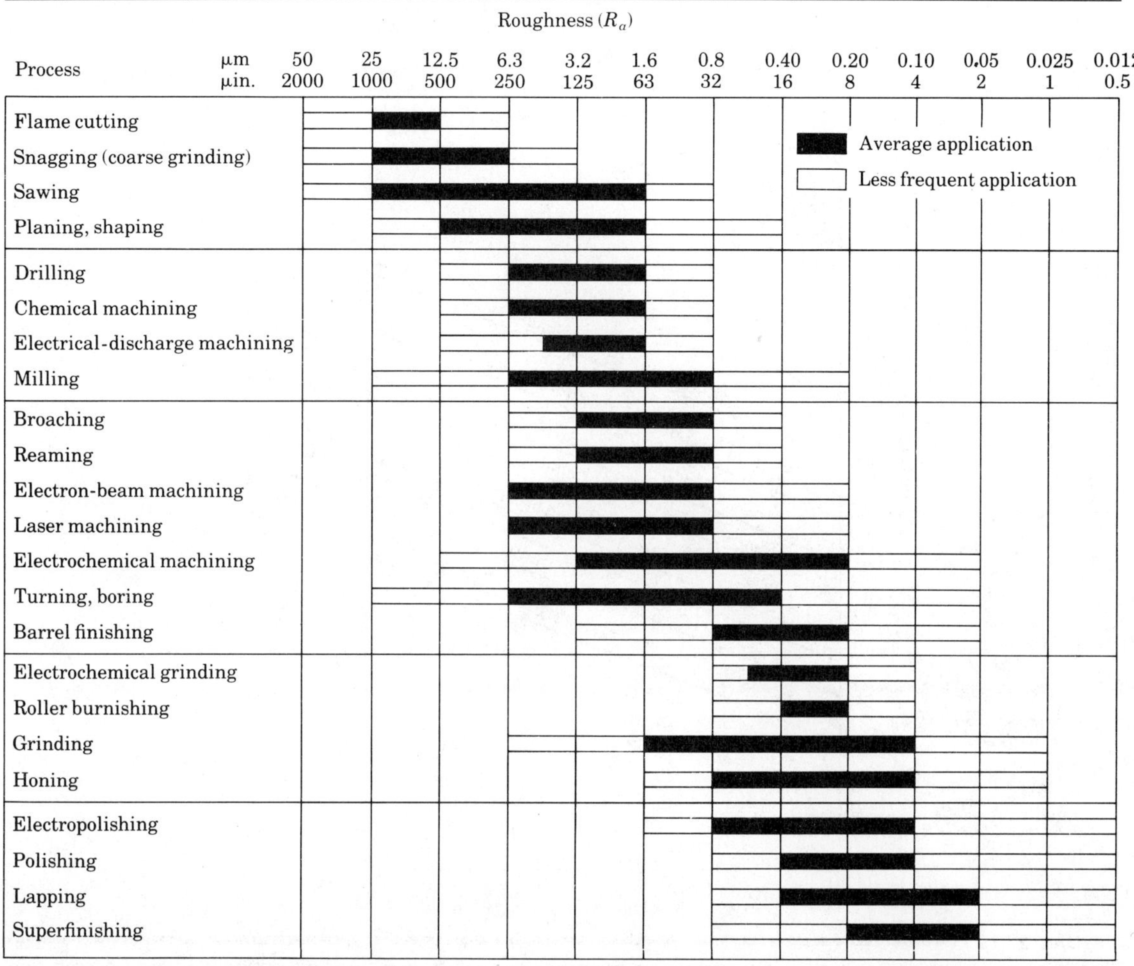

FIGURE 22.14
Range of surface roughnesses obtained in various machining processes. Note the wide range within each group. See also Fig. 26.4.

the parameters listed. In addition to the various recommendations concerning tools and process parameters that we have described thus far, an important consideration is the presence of vibration and chatter. Vibration during cutting can cause poor surface finish, poor dimensional accuracy, and premature tool wear and failure. Although we discuss this subject in greater detail in Section 24.3, we briefly outline here the generally accepted guidelines. Because of the complexity of the problem, however, some of the guidelines have to be implemented on a trial-and-error basis.

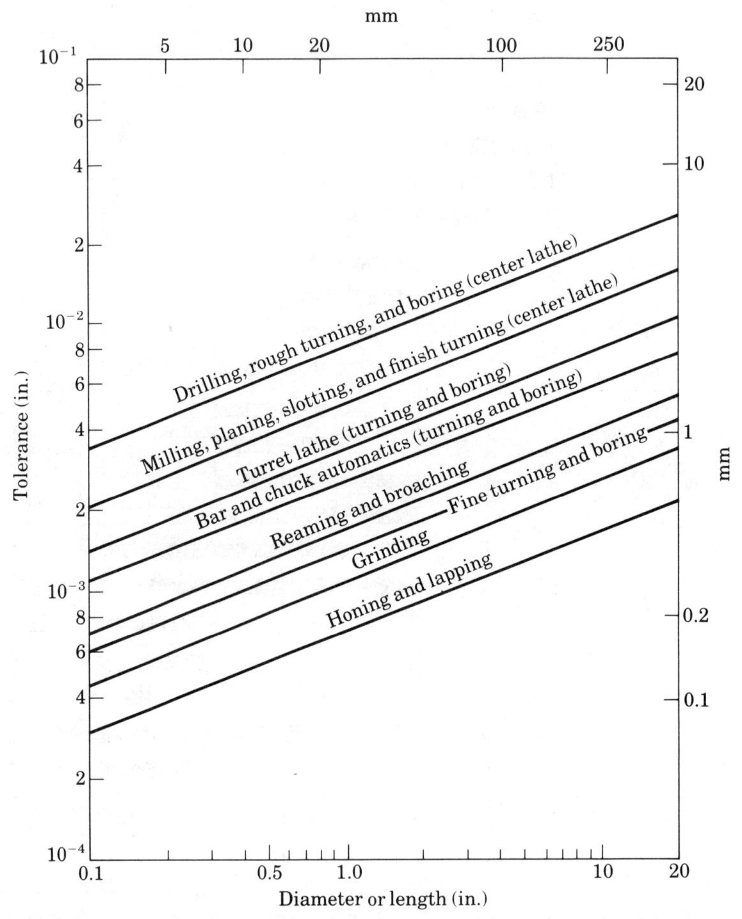

FIGURE 22.15 Range of tolerances obtained in various machining processes as a function of workpiece size. *Source:* Adapted from *Manufacturing Planning and Estimating Handbook,* McGraw-Hill, 1963.

TABLE 22.7
GENERAL TROUBLESHOOTING GUIDE FOR TURNING OPERATIONS

PROBLEM	PROBABLE CAUSES
Tool breakage	Tool material lacks toughness; improper tool angles; machine tool lacks stiffness; worn bearings and machine components; cutting parameters too high.
Excessive tool wear	Cutting parameters too high; improper tool material; ineffective cutting fluid; improper tool angles.
Rough surface finish	Built-up edge on tool; feed too high; tool too sharp, chipped or worn; vibration and chatter.
Dimensional variability	Lack of stiffness; excessive temperature rise; tool wear.
Tool chatter	Lack of stiffness; workpiece not supported rigidly; excessive tool overhang.

The guidelines are:

a) Minimize tool overhang.
b) Support workpiece rigidly.
c) Use machine tools with high stiffness and damping capacity.
d) When tools begin to vibrate and chatter, modify one or more of the process parameters, such as tool shape, cutting speed, feed rate, depth of cut, and cutting fluid.

22.4 Cutting Screw Threads

Screw threads are among the most important machine elements, as you can note by observing screws, bolts, and other threaded components in machines and various products. A **screw thread** may be defined as a ridge of uniform cross-section that follows a spiral or helical path on the outside or inside of a cylindrical (*straight thread*) or tapered or conical surface (*tapered thread*). Machine screws, bolts, and nuts have straight threads, as have threaded rods for applications such as the lead screw in lathes. Threads may be right-handed or left-handed. Tapered threads are commonly used for water or gas pipes and plumbing supplies so as to develop a watertight or airtight connection.

Threads may be produced basically by forming (*thread rolling*; see Section 13.5), which constitutes the largest quantity of threaded parts produced, or cutting. Casting threaded parts is also possible, although dimensional accuracy and production rate are not as high as those obtained in other processes. As we show in Fig. 22.1(k), turning operations are capable of producing threads on round bar stock. When threads are produced externally or internally by cutting with a lathe-type tool, the process is called *thread cutting* or *threading*. When cut internally with a special threaded tool (*tap*), it is called *tapping*, which we describe in Section 22.9. External threads may also be cut with a die or by milling. Although it adds considerably to the cost, threads may be ground for improved accuracy and surface finish.

22.4.1 Screw-thread nomenclature

Standardization of screw threads began in the middle 1880s, and several thread forms have since been standardized. Figure 22.16 shows the nomenclature for screw threads. Unified screw-thread forms (Fig. 22.17a) based on the American National thread system were adopted in the United States, Canada, and the United Kingdom in 1948 to obtain interchangeable screw threads. This standard has been revised periodically, particularly with regard to tolerances. In 1969, the ISO general-purpose screw-thread form was developed, with a wide range of metric sizes, and has been adopted by many countries (Fig. 22.17b). Other types of threads are shown in Fig. 22.18.

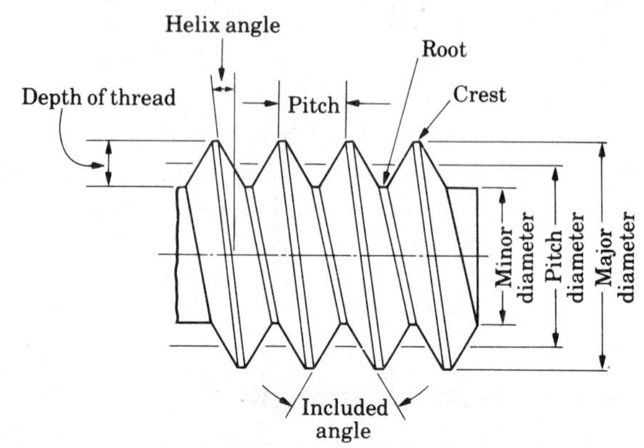

FIGURE 22.16
Standard nomenclature for screw threads.

In the Unified system, thread tolerances are specified as Class 1, 2, and 3, in decreasing order of looseness. In the ISO system, tolerance classes are a combination of tolerance grades and positions and are based on crest and pitch diameters. The *tolerance grade* is expressed by numerals ranging from 3 to 9, in order of increasing coarseness (loose tolerance). Comparatively, grade 6 in the ISO system is roughly equivalent to grade 2 in the Unified system. The letters represent *tolerance*

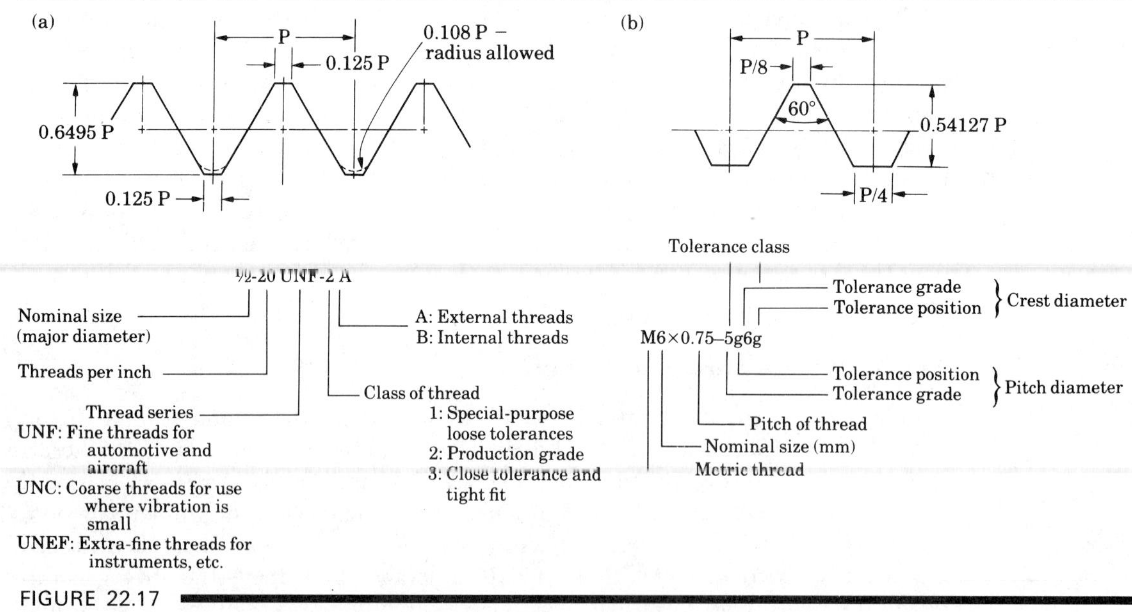

FIGURE 22.17
(a) Unified national thread and identification of threads. (b) ISO metric thread and identification of threads.

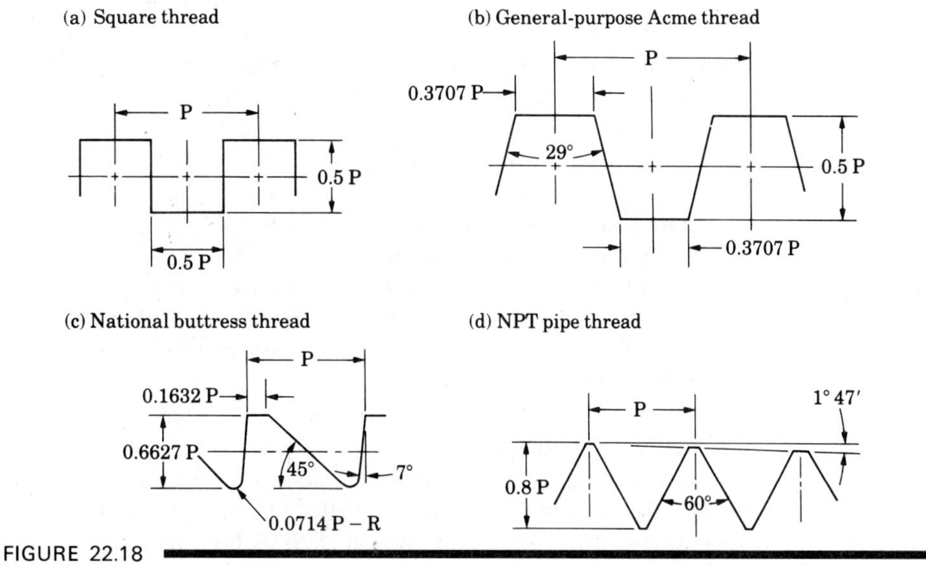

FIGURE 22.18
Various types of screw threads.

positions for the two diameters and indicate allowances (specified differences in size) according to the following system:

Allowance	External thread (bolt)	Internal thread (nut)
Large	e	—
Small	g	G
None	h	H

22.4.2 Screw-thread cutting on a lathe

A typical thread-cutting operation on a lathe is shown in Fig. 22.19(a). The cutting tool, whose shape depends on the type of thread to be cut, is mounted on a holder

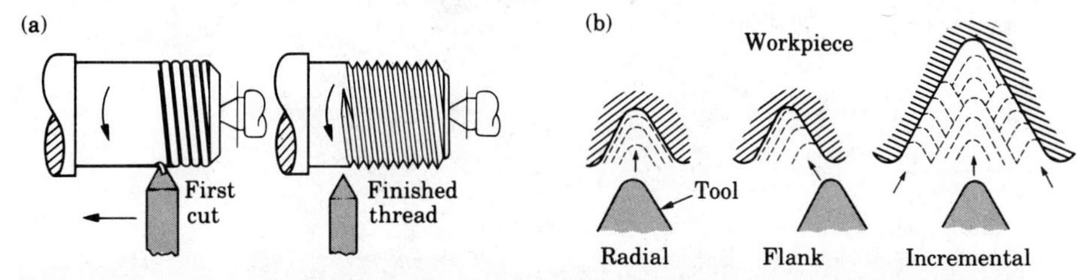

FIGURE 22.19
(a) Cutting screw threads on a lathe with a single-point cutting tool. (b) Cutting screw threads with a single-point tool in several passes. This process is normally utilized for large threads.

that is moved along the length of the workpiece by the lead screw (see Fig. 22.2). The axial movement of the tool in relation to the rotation of the workpiece determines the lead of the screw thread. Thus for a fixed spindle rpm, the slower the tool movement, the finer the thread will be. You may visualize threads as deep and coarse feed marks (see Fig. 20.26). In thread cutting, the cutting tool may be fed radially into the workpiece, thus cutting both sides of the thread at the same time as in form cutting. However, this method usually produces a poor surface finish. Generally, a number of passes in the sequence shown in Fig. 22.19(b) are required to produce good dimensional accuracy and surface finish. Although cutting threads on lathes is an old and versatile method, it requires considerable operator skill and is a slow, hence uneconomical, process. Consequently, except for small production runs, it has been largely replaced by other methods, such as thread rolling and automatic screw machining.

The production rate in cutting screw threads can be increased with tools called *die-head chasers* (Figs. 22.20a and b). These tools typically have four cutters with multiple teeth and can be adjusted radially. After threads are cut, the cutters open automatically (hence they are also called *self-opening die heads*) by rotating around their axes to allow the part to be removed. *Solid threading dies* (Fig. 22.20c) are also available for cutting straight or tapered screw threads. These dies are used mostly in threading ends of pipes and tubing and are not suitable for production work.

22.4.3 Automatic screw machines

Automatic screw machines are designed for high-production-rate machining of screws and similar threaded parts. Because they are capable of producing other components, they are now called **automatic bar machines.** All operations on these machines are performed automatically, with tools attached to a special turret. The bar stock is fed forward automatically after each screw or part is machined to finished dimensions and cut off. These machines may be equipped with single or multiple spindles. Capacities range from 3-mm to 150-mm ($\frac{1}{8}$-in. to 6-in.) diameter bar stock.

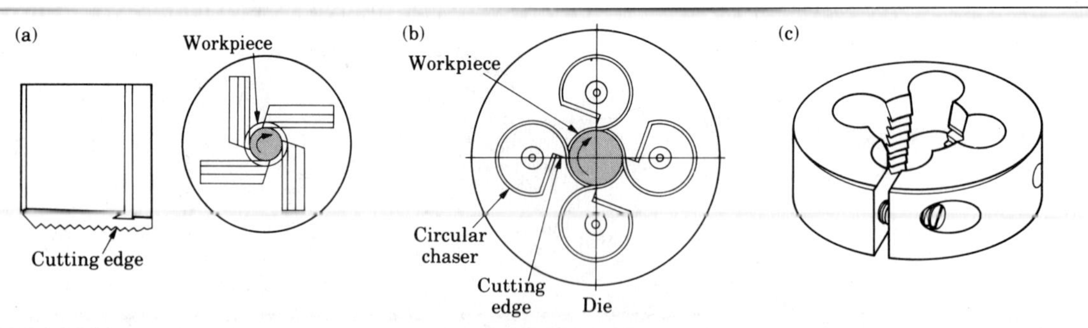

FIGURE 22.20
(a) Straight chasers for cutting threads on a lathe. (b) Circular chasers. (c) A solid threading die.

Single-spindle automatic machines are similar to turret lathes and are equipped with various cam-operated mechanisms. There are two types of single-spindle machines. In Swiss-type automatics (Fig. 22.21), the cylindrical surface of the rod is cut by a series of tools that move in radially, and in the same plane, toward the workpiece. The bar stock is clamped close to the headstock spindle, thus minimizing deflections. These machines are capable of high-precision machining of small-diameter parts. The other single-spindle machine is similar to a small automatic turret lathe. The turret is on a vertical plane, and all motions of the machine components are controlled by cams.

Multiple-spindle automatic machines have from four to eight spindles, arranged in a circle on a large drum, each carrying an individual workpiece. The cutting tools are arranged in various positions in the machine and move in both axial and radial directions. Each part is machined in stages as it moves from one station to the next. Because all operations are carried out simultaneously, cycle time per part is reduced.

22.4.4 Design considerations for screw-thread cutting

Certain design considerations must be taken into account in order to produce high-quality screw threads economically. The important factors are:

a) Designs should allow termination of threads before they reach a shoulder. Internal threads in blind holes should have an unthreaded length at the bottom.
b) Chamfers should be specified at the ends of threaded sections to minimize finlike threads with burrs.

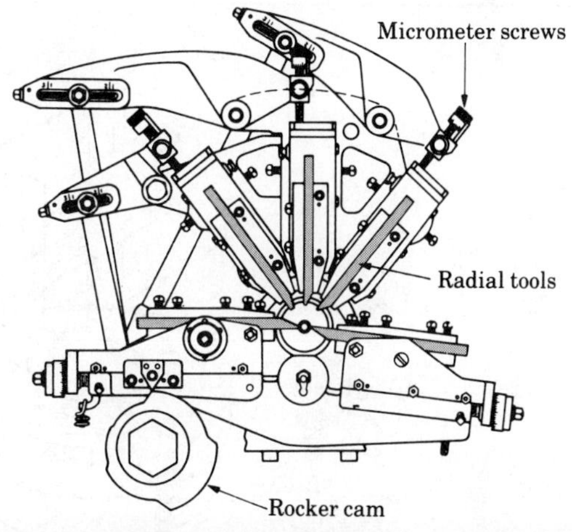

FIGURE 22.21 Schematic illustration of a Swiss-type automatic screw machine. *Source:* George Gorton Machine Company.

c) Threaded sections should not be interrupted with slots, holes, or other discontinuities.
d) Standard threading tooling and inserts should be used as much as possible.
e) Thin-walled parts should have sufficient thickness and strength to resist clamping and cutting forces.
f) Parts should be designed so that all cutting operations can be completed in one setup.

22.5 Boring and Boring Machines

We described the basic boring operation, as carried out on a lathe, in Fig. 22.1(h) and Section 22.2. Boring consists of producing circular internal profiles in hollow workpieces, or on a hole made by drilling or another process, and is done with cutting tools that are similar to those used in turning. However, because the boring bar has to reach the full length of the bore, tool deflection and hence maintaining dimensional accuracy can be a significant problem. The boring bar must be sufficiently stiff—that is, a material with high elastic modulus, such as carbides—to minimize deflection and avoid vibration and chatter. Boring bars have been designed with capabilities for damping vibration.

Although boring operations on relatively small workpieces can be carried out on a lathe, **boring mills** are used for large workpieces. These machines are either

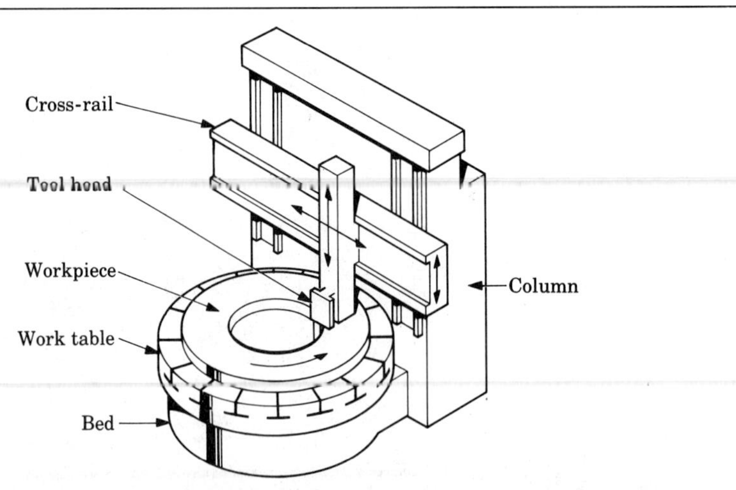

FIGURE 22.22
Schematic illustration of the components of a vertical boring mill.

vertical or horizontal and are capable of performing operations such as turning, facing, grooving, and chamfering. A *vertical boring machine* (Fig. 22.22) is similar to a lathe but with a vertical axis of workpiece rotation. The cutting tool, usually single point and made of M-2 and M-3 high-speed steel and C-7 and C-8 carbide, is mounted on the tool head, which is capable of vertical movement (for boring and turning) and radial movement (for facing), guided by the cross-rail. The head can be swiveled to produce tapered surfaces.

In *horizontal boring machines* (Fig. 22.23a), the workpiece is mounted on a table that can move horizontally in both the axial and radial directions. The cutting tool is mounted on a spindle that rotates in the headstock (Fig. 22.23b), which is capable of both vertical and longitudinal motions. Drills, reamers, taps, and milling cutters can also be mounted on the spindle.

Boring machines are available with a variety of features. Although workpiece diameters generally are 1–4 m (3–12 ft), workpieces as large as 20 m (60 ft) can be machined in some vertical boring machines. Machine capacities range up to 150 kW (200 hp). These machines are also available with computer numerical controls, whereby all movements are programmed, requiring little operator involvement and improving productivity. Cutting speeds and feeds for boring are similar to those for turning.

Jig borers are vertical boring machines with high-precision bearings. Although available in various sizes and used mainly in tool rooms for making jigs and fixtures, they are now being replaced by more versatile numerical control machines.

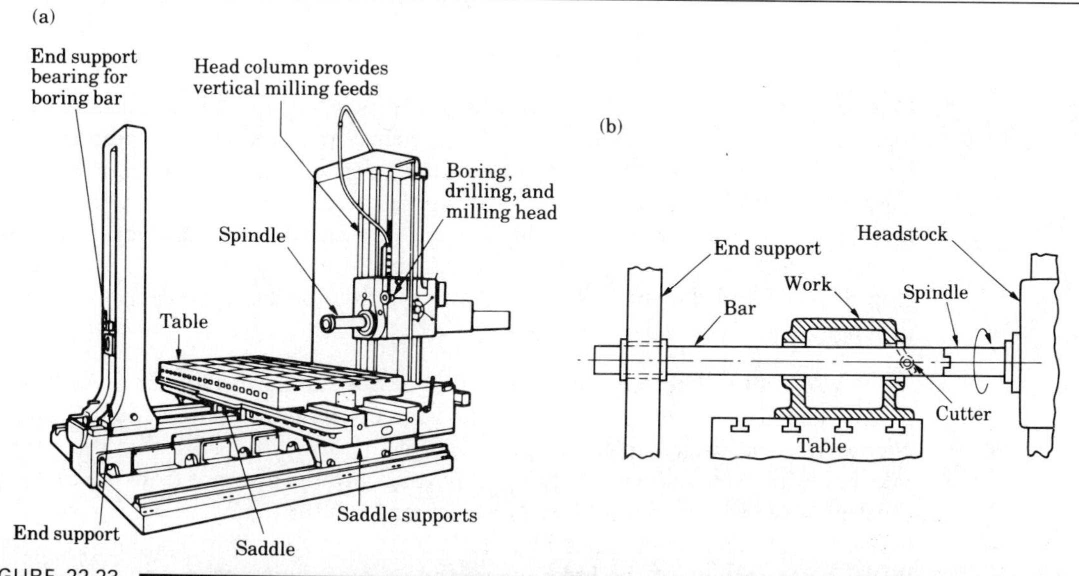

FIGURE 22.23
(a) Horizontal boring mill. *Source:* Giddings and Lewis, Inc. (b) Boring on a horizontal boring mill using an end support for the boring bar.

22.5.1 Design considerations for boring

Guidelines for efficient and economical boring operations are similar to those for turning operations. Additionally, the following factors should be considered:

a) Whenever possible, through holes rather than blind holes should be specified.
b) The greater the length-to-bore-diameter ratio, the more difficult it is to hold dimensions because of the deflections of the boring bar from cutting forces.
c) Interrupted internal surfaces should be avoided.

22.6 Drilling and Holemaking Operations

When inspecting various products around us, we realize that the vast majority have a variety of holes. Holes are used either for assembly with fasteners such as bolts, screws, and rivets or to provide access inside machinery or products. Note, for example, the number of rivets on an airplane's fuselage or the bolts in various components under the hood of an automobile, each rivet or bolt requiring a hole. **Holemaking** is thus among the most important operations in manufacturing. Two basic mechanical processes for producing holes are punching, which we described in Section 16.2, and drilling. We discuss various electrical, chemical, and high-energy beam methods of producing holes and similar cavities in Chapter 26.

22.6.1 Drills

One of the most common of all machining processes is drilling. **Drills** usually have a high length-to-diameter ratio (Fig. 22.24), hence are capable of producing deep holes. However, they are somewhat flexible, depending on their diameter, and should be used with care in order to drill holes accurately and to prevent the drill from breaking. Furthermore, the chips that are produced within the workpiece have to move in the direction opposite to the axial movement of the drill. Consequently, chip disposal and the effectiveness of cutting fluids can be significant problems in drilling.

Twist drill. The most common drill is the standard-point twist drill (Fig. 22.25). The main features of the drill point are a *point angle*, *lip-relief angle*, *chisel-edge angle*, and *helix angle*. The geometry of the drill tip is such that the normal rake angle and velocity of the cutting edge vary with the distance from the center of the drill. The drill–workpiece interface, viewed from the top of the workpiece as the hole is being drilled, is shown in Fig. 22.26. Note the two continuous chips produced in the outer regions of the hole, where the rake angle is positive and the cutting speed is high. In the center, however, the rake angle is highly negative and the speed is low, producing a highly deformed, coarse chip.

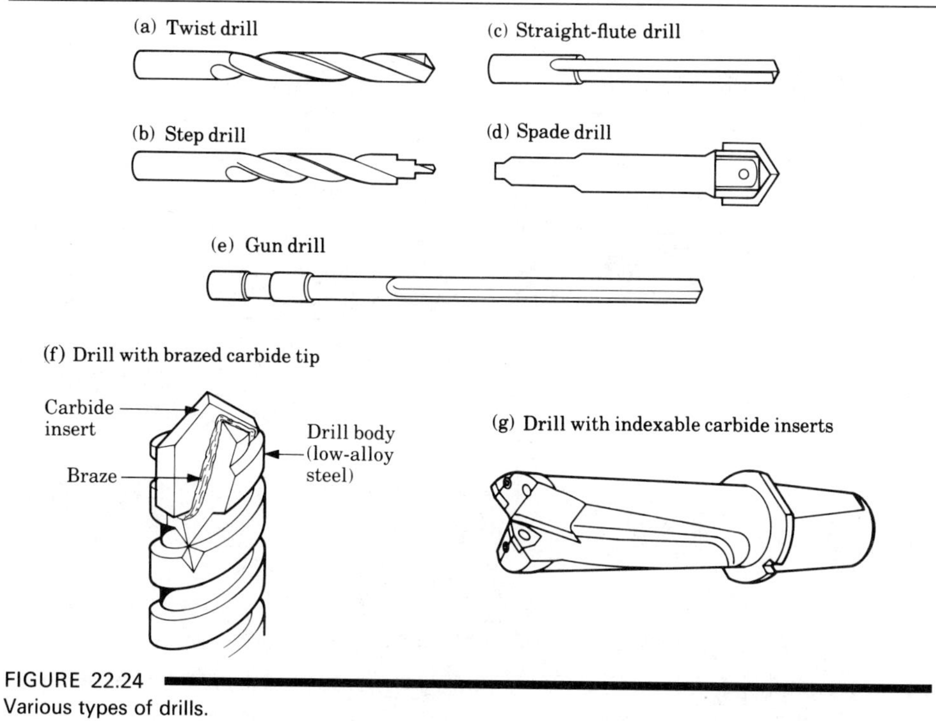

FIGURE 22.24
Various types of drills.

Generally, two spiral grooves (*flutes*) run the length of the drill, and the chips produced are guided upward through these grooves. The grooves also serve as passageways to enable the cutting fluid to reach the cutting edges. Drills are available with chip-breaker features ground along the cutting edges. This feature is important in drilling with automated machinery where disposal of long chips without operator interference is important. Drills are also provided with internal longitudinal holes through which cutting fluids are forced, thus improving lubrication and washing away the chips.

General recommendations for drill geometry for various workpiece materials are presented in Table 22.8 (p. 697). These angles are based on experience in drilling operations and are designed to produce accurate holes, minimize drilling forces and torque, and optimize drill life. Various other drill-point geometries have been developed to improve drill performance, but they require special drilling techniques and equipment. Small changes in drill geometry can have a significant influence on the drill's performance. For example, too small a lip relief angle (see Fig. 22.25) increases the thrust force, generates excessive heat, and increases wear. Conversely, too large an angle can cause chipping or breaking of the cutting edge.

Other types of drills. Several types of drills are shown in Figs. 22.24 and 22.27 (p. 698). A *step drill* produces holes of two or more different diameters. A *core drill*

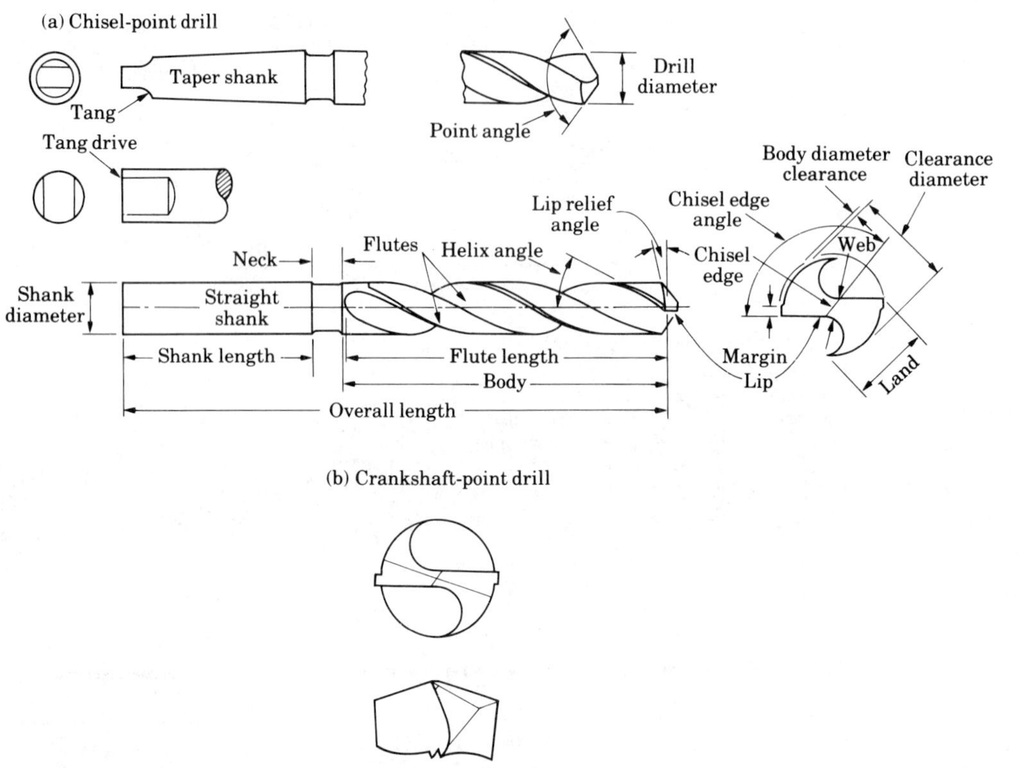

FIGURE 22.25
(a) Standard chisel-point drill indicating various features. (b) Crankshaft-point drill.

FIGURE 22.26
Top view of a hole being drilled with a standard-point drill, showing chip formation. *Source:* A. W. Haggerty.

TABLE 22.8
GENERAL RECOMMENDATIONS FOR DRILL GEOMETRY FOR HIGH-SPEED STEEL TWIST DRILLS

WORKPIECE MATERIAL	POINT ANGLE	LIP-RELIEF ANGLE	CHISEL-EDGE ANGLE	HELIX ANGLE	POINT
Aluminum alloys	90–118	12–15	125–135	24–48	Standard
Magnesium alloys	70–118	12–15	120–135	30–45	Standard
Copper alloys	118	12–15	125–135	10–30	Standard
Steels	118	10–15	125–135	24–32	Standard
High-strength steels	118–135	7–10	125–135	24–32	Crankshaft
Stainless steels, low strength	118	10–12	125–135	24–32	Standard
Stainless steels, high strength	118–135	7–10	120–130	24–32	Crankshaft
High-temp. alloys	118–135	9–12	125–135	15–30	Crankshaft
Refractory alloys	118	7–10	125–135	24–32	Standard
Titanium alloys	118–135	7–10	125–135	15–32	Crankshaft
Cast irons	118	8–12	125–135	24–32	Standard
Plastics	60–90	7	120–135	29	Standard

has a square end and is used to make an existing hole larger. *Counterboring* and *countersinking drills* produce depressions on the surface to accommodate the heads of screws and bolts. A *center drill* is a short drill and is used to help start a hole and guide the drill for regular drilling. *Spade drills* have a removable tip or bit and are used to produce large and deep holes. They have the advantages of higher stiffness (because of the absence of flutes in the body of the drill), ease of grinding the cutting edge, and lower cost. *Crankshaft drills* (Fig. 22.25b) have good centering ability, and because chips tend to break up easily these drills are suitable for drilling deep holes.

Gun drilling. Developed originally for drilling gun barrels, hence the word gun, *gun drilling* requires a special drill (see Figs. 22.24e and 22.27) and is used for drilling deep holes. Hole depth-to-diameter ratios can be 300 or higher. The thrust force (the radial force that tends to push the drill sideways) is balanced by bearing pads on the drill that slide along the inside surface of the hole (Fig. 22.28). Thus a gun drill is self-centering—an important factor in drilling straight, deep holes. The cutting fluid is forced under high pressure through a longitudinal hole in the body of the drill (Fig. 22.29). In addition to its lubricating and cooling functions, the fluid flushes out chips that otherwise would be trapped in the hole and interfere with the drilling operation. Cutting speeds are usually high and feeds are low.

Trepanning. In *trepanning* the cutting tool (Fig. 22.30a) produces a hole by removing a disk-shaped piece (core), usually from flat plates. Thus a hole is

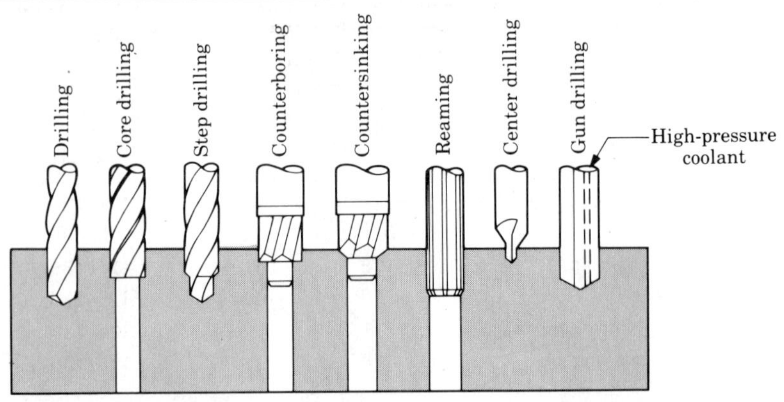

FIGURE 22.27
Various types of drills and drilling operations.

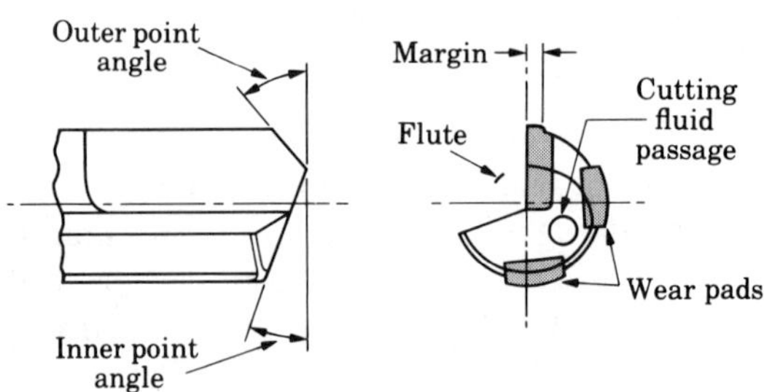

FIGURE 22.28
A gun drill showing various features.

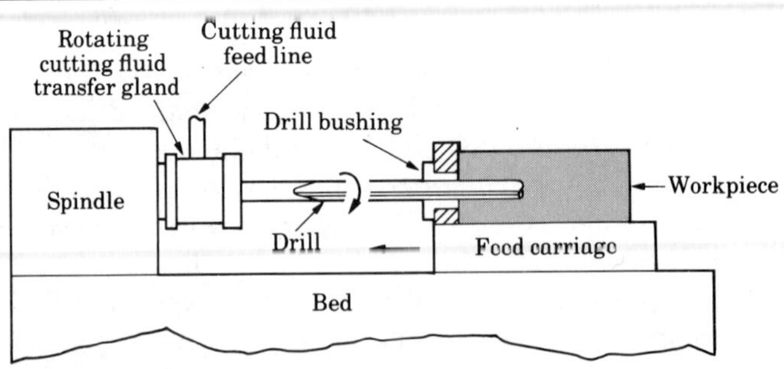

FIGURE 22.29
Method of gun drilling. *Source:* Eldorado Tool and Manufacturing Corporation.

produced without reducing all the material to chips. The process can be used to make disks up to 150 mm (6 in.) in diameter from flat sheet or plate. Trepanning can also be used to make circular grooves in which O-rings are to be placed (see also Fig. 22.1f). Trepanning can be done on lathes, drill presses (Fig. 22.30b), or other machines, using single-point or multipoint tools. A variation of trepanning is *gun-trepanning*, which uses a cutting tool similar to a gun drill except that the tool has a central hole in it.

22.6.2 Drill materials and sizes

Drills are usually made of high-speed steels (M1, M7, and M10) and can be coated with titanium nitride for increased wear resistance. Carbide-tipped (see Figs. 22.24f and g) or solid-carbide (C-2) drills are available for cast irons, steels, hard, high-temperature metals and abrasive materials, such as concrete and brick (*masonry drills*), and composite materials with abrasive fiber reinforcements, such as glass and graphite. Standard twist-drill sizes consist of the following series:

Numerical: No. 80 (0.0135 in.) to No. 1 (0.228 in.).

Letter: A (0.234 in.) to Z (0.413 in.).

Fractional: Straight shank from $\frac{1}{64}$ in. to $1\frac{1}{4}$ in. (in $\frac{1}{64}$ in. increments) to $1\frac{1}{2}$ in. (in $\frac{1}{32}$ in. increments), and larger drills in larger increments. Taper shank from $\frac{1}{8}$ in. to $1\frac{3}{4}$ in. (in $\frac{1}{64}$ in. increments) to $3\frac{1}{2}$ in. in $\frac{1}{16}$ in. increments.

Millimeter: In increments of 0.01–0.50 mm.

Typical ranges of drill diameters for various types are shown in Table 22.9.

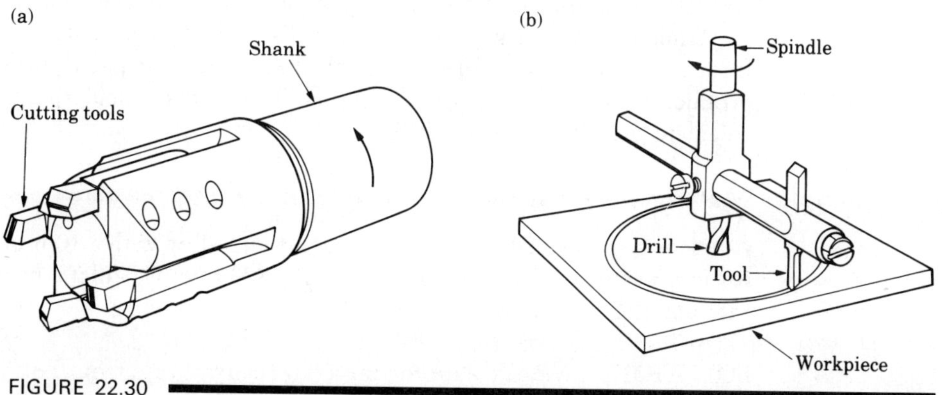

FIGURE 22.30
(a) Trepanning tool. (b) Trepanning with a drill-mounted single cutter.

TABLE 22.9
CAPABILITIES OF DRILLING AND BORING OPERATIONS

TOOL TYPE	DIAMETER RANGE (mm)	HOLE DEPTH/DIAMETER	
		TYPICAL	*MAXIMUM*
Twist	0.5–150	8	50
Spade	25–150	30	100
Gun	2–50	100	300
Trepanning	40–250	10	100
Boring	3–1200	5	8

22.6.3 Material removal rate

The material removal rate (MRR) in drilling is the ratio of volume removed to time. Thus,

$$\text{MRR} = \frac{\pi D^2}{4} fN, \tag{22.3}$$

where D is the drill diameter, f is the feed, and N is the rpm of the drill.

22.6.4 Thrust force and torque

The thrust force in drilling acts perpendicular to the hole axis. If this force is excessive, it can cause the drill to break or bend. The thrust force depends on factors such as the strength of the workpiece material, feed, rotational speed, cutting fluids, drill diameter, and drill geometry. Although some attempts have been made, accurate calculation of the thrust force on the drill has proven to be difficult. Experimental data are available as an aid in designing and using drills and drilling equipment. Thrust forces in drilling range from a few newtons for small drills to as high as 100 kN (22.5 klb) in drilling high-strength materials with large drills. Similarly, drill torque can range as high as 4000 N · m (3000 lb-ft).

Similarly, the torque during drilling is difficult to calculate. We can obtain it from the data in Table 20.1 by noting that power dissipated during drilling is the product of torque and rotational speed. Thus by first calculating the MRR, we can calculate the torque on the drill.

• Example: Material removal rate and torque in drilling.

A hole is being drilled in a block of magnesium alloy with a 10-mm drill at a feed of 0.2 mm/rev. The spindle is running at $N = 800$ rpm. Calculate the material removal rate and the torque on the drill.

SOLUTION. We calculate the material removal rate from Eq. (22.3). Thus,

$$\text{MRR} = [(\pi)(10)^2/4](0.2)(800) = 12{,}570 \text{ mm}^3/\text{min} = 210 \text{ mm}^3/\text{s}.$$

Referring to Table 20.1, we take an average unit power of 0.5 $W \cdot s/mm^3$ for magnesium alloys. Hence the power dissipated is

$$\text{Power} = (210)(0.5) = 105 \text{ W}.$$

Power is the product of the torque on the drill and the rotational speed in radians per second, which in this case is $(800)(2\pi)/60 = 83.8$. Noting that $W = J/s$ and $J = N \cdot m$, we have

$$T = 105/83.8 = 1.25 \text{ N} \cdot \text{m}.$$

22.6.5 Drilling practice

Drills and similar holemaking tools are usually held in *drill chucks* which may be tightened with or without keys (Fig. 22.31). Special chucks and collets, with various quick-change features that do not require stopping the spindle, are available for use on automated machinery. Because a drill doesn't have a centering action, it tends to "walk" on the workpiece surface at the beginning of the operation. This problem is particularly severe with small-diameter drills. To start a hole properly, the drill should be guided to keep it from deflecting sideways. A small starting hole can be made with a center drill, fixtures (such as a bushing) can be used, or the drill point may be ground to an S shape (*spiral point*). This shape's self-centering characteristic eliminates center drilling, produces accurate holes, and improves drill life. These factors are particularly important in automated production with computer numerical control machines.

Chip removal during drilling can be difficult, especially for deep holes in soft and ductile materials. The drill should be retracted periodically to remove chips that may have accumulated along the flutes; otherwise the drill may break because of excessive torque.

Because of its rotary motion, drilling produces holes with walls that have circumferential marks. In contrast, punched holes have longitudinal lines (see Fig. 16.5a). This difference is significant in terms of the hole's fatigue properties, which we discuss in Section 30.5. Drills generally leave a burr on the bottom surface upon breakthrough, thus necessitating deburring operations (see Section 25.10).

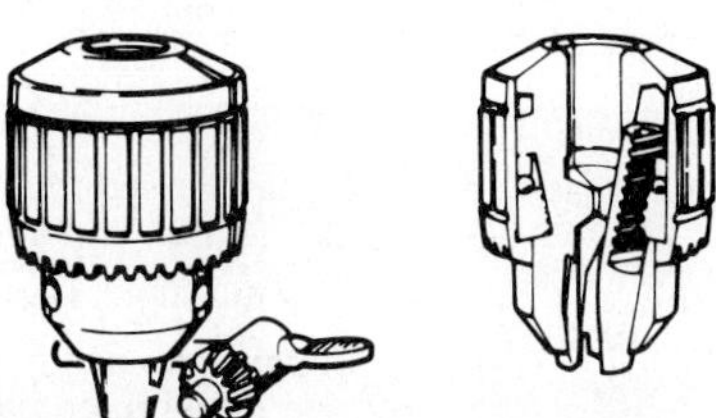

FIGURE 22.31
Cross-sectional view of a key-tightening type drill chuck. *Source:* Jacobs Manufacturing Company.

TABLE 22.10
GENERAL RECOMMENDATIONS FOR SPEEDS AND FEEDS IN DRILLING

WORKPIECE MATERIAL	SPEED (m/s)	FEED (mm/rev) DRILL DIAMETER (mm) 1.5	50
Aluminum alloys	0.5–2	0.025	0.75
Magnesium alloys	0.75–2	0.025	0.75
Copper alloys	0.25–1	0.025	0.65
Steels	0.3–0.5	0.025	0.75
Stainless steels	0.2–0.3	0.025	0.45
Titanium alloys	0.1–0.3	0.01	0.3
Cast irons	0.3–1	0.025	0.75
Thermoplastics	0.5–1	0.025	0.3
Thermosets	0.3–1	0.025	0.4

Note: As hole depth increases, speeds and feeds should be reduced.

Rotational speeds in drilling can range up to 30,000 rpm for drills less than 1 mm in diameter. Recommended ranges of drilling speed and feeds are given in Table 22.10. A general guide to the probable causes of problems in drilling operations is presented in Table 22.11.

Drill reconditioning. Drills are reconditioned by grinding them either manually or using special fixtures. Proper reconditioning of drills is important, particularly in automated manufacturing on computer numerical control machines. Hand grinding is difficult and requires considerable skill in order to produce symmetric and accurate cutting edges. Grinding on fixtures is accurate and is done on special computer-controlled grinders. Drills coated with titanium nitride can be recoated during reconditioning.

TABLE 22.11
GENERAL TROUBLESHOOTING GUIDE FOR DRILLING OPERATIONS

PROBLEM	PROBABLE CAUSES
Drill breakage	Dull drill; drill seizing in hole because of chips clogging flutes; feed too high; lip relief angle too small.
Excessive drill wear	Cutting speed too high; ineffective cutting fluid; rake angle too high; drill burned and strength lost when sharpened.
Tapered hole	Drill misaligned or bent; lips not equal; web not central.
Oversize hole	Same as above; machine spindle loose; chisel edge not central; side pressure on workpiece.
Poor hole surface finish	Dull drill; ineffective cutting fluid; welding of workpiece material on drill margin; improperly ground drill; improper alignment.

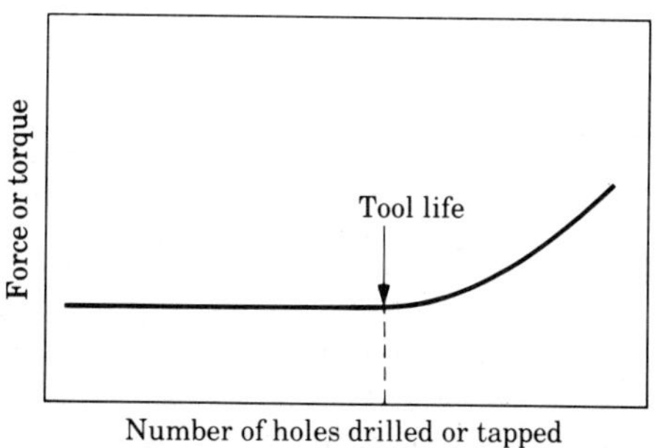

FIGURE 22.32
Determination of drill life by monitoring the rise in force or torque as a function of the number of holes drilled. This test is also used for determining tap life.

22.6.6 Measuring drill life

Dulling of drills and resharpening or replacing them are important, particularly in automated production. Use of dull drills increases forces, causes surface damage, and produces inaccurate holes. Life of drills, as well as taps, is usually measured by the number of holes drilled before they become dull.

The test procedure consists of clamping a block of material on a suitable dynamometer or force transducer and drilling a number of holes while recording and monitoring the torque or force during each successive operation. After a number of holes have been drilled, the torque and force begin to increase because the tool is becoming dull (Fig. 22.32). *Drill life* is defined as the number of holes drilled until this transition begins. Other techniques, such as monitoring vibration and acoustic emissions, may also be used to determine drill life. These techniques are particularly effective in computer-controlled operations.

22.7 Drilling Machines

Drilling machines are used for drilling holes, tapping, reaming, and other general-purpose, small-diameter boring operations. These machine tools are generally vertical. The most common vertical type is the **drill press**, the major components of which are shown in Fig. 22.33. The workpiece is placed on an adjustable table, either by clamping it directly into the slots and holes on the table or by using a vise, which in turn can be clamped to the table. The workpiece should be properly clamped, both for safety and accuracy, because the drilling torque can be high enough to rotate the workpiece. The drill is lowered manually by hand wheel or by power feed at preset rates. Manual feeding requires some skill in judging the

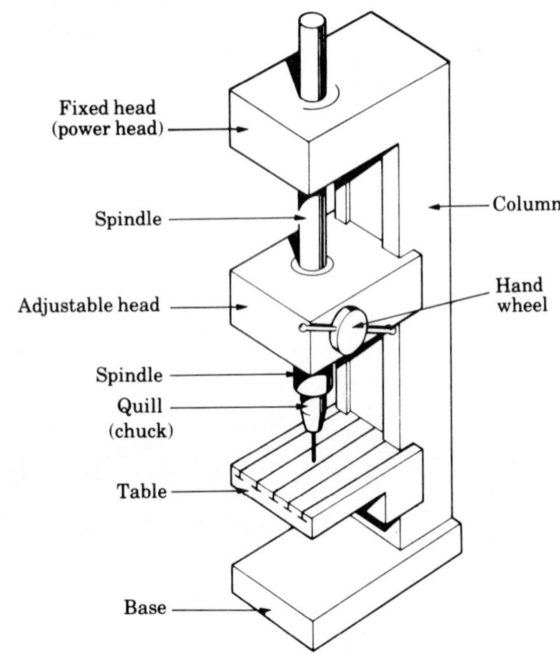

FIGURE 22.33
Schematic illustration of the components of a vertical drill press.

appropriate feed rate. In order to maintain proper cutting speeds at the cutting edges of drills, the spindle speed on drilling machines has to be adjustable to accommodate different sizes of drills. Adjustments are made by means of pulleys, gear boxes, or variable-speed motors. Drill presses are usually designated by the largest workpiece diameter that can be accommodated on the table. Sizes typically range from 150 mm to 1250 mm (6 in. to 50 in.).

Types of drilling machines range from simple *bench-type* units, used to drill small-diameter holes, to large *radial drills*, which can accommodate large workpieces (Fig. 22.34). The distance between the column and the spindle center can be as large as 3 m (10 ft). The drill head of *universal* drilling machines can be swiveled to drill holes at an angle. Recent developments in drilling machines include computer numerical control, three-axis machines in which various drilling operations are performed automatically and in the desired sequence with the use of a turret (Fig. 22.35). Note that the turret holds several different tools (see also Section 24.2).

Drilling machines with multiple spindles are used for high production-rate operations (*gang drilling*). These machines are capable of drilling as many as 50 holes of varying size, depth, and location in one step. Multiple-spindle machines are also used for reaming and counterboring operations.

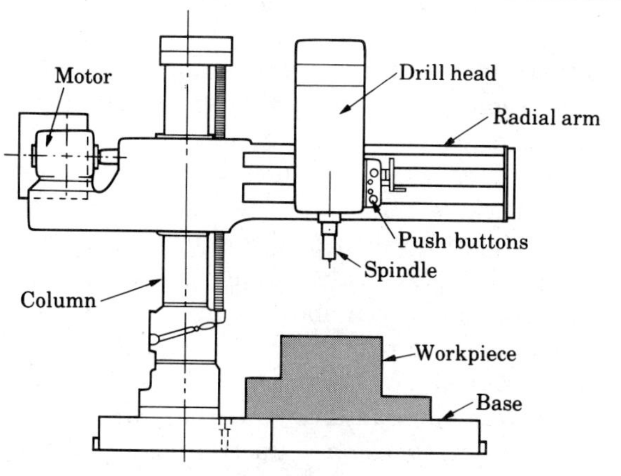

FIGURE 22.34
Schematic illustration of a radial drill press. Radial presses are designed to accommodate large workpieces.

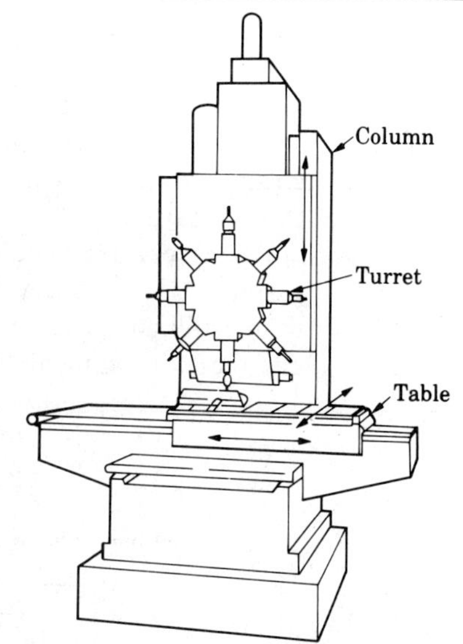

FIGURE 22.35
A three-axis computer numerical control drilling machine. The turret holds as much as eight different tools, such as drills, taps, and reamers.

22.8 Reaming and Reamers

Reaming is an operation used to make an existing hole dimensionally more accurate than can be obtained by drilling alone and to improve its surface finish. The most accurate holes are produced by the following sequence of operations:

- Centering
- Drilling
- Boring
- Reaming

For even better accuracy and surface finish, holes may be internally ground and honed.

A **reamer** (Fig. 22.36a) is a multiple-cutting-edge tool with straight or helically fluted edges, which removes very little material. The shanks may be straight or tapered, as in drills. The basic types of reamers are hand and machine.

Hand reamers are straight or have a tapered end in the first third of their length. Various *machine reamers*, also called chucking reamers because they are mounted in a chuck and operated by a machine, are available. There are two types of chucking reamers. *Rose* reamers have cutting edges with wide margins and no relief (see Fig. 22.36a). They remove considerable material and true up a hole for flute reaming. *Fluted* reamers have small margins and relief, with a rake angle of about 5°. They are usually used for light cuts of about 0.1 mm (0.004 in.) on the hole diameter. *Shell* reamers, which are hollow and mounted on an arbor, are generally used for holes larger than 20 mm (0.75 in.). *Expansion* reamers are adjustable for small variations in hole size and also to compensate for wear of the reamer's cutting edges. *Adjustable* reamers (Fig. 22.36b) can be set for specific hole diameters and are thus versatile.

Reamers are usually made of high-speed steels (M1, M2, and M7) or solid carbides (C-2)—or have carbide cutting edges. Reaming speeds are generally lower than drilling for the same material, but feeds are higher. Proper reamer maintenance and reconditioning are important for hole accuracy and surface finish. Reamers may

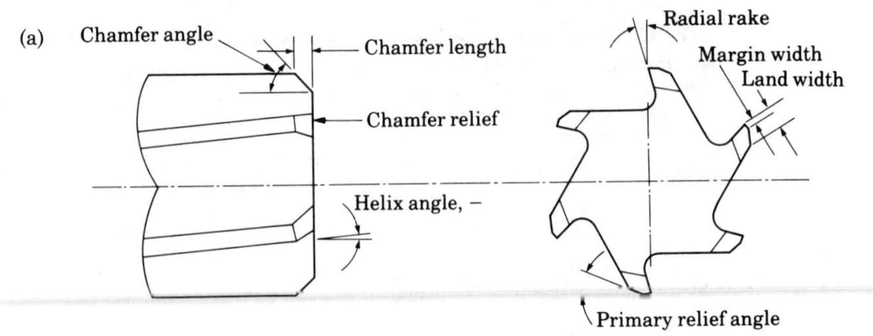

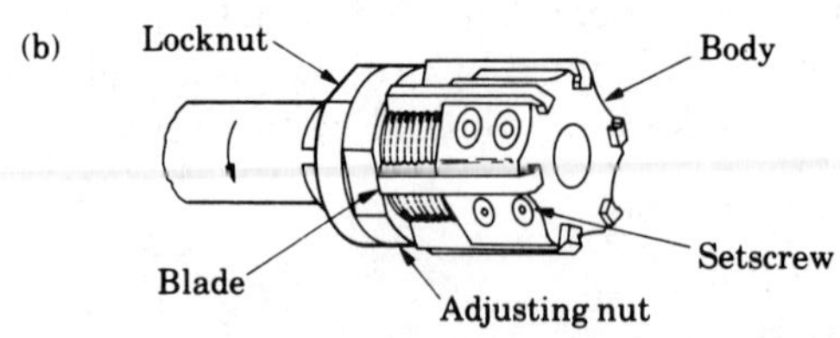

FIGURE 22.36 (a) Terminology for a helical reamer. (b) Inserted-blade adjustable reamer.

be held rigidly, as in a chuck, or they may float in their holding fixtures to ensure alignment or be piloted in guide bushings placed above and below the workpiece.

22.9 Tapping and Taps

Internal threads in workpieces can be produced by **tapping.** A **tap** is basically a threading tool with multiple cutting teeth (Fig. 22.37a). Taps are generally available with three or four flutes. Three-fluted taps are stronger because of the larger amount of material available in the flute. *Tapered taps* are designed to reduce the torque required for tapping through holes. *Bottoming taps* are for tapping blind holes to their full depth. *Collapsible taps* are for large diameter holes. After tapping has been completed, the tap is mechanically collapsed and, without having to rotate it, removed from the hole. Tap sizes range up to 100 mm (4 in.).

A recent development involves drilling and tapping in one stroke. The process, called *drapping,* uses a tool which has a drilling section followed by a tapping section. A hole is first drilled at a certain speed, then the speed is lowered for the tapping portion of the stroke. In this way, productivity is improved, particularly for through holes in parts that are subsequently fastened.

Chip removal can be a significant problem during tapping because of the small clearances involved. If chips aren't removed properly, the resulting excessive torque can break the tap. The use of a proper cutting fluid and periodic reversal and removal of the tap from the hole are effective means of chip removal and improving the quality of the tapped hole. The difficulty of tapping relative to other cutting operations can be appreciated by reviewing Fig. 21.13, where we note that tapping is among the most severe, requiring low cutting speeds and effective cutting fluids.

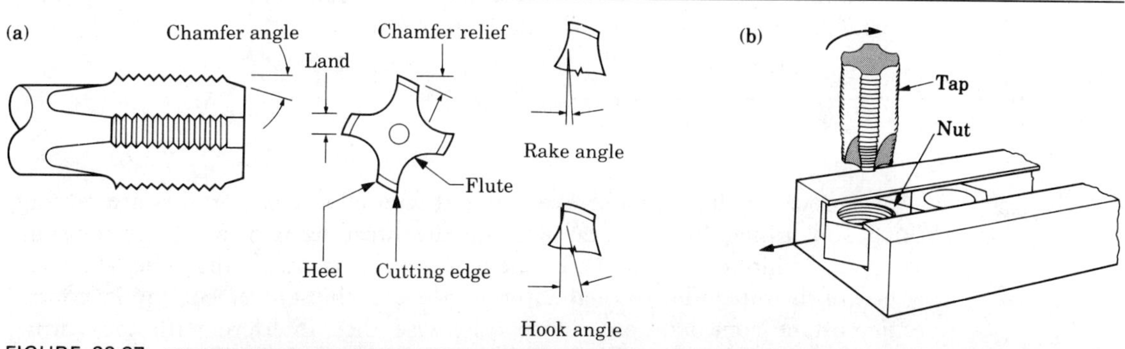

FIGURE 22.37
(a) Terminology for a tap. (b) Tapping of steel nuts in production.

Tapping may be done by hand or in drilling machines, lathes, or automatic screw machines, using tapping heads to hold the taps. Special tapping machines are also available with features for multiple tapping operations. The system for automatic tapping of nuts shown in Fig. 22.37(b) can achieve a production rate of 500 pieces per hour. With proper lubrication, tap life may be as high as 10,000 holes tapped. Tap life can be determined by the same technique used to measure drill life. Taps are usually made of carbon steels for light-duty applications and high-speed steels (M1, M2, M7, and M10) for production work.

22.10 Design Considerations for Drilling, Reaming, and Tapping

Some of the design guidelines for drilling, reaming, and tapping operations are as follows:

a) Designs should allow holes to be drilled on flat surfaces, perpendicular to the drill motion; otherwise the drill tends to deflect and the hole will not be located accurately. Exit surfaces for the drill should also be flat.
b) Interrupted hole surfaces should be avoided or minimized for better dimensional accuracy.
c) Hole bottoms should match standard drill-point angles (see Table 22.8). Flat bottoms or odd shapes should be avoided.
d) As in boring operations, through holes are preferred over blind holes. Hole depth should be minimized for good dimensional accuracy. If holes with large diameters are required, the workpiece should have a preexisting hole, preferably made during fabrication of the part by forming or casting.
e) Parts should be designed so that all drilling can be done with a minimum of fixturing or without repositioning the workpiece.
f) Reaming blind or intersecting holes may be difficult because of possible tool breakage. Extra hole depth should be provided.

SUMMARY

Cutting processes that produce external and internal circular profiles are turning, boring, and drilling. Reaming, tapping, and die threading are processes for finishing workpieces. Chip formation in all these processes is essentially the same. However, because of the three-dimensional nature of the cut, chip movement and its control are important considerations, since otherwise they interfere with the cutting operation. Chip removal can be a significant problem especially in drilling and tapping and can lead to tool breakage. Each process should be studied in order to understand the interrelationships of design parameters, such as dimensional accuracy, surface finish, and integrity, and process parameters, such as speed, feed, depth of cut, tool material and shape, and cutting fluids.

Design guidelines should be followed carefully to take full advantage of the capabilities of each process. Parts to be machined may have been produced by casting, forging, extrusion, powder metallurgy, and so on. The closer the blank to be machined to the final shape desired, the fewer the number and extent of machining processes required. Such net-shape manufacturing is of major significance in minimizing costs.

TRENDS

- The major trend in cutting processes is to optimize operations and minimize costs by automation, thus reducing labor.
- Sensors for forces, power, deflections, temperature, tool wear and fracture, and surface finish are under study as a means of on-line compensation to reduce defective parts and maintain quality.
- Machine tools are being designed and built such that setup time and idle time (such as time spent in non-cutting tool movements, retraction, acceleration and deceleration of tools, and turret indexing) are reduced.
- Machine tools are now capable of performing various simultaneous cutting operations on a workpiece, using two-spindle, four-axis, two-turret systems, as well as performing roughing and finishing operations in one setup.
- In-process gaging systems monitor the cutting operation and provide real-time control feedback. These include automatic tool wear and breakage monitoring and detection and implementation of inspection techniques for statistical process control.
- Computer-controlled turning machines are replacing screw machines in specialty fastener products.

KEY TERMS

Automatic bar machine	Form tool	Reamer
Boring	Holemaking	Reaming
Boring mill	Jig borer	Screw thread
Drill	Knurling	Tap
Drill press	Lathe	Tapping
Drilling	Material removal rate	Threading
Facing	Parting	Turning

BIBLIOGRAPHY

Boothroyd, G., and W.A. Knight, *Fundamentals of Machining and Machine Tools*, 2d ed. New York: Marcel Dekker, 1989.

Machining Data Handbook, 2 vols. 3d ed. Cincinnati: Machinability Data Center, 1980.

Metals Handbook, 8th ed., *Vol. 3: Machining*. Metals Park, Ohio: American Society for Metals, 1967.

Metals Handbook, 9th ed., Vol. 16: *Machining*. Metals Park, Ohio: ASM International, 1989.

Reshetov, D.N., and V.T. Portman, *Accuracy of Machine Tools*. New York: American Society of Mechanical Engineers, 1989.

Tool and Manufacturing Engineering Handbook, 4th ed., *Vol. 1: Machining*. Dearborn, Mich.: Society of Manufacturing Engineers, 1983.

Turning Handbook of High-Efficiency Cutting. Detroit: Carboloy Systems Department, General Electric Company, 1980.

Weck, M., *Handbook of Machine Tools*, 4 vols. New York: Wiley, 1984.

REVIEW QUESTIONS

22.1 Describe the type of machining operations that can be performed on a lathe.
22.2 Explain the functions of different angles on a single-point lathe cutting tool.
22.3 List the major components of a lathe and describe their functions.
22.4 What is the difference between feed rod and lead screw?
22.5 Why were power chucks developed?
22.6 How are lathes specified?
22.7 Why can boring on a lathe be a difficult operation?
22.8 Describe the differences between a tracer lathe and a turret lathe.
22.9 Why is there more than one turret in turret lathes?
22.10 Explain the reasoning behind the various design guidelines for turning.
22.11 List the features of automatic screw machines.
22.12 Explain the similarities and differences in the design guidelines for turning and boring.
22.13 Describe the differences in boring a workpiece on a lathe and on a boring mill.
22.14 List the different types of drills and explain their features.
22.15 Describe chip formation at the cutting edge of a drill.
22.16 Explain the consequence of drilling with a drill that was not sharpened properly.
22.17 Explain the gun-drilling process.
22.18 Describe various applications of trepanning.
22.19 How is drill life determined?
22.20 What units are used to describe feed in drilling?
22.21 List the advantages of spade drills.
22.22 Why are reaming operations performed?

22.23 Why is tapping a difficult operation?
22.24 What are the economic advantages of gang drilling?
22.25 Explain the reasoning behind the design guidelines for drilling.
22.26 What are the advantages of having a hollow spindle in the headstock of a lathe?
22.27 Explain how you would go about producing a tapered shape on a workpiece by turning.
22.28 Describe the difference between a steady rest and a follow rest. Give an application of each.
22.29 Assume that you are going to perform a boring operation on a large hollow workpiece. Would you use a horizontal or a vertical boring mill? Explain.
22.30 Ram-type turret lathes are used more commonly than saddle-type turret lathes. Why?
22.31 Explain the functions of the saddle on a lathe.
22.32 State whether you would set the height of the tool in a turning operation at the center of the workpiece, a little above it, or a little below it. Explain why.
22.33 We note in Table 22.8 that the helix angle for drills is different for different groups of materials. Why?
22.34 Explain the functions of the margin in a twist drill (see Fig. 22.25).
22.35 Explain why pipe threads are tapered.
22.36 Describe the relative advantages of self-opening and solid die heads for threading.
22.37 Explain the advantages and limitations of producing threads by forming and cutting.

QUESTIONS AND PROBLEMS

22.38 Explain why so many different kinds of cutting operations are performed on workpieces.
22.39 Explain why the sequence of drilling, boring, and reaming a series of holes in a workpiece at different locations is more accurate than just drilling and reaming.
22.40 State why machining operations may be necessary on cast, formed, or powder-metallurgy products.
22.41 As the diameter of a drill decreases, the recommended feed also becomes smaller. Explain why.
22.42 A badly oxidized and uneven round bar is being turned on a lathe. Would you recommend a small or large depth of cut? Explain your reasons.
22.43 Describe the problems, if any, that may be encountered in clamping a workpiece made of a soft metal in a three-jaw chuck.
22.44 A turning operation is being carried out on a long round bar at a constant depth of cut. Explain what changes, if any, may occur in the machined diameter from one end of the bar to the other. Give reasons for any changes that may occur.
22.45 Does the force or torque in drilling change as the hole depth increases? Explain.
22.46 A long conical workpiece with a large included angle is being turned on a lathe at a constant rpm and depth of cut. The tool travels at the same angle as the original workpiece. Discuss briefly whether the machined part will also be conical at the same angle or, if not, the shape that it will acquire.
22.47 Suggest remedies for the problems encountered in turning operations as listed in Table 22.7. Explain why you are making these suggestions.

22.48 Explain why the drilling problems listed in Table 22.11 have those particular causes. Suggest remedies and explain why you are making these suggestions.

22.49 Calculate the same quantities as in the example in Section 22.3.1 for high-strength cast iron and at $N = 500$ rpm.

22.50 Describe your observations concerning the contents of Tables 22.2 and 22.3, and explain why those particular recommendations have been made.

22.51 Estimate the machining time required in rough turning a 1-m long, annealed aluminum-alloy round bar, 100 mm in diameter, using a high-speed steel tool (see Table 22.4). Estimate the time for a carbide tool.

22.52 A titanium-alloy bar 4 in. in diameter is being turned on a lathe at a depth of cut $d = 0.05$ in. The lathe is equipped with a 10-hp electric motor and has a mechanical efficiency of 90 percent. The spindle speed is 250 rpm. What is the maximum feed that you can use before the lathe stalls?

22.53 A 0.5-in. diameter drill is used on a drill press operating at 200 rpm. If the feed is 0.005 in./rev, what is the MRR? What is the MRR if the drill diameter is tripled?

22.54 In the example in Section 22.6.4, assume that the workpiece material is high-strength copper alloy and the spindle is running at $N = 600$ rpm. Estimate the torque required in drilling.

22.55 We have seen that cutting speed, feed, and depth of cut are the main parameters in a turning operation. In relative terms, at what should these parameters be set for (a) a finishing operation and (b) a roughing operation?

22.56 Specify conditions under which a boring operation can be performed on (a) a lathe and (b) a boring mill. Explain why.

22.57 Explain the economic justification for purchasing a turret lathe instead of a conventional lathe.

22.58 The accompanying illustration shows a round, type 416 stainless steel (see Section 5.6) part machined from a long round bar stock. Suggest the type of machining operations required, their sequence, the machine tool(s) needed, the type of fixturing required, and the cutting speeds and feeds.

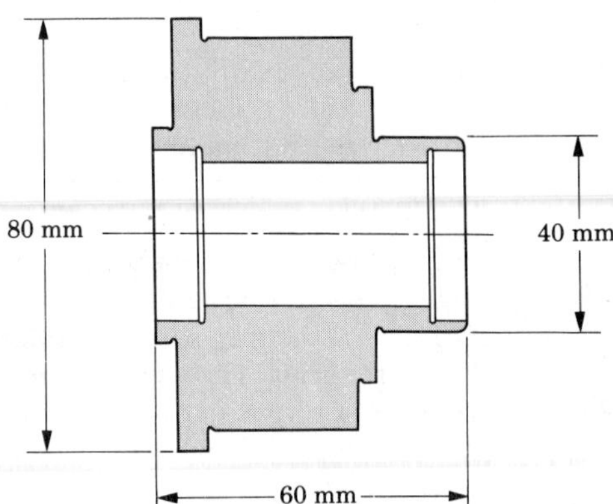

22.59 The accompanying illustration shows a cast-steel valve body (left) that is machined to the shape shown on the right. What type of machine tool would be suitable to machine this part? What type of machining operations are involved and what is the sequence of these operations? Note that not all surfaces are to be machined.

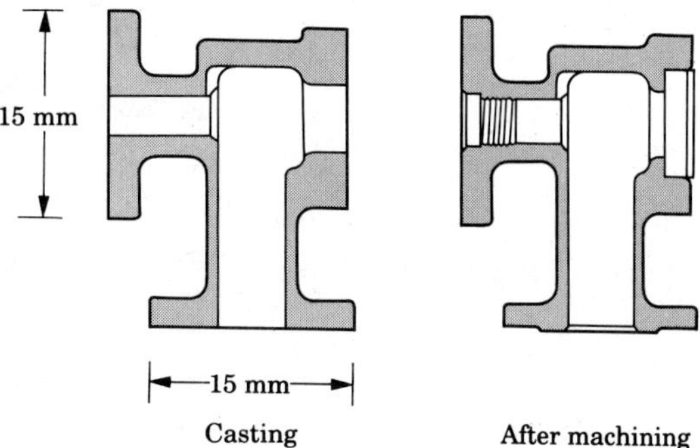

Casting After machining

22.60 Could the part shown on the left in Problem 22.59 be made by processes other than casting? Explain. How does the number of parts required influence your answer? Would you still need machining operations to complete the part, as shown on the right?

23

Cutting Processes for Producing Various Shapes

23.1 Introduction

In Chapter 22, we described cutting processes and machine tools that produce parts having either external or internal round profiles. Many other parts are manufactured by various cutting operations (Fig. 23.1). In this chapter, we consider several cutting processes and machine tools that are capable of producing complex shapes with the use of multitooth, as well as single-point, cutting tools.

We begin with one of the most versatile processes, called milling, in which a multitooth cutter rotates along various axes with respect to the workpiece. We then cover planing, shaping, and broaching, in which flat and shaped surfaces are produced, and the tool or workpiece travels along a straight path. Next, we describe sawing processes, which are generally used to prepare blanks, usually from rods, flat plate, and sheet, for further operations by forming, machining, and welding. We briefly cover filing, which is used to remove small amounts of material, usually from edges and corners.

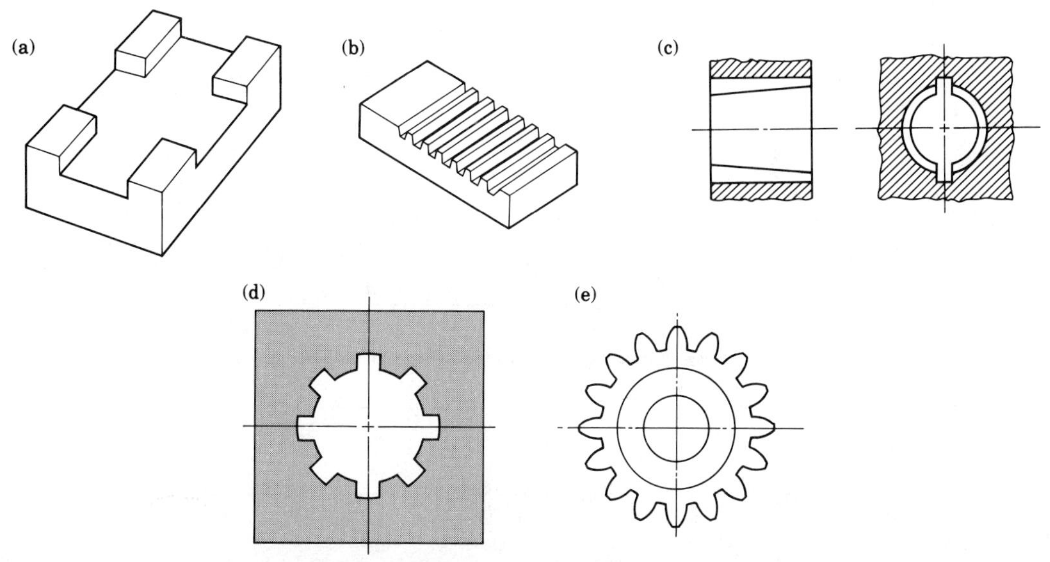

FIGURE 23.1
Typical parts and shapes produced by the cutting processes described in this chapter.

Recall that in Part III we described gear manufacturing by forming processes, in which gear teeth are produced by deformation of the metal. We also stated that gears can be manufactured by casting and powder-metallurgy techniques. In this chapter, we describe gear production by machining processes that use special cutters. Because each gear-manufacturing process has its own characteristics, we will compare the quality and properties of gears made by all these processes.

23.2 Milling Operations

Milling includes a number of versatile machining operations, which are capable of producing a variety of configurations (Fig. 23.2). A *milling cutter* is a multitooth tool that produces a number of chips in one revolution. Parts such as the one shown in Fig. 23.3 can be machined, with various milling cutters, efficiently and repeatedly.

23.2.1 Slab milling

In *slab milling*, also called *peripheral* milling, the axis of cutter rotation is parallel to the workpiece surface to be machined (Fig. 23.2a). The cutter has a number of teeth

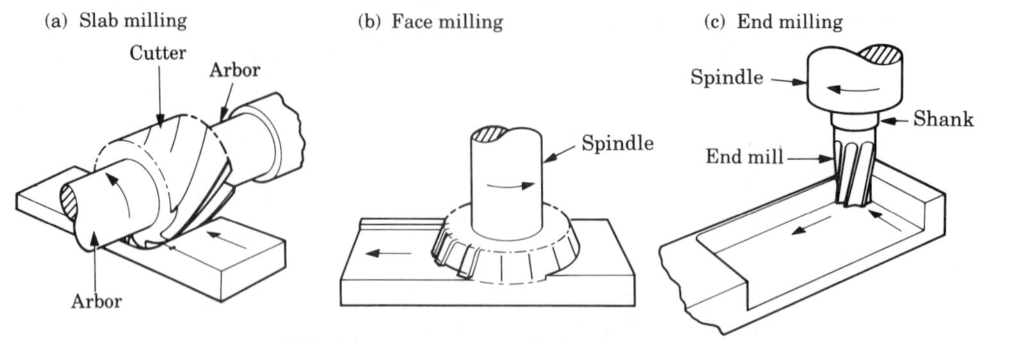

FIGURE 23.2
Basic types of milling cutters and operations.

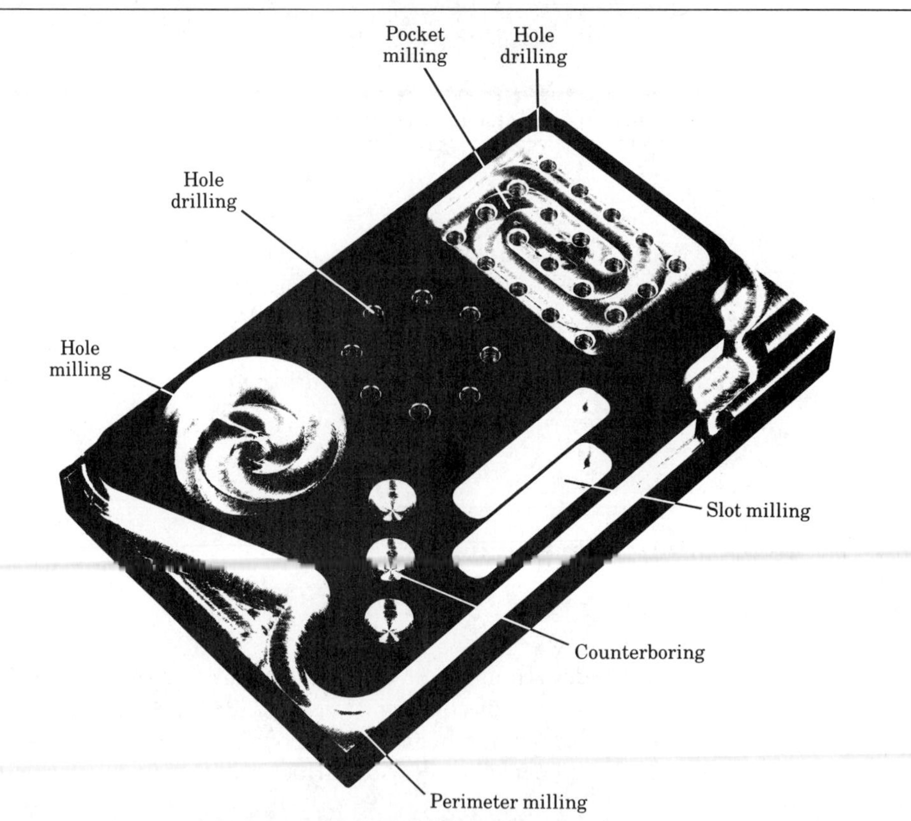

FIGURE 23.3
A typical part produced on a milling machine equipped with computer controls. Such parts can be made efficiently and repetitively on computer numerical control (CNC) machines, using little labor. *Source:* Bridgeport Machines Division, Textron Inc.

along its circumference, each tooth acting like a single-point cutting tool. The cutting speed is the peripheral speed of the cutter, or

$$V = \pi DN, \tag{23.1}$$

where D is the cutter diameter and N is the rotational speed of the cutter.

Note that the thickness of the chip in slab milling (Fig. 23.4a) varies along its length because of the relative longitudinal motion between cutter and workpiece. We can determine the approximate undeformed chip thickness t_c (*chip depth of cut*) from the equation

$$t_c = 2f\sqrt{\frac{d}{D}}, \tag{23.2}$$

where f is the feed per tooth of the cutter, measured along the workpiece surface, and d is the depth of cut. As the value of t_c becomes greater, the force on the cutter tooth increases.

Cutters may have *straight* or *helical* teeth. The helical teeth on the cutter shown in Fig. 23.2(a) are preferred over straight teeth because the contact load is less, resulting in a smoother operation and reducing tool forces and chatter (see also Fig. 20.9).

Feed per tooth is determined from the equation

$$f = v/Nn, \tag{23.3}$$

where v is the workpiece speed and n is the number of teeth on the cutter periphery.

The cutting time, t, is given by the expression

$$t = (l + l_c)/v, \tag{23.4}$$

where l is the length of the workpiece (Fig. 23.4b) and l_c is the extent of the cutter's first contact with the workpiece. Based on the assumption that $l_c \ll l$, the material

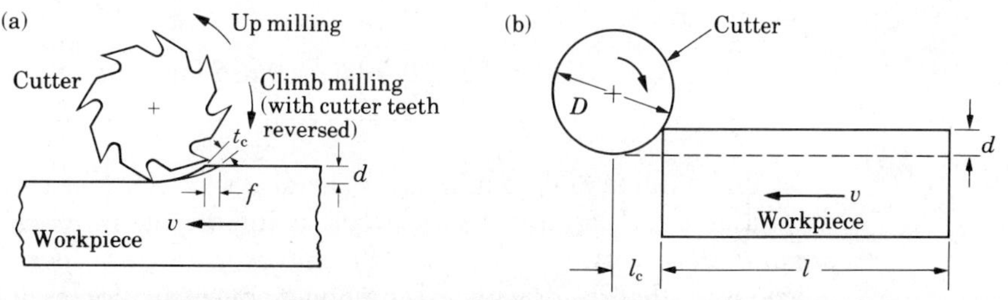

FIGURE 23.4
(a) Slab (peripheral) milling operation, showing depth of cut d, feed per tooth f, chip depth of cut t, and workpiece speed v. (b) Schematic illustration of cutter travel distance l_c to reach full depth of cut.

removal rate is

$$\text{MRR} = lwd/t = wdv, \tag{23.5}$$

where w is the width of the cut, which for a workpiece narrower than the length of the cutter, is the same as the width of the workpiece. As we stated in Section 22.3.1, the distances that the cutter travels in noncutting cycles of the milling operation are important economic considerations and should be minimized.

• Example: Calculation of material removal rate, power required, and cutting time in slab milling.

A slab-milling operation is being carried out on a 12-in. long, 4-in. wide annealed mild-steel block at a feed $f = 0.01$ in./tooth and depth of cut $d = \frac{1}{8}$ in. The cutter is $D = 2$ in. in diameter, has 20 straight teeth, rotates at $N = 100$ rpm, and is wider than the block to be machined. Calculate the material removal rate, estimate the power required for this operation, and calculate the cutting time.

SOLUTION. From the information given, we can calculate the workpiece speed v from Eq. (23.3). Thus,

$$v = fNn = (0.01)(100)(20) = 20 \text{ in./min.}$$

If we use Eq. (23.5), the material removal rate is

$$\text{MRR} = (4)(\tfrac{1}{8})(20) = 10 \text{ in}^3\text{/min.}$$

Since the workpiece is annealed mild steel, we can estimate unit power from Table 20.1 as $3 \text{ W} \cdot \text{s/mm}^3 = 1.1 \text{ hp} \cdot \text{min/in}^3$. Hence the power required in this operation can be estimated to be

$$\text{Power} = (1.1)(10) = 11 \text{ hp.}$$

The cutting time is given by Eq. (23.4) in which the quantity l_c can be shown, from simple geometric relationships and for $D \gg d$, to approximate

$$l_c = \sqrt{Dd} = \sqrt{(2)(\tfrac{1}{8})} = 0.5 \text{ in.}$$

Thus the cutting time is

$$t = (12 + 0.5)/20 = 0.625 \text{ min} = 37.5 \text{ s.}$$

•

Climb milling and up milling. The direction of cutter rotation in slab milling, as well as in other milling operations, is important. In *climb milling*, also called *down milling*, cutting starts with the chip at its thickest location (Fig. 23.4a). The advantage is that the downward component of cutting forces holds the workpiece in place, particularly slender parts. Because of the resulting high impact forces, however, this operation must have a rigid setup, and backlash must be eliminated in the gear mechanisms that rotate the cutter. Climb milling is not suitable for machining workpieces having surface scale, such as hot-worked metals and castings. The scale is hard and abrasive and causes excessive wear and damage to the cutter teeth during processing. For these reasons, tool life can be short.

In *up milling*, also called *conventional milling*, the maximum thickness of the chip is at the end of the cut. Advantages are that tooth engagement is not a function of workpiece surface characteristics, and contamination or scale on the surface does not affect tool life. The cut is smoother, provided that the cutter teeth are sharp, although there is a greater tendency for tool chatter than in climb milling. In up milling the workpiece has a tendency to be pulled upward, so proper clamping is important.

23.2.2 Face milling

In *face milling* the cutter is mounted on a spindle having an axis of rotation perpendicular to the workpiece surface (Fig. 23.2b). The cutter rotates at a speed N and the workpiece moves along a straight path at a speed v. When the cutter rotates as shown in Fig. 23.5(a), the operation is climb milling; when it rotates in the opposite direction (Fig. 23.5b), the operation is up milling. The cutting tools are usually carbide or high-speed steel inserts and are mounted on the cutter body as shown in Figs. 23.6 and 23.7. The terminology for face-milling cutters is shown in Fig. 23.8. The calculations for feed, cutting time, and material removal rate in face milling are similar to those for slab milling.

Because of the relative motion between the cutting teeth and the workpiece, a face-milling cutter leaves feed marks on the machined surface (Fig. 23.9a), much as in turning operations (see Figs. 20.26 and 20.27). Note that surface roughness depends on insert sharpness and feed per tooth, f. Thus the positioning of the cutter

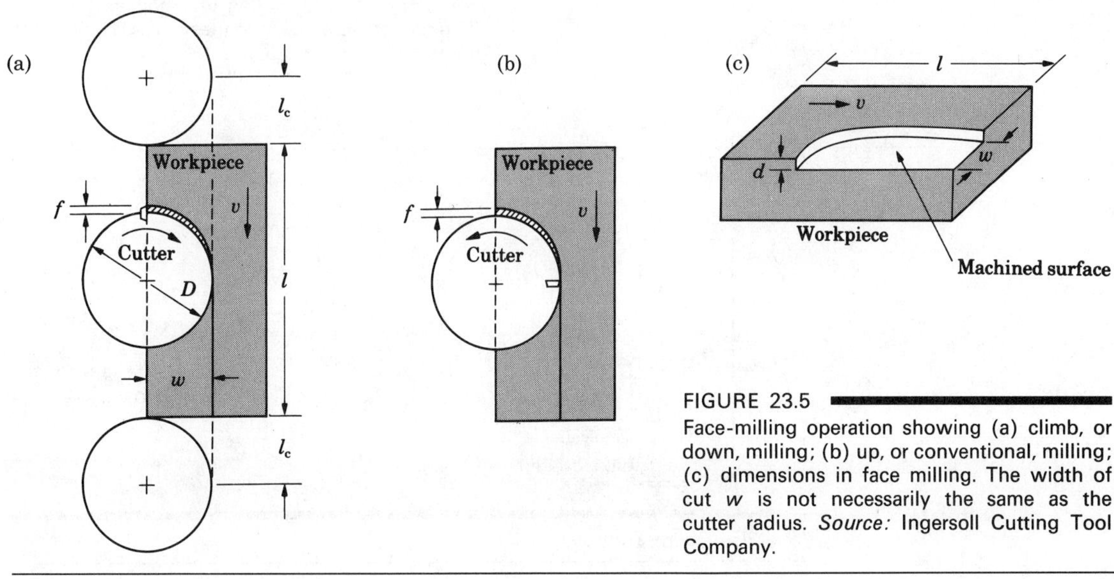

FIGURE 23.5 Face-milling operation showing (a) climb, or down, milling; (b) up, or conventional, milling; (c) dimensions in face milling. The width of cut w is not necessarily the same as the cutter radius. *Source:* Ingersoll Cutting Tool Company.

FIGURE 23.6
Face-milling cutter with indexable inserts. *Source:* Courtesy of Ingersoll Cutting Tool Company.

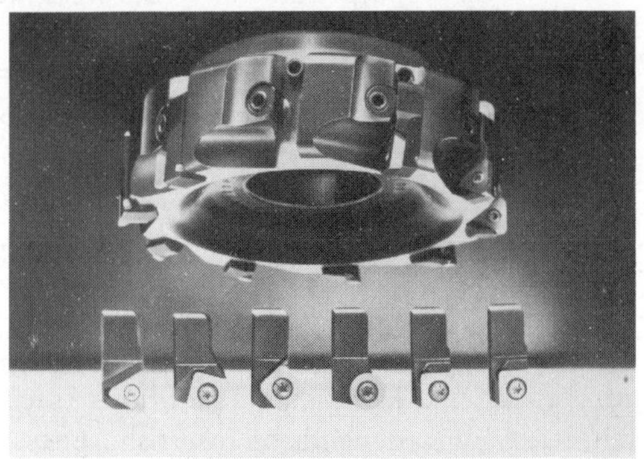

FIGURE 23.7
Types of inserts and cutting-tool holders for face-milling cutters. *Source:* Courtesy of GTE Valenite Corporation.

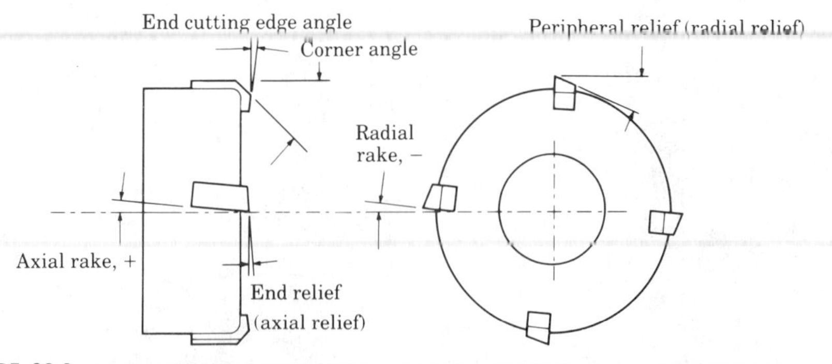

FIGURE 23.8
Terminology for a face-milling cutter.

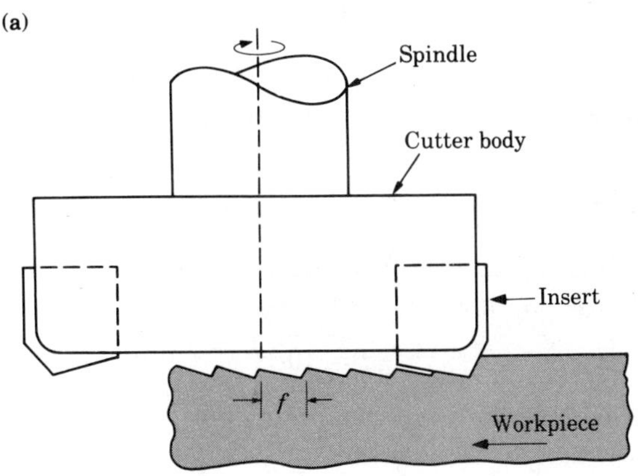

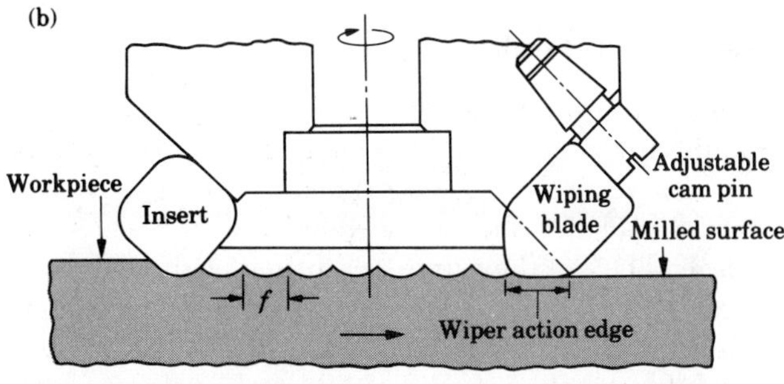

FIGURE 23.9
(a) Feed marks in face milling. (b) Wiper blade for smoothing feed marks in face milling.

and the wiping blade (Fig. 23.9b), whose function is to remove the peaks of roughness, is important.

• Example: Calculation of material removal rate, power required, and cutting time in face milling.

Refer to Fig. 23.5 and assume that $D = 150$ mm, $w = 60$ mm, $l = 500$ mm, $d = 3$ mm, $v = 0.01$ m/s, and $N = 100$ rpm. The cutter has 10 inserts, and the workpiece material is a high-strength aluminum alloy. Calculate the material removal rate, cutting time, and feed per tooth, and estimate the power required.

SOLUTION. The cross-section of the cut is $(w)(d) = (60)(3) = 180\text{ mm}^2$. Noting that the workpiece speed is 0.01 m/s = 10 mm/s, we calculate the material removal rate as

$$\text{MRR} = (180)(10) = 1800\text{ mm}^3/\text{s}.$$

The cutting time is given by

$$t = (l + 2l_c)/v.$$

We note from Fig. 23.5 that for this problem $l_c = D/2$, or 75 mm. Thus the cutting time is

$$t = (500 + 150)/10 = 65 \text{ s} = 1.08 \text{ min}.$$

We obtain the feed per tooth from Eq. (23.3). Noting that $N = 100$ rpm $=$ 1.67 rev/s, we find

$$f = 10/(1.67)(10) = 0.6 \text{ mm/tooth}.$$

For this material we estimate the unit power from Table 20.1 to be 1.1 $\text{W} \cdot \text{s/mm}^3$. Hence we estimate the power to be

$$\text{Power} = (1.1)(1800) = 1980 \text{ W} = 1.98 \text{ kW}.$$

23.2.3 End milling

The cutter in *end milling* is shown in Fig. 23.2(c) and has either straight or tapered shanks for smaller and larger cutter sizes, respectively. The cutter usually rotates on an axis vertical to the workpiece, although it can be tilted to machine tapered surfaces. Flat surfaces as well as various profiles can be produced by end milling. The end face of the cutter has cutting teeth, and thus it can be used as a drill to start a cavity. End mills are also available with hemispherical ends for producing curved surfaces, as in making dies. *Hollow end mills* have internal cutting teeth and are used for machining the cylindrical surface of solid round workpieces, as in preparing stock with accurate diameters for automatic screw machines.

23.2.4 Other milling operations and cutters

Several other types of milling operations and cutters are used to machine various surfaces (Fig. 23.10). In *straddle milling*, two or more cutters are mounted on an arbor and are used to machine two parallel surfaces on the workpiece (Fig. 23.11a). *Form milling* is used to produce curved profiles, with cutters that have specially shaped teeth (Fig. 23.11b). Such cutters are also used for cutting gear teeth (see Section 23.9).

Circular cutters for slotting and slitting are shown in Figs. 23.12(a) and (b), respectively. The teeth may be staggered slightly, as in a sawblade (see Section 23.7), to provide clearance for the cutter in making deep slots. *Slitting saws* are relatively thin: usually less than 5 mm ($\frac{3}{16}$ in.). *T-slot cutters* are used to mill T-slots (Fig. 23.13a), such as those in machine-tool work tables for clamping workpieces. A slot is first milled with an end mill. The T-slot cutter then cuts the complete profile of the slot in one pass. *Key seat cutters* are used to make the semicylindrical key seats (Woodruff) for shafts. *Angle milling cutters* with a single angle or double angles are used to produce tapered surfaces with various angles.

FIGURE 23.10 Various types of milling cutters with inserts. *Source:* Courtesy of Ingersoll Cutting Tool Company.

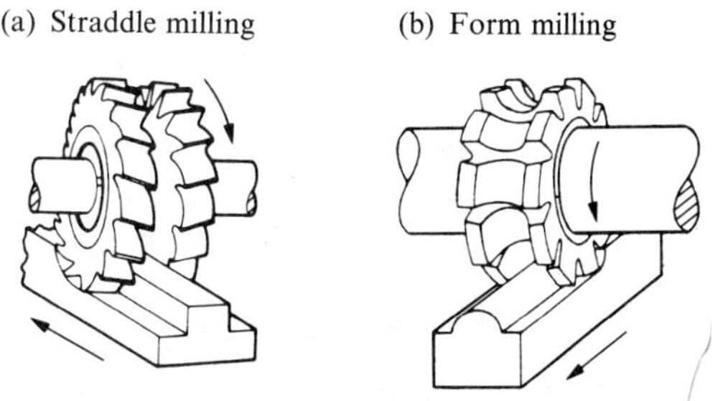

FIGURE 23.11 Cutters for (a) straddle milling and (b) form milling.

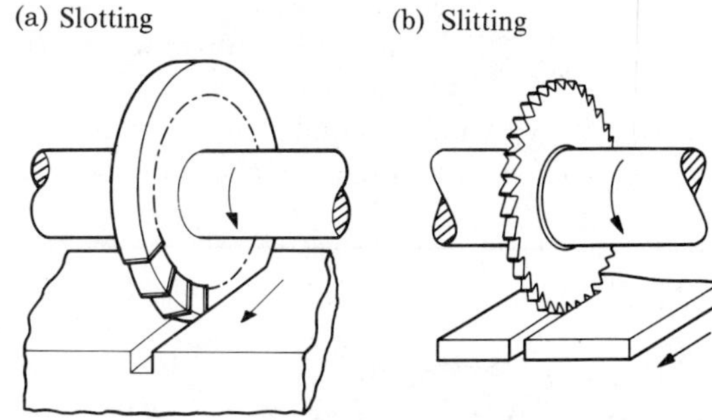

FIGURE 23.12 (a) Slotting and (b) slitting with milling cutters.

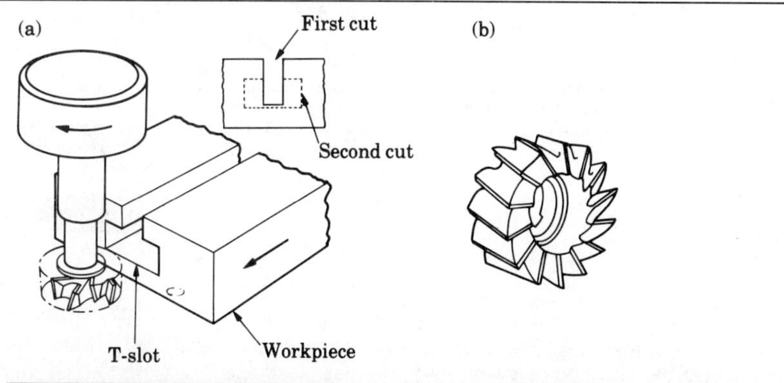

FIGURE 23.13 (a) T-slot cutting with a milling cutter. (b) Shell mill.

Shell mills (Fig. 23.13b) are hollow inside and are mounted on a shank, thus allowing the same shank to be used for different sized cutters. The use of shell mills is similar to that of end mills. Milling with a single cutting tooth mounted on a high-speed spindle is known as *fly cutting* and is generally used in simple face-milling and boring operations. The tool can be shaped as a single-point cutting tool and can be placed in various radial positions on the spindle.

23.2.5 Tool holders

Milling cutters are classified as arbor cutters or shank cutters. *Arbor cutters* are mounted on an arbor (Fig. 23.14), such as those for slab, face, straddle, and form milling. In *shank cutters*, the cutter and the shank are one piece. The most common examples of shank cutters are end mills. Whereas small end mills have straight shanks, larger end mills have tapered shanks for better clamping to resist the higher forces and torque involved. Cutters with straight shanks are mounted in chucks; those with tapered shanks are mounted in tapered tool holders (Fig. 23.15a).

The stiffness of cutters and tool holders is important for surface quality and reducing vibration and chatter in milling operations. Conventional tapered tool holders have a tendency to wear and bell mouth under the radial forces in milling. A new design has a spiral gearlike spindle and adapter (Fig. 23.15b), which distribute the radial forces evenly over all the gear teeth.

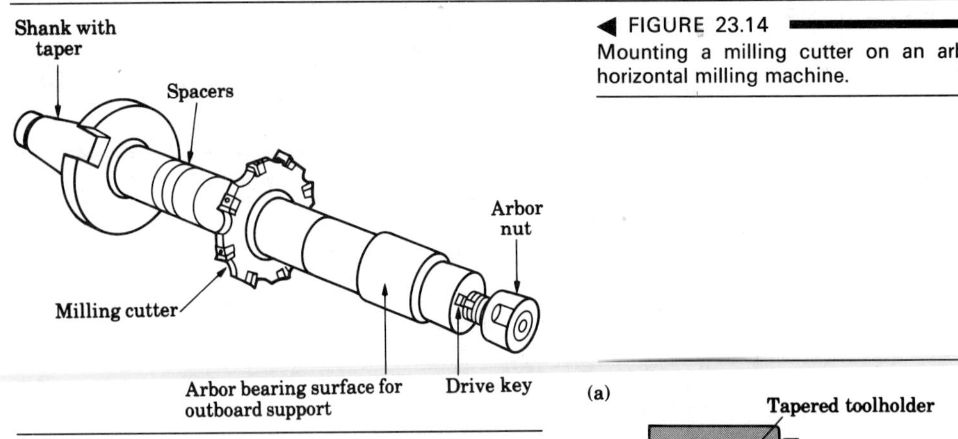

FIGURE 23.14 Mounting a milling cutter on an arbor for use on a horizontal milling machine.

FIGURE 23.15 (a) Conventional tapered tool holder, showing bell-mouthing. (b) New design with spiral gear teeth, which distributes the radial force evenly over the gear teeth; LD = Locating diameter. *Source:* Ingersoll Cutting Tool Company.

TABLE 23.1
TYPICAL CAPACITIES AND MAXIMUM WORKPIECE DIMENSIONS FOR SOME MACHINE TOOLS

MACHINE TOOL	MAXIMUM DIMENSION (m)	POWER (kW)	MAXIMUM SPEED
Milling machines (table travel)			
Knee-and-column	1.4	20	4000 rpm
Bed	4.3		
Numerical control	5		
Planers (table travel)	10	100	1.7 m/s
Shapers (stroke)			
Horizontal	0.7	7	1 m/s
Vertical	1.9		
Broaching machines (length)	2	0.9 MN	
Gear cutting (gear diameter)	5		

Note: Larger capacities are available for special applications.

23.3 Milling Machines

Because of their capabilities to perform a variety of cutting operations, milling machines (first built in 1876) are among the most versatile and useful machine tools. A wide selection of milling machines with numerous features is available (Table 23.1).

23.3.1 Column-and-knee type machines

Used for general-purpose milling operations, *column-and-knee type machines* are the most common milling machines. The spindle to which the milling cutter is attached may be horizontal (Fig. 23.16) for slab milling or vertical for face and end milling, boring, and drilling operations (Fig. 23.17). The various components are moved manually or by power. The basic components of these machines are:

a) *Work table*, on which the workpiece is clamped, using the T-slots. The table moves longitudinally with respect to the saddle.
b) *Saddle*, which supports the table and can move transversely.
c) *Knee*, which supports the saddle and gives the table vertical movement for adjusting the depth of cut.
d) *Overarm* in horizontal machines, which is adjustable to accommodate different arbor lengths.
e) *Head*, which contains the spindle and cutter holders. In vertical machines the head may be fixed or vertically adjustable and can be swiveled in a vertical plane on the column for milling tapered surfaces.

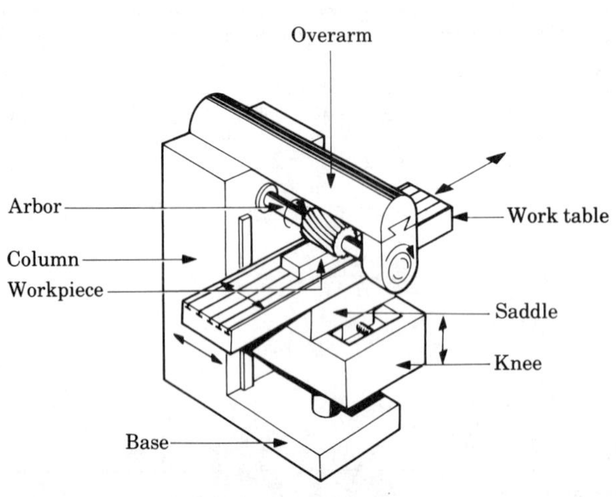

FIGURE 23.16 Schematic illustration of a horizontal-spindle column-and-knee type milling machine. *Source:* G. Boothroyd.

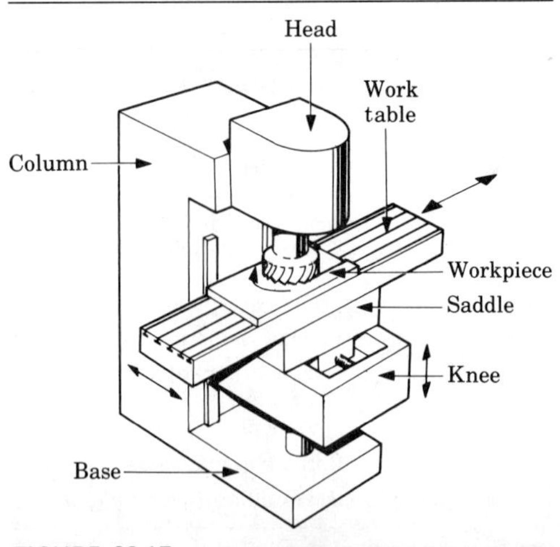

FIGURE 23.17 Schematic illustration of a vertical-spindle column-and-knee type milling machine (also called a *knee miller*). *Source:* G. Boothroyd.

Plain milling machines such as those we decribed have three axes of movement, which are usually imparted both manually and by power. In *universal column-and-knee milling machines*, the table can be swiveled on a horizontal plane. In this way complex shapes, such as helical grooves at various angles, can be produced on drills, taps, cutters, and gears by milling.

23.3.2 Bed-type machines

In *bed-type machines*, the work table is mounted directly on the bed, which replaces the knee, and can move only longitudinally (Fig. 23.18). These milling machines are not as versatile as others, but have high stiffness and are used for high-production-

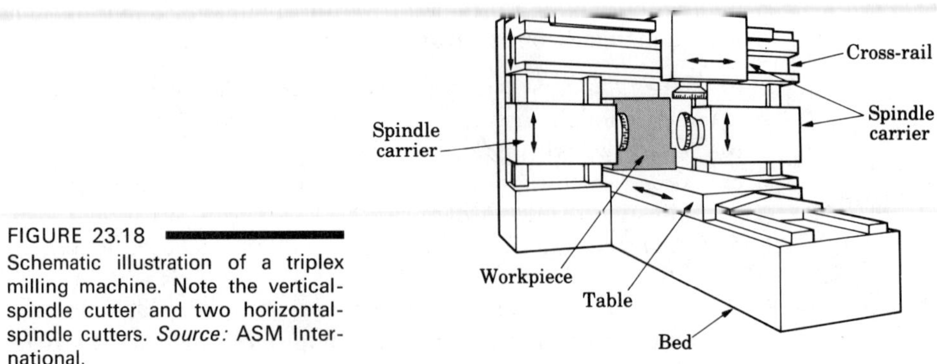

FIGURE 23.18 Schematic illustration of a triplex milling machine. Note the vertical-spindle cutter and two horizontal-spindle cutters. *Source:* ASM International.

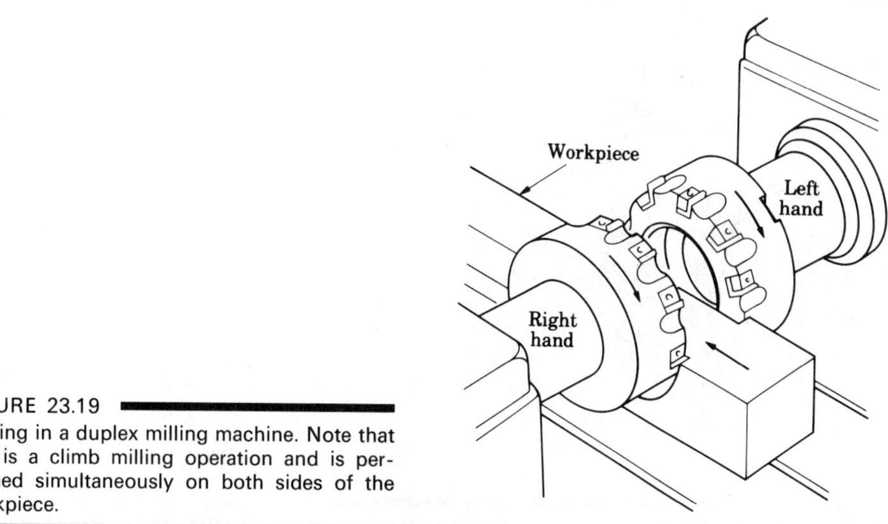

FIGURE 23.19 Cutting in a duplex milling machine. Note that this is a climb milling operation and is performed simultaneously on both sides of the workpiece.

rate work. The spindles may be horizontal or vertical and of duplex or triplex types, that is, with two (Fig. 23.19) or three spindles for simultaneously machining two or three workpiece surfaces.

23.3.3 Other types of milling machines

Several other types of milling machines are available. *Planer-type milling machines*, which are similar to bed-type machines, are equipped with several heads and cutters to mill various surfaces. They are used for heavy workpieces and are more efficient than planers (see Section 23.4) when used for similar purposes. *Rotary-table machines* are similar to vertical milling machines and are equipped with one or more heads for face-milling operations. Using tracer fingers, *duplicating machines* (copy milling machines) reproduce parts from a master model. They are used in the automotive and aerospace industries for machining complex parts and dies (die sinking).

Various milling-machine components are being replaced rapidly with computer numerical control (CNC) machines. These machine tools are versatile and capable of milling, drilling, boring, and tapping with repetitive accuracy (Fig. 23.20). Other developments include *profile milling machines*, with five-axis movements (Fig. 23.21). Note the three linear and two angular movements of machine components. A large five-axis profiling machine with computer numerical control is shown in Fig. 23.22.

23.3.4 Accessories

Accessories for milling machines include various fixtures and attachments for machine head and work table to adapt them to various milling operations. The accessory that has been used most commonly in the past is the *universal dividing*

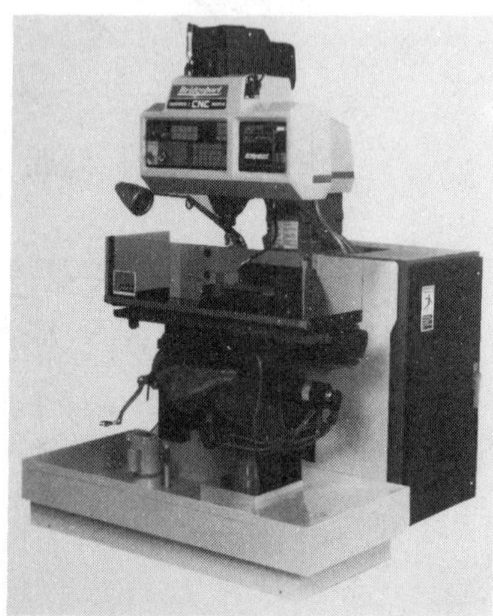

FIGURE 23.20
A computer numerical control, vertical-spindle milling machine. This machine is one of the most versatile machine tools. *Source:* Courtesy of Bridgeport Machines Division, Textron Inc.

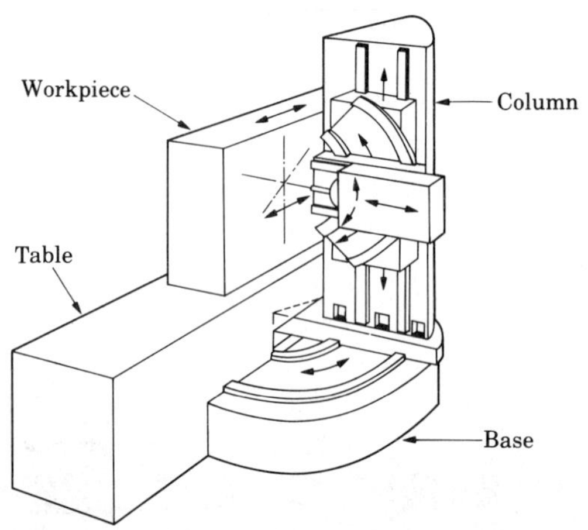

FIGURE 23.21
Schematic illustration of a five-axis profile milling machine. Note that there are three principal linear and two angular movements of machine components.

FIGURE 23.22
A large five-axis, computer numerical control milling machine (also called a profiling machine). *Source:* Courtesy of Cincinnati Milacron, Inc.

(*index*) *head*. Manually operated, this fixture rotates (indexes) the workpiece to specified angles between individual machining steps. It is typically used for milling parts with polygonal surfaces and in machining gear teeth. Dividing heads are now being replaced by automated and computer-controlled machinery, particularly for high production rates.

TABLE 23.2
RECOMMENDATIONS FOR CUTTING SPEEDS IN MILLING

WORKPIECE MATERIAL	CUTTING SPEED (m/s)	
	HSS	*WC*
Aluminum alloys	1.5–6	10 +
Magnesium alloys	3–5	12 +
Copper alloys	0.3–1.5	1.5–7
Steels	0.1–0.7	0.5–4
Stainless steels	0.2–1	1–2
High-temperature alloys	0.05–0.1	0.2–0.3
Titanium alloys	0.1–1	0.5–2
Cast irons	0.2–0.6	0.5–2

Note:
(a) Feed per tooth usually ranges from 0.1 mm to 0.5 mm.
(b) Cutting speeds are for uncoated tools. Speeds for coated tools are 25–75 percent higher.
(c) Speeds for free-machining metals are higher than those indicated.
(d) As hardness increases, cutting speed and feed should be decreased.

TABLE 23.3
GENERAL TROUBLESHOOTING GUIDE FOR MILLING OPERATIONS

PROBLEM	PROBABLE CAUSES
Tool breakage	Tool material lacks toughness; improper tool angles; cutting parameters too high.
Tool wear excessive	Cutting parameters too high; improper tool material; improper tool angles; improper cutting fluid.
Rough surface finish	Feed too high; spindle speed too low; too few teeth on cutter; tool chipped or worn; built-up edge; vibration and chatter.
Tolerances too broad	Lack of spindle stiffness; excessive temperature rise; dull tool; chips clogging cutter.
Workpiece surface burnished	Dull tool; depth of cut too low; radial relief angle too small.

23.3.5 Selection of cutting parameters

The conventional ranges of cutting speeds for milling are given in Table 23.2. Depths of cut are usually 1–8 mm (0.04–0.3 in.). Feeds range from about 0.1 mm/tooth (0.004 in./tooth) to 0.5 mm/tooth (0.02 in./tooth). The higher the strength and the lower the machinability of the material, the lower these values become. A general troubleshooting guide for milling operations is given in Table 23.3. Refer back to Table 22.5 for recommendations about cutting fluids.

23.3.6 Design and operating guidelines for milling

Many of the guidelines for turning and boring (see Chapter 22) are applicable to milling operations. Additional factors include:

a) Standard milling cutters should be used, and costly special cutters should be avoided. Design features include shape, size, depth, width, and corner radii.
b) Chamfers are preferable to radii because of the difficulty of matching various intersecting surfaces smoothly.
c) Workpieces should be sufficiently rigid to minimize deflections resulting from clamping and milling forces.

Guidelines to avoid vibration and chatter in milling are similar to those for turning (Section 22.3.4). In addition, the following should be considered:

a) Cutters should be mounted as close to the spindle base as possible to reduce deflections.
b) In case of vibration and chatter, tool shape and process condition should be modified, and cutters with fewer cutting teeth or with random tooth spacing should be used.

23.4 Planing and Planers

Planing is a relatively simple cutting process by which flat surfaces, as well as various cross-sections with grooves and notches, are produced along the length of the workpiece (Fig. 23.23). Planing is usually done on large workpieces—as large as 25 m × 15 m (75 ft × 40 ft).

In a *planer* (Fig. 23.24), the workpiece is mounted on a table that travels along a straight path. A horizontal cross-rail, which can be moved vertically along the ways in the column, is equipped with one or more tool heads. The cutting tools are attached to the heads, and machining is done along a straight path. Because of the reciprocating motion of the workpiece, elapsed noncutting time during the return

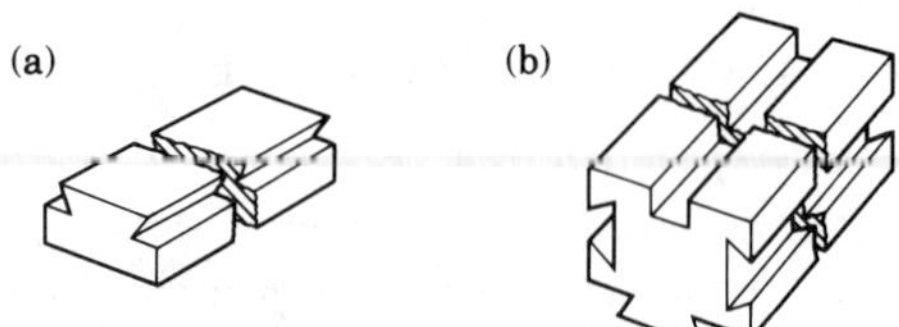

FIGURE 23.23
Typical parts made on a planar.

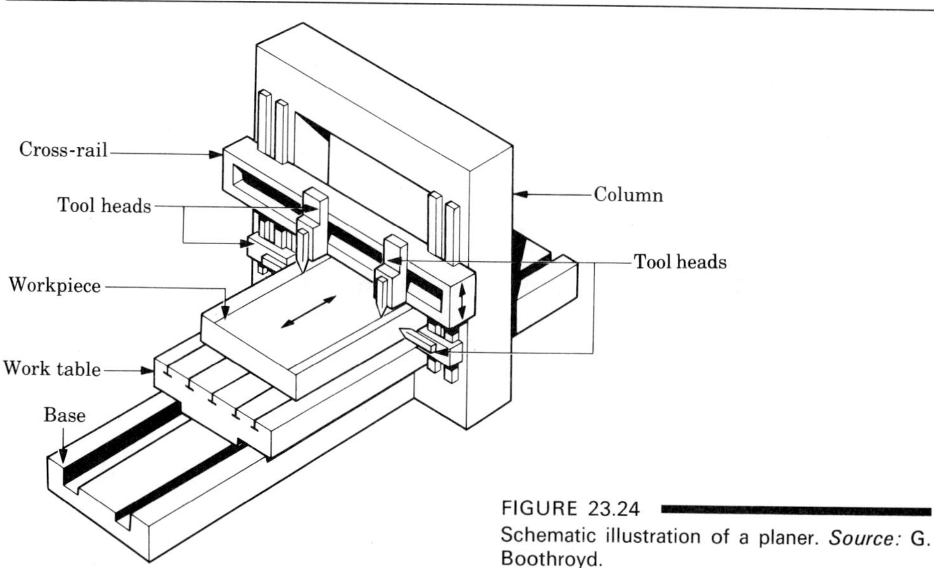

FIGURE 23.24 Schematic illustration of a planer. *Source:* G. Boothroyd.

stroke is significant both in planing and in shaping. Consequently, these operations are not efficient or economical, except for low-quantity production. Efficiency of the operation can be improved by equipping planers with tool holders and tools that cut in both directions of table travel.

In order to prevent chipping of tool cutting edges by rubbing along the workpiece during the return stroke, tools are either tilted away or lifted mechanically or hydraulically. Because of the length of the workpiece, equipping cutting tools with chip breakers is essential. Otherwise, the chips produced can be very long and interfere with the operation, as well as become a safety hazard.

The size of a planer is indicated by the width of the table or the distance between the columns, the maximum clearance under the cross-rail, and the length of the table. The width and height are usually equal. The length of the machine base (bed) is generally twice the length of the table to allow full table travel.

In order to accommodate large workpieces, planers are also available with single columns (open-side planer). However, unlike the two-column design shown in Fig. 23.24, these machines are not as rigid. Planers are among the largest tools, with capacities up to 110 kW (150 hp). However, because they are not efficient and require skilled labor, planers are now being replaced with planer-type milling machines similar to the machine shown in Fig. 23.18.

A planer can be equipped with tracer-control mechanisms which allows machining complex shapes, such as propeller blades, using patterns. The tracer finger guides the cutting tool of the planer by means of a hydraulic mechanism, duplicating the shape of the pattern. Such operations can now be performed more efficiently with computer numerical control machine tools.

23.4.1 Process parameters

Cutting speeds in planers range up to 2 m/s (400 ft/min) with capacities of up to 110 kW (150 hp). Recommended speeds are in the range of 0.05–0.1 m/s (10–20 ft/min) for cast irons and stainless steels, and up to 1.5 m/s (300 ft/min) for aluminum and magnesium alloys. Feeds usually are in the range of 0.5–3 mm/stroke (0.02–0.125 in./stroke). The most common tool materials are M2 and M3 high-speed steels and C-2 and C-6 carbides.

23.4.2 Design considerations

The basic design considerations for planing operations include:

a) Designs with nonparallel surfaces and contours should be avoided because all motions of the cutting tool and the workpiece are generally linear and reciprocating.
b) Designs should allow machining of all surfaces without having to reposition and reclamp the workpiece.
c) Workpieces should be sufficiently rigid to withstand the clamping and machining forces, particularly long pieces.

23.5 Shaping and Shapers

Shaping is used to machine parts much like planing does, except that the parts are smaller. Cutting by shaping is basically the same as in planing. In a *horizontal shaper*, the tool travels along a straight path, and the workpiece is stationary (Fig. 23.25). The cutting tool is attached to the tool head, which is mounted on the ram. The ram has a reciprocating motion, and in most machines cutting is done during the forward movement of the ram (*push cut*); in others it is done during the return stroke of the ram (*draw cut*).

Vertical shapers (*slotters*) are used for machining notches, keyways, and dies. Shapers are also capable of producing complex shapes, such as cutting a helical impeller, in which the workpiece is rotated during the cut through a master cam. Because of low production rate, shapers are generally used in toolrooms, job shops, and repair work.

23.5.1 Process parameters

Shapers are available with strokes of up to 1.2 m (48 in.). Cutting speeds are usually about 0.3 m/s (70 ft/min), with return strokes twice as fast. Some machines are capable of cutting speeds up to 2 m/s (400 ft/min). Ram reciprocations range up to 150 per min. Because of the small workpieces involved, shaper capacities are usually less than 15 kW (20 hp).

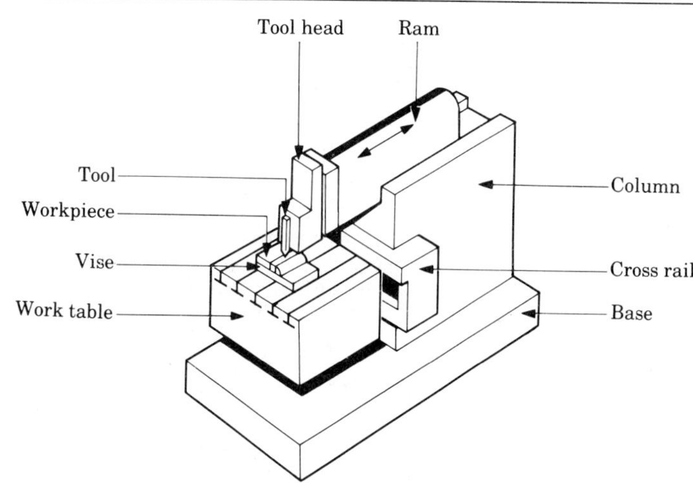

FIGURE 23.25 Schematic illustration of a horizontal shaper. These machines may also be vertical.

23.5.2 Design considerations

The basic design considerations for shaping are generally the same as those for planing (see Section 23.4.2).

23.6 Broaching and Broaching Machines

The **broaching** operation is similar to shaping with multiple teeth and is used to machine internal and external surfaces, such as holes of circular, square, or irregular section, keyways, teeth of internal gears, multiple spline holes, and flat surfaces (Fig. 23.26). A *broach* is, in effect, a long multitooth cutting tool (Fig. 23.27a) with successively deeper cuts. Thus the total depth of material removed in one stroke is the sum of the depths of cut of each tooth. A broach can remove material as deep as 6 mm (0.25 in.) in one stroke. Broaching can produce parts with good surface finish and dimensional accuracy; hence it competes favorably with other processes, such as boring, milling, shaping, and reaming, to produce similar shapes. Although broaches can be expensive, the cost is justified because of their use for high-quantity production runs.

The terminology for a broach is given in Fig. 23.27(b). The rake (hook) angle depends on the material cut, as in turning and other cutting operations, and usually ranges between 0° and 20°. The clearance angle is usually 1°–4°; finishing teeth have smaller angles. Too small a clearance angle causes rubbing of the teeth against the broached surface. The pitch of the teeth depends on factors such as length of the

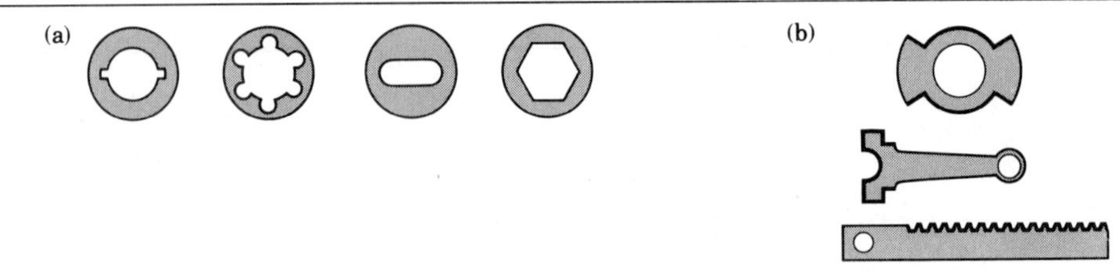

FIGURE 23.26
(a) Typical parts made by internal broaching. (b) Parts made by surface broaching. Heavy lines indicate broached surfaces. *Source:* General Broach and Engineering Company.

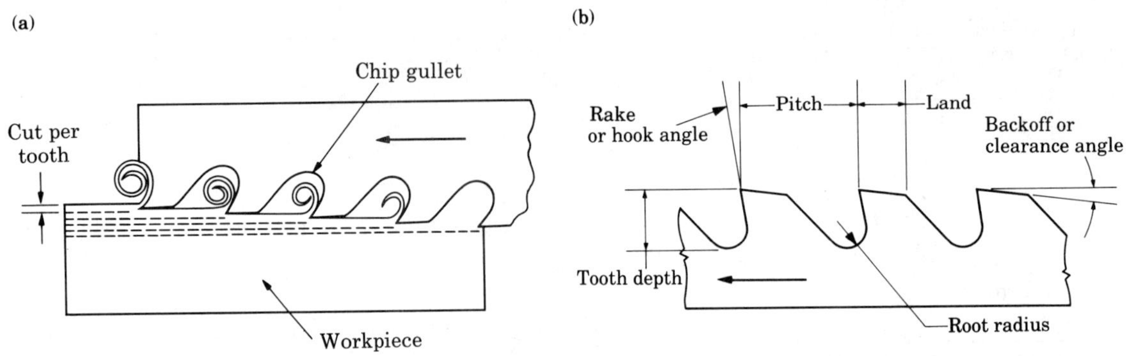

FIGURE 23.27
(a) Cutting action of a broach, showing various features. (b) Terminology for a broach.

workpiece (length of cut), tooth strength, and size and shape of chips. The tooth depth and pitch must be sufficiently large to accommodate the chips produced during broaching, particularly for long workpieces, but at least two teeth should be in contact with the workpiece at all times. The following formula may be used to obtain the pitch for a broach to cut a surface of length l:

$$\text{Pitch} = k\sqrt{l}, \tag{23.6}$$

where $k = 1.76$ for l in mm, and $k = 0.35$ for l in inches. An average pitch for small broaches is in the range of 3.2–6.4 mm (0.125–0.25 in.) and for large ones in the range of 12.7–25 mm (0.5–1 in.). The cut per tooth depends on the workpiece material and the surface finish desired. It is usually in the range of 0.025–0.075 mm (0.001–0.003 in.) for medium-size broaches and can be larger than 0.25 mm (0.01 in.) for larger broaches.

Broaches are available with various tooth profiles, including some with chip breakers (Fig. 23.28). They are also made round with circular cutting teeth and are used to enlarge holes (Fig. 23.29). Note that the cutting teeth on the broach have three regions: roughing, semifinishing, and finishing. A variety of broaches are made for producing various internal and external shapes. Irregular internal shapes

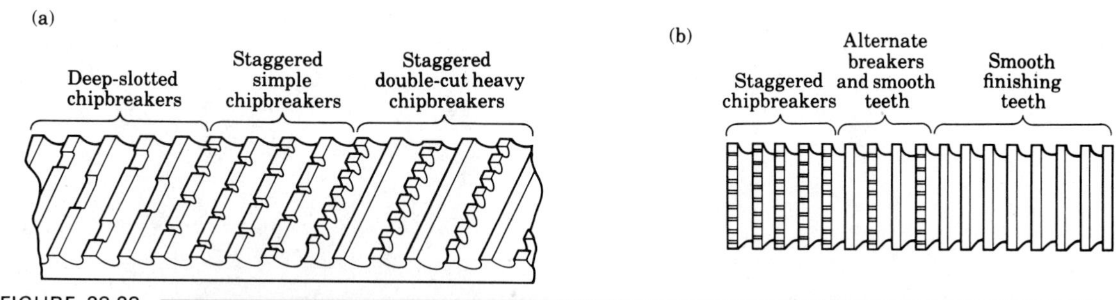

FIGURE 23.28
Chipbreaker features on (a) a flat broach and (b) a round broach.

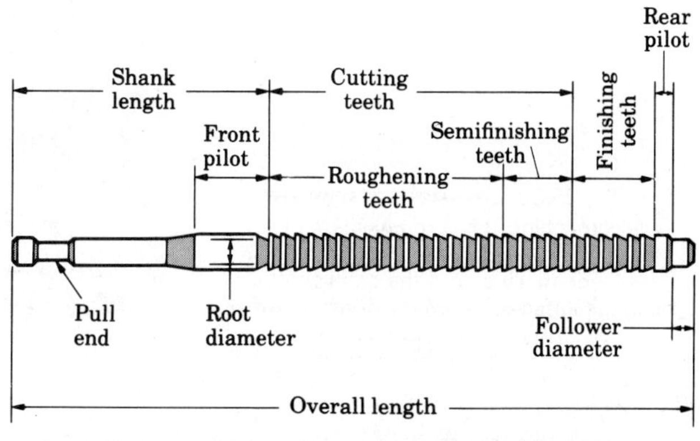

FIGURE 23.29
Terminology for a pull-type internal broach used for enlarging long holes.

are usually broached by starting with a round hole in the workpiece, produced by drilling or boring.

A recent advance in broaching technology is *turn broaching* of crankshafts (Fig. 23.30). The crankshaft rotates between centers and the broach, which is equipped with multiple inserts and passes tangentially across the part. The process is thus a combination of broaching and skiving. Straight as well as circular broaches have been used successfully for such applications. Machines that broach a number of crankshafts simultaneously have been built. Main bearings for engines are also broached in this way.

23.6.1 Broaching machines

Broaching machines either pull or push the broaches and are made horizontal or vertical. *Push broaches* are usually shorter, generally in the range of 150–350 mm (6–14 in.). *Pull broaches* tend to straighten the hole, whereas pushing permits the broach to follow any irregularity of the leader hole. Horizontal machines are capable of longer strokes. Broaching machines are relatively simple in construction,

FIGURE 23.30 Turn broaching of a crankshaft. The crankshaft rotates while the broaches pass tangentially across the crankshaft's bearing surfaces. *Source:* Courtesy of Ingersoll Cutting Tool Company.

have only linear motions, and are usually actuated hydraulically, although some are moved by crank, screw, or rack. Many styles of broaching machines are manufactured, some with multiple heads, allowing a variety of shapes and parts to be produced, including helical splines and rifled gun barrels. Sizes range from machines for making needlelike parts to those used for broaching gun barrels. The force required to pull or push the broach depends on the strength of the workpiece material, total depth of cut, and width of cut. Tooth profile and cutting fluids also affect this force. Capacities of broaching machines are as high as 0.9 MN (100 tons) of pulling force.

23.6.2 Process parameters

Cutting speeds for broaching may range from 0.025 m/s (5 ft/min) for high-strength alloys to as high as 0.25 m/s (50 ft/min) for aluminum and magnesium alloys. The most common tool materials are M2 and M7 high-speed steels. Cutting fluids are generally recommended (see Table 22.5).

23.6.3 Design considerations for broaching

As in other machining processes, broaching requires that certain guidelines be followed in order to obtain economical and high-quality production. The major requirements are:

a) Parts should be designed so that they can be clamped securely in broaching machines. Parts should have sufficient structural strength and stiffness to withstand cutting forces during broaching.
b) Blind holes, sharp corners, dovetail splines, and large surfaces should be avoided.
c) Chamfers are preferable to round corners.

Example: Broaching internal splines.

The part shown in the accompanying figure was made of nodular iron (65-45-15; see Section 12.3.2) with internal splines, each 50 mm (2 in.) long. The splines had 19 involute teeth with a pitch diameter of 63.52 mm (2.5009 in.). An M-2 high-speed-steel broach with 63 teeth, a length of 1.448 m (57 in.), and a diameter the same as the pitch diameter was used to produce the splines. The cut per tooth was 0.116 mm (0.00458 in.). The production rate was 63 pieces per hour. The number of parts per grind was 400, with a total broach life of about 6000 parts. *Source*: ASM International.

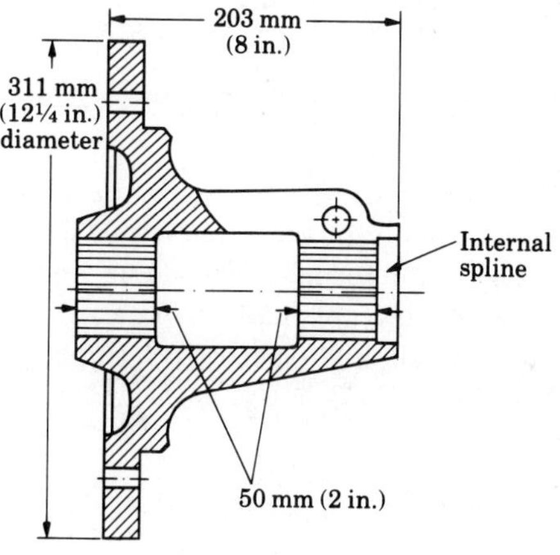

23.7 Sawing

Sawing is a cutting operation in which the cutting tool is a blade (*saw*) having a series of small teeth, with each tooth removing a small amount of material. This process is used for all metallic and nonmetallic materials that are machinable by other cutting processes and is capable of producing various shapes (Fig. 23.31). The

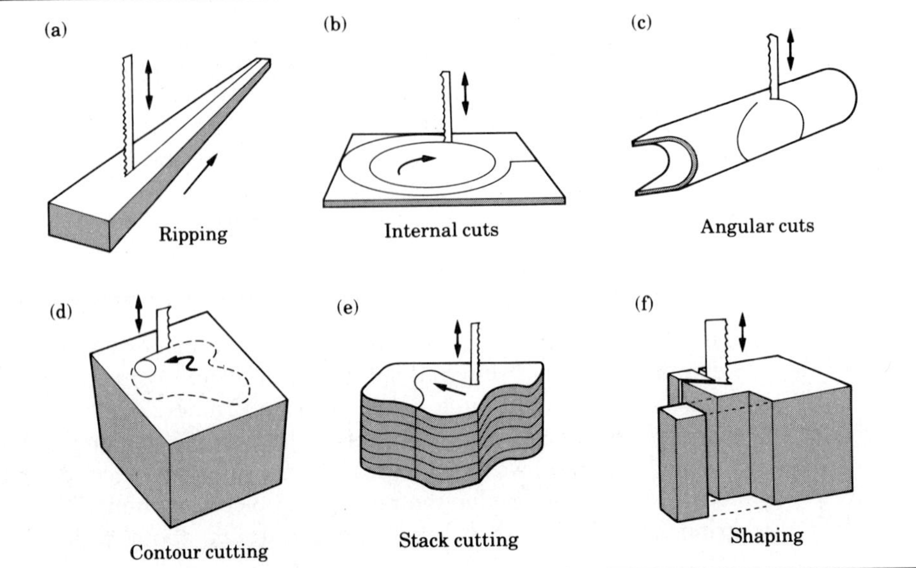

FIGURE 23.31
Examples of sawing operations. *Source:* DoALL Company.

width of cut (*kerf*) in sawing is usually narrow, and thus sawing wastes little material.

Typical saw-tooth and saw-blade configurations are shown in Fig. 23.32. Tooth spacing is usually in the range of 0.08–1.25 teeth per mm (2–32 per in.). A wide variety of tooth forms and spacing and blade thicknesses, widths, and sizes are available. Blades are made from carbon and high-speed steels (M-2 and M-7). Carbide-steel or high-speed-steel tipped steel blades are used for sawing harder materials (Fig. 23.33).

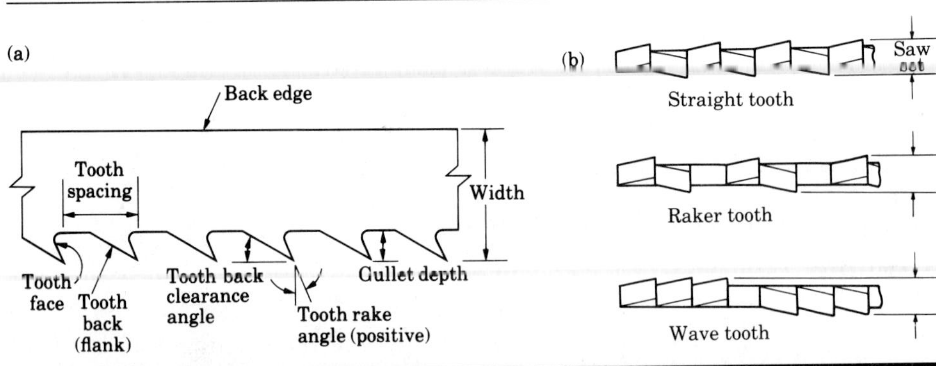

FIGURE 23.32
(a) Terminology for saw teeth. (b) Types of saw teeth, staggered to provide clearance for the saw blade to prevent binding during sawing.

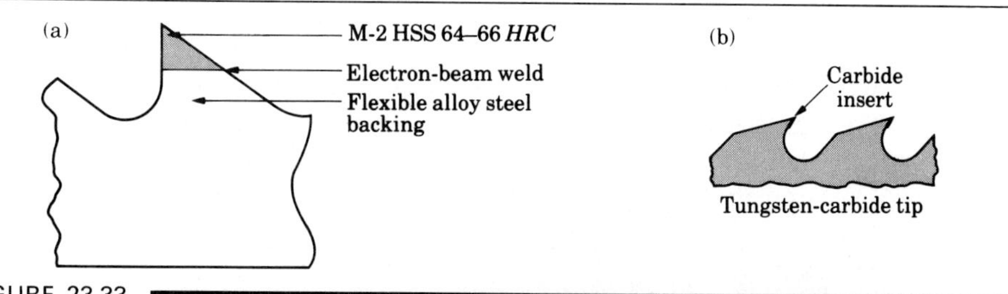

FIGURE 23.33
(a) High-speed-steel teeth welded on steel blade. (b) Carbide inserts brazed to blade teeth.

In order to prevent the saw from binding and rubbing during cutting, the teeth are alternately set in opposite directions so that the kerf is wider than the blade (Fig. 23.32b). At least two or three teeth should always be engaged with the workpiece in order to prevent snagging (catching of the saw tooth on the workpiece). This is why sawing thin materials satisfactorily is difficult. The thinner the stock, the finer the saw teeth should be—and the greater the number of teeth per unit length. Cutting speed in sawing usually ranges up to 1.5 m/s (300 ft/min), with lower speeds for high-strength metals. Cutting fluids are generally used to improve the quality of cut and the life of the saw.

23.7.1 Types of saws

Hacksaws have straight blades and reciprocating motions. Developed in the 1650s, they are used generally for cutting off bars, rods, and structural shapes. They may be manually or power operated. Because cutting takes place during only one of the two reciprocating strokes, hacksaws are not as efficient as band saws. Power hacksaw blades are usually 1.2–2.5 mm (0.05–0.10 in.) thick and up to 610 mm (24 in.) long. Strokes per minute range from 30 for high-strength alloys to 180 for carbon steels. The hacksaw frame in power hacksaws is weighted by various mechanisms, applying as much as 1.3 kN (300 lb) force on the workpiece to improve the cutting rate. Hand hacksaw blades are thinner and shorter than power hacksaw blades, with as many as 1.2 teeth per mm (32 per in.) for sawing sheet metal and thin tubing.

Circular saws, also called cold saws in cutting metal, are generally used for high-production-rate sawing (cutting off) of large cross-sections, such as products from rolling mills. They are available with a variety of saw tooth profiles and sizes.

Band saws have long, flexible, and continuous blades and have a continuous cutting action. Vertical band saws (direction of blade travel is vertical) are used for straight as well as contour cutting of flat sheets and other parts, which are supported on a horizontal table. Also available are computer control band saws with the capability of guiding the contour path automatically. Horizontal band saws have higher productivity than power hacksaws. With high-speed steel blades, cutting speeds for sawing are about 0.15 m/s (30 ft/min) for high-strength alloys and 2 m/s (400 ft/min) for carbon steels. With high-carbon-steel blades, cutting speeds range up to 6.7 m/s (1300 ft/min) for aluminum and magnesium alloys.

Blades and high-strength wire can be coated with diamond powder (*diamond-edged blades* and *diamond wire saws*) in which the diamond particles act as cutting teeth. They are suitable for sawing hard metallic and nonmetallic and composite materials. Wire diameters range from 13 mm (0.5 in.) for use in the rock industry to 0.08 mm (0.003 in.) for precision cutting. Hard materials can also be sawed with thin abrasive disks or by electrical and chemical processes (see Chapters 25 and 26).

23.7.2 Friction sawing

Friction sawing is a process in which a mild-steel blade or disk rubs against the workpiece at speeds of up to 125 m/s (25,000 ft/min). The frictional energy is converted into heat, which rapidly softens a narrow zone in the workpiece. The action of the blade or disk, which is sometimes provided with teeth or notches, pulls and ejects the softened metal from the cutting zone. The heat generated in the workpiece produces a heat-affected zone on the cut surfaces; thus properties can be adversely affected by the sawing process. Because only a small portion of the blade is engaged with the workpiece at any time, the blade cools rapidly as it passes through the air.

The friction-sawing process is suitable for hard ferrous metals and reinforced plastics but not for nonferrous metals because they have a tendency to stick to the blade. Disks for friction sawing as large as 1.8 m (6 ft) in diameter are used to cut off large steel sections in rolling mills. Friction sawing can also be used to remove flash from castings.

23.8 Filing

Filing is small-scale removal of material from a surface, corner, or hole. First developed in about 1000 B.C., files are usually made of hardened steels and are available in a variety of cross-sections, including flat, round, half round, square, and triangular. Files have many tooth forms and grades of coarseness, such as smooth-cut, second-cut, and bastard-cut. Although filing is usually done by hand, various machines with automatic features are available for high production rates, with files reciprocating at up to 500 strokes/min. *Band files* consist of file segments, each about 75 mm (3 in.) long, that are riveted to flexible steel bands and used in a manner similar to band saws. Disk type files are also available.

Rotary files and *burs* (Fig. 23.34) are available for special applications. These cutters are usually conical, cylindrical, or spherical in shape and have various tooth profiles. Similar to that of reamers, their cutting action removes small amounts of material. The rotational speeds range from 1500 rpm for cutting steel with large burs to as high as 45,000 rpm for magnesium with small burs.

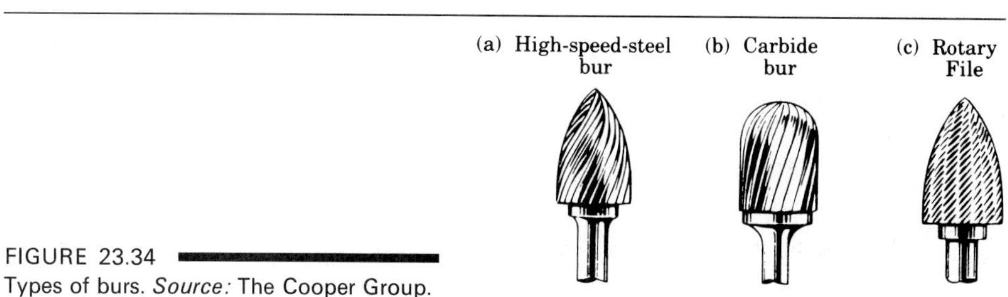

FIGURE 23.34
Types of burs. *Source:* The Cooper Group.

23.9 Gear Manufacturing by Machining

Because of their capability for transmitting motion and power, gears are among the most important of all mechanical components. In various sections in Parts II and III, we discussed several processes for making gears or producing gear teeth on various components. We have shown that gears can be manufactured by casting, forging, extrusion, drawing, thread rolling, powder metallurgy, and blanking (for making thin gears such as those used in watches and small clocks). Nonmetallic gears can be made by injection molding and casting.

The standard nomenclature for an involute spur gear is shown in Fig. 23.35. Gears may be as small as those used in watches and as large as 9 m (30 ft) in diameter. The dimensional accuracy and surface finish required in gear teeth depend

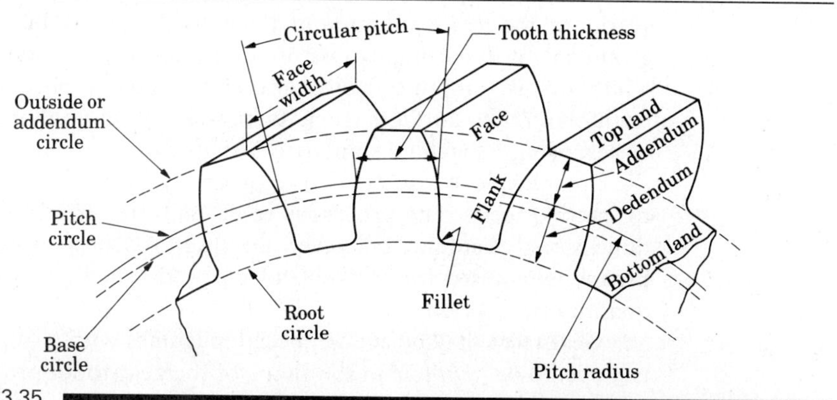

FIGURE 23.35
Nomenclature for an involute spur gear.

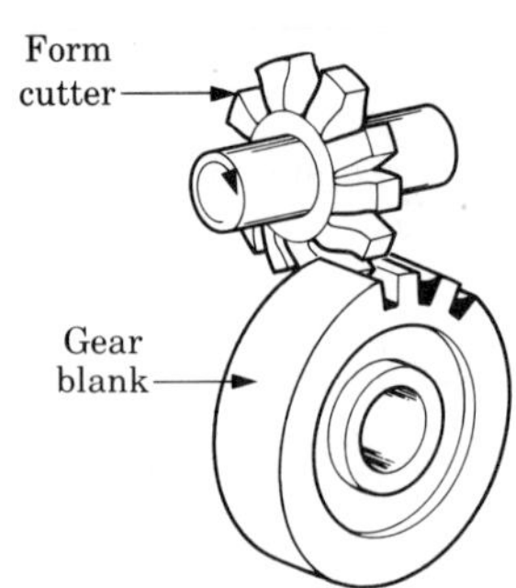

FIGURE 23.36
Producing gear teeth on a blank by form cutting.

on the gear's intended use. Poor gear-tooth quality contributes to inefficient energy transmission and noise and adversely affects the gear's frictional and wear characteristics. Submarine gears, for example, have to be of extremely high quality so as to reduce noise levels to avoid detection.

In this section we describe the two basic gear manufacturing methods by machining of a wrought or cast gear blank: form cutting and generating.

23.9.1 Form cutting

In **form cutting**, the cutting tool is similar to a form-milling cutter in the shape of the space between the gear teeth (Fig. 23.36). The gear-tooth shape is reproduced by cutting the gear blank around its periphery. The cutter travels axially along the length of the gear tooth at the appropriate depth to produce the gear tooth. After each tooth is cut, the cutter is withdrawn, the gear blank is rotated (indexed), and the cutter proceeds to cut another tooth. The process continues until all teeth are cut. Table 23.4 shows standard involute-gear cutter capabilities. Each cutter is designed to cut a range of tooth numbers. The precision of the form-cut tooth profile depends on the accuracy of the cutter and the machine and its stiffness.

Form cutting can be done on milling machines, with the cutter mounted on an arbor and the gear blank mounted in a dividing head. Because the cutter has a fixed geometry, form cutting can be used only to produce gear teeth that have constant width, that is, on spur or helical gears but not on bevel gears. Internal gears and gear teeth on straight surfaces, such as in rack and pinion, are form cut with a shaped cutter, using a machine similar to a shaper.

Broaching can also be used to produce gear teeth and is particularly applicable to internal teeth. The process is rapid and produces fine surface finish with high dimensional accuracy. However, because broaches are expensive—and a separate broach is required for each size of gear—this method is suitable mainly for high-quantity production.

Gear teeth may be cut on special machines with a single-point cutting tool that is guided by a *template* in the shape of the gear-tooth profile. Because the template can be made much larger than the gear tooth, dimensional accuracy is improved.

Form cutting is a relatively simple process and can be used for cutting gear teeth

TABLE 23.4
STANDARD CUTTERS FOR FORM CUTTING OF GEARS

STANDARD		HIGHER ACCURACY	
CUTTER NUMBER	*NUMBER OF TEETH*	*CUTTER NUMBER*	*NUMBER OF TEETH*
1	135 to infinite	1	135 to infinite
2	55–134	$1\frac{1}{2}$	80–134
3	35–54	2	55–79
4	26–34	$2\frac{1}{2}$	42–54
5	21–25	3	35–41
6	17–20	$3\frac{1}{2}$	30–34
7	14–16	4	26–29
8	12–13	$4\frac{1}{2}$	23–25
		5	21–22
		$5\frac{1}{2}$	19–20
		6	17–18
		$6\frac{1}{2}$	15–16
		7	14
		$7\frac{1}{2}$	13
		8	12

with various profiles; however, it is a slow operation, and certain types of machines require skilled labor. Consequently, it is suitable only for low-quantity production. Machines with semiautomatic features can be used economically for form cutting on a limited production basis.

23.9.2 Gear generating

In **gear generating,** the tool may be a (a) pinion-shaped cutter, (b) rack-shaped straight cutter, or (c) hob. The pinion-shaped cutter can be considered as one of the gears in a conjugate pair and the other as the gear blank (Fig. 23.37a). This type of cutter is used in gear generating on machines called *gear shapers* (Fig. 23.37b). The cutter has an axis parallel to that of the gear blank and rotates slowly with the blank at the same pitch-circle velocity with an axial reciprocating motion. A train of gears provides the required relative motion between the cutter shaft and the gear-blank shaft. Cutting may take place either at the downstroke or upstroke of the machine. Because the clearance required for cutter travel is small, gear shaping is suitable for gears that are located close to obstructing surfaces such as flanges (as in the gear blank in Fig. 23.37b). The process can be used for low-quantity as well as high-quantity production.

On a *rack shaper,* the generating tool is a segment of a rack (Fig. 23.37c), which reciprocates parallel to the axis of the gear blank. Because it is not practical to have more than 6–12 teeth on a rack cutter, the cutter must be disengaged at suitable

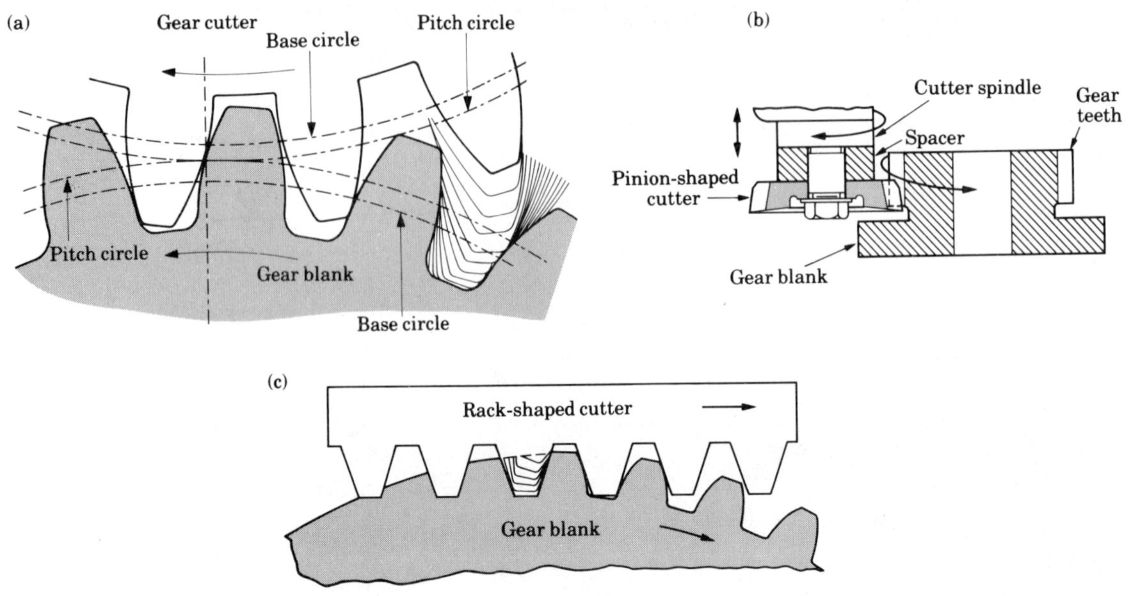

FIGURE 23.37 (a) Schematic illustration of gear generating with a pinion-shaped gear cutter. (b) Schematic illustration of gear generating in a gear shaper using a pinion-shaped cutter. Note that the cutter reciprocates vertically. (c) Gear generating with rack-shaped cutter.

intervals and returned to the starting point, the gear blank meanwhile remaining fixed.

A gear-cutting *hob* (Fig. 23.38) is basically a worm, or screw, made into a gear-generating tool by machining a series of longitudinal slots or gashes into it to form cutting teeth. When hobbing a spur gear, the angle between the hob and gear blank axes is 90° minus the lead angle at the hob threads. All motions in hobbing are rotary, and the hob and gear blank rotate continuously as in two gears meshing until all teeth are cut.

Hobs are available with one, two, or three threads. If the hob has a single thread and the gear is to have 40 teeth, for example, the hob and gear spindle must be geared together so that the hob makes 40 revolutions while the gear blank makes 1 revolution. Similarly, if a double-threaded hob is used, the hob would make 20 revolutions to 1 for the gear blank. In addition, the hob must be fed parallel to the gear axis for a distance greater than the face width of the gear tooth (see Fig. 23.35) in order to produce straight teeth on spur gears. The same hobs and machines can be used to cut helical gears by tilting the axis of the hob spindle.

Because it produces a variety of gears rapidly and with good dimensional accuracy, gear hobbing is used extensively. Although the process is suitable for low-quantity production, it is most economical for medium- to high-quantity production.

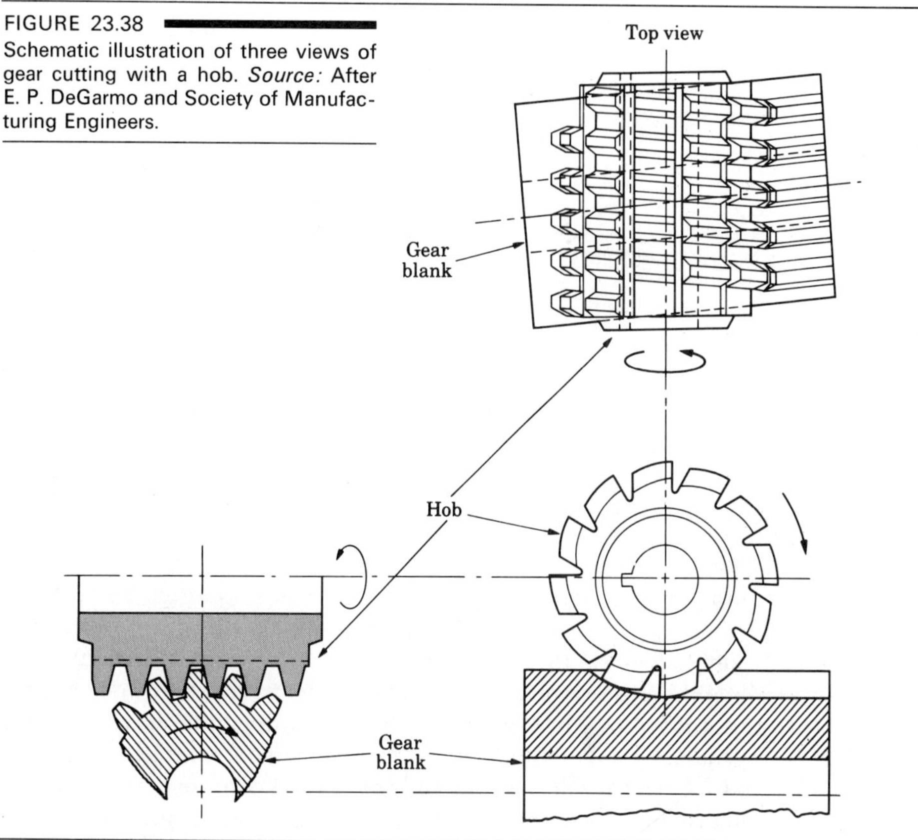

FIGURE 23.38 Schematic illustration of three views of gear cutting with a hob. *Source:* After E. P. DeGarmo and Society of Manufacturing Engineers.

23.9.3 Cutting bevel gears

Straight bevel gears are generally roughed out in one cut with a form cutter on machines that index automatically. The gear is then finished to the proper shape on a gear generator. The generating method is analogous to the rack-generating method that we described. The cutters reciprocate across the face of the bevel gear as does the tool on a shaper (Fig. 23.39a). The machines for spiral bevel gears operate on essentially the same principle. The spiral cutter is basically a face-milling cutter that has a number of straight-sided cutting blades protruding from its periphery (Fig. 23.39b).

23.9.4 Computer-controlled gear generating machines

As in most other machine tools, modern gear-generating machines are computer controlled. An 8-axis computer-controlled machine is shown in Fig. 23.40(a) (see p. 747) that has capabilities of generating a wide variety and sizes of gears using two indexable milling cutters. The eight motions of the machine components are shown

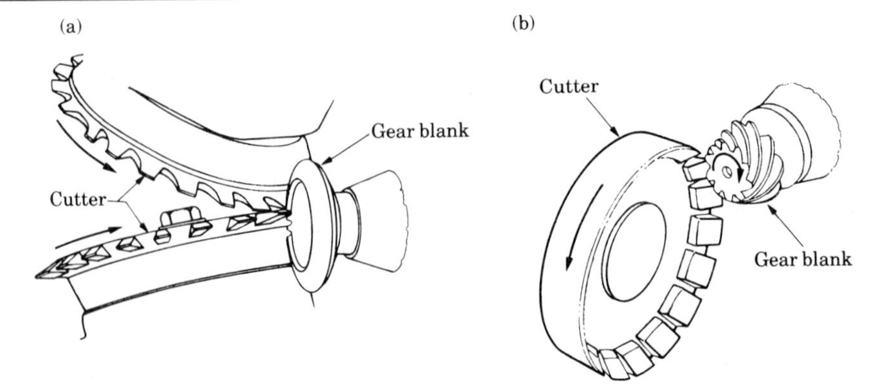

FIGURE 23.39
(a) Cutting a straight bevel-gear blank with two cutters. (b) Cutting a spiral bevel gear with a single cutter. *Source:* ASM International.

in the figure. Successive steps in generating a spur gear are shown in Fig. 23.40(b) where two cutters, one on each flank of the gear tooth (see Fig. 23.35), act simultaneously to produce accurate tooth profiles. With cutters rotating, the gear blank is fed forward (to the right in the figure) until one cutter is plunged to full tooth depth. Then with the blank rotating, the cutters are fed tangentially to generate involute tooth flanks. As each flank is completed, the gear blank is indexed to the next tooth position and the cutting process is repeated. Gears up to 12.5 m (41 ft) in diameter can be produced, and arc segments of gears (shown in the figure) can be made in even larger sizes.

23.9.5 Gear-finishing processes

As produced by any of the processes that we have described thus far, the surface finish and dimensional accuracy of gear teeth may not be accurate enough for certain applications. Moreover, the gears may be noisy or their mechanical properties, such as fatigue life, may not be sufficiently high. Several finishing processes are available for improving the surface quality of gears. The choice of process is dictated by the method of gear manufacture and whether the gears have been hardened by heat treatment. As we described in Chapter 4, heat treating can cause distortion of parts. Consequently, for precise gear-tooth profile, heat-treated gears should be finished by some means.

Shaving. In gear shaving, a cutter in the exact shape of the finished tooth profile removes small amounts of metal from the gear teeth. The cutter teeth are slotted or gashed at several points along its width, thus making the process similar to fine broaching. The motion of the cutter is reciprocating. Shaving and burnishing can only be performed on gears with a hardness of 40 HRC or lower. Although the

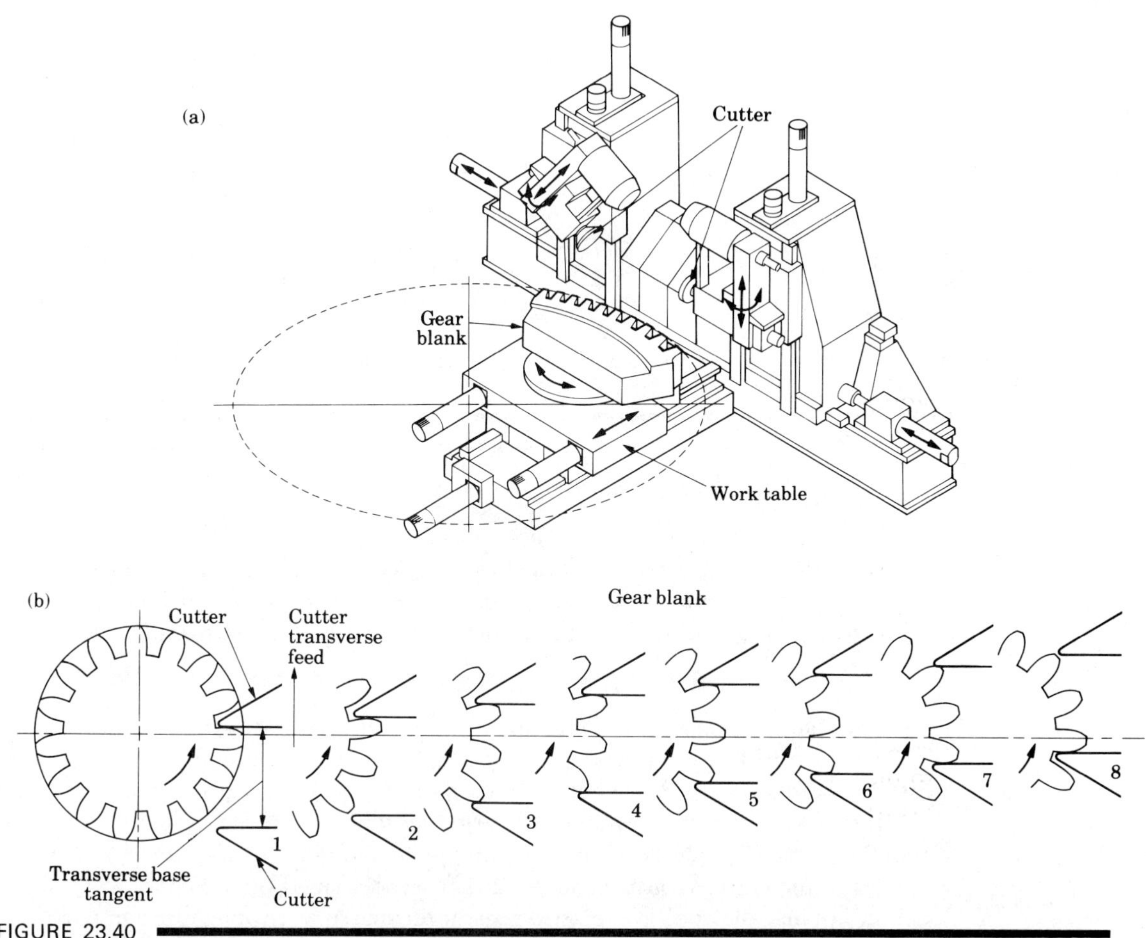

FIGURE 23.40 (a) Schematic illustration of an 8-axis, computer-controlled gear generating machine. Gears up to 12.5 m (41 ft) in diameter can be produced on this machine. (b) Steps in cutting each flank of the gear tooth. *Source:* The Ingersoll Milling Machine Company.

tools are expensive and special machines are necessary, shaving is a rapid and the most commonly used process for gear finishing. It produces gear teeth with improved surface finish and accuracy of tooth profile. Shaved gears may subsequently be heat treated and ground for improved hardness, wear resistance, and accurate tooth profile.

Burnishing. The surface finish of gear teeth can also be improved by burnishing. Introduced in the 1960s, this is basically a surface plastic deformation process, using a special hardened gear-shaped burnishing die that subjects the tooth surfaces to a surface-rolling action (gear rolling). Cold working of tooth surfaces

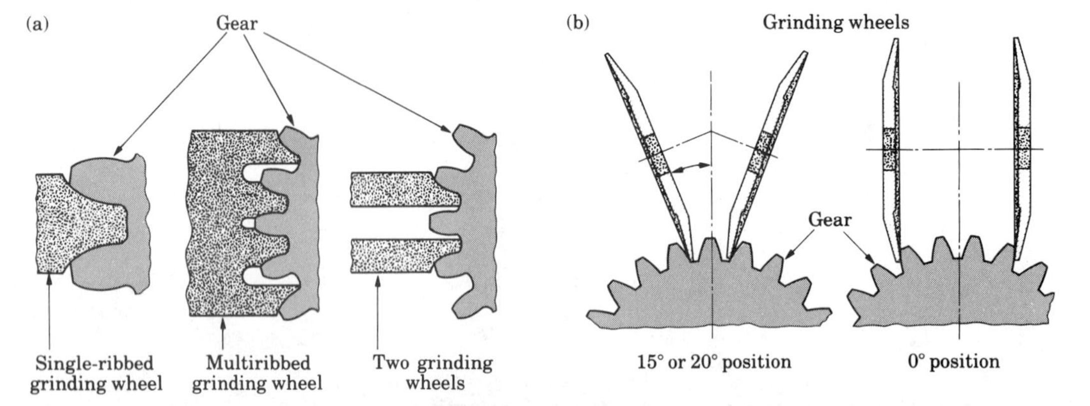

FIGURE 23.41
Finishing gears by grinding: (a) form grinding with shaped grinding wheels; (b) grinding by generating with two wheels.

improves the surface finish and induces surface compressive residual stresses on the gear teeth, thus improving their fatigue life. However, burnishing does not significantly improve gear-tooth accuracy. We describe burnishing further in Section 33.2.

Grinding, honing, and lapping. For highest dimensional accuracy, tooth spacing and form, and surface finish, gear teeth may subsequently be ground. Specially dressed grinding wheels are used either for forming or generating gear-tooth surfaces. In form grinding, the grinding wheel is in the exact shape of the tooth spacing (Fig. 23.41a). In generating, the grinding wheel acts in a manner similar to the cutter in gear generating that we described (Fig. 23.41b).

In honing, the tool is a plastic gear impregnated with fine abrasive particles. The process is faster than grinding and is used to improve surface finish. To further improve the surface finish, ground gear teeth are lapped using abrasive compounds with either a gear-shaped lapping tool (made of cast iron or bronze) or a pair of mating gears that are run together. Although costs are significantly higher and production rates are low, these finishing processes are particularly suitable for producing hardened gears of very high quality, long life, and quiet operation. We present the details of grinding, honing, and lapping processes in Chapter 25.

23.9.6 Economics of gear production

As in all cutting operations, the cost of gears increases rapidly with improved surface finish and quality. Figure 23.42 shows the relative manufacturing cost of gears as a function of quality, as specified by AGMA (American Gear Manufacturers Association) and DIN (Deutsches Institut für Normung) numbers. The higher the number, the smaller is the tolerance of gear teeth. Note how the cost can vary by an order of magnitude.

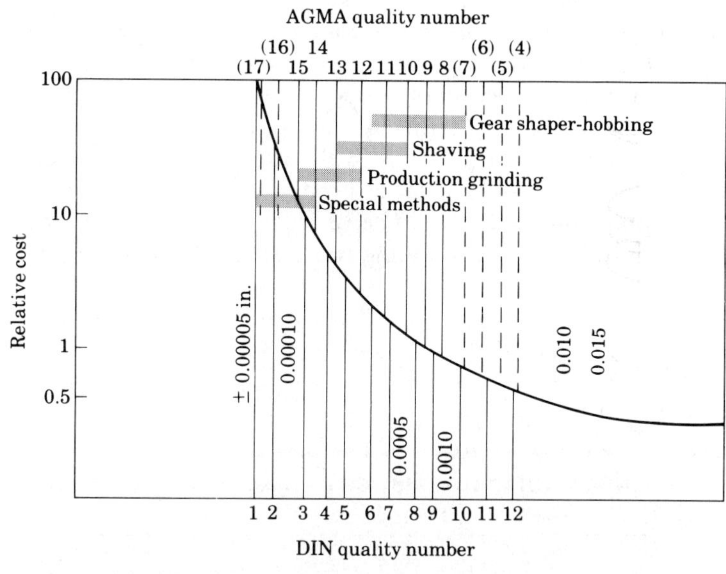

FIGURE 23.42 Gear manufacturing cost as a function of gear quality. The numbers along the vertical lines indicate tolerances. *Source:* Society of Manufacturing Engineers.

23.9.7 Design considerations for gear machining

Many of the design considerations for machining processes that we have described thus far also apply to gear cutting operations. Special considerations are:

a) Gear blanks should be designed properly, especially for complex gear teeth, and provision should be made for clamping blanks securely to the machine.
b) The specified dimensional accuracy and surface finish on gear teeth should be as broad as possible.
c) Wide gears are more difficult to machine than narrow ones.
d) Preferably, gears should be machined prior to their assembly on shafts.
e) Sufficient clearance should be provided between the gear teeth and flanges, shoulders, and other features so that the cutting tool can function without any interference.
f) Provision should be made for the use of standard cutters wherever possible.

SUMMARY

Some of the most versatile machining processes are milling, planing, shaping, broaching, and sawing. Milling is one of the most useful processes because of its capability to produce a variety of shapes from workpieces. Although there are similarities with processes such as turning, drilling, and boring, most of these processes utilize multitooth tools and cutters at various axes with respect to the workpiece. The machine tools employed have various features, attachments, and considerable flexibility in operation.

Gear manufacturing by various forming and shaping processes was described in previous chapters. Gears are also produced by machining, either by form cutting or generating, with the latter producing gears with better surface finish and greater dimensional accuracy. The surface finish and accuracy of tooth profile are further improved by gear finishing processes such as shaving, burnishing, and grinding.

SUMMARY TABLE 23.1

GENERAL CHARACTERISTICS OF MACHINING PROCESSES DESCRIBED IN CHAPTERS 22 AND 23

PROCESS	CHARACTERISTICS	COMMERCIAL TOLERANCES (± mm)
Turning	Turning and facing operations on all types of materials; uses single-point or form tools; requires skilled labor; production rate low, but medium to high with turret lathes and automatic machines, requiring less-skilled labor.	Fine: 0.05–0.13 Rough: 0.13 Skiving: 0.025–0.05
Boring	Internal surfaces or profiles, with characteristics similar to turning; stiffness of boring bar important to avoid chatter.	0.025
Drilling	Round holes of various sizes and depths; requires boring and reaming for improved accuracy; high production rate; labor skill required depends on hole location and accuracy specified.	0.075
Milling	Variety of shapes involving contours, flat surfaces and slots; wide variety of tooling; versatile; low to medium production rate; requires skilled labor.	0.13–0.25
Planing	Flat surfaces and straight contour profiles on large surfaces; suitable for low-quantity production; labor skill required depends on part shape.	0.08–0.13
Shaping	Flat surfaces and straight contour profiles on relatively small workpieces; suitable for low-quantity production; labor skill required depends on part shape.	0.05–0.13
Broaching	External and internal flat surfaces, slots, and contours with good surface finish; costly tooling; high production rate; labor skill required depends on part shape.	0.025–0.15
Sawing	Straight and contour cuts on flat or structural shapes; not suitable for hard materials unless saw has carbide teeth or coated with diamond; low production rate; requires only low labor skill.	0.8

TRENDS

- Machine tools for milling, broaching, sawing, and gear cutting are being designed and constructed with more computer-control features, thus improving productivity and product consistency and reducing the need for skilled labor.
- Milling machines are being designed and built such that vertical and horizontal spindles are combined in one machine.
- Machine tools are being made with greater power, cutting speed, and stiffness.
- The use of planers and shapers is now limited. Planers are being replaced with planer-type computer-controlled milling and similar machines.

KEY TERMS

Broaching
Filing
Form cutting
Gear generating
Milling
Planing
Sawing
Shaping

BIBLIOGRAPHY

Boothroyd, G., and W.A. Knight, *Fundamentals of Machining and Machine Tools*, 2d ed. New York: Marcel Dekker, 1989.

Gear Processing and Manufacturing, 2d ed. Dearborn, Mich.: Society of Manufacturing Engineers, 1984.

Lambert, B. (ed.), *Milling: Methods and Machines*. Dearborn, Mich.: Society of Manufacturing Engineers, 1980.

Machining Data Handbook, 2 vols., 3d ed. Cincinnati: Machinability Data Center, 1980.

Metals Handbook, 8th ed., Vol. 3: *Machining*. Metals Park, Ohio: American Society for Metals, 1967.

Metals Handbook, 9th ed., Vol 16: *Machining*. Metals Park, Ohio: ASM International, 1989.

Milling Handbook of High-Efficiency Metal Cutting. Detroit, Mich.: Carboloy Systems Department, General Electric Company, 1980.

Reshetov, D.N., and V.T. Portman, *Accuracy of Machine Tools*. New York: American Society of Mechanical Engineers, 1989.

Tool and Manufacturing Engineering Handbook, 4th ed., Vol. 1: *Machining*. Dearborn, Mich.: Society of Manufacturing Engineers, 1983.

Weck, M., *Handbook of Machine Tools*, 4 vols. New York: Wiley, 1984.

REVIEW QUESTIONS

23.1 Why is milling a versatile machining process?
23.2 Describe different types of cutters used in milling operations, and cite an application of each.
23.3 What are the advantages of helical teeth over straight teeth on cutters for slab milling?
23.4 Describe the relative characteristics of climb milling and up milling.
23.5 What are dividing heads used for?
23.6 Describe the features of different types of milling machines. What are duplex and triplex machines?
23.7 What is fly cutting? Name typical applications.
23.8 What are the differences in planing and shaping operations and their applications?
23.9 Explain why broaching is a commonly used process. Give some typical applications.
23.10 Describe the features of a broach and explain their functions.
23.11 Explain the methods of gear manufacturing by machining.
23.12 Why do machined gears have to undergo finishing operations? Which of the finishing processes are not suitable for hardened gear teeth? Why?
23.13 Why is sawing a commonly used process? Does it have any limitations? Explain.
23.14 Explain why hacksaws are not as productive as horizontal band saws.
23.15 Why do some saw blades have staggered teeth?
23.16 Why are some saw blades equipped with carbide or high-speed-steel cutting teeth?
23.17 Explain the principle of friction sawing.
23.18 Give some reasons for developing files as tools.
23.19 Explain the difference between shaving and burnishing.
23.20 Describe the methods of cutting bevel gears.
23.21 How is feed specified in planing and shaping?
23.22 What advantages do bed-type milling machines have over column-and-knee type machines for production operations?
23.23 Why is the axis of the hob tilted with respect to the axis of the gear blank?
23.24 Describe the difference between finishing by form grinding and generating.
23.25 Why does the cost of a machined gear rise so rapidly with its quality?

QUESTIONS AND PROBLEMS

23.26 You are performing a slab-milling operation at a certain cutting speed (surface speed of the cutter) and feed per tooth. Explain the procedure for determining the table speed required.
23.27 Would you use planers and shapers for high-production-rate work? Explain.
23.28 Explain why broaching crankshaft bearings is an attractive alternative to other machining processes.
23.29 Several guidelines were presented in this chapter for various cutting operations. Discuss the reasoning behind these guidelines.
23.30 In milling operations with horizontal- and vertical-spindle machines, which one is likely to hold dimensional accuracy better? Why?
23.31 What similarities and differences are there in slitting with a milling cutter and sawing?

23.32 Could the part shown in Fig. 23.3 be machined by any of the processes described in Chapter 22? Explain.

23.33 Explain the functions of the angles shown in Fig. 23.8.

23.34 How would you reduce the surface roughness shown in Fig. 23.9(a)?

23.35 State your opinion of the design change shown in Fig. 23.15.

23.36 Why were the two types of milling machines shown in Figs. 23.16 and 23.17 developed, instead of just one?

23.37 Why are machines such as the one shown in Fig. 23.20 so useful?

23.38 Describe how the choice of a gear-cutting process would depend on the quantity of gears to be produced.

23.39 Describe a type of product that could be produced efficiently on the type of machine shown in Fig. 23.21.

23.40 Comment on your observations concerning the designs shown in Fig. 23.26 and the usefulness of broaching operations.

23.41 Explain why push broaches are shorter than pull broaches.

23.42 Explain how contour cutting would be started in a band saw, as shown in Fig. 23.31(d).

23.43 In Fig. 23.33(a), high-speed-steel cutting teeth are welded to a steel blade. Would you recommend that the whole blade be made of high-speed steel? Explain your reasons.

23.44 Review reference books on gear manufacturing methods and describe how herring-bone gears are machined.

23.45 Show that the distance l_c in slab milling is approximately equal to $\sqrt{Dd}$ for situations where $D \gg d$. See Fig. 23.4(b).

23.46 Describe your observations concerning Table 23.2, and explain why those recommendations have been made.

23.47 In the example in Section 23.2.1, which of the quantities will be affected when the feed is increased to $f = 0.02$ in./tooth?

23.48 Calculate the chip depth of cut t_c and the torque in the example in Section 23.2.1.

23.49 Estimate the time required for face milling an 8-in. long, 2-in. wide brass block with an 8-in. diameter cutter with 10 high-speed-steel inserts.

23.50 Describe the factors that contribute to broaching force and explain why they do so.

23.51 A 10-in. long, 1-in. thick plate is being cut on a band saw at 100 ft/min. The saw has 12 teeth per in. If the feed per tooth is 0.003 in., how long will it take to saw the plate along its length?

23.52 A single-thread hob is used to cut 40 teeth on a spur gear. The cutting speed is 100 ft/min and the hob is 4 in. in diameter. Calculate the rotational speed of the spur gear.

23.53 Describe the conditions under which broaching would be the preferred method of machining over other methods.

23.54 How would you go about determining the proper direction of rotation of the cutter in a milling operation? What would be the proper direction in milling a hot-rolled steel block? Why?

23.55 List and explain the factors involved in the cost of producing gears by machining operations. Consider also the number of gears to be produced.

23.56 How would you go about designing the wiping blade for the milling cutter shown in Fig. 23.9?

23.57 Assume that in the face-milling operation shown in Fig. 23.5, the workpiece dimensions are 5 in. by 10 in. The cutter is 6 in. in diameter, has 6 teeth, and rotates at 400 rpm. The depth of cut is 0.125 in. and the feed is 0.005 in./tooth. Assume that the specific energy requirements for this material is 1.6 hp.min/in^3 (see Table 20.1) and that only 75 percent of the cutter diameter is engaged during cutting. Calculate (a) the power required and (b) the material removal rate.

23.58 Allowing an approach of 0.1 in. for both entry and exit of the cutter into the workpiece, calculate the cutting time needed for milling the whole surface in Problem 23.57.

23.59 The accompanying illustration shows a part that is machined from a rectangular blank. Suggest the machine tool(s) required, the fixturing needed, and the types of operations to be performed and their sequence.

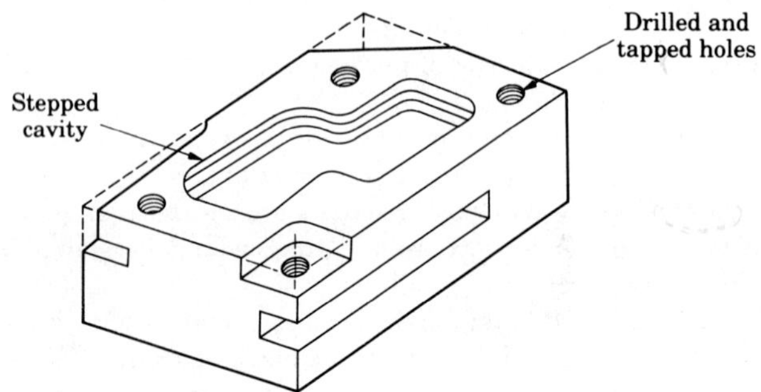

23.60 Referring to the illustration in Problem 23.59, would you prefer to machine this part from a preformed blank rather than a rectangular blank? If so, how would you prepare such a blank? How would the number of parts required influence your answer?

24

Machining Centers, Machine-Tool Structures, and Machining Economics

24.1 Introduction

In Chapters 22 and 23, we described machining processes used to produce a wide variety of parts. In this chapter, we first describe important developments in the design and functions of machine tools in general. Known as machining centers, these machines have a flexibility and versatility that other machine tools do not have. Consequently, they have a major economic impact in manufacturing operations. It is projected that by the year 1995, machining centers will constitute a significant portion of all metal-cutting type machine tools in the United States, as we can see from the following table showing market projections:

Turning centers	25 %
Machining centers	20
Grinders	18
Milling machines	9
Boring mills	3
Drilling machines	2.5
Gear cutting machines	2.5
Others	20

We also present the material and design aspects of machine tools as structures and review new developments in the use of various materials and composites in their construction. Included in these developments is an improved understanding of machine-tool performance, particularly with regard to their stiffness and vibration and damping characteristics. These are important considerations not only for dimensional accuracy and the quality of surfaces produced but also because of their influence on tool life.

In the final section of this chapter, we discuss the economic aspects of machining operations and identify the factors that contribute to machining costs. We also compare costs among machining and other methods of manufacturing a part.

24.2 Machining Centers

In describing machining processes and machinery in the preceding chapters, we noted that each machine tool, regardless of how well it is automated, is traditionally designed to perform basically one type of operation. We have also shown that in manufacturing operations most parts require a number of different cutting operations on their various surfaces.

Let's, for example, briefly review the diesel-engine block and heads depicted in Fig. 24.1. All surfaces on this engine block require different types of machining, such as turning, facing, milling, drilling, boring, and threading. Traditionally, these operations would be performed by moving the part from one machine tool to another until all machining is completed. Obviously, this procedure is not efficient

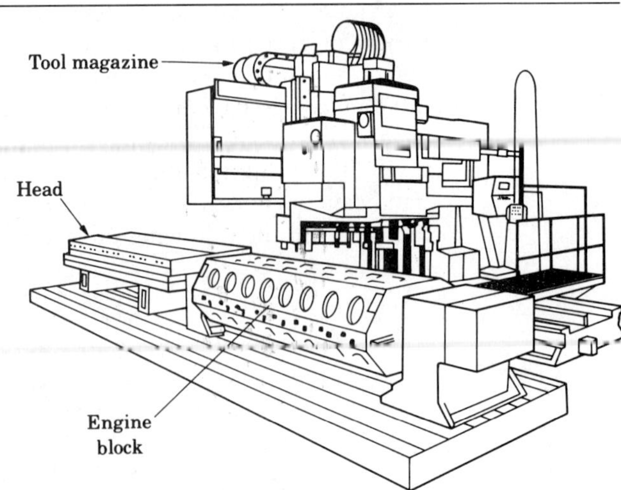

FIGURE 24.1 Machining of a diesel-engine block and heads on a machining center. Machining centers such as this one are capable of performing a variety of machining operations on various surfaces of workpieces. *Source:* Courtesy of Giddings and Lewis Machine Tool Company.

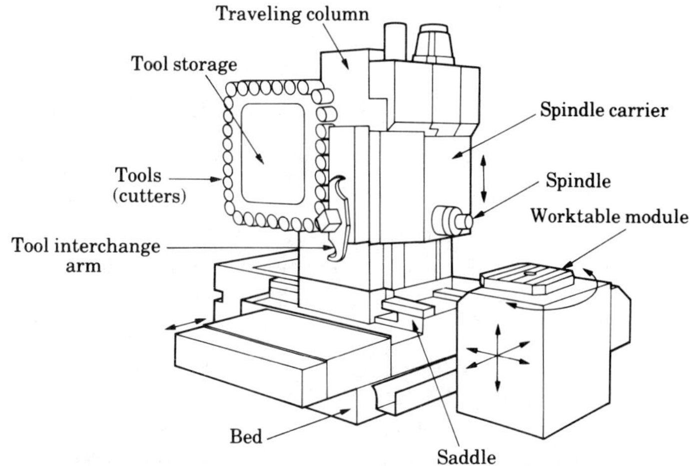

FIGURE 24.2 Schematic illustration of a horizontal-spindle machining center, equipped with an automatic tool changer. Tool magazines can store 120 cutting tools or more. *Source:* Courtesy of Cincinnati Milacron, Inc.

FIGURE 24.3 A computer-numerical-control horizontal machining center, with two work table modules. *Source:* Courtesy of Cincinnati Milacron, Inc.

because considerable time is wasted in moving parts from machine to machine, clamping them, and then removing them from machine tables.

An important concept, developed in the late 1950s, is **machining centers** (Figs. 24.2 and 24.3). A machining center is a computer-controlled machine tool (see Chapters 38 and 39) with automatic tool-changing capability. The machining center is designed to perform a variety of cutting operations on different surfaces of the workpiece, which is placed on a *module* that can be oriented in various directions. Thus, after a particular cutting operation, say milling, has been completed, the workpiece does not have to be moved to another machine for additional operations, say drilling, reaming, and threading. In other words, the tools and the machine are brought to the workpiece.

Note in Fig. 24.2 that the work module on which the workpiece is placed is capable of moving in three principal linear directions, as well as rotating around a vertical axis. Up to 120 cutting tools can be stored in a *magazine* (tool storage). Auxiliary tool storage is available on some machining centers for up to 480 tools. After all cutting operations have been completed, the module automatically moves away with the finished workpiece, and a new module containing another workpiece to be machined moves into its position.

24.2.1 Characteristics of machining centers

The major characteristics of machining centers are:

a) They are capable of handling a variety of part sizes and shapes efficiently, economically, and with repetitively high dimensional accuracy with tolerances on the order of ± 0.0025 mm (0.0001 in.).
b) The machines are versatile, having many axes of linear and angular movements and the capability of quick changeover from one type of product to another. Thus the need for a variety of machine tools and floor space is reduced significantly.
c) The time required for loading and unloading workpieces, changing tools, gaging, and troubleshooting is reduced, thus improving productivity, reducing labor (particularly skilled-labor) requirements, and minimizing machining costs.
d) Machining centers are highly automated, so that one operator can attend two or more machines at the same time.

24.2.2 Types of machining centers

The two major types of machining centers are vertical spindle and horizontal spindle, although many machines have the capability of using both axes. Machining centers are available in a variety of designs and sizes. Capacities range up to 75 kW (100 hp)—and even higher—and spindle speeds usually run 4000–8000 rpm—but some as high as 75,000 rpm.

Vertical-spindle machines are suitable for machining flat surfaces with deep cavities, such as in mold and die making. A small vertical-spindle machining center similar to a vertical-spindle milling machine is shown in Fig. 24.4. The tool magazine is on the left of the machine. All operations and movements in the machine are directed and modified through the computer-control panel on the right. Because the thrust forces in vertical machining are directed downward, such machines have high stiffness and produce parts with good dimensional accuracy. These machines are generally less expensive than horizontal-spindle machines.

Horizontal-spindle machines (see Fig. 24.2) are suitable for large, including tall, workpieces requiring machining on a number of their surfaces, with the use of a work table module (with workpieces mounted on a *pallet*) that can be rotated to various angular positions.

Universal machining centers have a variety of features and are capable of machining all surfaces of a workpiece. New developments include turret-type machines capable of performing turning operations. In-process and post-process gaging and inspection of machined workpieces are also new features of machining centers. The stiffness of machining centers, hence dimensional accuracy, is being improved continually. Many machines are being constructed on a modular basis, so

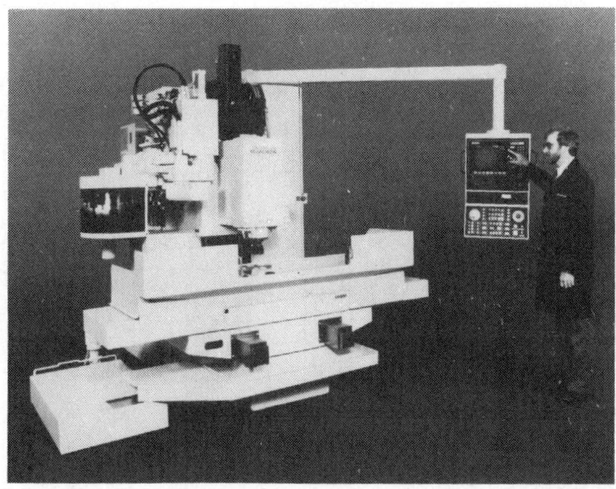

FIGURE 24.4
A vertical-spindle machining center. The tool magazine is on the left of the machine. *Source:* Courtesy of Cincinnati Milacron, Inc.

that various peripheral equipment can be installed and modified as the demand for and type of products change.

Machining centers come in a wide variety of sizes. The smallest machine is designed for parts up to 100 mm (4 in.), a spindle speed of up to 12,000 rpm, and capacity of 2.2 kW (3 hp). One of the largest machining centers built to date (Fig. 24.5) is capable of machining workpieces 11.5 m × 5.6 m × 5.4 m (38 ft × 18.5 ft × 18 ft), and weighing up to 200 tons. The structure consists of double columns. The master head in which the spindle is housed is mounted on a cross-rail and can be set to vertical and horizontal positions. Auxiliary spindle units are transported by robots.

Tooling systems. Machining centers are equipped with automatic tool changers. Tools are usually stored in a drum, magazine, or chain. Tools are automatically selected with random access for the shortest route to the spindle. The tool-interchange arm shown in Fig. 24.6 (see also Fig. 24.2) is a common design: It swings around to pick up the tool and place it in the spindle. Tool changing times are on the order of a few seconds. Tools are identified by coded tags, bar coding, or memory chips applied to the toolholders. Machining centers are also equipped with a tool-checking station that feeds information to the computer-numerical control to compensate for any variations in tool settings.

24.2.3 Economics and machine selection

Machining centers require significant capital expenditures. Thus to be cost effective, they have to be used for at least two shifts per day. Consequently, there must be sufficient and continued demand for products made in machining centers.

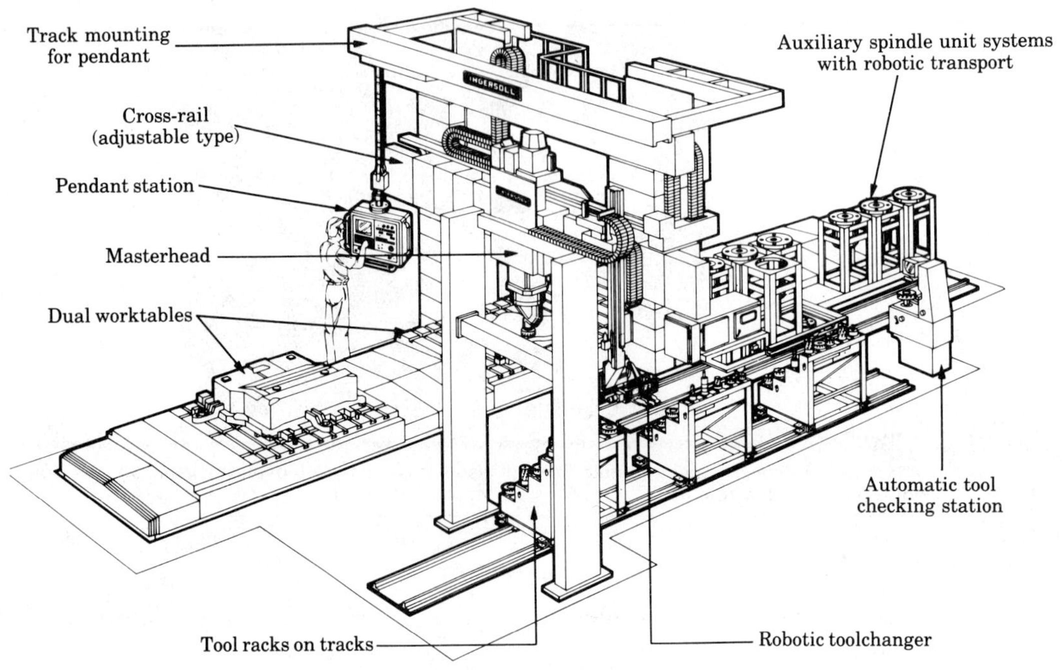

FIGURE 24.5 Schematic illustration of a large machining center, showing various components. *Source:* The Ingersoll Milling Machine Company.

The selection of a particular type and size of machining center depends on several factors, among which are:

a) Type of products, their size, and shape complexity.
b) Type of machining operations to be performed and type and number of tools required.

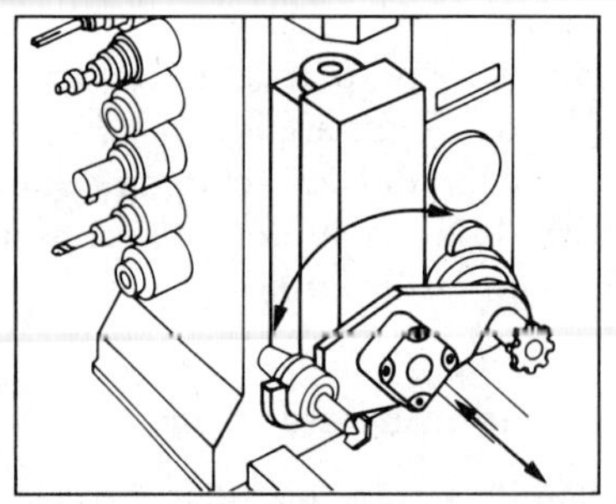

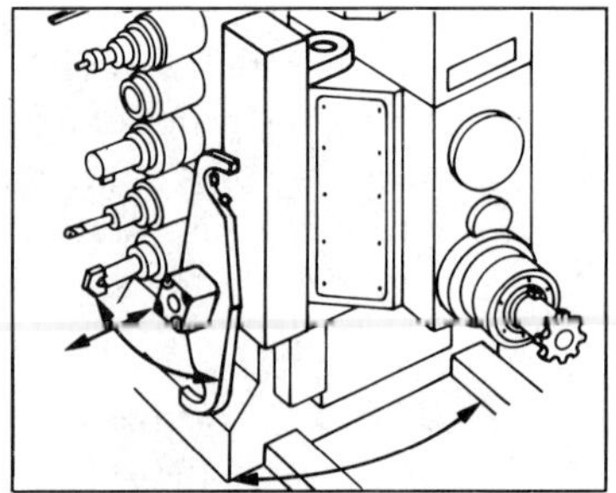

FIGURE 24.6 Swing-around tool changer (see also Fig. 24.2). *Source:* Cincinnati Milacron, Inc.

c) Dimensional accuracy required.
d) Production rate required.

These considerations must be weighed carefully against the high capital investment required for machining centers and compared to that of manufacturing the same products by using a number of the more traditional machine tools.

24.3 Vibration and Chatter

In describing cutting processes and machine tools, we pointed out the importance of machine stiffness in controlling dimensional accuracy and surface finish of parts. In this section we describe the adverse effects of low stiffness on product quality and machining operations. Low stiffness affects the level of vibration and chatter in tools and machines. If uncontrolled, vibration and chatter can result in the following:

- Poor surface finish (right central region in Fig. 24.7).
- Loss of dimensional accuracy of the workpiece.
- Premature wear, chipping, and failure of the cutting tool, which is crucial with brittle tool materials, such as ceramics, some carbides, and diamond (see Chapter 21).
- Damage to machine-tool components from excessive vibrations.
- Objectionable noise generated, particularly if it is of high frequency, such as the squeal heard in turning brass on a lathe.

Extensive studies over the past 40 years have shown that vibration and chatter in cutting are complex phenomena. Cutting operations cause two basic types of vibration: forced vibration and self-excited vibration.

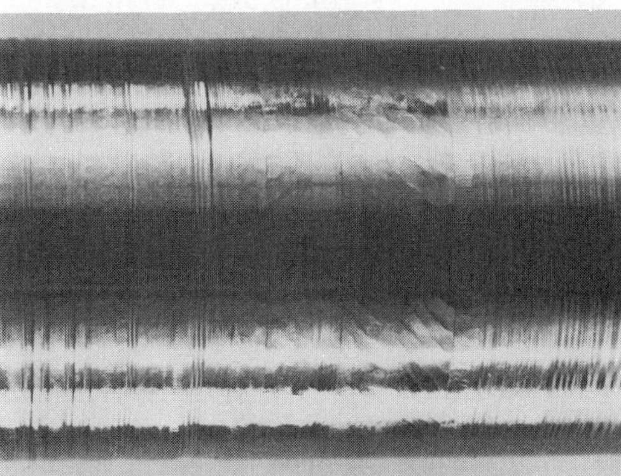

FIGURE 24.7 Chatter marks (right of center of photograph) on the surface of a turned part. *Source:* General Electric Company.

24.3.1 Forced vibrations

Forced vibration is generally caused by some periodic applied force present in the machine tool, such as from gear drives, imbalance of the machine-tool components, misalignment, or motors and pumps. In processes such as milling or turning a splined shaft or a shaft with a keyway, forced vibrations are caused by the periodic engagement of the cutting tool with and exit from the workpiece surface.

The basic solution to forced vibrations is to isolate or remove the forcing element. If the forcing frequency is at or near the natural frequency of a component of the machine-tool system, one of the frequencies may be raised or lowered. The amplitude of vibration can be reduced by increasing the stiffness or damping the system (see the discussion below). Although changing the cutting parameters generally does not appear to greatly influence forced vibrations, changing the cutting speed and the tool geometry can be helpful.

24.3.2 Self-excited vibrations

Generally called **chatter, self-excited vibration** is caused by the interaction of the chip-removal process (see Chapter 20) and the structure of the machine tool. Self-excited vibrations usually have a very high amplitude. Chatter typically begins with a disturbance in the cutting zone. Such disturbances include lack of homogeneity in the workpiece material or its surface condition, changes in type of chips produced, or a change in frictional conditions at the tool–chip interface as also influenced by cutting fluids and their effectiveness.

Regenerative chatter. The most important type of self-excited vibration is **regenerative chatter**. It is caused when a tool cuts a surface that has a roughness or disturbances from the previous cut (see, for example, Figs. 20.26 and 20.27). Because the depth of cut varies, the resulting variations in the cutting force subject the tool to vibrations and the process continues repeatedly—hence the term regenerative. You may observe this type of vibration while driving over a rough road (the so-called washboard effect).

Self-excited vibrations can generally be controlled by increasing the dynamic stiffness of the system and by damping. We define **dynamic stiffness** as the ratio of the amplitude of the force applied to the amplitude of vibration. Since a machine tool has different stiffnesses at different frequencies, changes in cutting parameters, such as cutting speed, can influence chatter.

24.3.3 Damping

We define **damping** as the rate at which vibrations decay. This effect is like testing your car's shock absorbers by pushing down on the car's front or rear end and observing how soon the motion stops. Damping is an important factor in controlling machine-tool vibration and chatter.

Internal damping of structural materials. Internal damping results from the energy loss in materials during vibration. Thus, for example, steel has less damping than gray cast iron, and composite materials (see Section 24.4) have more damping

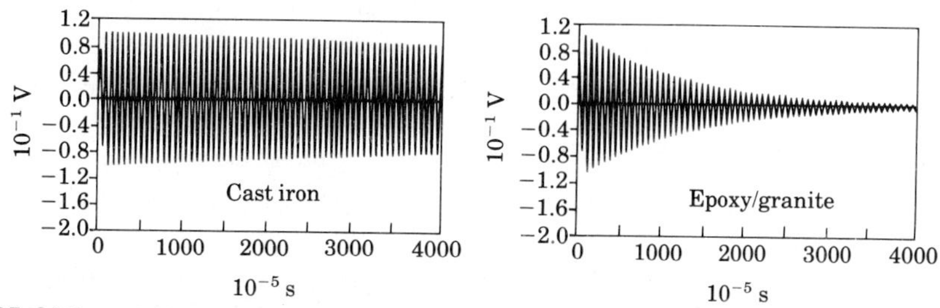

FIGURE 24.8

Relative damping capacity of gray cast iron and epoxy–granite composite material. The vertical scale is the amplitude of vibration and the horizontal scale is time. *Source:* Cincinnati Milacron, Inc.

than gray iron (Fig. 24.8). You can observe this difference in the damping capacity of materials by striking them with a gavel and listening to the sound. Try, for example, striking pieces of steel, concrete, and wood.

Joints in the machine-tool structure. Although less significant than internal damping, bolted joints in the structure of a machine tool are also a source of damping. Because friction dissipates energy, small relative movements along dry (unlubricated) joints dissipate energy and thus improve damping. In joints where oil is present, the internal friction of the oil layers also dissipates energy, thus contributing to damping. Note, for example, that stirring oil with a stick requires force; thus energy is dissipated in stirring it.

In describing the machine tools for various cutting operations, we noted that all machines consist of a number of large and small components, assembled into a structure. Consequently, this type of damping is cumulative, owing to the presence of a number of joints in a machine tool. Note in Fig. 24.9 how damping increases as the number of components on a lathe and their contact area increase. Thus the more joints, the greater is the amount of energy dissipated and the higher is the damping.

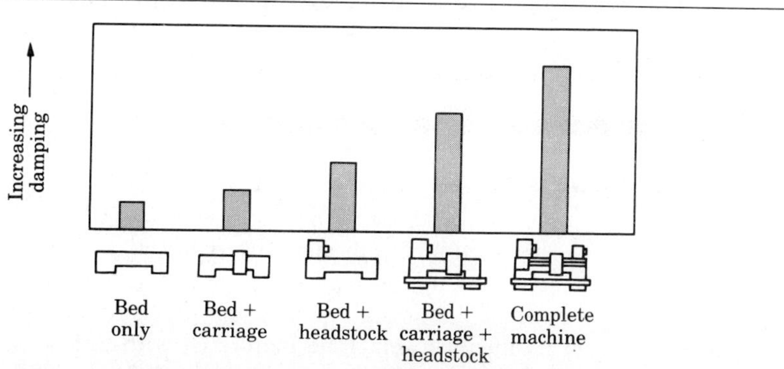

FIGURE 24.9

Damping of vibrations as a function of the number of components on a lathe. Joints dissipate energy; thus the greater the number of joints, the higher the damping will be. *Source:* J. Peters.

External damping. External damping is accomplished with external dampers, which are similar to shock absorbers on automobiles. Special vibration absorbers have been developed and installed on machine tools for this purpose.

24.3.4 Factors influencing chatter

Studies have indicated that the tendency for a particular workpiece to chatter during cutting is proportional to the cutting forces and the depth and width of cut. Consequently, because cutting forces increase with strength (hardness), the tendency to chatter generally increases as the hardness of the workpiece material increases. Thus aluminum and magnesium alloys have less tendency to chatter than do martensitic and precipitation-hardening stainless steels, nickel alloys, and high-temperature and refractory alloys.

An important factor in chatter is the type of chip produced during cutting operations. As you have seen, continuous chips involve steady cutting forces. Consequently, they generally do not cause chatter. Discontinuous chips and serrated chips, on the other hand, may do so. These chips are produced periodically, and the resulting variations in force during cutting can cause chatter.

24.3.5 Guidelines for reducing vibration and chatter

We have presented general guidelines for reducing vibration and chatter in machining operations in various sections of Chapters 22 and 23. We can summarize the basic guidelines as follows:

a) Minimize tool overhang.
b) Support workpiece rigidly.
c) Modify tool and cutter geometry.
d) Change process parameters, such as cutting speed, feed, depth of cut, and cutting fluids.
e) Increase the stiffness of the machine tool and its components by improving design, using larger cross-sections, and using materials with higher elastic modulus.
f) Improve the damping capacity of the machine tool.

24.4 Machine-Tool Structures

In this section we discuss the *material* and *design* aspects of machine tools as structures that have certain desired characteristics. Today's markets have stringent requirements for high quality and precision in manufactured products, often made with difficult-to-machine materials, with precise specifications concerning dimensional accuracy and surface finish and integrity. Consequently, the design and construction of machine tools are important aspects of manufacturing engineering.

The proper design of machine-tool structures requires a thorough knowledge of the materials available for construction, their forms and properties, the dynamics of the particular machining process, and the forces involved. As we stated in Section 24.3, *stiffness* and *damping* are important factors in machine-tool structures. Stiffness involves both the dimensions of the structural components and the elastic modulus of the materials used. Damping, as you have seen, involves the type of materials used, as well as the number and nature of the joints in the structure.

24.4.1 Materials and design

Traditionally, the base and some of the major components of machine tools have been made of gray or nodular cast iron, which has the advantages of low cost and good damping capacity, but are heavy. Lightweight designs are desirable because of ease of transportation, higher natural frequencies, and lower inertial forces of moving members. Lightweight designs and design flexibility require fabrication processes such as (a) mechanical fastening (bolts and nuts) of individual components and (b) welding. However, this approach to fabrication increases labor, as well as material, costs because of the preparations involved.

Wrought steels are the likely choice for such lightweight structures because of their low cost, availability in various section sizes and shapes (such as channels, angles, and tubes), desirable mechanical properties, and favorable characteristics (such as formability, machinability, and weldability). Tubes, for example, have high stiffness-to-weight ratios. On the other hand, the benefit of the higher damping capacity of castings and composites isn't available with steels.

In addition to stiffness, another factor that contributes to lack of precision of the machine tool is the thermal expansion of its components, causing distortion. The source of heat may be internal, such as bearings, ways, motors, and heat generated from the cutting zone, or external, such as nearby furnaces, heaters, sunlight, and fluctuations in cutting fluid and ambient temperatures.

Equally important in machine-tool precision are foundations, their mass, and how they are installed in a plant. For example, in one grinder for grinding 2.75-m (9-ft) diameter marine-propulsion gears with high precision, the concrete foundation is 6.7 m (22 ft deep). The large mass of concrete and the machine base reduces the amplitude of vibrations and its adverse effects.

24.4.2 New developments

Significant developments concerning the materials used for machine-tool bases and components have taken place. These developments are summarized in the following paragraphs.

Concrete. Concrete has been used for machine-tool bases since the early 1970s. Compared to cast iron, concrete is less expensive, its curing time is about three weeks (much shorter than for making castings), and it has good damping capacity. However, concrete is brittle and thus not suitable for applications involving impact loads. Concrete can also be poured into cast-iron base structures to increase their mass and improve machine-tool damping capacity. Filling the cavities of bases with loose sand is also an effective means of improving damping capacity.

Acrylic concretes. A mixture of concrete and plastic (polymethylmethacrylate; see Chapter 7), acrylic concretes can easily be cast into desired shapes for machine bases and various components. They were first introduced in the 1980s, and several new compositions are being developed. These materials can also be used for sandwich construction with cast irons, thus combining the advantages of each type of material.

Granite–epoxy composite. A castable composite has been developed, with a composition of about 93 percent crushed granite and 7 percent epoxy binder. First used in precision grinders in the early 1980s, this composite material has several favorable properties: (a) good castability, which allows design versatility in machine tools; (b) high stiffness-to-weight ratio; (c) thermal stability; (d) resistance to environmental degradation; and (e) good damping capacity (see Fig. 24.8).

Ceramics. Ceramic components are being used in machine tools for their high stiffness and good thermal stability. Spindles and bearings are being made of silicon nitride, which has better friction and wear characteristics than traditional metallic materials. Furthermore, the low density of ceramics makes them suitable for components in high-speed machinery that undergo rapid reciprocal movements, in which low inertial forces are desirable to maintain the system's stability.

Assembly techniques. Resin bonding is being used to assemble machine tools, which normally involved mechanical fastening and welding. These adhesives have favorable characteristics for machine-tool construction, do not require special preparation, and are suitable for assembling nonmetallic machine-tool components (see also Section 30.4).

24.5 Machining Economics

In the Introduction to Part IV, we stated that the limitations of machining processes include the amount of time required to machine a part and the amount of material wasted. On the other hand, machining is indispensable in many situations, particularly for complex shapes and obtaining good dimensional accuracy and surface finish, as you have seen from our discussion.

You have also seen that machining involves a number of material and process variables and that the proper choice of these variables is important in the productivity of machining operations. In this section we describe methods of economic analysis aimed at minimizing the cost of machining.

In machining a part, the total machining cost per piece, C, consists of:

$$C = C_1 + C_2 + C_3 + C_4, \tag{24.1}$$

where

C_1 = *nonproductive cost*: labor, overhead, and machine-tool costs involved in setting up for machining, mounting the cutting tool, preparing the fixtures and the machine, advancing and retracting the tool, and so on.

C_2 = *machining cost*: labor, overhead, and machine-tool costs while the cutting operation is taking place.

C_3 = *tool-change cost*: labor, overhead, and machine-tool costs during tool change.

C_4 = cost of *cutting tool.*

The costs of several cutting tools are listed in Tables 24.1 and 24.2.

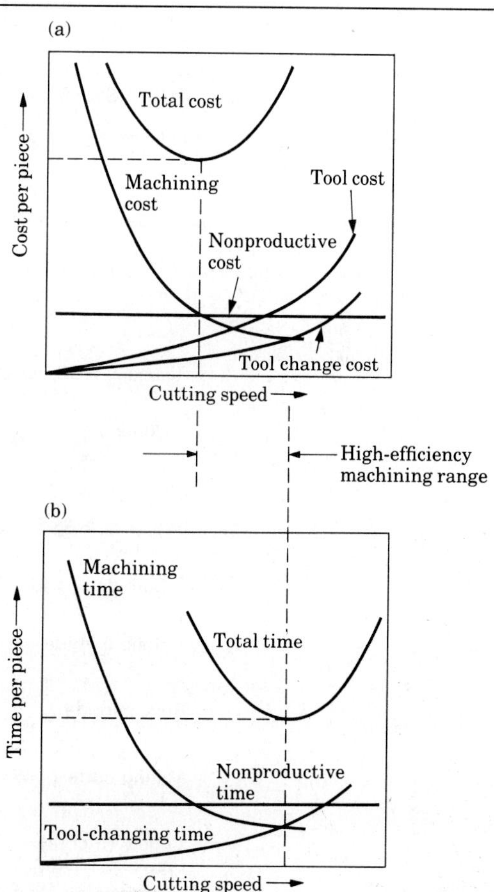

FIGURE 24.10 Graphs showing (a) cost per piece and (b) time per piece in machining. Note the optimum speeds for both cost and time. The range between the two is known as the high-efficiency machining range.

TABLE 24.1
COST OF SELECTED CUTTING TOOLS

TOOL	SIZE (in.)	COST ($)
High-speed steel tool bits	$\frac{1}{4}$ sq. × $2\frac{1}{2}$ long	1–2
	$\frac{1}{2}$ sq. × 4	3–7
Carbide-tipped (brazed) tools for turning	$\frac{1}{4}$ sq.	2
	$\frac{3}{4}$ sq.	5
Carbide inserts, square $\frac{3}{16}$" thick		
Plain	$\frac{1}{2}$ inscribed circle	2–5
Coated		3–6
Ceramic inserts, square	$\frac{1}{2}$ inscribed circle	5–8
Cubic boron-nitride inserts, square	$\frac{1}{2}$ inscribed circle	60–80
Diamond-tipped inserts (polycrystalline)	$\frac{1}{2}$ inscribed circle	75–100
Tool holders	1 × 1 × 6	50

TABLE 24.2
COST OF SELECTED TOOLS FOR MACHINING*

TOOL	SIZE (in.)	COST ($)
Drills, HSS, straight shank	$\frac{1}{4}$	1.00–1.70
	$\frac{1}{2}$	3.00–6.00
Coated (TiN)	$\frac{1}{4}$	2.60–3.00
	$\frac{1}{2}$	7–12
Tapered shank	$\frac{1}{4}$	2.50–7.00
	1	16–42
	2	80–85
	3	250
	4	950
Reamers, HSS, hand	$\frac{1}{4}$	8–10
	$\frac{1}{2}$	12–14
Chucking	$\frac{1}{2}$	7–9
	1	20–25
	$1\frac{1}{2}$	40–54
End mills, HSS	$\frac{1}{2}$	6–15
	1	18–27
Carbide tipped	$\frac{1}{2}$	30–33
	1	45–60
Solid carbide	$\frac{1}{2}$	30–70
	1	180
Burs, carbide	$\frac{1}{2}$	12–18
	1	50–60
Milling cutters, HSS, staggered tooth, $\frac{3}{8}$" wide	4	35–75
	8	130–260
Collets (5 core)	1	7–18

* Cost depends on the particular type of material and shape of tool.

Three of the four cost variables depend on cutting speed, as shown in Fig. 24.10. Note that as cutting speed increases, the machining time (hence cost per piece) decreases. However, as we have shown in Section 20.7, tool life decreases with cutting speed. Hence the tool cost increases, as does the tool-changing cost, since tools have to be changed more frequently. Note that the nonproductive cost C_1 does not vary with the cutting speed.

When we add these costs to obtain the total cost per piece, we find one optimum cutting speed for *minimum cost per piece* and another optimum speed for *minimum time per piece* (high production rate). The range between the two optimum speeds is sometimes called the **high-efficiency machining range**.

This analysis indicates the importance of identifying all relevant parameters in a machining operation, determining various cost factors, obtaining relevant tool-life curves for the particular operation, and properly measuring the various time intervals involved in the overall operation. The importance of obtaining accurate data is clearly shown in Fig. 24.10: Small changes in cutting speed can have a significant effect on the minimum cost or time per piece.

Example: Cost in end milling.

The results of an economic study involving end milling are shown in the following table. The individual factors were tool cost (highest for carbide), relative material-removal rate (carbides can be used for the highest cutting speed), and tool life (carbide has the longest tool life). The relative machining costs for three different types of cutting-tool materials are shown. Note that each tool material has different cost and machining characteristics. In the final analysis, however, the brazed carbide tool has the lowest cost per unit volume of material machined.

CUTTER	RELATIVE PURCHASE COST	RELATIVE MATERIAL-REMOVAL RATE	RELATIVE TOOL LIFE	RELATIVE MACHINING COST/UNIT VOLUME MACHINED
HSS	1	1	1	1.0
Carbide	6	3	3	0.35
Brazed carbide	3	2.5	2.5	0.29

Source: After V. A. Tipnis.

Example: Manufacturing bolts by forming and machining.

We have previously described and given various examples of making a particular part by different manufacturing processes. The threaded bolt shown in the accompanying figure, for example, can be made by two different methods. The bolt can be completely machined from a round bar stock, or the bolt can be made by the sequence of cold-forming operations shown and the threads produced by thread rolling.

Note that for low quantities the cost per bolt, including tooling cost, is low when the part is completely machined. The reason is that machining does not require special tooling or dies, other than standard turning and threading tools. However, as you have seen, machining this part takes considerable time.

The two curves intersect at 16 bolts. Thus if more than 16 bolts are required, making this bolt by forming is more economical. Although forming requires special dies, which can be costly and time-consuming to make, the cost of dies per piece decreases as the number of parts made with the same set of dies increases. Note that a similar situation exists in forging (see Fig. 14.35).

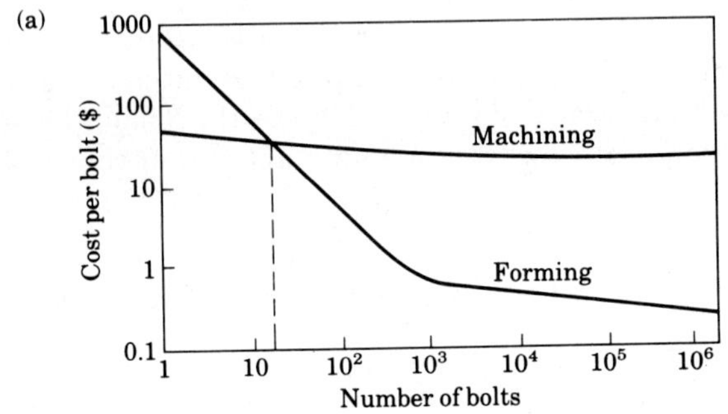

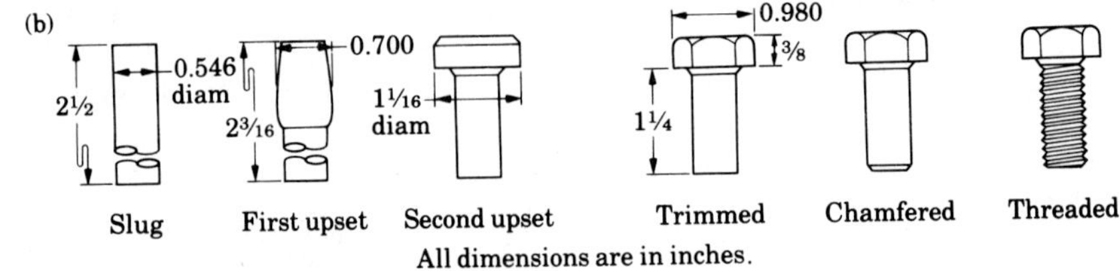

SUMMARY

Because of their versatility and capability of performing a variety of cutting operations on small and large workpieces, machining centers have become one of the most important developments in machine tools. Various types of machining centers are available, and their selection depends on factors such as part complexity, the number and type of cutting operations involved, dimensional accuracy, and production rate required.

Vibration and chatter in machining are important considerations for workpiece dimensional accuracy and surface finish, as well as tool life. Stiffness and damping capacity of machine tools are important factors in controlling vibration and chatter. New materials are being developed and used for constructing machine-tool structures.

The economics of machining processes depend on nonproductive, machining, tool-change, and tool costs. Optimum cutting speeds can be obtained for minimum machining time per piece and minimum cost per piece, respectively.

TRENDS

- Machining centers and computer controls on all types of machine tools have progressed rapidly and continue to do so. Improvements are constantly being made in the power, speed, and stiffness of these machines.
- In many cases, ordinary machine tools, described in Chapters 22 and 23, are now becoming machining centers with a variety of features and computer controls.
- Machining centers are being constructed with multitool heads and rapid traverses in all axes to reduce idle time, with traverse speeds reaching 4 m/s (100 in./min).
- The design of machining centers is being reviewed and modified to make them fit and integrate better when placed in line with other machines in computer-integrated manufacturing systems. These include modular machining centers.
- Proper fixturing of workpieces on machine tools to eliminate or reduce part distortion, due to cutting forces or heat generated during machining, continue to be an important aspect of fixture design.
- The characteristics of machine-tool structures and their construction are constantly being investigated in order to minimize distortions and deviations during cutting operations and improve surface finish and dimensional accuracy.
- New materials are being developed for machine tool bases and components. Among these materials are concrete, acrylic concrete mixes, epoxy–granite composites, ceramics, and various combinations of them.

KEY TERMS

Chatter
Damping
Dynamic stiffness
Forced vibration
High-efficiency machining range
Machining centers
Regenerative chatter
Self-excited vibration

BIBLIOGRAPHY

Boothroyd, G., and W.A. Knight, *Fundamentals of Machining and Machine Tools*, 2d ed. New York: Marcel Dekker, 1989.

Koenigsberger, F., and J. Tlusty, *Machine Tool Structures*. New York: Pergamon, 1970.

Metals Handbook, 8th ed., Vol. 3: *Machining*. Metals Park, Ohio: American Society for Metals, 1967.

Metals Handbook, 9th ed., Vol. 16: *Machining*. Materials Park, Ohio: ASM International, 1989.

Reshetov, D.N., and V.T. Portman, *Accuracy of Machine Tools*. New York: American Society of Mechanical Engineers, 1989.

Tool and Manufacturing Engineering Handbook, 4th ed., Vol. 1: *Machining*. Dearborn, Mich.: Society of Manufacturing Engineers, 1983.

Weck, M., *Handbook of Machine Tools*, 4 vols. New York: Wiley, 1984.

REVIEW QUESTIONS

24.1 What are the distinctive features of machining centers? Why are they so versatile?

24.2 What are the approximate size and weight of the workpiece that can presently be machined in a machining center?

24.3 Explain the features of basic types of machining centers.

24.4 What advantages do gantry-type machining centers have over other types?

24.5 Explain the tooling system in a machining center and how it operates.

24.6 Describe the economic considerations in selecting machining centers.

24.7 What are forced vibrations? Self-excited vibrations? What is regenerative chatter? How would you eliminate these vibrations?

24.8 Describe the adverse effects of vibrations and chatter in machining.

24.9 Why is damping of machine tools important? How is it accomplished?

24.10 List the factors that contribute to chatter in machining.

24.11 Explain the trends in materials used for machine-tool structures.

24.12 Why is thermal expansion of machine-tool components important?

24.13 What factors contribute to costs in machining operations?

24.14 What is meant by high-efficiency machining range?

QUESTIONS AND PROBLEMS

24.15 Explain the technical requirements that led to the development of machining centers.

24.16 Would it be possible to design and build machining centers without the use of computer controls? Explain.

24.17 If you were the chief engineer in charge of the design of machining centers, what changes and improvements would you make on existing models?

24.18 Why do spindle speeds in machining centers vary over a wide range?

24.19 Why is the stiffness of machine tools so important in machining operations?

24.20 Outline the trends in the design of machine tools.

24.21 Explain the concept of machining centers in terms that someone who does not know much about machine tools can understand.

24.22 Are there cutting operations that cannot be performed in machining centers? Why or why not?

24.23 Is the control of cutting-fluid temperature important? In what way?

24.24 On the basis of the information given in Part I of this text and other sources, make a list of the elastic modulus of the materials that are now being used for advanced machine tools.

24.25 Make a list of components of machine tools that could be made of ceramics and explain why ceramics would be suitable.

24.26 Explain how you would go about reducing each of the cost factors in machining operations. What difficulties would you encounter?

24.27 Review the various parts made by the processes described in Part III, and identify those that can be made economically by machining. Explain your reasons.

24.28 In a drilling operation, it has been shown that a $\frac{1}{2}$-in. diameter coated high-speed steel drill has three times the life of an uncoated drill made of the same material. On the basis of the discussions in Section 24.5 and Table 24.2, which drill would you recommend, and why?

24.29 We know that we may be able to decrease machining costs by increasing the cutting speed. Explain which costs are likely to change, and how, as the cutting speed increases.

24.30 In addition to the number of joints in a machine tool, what other factors influence the rate at which damping increases, as shown in Fig. 24.9?

24.31 Would the parts shown in Problems 22.59 and 23.59 be suitable for machining on a machining center? Explain.

24.32 Comment on the cost of tools as a function of their size, as shown in Tables 24.1 and 24.2.

25

Abrasive Processes and Finishing Operations

25.1 Introduction

In many cases the surface finish and dimensional accuracy requirements for a part are too fine, the workpiece material is too hard, or the workpiece material is too brittle to produce the part solely by any of the processes that we have described so far. For example, ball and roller bearings, pistons, valves, cylinders, some cutting tools and dies, cams, gears, and precision components for instrumentation generally require high dimensional accuracy and fine surface finish.

One of the best methods for producing such parts is to use **abrasives**. An abrasive is a small, hard particle having sharp edges and an irregular shape, unlike the cutting tools we described earlier. Abrasives are capable of removing small amounts of material from a surface by a cutting process that produces tiny chips. Most of us are familiar with using bonded abrasives (grinding wheels) to sharpen

FIGURE 25.1
A variety of bonded abrasives. *Source:* Courtesy of Norton Company.

knives and tools (Fig. 25.1) and using sandpaper to smoothen surfaces and sharp corners. Abrasives are also used to hone, lap, buff, and polish workpieces.

Abrasive processes are generally among the last operations performed on manufactured products. These processes, however, are not necessarily confined to fine or small-scale material removal. They are also used for large-scale removal operations and can indeed compete economically with some machining processes, such as milling and turning.

Because they are hard, abrasives are also used in finishing very hard or heat-treated parts; shaping hard nonmetallic materials, such as ceramics and glasses; removing unwanted weld beads; cutting off lengths of bars, structural shapes, masonry, and concrete; and cleaning surfaces with jets of air or water containing abrasive particles.

In this chapter we first present the characteristics of abrasives and how they are used in various removal processes. As we have done with cutting operations, we will describe the mechanics of these operations. Knowing the mechanics allows us to establish the interrelationships of material and process variables and the quality and dimensional accuracy of surfaces produced. Finally, we discuss grinding and related processes, such as honing and lapping, as well as surface finishing operations.

25.2 Abrasives

Abrasives commonly used in material removal processes in manufacturing operations are:

a) Aluminum oxide (Al_2O_3)
b) Silicon carbide (SiC)
c) Cubic boron nitride (CBN)
d) Diamond

These abrasives (see also Chapter 8) are considerably harder than cutting-tool materials (Table 25.1). Because of their extreme hardness, diamond and CBN are generally called *superabrasives*. As used in manufacturing processes, abrasives are generally very small in size as compared to cutting tools or inserts, and have sharp edges, allowing the removal of very small quantities of material from the workpiece surface. Consequently, very fine surface finish and dimensional accuracy can be obtained. See, for example, Figs. 22.14 and 26.4.

The size of an abrasive *grain* (or grit) is identified by a number (Fig. 25.2). The smaller the grain size, the larger the number. For example, grain size 10 is regarded as very coarse, 100 as fine, and 500 as very fine. Sandpaper or emery cloth, for example, are also identified in this manner, with the grain size printed on the back of the paper or cloth.

Abrasives found in nature are emery, corundum (alumina, aluminum oxide), quartz, garnet, and diamond. However, natural abrasives contain unknown amounts of impurities and possess nonuniform properties. Consequently, their performance is unreliable. As a result, the aluminum oxide and silicon carbide used are now almost totally synthetic. Synthetic aluminum oxide, first made in 1893, is obtained by fusing bauxite, iron filings, and coke. Silicon carbide (1891) is made with silica sand, petroleum coke, and small amounts of sodium chloride and sawdust. First molten, then fused in electric furnaces, and later cooled, the abrasives are finally crushed and graded into various sizes by passing them through standard screens.

TABLE 25.1
KNOOP HARDNESS FOR VARIOUS MATERIALS AND ABRASIVES

Common glass	350–500	Titanium nitride	2000
Flint, quartz	800–1100	Titanium carbide	1800–3200
Zirconium oxide	1000	Silicon carbide	2100–3000
Hardened steels	700–1300	Boron carbide	2800
Tungsten carbide	1800–2400	Cubic boron nitride	4000–5000
Aluminum oxide	2000–3000	Diamond	7000–8000

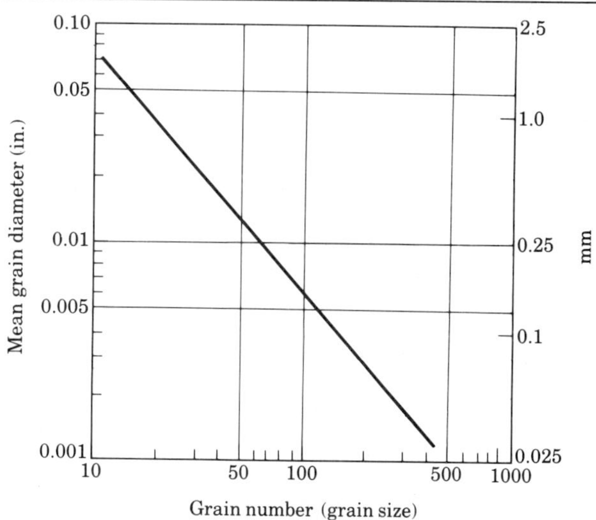

FIGURE 25.2 Relationship between abrasive grain size (number) and its mean diameter.

A recent advance in abrasives is the development of *seeded gel* aluminum-oxide abrasives, made by a nonfusion sintering process. The size of these abrasive particles typically is on the order of 0.2 μm, which is much smaller than commonly used abrasive particles. **Seeded gel** (SG) **abrasives** are free of flaws found in abrasives made by other processes. They do not dull as readily and, because of their unique crystal structure, maintain their sharpness during use by fracturing in a manner that leaves sharp cutting edges.

25.3 Bonded Abrasives (Grinding Wheels)

Because each abrasive grain usually removes only a very small amount of material at a time, high rates of material removal can be obtained only if a large number of these grains act together. This is done by using **bonded abrasives**, typically in the form of a **grinding wheel**. A simple grinding wheel is shown schematically in Fig. 25.3. The abrasive grains are spaced at some distance from each other and are held together by bonds, which act as supporting posts or braces between the grains.

Some of the commonly used types of grinding wheel are shown in Fig. 25.4, with their grinding surfaces indicated by arrows. An estimated 250,000 different types and sizes of abrasive wheels are made today.

Bonded abrasives are marked with a standardized system of letters and numbers, indicating the type of abrasive, grain size, grade, structure, and bond type.

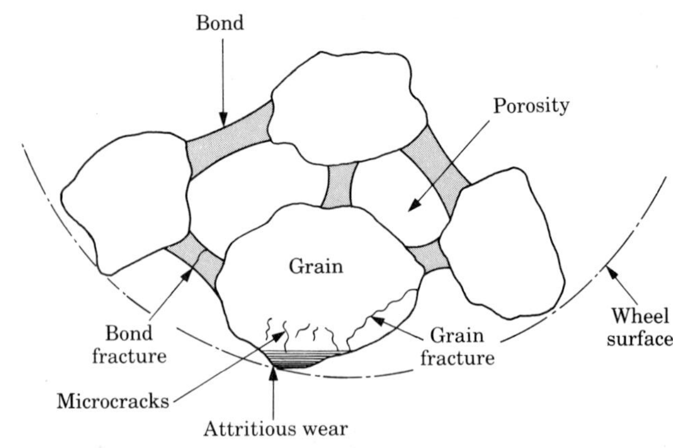

FIGURE 25.3 Schematic illustration of a physical model of a grinding wheel, showing its structure and wear and fracture patterns.

Figure 25.5 shows the marking system for aluminum-oxide and silicon-carbide bonded abrasives. Figure 25.6 (p. 780) shows the marking system for diamond and cubic boron nitride bonded abrasives.

25.3.1 Bond types

Vitrified. Essentially a glass, *vitrified bond* is also called a ceramic bond, particularly outside the United States. It is the most common and widely used bond. The raw materials consist of feldspar (a crystalline mineral) and clays. They are mixed with the abrasives, moistened, and molded under pressure into the shape of

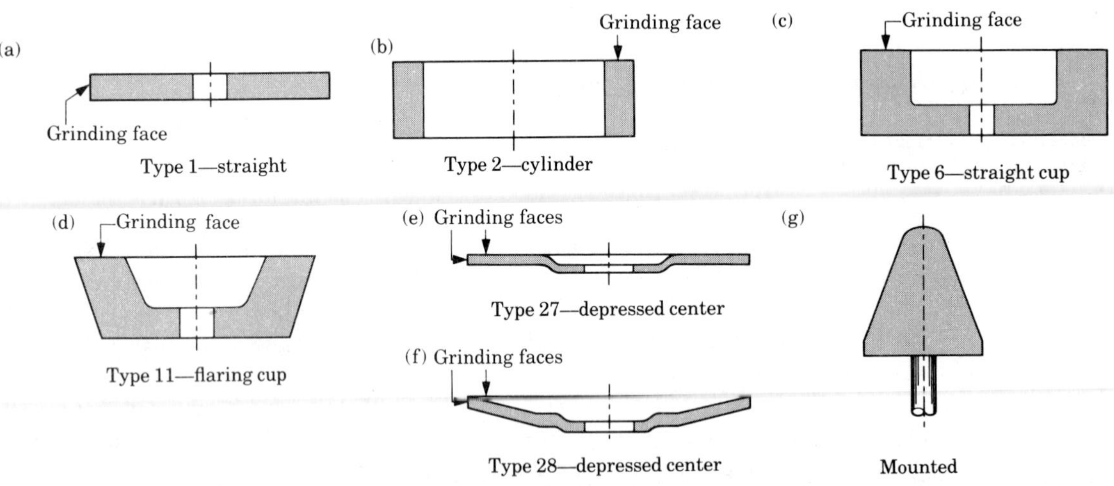

FIGURE 25.4 Some common types of grinding wheels. Note that each wheel has a specific grinding face. Grinding on other surfaces is improper and unsafe.

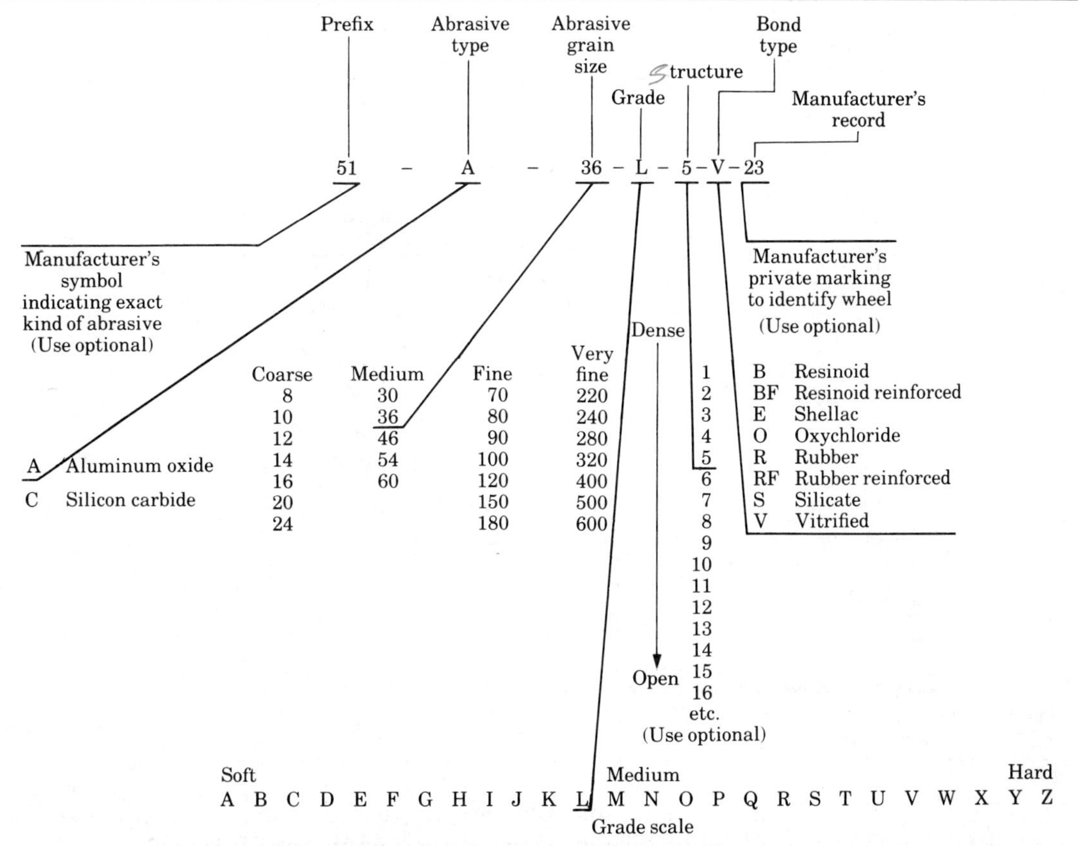

FIGURE 25.5
Standard marking system for aluminum-oxide and silicon-carbide bonded abrasives.

grinding wheels. These "green" products, which are similar to powder-metallurgy parts, are then fired slowly, up to a temperature of about 1250 °C (2300 °F), to fuse the glass and develop structural strength. The wheels are then cooled slowly to avoid thermal cracking, finished to size, inspected for quality and dimensional accuracy, and tested for defects.

Vitrified bonds produce wheels that are strong, stiff, porous, and resistant to oil, acids, and water. Because they are brittle, they lack resistance to mechanical and thermal shock. However, vitrified wheels are also available with steel backing plates or cups for better structural support during their use.

Resinoid. Resinoid bonding materials are thermosetting resins, and are available in a wide range of compositions and properties. Because the bond is an organic compound, wheels with resinoid bonds are also called *organic wheels*. The basic manufacturing technique consists of mixing the abrasive with liquid or

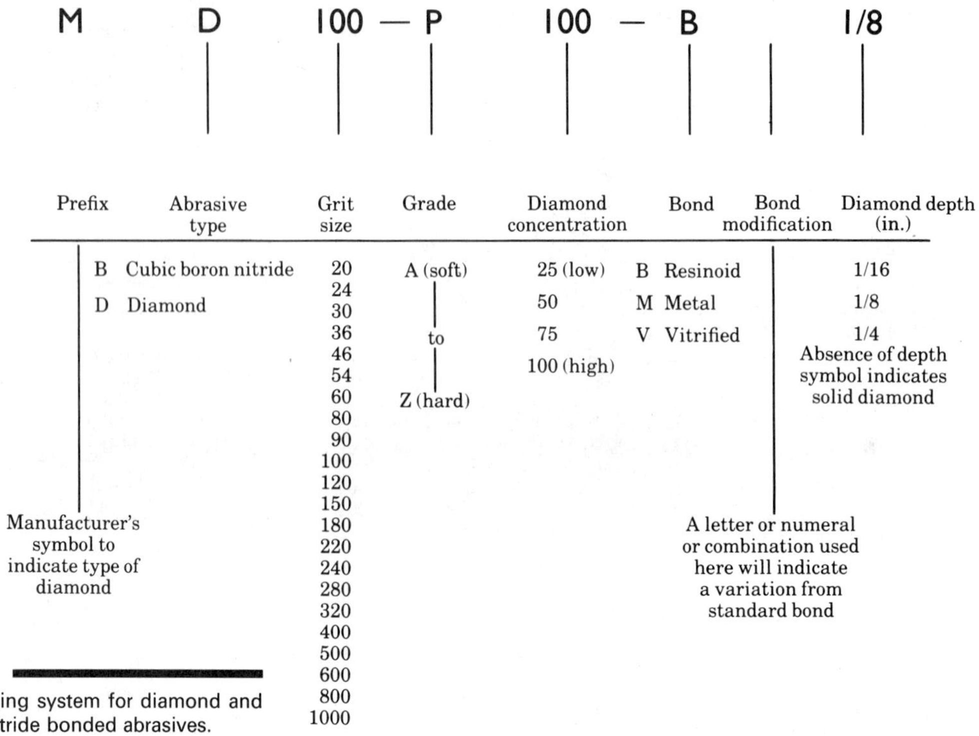

M	D	100 —	P	100 —	B		1/8
Prefix	Abrasive type	Grit size	Grade	Diamond concentration	Bond	Bond modification	Diamond depth (in.)
Manufacturer's symbol to indicate type of diamond	B Cubic boron nitride D Diamond	20 24 30 36 46 54 60 80 90 100 120 150 180 220 240 280 320 400 500 600 800 1000	A (soft) to Z (hard)	25 (low) 50 75 100 (high)	B Resinoid M Metal V Vitrified	A letter or numeral or combination used here will indicate a variation from standard bond	1/16 1/8 1/4 Absence of depth symbol indicates solid diamond

FIGURE 25.6 Standard marking system for diamond and cubic boron nitride bonded abrasives.

powdered phenolic resins and additives, pressing the mixture into the shape of a grinding wheel, and curing it at temperatures of about 175 °C (350 °F).

Because the elastic modulus of thermosetting resins is lower than that of glasses, resinoid wheels are more flexible than vitrified wheels. *Reinforced wheels* are available. One or more layers of fiberglass mats of various mesh sizes provide the reinforcement. Its purpose is to retard the disintegration of the wheel should it break for some reason, rather than to improve its strength. Large-diameter resinoid wheels can be supported additionally with one or more internal steel rings, which are inserted during molding of the wheel.

Rubber. The most flexible bond used in abrasive wheels is rubber. The manufacturing process consists of mixing crude rubber, sulfur, and the abrasive grains together, rolling the mixture into sheets, cutting out circles, and heating them under pressure to vulcanize the rubber. Thin wheels can be made in this manner and are used like saws for cutting-off operations.

Metal bonds. Using powder-metallurgy techniques, the abrasive grains, which are usually diamond or cubic boron nitride, are bonded to the periphery of a metal wheel, to depths of 6 mm (0.25 in.) or less. Bonding is carried out under high pressure and temperature, without the use of bonding materials. The wheel itself may be made of steel or bronze, although plastics can also be used.

Other bonds. In addition to those we described above, other bonds include silicate, shellac, and oxychloride bonds. However, they have limited uses, and we won't discuss them further here. A new development is the use of polyimide as a substitute for the phenolic in resinoid wheels. It is tough and has resistance to high temperatures.

25.3.2 Wheel grade and structure

The **grade** of a bonded abrasive is a measure of the bond's strength. Thus it includes both the type and the amount of bond in the wheel. Because strength and hardness are directly related, the grade is also referred to as the *hardness* of a bonded abrasive. Thus a hard wheel has a stronger bond and/or a larger amount of bonding material between the grains than a soft wheel.

The **structure** is a measure of the *porosity* (spacing between the grains in Fig. 25.3) of the bonded abrasive. Some porosity is essential to provide clearance for the grinding chips; otherwise they would interfere with the grinding process. The structure of bonded abrasives ranges from dense to open.

25.4 The Grinding Process

Grinding is a chip-removal process, and the cutting tool is an individual abrasive grain. The major differences between grain and single-point cutting tool actions are that individual grains have irregular shapes and are spaced randomly along the periphery of the wheel (Fig. 25.7). The average rake angle of the grains is highly negative (Fig. 25.8).

FIGURE 25.7 The surface of a grinding wheel (A46-J8V) showing grains, porosity, wear flats on grains, and metal chips from the workpiece adhering to the grains. Note the random distribution and shape of abrasive grains. Magnification: 50×.

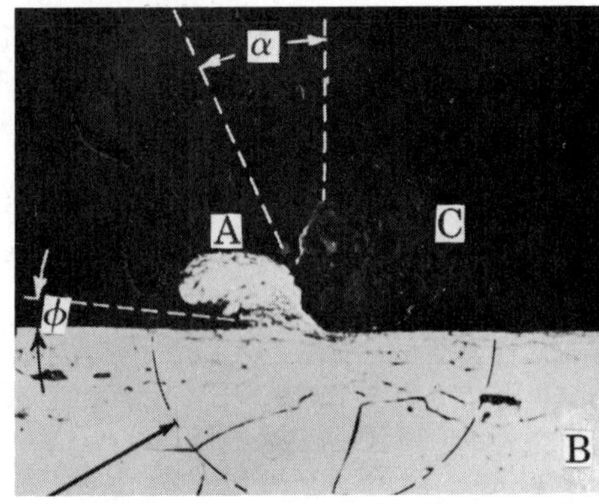

FIGURE 25.8 Grinding chip being produced by a single abrasive grain. (A) chip, (B) workpiece, (C) abrasive grain. Note the large negative rake angle of the grain. The inscribed circle is 0.065 mm (0.0025 in.) in diameter. *Source:* M. E. Merchant.

The grinding process and its parameters can best be observed in the surface grinding operation, shown schematically in Fig. 25.9. A straight grinding wheel (see Fig. 25.4a) of diameter D removes a layer of metal at a depth d (*wheel depth of cut*). An individual grain on the periphery of the wheel moves at a tangential velocity V, while the workpiece moves at a velocity v. Each abrasive grain removes a small chip whose undeformed thickness (*grain depth of cut*) is t and the undeformed length is l.

• Example: Chip dimensions in surface grinding.

We can show that the undeformed chip length l in surface grinding (see Fig. 25.9) is approximately given by the expression

$$l = \sqrt{Dd}$$

and the undeformed chip thickness t by the expression

$$t = \sqrt{\frac{4v}{VCr}\sqrt{\frac{d}{D}}},$$

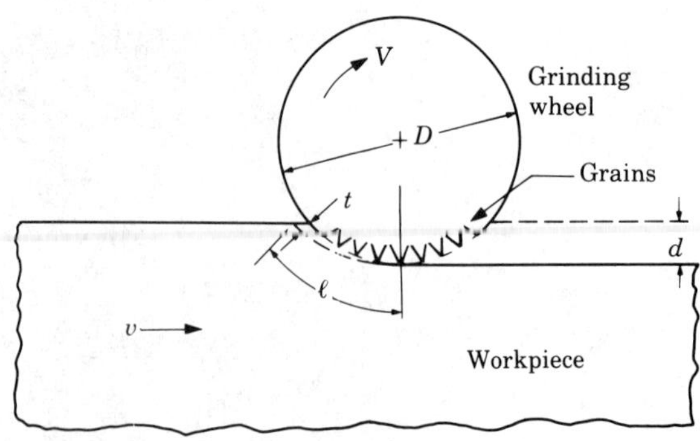

FIGURE 25.9 Schematic illustration of the surface grinding process, showing various process variables. The figure depicts up grinding. See also Fig. 23.4(a).

where C is the number of cutting points per unit area of the periphery of the wheel and is estimated to be in the range of 0.1–10 per mm^2 (10^2–10^3 per in^2). The quantity r is the ratio of chip width to average undeformed chip thickness and has an approximate value of between 10 and 20. Calculate l and t for the following process parameters: $D = 200$ mm, $d = 0.05$ mm, $v = 0.5$ m/s, and $V = 30$ m/s.

Using the above formulas, we find that

$$l = \sqrt{(200)(0.05)} = 3.2 \text{ mm} = 0.13 \text{ in.}$$

Assuming that $C = 2$ per mm^2 and that $r = 15$, we obtain

$$t = \sqrt{\frac{(4)(0.5)}{(30)(2)(15)}\sqrt{\frac{0.05}{200}}} = 0.006 \text{ mm} = 0.00023 \text{ in.}$$

Note that these chips are much smaller than those obtained in metal cutting operations in general. Furthermore, because of deformation, the actual chip will be shorter and thicker than the calculated values.

●

25.4.1 Forces

A knowledge of grinding forces is necessary not only in the design of grinding machines and estimating power requirements, but also in determining the deflections that the workpiece will undergo. Deflections, in turn, adversely influence dimensional accuracy and are critical in precision grinding.

If we assume that the cutting force on the grain is proportional to the cross-sectional area of the undeformed chip, we can show that the grain force (tangential force on the wheel) is proportional to the process variables:

$$\text{Grain force} \propto \left(\frac{v}{V}\sqrt{\frac{d}{D}}\right)(\text{strength of the material}). \tag{25.1}$$

Because of the small dimensions involved, forces in grinding are much smaller than those in cutting operations (see Chapters 22 and 23). Grinding forces should be kept low in order to avoid distortion and to maintain dimensional accuracy of the workpiece.

The energy dissipated in producing a grinding chip consists of the energy required for (a) chip formation, (b) ploughing (as shown by the ridges in Fig. 25.10), and (c) friction caused by rubbing of the grain along the surface. The grain develops a **wear flat** (Fig. 25.11)—similar to flank wear in cutting tools—as a result of the grinding operation. The wear flat rubs along the ground surface and, because of friction, dissipates energy.

Typical *specific energy* requirements (energy per unit volume of material removed) in grinding are shown in Table 25.2. Note that these energy levels are much higher than those in machining operations with cutting tools (see Table 20.1). This difference can be attributed to factors such as the presence of a wear flat and chips produced with a high neg ıtive rake angle. Thus the chips undergo much larger deformation and consequently equire higher energies.

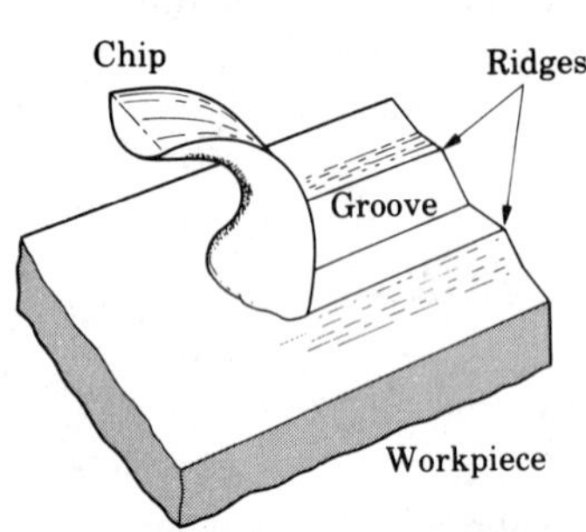

FIGURE 25.10 Chip formation and plowing of the workpiece surface by an abrasive grain.

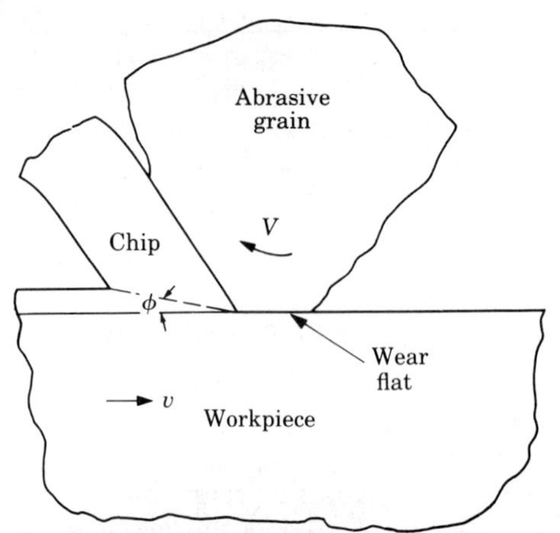

FIGURE 25.11 Schematic illustration of chip formation by an abrasive grain with a wear flat. Note the negative rake angle and small shear angle. Wear flat is similar to flank wear in cutting tools (see Fig. 20.17a).

25.4.2 Temperature

Temperature rise in grinding is an important consideration because it can adversely affect the surface properties and cause residual stresses on the workpiece. Furthermore, temperature gradients in the workpiece cause distortions by differential thermal expansion and contraction. When a portion of the heat generated is conducted into the workpiece, it expands the part being ground, thus making it difficult to control dimensional accuracy.

The surface temperature in grinding is related to process variables by the following expression:

$$\text{Temperature rise} \propto D^{1/4} d^{3/4} \left(\frac{V}{v}\right)^{1/2}. \tag{25.2}$$

TABLE 25.2
APPROXIMATE UNIT POWER REQUIREMENTS FOR SURFACE GRINDING

WORKPIECE MATERIAL	HARDNESS	SPECIFIC ENERGY* ($W \cdot s/mm^3$)
Aluminum	150 *HB*	6.8–27
Cast iron (class 40)	215 *HB*	12–60
Low-carbon steel (1020)	110 *HB*	13.7–68
Titanium alloy	300 *HB*	16.4–55
Tool steel (T15)	67 *HRC*	17.7–82

* Divide by 2.73 to obtain $hp \cdot min/in^3$.

Thus temperature increases with increasing depth of cut, wheel diameter, and wheel speed, and decreases with increasing work speed. Note that the depth of cut d has the greatest influence on temperature.

Peak temperatures during grinding may reach 1600 °C (3000 °F). However, the time involved in producing a chip is extremely short (microseconds), so the chip may or may not melt. Because the chips carry away much of the heat generated, as with chips in cutting with tools (see Fig. 20.16), only a fraction of the heat produced in grinding is conducted to the workpiece.

Sparks. The sparks produced during grinding metals are actually chips that glow and result from the exothermic (heat producing) reaction of the hot chips with oxygen in the atmosphere. Sparks do not occur when any metal is ground in an oxygen-free environment. The color, intensity, and shape of the sparks depend on the composition of the metal being ground. Thus by observing sparks, we can identify the type of metal being ground. If the heat generated due to exothermic reaction is sufficiently high, chips can melt and, owing to surface tension, acquire a spherical shape and solidify as spherical metal particles.

Tempering. Excessive temperature rise in grinding can cause tempering and softening of the workpiece surface. Process variables must therefore be selected carefully in order to avoid excessive temperature rise. The use of grinding fluids is an effective means of controlling temperature.

Burning. Excessive temperature during grinding may burn the ground surface. A *burn* is characterized by a bluish color on ground steel surfaces, an indication that high temperature caused oxidation. A burn may not be objectionable in itself. However, the surface layers may undergo phase transformations, with martensite forming in higher carbon steels from rapid cooling (*metallurgical burn*). This condition will influence the surface properties of ground parts, reducing surface ductility and toughness.

Heat checking. High temperatures in grinding may cause the workpiece surface to crack (*heat checking*). These cracks are usually perpendicular to the grinding direction; however, under severe grinding conditions, parallel cracks may also appear. Such a surface has low toughness and low fatigue and corrosion resistance.

25.4.3 Residual stresses

Temperature gradients within the workpiece during grinding are primarily responsible for residual stresses. Grinding fluids and method of application, as well as process parameters such as depth of cut and speeds, significantly influence the magnitude and type (tension or compression) of residual stresses developed.

Because of the deleterious effect of tensile residual stresses on fatigue strength, process variables should be carefully selected. Residual stresses can usually be reduced by lowering wheel speed and increasing work speed (*low-stress*, or *gentle*, *grinding*). Softer grade wheels, known as *free-cutting wheels*, may also be used.

25.5 Grinding Wheel Wear

Grinding wheel wear is an important consideration because it adversely affects the shape and accuracy of ground surfaces, as is the case with cutting tools. Grinding wheels wear by three different mechanisms: attritious grain wear, grain fracture, and bond fracture.

25.5.1 Attritious wear

In **attritious wear**, the cutting edges of a sharp grain become dull by attrition, developing a wear flat that is similar to flank wear in cutting tools. Wear is caused by the interaction of the grain with the workpiece material, involving both physical and chemical reactions. These reactions are complex and involve diffusion, chemical degradation or decomposition of the grain, fracture at a microscopic scale, plastic deformation, and melting.

Attritious wear is low when the two materials are chemically inert with respect to each other, much like with the use of cutting tools. The more inert the materials, the lower will be the tendency for reaction and adhesion to occur between the grain and the workpiece. For example, because aluminum oxide is relatively inert with respect to iron, its rate of attritious wear when it is used to grind steels is much lower than that for silicon carbide and diamond. On the other hand, silicon carbide can dissolve in iron, and hence it is not suitable for grinding steels. Cubic boron nitride has a higher inertness with respect to steels, and thus it is suitable for use as an abrasive.

The selection of the type of abrasive for low attritious wear is therefore based on the reactivity of the grain and the workpiece and their relative mechanical properties, such as hardness and toughness. The environment and the type of grinding fluid used also have an influence on grain–workpiece material interactions.

25.5.2 Grain fracture

Because abrasive grains are brittle, their fracture characteristics in grinding are important. If the wear flat caused by attritious wear is excessive, the grain becomes dull and grinding becomes inefficient and produces undesirably high temperatures. Optimally, the grain should fracture or fragment at a moderate rate, so that new sharp cutting edges are produced continuously during grinding. This is equivalent to breaking a dull piece of chalk or resharpening a pencil in order to draw fine lines.

Friability. The ability of grains to fracture gives abrasive wheels self-sharpening characteristics. The parameter describing the fracture behavior of abrasives is called **friability**. High friability indicates low strength or low fracture resistance of the grain. Thus a highly friable grain fragments more rapidly under dynamic forces in grinding than a grain with low friability. Aluminum oxide has lower friability, hence less tendency to fragment, than silicon carbide.

The shape and size of the grain also affect its friability. Blocky grains, for example, are less friable than platelike grains. Also, because the probability of defects in small grains is lower, they are stronger and less friable than larger grains (size effect).

The selection of grain type and size for a particular application also depends on the attritious wear rate. A grain–workpiece material combination with high attritious wear and low friability dulls grains and develops a large wear flat. Grinding then becomes inefficient, and surface damage is likely to occur.

25.5.3 Bond fracture

The strength of the bond (grade) is a significant parameter in grinding. If the bond is too strong, dull grains cannot be dislodged and grinding becomes inefficient. On the other hand, if the bond is too weak, the grains are dislodged easily and the wear rate of the wheel increases. Consequently, maintaining dimensional accuracy becomes difficult. In general, softer bonds are recommended for harder materials and for reducing residual stresses and thermal damage to the workpiece. Hard-grade wheels are used for softer materials and for removing large amounts of material at high rates.

25.5.4 Dressing

Dressing is the process of conditioning worn grains on the surface of a grinding wheel in order to produce sharp new grains. Dressing is necessary when excessive attritious wear dulls the wheel—called **glazing** because of the shiny appearance of the wheel surface—or when the wheel becomes loaded. **Loading** occurs when the porosities on the grinding surfaces of the wheel (see Fig. 25.7) become filled or clogged with chips. A loaded wheel cuts very inefficiently, generating much frictional heat, causing surface damage and loss of dimensional accuracy.

Dressing is done by various techniques. In one method, a specially shaped diamond or diamond cluster is moved across the width of the grinding face of a rotating wheel and removes a small layer from the wheel surface with each pass (Fig. 25.12).This method can be either dry or wet, depending on whether the wheel is to be used dry or wet, respectively. In another dressing method, a set of star-

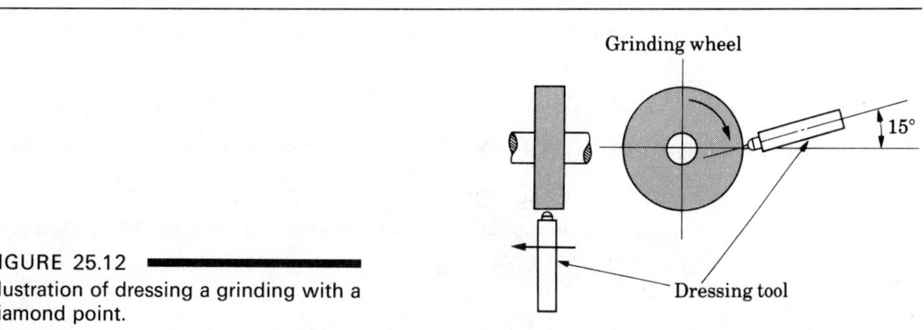

FIGURE 25.12 Illustration of dressing a grinding with a diamond point.

shaped steel disks is pressed against the wheel. Material is removed from the wheel surface by crushing the grains (*crush dressing*). This method produces a coarse grinding surface on the wheel and is used only for rough grinding operations. Abrasive sticks may also be used to dress grinding wheels.

Dressing techniques and how frequently the wheel surface is dressed are significant, affecting grinding forces and surface finish. Modern grinders are equipped with automatic dressing features, which dress the wheel as grinding continues. Dressing can also be done to generate a certain shape or form on a grinding wheel for the purpose of grinding profiles on workpieces (see Section 25.6.2).

Truing is a dressing operation by which a wheel is restored to its original shape. Thus a round wheel is dressed to make its circumference a true cricle (hence the word truing).

25.5.5 Grinding ratio

Grinding wheel wear is generally correlated with the amount of material ground by a parameter called the **grinding ratio**, G, which is defined as

$$G = \frac{\text{Volume of material removed}}{\text{Volume of wheel wear}}. \tag{25.3}$$

In practice, grinding ratios vary widely, ranging from 2 to 200 and higher, depending on the type of wheel, workpiece material, the grinding fluid, and process parameters (such as depth of cut and speeds of wheel and workpiece).

During a grinding operation, a particular wheel may *act soft* (wear is high) or *hard* (wear is low), regardless of its grade. Note, for example, that an ordinary pencil acts soft when you write on rough paper and acts hard when you write on soft paper. This behavior is a function of the force on the grain. The greater the force, the greater the tendency for the grains to fracture or be dislodged from the wheel surface, thus the higher the wheel wear and the lower the grinding ratio. Equation (25.1) shows that the grain force increases with the strength of the workpiece material, work speed, and depth of cut, and decreases with increasing wheel speed and wheel diameter. Thus a wheel acts soft when v and d increase or when V and D decrease.

Attempting to obtain a high grinding ratio in practice isn't always desirable because high ratios could indicate grain dulling and possible surface damage. A lower ratio may be quite acceptable when an overall economic analysis justifies it.

• Example: Action of a grinding wheel.

A surface-grinding operation is being carried out with the wheel running at a constant spindle speed. Will the wheel act soft or hard as the wheel wears down over a period of time?

SOLUTION. Referring to Eq. (25.1), we note that the parameters that change by time in this operation are the wheel diameter D (assuming that d remains constant and the wheel is dressed periodically) and V. Hence, as D becomes smaller, the relative grain force increases. Therefore the wheel acts softer. Some grinding machines are equipped with variable-speed spindle motors to accommodate these changes, and also to make provision for different diameter wheels.

25.5.6 Grinding wheel selection

Selection of a grinding wheel for a certain application greatly influences the quality of surfaces produced, as well as the economics of the operation. The selection involves not only the shape of the wheel with respect to the shape of the part to be produced, but the characteristics of the workpiece material as well. On the basis of our discussion thus far, you can see the importance of the physical and mechanical properties of the workpiece material in the selection of type of abrasive and bond. General recommendations for the types of grinding wheels to be used for various materials and applications are given in Table 25.3.

TABLE 25.3
GENERAL RECOMMENDATIONS FOR GRINDING WHEELS FOR USE WITH VARIOUS MATERIALS AND APPLICATIONS

MATERIAL	TYPE OF GRINDING WHEEL
Aluminum	C46-K6V
Brass	C46-K6V
Bronze	A54-K6V
Cast iron	C60-L6V, A60-M6V
Carbides	C60-I9V, D150-R75B
Ceramics	D150-N50M
Copper	C60-J8V
Nickel alloys	B150H100V
Nylon	A36-L8V
Steels	A60-M6V
Titanium	A60-K8V
Tool steels (> 50 HRC)	B120WB
APPLICATION	
Ball bearings	A180-Z6V
Crankshafts	A60-N6V
Cutters (milling, drills)	A46-K8V
Knives (sharpening)	A120-G8V
Razor blades (finishing)	A800-K7B
Saws	A60-N6V
Skates	A60-N6V
Valves (automotive, refacing)	A80-J5V

Note: These recommendations vary significantly, depending on material composition, the particular grinding operation, and grinding fluids used.

25.6 Grinding Operations and Machines

Grinding operations are carried out with a variety of wheel–workpiece configurations. The selection of a grinding process for a particular application depends on part shape, part size, ease of fixturing, and the production rate required.

The basic types of grinding operations are surface, cylindrical, internal, and centerless grinding. The relative movement of the wheel may be along the surface of the workpiece (*traverse* grinding, *through feed* grinding, *cross-feeding*), or it may be radially into the workpiece (*plunge* grinding). Surface grinders comprise the largest percentage of grinders in use in industry, followed by bench grinders (usually with two wheels), cylindrical grinders, and tool and cutter grinders. The least used are internal grinders.

Grinding machines are available for various workpiece geometries and sizes. Special machines have been built with features for automatic workpiece loading, clamping, cycling, and gaging and wheel dressing.

25.6.1 Surface grinding

Surface grinding involves grinding flat surfaces and is one of the most common grinding operations (Fig. 25.13). Typically, the workpiece is secured on a magnetic chuck attached to the work table of a *surface grinder* (Fig. 25.14). Nonmagnetic materials generally are held by vices, special fixtures, vacuum chucks, or double-sided adhesive tapes. A straight wheel is mounted on the horizontal spindle of the grinder. Grinding is done as the table reciprocates longitudinally and feeds laterally

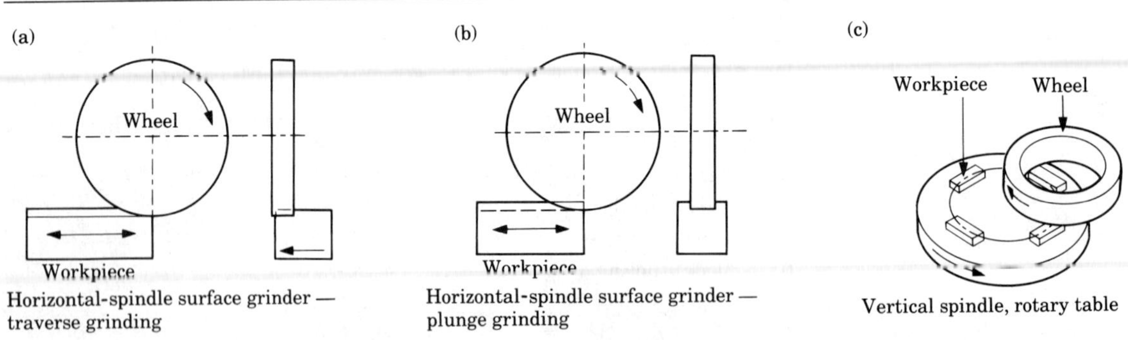

FIGURE 25.13
Schematic illustrations of surface grinding operations.

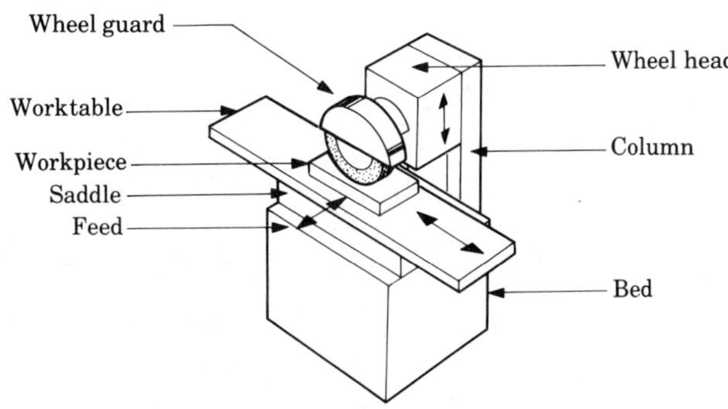

FIGURE 25.14
Schematic illustration of a horizontal-spindle surface grinder. The majority of grinding operations are done on such machines.

after each stroke. In plunge grinding, the wheel is moved radially into the workpiece, as in grinding a groove (Fig. 25.13b).

The size of a surface grinder is determined by the surface dimensions that can be ground on the machine (see Summary Table 25.1). In addition to the grinder shown in Fig. 25.14, other types have vertical spindles and rotary tables (Fig. 25.13c). These configurations allow a number of pieces to be ground in one setup. Steel balls for ball bearings are ground in special setups at high production rates (Fig. 25.15).

25.6.2 Cylindrical grinding

In *cylindrical grinding* (also called *center-type* grinding; Fig. 25.16), the workpiece's external cylindrical surfaces and shoulders are ground. Typical applications include crankshaft bearings, spindles, pins, bearing rings, and rolls for rolling mills. The rotating cylindrical workpiece reciprocates laterally along its axis, although in grinders used for large and long workpieces the grinding wheel reciprocates. The latter design is called a *roll grinder* and is capable of grinding rolls as large as 1.8 m (72 in.) in diameter.

The workpiece in cylindrical grinding is held between centers or in a chuck, or it is mounted on a faceplate in the headstock of the grinder. For straight cylindrical surfaces, the axes of rotation of the wheel and workpiece are parallel. Separate motors drive the wheel and workpiece at different speeds. Long workpieces with two or more diameters (Fig. 25.17) are ground on cylindrical grinders. Cylindrical grinding can produce shapes (form grinding and plunge grinding) in which the wheel is dressed to the form to be ground (Fig. 25.18), using a template to dress the wheel (Fig. 25.19).

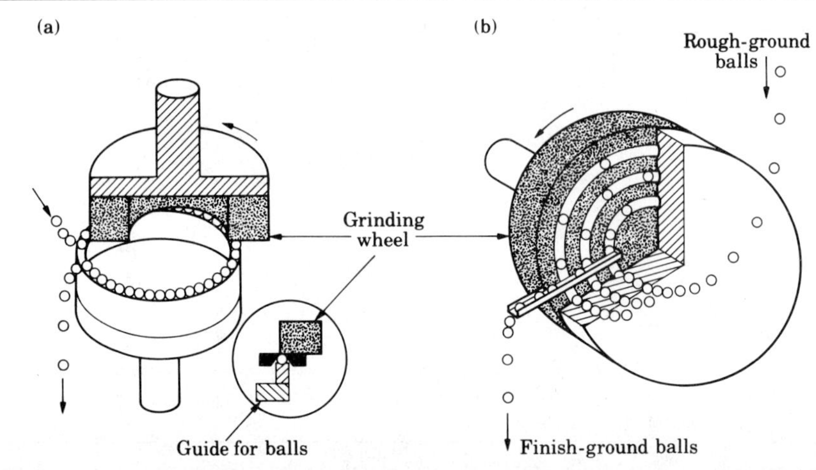

FIGURE 25.15
(a) Rough grinding of steel balls on a vertical-spindle grinder, guided by a special rotary fixture. (b) Finish grinding of balls in a multiple-groove fixture. The balls are ground to within 0.013 mm (0.0005 in.) of their final size. *Source: American Machinist.*

Cylindrical grinders are identified by the maximum diameter and length of the workpiece that can be ground, similar to engine lathes. In Universal grinders, both the workpiece and the wheel axes can be moved and swiveled around a horizontal plane (Fig. 25.20), thus permitting the grinding of tapers and other shapes. These machines are equipped with computer controls, reducing labor and producing parts accurately and repetitively.

Thread grinding is done on cylindrical grinders, as well as centerless grinders, with specially dressed wheels matching the shape of the threads (Fig. 25.21). Threads produced by grinding are the most accurate of any manufacturing process and have a very fine surface finish. The workpiece and wheel movements are synchronized to produce the pitch of the thread, usually in about six passes.

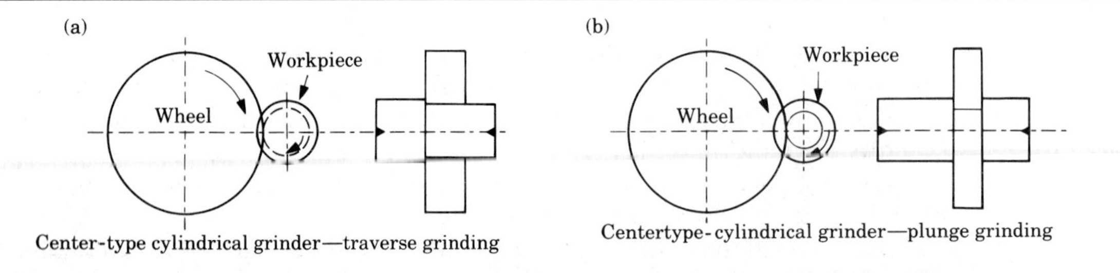

FIGURE 25.16
Schematic illustrations of cylindrical grinding operations.

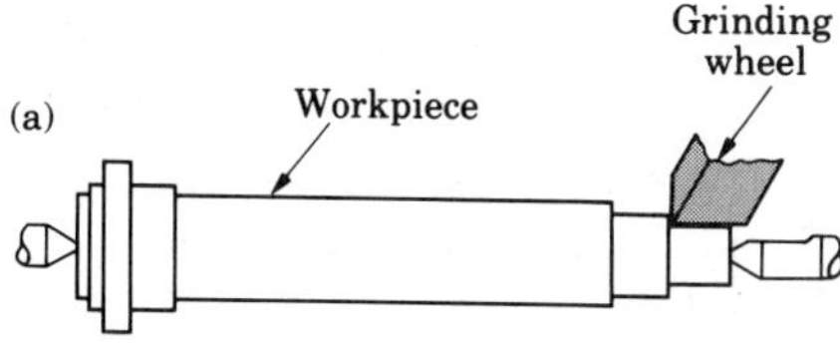

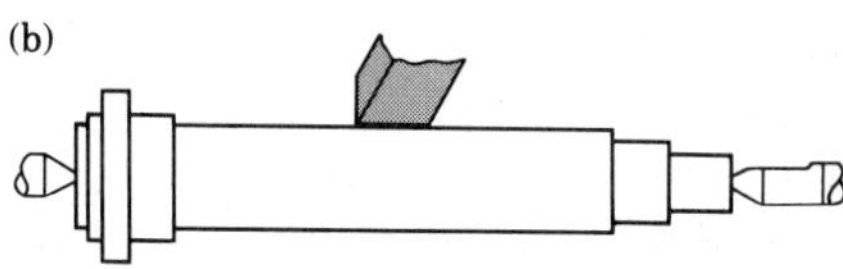

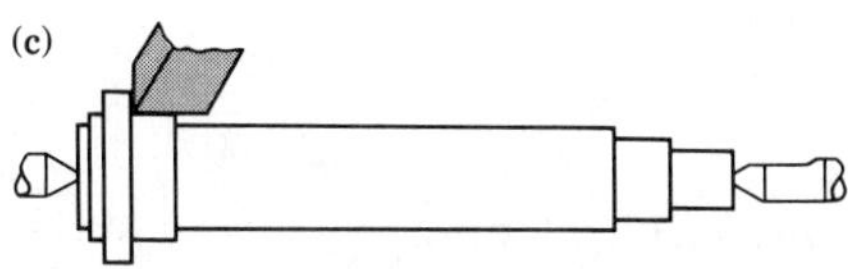

FIGURE 25.17

Multiple-diameter grinding of a cylindrical workpiece on a cylindrical grinder: (a) plunge grinding, (b) traverse grinding, and (c) shoulder grinding. The diameters at the ends of the part are produced by plunge grinding. Note that in these regions the lengths ground are narrower than the face of the grinding wheel.

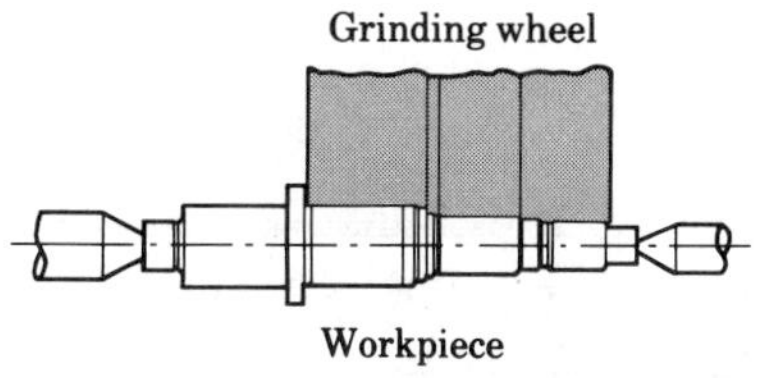

FIGURE 25.18

Plunge grinding of a workpiece on a cylindrical grinder with the wheel dressed to a stepped shape.

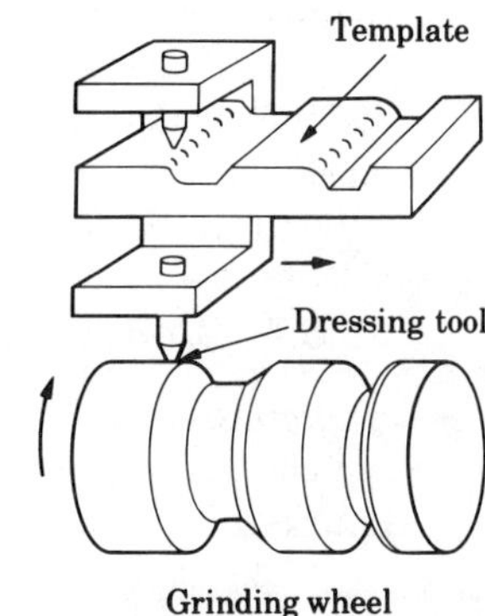

FIGURE 25.19

Method of shaping a grinding wheel with a template, to be used for plunge grinding.

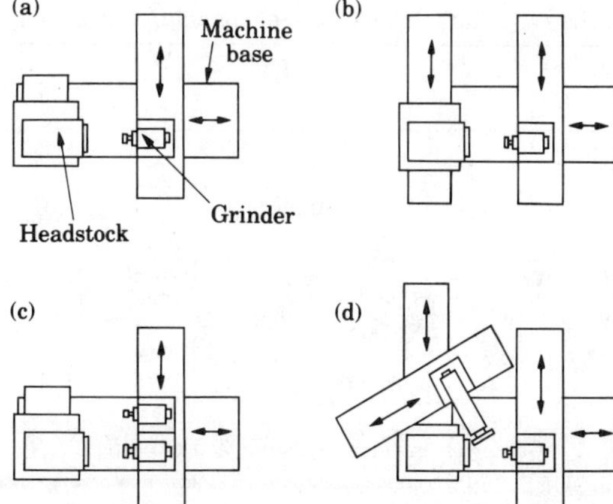

FIGURE 25.20

Schematic illustrations of various types of universal grinders. Note the movements of the headstock and the grinding head that carries the wheel. (a) Compounds slide under wheel (2-axis machine). (b) Compound slide under wheel with headstock cross slide (3 axes). (c) Compound slide with two wheels (2 axes). (d) Compound slide under two wheels, one for outside surfaces and one for internal grinding (4 axes). *Source:* Bryant Grinder Corporation.

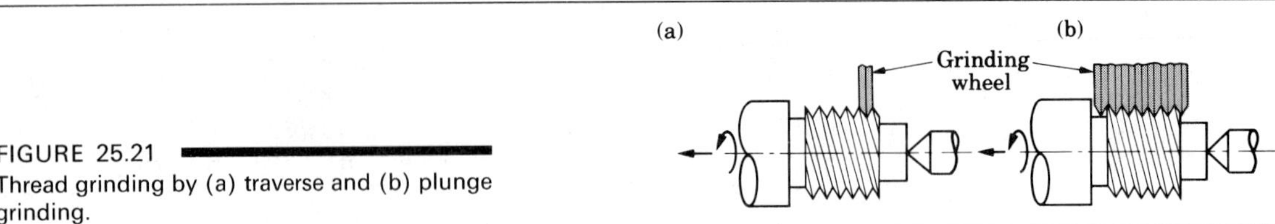

FIGURE 25.21
Thread grinding by (a) traverse and (b) plunge grinding.

25.6.3 Internal grinding

In *internal grinding* (Figs. 25.22a and b), a small wheel is used to grind the inside diameter of the part, such as bushings and bearing races. The workpiece is held in a rotating chuck and the wheel rotates at 30,000 rpm or higher. Internal profiles can also be ground with profile-dressed wheels that move radially into the workpiece. The headstock of internal grinders can be swiveled on a horizontal plane to grind tapered holes, similar to those shown in Fig. 25.20.

25.6.4 Centerless grinding

Centerless grinding is a process for continuously grinding cylindrical surfaces in which the workpiece is supported not by centers (hence the term centerless) or chucks but by a blade (Figs. 25.23 and 25.24). Typical parts made by centerless

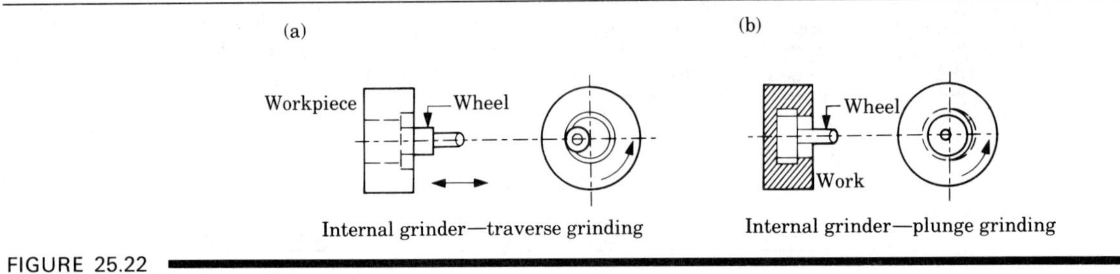

FIGURE 25.22
Schematic illustrations of internal grinding operations.

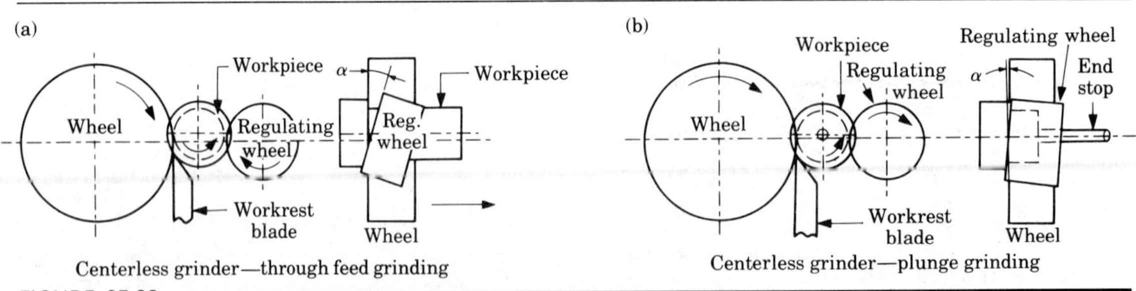

FIGURE 25.23
Schematic illustrations of centerless grinding operations.

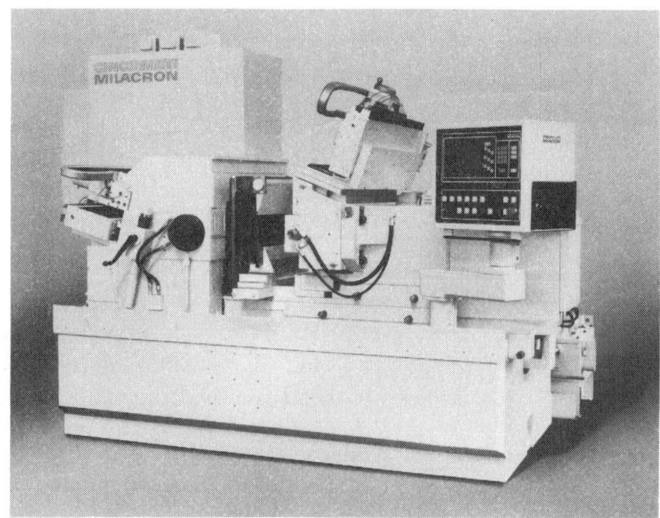

FIGURE 25.24 A computer-numerical-control cylindrical grinding machine. The movement of the workpiece is perpendicular to the page. *Source:* Courtesy of Cincinnati Milacron, Inc.

grinding are roller bearings, piston pins, shafts, and similar components. This continuous production process requires little operator skill. Parts with diameters as small as 0.1 mm (0.004 in.) can be ground.

In *through-feed grinding* the workpiece is supported on a workrest blade and is ground between two wheels. Grinding is done by the larger wheel, while the smaller wheel regulates the axial movement of the workpiece. The *regulating wheel*, which is rubber bonded, is tilted and runs at speeds of only about 1/20 those of the grinding wheel.

Parts with variable diameters, such as bolts, valve tappets, and distributor shafts, can be ground by centerless grinding. Called *infeed*, or plunge, *grinding* (Fig. 25.23b), the process is similar to plunge or form grinding with cylindrical grinders. Tapered pieces are centerless ground by *end-feed grinding*. High production rate thread grinding can be done with centerless grinders using specially dressed wheels. In *internal centerless grinding*, the workpiece is supported between three rolls and is internally ground. Typical applications are sleeve-shaped parts and rings.

25.6.5 Other grinders

A variety of special-purpose grinders are available. *Bench grinders* are used for routine offhand grinding of tools and small parts. They are usually equipped with two wheels mounted on the two ends of the shaft of an electric motor. *Pedestal grinders* are placed on the floor and used similarly to bench grinders.

Universal tool and cutter grinders are used for grinding single-point or multipoint tools and cutters. They are equipped with special workholding devices for accurate positioning of the tools to be ground. *Tool-post grinders* are self-contained units and are usually attached to the tool post of a lathe. The workpiece is mounted on the headstock and is ground by moving the tool post. These grinders are versatile, but the lathe should be protected from abrasive debris.

Swing-frame grinders are used in foundries for grinding large castings. Rough grinding of castings is called **snagging**, and is usually done on floorstand grinders using wheels as large as 0.9 m (36 in.) in diameter. *Portable grinders*, either air or electrically driven, or with a flexible shaft connected to an electric motor or gasoline engine are available for operations such as grinding off weld beads and cutting-off operations, usually on large workpieces or structures.

25.6.6 Creep-feed grinding

Grinding has traditionally been associated with small rates of material removal (Table 25.4) and fine finishing operations. However, grinding can also be used for large-scale metal removal operations similar to milling, shaping, and planing. In *creep-feed grinding*, developed in the late 1950s, the wheel depth of cut d is as much as 6 mm (0.25 in.), and the workpiece speed is low (Fig. 25.25a). The wheels are mostly softer grade resin bonded with open structure to keep temperatures low and improve surface finish. Grinders with capabilities for continuously dressing the grinding wheel with a diamond roll are now available. The machines used for creep-feed grinding have special features, such as high power—up to 225 kW (300 hp)—high stiffness and damping capacity, variable and well-controlled spindle and work-table speeds, and ample capacity for grinding fluids.

Its overall economics and competitive position with other material-removal processes indicate that creep-feed grinding can be economical for specific applications, such as in grinding shaped punches, key seats, twist-drill flutes, and the roots of turbine blades (Fig. 25.25b). The wheel is dressed to the shape of the workpiece to be produced. Consequently, the workpiece does not have to be previously milled, shaped, or broached. Thus near-net shape castings and forgings are suitable parts for creep-feed grinding.

25.6.7 Grinding chatter

Chatter is particularly bothersome in grinding because it adversely affects surface finish and wheel performance. Vibrations during grinding may be caused by bearings, spindles, and unbalanced wheels, as well as external sources, such as from nearby machinery. The grinding process can itself cause regenerative chatter. The analysis of chatter in grinding is similar to that for machining operations (see

TABLE 25.4
TYPICAL RANGE OF SPEEDS AND FEEDS FOR ABRASIVE PROCESSES

PROCESS VARIABLE	CONVENTIONAL GRINDING	CREEP-FEED GRINDING	BUFFING	POLISHING
Wheel speed (m/s)	25–50	25–50	30–60	25–40
Work speed (m/s)	0.2–1	0.002–0.02	—	—
Feed (mm/pass)	0.01–0.05	1–6	—	—

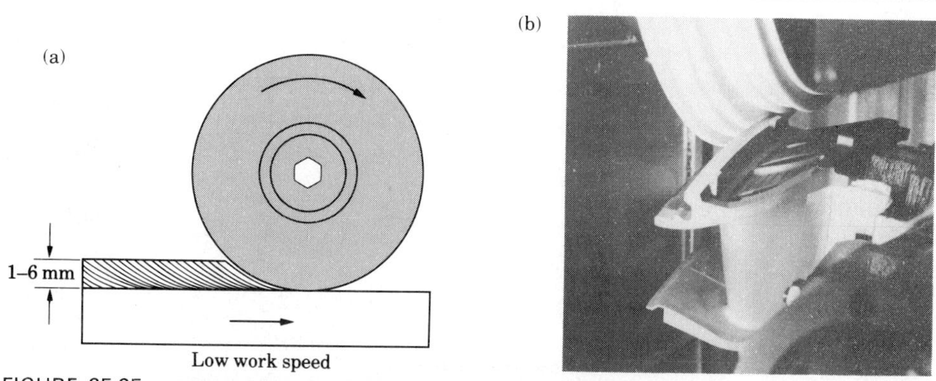

FIGURE 25.25 (a) Schematic illustration of the creep-feed grinding process. Note the large wheel depth of cut. (b) An example of creep-feed grinding with a shaped wheel. *Source*: Courtesy of Blohm, Inc., and *Manufacturing Engineering Magazine*, Society of Manufacturing Engineers. This operation can also be performed by some of the processes described in Chapter 26.

Section 24.3). Thus the important variables are stiffness of the machine tool and damping. Additional factors that are unique to grinding chatter are nonuniformities in the grinding wheel, dressing techniques used, and uneven wheel wear.

Because these variables produce characteristic chatter marks on ground surfaces, careful study of these marks can often lead to the source of the problem. General guidelines have been established to reduce the tendency for chatter in grinding, such as using soft-grade wheels, dressing the wheel frequently, changing dressing techniques, reducing the material-removal rate, and supporting the workpiece rigidly.

25.6.8 Safety

Because grinding wheels are brittle and are operated at high speeds, certain procedures must be carefully followed in their handling, storage, and use. Failure to follow these rules—and the instructions and warnings printed on individual wheel labels—may result in serious injury or death. Grinding wheels should be stored properly and protected from environmental extremes. They should be visually inspected for cracks and damage prior to installing them on grinders.

Wheels should be mounted on spindles of proper size, so that they are neither forced, which may fracture the wheel at its center, nor loose, which can cause unbalance. Flanges should be of appropriate design and dimensions. Wheels should be balanced because otherwise the surface produced will be wavy and the wheel will cause vibrations, possibly leading to wheel fracture. Some machine spindles and flanges provide for balancing wheels (Fig. 25.26). Grinding wheels should be used according to their specifications and maximum operating speeds and should not be dropped or abused. Wheel guarding, operator protection, and bystander safety are all important.

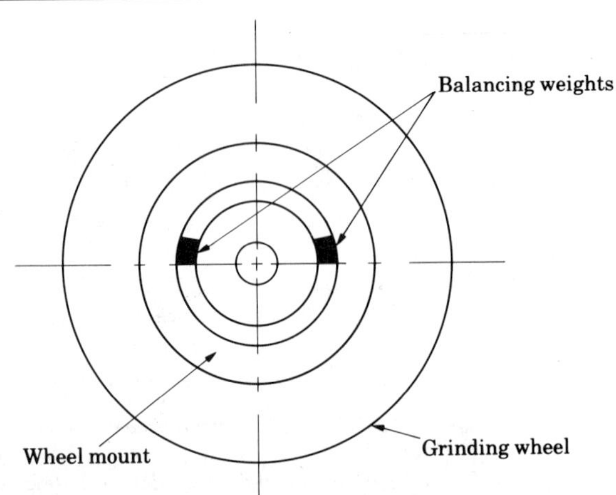

FIGURE 25.26 Balancing of a grinding wheel by shifting the weights on the wheel mount. The procedure is similar to balancing automobile wheels. Some wheel mounts have four weights.

25.7 Design Considerations

Design considerations for grinding operations are similar to those for machining processes. However, specific attention should be given to the following:

a) Parts should be designed so that they can be held securely, either in chucks, magnetic tables, or suitable fixtures and workholding devices. Otherwise, thin straight or tubular workpieces may distort during grinding.
b) Interrupted surfaces, such as holes and keyways, should be avoided if high dimensional accuracy is needed because such interruptions can cause vibrations.
c) In cylindrical grinding, parts should be balanced, and long slender designs should be avoided to minimize deflections. Fillets and corner radii should be as large as possible, or relief should be provided for by prior machining.
d) In centerless grinding, short pieces may be difficult to grind accurately because of their lack of support on the blade. In through-feed grinding, only the largest diameters can be ground. Small-diameter sections can be ground by plunge grinding.
e) Designs requiring accurate form grinding should be kept simple to avoid frequent wheel dressing.
f) Deep and small holes, and blind holes requiring internal grinding, should be avoided, or should include a relief.

In general, design should require that a minimum amount of material be removed by grinding. Moreover, designs should allow, insofar as possible, all grinding to be done without having to reposition the workpiece.

25.8 Ultrasonic and Abrasive-Jet Machining

In **ultrasonic machining,** material is removed from a surface by microchipping or erosion with fine abrasive grains. The tip of the tool (Fig. 25.27a) vibrates at a frequency of 20 kHz and a low amplitude (0.05–0.125 mm; 0.002–0.005 in.). This vibration, in turn, transmits a high velocity to abrasive grains between the tool and the workpiece. The stress produced by the abrasive particles hitting the workpiece surface are high because (a) the time of contact between the particle and the surface is very short (10–100 μs), and (b) the area of contact is very small. In brittle materials, impact stresses are high enough to cause microchipping and erosion of the workpiece surface.

The tip of the tool, which is attached to a transducer through the tool holder, is usually made of mild steel and undergoes wear. The grains are usually boron carbide, although aluminum oxide or silicon carbide are also used. Grain sizes range from 100 for roughing to 1000 for finishing operations. The grains are carried in a water slurry, with concentrations of 20–60 percent by volume. The slurry also carries the debris away from the cutting zone.

Ultrasonic machining is best suited for materials that are hard and brittle, such as ceramics, carbides, precious stones, and hardened steels. Two applications of ultrasonic machining are shown in Figs. 25.27(b) and (c).

In **abrasive-jet machining,** a jet of air or carbon dioxide containing abrasive particles is aimed at the workpiece surface under controlled conditions. The impact of the particles is capable of cutting holes or slots in very hard metallic and nonmetallic materials. Because the flow of free abrasives tends to round off corners, designs should avoid sharp corners.

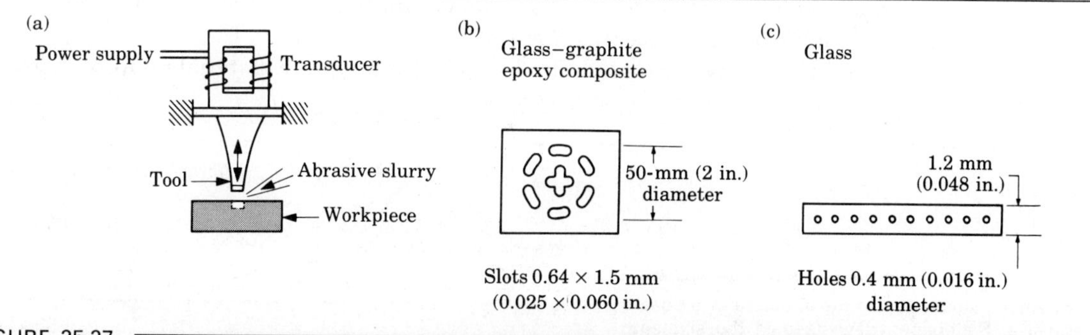

FIGURE 25.27 (a) Schematic illustration of the ultrasonic machining process. (b) and (c) Types of parts made by this process. Note the small size of holes produced.

25.8.1 Design considerations

Design guidelines for ultrasonic machining include:

a) Designs should avoid sharp profiles, corners, and radii because the abrasive slurry erodes away sharp corners.
b) Some taper is to be expected for holes made by this process.
c) In order to avoid chipping of brittle materials in producing through holes, the bottom of the parts should be supported with a backup plate.

25.9 Finishing Operations

In addition to those described thus far, several processes are generally used on workpieces as the final finishing operation. These processes mainly utilize abrasive grains. Commonly used finishing operations are described in this section in the order of improved surface finish produced. Finishing operations can contribute significantly to production time and product cost. Thus they should be specified with due consideration to their costs and benefits.

25.9.1 Coated abrasives

Typical examples of **coated abrasives** are sandpaper and emery cloth. The grains used in coated abrasives are more pointed than those used for grinding wheels. The grains are electrostatically deposited on flexible backing materials, such as paper or cloth (Fig. 25.28), with their long axes perpendicular to the plane of the backing. The matrix (coating) is made of resins.

Coated abrasives are available as sheets and belts and usually have a much more open structure than the abrasives on grinding wheels. Coated abrasives are used extensively in finishing flat or curved surfaces on metallic and nonmetallic parts, of metallographic specimens, and in woodworking. The precision of surface finish obtained depends primarily on grain size.

Coated abrasives are also used as belts for high-rate material removal. *Abrasive*

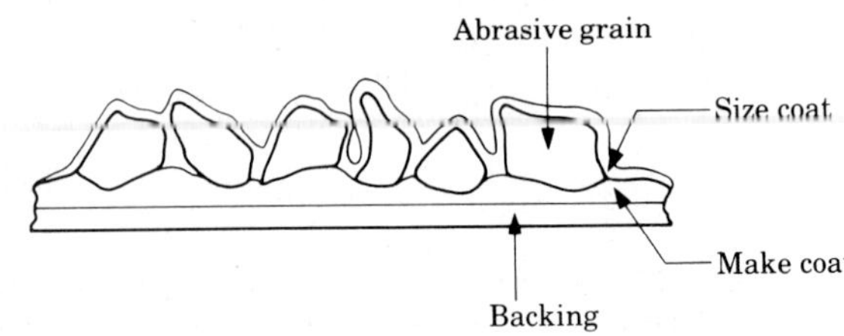

FIGURE 25.28 Schematic illustration of the structure of a coated abrasive. Sandpaper, developed in the sixteenth century, and emery cloth are common examples of coated abrasives.

belts have become an important production process, in some cases replacing conventional grinding operations. Belt speeds range between 12 m/s and 30 m/s (2500 ft/min and 6000 ft/min). Machines for abrasive-belt operations require proper belt support and rigid construction to minimize vibrations.

Example: Belt grinding of turbine nozzle vanes.

The turbine nozzle vane shown in the accompanying figure was investment cast from a cobalt-base superalloy. To remove a thin diffusion layer from the root skirt and tip skirt sections, this part was ground on a cloth-backed abrasive belt (aluminum oxide, 60 grain size). The vanes were mounted on a fixture and ground at a belt speed of 30 m/s (6000 ft/min) without a grinding fluid. Production rate was 93 s per piece. Each vane weighed 21.65 g before belt grinding and 20.25 g after. (*Source:* ASM International.)

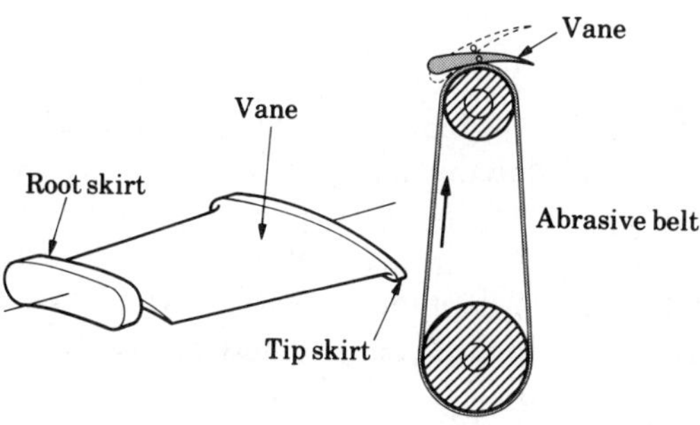

25.9.2 Wire brushing

In the **wire brushing** process, the workpiece is held against a circular wire brush that rotates at high speed. The tips of the wire produce longitudinal scratches on the workpiece surface. This process is used to produce a fine surface texture.

25.9.3 Honing

Honing is an operation used primarily to give holes a fine surface finish. The honing tool (Fig. 25.29) consists of a set of aluminum-oxide or silicon-carbide sticks. They are mounted on a mandrel that rotates in the hole, applying a radial force with a reciprocating axial motion, thus producing a cross-hatched pattern. The sticks can be adjusted radially for different hole sizes. The fineness of surface finish can be controlled by the type and size of abrasive used, the speed of rotation, and the pressure applied. A fluid is used to remove chips and to keep temperatures low. Honing is also done on external cylindrical and flat surfaces.

Superfinishing is similar to honing, but the pressure applied is very light and the motion of the hone has a short stroke. The process is controlled so that the grains do not travel along the same path along the surface of the workpiece. An example of external superfinishing of a round part is shown in Fig. 25.30.

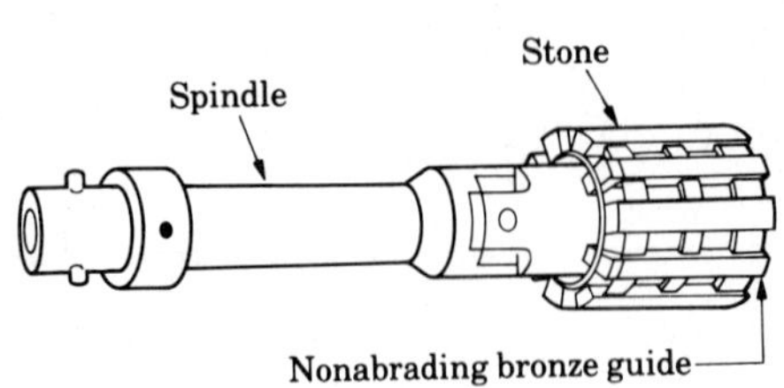

FIGURE 25.29 Schematic illustration of a honing tool to improve the surface finish of bored or ground holes.

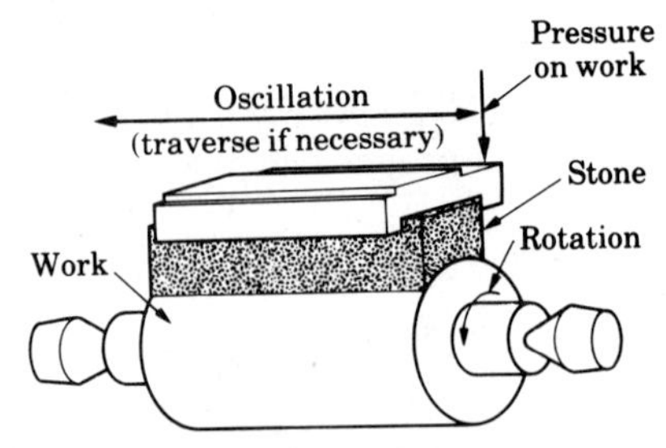

FIGURE 25.30 Schematic illustration of the superfinishing process for a cylindrical part.

25.9.4 Lapping

Lapping is a finishing operation used on flat or cylindrical surfaces. The lap (Fig. 25.31a) is usually made of cast iron, copper, leather, or cloth. The abrasive particles are embedded in the lap, or they may be carried through a slurry. Tolerances on the order of ± 0.0004 mm (0.000015 in.) can be obtained with the use of fine abrasives—up to size 900. Surface finish can be as smooth as 0.025–0.1 μm (1–4 μin.).

Production lapping on flat or cylindrical pieces is done on machines such as those shown in Figs. 25.31(b) and (c). Lapping is also done on curved surfaces, such as spherical objects and glass lenses, using specially shaped laps. Running-in of mating gears can be done by lapping. Depending on the hardness of the workpiece, lapping pressures range from 7–140 kPa (1–20 psi).

25.9.5 Polishing and buffing

Polishing is a process that produces a smooth, lustrous surface finish. Two basic mechanisms are involved in the polishing process: (a) fine-scale abrasive removal,

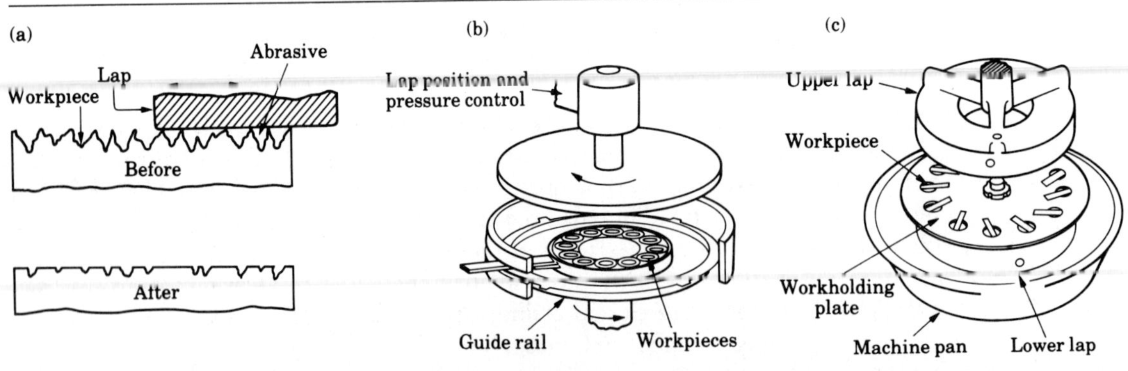

FIGURE 25.31 (a) Schematic illustration of the lapping process. (b) Production lapping on flat surfaces. (c) Production lapping on cylindrical surfaces.

and (b) softening and smearing of surface layers by frictional heating during polishing. The shiny appearance of polished surfaces results from the smearing action.

Polishing is done with disks or belts made of fabric, leather, or felt and coated with fine powders of aluminum oxide or diamond. Parts with irregular shapes, sharp corners, deep recesses, and sharp projections are difficult to polish.

Buffing is similar to polishing, with the exception that very fine abrasives are used on soft disks made of cloth or hide. The abrasive is supplied externally from a stick of abrasive compound. Polished parts may be buffed to obtain an even finer surface finish.

25.9.6 Electropolishing

Mirror-like finishes can be obtained on metal surfaces by **electropolishing**, a process which is the reverse of electroplating (see Section 33.7). Because there is no mechanical contact with the workpiece, this process is particularly suitable for polishing irregular shapes. The electrolyte attacks projections and peaks on the workpiece surface at a higher rate than the rest, thus producing a smooth surface. Electropolishing is also used for deburring operations.

25.10 Deburring

Burrs are thin ridges, usually triangular in shape, that develop along the edges of a workpiece from shearing sheet materials, trimming forgings and castings, and machining. Burrs may interfere with the assembly of parts and can cause jamming of parts, misalignment, and short circuits in electrical components. Furthermore, burrs may reduce the fatigue life of components. Because they are usually sharp, they can be a safety hazard to personnel. On the other hand, burrs on thin drilled or tapped components, such as tiny parts in watches, can provide extra thickness and, thus, improve the holding torque of screws.

Several **deburring** processes are available. Burrs may be removed manually with files or mechanically by cutting, wire brushing (including the use of rotary nylon brushes with filaments containing silicon-carbide grits), abrasive belts, tumbling, vibratory finishing, abrasive flow, ultrasonics, and abrasive jets or water jets. The need for deburring may be reduced by adding chamfers to sharp edges on parts.

Vibratory and **barrel finishing** processes are used to improve the surface finish and remove burrs from large numbers of relatively small workpieces. In this batch-type operation, specially shaped abrasive pellets (Fig. 25.32) are placed in a container along with the parts to be deburred. The container is either vibrated or tumbled. The impact of individual abrasives and metal particles removes sharp edges and burrs from the parts.

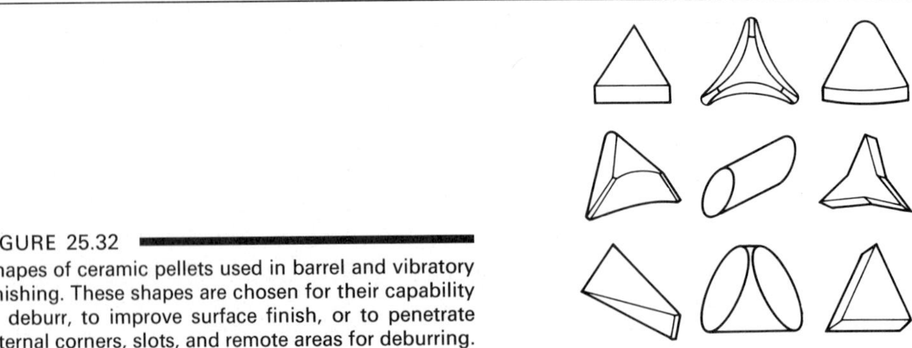

FIGURE 25.32 Shapes of ceramic pellets used in barrel and vibratory finishing. These shapes are chosen for their capability to deburr, to improve surface finish, or to penetrate internal corners, slots, and remote areas for deburring.

In **shot blasting** (also called *grit blasting*), abrasive particles (usually sand) are propelled by a high-velocity jet of air, or by a rotating wheel, onto the surface of the workpiece. Shot blasting is particularly useful in deburring metallic and nonmetallic materials and stripping, cleaning, and removing surface oxides. The surface produced has a matte finish. Small-scale polishing and etching can also be done by this process on bench-type units (*microabrasive blasting*).

In **abrasive-flow machining**, abrasive grains, such as silicon carbide or diamond, are mixed in a viscous matrix, which is then forced back and forth through the openings and passageways in the workpiece. The movement of the abrasive matrix under pressure erodes away burrs and sharp corners and polishes the part. The process is particularly suitable for workpieces with internal cavities that are inaccessible by other means. Pressures applied range from 0.7 MPa to 11 MPa (100 psi to 1600 psi).

Deburring and flash removal from castings and forgings are now being performed increasingly by programmable robots (see Section 38.7), thus eliminating tedious manual labor and resulting in more consistent deburring. One example of deburring die-cast parts for an outdoor motor housing is shown in Fig. 25.33.

25.11 Grinding Fluids

The functions of grinding fluids are similar to those for cutting fluids (Section 21.12). Although grinding and other abrasive-removal processes can be performed dry, the use of a fluid is important. It prevents temperature rise in the workpiece and improves the part's surface finish and dimensional accuracy. Fluids also improve the efficiency of the operation by reducing wheel loading and wear and lowering power consumption.

Grinding fluids are typically water-base emulsions for general grinding and oils for thread grinding (Table 25.5). They may be applied as a stream (flood) or as mist, which is a mixture of fluid and air. Because of the high surface speeds involved, an

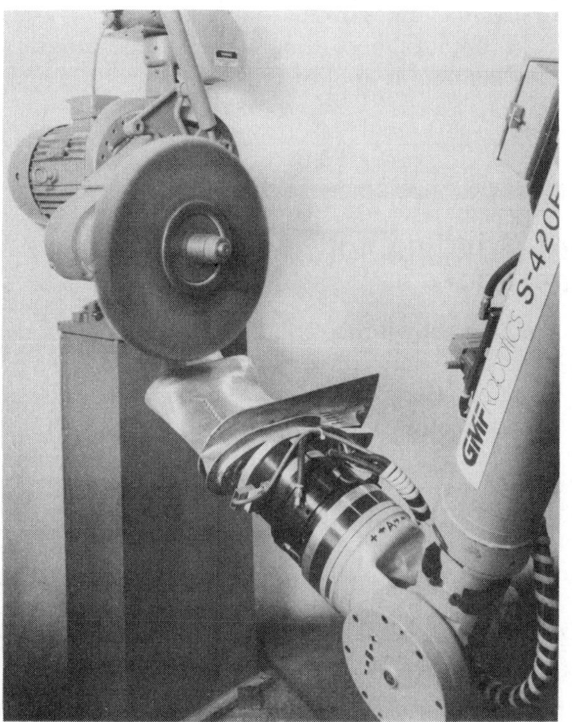

FIGURE 25.33

A deburring operation on a robot-held die-cast part for an outboard motor housing, using a grinding wheel. Abrasive belts can also be used for such operations. *Source:* Courtesy of Acme Manufacturing Company and *Manufacturing Engineering Magazine*, Society of Manufacturing Engineers.

TABLE 25.5

GENERAL RECOMMENDATIONS FOR GRINDING FLUIDS

MATERIAL	GRINDING FLUID
Aluminum	E, EP
Beryllium	D, E, CSN
Copper	CSN, E, MO + FO
Magnesium	D, MO
Nickel	CSN, EP
Refractory metals	EP
Steels	CSN, E
Titanium	CSN, E

D: dry; E: emulsion; EP: extreme pressure; CSN: chemicals and synthetics; MO: mineral oil; FO: fatty oil.

airstream or air blanket around the periphery of the wheel usually prevents the fluid from reaching the cutting zone. Special nozzles have been designed in which the grinding fluid is applied under high pressure and effectively.

25.12 Economics of Grinding and Finishing Operations

We have shown that grinding may be used both as a finishing operation and as a large-scale removal operation (as in creep-feed grinding). The use of grinding as a finishing operation is often necessary because forming and machining processes alone usually cannot produce parts with the desired dimensional accuracy and surface finish. However, because it is an additional operation, grinding contributes significantly to product cost. Creep-feed grinding, on the other hand, has proved to be an economical alternative to machining operations such as milling, even though wheel wear is high.

All finishing operations contribute to product cost. On the basis of the discussion thus far, you can see that as the surface finish improves, more operations are required, and hence the cost increases. Note in Fig. 25.34 how rapidly cost increases as surface finish is improved by processes such as grinding and honing.

Much progress has been made in automating the equipment involved in finishing operations, including computer controls. Consequently, labor costs and production times have been reduced, even though such machinery may require significant capital investment. If finishing is likely to be an important factor in manu-

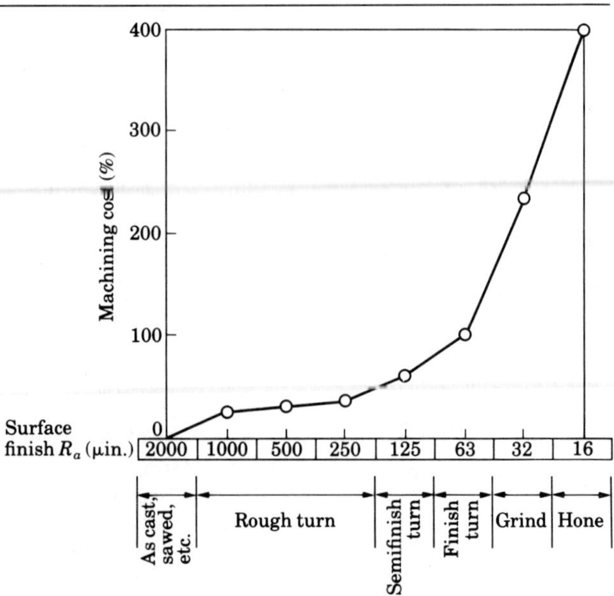

FIGURE 25.34 Increase in the cost of the machining and finishing a part as a function of the surface finish required.

facturing a particular product, the conceptual and design stages should involve an analysis of the degree of surface finish and dimensional accuracy required.

Furthermore, all processes that precede finishing operations should be analyzed for their capability to produce a more acceptable surface finish and dimensional accuracy. As you have seen, this can be accomplished through proper selection of tools and process parameters and the characteristics of the machine tools involved.

SUMMARY

Grinding and various abrasive-removal processes are capable of producing the finest accuracy and surface finish in maufactured products. The majority of abrasive processes are basically finishing operations that are usually performed on machined or cold-worked parts. However, abrasives are also used for large-scale material-removal processes, such as creep-feed grinding and snagging in foundries.

A variety of abrasive processes and machinery are available for surface, external, and internal grinding. The selection of abrasives and process variables in these operations must be controlled in order to obtain the desired surface finish and dimensional accuracy. Otherwise, damage to surfaces such as burning, heat checking, and harmful residual stresses may develop. Several finishing operations are available for deburring. Because they contribute significantly to product cost, proper selection and implementation of finishing operations are important.

SUMMARY TABLE 25.1

GENERAL CHARACTERISTICS OF ABRASIVE PROCESSES AND MACHINES

PROCESS	CHARACTERISTICS	MAXIMUM DIMENSION (m)*
Surface	Flat surfaces on most materials; production rate depends on table size and automation; labor skill depends on part; production rate is high on vertical-spindle rotary-table type.	Reciprocating table L: 6 Rotary table D: 3
Cylindrical	Round workpieces with stepped diameters; low production rate unless automated; labor skill depends on part shape.	Workpiece D: 0.8 Roll grinders D: 1.8 Universal grinders D: 2.5
Centerless	Round workpieces; high production rate; low to medium labor skill.	Workpiece D: 0.8
Internal	Bores in workpiece; low production rate; low to medium labor skill.	Hole D: 2
Honing	Bores and holes in workpiece; low production rate; low labor skill.	Spindle D: 1.2
Lapping	Flat surfaces; high production rate; low labor skill.	Table D: 3.7
Ultrasonic machining	Holes and cavities of various shapes, particularly in hard and brittle nonconducting materials.	—

* Larger capacities are available for special applications. L = length; D = diameter.

TRENDS

- As with other machine tools, the trend for abrasive operations is for greater automation and computer control of machine tools.
- Advanced controls on grinders are now capable of calculating when and how long a grinding wheel should be dressed, as well as automatically and continuously compensating for wheel-diameter reduction, feeds, and wheel surface speeds.
- Manual tasks in abrasive processes are being eliminated that can otherwise cause carpal tunnel syndrome and other wrist-related injuries to the operator due to repetitive motions of the hands.
- Cooling the workpiece in some grinding operations, such as creep feed, is requiring mist collecting (from grinding fluids) and automatic fire-extinguishing systems.
- Grinding machines are now capable of competing with machining operations and perform roughing and finishing operations in one setup of the workpiece. Creep-feed grinding is beginning to compete with milling and broaching processes.
- Developments in grinding wheels include better control of manufacturing parameters and greater uniformity of properties.
- Operating speeds of wheels are being increased for higher productivity. The design, guarding, and safe operation of these wheels is an important consideration.

KEY TERMS

Abrasive-flow machining
Abrasive-jet machining
Abrasives
Attritious wear
Barrel finishing
Bonded abrasives
Buffing
Coated abrasives
Deburring
Dressing
Electropolishing
Friability
Glazing
Grade
Grinding
Grinding ratio
Grinding wheel
Honing
Lapping
Loading
Polishing
Seeded gel abrasive
Shot blasting
Snagging
Structure
Superfinishing
Truing
Ultrasonic machining
Vibratory finishing
Wear flat
Wire brushing

BIBLIOGRAPHY

Andrew, C., T.D. Howes, and T.R.A. Pearce, *Creep Feed Grinding*. New York: Industrial Press, 1985.

Farago, F.T., *Abrasive Methods Engineering*, Vol. 1, 1976; Vol. 2, 1980. New York: Industrial Press.

Gillespie, L.K., *Deburring Technology for Improved Manufacturing*. Dearborn, Mich.: Society of Manufacturing Engineers, 1981.

King, R.I., and R.S. Hahn, *Handbook of Modern Grinding Technology*. New York: Chapman and Hall/Methuen, 1987.

Krar, S., and E. Ratterman, *Superabrasives: Grinding and Machining with CBN and Diamond*. New York: McGraw-Hill, 1990.

Lewis, K.B., and W.F. Schleicher, *The Grinding Wheel: A Textbook of Modern Grinding Practice*, 3d ed. Cleveland: The Grinding Wheel Institute, 1976.

Machinery's Handbook, revised periodically. New York: Industrial Press.

Machining Data Handbook, 3d ed., 2 vols. Cincinnati: Machinability Data Center, 1980.

Malkin, S., *Grinding Technology: Theory and Applications of Machining with Abrasives*. New York: Wiley, 1989.

McKee, R.L., *Machining with Abrasives*. New York: Van Nostrand Reinhold, 1982.

Metals Handbook, 8th ed., Vol. 3: *Machining*. Metals Park, Ohio: American Society for Metals, 1967.

Metals Handbook, 9th ed., Vol. 16: *Machining*. Metals Park, Ohio: ASM International, 1989.

Metzger, J.L., *Superabrasive Grinding*. Stoneham, Mass.: Butterworths, 1986.

Tool and Manufacturing Engineers Handbook, 4th ed., Vol. 1: *Machining*. Dearborn, Mich.: Society of Manufacturing Engineers, 1983.

REVIEW QUESTIONS

25.1 What is an abrasive? What are superabrasives?

25.2 How is the size of an abrasive grain related to its number?

25.3 Name the abrasives commonly used in grinding operations.

25.4 Why are most abrasives now made synthetically?

25.5 Describe the structure of a grinding wheel.

25.6 Explain the characteristics of each type of bond used in bonded abrasives.

25.7 Describe the (a) grade and (b) structure of bonded abrasives.

25.8 List the variables that influence the (a) grain force and (b) temperature in grinding.

25.9 What is the consequence of allowing the temperature to rise in grinding?

25.10 What are grinding sparks caused by?

25.11 Define metallurgical burn.

25.12 Explain the mechanisms by which grinding wheels wear.

25.13 Define (a) friability, (b) wear flat, (c) grinding ratio, (d) truing, and (e) dressing.

25.14 Explain what is meant by a grinding wheel acting soft or hard.

25.15 List the type of grinding operations commonly used in manufacturing.

25.16 What is creep-feed grinding and what are its advantages?

25.17 List the factors that may cause chatter in grinding.

25.18 Explain the major design guidelines for grinding.

25.19 What is the principle of ultrasonic machining? Why is it not suitable for ductile materials?

25.20 List the finishing operations commonly used in manufacturing. Why are they necessary? Explain why they should be minimized.

25.21 Describe the structure of a coated abrasive.

25.22 What are the differences between lapping, polishing, and buffing?

25.23 What applications does electropolishing have that conventional polishing does not?

25.24 Name the major deburring operations and describe briefly their principles.

25.25 What is the function of grinding fluids?

QUESTIONS AND PROBLEMS

25.26 Why are grinding operations necessary for components that have been machined by the processes described in Chapters 22 and 23?

25.27 Explain why there are so many different types and sizes of grinding wheels.

25.28 Explain the reasons for the large difference between the specific energies involved in machining (Table 20.1) and grinding (Table 25.3).

25.29 Comment on the selection of grinding wheels for the applications shown in Table 25.3.

25.30 What precautions would you take when grinding with high precision? Make comments about the machine, process parameters, the grinding wheel, and grinding fluids.

25.31 Explain why the same grinding wheel may act soft or hard.

25.32 What factors could contribute to chatter in grinding?

25.33 The grinding ratio *G* depends on the following: type of grinding wheel, workpiece hardness, wheel depth of cut, wheel and workpiece speeds, and type of grinding fluid. Explain why.

25.34 It is generally recommended that in grinding hardened steels, the grinding wheel be of a softer grade. Explain the reason.

25.35 In Fig. 25.4, the proper grinding surfaces are shown for each type of wheel with an arrow. Explain why the other surfaces of the wheels should not be used for grinding.

25.36 Explain the factors involved in selecting the appropriate type of abrasive for a particular grinding operation.

25.37 What are the effects of a wear flat on the grinding process?

25.38 Describe the methods you would use to determine the number of active cutting points per unit surface area on the periphery of a straight (Type 1) grinding wheel. What would be the significance of this number?

25.39 Would you encounter any difficulties in grinding thermoplastics? If so, what precautions would you take?

25.40 Calculate the chip dimensions for the example problem in Section 25.4 for the following process variables: $D = 8$ in., $d = 0.001$ in., $v = 100$ ft/min, $V = 6000$ ft/min, $C = 500$ per in^2, and $r = 20$.

25.41 If the strength of the workpiece material is doubled, what should be the percentage decrease in the wheel depth of cut d in order to maintain the same grain force, all other variables being the same?

25.42 Derive a formula for the material removal rate (MRR) in surface grinding in terms of process parameters.

25.43 Assume that a surface grinding operation is being carried out under the following conditions: $D = 200$ mm, $d = 0.1$ mm, $v = 0.4$ m/s, and $V = 30$ m/s. These conditions are then changed to the following: $D = 150$ mm, $d = 0.1$ mm, $v = 0.3$ m/s, and $V = 25$ m/s. How different is the temperature rise from the initial conditions?

25.44 Is the grinding ratio G important in evaluating the economics of a grinding operation? Explain.

25.45 We know that grinding can produce a very fine surface finish on a workpiece. Is this necessarily an indication of the quality of a part? Explain.

25.46 Estimate the percent increase in cost if the specification for the surface finish of a part is changed from 63 μin. to 16 μin.

25.47 What costs would be associated with dressing a grinding wheel? Why?

25.48 Assume that the energy cost for grinding an aluminum part, with a specific energy requirement of 7 $\text{W} \cdot \text{s/mm}^3$, is \$0.80 per piece. What would be the energy cost of carrying out the same operation if the workpiece material is T15 tool steel?

25.49 On the basis of the information given in Chapters 22 and 26, estimate the time involved in producing a 2-mm hole 8 mm deep in a titanium alloy by (a) conventional drilling and (b) laser-beam machining.

25.50 What other methods could you use in producing the hole in Problem 25.49? Outline the factors involved in the final selection of a specific process. Could you use a combination of processes?. Would the diameter and depth of the hole be a factor in the final selection? Explain.

26

Nontraditional Machining Processes

26.1 Introduction

The machining processes that we have described so far remove material by chip formation, abrasion, or microchipping. There are situations, however, where these processes are not satisfactory or economical for the following reasons:

- The hardness and strength of the material is very high, typically above 400 HB.
- The workpiece is too flexible, slender, or delicate to withstand the cutting or grinding forces, or parts are difficult to clamp in workholding devices.
- The shape of the part is complex (Fig. 26.1a), such as internal and external profiles, or small-diameter holes, such as in fuel-injection nozzles (Fig. 26.1b).
- Surface finish and tolerances better than those obtainable by other processes are required.
- Temperature rise or residual stresses in the workpiece are undesirable or unacceptable.

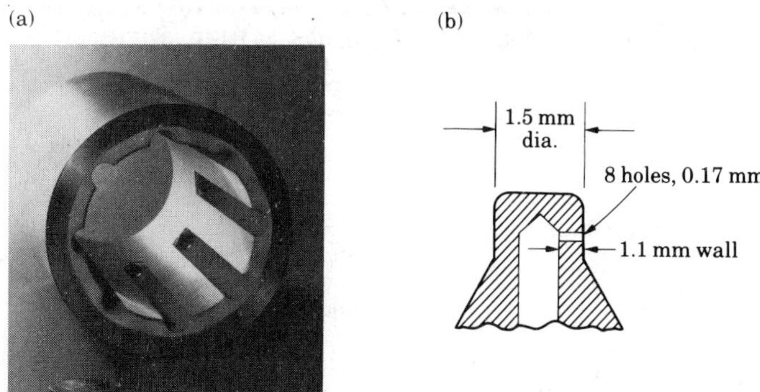

FIGURE 26.1

Examples of parts made by nontraditional machining processes. These parts would be difficult or uneconomical to manufacture by conventional processes. (a) Internal splines in a bushing produced by electrochemical machining. (b) Holes in fuel-injection nozzle made by electrical-discharge machining. Material: Heat-treated steel.

These requirements led to the development of chemical, electrical, and other means of material removal in the 1940s. These methods are called *nontraditional* or *unconventional machining* (Table 26.1 on p. 814). When selected and applied properly, these processes offer significant economic and technical advantages over the traditional machining methods described in the preceding chapters. We also describe their typical applications and limitations and considerations of quality, dimensional accuracy, properties of surfaces produced, and economics. Because many of the processes described in this chapter are also used to make dies, die manufacturing methods are also discussed.

26.2 Chemical Machining

We know that certain chemicals attack metals and etch them, thereby removing small amounts of material from the surface. Thus **chemical machining** (CM) was developed, whereby material is removed from a surface by chemical dissolution, using chemical **reagents**, or **etchants**, such as acids and alkaline solutions. Chemical machining is the oldest of the nontraditional machining processes, and has been used for many years for engraving metals and hard stones, as well as in the production of printed-circuit boards and microprocessor chips. Parts can also be deburred by chemical means.

26.2.1 Chemical milling

In **chemical milling**, shallow cavities are produced on plates, sheets, forgings, and extrusions for overall reduction of weight (Fig. 26.2 on p. 815). Chemical milling

TABLE 26.1
GENERAL CHARACTERISTICS OF NONTRADITIONAL MACHINING PROCESSES

PROCESS	CHARACTERISTICS	PROCESS PARAMETERS AND TYPICAL MATERIAL REMOVAL RATE OR CUTTING SPEED
Chemical machining (CM)	Shallow removal (up to 12 mm) on large flat or curved surfaces; blanking of thin sheets; low tooling and equipment cost; suitable for low production runs.	0.0025–0.1 mm/min
Electrochemical machining (ECM)	Complex shapes with deep cavities; highest rate of material removal among nontraditional processes; expensive tooling and equipment; high power consumption; medium to high production quantity.	V: 5–25 dc; A: 1.5–8 A/mm^2; 2.5–12 mm/min, depending on current density.
Electrochemical grinding (ECG)	Cutting off and sharpening hard materials, such as tungsten-carbide tools; also used as a honing process; higher removal rate than grinding.	A: 1–3 A/mm^2; Typically 1.5 cm^3/min per 1000 A.
Electrical-discharge machining (EDM)	Shaping and cutting complex parts made of hard materials; some surface damage may result; also used as a grinding and cutting process with traveling wire; expensive tooling and equipment.	V: 50–300; A: 0.1–500; Typically 0.15 cm^3/min; 0.1–0.25 cm^3/min for grinding.
Laser-beam machining (LBM)	Cutting and holemaking on thin materials; heat-affected zone; does not require a vacuum; expensive equipment; consumes much energy.	50–750 cm/min
Electron-beam machining (EBM)	Cutting and holemaking on thin materials; very small holes and slots; heat-affected zone; requires a vacuum; expensive equipment.	0.0008–0.002 cm^3/min
Hydrodynamic machining (HDM)	Cutting all types of nonmetallic materials to 25 mm and greater in thickness; suitable for contour cutting of flexible materials; no thermal damage; noisy.	Varies considerably with material.

has been used on a wide variety of metals, with depths of removal to as much as 12 mm (0.5 in.) Selective attack by the chemical reagent on different areas of the workpiece surfaces is controlled by removable *masking* (Fig. 26.3a), or by partial immersion in the reagent.

The procedure for chemical milling consists of the following steps:

1. If the part to be machined has residual stresses from prior processing, the stresses should be relieved in order to prevent warping after chemical milling.
2. The surfaces are thoroughly degreased and cleaned (see Section 33.15) to

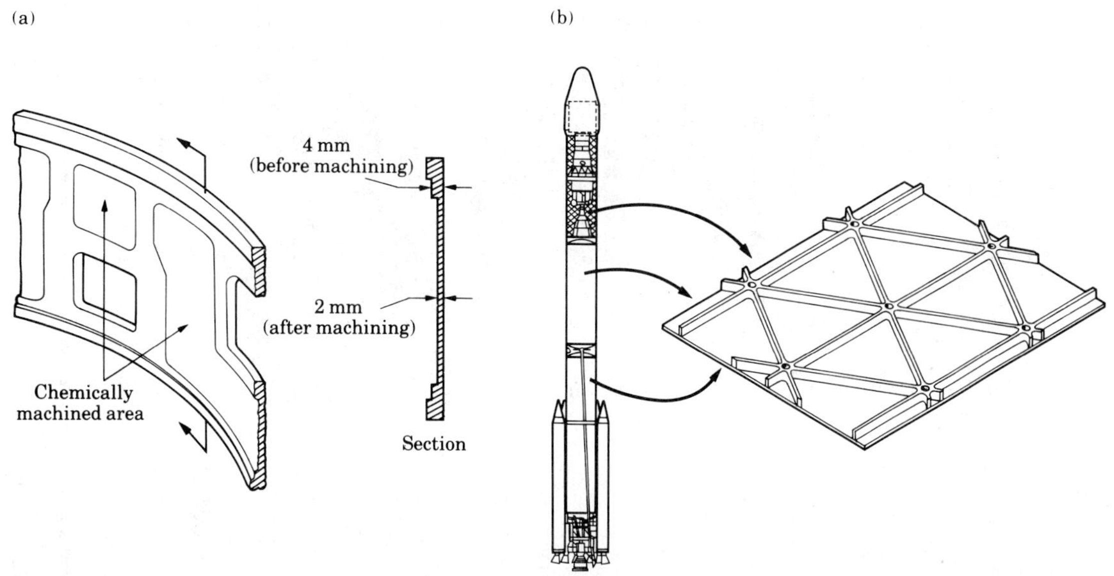

FIGURE 26.2
(a) Missile skin-panel section contoured by chemical milling to improve the stiffness-to-weight ratio of the part. (b) Weight reduction of space launch vehicles by chemical milling aluminum-alloy plates. These panels are milled after the plates have first been formed into shape, such as by roll forming or stretch forming. The design of the chemically machined rib patterns can be modified readily at minimal cost. *Source*: *Advanced Materials and Processes*, p. 43, December 1990. ASM International.

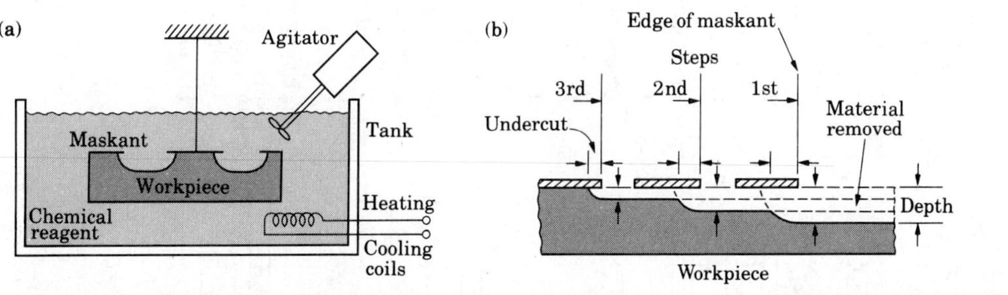

FIGURE 26.3
(a) Schematic illustration of the chemical machining process. Note that no forces or machine tools are involved in this process. (b) Stages in producing a profiled cavity by chemical machining.

ensure good adhesion of the masking material and uniform material removal. Scale from heat treatment should be removed.

3. The masking material is applied. Masking with tapes or paints (**maskants**) is a common practice, although elastomers (rubber and neoprene) and plastics (polyvinyl chloride, polyethylene, and polystyrene) are also used as maskants. The maskant material must be such that it adheres to the workpiece surface and is not reactive with the chemical reagent.

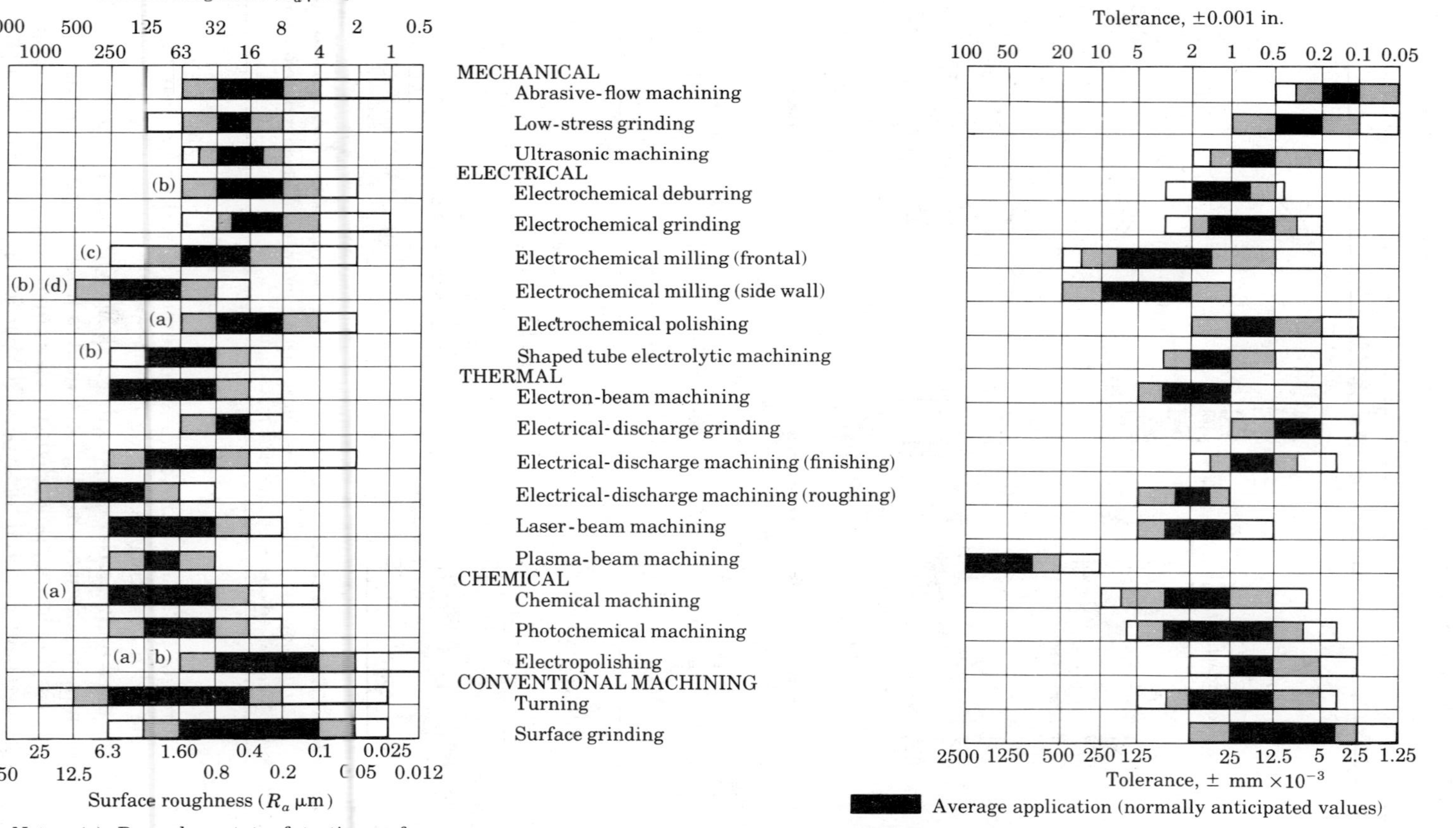

FIGURE 26.4

Surface roughness and tolerances obtained in various machining processes. Note the wide range within each process (see also Fig. 22.14). *Source: Machining Data Handbook,* 3rd ed. Copyright © 1980. Used by permission of Metcut Research Associates, Inc.

4. The masking covering regions that require etching is then peeled off by the scribe-and-peel technique.
5. The exposed surfaces are etched. Common etchants are sodium hydroxide for aluminum, solutions of hydrochloric and nitric acids for steels, and iron chloride for stainless steels. Temperature control and stirring during chemical milling is important in order to obtain a uniform depth of material removed.
6. After machining, the parts should be washed thoroughly to prevent further reactions with any etchant residues.
7. Remove the rest of the masking material and clean and inspect the part.
8. Additional finishing operations may be performed on chemically milled parts.

This sequence of operations can be repeated to produce stepped cavities and contours (Fig. 26.3b on p. 815). This process is used in the aerospace industry for removing shallow layers of material from large aircraft, missile skin panels, and extruded parts for airframes. Tank capacities for reagents are as large as 3.7 m × 15 m (12 ft × 50 ft). The process is also used to fabricate microelectronic devices (Chapter 34). The range of surface finish and tolerances obtained by chemical machining and other machining processes is shown in Fig. 26.4.

Some surface damage may result from chemical milling because of preferential etching and intergranular attack, which adversely affect surface properties. Chemical milling of welded and brazed structures may produce uneven material removal. Chemical milling of castings may result in uneven surfaces caused by porosity and nonuniformity of structure.

26.2.2 Chemical blanking

Chemical blanking is similar to blanking of sheet metal (see Fig. 16.4), with the exception that material is removed by chemical dissolution rather than by shearing. Typical applications for chemical blanking are burr-free etching of printed circuit boards, decorative panels, thin sheet-metal stampings, and the production of complex or small shapes.

26.2.3 Photochemical blanking

Also called photoetching, **photochemical blanking** is a modification of chemical milling. Material is removed, usually from flat thin sheet, by photographic techniques. Complex burr-free shapes can be blanked (Fig. 26.5) on metals as thin as 0.0025 mm (0.0001 in.). Sometimes called *photochemical machining*, it is also used for etching. The procedure consists of the following steps (see also Sections 34.6 and 34.7):

1. The design of the part to be blanked is prepared at a magnification of up to 100×. A photographic negative is then made and reduced to the size of the finished part. The reduced negative of the design is called *artwork*. The original (enlarged) drawing allows inherent design errors to be reduced by the amount of reduction (such as 100×) for the final artwork image.

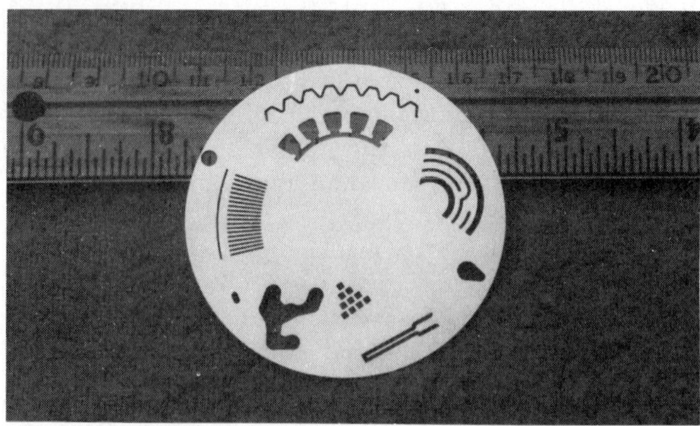

FIGURE 26.5 An example of photochemical blanking of a 0.1-mm (0.004-in.) thick, 60-mm ($2\frac{3}{8}$-in.) diameter steel sheet with various blanked profiles. Note the fine detail. *Source:* Newcut.

2. The sheet blank is coated with a photosensitive material (*photoresist*) by dipping, spraying, or roller coating, and dried in an oven. This coating is often called the *emulsion*.
3. The negative is placed over the coated blank and exposed to ultraviolet light, which hardens the exposed areas.
4. The blank is developed, which dissolves the unexposed areas.
5. The blank is immersed into a bath of reagent, as in chemical milling, or sprayed with the reagent, which etches away exposed areas. Finally, the part is thoroughly washed.

Typical applications for photochemical blanking are fine screens, printed-circuit cards, electric-motor laminations, flat springs, and masks for color television. Although skilled labor is required, tooling costs are low, and the process can be automated.

Photochemical blanking is capable of forming very small parts where traditional blanking dies (Section 16.2) are difficult to make. Also, the process is effective for blanking fragile workpieces and materials.

Handling of chemical reagents requires precautions and special safety considerations to protect the workers against both liquid chemicals and volatile chemical exposure. Futhermore, disposal of chemical by-products from this process is a major consideration, although some by-products can be recycled.

26.2.4 Design considerations

Because the etchant attacks all exposed surfaces continuously, designs involving sharp corners, deep and narrow cavities, severe tapers, folded seams, and porous workpiece materials should be avoided. Moreover, the etchant attacks the material in both vertical and horizontal directions, so undercuts may develop, as shown by the areas under the edges of the maskant in Fig. 26.3. In order to improve the production rate, the bulk of the workpiece should be shaped by other processes prior to chemical machining.

Dimensional variations can occur due to size changes in artwork by humidity and temperature. This variation can be minimized by careful selection of artwork media and controlling the environment in the artwork generation and production area in the plant. Many product designs are now made with computer-aided design systems. However, product drawings must be translated into a protocol that is compatible with the equipment for photochemical artwork generation.

26.3 Electrochemical Machining

Electrochemical machining (ECM) is basically the reverse of electroplating. Electrolytes dissolve the reaction products formed on the workpiece (*anode*) by electrochemical action, thus removing material from the surface and producing a cavity (Fig. 26.6a). Note that the cavity produced is the female mating image of the tool (*cathode*). Modifications of this process are used for turning, facing, slotting, trepanning, and profiling operations in which the electrode becomes the cutting tool.

The shaped tool is generally made of brass, copper, bronze, or stainless steel. The electrolyte is a highly conductive inorganic salt solution such as sodium chloride mixed in water or sodium nitrate. It is pumped at a high rate through the passages in the tool. A dc power supply in the range of 5–25 V maintains current

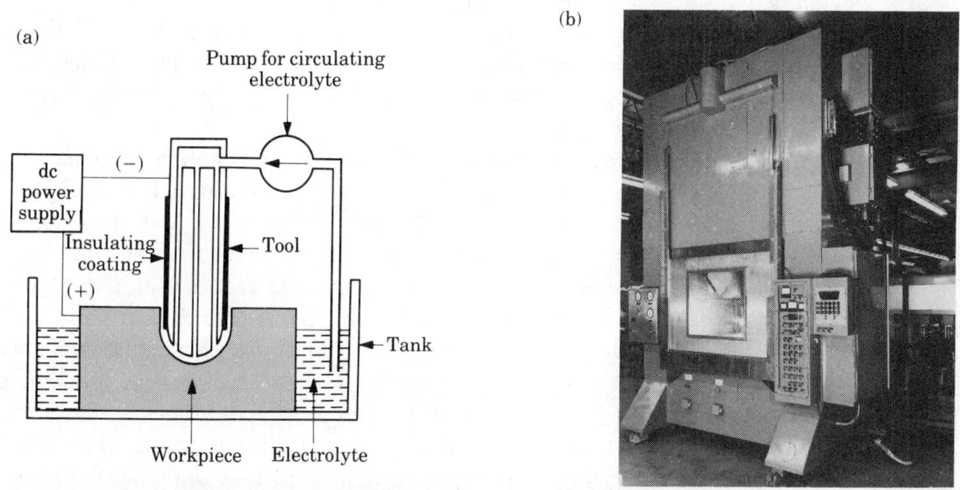

FIGURE 26.6
(a) Schematic illustration of the electrochemical-machining process. This process is the reverse of electroplating described in Section 33.7. (b) An ECM machine. *Source:* Courtesy of Anocut, Inc.

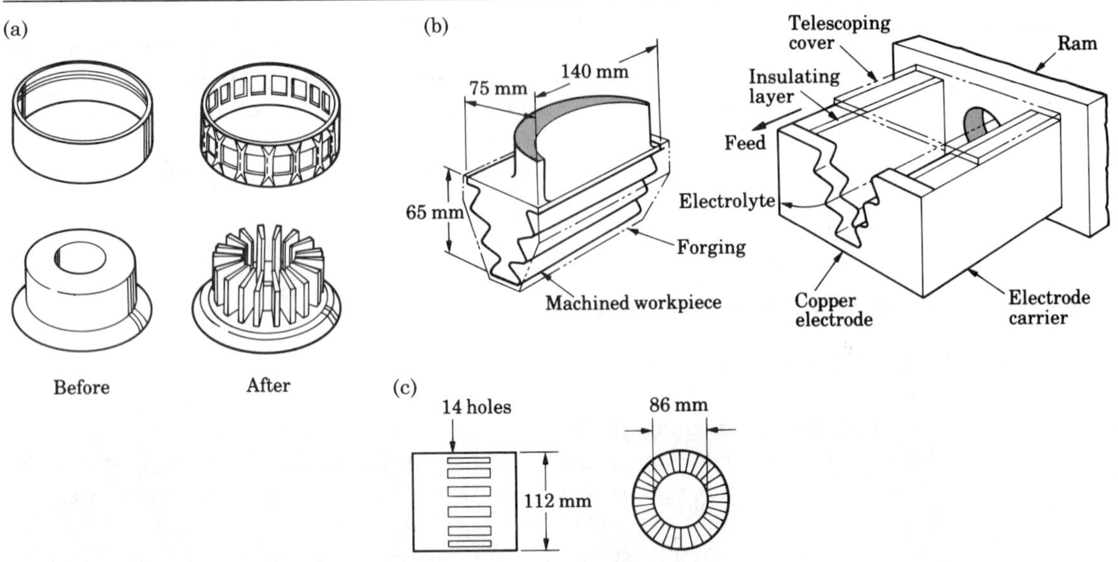

FIGURE 26.7 Typical parts made by electrochemical machining. (a) Parts with various shapes. (b) Electrochemically machined turbine blade made of a nickel alloy, 360 HB. *Source:* ASM International. (c) Thin slots on a 4340-steel roller-bearing cage, produced by ECM.

densities, which for most applications are 1.5–8 A/mm^2 (1000–5000 A/in^2) of active machined surface. Machines having current capacities as high as 40,000 A and as small as 5 A are available (Fig. 26.6b). The penetration rate of the tool is proportional to the current density. Because the metal removal rate is only a function of ion exchange rate, it is not affected by the strength, hardness, or toughness of the workpiece.

Electrochemical machining is generally used for machining complex cavities in high-strength materials, particularly in the aerospace industry for mass production of turbine blades, jet-engine parts, and nozzles (Figs. 26.7a and b). It is also used for machining forging-die cavities and producing small holes (Fig. 26.7c).

The ECM process leaves a burr-free surface; in fact, it can also be used as a deburring process. It does not cause any thermal damage to the part, and the lack of tool forces prevents distortion of the part. Furthermore, there is no tool wear, and the process is capable of producing complex shapes as well as machining hard materials. However, the mechanical properties of components made by ECM should be compared carefully to those of other material-removal methods (Fig. 26.8). Note the lower fatigue strength compared to that obtained for ground and polished surfaces. Recent trends in electrochemical machining involve numerically controlled machining centers, with the capability for higher production rates, high flexibility, and maintenance of close tolerances.

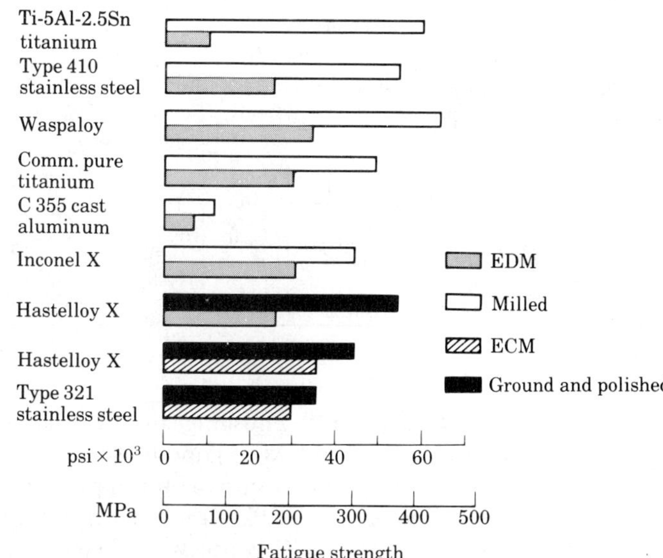

FIGURE 26.8 Effect of machining process on the fatigue strength of metal alloys. Note the severe reduction in strength in some operations. Fatigue strength can be improved with proper control of processing parameters and subsequent treatments, such as shot peening, surface rolling, and polishing. *Source:* Pratt and Whitney.

26.3.1 Design considerations

Because of the tendency for the electrolyte to erode away sharp profiles, electrochemical machining is not suited for producing sharp square corners or flat bottoms. Moreover, controlling the electrolyte flow is difficult, and irregular cavities may not be produced to the desired shape with acceptable dimensional accuracy. Designs should make provision for a small taper for holes and cavities.

26.4 Electrochemical Grinding

Electrochemical grinding (ECG) combines electrochemical machining with conventional grinding. The equipment used in electrochemical grinding is similar to a grinder, except that the wheel is a rotating cathode with abrasive particles (Fig. 26.9a). The wheel is metal-bonded with diamond or aluminum-oxide abrasives, and rotates at a surface speed of 20–35 m/s (4000–7000 ft/min). The abrasives serve as insulators between the wheel and the workpiece and mechanically remove electrolytic products from the working area. A flow of electrolyte, usually sodium nitrate, is provided for the electrochemical machining phase of the operation.

The majority of metal removal in ECG is by electrolytic action, and typically less than 5 percent of metal is removed by the abrasive action of the wheel. Therefore, wheel wear is very low, and current densities range from 1–3 A/mm^2

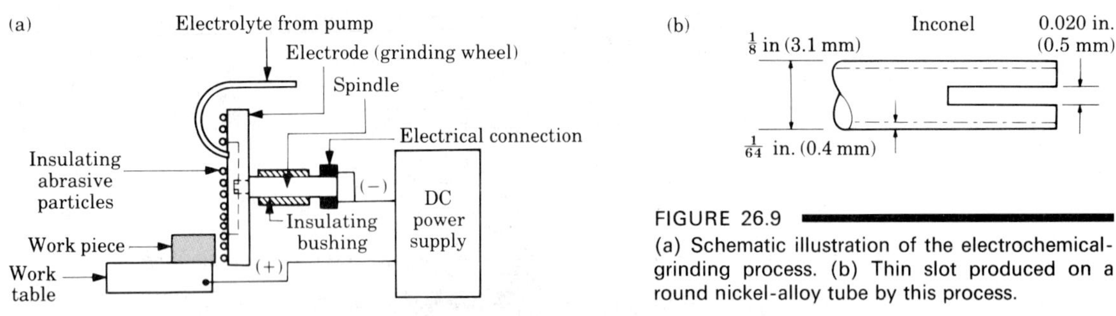

FIGURE 26.9 (a) Schematic illustration of the electrochemical-grinding process. (b) Thin slot produced on a round nickel-alloy tube by this process.

(500–2000 A/in^2). Finishing cuts are usually made by the grinding action but only to produce a surface with good finish and dimensional accuracy.

This process is suitable for applications similar to those for milling, grinding, and sawing (Fig. 26.9b). It is not adaptable to cavity-sinking operations, such as die making. The ECG process has been successfully applied to carbides and high-strength alloys. This process offers a distinct advantage over traditional diamond-wheel grinding when processing very hard materials, where wheel wear can be high. ECG machines are now available with numerical controls, thus improving accuracy, repeatability, and increased productivity.

Electrochemical honing combines the fine abrasive action of honing with electrochemical action. Although the equipment is costly, the process is as much as five times faster than conventional honing. It is used primarily for finishing internal cylindrical surfaces.

26.4.1 Design considerations

In addition to those already listed for electrochemical machining, ECG requires the following additional design considerations:

a) Designs should avoid sharp inside radii.
b) If flat surfaces are to be produced, the electrochemically ground surface should preferably be narrower than the width of the grinding wheel.

• Example: Electrochemical grinding of thin-walled tube.

Three slots on a 3.18-mm (1/8-in.) diameter, 0.25-mm (0.010-in.) thick tube made of Type 316 stainless steel were produced by electrochemical grinding (see the accompanying figure). The slots were ground in one pass using a metal-bonded aluminum oxide wheel running at 30 m/s (6000 ft/min). The ground parts were free from distortion and the ground edges were free of burrs. Production rate was 12 pieces per hour. (*Source*: ASM International.)

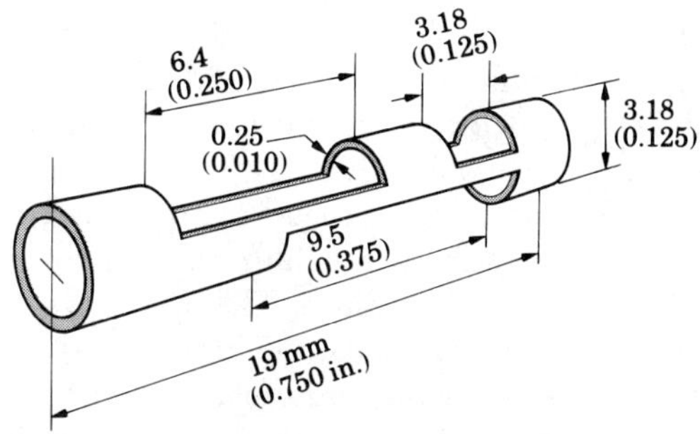

26.5 Electrical-Discharge Machining

The principle of **electrical-discharge machining** (EDM), also called *electrodischarge* or *spark-erosion* machining, is based on erosion of metals by spark discharges. We know that when two current-conducting wires are allowed to touch each other, an arc is produced. If we look closely at the point of contact between the two wires, we note that a small portion of the metal has been eroded away, leaving a small crater. In other words, we have removed a small amount of material from the surface of the wire. Although this phenomenon has been known since the discovery of electricity, it was not until the 1940s that a machining process based on this principle was developed.

The EDM system consists of a shaped tool (*electrode*) and the workpiece, connected to a dc power supply and placed in a *dielectric fluid* (Fig. 26.10). When the potential difference between the tool and the workpiece is sufficiently high, a

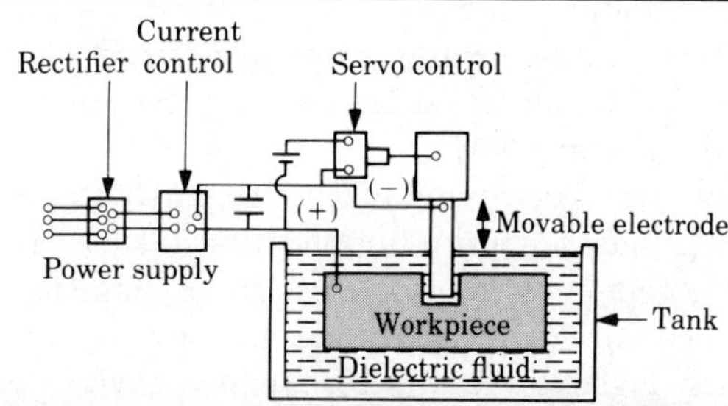

FIGURE 26.10 Schematic illustration of the electrical-discharge machining process.

transient spark discharges through the fluid, removing a very small amount of metal from the workpiece surface. The discharge is repeated at rates of between 50 kHz and 500 kHz, with voltages usually ranging between 50 V and 300 V, and currents from 0.1 A to 500 A.

The dielectric fluid (a) acts as an insulator until the potential is sufficiently high, (b) acts as a flushing medium and carries away the debris in the gap, and (c) provides a cooling medium. The gap between the tool and the workpiece (called *overcut*) is critical; thus the downward feed of the tool is controlled by a servomechanism, which automatically maintains a constant gap. The most common dielectric fluids are mineral oils, although kerosene and distilled and deionized water may be used in specialized applications.

The workpiece is fixtured within the tank containing the dielectric fluid, and its movements are controlled by numerically controlled systems. The machines are equipped with a pump and filtering system for the dielectric fluid.

The EDM process can be used on any material that is an electrical conductor. The melting point and latent heat of melting are important physical properties that determine the volume of metal removed per discharge. As these values increase, the rate of material removal slows. The volume of material removed per discharge is typically in the range of 10^{-6}–10^{-4} mm^3 (10^{-10}–10^{-8} in^3). Since the process doesn't involve mechanical energy, hardness, strength, and toughness of the workpiece material don't necessarily influence the removal rate.

The frequency of discharge or the energy per discharge is usually varied to control the removal rate, as are the voltage and current. The rate and surface roughness increase with increasing current density and decreasing frequency of sparks. Metal removal rates usually range from 0.1–25 cm^3/h (0.005–1.5 in^3/h). Higher rates are possible but they produce a very rough finish, having a molten and resolidified (recast) structure with poor surface integrity and low fatigue properties (see Fig. 26.8). Thus finishing cuts are made at low removal rates, or the recast layer is removed later by finishing operations.

Electrodes for EDM are usually made of graphite, although brass, copper, or copper–tungsten alloy may be used. The tools are shaped by forming, casting, powder metallurgy, or machining. Tool wear is an important factor since it affects dimensional accuracy and the shape produced. Tool wear in EDM is related to the melting points of the materials involved: the lower the melting point, the higher the wear rate. Consequently, graphite electrodes have the highest wear resistance. Tool wear can be minimized by reversing the polarity and using copper tools, a process called **no-wear EDM.**

Electrical-discharge machining has become an important manufacturing process. It has numerous applications, such as producing die cavities for large automotive body components, small-diameter deep holes (see Fig. 26.1b) using tungsten wire as the electrode, narrow slots, turbine blades, and various intricate shapes (Figs. 26.11a and b). Stepped cavities can be produced by controlling the relative movements of the workpiece in relation to the electrode (Fig. 26.12). In another setup, internal cavities are produced by a rotating electrode with a movable tip. The electrode is rotated mechanically during machining (Fig. 26.13).

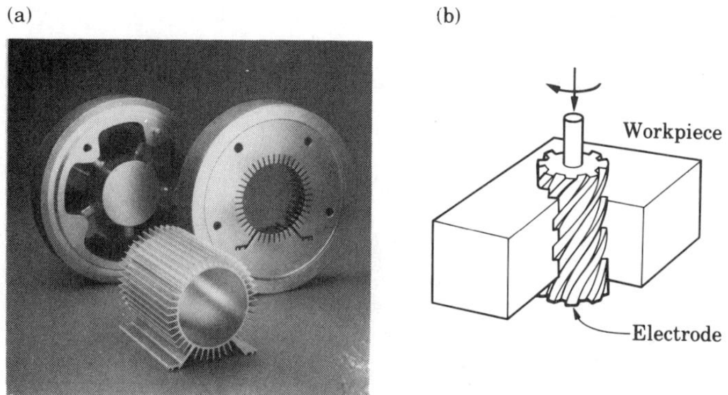

FIGURE 26.11 (a) Examples of cavities produced by the electrical-discharge machining process, using shaped electrodes. Two round parts (rear) are the set of dies for extruding the aluminum piece shown in front (see also Fig. 15.9b). *Source:* Courtesy of AGIE USA Ltd. (b) A spiral cavity produced by a rotating electrode. *Source: American Machinist.*

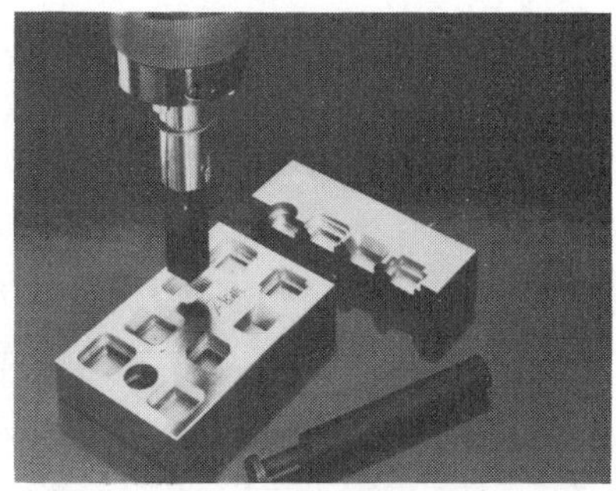

FIGURE 26.12 Stepped cavities produced with a square electrode by EDM. The workpiece moves in the two principal horizontal directions, and its motion is synchronized with the downward movement of the electrode to produce various cavities. Also shown is a round electrode capable of producing round or elliptical cavities. *Source:* Courtesy of AGIE USA Ltd.

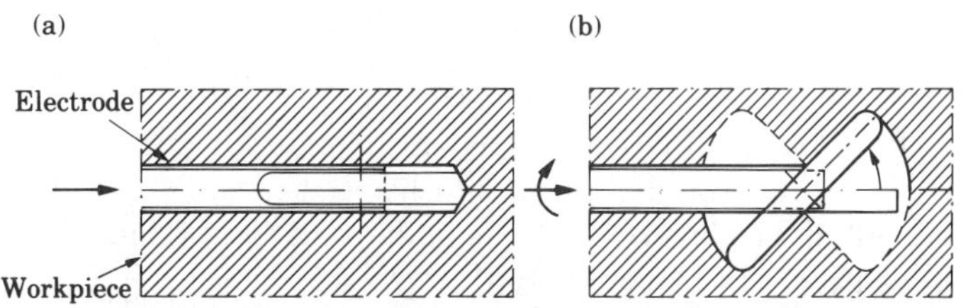

FIGURE 26.13 Schematic illustration of producing an inner cavity by EDM, using a specially designed electrode with a hinged tip, which is rotated along two axes to produce the large cavity. *Source:* Luziesa France.

26.5.1 Electrical-discharge grinding

The grinding wheel in *electrical-discharge grinding* (EDG) is made of graphite or brass and contains no abrasives. Material is removed from the surface of the workpiece by repetitive spark discharges between the rotating wheel and the workpiece. An application of electrical-discharge grinding is shown in Fig. 26.14, which shows longitudinal grooves being produced on the workpiece using a shaped wheel.

The EDG process can be combined with electrochemical grinding. The process is then called *electrochemical-discharge grinding* (ECDG). Material is removed by chemical action, with the electrical discharges from the graphite wheel breaking up the oxide film, and is washed away by the electrolyte flow. The process is used primarily for grinding carbide tools and dies but can also be used for fragile parts,

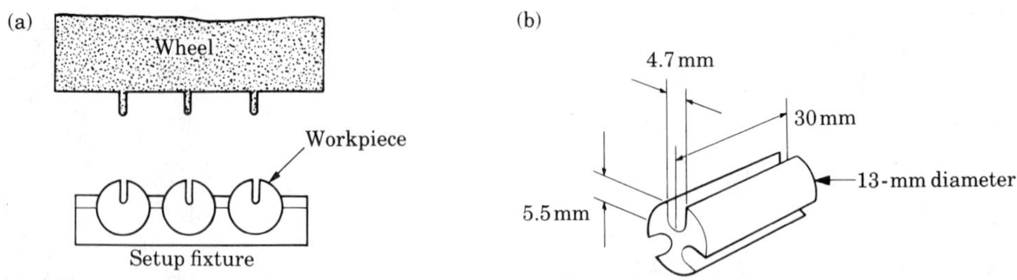

FIGURE 26.14 (a) Longitudinal grooves produced in a round rod by the electrical-discharge grinding process. (b) Sample dimensions. Note that the workpieces are indexed by 120° to produce each set of grooves.

such as surgical needles, thin-walled tubes, and honeycomb structures. The ECDG process is faster than EDG, but power consumption is higher.

In *sawing* with EDM, a setup similar to a band or circular saw (but without any teeth) is used with the same electrical circuit as in EDM. Narrow cuts can be made at high rates of metal removal. Because cutting forces are negligible, the process can be used on slender components.

26.5.2 Design considerations

Parts should be designed so that the required electrodes can be shaped properly and economically. Deep slots and narrow openings should be avoided. For economic production, the surface finish specified should not be too fine. In order to achieve a high production rate, the bulk of material removal should be done by conventional processes (roughing out).

26.6 Traveling-Wire EDM

A variation of EDM is **traveling-wire EDM** (Figs. 26.15a and b), or *electrical-discharge wire cutting*. In this process, which is similar to contour cutting with a band saw, a slowly moving wire travels along a prescribed path, cutting the workpiece, with the discharge sparks acting like cutting teeth. This process is used to cut plates as thick as 150 mm (6 in.), and for making punches, tools, and dies from hard metals. It can also cut intricate components for the electronics industry. Machines are equipped with computer controls to control the cutting path of the wire (Fig. 26.15c). They are also equipped with automatic self-threading features, multiheads for cutting two parts at the same time, and features such as controls that prevent wire breakage. New developments include a two-axis computer-controlled

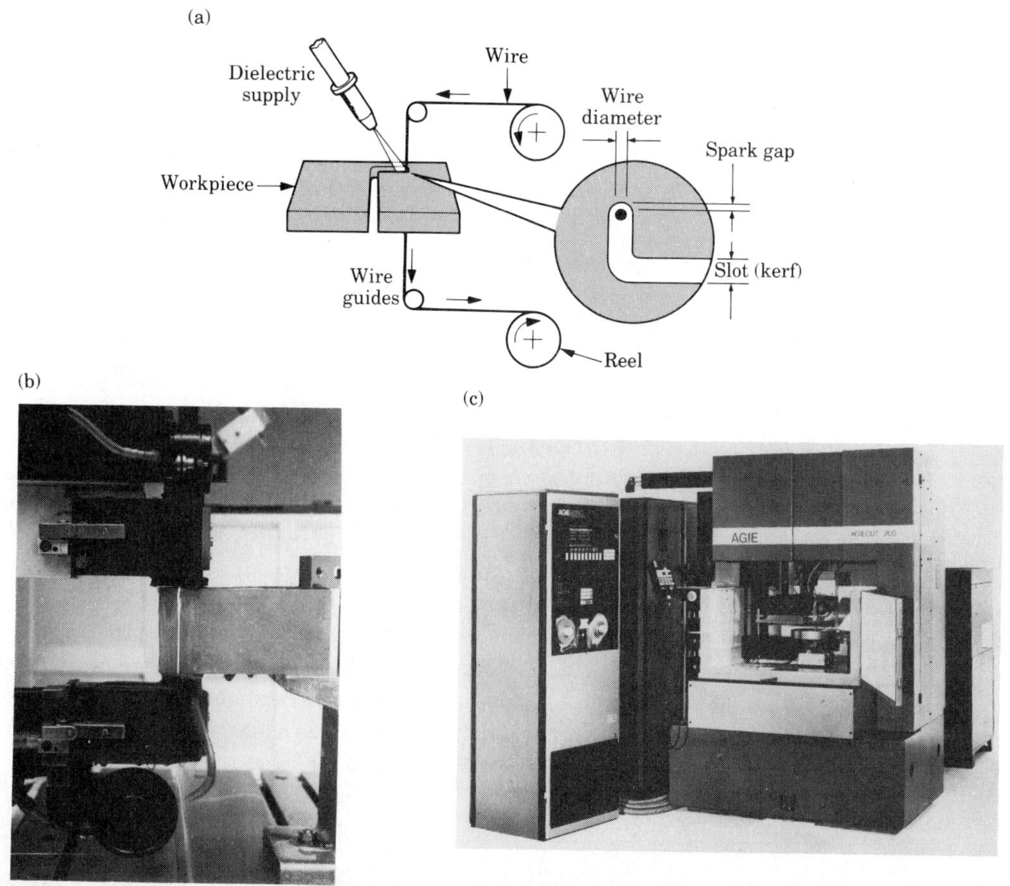

FIGURE 26.15 (a) Schematic illustration of the traveling-wire EDM process. As much as 50 hours of machining can be performed with one reel of wire, which is then discarded. (b) Cutting a thick plate with wire EDM. (c) A computer-controlled wire EDM machine. *Source:* Courtesy of AGIE USA Ltd.

machine that can produce cylindrical shapes in a manner similar to a turning operation or cylindrical grinding.

The wire is usually made of brass, copper, or tungsten and is typically about 0.25 mm (0.01 in.) in diameter, making narrow cuts possible. Coated and multi-coated wires are also used. The wire should have sufficient tensile strength and fracture toughness, as well as high electrical conductivity and capacity fo flush away the debris produced during cutting. The wire is generally used only once, as it is relatively inexpensive. It travels at sufficiently high and constant velocity, 2.5–150 mm/s (0.1–6 in./s), and a constant gap (kerf) is maintained during the cut.

26.7 Laser-Beam Machining

In **laser-beam machining** (LBM), the source of energy is a laser (an acronym for *L*ight *A*mplification by *S*timulated *E*mission of *R*adiation), which focuses optical energy on the surface of the workpiece (Fig. 26.16a). The highly focused, high-density energy melts and evaporates portions of the workpiece in a controlled manner.

This process, which does not require a vacuum, is used to machine a variety of metallic and nonmetallic materials. It is also used for small-scale cutting operations, such as slitting, and drilling holes (Figs. 26.16b and c) as small as 0.005 mm (0.0002 in.), with hole depth to diameter ratios of 50 to 1. The cooling passages in the first-stage vanes of the Boeing 747 turbine engines are produced by this process. Machines are now being designed and built utilizing the laser-machining process. A five-axis laser machine tool with a capacity of up to 750 W carbon-dioxide laser is shown in Fig. 26.17. A mold for a tire tread made by this machine is shown in the inset to the figure. As can be seen, the mold has sharp edges and corners, with web dimensions on the order of 1–3 mm (0.04–0.12 in.).

Important physical parameters in LBM are the reflectivity and thermal conductivity of the workpiece surface and its specific heat and latent heats of melting and evaporation. The lower these quantities, the more efficient the process is. The surface produced by LBM is usually rough and has a heat-affected zone which, in critical applications, may have to be removed or heat treated. Kerf width is an

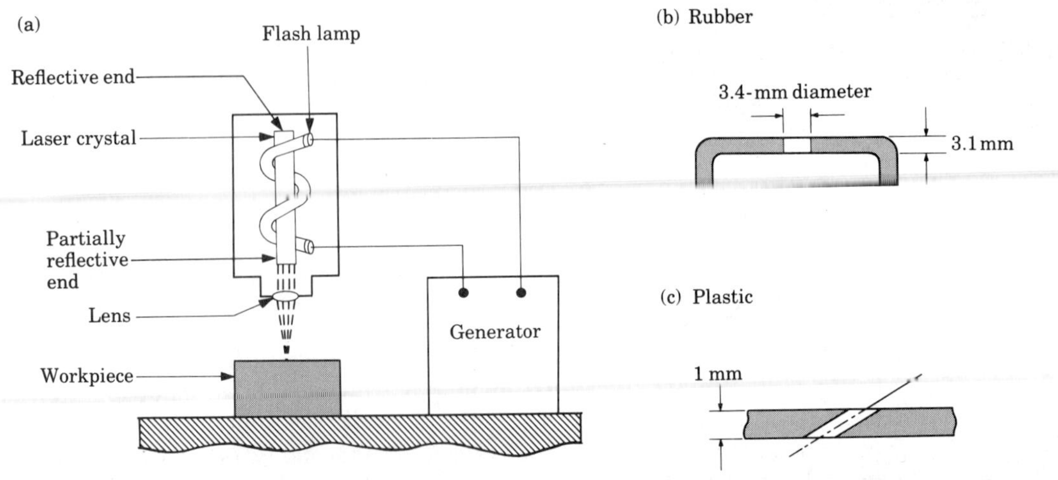

FIGURE 26.16
(a) Schematic illustration of the laser-beam machining process. (b) and (c) Examples of holes produced in nonmetallic parts by LBM.

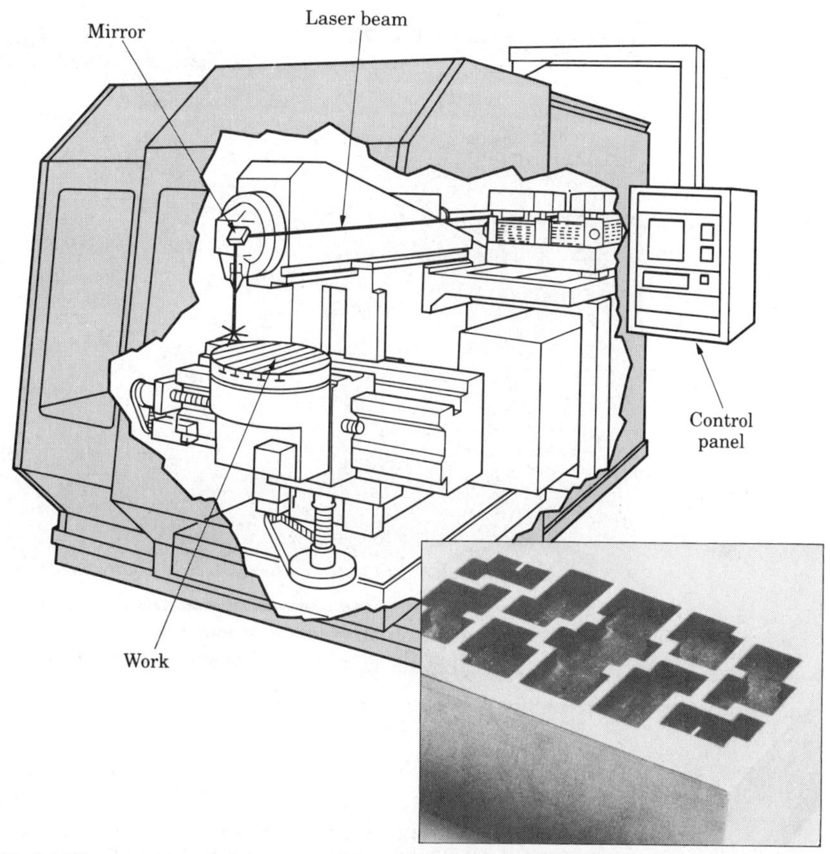

FIGURE 26.17

Schematic illustration of a five-axis laser machine tool. Inset shows a mold for tire treads. The process is numerically controlled by a multiprocessor system that can be programmed on the shop floor. *Source:* Courtesy of Maho Machine Tool Corporation, Naugatuck, CT.

important consideration, as it is in other cutting processes such as sawing, wire EDM, and electron-beam machining.

Laser beams may be used in combination with a gas stream, such as oxygen, nitrogen, or argon (*laser-beam torch*), for cutting thin sheet materials. High-pressure inert-gas (nitrogen)–assisted laser cutting is used for stainless steel and aluminum; it leaves an oxide-free edge that can improve weldability. Gas streams also have the important function of blowing away molten and vaporized material from the workpiece surface. Laser-beam machining is used widely in drilling and cutting composite materials, especially for the electronics industry. The abrasive nature of composites and the cleanliness of the operation make laser-beam machining an attractive alternative to traditional machining methods. Laser beams are also used for small-scale heat-treating and welding operations.

FIGURE 26.18
Lubrication hole being drilled in an automotive transmission hub, using a 250 W, Nd:YAG laser. *Source:* Courtesy of Coherent General, Inc.; Dr. David Roessler, General Motors Research Laboratory; and *Mechanical Engineering*, April 1990, p. 41.

Laser-beam machining is being used increasingly in the automotive industries. Bleeder holes for fuel-pump covers and lubrication holes in transmission hubs, for example, are being drilled with lasers (Fig. 26.18). The part shown, with a hardness of 60 HRC, requires drilling three holes, 1.5 mm in diameter and 9.5 mm deep (0.060 in. and 0.38 in.), and at an angle of 35° to the curved surface. It takes 3.5 sec to drill each hole. Significant cost savings have been achieved by laser-beam machining, a process that is beginning to compete with electrical-discharge machining.

Extreme caution should be exercised with lasers, as even low-power lasers can cause damage to the retina of the eye without the person being aware of it at first.

26.7.1 Design considerations

Reflectivity of the workpiece surface is an important consideration, and because they reflect less, dull and unpolished surfaces are preferable. Designs with sharp corners should be avoided since they are difficult to produce. Deep cuts produce tapers. Any adverse effects on the properties of the machined materials caused by the high local temperatures involved should be investigated.

26.8 Electron-Beam Machining

The source of energy in **electron-beam machining** (EBM) is high-velocity electrons, which strike the surface of the workpiece (Fig. 26.19). Its applications are similar to those of laser-beam machining, except that EBM requires a vacuum. These machines utilize voltages in the range of 50–200 kV to accelerate the electrons to speeds of 50–80 percent of the speed of light. The interaction of the electron beam with the workpiece surface produces hazardous x-rays; consequently, the equipment should be used only by highly trained personnel.

Plasma (ionized gas) **beams** are also used to cut sheet and plate at high speeds. Material-removal rates in this process are much higher than in the EDM and LBM processes, and parts can be machined with good reproducibility.

26.8.1 Design considerations

The guidelines for LBM generally apply to EBM as well. Because vacuum chambers have limited capacity, part sizes should be as small as possible for a higher production rate per cycle. If a product requires electron-beam machining on only a small portion of the workpiece, consideration should be given to manufacturing the product as a number of smaller components and assembling them following electron-beam machining.

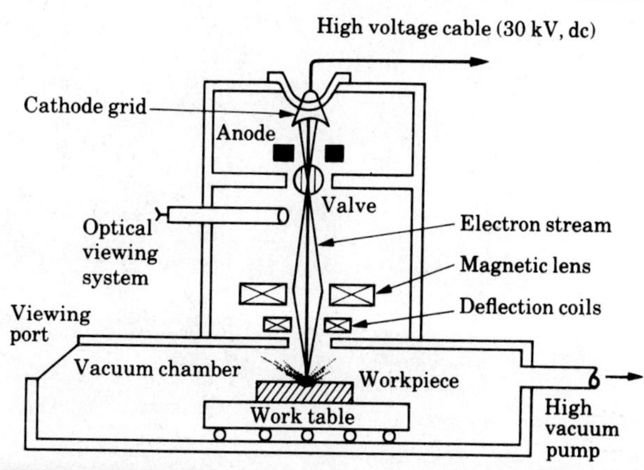

FIGURE 26.19 Schematic illustration of the electron-beam machining process. Unlike LBM, this process requires a vacuum, hence workpiece size is limited.

26.9 Hydrodynamic Machining

We know that when we put our hand across a jet of water or air, we feel a considerable force acting on it. This force results from the momentum change of the stream—and in fact is the principle on which the operation of water or gas turbines is based. In **hydrodynamic machining** (HDM), also called *water-jet machining* (Fig. 26.20), this force is utilized in cutting and deburring operations.

The water jet acts like a saw and cuts a narrow groove in the material. Although

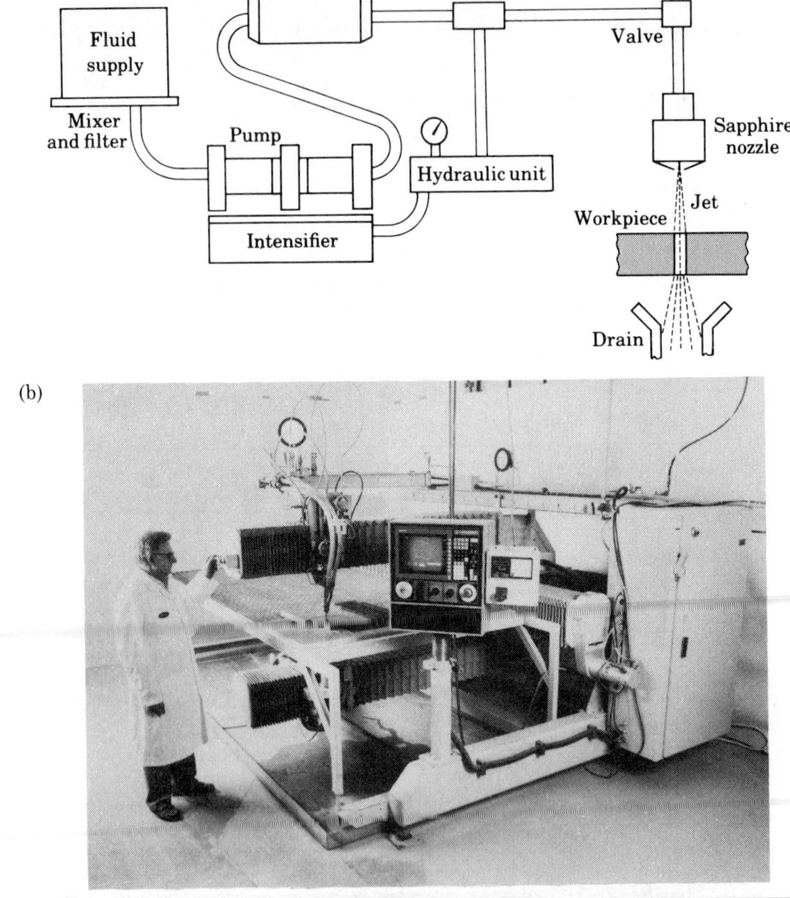

FIGURE 26.20 (a) Schematic illustration of the hydrodynamic machining process, also called water-jet maching. (b) A computer-controlled, water-jet cutting machine cutting a granite plate. *Source:* Courtesy of Possis Corporation.

FIGURE 26.21
Example of various nonmetallic parts cut by a water-jet machine. *Source:* Courtesy of Possis Corporation.

pressures as high as 1400 MPa (200 ksi) can be generated, a pressure level of about 400 MPa (60 ksi) is generally used for efficient operation. Jet-nozzle diameters usually range between 0.05 mm and 1 mm (0.002 in. and 0.040 in.). Long-chain polymers are often added to the water to improve fluid coherency and eliminate the misting effects of the water jet at the exit of the nozzle.

A water-jet cutting machine and its operation are shown in Fig. 26.20(b). A variety of materials can be cut with this technique, including plastics, fabrics, rubber, wood products, paper, leather, insulating materials, brick, and composite materials (Fig. 26.21). Thicknesses range up to 25 mm (1 in.) and higher. Vinyl and foam coverings for some automobile dashboards, for example, are being cut using multiaxis, robot-guided hydrodynamic machining equipment. The advantages of this process are that cuts can be started at any location without the need for predrilled holes, no heat is produced, no deflection of the rest of the workpiece takes place (hence the process is suitable for flexible materials), little wetting of the workpiece takes place, and the burr produced is minimal.

In *abrasive water-jet* machining, the water jet contains abrasive particles such as silicon carbide or aluminum oxide, thus increasing the material-removal rate. Metallic and nonmetallic materials of various thicknesses can be cut in single or multilayers. Cutting speeds are as high as 7.5 m/min (25 ft/min) for reinforced plastics, but much lower for metals. Carbide-based composite materials have been developed for nozzles, which otherwise have a limited life due to wear caused by the abrasive particles.

26.10 Economics of Nontraditional Machining Processes

We have shown that nontraditional processes have unique applications, particularly for difficult-to-machine materials and complex internal and external profiles. The economic production run for a particular process depends on the cost of tooling and equipment, the material-removal rate, operating costs, and the level of operator skill required, as well as secondary and finishing operations that may be necessary. In chemical machining, the costs of reagents, maskants, and disposal, together with the cost of cleaning the parts, are important factors. In electrical-discharge machining, the cost of electrodes and the need to replace them periodically are significant.

The rate of material removal, hence production rate, can vary significantly in these processes (see Table 26.1). The cost of tooling and equipment also varies significantly, as does the operator skill required. The high capital investment for machines such as electrical and high-energy beam machining should be justified in terms of the production runs and the feasibility of manufacturing the same part by other means, if at all possible.

26.11 Die Manufacturing Methods

Various manufacturing methods, either singly or in combination, are used in making dies. These processes include casting, forging, machining, grinding, electrical, and electrochemical methods of die sinking, and finishing operations such as honing and polishing. The choice of a die manufacturing method depends on the particular manufacturing process in which the die is to be used and the size and shape of the die. Cost often dictates the process selected because, as we have shown, tool and die costs can be significant in manufacturing operations. For example, the cost of a set of dies for automotive body panels may run $2 million. Even small and relatively simple dies can cost hundreds of dollars. On the other hand, because a large number of parts are usually made from the same die, die cost per piece is generally only a small portion of a part's manufacturing cost.

Dies of various sizes and shapes can be cast from steels, cast irons, and nonferrous alloys. The processes used may be sand casting for large dies weighing many tons to shell molding for small dies. Cast steels are generally preferred as die materials because of their strength and toughness and the ease with which their composition, grain size, and properties can be controlled and modified.

Most commonly, dies are machined from die blocks by processes such as milling, turning, grinding, electrical and electrochemical machining, and polishing. Typically, a die for hot-working operations is machined by milling on an automatic copy machine that traces a master model (pattern), or on computer-controlled machine tools that use various software. The patterns may be made of wood, epoxy resins, aluminum, or gypsum. Conventional machining can be difficult for high-strength and wear-resistant die materials that are hard or are heat treated. These operations can be very time-consuming even with copy milling machines. As a result, nontraditional machining processes are used extensively, particularly for small- or medium-sized dies. These processes are generally faster and more economical, and the dies usually do not require additional finishing. Diamond dies for drawing fine wire are manufactured by producing holes with a thin rotating needle, coated with diamond dust suspended in oil.

Failure of dies in manufacturing operations generally result from one or more of the following causes: improper design, defective material, improper heat treatment and finishing operations, overheating and heat checking (cracking caused by temperature cycling), excessive wear, overloading, misuse, and improper handling. The proper design of dies is as important as the proper selection of die materials. Sharp corners, radii, and fillets, as well as sudden changes in cross-section, act as stress raisers and can have detrimental effects on die life. Dies may be made in segments and prestressed during assembly for improved strength.

To obtain improved hardness, wear resistance, and strength, die steels are usually heat treated. Improper heat treatment is one of the most common causes of die failure. Dies may be subjected to various surface treatments for improved frictional and wear characteristics (see Chapter 33).

SUMMARY

Machining processes involve not only single-point or multipoint tools, but also other methods using chemical, electrical, and high-energy-beam sources of energy. The mechanical properties of the workpiece material are not significant because these processes rely on mechanisms that do not involve the strength, hardness, ductility, or toughness of the material. Rather, they involve physical, chemical, and electrical properties.

Chemical and electrical methods of machining are particularly suitable for hard materials and complex shapes. They do not produce forces (hence can be used for slender and flexible workpieces), significant temperatures, or residual stresses. However, the effects of these processes on surface integrity must be understood, as they can damage surfaces considerably, thus reducing fatigue life.

TRENDS

- The need for economical methods of material removal will increase further because of the development of new materials and composites that will be difficult to machine with traditional processes.
- In spite of their advantages, the effects of these processes on the properties and service life of workpieces are important considerations, particularly for critical applications.
- Computer-aided design and manufacturing of dies are important developments, for economic reasons as well as for improving the quality of products.
- Laser-beam machining of automotive components is being implemented at an increasing rate.

KEY TERMS

Chemical blanking
Chemical machining
Chemical milling
Electrical-discharge machining
Electrochemical grinding
Electrochemical machining
Electron-beam machining
Etchant
Hydrodynamic machining
Laser-beam machining
Maskant
No-wear EDM
Photochemical blanking
Plasma beams
Reagent
Traveling-wire EDM

BIBLIOGRAPHY

Benedict, G.F., *Nontraditional Manufacturing Processes*. New York: Marcel Dekker, 1987.

Harris, W.T., *Chemical Milling: The Technology of Cutting Materials by Etching*. New York: Oxford, 1976.

Kalpakjian, S. (ed.), *Tool and Die Failures: Source Book*. Metals Park, Ohio: American Society for Metals, 1982.

Lange, K. (ed.), *Handbook of Metal Forming* (Chapter 32, Die Manufacture). New York: McGraw-Hill, 1985.

Machinery's Handbook, revised periodically. New York: Industrial Press.

Machining Data Handbook, 3d ed., 2 Vols. Cincinnati: Machinability Data Center, 1980.

Metals Handbook, 8th ed., Vol. 3: *Machining*. Metals Park, Ohio: American Society for Metals, 1967.

Metals Handbook, 9th ed., Vol. 16: *Machining*. Metals Park, Ohio: ASM International, 1989.

McGeough, J.A., *Principles of Electrochemical Machining*. London: Chapman and Hall, 1974.

——, *Advanced Methods of Machining*. London: Chapman and Hall, 1988.

Source Book on Applications of the Laser in Metalworking. Metals Park, Ohio: American Society for Metals, 1979.

Tool and Manufacturing Engineers Handbook, 4th ed., Vol. 1: *Machining*. Dearborn, Mich.: Society of Manufacturing Engineers, 1983.

REVIEW QUESTIONS

26.1 List the reasons for development of unconventional machining processes.

26.2 Name the processes involved in chemical machining. Describe briefly their principles.

26.3 What should be the properties of maskants?

26.4 Describe chemical blanking and compare it with conventional blanking with dies.

26.5 Explain the difference between chemical machining and electrochemical machining.

26.6 What is the underlying principle of electrochemical grinding?

26.7 Why has electrical-discharge machining become so widely used?
26.8 Explain how EDM is capable of producing complex shapes.
26.9 What are the capabilities of traveling-wire EDM? Could this process be used to make tapered pieces? Explain.
26.10 List the functions of the dielectric fluid in EDM.
26.11 What is the difference between laser-beam and electron-beam machining?
26.12 What is a plasma?
26.13 Describe the advantages of hydrodynamic machining.
26.14 Why is preshaping or premachining of parts sometimes desirable in the processes described in this chapter?

QUESTIONS AND PROBLEMS

26.15 Give possible technical and economic reasons why the processes described in this chapter might be preferred, or even necessary, over those described in the preceding chapters.
26.16 On the basis of the discussion in this chapter, what precautions would you take in chemical machining and why?
26.17 Which of the processes described in Part IV would you recommend for die sinking in a steel die block?
26.18 Explain why the mechanical properties of workpiece materials are not significant in the processes described in this chapter.
26.19 Why has the traveling-wire EDM process become so widely accepted in industry?
26.20 Why do material-removal processes affect the fatigue strength of materials to different degrees?
26.21 Make a list of machining processes that may be suitable for the following materials: (a) ceramics, (b) cast iron, (c) thermoplastics, (d) thermosets, (e) diamond, and (f) annealed copper.
26.22 How would you manufacture a very thin, large-diameter round disk with a thickness that decreases nonlinearly from the center?
26.23 Explain why producing sharp profiles and corners with some of the processes described in this chapter is difficult.
26.24 Which of the nontraditional machining processes causes thermal damage? What is the consequence of such damage to workpieces?
26.25 Describe the similarities and differences among the various design guidelines presented in this chapter.
26.26 A 40-mm-deep hole 20 mm in diameter is being produced by electrochemical machining. High production rate is more important than machined surface quality. Estimate the maximum current and the time required to perform this operation.
26.27 If the operation in Problem 26.26 were performed on an electrical-discharge machine, what would be the estimated machining time?
26.28 A cutting off operation is being performed with a laser beam. The workpiece being cut is 1/2 in. thick and 2 in. wide. If the kerf is 1/16 in. wide, estimate the time required to perform this operation.
26.29 What factors are involved in the material removal rate in chemical milling, and why?

26.30 Material removal rate in electrical-discharge machining can be increased by increasing the current density and decreasing the frequency of the sparks. Are there any disadvantages in doing so? Explain.

26.31 Would you consider designing a machine tool that combines in one machine two or more of the processes described in this chapter? Explain. For what types of parts would such a machine be useful? Give a preliminary sketch for such a machine.

26.32 Same as Problem 26.31 but combining processes described in (a) Chapters 13–16, (b) Chapters 22 and 23, and (c) Chapters 25 and 26. Give a preliminary sketch of a machine for each of the three groups. How would you convince a prospective customer of the merits of such machines?

PART V

Joining Processes and Equipment

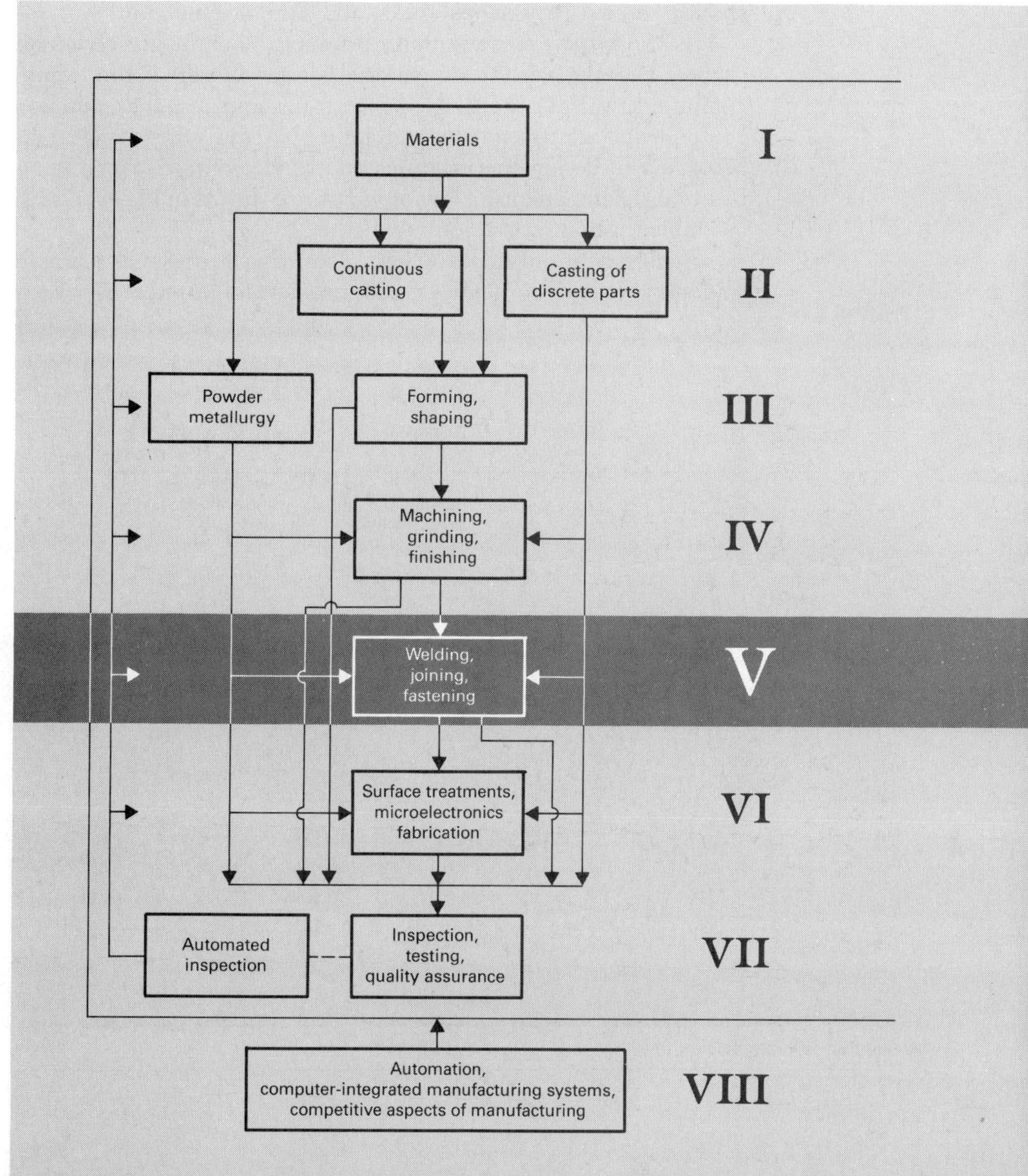
Materials
I
Continuous casting
Casting of discrete parts
II
Powder metallurgy
Forming, shaping
III
Machining, grinding, finishing
IV
Welding, joining, fastening
V
Surface treatments, microelectronics fabrication
VI
Automated inspection
Inspection, testing, quality assurance
VII
Automation, computer-integrated manufacturing systems, competitive aspects of manufacturing
VIII

• INTRODUCTION •

When we inspect the vast numbers of products around us, we soon realize that almost all of them are assemblages of components that were manufactured as individual parts. Even relatively simple products consist of at least two different parts joined by various means. Some kitchen knives, for example, have wooden handles that are attached to the knife blade with metal fasteners. Cooking pots and pans have plastic or wooden handles and knobs that are attached by various means. The eraser of an ordinary pencil is attached with a brass sleeve. However, some products are made of only one component: bolts, nails, steel balls for bearings, staples, screws, paper clips, forks, and similar items.

As you inspect more complex products, note the greater number of parts, their shapes, and the variety of materials used. Observe, for example, motorcycles, computers, washing machines, power tools, and airplanes, and how their numerous components are assembled and joined so that they can function reliably and according to design specifications and service requirements. A typical automobile has 15,000 components, a few of which are shown in Fig. V.1, all of which must be assembled, using several joining methods.

Joining is an all inclusive term, covering processes such as welding, brazing, soldering, adhesive bonding, and mechanical joining. These processes are an

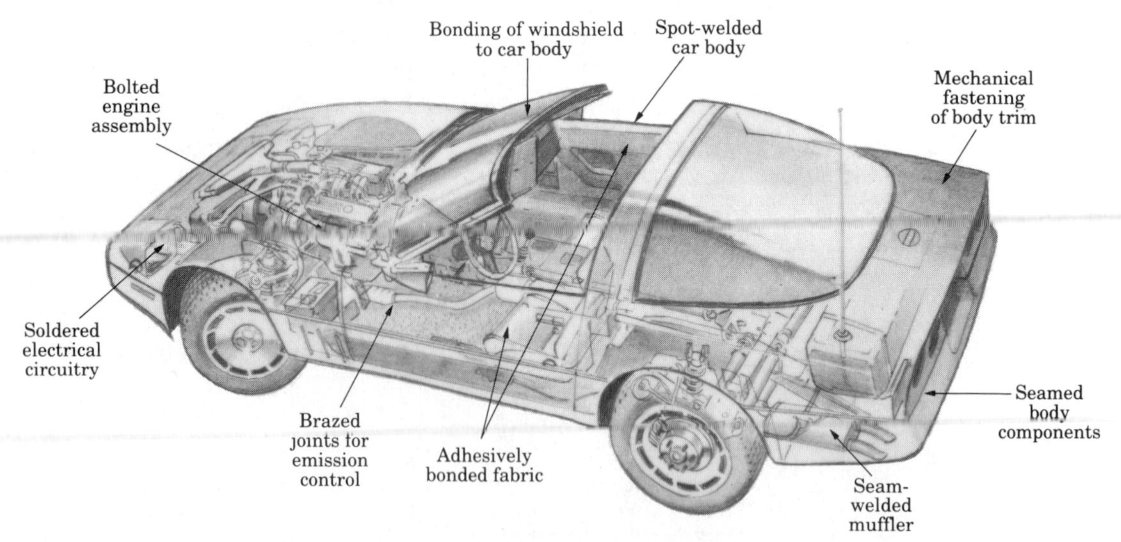

FIGURE V.1
Welded parts in a typical automobile.

important and necessary aspect of manufacturing operations for the following reasons:

- The product is impossible to manufacture as a single piece. Consider, for example, the tubular part shown in Fig. V.2. Assume that each of the arms of this product is 5 m (15 ft) long, the tubes are 100 mm (4 in.) in diameter, and their wall thickness is 1 mm (0.04 in.). Which of the processes that we have described thus far can be used to produce such a part? After reviewing all those manufacturing processes, you would soon conclude that manufacturing this part in one piece would be impossible.
- The product—such as a cooking pot with a handle—is easier and more economical to manufacture as individual components, which are then assembled.
- Products such as automobile engines, hair dryers, printers, and soldered-joint electronic devices may have to be taken apart for repair or maintenance during their service lives.
- Different properties may be desirable for functional purposes of the product. Surfaces subjected to friction and wear, or corrosion and environmental attack, generally require characteristics different from those of the component's bulk. Examples are carbide cutting tips brazed to the shank of a drill and brake shoes bonded to a metal backing.
- Transporting the product in individual components and assembling them at home or at the customer's plant may be easier and less costly. Some bicycles and toys, most machine tools, and mechanical or hydraulic presses are assembled after the components have been transported to the appropriate site.

As you can see in Fig. V.3 on the following page, numerous joining processes have been developed. As we describe in this Part, the choice of a particular process depends on factors such as joint design and its application, the materials to be joined, and the size, shape, and thickness of the components. Other factors are workplace location, equipment cost, number of parts to be joined, and operator skill required.

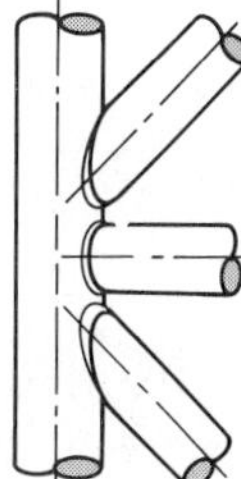

FIGURE V.2
A tubular part fabricated by joining. See text for details.

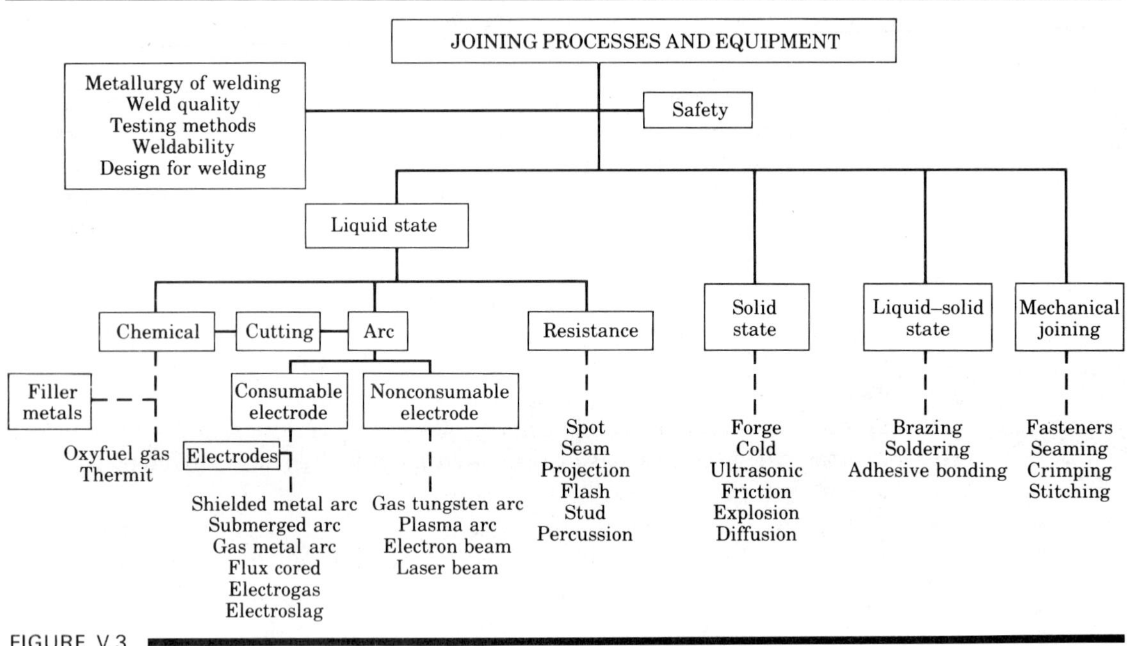

FIGURE V.3
Outline of topics described in Part V.

27

Oxyfuel Gas, Arc, and Resistance Welding and Cutting Processes

27.1 Introduction

The welding processes we describe in this chapter involve partial melting and fusion of the joint between two members. A variety of typical joints is shown in Fig. 27.1, in which the broken lines represent the workpiece shapes before welding. We define *fusion* as melting together and coalescing materials by means of heat. The thermal energy required for these welding operations is usually supplied by chemical and electrical means. Filler metals, which are metals added to the weld area during welding of the joint, may or may not be used. Fusion welds made without the addition of filler metals are known as *autogenous* welds.

In this chapter we describe three major classes of welding processes: oxyfuel gas, arc, and resistance welding. We present the basic principles of each welding process and the equipment used, their relative advantages, limitations, and costs, and economic considerations in their selection. Various other welding processes and equipment are similarly described in Chapter 28. Heating the weld joint to a sufficiently high temperature for fusion to occur involves important metallurgical

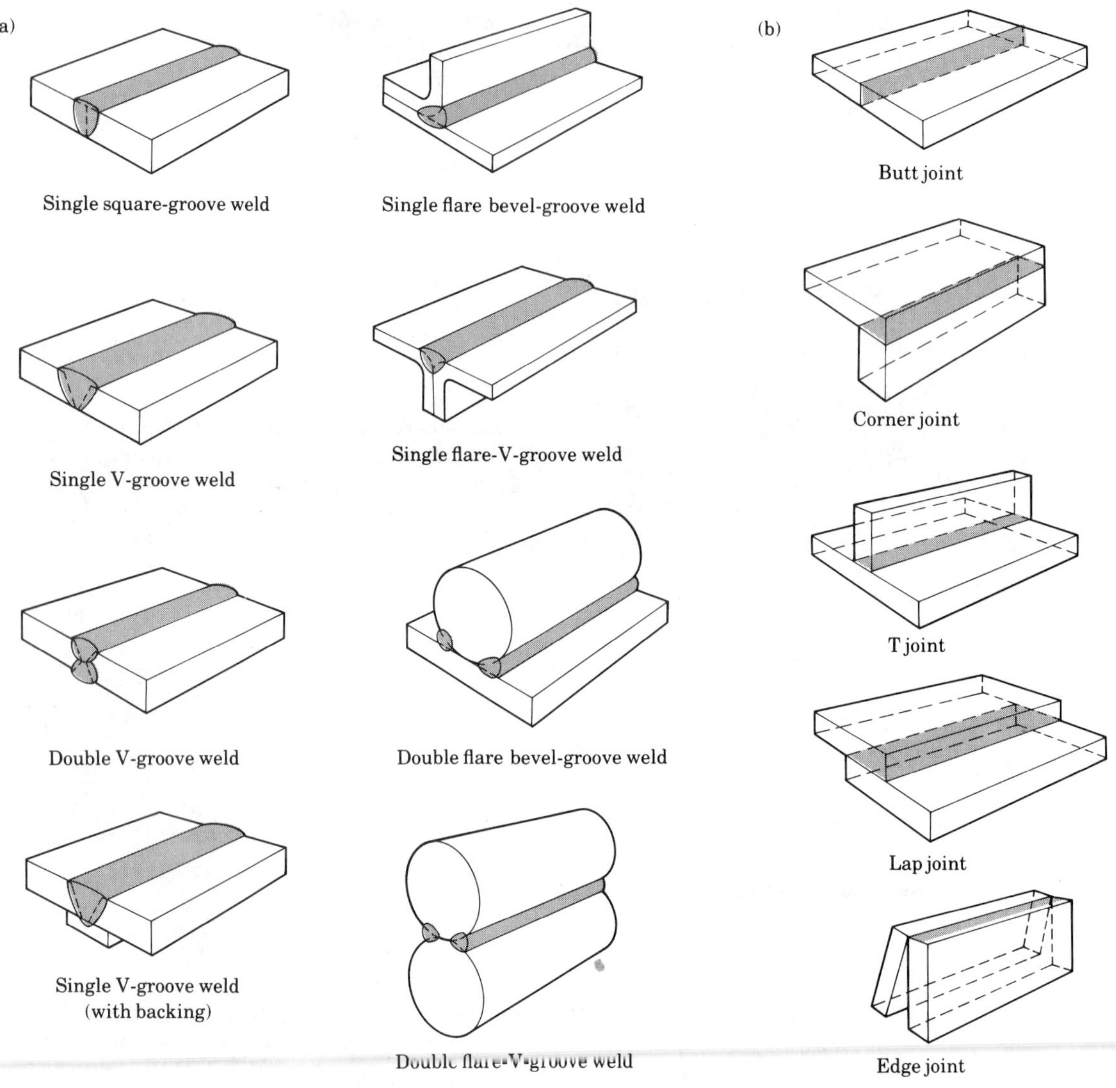

FIGURE 27.1 Types of (a) grooves and (b) joints commonly used in fusion welds. The broken lines represent the shape of the workpieces before welding. The shaded volumes are the welds.

and physical changes. We describe this subject and the weldability of metals in Chapter 29. We summarize joining processes in Table 27.1 as a guide for the capabilities of these processes, as well as to serve as an outline for the topics that we cover in these chapters.

TABLE 27.1
OVERVIEW OF COMMERCIAL JOINING PROCESSES*

MATERIAL	THICKNESS	JOINING PROCESS																						
																	B							
		SMAW	SAW	GMAW	FCAW	GTAW	PAW	ESW	EGW	RW	FW	OFW	DFW	FRW	EBW	LBW	TB	FB	IB	RB	DB	IRB	DFB	S
Carbon steel	S	X	X	X		X				X	X	X			X	X	X	X	X	X	X	X	X	X
	I	X	X	X	X	X				X	X	X		X	X	X	X	X	X	X	X		X	X
	M	X	X	X	X					X	X	X		X	X	X	X	X	X				X	
	T	X	X	X	X			X	X		X	X		X	X			X					X	
Low-alloy steel	S	X	X	X		X				X	X	X	X		X	X	X	X	X	X	X	X	X	X
	I	X	X	X	X	X				X	X		X	X	X	X	X	X	X				X	X
	M	X	X	X	X						X		X	X	X	X	X	X	X				X	
	T	X	X	X	X			X		X	X		X	X	X		X					X		
Stainless steel	S	X	X	X		X	X			X	X	X	X		X	X	X	X	X	X	X	X	X	X
	I	X	X	X	X	X	X			X	X		X	X	X	X	X	X	X				X	X
	M	X	X	X	X		X				X		X	X	X	X	X	X	X				X	
	T	X	X	X	X			X			X		X	X	X			X					X	
Cast iron	I	X										X					X	X	X				X	X
	M	X	X	X	X							X					X	X	X				X	X
	T	X	X	X	X							X						X					X	
Nickel and alloys	S	X		X		X	X			X	X	X			X	X	X	X	X	X	X	X	X	X
	I	X	X	X		X	X			X	X			X	X	X	X	X	X				X	X
	M	X	X	X			X				X			X	X	X	X	X					X	
	T	X		X				X			X			X	X			X					X	
Aluminum and alloys	S	X		X		X	X			X	X	X	X	X	X	X	X	X	X	X	X	X	X	X
	I	X		X		X				X	X		X	X	X	X	X	X			X		X	X
	M	X		X		X					X			X	X		X	X			X		X	
	T	X		X				X	X		X				X			X					X	
Titanium and alloys	S			X		X	X			X	X		X		X	X		X	X			X	X	
	I			X		X	X				X		X	X	X	X		X					X	
	M			X		X	X				X		X	X	X	X		X					X	
	T			X							X		X		X		X	X					X	

(continued)

TABLE 27.1
OVERVIEW OF COMMERCIAL JOINING PROCESSES* *(continued)*

MATERIAL	THICKNESS	JOINING PROCESS																						
																	B							
		SMAW	SAW	GMAW	FCAW	GTAW	PAW	ESW	EGW	RW	FW	OFW	DFW	FRW	EBW	LBW	TB	FB	IB	RB	DB	IRB	DFB	S
Copper and alloys	S			X		X	X				X				X		X	X	X	X			X	X
	I			X			X				X			X	X		X	X		X			X	X
	M			X							X			X	X		X	X					X	
	T			X							X				X			X					X	
Magnesium and alloys	S			X		X				X					X	X	X	X			X		X	
	I			X		X				X	X			X	X	X	X	X			X		X	
	M			X							X			X	X	X		X					X	
	T			X							X				X									
Refractory alloys	S			X		X	X			X	X				X		X	X	X	X		X	X	
	I			X			X				X				X		X	X					X	
	M									X	X													
	T																							

* This table presented as a general survey only. In selecting processes to be used with specific alloys, the reader should refer to other appropriate sources of information. *Source:* Courtesy of the American Welding Society.

LEGEND

PROCESS CODE

SMAW—Shielded Metal-Arc Welding
SAW—Submerged Arc Welding
GMAW—Gas Metal-Arc Welding
FCAW—Flux-Cored Arc Welding
GTAW—Gas Tungsten-Arc Welding
PAW—Plasma Arc Welding
ESW—Electroslag Welding
EGW—Electrogas Welding
RW—Resistance Welding
FW—Flash Welding
OFW—Oxyfuel Gas Welding
DFW—Diffusion Welding

FRW—Friction Welding
EBW—Electron Beam Welding
LBW—Laser Beam Welding
B—Brazing
  TB—Torch Brazing
  FB—Furnace Brazing
  IB—Induction Brazing
  RB—Resistance Brazing
  DB—Dip Brazing
  IRB—Infrared Brazing
  DFB—Diffusion Brazing
S—Soldering

THICKNESS

S—Sheet: up to 3 mm ($\frac{1}{8}$ in.)
I—Intermediate: 3 to 6 mm ($\frac{1}{8}$ to $\frac{1}{4}$ in.)
M—Medium: 6 to 19 mm ($\frac{1}{4}$ to $\frac{3}{4}$ in.)
T—Thick: 19 mm ($\frac{3}{4}$ in.) and up

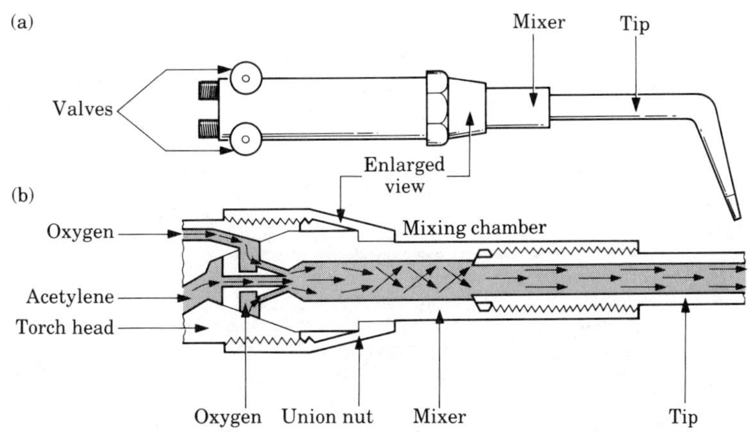

FIGURE 27.2
(a) General view and (b) cross-section of a torch used in oxyacetylene welding. The acetylene valve is opened first, the gas is lit with a spark lighter or a pilot light, and then the oxygen valve is opened and the flame adjusted.

27.2 Oxyfuel Gas Welding

Oxyfuel gas welding (OFW) is a general term used to describe any welding process that uses a *fuel gas* combined with oxygen to produce a flame. This flame is used as the source of heat to melt the metals at the joint. The most common gas welding process uses *acetylene fuel*, and is known as *oxyacetylene welding*. Developed in the early 1900s, this process utilizes the heat generated by the combustion of acetylene gas (C_2H_2) in a mixture with oxygen in a **torch** (Fig. 27.2).

The heat is generated in accordance with the following chemical reactions. The primary combustion process, which occurs in the inner core of the flame (Fig. 27.3), is

$$C_2H_2 + O_2 \rightarrow 2CO + H_2 + \text{heat}. \tag{27.1}$$

This reaction dissociates the acetylene into carbon monoxide and hydrogen and produces about one-third of the total heat generated in the flame. The second reaction is

$$2CO + H_2 + 1.5O_2 \rightarrow 2CO_2 + H_2O + \text{heat}, \tag{27.2}$$

which results in burning of the hydrogen and combustion of the carbon monoxide, producing about two-thirds of the total heat. The temperatures developed in the flame as a result of these reactions can reach 3300 °C (6000 °F). The reaction of hydrogen with oxygen produces water vapor.

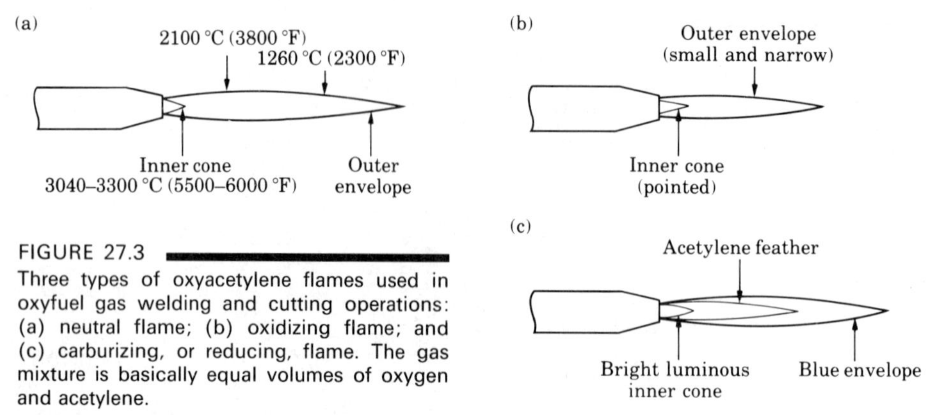

FIGURE 27.3 Three types of oxyacetylene flames used in oxyfuel gas welding and cutting operations: (a) neutral flame; (b) oxidizing flame; and (c) carburizing, or reducing, flame. The gas mixture is basically equal volumes of oxygen and acetylene.

27.2.1 Types of flames

The proportions of acetylene and oxygen in the gas mixture are an important factor in oxyfuel gas welding. At a ratio of 1 : 1, that is, when there is no excess oxygen, it is considered to be a **neutral flame**. With a greater oxygen supply, it becomes an **oxidizing flame**. This flame is harmful, especially for steels, because it oxidizes the steel. Only in copper and copper-base alloys is an oxidizing flame desirable because a thin protective layer of slag forms over the molten metal. If the supply of oxygen is lowered, it becomes a *reducing* or **carburizing flame**. The temperature of a reducing, or excess-acetylene, flame is lower. Hence it is suitable for applications requiring low heat, such as brazing, soldering, and flame hardening.

Other fuel gases such as hydrogen and methylacetylene propadiene can be used in oxyfuel gas welding. However, the temperatures developed are low, and hence they are used for welding metals with low melting points, such as lead, and parts that are thin and small. The flame with hydrogen gas is colorless, making it difficult to adjust the flame by eyesight. Other gases, such as natural gases, propane, and butane, are not suitable for oxyfuel welding because of the low heat output or because the flame is oxidizing.

27.2.2 Filler metals

Filler metals are used to supply additional material to the weld zone during welding. They are available as rods or wire, and are made of metals similar to those to be welded. These consumable **filler rods** may be bare, or they may be coated with *flux*. The purpose of the flux is to retard oxidation of the surfaces of the parts being welded, by generating a gaseous shield around the weld zone. The flux also helps dissolve and remove oxides and other substances from the workpiece, resulting in a stronger joint. The slag developed protects the molten puddle of metal against oxidation as it cools.

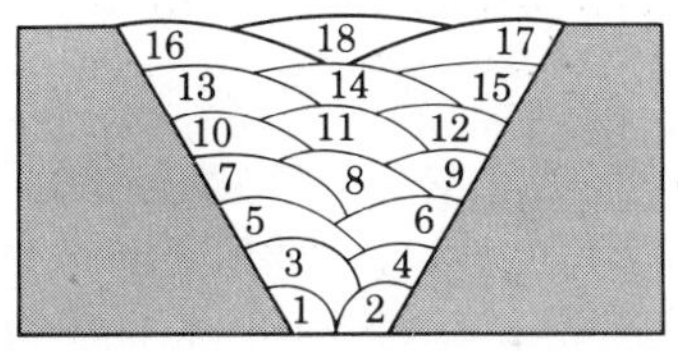

FIGURE 27.4 A weld zone showing the buildup sequence of individual weld beads in deep welds.

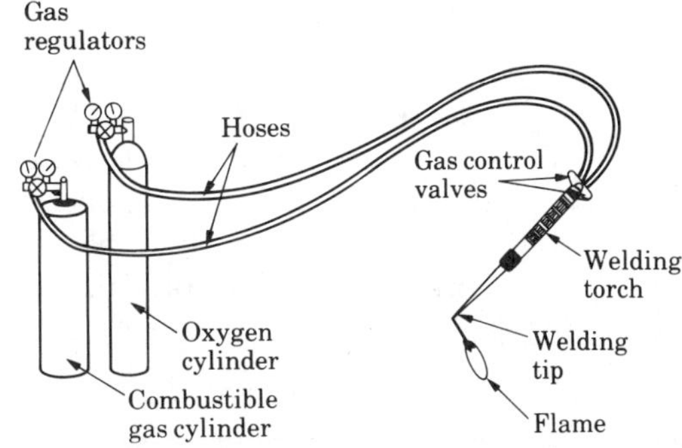

FIGURE 27.5 Basic equipment used in oxyfuel gas welding. Other equipment includes safety shields, goggles, gloves, and protective clothing. To ensure correct connections, all threads on acetylene fittings are left-handed, and those for oxygen are right-handed. Oxygen regulators are usually painted green, and acetylene regulators red.

27.2.3 Welding practice and equipment

Oxyfuel gas welding can be used with most ferrous and nonferrous metals for any thickness of workpiece, but the relatively low heat input limits the process economically to less than 6 mm (0.25 in.). A variety of joints can be produced by this method. The details of welding procedures are given in various texts, so we only summarize the basic steps as follows:

1. Prepare the edges to be joined, and maintain their proper position using suitable clamps and fixtures.
2. Open the acetylene valve and ignite the gas at the tip of the torch. Open the oxygen valve and adjust the flame for the particular operation.
3. Hold the torch at about 45° from the plane of the workpiece, with the inner flame near the workpiece and the filler rod at about 30°–40°.
4. Touch the filler rod to the joint and control its movement along the joint length by observing the rate of melting and filling of the joint.

Small joints may consist of a single weld bead, as shown in Fig. 27.1. Deep V-groove joints are made in multiple passes (Fig. 27.4). Cleaning the surface of each weld bead prior to depositing a second layer is important for joint strength and avoiding defects (see Chapter 29). Hand or power wire brushes may be used for this purpose.

The equipment for oxyfuel gas welding basically consists of a welding torch, which is available in various sizes and shapes, connected by hoses to high-pressure gas cylinders and equipped with pressure gauges and regulators (Fig. 27.5). The use of safety equipment such as goggles with shaded lenses, face shields, gloves, and protective clothing is important. Proper connection of hoses to the cylinders is also an important factor in safety. Oxygen and acetylene cylinders have different threads, so hoses cannot be connected to the wrong cylinders. Gas cylinders should be anchored securely and should not be dropped or mishandled.

Process capabilities. The low cost of the equipment, usually less than $500 for smaller units, is an attractive feature of oxyfuel gas welding. Although it can be mechanized, this welding operation is essentially manual and hence slow. It has the advantages of being portable, versatile, and economical for low-quantity and repair work. Proper operator training and skill are essential.

27.3 Arc-Welding Processes: Consumable Electrode

In **arc welding**, developed in the mid-1800s, the heat required is obtained through electrical energy. Using either a *consumable* or *nonconsumable electrode* (rod or wire), an arc is produced between the tip of the electrode and the workpiece to be welded, using ac or dc power supplies. This arc produces temperatures in the range of 5000–30,000 °C (9000–54,000 °F), which are much higher than those developed in oxyfuel gas welding. The arc also produces radiation, which may dissipate as much as 20 percent of the total energy. Arc welding includes various welding processes, which we describe below.

27.3.1 Shielded metal-arc welding

Shielded metal-arc welding (SMAW) is one of the oldest, simplest, and most versatile joining processes. Currently, about 50 percent of all industrial and maintenance welding is performed by this process. The electric arc is generated by touching the tip of a coated electrode against the workpiece and then withdrawing it quickly to a distance sufficient to maintain the arc (Fig. 27.6). The electrodes are in the shape of thin, long sticks (see Section 27.4), so this process is also known as stick welding.

The heat generated melts a portion of the tip of the electrode, its coating, and the base metal in the immediate area of the arc. A weld forms after the molten metal—a mixture of the base metal (workpiece), electrode metal, and substances from the coating on the electrode—solidifies in the weld area.

A bare section at the end of the electrode is clamped in an electrode holder. The holder is connected to one terminal of the power source, while the other terminal is connected to the workpiece being welded (Fig. 27.7). The current usually ranges between 50 A and 300 A, with power requirements generally less than 10 kW. The current may be ac or dc, and the polarity of the electrode may be positive (reverse polarity) or negative (straight polarity). The choice depends on the type of electrode, type of metals to be welded, arc atmosphere, and depth of the heated

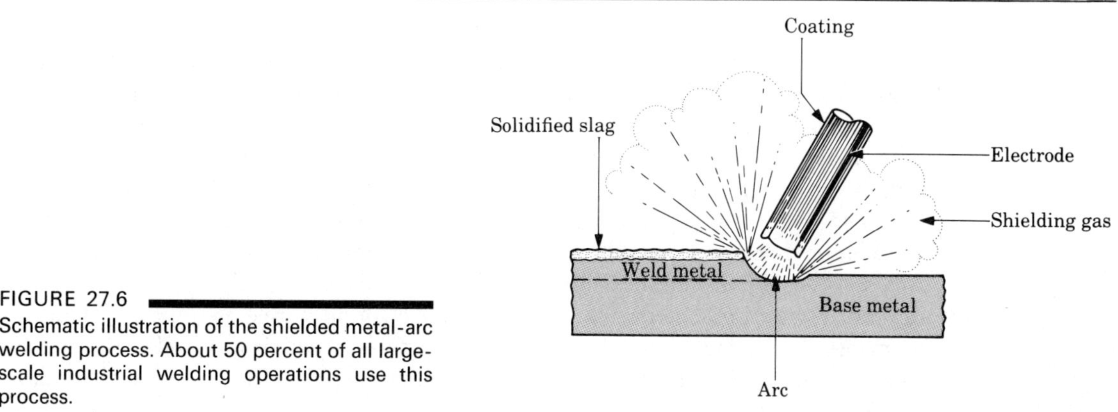

FIGURE 27.6 Schematic illustration of the shielded metal-arc welding process. About 50 percent of all large-scale industrial welding operations use this process.

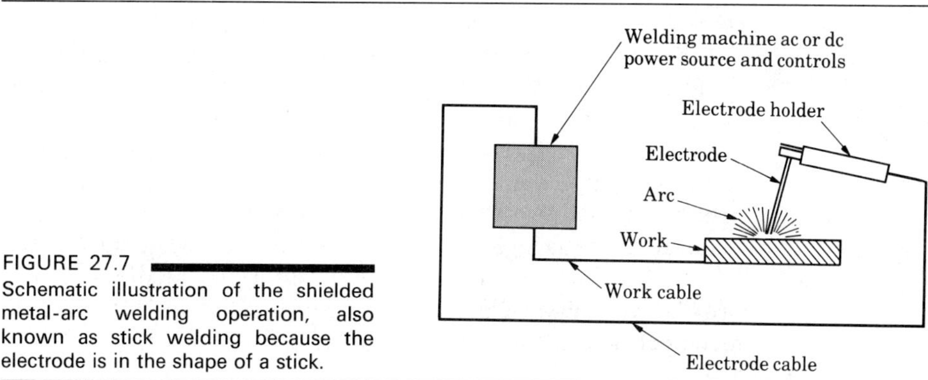

FIGURE 27.7 Schematic illustration of the shielded metal-arc welding operation, also known as stick welding because the electrode is in the shape of a stick.

zone. Too low a current causes incomplete fusion, and too high a current can damage the electrode coating and reduce its effectiveness.

Process capabilities. The SMAW process has the advantage of being relatively simple and versatile, requiring a relatively small variety of electrodes. The equipment consists of a power supply, power cables, and electrode holder, and the total cost of equipment is typically below $1500. The use of safety equipment, similar to that used with oxyfuel gas welding, is essential.

This process is commonly used in general construction, shipbuilding, and pipelines, as well as for maintenance work, since the equipment is portable and can be easily maintained. It is especially useful for work in remote areas where portable fuel-powered generators can be used as the power supply. The SMAW process is best suited for workpiece thicknesses of 3–19 mm (0.12–0.75 in.), although this range can be easily extended using special techniques and highly skilled operators. This process requires that slag be cleaned after each weld bead. Thus labor costs are high, as are material costs.

27.3.2 Submerged arc welding

In *submerged arc welding* (SAW), the weld arc is shielded by granular flux, consisting of lime, silica, manganese oxide, calcium fluoride, and other elements. The flux is fed into the weld zone by gravity flow through a nozzle (Fig. 27.8). The thick layer of flux completely covers the molten metal and prevents spatter and sparks—and without the intense radiation and fumes of the SMAW process. The flux also acts as a thermal insulator, allowing deep penetration of heat into the workpiece. The welder must wear gloves, but other than tinted safety glasses, face shields generally are unnecessary.

The consumable electrode is a coil of bare round wire 1.5–10 mm ($\frac{1}{16}$–$\frac{3}{8}$ in.) in diameter, and is fed automatically through a tube (welding gun). Electric currents usually range between 300 A and 2000 A, from either ac or dc power sources, at up to 440 V.

Process capabilities. Because the flux is fed by gravity, the SAW process is somewhat limited to welds in a flat or horizontal position with backup piece. Circular welds can be made on pipes, provided that they are rotated during welding. As Fig. 27.8 shows, the unfused flux can be recovered, treated, and reused.

Developed in the 1940s, the SAW process can be automated for greater economy. Total cost for a welding system usually ranges from $2000 to $10,000, but it can be considerably higher for larger systems with multiple electrodes. This process is used to weld a variety of carbon and alloy steel and stainless steel sheet or plate, at speeds as high as 85 mm/s (17 ft/min). The quality of the weld is very high, with good toughness, ductility, and uniformity of properties. The SAW process provides very high welding productivity, depositing 4–10 times the amount of weld metal per hour as the SMAW process.

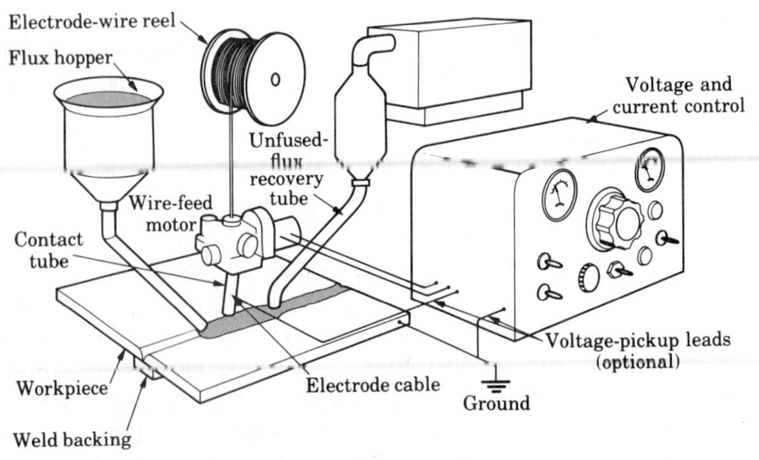

FIGURE 27.8 Schematic illustration of the submerged-arc welding process and equipment. Unfused flux is recovered and reused. *Source:* American Welding Society.

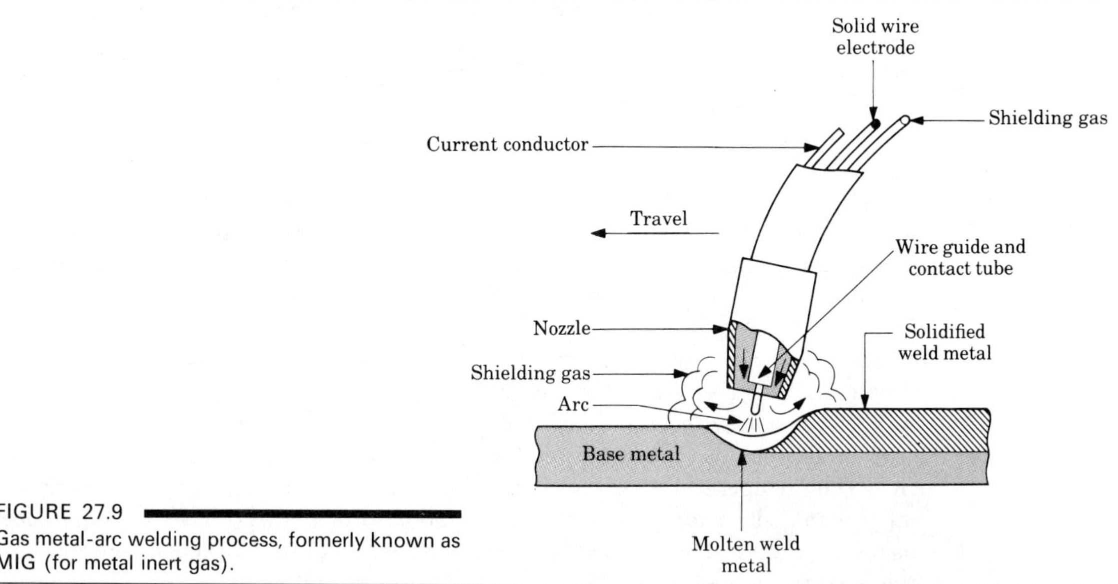

FIGURE 27.9 Gas metal-arc welding process, formerly known as MIG (for metal inert gas).

27.3.3 Gas metal-arc welding

In *gas metal-arc welding* (GMAW), the weld area is shielded by an external source, such as argon, helium, carbon dioxide, or various other gas mixtures (Fig. 27.9). The consumable bare wire is fed automatically through a nozzle into the weld arc (Fig. 27.10). In addition to the use of inert shielding gases, deoxidizers are usually present in the electrode metal itself, in order to prevent oxidation of the molten weld puddle. The welds made by this process are thus relatively free of slag, and hence multiple weld layers can be deposited at the joint without the necessity for intermediate cleaning of slag.

Metal can be transferred three ways in the GMAW process: spray, globular, and short circuiting. In *spray transfer*, small droplets of molten metal from the

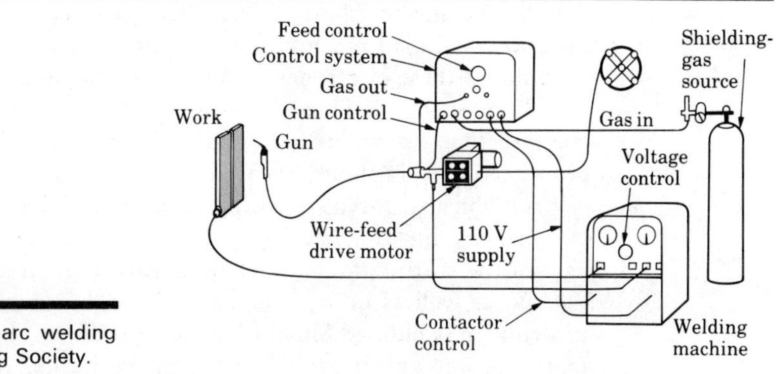

FIGURE 27.10 Basic equipment used in gas metal-arc welding operations. *Source:* American Welding Society.

electrode are transferred to the weld area at rates of several hundred droplets per second. The transfer is spatter-free and very stable. High dc current and voltages and large-diameter electrodes are used, with argon or argon-rich gas mixtures used as the shielding gas. The average current required in this process can be reduced by pulsed arcs, which are high-amplitude pulses superimposed over a low, steady current.

In *globular transfer*, carbon-dioxide rich gases are utilized, and globules propelled by the forces of the electric arc transfer the metal, resulting in considerable spatter. High welding currents are used, with greater weld penetration and welding speed than in spray transfer. Heavier sections are commonly joined by this method.

In *short circuiting*, the metal is transferred in individual droplets, at rates of more than 50 per second, as the electrode tip touches the molten weld metal and short circuits. Low currents and voltages are utilized, with carbon-dioxide rich gases and electrodes made of small-diameter wire. The power required is about 2 kW. The temperatures involved are relatively low. Thus this method is suitable only for thin sheets and sections (less than 6 mm; 0.25 in.); otherwise, incomplete fusion may occur. This process is very easy to use and may be the most popular for welding ferrous metals in thin sections. However, pulse-arc systems are gaining wide usage for thin ferrous and nonferrous metals.

Process capabilities. The GMAW process was developed in the 1950s and was formerly called *metal inert-gas* (MIG) welding. It is suitable for welding a variety of ferrous and nonferrous metals and is used extensively in the metal-fabrication industry. Because of the relatively simple nature of the process, training operators is easy. This process is rapid, versatile, and economical; welding productivity is double that of the SMAW process. The cost of equipment usually ranges from $1000 to $3000. The GMAW process can easily be automated and lends itself readily to flexible manufacturing systems and robotics.

27.3.4 Flux-cored arc welding

The *flux-cored arc welding* process (FCAW, Fig. 27.11) is similar to gas metal-arc welding, with the exception that the electrode is tubular in shape and is filled with flux (hence the term flux cored). Cored electrodes produce a more stable arc, improve weld contour, and improve the mechanical properties of the weld metal. The flux in these electrodes is much more flexible than the brittle coating used on SMAW electrodes. Thus the tubular electrode can be provided in long coiled lengths. The electrodes are usually 0.5–4 mm (0.020–0.15 in.) in diameter. The power required is about 20 kW.

Self-shielded cored electrodes are also available. These electrodes do not require external gas shielding because they contain emissive fluxes that shield the weld area against the surrounding atmosphere. Advances in manufacturing of electrodes for FCAW, as well as in the chemistry of the flux, have made this process the fastest growing in welding. Small-diameter electrodes have made welding of thinner materials not only possible but often desirable. Also, small-diameter electrodes

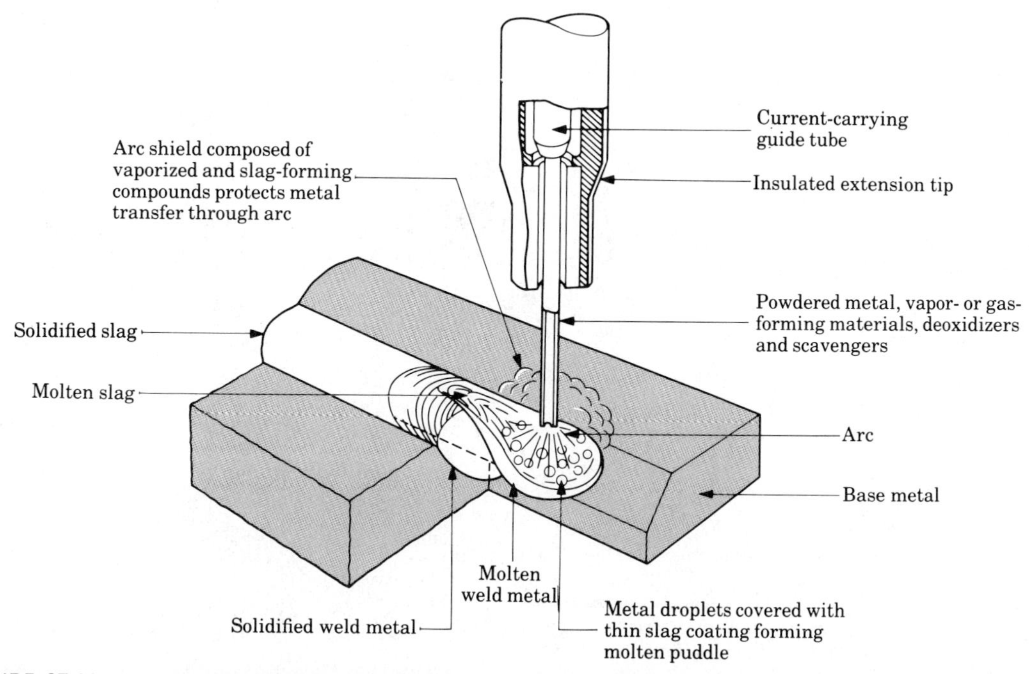

FIGURE 27.11 Schematic illustration of the flux-cored arc welding process. This operation is similar to gas metal-arc welding shown in Fig. 27.9.

make it relatively easy to weld parts out of position, and the flux chemistry enables welding of many base metals.

Process capabilities. The flux-cored arc welding process combines the versatility of SMAW with the continuous and automatic electrode-feeding feature of GMAW. It is economical and is used for welding a variety of joints, mainly with steels, stainless steels, and nickel alloys. The higher weld metal deposition rate of the FCAW process, over that of GMAW, has led to its use in joining sections of all thicknesses. Recent development of tubular electrodes with very small diameters has extended the use of this process to smaller workpiece section sizes.

A major advantage of FCAW is the ease with which specific weld metal chemistries can be developed. By adding alloys to the flux core, virtually any alloy composition can be developed. This process is easy to automate and is readily adaptable to flexible manufacturing systems and robotics. The cost of equipment is generally in the range of $1000 and $3000.

27.3.5 Electrogas welding

Electrogas welding (EGW) is used primarily for welding the edges of sections vertically in one pass with the pieces placed edge to edge (butt) (Fig. 27.12). It is

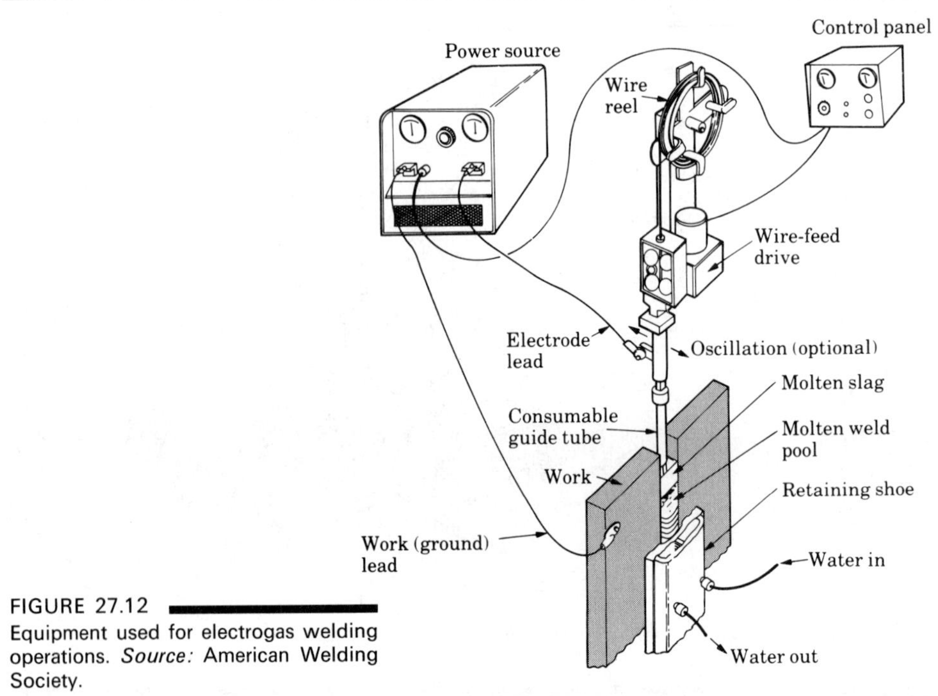

FIGURE 27.12 Equipment used for electrogas welding operations. *Source:* American Welding Society.

classified as a machine-welding process because it requires special equipment. The weld metal is deposited into a weld cavity between the two pieces to be joined. The space is enclosed by two water-cooled copper dams (shoes) to prevent the molten slag from running off. Mechanical drives move the shoes upward. Circumferential welds are also possible, with the workpiece rotating.

Single or multiple electrodes are fed through a guide tube and a continuous arc is maintained, using flux-cored electrodes at up to 750 A, or solid electrodes at 400 A. Power requirements are about 20 kW. Shielding is by inert gas, such as carbon dioxide, argon, or helium, depending on the type of material being welded. The gas may be provided from an external source, or it may be produced from a flux-cored electrode, or both.

Process capabilities. The equipment for electrogas welding is reliable, and training operators is relatively simple. Weld thickness ranges from 12 mm to 75 mm (0.5 in. to 3 in.) on steels, titanium, and aluminum alloys. The cost of machines typically ranges from $15,000 to $25,000, although portable machines with less power cost as little as $5000.

27.3.6 Electroslag welding

Developed in the 1950s, *electroslag welding* (ESW) and its applications are similar to electrogas welding. The main difference is that the arc is started between the electrode tip and the bottom of the part to be welded. Flux is added and melted by the heat of the arc. After the molten slag reaches the tip of the electrode, the arc is extinguished. Energy is supplied continuously through the electrical resistance of the molten slag. Thus because the arc is extinguished, ESW is not strictly an arc welding process. Single or multiple solid as well as flux-cored electrodes may be used. The guide may be nonconsumable (conventional method) or consumable.

Process capabilities. Electroslag welding is capable of welding plates with thicknesses ranging from 50 mm to more than 900 mm (2 in. to more than 36 in.). Welding is done in one pass. The current required is about 600 A at 40–50 V, although higher currents are used for thick plates. Travel speed of the weld is 0.2–0.6 mm/s (0.5–1.5 in./min). The weld quality is good and the process is used for heavy structural steel sections, such as heavy machinery and nuclear-reactor vessels. The cost of a typical ESW system ranges from $15,000 to $25,000, and higher for multiple-electrode units.

27.4 Electrodes

Electrodes for the consumable arc welding processes described are classified according to the strength of the deposited weld metal, current (ac or $\pm$ dc), and the type of coating. Electrodes are identified by numbers and letters (Table 27.2) or by color code, particularly if they are too small to imprint with identification. Typical coated electrode dimensions are 150–460 mm (6–18 in.) long and 1.5–8 mm ($\frac{1}{16}$–$\frac{5}{16}$ in.) in diameter. The thinner the sections to be welded and the lower the current required, the smaller the diameter of the electrode should be.

The specifications for electrodes and filler metals, including tolerances, quality control procedures, and processes, are stated by the American Welding Society (AWS) and the American National Standards Institute (ANSI), and in the Aerospace Materials Specifications (AMS) by the Society of Automotive Engineers (SAE). Among others, the specifications require that the wire diameter not vary more than 0.05 mm (0.002 in.) from nominal size, and that the coatings be concentric with the wire. Electrodes are sold by weight and are available in a wide variety of sizes and specifications.

27.4.1 Electrode coatings

Electrodes are coated with claylike materials that include silicate binders and powdered materials such as oxides, carbonates, fluorides, metal alloys, and cellulose (cotton cellulose and wood flour). The coating, which is brittle and has complex interactions during welding, has the following basic functions:

- Stabilize the arc.

TABLE 27.2
DESIGNATIONS FOR MANUAL ELECTRODES

The prefix "E" designates arc welding electrode.

The first two digits of four-digit numbers and the first three digits of five-digit numbers indicate minimum tensile strength:

E60XX	60,000 psi minimum tensile strength
E70XX	70,000 psi minimum tensile strength
E110XX	110,000 psi minimum tensile strength

The next-to-last digit indicates position:

EXX1X	All positions
EXX2X	Flat position and horizontal fillets

The suffix (Example: EXXXX-A1) indicates the approximate alloy in the weld deposit:

—A1	0.5% Mo
—B1	0.5% Cr, 0.5% Mo
—B2	1.25% Cr, 0.5% Mo
—B3	2.25% Cr, 1% Mo
—B4	2% Cr, 0.5% Mo
—B5	0.5% Cr, 1% Mo
—C1	2.5% Ni
—C2	3.25% Ni
—C3	1% Ni, 0.35% Mo, 0.15% Cr
—D1 and D2	0.25–0.45% Mo, 1.75% Mn
—G	0.5% min. Ni, 0.3% min. Cr, 0.2% min. Mo, 0.1% min. V, 1% min. Mn (only one element required)

- Generate gases to act as a shield against the surrounding atmosphere. The gases produced are carbon dioxide and water vapor, and carbon monoxide and hydrogen in small amounts.
- Control the rate at which the electrode melts.
- Act as a flux to protect the weld against formation of oxides, nitrides, and other inclusions, and with the resulting slag, protect the molten weld pool.
- Add alloying elements to the weld zone to enhance the properties of the weld, including deoxidizers to prevent the weld from becoming brittle.

The deposited electrode coating or slag must be removed after each pass in order to ensure a good weld. A manual or power wire brush can be used for this purpose. Bare electrodes and wire, made of stainless steels and aluminum alloys, are also available. They are used as filler metals in various welding operations.

27.5 Arc-Welding Processes: Nonconsumable Electrode

Unlike the arc-welding processes that use consumable electrodes, which we described in Section 27.3, nonconsumable-electrode processes typically use a tungsten electrode. As one pole of the arc, it generates the heat required for welding. A shielding gas is supplied from an external source. We describe below the advantages, limitations, and typical applications of these processes.

27.5.1 Gas tungsten-arc welding

In *gas tungsten-arc welding* (GTAW), formerly known as TIG welding (for tungsten inert gas), the filler metal is supplied from a filler wire (Fig. 27.13). Because the tungsten electrode is not consumed in this operation, a constant and stable arc gap is maintained at a constant current level. The filler metals are similar to the metals to be welded, and flux is not used. The shielding gas is usually argon or helium, or a mixture of the two. Welding with GTAW may be done without filler metals, as in welding close-fit joints.

The power supply (Fig. 27.14) is either dc at 200 A, or ac at 500 A, depending on the metals to be welded. In general, ac is preferred for aluminum and magnesium

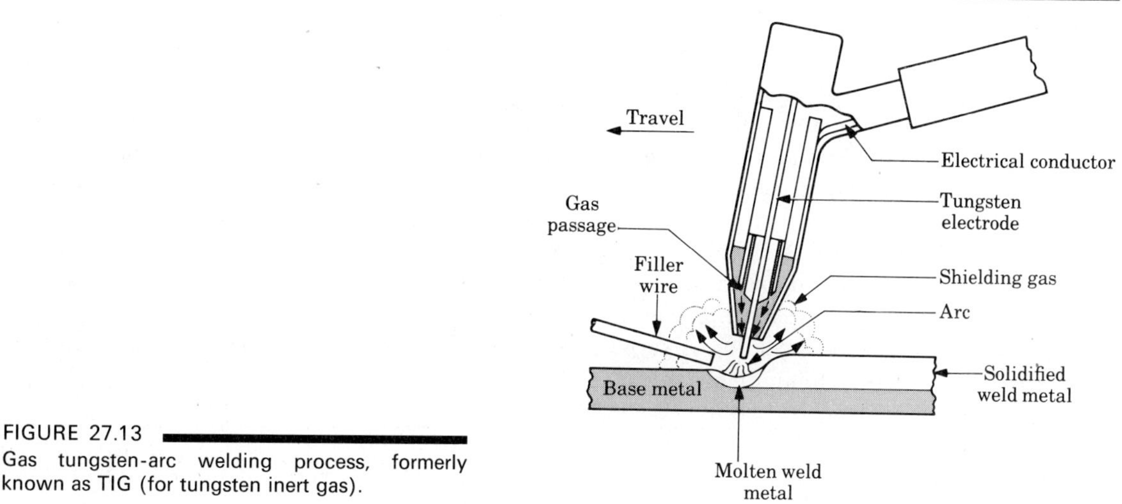

FIGURE 27.13 Gas tungsten-arc welding process, formerly known as TIG (for tungsten inert gas).

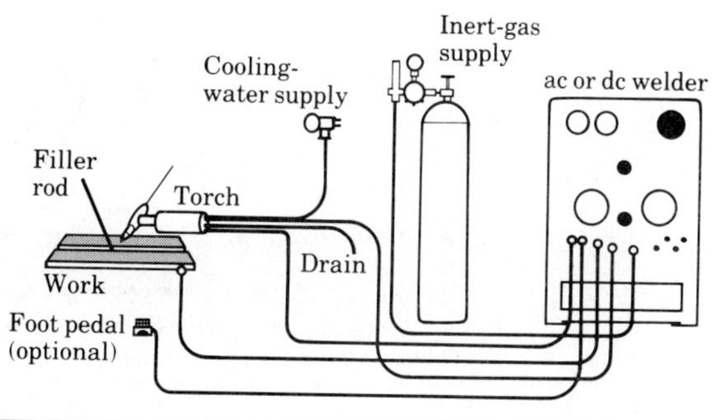

FIGURE 27.14 Equipment for gas tungsten-arc welding operations. *Source:* American Welding Society.

because the cleaning action of ac removes oxides and improves weld quality. Thorium or zirconium may be used in the tungsten electrodes to improve their electron emission characteristics. Power requirements range from 8 kW to 20 kW.

Process capabilities. The GTAW process is used for a wide variety of metals, particularly aluminum, magnesium, titanium, and refractory metals. It is especially suitable for thin metals. The cost of the inert gas makes this process more expensive than SMAW, but it provides welds with very high quality and surface finish. It is used in a variety of critical applications with a wide range of workpiece thicknesses and shapes. The equipment is portable and typically costs from $1000 to $5000.

27.5.2 Atomic hydrogen welding

Atomic hydrogen welding (AHW) uses an arc in a shielding atmosphere of hydrogen. The arc is between two tungsten or carbon electrodes. Thus the workpiece is not part of the electrical circuit, as it is in GTAW. The hydrogen gas also cools the electrodes.

27.5.3 Plasma-arc welding

In *plasma-arc welding* (PAW), developed in the 1960s, a concentrated plasma arc is produced and aimed at the weld area. The arc is stable and reaches temperatures as high as 33,000 °C (60,000 °F). A *plasma* is ionized hot gas, composed of nearly equal numbers of electrons and ions. The plasma is initiated between the tungsten electrode and the orifice, using a low-current pilot arc. Unlike other processes, the plasma arc is concentrated because it is forced through a relatively small orifice. Operating currents are usually below 100 A, but they can be higher for special applications. When a filler metal is used, it is fed into the arc, as in GTAW. Arc and weld-zone shielding is supplied through an outer shielding ring by gases such as argon, helium, or mixtures.

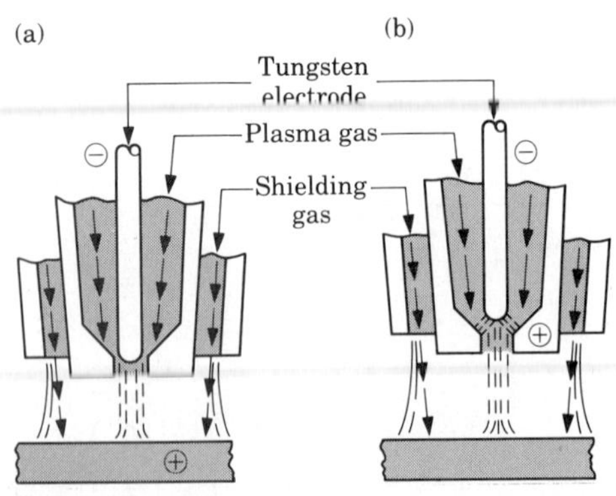

FIGURE 27.15
Two types of plasma-arc welding processes: (a) transferred and (b) nontransferred. Deep and narrow welds are made by this process at high welding speeds.

There are two methods of plasma-arc welding. In the *transferred-arc* method (Fig. 27.15a), the workpiece being welded is part of the electrical circuit. The arc thus transfers from the electrode to the workpiece—hence the term transferred. In the *nontransferred* method (Fig. 27.15b), the arc is between the electrode and the nozzle, and the heat is carried to the workpiece by the plasma gas.

Process capabilities. Compared to other arc-welding processes, plasma-arc welding has greater energy concentration (hence deeper and narrower welds can be made), better arc stability, and higher welding speeds, such as 2–16 mm/s (5–40 in./min). A variety of metals can be welded, with part thicknesses generally less than 6 mm (0.25 in.). The high heat concentration can penetrate completely through the joint (*keyhole* technique), with thicknesses as much as 20 mm (0.75 in.) for some titanium and aluminum alloys.

Plasma-arc welding, rather than the GTAW process, is often used for butt and lap joints because of higher energy concentration, better arc stability, and higher welding speeds. Equipment costs are typically in the range of $3000–$6000. Proper training and skill are essential for operators who use this equipment. Safety considerations include protection against glare, spatter, and noise from the plasma arc.

27.6 Resistance Welding Processes

Resistance welding (RW) covers a number of processes in which the heat required for welding is produced by means of the electrical resistance between the two members to be joined. These processes have major advantages, such as not requiring consumable electrodes, shielding gases, or flux.

The heat generated in resistance welding is given by the general expression

$$H = I^2 R t, \tag{27.3}$$

where

H = heat generated, in joules (watt-seconds);
I = current, in amperes;
R = resistance, in ohms; and
t = time of current flow, in seconds.

The total resistance in these processes, such as in the resistance spot welding shown in Fig. 27.16, is the sum of the following:

a) Resistance of the electrodes.
b) Electrode–workpiece contact resistances.
c) Resistances of the individual parts to be welded.
d) Workpiece–workpiece contact resistances (**faying surfaces**).

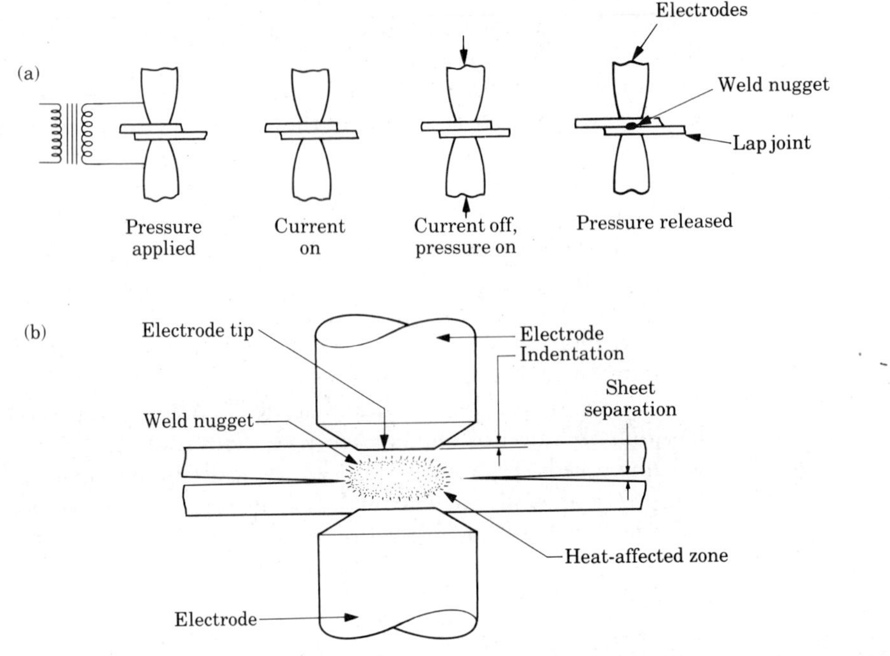

FIGURE 27.16 (a) Sequence in the resistance spot welding process. (b) Cross-section of a spot weld, showing weld nugget and light indentation by the electrode on sheet surfaces. This is one of the most common processes used in sheet-metal fabrication and automotive-body assembly.

For high heat generation at the junction, workpiece–workpiece contact resistances should be kept high, and the rest should be kept low. However, because these resistances are generally low (typically 100 $\mu\Omega$), the current required is high, as you can see in Eq. (27.3). The actual temperature rise at the joint depends on the specific heat and thermal conductivity of the metals to be joined. Thus because they have high thermal conductivity, metals such as aluminum and copper require high heat concentrations. Similar as well as dissimilar metals can be joined by resistance welding.

The workpieces are generally in the secondary circuit of a transformer. The transformer converts high-voltage, low-current power to low-voltage, high-current power. The magnitude of the current in resistance welding operations may be as high as 100,000 A, although the voltage is typically only 0.5–10 V.

Developed in the early 1900s, resistance welding processes require specialized machinery. Many are now operated by programmable computer control. Cost of the total system typically ranges from $20,000 to more than $50,000. The machinery is generally not portable, and the process is more suitable for use in manufacturing plants and machine shops. Operator skill required is minimal, particularly with modern machinery. Safety precautions are similar to those for other welding operations.

There are five basic methods of resistance welding: spot, seam, projection, flash, and upset welding. Lap joints are used in the first three processes, and butt joints in the last two.

27.6.1 Resistance spot welding

In *resistance spot welding* (RSW), the tips of two opposing solid cylindrical electrodes contact the lap joint of two sheet metals, and resistance heating produces a spot weld (Fig. 27.16a). In order to obtain a good bond in the weld nugget, pressure is also applied until the current is turned off. Accurate control and timing of the electric current and pressure are essential in resistance welding. The strength of the bond depends on surface roughness and the cleanliness of the mating surfaces. Thus oil, paint, or thick oxide layers should be removed before welding. The presence of uniform, thin oxide layers and other contaminants is not critical.

The **weld nugget** (Fig. 27.16b) is generally 6–10 mm (0.25–0.375 in.) in diameter. The surface of the weld spot has a slightly discolored indentation. Currents range from 3000 A to 40,000 A, depending on the materials being welded and their thickness.

Process capabilities. Spot welding is the simplest and most commonly used resistance welding process. Welding may be performed by means of single (most common) or multiple electrodes, and the required pressure is supplied through mechanical or pneumatic means. The rocker-arm type spot welding machines (Fig. 27.17) are normally used for smaller parts, with press-type machines being used for

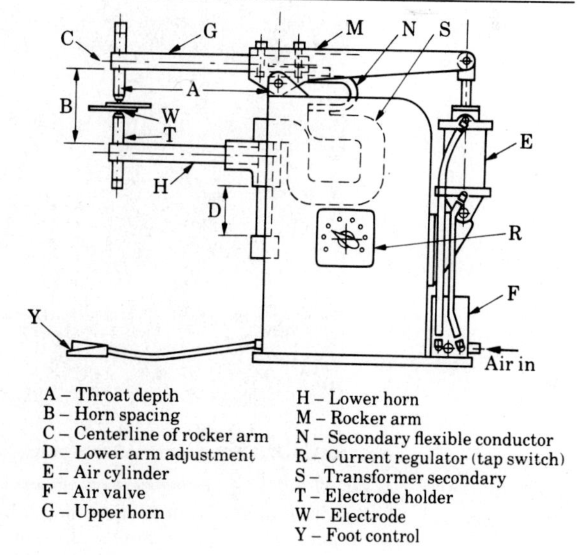

FIGURE 27.17 Schematic illustration of an air-operated rocker-arm spot welding machine. *Source:* American Welding Society.

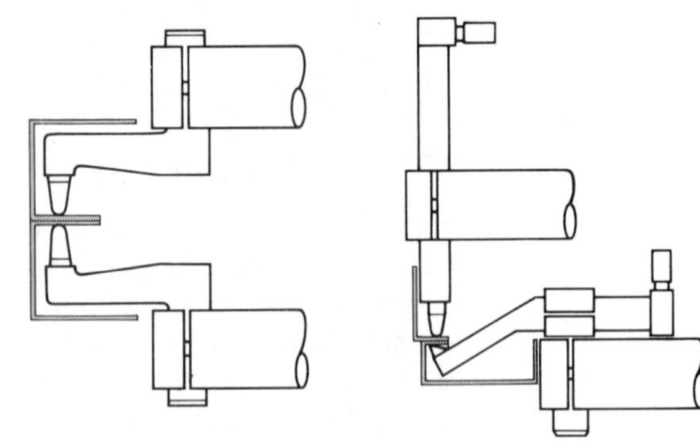

FIGURE 27.18 Types of special electrodes designed for easy access in spot welding operations for complex shapes.

larger workpieces. The shape and surface condition of the electrode tip and accessibility are important factors in spot welding. A variety of electrode shapes are used to spot weld areas that are difficult to reach (Fig. 27.18).

Spot welding is widely used for fabricating sheet metal. Examples range from attaching handles to stainless-steel cookware, to rapid spot welding of automobile bodies with multiple electrodes (Fig. 27.19). An automobile may have as many as 10,000 spot welds. Modern equipment used for spot welding is computer controlled for optimum timing of current and pressure (Fig. 27.20), and the spot-welding guns are manipulated by programmable robots.

FIGURE 27.19 Robots equipped with spot welding guns and operated by computer controls in mass-production line for automotive bodies. *Source:* Courtesy of Cincinnati Milacron, Inc.

FIGURE 27.20
An automated spot welding machine with programmable robot. The welding tip can move in three principal directions. Sheets as large as 2.2 m × 0.55 m (88 in. × 22 in.) can be accommodated in this machine. *Source:* Courtesy of Taylor-Winfield Corporation.

Example: Heat generated in resistance spot welding.

Assume that two 1-mm (0.04-in.) thick steel sheets are being spot welded at a current of 5000 A and current flow time $t = 0.1$ s, and using electrodes 5 mm (0.2 in.) in diameter. Estimate the heat generated and its distribution in the weld zone.

SOLUTION. Let's assume that the effective resistance in this operation is 200 $\mu\Omega$. Then, according to Eq. (27.3),

$$\text{Heat} = (5000)^2(0.0002)(0.1) = 500 \text{ J}.$$

From the information given, we estimate the weld nugget volume to be 30 mm^3 (0.0018 in^3). If we assume the density for steel to be 8000 kg/m^3 (0.008 g/mm^3), the weld nugget has a mass of 0.24 g. Since the heat required to melt 1 g of steel is about 1400 J, the heat required to melt the weld nugget is (1400)(0.24) = 336 J. Consequently, the remaining heat (164 J) is dissipated into the metal surrounding the nugget.

27.6.2 Resistance seam welding

Resistance seam welding (RSEW) is a modification of resistance spot welding, wherein the electrodes are replaced by rotating wheels or rollers (Fig. 27.21). With continuous ac power supply, the electrically conducting rollers produce continuous spot welds whenever the current reaches a sufficiently high level in the ac cycle. These are actually overlapping spot welds and produce a joint that is liquid tight and gas tight (Fig. 27.21b). With intermittent application of current to the rollers, a

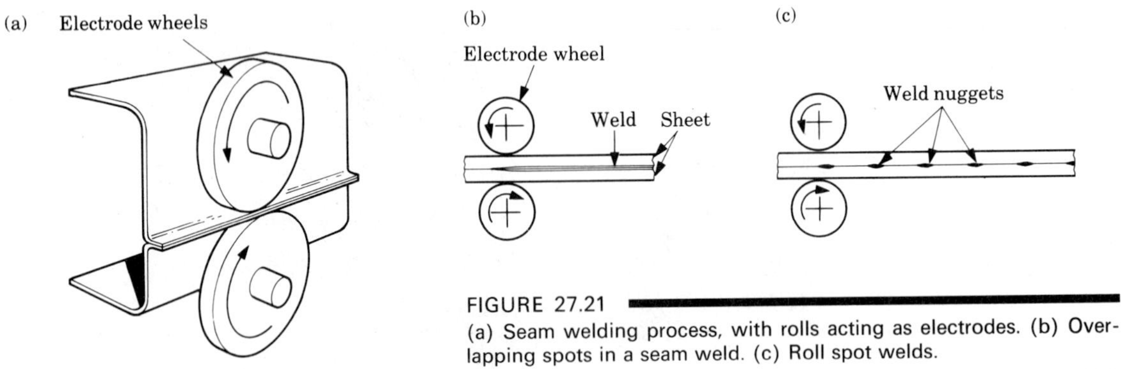

FIGURE 27.21
(a) Seam welding process, with rolls acting as electrodes. (b) Overlapping spots in a seam weld. (c) Roll spot welds.

series of spot welds at various intervals can be made along the length of the seam (Fig. 27.21c). This procedure is called roll spot welding. The RSEW process is used to make the longitudinal (side) seam of cans for household products, mufflers, gasoline tanks, and other containers. The typical welding speed is 25 mm/s (60 in./min) for thin sheet.

27.6.3 High-frequency resistance welding

High-frequency resistance welding (HFRW) is similar to seam welding, except that high frequency current (up to 450 kHz) is employed. A typical application is making butt-welded tubing. In one method, the current is conducted through two sliding contacts (Fig. 27.22a) to the edges of roll-formed tubes. The heated edges are then pressed together by passing the tube through a pair of squeeze rolls.

In another method, the roll-formed tube is subjected to high-frequency induction heating (Fig. 27.22b). Structural sections can also be fabricated by HFRW (Fig. 27.23). Spiral pipe and tubing, and finned tubes (for heat exchangers), may be made by these techniques.

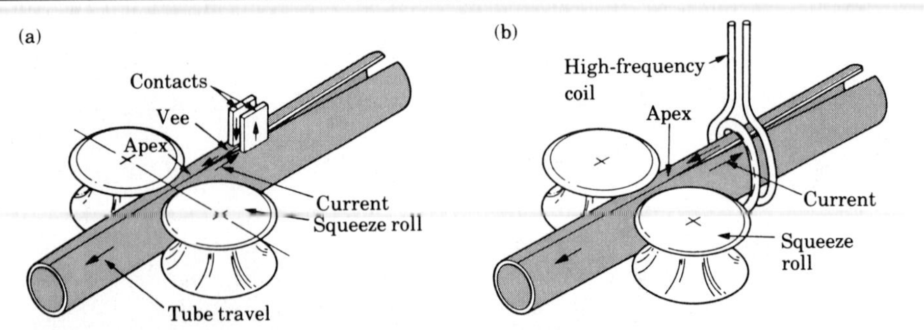

FIGURE 27.22
Methods of high-frequency butt welding of tubes.

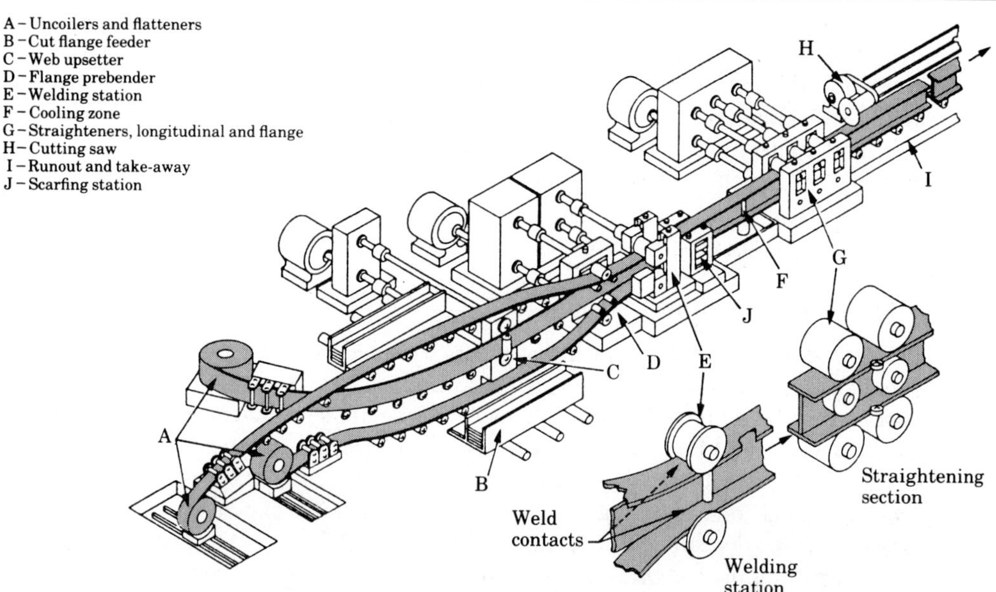

FIGURE 27.23 Continuous welding of fabricated I-beams in an automated facility. *Source:* American Welding Society.

27.6.4 Resistance projection welding

In *resistance projection welding* (RPW), high electrical resistance at the joint is developed by embossing one or more projections (dimples) on one of the surfaces to be welded (Fig. 27.24). High localized temperatures are generated at the projections, which are in contact with the flat mating part. The electrodes—made of copper-base alloys and water cooled to keep their temperature low—are large and flat. Weld nuggets, similar to those in spot welding, are formed as the electrodes exert pressure to compress the projections. The projections may be round or oval for design or strength purposes.

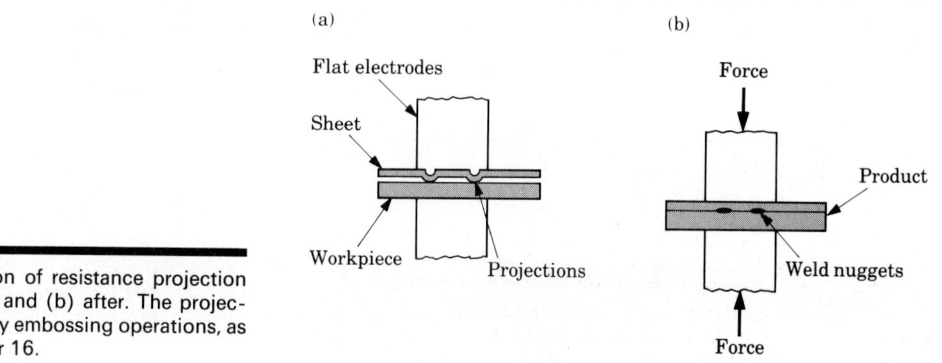

FIGURE 27.24 Schematic illustration of resistance projection welding: (a) before and (b) after. The projections are produced by embossing operations, as described in Chapter 16.

Process capabilities. Spot welding equipment can be used for RPW by modifying the electrodes. Although embossing workpieces is an added expense, this process produces a number of welds in one stroke, extends electrode life, and is capable of welding metals of different thicknesses. Nuts and bolts are also welded to sheet and plate by this process (Figs. 27.25a and b), with projections that are produced by machining or forging. Joining a network of wires such as metal baskets, grills, oven racks, and shopping carts is also considered resistance projection welding because of the small contact area between crossing wires (grids).

27.6.5 Flash welding

In *flash welding* (FW), also called flash butt welding, heat is generated from the arc as the ends of the two members begin to make contact, developing an electrical resistance at the joint (Fig. 27.26a). Because of the arc's presence, this process is also classified as arc welding. After the proper temperature is reached and the interface begins to melt, an axial force is applied at a controlled rate, and a weld is formed by plastic deformation (upsetting) of the joint. Some metal is expelled from the joint as a shower of sparks during the flashing process.

Because impurities and contaminants are squeezed out during this operation, the quality of the weld is good, although a significant amount of material may be burned off during the welding process. The joint may later be machined to improve its appearance. The machines for FW are usually automated and large, with a variety of power supplies ranging from 10 to 1500 kVA.

Process capabilities. The flash welding process is suitable for end-to-end or edge-to-edge joining of similar or dissimilar metals 1–75 mm (0.05–3 in.) in diameter and sheet and bars 0.02–25 mm (0.01–1 in.) thick. Thinner sections have a tendency to buckle under the axial force applied during welding. This process is also used to repair broken band-saw blades, with fixtures that are attached to the band-saw frame.

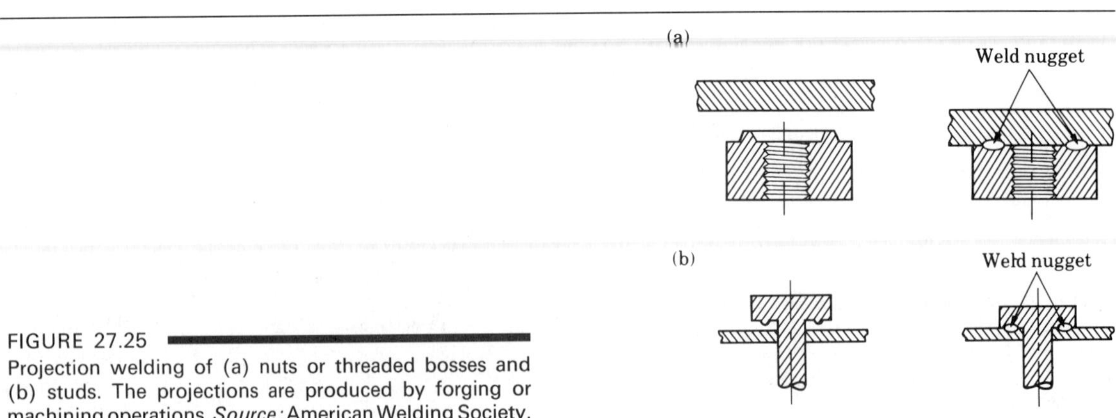

FIGURE 27.25 Projection welding of (a) nuts or threaded bosses and (b) studs. The projections are produced by forging or machining operations. *Source:* American Welding Society.

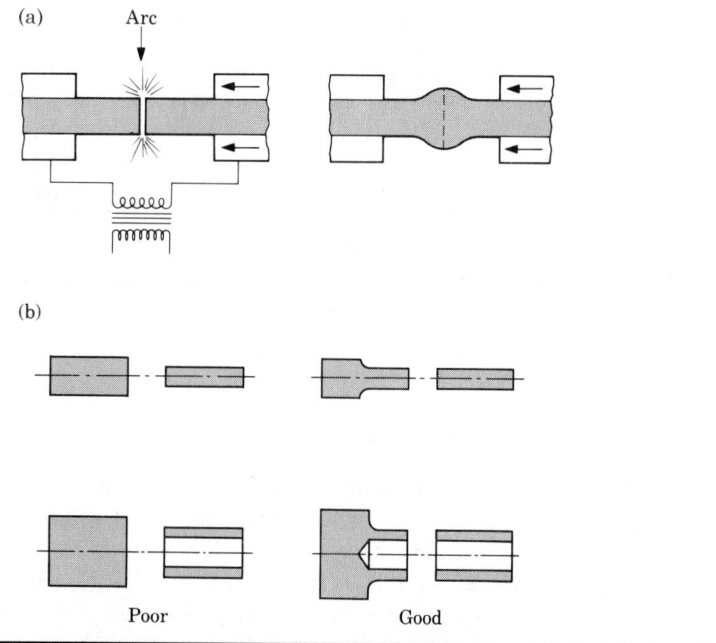

FIGURE 27.26
(a) Flash welding process for end-to-end welding of solid rods or tubular parts. (b) Design guidelines for flash welding.

This process can be automated for reproducible welding operations. Typical FW applications are joining pipe and tubular shapes for metal furniture and windows. It is also used for welding the ends of coils of sheet or wire in continuously operating rolling mills and wire-drawing equipment. Once the appropriate process parameters are established, the required operator skill is minimal. Prices of FW machines range from about $5000 to as much as $1 million for the large machines used in steel mills.

Design guidelines for mating surfaces in flash welding are shown in Fig. 27.26(b). Note the importance of uniform cross-sections at the joint.

27.6.6 Stud welding

Stud welding (SW) is similar to flash welding and is also called stud arc welding. The stud, such as a small part or a threaded rod or hanger, serves as one of the electrodes while being joined to another member, which is usually a flat plate (Fig. 27.27). In order to concentrate the heat generated, prevent oxidation, and retain the molten metal in the weld zone, a disposable ceramic ring (ferrule) is placed around the joint. The equipment for stud welding can be automated, with various controls for arcing and applying pressure. Portable stud welding equipment is also available.

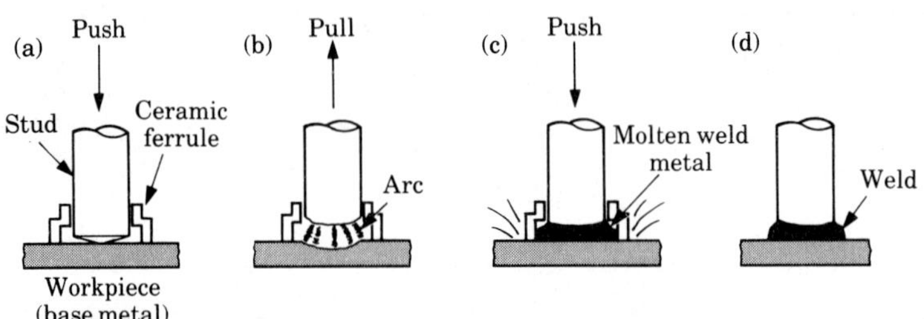

FIGURE 27.27
Sequence of operations in stud welding, which is used for welding bars, threaded rods, and various fasteners on metal plates.

27.6.7 Percussion welding

The resistance welding processes we have described require a transformer to meet power requirements. However, the electrical energy for welding may be stored in a condenser. *Percussion welding* (PEW) utilizes this technique, in which the power is discharged in a very short time (1–10 ms), developing high localized heat at the joint. This process is useful where heating of the components adjacent to the joint is to be avoided, such as in electronic components.

27.7 Cutting

A piece of metal can be separated into two or more pieces, or into various contours, not only by mechanical means such as sawing (see Fig. 23.31), but also by a source of heat that removes a narrow zone in the workpiece. The sources of heat that we have described thus far, namely, torches and electric arcs, can be used for this purpose.

27.7.1 Oxyfuel gas cutting

Oxyfuel gas cutting (OFC) is similar to oxyfuel welding, but the heat source is now used to remove a narrow zone from a metal plate or sheet (Fig. 27.28a). This process is particularly suitable for steels. The basic reactions with steel are:

$$\mathrm{Fe} + \mathrm{O} \rightarrow \mathrm{FeO} + \text{heat}; \tag{27.4}$$

$$3\mathrm{Fe} + 2\mathrm{O}_2 \rightarrow \mathrm{Fe}_3\mathrm{O}_4 + \text{heat}; \tag{27.5}$$

$$4\mathrm{Fe} + 3\mathrm{O}_2 \rightarrow 2\mathrm{Fe}_2\mathrm{O}_3 + \text{heat}. \tag{27.6}$$

The highest heat is generated in the second reaction, resulting in a temperature rise of about 870 °C (1600 °F). This temperature is not sufficiently high to cut steels,

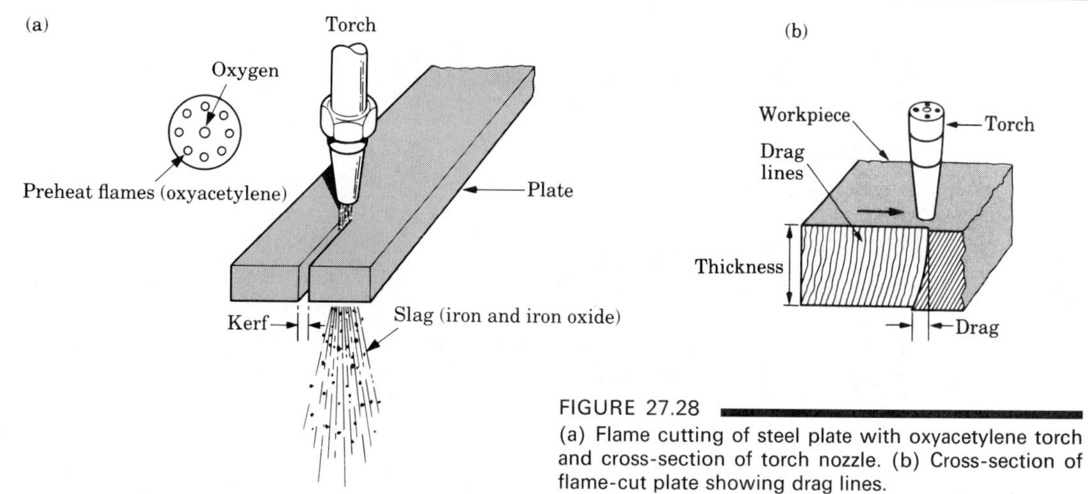

FIGURE 27.28 (a) Flame cutting of steel plate with oxyacetylene torch and cross-section of torch nozzle. (b) Cross-section of flame-cut plate showing drag lines.

so the workpiece is preheated with fuel gas, and oxygen is introduced later (see nozzle cross-section in Fig. 27.28a). The higher the carbon content of the steel, the higher the preheating temperature must be. Cutting occurs mainly by oxidation and burning of the steel, with some melting taking place. Cast irons and steel castings can also be cut by this method. The process generates a kerf, similar to that produced by sawing with a saw blade.

For metals and alloys that do not oxidize as readily or have high thermal conductivity, such as stainless steels and copper, iron powders or fluxes are introduced into the flame. These elements increase the local temperature and help remove the oxides in the cutting zone, which otherwise would obstruct the heat transfer from the torch to the metal.

Process capabilities. The maximum thickness that can be cut by OFC depends mainly on the gases used. With oxyacetylene gas, the maximum thickness is about 300–350 mm (12–14 in.); with oxyhydrogen, about 600 mm (24 in.). Kerf widths range from about 1.5 mm to 10 mm (0.06 to 0.4 in.), with reasonably good control of tolerances. The flame leaves drag lines on the cut surface (Fig. 27.28b), which is rougher than surfaces produced by sawing, blanking, or other operations using cutting tools. Distortion caused by uneven temperature distribution can be a problem in OFC.

Although long used for salvage and repair work, oxyfuel gas cutting has become an important manufacturing process. Torches may be guided along various paths manually, mechanically, or by automatic machines using programmable controllers and robots. Also, two or more layers of flat sheet can be cut with OFC (stack cutting), thus improving productivity and reducing cutting cost. Underwater cutting is done with specially designed torches that produce a blanket of compressed air between the flame and the surrounding water.

27.7.2 Arc cutting

Arc-cutting processes are based on the same principles as arc-welding processes. A variety of materials can be cut at high speeds by arc cutting.

In *air carbon-arc cutting* (CAC-A), a carbon electrode is used, and the molten metal is blown away by a high-velocity air jet. Thus the metal being cut doesn't have to oxidize. This process is used especially for gouging and scarfing (removal of metal from a surface). The AAC process is noisy, and the molten metal can be blown substantial distances, causing safety hazards.

Plasma-arc cutting (PAC) produces the highest temperatures. It is used for rapid cutting of nonferrous and stainless-steel plates. The cutting productivity of this process is higher than that of oxyfuel gas methods. It produces good surface finish and narrow kerfs. It also is the most popular cutting process utilizing programmable controls that is used in manufacturing today.

Lasers and *electron beams* (see Sections 26.7 and 26.8) are used for very accurately cutting a wide variety of metals. Surface finish is better and kerf width is narrower than that for other thermal cutting processes. Proper safety precautions are important.

SUMMARY

Oxyfuel gas, arc, and resistance welding are among the most commonly used joining operations. Gas welding uses chemical energy, whereas arc and resistance welding use electrical energy to supply the necessary heat for welding. In all these processes, heat is used to bring the joint being welded to a liquid state. Shielding gases are used to protect the molten weld pool and weld area against oxidation. Filler rods may or may not be used in oxyfuel gas and arc welding to fill the weld area. Resistance welding operations do not require filler rods.

The selection of a welding process for a particular operation depends on the workpiece material, its thickness and size, shape complexity, type of joint, strength required, and change in product appearance caused by welding. A variety of equipment is available, much of which is now computer controlled with programmable features.

Cutting metals is also done with processes whose principles are based on oxyfuel gas and arc welding. Cutting of steels occurs mainly by oxidation and burning. The highest temperatures for cutting are obtained by plasma-arc cutting.

TRENDS

- The trend is to automate welding processes in order to control process variables accurately and repeatedly and reduce the need for skilled labor.

(*continued*)

TRENDS *(continued)*

- Computer controls and programmable robots are being used extensively for many welding operations. These developments are being further enhanced by the use of appropriate sensors, such as infrared. They monitor the conditions during welding and with feedback controls, make necessary adjustments to maintain weld quality and integrity.
- Electrodes are being developed that have nonsticking characteristics and can be used in any position in arc-welding operations.
- Due to advances made during the past few years, the flux-cored arc welding process has advanced far beyond gas metal-arc welding in replacing shielded-metal-arc welding.
- Pulsed-arc gas metal-arc welding is gaining wide usage for thin ferrous and nonferrous metals.

KEY TERMS

Arc cutting
Arc welding
Carburizing flame
Faying surfaces
Filler rod
Neutral flame
Oxidizing flame
Oxyfuel gas cutting
Oxyfuel gas welding
Resistance welding
Shielded metal-arc welding
Torch
Weld nugget

BIBLIOGRAPHY

Cary, H.B., *Modern Welding Technology*. Englewood Cliffs, N.J.: Prentice-Hall, 1979.

Davies, A.C., *The Science and Practice of Welding*, 7th ed. Cambridge, England: Cambridge University Press, 1977.

——, *The Science and Practice of Welding*, 2 vols. New York: Cambridge University Press, 1989.

Galyen, J., G. Sear, and C. Tuttle, *Welding: Fundamentals and Procedures*. New York: Wiley, 1984.

Gray, T.G.F., J. Spence, and T.H. North, *Rational Welding Design*. New York: Butterworths, 1975.

Houldcroft, P.T., *Welding Process Technology*. Cambridge, England: Cambridge University Press, 1977.

Metals Handbook, 9th ed., *Vol. 6: Welding, Brazing, and Soldering*. Metals Park, Ohio: American Society for Metals, 1983.

Mohler, R., *Practical Welding Technology*. New York: Industrial Press, 1983.

Principles of Industrial Welding. Cleveland: The James F. Lincoln Arc Welding Foundation, 1978.

Tool and Manufacturing Engineers Handbook, Vol. 4, *Quality Control and Assembly*. Dearborn, Mich.: Society of Manufacturing Engineers, 1986.

Welding Handbook, 8th ed., 3 vols. Miami: American Welding Society, 1987.

REVIEW QUESTIONS

27.1 Explain fusion as it relates to welding operations.

27.2 Describe the reactions that take place in an oxyfuel gas torch. What is the level of temperatures generated?

27.3 Explain the features of neutral, reducing, and oxidizing flames. Why is it called reducing?

27.4 Why is an oxidizing flame desirable in welding copper alloys?

27.5 Describe the procedure to be followed in an oxyfuel gas welding operation.

27.6 Explain the basic principles of arc welding processes.

27.7 Why is shielded metal-arc welding a commonly used process? Why is it also called stick welding?

27.8 Why is the quality of submerged arc welding very good?

27.9 Describe the features of three types of arcs in gas metal-arc welding. Why has it been called MIG welding?

27.10 Describe the functions and characteristics of electrodes. What functions do coatings have? How are electrodes classified?

27.11 What are the similarities and differences between consumable and nonconsumable electrodes?

27.12 Name the types of shielding gas used in welding.

27.13 What advantages do resistance welding processes have over others described in this chapter?

27.14 Describe the features of a weld nugget. What does its strength depend on?

27.15 Make a list of products that can be fabricated by resistance welding processes.

27.16 What is the difference between spot welding and projection welding?

27.17 Name some applications for (a) flash welding, (b) stud welding, and (c) percussion welding.

27.18 Explain the functions of filler metals.

27.19 Explain how cutting takes place using an oxyfuel gas torch. How is underwater cutting done?

27.20 What is the principle of arc cutting?

QUESTIONS AND PROBLEMS

27.21 Explain why so many different welding processes have been developed.

27.22 What is the effect of the thermal conductivity of the workpiece on kerf width in oxyfuel gas cutting?

27.23 Describe the differences between oxyfuel gas cutting of ferrous and nonferrous alloys.

27.24 Give reasons why spot welding is so commonly used in automotive bodies and home appliances.

27.25 Could you use oxyfuel gas cutting for a stack of sheet metals (stack cutting)?

27.26 Discuss the need and role of fixtures in holding workpieces in welding operations.

27.27 Discuss your observations concerning the welding design guidelines illustrated in Fig. 27.26.

27.28 Could plasma arc cutting be used for nonmetallic materials? If so, would you select a transferred or nontransferred type of arc? Explain.

27.29 Explain the significance of the magnitude of the pressure applied through the electrodes during resistance welding operations.

27.30 Describe all factors that contribute to the total resistance in resistance welding processes. Which ones should be minimized?

27.31 Give several examples of how you would apply the types of joints shown in Figs. 27.1(a) and (b).

27.32 Discuss your observations concerning Table 27.1.

27.33 What factors influence the shape of the upset joint in flash welding, as shown in Fig. 27.26(a)?

27.34 Two 1.5-mm (0.06-in.) thick copper sheets are being spot welded using a current of 8000 A and a current flow time of $t = 0.2$ s. The electrodes are 4 mm (0.16 in.) in diameter. Estimate the heat generated in the weld zone.

27.35 Design a product or structure that could utilize at least three of the joints shown in Fig. 27.1 in one unit.

27.36 Make a summary table outlining the processes described in this chapter, together with their characteristics and applications.

27.37 Comment on the position of Part V in the diagram on the opening page of this Part. Give an application of this chapter showing its relationship to other parts preceding it.

28

Solid-State and Other Welding Processes

28.1 Introduction

In this chapter we describe processes in which joining takes place without fusion of the workpieces. Unlike the oxyfuel gas, arc, and resistance welding processes we described in Chapter 27, no liquid (molten) phase is present in the joint. We also describe welding processes that use energy sources other than those discussed thus far, including laser and electron beams.

We can best demonstrate the principle of **solid-state bonding** with the following example. If two clean surfaces are brought into atomic contact with each other under sufficient pressure—and in the absence of oxide films and other contaminants (see Section 31.2)—they form bonds and produce a strong joint. Heat and some movement of the mating surfaces by plastic deformation (as in forge and cold welding and roll bonding) may be employed to improve the strength of the joint.

Applying external heat improves the bond by diffusion (as in diffusion bonding). Small interfacial movements on the faying surfaces disturb the surfaces, breaking up oxide films and generating new and clean surfaces, thus improving the strength of the bond (as in ultrasonic welding). Heat may also be generated by friction, which is utilized in friction welding. Very high contact pressures are also utilized in joining processes, as in explosion welding.

In other processes covered in this chapter, we describe the use of high-energy beams as the source of energy. Electron-beam and laser-beam welding processes are based on this principle and have important and unique applications in manufacturing. With this chapter, we conclude our coverage of welding processes (see also Table 27.1). Other joining processes, such as brazing, soldering, adhesive bonding, and mechanical joining methods, are covered in Chapter 30.

28.2 Forge Welding

In forge welding (FOW), both elevated temperature and pressure are applied to obtain a strong bond between the members being joined. The components are heated and pressed or hammered together with suitable tools, dies, or rollers. Local plastic deformation at the interface breaks up the oxide films and other contaminants, thus improving bond strength. However, compared to other welding processes, the resulting joint does not have particularly high load-bearing strength.

Forge welding is an ancient process, dating back to the period 1000–1 B.C. It has been practiced widely by blacksmiths using anvils and hammers, with charcoal as the heat source, making iron and steel chain links and medieval swords. Because of the difficulties involved in controlling the process—and the considerable skill and labor involved—this process has been largely replaced by various other joining methods.

28.3 Cold Welding

In **cold welding** (CW), pressure is applied to the workpieces, either through dies or rolls. Because of the plastic deformation involved, it is necessary that at least one, but preferably both, of the mating parts be ductile. The interface is usually cleaned by wire brushing prior to welding.

However, in bonding two dissimilar metals that are mutually soluble, brittle intermetallic compounds may form (see Section 4.2), resulting in a weak and brittle joint. An example is the bonding of aluminum and steel, where a brittle intermetallic compound is formed at the interface. The best bond strength and ductility is obtained with two similar materials. Cold welding can be used to join small

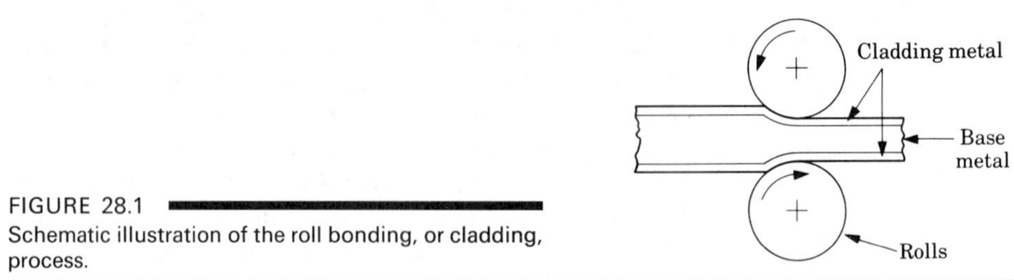

FIGURE 28.1
Schematic illustration of the roll bonding, or cladding, process.

workpieces made of soft, ductile metals. Applications include electrical connections and welding wire stock.

28.3.1 Roll bonding

The pressure required for cold welding can be applied through a pair of rolls (Fig. 28.1). Hence the process is called *roll bonding*. Developed in the 1960s, roll bonding is used for manufacturing certain U.S. coins. The process can be carried out at elevated temperatures (hot roll bonding). Typical examples are cladding pure aluminum over aluminum-alloy sheet and stainless steel over mild steel for corrosion resistance.

Example: Roll bonding of the U.S. quarter.

The technique used for manufacturing composite U.S. quarters is roll bonding of two outer layers of 75 percent copper–25 percent nickel (cupronickel), each 1.2 mm (0.048 in.) thick, with an inner layer of pure copper 5.1 mm (0.20 in.) thick. To obtain good bond strength, the faying surfaces are chemically cleaned and wire brushed. The strips are first rolled to a thickness of 2.29 mm (0.090 in.). A second rolling operation reduces the final thickness to 1.36 mm (0.0535 in.). The strips thus undergo a total reduction in thickness of 82 percent. Since volume constancy is maintained in plastic deformation, there is a major increase in the surface area between the layers, thus generating clean interfacial surfaces. This extension in surface area under the high pressure of the rolls, combined with the solid solubility of nickel in copper (see Section 4.2.1), produces a strong bond.

28.4 Ultrasonic Welding

In **ultrasonic welding** (USW), the faying surfaces of the two members are subjected to a static normal force and oscillating shearing (tangential) stresses. The shearing

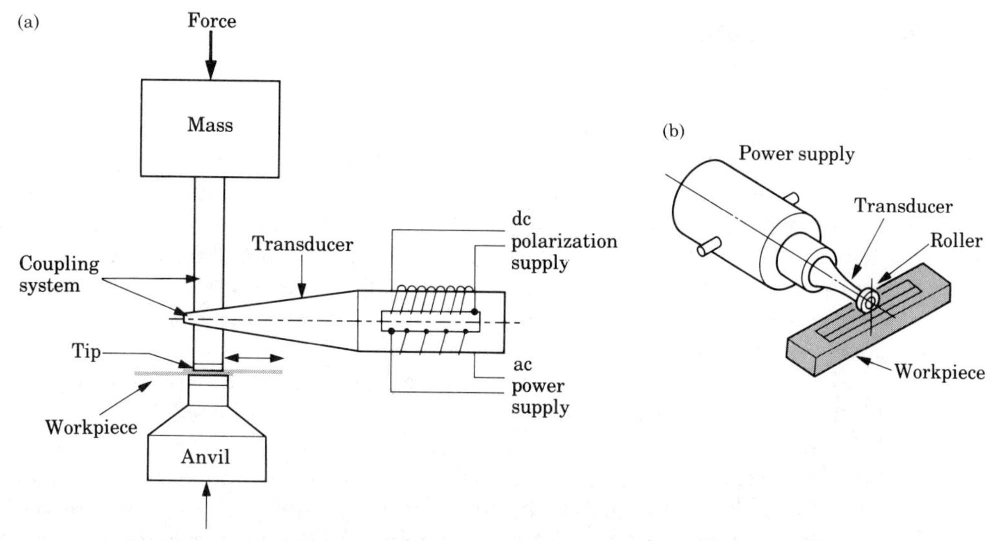

FIGURE 28.2
(a) Components of an ultrasonic welding machine for lap welds. The lateral vibrations of the tool tip cause plastic deformation and bonding at the interface of the workpieces. (b) Ultrasonic seam welding using a roller.

stresses are applied by the tip of a transducer (Fig. 28.2a), similar to that used for ultrasonic machining (see Fig. 25.27a). The frequency of oscillation generally ranges from 10 kHz to 75 kHz, although both lower and higher frequencies than these can be employed. The energy required increases with the thickness and hardness of the materials being joined. Proper coupling between the transducer and the tip (called *sonotrode*, from the word *sonic*, as contrasted to electrode) is important for efficient operation. These units are now operated with solid-state frequency converters.

The shearing stresses cause lateral movement and plastic deformation at the workpiece interfaces, breaking up oxide films and contaminants and thus allowing good contact and producing a strong bond. Temperatures generated in the weld zone are usually in the range of one-third to one-half the melting point (absolute scale) of the metals joined. Therefore no melting and fusion take place. In certain situations, however, temperatures can be sufficiently high to cause metallurgical changes in the weld zone.

The ultrasonic welding process is reliable and versatile. It can be used with a wide variety of metallic and nonmetallic materials, including dissimilar metals (bimetallic strips). It is used extensively in the electronics industry for lap welding of sheet, foil, and thin wire and in packaging with foils. The welding tip can be replaced with rotating disks (Fig. 28.2b) for seam welding structures, similar to those shown in Fig. 27.21 for resistance seam welding, one component of which is a sheet or foil. Moderate skill is required to operate the equipment.

28.5 Friction Welding

In the joining processes that we have described so far, the energy required for welding such as chemical, electrical, and ultrasonic, is supplied externally. In **friction welding** (FRW), the required heat for welding is, as the name implies, generated through friction at the interface of the two members being joined. Thus the source of energy is mechanical. You can demonstrate the significant rise in temperature from friction by rubbing your hands together fast or sliding down a rope rapidly.

In friction welding, one of the members remains stationary while the other is placed in a chuck or collet and rotated at a high constant speed. The two members to be joined are then brought into contact under an axial force (Fig. 28.3). After sufficient contact is established, the rotating member is brought to a quick stop, so that the weld is not destroyed by shearing, while the axial force is increased. The rotating member must be clamped securely to the chuck or collet to resist both torque and axial forces without slipping.

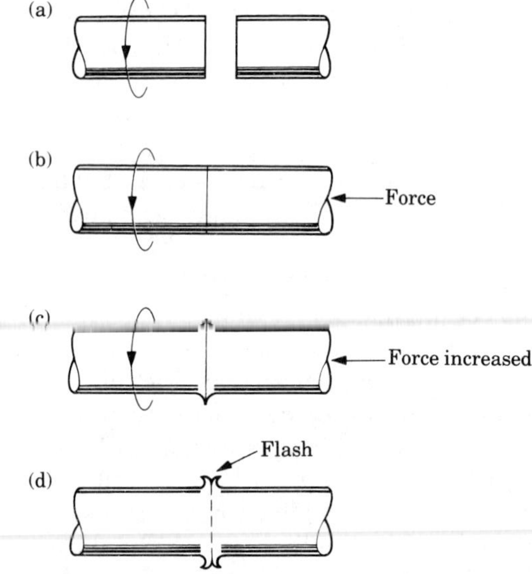

FIGURE 28.3 Sequence of operations in the friction welding process. (a) Left part is rotated at high speed. (b) Right part is brought into contact under an axial force. (c) Axial force is increased; flash begins to form. (d) Left part stops rotating. Weld is completed. Flash can be removed by machining.

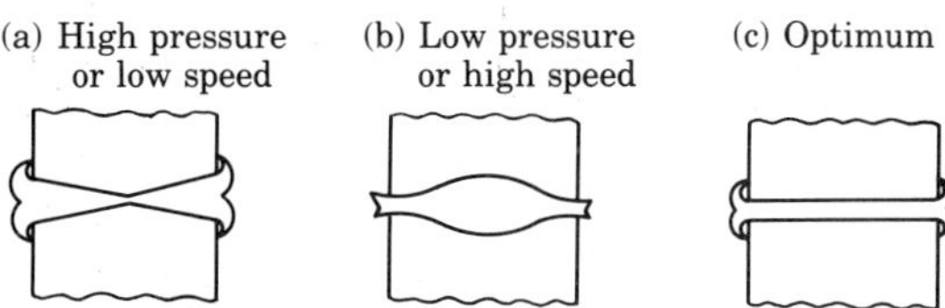

FIGURE 28.4
Shape of fusion zone in friction welding, as a function of force applied and rotational speed.

The pressure at the interface and the resulting friction produce sufficient heat for melting and fusion to take place. The weld zone is usually confined to a narrow region, depending on (a) the amount of heat generated, (b) the thermal conductivity of the materials, and (c) the materials' mechanical properties at elevated temperatures. The shape of the welded joint depends on the rotational speed and the axial pressure applied (Fig. 28.4). These factors must be controlled to obtain a uniformly strong joint. Oxides and other contaminants at the interface are removed by the radial movement of the hot metal at the interface (Fig. 28.3d).

28.5.1 Process capabilities

Developed in the 1940s, friction welding can be used to join a wide variety of materials, provided that one of the components has some rotational symmetry. Solid, as well as tubular parts, can be joined by this method, with good joint strength. Solid steel bars up to 100 mm (4 in.) in diameter and pipes up to 250 mm (10 in.) outside diameter have been welded successfully by friction. The surface speed of the rotating member may be as high as 15 m/s (3000 ft/min). Because of the combined heat and pressure, the interface in FRW develops a flash by plastic deformation (upsetting) of the heated zone. This flash, if objectionable, can easily be removed by machining.

Friction welding machines generally cost between $75,000 and $300,000, depending on their size and capacity. They are fully automated, and operator skill required is minimal if individual cycle times for the complete process are set properly.

Example: Friction welding of bolt head.

The square-head bolt shown in the accompanying figure was originally made by machining from a 64-mm (2.5-in.) square AISI 4140 steel bar. Because of the cost involved, inertia friction welding was considered as an alternative process. The shank of the bolt was machined and threaded from a 32-mm (1.25-in.) diameter bar, and the square head was recessed to a matching contour, as shown in the figure. The shank and the head were then inertia welded, the completed bolt was normalized, and the flash was removed by machining. Inspection of the joint indicated that it was a sound weld. The cost of the inertia-welded bolt was less than one half of the original method of manufacturing.

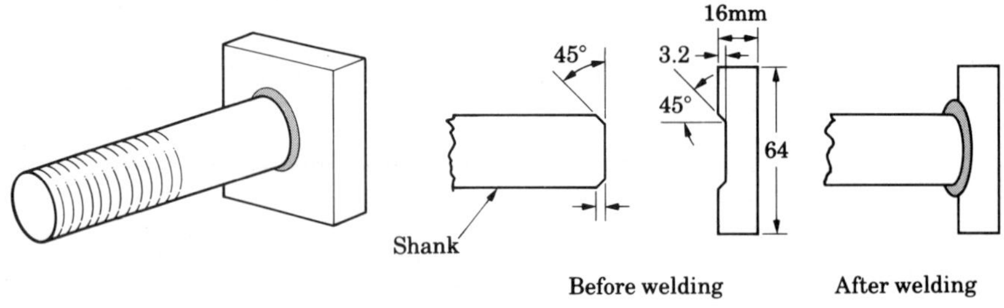

28.5.2 Inertia friction welding

Inertia friction welding is a modification of FRW. The energy required for frictional heating in inertia friction welding is supplied through the kinetic energy of a flywheel. The flywheel is accelerated to the proper speed, the two members are brought into contact, and an axial force is applied. As friction at the interface slows the flywheel, the axial force is increased. The weld is completed when the flywheel comes to a stop. The timing of this sequence is important. If the timing is not properly controlled, weld quality will be poor. The rotating mass of inertia friction welding machines can be adjusted for applications requiring different levels of energy, which depend on workpiece size and properties.

28.6 Explosion Welding

In **explosion welding** (EXW), pressure is applied by detonating a layer of explosive that has been placed over one of the members being joined, called the *flyer plate* (Figs. 28.5a and b). The contact pressures developed are extremely high, and the plate's kinetic energy striking the mating member produces a wavy interface. This impact mechanically interlocks the two surfaces (Figs. 28.6a and b). Cold pressure welding by plastic deformation also takes place. The flyer plate is placed at an angle, and any oxide films present at the interface are broken up and propelled from the interface. As a result, bond strength in explosion welding is very high.

The explosive may be flexible plastic sheet, cord, or granulated or liquid, which is cast or pressed onto the flyer plate. Detonation speeds are usually 2400–3600 m/s (8000–12,000 ft/s), depending on the type of explosive, thickness of the explosive layer, and its packing density. There is a minimum detonation speed for welding to occur in this process. Detonation is carried out using a standard commercial blasting cap.

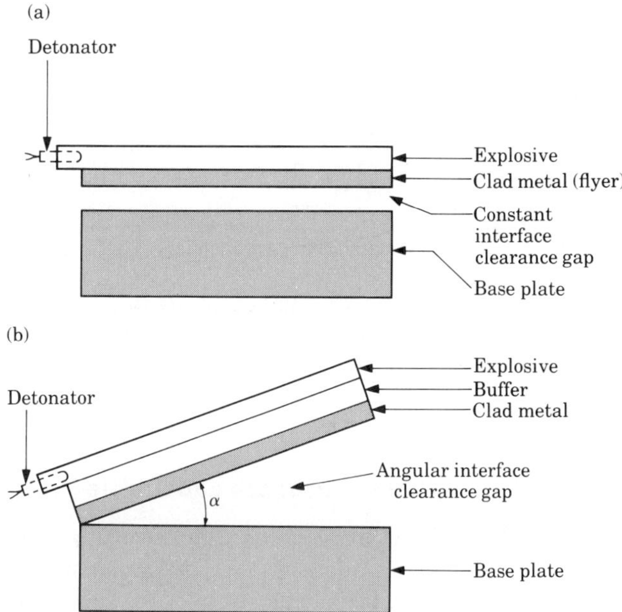

FIGURE 28.5 Schematic illustration of the explosion welding process: (a) constant interface clearance gap and (b) angular interface clearance gap.

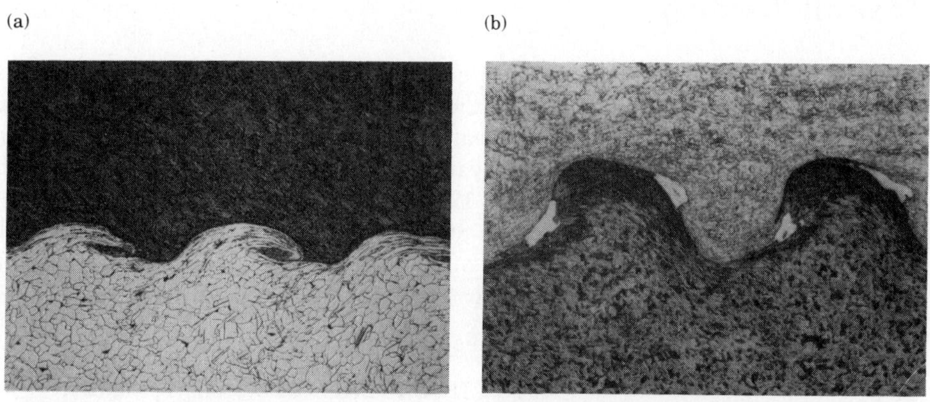

FIGURE 28.6 Cross-sections of explosion-welded joints: (a) titanium (top) on low-carbon steel (bottom) and (b) Incoloy 800 (iron–nickel base alloy) on low-carbon steel. *Source:* Courtesy of E. I. du Pont de Nemours & Co.

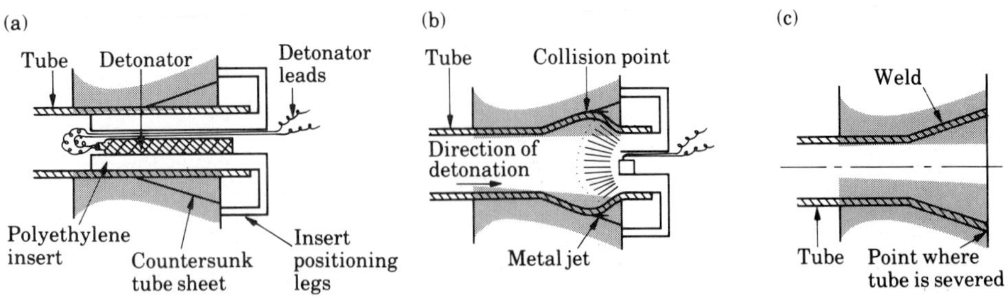

FIGURE 28.7
Explosion welding of tube on head plate for heat exchangers and boilers: (a) before welding, (b) during welding, and (c) completed weld. The tube ends are expanded by placing and detonating the explosives inside the tube. *Source:* American Welding Society.

28.6.1 Process capabilities

This process was developed in the 1950s. It is particularly suitable for cladding plates and slab with dissimilar metals. Plates as large as 6 m × 2 m (20 ft × 7 ft) have been explosively clad. These plates may then be rolled into thinner sections. Tube and pipe are often joined to the holes in head plates of boilers and heat exchangers by placing the explosive inside the tube; explosion expands the tube (Fig. 28.7). Explosion welding is inherently dangerous and requires safe handling by well-trained and experienced personnel.

28.7 Thermit Welding

Thermit welding (TW) gets its name from *thermite*, which is based on the word *therm*, meaning heat; the word Thermit is a registered trademark. The process involves exothermic (heat producing) reactions between metal oxides and metallic reducing agents. The heat of the reaction is then utilized in welding. This process dates back to the early 1900s.

The most common mixture of materials used in welding steel and cast iron is finely divided particles of iron oxide (Fe_3O_4), aluminum oxide (Al_2O_3), iron, and aluminum. The basic reactions are

$$\tfrac{3}{4}Fe_3O_4 + 2Al \rightarrow \tfrac{9}{4}Fe + Al_2O_3 + \text{heat}; \qquad (28.1)$$

$$3FeO + 2Al \rightarrow 3Fe + Al_2O_3 + \text{heat}; \qquad (28.2)$$

$$Fe_2O_3 + 2Al \rightarrow 2Fe + Al_2O_3 + \text{heat}. \qquad (28.3)$$

This nonexplosive mixture produces a maximum theoretical temperature of 3200 °C (5800 °F) within less than a minute. In practice, however, this temperature is only about 2200–2400 °C (4000–4350 °F). The mixture may also contain other materials to impart special properties to the weld. The reaction is started by applying a magnesium fuse to special compounds of peroxides, chlorates, or chromates, known as oxidizing agents, with an ignition temperature of about 1200 °C (2200 °F).

Welding copper, brasses, and bronzes, and copper alloys to steels, involves the following reactions:

$$3CuO + 2Al \rightarrow 3Cu + Al_2O_3 + \text{heat}; \tag{28.4}$$

$$3Cu_2O + 2Al \rightarrow 6Cu + Al_2O_3 + \text{heat}. \tag{28.5}$$

Oxides of copper, nickel, chromium, and manganese are also used in Thermit welding, resulting in temperatures ranging up to 5000 °C (9000 °F).

28.7.1 Process capabilities

Thermit welding involves aligning the parts to be joined but with a gap between them (usually filled with wax), around which a sand or ceramic mold is built. If the parts are thick, the mold cavity may be preheated to improve welding and to dry the mold. Drying the mold is very important; otherwise superheated steam trapped in the mold can cause explosions. The superheated products of the reaction are allowed to flow into the gap, melting the edges of the parts being joined. After the weld cools, excess material is removed by machining, grinding, or other methods. Thermit welding is suitable for welding and repairing large forgings and castings. It can also be used to weld thick steel structural sections, railroad rails, and pipe.

28.8 Electron-Beam Welding

In **electron-beam welding** (EBW), heat is generated by high-velocity narrow-beam electrons. The kinetic energy of the electrons is converted into heat as they strike the workpiece. This process requires special equipment to focus the beam on the workpiece in a vacuum. The higher the vacuum, the more the beam penetrates and the greater the depth-to-width ratio is. Almost any metal can be welded by EBW, with workpiece thicknesses ranging from foil to plate. The intense energy is also capable of producing holes in the workpiece (keyhole). Generally, no shielding gas, flux, or filler metal is required. Capacities of electron beam guns range up to 175 kV and 1000 milliamps.

28.8.1 Process capabilities

Developed in the 1960s, EBW has the capability to make high-quality welds that are almost parallel sided, are deep and narrow, and have small heat-affected zones. Depth-to-width ratios range between 10:1 and 30:1. The size of welds made by

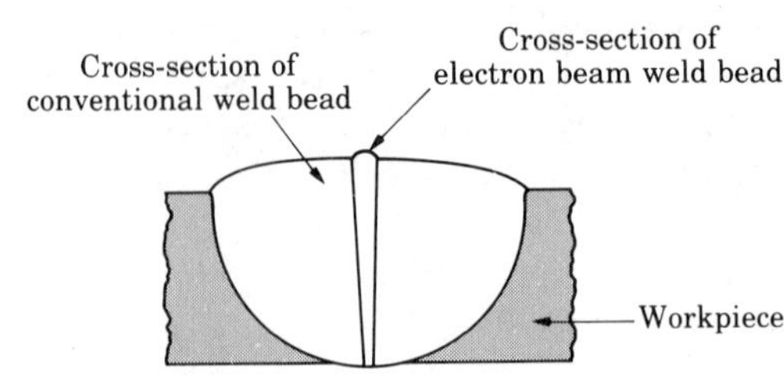

FIGURE 28.8
Comparison of the size of weld beads in conventional (tungsten-arc) and electron-beam welding.

EBW and conventional processes are compared in Fig. 28.8. Using servo controls, parameters can be controlled accurately, with welding speeds as high as 200 mm/s (40 ft/min). The electron beam can be projected a distance of several meters, thus making welding at otherwise inaccessible locations possible. Almost any metal can be butt or lap welded with this process, with thicknesses to as much as 150 mm (6 in.).

Distortion and shrinkage in the weld area is minimal, although the weld has a propensity to crack along its centerline. Precisely controlling the welding parameters can usually eliminate this situation. Electron-beam welding equipment generates x-rays, and hence proper monitoring and periodic maintenance are important. The cost of equipment ranges from about $75,000 to over $1 million, depending on capacity.

28.9 Laser-Beam Welding

Laser-beam welding (LBW) utilizes a focused high-power coherent monochromatic (one wavelength) light beam as the source of heat (Fig. 26.16). Because the beam has high energy density, it has deep penetrating power. It can be directed, shaped, and focused precisely on the workpiece. Consequently, this process is particularly suitable for welding narrow and deep joints. In automotive applications, for example, lubrication holes in transmission hubs are drilled using Nd:YAG lasers, and transmission gears are welded using CO_2 lasers. Welding of transmission components with high-powered lasers is the most widespread application in the automotive industry. Other applications include cutting mounting holes on the Ford Aerostar van that are ordered with roof racks, using a 500-W CO_2 laser, and marking (scribing) bar codes and alphanumeric data on vehicle identification tags for General Motors cars, using a 50-W Nd:YAG laser.

28.9.1 Process capabilities

Laser-beam welding can be used successfully on a variety of materials with thicknesses of up to 25 mm (1 in.) and is particularly effective on thin workpieces. Welding speeds range from 40 mm/s (8 ft/min), to as high as 1.3 m/s (250 ft/min) for thin metals. Because of the nature of the process, welding can be done in otherwise inaccessible locations. Laser-beam welding produces welds of good quality, with minimum shrinkage and distortion, and with depth-to-width ratios as high as 30:1. However, as in EBW, this high ratio can cause centerline cracking in the weld.

The major advantages of LBW over EBW are:

- The beam can be transmitted through air, so a vacuum is not required.
- Because laser beams can be shaped, directed, and focused optically, the process can easily be automated.
- The beams do not generate x-rays.
- The quality of the weld is better, with less tendency for incomplete fusion, spatter, and porosity.

The cost of equipment for LBM usually ranges from $40,000 to almost $1 million. As in other, similar automated welding systems, the operator skill required is minimal.

• Example: Laser welding of razor blades.

The accompanying photograph shows a closeup of the Gillette Sensor razor cartridge. Each of the two narrow, high-strength blades has 13 pinpoint welds, 11 of which can be seen (as darker spots, about 0.5 mm in diameter) on each blade in the photograph. The welds are made with a Nd:YAG laser equipped with fiber-optic

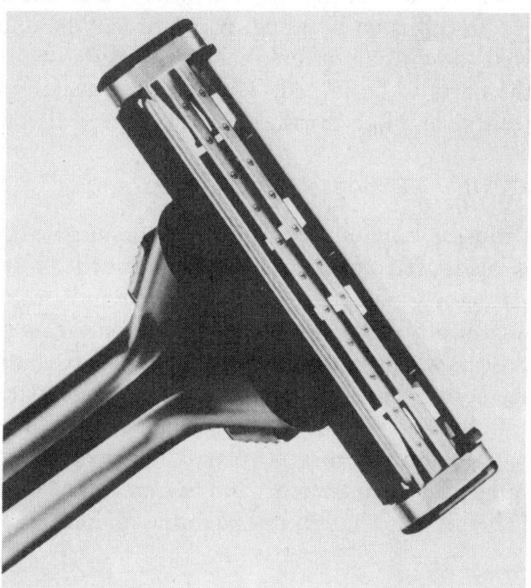

delivery, thus providing very flexible beam manipulation to the exact locations along the length of the blade. Production is at a rate of 3 million welds per hour, with accurate and consistent weld quality. *Note*: The student is encouraged to inspect the welds on these blades with a magnifying glass. *Source*: Courtesy of Lumonics Corporation, Industrial Products Division.

28.10 Diffusion Bonding

Diffusion bonding, or diffusion welding (DFW), is a solid-state joining process in which the strength of the joint results primarily from diffusion and, to a lesser extent, small plastic deformation of the faying surfaces. This process requires temperatures of about $0.5T_m$ (where T_m is the melting point of the metal on the absolute scale) in order to have a sufficiently high diffusion rate between the parts being joined. The bonded interface in DFW (Fig. 28.9) has essentially the same physical and mechanical properties as the base metal. Its strength depends on pressure, temperature, time of contact, and the cleanliness of the faying surfaces. These requirements can be lowered by using filler metal at the interfaces.

Although developed in the 1970s as a modern welding technology, the principle of diffusion bonding actually dates back centuries, when goldsmiths bonded gold over copper. Called filled gold, a thin layer of gold foil is obtained by hammering; the gold is then placed over copper, and a weight is placed on top of it. The assembly is then placed in a furnace and left until a good bond is obtained.

In diffusion bonding, pressure may be applied by dead weights or a press, as well as by using differential gas pressure or utilizing relative thermal expansion of the parts to be joined. The parts are usually heated in a furnace or by electrical resistance. High-pressure autoclaves are also used for bonding complex parts.

28.10.1 Process capabilities

Diffusion bonding is generally most suitable for dissimilar metal pairs. However, it is also used for reactive metals, such as titanium, beryllium, zirconium, and refractory metal alloys, and for composite materials. Because diffusion involves migration of the atoms across the joint, this process is slower than other welding processes. Although DFW is used for fabricating complex parts in low demand for the aerospace, nuclear, and electronics industries, it has been automated to make it suitable and economical for moderate-volume production.

Equipment cost is related approximately to the diffusion bonded area and ranges between $3/mm^2 and $6/mm^2 ($2000/in^2 and $4000/in^2). Unless the process is highly automated, considerable operator training and skill is required.

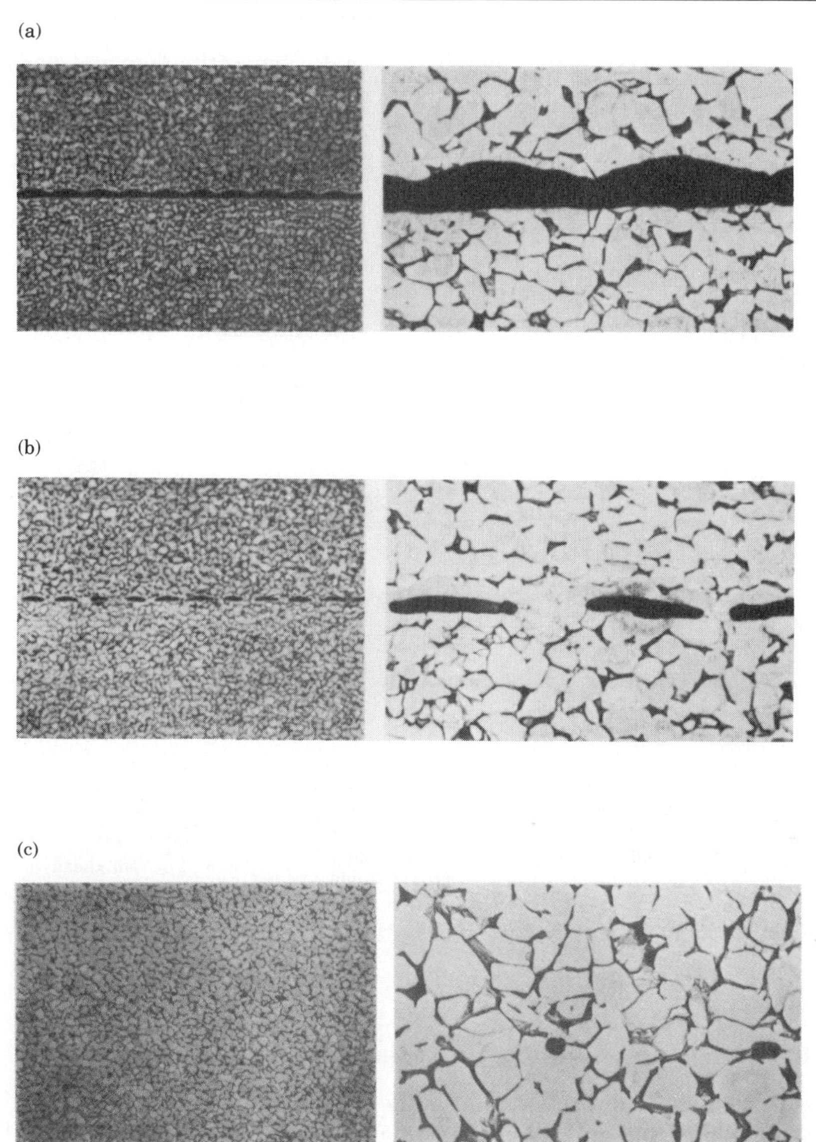

FIGURE 28.9 Sequence of diffusion bonding between two titanium-alloy sheets. Temperature: 925 °C (1700 °F); time: 1 hr; pressure: 700–3500 kPa (100–500 psi). Magnification: 500 ×. *Source:* Courtesy of Rockwell International Science Center.

Example: Diffusion bonding applications.

Diffusion bonding is especially suitable for metals such as titanium and superalloys used in military aircraft. Design possibilities allow the conservation of expensive strategic materials and reduce manufacturing costs. The military aircraft illustrated below has more than 100 diffusion-bonded parts, some of which are shown in the figure.

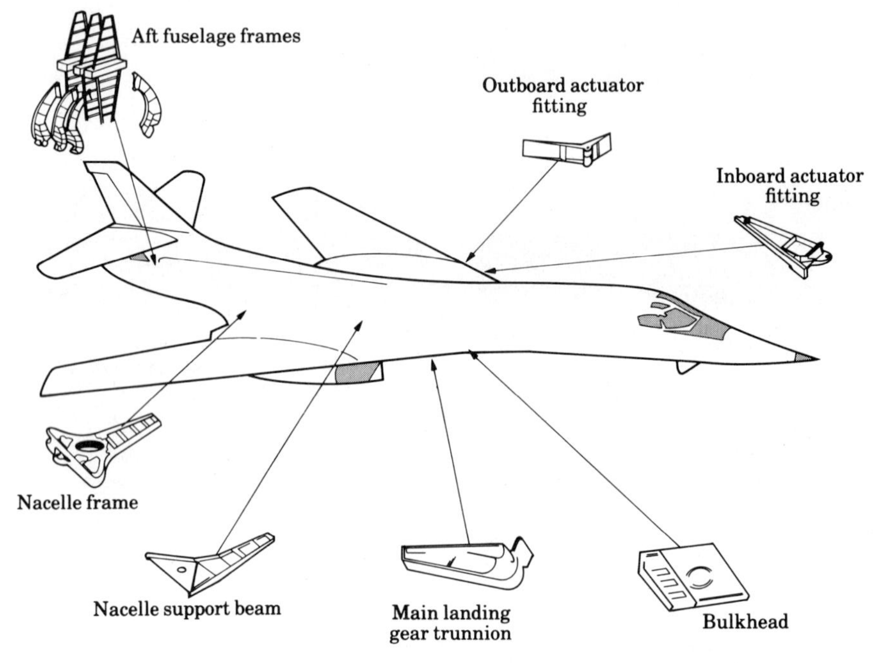

28.10.2 Diffusion bonding/superplastic forming

An important development is the ability to fabricate sheet-metal structures by combining diffusion bonding with superplastic forming (see Section 16.12). Typical structures in which flat sheets are diffusion bonded and formed are shown in Fig. 28.10. After diffusion bonding of selected locations of the sheets, the unbonded regions are expanded into a mold by air pressure. These structures are thin and have high stiffness-to-weight ratios. Hence they are particularly important in aircraft and aerospace applications.

This process improves productivity by eliminating mechanical fasteners and produces parts with good dimensional accuracy and low residual stresses. The technology is now well advanced for titanium structures (typically Ti-6Al-4V alloy) for aerospace applications. Structures made of 7475-T6 aluminum alloy are also being developed using this technique.

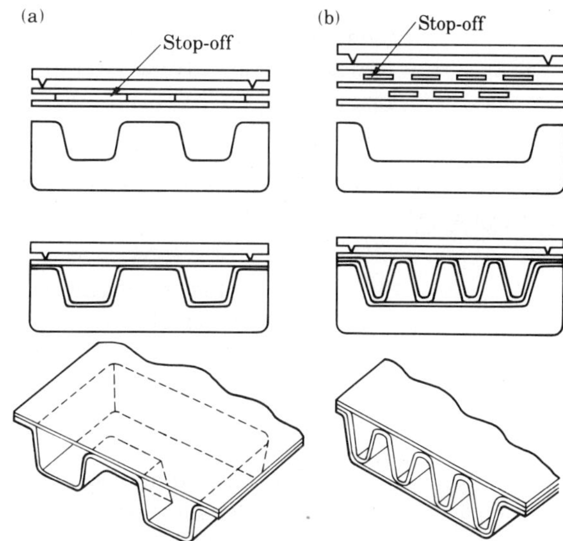

FIGURE 28.10 Two types of structures made by diffusion bonding and superplastic forming of sheet metal. Such structures have a high stiffness-to-weight ratio. *Source*: Rockwell International Corp.

28.11 Safe Welding Practices

As in all manufacturing operations, there are certain hazards in welding and cutting. Although we have referred to safety considerations throughout Chapters 27 and 28, we need to summarize the major hazards that are present where welding is done and safe welding practices to counteract them. Some of the hazards are related to the machinery and equipment involved, and others are related to the processes themselves (see also Chapter 37).

Because of the heat sources, such as open flames, arcs, sparks, and hot metal, used in welding and related operations, fire and explosion hazards are always present in the work area. Thus welding processes should be carried out away from all combustible materials. They include flammable fluids, vapors, gases, fuel, wood, and textiles. Floors, partitions, platforms, and ceilings may also have been made of flammable materials and thus present potential safety hazards.

Fires and explosions can cause serious injury and even fatality. Protection of the operator's eyes, face, and body against sparks and spatter, as well as infrared and ultraviolet radiation, is essential. Several types of safety equipment and protective clothing are available and should be used.

Excessive and prolonged noise generated by welding or cutting operations can cause temporary or permanent hearing loss. Ear protection devices should be used.

Welding and related methods and machinery that use electricity as a source of energy also present hazards. Proper installation and maintenance of equipment and training of personnel is essential.

In addition to the use of various shielding gases during welding operations, many processes emit fumes and gases. Overexposure to these, some of which are toxic, can be hazardous to health. Proper ventilation systems must be installed and maintained.

SUMMARY

In addition to the traditional joining processes of oxyfuel gas, arc, and resistance welding, a number of other joining processes that are based on producing a strong joint under pressure and/or heat are available. Surface preparation and cleanliness are important in some of these processes. Pressure is applied mechanically or by explosives. Heat may be supplied externally, including high-energy beams, or generated internally, as in friction welding.

Among important developments is the combining of the diffusion bonding and superplastic forming processes. Productivity is improved, as is the capability to make complex parts economically.

As in all manufacturing operations, certain hazards are inherent in welding operations. Some relate to the machinery and equipment used and others to the nature of the process. Proper safety precautions must always be taken in work areas where welding is done.

SUMMARY TABLE 28.1

GENERAL CHARACTERISTICS OF JOINING PROCESSES

PROCESS	OPERATION	ADVANTAGE	SKILL LEVEL REQUIRED	WELDING POSITION	CURRENT TYPE	DISTORTION*	COST OF EQUIPMENT
SMAW	Manual	Portable and flexible	High	All	ac,dc	1 to 2	Low
SAW	Automatic	High deposition	Medium to low	Flat and horizontal	ac, dc	1 to 2	Medium
GMAW	Semiautomatic or automatic	Most metals	High to low	All	dc	2 to 3	Medium to high
GTAW	Manual or automatic	Most metals	High to low	All	ac, dc	2 to 3	Medium
FCAW	Semiautomatic or automatic	High deposition	High to low	All	dc	1 to 3	Medium
OFW	Manual	Portable and flexible	High	All	—	2 to 4	Low
EBW, LBW	Semiautomatic or automatic	Most metals	Medium to high	All	—	3 to 5	High

*1, highest; 5, lowest.

TRENDS

- As in other joining processes, the trend in the processes described in this chapter is toward automation, utilizing computer controls and programmable robots even more extensively.
- High-energy beam welding processes will continue to be prominent in joining operations. Efforts are continuing in improving the strength of welded joints.
- Laser welding of flat sheet metal for automotive bodies (butt welding) prior to stamping is being studied extensively.
- Diffusion bonding combined with superplastic forming is becoming an important manufacturing process.

KEY TERMS

Cold welding
Diffusion bonding
Electron-beam welding
Explosion welding
Forge welding
Friction welding
Laser-beam welding
Solid-state bonding
Thermit welding
Ultrasonic welding

BIBLIOGRAPHY

Arata, Y. (ed.), *Plasma, Electron, and Laser Beam Technology*. Metals Park, Ohio: American Society for Metals, 1986.

Cary, H.B., *Modern Welding Technology*. Englewood Cliffs, N.J.: Prentice-Hall, 1979.

Davies, A.C., *The Science and Practice of Welding*, 7th ed. Cambridge, England: Cambridge University Press, 1977.

——, *The Science and Practice of Welding*, 2 vols. New York: Cambridge University Press, 1989.

Metals Handbook, 9th ed., *Vol. 6: Welding, Brazing, and Soldering*. Metals Park, Ohio: American Society for Metals, 1983.

Mohler, R., *Practical Welding Technology*. New York: Industrial Press, 1983.

Schwartz, M.M. (ed.), *Source Book on Innovative Welding Processes*. Metals Park, Ohio: American Society for Metals, 1981.

Tool and Manufacturing Engineers Handbook, Vol. 4: Quality Control and Assembly. Dearborn, Mich.: Society of Manufacturing Engineers, 1986.

Welding Handbook, 8th ed., 3 vols. Miami: American Welding Society, 1987.

REVIEW QUESTIONS

28.1 Explain what is meant by solid-state welding.
28.2 What is the difference between forge welding and cold welding?
28.3 What are exothermic chemical reactions as applied to welding?
28.4 Describe the principle of ultrasonic welding.
28.5 What advantages does friction welding have over other methods described in this chapter?
28.6 Explain the difference between friction welding and inertia friction welding.
28.7 Describe the advantages and limitations of explosion welding.
28.8 Discuss the chemical reactions that take place in Thermit welding.
28.9 Explain the similarities and differences between electron-beam and laser-beam welding. Give typical applications for each.
28.10 Describe the mechanism of diffusion bonding.
28.11 Why is diffusion bonding when combined with superplastic forming of sheet metals an attractive fabrication process? Does it have any limitations?
28.12 Can roll bonding be applied to a variety of part configurations? Explain.
28.13 Discuss your observations concerning Summary Table 28.1.

QUESTIONS AND PROBLEMS

28.14 Explain the similarities and differences between the joining processes described in this chapter and Chapter 27.
28.15 Why were the processes described in this chapter developed?
28.16 Discuss various other processes that can be used in attaching tubes to head plates, as shown in Fig. 28.7.
28.17 What factors influence the size of the weld beads shown in Fig. 28.8?
28.18 Describe what you observe in Fig. 28.9.
28.19 Make a list of safety precautions necessary in applying the joining processes described thus far.
28.20 Discuss the factors that influence the strength of a (a) diffusion-bonded and (b) cold-welded component.
28.21 Describe designs that cannot be joined by friction welding processes.
28.22 Explain the sources of heat for the processes described in this chapter.
28.23 Describe the difficulties involved in making deep narrow welds.
28.24 Name some applications for ultrasonic seam welding, shown in Fig. 28.2(b).
28.25 Explain how you would fabricate the structures shown in Fig. 28.10 with methods other than diffusion bonding and superplastic forming.
28.26 Prepare an extended version of Summary Table 28.1 and include additional process characteristics and applications.
28.27 Describe the difficulties you might encounter in applying explosion welding in a factory environment.
28.28 Inspect the edges of a U.S. quarter and comment on your observations. Is the cross-section symmetrical? Explain.
28.29 Design a machine that can perform friction welding of two cylindrical pieces as well as removing the flash from the welded joint (see Fig. 28.3).
28.30 How would you modify your design in Problem 28.29 if one of the pieces to be welded is noncircular?

29

The Metallurgy of Welding; Welding Design and Process Selection

29.1 Introduction

In Chapters 27 and 28, we presented the basic principles of the welding processes that utilize chemical, electrical, thermal, and mechanical sources of energy. By now you can appreciate that heating the workpieces to a temperature sufficiently high to produce a weld involves important metallurgical and physical changes in the materials being welded. In this chapter, we describe the following aspects of welding processes:

a) The nature of the welded joint.
b) The quality and properties of the welded joint.
c) Weldability of metals.
d) Methods for testing welds.
e) Welding design.
f) Welding process selection.

The strength, toughness, and ductility of a welded joint depend on many factors. For example, the rate of heat application and the thermal properties of metals are important, in that they control the magnitude and distribution of temperature in a joint during welding. The grain structure of the welded joint depends on the magnitude of heat applied and temperature rise, the degree of prior cold work of the metals, and the rate of cooling after the weld is made.

Weld quality depends on factors such as the geometry of the weld bead and the presence of cracks, residual stresses, inclusions, and oxide films. Their control is essential to reliable welds that have acceptable mechanical properties. Finally, we present general guidelines for proper weld design, as we have done for other manufacturing processes, and factors involved in selecting the appropriate welding process for various applications.

29.2 The Welded Joint

A typical fusion weld joint is shown in Fig. 29.1, where three distinct zones can be identified:

- The **base metal**, that is, the metal to be welded.
- The **heat-affected zone** (HAZ).
- The **weld metal**, that is, the region that has melted during welding.

The metallurgy and properties of the second and third zones depend strongly on the metals joined, the welding process, filler metals used, if any, and process

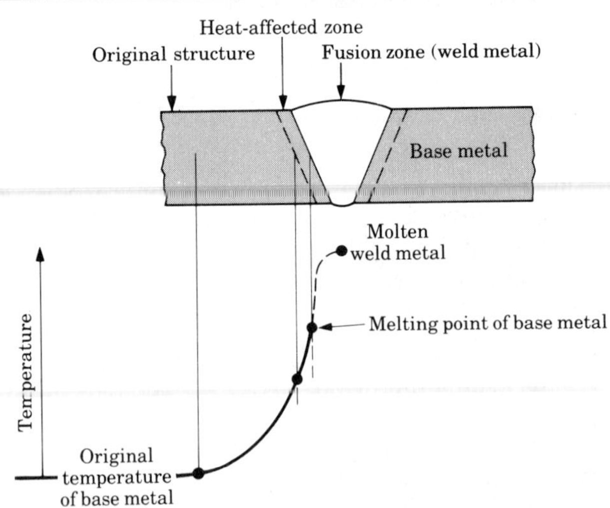

FIGURE 29.1
Characteristics of a typical fusion weld zone in oxyfuel gas and arc welding.

variables. A joint produced without a filler metal is called *autogenous*, and the weld zone is composed of the molten and resolidified base metal. A joint made with a filler metal has a central zone called the weld metal and is composed of a mixture of the base and weld metals.

The study of a weld joint requires an understanding of metal and alloy solidification and phase diagrams, which we discussed in Chapters 1, 4, and 10. This subject is complex and we merely introduce it here.

29.2.1 Solidification of the weld metal

After applying heat and introducing filler metal, if any, into the weld area, the molten weld joint is allowed to cool naturally to ambient temperature. The solidification process is similar to that in casting and begins with the formation of columnar (dendritic) grains. These grains are relatively long and form parallel to the heat flow (see Fig. 10.2). Because metals are much better heat conductors than the surrounding air, the grains lie parallel to the plane of the two plates or sheets being welded (Fig. 29.2a). The grains in a shallow weld are shown in Figs. 29.2(b) and 29.3. Grain structure and size depend on the specific alloy, the welding process, and the filler metal used.

The weld metal is basically a cast structure and, because it has cooled slowly, it generally has coarse grains. Consequently, this structure has generally low strength, toughness, and ductility. However, the proper selection of filler-metal composition or heat treatments following welding can improve the joint's mechanical properties. The results depend on the particular alloy, its composition, and the thermal cycling to which the joint is subjected. Cooling rates may, for example, be controlled and reduced by preheating the general weld area prior to welding. Preheating is particularly important for metals with high thermal conductivity, such as aluminum and copper; otherwise, the heat during welding dissipates rapidly.

29.2.2 Heat-affected zone

The **heat-affected zone** is within the base metal itself. It has a microstructure different from that of the base metal before welding, because it has been subjected to elevated temperatures for a period of time during welding. The portions of the base metal that are far enough away from the heat source do not undergo any structural changes during welding.

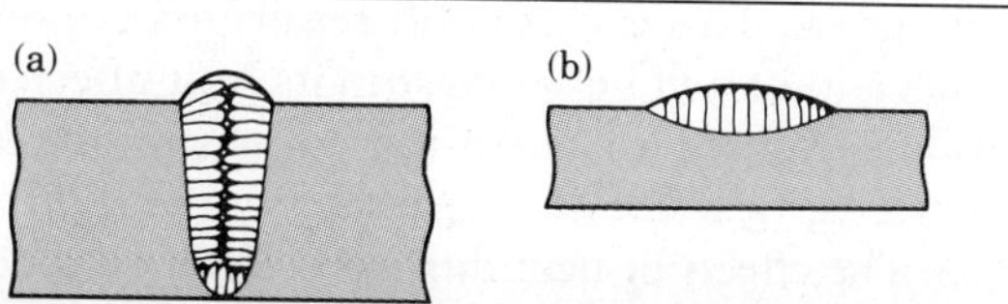

FIGURE 29.2 Grain structure in (a) a deep and (b) a shallow weld. Note that the grains in the solidified weld metal are perpendicular to the surface of the base metal.

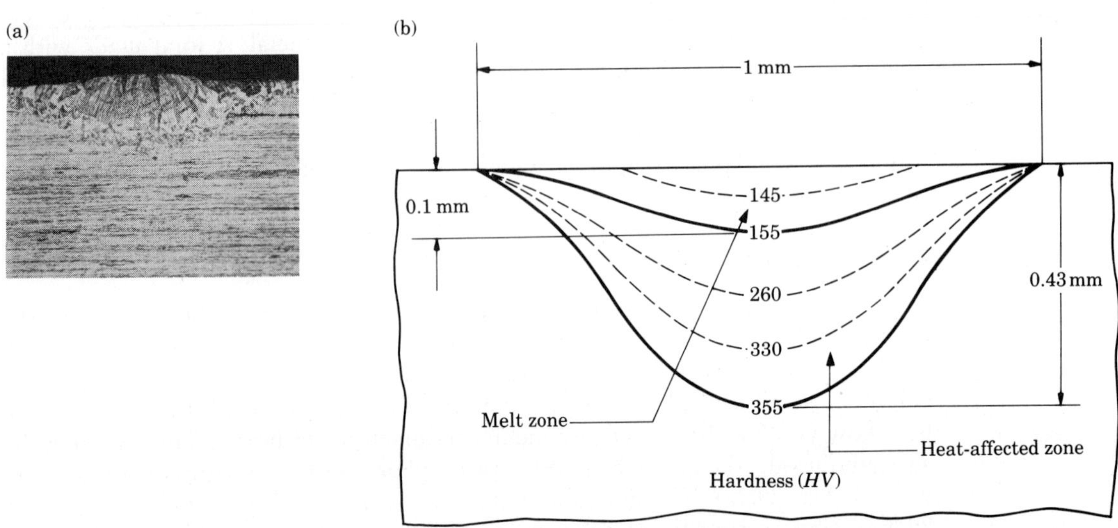

FIGURE 29.3 (a) Weld bead on a cold-rolled nickel strip produced by a laser beam. (b) Microhardness profile across the weld bead. Note the softer condition of the weld bead compared to the base metal. *Source:* IIT Research Institute.

The properties and microstructure of HAZ depend on (a) the rate of heat input and cooling; and (b) the temperature to which this zone was raised. The HAZ and the corresponding phase diagram for 0.3 percent carbon steel are shown in Fig. 29.4. In addition to metallurgical factors (such as original grain size, grain orientation, and degree of prior cold work), the specific heat and thermal conductivity of the metals influence the HAZ's size and characteristics.

The strength and hardness of the heat-affected zone depend partly on how the original strength and hardness of the particular alloy was developed prior to welding. As we described in Chapters 2 and 4, they may have been developed by cold working, solid-solution strengthening, or precipitation hardening or by various heat treatments. Of these strengthening methods, the simplest to analyze is base metal that has been cold worked, say, by cold rolling or forging.

The heat applied during welding recrystallizes the elongated grains (preferred orientation) of the cold-worked base metal. Grains that are away from the weld metal will recrystallize into fine equiaxed grains. However, grains close to the weld metal, having been subjected to elevated temperatures for a longer period of time, will grow. This growth will result in a region that is softer and has less strength. Such a joint will be weakest in its heat-affected zone. The grain structure of such a weld—exposed to corrosion by chemical reaction—is shown in Fig. 29.5. The center vertical line is where the two workpieces meet.

The effects of heat during welding on the HAZ for joints made with dissimilar metals, and for alloys strengthened by other methods, are complex and beyond the scope of this text. Details can be found in the more specialized texts listed in the bibliography at the end of this chapter.

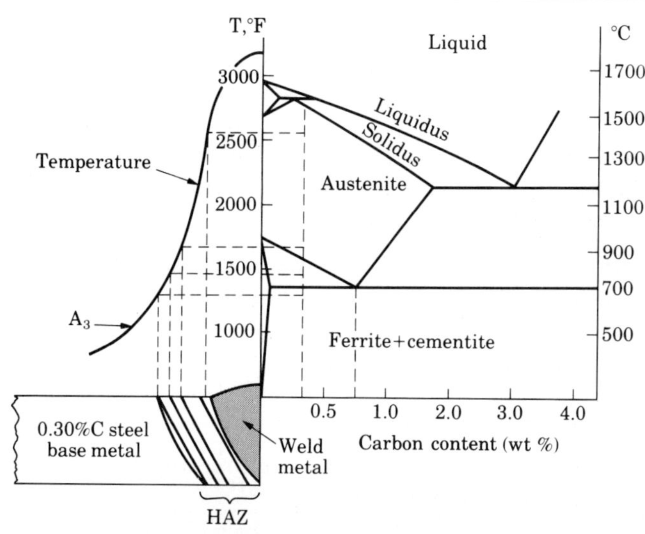

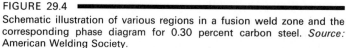

FIGURE 29.4 Schematic illustration of various regions in a fusion weld zone and the corresponding phase diagram for 0.30 percent carbon steel. *Source:* American Welding Society.

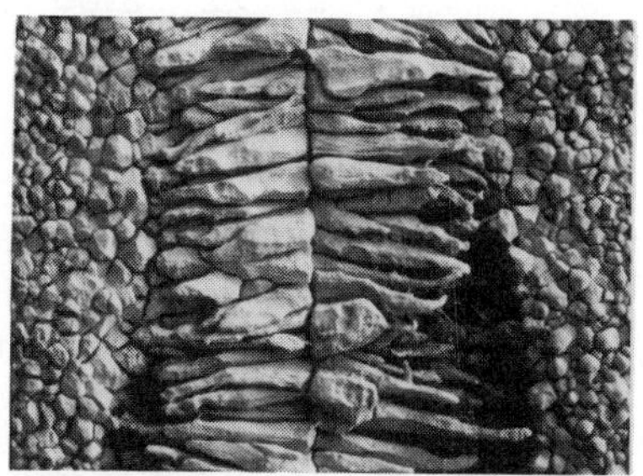

FIGURE 29.5 Intergranular corrosion of a 310 stainless steel welded tube after exposure to a caustic solution. The weld line is at the center of the photograph. Scanning electron micrograph at 20 ×. *Source*: Courtesy of B. R. Jack, Allegheny Ludlum Steel Corp.

29.3 Weld Quality

Because of its history of thermal cycling and attendant microstructural changes, a welded joint may develop certain imperfections and discontinuities. It should be pointed out, however, that many of these defects can also be caused by poor or careless application of welding techniques or poor training of the operator.

29.3.1 Porosity

Porosity in welds is caused by trapped gases released during solidification of the weld area, chemical reactions during welding, or contaminants. Most welded joints contain some porosity, which is generally spherical in shape or in the form of elongated pockets. The distribution of porosity in the weld zone may be random, or it may be concentrated in a certain region.

Porosity in welds can be reduced by the following methods:

a) Proper selection of electrodes and filler metals.
b) Improving welding techniques, such as preheating the weld area or increasing the rate of heat input.
c) Proper cleaning and preventing contaminants from entering the weld zone.

29.3.2 Slag inclusions

Slag inclusions are compounds such as oxides, fluxes, and electrode-coating materials that are trapped in the weld zone. If shielding gases are not effective during welding, contamination from the environment may also contribute to such inclusions. Welding conditions are important, and with proper techniques the molten slag will float to the surface of the molten weld metal and not be entrapped. Slag inclusions may be prevented by:

a) Cleaning the weld-bead surface before the next layer is deposited by using a hand or power wire brush (see Fig. 27.4).
b) Providing adequate shielding gas.
c) Changing the type of electrode.

29.3.3 Incomplete fusion or penetration

Incomplete fusion (or lack of fusion) produces poor weld beads, such as those shown in Fig. 29.6. A better weld can be obtained by:

a) Raising the temperature of the base metal.
b) Cleaning the weld area prior to welding.
c) Changing the joint design and type of electrode.
d) Providing adequate shielding gas.

Incomplete penetration occurs when the depth of the welded joint is insufficient. Penetration can be improved by:

a) Increasing the heat input.
b) Lowering travel speed during welding.
c) Changing the joint design.

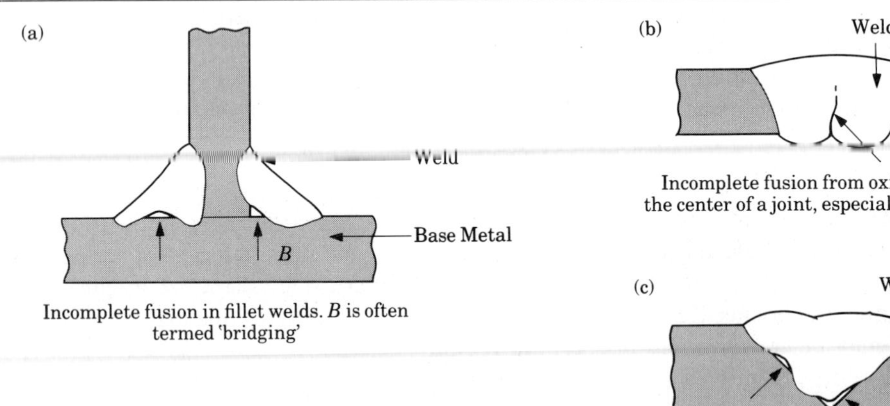

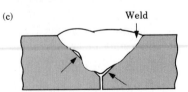

FIGURE 29.6
Examples of various defects in fusion welds. *Source:* American Welding Society.

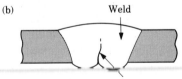

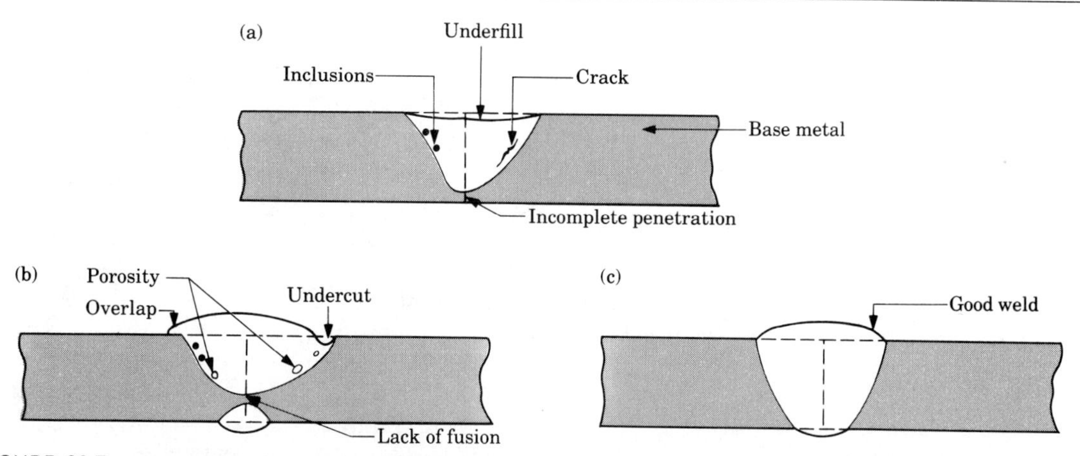

FIGURE 29.7
Examples of various defects in fusion welds. *Source:* American Welding Society.

29.3.4 Weld profile

Weld profile is important not only because of its effects on the strength and appearance of the weld, but also because it can indicate incomplete fusion or the presence of slag inclusions in multiple-layer welds (see Fig. 27.4). **Underfilling** results when the joint is not filled with the proper amount of weld metal (Fig. 29.7a). **Undercutting** (Fig. 29.7b) results from melting away the base metal and subsequently generating a groove in the shape of a sharp recess or notch. Unless it is not deep or sharp, an undercut can act as a stress raiser and reduce the fatigue strength of the joint—and may lead to premature failure. **Overlap** (Fig. 29.7b) is a surface discontinuity generally caused by poor welding practice and selection of the wrong materials. A proper weld is shown in Fig. 29.7(c).

29.3.5 Cracks

Cracks may occur in various locations and directions in the weld area. The types of cracks are typically longitudinal, transverse, crater, underbead, and toe cracks (Fig. 29.8). These cracks generally result from a combination of the following factors:

a) Temperature gradients that cause thermal stresses in the weld zone.
b) Variations in the composition of the weld zone that cause different contractions.
c) Embrittlement of grain boundaries by segregation of elements, such as sulfur, to the grain boundaries as the solid–liquid boundary moves when the weld metal begins to solidify.
d) Hydrogen embrittlement.
e) Inability of the weld metal to contract during cooling (Fig. 29.9)—a situation similar to hot tears that develop in castings (see Fig. 10.11).

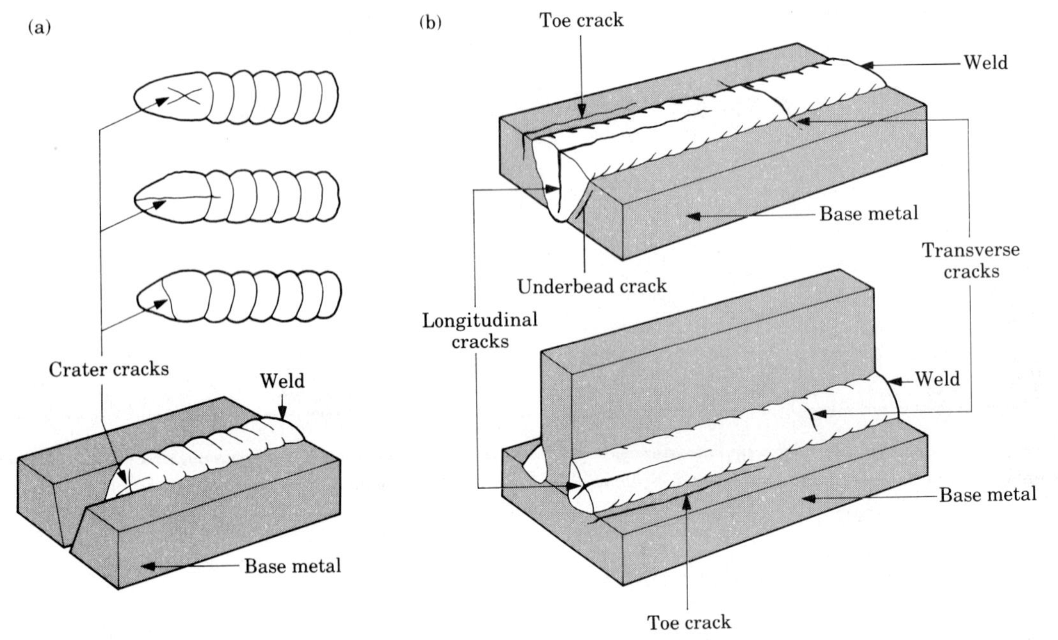

FIGURE 29.8 Types of cracks in welded joints caused by thermal stresses that develop during solidification and contraction of the weld bead and the welded structure. (a) Crater cracks. (b) Various types of cracks in butt and T joints.

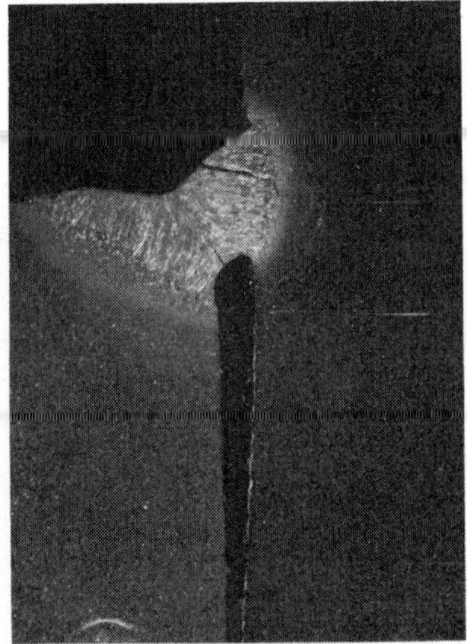

FIGURE 29.9 Crack in a weld bead. The two components were not allowed to contract after the weld was completed. *Source:* S. L. Meiley, Packer Engineering Associates, Inc.

Cracks are classified as *hot* or *cold cracks*. Hot cracks occur while the joint is still at elevated temperatures. Cold cracks develop after the weld metal has solidified. Some crack-prevention measures are:

a) Change the weld design to minimize stresses from shrinkage during cooling.
b) Change welding-process parameters, procedures, and sequence.
c) Preheat components being welded.
d) Avoid rapid cooling of the components after welding.

29.3.6 Lamellar tears

In describing the anisotropy of plastically deformed metals (see Section 1.5.1), we stated that because of the alignment of nonmetallic impurities and inclusions (stringers) the workpiece is weaker when tested in its thickness direction. This condition is particularly evident in rolled plates and structural shapes. In welding such components, *lamellar tears* may develop because of shrinkage of the restrained members in the structure during cooling (Fig. 29.10). Such tears can be avoided by providing for shrinkage of the members or by changing the joint design to make the weld bead penetrate the weaker member farther.

29.3.7 Surface damage and irregularities

During welding, some of the metal may spatter and be deposited as small droplets on adjacent surfaces. In arc welding processes, the electrode may inadvertently touch the parts being welded at places not in the weld zone (arc strikes). Such surface defects may be objectionable for reasons of appearance or subsequent use of the welded part. If severe, these defects may adversely affect the properties of the welded structure, particularly for notch-sensitive metals. Using proper welding techniques and procedures is important in avoiding surface damage.

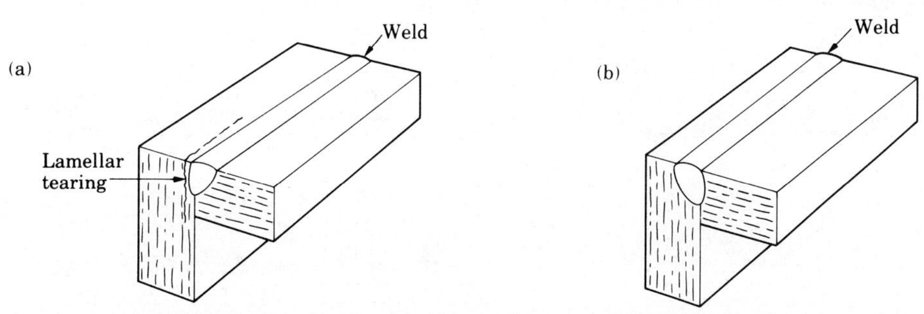

FIGURE 29.10 (a) Lamellar tears in heavy welded construction, showing lack of ductility of the vertical plate in its thickness direction, resulting from the presence of stringers (mechanical fibering). (b) Avoiding tears by changing the shape of the weld groove, with deeper penetration into the vertical plate.

29.3.8 Residual stresses

Because of localized heating and cooling during welding, expansion and contraction of the weld area causes residual stresses in the workpiece. Residual stresses can cause:

a) Distortion, warping, and buckling of the welded parts (Fig. 29.11).
b) Stress-corrosion cracking.
c) Further distortion if a portion of the welded structure is subsequently removed, say, by machining or sawing.
d) Reduced fatigue life.

We can best describe the type and distribution of residual stresses in welds by reference to Fig. 29.12(a). When two plates are being welded, a long narrow region is subjected to elevated temperatures, whereas the plates as a whole are essentially at ambient temperature. As the weld is completed and time elapses, the heat from the weld area dissipates laterally to the plates as the weld area cools. The plates thus begin to expand longitudinally while the welded length begins to contract.

These two opposing effects cause residual stresses that are typically distributed as shown in Fig. 29.12(b). Note that the magnitude of compressive residual stresses in the plates diminishes to zero at a point away from the weld area. Because no external forces are acting on the welded plates, the tensile and compressive forces represented by these residual stresses must balance each other.

In complex welded structures, residual stress distributions are three dimensional and difficult to analyze. The preceding example involves two plates that are not restrained from movement. In other words, the plates are not an integral part of a larger structure. If they are restrained, reaction stresses will be generated because the plates are not free to expand or contract. This situation arises particularly in structures with high stiffness.

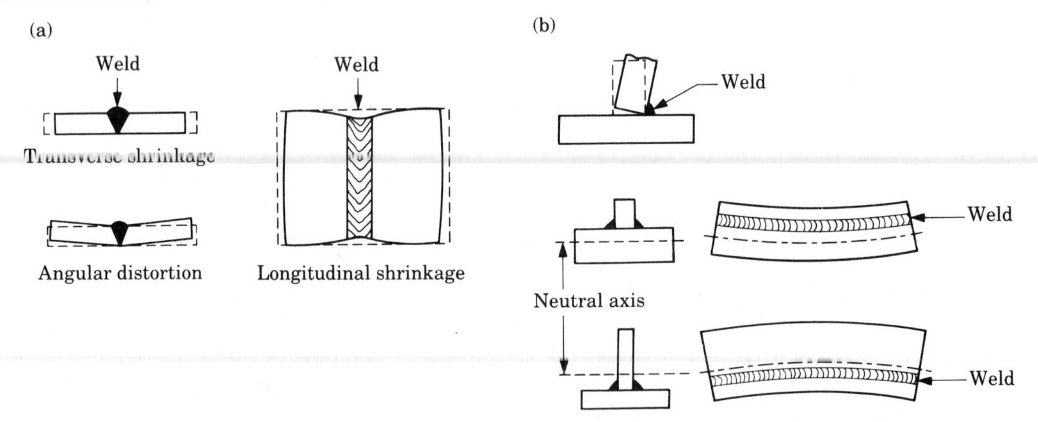

FIGURE 29.11
Distortion of parts after welding: (a) butt joints and (b) fillet welds. Distortion is caused by differential thermal expansion and contraction of different parts of the welded assembly. Warping can be reduced or eliminated by proper fixturing of the parts prior to welding.

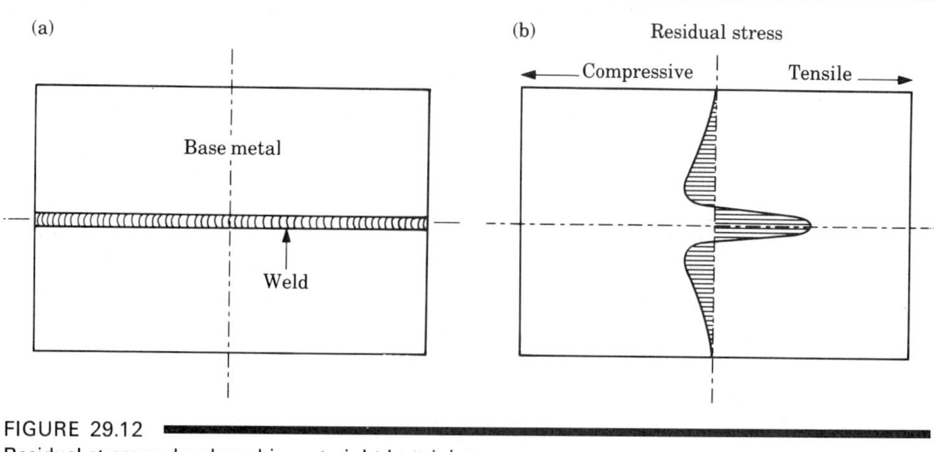

FIGURE 29.12
Residual stresses developed in a straight butt joint.

29.3.9 Stress relieving of welds

The problems caused by residual stresses, such as distortion, buckling, or cracking, can be reduced by preheating the base metal or the parts to be welded. Preheating reduces distortion by reducing the cooling rate and the level of thermal stresses (by reducing the elastic modulus). This technique also reduces shrinkage and possible cracking of the joint. The workpieces may be heated in a furnace or electrically or inductively, and for thin sections, by radiant lamps or hot-air blast. For optimum results, preheating temperatures and cooling rates must be controlled carefully in order to maintain acceptable strength and toughness in the welded structure.

Residual stresses can be reduced by *stress relieving* the welded structure. The temperature and time required for stress relieving depend on the type of material and magnitude of the residual stresses developed. These parameters must be selected carefully in order to retain the base metal's original properties of strength and toughness. For structures that are too large to be stress relieved in a furnace, the technique of in-situ (meaning "in the original shape") stress relieving is utilized. The structure (such as a vessel) is completely insulated, and the entire structure is used as a furnace to bring it up to the required temperature. The technique of partial stress relief may also be employed. In this case, only the immediate areas surrounding the welds are subjected to the stress-relief cycle.

Other methods of stress relieving include peening, hammering, or surface rolling the weld bead area (see Section 33.2). These processes induce compressive residual stresses, thus reducing or eliminating tensile residual stresses in the weld. For multilayer welds, the first and last layers should not be peened in order to protect them against possible peening damage.

Residual stresses can also be relieved, or reduced, by plastically deforming of the structure by a small amount. This technique can be used in some welded structures, such as pressure vessels, by pressurizing the vessels internally (proof-stressing). In order to reduce the possibility of sudden fracture under high internal pressure, the weld must be made properly and be free from notches and discontinuities, which could act as points of stress concentration.

Vibratory means of stress relieving in which the welded part is vibrated at one of its resonant frequencies may also be used. This technique is relatively new, and its results are controversial. The metallurgical structure and hardness of the weld area is not affected by this process.

In addition to stress relieving, welds may also be heat treated by various techniques in order to modify their properties (see Chapter 4). These techniques include annealing, normalizing, or quenching and tempering of steels and solution treatment and aging of various alloys.

29.4 Weldability

We may define **weldability** of a metal as its capacity to be welded into a specific structure that has certain properties and characteristics and that will satisfactorily meet its service requirements. Weldability involves a large number of variables, making generalizations difficult. As you have seen, the material characteristics —such as alloying elements, impurities, inclusions, grain structure, and processing history—of the base metal and filler metal are important.

Because of the melting, solidification, and microstructural changes involved, a thorough knowledge of the phase diagram and the response of the metal or alloy to elevated temperatures over a period of time is essential. Also influencing weldability are the mechanical and physical properties of strength, toughness, ductility, notch sensitivity, elastic modulus, specific heat, melting point, thermal expansion, surface-tension characteristics of the molten metal, and corrosion.

Preparation of surfaces for welding is important, as are the nature and properties of surface oxide films and adsorbed gases. The welding process employed significantly affects the temperatures developed and their distribution in the weld zone. Other factors are shielding gases, fluxes, moisture content of the coatings on electrodes, welding speed, welding position, cooling rate, preheating, and post-welding techniques (such as stress relieving and heat treating).

The following list states generally the weldability of specific metals, which can vary if special welding techniques are used.

a) *Plain-carbon steels*: Excellent for low-carbon steels; fair to good for medium-carbon steels; poor for high-carbon steels.
b) *Low-alloy steels*: Similar to that for medium-carbon steels.
c) *High-alloy steels*: Generally good under well-controlled conditions.
d) *Stainless steels*: Weldable by various processes.

e) *Aluminum alloys*: Weldable at a high rate of heat input. Alloys containing zinc or copper generally are considered unweldable.
f) *Copper alloys*: Similar to that of aluminum alloys.
g) *Magnesium alloys*: Weldable with the use of protective shielding gas and fluxes.
h) *Nickel alloys*: Similar to that of stainless steels.
i) *Titanium alloys*: Weldable with the proper use of shielding gases.
j) *Tantalum*: Similar to that of titanium.
k) *Tungsten*: Weldable under well-controlled conditions.
l) *Molybdenum*: Similar to that of tungsten.
m) *Columbium*: Good.
n) *Beryllium*: Weldable under well-controlled conditions.

29.5 Testing Welded Joints

As in all manufacturing processes, the quality of a welded joint is established by testing. Several standardized tests and test procedures have been established and are available from organizations such as ASTM, AWS, ANSI, ASME, ASCE, and federal agencies.

Welded joints may be tested either destructively or nondestructively (see Sections 36.2 and 36.3). Each technique has certain capabilities, sensitivity, limitations, reliability, and need for special equipment and operator skill.

29.5.1 Destructive techniques

Five methods of destructively testing welded joints are commonly used. We look at each method briefly below.

Tension test. Longitudinal and transverse tension tests are performed on specimens removed from actual welded joints and from the weld metal area (Fig. 29.13). Stress–strain curves are obtained by the procedures we described in Chapter 2. These curves indicate the yield strength (Y), ultimate tensile strength (UTS), and ductility of the welded joint in different locations and directions. Ductility is measured in terms of percentage elongation and percentage reduction of area.

Tension-shear test. The specimens in the tension-shear test (Figs. 29.14a and b) are specially prepared to simulate actual welded joints and procedures. The specimens are subjected to tension, and the shear strength of the weld metal and the location of fracture are determined.

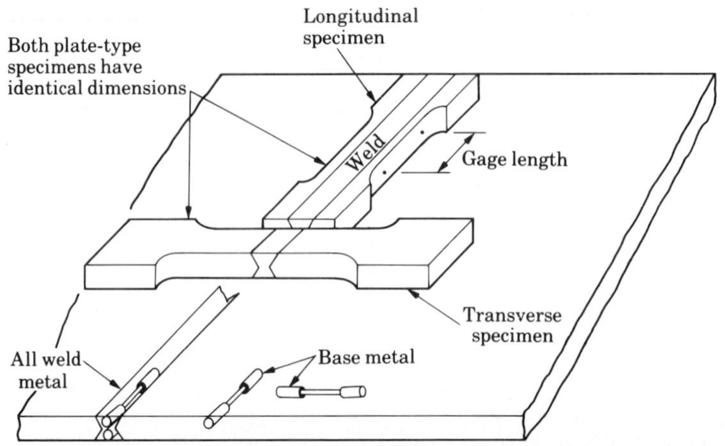

FIGURE 29.13
Typical tension-test specimens removed from welded joints for evaluating weld strength.

Bend test. Several bend tests have been developed to determine the ductility and strength of welded joints. In one test, the welded specimen is bent around a fixture (wrap-around bend test; Fig. 29.15a). In another test, the specimens are tested in three-point transverse bending (Fig. 29.15b). These tests help establish the relative ductility and strength of welded joints.

Fracture toughness test. Fracture toughness tests commonly utilize the impact testing techniques we described in Section 2.9. Charpy V-notch specimens are prepared and tested for toughness. Other toughness tests include the drop-weight test in which the energy is supplied by a falling weight.

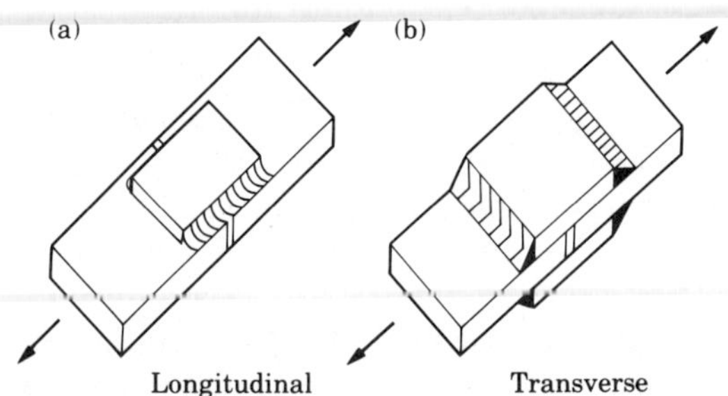

FIGURE 29.14
Types of specimens for tension-shear testing of welds.

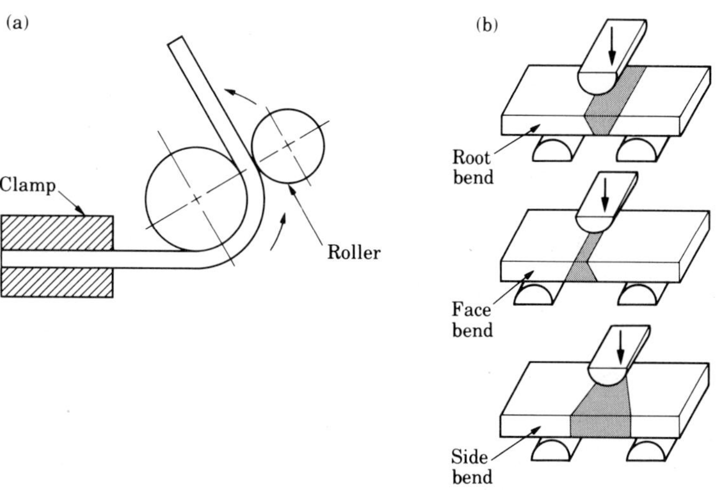

FIGURE 29.15
(a) Wrap-around bend test method. (b) Three-point bending of welded specimens (see also Fig. 2.14).

Corrosion and creep tests. In addition to mechanical tests, welded joints may be tested for corrosion and creep resistance. Because of the difference in the composition and microstructure of the materials in the weld zone, preferential corrosion may take place in there (see Fig. 29.5). Creep tests are important in determining the behavior of welded joints at elevated temperatures. Testing weld hardness may also be used to indicate weld strength and microstructural changes in the weld zone.

Testing of spot welds. Spot welded joints may be tested for weld-nugget strength using the (a) tension-shear, (b) cross-tension, (c) twist, and (d) peel tests (Fig. 29.16). Because they are easy to perform and inexpensive, tension-shear tests are commonly used in fabricating facilities. The cross-tension and twist tests are capable of revealing flaws, cracks, and porosity in the weld area. The peel test is commonly used for thin sheets. After bending and peeling the joint, the shape and size of the torn-out weld nugget is observed.

29.5.2 Nondestructive techniques

Welded structures often have to be tested nondestructively, particularly for critical applications where weld failure can be catastrophic, such as in pressure vessels, load-bearing structural members, and power plants. Nondestructive testing techniques for welded joints usually consist of visual, radiographic, magnetic particle, liquid penetrant, and ultrasonic testing methods.

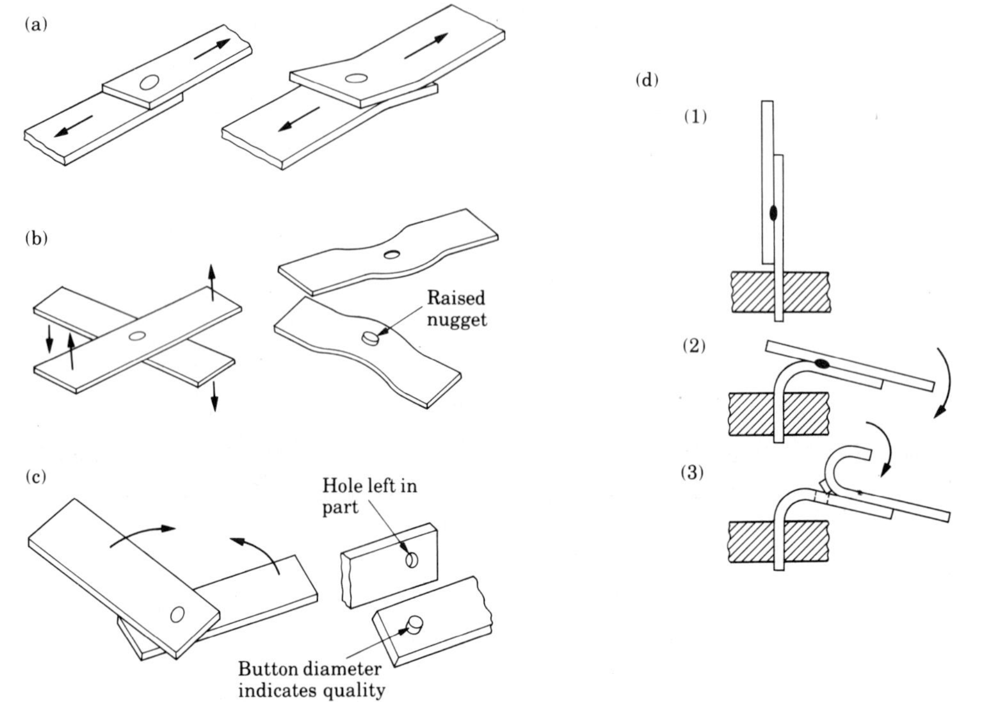

FIGURE 29.16
(a) Tension-shear test for spot welds. (b) Cross-tension test. (c) Twist test. (d) Peel test.

29.6 Welding Design and Process Selection

We have already shown some of the basic types of joint and weld design (see Fig. 27.1). In addition to the material characteristics that we have described thus far, selection of a joint and a welding process involves the following considerations:

a) Configuration of the parts or structure to be welded and their thickness and size.
b) The methods used to manufacture component parts.
c) Service requirements, such as the type of loading and stresses generated.
d) Location, accessibility, and ease of welding.
e) Effects of distortion and discoloration.

f) Appearance.
g) Costs involved in edge preparation, welding, and post-processing of the weld, including machining and finishing operations.

As in all manufacturing processes, the optimum choice is the one that meets all design and service requirements at minimum cost. Some examples of weld characteristics are shown in Fig. 29.17, emphasizing the need for careful consideration of some of the factors just identified. We may summarize general design

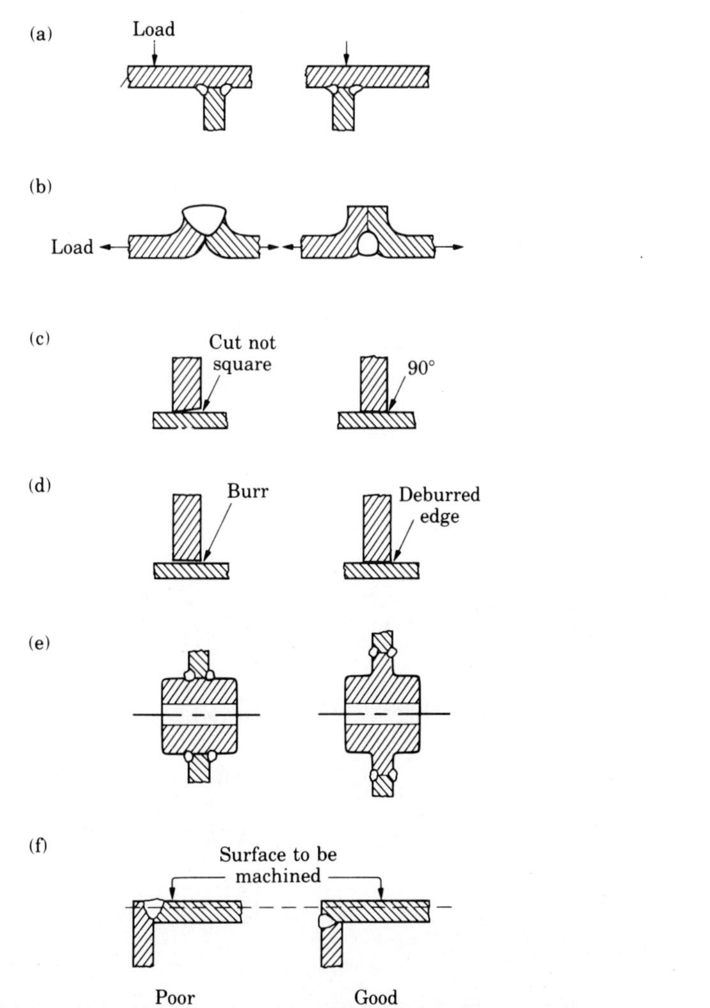

FIGURE 29.17
Design guidelines for welding. *Source:* J. G. Bralla (ed.), *Handbook of Product Design for Manufacturing.* Copyright © 1986, McGraw-Hill Publishing Company. Used with permission.

guidelines as follows:

a) Product design should minimize the number of welds, as welding can be costly unless automated. The weld location should be selected to avoid excessive stresses or stress concentrations in the welded structure and for appearance.
b) Parts should fit properly before welding. Thus the method used to produce edges (sawing, machining, shearing, and flame cutting) can affect weld quality.
c) Some designs can avoid the need for edge preparation.
d) Weld-bead size should be kept to a minimum to conserve weld metal. Weld location should be selected so as not to interfere with further processing of the part or with its intended use and appearance.

Standardized symbols used in engineering drawings to describe the type of weld and its characteristics are shown in Fig. 29.18. These symbols identify the type of

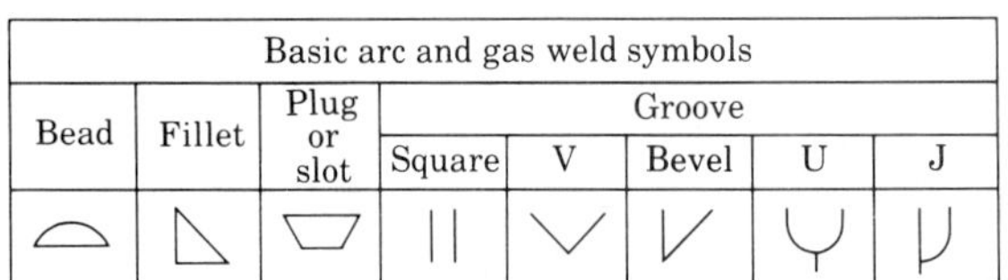

Basic arc and gas weld symbols							
Bead	Fillet	Plug or slot	Groove				
			Square	V	Bevel	U	J

Basic resistance weld symbols			
Spot	Projection	Seam	Flash or upset

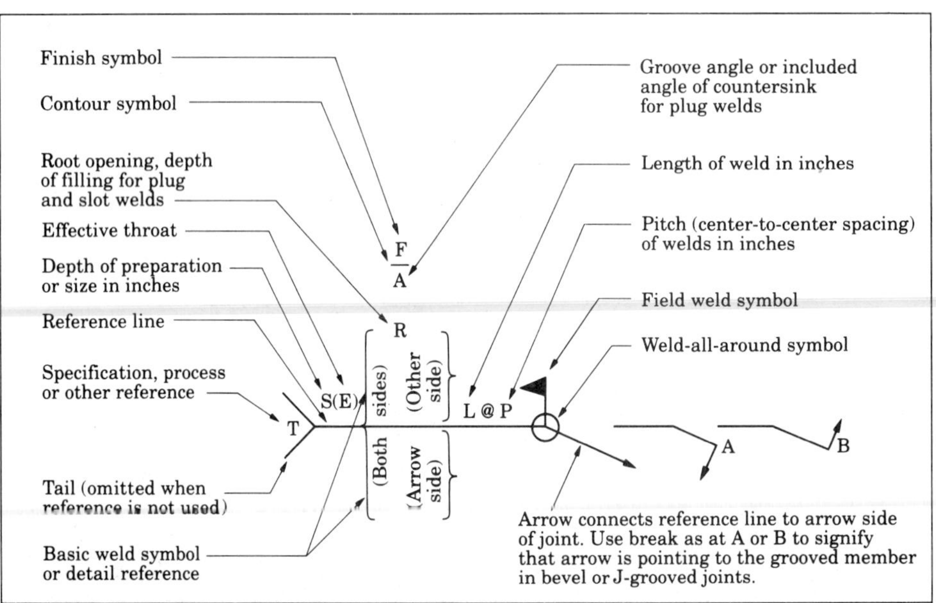

FIGURE 29.18
Standard identification and symbols for welds.

weld, groove design, weld size and length, welding process, sequence of operations, and various other information. Further information about these symbols can be found in the references cited at the end of this chapter.

Example: Weld design selection.

The figures below show three different types of weld design. In example (a), the two vertical joints can be welded externally or internally. Full-length external welding takes considerable time and requires more weld material than the alternative design, which consists of intermittent internal welds. Moreover, in the alternative method, the appearance of the structure is improved and distortion is reduced. In example (b), although both designs require the same amount of weld material and welding time, it can be shown that the design on the right can carry three times the moment *M* of the one on the left.

In example (c), the weld on the left requires about twice the amount of weld material than the design on the right. Note also that because more material must be machined, the design on the left will require more time for edge preparation and more base metal will be wasted.

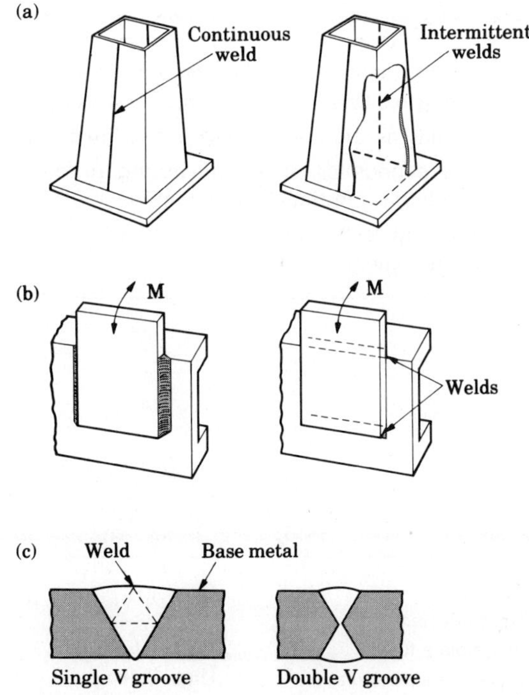

SUMMARY

The metallurgy of the welded joint is an important aspect of all welding processes because it determines the strength and toughness of the joint. The welded joint consists of solidified metal and a heat-affected zone, with a wide variation in microstructure and properties, depending on the metals joined and the filler metals. Because of severe thermal gradients in the weld zone, distortion, residual stresses, and cracking can be a significant problem.

Metals and alloys can be welded and joined by a variety of processes. Their weldability depends greatly on their composition, the type of welding operation involved, and the control of welding parameters. Joint design, as well as surface preparation, protective atmospheres, the appearance and quality of the weld, and subsequent testing of welded joints for safety and reliability are important. General guidelines are available to help in the initial selection of suitable and economical welding processes for a particular application.

TRENDS

- The characteristics and properties of the welded joint for newly developed high-strength alloys are under continued study.
- Quality assurance and weld reliability continue to be important topics.
- The economics of manufacturing products by various methods other than welding is continually under study.
- Welding techniques for aluminum aerospace structures are under continued development.

KEY TERMS

Base metal
Heat-affected zone
Incomplete fusion
Incomplete penetration
Overlap
Porosity
Slag inclusions
Undercutting
Underfilling
Weld metal
Weldability

BIBLIOGRAPHY

Cary, H.B., *Modern Welding Technology*. Englewood Cliffs, N.J.: Prentice-Hall, 1979.

Davies, A.C., *The Science and Practice of Welding*, 7th ed. Cambridge, England: Cambridge University Press, 1977.

——, *The Science and Practice of Welding*, 2 vols. New York: Cambridge University Press, 1989.

Esterling, K.E., *Introduction to the Physical Metallurgy of Welding*. Stoneham, Mass.: Butterworths, 1983.

Galyen, J., G. Sear, and C. Tuttle, *Welding: Fundamentals and Procedures*. New York: Wiley, 1985.

Gray, T.G.F., J. Spence, and T.H. North, *Rational Welding Design*. New York: Butterworths, 1975.

Horwitz, H., *Welding: Principles and Practices*. Boston: Houghton Mifflin, 1979.

Lancaster, J.F., *Metallurgy of Welding*, 3d ed. London: George Allen and Unwin, 1980.

Metals Handbook, 9th ed., *Vol. 6: Welding, Brazing, and Soldering*. Metals Park, Ohio: American Society for Metals, 1983.

Mohler, R., *Practical Welding Technology*. New York: Industrial Press, 1983.

Principles of Industrial Welding. Cleveland, Ohio: The James F. Lincoln Arc Welding Foundation, 1978.

Schwartz, M.M., *Metals Joining Manual*. New York: McGraw-Hill, 1979.

Welding Handbook, 8th ed., 3 vols. Miami: American Welding Society, 1987.

Welding Inspection. Miami: American Welding Society, 1980.

REVIEW QUESTIONS

29.1 Describe the features of a fusion weld and identify the different regions.

29.2 Describe the effects of rising temperature in the weld zone, as shown in Fig. 29.1.

29.3 Discuss factors that contribute to the shape of grains in a weld zone.

29.4 What are the characteristics of the heat-affected zone?

29.5 Describe your observations concerning Figs. 29.3 and 29.4.

29.6 What is meant by weld quality?

29.7 Discuss the factors that influence weld quality.

29.8 Describe the common types of defects in welds.

29.9 Why are residual stresses important in welded components?

29.10 Describe the methods used for relieving or reducing residual stresses in welds.

29.11 Explain how the residual stress distribution shown in Fig. 29.12 develops during welding.

29.12 What is weldability?

29.13 Explain why aluminum can be difficult to weld.

29.14 How does the weldability of steel change as its carbon content increases?

29.15 Explain why some joints may have to be preheated prior to welding.

29.16 Explain the common methods used for testing welded joints.

29.17 Discuss the basic concepts for good weld design practice.

29.18 What is the significance of structures that undergo distortion during welding, as shown in Fig. 29.11?

29.19 List the rules that must be followed to avoid cracking in welded joints.

29.20 Describe your observations concerning welding design guidelines illustrated in Fig. 29.17.

QUESTIONS AND PROBLEMS

29.21 What are the similarities and differences between casting of metals and fusion welds?

29.22 Are there common factors among the weldability, castability, formability, and machinability of metals? Explain with appropriate examples.

29.23 Explain the role of the stiffness of various components to be welded on weld defects.

29.24 Discuss the weldability of several metals and explain why some metals are easier to weld than others. Cast iron is generally difficult to weld. Why?

29.25 Must the filler metal be of the same composition as the base metal to be welded? Explain.

29.26 Comment on your observations concerning Fig. 29.16.

29.27 Describe the factors that contribute to the change in properties across a welded joint.

29.28 Describe the factors that contribute to the cost of welding. Which costs would be difficult to minimize?

29.29 Assume that you are asked to inspect a weld for a critical application. Describe the procedure that you would follow during your inspection. Would the size of the part or structure you are inspecting have an influence on your methodology?

29.30 If you find a flaw in a welded joint during inspection, how would you go about determining whether or not this flaw is important?

30

Brazing, Soldering, Adhesive Bonding, and Mechanical Joining Processes

30.1 Introduction

In almost all the joining processes we described in Chapters 27–29, the metals to be joined were heated to elevated temperatures by various means to cause fusion or bonding at the joint. But what if you want to join materials that cannot withstand high temperatures, such as electronic components? What if the parts to be joined are delicate or intricate or are made of two or more materials with very different characteristics, properties, thicknesses, and cross-sections?

In this chapter, we first describe two joining processes, namely, brazing and soldering, that permit lower temperatures than those required for welding. In both brazing and soldering, filler metals are placed in or supplied to the joint. They are then melted using an external source of heat and, upon solidification, a strong joint results. Soldering temperatures are lower than those of brazing, and the strength of a soldered joint is not high. Thus brazing and soldering are arbitrarily distinguished by temperature.

We also describe adhesive bonding techniques. The ancient method of sticking parts with animal glue, as is done in labeling, packaging, and bookbinding, has now developed into an important technology with wide applications in the aerospace and various other industries, using a wide variety of adhesives.

All the joints that we have described so far are of a permanent nature. In many applications, there are situations where joined parts have to be taken apart for replacement, maintenance, repair, or adjustment. How do we take apart a product without destroying the joint? If we need joints that are truly nonpermanent—but are as strong as welded joints—the solution obviously is to use mechanical means, such as bolts, screws, nuts, and a variety of similar fasteners. Thus in this chapter, we also discuss the advantages and limitations of mechanical joining techniques. With this chapter, we conclude our description of all commonly used methods of joining processes.

30.2 Brazing

Brazing is a joining process in which a filler metal is placed at or between the faying surfaces to be joined, and the temperature is raised to melt the filler metal but not the workpieces (Fig. 30.1a). The molten metal fills the closely fitting space by capillary action. Upon cooling and solidification of the filler metal, a strong joint is obtained. Brazing comes from the word *brass*, an archaic word meaning to harden, and was first used as far back as 3000–2000 B.C. Actually, there are two types of brazing processes: (a) that which we have already described, and (b) **braze welding** (Fig. 30.1b), in which the filler metal is deposited at the joint with a technique similar to oxyfuel gas welding.

Filler metals used for brazing melt above 450 °C (840 °F). The temperatures employed in brazing are below the melting point (solidus temperature) of the metals to be joined. Thus this process is unlike liquid-state welding processes in which the

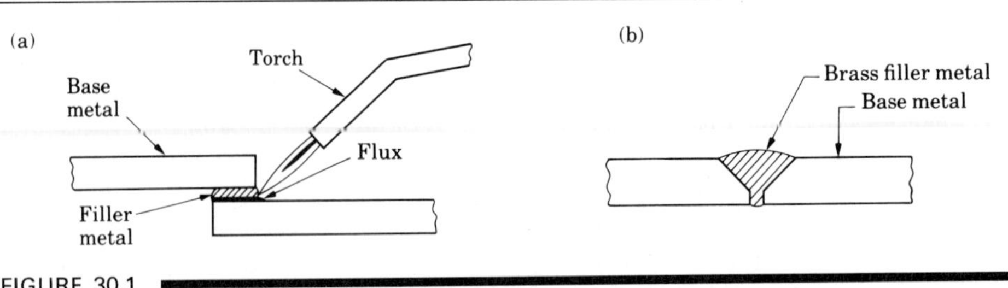

FIGURE 30.1
(a) Brazing and (b) braze welding operations.

workpieces must melt in the weld area for fusion to occur. Problems associated with heat-affected zones, warping, and residual stresses (see Chapter 29) are therefore reduced in brazing.

The strength of the brazed joint depends on (a) joint design and (b) the adhesion at the interfaces of the workpiece and filler metal. Consequently, the surfaces to be brazed should be chemically or mechanically cleaned to ensure full capillary action; hence the use of a flux is important.

30.2.1 Filler metals

Several **filler metals** (*braze metals*) are available and have a range of brazing temperatures (Table 30.1). They come in a variety of shapes, such as wire, rings, shims, and filings. The choice of filler metal and its composition are important in order to avoid embrittlement of the joint (by grain boundary penetration of liquid metal), formation of brittle intermetallic compounds at the joint, and galvanic corrosion in the joint. Thus studying the relevant phase diagrams prior to the final selection of a filler metal for a particular application is essential. Note that filler metals for brazing, unlike other welding operations, generally have significantly different compositions than the metals to be joined.

Because of diffusion between the filler metal and the base metal, mechanical and metallurgical properties of joints can change in subsequent processing or during the service life of brazed components. For example, when titanium is brazed with pure tin filler metal, it is possible for the tin to completely diffuse into the titanium base metal by subsequent aging or heat treatment. When that happens, the joint no longer exists.

30.2.2 Fluxes

The use of a flux is essential in brazing in order to prevent oxidation and to remove oxide films from workpiece surfaces. Brazing fluxes are generally made of borax, boric acid, borates, fluorides, and chlorides. Wetting agents may also be added to

TABLE 30.1
TYPICAL FILLER METALS FOR BRAZING VARIOUS METALS AND ALLOYS

BASE METAL	FILLER METAL	BRAZING TEMPERATURE, (°C)
Aluminum and its alloys	Aluminum–silicon	570–620
Magnesium alloys	Magnesium–aluminum	580–625
Copper and its alloys	Copper–phosphorus	700–925
Ferrous and nonferrous (except aluminum and magnesium)	Silver and copper alloys, copper–phosphorus	620–1150
Iron-, nickel-, and cobalt-base alloys	Gold	900–1100
Stainless steels, nickel- and cobalt-base alloys	Nickel–silver	925–1200

improve both the wetting characteristics of the molten filler metal and capillary action.

Surfaces to be brazed must be clean and free from rust, oil, and other contaminants. Clean surfaces are essential to obtain the proper wetting and spreading characteristics of the molten filler metal in the joint and maximum bond strength. Sand blasting may also be used to improve surface finish of faying surfaces. Because they are corrosive, fluxes should be removed after brazing has been completed. Fluxes can usually be removed by washing with hot water.

30.2.3 Brazing methods

The heating methods used also identify the various brazing processes. We describe these processes below.

Torch brazing. The heat source in *torch brazing* (TB) is oxyfuel gas with a carburizing flame. Brazing is performed by first heating the joint with the torch, then depositing the brazing rod or wire in the joint. Suitable part thicknesses are usually in the range of 0.25–6 mm (0.01–0.25 in.). More than one torch may be used in this process. Although it can be automated as a production process, torch brazing is difficult to control and requires skilled labor. This process can also be used for repair work. The basic equipment for manual brazing costs about \$300 but can run more than \$50,000 for automated systems.

Furnace brazing. As the name suggests, *furnace brazing* (FB) is carried out in a furnace. The parts are precleaned and preloaded with brazing metal in appropriate configurations before being placed in the furnace (Fig. 30.2). Furnaces may be batch type for complex shapes or continuous type for high production runs, especially for small parts with simple joint designs. Vacuum furnaces or neutral atmospheres are used for metals that react with the environment. This process is similar to using shielding gas in welding operations. Skilled labor is not required, and complex shapes can be brazed since the whole assembly is heated uniformly in the furnace.

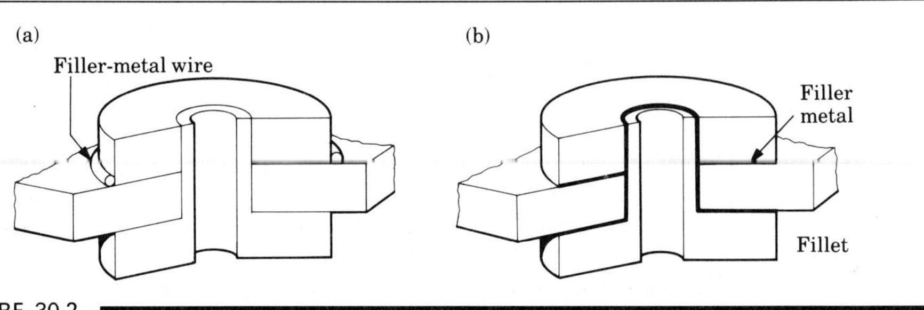

FIGURE 30.2
An example of brazing: (a) before and (b) after. Note that the filler metal is a shaped wire.

The cost of furnaces varies widely, ranging from about $2,000 for simple batch furnaces to more than $300,000 for continuous vacuum furnaces.

Induction brazing. The source of heat in *induction brazing* (IB) is induction heating by high-frequency ac current. Parts are preloaded with filler metal and are placed near the induction coils for rapid heating. Unless a protective atmosphere is utilized, fluxes are generally used. Part thicknesses are usually less than 3 mm (0.125 in.). Induction brazing is particularly suitable for continuously brazing many parts. The cost for small units is about $10,000.

Resistance brazing. In *resistance brazing* (RB), the source of heat is through electrical resistance of the components to be brazed. Electrodes are utilized for this purpose, as in resistance welding. Parts are either preloaded with filler metal, or it is supplied externally during brazing. Parts that are commonly brazed by this process have thicknesses of 0.1–12 mm (0.004–0.5 in.). As in induction brazing, the process is rapid, heating zones can be confined to very small areas, and the process can be automated to produce uniform quality. Equipment costs range from $1,000 for simple units to more than $10,000 for larger, more complex units.

Dip brazing. *Dip brazing* (DB) is carried out by dipping the assemblies to be brazed into either a molten filler-metal bath or a molten salt bath (at a temperature just above the melting point of the filler metal), which serves as the heat source. All workpiece surfaces are thus coated with the filler metal. Consequently, dip brazing in metal baths is used only for small parts, such as sheet, wire, and fittings, usually of less than 5 mm (0.2 in.) in thickness or diameter. Molten salt baths, which also act as fluxes, are used for complex assemblies of various thicknesses. Depending on the size of the parts and the bath, as many as 1000 joints can be made at one time by dip brazing. Cost of equipment varies widely: from about $2,000 to more than $200,000; the more expensive equipment comes with various computer controls.

Infrared brazing. The heat source in *infrared brazing* (IRB) is a high-intensity quartz lamp. This process is particularly suitable for brazing very thin components, usually less than 1 mm (0.04 in.) thick, including honeycomb structures. The radiant energy is focused on the joint, and the process can be carried out in a vacuum. Equipment cost ranges from $500 to $30,000.

Diffusion brazing. *Diffusion brazing* (DFB) is carried out in a furnace where—with proper control of temperature and time—the filler metal diffuses into the faying surfaces of the components to be joined. The brazing time required may range from 30 minutes to 24 hours. Diffusion brazing is used for strong lap or butt joints and for difficult joining operations. Because the rate of diffusion at the interface does not depend on the thickness of the components, part thicknesses may range from foil to as much as 50 mm (2 in.). The cost of equipment ranges from about $50,000 to $300,000.

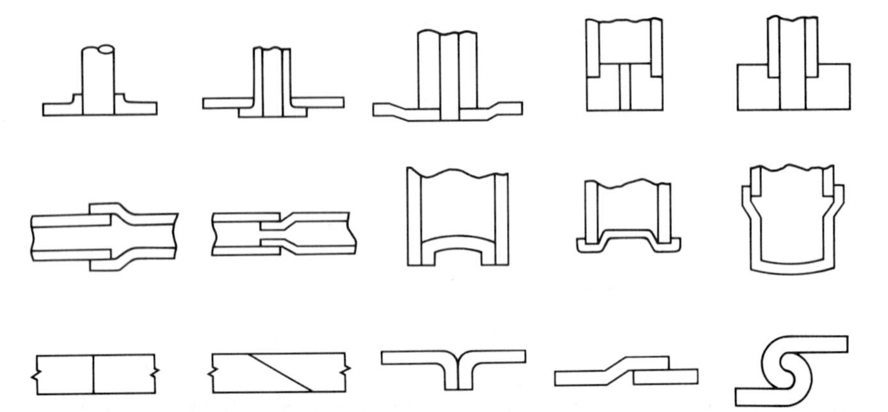

FIGURE 30.3
Joint designs commonly used in brazing operations.

30.2.4 Braze welding

The joint in braze welding is prepared as in fusion welding (see Chapter 27). Using an oxyacetylene torch with an oxidizing flame, filler metal is deposited at the joint (see Fig. 30.1b), rather than by capillary action as in brazing. Thus considerably more filler metal is used, compared to brazing. However, temperatures in braze welding are generally lower than in fusion welding, and part distortion is minimal. The use of a flux is essential in this process. The principal use of braze welding is to maintain and repair parts, such as ferrous castings and steel components.

30.2.5 Brazing process capabilities

Examples of typical brazed joints are shown in Fig. 30.3. In general, dissimilar metals can be assembled with good joint strength, including carbide drill bits or carbide inserts on steel shanks (see Fig. 22.24f). The shear strength of brazed joints can reach 800 MPa (120 ksi) using brazing alloys containing silver (silver solder). Intricate, lightweight shapes can be joined rapidly and with little distortion. Brazing can be automated and used for mass production.

30.2.6 Design for brazing

As in all joining processes, joint design is important in brazing. Some design guidelines are given in Fig. 30.4. Note that strong joints require greater contact area for brazing than for welding. The typical joint clearance in brazing ranges from 0.025 mm to 0.2 mm (0.001 in. to 0.008 in.). The clearances must fit within a very small tolerance range because larger clearances reduce the strength of the brazed joint. A variety of special fixtures may be used to hold the parts together, some with provision for thermal expansion and contraction, during brazing.

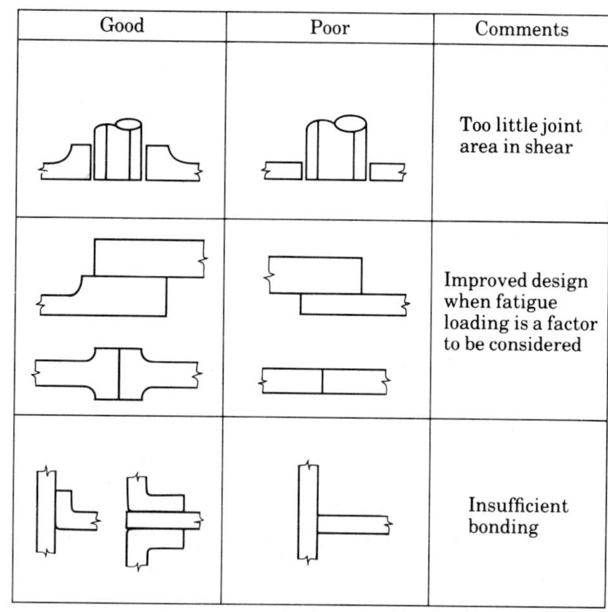

Good
Poor
Comments
Too little joint area in shear
Improved design when fatigue loading is a factor to be considered
Insufficient bonding

FIGURE 30.4
Examples of good and poor design for brazing. *Source:* American Welding Society.

30.3 Soldering

In **soldering**, the filler metal, called *solder*, melts below 450 °C (840 °F). As in brazing, the solder fills the joint by capillary action between closely fitting or closely placed components. Heat sources for soldering are usually soldering irons, torches, or ovens. Soldering with copper–gold and tin–lead alloys was first practiced as far back as 4000–3000 B.C.

There are several soldering methods, which are similar to brazing methods:

a) Torch soldering (TS).
b) Furnace soldering (FS).
c) Iron soldering (INS), using a soldering iron.
d) Induction soldering (IS).
e) Resistance soldering (RS).
f) Dip soldering (DS).
g) Infrared soldering (IRS).
h) Wave soldering (WS), used for automated soldering of printed circuit boards.

TABLE 30.2
TYPICAL SOLDERS AND THEIR APPLICATIONS

Tin–lead	General purpose
Tin–zinc	Aluminum
Lead–silver	Strength at higher than room temperature
Cadmium–silver	Strength at high temperatures
Zinc–aluminum	Aluminum; corrosion resistance

i) Ultrasonic soldering, in which a transducer subjects the molten solder to ultrasonic cavitation, which removes the oxide films from the surfaces to be joined. The need for a flux is thus eliminated.

30.3.1 Types of solders and fluxes

Solders (from the Latin *solidare*, meaning to make solid) are usually tin–lead alloys in various proportions. For better joint strength and special applications, other solder compositions that can be used are tin–zinc, lead–silver, cadmium–silver, and zinc–aluminum alloys (Table 30.2).

In soldering, fluxes are used as in welding and brazing and for the same purposes. Fluxes are generally of two types:

a) Inorganic acids or salts, such as zinc ammonium chloride solutions, which clean the surface rapidly. After soldering, the flux residues should be removed by washing thoroughly with water to avoid corrosion.
b) Noncorrosive resin-based fluxes, used in electrical applications.

30.3.2 Process capabilities

Soldering is used extensively in the electronics industry and in making containers for liquid- or air-tight joints (lock-seam cans; Fig. 30.5). Unlike brazing, soldering temperatures are relatively low. Thus a soldered joint has very limited use at elevated temperatures. Moreover, because solders do not generally have much strength, they are not used for load-bearing structural members. Because of the small faying surfaces, butt joints are rarely made with solders. In other situations, joint strength is improved by mechanical interlocking of the joint.

Copper and precious metals, such as silver and gold, are easy to solder, as is tinplate for food containers. Aluminum and stainless steels are difficult to solder because of their strong, thin oxide film (see Section 31.2). However, these and other metals can be soldered using special fluxes that modify surfaces.

Soldering can be used to join various metals and thicknesses. Although manual operations require skill and are time-consuming, soldering speeds can be high with automated equipment. The cost of soldering equipment depends on its complexity and the level of automation. It ranges from less than $100 for industrial soldering irons to more than $50,000 for automated equipment.

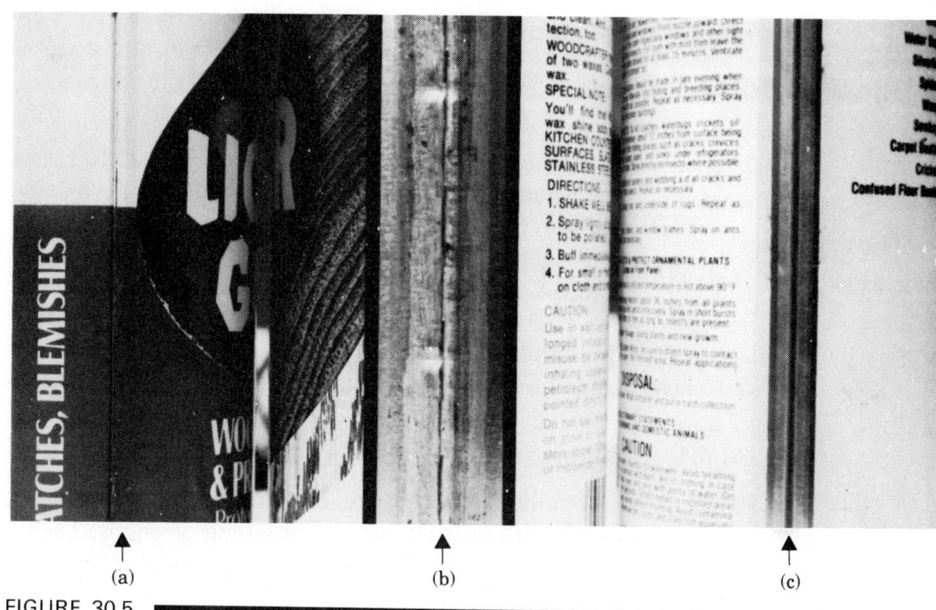

FIGURE 30.5

Three types of joining processes used on cans for household products: (a) seam lock, (b) soldered joint, and (c) resistance seam welded joint.

30.3.3 Design for soldering

Design guidelines for soldering are similar to those for brazing. Some frequently used joint designs are shown in Fig. 30.6. Note again the importance of large contact surfaces to develop sufficient joint strength in soldered products.

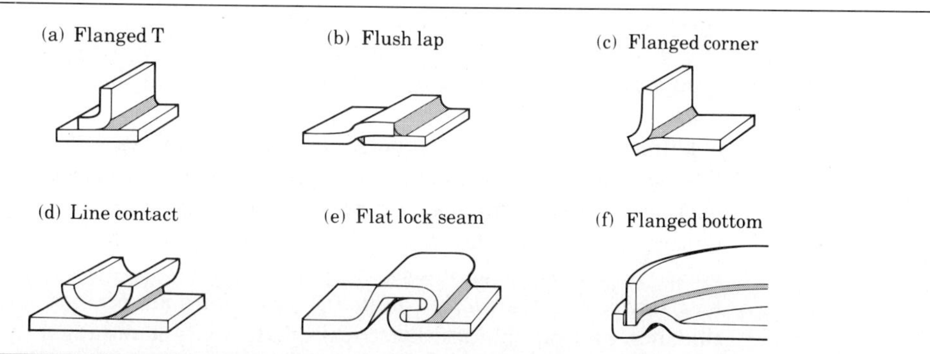

FIGURE 30.6

Joint designs commonly used for soldering. *Source:* American Welding Society.

30.4 Adhesive Bonding

Numerous components and products can be joined and assembled using an *adhesive*, rather than by any of the joining methods we have described thus far. For a long time, **adhesive bonding** has been a common method of joining and assembly for applications such as labeling, packaging, bookbinding, home furnishings, and footware. Plywood, developed in 1905, is a typical example of adhesive bonding of several layers of wood with glue. Adhesive bonding has been gaining increased acceptance in manufacturing ever since its first use on a large scale in assembling load-bearing components in aircraft during World War II (1939–1945).

30.4.1 Adhesives

Many types of adhesives are available—and continue to be developed—which provide adequate joint strength, including fatigue strength. The three basic types of adhesives are:

- *Natural adhesives*, such as starch, dextrin (a gummy substance obtained from starch), soya flour, and animal products.
- *Inorganic adhesives*, such as sodium silicate and magnesium oxychloride.
- *Synthetic organic adhesives*, which may be thermoplastics (used for non-structural and some structural bonding) or thermosetting polymers (used primarily for structural bonding).

Because of their strength, synthetic organic adhesives are the most important in manufacturing processes, particularly for load-bearing applications. They are classified as

a) *Chemically reactive*; examples are polyurethanes, silicones, epoxies, cyanoacrylates, modified acrylics, phenolics, polyimides, and anaerobics.
b) *Pressure sensitive*; examples are natural rubber, styrene-butadiene rubber, butyl rubber, nitrile rubber, and polyacrylates.
c) *Hot melt*; which are thermoplastics such as ethylene-vinyl acetate copolymers, polyolefins, polyamides, polyester, and thermoplastic elastomers.
d) *Evaporative* or *diffusion*; examples are vinyls, acrylics, phenolics, polyurethanes, synthetic rubbers, and natural rubbers.
e) *Film* and *tape*; examples are nylon-epoxies, elastomer-epoxies, nitrile-phenolics, vinyl-phenolics, and polyimides.
f) *Delayed tack*; examples are styrene-butadiene copolymers, polyvinyl acetates, polystyrenes, and polyamides.
g) *Electrically* and *thermally conductive*; examples are epoxies, polyurethanes, silicones, and polyimides. Electrical conductivity is obtained by addition of fillers such as silver (used most commonly), copper, aluminum, and gold.

Fillers that improve the electrical conductivity of adhesives generally also improve their thermal conductivity.

Least expensive of adhesives are epoxies and phenolics, followed by polyurethanes, acrylics, silicones, and cyanoacrylates. Adhesives for high-temperature applications in a range up to about 260 °C (500 °F), such as polyimides and polybenzimidazoles, are generally the most expensive.

Adhesives are available in various forms, such as liquids, pastes, solutions, emulsions, powder, tape, and film. When applied, adhesives generally are about 0.1 mm (0.004 in.) thick. Depending on the particular application, an adhesive must have one or more of the following properties (Table 30.3 on p. 928):

a) Strength (shear and peel).
b) Toughness.
c) Resistance to various fluids and chemicals.
d) Resistance to environmental degradation, including heat and moisture.
e) Capability to wet the interface to be bonded.

Adhesive joints are designed to withstand shear, compressive, and tensile forces, but they should not be subjected to peeling forces (Fig. 30.7 on p. 929). Note, for example, how easily you can peel adhesive tape from a surface. During peeling the behavior of an adhesive may be brittle, or it may be ductile and tough, requiring large forces to peel it.

30.4.2 Surface preparation and application

Surface preparation is very important in adhesive bonding. Joint strength depends greatly on the absence of dirt, dust, oil, and various other contaminants. You have no doubt tried to apply an adhesive tape over a dusty or oily surface with little success. Contaminants also affect the wetting ability of the adhesive, and they can prevent spreading of the adhesive evenly over the interface. Thick, weak, or loose oxide films on workpieces are detrimental to adhesive bonding. On the other hand, a porous or thin, strong oxide film may be desirable, particularly one with some surface roughness to improve adhesion. Various compounds and primers are available to modify surfaces to increase adhesive bond strength. Liquid adhesives may be applied by brushing, spraying, and rollers.

30.4.3 Process capabilities

A wide variety of similar and dissimilar metallic and nonmetallic materials and components with different shapes, sizes, and thicknesses can be bonded to each other by adhesives. Adhesive bonding can be combined with mechanical joining methods (see Section 30.5) to further improve the strength of the bond.

An important consideration in the use of adhesives in production is curing time, which can range from a few seconds at high temperatures to many hours at room temperature, particularly for thermosetting adhesives. (In using various glues, you undoubtedly have read instructions on the labels to this effect.) Thus production

TABLE 30.3
TYPICAL PROPERTIES AND CHARACTERISTICS OF CHEMICALLY REACTIVE STRUCTURAL ADHESIVES

	EPOXY	POLYURETHANE	MODIFIED ACRYLIC	CYANOACRYLATE	ANAEROBIC
Impact resistance	Poor	Excellent	Good	Poor	Fair
Tension–shear strength, MPa (10^3 psi)	15.4 (2.2)	15.4 (2.2)	25.9 (3.7)	18.9 (2.7)	17.5 (2.5)
Peel strength, N/m (lbf/in.)	< 525 (3)	14,000 (80)	5250 (30)	< 525 (3)	1750 (10)
Substrates bonded	Most	Most smooth, nonporous	Most smooth, nonporous	Most nonporous metals or plastics	Metals, glass, thermosets
Service temperature range, °C (°F)	– 55 to 120 (– 70 to 250)	– 160 to 80 (– 250 to 175)	– 70 to 120 (– 100 to 250)	– 55 to 80 (– 70 to 175)	– 55 to 150 (– 70 to 300)
Heat cure or mixing required	Yes	Yes	No	No	No
Solvent resistance	Excellent	Good	Good	Good	Excellent
Moisture resistance	Excellent	Fair	Good	Poor	Good
Gap limitation, mm (in.)	None	None	0.75 (0.03)	0.25 (0.01)	0.60 (0.025)
Odor	Mild	Mild	Strong	Moderate	Mild
Toxicity	Moderate	Moderate	Moderate	Low	Low
Flammability	Low	Low	High	Low	Low

Source: Advanced Materials & Processes July 1990, ASM International.

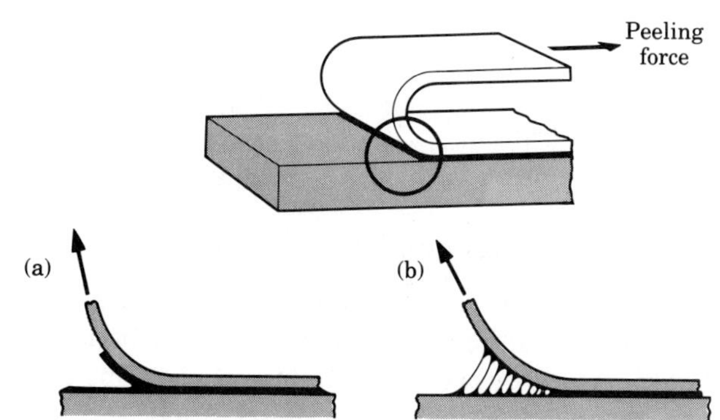

FIGURE 30.7 Characteristic behavior of (a) brittle and (b) tough adhesives in a peeling test. This test is similar to peeling adhesive tape from a solid surface. Adhesive joints should not be subjected to this type of loading in service.

rates may be low. Furthermore, adhesive bonds for structural applications are rarely suitable for service above 250 °C (500 °F). Joint design and bonding methods require care and skill. Special equipment, such as fixtures, presses, tooling, and autoclaves and ovens for curing, is usually needed.

Nondestructive inspection of the quality and strength of adhesively bonded components can be difficult. Some of the techniques that we describe in Section 36.2, such as acoustic impact (tapping), holography, infrared detection, and ultrasonic testing, are effective nondestructive testing methods.

Major industries that use adhesive bonding extensively are aerospace, automotive, appliances, and building products. Applications include attaching rear-view mirrors to windshields, automotive brake-lining assemblies, laminated windshield glass, appliances, helicopter blades, honeycomb structures, and aircraft bodies and control surfaces.

The major advantages of adhesive bonding are:

a) It provides a bond at the interface either for structural strength or for nonstructural applications such as sealing, insulating, preventing electrochemical corrosion between dissimilar metals, and reducing vibration and noise through internal damping at the joints.
b) It distributes the load at an interface, thus eliminating localized stresses that result from joining the components with welds or mechanical fasteners, such as bolts and screws. Moreover, structural integrity of the sections is maintained, since no holes are required, and the appearance of the components is generally improved.
c) Very thin and fragile components can be bonded with adhesives without contributing significantly to weight.
d) Porous materials and materials of very different properties and sizes can be joined.
e) Because it is usually carried out at a temperature between room temperature and about 200 °C (400 °F), there is no significant distortion of the components or change in their original properties. This is particularly important for materials that are heat sensitive.

The major limitations of adhesive bonding are:

a) Limited service temperatures.
b) Possibly long bonding time.
c) Need for great care in surface preparation.
d) Difficulty in testing bonded joints nondestructively, particularly for large structures.
e) Reliability of adhesively bonded structures during their service life.

The cost of adhesive bonding depends on the particular operation. However, in many applications the overall economics of the process makes adhesive bonding an attractive, and sometimes the only feasible or practical, joining process. The cost of equipment varies greatly, depending on the size and type of operation.

30.4.4 Design for adhesive bonding

Designs for adhesive bonding should ensure that joints are subjected to compressive, tensile, and shear forces, but not peeling or cleavage (see Fig. 30.7 on p. 929). Several joint designs for adhesive bonding are shown in Figs. 30.8 (p. 931) and 30.9 (p. 932). They vary considerably in strength; hence selection of the appropriate design is important and should include considerations such as type of loading and the environment.

Butt joints require large bonding surfaces. Simple lap joints tend to distort under tension because of the force couple at the joint. The coefficients of expansion of the components to be bonded should preferably be close in order to avoid internal stresses during adhesive bonding.

30.5 Mechanical Joining

Two or more components may have to be joined or fastened in such a way that they can be taken apart sometime during the product's service life. Numerous objects, including mechanical pencils, caps and lids on containers, mechanical watches, engines, and bicycles, have components that are fastened mechanically. **Mechanical joining** may be preferred for the following reasons:

a) Ease of manufacturing.
b) Ease of assembly and transportation.
c) Ease of parts replacement, maintenance, or repair.
d) Designs requiring movable joints, such as hinges, sliding mechanisms for drawers and doors, and adjustable components and fixtures.
e) Lower overall cost of manufacturing the product.

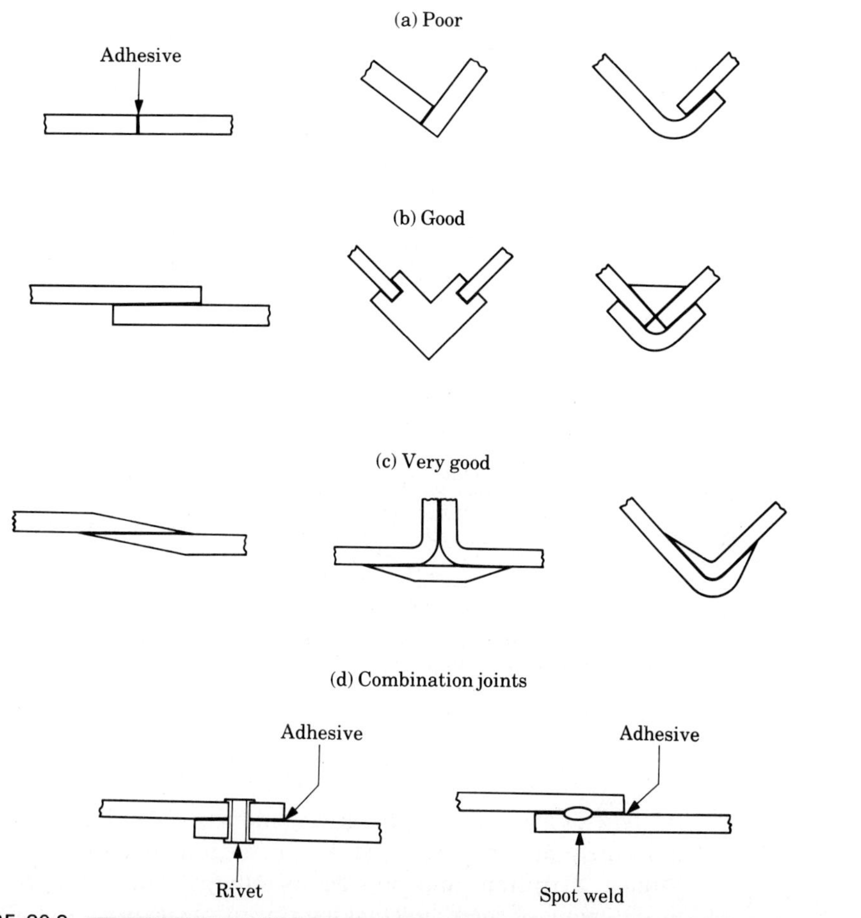

FIGURE 30.8
Various joint designs in adhesive bonding. Note that good design requires large contact areas between the members to be joined.

The most common method of mechanical joining is by fastening, using bolts, nuts, screws, pins, and a variety of other **fasteners.** These processes are also known as mechanical assembly.

Mechanical joining generally requires that the components have holes through which fasteners are inserted. These joints may be subjected to both shear and tensile stresses and should be designed to resist these forces.

30.5.1 Hole preparation

Hole preparation is an important aspect of mechanical joining. As we described in Chapters 16, 22, and 26, a hole in a solid body can be produced by punching, drilling, chemical, and electrical means, and by high-energy beams, depending on

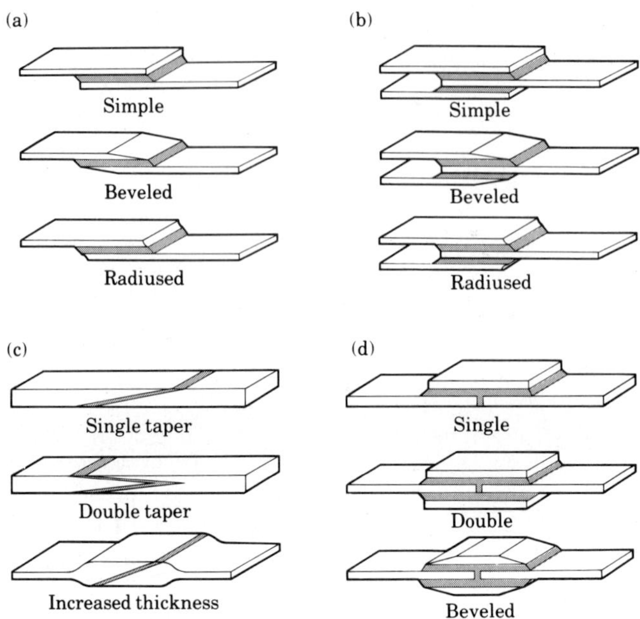

FIGURE 30.9
Various configurations for adhesively bonded joints: (a) single lap, (b) double lap, (c) scarf, and (d) strap.

the type of material, its properties, and thickness. Recall from Parts II and III that holes may also be produced as an integral part of the product during casting, forging, extrusion, and powder metallurgy. For improved accuracy and surface finish, many of these holemaking operations may be followed by finishing operations, such as shaving, deburring, reaming, and honing, which we described in Part IV.

Because of their fundamental differences, each holemaking operation produces a hole with different surface finish and properties and dimensional characteristics. The most significant influence of a hole in a solid body is its tendency to reduce the component's fatigue life (stress concentration). Fatigue life can best be improved by inducing compressive residual stresses on the cylindrical surface of the hole (Fig. 30.10). These stresses are usually developed by pushing a round rod (drift pin) through the hole and expanding it by a very small amount (see insert in Fig. 30.10). This process plastically deforms the surface layers of the hole, in a manner similar to shot peening or roller burnishing (Section 33.2).

30.5.2 Threaded fasteners

Bolts, screws, and nuts are among the most commonly used threaded fasteners. Texts on machine design describe numerous standards and specifications, including

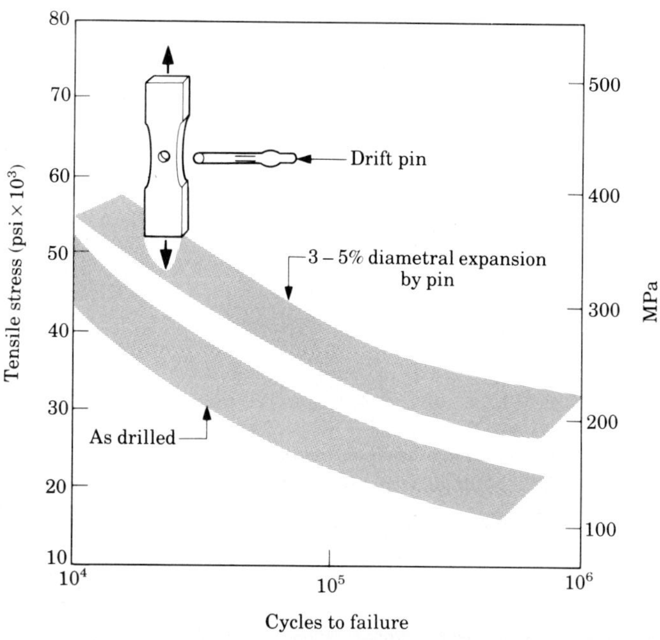

FIGURE 30.10 Effect of hole preparation on the fatigue life of 7075-T6 aluminum in tension (see also Section 2.7).

thread dimensions, tolerances, pitch, strength, and the quality of materials used to make these fasteners (see also Section 22.4.1).

Bolts and screws may be secured with nuts (machine screws), or they may be self-tapping, whereby the screw either cuts or forms the thread into the part to be fastened. This method is particularly effective and economical in plastic products. In this way, fastening does not require a tapped hole and the need for—and cost of—a nut.

If the joint is to be subjected to vibration, such as in aircraft and various types of engines and high-speed machinery, several specially designed nuts and lock washers are available. They increase the frictional resistance in the torsional direction, thus preventing vibrational loosening of the fasteners.

30.5.3 Rivets

The most common method of permanent or semipermanent mechanical joining is by *riveting* (Fig. 30.11). Hundreds of thousands of rivets may be used in the construction and assembly of large commercial aircraft. Holes for the rivets are required in the components to be joined. Installing a rivet consists of placing the rivet in the hole and deforming the end of its shank by upsetting (heading). Riveting may be done either at room temperature or hot, using special tools or explosives in the rivet cavity. Some design guidelines for riveting are illustrated in Fig. 30.12.

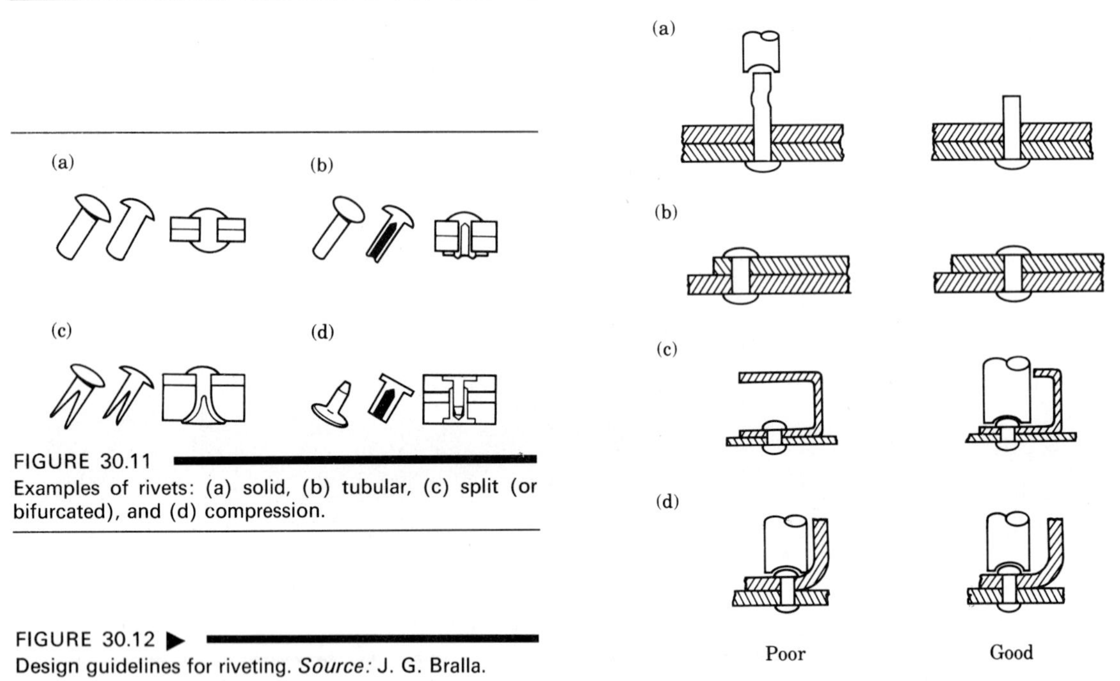

FIGURE 30.11
Examples of rivets: (a) solid, (b) tubular, (c) split (or bifurcated), and (d) compression.

FIGURE 30.12
Design guidelines for riveting. *Source:* J. G. Bralla.

30.5.4 Design for mechanical joining

The design of mechanical joints requires consideration of the type of loading, such as shear and tension, to which the structure will be subjected and the size and spacing of holes.

Compatibility of the fastener material with that of the components to be joined is important. Incompatibility may lead to galvanic corrosion, also known as *crevice corrosion* (Fig. 30.13). In a system where, for example, a steel rivet is used to fasten

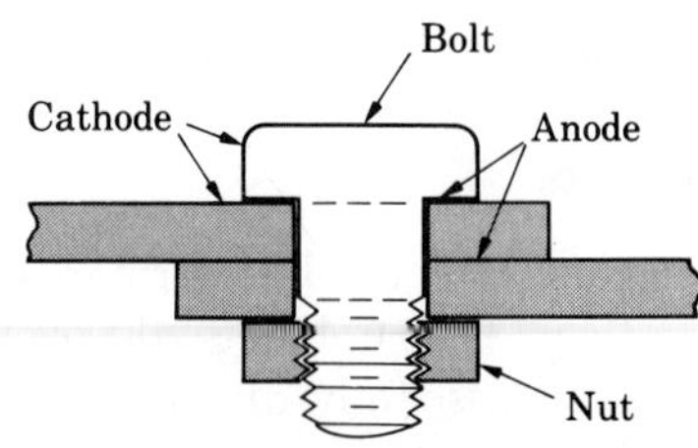

FIGURE 30.13
Galvanic corrosion of a bolted assembly. The areas that are oxygen deficient serve as the anode, and areas exposed to the environment serve as the cathode. Dirt and other contaminants increase oxygen deficiency.

copper sheets, the rivet is anodic and the copper plate cathodic, thus leading to rapid corrosion and loss of joint strength. Aluminum or zinc fasteners on copper products react similarly.

The effect of holes on the strength of the joint must be carefully evaluated. Because of the hole and its effect as a stress raiser, fatigue failure is a common problem in mechanically fastened joints.

Other general design guidelines for mechanical joining include (see also Section 38.10):

a) Using fewer but larger fasteners is generally less costly than using a large number of smaller ones.
b) Part assembly should be accomplished with a minimum number of fasteners.
c) The fit between parts to be joined should be as loose as possible to reduce costs and facilitate the assembly process.
d) Fasteners of standard size should be used whenever possible.
e) Holes should not be too close to edges or corners to avoid tearing the material when it is subjected to forces.

30.5.5 Other methods of fastening

Many types of fasteners are used in numerous joining and assembly applications. We describe the most common types below.

Metal stitching or stapling. The process of *metal stitching* or *stapling* (Fig. 30.14) is much like that of ordinary stapling of papers. This operation is fast and particularly suitable for joining thin metallic and nonmetallic materials, and it does not require holes in the components. A common example is the stapling of cardboard containers for appliances and other consumer products.

Seaming. *Seaming* is based on the simple principle of folding two thin pieces of material together. Seaming is much like joining two pieces of paper, in the absence of a paper clip, by folding them at the corner. The most common examples

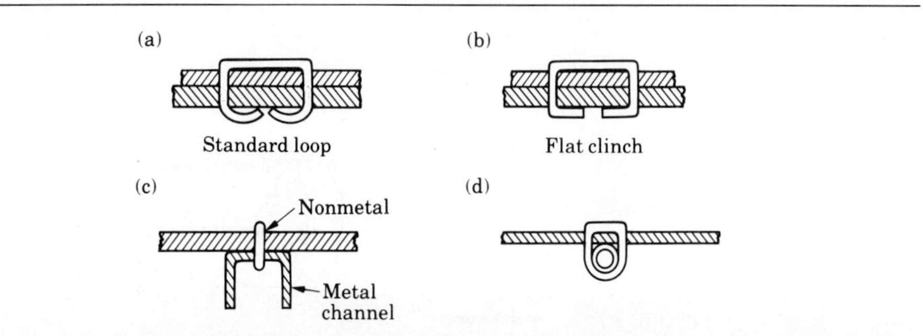

FIGURE 30.14
Examples of metal stitching.

of seaming are the tops of beverage cans and containers for food and household products (Fig. 30.15).

In seaming, the materials should be capable of undergoing bending and folding at very small radii. Otherwise, they will crack and the seams will not be airtight or watertight. The performance of seams may be improved with adhesives, coatings, seals, or by soldering.

Crimping. *Crimping* is a method of joining without using fasteners. It can be done with beads or dimples (Fig. 30.16), which can be produced by shrinking or swaging operations. Crimping can be used on both tubular and flat parts, provided that the materials are thin and ductile enough to withstand the large localized deformations. Caps on glass bottles are attached by crimping, as are connectors for electrical wiring.

Snap-in fasteners. Various spring and snap-in fasteners are shown in Fig. 30.17. Such fasteners are widely used in automotive bodies and household appliances. They are economical and permit easy and rapid component assembly.

Shrink and press fits. Components may also be assembled by shrink fitting and press fitting. *Shrink fitting* is based on the principle of differential thermal expansion and contraction of two components. Typical applications are the assembly of die components and mounting gears and cams on shafts. In *press fitting*, one component is forced over another, resulting in high joint strength.

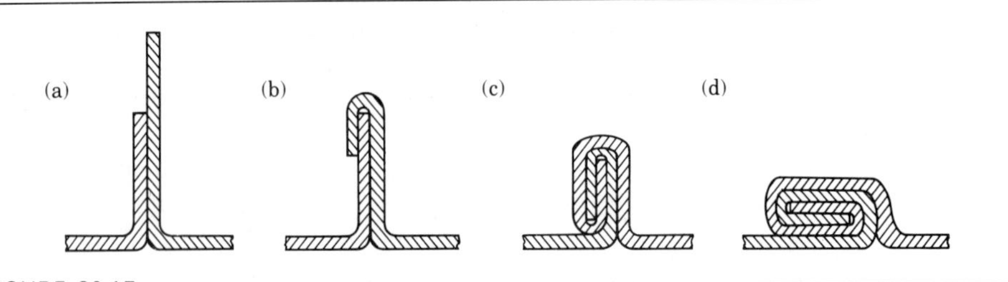

FIGURE 30.15
Two examples of mechanical joining by crimping.

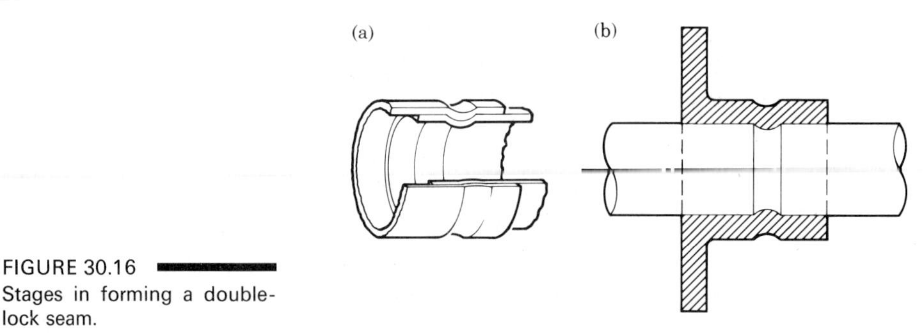

FIGURE 30.16
Stages in forming a double-lock seam.

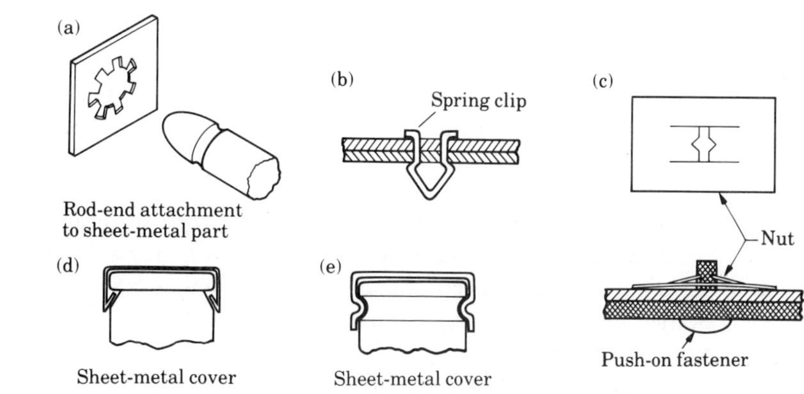

FIGURE 30.17
Examples of spring and snap-in fasteners to facilitate assembly.

30.6 Joining Plastics

Plastics can be joined by many of the methods we have described for joining metals and nonmetallic materials, including:

a) Fusion welding with heat.
b) Ultrasonic welding.
c) Friction welding.
d) Adhesive bonding.
e) Bonding with solvents.
f) Mechanical joining, particularly with self-tapping screws. Mechanical joining is particularly effective with most plastics because of their inherent toughness and resilience.

30.6.1 Thermoplastics

Because thermoplastics soften and melt as temperature is increased, they can be joined by the fusion-welding techniques we described in Chapters 27 and 28, but at much lower temperatures. This method is particularly effective with plastics that cannot be bonded easily with adhesives. Plastics such as polyvinyl chloride, polyethylene, polypropylene, acrylics, and acrylonitrile butadiene styrene (ABS) can be joined in this manner. For example, specially designed portable fusion sealing systems have been developed to allow in-field joining of pipe, which is usually made of polyethylene, used for natural-gas delivery.

The heat source in fusion welding of thermoplastics is usually hot air or other gases. The heat melts the joint and, with the application of pressure to ensure a good bond, allows fusion to take place at the interface. Filler materials of the same type of polymer may be used. Other heating methods may also be utilized in joining thermoplastics: (1) heated tools and dies (for joining thin materials, since plastics are poor thermal conductors), (2) electrical-resistance wire, and (3) high-frequency heating.

Coextruded multiple food wrappings consist of different types of films, which are bonded by heat during extrusion. Each film has a different function, such as to keep out moisture, keep out oxygen, and facilitate heat sealing during packaging. Some wrappings have as many as seven layers, all bonded together during production of the film.

Oxidation can be a problem in joining some polymers, such as polyethylene, causing degradation. In these cases an inert shielding gas such as nitrogen is used to prevent oxidation. Because of low thermal conductivity of thermoplastics, the heat source may burn or char the surfaces of the components if applied at a high rate, possibly causing difficulties in obtaining sufficiently deep fusion.

Adhesive bonding of plastics is best illustrated in joining sections of polyvinyl chloride (PVC) pipe, used extensively in home irrigation systems, and ABS pipe, used in drain, waste, and vent systems. The adhesive is applied to the connecting sleeve and pipe surfaces, using a primer to improve adhesion (much like using primers in painting), and the pieces are pushed together. Adhesive bonding of polyethylene, polypropylene, and polytetrafluoroethylene (Teflon) can be difficult because adhesives do not stick readily to their surfaces. Their surfaces usually have to be treated chemically to improve bonding. The use of adhesive primers and double-sided adhesive tapes are also effective.

30.6.2 Thermosets

Because they do not soften or melt with temperature, thermosetting plastics, such as epoxy and phenolics, are usually joined by using (1) threaded or other molded-in inserts, (2) mechanical fasteners, and (3) solvent bonding.

The basic process of thermoset bonding with solvents involves:

1. Roughening the surfaces with an abrasive.
2. Wiping the surfaces with a solvent.
3. Pressing the surfaces together and holding until sufficient bond strength is developed.

SUMMARY

Joining processes that do not rely on fusion or pressure at the interfaces include brazing and soldering. Instead, these processes utilize filler material that requires some temperature rise in the joint. They can be used to join dissimilar metals of intricate shapes and various thicknesses.

Adhesive bonding has gained increased acceptance in major industries, such as the aerospace and automotive. In addition to good bond strength, adhesives have other favorable characteristics, such as sealing, insulating, preventing electrochemical corrosion between dissimilar metals, and reducing vibration and noise through internal damping in the bond. Surface preparation and joint design are important factors in adhesive bonding.

Mechanical joining is one of the oldest and most common joining methods. Bolts, screws, and nuts are common fasteners for machine components and structures, which are likely to be taken apart for maintenance, ease of transportation, and various other reasons. Rivets are semipermanent or permanent fasteners used in buildings, bridges, and transportation equipment. A wide variety of other fasteners and fastening techniques is available for numerous permanent or semipermanent applications.

Thermoplastics can be joined by fusion-welding techniques, adhesive bonding, and mechanical fastening. Thermosets are usually joined by mechanical means, such as molded-in inserts and fasteners, as well as solvent bonding.

SUMMARY TABLE 30.1

COMPARISON OF VARIOUS JOINING METHODS

	CHARACTERISTICS								
METHOD	*STRENGTH*	*DESIGN VARIABILITY*	*SMALL PARTS*	*LARGE PARTS*	*TOLERANCES*	*RELIABILITY*	*EASE OF MAINTENANCE*	*VISUAL INSPECTION*	*COST*
Arc welding	1	2	3	1	3	1	2	2	2
Resistance welding	1	2	1	1	3	3	3	3	1
Brazing	1	1	1	1	3	1	3	2	3
Bolts and nuts	1	2	3	1	2	1	1	1	3
Riveting	1	2	3	1	1	1	3	1	2
Fasteners	2	3	3	1	2	2	2	1	3
Seaming, crimping	2	2	1	3	3	1	3	1	1
Adhesive bonding	3	1	1	2	3	2	3	3	2

Note: 1, very good; 2, good; 3, fair.

TRENDS

- Automation and the consequent reduction in labor requirements for many of the joining processes described is continuing.
- Adhesive bond strength and reliability under various environmental conditions are under constant study.
- A wide variety of metallic and nonmetallic fasteners are being developed, primarily for economic assembly of parts manufactured in computer-integrated manufacturing systems.

KEY TERMS

Adhesive bonding
Braze welding
Brazing
Fasteners
Filler metals
Hole preparation
Mechanical joining
Soldering

BIBLIOGRAPHY

Blake, A., *What Every Engineer Should Know about Threaded Fasteners.* New York: Marcel Dekker, 1986.

Brazing Handbook, 4th ed. Miami: American Welding Society, 1991.

DeFryne, G. (ed.), *High Performance Adhesive Bonding.* Dearborn, Mich.: Society of Manufacturing Engineers, 1982.

Engineered Materials Handbook, Vol. 3: *Adhesives and Sealants.* Metals Park, Ohio: ASM International, 1991.

Haviland, G.S., *Machinery Adhesives for Locking, Retaining, and Sealing.* New York: Marcel Dekker, 1986.

Landrock, A.H., *Adhesives Technology Handbook.* Park Ridge, N.J.: Noyes, 1985.

Lincoln, B., K.J. Gomes, and J.F. Braden, *Mechanical Fastening of Plastics.* New York: Marcel Dekker, 1983.

Metals Handbook, 9th ed., Vol 6: *Welding, Brazing, and Soldering.* Metals Park, Ohio: American Society for Metals, 1983.

Parmley, R.O. (ed.), *Standard Handbook of Fastening and Joining*, 2d ed. New York: McGraw-Hill, 1989.

Principles of Industrial Welding. Cleveland, Ohio: The James F. Lincoln Arc Welding Foundation, 1978.

Schneberger, G.L. (ed.), *Adhesives in Manufacturing.* New York: Marcel Dekker, 1983.

Schwartz, M.M., *Brazing*. Metals Park, Ohio: ASM International, 1987.
——, *Ceramic Joining*. Metals Park, Ohio: ASM International, 1990.
Shields, J., *Adhesives Handbook*, 3d ed. London: Butterworths, 1984.
Soldering Manual. Miami: American Welding Society, 1978.
Solders and Soldering. New York: Lead Industries Association, 1982.
Source Book on Brazing and Brazing Technology. Metals Park, Ohio: American Society for Metals, 1980.
Thwaites, C.J., *Capillary Joining—Brazing and Soft-Soldering*. New York: Wiley, 1982.
Welding Handbook, 8th ed., 3 vols. Miami: American Welding Society, 1987.

REVIEW QUESTIONS

30.1 Describe the similarities and differences between the processes described in this chapter and those in Chapters 27 and 28.
30.2 Explain the principle of brazing.
30.3 What is the difference between brazing and braze welding?
30.4 What are the relative advantages of braze welding and fusion welding?
30.5 Are fluxes necessary in brazing? If so, why?
30.6 Describe various brazing processes, giving applications for each method.
30.7 What type of materials and part configurations would be difficult to join by brazing?
30.8 Do you think it is acceptable to differentiate brazing and soldering arbitrarily by temperature of application? Comment.
30.9 Describe the types of fluxes used in soldering and their applications.
30.10 How different is adhesive bonding from other joining methods? What limitations does it have?
30.11 Describe the basic types of adhesives.
30.12 Why is surface preparation important in adhesive bonding?
30.13 Why have mechanical joining methods been developed? Give specific examples of their applications.
30.14 Explain why hole preparation may be important in mechanical joining.
30.15 Describe the similarities and differences in the functions of a bolt and a rivet.
30.16 Give examples where rivets in a structure or assembly could be removed and riveted again.
30.17 What precautions should be taken in mechanical joining of dissimilar metals?
30.18 Explain the principles of various types of mechanical joining and fastening methods.
30.19 What difficulties are involved in joining plastics? Why?

QUESTIONS AND PROBLEMS

30.20 Comment on your observations concerning the joints shown in Figs. 30.3 and 30.9.
30.21 Give several applications for fasteners in various products, and explain why other joining methods have not been used.
30.22 In Fig. 30.12, why are the designs on the left labeled poor and the ones on the right good?

30.23 Inspect various household products and describe how they are joined and assembled. Explain why those particular processes were used.

30.24 Name several products that have been assembled by (a) seaming, (b) stitching, and (c) soldering.

30.25 You will note in Table 27.1 that soldering is generally applied to thinner components. Explain why.

30.26 Do you think the strength of an adhesively bonded structure is as high as that obtained by diffusion bonding? Explain.

30.27 Explain why adhesively bonded joints tend to be weak in peeling.

30.28 Suggest methods of attaching a round bar (made of thermosetting plastic) perpendicular to a flat metal plate.

30.29 Describe the tooling and equipment that is necessary to perform the double-lock seaming operation shown in Fig. 30.15, starting with flat sheet.

30.30 What joining methods would be suitable to assemble a thermoplastic cover over a metal frame? Assume that the cover is removed periodically.

30.31 Same as Problem 30.30, but for a cover made of (a) a thermoset, (b) metal, and (c) ceramic. Describe the factors involved in your selection of methods.

30.32 Write brief paragraphs on the soldering processes listed in Section 30.3, giving an application for each.

30.33 Prepare a list of guidelines for design for joining by the processes described in this chapter. Would these guidelines be common to most processes? Explain.

30.34 Inspect your automobile and identify various components that have been joined by the processes described in this chapter.

30.35 In your inspection in Problem 30.34, would you join these components by different methods, and if so, what would be the advantages?

30.36 Describe the various costs involved in using the joining processes described in Part V. Identify those costs that are likely to be the highest in each group and those that are lowest.

PART VI

Surface Technology

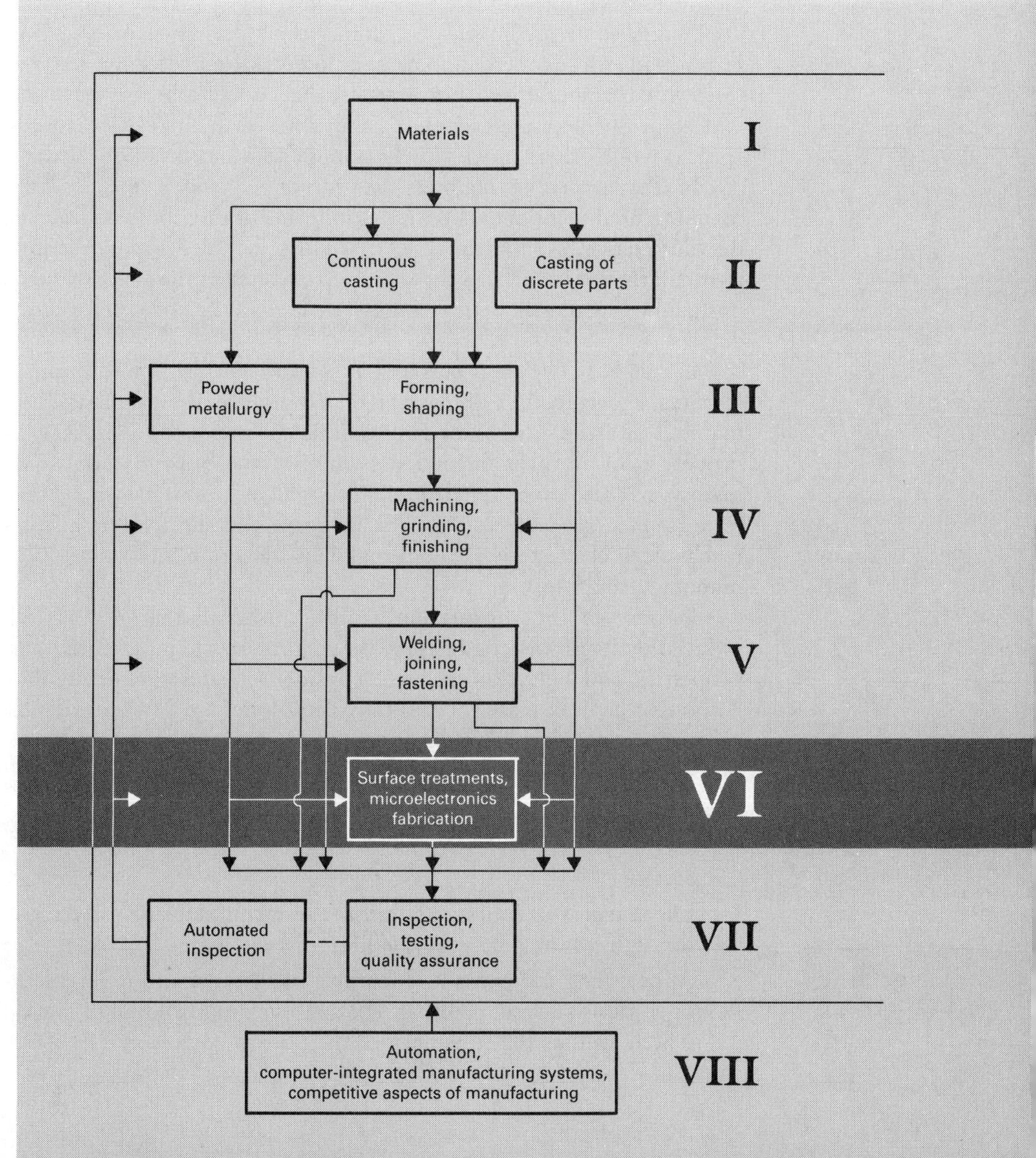

• INTRODUCTION •

You may not realize it, but your first visual or tactile contact with objects is through their *surfaces*. You can see or feel surface roughness, waviness, color, reflectivity, texture, and other features such as scratches, nicks, and depressions. Prior to purchasing a household appliance, kitchen utensil, or automobile, you inspect its surfaces.

Machinery and equipment, including a variety of manufacturing equipment such as presses and machine tools, often have members that slide against each other: pistons and cylinders, slideways, bearings, and cutting and forming tools and dies. Close examination reveals that some of these surfaces are smooth while others are rough, some are lubricated while others are dry, and some are subjected to heavy loads while others support light loads. Moreover, some surfaces are subjected to elevated temperatures (hot-working dies) while others are at room temperature, and some surfaces slide against each other at high relative speeds (high cutting speeds) while others move slowly (saddle on a machine tool bed).

In the preceding chapters, we described the properties of materials and manufactured components basically in terms of their bulk characteristics: strength, ductility, hardness, and toughness. We also included some description of surfaces, particularly if it influenced these properties, such as the effect of surface preparation on fatigue life and the sensitivity of brittle materials to surface scratches and other defects.

In this Part, we treat surfaces as distinct entities. In addition to its geometric features, a surface is a thin layer of material at the boundary between the bulk material and the environment. Physical, chemical, metallurgical, and mechanical properties of surfaces depend not only on the bulk material and its processing history, but also on the environment to which the surfaces are exposed. We use the term *surface integrity* to describe not only the surface but also its physical and mechanical characteristics, especially as they are influenced by the particular manufacturing process.

Because of the various mechanical, physical, thermal, and chemical effects induced by its processing history, the surface of a manufactured part generally has properties and behavior that are considerably different from those of its bulk. Although the bulk material generally determines the component's overall mechanical properties, the component's surface directly influences several important properties and characteristics of the manufactured part (Fig. VI.1):

- Friction and wear properties of the part during subsequent processing when it comes directly into contact with tools and dies and when it is placed in service.
- Effectiveness of lubricants during the manufacturing processes, as well as throughout the part's service life.
- Appearance and geometric features of the part and their role in subsequent operations such as painting, coating, welding, soldering, and adhesive bonding, as well as the resistance of the part to corrosion.

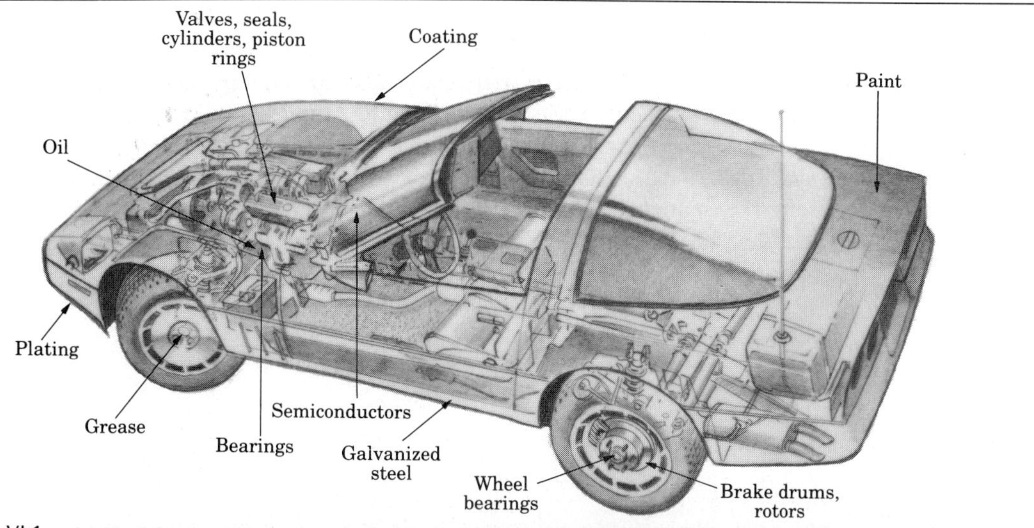

FIGURE VI.1
Components of a typical automobile that are related to the topics described in Part VI.

- Initiation of cracks because of surface defects, such as roughness, scratches, seams, and heat-affected zones, which could lead to weakening and premature failure of the part by fatigue or other fracture mechanisms.
- Thermal and electrical conductivity of contacting bodies, with rough surfaces, for example, having higher thermal and electrical resistance than smooth surfaces.

Following the outline shown in Fig. VI.2, we present surface characteristics in terms of their structure and topography. We then discuss the material and process

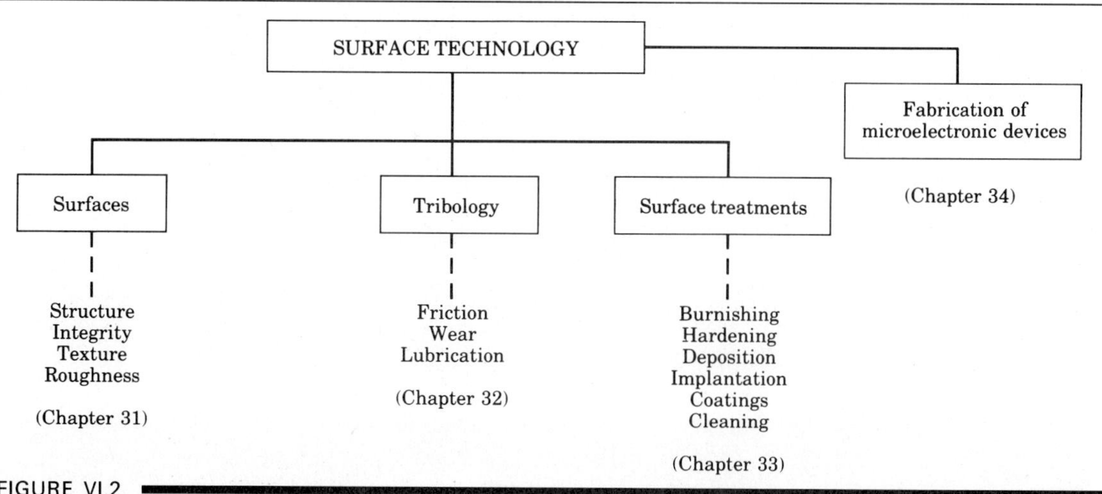

FIGURE VI.2
An outline of topics covered in Part VI.

variables that influence friction and wear of materials. We also describe several mechanical, thermal, electrical, and chemical methods that can be used to modify surfaces of parts to improve frictional behavior, effectiveness of lubricants, resistance to wear and corrosion, and surface finish and appearance.

Because much of the technology involved relies on the topics covered thus far, we also describe the manufacturing of microelectronic devices in this part. We present the sequence of operations involved in manufacturing semiconductors, which have been such a critical element in integrated-circuit technology and their extensive use in computer-integrated manufacturing operations.

31

Surfaces: Their Nature, Roughness, and Measurement

31.1 Introduction

In the Introduction to Part VI, we described the significance of surfaces as they influence manufacturing operations and are, in turn, influenced by them. In this chapter we discuss the nature of surfaces and their structure, both on workpieces and tools and dies, as they influence manufacturing operations and the service life of manufactured products. Our discussion includes the effects of various flaws typically found in manufactured products.

In order to define surface roughness and communicate it to others, we have to know the methods by which surface roughness is measured in engineering practice, including the instrumentation involved. We also present typical surface roughness ranges encountered in engineering practice.

31.2 Surface Structure and Properties

Upon close examination of the surface of a piece of metal, you will find that it generally consists of several layers (Fig. 31.1). The bulk metal, also known as the metal *substrate*, has a structure that depends on the composition and processing history of the metal. Above this bulk metal is a layer that usually has been plastically deformed and work hardened during the manufacturing process.

The depth and properties of the work-hardened layer—the **surface structure**—depend on factors such as the processing method used and frictional sliding to which the surface was subjected. Sharp tools and selection of proper process parameters produce surfaces with little or no disturbance. For example, if the surface is produced by machining with a dull tool or under poor cutting conditions—or is ground with a dull grinding wheel—this layer will be relatively thick. Also, nonuniform surface deformation or severe temperature gradients during manufacturing operations usually cause residual stresses in this work-hardened layer.

Unless the metal is processed and kept in an inert (oxygen-free) environment—or it is a noble metal, such as gold or platinum—an **oxide layer** usually lies on top of the work-hardened layer. Some examples are:

a) Iron has an oxide structure with FeO adjacent to the bulk metal, followed by a layer of Fe_3O_4 and then a layer of Fe_2O_3, which is exposed to the environment.
b) Aluminum has a dense amorphous (without crystalline structure) layer of Al_2O_3 with a thick, porous hydrated aluminum-oxide layer over it.
c) Copper has a bright shiny surface when freshly scratched or machined. Soon after, however, it develops a Cu_2O layer, which is then covered with a layer of CuO. This gives copper its somewhat dull color, such as we see in kitchen utensils.

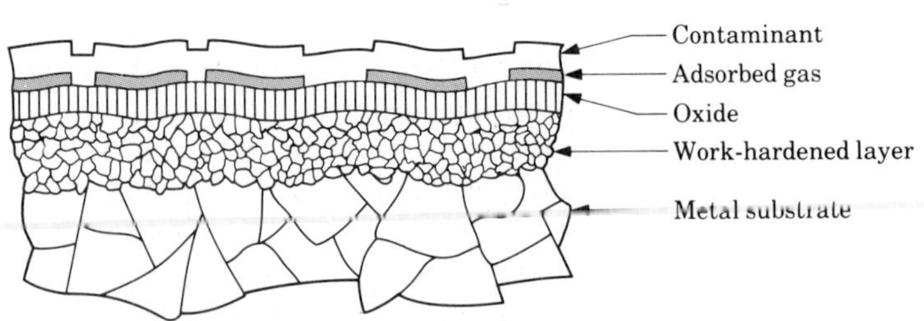

FIGURE 31.1 Schematic illustration of the cross-section of the surface structure of metals. The thickness of the individual layers depends on processing conditions and the environment.

TABLE 31.1
HARDNESS RATIO OF OXIDE TO BASE METAL FOR SEVERAL METALS

METAL	RATIO
Molybdenum	0.3
Tantalum	0.6
Copper	1.6
Nickel	2
Tungsten	5
Iron	5
Lead	20
Aluminum	70
Tin	100

d) Stainless steels are "stainless" because they develop a protective layer of chromium oxide (passivation), as we described in Section 3.8.

Under normal environmental conditions, surface oxide layers are generally covered with adsorbed layers of gas and moisture (see Fig. 31.1). Finally, the outermost surface of the metal may be covered with contaminants, such as dirt, dust, grease, lubricant residues, cleaning-compound residues, and pollutants from the environment.

Thus surfaces generally have properties that are very different from those of the substrate. The oxide on a metal's surface, for example, is generally much harder than the base metal (Table 31.1). Hence oxides tend to be brittle and abrasive. As you will see throughout the rest of this and the next two chapters, this surface characteristic has several important effects on friction, wear, and lubrication in materials processing and coatings on products.

The factors involved in the surface structure of metals are also relevant to a great extent to the surface structure of plastics and ceramics. The surface texture of these materials depends, as with metals, on the method of production (see Chapters 18 and 19). Environmental conditions also influence the surface characteristics of these materials.

31.3 Surface Integrity

Surface integrity describes not only the topological (geometric) aspects of surfaces, but also their mechanical and metallurgical properties and characteristics. Surface integrity is an important consideration in manufacturing operations because it

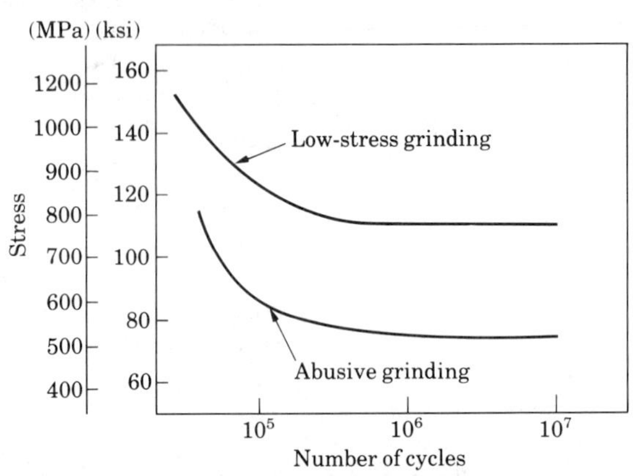

FIGURE 31.2
Fatigue curve for surface-ground 4340 steel, quenched and tempered, 51 HRC. Note the severe reduction in fatigue strength under abusive grinding conditions. (See also Fig. 2.34.)

influences the properties of the product, such as its fatigue strength and resistance to corrosion, and its service life. The detrimental effect of improper (abusive) grinding, for example, on the fatigue life of an alloy steel is shown in Fig. 31.2 (see also Fig. 2.34).

Several defects caused by and produced during component manufacturing can be responsible for lack of surface integrity. These defects are usually caused by a combination of factors, such as defects in the original material, the method by which the surface is produced, and lack of proper control of process parameters that can result in excessive stresses and temperatures. The following are definitions (listed in alphabetical order) of the major surface defects found in practice.

Cracks are external or internal separations with sharp outlines. Cracks requiring a magnification of 10 × to be seen by the naked eye are called microcracks.

Craters are shallow depressions.

Folds are the same as seams.

Heat-affected zone is the portion of a metal subjected to thermal cycling without melting.

Inclusions are small, nonmetallic elements or compounds in the metal.

Intergranular attack is the weakening of grain boundaries by liquid-metal embrittlement and corrosion.

Laps are the same as seams.

Metallurgical transformation involves microstructural changes caused by temperature and pressure cycling. Included are phase transformations, recrystallization, alloy depletion, decarburization, and molten and recast, resolidified, or redeposited metal, as in electrical-discharge machining.

Pits are shallow surface depressions, usually the result of chemical or physical attack.

Plastic deformation is a severe surface deformation caused by high friction, tool and die geometry, worn tools, and method of processing.

Residual stresses are surface stresses (tension or compression) caused by nonuniform deformation and temperature distribution.

Seams are surface defects resulting from overlapping of the material during processing.

Splatter consists of small resolidified molten metal particles deposited on a surface, as from welding.

31.3.1 Techniques for testing surface integrity

One of the most commonly used techniques for testing surface integrity is *metallography*. Samples are removed from the workpiece, polished, etched, and observed under an optical or electron microscope. The test samples are usually much smaller than the part or component being analyzed, so they must be taken from appropriate locations in the workpiece. Several nondestructive techniques used to observe and test surfaces are described in Section 36.2.

31.4 Surface Texture

Regardless of the method of production, all surfaces have their own characteristics, which we refer to as **surface texture**, roughness, and finish. The description of surface texture as a geometrical property is complex. However, certain guidelines have been established for identifying surface texture in terms of well-defined and measurable quantities (Fig. 31.3).

Flaws, or *defects*, are random irregularities, such as scratches, cracks, holes, depressions, seams, tears, and inclusions (see Section 31.3).

Lay, or *directionality*, is the direction of the predominant surface pattern and is usually visible to the naked eye.

Roughness consists of closely spaced, irregular deviations on a scale smaller than that for waviness. Roughness may be superimposed on waviness. Roughness is expressed in terms of its height, its width, and the distance on the surface along which it is measured.

Waviness is a recurrent deviation from a flat surface, much like waves on the surface of water. It is measured and described in terms of the space between adjacent crests of the waves (*waviness width*) and height between the crests and valleys of the waves (*waviness height*). Waviness may be caused by deflections of

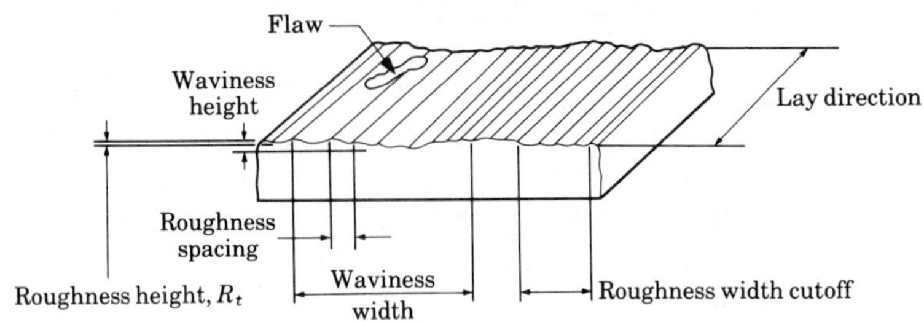

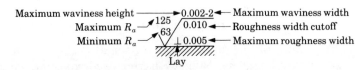

FIGURE 31.3
Standard terminology and symbols to describe surface finish. The quantities are given in μin.

tools, dies, and the workpiece, warping from forces or temperature, uneven lubrication, and vibration or any periodic mechanical or thermal variations in the system during the manufacturing process.

31.5 Surface Roughness

We generally describe surface roughness by two methods: arithmetic mean value (R_a); and root-mean-square average (R_q). The **arithmetic mean value** (R_a; formerly identified as *AA* for *arithmetic average* or *CLA* for *center-line average*) is based on the schematic illustration of a rough surface shown in Fig. 31.4. The arithmetic mean value R_a is defined as

$$R_a = \frac{a + b + c + d + \cdots}{n}, \tag{31.1}$$

where all ordinates, a, b, c, . . . , are absolute values.

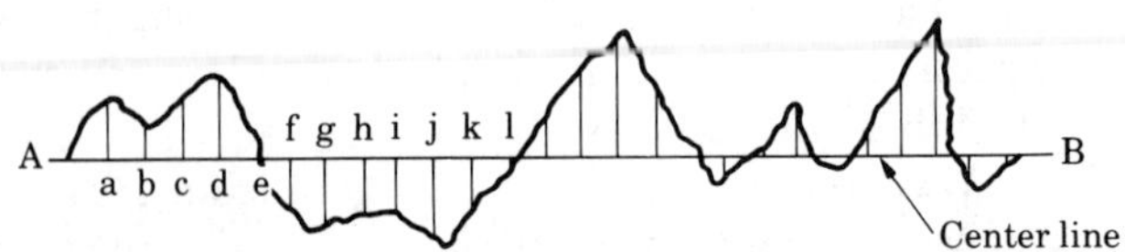

FIGURE 31.4
Coordinates used for surface-roughness measurement, using Eqs. (31.1) and (31.2).

The **root-mean-square average** (R_q; formerly identified as *RMS*) is defined as

$$R_q = \sqrt{\frac{a^2 + b^2 + c^2 + d^2 + \cdots}{n}}. \tag{31.2}$$

The datum line AB in Fig. 31.4 is located so that the sum of the areas above the line is equal to the sum of the areas below the line. The units generally used for surface roughness are μm (micrometer, or micron) or μin. (microinch), where 1 μm = 40 μin. and 1 μin. = 0.025 μm.

Additionally, we may also use the *maximum roughness height* (R_t) as a measure of roughness. It is defined as the height from the deepest trough to the highest peak (see Fig. 31.4). It indicates the amount of material that has to be removed to obtain a smooth surface by polishing or other means.

Because of its simplicity, the arithmetic mean value R_a was adopted internationally in the mid-1950s and is widely used in engineering practice. Equations (31.1) and (31.2) show that there is a relationship between R_q and R_a. For a surface roughness in the shape of a sine curve, R_q is larger than R_a by a factor of 1.11. This factor is 1.1 for most machining processes by cutting, 1.2 for grinding, and 1.4 for lapping and honing.

In general, we cannot adequately describe a surface by its R_a or R_q value alone, since these values are averages. Two surfaces may have the same roughness value but their actual topography may be quite different (Fig. 31.5). A few deep troughs, for example, affect the roughness values insignificantly. However, differences in the surface profile can be significant in terms of fatigue, friction, and the wear characteristics of a manufactured product.

31.5.1 Symbols for surface roughness

Acceptable limits for surface roughness are specified on technical drawings by the symbols shown around the check mark in the lower portion of Fig. 31.3, and their values are placed to the left of the check mark. The symbols and their meaning

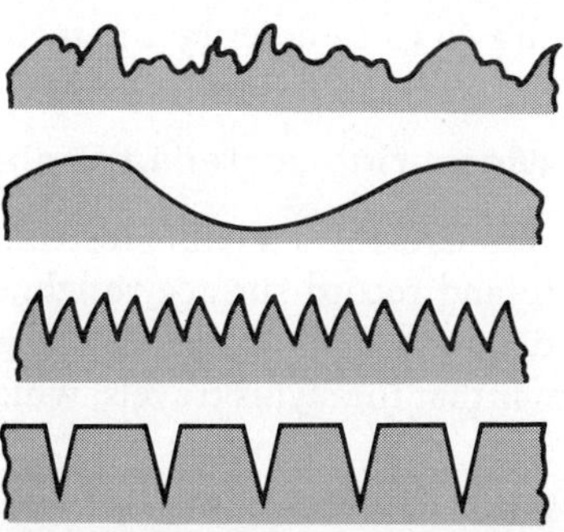

FIGURE 31.5
Various surface profiles that have the same R_a and R_q values, illustrating the difficulty in describing surface profiles by roughness values only.

Lay symbol	Interpretation	Examples
=	Lay parallel to the line representing the surface to which the symbol is applied	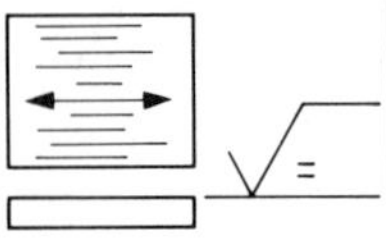 =
⊥	Lay perpendicular to the line representing the surface to which the symbol is applied	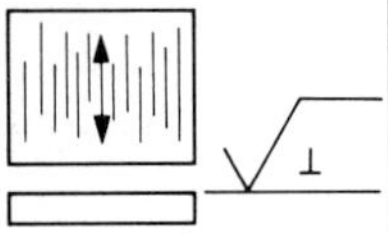 ⊥
X	Lay angular in both directions to line representing the surface to which symbol is applied	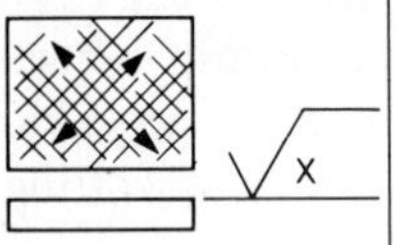 X
M	Lay multidirectional	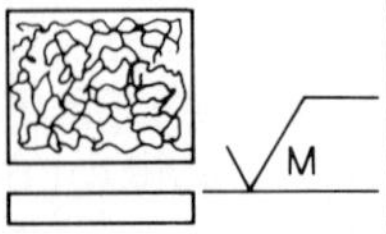 M

Lay symbol	Interpretation	Examples
C	Lay approximately circular relative to the center of the surface to which the symbol is applied	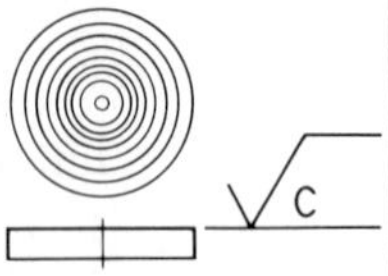 C
R	Lay approximately radial relative to the center of the surface to which the symbol is applied	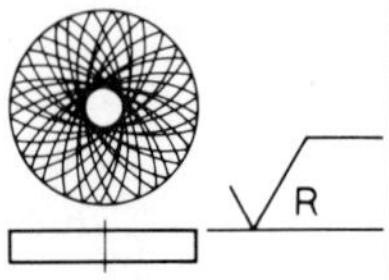 R
P	Pitted, protuberant, porous, or particulate nondirectional lay	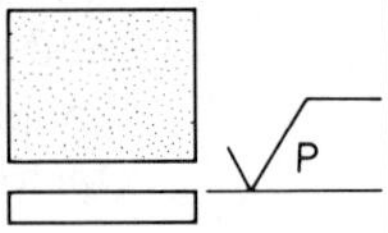 P

FIGURE 31.6
Standard lay symbols for engineering surfaces.

concerning lay are given in Fig. 31.6. Note that the symbol for lay is placed at the lower right of the check mark.

Symbols used to describe a surface only specify the roughness, waviness, and lay; they do not include flaws. Whenever important, a special note is included in technical drawings to describe the method to be used to inspect for surface flaws.

31.5.2 Measuring surface roughness

Various commercially available instruments, called **surface profilometers**, are used to measure and record surface roughness. The most commonly used instruments feature a diamond stylus traveling along a straight line over the surface (Fig. 31.7a). The distance that the stylus travels, which can be varied, is called the *cut off* (see Fig. 31.3).

To highlight the roughness, profilometer traces are recorded on an exaggerated vertical scale (a few orders of magnitude greater than the horizontal scale; Fig. 31.8), called *gain* on the recording instrument. Thus the recorded profile is significantly distorted, and the surface appears to be much rougher than it actually

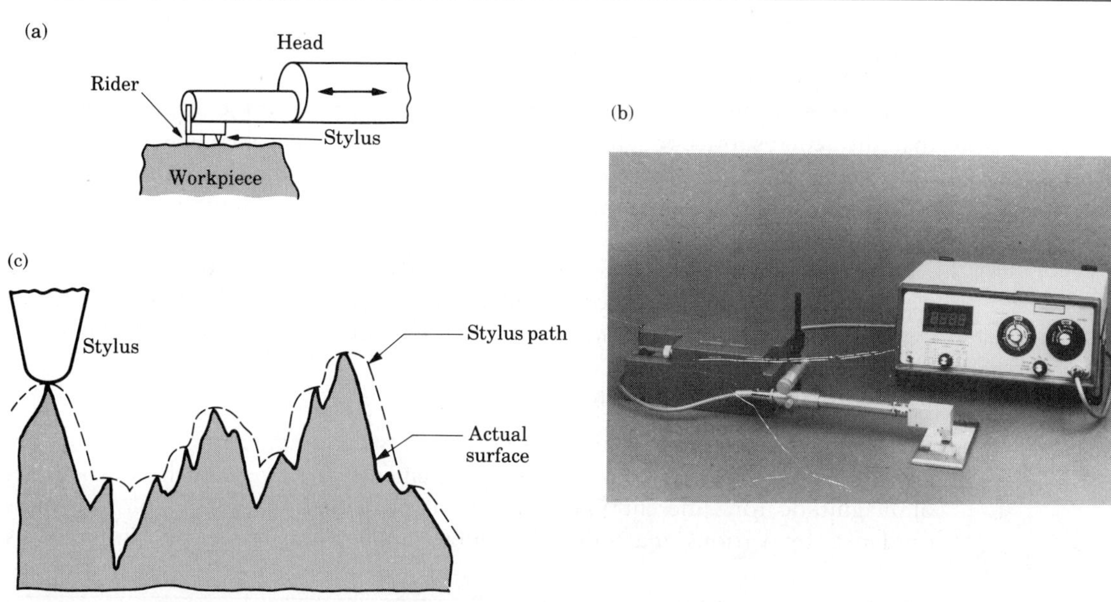

FIGURE 31.7

(a) Measuring surface roughness with a stylus. The rider supports the stylus and guards against damage. (b) Surface measuring instrument. *Source:* Sheffield Measurement Division of Warner & Swasey Co. (c) Path of stylus in surface roughness measurements (broken line) compared to actual roughness profile. Note that the profile of the stylus path is smoother than the actual surface profile. *Source:* D. H. Buckley.

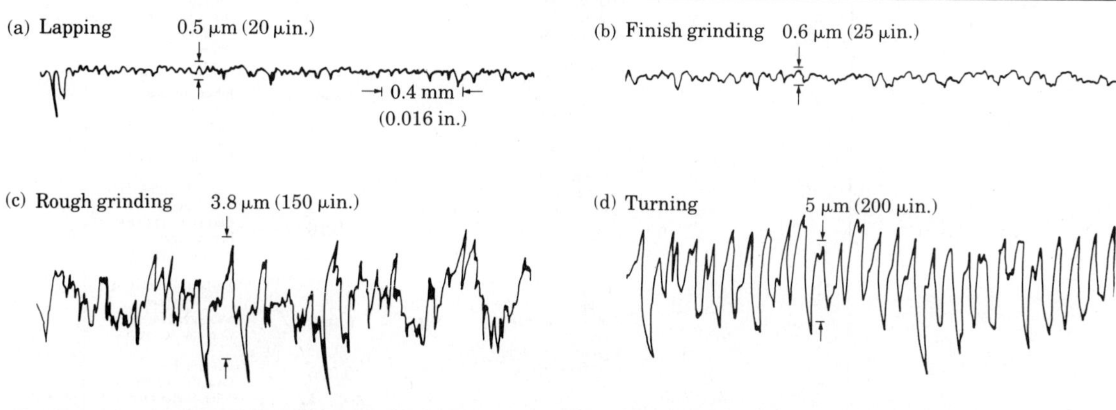

FIGURE 31.8

Typical surface profiles produced by various machining and surface-finishing processes. Note the difference between the vertical and horizontal scales. *Source:* D. B. Dallas (ed.), *Tool and Manufacturing Engineers Handbook,* 3d ed. Copyright © 1976, McGraw-Hill Publishing Company. Used with permission.

is. The recording instrument compensates for any surface waviness and indicates only roughness. A record of the surface profile is made by mechanical and electronic means (Fig. 31.7b).

Because of the finite radius of the stylus tip, the path of the stylus is smoother than the actual surface roughness (note the path with the broken line in Fig. 31.7c). The smaller the tip radius and the smoother the surface, the closer the path of the stylus will represent the actual surface profile. The most commonly used stylus diameter is 10 μm (400 μin.).

We can observe surface roughness directly through an optical or scanning electron microscope. Stereoscopic photographs are particularly useful for three-dimensional views of surfaces, and we can use them to measure surface roughness.

31.5.3 Surface roughness in practice

Surface roughness design requirements for some typical engineering applications are given in Fig. 31.9. Note that roughness requirements vary by as much as two orders of magnitude for different parts. We have presented the ranges of roughness produced by various machining, forming, and casting processes in previous

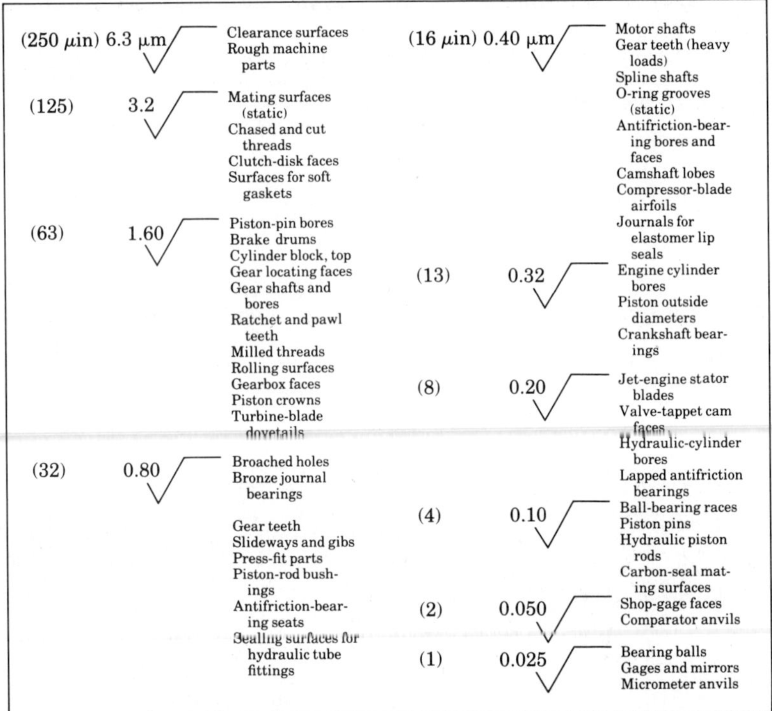

FIGURE 31.9
Typical surface roughness design requirements in engineering components. Note that the range of roughness in these applications is two orders of magnitude.

chapters. Because of the many material and process variables involved, the range of roughness produced, even within a particular process, can be great.

Why should surface roughness design requirements vary so much in the applications shown in Fig. 31.9? Reasons and considerations for these differences include:

a) *Accuracy required on mating surfaces*, such as seals, fittings, gaskets, tools, and dies. Note in Fig. 31.9, for example, that ball bearings and gages require very smooth surfaces, whereas surfaces for gaskets and brake drums can be quite rough.
b) *Frictional considerations*, that is, the effect of roughness on friction, wear, and lubrication.
c) *Fatigue and notch sensitivity*, because the rougher the surface is, the shorter the fatigue life will be (see Fig. 2.34).
d) *Electrical and thermal contact resistance*, because the rougher the surface is, the higher the resistance will be.
e) *Corrosion resistance*, because the rougher the surface is, the greater will be the possibility of entrapped corrosive media.
f) *Subsequent processing*, such as painting and coating, in which a certain roughness can result in better bonding.
g) *Appearance*.
h) *Cost considerations*, because the finer the finish is, the higher the cost will be (see Fig. 25.33).

All these factors should be carefully considered before a decision is made as to the recommendation about surface roughness for a certain product. As in all manufacturing processes, the cost involved in the selection should also be a major consideration.

SUMMARY

In manufacturing processes, surfaces and their properties are as important as the bulk properties of the materials. A surface not only has a particular shape, roughness, and appearance, but it also is a layer having properties that can differ significantly from those of the bulk material.

Surfaces are exposed to the environment and thus are subject to environmental attack. They also may come into contact with tools and dies (during processing) or other components (during their service life). Consequently, their geometric and material properties can significantly affect friction, wear, fatigue, corrosion, and electrical and thermal conductivity.

Measuring and describing surface features and their characteristics are among the most important aspects of manufacturing processes. The most common surface roughness measurement is the arithmetic mean value. Profilometers are commonly used for measuring surface roughness.

TRENDS

- Efforts are continually being made to accurately define and describe surfaces and their characteristics.
- In-process measurement of surface roughness as parts are being produced is an important trend in manufacturing quality control.
- Measuring instruments are being developed with better sensitivity and accuracy and for use in computer-controlled manufacturing systems.

KEY TERMS

Arithmetic mean value
Flaw
Lay
Oxide layer
Root-mean-square average
Roughness
Surface integrity
Surface profilometer
Surface structure
Surface texture
Waviness

BIBLIOGRAPHY

Bhushan, B., and B.K. Gupta, *Handbook of Tribology: Materials, Coatings, and Surface Treatments*. New York: McGraw-Hill, 1991.

Buck, O., and S.M. Wolf, *Nondestructive Evaluation: Application to Materials Processing*. Metals Park, Ohio: American Society for Metals, 1984.

Farago, F.T., *Handbook of Dimensional Measurement*, 2d ed. New York: Industrial Press, 1982.

Halling, J., *Principles of Tribology*. New York: Macmillan, 1975.

Machining Data Handbook, 3d ed., 2 vols. Cincinnati: Machinability Data Center, 1980.

Metals Handbook, 9th ed., *Vol. 8: Mechanical Testing*, 1985; *Vol. 10: Materials Characterization*, 1986. Metals Park, Ohio: American Society for Metals.

Moore, D.F., *Principles and Applications of Tribology*. New York: Pergamon, 1975.

Peterson, M.R., and W.O. Winer (eds.), *Wear Control Handbook*. New York: American Society of Mechanical Engineers, 1980.

Schey, J.A., *Tribology in Metalworking—Friction, Lubrication and Wear*. Metals Park, Ohio: American Society for Metals, 1983.

REVIEW QUESTIONS

31.1 Explain the importance of studying surfaces of workpieces, tools, and dies.
31.2 What is a substrate?
31.3 Describe the various layers usually present on the surface of a metal.
31.4 What is meant by surface integrity? Surface texture?
31.5 List the types of defects found on surfaces.
31.6 Explain the terms (a) roughness, (b) waviness, and (c) lay.
31.7 How is surface roughness generally measured?
31.8 What is cutoff?
31.9 Why are surface roughness design requirements in engineering so broad?

QUESTIONS AND PROBLEMS

31.10 Explain the presence of various layers on the surface of a metal workpiece. What influences the thickness of these layers?
31.11 What is the consequence of the hardness of oxides of metals being generally much higher than the base metals?
31.12 Why does the method of grinding have a major influence on fatigue life?
31.13 Describe the effects of various surface defects on the performance of products in service.
31.14 What factors would you consider in specifying the lay of a surface?
31.15 Explain why the same surface roughness values do not necessarily represent the same type of surface.
31.16 In using a surface roughness measuring instrument, how would you go about determining the cutoff value?
31.17 Explain the trends that you see in surface roughness limitations given in Fig. 31.9.
31.18 What is the significance of the fact that the stylus path and the actual surface roughness are not necessarily the same?
31.19 How would you utilize the information given in Table 31.1? Give specific examples.
31.20 What processes could be used to produce the surfaces shown in Fig. 31.6? What would be their effect on finishing costs?
31.21 Give two examples each in which waviness of a surface would be (a) desirable and (b) undesirable.
31.22 Same as Problem 31.21, but for surface roughness.
31.23 Would it be desirable to integrate the surface measuring instruments described in this chapter into the machine tools described in Parts III and IV? How would you go about doing so, particularly considering the factory environment in which they are to be used? Make some preliminary sketches.
31.24 In Section 31.3, we listed the major surface defects. How would you go about determining whether or not each of these defects is important for a particular application?

32

Tribology: Friction, Wear, and Lubrication

32.1 Introduction

Throughout the text so far, we have emphasized that friction, wear, and lubrication significantly influence the technology and economics of manufacturing operations. We have noted several times the nature and extent of wear that tools and dies undergo, eventually requiring reconditioning and replacement. In the United States alone, the total annual cost of replacing parts because of wear is estimated to be more than $100 billion.

Although we have covered their general effects in manufacturing processes, we have not yet described the fundamental mechanisms by which friction and wear occur during the interaction of workpieces, dies, and tools. Only after understanding these interactions can we begin to recommend appropriate materials and lubricants to meet specific requirements for friction and wear.

In this chapter we discuss those aspects of tribology that are relevant to manufacturing processes. After establishing the technological basis of friction and wear, we present the fundamentals of lubrication and the metalworking fluids commonly used in manufacturing operations.

32.2 Friction in Metals

Friction is defined as the resistance to relative sliding between two bodies in contact under a normal load. Friction plays an important role in all metalworking and manufacturing processes because of the relative motion and forces that are always acting on tools, dies, and workpieces.

Friction dissipates energy, thus generating heat, which can have detrimental effects on an operation. Furthermore, because it impedes free movement at interfaces, friction can significantly affect the flow and deformation of materials in metalworking processes. However, friction is not always undesirable. Without friction, for example, rolling metals, clamping workpieces on machines, or holding drills in chucks would be impossible.

The most commonly accepted theory of friction is the *adhesion theory*. It is based on the observation that two clean dry surfaces, regardless of how smooth they are, contact each other (junction) at only a fraction of their apparent area of contact (Fig. 32.1). The slope of the hills on these surfaces has been shown to range typically between 5° and 15°. In such a situation, the normal (contact) load N is supported by the minute **asperities** (small projections from the surface) in contact with each other. The normal stresses at these asperities therefore are high, causing plastic deformation at the junctions. This contact creates an adhesive bond between the asperities. In other words, the asperities form *microwelds*. Cold pressure welding (see Section 28.3) is based on this principle.

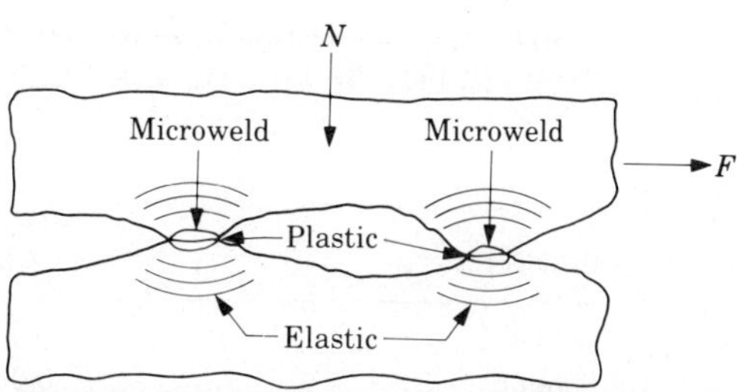

FIGURE 32.1 Schematic illustration of the interface of two bodies in contact, showing real areas of contact at the asperities. In engineering surfaces, the ratio of the apparent to real areas of contact can be as high as 4–5 orders of magnitude.

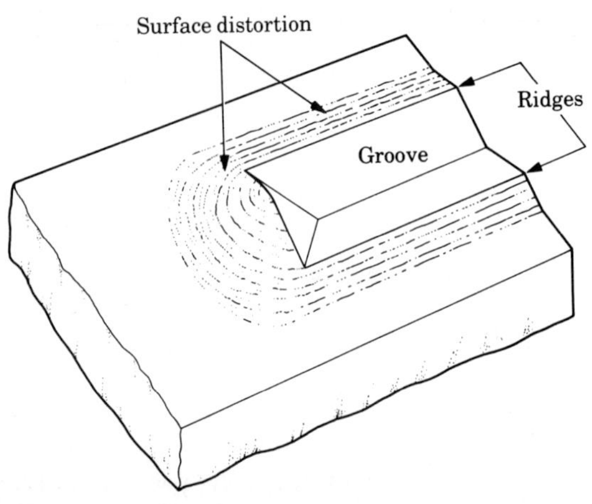

FIGURE 32.2 Schematic illustration of ploughing, in which the material on the surface is displaced sideways, as in scraping a clay surface with a pencil.

Sliding motion between two bodies having such an interface is possible only if a tangential force is applied. This tangential force is the force required to *shear* the junctions and is called the **friction force** (F). The ratio of F to N (see Fig. 32.1) is called the **coefficient of friction** (μ). In addition to the force required for breaking these junctions by shearing, a *ploughing force* can also be present if one surface scratches the other (Fig. 32.2). This force can contribute significantly to friction at the interface. Ploughing may either cause displacement of the material, as shown in Fig. 32.2, or produce small chips or slivers, as in cutting and abrasive processes. Depending on the materials and the process variables involved, coefficients of friction in manufacturing vary greatly, as you can see in Table 32.1.

Almost all the energy dissipated in overcoming friction is converted into heat, which raises the interface temperature. Temperature increases with friction, speed, and low thermal conductivity and specific heat of the sliding materials. The interface temperature may be high enough to soften and even melt the surfaces, as well as cause structural changes. These phenomena, in turn, adversely affect the operations involved, causing surface damage. Temperature also affects the viscosity

TABLE 32.1
COEFFICIENT OF FRICTION IN METALWORKING PROCESSES

PROCESS	COEFFICIENT OF FRICTION (μ)	
	COLD	*HOT*
Rolling	0.05–0.1	0.2–0.7
Forging	0.05–0.1	0.1–0.2
Drawing	0.03–0.1	—
Sheet-metal forming	0.05–0.1	0.1–0.2
Machining	0.5–2	—

and other properties of lubricants, causing their breakdown. Note, for example, how butter and oils burn and degrade when temperatures are excessive.

32.3 Friction in Plastics and Ceramics

Although their strength is low compared to metals, plastics generally possess low frictional characteristics. This property makes polymers attractive for bearings, gears, seals, prosthetic joints, and general friction-reducing applications. In fact, polymers are sometimes called self-lubricating. The factors involved in friction and metal wear are also generally applicable to polymers. In sliding, the ploughing component of friction in thermoplastics and elastomers is a significant factor because of their viscoelastic behavior (exhibiting both viscous and elastic behavior) and subsequent hysteresis loss (see Fig. 7.11). You can simulate this condition by dragging a dull nail across the surface of a rubber tire.

An important factor in plastics applications is the effect of temperature rise at the sliding interfaces caused by friction. As we described in Chapter 7, thermoplastics lose their strength and become soft as temperature increases. Their low thermal conductivity and low melting points are significant in terms of heat generation by friction. If the temperature rise is not controlled, sliding surfaces can undergo deformation and thermal degradation.

The frictional behavior of various polymers relative to metals is similar to that for metals on metals. The well-known low friction of PTFE (Teflon) is attributed to its molecular structure, in that no reaction takes place between PTFE and metals. Thus adhesion is poor and friction is low.

In view of the importance of ceramics, their frictional behavior is now being studied intensively. Initial investigations indicate that the origin of friction in ceramics is similar to that in metals. Thus adhesion and ploughing at interfaces contribute to the friction force in ceramics.

32.4 Reducing Friction

Friction can be reduced by selecting materials that have low adhesion, such as carbides and ceramics, and by using surface films and coatings (see the remaining sections of Chapter 32 and Chapter 33).

Lubricants, such as oils, or solid films, such as graphite, interpose an adherent film between tool, die, and workpiece. The film minimizes adhesion and interactions of one surface to the other, thus reducing friction.

Friction can also be reduced significantly by subjecting the die–workpiece interface to **ultrasonic vibrations**, generally at 20 kHz. These vibrations momentarily separate die and workpiece, thus allowing the lubricant to flow more freely into the interface. An additional factor is the high-frequency variation of the relative velocity between die and workpiece, thus reducing the friction at the interface.

32.5 Measuring Friction

The coefficient of friction is usually determined experimentally, either during actual manufacturing processes or in simulated tests using small-scale specimens of various shapes. The techniques used generally involve measurement of either forces or dimensional changes in the specimen.

Because of difficulties involved in full-scale experimentation on production equipment, particularly the high cost of interrupting production, small-scale tests simulating actual production conditions have been developed and used extensively. Unfortunately, many of these tests do not duplicate the exact conditions of the actual metalworking process, such as size of workpiece and its surface condition, the forces involved, and operating speed and temperature. They can, however, be used to compare different materials and lubricants.

32.5.1 Ring compression test

A test that has gained wide acceptance, particularly for bulk deformation processes such as forging, is the **ring compression test**. A flat ring is compressed plastically between two flat platens (Fig. 32.3a). As its height is reduced, the ring expands radially outward. If friction at the interfaces is zero, both the inner and outer diameters of the ring expand as if it were a solid disk. With increasing friction, the inner diameter becomes smaller.

For a particular reduction in height, there is a critical friction value at which the internal diameter increases (from the original) if μ is low and decreases if μ is high (Fig. 32.3b). By measuring the change in the specimen's internal diameter, and using the curves shown in Fig. 32.4, which are obtained through theoretical analyses, we can determine the coefficient of friction.

Each ring geometry has its own specific set of curves. The most common geometry has outer diameter to inner diameter to height proportions of the specimen of 6:3:2. The actual size of the specimen usually is not relevant in these tests. Thus, once you know the percentage reductions in internal diameter and height, you can determine μ from the appropriate chart.

The major advantages of the ring compression test are that it does not require any force measurements and that it involves large-scale deformation of the workpiece material. This test can also be used for rating different metalworking fluids (also known as lubricants and coolants).

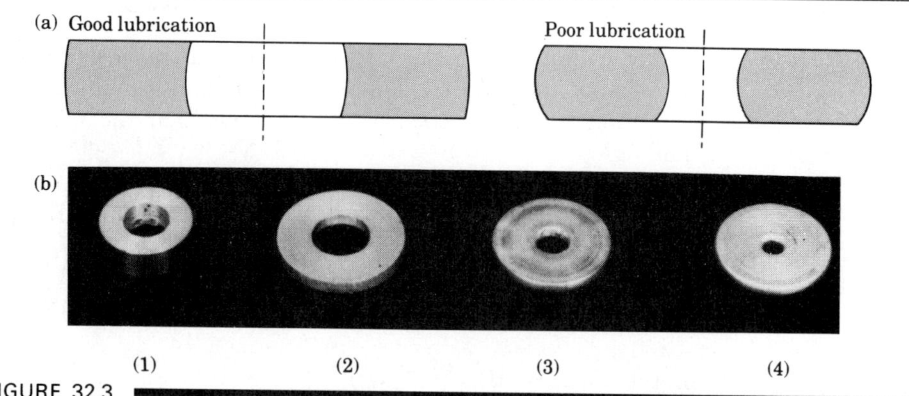

FIGURE 32.3

Ring compression test between flat dies. (a) Effect of lubrication on type of ring specimen barreling. (b) Test results: (1) original specimen and (2)–(4) increasing friction. *Source:* A. T. Male and M. G. Cockcroft.

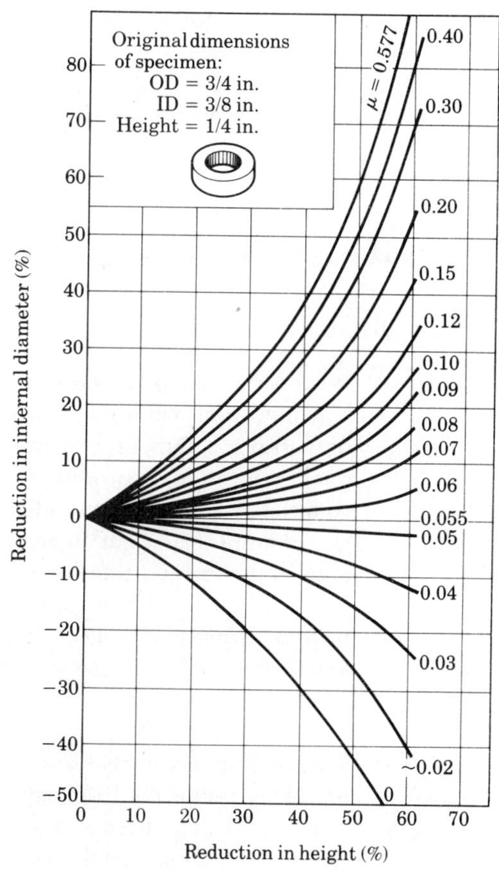

FIGURE 32.4

Chart to determine friction coefficient from ring compression test. Reduction in height and change in internal diameter of the ring are measured; then μ is read directly from this chart. Example: If the ring specimen is reduced in height by 40 percent and its internal diameter decreases by 10 percent, the coefficient of friction is 0.10.

• Example: Determination of coefficient of friction.

In a ring compression test, a specimen 10 mm in height with outside diameter OD = 30 mm and inside diameter ID = 15 mm is reduced in thickness by 50 percent. Determine the coefficient of friction μ if the OD after deformation is 38 mm.

SOLUTION. It is first necessary to determine the new ID. It is obtained from volume constancy as follows:

$$\text{Volume} = \frac{\pi}{4}(30^2 - 15^2)10 = \frac{\pi}{4}(38^2 - \text{ID}^2)5.$$

From this, the new ID = 9.7 mm.

Thus the change in internal diameter is

$$\text{Change in ID} = \frac{15 - 9.7}{15} \times 100 = 35 \text{ percent (decrease).}$$

For a 50-percent reduction in height and a reduction in internal diameter of 37 percent, the following value is obtained from Fig. 32.4:

$$\mu = 0.21.$$

32.6 Wear

Wear is defined as the progressive loss or removal of material from a surface. Wear has important technologic and economic significance because it changes the shape of workpiece, tool, and die interfaces. By doing so, it affects the process and size and quality of the parts produced. The magnitude of the wear problem is evident in the countless parts and components that continually have to be replaced or repaired.

Examples of wear in manufacturing processes are dull drills that have to be reground, worn cutting tools that have to be indexed or resharpened, and forming tools and dies that have to be repaired or replaced. Important components in some metalworking machinery are *wear plates* (Fig. 32.5), which are subjected to high loads. These plates, also known as *wear parts* because they are expected to wear, can be replaced easily.

Although wear generally alters the surface topography and may result in severe surface damage, it also has a beneficial effect: It can reduce surface roughness by removing the peaks from asperities (Fig. 32.6). Thus, under controlled conditions, wear may be regarded as a kind of smoothing or polishing process. The *running-in* period for various machines and engines produces this type of wear.

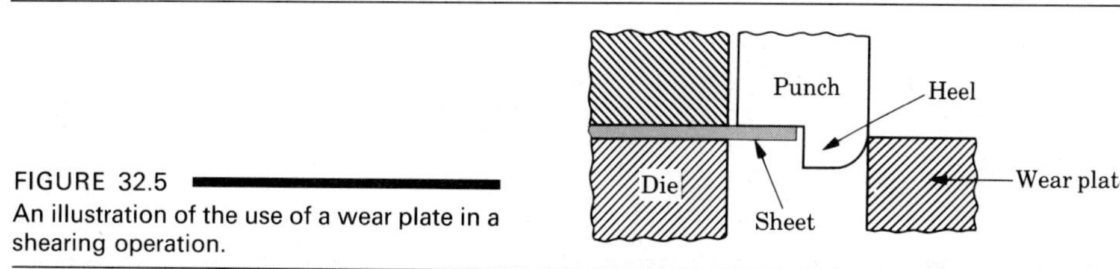

FIGURE 32.5
An illustration of the use of a wear plate in a shearing operation.

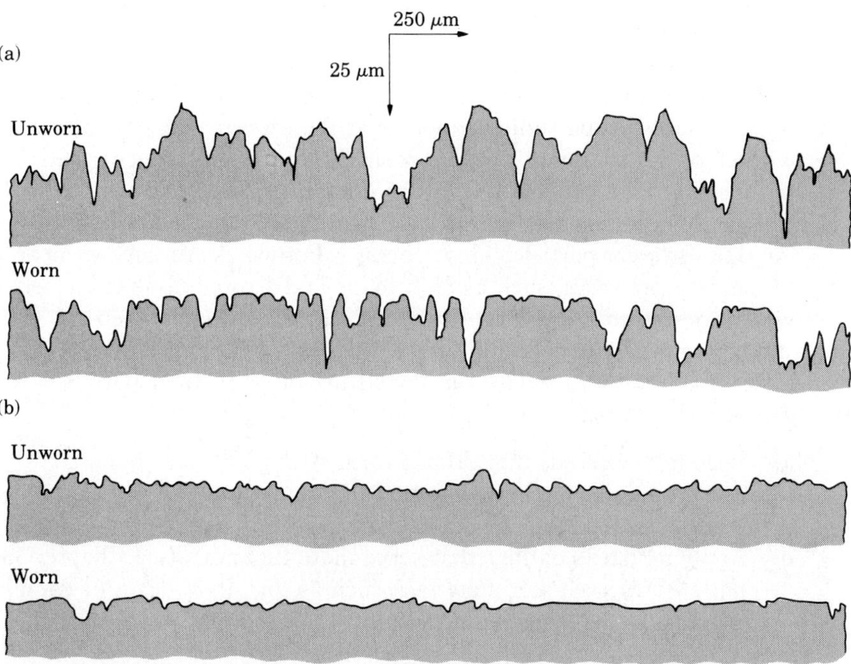

FIGURE 32.6
Changes in originally (a) wire-brushed and (b) ground-surface profiles after wear. *Source:* E. Wild and K. J. Mack.

Wear is usually classified as adhesive, abrasive, corrosive, fatigue, erosion, fretting, and impact. We describe below the major types of wear relevant to manufacturing operations.

32.6.1 Adhesive wear

If a tangential force is applied to the model shown in Fig. 32.1, shearing can take place either at the original interface or along a path below or above it (Fig. 32.7), causing **adhesive wear.** The fracture path depends on whether or not the strength of

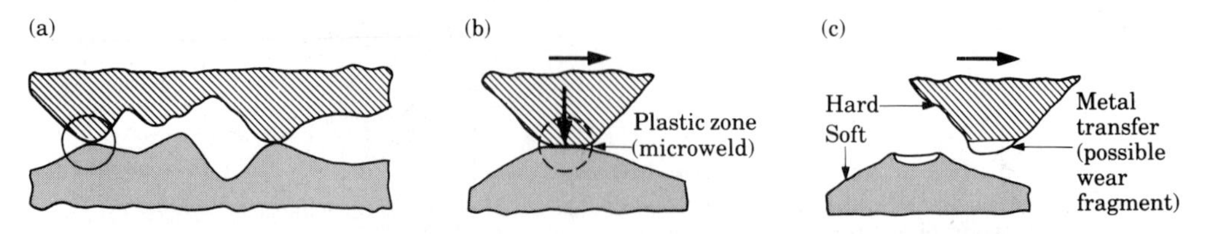

FIGURE 32.7
Schematic illustration of (a) two asperities contacting, (b) adhesion between two asperities, and (c) the formation of a wear particle.

the adhesive bond of the asperities is greater than the cohesive strength of either of the two sliding bodies.

Because of factors such as strain hardening at the asperity contact, diffusion, and mutual solid solubility, the adhesive bonds are often stronger than the base metals. Thus during sliding, fracture at the asperity usually follows a path in the weaker or softer component. A wear fragment is then generated. Although this fragment is attached to the harder component (upper member in Fig. 32.7), it eventually becomes detached during further rubbing at the interface and develops into a loose wear particle. This process is known as adhesive wear or sliding wear. In more severe cases, such as high loads and strongly bonded asperities, adhesive wear is described as scuffing, smearing, tearing, galling, or seizure. Oxide layers on surfaces greatly influence adhesive wear. They can act as a protective film, resulting in what is known as *mild wear*, consisting of small wear particles. Adhesive wear can be reduced by:

a) Selecting materials that do not form strong adhesive bonds.
b) Using a harder material as one of the pair.
c) Using materials that oxidize more easily.
d) Applying hard coatings that serve these functions (see Chapter 33). Coating one surface with soft materials, such as tin, silver, lead, or cadmium, is also effective in reducing sliding wear.

32.6.2 Abrasive wear

Abrasive wear is caused by a hard and rough surface—or a surface containing hard, protruding particles—sliding across a surface. This type of wear removes particles by forming microchips or slivers, thereby producing grooves or scratches on the softer surface (Fig. 32.8). In fact, the abrasive processes that we described in Chapter 25, such as grinding, ultrasonic machining, and abrasive-jet machining, act in this manner. The difference is that in those operations we control the process parameters to produce desired shapes and surfaces, whereas abrasive wear is unintended and unwanted.

The abrasive wear resistance of pure metals and ceramics is directly proportional to their hardness (Fig. 32.9). Abrasive wear can thus be reduced by increasing the

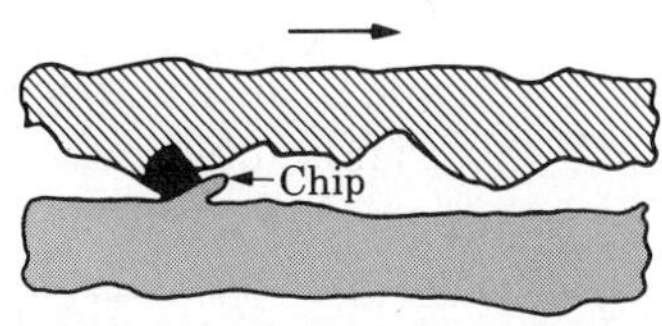

FIGURE 32.8 Schematic illustration of abrasive wear in sliding. Longitudinal scratches on a surface usually indicate abrasive wear.

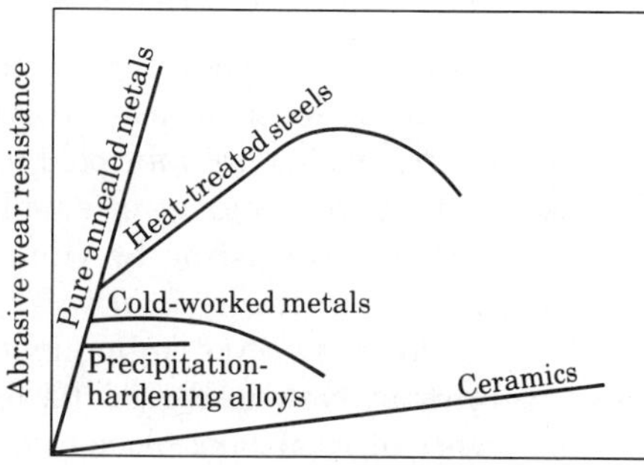

FIGURE 32.9 Abrasive wear resistance as a function of the hardness of several classes of materials.

hardness of materials (such as by heat treating and microstructural changes) or by reducing the normal load.

Other materials that resist abrasive wear are elastomers and rubbers, because they deform elastically and then recover when abrasive particles cross over their surfaces. The best example is automobile tires, which have long lives even though they are operated on abrasive road surfaces. Even hardened steels would not last long under such conditions.

32.6.3 Corrosive wear

Corrosive wear, also known as *oxidation*, or *chemical*, wear, is caused by chemical or electrochemical reactions between the surface and the environment. The fine corrosive products on the surface constitute the wear particles. When the corrosive layer is destroyed or removed, as by sliding or abrasion, another layer begins to form, and the process of removal and corrosive-layer formation is repeated.

Among corrosive media are water, seawater, oxygen, acids and chemicals, and atmospheric hydrogen sulfide and sulfur dioxide. We can reduce corrosive wear by selecting materials that will resist environmental attack, controlling the environment, and reducing operating temperatures to lower the rate of chemical reaction.

32.6.4 Fatigue wear

Fatigue wear, also called *surface fatigue* or surface-fracture wear, is caused when the surface of a material is subjected to cyclic loading, such as in rolling contact in bearings. The wear particles are usually formed by spalling or pitting. Another type of wear is by *thermal fatigue*. Cracks on the surface are generated by thermal stresses from thermal cycling, such as a cool die repeatedly contacting hot workpieces (heat checking). These cracks then join, and the surface begins to spall, producing fatigue wear. This type of wear usually occurs in hot-working and die-casting dies. We can reduce fatigue wear by lowering contact stresses, reducing thermal cycling, and improving the quality of materials by removing impurities, inclusions, and various other flaws that may act as local points for crack initiation.

32.6.5 Other types of wear

Several other types of wear are important in manufacturing processes. *Erosion* is caused by loose abrasive particles abrading a surface. *Fretting corrosion* occurs at interfaces that are subjected to very small movements, such as in machinery. *Impact wear* is the removal of small amounts of materials from a surface by impacting particles. Deburring by vibratory finishing and tumbling (see Section 25.10) is an example of this type of wear.

In many cases component wear is the result of a combination of different types of wear. Note in Fig. 32.10, for example, that even in the same forging die, various types of wear take place in different locations. A similar situation exists in cutting tools (see Fig. 20.20).

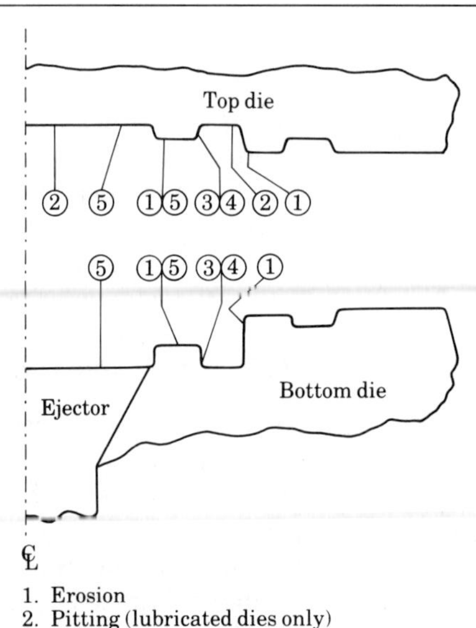

FIGURE 32.10 Types of wear observed in a single die used for hot forging. *Source:* T. A. Dean.

32.7 Wear of Plastics and Ceramics

The wear behavior of plastics is similar to that of metals. Thus wear may occur in ways similar to those described in Section 32.6. Abrasive wear behavior depends partly on the ability of the polymer to deform and recover elastically, similar to the behavior of elastomers. There is evidence that the parameter describing this behavior may be the ratio of hardness to elastic modulus. Thus the abrasive wear resistance of the polymer increases as this ratio increases.

Typical polymers with good wear resistance are polyimides, nylons, polycarbonate, polypropylene, acetals, and high-density polyethylene. These polymers are molded or machined to make gears, pulleys, sprockets, and similar mechanical components. Because plastics can be made with a wide variety of compositions, they can also be blended with internal lubricants, such as PTFE, silicon, graphite, and molybdenum disulfide, and with rubber particles that are interspersed within the polymer matrix.

32.7.1 Wear of reinforced plastics

Wear resistance of reinforced plastics depends on the type, amount, and direction of reinforcement in the polymer matrix. Carbon, glass, and aramid fibers all improve wear resistance. Wear takes place when fibers are pulled out of the matrix (fiber pullout). Wear is greatest when the sliding direction is parallel to the fibers because they can be pulled out more easily. Long fibers increase the wear resistance of composites because they are more difficult to pull out, and cracks in the matrix cannot propagate to the surface as easily.

32.7.2 Wear of ceramics

When ceramics slide against metals, wear is caused by small-scale plastic deformation and brittle surface fracture, surface chemical reactions, ploughing, and possibly some fatigue. Metals can be transferred to the oxide-type ceramic surfaces, forming metal oxides. Thus sliding actually takes place between the metal and metal-oxide surface. Conventional lubricants do not appear to influence ceramics wear significantly.

32.8 Measuring Wear

Several methods can be used to observe and measure wear. The choice of a particular method depends on the accuracy desired and the physical constraints of the system, such as specimen or part size and difficulty of disassembly.

Although not quantitative, the simplest method is visual and tactile (touching) inspection. Measuring dimensional changes, gaging the worn component, profilometry, and weighing are more accurate methods. However, for large workpieces or tools and dies, the weighing method is not accurate because the amount of wear is usually very small compared to the overall weight of the components involved. Performance and noise level can be monitored. Worn machinery components emit more noise than new parts.

Radiography is a method in which wear particles from an irradiated surface are transferred to the mating surface, which is then measured for the amount of radiation. An example is the transfer of wear particles from irradiated cutting tools to the back side of chips. In other situations, the lubricant can be analyzed for wear particles (spectroscopy). This is a precision method and is used widely for applications such as checking jet-engine component wear.

32.9 Lubrication

We have noted that the surfaces of tools, dies, and workpieces are usually subjected to a wide range of process parameters, such as contact pressures (ranging from very low values to multiples of the yield stress of the workpiece material), relative speeds (from very low to very high), and temperatures (ranging from ambient to melting).

In addition to selecting appropriate materials and controlling process parameters, we can also select metalworking fluids to effectively reduce friction and wear. Lubricants are used extensively in all types of machinery and engines to reduce friction and wear. Although we generally think of lubricants as fluids, they can be effective in solid or semisolid forms as well.

32.9.1 Types of lubrication

In **thick film lubrication**, the surfaces are completely separated, and lubricant viscosity is the important factor. Such films can develop in some regions of the workpiece in high-speed operations, and with high-viscosity lubricants that become trapped at die–workpiece interfaces. A thick lubricant film generates a dull, grainy surface appearance on the workpiece, the degree of roughness depending on grain size (Fig. 32.11). In operations such as coining and precision forging, trapped lubricants are undesirable because they prevent accurate shape generation and produce rough surfaces.

The lubricant film becomes thinner as the load between the die and workpiece increases or as the speed and viscosity of the metalworking fluid decreases. This condition increases friction and leads to slight wear. In **mixed lubrication**, a significant portion of the load is carried by the metal-to-metal contact of the asperities and the rest by the fluid film trapped in pockets, such as the valleys of asperities.

FIGURE 32.11
Rough surface developed on an aluminum compression specimen by the presence of a high-viscosity lubricant and high compression speed. The coarser the grain size, the rougher the surface will be. *Source:* A. Mulc and S. Kalpakjian.

In **boundary lubrication**, the load is supported by contacting surfaces covered with a boundary layer of lubricant, typically natural oils, fats, fatty acids, and soaps, thus preventing direct metal-to-metal contact and reducing wear. However, boundary films can break down because of desorption caused by high temperatures at the interface or can be rubbed off during sliding. Deprived of this protective film, the metal surfaces may then wear and score severely.

32.9.2 Other considerations

The valleys in the surface roughnesses of the contacting bodies can serve as local reservoirs or pockets for lubricants, supporting a substantial portion of the load. The workpiece, not the die, should have the rougher surface; otherwise, the rougher and harder die surface may act like a file, damaging the workpiece surface. The recommended surface roughness on most dies is about 0.4 μm (15 μin.).

The overall geometry of the interacting bodies is also an important consideration in ensuring proper lubrication. The movement of the workpiece into the deformation zone, such as in wire drawing, extrusion, and rolling, should allow a supply of lubricant to be carried into the die–workpiece interface. With proper selection of process parameters, a relatively thick lubricant film can be entrained and maintained.

32.10 Metalworking Fluids

On the basis of our previous discussions, we can summarize the functions of metalworking fluids as follows:

- Reduce friction, thus reducing force and energy requirements and temperature rise.

- Reduce wear, seizure, and galling.
- Improve material flow in tools, dies, and molds.
- Act as a thermal barrier between the workpiece and tool and die surfaces, thus preventing workpiece cooling in hot-working processes.
- Act as a release or parting agent to help in the removal or ejection of parts from dies and molds.

Many types of metalworking fluids are now available to fulfill these requirements. Because of their diverse chemistries, properties, and characteristics, the behavior and performance of lubricants can be complex. In this section we describe the general properties of only the most commonly used lubricants.

32.10.1 Oils

Oils have high film strength on the surface of a metal, as you well know if you have ever tried to clean an oily surface. Although they are very effective in reducing friction and wear, oils have low thermal conductivity and specific heat. Thus they are not effective in conducting away the heat generated by friction and plastic deformation in metalworking operations. Oils also are difficult to remove from component surfaces that subsequently are to be painted or welded, and they are difficult to dispose of.

The sources of oils are mineral (petroleum), animal, vegetable, and fish. Oils may be *compounded* with a variety of additives or with other oils to impart special properties, such as their viscosity–temperature behavior and surface tension, heat resistance, and boundary layer characteristics. Mineral oils with or without fillers, and used undiluted, as known as *neat oils*.

Additives. Metalworking fluids are usually blended with several additives, such as oxidation inhibitors, rust preventatives, odor control agents, antiseptics, and foam inhibitors. Important additives in oils are sulfur, chlorine, and phosphorus. Known as **extreme-pressure (EP) additives** and used singly or in combination, they react chemically with metal surfaces and form adherent surface films of metallic sulfides and chlorides. These films have low shear strength and good antiweld properties and thus effectively reduce friction and wear. While EP additives are important in boundary lubrication, these lubricants may preferentially attack the cobalt binder in tungsten carbide tools and dies, causing changes in their surface roughness and integrity.

32.10.2 Emulsions

An **emulsion** is a mixture of two immiscible liquids, usually mixtures of oil and water in various proportions, along with additives. Emulsions, also known as *water-soluble fluids*, are of two types: direct and indirect. In a direct emulsion, mineral oil is dispersed in water as very small droplets. In an indirect emulsion, water droplets are dispersed in the oil.

Direct emulsions are important fluids because the presence of water gives them high cooling capacity. They are particularly effective in high-speed metal machining where temperature rise has detrimental effects on tool life, workpiece surface integrity, and dimensional accuracy.

32.10.3 Synthetic solutions

Synthetic solutions are chemical fluids containing inorganic and other chemicals dissolved in water. Various chemical agents are added to impart different properties. *Semisynthetic* solutions are basically synthetic solutions to which small amounts of emulsifiable oils have been added.

32.10.4 Soaps, greases, and waxes

Soaps are generally reaction products of sodium or potassium salts with fatty acids. Alkali soaps are soluble in water, but other metal soaps are generally insoluble. They are effective boundary lubricants and can also form thick film layers at die–workpiece interfaces, particularly when applied on conversion coatings for cold metalworking applications (Section 32.12).

Greases are solid or semisolid lubricants and generally consist of soaps, mineral oil, and various additives. They are highly viscous and adhere well to metal surfaces. Although used extensively in machinery, greases have limited use in manufacturing processes.

Waxes may be of animal or plant (paraffin) origin and have complex structures. Compared to greases, waxes are less "greasy" and are more brittle. They have limited use in metalworking operations, except for copper and, as chlorinated paraffin, for stainless steels and high-temperature alloys.

32.11 Solid Lubricants

Because of their unique properties and characteristics, several solid materials are used as lubricants in manufacturing operations. We describe below four of the most commonly used **solid lubricants**.

32.11.1 Graphite

The general properties of graphite are described in Section 8.6. Graphite is weak in shear along its layers and has a low coefficient of friction in that direction. Thus it can be a good solid lubricant, particularly at elevated temperatures. However, the graphite friction is low only in the presence of air or moisture. In a vacuum or an inert gas atmosphere, friction is very high; in fact, graphite can be quite abrasive. Graphite may be applied either by rubbing it on surfaces or as a *colloidal* (dispersion of small particles) suspension in liquid carriers such as water, oil, or alcohols.

32.11.2 Molybdenum disulfide

This is another widely used lamellar solid lubricant. It is somewhat similar in appearance to graphite. However, unlike graphite it has a high friction coefficient in ambient environment. Oils are commonly used as carriers for molybdenum disulfide (MoS_2) and are used as a lubricant at room temperature. Molybdenum disulfide can also be rubbed onto the surfaces of a workpiece.

32.11.3 Metallic and polymeric films

Because of their low strength, thin layers of soft metals and polymer coatings are also used as solid lubricants. Suitable metals are lead, indium, cadmium, tin, silver, and polymers such as PTFE, polyethylene, and methacrylates. However, these coatings have limited applications because of their lack of strength under high stresses and at elevated temperatures.

Soft metals are used to coat high-strength metals such as steels, stainless steels, and high-temperature alloys. Copper or tin, for example, is chemically deposited on the surface of the metal before it is processed. If the oxide of a particular metal has low friction and is sufficiently thin, the oxide layer can serve as a solid lubricant, particularly at elevated temperatures.

32.11.4 Glasses

Although a solid material, glass becomes viscous at elevated temperatures and hence can serve as a liquid lubricant. Viscosity is a function of temperature, but not of pressure, and depends on the type of glass. Poor thermal conductivity also makes glass attractive, since it acts as a thermal barrier between hot workpieces and relatively cool dies. Typical glass lubrication applications are in hot extrusion and forging.

32.12 Conversion Coatings

Lubricants may not always adhere properly to workpiece surfaces, particularly under high normal and shearing stresses. This condition is a special problem in forging, extrusion, and wire drawing of steels, stainless steels, and high-temperature alloys.

For these applications, acids transform the workpiece surface by chemically reacting with it, leaving a somewhat rough and spongy surface that acts as a carrier for the lubricant. After treatment, borax or lime is used to remove any excess acid from the surfaces. A liquid lubricant, such as a soap, is then applied to the coated surface. The lubricant film adheres to the surface and cannot be scraped off easily. *Zinc phosphate* conversion coatings are often used on carbon and low-alloy steels. *Oxalate* coatings are used for stainless steels and high-temperature alloys.

32.13 Lubricant Selection

Selecting a lubricant for a particular process and workpiece material involves consideration of several factors:

a) The particular manufacturing process.
b) Compatibility of the lubricant with the workpiece and tool and die materials.
c) Surface preparation required.
d) Method of lubricant application.
e) Removal of lubricant after processing.
f) Contamination of the lubricant by other lubricants, such as those used to lubricate the machinery.
g) Treatment of waste lubricant.
h) Storage and maintenance of lubricants.
i) Biological and ecological considerations.

In selecting an oil as a lubricant, we should recognize the importance of its viscosity–temperature–pressure characteristics. Low viscosity can have a significant detrimental effect on friction and wear. The different functions of a metalworking fluid, whether primarily a lubricant or a coolant, must also be taken into account. Water-base fluids are very effective coolants but as lubricants are not as effective as oils.

Metalworking fluids should not leave any harmful residues that could interfere with machinery operations. The fluids should not stain or corrode workpiece or equipment. The fluids should be checked periodically for deterioration caused by bacterial growth, accumulation of oxides, chips, and wear debris—and also general degradation and breakdown because of temperature and time. A lubricant may carry with it wear particles and cause damage to the system, so proper inspection and filtering of metalworking fluids are important. These precautions are necessary for all types of machinery, including internal combustion engines and jet engines.

After completion of manufacturing operations, metal surfaces are usually covered with lubricant residues, which should be removed prior to further workpiece processing, such as welding or painting. Various cleaning solutions and techniques can be used for this purpose (Section 33.15).

Biological and ecological considerations, with their accompanying legal ramifications, are also important considerations. Potential health hazards may be involved in contacting or inhaling some metalworking fluids. Recycling of waste fluids and their disposal are additional important factors to be considered. Laws and regulations concerning the manufacture, transportation, use, and disposal of metalworking fluids are promulgated by the Occupational Safety and Health Act (OSHA), the National Institute for Occupational Safety and Health (NIOSH), and the Environmental Protection Agency (EPA) of the United States. We discuss these issues in Section 37.4.

SUMMARY

Friction and wear are among the most significant factors in processing materials. Much progress has been made in understanding these phenomena and identifying the factors that govern them. Among these factors are the affinity and solid solubility of the two materials in contact, the nature of surface films, the presence of contaminants, and process parameters, such as load, speed, and temperature.

A wide variety of metalworking fluids is available for specific applications, including oils, emulsions, synthetic solutions, and solid lubricants. Their selection and use require careful consideration of many factors in regard to workpiece and die materials and the particular manufacturing process. These fluids have various lubricating and cooling characteristics. Biological and ecological considerations are also important factors.

TRENDS

- Because of their major economic impact, friction and wear are topics of continued study, particularly for new metal alloys, ceramics, and composite materials.
- Biological and ecological aspects of lubricants and lubrication practices are being studied extensively. Ecologically acceptable metalworking fluids are under constant development.
- Studies are continuing on the development of test methods to determine friction and wear under various processing conditions.
- Work is in progress concerning lubricant film thickness and lubrication regimes in metalworking processes.
- Surface alterations and development of various synthetic solutions as metalworking fluids are under continuing investigation.
- Improvements in the wear resistance of ceramics continue to be made by controlling their grain size, porosity, and phase structure; of polymers, by reinforcing fibers and lubricating additives.

KEY TERMS

Abrasive wear
Adhesive wear
Asperities
Boundary lubrication
Coefficient of friction
Corrosive wear
Emulsion
Extreme-pressure additives
Fatigue wear
Friction
Friction force
Greases
Lubrication
Mixed lubrication
Oils
Ring compression test
Soaps
Solid lubricants
Thick-film lubrication
Ultrasonic vibrations
Waxes
Wear

BIBLIOGRAPHY

Bhushan, B., and B.K. Gupta, *Handbook of Tribology: Materials, Coatings, and Surface Treatments.* New York: McGraw-Hill, 1991.

Friction and Wear Devices, 2d ed. Park Ridge, Ill.: American Society of Lubrication Engineers, 1976.

Halling, J., *Principles of Tribology.* New York: Macmillan, 1975.

Lansdown, A.R., *Lubrication: A Practical Guide to Lubricant Selection.* Oxford: Pergamon, 1982.

——, and A.L. Price, *Materials to Resist Wear: A Guide to Their Selection and Use.* Oxford: Pergamon Press, 1986.

Moore, D.E., *Principles and Applications of Tribology.* New York: Pergamon, 1975.

Nachtman, E.S., and S. Kalpakjian, *Lubricants and Lubrication in Metalworking Operations.* New York: Marcel Dekker, 1985.

Olds, N.J., *Lubricants, Cutting Fluids and Coolants.* Boston: Cahners, 1973.

Peterson, M.B., and W.O. Winer (eds.), *Wear Control Handbook.* New York: American Society of Mechanical Engineers, 1980.

Rabinowicz, E., *Friction and Wear of Materials.* New York: Wiley, 1965.

Rigney, D.A. (ed.), *Fundamentals of Friction and Wear of Materials.* Metals Park, Ohio: American Society for Metals, 1981.

Schey, J.A., *Tribology in Metalworking—Friction, Lubrication and Wear.* Metals Park, Ohio: American Society for Metals, 1983.

Source Book on Wear Control Technology. Metals Park, Ohio: American Society for Metals, 1978.

REVIEW QUESTIONS

32.1 Explain the importance of friction in manufacturing processes.

32.2 Explain the adhesion theory of friction. What is the nature of the friction force?

32.3 What is ploughing force?

32.4 What is the significance of surface temperature rise resulting from friction?
32.5 Why does surface temperature rise with increasing sliding speed?
32.6 List the methods that can be used to reduce friction in metalworking processes.
32.7 How are ultrasonic vibrations effective in reducing friction? What frequency of vibration is commonly used?
32.8 Describe the features of the ring compression test. Does it require the measurement of forces?
32.9 Explain what is meant by wear.
32.10 List the types of wear generally observed in engineering practice.
32.11 How can adhesive wear be reduced? Abrasive wear? Fatigue wear?
32.12 List the methods that can be used to measure wear.
32.13 What is a lubricant? What functions should it perform in manufacturing processes?
32.14 Explain the characteristics of various types of lubricating methods.
32.15 List the different types of fluid and solid lubricants used in metalworking.
32.16 What is the difference between a direct and an indirect emulsion?
32.17 What are conversion coatings used for?
32.18 Describe the factors involved in lubricant selection.

QUESTIONS AND PROBLEMS

32.19 Describe the tribological differences between ordinary machine elements (such as gears and bearings) and metalworking processes involving tools and dies. Consider factors such as load, speed, and temperature.
32.20 Explain why the bearings of skateboard wheels (if not sealed) should not be lubricated.
32.21 Explain why the specimen in a ring compression test may become oval.
32.22 Describe the applicability of simulated friction and wear tests to actual manufacturing operations.
32.23 Can the temperature rise at a sliding interface exceed the melting point of the metals? Explain.
32.24 Describe the changes that occur in the model for adhesion theory of friction if the surfaces are similar to that shown in Fig. 31.1.
32.25 Describe conditions that lead to ploughing.
32.26 Explain what is involved in the running-in process.
32.27 It has been stated that as the normal load decreases, abrasive wear is reduced. Explain why this is so.
32.28 Explain why the particular types of wear in Fig. 32.10 occur in those particular locations in the forging die.
32.29 Why is the abrasive-wear resistance of a material a function of its hardness?
32.30 Explain the similarities and differences between the friction of metals and polymers.
32.31 Would ceramics be suitable for sliding at high speeds? Explain.
32.32 Using Fig. 32.4, make a plot of the coefficient of friction versus change in internal diameter for a constant reduction in height of 50 percent.
32.33 Assume that in the example problem in Section 32.5.1 the coefficient of friction is 0.15. If all other parameters remain the same, what is the new internal diameter of the specimen?

32.34 We have seen that wear can have detrimental effects in manufacturing operations. Can you visualize situations in which wear could be beneficial? Give examples.

32.35 Write a brief paper on why Teflon has such a low friction coefficient.

32.36 Explain the similarities and differences between wear of metals and of polymers.

32.37 On the basis of the topics discussed in this chapter, do you think there is a direct correlation between friction and wear of materials? Explain.

32.38 You have undoubtedly replaced parts in various appliances and automobiles because they were worn. Describe the methodology you would follow in determining the type(s) of wear these components have undergone.

32.39 Following your observations and opinions regarding Problem 32.38, state how you would go about changing materials or designs to reduce the wear of these components.

32.40 As we stated in Section 32.2, in the ploughing process (Fig. 32.2) it is possible that material is removed from the surface in the shape of slivers or chips. Describe the conditions under which this can happen.

33

Surface Treatment, Coating, and Cleaning

33.1 Introduction

After a component is manufactured, all or parts of its surfaces may have to be processed further in order to impart certain properties and characteristics (Table 33.1). Surface treatment may be necessary to:

- Improve resistance to wear, erosion, and indentation (slideways in machine tools, wear surfaces of machinery, and shafts, rolls, cams, and gears).
- Control friction (sliding surfaces on tools, dies, bearings, and machine ways).
- Reduce adhesion (electrical contacts).
- Improve lubrication (surface modification to retain lubricants).
- Improve corrosion and oxidation resistance (sheet metals for automotive or other outdoor uses, gas turbine components, and medical devices).
- Improve fatigue resistance (bearings and multiple-diameter shafts with fillets).

TABLE 33.1
COATINGS USED FOR VARIOUS FUNCTIONS

FUNCTION	COATINGS
Reduce wear	Titanium carbide, titanium nitride, diamond
Reduce friction	PTFE, molybdenum disulfide
Increase friction	Titanium, bonded abrasives
Improve lubrication	Copper, lead
Increase temperature or load capacity	Electroless nickel
Prevent adhesion	Silver/gold plate
Imbed particles	Indium, lead
Reduce corrosive wear	Chromium plate or diffusion
Retain fluid lubricants	Phosphating, nylon
Rebuild surface	Steel hard surfacing
Reduce surface roughness	Silver plate
Prevent drop erosion	Polyurethane, neoprene
Prevent particle erosion	Cobalt alloy, molybdenum

Source: After S. Ramalingam.

TABLE 33.2
SUGGESTED SURFACE TREATMENTS FOR VARIOUS METALS

METAL	TREATMENT
Aluminum	Chrome plate; anodic coating, phosphate; chromate conversion coating
Beryllium	Anodic coating; chromate conversion coating
Cadmium	Phosphate; chromate conversion coating
Die steels	Boronizing; ion nitriding; liquid nitriding
High-temperature steels	Diffusion
Magnesium	Anodic coating; chromate conversion coating
Mild steel	Boronizing; phosphate; carburizing; liquid nitriding; carbonitriding; cyaniding
Molybdenum	Chrome plate
Nickel- and cobalt-base alloys	Boronizing; diffusion
Refractory metals	Boronizing
Stainless steel	Vapor deposition; ion nitriding; diffusion; liquid nitriding; nitriding
Steel	Vapor deposition; chrome plate; phosphate; ion nitriding; induction hardening; flame hardening; liquid nitriding
Titanium	Chrome plate; anodic coating; ion nitriding
Tool steel	Boronizing; ion nitriding; diffusion; nitriding; liquid nitriding
Zinc	Vapor deposition; anodic coating; phosphate; chromate chemical conversion coating

Source: After M. K. Gabel and D. M. Donovan in *Wear Control Handbook*, New York, ASME, 1980, p. 349.

- Rebuild surfaces on worn components (worn tools, dies, and machine components).
- Improve surface roughness (appearance, dimensional accuracy, and frictional characteristics).
- Impart decorative features, color, or special surface texture.

Several techniques that are suitable and applicable to certain groups of materials have been developed (Table 33.2). In this chapter we describe the methods used to modify the surface structure and its properties in order to impart these desirable characteristics. We begin with surface hardening techniques involving mechanical or thermal means and continue with different types of coatings that are applied by various means. Some of these techniques are also used in the manufacture of semiconductor devices (see Chapter 34). Finally, we discuss cleaning techniques for manufactured surfaces, particularly lubricant residues, before components are processed further, assembled, and the product is placed in service.

33.2 Mechanical Surface Treatment and Coating

Several techniques are available for mechanically improving the surface properties of finished components. We describe the more common ones below.

33.2.1 Shot peening

In **shot peening**, the workpiece surface is hit repeatedly with a large number of cast steel, glass, or ceramic shot (small balls), making overlapping indentations on the surface. This action causes plastic deformation of surfaces, to depths up to 1.25 mm (0.05 in.), using shot sizes ranging from 0.125 mm to 5 mm (0.005 in. to 0.2 in.) in diameter.

Because the plastic deformation is not uniform throughout the part's thickness, shot peening imparts compressive residual stresses on the surface, thus improving the fatigue life of the component. This process is used extensively on shafts, gears, springs, oil-well drilling equipment, and jet-engine parts (such as turbine and compressor blades).

33.2.2 Roller burnishing

In **roller burnishing**, also called *surface rolling*, the surface of the component is cold worked by a hard and highly polished roller or rollers. This process is used on various flat, cylindrical, or conical surfaces (Figs. 33.1 and 33.2). Roller burnishing

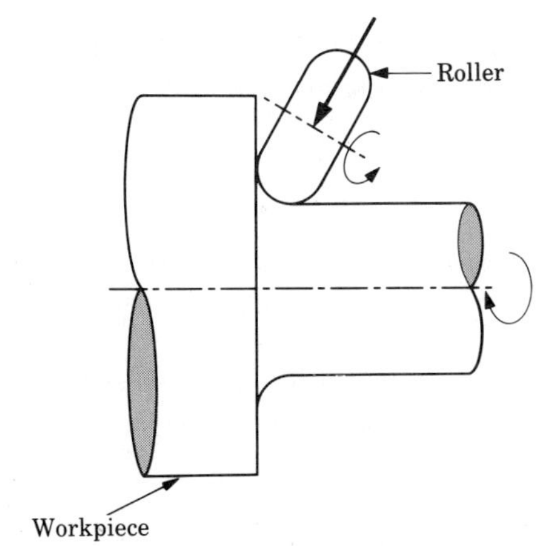

FIGURE 33.1 Roller burnishing of the fillet of a stepped shaft to induce compressive surface residual stresses for improved fatigue life.

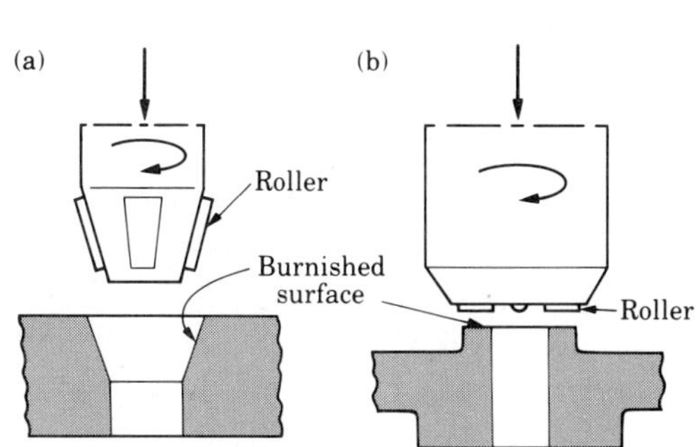

FIGURE 33.2 Examples of roller burnishing of (a) a conical surface and (b) a flat surface and the burnishing tools used. *Source:* Sandvik, Inc.

improves surface finish by removing scratches, tool marks, and pits. Consequently, corrosion resistance is also improved since corrosive products and residues cannot be entrapped. Internal cylindrical surfaces are burnished by a similar process, called *ballizing* or *ball burnishing*. A smooth ball, slightly larger than the bore diameter, is pushed through the length of the hole.

Roller burnishing is used to improve the mechanical properties of surfaces, as well as the shape and surface finish of components. It can be used either singly or in combination with other finishing processes, such as grinding, honing, and lapping. Soft and ductile, as well as very hard metals, can be roller burnished. Typical applications include hydraulic-system components, seals, valves, spindles, and fillets on shafts.

33.2.3 Explosive hardening

In **explosive hardening**, surfaces are subjected to high transient pressures by placing a layer of explosive sheet directly on the workpiece surface and detonating it. The contact pressures developed can be as high as 35 GPa (5×10^6 psi), lasting about 2–3 μs. Large increases in surface hardness can be obtained by this method, with very little change (less than 5 percent) in the shape of the component. Railroad rail surfaces can be hardened by this method.

33.2.4 Cladding (clad bonding)

In **cladding**, metals are bonded with a thin layer of corrosion-resistant metal by applying pressure with rolls or other means. A typical application is cladding of aluminum (*Alclad*) in which a corrosion-resistant layer of aluminum alloy is clad over pure aluminum. Other applications are steels clad with stainless steel or nickel alloys. The cladding material may also be applied through dies, as in cladding steel wire with copper, or by explosives. Multiple-layer cladding is also utilized for special applications (see also Fig. 28.1).

33.2.5 Mechanical plating

In **mechanical plating** (also called mechanical coating, impact plating, or peen plating), fine metal particles are compacted over the workpiece surfaces by impacting them with spherical glass, ceramic, or porcelain beads. The beads are propelled by rotary means. The process is used typically for hardened-steel parts for automobiles, with plating thickness usually less than 0.025 mm (0.001 in.).

33.3 Surface Hardening by Heat Treatment and Thermal Spraying

Surfaces may be hardened by thermal means in order to improve their frictional and wear properties, as well as resistance to indentation, erosion, abrasion, and corrosion. We describe the most common methods below.

33.3.1 Case hardening

We described traditional methods of **case hardening** (carburizing, carbonitriding, cyaniding, nitriding, flame hardening, and induction hardening) in Section 4.10 and summarized them in Table 4.1. In addition to the common heat sources of gas and electricity, laser beams are also used as a heat source in surface hardening of both metals and ceramics.

The hardness levels obtained in carbon and alloy steels by various heat-treating processes are shown in Fig. 33.3. It also includes hardnesses obtained in steels by cold and hot metalworking techniques. Case hardening, as well as some of the other surface-treatment processes, induce residual stresses on surfaces. The formation of martensite in case hardening causes compressive residual stresses on surfaces. Such stresses are desirable because they improve the fatigue life of components by delaying the initiation of fatigue cracks.

33.3.2 Hard facing

In **hard facing**, a relatively thick layer, edge, or point of wear-resistant hard metal is deposited on the surface by any of the welding techniques we described in Chapters 27 and 28. A number of layers are usually deposited (*weld overlay*). Hard coatings of tungsten carbide and chromium and molybdenum carbides can also be deposited

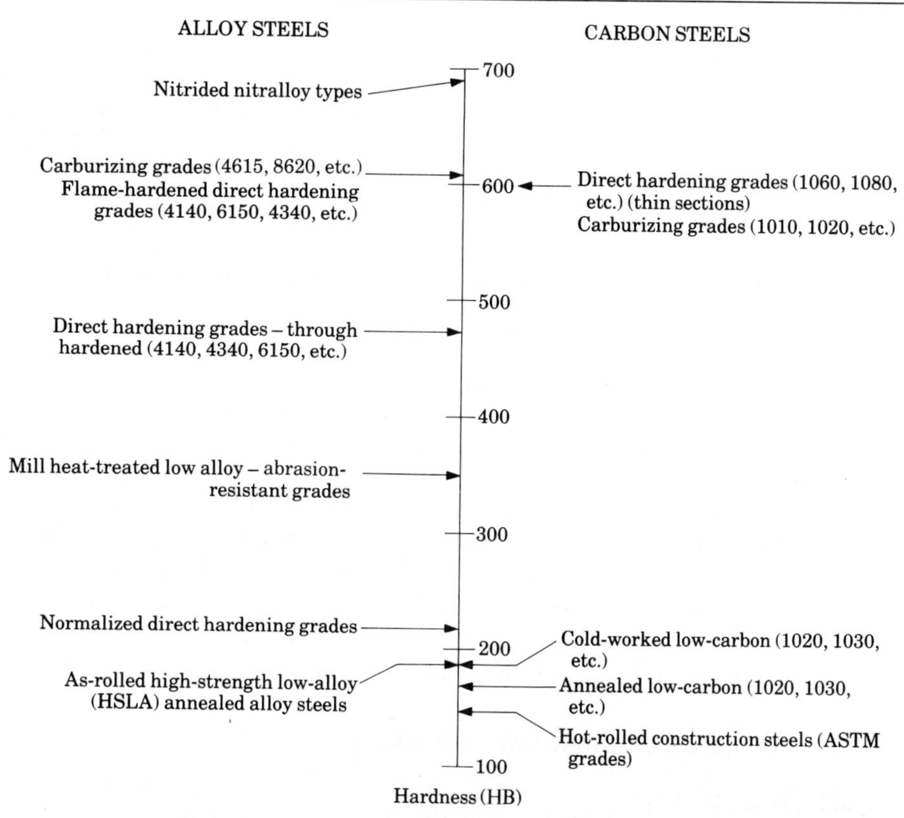

FIGURE 33.3 Hardness ranges obtainable by various thermal and mechanical treatment processes for carbon and alloy steels. *Source:* K. G. Budinski.

using an electric arc (*spark hardening*). Hard-facing alloys are available as electrodes, rod, wire, and powder. Typical applications for hard facing are valve seats, oil-well drilling tools, and dies for hot metalworking. Worn parts are also hard faced for extended use.

33.3.3 Thermal spraying

In **thermal spraying,** metal in the form of rod, wire, or powder is melted in a stream of oxyacetylene flame, electric arc, or plasma arc, and the droplets are sprayed on the preheated surface, at speeds up to 100 m/s (20,000 ft/min) with a compressed-air spray gun (Fig. 33.4). Surfaces to be sprayed should be cleaned and roughened to improve bond strength. Typical applications for thermal spraying, also called *metallizing*, are steel structures, storage tanks, and tank cars, sprayed with zinc or aluminum up to 0.25 mm (0.010 in.) in thickness. The deposited metal layer has a porosity of 5–20 percent.

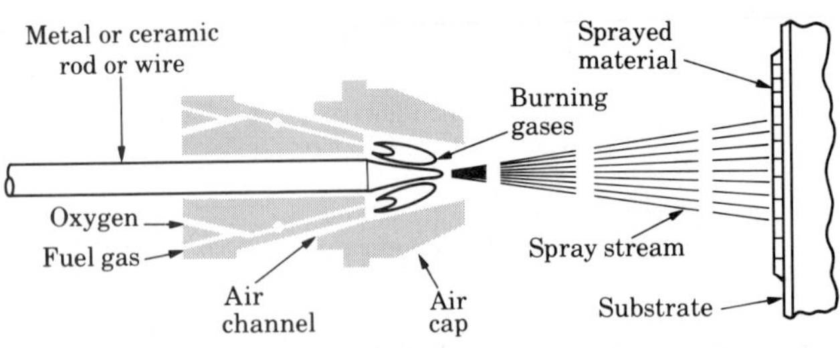

FIGURE 33.4
Schematic illustration of the cross-section of a thermal spray gun.

Techniques have also been developed for spraying *ceramic coatings* for high-temperature and electrical-resistance applications, such as to withstand repeated arcing. Powders of hard metals and ceramics are also used as spraying materials. Plasma-arc temperatures may reach 15,000 °C (27,000 °F), which is much higher than that obtained with flames. Typical applications are nozzles for rocket motors and wear-resistant parts. Another method of deposition is *detonation spraying* or plating. Explosive gases are ignited in a steel tube, and the very high velocities developed propel the powder particles onto the surface being coated.

33.4 Vapor Deposition

Vapor deposition is a process in which the substrate (workpiece surface) is subjected to chemical reactions by gases that contain chemical compounds of the materials to be deposited. The coating thickness is usually a few μm, which is much less than the thicknesses provided by the techniques described in Sections 33.2 and 33.3. The deposited materials may consist of metals, alloys, carbides, nitrides, borides, ceramics, or various oxides. The substrate may be metal, plastic, glass, or paper. Typical applications are coating cutting tools, drills, reamers, milling cutters, punches, dies, and wear surfaces.

There are two major deposition processes: physical vapor deposition and chemical vapor deposition. These techniques allow effective control of coating composition, thickness, and porosity.

33.4.1 Physical vapor deposition

The three basic types of **physical vapor deposition** (PVD) processes are vacuum or arc evaporation (PV/ARC), sputtering, and ion plating. These processes are carried out in a high vacuum at temperatures in the range of 200–500 °C (400–900 °F). In physical vapor deposition, the particles to be deposited are carried physically to the workpiece, rather than by chemical reactions as in chemical vapor deposition.

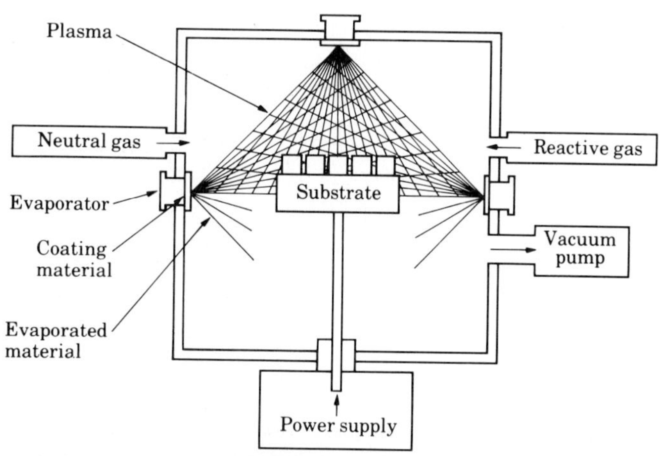

FIGURE 33.5
Schematic illustration of the physical deposition process. *Source: Cutting Tool Engineering.*

Vacuum evaporation. In *vacuum evaporation*, the metal to be deposited is evaporated at high temperatures in a vacuum and is deposited on the substrate, which is usually at room temperature or slightly higher. Uniform coatings can be obtained on complex shapes.

In PV/ARC, which was developed recently, the coating material (cathode) is evaporated by a number of arc evaporators (three are shown in Fig. 33.5), using highly localized electric arcs. The arcs produce a highly reactive plasma consisting of ionized vapor of the coating material. The vapor condenses on the substrate (anode) and coats it. Applications for this process may be functional (oxidation-resistant coatings for high temperature applications, electronics, and optics) or decorative (hardware, appliances, and jewelry).

Sputtering. In *sputtering*, an electric field ionizes an inert gas (usually argon). The positive ions bombard the coating material (cathode) and cause sputtering (ejecting) of its atoms. These atoms then condense on the workpiece, which is heated to improve bonding (Fig. 33.6). In *reactive sputtering*, the inert gas is replaced by a reactive gas, such as oxygen, in which case the atoms are oxidized and the oxides are deposited. Carbides and nitrides are also deposited by reactive sputtering. Very thin polymer coatings can be deposited on metal and polymeric substrates with a reactive gas, causing polymerization of the plasma. *Radio-frequency* (RF) *sputtering* is used for nonconductive materials such as electrical insulators and semiconductor devices.

Ion plating. *Ion plating* is a generic term describing the combined processes of sputtering and vacuum evaporation. An electric field causes a glow discharge, generating a plasma (Fig. 33.7). The vaporized atoms in this process are only partially ionized.

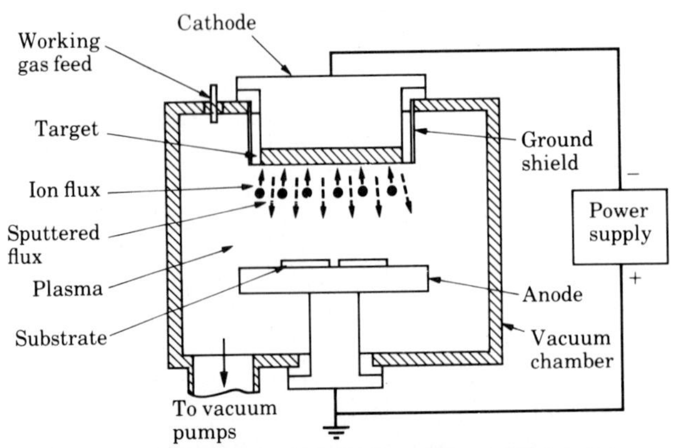

FIGURE 33.6
Schematic illustration of the sputtering process. *Source:* ASM International.

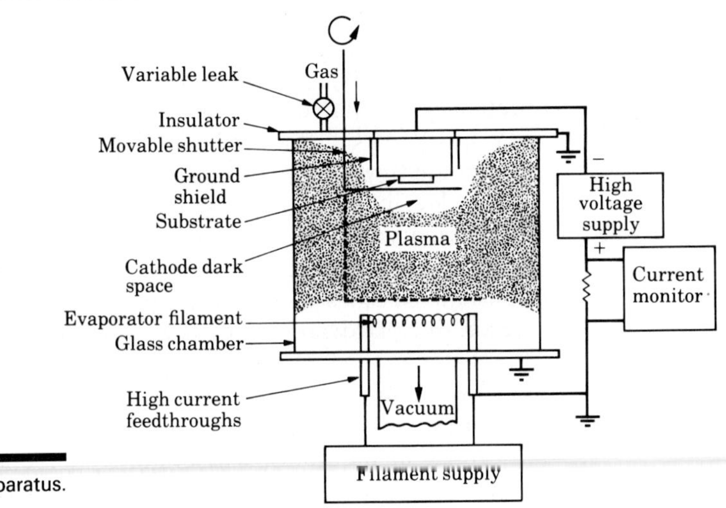

FIGURE 33.7
Schematic illustration of an ion-plating apparatus.
Source: ASM International.

33.4.2 Chemical vapor deposition

Chemical vapor deposition (CVD) is a thermochemical process (Fig. 33.8). In a typical application, such as for coating cutting tools with titanium nitride (TiN), the tools are placed on a graphite tray and heated to 950–1050 °C (1740–1920 °F) at atmospheric pressure in an inert atmosphere. Titanium tetrachloride (a vapor), hydrogen, and nitrogen are then introduced into the chamber. The chemical reactions form titanium nitride on the tool surfaces. For coating with titanium carbide, methane is substituted for the gases.

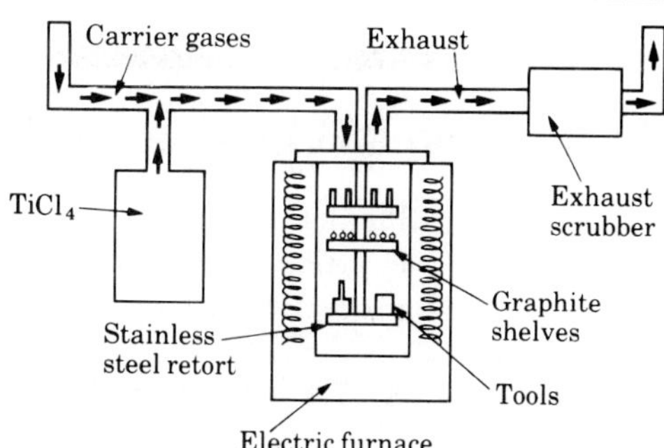

FIGURE 33.8
Schematic illustration of the chemical vapor deposition process.

Chemical vapor deposition coatings are usually thicker than those obtained from PVD. A typical cycle for CVD is long, consisting of 3 hours of heating, 4 hours of coating, and 6–8 hours of cooling to room temperature. The thickness of the coating depends on the flow rates of the gases used, time, and temperature.

33.5 Ion Implantation

In **ion implantation**, ions are introduced into the surface of the workpiece material. The ions are accelerated in a vacuum to such an extent that they penetrate the substrate to a depth of a few μm. Ion implantation (not to be confused with ion plating) modifies surface properties by increasing surface hardness and improving friction, wear, and corrosion resistance. This process can be controlled accurately, and the surface can be masked to prevent ion implantation in unwanted places. When used in specific applications, such as semiconductors, this process is called *doping* (meaning alloying with small amounts of various elements).

33.6 Diffusion Coating

Diffusion coating is a process in which an alloying element is diffused into the surface of the substrate, thus altering its properties. Such elements can be supplied in solid, liquid, or gaseous states. This process acquires different names, depending on the diffused element, as you can see in Table 4.1, which describes diffusion processes such as carburizing, nitriding, and boronizing.

33.7 Electrochemical Plating

Plating, as in other coating processes, imparts resistance to wear and corrosion, high electrical conductivity, better appearance and reflectivity, and similar desirable properties. We describe the basic types of **electrochemical plating** below.

33.7.1 Electroplating

In **electroplating,** the workpiece (cathode) is plated with a different metal (anode), while both are suspended in a bath containing a water-base electrolyte solution. Although the plating process involves a number of reactions, basically the metal ions from the anode are discharged under the potential from the external source of electricity, combine with the ions in the solution, and are deposited on the cathode.

All metals can be electroplated, with thicknesses ranging from a few atomic layers to a maximum of about 0.05 mm (0.002 in.). Complex shapes may have varying plating thicknesses. Some design guidelines for electroplating are shown in Fig. 33.9. Chemical cleaning and degreasing and thorough rinsing of the workpiece prior to plating are essential. The parts are placed on racks or in a barrel (bulk plating) and lowered into the plating bath.

Common plating materials are chromium, nickel, cadmium, copper, zinc, and tin. Chromium plating is carried out by first plating the metal with copper, then with nickel, and finally with chromium. *Hard chromium plating* is done directly on the base metal and has a hardness up to 70 HRC. This method is used to improve wear and corrosion resistance of tools, valve stems, hydraulic shafts, and diesel- and aircraft-engine cylinder liners—and also for rebuilding worn parts.

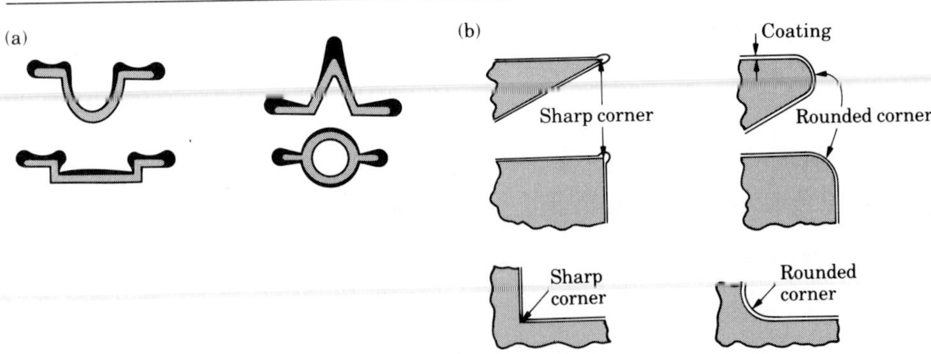

FIGURE 33.9 (a) Schematic illustration of nonuniform coatings (exaggerated) in electroplated parts. (b) Design guidelines for electroplating. Note that sharp external and internal corners should be avoided for uniform plating thickness. *Source:* ASM International.

Typical electroplating applications are copper plating aluminum wire and phenolic boards for printed circuits, chrome plating hardware, tin plating copper electrical terminals for ease of soldering, and components requiring resistance to wear and corrosion and good appearance. Because they do not develop oxide films, noble metals (such as gold, silver, and platinum) are important electroplating materials for the electronics and jewelry industries.

Plastics such as ABS, polypropylene, polysulfone, polycarbonate, polyester, and nylon also can be electroplated. Because they are not electrically conductive, plastics must be preplated by such processes as electroless nickel plating (see Section 33.8). Parts to be coated may be simple or complex, and size is not a limitation.

33.7.2 Electroforming

A variation of electroplating is **electroforming**, which actually is a metal fabricating process. Metal is electrodeposited on a mandrel (also called mold or matrix), which is then removed. Thus the coating itself becomes the product. Simple and complex shapes can be produced by electroforming, with wall thicknesses as small as 0.025 mm (0.001 in.). Parts may weigh from a few grams to as much as 270 kg (600 lb).

Mandrels are made from a variety of metallic (such as zinc or aluminum) or nonmetallic materials, which can be made electrically conductive with proper coatings. Mandrels should be physically removable without damaging the electroformed part. They may also be made of low-melting alloys, wax, or plastics, which can be melted away or dissolved with suitable chemicals.

The electroforming process is particularly suitable for low production quantities or intricate parts (such as molds, dies, waveguides, nozzles, and bellows) made of nickel, copper, gold, and silver. It is also suitable for aerospace, electronics, and electrooptics applications. Production rates can be increased with multiple mandrels.

33.8 Electroless Plating

Electroless plating is carried out by chemical reactions, without the use of an external source of electricity. The most common application utilizes nickel, although copper is also used. In electroless nickel plating, nickel chloride (a metallic salt) is reduced—with sodium hypophosphite as the reducing agent—to nickel metal, which is then deposited on the workpiece. The hardness of nickel plating ranges between 425 HV and 575 HV, and can be heat treated to 1000 HV. The coating has excellent wear and corrosion resistance.

Cavities, recesses, and the inner surface of tubes can be plated successfully. This process can also be used with nonconductive materials, such as plastics and ceramics. Electroless plating is more expensive than electroplating. However, unlike electroplating, the coating thickness in electroless plating is uniform (see Fig. 33.9).

33.9 Anodizing

Anodizing is an oxidation process (*anodic oxidation*) in which the workpiece surfaces are converted to a hard and porous oxide layer that provides corrosion resistance and a decorative finish. The workpiece is the anode in an electrolytic cell immersed in an acid bath, resulting in chemical adsorption of oxygen from the bath. Organic dyes of various colors (typically black, red, bronze, gold, gray) can be used to produce stable, durable surface films.

Typical applications for anodizing are aluminum furniture and utensils, architectural shapes, automobile trim, picture frames, keys, and sporting goods. Anodized surfaces also serve as a good base for painting, especially for aluminum, which otherwise is difficult to paint.

33.10 Conversion Coating

Conversion coating, also called *chemical reaction priming*, is a coating that forms on metal surfaces as a result of chemical or electrochemical reactions. Various metals, particularly steel, aluminum, and zinc, can be conversion coated. Oxides that naturally form on their surfaces are a form of conversion coating. Phosphates, chromates, and oxalates are used to produce conversion coatings. These coatings are for purposes such as corrosion protection, prepainting, and decorative finish. An important application is in conversion coating of workpieces as a lubricant carrier in cold forming operations (see Section 32.12). The two common methods of coating are immersion and spraying. The equipment involved depends on the method of application, the type of product, and considerations of quality.

As the name implies, *coloring* involves processes that alter the color of metals, alloys, and ceramics. It is caused by the conversion of surfaces (by chemical, electrochemical, or thermal processes) into chemical compounds, such as oxides, chromates, and phosphates. Note, for example, how the color of iron and steel changes as it oxidizes (rust color).

33.11 Hot Dipping

In **hot dipping**, the workpiece, usually steel or iron, is dipped into a bath of molten metal, such as zinc (for galvanized-steel sheet and plumbing supplies), tin (for tinplate and tin cans for food containers), aluminum (aluminizing), and terne (lead alloyed with 10–20 percent tin). Hot-dipped coatings on discrete parts or sheet metal provide long-term corrosion resistance to galvanized pipe, plumbing supplies, and many other products.

A typical continuous hot-dipped galvanizing line for steel sheet is shown in Fig. 33.10. The rolled sheet is first cleaned electrolytically and scrubbed by brushing. The sheet is then annealed in a continuous furnace with controlled atmosphere and temperature and dipped in molten zinc at about 450 °C (840 °F). The thickness of the zinc coating is controlled by a wiping action from a stream of air or steam, called *air knife* (similar to air-drying cars in car washes). The coating thickness is usually given in terms of coating weight per unit surface area of the sheet, typically 150–900 g/m^2 (0.5–3 oz/ft^2). Service life depends on the thickness of the zinc coating and the environment to which it is exposed (Fig. 33.11). Various precoated sheet steels are used extensively in automobile bodies. Proper draining to remove excess coating materials is an important consideration.

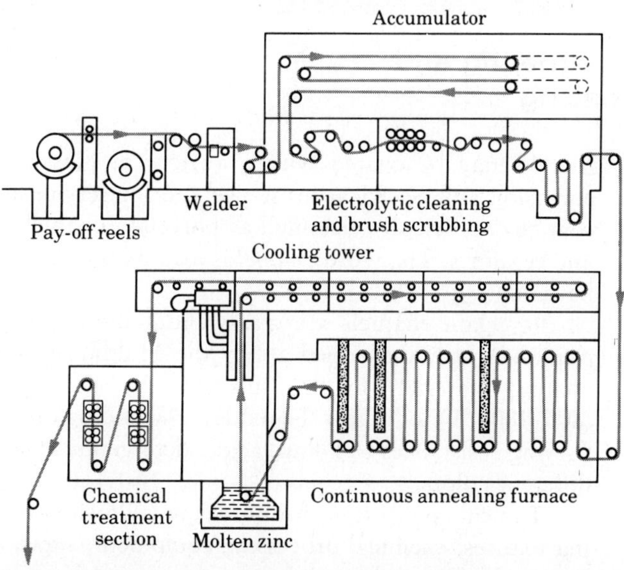

FIGURE 33.10 Flowline for continuous hot-dip galvanizing of sheet steel. The welder (upper left) is used to weld the ends of coils to maintain continuous material flow. *Source:* American Iron and Steel Institute.

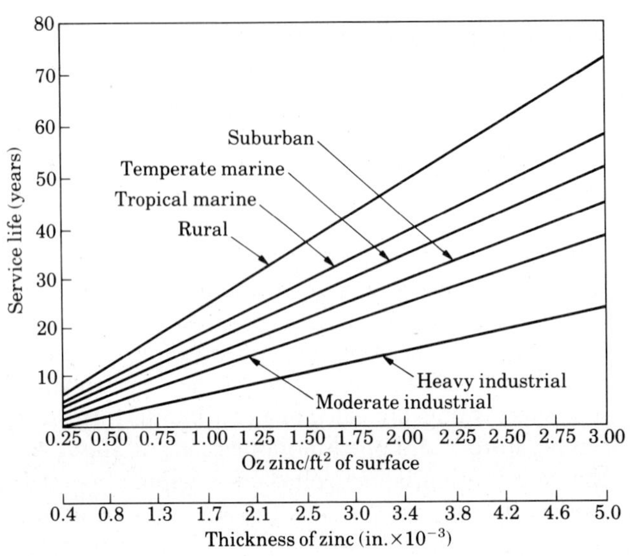

FIGURE 33.11 Typical life of zinc coatings in various environments. *Source:* American Hot Dip Galvanizers Association. Reprinted with permission.

33.12 Porcelain Enameling and Ceramic Coating

Metals may be coated with a variety of glassy (vitreous) coatings to provide corrosion and electrical resistance and for service at elevated temperatures. These coatings are usually classified as porcelain enamels and generally include enamels and ceramics. The word *enamel* is also used for glossy paints, indicating a smooth, hard coating.

Porcelain enamels are glassy inorganic coatings consisting of various metal oxides. A fully developed art by the Middle Ages, **enameling** involves fusing the coating material on the substrate by heating them both to 425–1000 °C (800–1800 °F) to liquefy the oxides. Depending on their composition, enamels have varying resistances to alkali, acids, detergents, cleansers, and water—and come in different colors.

Typical applications for porcelain enameling are household appliances, plumbing fixtures, chemical processing equipment, signs, cookware, and jewelry. Porcelain enamels are also used as protective coatings on jet-engine components. The coating may be applied by dipping, spraying, or electrodeposition, and thicknesses are usually 0.05–0.6 mm (0.002–0.025 in.). Metals that are coated are typically steels, cast iron, and aluminum. Glasses are used as lining for chemical resistance, and the thickness is much greater than in enameling. **Glazing** is the application of glassy coatings on ceramic wares to give them decorative finishes and to make them impervious to moisture.

Ceramic coatings such as aluminum oxide or zirconium oxide are applied, with the use of binders, to the substrate at room temperature. Such coatings have been used in hot extrusion dies to extend their life.

Example: Ceramic coatings for high temperature applications.

Characteristics such as wear resistance and thermal and electrical insulation, particularly at elevated temperatures, can be imparted on products by ceramic coatings rather than imparting these properties to the base metals or materials themselves. Selecting materials with such bulk properties can be expensive or may not meet the structural strength requirements in a particular application. Thus, for example, a wear-resistance component does not have to be made completely from a wear-resistant material, since the properties of only a thin layer on the component's surface are relevant for wear. Consequently, coatings have important applications. The table below shows various ceramic coatings and typical applications at elevated temperatures. These coatings may be applied either singly or in layers, as is done in multiple-layer coated cutting tools (see Section 21.6).

PROPERTY	TYPE OF CERAMIC	APPLICATIONS
Wear resistance	Chromium oxide Aluminum oxide Aluminum titania	Pumps, turbine shafts, seals, compressor rods for the petroleum industry; plastics extruder barrels; extrusion dies.
Thermal insulation	Zirconium oxide (yttria stabilized) Zirconium oxide (calcia stabilized) Magnesium zirconate	Fan blades, compressor blades, and seals for gas turbines; valves, pistons, and combustion heads for automotive engines.
Electrical insulation	Magnesium aluminate Aluminum oxide	Induction coils, brazing fixtures, general electrical applications.

33.12.1 Other coatings for sheet metal

Sheet metals also are precoated with a variety of organic coatings, films, and laminates to improve appearance, eye appeal, and corrosion resistance. Coatings are applied to the coil stock on continuous lines, with thicknesses generally of 0.0025–0.2 mm (0.0001–0.008 in.).

Such coatings have a wide range of properties: flexibility, durability, hardness, resistance to abrasion and chemicals, color, texture, and gloss. Coated sheet metal is subsequently formed into various products, such as TV cabinets, appliance housings, paneling, shelving, residential building siding, gutters, and metal furniture.

33.13 Diamond Coating

We described the properties of diamond that are relevant to manufacturing engineering in Section 8.7. Important advances have been made in diamond coating of metals, glass, ceramics, and plastics, using various vapor-deposition processes. Techniques have also been developed to produce free-standing diamond films on the order of 1 mm (0.040 in.) thick and up to 125 mm (5 in.) in diameter, including smooth and optically clear diamond film (unlike the hazy gray diamond film formerly produced). Development of these techniques, combined with important properties of diamond such as hardness, wear resistance, high thermal conductivity, and transparency to ultraviolet light and microwave frequencies, have enabled the production of various aerospace and electronic parts and components.

Examples of diamond coated products are scratchproof windows (such as for aircraft and missile sensors to protect against sandstorms), sunglasses, cutting tools, surgical knives, razors, electronic and infrared heat seekers and sensors, light-emitting diodes, diamond-coated speakers for stereo systems, turbine blades, and fuel-injection nozzles. An important application is in making computer chips. Diamond can be doped to form *p*- and *n*-type ends on semiconductors to make transistors (see Chapter 34), and its high thermal conductivity allows closer packing of chips than silicon or gallium-arsenide chips, thus significantly increasing the speed of computers.

33.14 Painting

Because of its decorative and functional properties (such as environmental protection), low cost, relative ease of application, and the range of available colors, paint is widely used as a surface coating. Engineering applications of **painting** range from all types of machinery to automobile bodies. Paints are classified as *enamels*, which produce a smooth coat and dry with a glossy or semiglossy appearance; *lacquers*, which form a film by evaporation of a solvent; and *water-base paints*, which are easily applied, but have a porous surface, absorb water, and are not as easily cleaned as other paints. Paints are now available with good resistance to abrasion, fading, and temperature extremes; they are easy to apply and dry quickly.

Selection of a particular paint depends on specific requirements. Among these are resistance to mechanical actions (abrasion, marring, impact, and flexing) or to chemical actions (acids, solvents, detergents, alkalis, fuels, staining, and general environmental attack).

Common methods of applying paint are dipping, brushing, and spraying (Fig. 33.12). In electrocoating or electrostatic spraying, paint particles are charged electrostatically and are attracted to the workpiece surfaces, producing a uniformly adherent coating. Unlike conventional spraying, in which as much as 70 percent of

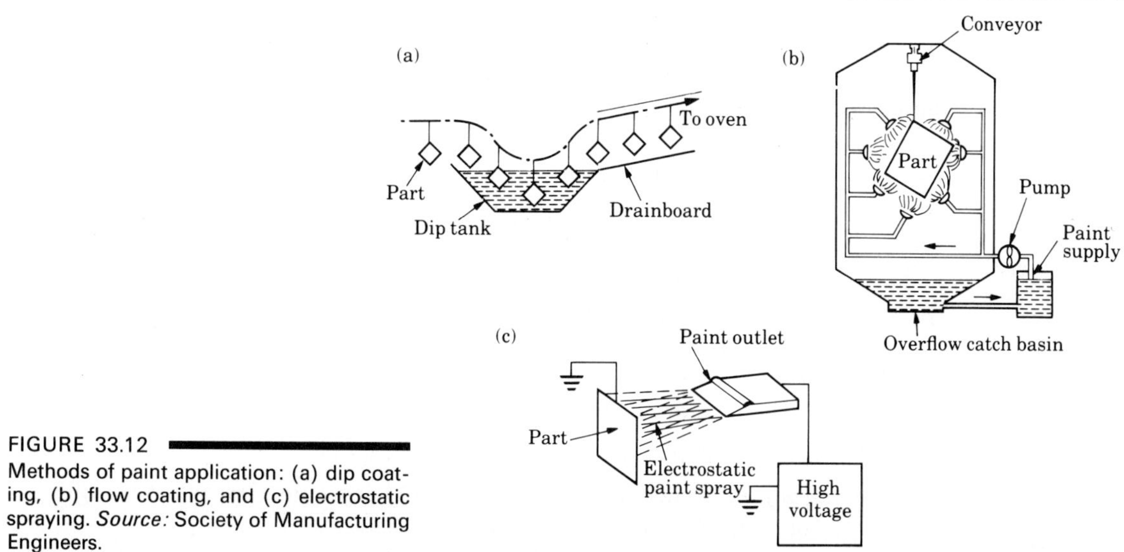

FIGURE 33.12 Methods of paint application: (a) dip coating, (b) flow coating, and (c) electrostatic spraying. *Source:* Society of Manufacturing Engineers.

the paint may be lost, in electrostatic spraying the loss can be as low as 10 percent. However, deep recesses and corners are difficult to coat by this method.

33.15 Cleaning Surfaces

We have stressed the importance of surfaces and the influence of deposited or adsorbed layers of various elements and contaminants on surfaces. A clean surface can have both beneficial and detrimental effects. Although an unclean surface may reduce the tendency for adhesion and galling, in general cleanliness is essential for more effective application of metalworking fluids, coating and painting, adhesive bonding, welding, brazing, soldering, reliable functioning of manufactured parts in machinery, food and beverage containers, storage, and in assembly operations.

Cleaning involves removal of solid, semisolid, or liquid contaminants from a surface, and it is an important part of manufacturing operations and the economics of production. The word *clean*, or the degree of cleanliness of a surface, is somewhat difficult to define. How, for example, would you test the cleanliness of a fork or dinner plate? Two simple and common tests are based on:

1. Wiping with a clean cloth and observing any residues on the cloth, as we all have done at one time or another.
2. Observing whether water continuously coats the surface. If water collects as individual droplets, the surface is not clean (*waterbreak test*). Test this phenomenon yourself by wetting dinner plates that have been cleaned to varying degrees.

The type of cleaning process required depends on the type of contaminants to be removed. Contaminants, also called *soils*, may consist of rust, scale, chips and other metallic and nonmetallic debris, metalworking fluids, solid lubricants, pigments, polishing and lapping compounds, and general environmental elements.

33.15.1 Cleaning processes

Basically there are two types of cleaning methods: mechanical and chemical. *Mechanical* methods consist of physically disturbing the contaminants, as with wire or fiber brushing, dry or wet abrasive blasting, tumbling, and steam jets. Many of these processes are particularly effective in removing rust, scale, and other solid contaminants. Ultrasonic cleaning may also be placed in this category.

Chemical cleaning usually involves the removal of oil and grease from surfaces. It consists of one or more of the following processes:

- *Solution.* The soil dissolves in the cleaning solution.
- *Saponification.* A chemical reaction that converts animal or vegetable oils into a soap that is soluble in water.
- *Emulsification.* The cleaning solution reacts with the soil or lubricant deposits and forms an emulsion. The soil and the emulsifier then become suspended in the emulsion.
- *Dispersion.* The concentration of soil on the surface is decreased by surface-active materials in the cleaning solution.
- *Aggregation.* Lubricants are removed from the surface by various agents in the cleaner and collect as large dirt particles.

Some common *cleaning fluids* are used in conjunction with electrochemical processes for more effective cleaning. These fluids include:

- *Alkaline solutions* are a complex combination of water-soluble chemicals. They are the least expensive and most widely used in manufacturing operations. Small parts may be cleaned in rotating drums or barrels. Most parts are cleaned on continuous conveyors by spraying them with the solution and then rinsing them with water.
- *Emulsions* generally consist of kerosene and oil in water and various types of emulsifiers.
- The most common *solvents* are petroleum solvents, chlorinated hydrocarbons, and mineral spirits. Solvents are generally used for short runs; fire and toxicity are major hazards.
- Parts are subjected to *hot vapors* of chlorinated solvents to remove oil, greases, and wax. The solvent is boiled in a container and then condensed. The process is simple and the cleaned parts are dry.
- Various *acids*, *salts*, and *organic compound mixtures* are effective in cleaning parts covered with heavy paste or oily deposits and rust.

Cleaning discrete parts having complex shapes can be difficult. Design engineers should be aware of this difficulty and provide alternative designs, such as avoiding deep blind holes, making several smaller components instead of one large component that may be difficult to clean, and provide appropriate drain holes in the part to be cleaned.

SUMMARY

Surface treatment is an important aspect of all manufacturing processes. It is used to impart certain physical and mechanical properties, such as appearance and corrosion, friction, wear, and fatigue resistance. Several techniques are available for modifying surfaces. They include mechanical working and coating of surfaces, heat treatment, deposition, plating, and coatings such as enamels, nonmetallic materials, and paints.

Clean surfaces are important to further processing of the product, such as in coating, painting, and welding. Cleaning can have a significant economic impact on manufacturing operations. Various mechanical and chemical cleaning methods may be utilized.

TRENDS

- The structure and properties of coatings, bonding to the substrate, surface preparation and modification, porosity and coating densification, and coating integrity are major areas of study.
- Coatings continue to be an important application for cutting tool materials to improve their friction and wear characteristics.
- Coatings for corrosion resistance for high-temperature applications are under continued development. Work on aluminizing steels for service to 1000 °C (1800 °F) is continuing. Gas turbine blades are being coated with alloys of chromium, nickel, aluminum, iron, cobalt, and yttrium.
- Ceramics, tungsten carbide, and silicon-based coatings are being developed for wear resistance at elevated temperatures.
- Techniques are being developed to coat various metallic and nonmetallic materials with diamond, as well as preparing diamond film.
- Solar energy is being used as an energy source for surface hardening of ferrous alloys, post-treatment of coatings and films, and electronic materials processing. High-power arc lamps are also being used for surface hardening of alloys.
- Newly developed is a hybrid coating technique called dual ion-beam assisted (or enhanced) deposition, which combines physical vapor deposition with simultaneous ion-beam bombardment, with good adhesion on metals, ceramics, and polymers. Ceramic bearings and dental instruments are examples of applications of this technique.
- Chemical vapor deposition is being modified incorporating a fluidized bed, using a bed of powder; parts with various shapes and with close dimensional control have been fabricated by this method.

KEY TERMS

Anodizing	Electroless plating	Mechanical plating
Case hardening	Electroplating	Painting
Chemical vapor deposition	Enameling	Physical vapor deposition
Cladding	Explosive hardening	Roller burnishing
Conversion coating	Glazing	Shot peening
Diffusion coating	Hard facing	Thermal spraying
Electrochemical plating	Hot dipping	Vapor deposition
Electroforming	Ion implantation	

BIBLIOGRAPHY

Auciello, O., and R. Kelly (eds.), *Ion Bombardment Modification of Surfaces*. Amsterdam: Elsevier, 1984.

Banov, A. (ed.), *Paints and Coatings Handbook*. New York: McGraw-Hill, 1982.

Bhushan, B., and B.K. Gupta, *Handbook of Tribology: Materials, Coatings, and Surface Treatments*. New York: McGraw-Hill, 1991.

Bunshah, R.F., et al., *Deposition Technologies for Films and Coatings*. Park Ridge, N.J.: Noyes Publications, 1982.

Metals Handbook, 9th ed., Vol. 5: *Surface Cleaning, Finishing, and Coating*. Metals Park, Ohio: American Society for Metals, 1982.

Nachtman, E.S., and S. Kalpakjian, *Lubricants and Lubrication in Metalworking Operations*. New York: Marcel Dekker, 1985.

Peterson, M.B., and W.O. Winer (eds.), *Wear Control Handbook*. New York: ASME, 1980.

Schey, J.A., *Tribology in Metalworking—Friction, Lubrication and Wear*. Metals Park, Ohio: American Society for Metals, 1983.

Source Book on Wear Control Technology. Metals Park, Ohio: American Society for Metals, 1978.

Strafford, K.N., P.K. Datta, and C.G. Googan (eds.), *Coatings and Surface Treatments for Corrosive and Wear Resistance*. New York: Wiley, 1984.

Stuart, R.V., *Vacuum Technology, Thin Films, and Sputtering*. New York: Academic Press, 1983.

Tool and Manufacturing Engineers Handbook, 4th ed., *Vol. 3: Materials, Finishing and Coating*. Dearborn, Mich.: Society of Manufacturing Engineers, 1985.

REVIEW QUESTIONS

33.1 Explain why surface treatment of manufactured products may be necessary.

33.2 Which surface treatments are functional and which are decorative? Name some that serve both functions.

33.3 Give some applications of mechanical surface treatment.
33.4 What are the advantages of cladding?
33.5 Explain the difference between case hardening and hard facing.
33.6 State why you might want to coat parts with ceramics.
33.7 Explain the principles of physical and chemical vapor deposition. What applications do they have?
33.8 What is doping? Why is it used?
33.9 Name some parts that could be electroplated.
33.10 What is the principle of electroforming? What are its advantages?
33.11 Explain the difference between electroplating and electroless plating.
33.12 Give some applications of conversion coatings. How are they used in metalworking operations?
33.13 How is hot dipping performed?
33.14 What is an air knife?
33.15 List several applications of coated sheet metal.
33.16 What tests are there to determine the cleanliness of surfaces?
33.17 What are soils?
33.18 Explain the common methods of cleaning and the solutions used for manufactured products.

QUESTIONS AND PROBLEMS

33.19 Explain how roller burnishing processes induce residual stresses on the surfaces of parts.
33.20 How would you go about estimating the forces required for roller burnishing, based on the topics covered in Chapter 2?
33.21 Name some materials that approach the hardness levels shown in Fig. 33.3.
33.22 Explain the principles involved in various techniques for application of paints.
33.23 Give examples of part designs that are suitable for hot-dip galvanizing.
33.24 Repeat Question 33.23, but for cleaning.
33.25 Refer to Table 33.2 and name two surface-treatment processes that are the most common. Why are they common?
33.26 Outline the reasons why the topics discussed in this chapter are important in manufacturing processes and operations.
33.27 List several products or components that could not be made properly, or function effectively in service, without implementing the knowledge involved in this chapter.
33.28 Describe your observations concerning Table 33.1.
33.29 Write a brief paper on which the processes described in this chapter are used in improving the corrosion resistance of an automobile.
33.30 We stated in the Trends section in this chapter that solar energy is being used as an energy source for surface hardening and post-treatment of coatings and films. Make a sketch of an installation that could use solar energy for this purpose.

34

Fabrication of Microelectronic Devices

by Kent M. Kalpakjian

34.1 Introduction

Although semiconducting materials have been used in electronics since the early decades of this century, it was the invention of the transistor in 1948 that set the stage for what would become one of the greatest technological advancements in all of history. Microelectronics has played an ever-increasing role in our lives since **integrated circuit** (IC) technology became the foundation for calculators, wrist watches, control of home appliances, information systems, telecommunications, automotive control, robotics, space travel, military weaponry, and personal computers.

The major advantages of today's ICs are their reduced size and cost. As fabrication technology becomes more advanced, the size of devices decreases. Consequently, more components can be put onto a **chip** (a small slice of semiconducting material on which the circuit is fabricated), thereby reducing cost. In addition, mass processing (also known as batch processing) and process automation have helped to reduce the cost of each completed circuit. The components fabricated include transistors, diodes, resistors, and capacitors. Typical chip sizes range from 3 mm × 3 mm to 15 mm × 15 mm, respectively. In the past no more than 100 devices could be fabricated on a single chip, but technology now allows densities in hundreds of thousands of devices per chip (Fig. 34.1). This scale of integration has been termed *very large scale integration* (VLSI). Some of the most advanced ICs may contain more than 10 million devices.

Because of the minute scale of microelectronic devices, all fabrication must take place in an extremely clean environment. Clean rooms are used for this purpose and are given ratings that refer to the maximum number of 0.5 μm particles per cubic foot. These ratings range from 10,000 down to 10—or even to 1 in some of the most advanced fabrication facilities. For comparison, the dust level in modern hospitals is on the order of 10,000 particles per cubic foot.

In this chapter we describe the current processes used in fabricating microelectronic devices and integrated circuits, following the outline shown in Fig. 34.2. The major steps in fabricating a **metal-oxide-semiconductor field effect transistor**, or **MOSFET**, which is one of the dominant devices used in IC technology, are shown in Fig. 34.3. We first introduce the basic properties of semiconductors and the material properties of silicon, followed by a discussion of each of the major fabrication steps. Finally, we look at trends and expectations of the microelectronics industry.

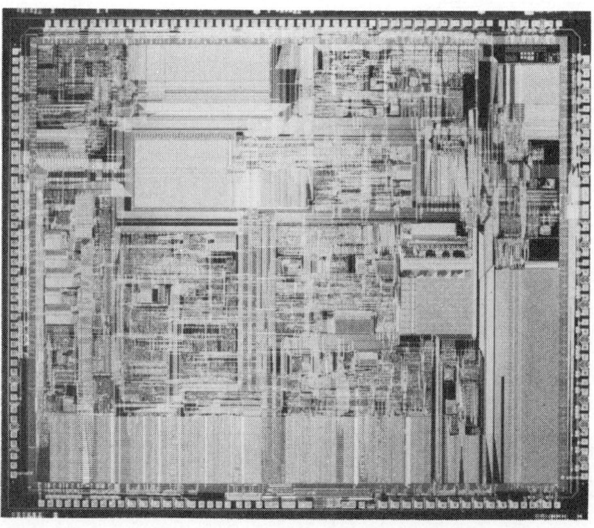

FIGURE 34.1

A 32-bit microprocessor chip capable of executing 15 million instructions per second. It contains 370,000 transistors with a minimum feature width of 1.25 μm and measures 13.5 mm by 11.2 mm. *Source:* National Semiconductor Corporation.

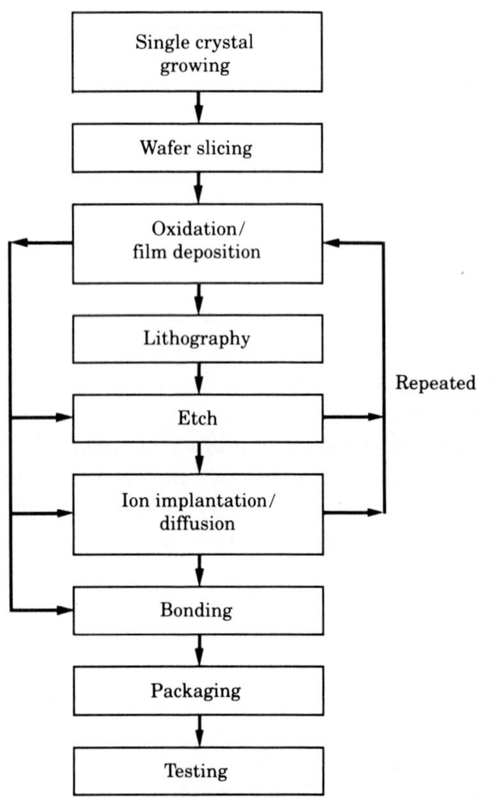

FIGURE 34.2
General fabrication sequence for integrated circuits.

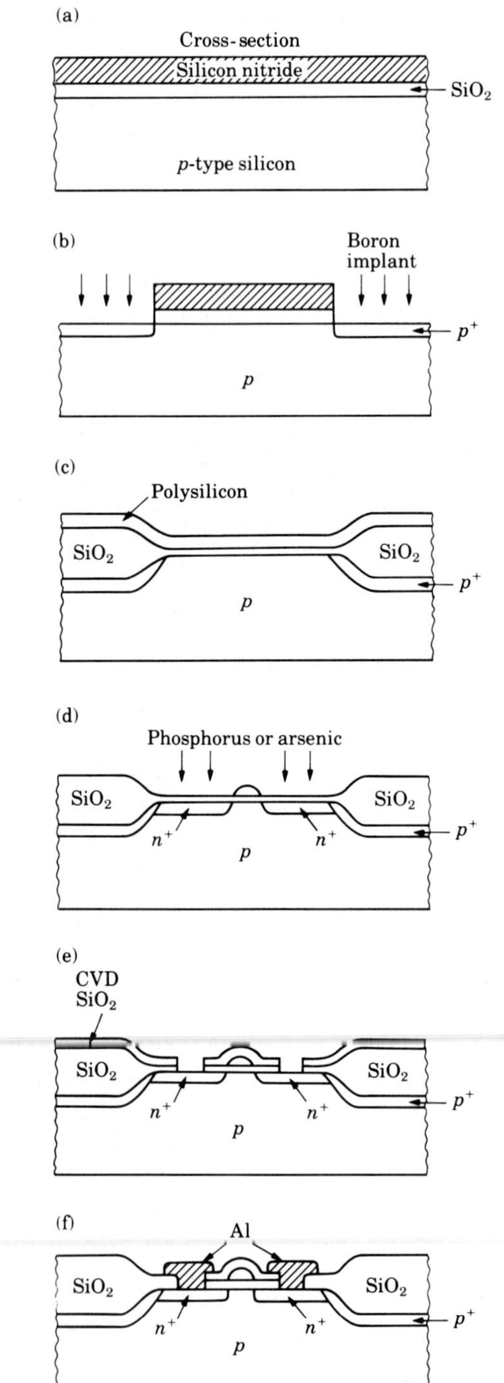

FIGURE 34.3 ▶
Cross-sectional views of the fabrication of a MOS transistor. *Source:* R. C. Jaeger.

34.2 Semiconductors and Silicon

As the name suggests, **semiconductor** materials have electrical properties that lie between those of conductors and insulators and exhibit resistivities between 10^{-3} Ω-cm and 10^{8} Ω-cm. Semiconductors have become the foundation for electronic devices because their electrical properties can be altered by adding controlled amounts of selected impurity atoms into their crystal structures (Fig. 34.4). These impurity atoms, also known as *dopants*, either have one more valence electron (*n*-type dopant) or one less valence electron (*p*-type dopant) than the atoms in the semiconductor lattice, where *n* and *p* refer to negative and positive. For silicon, which is a group IV element, typical *n*-type and *p*-type dopants include phosphorous (group V) and boron (group III), respectively. The electrical operation of semiconductor devices is controlled by creating regions of different doping types and concentrations.

Although the earliest electronic devices were fabricated on germanium, *silicon* has without a doubt become the industry standard. The abundance of silicon in its alternative forms is second only to that of oxygen, making it economically attractive. Silicon's main advantage over germanium is its larger energy gap (1.1 eV) compared to that of germanium (0.66 eV). This allows silicon-based devices to

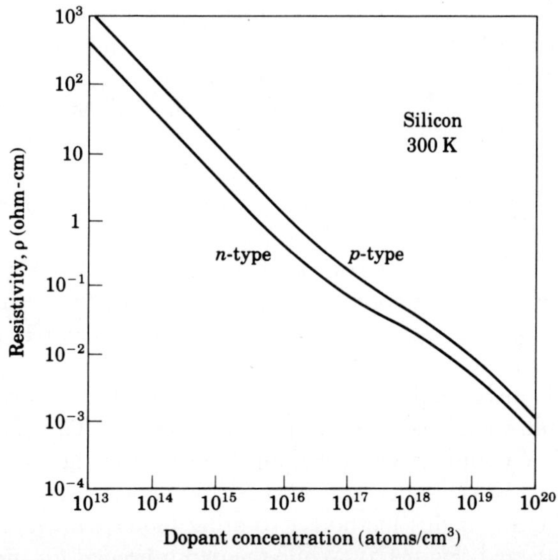

FIGURE 34.4

Plot of resistivity of silicon at room temperature as a function of dopant type and concentration. *Source:* After W. R. Turber, R. C. Mattis, and Y. M. Liu.

operate at temperatures (≈150 °C; 270 °F) higher than devices fabricated on germanium (≈100 °C; 180 °F). Silicon's important processing advantage is that its oxide, silicon dioxide, is an excellent insulator and is used for isolation and passivation purposes. Conversely, germanium oxide is water soluble and unsuitable for electronic devices.

However, silicon has some limitations, which has encouraged the development of compound semiconductors. The most advanced and researched compound semiconductor to date is *gallium arsenide*. Its major advantage over silicon is its capability for light emission (allowing fabrication of devices such as lasers and light-emitting diodes, LEDs), in addition to having a larger energy gap (1.43 eV) and therefore a higher maximum operating temperature (≈200 °C; 400 °F). Devices fabricated on gallium arsenide also have much higher operating speeds than those on silicon. Some of gallium arsenide's disadvantages include its considerably higher cost, greater processing complications, and the difficulty of growing high-quality oxide layers, the need for which we emphasize throughout this chapter.

34.3 Crystal Growing and Wafer Preparation

Silicon occurs naturally in the forms of silicon dioxide and various silicates. They must undergo a series of purification steps in order to become the high-quality, defect-free, and single-crystal material needed for semiconductor device fabrication. The process begins by heating silica and carbon together in an electric furnace, resulting in 95–98% pure polycrystalline silicon. This material is converted to an alternative form, commonly trichlorosilane, which in turn is purified and decomposed in a high-temperature hydrogen atmosphere. The result is an extremely high-quality electronic-grade silicon (EGS).

Single-crystal silicon is almost always obtained by using the **Czochralski process**. This method utilizes a seed crystal that is dipped into a silicon melt and then slowly pulled out while being rotated (see Fig. 11.31a). At this point, controlled amounts of impurities can be added to the system to obtain a uniformly doped crystal. Typical pull rates are on the order of 10 μm/s. The result of this growing technique is a cylindrical single-crystal ingot, typically 50–200 mm (2–8 in.) in diameter and over 1 m (40 in.) in length. Unfortunately, this technique does not allow exact control of the ingot diameter. Therefore, ingots are commonly grown a few millimeters larger than required and then ground to a precise diameter.

Next, the crystal is sliced into individual **wafers** by using an inner diameter blade. In this method a rotating blade with its cutting edge on the inner ring is utilized (Fig. 34.5). While the depth needed for most electronics devices is no more than several microns, wafers are cut to a thickness of about 0.5 mm (0.02 in.). This thickness provides the necessary physical support to absorb temperature changes

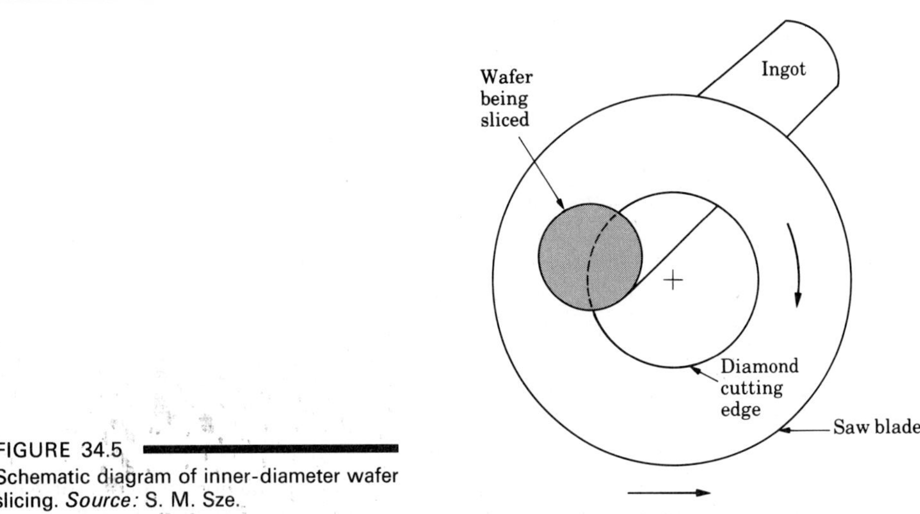

FIGURE 34.5
Schematic diagram of inner-diameter wafer slicing. *Source:* S. M. Sze.

and strains during subsequent fabrication. One concern about the process is wastage. Since typical blade thicknesses range from 250 to 300 microns, approximately one third of the silicon crystal is wasted as sawdust in the wafer-cutting process. Finally, the wafers must be polished and cleaned to remove surface damage caused by the cutting process.

Device fabrication takes place over the entire wafer surface, and many identical circuits are generated at the same time. Because of decreasing device sizes and larger wafer diameters, hundreds of individual circuits can be put on one wafer. Once processing is finished, the wafer is sliced into individual "chips," each containing one complete integrated circuit.

34.4 Film Deposition

Films of many different types, particularly insulating and conducting types, are used extensively in microelectronic device processing. Common deposition films include polysilicon, silicon nitride, silicon dioxide, and conductive metals. In some instances, the wafers serve merely as a mechanical support on which custom layers are grown. The devices are then fabricated on the layers. Among the advantages of processing on these deposited films, instead of the actual wafers, include fewer impurities (notably carbon and oxygen), improved device performance, and tailored material properties not obtainable on the wafers themselves.

Some of the major functions of deposited films are *masking* for diffusion or implants and protection of the semiconductor surface. In masking applications, the film must effectively inhibit the passage of dopants while also displaying an ability to be etched into patterns of high resolution. Upon completion of device fabrication, films are applied to protect the underlying circuitry. Films used for masking and protection include silicon dioxide, phosphosilicate glass (PSG), and silicon nitride. Each of these materials has distinct advantages, and they are often used in combination.

Other films contain dopant impurities and are used as doping sources for the underlying substrate. Conductive films are used primarily for device interconnection. These films must be highly conductive, capable of carrying large currents, and able to connect to terminal packaging leads while maintaining their physical integrity. Generally, gold and aluminum are used for this purpose. Increasing circuit complexity has required multiple levels of conductive layers, which must be separated by insulating films.

Films may be deposited by a number of techniques, which involve a variety of pressures, temperatures, and vacuum systems. One of the simplest and oldest methods is *evaporation*, which is used primarily for depositing metal films. In this process the metal is heated to its point of vaporization in a vacuum. Upon evaporation, the metal forms a thin layer on the substrate surface. The heat of evaporation is usually provided by a heating filament or electron beam.

Another method of depositing metals is *sputtering* and involves bombarding a target with high-energy ions, usually argon (Ar^+), in a vacuum. Sputtering systems usually include a dc power source to obtain the energized ions. As the ions impinge on the target, atoms are knocked off and subsequently deposited on wafers mounted within the system. Although some argon may be trapped within the film, this technique provides very uniform coverage. Advances in this field include using a radio-frequency power source (*RF sputtering*) and introducing magnetic fields (*magnetron sputtering*).

In one of the most common techniques, **chemical vapor deposition** (**CVD**), film deposition is achieved by the reaction and/or decomposition of gaseous compounds (see Section 33.4). Using this technique, silicon dioxide is routinely deposited by the oxidation of silane or a chlorosilane. Figure 34.6(a) shows a continuous CVD reactor that operates at atmospheric pressure. A similar method that operates at lower pressures, referred to as *low-pressure chemical vapor deposition* (LPCVD), is shown in Fig. 34.6(b). Capable of coating hundreds of wafers at a time, this method has a much higher production rate than atmospheric-pressure CVD and provides superior film uniformity with less consumption of carrier gases. This technique is commonly used for depositing polysilicon, silicon nitride, and silicon dioxide. *Plasma-enhanced chemical vapor deposition* (PECVD) involves placing wafers in an RF plasma containing the source gases and offers the advantage of maintaining low wafer temperature during deposition.

Other deposition methods are categorized as **epitaxy**, in which the crystalline layer is formed using the substrate as a seed crystal. This is a very common method for growing layers of silicon. If the silicon is deposited from the gaseous phase, the process is known as *vapor-phase epitaxy* (VPE). In another variation, the heated substrate is brought into contact with a liquid solution containing the material to be

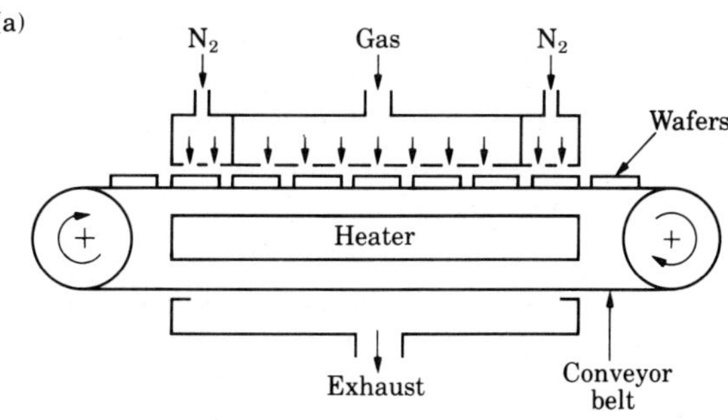

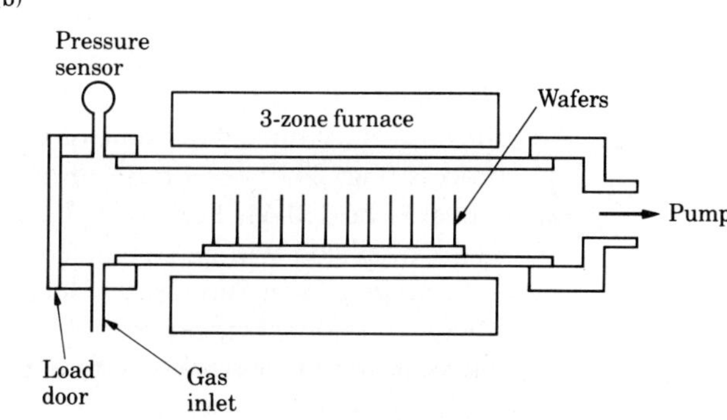

FIGURE 34.6 Schematic diagrams of (a) continuous, atmospheric-pressure CVD reactor and (b) low-pressure CVD. *Source:* S. M. Sze.

deposited (*liquid-phase epitaxy*, or LPE). Another high-vacuum process utilizes evaporation to produce a thermal beam of molecules that deposit on the heated substrate. This process, called *molecular beam epitaxy* (MBE), offers a very high degree of purity. In addition, since the films are grown one atomic layer at a time, excellent control of doping profiles is achieved, which is especially important in gallium arsenide technology. Unfortunately, MBE suffers from relatively slow growth rates and a lower production rate than other conventional film-deposition techniques.

34.5 Oxidation

Recall that the term *oxidation* refers to the growth of an oxide layer by the reaction of oxygen with the substrate material. Oxide films can also be formed by the previously described deposition techniques. The thermally grown oxides described in this section display a higher level of purity than deposited oxides because they are grown directly from the high-quality substrate. However, deposition methods must be used if the composition of the desired film is different from that of the substrate material. Silicon dioxide is the most widely used oxide in IC technology today, and its excellent characteristics are one of the major reasons for the widespread use of silicon. In addition to its functions of dopant masking, device isolation, and surface passivation, silicon dioxide is even used as an actual device component in some technologies.

Silicon surfaces have an extremely high affinity for oxygen, and a freshly sawed slice of silicon will instantly acquire a thin layer of oxide of no more than 40 Å thick. Modern IC technology requires oxide thicknesses in the hundreds and thousands of angstroms. *Dry oxidation* is a relatively simple process and is accomplished by elevating the substrate temperature typically to 900–1200 °C (1650–2200 °F), in an oxygen-rich environment.

As a layer of oxide forms, the oxidizing agents must be able to pass through the oxide and reach the silicon surface where the actual reaction takes place. Thus an oxide layer does not continue to grow on top of itself, but rather it grows from the silicon surface outward. Some of the silicon substrate is consumed in the oxidation process (Fig. 34.7). The ratio of oxide thickness to amount of silicon consumed is found to be 1:0.44. For example, to obtain an oxide layer 1000 Å thick, roughly 440 Å of silicon will be consumed. This does not present a problem, as substrates are always grown sufficiently thick. One important effect of this consumption of silicon is the rearrangement of dopants in the substrate near the interface. As different impurities have different mobilities in silicon dioxide, some dopants deplete away from the oxide interface while others pile up, and processing parameters have to be adjusted to compensate for this effect.

Another oxidizing technique utilizes a water-vapor atmosphere as the agent and is appropriately called *wet oxidation*. This method offers a considerably higher growth rate than that of dry oxidation, but it suffers from a lower oxide density and therefore a lower dielectric strength. Common practice is to combine both dry and wet oxidation methods and grow an oxide in a three-part layer: dry, wet, dry. This approach combines the advantages of wet oxidation's much higher growth rate and dry oxidation's high quality.

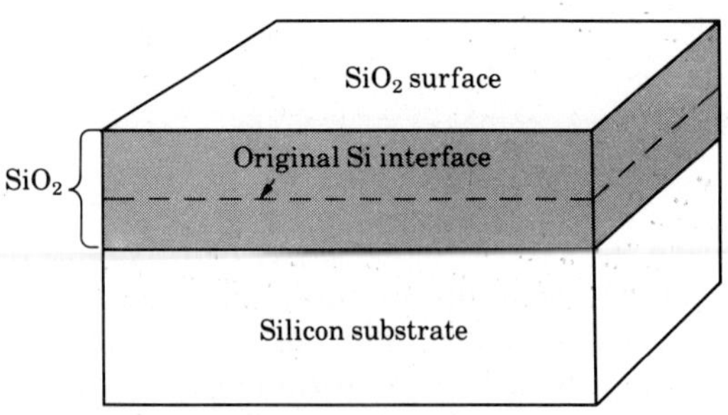

FIGURE 34.7
Growth of silicon dioxide, showing consumption of silicon. *Source:* S. M. Sze.

These oxidation methods are useful primarily for coating the entire silicon surface with oxide, but it is also necessary to oxidize only certain portions of the substrate surface. This procedure is termed *selective oxidation* and uses silicon nitride, which inhibits the passage of oxygen and water vapor. Thus by masking certain areas with silicon nitride, the silicon under these areas remains unaffected while the uncovered areas are oxidized.

34.6 Lithography

Lithography is the process by which the geometric patterns that define the devices are transferred from a **mask** to the substrate surface. In current practice, the lithographic process is applied to microelectronic circuits several times, each time using a different mask to define the different areas of the working devices. Typically designed at 100–2000 times their final size, mask patterns then go through a series of reductions before being applied to a permanent glass plate. Computer-aided design (CAD) has had a major impact on mask design and generation. Cleanliness is especially important in lithography, and many manufacturers are turning to robotics and specialized wafer-handling apparatus in order to minimize dust and dirt contamination.

Once the film deposition process is completed and the desired masking patterns have been generated, the wafer is cleaned and coated with an emulsion, called *photoresist* (PR), which is sensitive to ultraviolet (UV) light. Photoresist layers of 0.5–2.5 μm (20–100 μin.) thick are obtained by applying the PR to the substrate in liquid form and then spinning it at several thousand rpm for 30 or 60 seconds to give uniform coverage.

The next step in lithography is *prebaking* the wafer to remove the solvent from the PR and harden it. This step is carried out in an oven at around 100 °C for 10–30 min. The wafer is then aligned under the desired mask in a mask aligner. In this crucial step, called **registration,** the mask must be aligned correctly with the previous layer on the wafer. Once the wafer and mask are aligned, they are subjected to UV radiation. Upon development and removal of the exposed PR, a duplicate mask pattern will appear in the PR layer.

Following the exposure and development sequence, *postbaking* the wafer toughens and improves the adhesion of the remaining resist. In addition, a deep UV treatment (baking the wafer to 150–200 °C in ultraviolet light) can also be used to further toughen the resist against high-energy implants and dry etches. The underlying film not covered by the PR is then etched away (Section 34.7). Finally, the PR is stripped by dipping the wafer in a solvent solution (Fig. 34.8). The lithography process may be repeated as many as 25 times in the fabrication of the most advanced ICs.

One of the major issues in the area of lithography is **linewidth,** which refers to

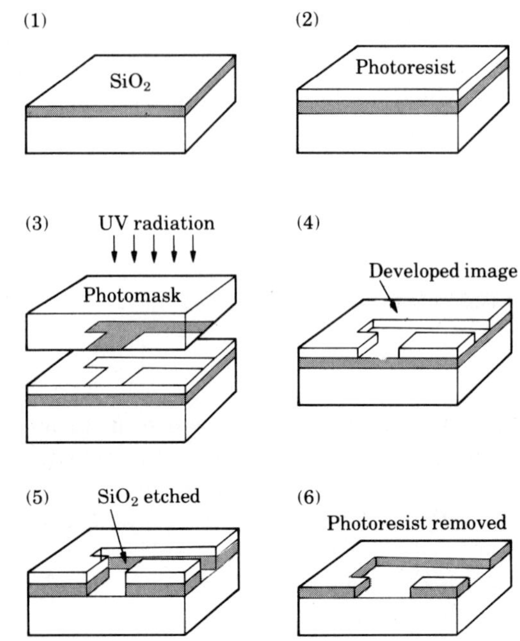

FIGURE 34.8
Pattern transfer by lithography. *Source:* After W. C. Till and J. T. Luxon.

the width of the smallest feature obtainable on the silicon surface. As circuit densities have escalated over the years, device sizes and features have become smaller and smaller. Today, minimum commercially feasible linewidths are 0.8 μm (40 μin.), but considerable research is being done concerning 0.5 and 0.3 μm structures. However, pattern resolution—and therefore device miniaturization—is limited by the wavelength of the radiation source used. Thus the need has arisen to move to wavelengths shorter than those in the ultraviolet range. This problem has been solved through the use of shorter or "deep" UV wavelengths, electron beams, and x-rays. In these technologies, the photoresist is replaced by a similar resist that is sensitive to a specific range of shorter wavelengths.

34.7 Etching

Etching is the process by which entire films or particular sections of films are removed, and it plays an important role in the fabrication sequence. One of the most important criteria in choosing an etchant is its *selectivity*, which refers to its ability to etch one material without etching another. In silicon technology, an

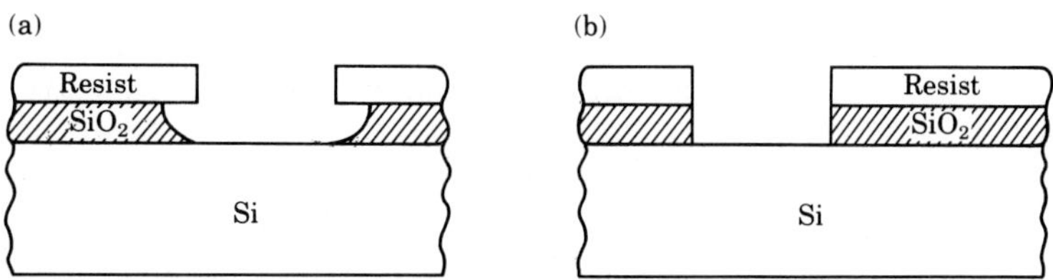

FIGURE 34.9
Etching profiles resulting from (a) isotropic wet etching and (b) anisotropic dry etching. *Source:* R. C. Jaeger.

etching process must effectively etch the silicon dioxide layer with minimal removal of the underlying silicon or the resist material. In addition, polysilicon and metals must be etched into high-resolution lines with vertical wall profiles and with minimal removal of the underlying insulating film. Typical etch rates range from hundreds to several thousands of angstroms per minute, and selectivities (defined as the ratio of the etch rates of the two films) can range from 1:1 to 100:1.

One method involves immersing the wafers in a liquid solution (*wet etching*). If silicon dioxide is to be etched, this solution contains hydrofluoric acid, which etches silicon very slowly. The main drawback of this etching technique is that it is isotropic, meaning that the etch occurs equally in all directions. This condition leads to undercutting (Fig. 34.9a), which in turn prohibits the transfer of very high resolution patterns.

The solution to this problem is *dry etching*, which involves the use of chemical reactants in a low-pressure system. In contrast to the wet process, dry etching has a high degree of directionality, resulting in highly anisotropic etches (Fig. 34.9b). The dry process also requires only small amounts of the reactant gases, whereas the aqueous solutions used in the wet process need to be refreshed periodically. However, a complete dry-etching system can cost in excess of one million dollars. The most widely used dry-etching techniques include *sputter etching*, which removes material by bombarding it with noble gas ions, usually Ar^+, and *plasma etching*, which utilizes a gaseous plasma of chlorine or fluorine ions generated by RF excitation. *Reactive ion etching* combines these two processes, using both momentum transfer and chemical reaction to remove material.

34.8 Diffusion and Ion Implantation

We should mention again that the electrical operation of microelectronic devices depends on regions of different doping types and concentrations. The electrical character of these regions is altered by introducing dopants into the substrate, accomplished by the *diffusion* and *ion implantation* processes. This step in the fabrication sequence is repeated several times, since many different regions of microelectronic devices must be defined.

In the diffusion process, the movement of atoms results from thermal excitation. Dopants can be introduced to the substrate surface in the form of a deposited film, or the substrate can be placed in a vapor containing the dopant source. The process takes place at elevated temperatures, usually 800–1200 °C (1500–2200 °F). Dopant movement within the substrate is strictly a function of temperature, time, and the diffusion coefficient (or diffusivity) of the dopant species, as well as the type and quality of the substrate material.

Because of the nature of diffusion, the dopant concentration is very high at the substrate surface and drops off sharply away from the surface. To obtain a more uniform concentration within the substrate, the wafer is heated further to drive in the dopants, which is called a *drive-in diffusion.* The fact that diffusion, desired or undesired, will always occur at high temperatures is always taken into account during subsequent processing steps. Although the diffusion process is relatively inexpensive, it is highly isotropic.

Ion implantation is a much more extensive process and requires specialized equipment (Fig. 34.10). Implantation is accomplished by accelerating ions through a high-voltage field of as much as one million electron-volts and then choosing the desired dopant by means of a mass separator. In a manner similar to that in cathode-ray tubes, the beam is swept across the wafer by sets of deflection plates, thus ensuring uniform coverage of the substrate. The complete implantation system must be operated in a vacuum.

The high-velocity impact of ions on the silicon surface damages the lattice structure, resulting in lower electron mobilities. This condition is undesirable, but the damage can be fixed somewhat by an annealing step, which involves heating the substrate to relatively low temperatures, usually 400–800 °C (750–1500 °F), for 15–30 min. This provides the energy that the silicon lattice needs to rearrange and mend itself. Another important function of annealing is to drive in the implanted dopants. Implantation alone imbeds the dopants less than half a micron below the silicon surface, and the annealing step enables the dopants to diffuse to a more desirable depth of a few microns.

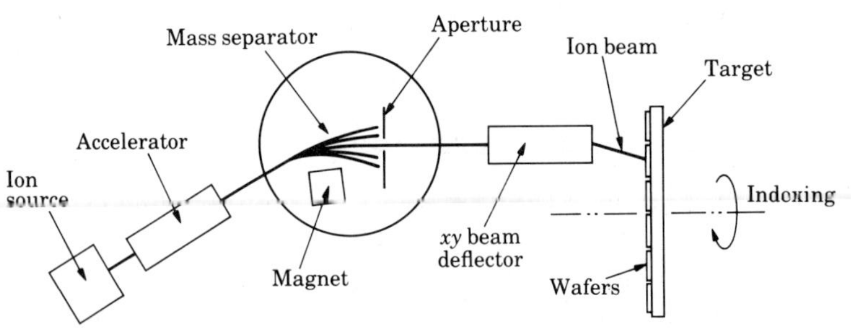

FIGURE 34.10
Apparatus for ion implantation. *Source:* J. A. Schey.

The major advantages of ion implantation include a high degree of anisotropy and wider range of dopant species. Although a complete ion implantation system may cost in excess of $1 million, its flexibility has made it the dominant technique in modern IC technology.

• Example: Processing of a *p*-type region in *n*-type silicon.

We wish to create a *p*-type region within a sample of *n*-type silicon. Draw cross-sections of the sample at each processing step in order to accomplish this.

SOLUTION

CROSS-SECTION	DESCRIPTION
a) 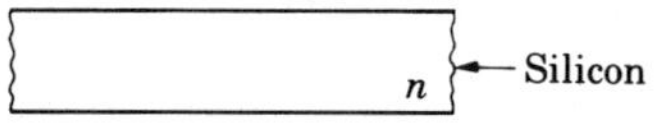 *n* — Silicon	Sample of *n*-type silicon
b) 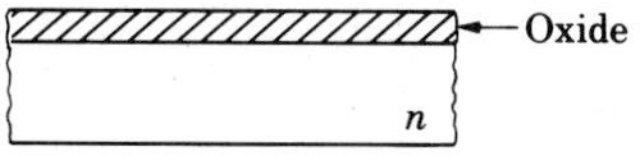 *n* — Oxide	Grow silicon dioxide by oxidation.
c) 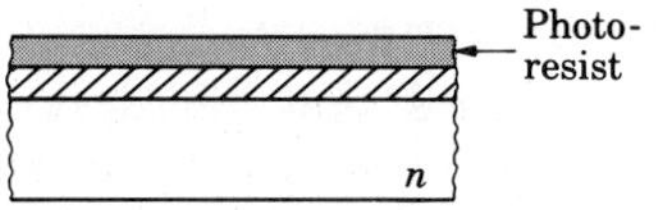 *n* — Photo-resist	Apply photoresist.
d) 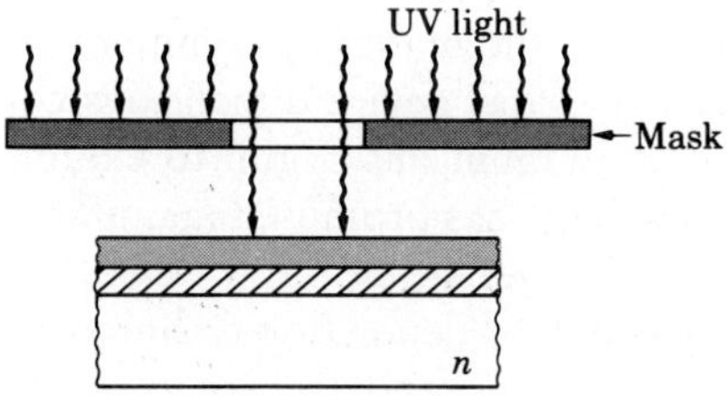 UV light — Mask — *n*	Expose photoresist using appropriate lithographic mask.
e) 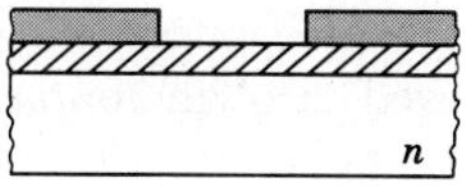 *n*	Develop photoresist.
f) 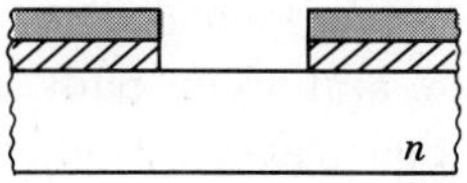 *n*	Etch silicon dioxide.
g) 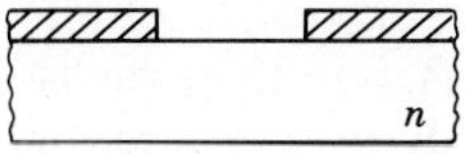 *n*	Remove photoresist.

(*continued*)

CROSS-SECTION	DESCRIPTION
h) p n	Implant boron.
i) p n	Remove silicon dioxide.

This simple device is known as a *pn junction diode*, and the physics of its operation is the foundation for most semiconductor devices.

34.9 Metallization and Testing

In the preceding sections we focused only on device fabrication. However, generating a complete and functional integrated circuit requires these devices to be interconnected. Interconnections are made by metals that exhibit low electrical resistance and good adhesion to dielectric insulator surfaces. Aluminum and aluminum-silicon-copper alloys are the materials commonly used for this purpose in VLSI technology today. However, as device dimensions continue to shrink and circuit geometries get tighter, aluminum has begun to exhibit poor step coverage over very extreme topographies. This has prompted manufacturers to use tungsten, which displays excellent step coverage.

The metal is deposited by standard deposition techniques, and interconnection patterns are generated by lithographic and etching processes. Modern ICs can typically have 1–4 layers of metallization, in which case each layer of metal is insulated by a dielectric. Layers of metal are connected together by *vias* and access to the devices on the substrate is achieved through *contacts* (Fig. 34.11).

Wafer processing is complete upon application of a passivation layer, and the next step is to test each of the individual circuits on the wafer. Each chip, also referred to as a *die*, is tested with a computer-controlled probe platform that contains needlelike probes to access the aluminum pads on the die. The platform steps across the wafer, testing whether each circuit functions properly with computer-generated simulations. If a defective chip is encountered, it is marked with a drop of ink.

After completion of this preliminary testing, each die is separated from the wafer. Diamond sawing is a commonly used separation technique and results in

(a)

(b)

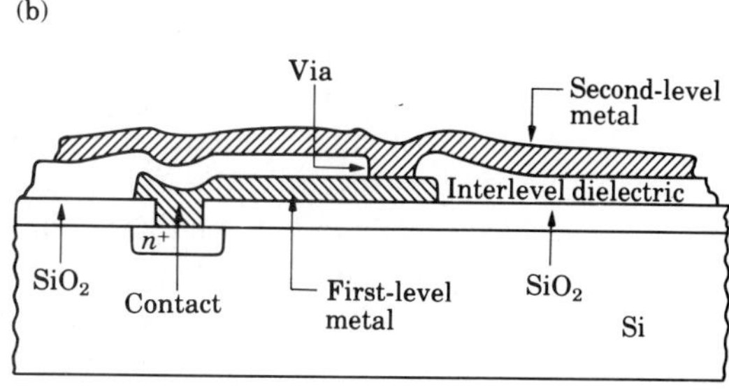

FIGURE 34.11 (a) Scanning electron microscope photograph of a two-level metal interconnect. Note the varying surface topography. *Source:* National Semiconductor Corporation. (b) Schematic drawing of a two-level metal interconnect structure. *Source:* R. C. Jaeger.

very straight edges, with minimal chipping and cracking damage. Another method cuts the wafer only partially and then separates the chips by applying pressure to the scribed lines, which crack and break along the crystallographic planes of the crystal. The chips are then sorted, with those damaged during sawing and those with ink dots being discarded.

34.10 Bonding and Packaging

The working dies must be attached to a more rugged foundation to ensure reliability. One simple method is to fasten a die to its packaging material with an epoxy cement. Another method uses a eutectic bond, made by heating metal-alloy systems. One widely used mixture is 96.4 percent gold and 3.6 percent silicon, which has a eutectic point at 370 °C (700 °F).

When the chip has been attached to the substrate, it must be accessible to electrical connections from the package leads. Bonding pads are located around the perimeter of the die and are typically 100–125 μm (0.004–0.005 in.) on a side. The most widely used method of attaching these pads is by **wire bonding**, which utilizes very thin (25 μm diameter; 0.001 in.) gold or aluminum wire. The pads are attached by thermocompression, ultrasonic, or thermosonic techniques.

The connected circuit is now ready for final packaging. The *packaging* process largely determines the overall cost of each completed IC, since the circuits are mass produced on the wafer but then packaged individually. Packages are available in a variety of styles, and selecting the appropriate one must reflect operating requirements. Consideration of a circuit's package includes chip size, number of external leads, operating environment, heat dissipation, and power requirements. For example, ICs that are used for military and industrial applications need packages of particularly high strength and toughness, as well as high temperature resistance.

The most common style used today is the **dual-in-line package (DIP)**, shown schematically in Fig. 34.12(a). Characterized by low cost and ease of handling, DIP packages are made of thermoplastic, epoxy, or ceramic and can have from 2 to 500 external leads. Ceramic packages are designed for use over a broader temperature range and in high-reliability and high-performance situations, although they are considerably more expensive than plastic packages.

Another packaging style is the metal can shown in Fig. 34.12(b), which is most useful for its excellent heat dissipation and its ability to shield the enclosed circuit from external electromagnetic interference. Figure 34.12(c) shows a flat ceramic package in which the package and all the leads are in the same plane. This package style does not offer the ease of handling nor the modular design of the DIP package. Thus it is usually permanently affixed to a multiple-level circuit board in which the low profile of the flat pack is necessary.

After the chip has been sealed in the package, it undergoes final testing. Because one of the main purposes of packaging is isolation from the environment, testing at this stage usually encompasses heat, humidity, mechanical shock, corrosion, and vibration. Destructive tests are also performed to investigate the effectiveness of sealing.

34.11 Reliability and Yield

The major concern about completed ICs is their reliability and failure rate, since no device has an infinite lifetime. Statistical methods are used to characterize the expected lifetimes and failure rates of microelectronic devices. The unit for failure rate is the FIT, defined as 1 failure per 1 billion device-hours. However, complete systems may have millions of devices, so the overall failure rate in entire systems is correspondingly higher. Failure rates greater than 100 FIT are generally unacceptable.

Equally important in failure analysis is determining the failure mechanism, that is, the actual process that causes the device to fail. Common failures due to processing involve diffusion regions (nonuniform current flow and junction breakdown), oxide layers (dielectric breakdown and accumulation of surface charge), lithography (uneven definition of features and mask misalignment), and metal

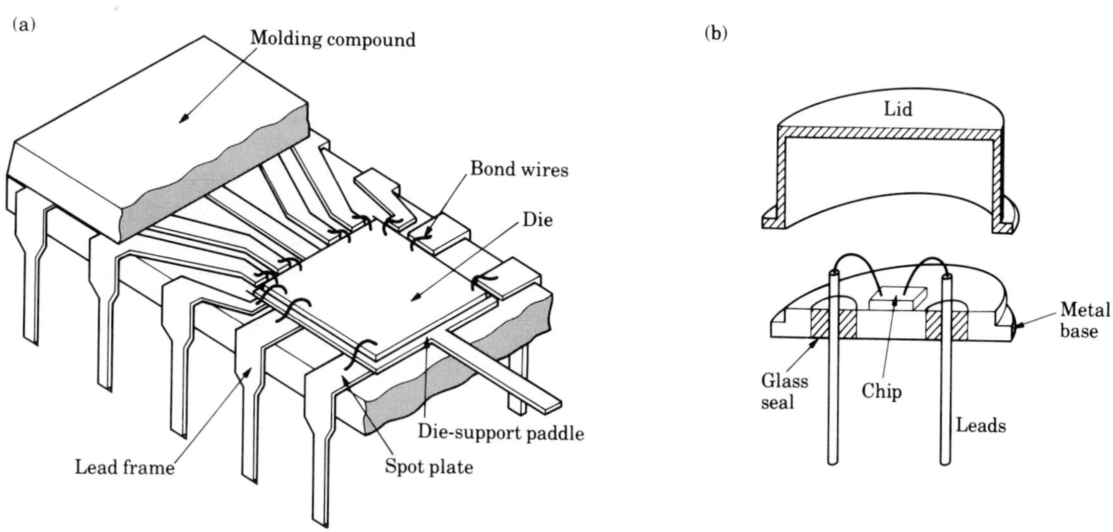

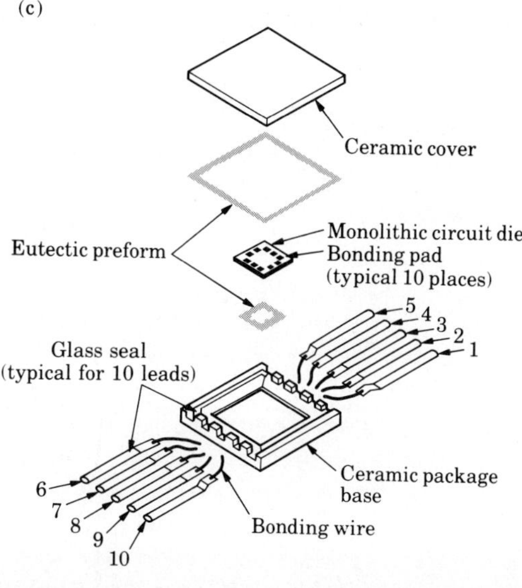

FIGURE 34.12

Schematic illustrations of different IC packages: (a) dual-in-line (DIP), (b) metal can, and (c) ceramic flat pack. *Sources:* R. C. Jaeger and A. B. Glaser; G. E. Subak-Sharpe.

layers (poor contact and electromigration resulting from high current densities). Other failures can originate from improper chip mounting, degradation of wire bonds, and loss of package hermeticity. A distribution of failure mechanisms in integrated circuits is presented in Fig. 34.13.

Because device lifetimes are very long, it is impractical to study device failure under normal operating conditions. One method of studying failures efficiently is by **accelerated life testing,** which involves accelerating the conditions whose effects cause device breakdown. Cyclic variations in temperature, humidity, voltage, and current are used to stress the components. Statistical data taken from these tests are then used to predict device failure nodes and device life. Chip mounting and packaging is strained by cyclical temperature variations.

Yield is defined as the ratio of good chips to the total number of chips produced on a wafer. Yields can range from only a few percent up to almost 100 percent. Commonly the wafer is separated into regions of good and bad chips. Sometimes imperfections are caused by point defects, such as oxide pinholes, film contamination, and crystal dislocations, and others are caused by area defects, such as uneven film deposition and etch nonuniformity.

In addition to the metal-oxide semiconductor structure introduced at the beginning of this chapter, the **bipolar junction transistor (BJT)** is also widely used. The fabrication sequence for making bipolar devices is outlined in Fig. 34.14. While the actual fabrication steps are very similar to those of both the MOSFET and BJT technologies, their circuit applications are different. Memory circuits, such as RAMs and ROMs, consist primarily of MOS devices, whereas linear circuits, such as amplifiers and filters, contain mostly bipolar transistors. Other differences between these two devices include the faster operating speeds of the BJT and the smaller size (and therefore greater circuit density) of the MOSFET.

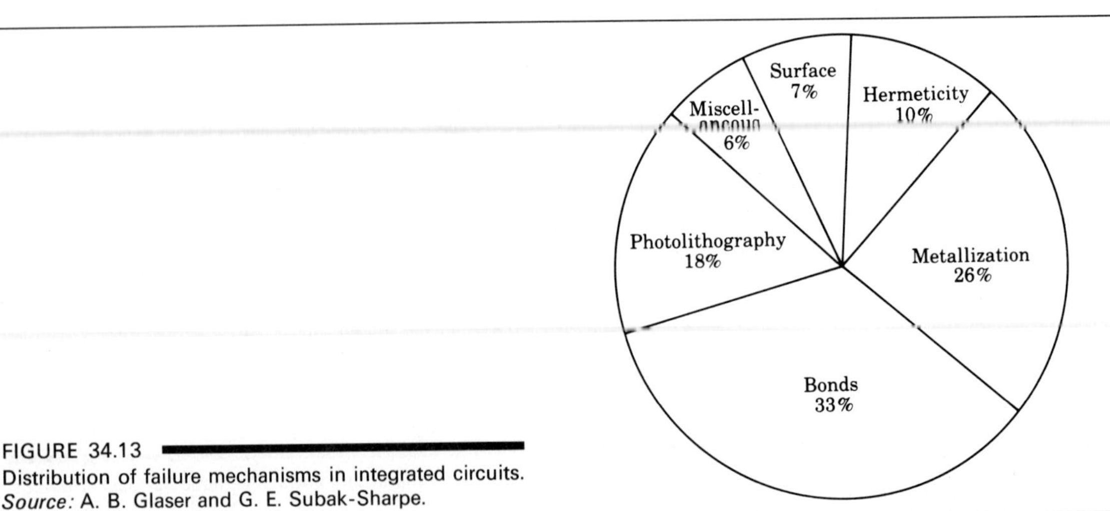

FIGURE 34.13
Distribution of failure mechanisms in integrated circuits.
Source: A. B. Glaser and G. E. Subak-Sharpe.

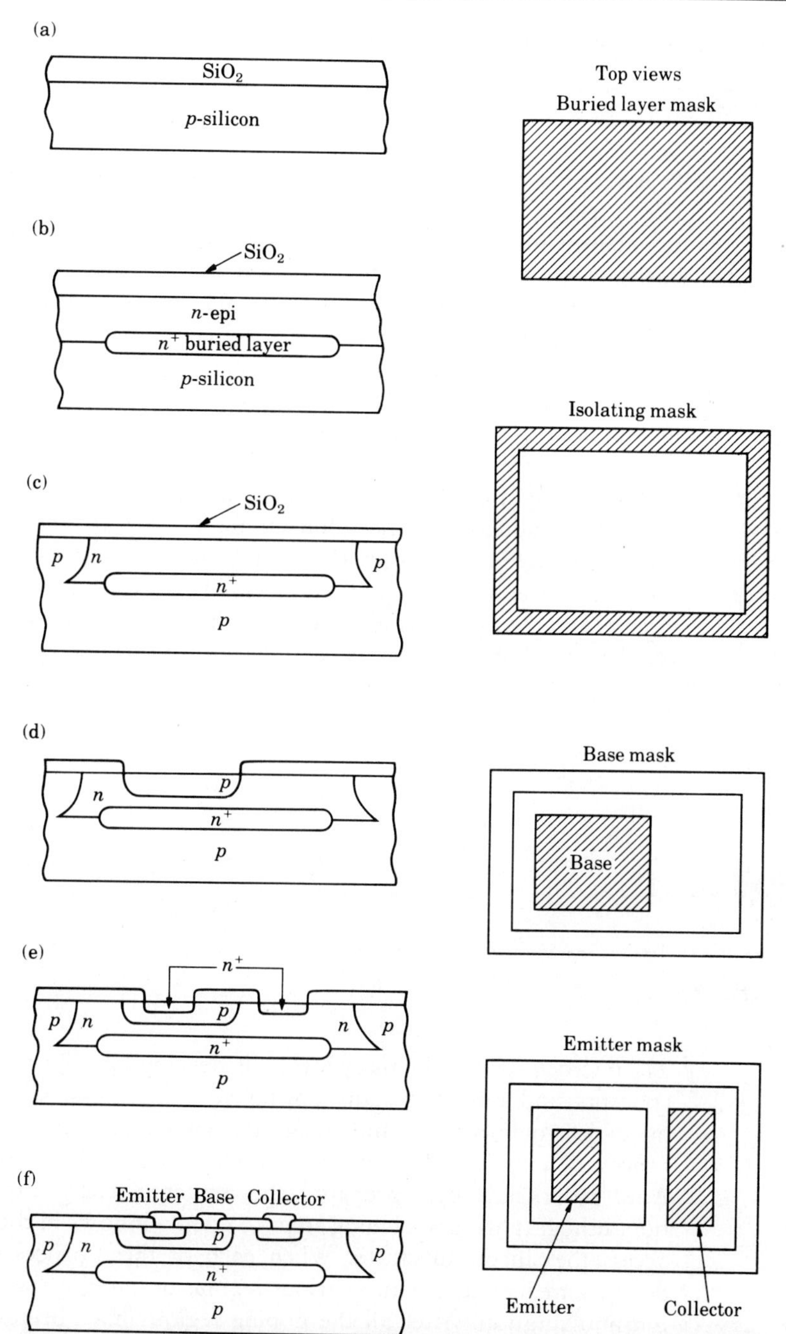

FIGURE 34.14

Cross-sectional views and masking levels of the major processing steps for a bipolar transistor. The n^+ buried layer allows a more uniform current flow within the transistor. *Source:* R. C. Jaeger.

34.12 Printed Circuit Boards

Packaged ICs are seldom used alone. Rather, they are usually combined with other ICs to serve as building blocks of a yet larger circuit. A **printed circuit board** is the foundation for the final interconnections among all the completed circuits and basically serves as the communication link between the outside world and the microelectronic circuitry within each packaged IC. In addition to the ICs, circuit boards also usually contain discrete circuit components, such as resistors and capacitors, which would take up too much "real estate" on the silicon surface or have special power dissipation requirements. Other common discrete components are inductors, which cannot be integrated onto the silicon surface, and high-performance transistors.

A printed circuit board is basically a plastic resin material coated with copper foil. The conductive patterns on circuit boards are defined by lithography and selectively etching the copper. The ICs and other discrete components are then fastened to the board by soldering. This is the final step in making the integrated circuits and the microelectronic devices they contain accessible. Such circuit boards are the hearts of computers and other large electronic systems.

SUMMARY

The microelectronics industry is rapidly developing. The possibilities for new device concepts and circuit designs appear to be endless. Clearly the fabrication of microelectronic devices and integrated circuits involves many different types of processes, most of which have been adapted from other fields of manufacturing. After wafers have been prepared, they undergo repeated oxidation or film deposition, lithographic, and etching steps to open windows in the oxide layer in order to access the silicon substrate. When each of these processing cycles is complete, dopants are introduced into various regions of the silicon structure by diffusion and ion implantation. After all the doping regions have been established, devices are interconnected by metal layers, and the completed circuit is packaged and made accessible through electrical connections. Finally, the packaged circuit and other discrete devices are soldered to a printed circuit board for final installation.

TRENDS

- Progress is being made in miniaturizing devices to below 0.5-μm linewidths and integration scales reaching the hundreds of millions of components per chip.
- Stacked and three-dimensional device and circuit structures are being developed.
- Gallium arsenide layers are being grown on silicon substrates to combine the optical capabilities of gallium arsenide with the established electrical properties and superior strength and heat dissipation of silicon.
- Research is being conducted on optical integration of devices (transmission by light pulses, not electrical signals) to yield faster operating speeds.
- High-temperature superconductors are being investigated as a means of circuit interconnection and also as a substrate material for a new generation of devices that operate on the storage of magnetic flux, not electric charge.
- Research is being conducted on developing "vacuum transistors," in which electrons travel between device terminals through a vacuum, not through the host material, resulting in much higher electron mobilities and increased temperature and radiation resistance.

KEY TERMS

Accelerated life testing
Bipolar junction transistor
Chemical vapor deposition
Chip
Czochralski process
Dual-in-line package (DIP)
Epitaxy
Etching
Integrated circuit
Linewidth
Mask
Metal-oxide-semiconductor field effect transistor
Printed circuit board
Registration
Semiconductor
Wafer
Wire bonding
Yield

BIBLIOGRAPHY

Colclaser, R.A., *Microelectronics: Processing and Device Design*. New York: Wiley, 1980.

Ghandhi, S.K., *VLSI Fabrication Principles*. New York: Wiley, 1983.

Glaser, A.B., and G.E. Subak-Sharpe, *Integrated Circuit Engineering*. Reading, Mass.: Addison-Wesley, 1977.

Harper, C.A. (ed.), *Handbook of Materials and Processes for Electronics.* New York: McGraw-Hill, 1970.
Jaeger, R.C., *Introduction to Microelectronic Fabrication.* Reading, Mass.: Addison-Wesley, 1988.
Keyes, R.W., *The Physics of VLSI Systems.* Reading, Mass.: Addison-Wesley, 1987.
Ruska, W.S. *Microelectronic Processing.* New York: McGraw-Hill, 1987.
Streetman, B.G., *Solid State Electronic Devices.* Englewood Cliffs, N.J.: Prentice-Hall, 1980.
Sze, S.M. (ed.), *VLSI Technology.* New York: McGraw-Hill, 1983.
———, *Physics of Semiconductor Devices.* New York: Wiley, 1981.
Till, W.C., and J.T. Luxon, *Integrated Circuits: Materials, Devices, and Fabrication.* Englewood Cliffs, N.J.: Prentice-Hall, 1982.
Wolf, S., and R.N. Tauber, *Silicon Processing for the VSLI Era,* Vol. 1: *Process Technology.* Sunset Beach, CA: Lattice Press, 1986.

REVIEW QUESTIONS

34.1 Define wafer, chip, device, and integrated circuit.
34.2 Why is silicon the semiconductor most used in IC technology?
34.3 What do VLSI, IC, CVD, and DIP stand for?
34.4 How do *n*-type and *p*-type dopants differ?
34.5 How is epitaxy different from other forms of film deposition?
34.6 Compare wet and dry oxidation.
34.7 How is silicon nitride used in oxidation?
34.8 What are the purposes for prebaking and postbaking in lithography?
34.9 Define selectivity and isotropy and their importance in relation to etching.
34.10 What do the terms linewidth and registration refer to?
34.11 Compare diffusion and ion implantation.
34.12 What is the difference between evaporation and sputtering?
34.13 What is the definition of yield?
34.14 What is accelerated life testing?
34.15 What do BJT and MOSFET stand for?

QUESTIONS AND PROBLEMS

34.16 A certain wafer manufacturer produces two equal-sized wafers, one containing 100 chips and the other containing 70 chips. After testing, it is observed that 20 chips on each wafer are bad. What are the yields of these two wafers? Can any relationship be drawn between chip size and yield?

34.17 A chlorine-based polysilicon etch process displays a polysilicon:resist selectivity of 4:1 and a polysilicon:oxide selectivity of 50:1. How much resist and exposed oxide will be consumed in etching 3500 Å of polysilicon? What would the polysilicon:oxide selectivity have to be in order to lose only 40 Å of exposed oxide?

34.18 In a horizontal epitaxial reactor (see the accompanying figure), the wafers are placed on a stage (susceptor) that is tilted by a small amount, usually 1–3°. Why is this done?

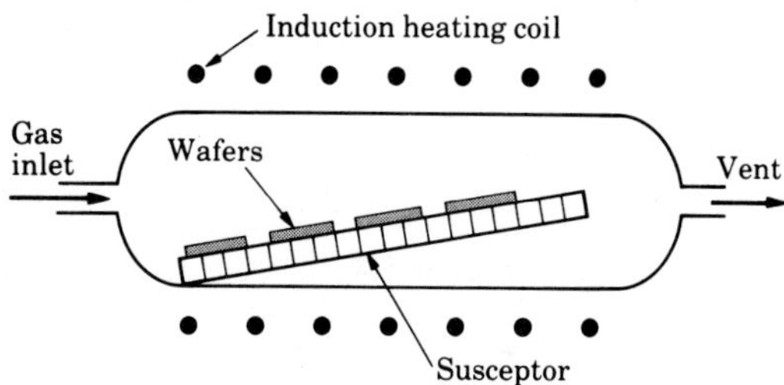

34.19 During a processing sequence, three silicon dioxide layers are grown by oxidation: 2500 Å, 4000 Å, and 1500 Å. How much of the silicon substrate is consumed?

34.20 A common problem in ion implantation is channeling, in which the high velocity ions travel deep into the material via channels along the crystallographic planes before finally being stopped. What is one simple way to stop this effect?

34.21 The accompanying figure shows the cross-section of a simple *npn* bipolar transistor. Develop a process flowchart to fabricate this device.

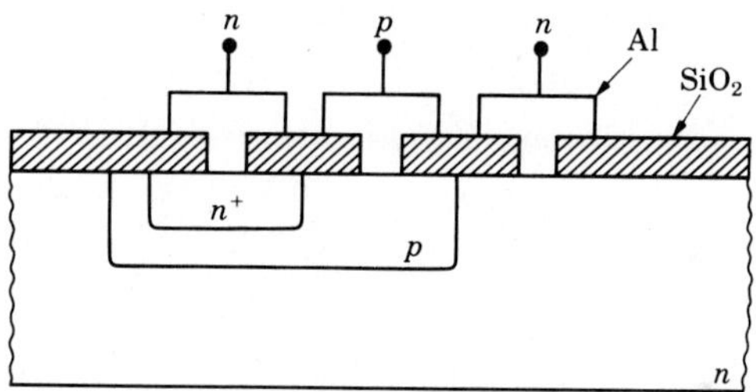

34.22 What is the resistivity of silicon (at room temperature) that has been doped with boron at 1×10^{16} atoms/cm^3? arsenic at 1×10^{15} atoms/cm^3?

34.23 A certain design rule calls for metal lines to be no less than 2 microns wide. If a 1-micron thick metal layer is to be wet-etched, what is the minimum photoresist width allowed (assuming that the wet etch is perfectly isotropic)? What would be the minimum photoresist width if a perfectly anisotropic dry-etch process is used?

PART VII

Common Aspects of Manufacturing

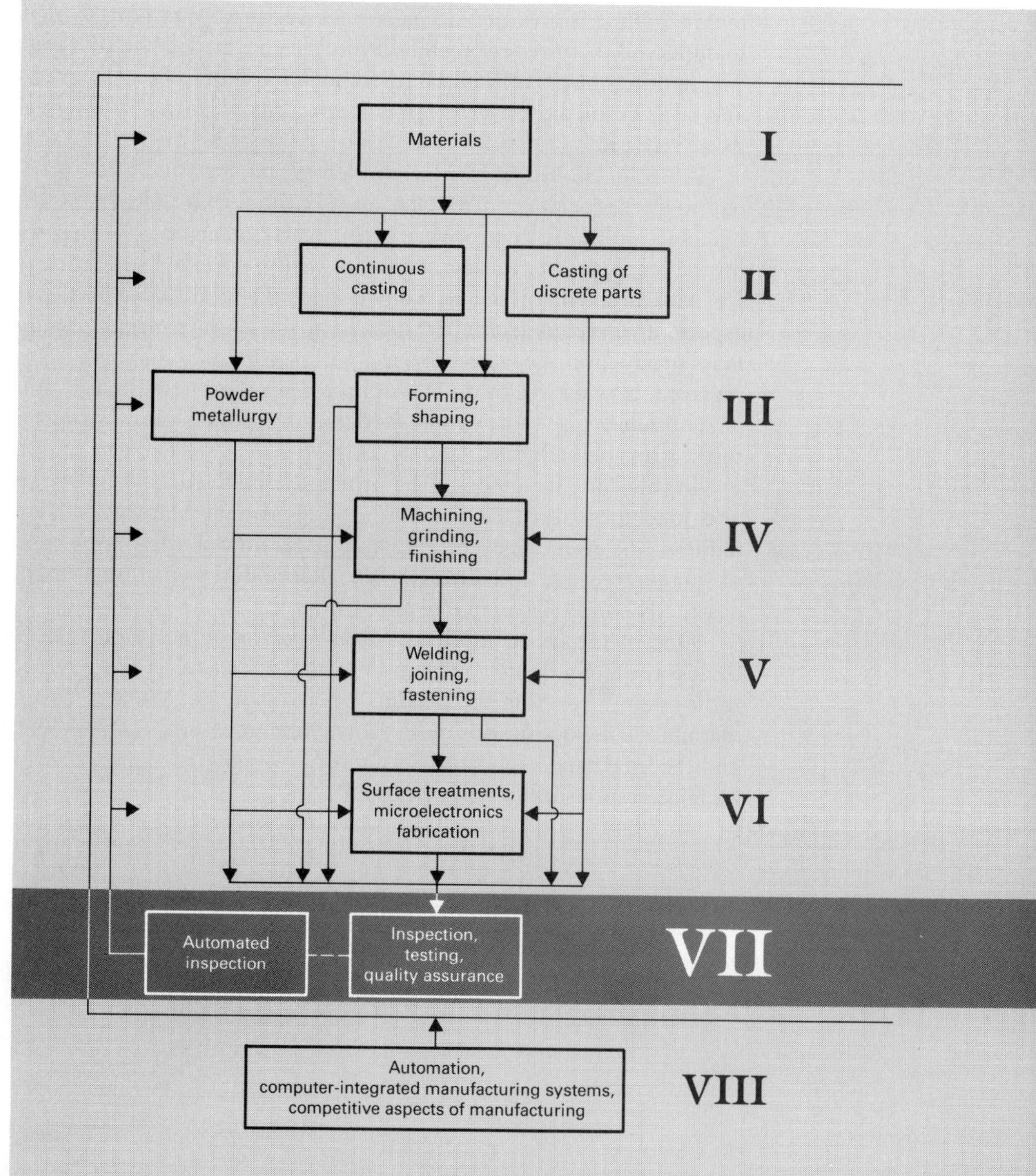

• INTRODUCTION •

Thus far we have described basic manufacturing processes for making a wide variety of industrial and consumer products. We have presented the techniques used to modify surfaces of components and products to obtain certain desirable properties, showing the advantages and limitations of each technique. Although we have discussed dimensional accuracies obtained in specific manufacturing processes, we have not yet described how parts are measured and inspected before they are marketed.

Why must we measure dimensions of parts accurately and inspect them? We measure dimensions and other surface features of a part to make sure that it is manufactured consistently and within the specified range of dimensional accuracy. The majority of parts manufactured are components of a product, and they must fit and be assembled properly so that the product performs its intended purpose during its service life.

A piston should fit into a cylinder within specified tolerances. A turbine blade has to fit properly into its slot on a turbine disk. Similarly, the slideways of a machine tool must be produced with a certain accuracy so that the parts produced on the machine are, in turn, accurate within specified tolerances.

Measurement of the relevant dimensions and features of parts is an integral aspect of *interchangeable manufacture*, the basic concept of standardization and mass production. For example, if a ball bearing in a machine is worn and has to be replaced, all we have to do is purchase a similar one with the same specification or part number. The same can be said about products such as bolts and nuts, staples, fuses, tires, razor blades, and spark plugs.

In this Part, we describe the principles involved in and the various instruments and machines used for measuring dimensional features such as length, angle, flatness, and roundness. Testing and inspection of parts are also important aspects of manufacturing operations. We describe the methods used for both non-destructive and destructive testing of parts.

One of the most important aspects of manufacturing is product quality. We discuss the technologic and economic importance of *building quality into a product*, rather than inspecting the product after it is made. We also cover those aspects of manufacturing operations that involve human beings, such as safety, hazards, risks, and the legal aspects of product liability. Design and manufacturing engineers have a major responsibility in this regard.

35

Engineering Metrology

35.1 Introduction

In this chapter we present the principles, methods, and characteristics of the instruments used for various dimensional measurements. **Engineering metrology** is defined as the measurement of dimensions: length, thickness, diameter, taper, angle, flatness, profiles, and others. For example, consider the slideways for machine tools (Fig. 35.1). These slideways need to have specific dimensions, angles, and flatness in order for the machine to function properly and with the accuracy desired.

Traditionally, measurements have been made after the part has been produced, which is known as postprocess inspection. The current trend in manufacturing is to make measurements while the part is being produced on the machine, which is known as in-process, on-line, or real-time inspection. Here the term *inspection* means to check the dimensions of what we have produced or are producing and to see whether it complies with the specified dimensional accuracy. Much progress has been made in developing new and automated instruments that are highly precise and sensitive.

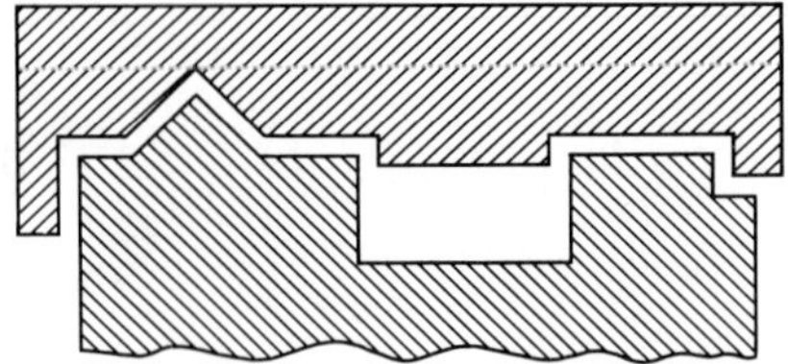

FIGURE 35.1
Cross-section of a machine tool slideway. The width, depth, angles, and other dimensions have to be produced and measured accurately for the machine tool to function as expected.

An important aspect of metrology in manufacturing processes is *dimensional tolerances*, that is, the permissible variation in the dimensions of a part. Tolerances are important not only for proper functioning of products, but they also have a major economic impact on manufacturing costs. The smaller we make the tolerance, the higher the production costs. We describe these and related aspects of tolerances and tolerancing in this chapter.

35.2 Measurement Standards

Our earliest experience with measurement was with a simple *ruler* to measure lengths (linear dimensions). A ruler is inscribed with lines that are a certain distance apart, typically 1 mm or $\frac{1}{16}$ in. Thus we use the ruler as a standard against which we measure dimensions. Traditionally, in English-speaking countries the units of *inch* and *foot*—the origins of which were based on parts of the human body—have been used. Consequently, different lengths of a one-foot measure in different areas of the country were not unusual.

However, in much of the world, the *meter* has been used as a standard (metric system). Originally, one meter was defined as $\frac{1}{10}$ of a millionth of the distance between the North Pole and the equator. The original meter length was standardized by the distance between two scratches on a platinum–iridium bar that is kept, under controlled conditions, in a building outside Paris. One meter equals 39.37 in.

In 1960 the meter was officially defined as 1,650,763.73 wavelengths, in vacuum, of the orange light given off by electrically excited krypton 86 (a rare gas). The precision of this measurement was set at 1 part in 10^9. The meter is now a Systéme International d'Unites, or SI, unit of length and is the international standard. The smallest dimensions are measured in nanometers (1 nm $= 10^{-9}$ m).

Numerous measuring instruments and devices are used in engineering metrology, each having its own application, sensitivity, and precision, as we describe throughout this chapter. Let's define two terms that are commonly used to describe the type and quality of an instrument (see also Section 35.10).

Sensitivity, also called *resolution*, is defined as the smallest difference in dimensions that the instrument can detect or distinguish. A wooden yardstick, for example, has far less sensitivity than a finely-graduated steel rule. **Precision** (sometimes incorrectly called accuracy) is defined as the degree to which the instrument gives repeated measurements of the same standard. A common wooden ruler, for example, will expand or contract, depending on the environment, thus giving different and hence unreliable measurements.

Temperature control is important, particularly for making fine measurements with precision instruments. In engineering metrology, the words instrument and gage are often used interchangeably. However, we discuss a specific category of gages in Section 35.7. The standard measuring temperature is 20 °C (68 °F), and all gages are calibrated at this temperature. Consequently, accurate measurements should be done in controlled environments, maintaining this temperature usually within ± 0.3 °C (0.5 °F).

• Example: Length measurements throughout history.

A variety of standards for length measurement have been developed during the past 6000 years. About 4000 B.C., a common standard in Egypt was the King's Elbow, which was equivalent to 0.4633 m. One *elbow* was equal to 1.5 feet, 2 handspans, 6 handwidths, or 24 finger-thicknesses. In A.D. 1101, King Henry I declared a new standard—the *yard* (0.9144 m)—which was the distance from his nose to the tip of his thumb. During the Middle Ages almost every kingdom and city established its own length standard, some with identical names. In 1528, the French physician J. Fernel proposed the distance between Paris and Amiens as the general length reference. During the seventeenth century, suggestions were made to use the length of a pendulum as a standard. In 1661, the British architect Sir Christopher Wren suggested the length of a pendulum, with a period of $\frac{1}{2}$ second. The Dutch mathematician C. Huygens suggested a pendulum length that was one third of that, with a period of 1 second.

To avoid the confusion and proliferation of length measurement, a decisive movement toward a definitive length standard began in 1790 in France, with the concept of a *métre*, from the Greek word *metron*, meaning measure. A gage block one meter long was made of pure platinum with a rectangular cross-section and was placed in the National Archives in Paris in 1799. Copies of this gage were made and distributed to other countries over the years. During 1870–1872, international committees met and decided on an international meter standard. The new bar was made of 90 percent platinum–10 percent iridium, with an X-shaped cross-section and 20 mm × 20 mm overall dimensions. Three marks were engraved at each end of the bar. The standard meter is the distance between the central marks at each end, measured at 0 °C. The latest trend in extremely accurate measurement is based on the speed of light in vacuum, obtained by measuring the wavelength and the frequency of the standardized infrared beam of a laser.

•

35.3

Line-Graduated Instruments

Line-graduated instruments are used for measuring length (linear measurements) or angles (angular measurements). *Graduated* means marked to indicate a certain quantity.

35.3.1 Linear measurements (direct reading)

Several commonly used linear-measurement instruments can be used to read dimensions directly.

Rules. The simplest and most commonly used instrument for making linear measurements is a steel rule (*machinist's rule*), bar, or tape, with fractional or decimal graduations (Fig. 35.2). Lengths are measured directly, to an accuracy that is limited to the nearest division, usually 1 mm or $\frac{1}{64}$ in. Rules may be rigid or flexible and may be equipped with a hook at one end for ease of measuring from an edge. Rule depth gages are similar to rules and slide along a special head, similar to the vertical instrument shown in Fig. 35.2.

Vernier calipers. Named for P. Vernier, who lived in the 1600s, **vernier calipers** have a graduated beam and a sliding jaw with a *vernier*. These instruments are also called *caliper gages* (Fig. 35.3a). The two jaws of the caliper contact the part being measured, and the dimension is read at the matching graduated lines (Fig. 35.3b). Vernier calipers can be used to measure inside or outside lengths. The vernier improves the sensitivity of a simple rule by indicating fractions of the

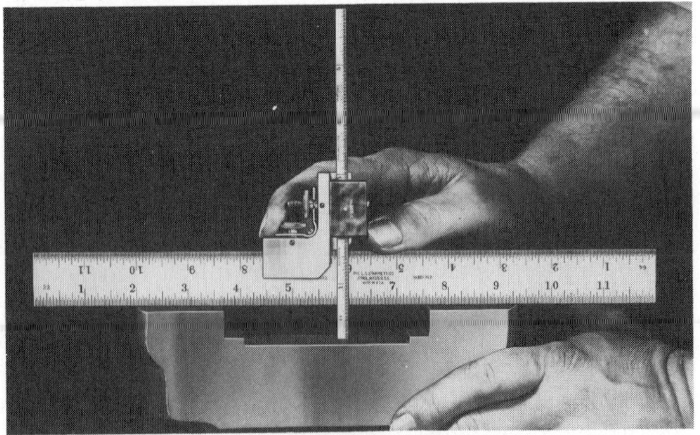

FIGURE 35.2
A steel rule (horizontal) and a rule depth gage (vertical).

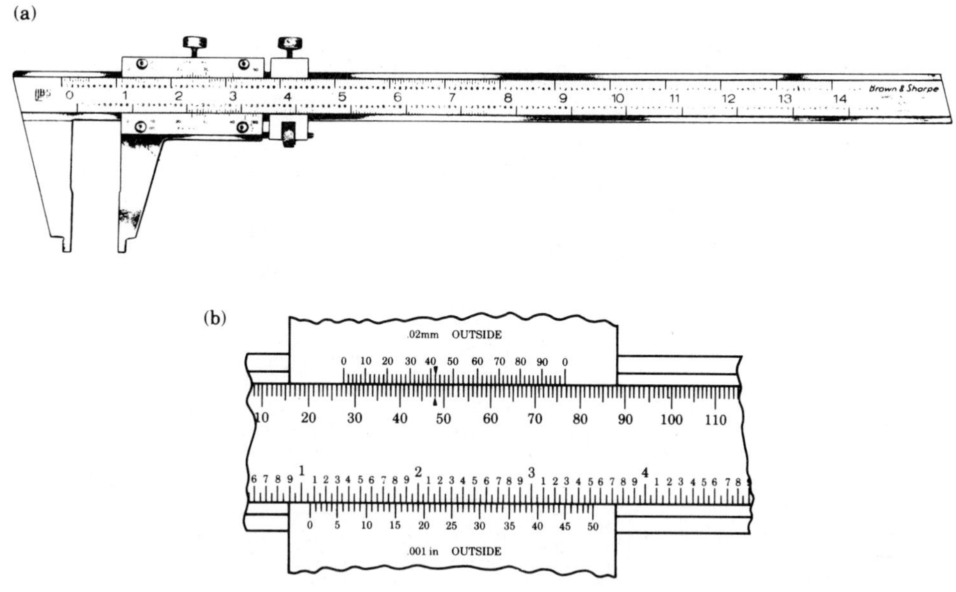

FIGURE 35.3

(a) A caliper gage with a vernier. (b) A vernier, reading 27.00 + 0.42 = 27.42 mm, or 1.000 + 0.050 + 0.029 = 1.079 in. We arrive at the last measurement as follows. First note that the two lowest scales pertain to the inch units. We next note that the 0 (zero) mark on the lower scale has passed the 1-in. mark on the upper scale. Thus we first record a distance of 1.000 in. Next we note that the 0 mark has also passed the first (shorter) mark on the upper scale. Noting that the 1-in. distance on the upper scale is divided into 20 segments, we have passed a distance of 0.050 in. Finally, note that the marks on the two scales coincide at the number 29. Each of the 50 graduations on the lower scale indicates 0.001 in. Consequently, we also have 0.029 in. Thus the total dimension is 1.000 in. + 0.050 in. + 0.029 in. = 1.079 in.

smallest division on the graduated beam, usually to 25 μm (0.001 in.). Vernier calipers are also equipped with digital readouts, easy to read, and less subject to human error than reading verniers. Vernier height gages are vernier calipers with setups similar to a depth gage and have similar sensitivity (see also electronic gages, Section 35.4.2).

Micrometers. Commonly used for measuring the thickness and inside or outside diameters of parts (Fig. 35.4a), the **micrometer** has a graduated, threaded spindle (Fig. 35.4b). Circumferential vernier readings to a sensitivity of 2.5 μm (0.0001 in.) can be obtained. Micrometers are also available for measuring depths (*micrometer depth gage*) and internal diameters (*inside micrometer*) with the same sensitivity. Micrometers are also equipped with digital readout to reduce errors in reading. The anvils on micrometers (flat, as shown) can be equipped with conical or ball contacts. They are used to measure inside recesses, threaded rod diameters, and wall thicknesses of tubes and curved sheets.

Diffraction gratings. **Diffraction gratings** consist of two flat optical glasses with closely spaced parallel lines scribed on their surfaces (Fig. 35.5). The grating on

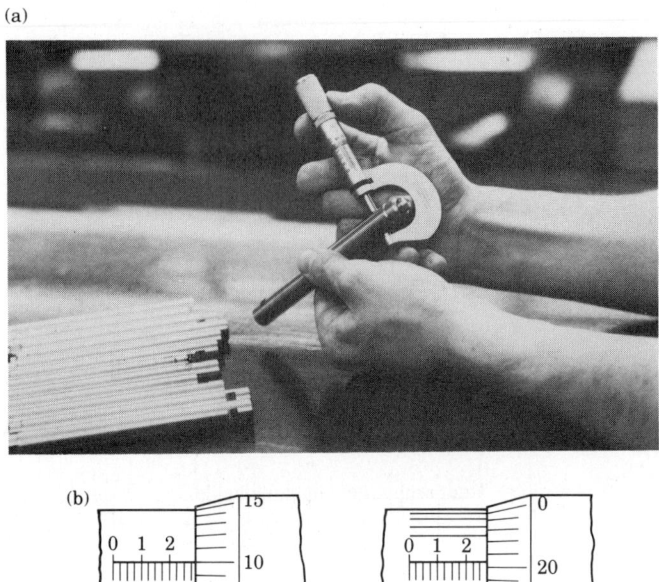

FIGURE 35.4
(a) A micrometer being used to measure the diameter of round rods. *Source:* L. S. Starrett Co. (b) Vernier on the sleeve and thimble of a micrometer. Left one reads 0.200 + 0.075 + 0.010 = 0.285 in.; right one reads 0.200 + 0.050 + 0.020 + 0.0003 = 0.2703 in. These dimensions are read in a manner similar to that described in the caption for Fig 35.3.

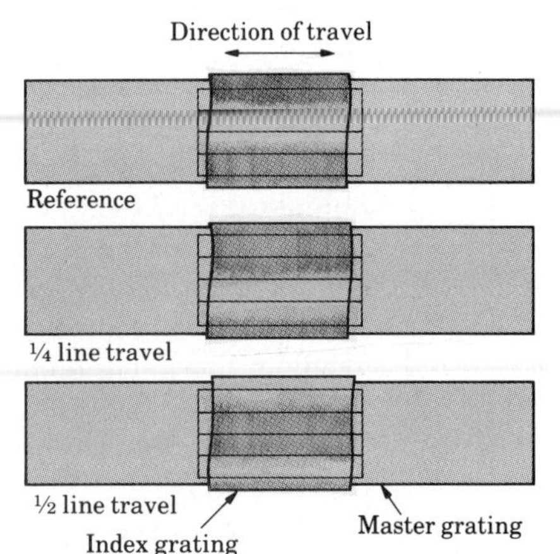

FIGURE 35.5
Measurement of very small linear displacements by fringes (moiré fringes). Note the shift in the fringe patterns. Gratings can typically be 40 lines/mm (1000 lines/in.).

the shorter glass is inclined slightly. As a result, interference fringes develop when it is viewed over the longer glass. The position of these fringes depends on the relative position of the two sets of glasses. With modern equipment, using electronic counters and photoelectric sensors, sensitivities of 2.5 μm (0.0001 in.) can be obtained with gratings having 40 lines/mm (1000 lines/in.).

35.3.2 Linear measurements (indirect reading)

Indirect-reading instruments are typically *calipers* and *dividers* without any graduated scales. They are used to transfer the size measured to a direct-reading instrument, such as a rule. After adjusting the legs to contact the part at the desired location, the instrument is held against a graduated rule, and the dimension is read. Because of the experience required in using them and their dependence on graduated scales, the accuracy of this type of indirect measurement is limited. *Telescoping gages* are available for indirect measurement of holes or cavities.

35.3.3 Angle-measuring instruments

Angles are measured by the methods described below, either in degrees, radians, or minutes and seconds of arc. Because of the geometry involved, angles are usually more difficult to measure than are linear dimensions.

Bevel protractor. A **bevel protractor** is a direct-reading instrument similar to a common protractor, except that it has a movable member (Fig. 35.6a). The two blades of the protractor are placed in contact with the part being measured, and the angle is read directly on the vernier scale. The sensitivity of the instrument depends on the graduations of the vernier (Fig. 35.6b). Another type of bevel protractor is the *combination square*, which is a steel rule equipped with devices for measuring 45° and 90° angles.

Sine bar. Measuring with a **sine bar** involves placing the part on an inclined bar or plate and adjusting the angle by placing gage blocks (Section 35.7) on a surface plate (Fig. 35.7). After the part is placed on the sine bar, a dial indicator (Section 35.4) is used to scan the top surface of the part. Gage blocks are added or removed as necessary until the top surface is parallel to the surface plate. The angle on the part is then calculated from geometric relationships.

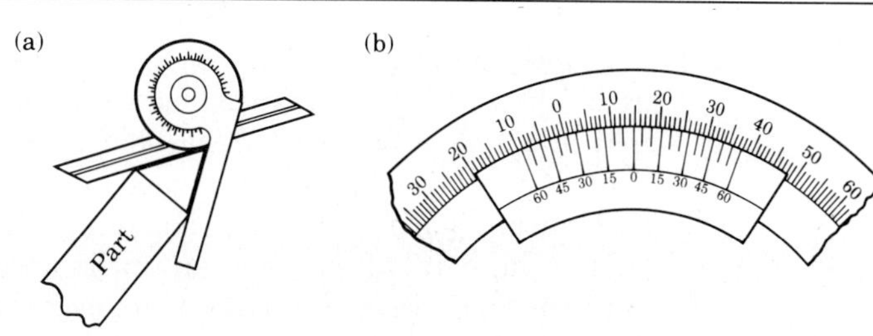

FIGURE 35.6 (a) Schematic illustration of a bevel protractor for measuring angles. (b) Vernier for angular measurement, indicating 14° 30′.

(a)

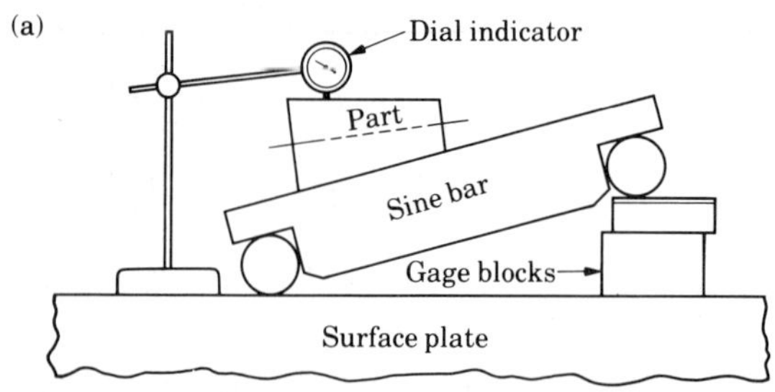

(b)

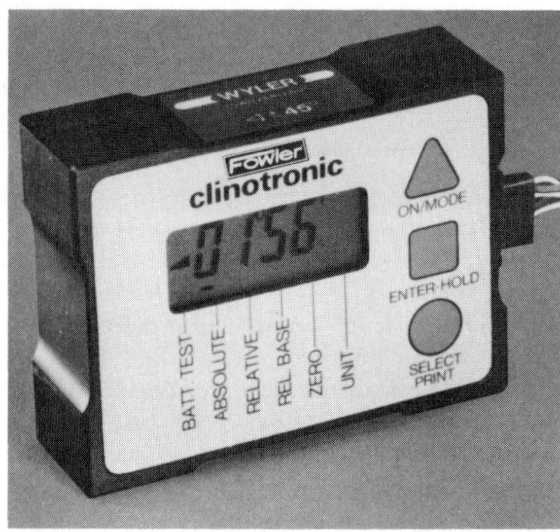

FIGURE 35.7
(a) Setup showing the use of a sine bar for precision measurement of workpiece angles. (b) Measuring inclination angles with a digital electronic instrument. *Source:* Fred V. Fowler Co., Inc., exclusive North American distributor of the Clinotronic 45.

Surface plates are made of cast iron or natural stones, such as granite, and are used extensively in engineering metrology. Granite surface plates have the desirable properties of resistance to corrosion, nonmagnetic, and low thermal expansion.

Other methods. Angles can also be measured using *angle gage blocks*. These are blocks with different tapers that can be assembled in various combinations and used in a manner similar to sine bars. Angles on small parts can be measured through microscopes, with graduated eyepieces, or with optical projectors (Section 35.5). Inclination angles can be measured with a digital electronic instrument (Fig. 35.7b). This particular instrument (inclinometer) has a measuring capacity of $\pm 45^\circ$ and an accuracy of <3 minutes of arc over the full range. The output socket on the right allows connection to a printer and directly to computers.

35.4 Comparative Length-Measuring Instruments

Unlike the instruments we have just described, instruments used for measuring comparative lengths, also called *deviation-type* instruments, amplify and measure variations or deviations in distance between two or more surfaces. These instruments compare dimensions, hence the word *comparative*. We describe below common types of instruments used for making comparative measurements.

35.4.1 Dial indicators

Dial indicators are simple mechanical devices that convert linear displacements of a pointer to rotation of an indicator on a circular dial (Fig. 35.8). The indicator is set to zero at a certain reference surface, and the instrument or the surface to be measured—either external or internal—is brought into contact with the pointer. The movement of the indicator is read directly on the circular dial—either plus or minus—to accuracies as high as 1 μm (40 μin.).

Dial indicators of several designs are available for use as portable or benchtop units. The basic design consists of a rack-and-pinion and a gear-train mechanism that convert linear motion to rotary motion with large amplifications. These instruments are also used for multiple-dimension gaging of parts (Fig. 35.8c). Instruments with electrical and fluidic amplification mechanisms and with digital readout are also available.

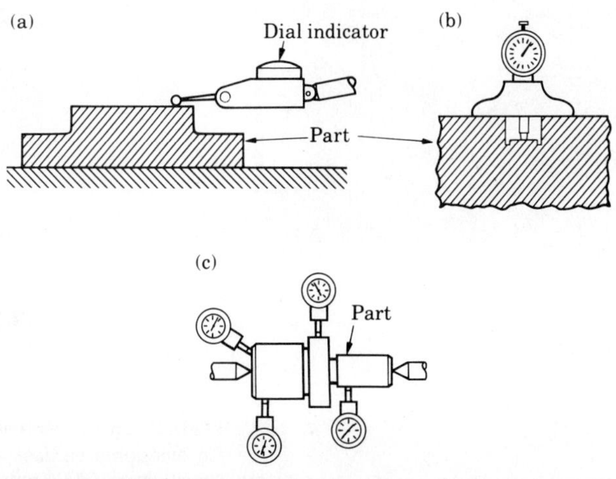

FIGURE 35.8
Three uses for dial indicators: (a) roundness, (b) depth, and (c) multiple-dimension gaging of a part.

35.4.2 Electronic gages

Unlike mechanical systems, **electronic gages** sense the movement of the contacting pointer through changes in the electrical resistance of a strain gage or through inductance or capacitance. The electrical signals are then converted and displayed as linear dimensions. A hand-held electronic gage for measuring bore diameters is shown in Fig. 35.9. The tool is inserted into the bore by squeezing the handle slightly, and the bore diameter is read directly (shown in millimeters in Fig. 35.9). A microprocessor-assisted electronic gage for measuring vertical length is shown in Fig. 35.10. A commonly used electronic gage is the *linear variable differential transformer* (LVDT), used extensively for measuring small displacements.

Although they are more expensive than other types, electronic gages have advantages such as ease of operation, rapid response, digital readout, less possibility of human error, versatility, flexibility, and the capability to be integrated into automated systems through microprocessors and computers (Section 35.9).

FIGURE 35.9 An electronic gage for measuring bore diameters. The measuring head is equipped with three carbide-tipped steel pins for wear resistance. The LED display reads 29.158 mm. *Source:* Courtesy of TESA SA.

FIGURE 35.10 An electronic vertical length measuring instrument, with a sensitivity of 0.001 mm (40 μin.). *Source:* Courtesy of TESA SA.

35.5 Measuring Straightness, Flatness, Roundness, and Profile

The geometric features of straightness, flatness, roundness, and profile are important aspects of engineering design and manufacturing. For example, piston rods, instrument components, and machine-tool slideways should all meet certain requirements with regard to these characteristics in order to function properly. Consequently, their accurate measurement is essential.

35.5.1 Straightness

Straightness can be checked with straight edges or with dial indicators (Figs. 35.11a and b). **Autocollimators**, resembling a telescope with a light beam that bounces back from the object, are used for accurately measuring small angular deviations on a flat surface. Optical means such as *transits* and *laser beams* are used for aligning individual machine elements in the assembly of machine components.

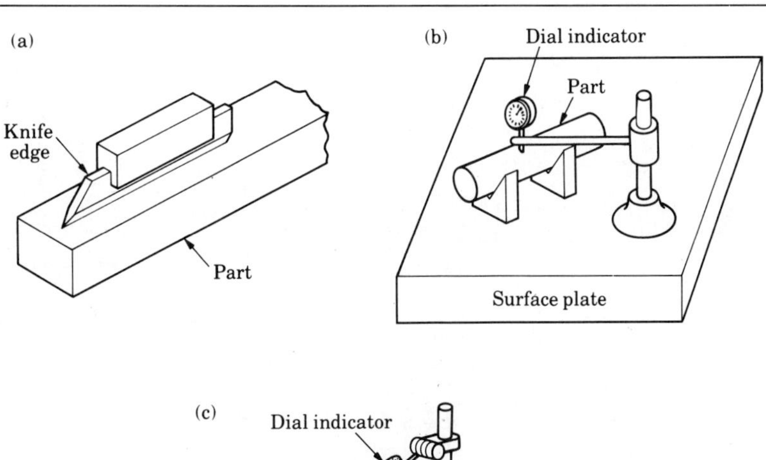

FIGURE 35.11 Measuring straightness with (a) a knife-edge rule and (b) a dial indicator attached to a movable stand resting on a surface plate. (c) Measuring flatness with a dial indicator attached to a movable stand resting on a surface plate. *Source:* After F. T. Farago.

35.5.2 Flatness

Flatness can be measured by mechanical means using a surface plate and a dial indicator (Fig. 35.11c). This method can be used for measuring perpendicularity, which can also be measured with the use of precision steel squares. The instrument shown in Fig. 35.10 can also be used for measuring perpendicularity.

Interferometry. Another method for measuring flatness is by interferometry, using an **optical flat**. The device—a glass or fused quartz disk with parallel flat surfaces—is placed on the surface of the workpiece (Fig. 35.12a). When a monochromatic (one wavelength) light beam is aimed at the surface at an angle, the optical flat splits it into two beams, appearing as light and dark bands to the naked eye (Fig. 35.12b).

The number of fringes that appear is related to the distance between the surface of the part and the bottom surface of the optical flat (Fig. 35.12c). Consequently, a truly flat workpiece surface (that is, when the angle between the two surfaces in Fig. 35.12a is zero) will not split the light beam and fringes will not appear. When surfaces are not flat, fringes are curved (Fig. 35.12d). The interferometry method is also used for observing surface textures and scratches (Fig. 35.12e) through microscopes for better visibility.

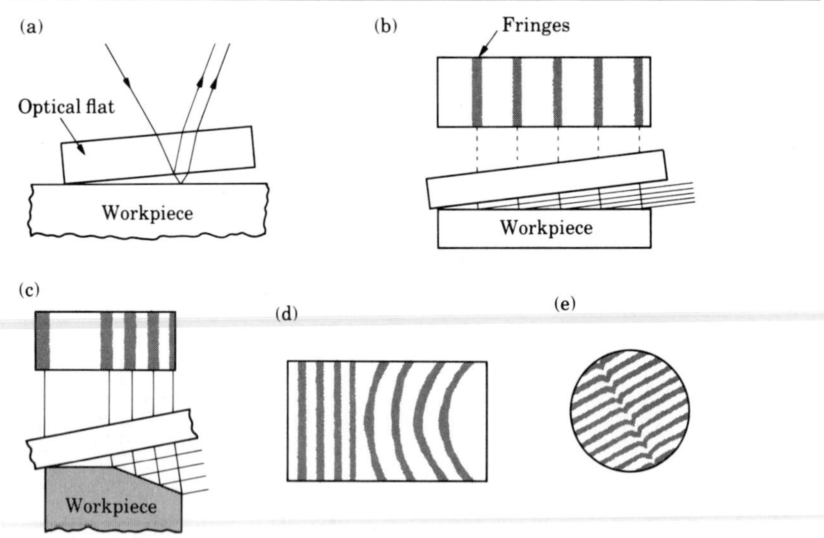

FIGURE 35.12
(a) Interferometry method for measuring flatness using an optical flat. (b) Fringes on a flat inclined surface. An optical flat resting on a perfectly flat workpiece surface will not split the light beam, and no fringes will be present. (c) Fringes on a surface with two inclinations. Note: the greater the incline, the closer the fringes. (d) Curved fringe patterns indicate curvatures on the workpiece surface. (e) Fringe pattern indicating a scratch on the surface.

35.5.3 Roundness

Roundness is usually described as deviations from true roundness (mathematically, a circle). The term *out of roundness* is actually more descriptive of the shape of the part (Fig. 35.13a). Roundness is very important to the proper functioning of rotating shafts, bearing races, pistons and cylinders, and steel balls in bearings.

The several methods of measuring roundness basically fall into two categories. In the first, the round part is placed on a V-block or between centers (Figs. 35.13b and c, respectively) and is rotated, with the point of a dial indicator in contact with the surface. After a full rotation of the workpiece, the difference between the maximum and minimum readings on the dial is noted. This difference is called the *total indicator reading* (TIR) or full indicator movement. This method is also used for measuring the straightness (squareness) of shaft end faces.

In the second method, called *circular tracing*, the part is placed on a platform, and its roundness is measured by rotating the platform (Fig. 35.13d). Conversely, the probe can be rotated around a stationary part to make the measurement.

35.5.4 Profile

Profile may be measured by several methods. In one method, a surface is compared with a template or profile gage to check shape conformity. Radii or fillets can be measured by this method (Fig. 35.14a). Profile may also be measured with a number

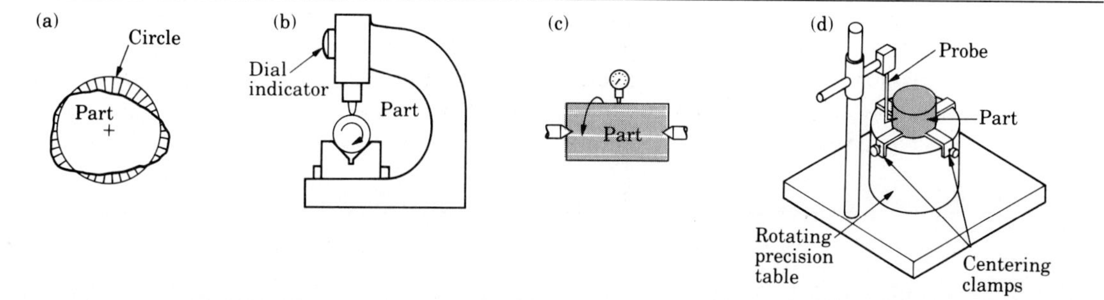

FIGURE 35.13
(a) Schematic illustration of out of roundness (exaggerated). Measuring roundness using (b) V-block and dial indicator, (c) part supported on centers and rotated, and (d) circular tracing, with part being rotated on a vertical axis. *Source:* After F. T. Farago.

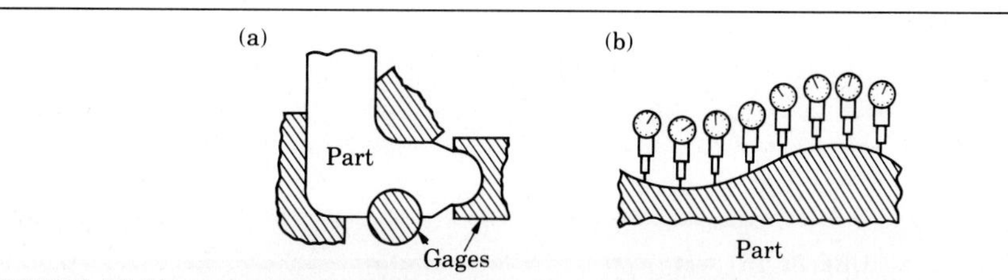

FIGURE 35.14
Measuring profiles with (a) radius gages and (b) dial indicators.

of dial indicators or similar instruments (Fig. 35.14b). Profile-tracing instruments are the latest development (Section 35.6).

35.5.5 Measuring screw threads and gear teeth

Threads and gear teeth have several features with specific dimensions and tolerances (see Figs. 22.17 and 23.35). These dimensions must be produced accurately for smooth operation of gears, reducing wear and noise level, and part interchangeability. These features are measured basically by means of thread gages of various designs that compare the thread produced against a standard thread. Some of the gages used are threaded plug gages, screw-pitch gages (similar to radius gages; see Fig. 35.14a), micrometers with cone-shaped points, and snap gages (Section 35.7) with anvils in the shape of threads.

Gear teeth are measured with instruments that are similar to dial indicators, with calipers (Fig. 35.15a), and with micrometers using pins or balls of various diameters (Fig. 35.15b). Special profile-measuring equipment is also available, including optical projectors.

35.5.6 Optical contour projectors

Optical contour projectors, also called *optical comparators*, were first developed in the 1940s to check the geometry of cutting tools for machining screw threads but are now used for checking all profiles (Fig. 35.16). The part is mounted on a table, or between centers, and the image is projected on a screen at magnifications up to 100 × or higher. Linear and angular measurements are made directly on the screen, which is equipped with reference lines and circles. The screen can be rotated to allow angular measurements as small as 1 min, using verniers such as that shown in Fig. 35.6(b).

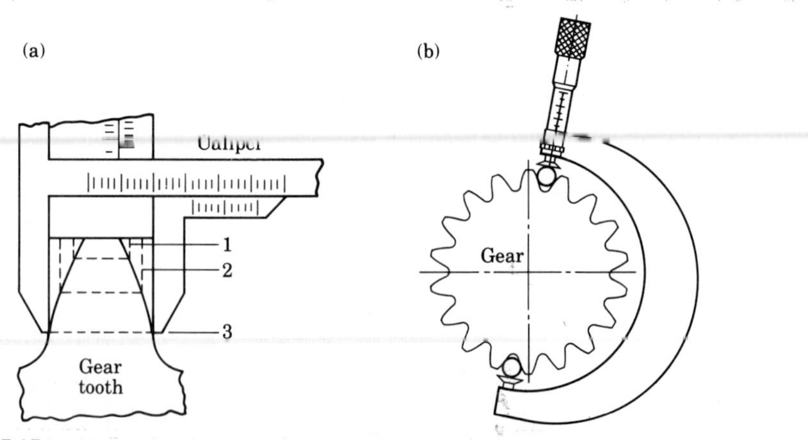

FIGURE 35.15
Measuring gear-tooth thickness and profile with (a) a gear-tooth caliper and (b) pins or balls and a micrometer. *Source:* American Gear Manufacturers Association.

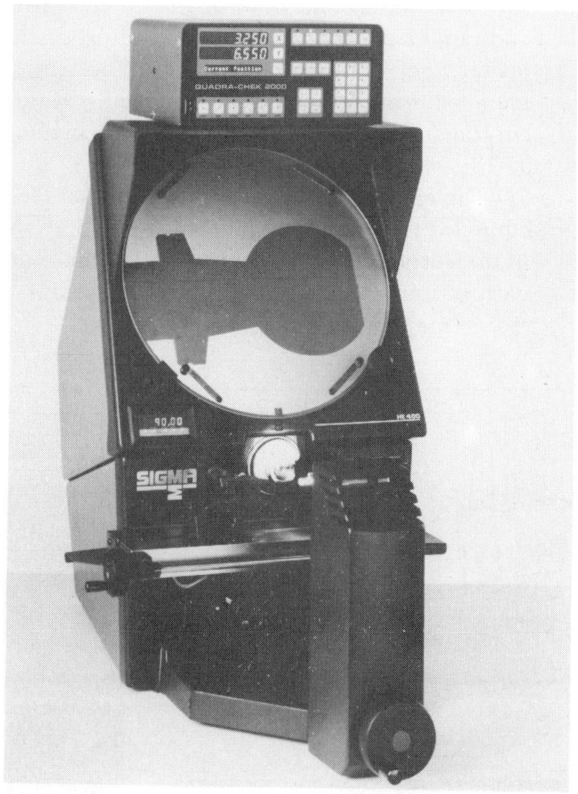

FIGURE 35.16
A bench model horizontal-beam contour projector with a 16 in.-diameter screen with 150-W tungsten halogen illumination. *Source*: Courtesy of L. S. Starrett Company, Precision Optical Division.

35.6 Coordinate Measuring and Layout Machines

Coordinate measuring and layout machines are the latest development in measurement technology. Basically, they consist of a platform on which the workpiece being measured is placed and moved linearly or rotated. A stylus, attached to a head capable of lateral and vertical movements, records all measurements (Fig. 35.17). **Coordinate measuring machines** (CMM), also called *measuring machines*, are versatile in their capability to record measurements of complex profiles with high sensitivity (0.25 μm; 10 μin.) and speed.

These machines are built rigidly and are very precise. They are equipped with digital readout or can be linked to computers for on-line inspection of parts (Section

35.9). These machines can be placed close to machine tools for efficient inspection and rapid feedback for correction of processing parameters before the next part is made. They are also being made more rugged to resist environmental effects in manufacturing plants, such as temperature variations, vibration, and dirt. A coordinate measuring machine for inspection of an engine block is shown in Fig. 35.18.

Dimensions of large parts are measured by **layout machines**, which are equipped with digital readout. Figure 35.19 shows a layout machine in the process of measuring the dimensions of an automobile door. These machines are equipped with scribing tools for marking dimensions on large parts, with an accuracy of ± 0.04 mm (0.0016 in.).

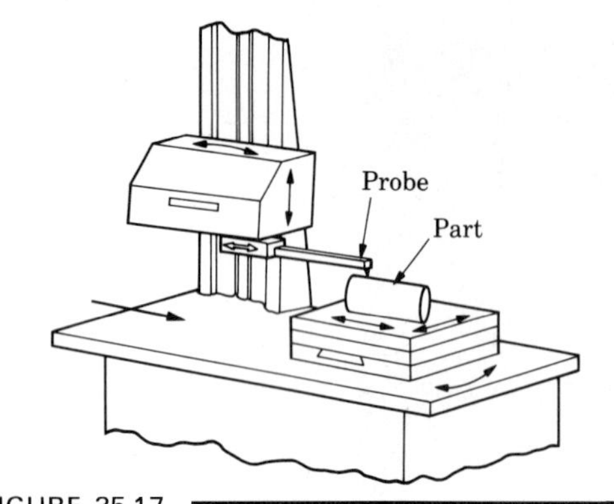

FIGURE 35.17 Schematic illustration of one type of coordinate measuring machine.

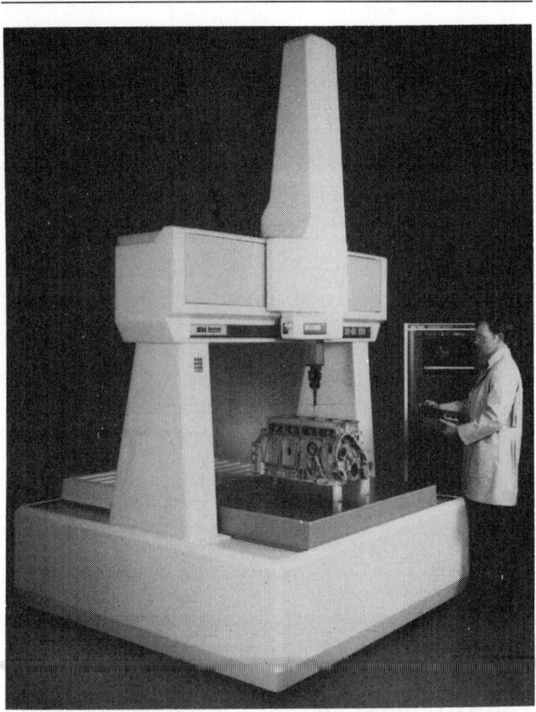

FIGURE 35.18 A coordinate measuring machine, measuring dimensions on an engine block. *Source:* Courtesy of Sheffield Measurement Division, Warner & Swasey Company.

FIGURE 35.19 A layout machine capable of 3-axis measurement. *Source:* Courtesy of Sheffield Measurement Division, Warner & Swasey Company.

• Example: Checking dimensions with CMM.

The accompanying illustration shows a process sheet, indicating the dimensions to be checked on a bearing seal for an aircraft jet engine. The function of the seal is to keep the lubricant in while keeping contaminants out of the bearing area. All the dimensions shown in the figure can be checked in about 2.5 minutes with a coordinate measuring machine. All measurements can be graphically represented. (Figure reprinted with permission from *Mechanical Engineering*, Dec. 1989, p. 55.)

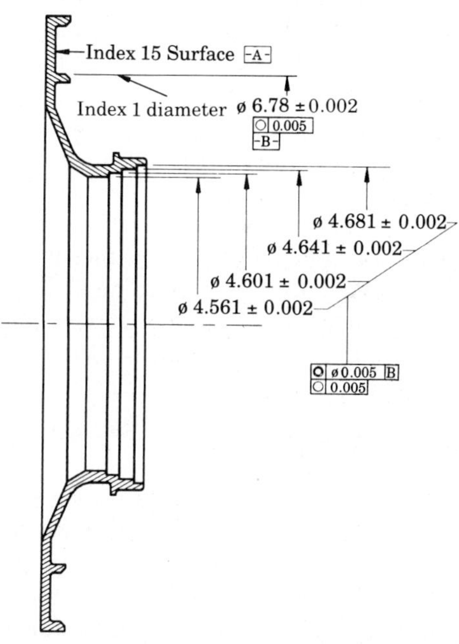

35.7 Gages

Thus far we have used the word *gage* to describe some types of measuring instruments, such as a caliper gage, depth gage, telescoping gage, electronic gage, strain gage, and radius gage. The reason is that the words instrument and gage (also spelled gauge) have traditionally been used interchangeably. However, gage has a variety of meanings, such as pressure gage, gage length of a tension-test specimen, and gages for sheet metal, wire, railroad rail, and the bore of shotguns. In this section we describe several common gages that have simple solid shapes and cannot be classified as instruments.

35.7.1 Gage blocks

Gage blocks are individual square, rectangular, or round metal blocks of various sizes (see the gage blocks in Fig. 35.7), made very precisely from heat-treated and stress-relieved alloy steels or from carbides. Their surfaces are lapped and are flat and parallel within a range of 0.02–0.12 μm (1–5 μin.). Zirconia ceramic gage blocks are also being made.

First developed by C. E. Johansson in the early 1900s, gage blocks are available in sets of various sizes, some sets containing almost a hundred gage blocks. The blocks can be assembled in many combinations to obtain desired lengths. Dimensional accuracy can be as high as 0.05 μm (2 μin.). Environmental temperature control is important in using gages for high-precision measurements.

The individual gage blocks are assembled by *wringing*, which is a sliding and twisting motion. The adsorbed films of moisture and oil between the gage blocks develop negative pressure at the interface, thus allowing the blocks to adhere to each other under external (atmospheric) pressure. This phenomenon is similar to the tendency for papers to stick together in a humid environment.

Although their use requires some skill, gage-block assemblies are commonly utilized in industry as an accurate reference length. Angle blocks are made similarly and are available for angular gaging. Worn or damaged gage blocks should not be used when highly accurate measurements are required, although they may still be used when less accuracy is needed. The four basic grades of gage blocks, in decreasing order of accuracy, are:

a) Grade 0.5 (formerly AAA)—reference gages, for very high precision work.
b) Grade 1 (AA)—laboratory grade, for calibration of instruments and other gages.
c) Grade 2 (A+)—precision grade, for toolrooms and inspection.
d) Grade 3 (A and B)—working grade, for use in production.

35.7.2 Fixed gages

Fixed gages are replicas of the shapes of the parts to be measured. **Plug gages** are commonly used for holes (Figs. 35.20a and b). The *GO gage* is smaller than the *NOT GO* (or *NO GO*) *gage* and slides into any hole whose smallest dimension is less than the diameter of the gage. The *NOT GO* gage must not go into the hole. Two gages are required for such measurements, although both may be on the same device, either at opposite ends or in two steps at one end (step-type gage). Plug gages are also available for measuring internal tapers (in which deviations between the gage and the part are indicated by the looseness of the gage), splines, and threads (in which the *GO* gage must screw into the threaded hole).

Ring gages (Fig. 35.20c) are used to measure shafts and similar round parts. Ring thread gages are used to measure external threads. The *GO* and *NOT GO* features on these gages are identified by the type of knurling on the outside diameters of the rings, as shown in the figure.

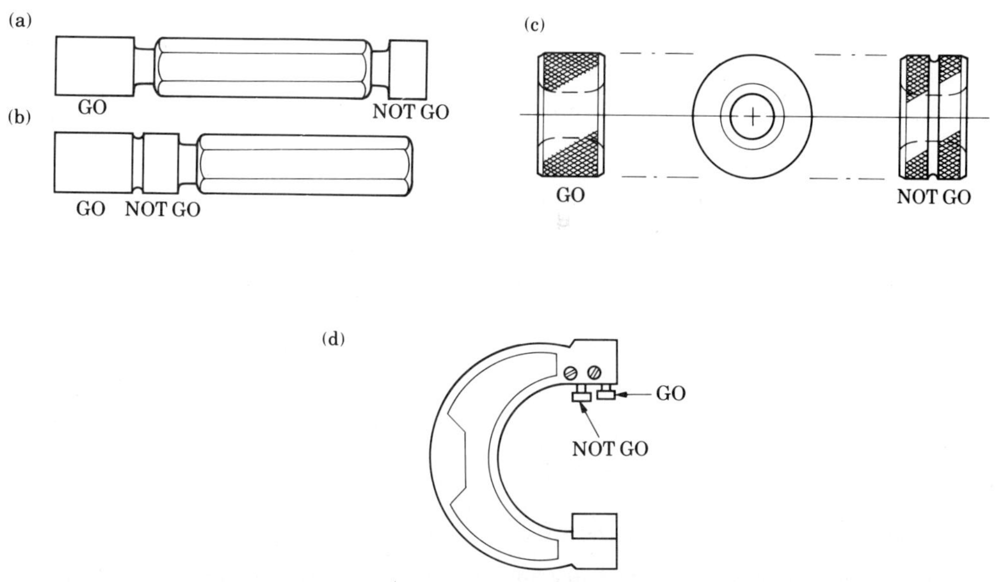

FIGURE 35.20
(a) Plug gage for holes, with GO–NOT GO on opposite ends. (b) Plug gage with GO–NOT GO on one end. (c) Plain ring gages for gaging round rods. Note the difference in knurled surfaces to identify the two gages. (d) Snap gage with adjustable anvils.

Snap gages (Fig. 35.20d) are commonly used to measure external dimensions. They are made with adjustable gaging surfaces for use with parts having different dimensions. One of the gaging surfaces may be set at a different gap from the other, thus making a one-unit *GO–NOT GO* gage.

Although fixed gages are easy to use and inexpensive, they only indicate whether a part is too small or too large, compared to an established standard. They do not measure actual dimensions.

35.7.3 Pneumatic gages

Although there are several types of **pneumatic gages**, also called *air gages*, their basic operation is shown in Fig. 35.21. The gage head has holes through which pressurized air, supplied by a constant-pressure line, escapes. The smaller the gap between the gage and the hole, the more difficult it is for the air to escape, and hence the back pressure is higher. The back pressure, sensed and indicated by a pressure gage, is calibrated to read dimensional variations of holes. You can observe the principle of air gages by blowing air through a soda straw while holding it at different distances from a perpendicular surface.

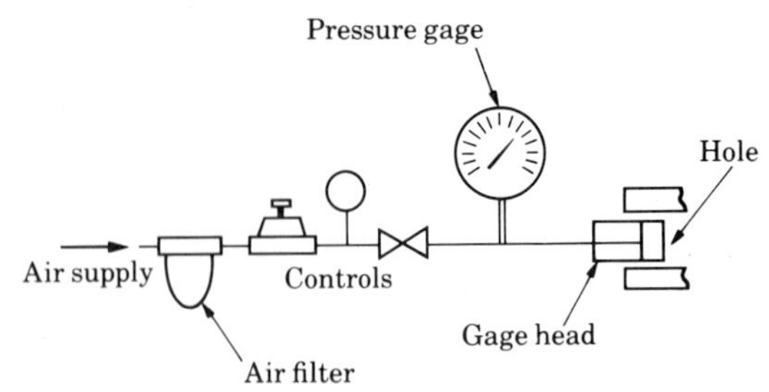

FIGURE 35.21
Schematic illustration of one type of pneumatic gage.

35.8 Microscopes

Microscopes are optical instruments used to view and measure very fine details, shapes, and dimensions on small and medium-sized tools, dies, and workpieces. The most common and versatile microscope used in tool rooms is the *toolmaker's microscope.* It is equipped with a stage that is movable in two principal directions and can be read to 2.5 μm (0.0001 in.). Several models of microscopes are available with various features for specialized inspection, including models with digital readout.

Light section microscope. The *light section microscope* is used to measure small surface details, such as scratches, and the thickness of deposited films and coatings. A thin light band is applied obliquely to the surface and the reflection is viewed at 90°, showing surface roughness, contours, and other features.

Scanning electron microscope. Unlike ordinary optical microscopes, the *scanning electron microscope* (SEM) has excellent depth of field. As a result, all regions of a complex part are in focus and can be viewed in and photographed to show extremely fine detail. This type of microscope is particularly useful for studying surface textures and fracture patterns. Although expensive ($50,000 and higher), such microscopes are capable of magnifications greater than 100,000 ×.

35.9 Automated Measurement

With increasing automation in all aspects of manufacturing processes and operations, the need for automated measurement (also called *automated inspection*; Section 36.4) has become much more apparent. Flexible manufacturing systems and manufacturing cells (Chapter 39) have led to the adoption of advanced measurement techniques and systems. In fact, installation and utilization of these systems is now a necessary—not an optional—manufacturing technology.

Traditionally, a batch of parts was manufactured and sent for measurement in a separate quality-control room, and if they passed measurement inspection, they were put into inventory. Automated inspection, however, is based on various on-line sensor systems that monitor the dimensions of parts being made and use these measurements to correct the process (see Section 38.8).

To appreciate the importance of on-line monitoring of dimensions, let's find the answer to the following question: If a machine has been producing a certain part with acceptable dimensions, what factors contribute to subsequent deviation in the dimensions of the same part produced by the same machine? The major factors are:

- Static and dynamic deflections of the machine because of vibrations and fluctuating forces, caused by variations such as in the properties and dimensions of the incoming material.
- Deformation of the machine because of thermal effects. These effects include changes in temperatures of the environment, metalworking fluids, and machine bearings and components.
- Wear of tools and dies, which, in turn, affects the dimensions of the parts produced.

As a result of these factors, the dimensions of parts produced will vary, necessitating monitoring of dimensions during production. In-process workpiece control is accomplished by special gaging and is used in a variety of applications, such as high-quantity machining and grinding.

35.10 General Characteristics and Selection of Measuring Instruments

The characteristics and quality of measuring instruments are generally described by certain specific terms. These terms, in alphabetical order, are defined as follows:

a) *Accuracy*. The degree of agreement of the measured dimension with its true magnitude.
b) *Amplification*. See Magnification.
c) *Calibration*. Adjusting or setting an instrument to give readings that are accurate within a reference standard.
d) *Drift*. See Stability.
e) *Linearity*. The accuracy of the readings of an instrument over its full working range.

f) *Magnification*. The ratio of instrument output to the input dimension.
g) *Precision*. Degree to which an instrument gives repeated measurement of the same standard.
h) *Repeat accuracy*. Same as accuracy, but repeated many times.
i) *Resolution*. Smallest dimension that can be read on an instrument.
j) *Rule of 10*. An instrument or gage should be 10 times more accurate than the dimensional tolerances of the part being measured.
k) *Sensitivity*. Smallest difference in dimension that an instrument can distinguish or detect.
l) *Speed of response*. How rapidly an instrument indicates the measurement, particularly when a number of parts are measured in rapid succession.
m) *Stability*. An instrument's capability to maintain its calibration over a period of time (also called *drift*).

Selection of an appropriate measuring instrument for a particular application depends on the foregoing factors. In addition, the size and type of parts to be measured, the environment (temperature, humidity, dust, and so on), operator skills required, and costs have to be considered in the purchase of such equipment.

35.11 Dimensional Tolerances

Individually manufactured parts and components are eventually assembled into products. We take it for granted that when, for example, a thousand lawnmowers are manufactured and assembled, each part of this product will mate properly with another component. For example, the wheels of the lawnmower will slip easily into their axles, or the pistons will fit properly into the cylinders, being neither too tight nor too loose.

Likewise, when we replace a broken or worn bolt on an old machine, all we have to do is to purchase an identical bolt. We are confident that, from similar experiences in the past, the new bolt will fit properly in the machine. The reason why we feel confident is that the bolt is manufactured according to certain standards and that the dimensions of all similar bolts vary by only a specified small amount. In other words, the bolts are manufactured within a certain range of dimensional **tolerance**. Thus all similar bolts are interchangeable.

We also expect that the new bolt, unless abused or misused, will function satisfactorily for a period of time. The reason is that bolts are tested periodically during their production to make sure that their quality is within specified ranges (see Chapter 36).

Dimensional tolerance is defined as the permissible or acceptable variation in the dimensions (height, width, depth, diameter, angles) of a part. The root of the word tolerance is the Latin *tolerare*, meaning to endure or put up with. Tolerances are unavoidable because it is virtually impossible (and unnecessary) to manufacture two parts that have precisely the same dimensions. Furthermore, because close tolerances substantially increase the product cost, a narrow tolerance range is undesirable economically.

35.11.1 Importance of tolerance control

Tolerances become important only when a part is to be assembled or mated with another part. Surfaces that are free and not functional do not need close tolerance control. Thus, for example, the accuracies of the holes and the distance between the holes for a connecting rod are far more critical than the rod's width and thickness at various locations along its length. By reviewing the figures throughout this text, you can determine which dimensions and features of the parts illustrated are more critical than others.

To illustrate the importance of dimensional tolerances, let's assemble a simple shaft (axle) and a wheel with a hole, assuming that we want the axle's diameter to be 1 in. (Fig. 35.22). We go to the hardware store and purchase a 1-in. round rod and a wheel with a 1-in. hole. Will the rod fit into the hole without forcing it, or will it be loose in the hole?

The 1-in. dimension is the *nominal* size of the shaft. If we purchase such a rod from different stores or at different times—or select one randomly from a lot of, say, 50 shafts—the chances are that each rod will have a slightly different diameter. Machines may, with the same setup, produce rods of slightly different diameters, depending on a number of factors, such as speed of operation, temperature, lubrication, variations in the incoming material, and similar variables.

If we now specify a range of diameters for both the rod and the hole of the wheel, we can predict correctly the type of fit that we will have after assembly.

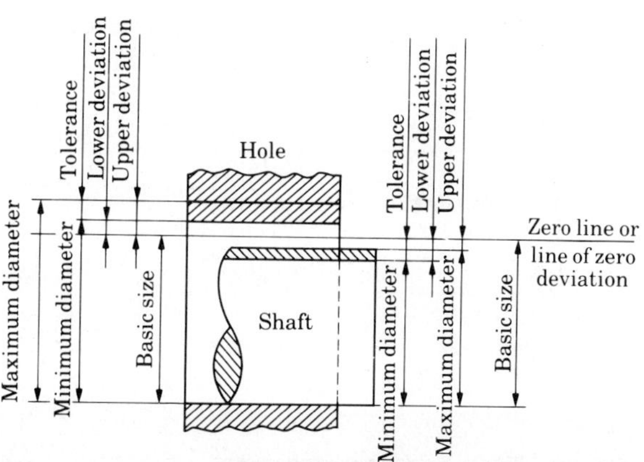

FIGURE 35.22
Basic size, deviation, and tolerance on a shaft, according to the ISO system.

Certain terminology has been established to clearly define these geometric quantities, such as the ISO system shown in Fig. 35.22. Note that both the shaft and the hole have minimum and maximum diameters, respectively, the difference being the tolerance for each member. A proper engineering drawing should specify these parameters with numerical values, as shown in Fig. 35.23.

The range of tolerances obtained in machining processes is given in Figs. 22.15 and 26.4 and Summary Table 23.1. See also various sections for tolerances obtained in other processes.

There is a general relationship between tolerances and surface finish of parts manufactured by various processes (Fig. 35.24). Note the wide range of tolerances and surface finishes obtained. Also, the larger the part, the greater the obtainable tolerance range becomes (see Fig. 22.15).

Experience has shown that dimensional inaccuracies of manufactured parts are approximately proportional to the cube root of the size of the part. Thus doubling the size of a part increases the inaccuracies by $\sqrt[3]{2} = 1.26$ times, or 26 percent.

35.11.2 Definitions

Several terms are used to describe features of dimensional relationships between mating parts. These terms, in alphabetical order, are defined as follows:

a) *Allowance.* The specified difference in dimensions between mating parts; also called *functional dimension* or *sum dimension.*
b) *Basic size.* Dimension from which limits of size are derived, using tolerances and allowances.
c) *Bilateral tolerance.* Deviation—plus or minus—from the basic size.
d) *Clearance.* The space between mating parts.
e) *Clearance fit.* Fit that allows for rotation or sliding between mating parts.
f) *Datum.* A theoretically exact axis, point, line, or plane.
g) *Feature.* Physically identifiable portion of a part, such as hole, slot, pin, or chamfer.
h) *Fit.* The amount of clearance or interference between mating parts.

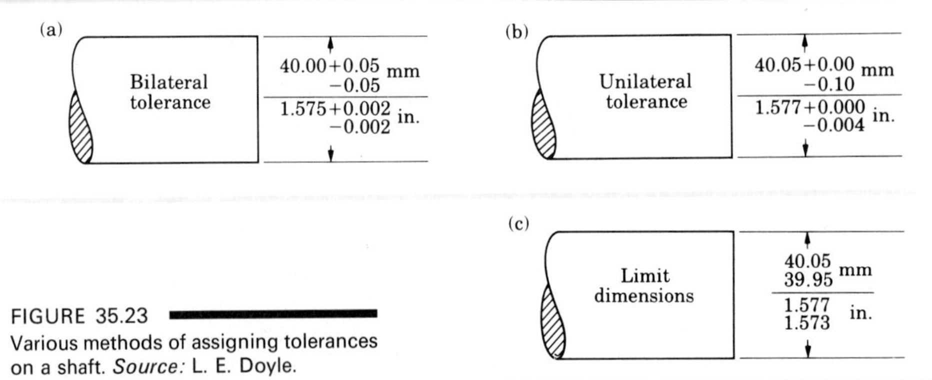

FIGURE 35.23 Various methods of assigning tolerances on a shaft. *Source:* L. E. Doyle.

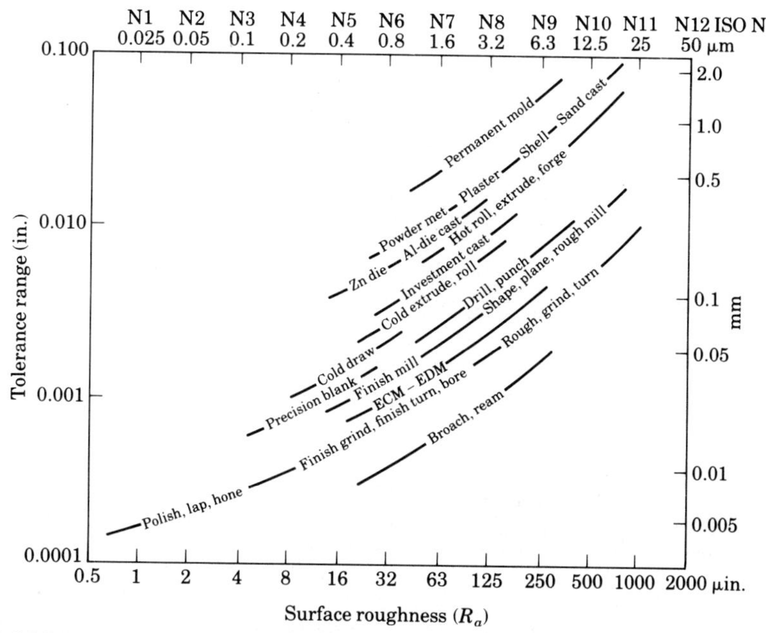

FIGURE 35.24
Tolerances and surface roughness obtained in various manufacturing processes. These tolerances apply to a 25-mm (1-in.) workpiece dimension. *Source:* J. A. Schey.

i) *Geometric tolerancing.* Tolerances that involve shape features of the part.
j) *Hole-basis system.* Tolerances based on a zero line on the hole; also called *standard hole practice* or *basic hole system.*
k) *Interference.* Negative clearance.
l) *Interference fit.* Fit that allows for alignment and stiffness of the mating parts.
m) *International tolerance grade (IT).* A group of tolerances that vary depending on the basic size, but provide the same relative level of accuracy within a grade.
n) *Limit dimensions.* The maximum and minimum dimensions of a part; also called *limits.*
o) *Maximum material condition.* A part whose dimensions are at the maximum limit dimensions.
p) *Nominal size.* Dimension that is used for the purpose of general identification.
q) *Positional tolerancing.* A system of specifying the true position, size, and form of the features of a part, including allowable variations.
r) *Shaft-basis system.* Tolerances based on a zero line on the shaft; also called *standard shaft practice* or *basic shaft system.*
s) *Standard size.* Nominal size in integers and common subdivisions of length.

(a)

Type of feature	Type of tolerance	Characteristic	Symbol
Individual (no datum reference)	Form	Flatness	▱
		Straightness	—
		Circularity (roundness)	○
		Cylindricity	⌭
Individual or related	Profile	Profile of a line	⌒
		Profile of a surface	⌓
Related (datum reference required)	Orientation	Perpendicularity	⊥
		Angularity	∠
		Parallelism	//
	Location	Position	⌖
		Concentricity	◎
	Runout	Circular runout	↗
		Total runout	⌰

(b)

[.605]
Basic, or exact, dimension

[-A-]
Datum feature symbol

Ⓜ
Maximum material condition

Ⓢ
Regardless of feature size

Ⓛ
Least material condition

Ⓟ
Projected tolerance zone

∅
Diametrical (cylindrical) tolerance zone or feature

[⌖|∅.005 Ⓜ|A]
Feature control frame

(A1)
Datum target symbol

FIGURE 35.25
Geometric characteristic symbols to be indicated on engineering drawings of parts to be manufactured. *Source:* The American Society of Mechanical Engineers.

t) *Transition fit.* Fit with small clearance or interference that allows for accurate location of mating parts.
u) *Unilateral tolerancing.* Deviation in one direction only from the nominal dimension.
v) *Zero line.* Reference line along the basic size from which a range of tolerances and deviations are specified.

Because the dimensions of holes are more difficult to control than those of shafts, the hole-basis system is commonly used for specifying tolerances in shaft and hole assemblies. The symbols commonly used to indicate geometric characteristics are shown in Figs. 35.25(a) and (b).

35.11.3 Limits and fits

Limits and fits are important in specifying dimensions for holes and shafts. There are two standards on limits and fits, as described by the American National Standards Institute (see ANSI B4.1, B4.2, and B4.3). One standard is based on the older inch unit; the other is based on the metric unit and has been developed in greater detail. In these standards, capital letters always refer to the hole and lowercase letters to the shaft.

In the *inch* units, **fits** are divided into the following general classifications; each of these classes has limits of clearances with the hole size as the basic size.

a) *Running and sliding fits*, which are subdivided into the following classes:

   Class RC1. Close-sliding fits, for accurate location of parts to be assembled without perceptible play.

   Class RC2. Sliding fits, for parts that turn and move easily.

   Class RC3. Precision-running fits, for precision work at low speeds and light pressures.

   Class RC4. Close-running fits, for accurate machinery with moderate speeds and pressures.

   Class RC5 and RC6. Medium-running fits, for higher speeds and high pressures.

   Class RC7. Free-running fits, where accuracy is not important.

   Class RC8 and RC9. Loose-running fits.

b) *Locational clearance fits*, for stationary parts to be freely assembled and disassembled.
c) *Locational transition fits*, where accuracy of location is important.
d) *Locational interference fits*, where accuracy of location is very important and for parts requiring rigidity and proper alignment.
e) *Force and shrink fits*, which are subdivided into the following classes:

   Class FN1. Light-drive fits, for assembly requiring light pressures.

   Class FN2. Medium-drive fits, for ordinary parts and shrink fits on light sections.

   Class FN3. Heavy-drive fits, for heavier parts and shrink fits on medium sections.

   Class FN4 and FN5. Force fits, for parts that are to be highly stressed.

In the *metric* system, fits are classified in a similar manner by the International Standards Organization (ISO) and are outlined below.

	ISO symbol	
	Hole basis	Shaft basis
a) *Clearance fits*		
Loose running	H11/c11	C11/h11
Free running	H9/d9	D9/h9
Close running	H8/f7	F8/h7
Sliding	H7/g6	G7/h6
Locational clearance	H7/h6	H7/h6
b) *Transition fits*		
Locational transition (accurate)	H7/k6	K7/h6
Locational transition (more accurate)	H7/n6	N7/h6
c) *Interference fits*		
Locational interference	H7/p6	P7/h6
Medium drive	H7/s6	S7/h6
Force fits	H7/u6	U7/h6

SUMMARY

In modern manufacturing technology, many parts are made with a high degree of precision, thus requiring measuring instrumentation with several features and characteristics. A variety of devices are available, from simple gage blocks to electronic gages with high sensitivity. The selection of a particular measuring instrument depends on factors such as the type of measurement, the environment, and the accuracy of measurement. Gages must be checked against a reliable standard. Humidity, heat, cold, vibration, and dirt can have adverse effects on the accuracy and reliability of measurements. Operator skill required and cost of instrumentation are also important considerations. Great advances have been made in automated measurement, linking them to microprocessors and computers for accurate in-process control of manufacturing operations. Reliable linking, monitoring, display, distribution, and manipulation of data obtained are important factors, as are the significant costs involved in doing so.

Dimensional tolerances and their selection are important factors in manufacturing. Tolerances not only affect the accuracy and operation of all types of machinery and equipment, but can also significantly influence product cost. The smaller the range of tolerances specified, known as tight tolerances, the greater becomes the cost of achieving it. Consequently, tolerances should be as broad as possible, while maintaining the operational requirements of the product.

SUMMARY TABLE 35.1

TYPES OF MEASUREMENT AND INSTRUMENTS USED

MEASUREMENT	INSTRUMENT	SENSITIVITY μm	SENSITIVITY μin.
Linear	Steel rule	0.5 mm	$\frac{1}{64}$ in.
	Vernier caliper	25	1000
	Micrometer, with vernier	2.5	100
	Diffraction grating	2.5	100
Angle	Bevel protractor, with vernier	5 min	
	Sine bar		
Comparative length	Dial indicator	1	40
	Electronic gage	1	40
	Gage blocks	0.05	2
Straightness	Autocollimator		
	Transit		
	Laser beam	2.5	100
Flatness	Interferometry	0.03	1
Roundness	Dial indicator		
	Circular tracing	1	40

(continued)

SUMMARY TABLE 35.1 (*continued*)

TYPES OF MEASUREMENT AND INSTRUMENTS USED

MEASUREMENT	INSTRUMENT	SENSITIVITY	
		μm	μin.
Profile	Radius or fillet gage		
	Dial indicator	1	40
	Optical comparator	125	5000
	Coordinate measuring machines	0.25	10
GO–NOT GO	Plug gage		
	Ring gage		
	Snap gage		
Microscopes	Toolmaker's	2.5	100
	Light Section		
	Scanning electron		

TRENDS

- Studies are continually being made in the accuracy, reliability, and speed of measuring instruments and coordinate measuring machines, used either individually or as elements in computer-integrated manufacturing systems.
- Automated measurement and inspection will continue to be an essential part of all manufacturing operations.
- Tolerances and their control during manufacturing processes continue to be an important activity in product quality and for reliable use and operation of products.

KEY TERMS

Autocollimator
Bevel protractor
Coordinate measuring machines
Dial indicator
Diffraction grating
Electronic gage
Engineering metrology
Fits
Gage block
Layout machine
Micrometer
Optical contour projector
Optical flat
Plug gage
Pneumatic gage
Precision
Ring gage
Sensitivity
Sine bar
Snap gage
Tolerance
Total indicator reading
Vernier caliper

BIBLIOGRAPHY

Bentley, J.P., *Principles of Measurement Systems*, 2d ed. New York: Wiley, 1988.

Farago, F.T., *Handbook of Dimensional Measurement*, 2d ed. New York: Industrial Press, 1982.

Kennedy, C.W., and E.G. Hoffman, *Inspection and Gaging*, 6th ed. New York: Industrial Press, 1987.

Lange, J.C., *Design Dimensioning with Computer Graphics Applications*. New York: Marcel Dekker, 1984.

Lenk, J.D., *Handbook of Controls and Instrumentation*. Englewood Cliffs, N.J.: Prentice-Hall, 1980.

Lowell, W.F., *Modern Geometric Dimensioning and Tolerancing*, 2d ed. Fort Washington, Md.: National Tooling and Machining Association, 1982.

Machinery's Handbook. New York: Industrial Press (revised periodically).

Murphy, S.D. (ed.), *In-Process Measurement and Control*. New York: Marcel Dekker, 1990.

Puncochar, D.E., *Interpretation of Geometric Dimensioning and Tolerancing*. New York: Industrial Press, 1990.

Spotts, M.F., *Dimensioning and Tolerancing for Quantity Production*. Englewood Cliffs, N.J.: Prentice-Hall, 1983.

Tool and Manufacturing Engineers Handbook, 4th ed., *Vol. 4: Quality Control and Assembly*. Dearborn, Mich.: Society of Manufacturing Engineers, 1987.

Warnecke, H.J., and W. Dutschke (eds.), *Metrology in Manufacturing Technology*. Berlin: Springer, 1984.

REVIEW QUESTIONS

35.1 Define engineering metrology.
35.2 Explain what is meant by standards for measurement.
35.3 How is the meter currently defined?
35.4 Why is it important to control temperature during measurement of dimensions?
35.5 Explain the difference between direct- and indirect-reading linear measurements. Name the instruments used in each category.
35.6 Describe the principle of a vernier.
35.7 Explain how diffraction grating works.
35.8 How does a telescoping gage work?
35.9 Describe how a sine bar is used to measure angles.
35.10 Explain what is meant by comparative length measurement.
35.11 Explain the principle of a mechanical dial indicator.
35.12 Describe the attributes of electronic gages.
35.13 Explain how flatness is measured. What is an optical flat?
35.14 Describe the principle of an optical comparator.
35.15 Why have coordinate measuring machines become important instruments?
35.16 Explain what wringing of gage blocks is.
35.17 Why are there different grades of gage blocks?
35.18 What is the difference between a plug gage and a ring gage?

35.19 How are snap gages used? Name some applications.
35.20 Explain the principle of an air gage. What advantages do they have over other types of gages?
35.21 Describe what is meant by automated inspection.
35.22 List and explain the general characteristics of measuring instruments.
35.23 What are dimensional tolerances? Why is their control important?
35.24 Explain the difference between tolerance and allowance.
35.25 Explain what is meant by fit of mating parts.

QUESTIONS AND PROBLEMS

35.26 Why are the words accuracy and precision so often incorrectly interchanged?
35.27 Assume that a steel rule expands by 2 percent because of an increase in environmental temperature. What will be the indicated diameter of a shaft whose diameter at room temperature was 2.000 in.?
35.28 Explain why an instrument may not have sufficient precision.
35.29 Explain how the presence of moisture and oil between gage blocks develops negative pressure.
35.30 Why do manufacturing processes produce parts with such a wide range of tolerances?
35.31 Explain the need for automated inspection.
35.32 Tolerances for nonmetallic stock are usually wider. Explain why.
35.33 Sketch a vernier similar to the one shown in Fig. 35.4(b) to read 0.106 in. for the top illustration and 0.3997 in. for the bottom illustration.
35.34 Sketch a vernier similar to the one shown in Fig. 35.6(b) to read (a) 17°24′ and (b) 1°56′.
35.35 Calculate the included angle of the part being measured in Fig. 35.7 if the height of the gage blocks is 4.1400 in. and the distance between the centers of the round bars under the sine bar is 5.00 in.
35.36 Would it be desirable to integrate the instruments and machines described in this chapter into the machine tools described in Parts III and IV? How would you go about doing so, particularly considering the factory environment in which they are to be used? Make some preliminary sketches of such machines.
35.37 Comment on your observations regarding Fig. 35.24. Why does tolerance increase with increasing surface roughness?
35.38 Can the gages shown in Fig. 35.20 be automated to be used in a high production facility? Give examples.
35.39 We stated in Section 35.7.1 that zirconia ceramic gage blocks are now being made. What would be the advantages and limitations of such gages?
35.40 How would you go about specifying tolerances in the layout machine application shown in Fig. 35.19 concerning an automobile door? What would be the consequences of exceeding these limits?

36

Testing, Inspection, and Quality Assurance

36.1 Introduction

Throughout this text we have noted that a manufactured product develops certain external and internal characteristics, which result in part from the production processes used. External characteristics involve surface finish and surface integrity, such as surface damage from cutting tools or friction during processing of the workpiece in dies. Internal characteristics include various defects, such as porosity, impurities, inclusions, phase transformations, embrittlement, cracks, debonding of laminations, and harmful residual stresses. Some of these defects may exist in the original stock, or they may be induced or introduced during the manufacturing operation.

Before they are marketed, manufactured parts and products are inspected for several characteristics. This inspection routine is particularly important for products or components whose failure or malfunction has potentially serious implications, such as bodily injury or fatality. Typical examples are cables breaking, switches malfunctioning, brakes failing, grinding wheels breaking, railroad wheels fracturing, turbine blades failing, pressure vessels bursting, and weld or joints failing. In this chapter we identify and describe the various methods that are commonly used to inspect manufactured products.

Product quality has always been one of the most important elements in manufacturing operations, and with increasing domestic and international competition, it has become even more important. Prevention of defects in products and on-line inspection are now major goals in all manufacturing activities. We again emphasize that quality must be built into a product and not merely checked after the product has been made. Thus close cooperation and communication between design and manufacturing engineers are essential. Important advances in quality engineering and productivity have been made, largely because of the efforts of quality experts such as Deming and Taguchi, which we describe in this chapter. We also review various statistical methods used to monitor and minimize the defect rate.

36.2 Nondestructive Testing

Nondestructive testing (NDT) is carried out in such a way that product integrity and surface texture remain unchanged. These techniques generally require considerable operator skill. Interpreting test results accurately may be difficult because test results can be quite subjective. However, the use of computer graphics and other enhancement techniques have reduced the likelihood of human error in nondestructive testing. We describe below the basic principles of the more commonly used nondestructive testing techniques.

36.2.1 Liquid penetrants

In the *liquid-penetrants technique*, fluids are applied to the surfaces of the part and allowed to penetrate into surface openings, cracks, seams, and porosity (Fig. 36.1). The penetrant can seep into cracks as small as 0.1 μm (4 μin.) in width. Two common types of liquids are (1) fluorescent penetrants with various sensitivities, which fluoresce under ultraviolet light; and (2) visible penetrants, using dyes usually red in color, which appear as bright outlines on the surface.

The surface to be inspected is first thoroughly cleaned and dried. The liquid is brushed or sprayed on the surface to be inspected and allowed to remain long enough to seep into surface openings. Excess penetrant is then wiped off or washed away with water or solvent. A developing agent is then added to allow the penetrant to seep back to the surface and spread to the edges of openings, thus magnifying the size of defects. The surface is then inspected for defects, either visually in the case of dye penetrants or with fluorescent lighting.

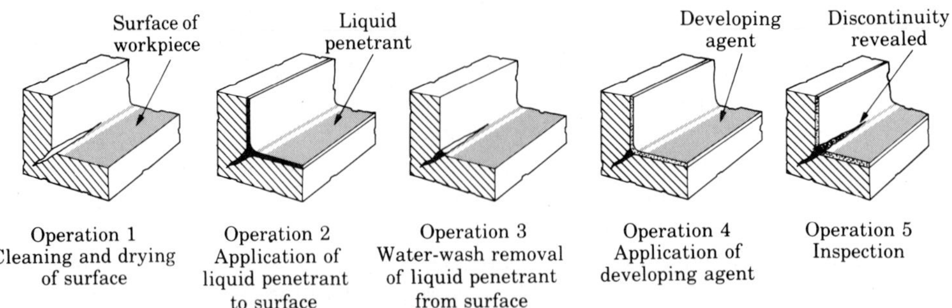

FIGURE 36.1
Sequence of operations for liquid-penetrant inspection to detect the presence of cracks and other flaws in a workpiece. *Source: Metals Handbook, Desk Edition.* Copyright © 1985, ASM International, Metals Park, Ohio. Used with permission.

This method is capable of detecting a variety of surface defects and is used extensively. The equipment is simple and easy to use, can be portable, and is less costly to operate than other methods. However, this method can only detect defects that are open to the surface, not internal defects.

36.2.2 Magnetic-particle inspection

The *magnetic-particle inspection technique* consists of placing fine ferromagnetic particles on the surface. The particles can be applied either dry or in a liquid carrier such as water or oil. When the part is magnetized with a magnetic field, a discontinuity (defect) on the surface causes the particles to gather visibly around it (Fig. 36.2). The collected particles generally take the shape and size of the defect. Subsurface defects can also be detected by this method, provided they are not deep. The ferromagnetic particles may be colored with pigments for better visibility on metal surfaces. Wet particles are used for detecting fine discontinuities, such as fatigue cracks.

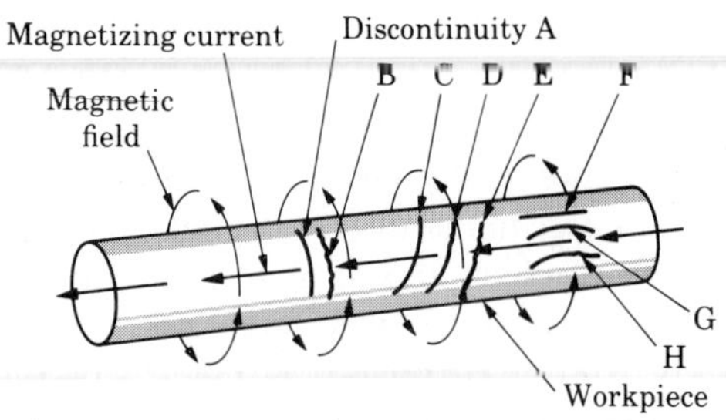

FIGURE 36.2
Schematic illustration of magnetic-particle inspection of a part with a defect in it. Cracks that are in a direction parallel to the magnetic field, such as in A, would not be detected, whereas the others shown would. Cracks F, G, and H are the easiest to detect. *Source: Metals Handbook, Desk Edition.* Copyright © 1985, ASM International, Metals Park, Ohio. Used with permission.

The magnetic fields can be generated either with direct current or alternating current, using yokes, bars, and coils. Subsurface defects can be detected best with direct current. The magnetic-particle method can be used on ferromagnetic materials only, and parts have to be demagnetized and cleaned. The equipment may be portable or stationary.

36.2.3 Ultrasonic inspection

In *ultrasonic inspection*, an ultrasonic beam travels through the part. An internal defect, such as a crack, interrupts the beam and reflects back a portion of the ultrasonic energy. The amplitude of the energy reflected and the time required for return indicates the presence and location of any flaws in the workpiece. The ultrasonic waves are generated by transducers, called search units or probes, of various types and shapes. They operate on the principle of piezoelectricity (see Section 3.7), using materials such as quartz, lithium sulfate, and various ceramics. Most inspections are carried out at a frequency range of 1–25 MHz. Couplants are used to transmit the ultrasonic waves from the transducer to the test piece. Typical couplants are water, oil, glycerin, and grease.

The ultrasonic inspection method has high penetrating power and sensitivity. It can be used to inspect flaws in large volumes of material, such as railroad wheels, pressure vessels, and die blocks, from various directions. Accuracy is higher than that of other nondestructive inspection methods. However, this method requires experienced personnel to carry out the inspection and interpret the results correctly.

36.2.4 Acoustic methods

The *acoustic-emission technique* detects signals (high-frequency stress waves) generated by the workpiece itself during plastic deformation, crack initiation and propagation, phase transformation, and sudden reorientation of grain boundaries. Bubble formation during boiling and friction and wear of sliding interfaces are other sources of acoustic signals.

Acoustic-emission inspection is typically performed by stressing elastically the part or structure, such as bending a beam, applying torque to a shaft, and pressurizing a vessel. Acoustic emissions are detected by sensors consisting of piezoelectric ceramic elements. This method is particularly effective for continuous surveillance of load-bearing structures.

The *acoustic-impact technique* consists of tapping the surface of an object and listening to and analyzing the signals to detect discontinuities and flaws. The principle is basically the same as tapping walls, desktops, or countertops in various locations with your fingers or a hammer and listening to the sound emitted. Vitrified grinding wheels are tested in a similar manner (ring test) to detect cracks in the wheel that may not be visible to the naked eye.

The acoustic-impact technique can be instrumented and automated and is easy to perform. However, the results depend on part geometry and mass, thus requiring a reference standard to identify flaws.

36.2.5 Radiography

Radiography involves x-ray inspection to detect internal flaws or density and thickness variations in the part. The radiation source is typically an x-ray tube, and a visible permanent image is made on an x-ray film or radiographic paper (Fig. 36.3a). Fluoroscopes are used to produce x-ray images very quickly—a real-time radiography technique that shows events as they are occurring. It does not require film handling and processing. Radiographic techniques require expensive equipment and proper interpretation of results and involve radiation hazard.

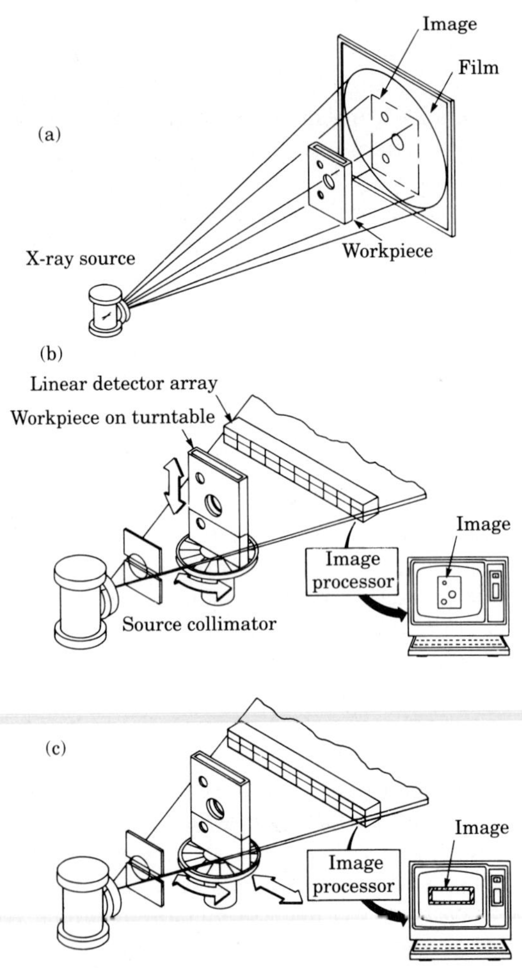

FIGURE 36.3
Three methods of radiographic inspection: (a) conventional radiography, (b) digital radiography, and (c) computed tomography. *Source:* Courtesy of *Advanced Materials and Processes*, p. 56, November 1990. ASM International.

In *digital radiography*, the film is replaced by a linear array of detectors (Fig. 36.3b). The x-ray beam is collimated into a fan beam (compare Figs. 36.3a and b), and the workpiece is moved vertically. The radiation is digitally sampled by the detectors and the data are stored in computer memory, which is then displayed on the CRT as a two-dimensional image of the workpiece.

In *computed tomography*, the same system as described above is used except that the workpiece is rotated along a vertical axis as well as being moved vertically (Fig. 36.3c), producing x-ray images of thin cross-sections of the workpiece. The translation and rotation of the workpiece provide several viewing angles of the object. From the data, the computer mathematically reconstructs and displays an image of the cross-section of the workpiece. The size, location, and distribution of flaws can thus be determined more reliably than is possible by ordinary radiography.

36.2.6 Eddy-current inspection

The *eddy-current inspection method* is based on the principle of electromagnetic induction. The part is placed in or adjacent to an electric coil through which alternating current (exciting current) flows at frequencies ranging from 60 Hz to 6 MHz. This current causes eddy currents to flow in the part. Defects in the part impede and change the direction of eddy currents (Fig. 36.4), causing changes in the electromagnetic field. These changes affect the exciting coil (inspection coil) whose voltage is monitored to determine the presence of flaws.

Inspection coils can be made in various sizes and shapes to suit the geometry of the part being inspected. Parts must be electrically conductive, and flaw depths detected are usually limited to 13 mm (0.5 in.). Moreover, the technique requires using a standard reference sample to set the sensitivity of the tester.

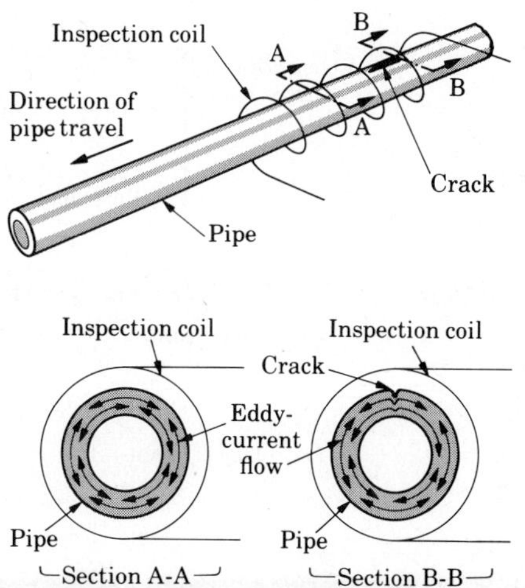

FIGURE 36.4

Changes in eddy-current flow in a workpiece caused by a defect. *Source: Metals Handbook, Desk Edition.* Copyright © 1985, ASM International, Metals Park, Ohio. Used with permission.

36.2.7 Thermal inspection

Thermal inspection involves observing temperature changes by contact- or noncontact-type heat-sensing devices. Defects in the workpiece, such as cracks, debonded regions in laminated structures, and poor joints, cause a change in temperature distribution. In *thermographic inspection*, materials such as heat-sensitive paints and papers, liquid crystals, and other coatings are applied to the surface. Any changes in their color or appearance indicate defects.

The most common method of noncontact thermographic inspection uses infrared detectors, such as infrared scanning microscopes and cameras, with high response time and sensitivities of 1 °C (2 °F). Thermometric inspection utilizes devices such as thermocouples, meltable materials such as waxlike crayons, radiometers, and pyrometers.

36.2.8 Holography

The *holography technique* creates a three-dimensional image of the part utilizing an optical system (Fig. 36.5). This technique is generally used on simple shapes and highly polished surfaces, and the image is recorded on a photographic film. Its use has been extended to inspection of parts (*holographic interferometry*) having various shapes and surface conditions. Using double- and multiple-exposure techniques while the part is being subjected to external forces or time-dependent variations, changes in the images reveal defects in the part.

In *acoustic holography*, information on internal defects is obtained directly from the image of the interior of the part. In *liquid-surface acoustical holography*,

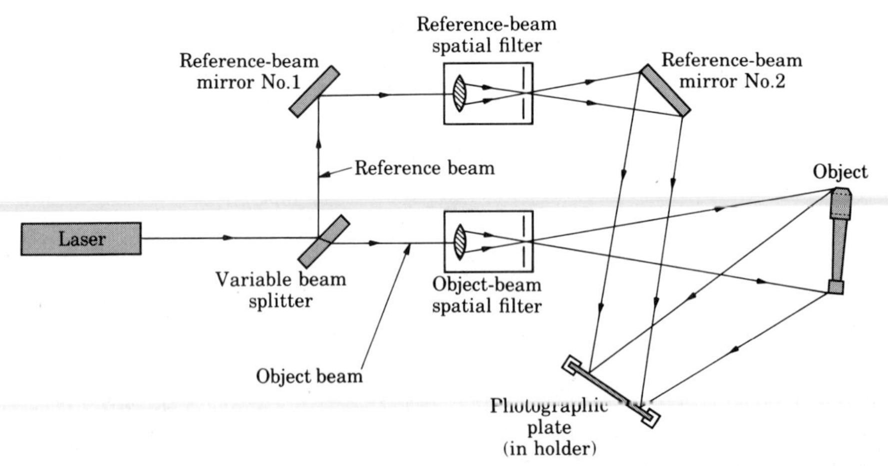

FIGURE 36.5 Schematic illustration of the basic optical system used in holography elements in radiography for detecting flaws in workpieces. *Source: Metals Handbook, Desk Edition.* Copyright © 1985, ASM International, Metals Park, Ohio. Used with permission.

the workpiece and two ultrasonic transducers (one for the object beam and the other for the reference beam) are immersed in a water-filled tank. The holographic image is then obtained from the ripples in the tank. In *scanning acoustical holography*, only one transducer is used and the hologram is produced by electronic-phase detection. This system is more sensitive, the equipment is usually portable, and very large workpieces can be accommodated by using a water column instead of a tank.

36.3 Destructive Testing

As the name suggests, the part or product tested using destructive testing methods no longer maintains its integrity, original shape, or surface texture. Mechanical test methods (see Chapter 2) are all destructive, in that a sample or specimen has to be removed from the product in order to test it (see, for example, Fig. 14.24). In addition to mechanical testing, other destructive tests include speed testing of grinding wheels to determine their bursting speed and high-pressure testing of pressure vessels to determine their bursting pressure. Destructive testing has several advantages and limitations, compared to nondestructive testing, which are outlined in Table 36.1.

Hardness tests leaving large impressions may be regarded as destructive testing. However, microhardness tests may be regarded as nondestructive because of the very small permanent indentations. This distinction is based on the assumption that the material is not notch sensitive. However, most glasses, highly heat-treated metals, and ceramics are notch sensitive; that is, the small indentation produced by the indenter may lower their strength and toughness.

36.4 Automated Inspection

Note that in all the preceding examples we discussed testing parts or products that had already been manufactured. Traditionally, individual parts and assemblies of parts have been manufactured in batches, sent to inspection in quality-control rooms, and if approved, put in inventory. If products do not pass the quality inspection, they are either scrapped or kept on the basis of a certain acceptable deviation from the standard. Obviously, such a system lacks flexibility, requires maintaining an inventory, and inevitably results in some defective parts going through the system. The traditional method actually counts the defects after they occur (*postprocess inspection*), and in no way attempts to prevent defects.

TABLE 36.1
ADVANTAGES AND LIMITATIONS OF NONDESTRUCTIVE AND DESTRUCTIVE TESTING

NONDESTRUCTIVE TESTING	DESTRUCTIVE TESTING
ADVANTAGES 1. Can be done directly on production items without regard to part cost or quantity available, and no scrap losses are incurred except for bad parts 2. Can be done on 100% of production or on representative samples 3. Can be used when variability is wide and unpredictable 4. Different tests can be applied to the same item simultaneously or sequentially 5. The same test can be repeated on the same item 6. May be performed on parts in service 7. Cumulative effect of service usage can be measured directly 8. May reveal failure mechanism 9. Little or no specimen preparation is required 10. Equipment is often portable for use in field 11. Labor costs are usually low, especially for repetitive testing of similar parts	*ADVANTAGES* 1. Can often directly and reliably measure response to service conditions 2. Measurements are quantitative, and usually valuable for design or standardization 3. Interpretation of results by a skilled technician usually not required 4. Correlation between tests and service usually direct, leaving little margin for disagreement among observers as to meaning and significance of test results
LIMITATIONS 1. Results often must be interpreted by a skilled, experienced technician 2. In absence of proven correlation, different observers may disagree on meaning and significance of test results 3. Properties are measured indirectly, and often only qualitative or comparative measurements can be made 4. Some nondestructive tests require large capital investments	*LIMITATIONS* 1. Can be applied only to a sample, and separate proof that the sample represents the population is required 2. Tested parts cannot be placed in service 3. Repeated tests of same item are often impossible, and different types of tests may require different samples 4. Extensive testing usually cannot be justified, because of large scrap losses 5. May be prohibited on parts with high material or fabrication costs, or on parts of limited availability 6. Cumulative effect of service usage cannot be measured directly, but only inferred from tests on parts used for different lengths of time 7. Difficult to apply to parts in service, and usually terminates their useful life 8. Extensive machining or other preparation of test specimens is often required 9. Capital investment and manpower costs are often high

Source: ASM Metals Reference Book, 2d ed., 1983.

In contrast, one of the important trends in modern manufacturing is **automated inspection.** This method uses a variety of sensor systems that monitor the relevant parameters *during* the manufacturing process (*on-line inspection*). Then, using these measurements, the process automatically corrects itself to produce acceptable parts. Thus further inspection of the part at another location in the plant is unnecessary. Parts may also be inspected immediately after they are produced (*in-process inspection*).

The use of accurate sensors and computer-control systems has integrated automated inspection into manufacturing operations. Such a system ensures that no part is moved from one manufacturing process to another (for example, a turning operation followed by cylindrical grinding) unless the part is made correctly and meets the standards of the first operation. Automated inspection is flexible and responsive to product design changes. Furthermore, because of automated equipment, less operator skill is required, productivity is increased, and parts have higher quality, reliability, and dimensional accuracy.

36.4.1 Sensors for automated inspection

Recent advances in sensor technology are making on-line or real-time monitoring of manufacturing processes feasible. Directly or indirectly, and with the use of various probes, sensors can detect dimensional accuracy, surface roughness, temperature, force, power, vibration, tool wear, and the presence of external or internal defects (see Section 38.8). Sensors operate on the principles of strain gages, inductance, capacitance, ultrasonics, acoustics, pneumatics, infrared radiation, optics, lasers, and various electronic gages. Sensors may be tactile (touching) or nontactile.

Sensors, in turn, are linked to microprocessors and computers for graphic data display. This capability allows rapid on-line adjustment of any processing parameters in order to produce parts that are consistently within specified standards of tolerance and quality. Such systems have already been implemented as standard equipment on many metal-cutting machine tools and grinding machines.

36.5 Product Quality

We have all used terms like "poor quality" or "high quality" to describe a certain product, a certain store, or the products of a certain company. What is quality? Although we may know it when we see or use a product, quality, unlike most technical terms, is difficult to define precisely.

Quality is a broadbased characteristic or property, and it not only consists of several well-defined technical considerations, but also can be quite subjective. A handle on a kitchen utensil that has been installed crooked, a product whose walls are so thin that it warps when subjected to small forces or temperature variations, and a machine tool that cannot maintain accuracy of the workpiece because of lack of stiffness or poor construction—all lead us to believe that the product is of low quality. A calculator or weighing scale that functions erratically and an arm rest on an automobile door that repeatedly comes loose are further examples of what we think of as low-quality products.

The public's perception is that a high-quality product is one that performs its functions reliably over a long period of time, without breaking down or requiring repairs. A few examples of this type of product are "good quality" refrigerators, washing machines, automobiles, bicycles, and kitchen knives. On the other hand, if the stem of a screwdriver bends, its handle discolors or cracks, or its tip wears off more rapidly than we had expected, we say that this screwdriver is of low quality.

Although the definition is somewhat outmoded, quality has been defined as (a) a product's fitness for use, and (b) the totality of features and characteristics that bear on a product's ability to satisfy a given need. More recently, several dimensions of quality have been identified, which include the product's performance, features, conformance, durability, reliability, serviceability, aesthetics, and perceived quality.

Note that, in describing good- or poor-quality products, we have not stated the lifetimes of products or any of their technical specifications. Throughout this text you have seen that design and manufacturing engineers have the freedom and responsibility to select and specify materials for the products to be made. Thus when selecting the metal for a screwdriver stem, you can specify materials that have high strength and high resistance to wear and corrosion. As a result, the screwdriver will perform better and last longer than one made of materials with inferior properties.

We must recognize, however, that materials possessing better properties also are generally more expensive and may be more difficult to process than those with poorer properties. Moreover, because the range of available materials and properties is so broad, manufacturers in the past have usually been forced to set some limit on expected useful product life. For example, automobile disk brakes under normal use are generally designed and manufactured to last an average of about 40,000 miles, mufflers 30,000 miles, batteries 4 years, and tires 40,000–60,000 miles. Similarly, a typical hot-water heater for homes is expected to last about 10 years, a dollar bill 18 months, and nuclear reactors 40 years.

The level of quality that a manufacturer chooses for its products may be market-dependent. For example, low-cost, low-quality tools have their own market niche. Even this sort of product, however, has its own degree of required quality performance.

Quality standards are essentially trade-offs among several considerations. As you will see in Chapter 40, the total product cost depends on several variables, including the level of automation in the manufacturing plant. Thus the engineer has many opportunities to review and modify overall product design and manufacturing processes in order to minimize costs without sacrificing quality.

Contrary to general public perception, quality products do not necessarily cost more. In fact, higher quality actually means lower cost. It is important to recognize that poor quality has significant built-in costs of customer dissatisfaction, difficulties in assembling and maintaining components, and need for in-field repair.

36.6 Quality Assurance

Quality assurance is the total effort by a manufacturer to ensure that its products conform to a detailed set of specifications and standards. These standards cover several parameters, such as dimensions, surface finish, tolerances, composition, color, and mechanical, physical, and chemical properties. In addition, standards are usually written to ensure proper assembly using interchangeable, defect-free components and a product that performs as intended by its designers.

Quality assurance is the responsibility of everyone involved with design and manufacturing. The often-repeated statement that quality must be built into a product reflects this important concept. Quality cannot be inspected into a finished product.

Although product quality has always been important, increased domestic and international competition has caused quality assurance to become even more important. Every aspect of design and manufacturing operation, such as material selection, production, and assembly, is now being analyzed in detail to ensure that quality is truly built into the final product.

36.6.1 Quality control

If you were in charge of product quality for a manufacturing plant, how would you make sure that the final product is of acceptable quality? The best method is to control materials and processes in such a manner that the products are made correctly in the first place. However, 100 percent inspection is usually too costly to maintain. Therefore several methods of inspecting smaller, statistically relevant sample lots have been devised. These methods all use statistics to determine the probability of defects occurring in the total production batch.

Inspection involves a series of steps:

1. Inspecting incoming materials to make sure that they meet certain property, dimension, and surface finish and integrity requirements.
2. Inspecting individual product components to make sure that they meet specifications.
3. Inspecting the product to make sure that individual parts have been assembled properly.
4. Testing the product to make sure that it functions as designed and intended.

Why must inspections be continued once a product of acceptable quality has been produced? Because there will always be variations in the dimensions and properties of incoming materials, variations in the performance of tools, dies, and machines used in various stages of manufacturing, possibilities of human error, and errors made during assembly of the product. As a result, no two products are ever made exactly alike.

Another important aspect of quality control is the capability to analyze defects and promptly eliminate them or reduce them to acceptable levels. In an even broader sense, quality control involves evaluating the product design and customer satisfaction. The sum total of all these activities is referred to as *total quality control*.

From the discussion so far, you should realize by now that in order to control quality you have to be able to (1) measure quantitatively the level of quality, and (2) identify all the material and process variables that can be controlled. The level of quality obtained during production can then be established by inspecting the product to determine whether it meets the specifications for tolerances, surface finish, defects, and other characteristics. The identification of material and process variables and their effect on product quality is now possible through the extensive knowledge gained from research and development activities in all aspects of manufacturing.

36.7 Statistical Methods of Quality Control

Statistics deals with the collection, analysis, interpretation, and presentation of large amounts of numerical data. The use of statistical techniques in modern manufacturing operations is necessary because of the large number of material and process variables involved. For example:

a) Cutting tools, dies, and molds wear. Thus part dimensions vary over a period of time.
b) Machinery performs differently depending on its age, condition, and maintenance. Thus older machines tend to vibrate, are difficult to adjust, and do not maintain tolerances as well as new machines do.
c) Metalworking fluids perform differently as they degrade. Thus surface finish of the workpiece, tool life, and forces are affected.
d) Environmental conditions, such as temperature, humidity, and air quality in the plant may change from one hour to the next, affecting machines, workspaces, and employees.
e) Different shipments of raw materials may have significantly different dimensions, properties, and surface characteristics.
f) Operator skill and attention may vary during the day, from machine to machine, or among operators.

In the preceding list, those events that occur randomly, that is, without any particular trend or pattern, are called *chance variations*. Those that can be traced to specific causes are called *assignable variations*. The existence of *variability* in production operations has been recognized for centuries, but Eli Whitney (1765–1825) first grasped its full significance when he found that interchangeable parts were indispensible to the mass production of firearms. Modern statistical concepts relevant to manufacturing engineering were first developed in the early 1900s, notably through the work of W. A. Shewhart.

36.7.1 Statistical quality control

To understand **statistical quality control** (SQC), we first need to define some of the terms that are commonly used in this field.

- **Sample size.** The number of parts to be inspected in a sample, whose properties are studied to gain information about the whole population.
- **Random sampling.** Taking a sample from a population or lot in which each item has an equal chance of being included in the sample. Thus when taking samples from a large bin, the inspector does not take only those that happen to be within reach.
- **Population.** The totality of individual parts of the same design from which samples are taken (also called the *universe*).
- **Lot size.** A subset of population. A lot or several lots can be considered subsets of the population and may be treated as representative of the population.

The sample is inspected for certain characteristics and features, such as tolerances, surface finish, and defects, with the instruments and techniques that we described in Chapter 35 and Sections 36.2 and 36.3. These characteristics fall into two categories: those that can be measured quantitatively (method of variables) and those that are qualitative (method of attributes).

The **method of variables** is the quantitative measurement of characteristics such as dimensions, tolerances, surface finish, or physical or mechanical properties. Such measurements are made for each of the units in the group under consideration, and the results are compared against specifications.

The **method of attributes** involves observing the presence or absence of qualitative characteristics, such as external or internal defects in machined, formed, or welded parts or dents in sheet-metal products, for each of the units in the group under consideration. Sample size for attributes-type data is generally larger than for variables-type data.

During the inspection process, measurement results will vary. For example, assume that you are measuring the diameter of turned shafts as they are produced on a lathe, using a micrometer as shown in Fig. 35.4(a). You soon note that their diameters vary, even though ideally you want all the shafts to be exactly the same size. Let's now turn to consideration of statistical quality-control techniques, which allow us to evaluate these variations and set limits for the acceptance of parts.

If we list the measured diameters of the turned shafts in a given population, we note that one or more parts have the smallest diameter, and one or more have the largest diameter. The majority of the turned shafts have diameters that lie between these extremes. If we group these diameters and plot them, the plot consists of a bar graph representing the number of parts in each diameter group (Fig. 36.6). The bars show a **distribution**, also called a *spread* or *dispersion*, of the shaft-diameter measurements.

The bell-shaped curve in Fig. 36.6 is called a **frequency distribution**, or the frequency with which parts within each diameter group are being produced.

Data from manufacturing processes often fit curves represented by a mathematically derived **normal distribution curve** (Fig. 36.7). These curves are also called *Gaussian*, after K. F. Gauss (1777–1855), who developed them on the basis of probability. The bell-shaped normal distribution curve fitted to the data in Fig. 36.6

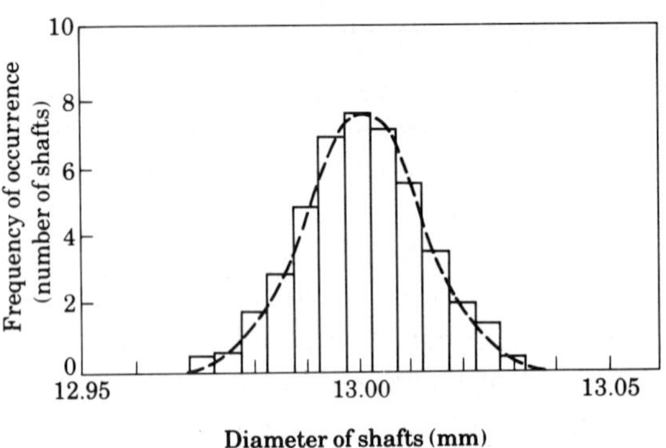

FIGURE 36.6
A plot of the number of shafts measured and their respective diameters. This type of curve is called a frequency distribution.

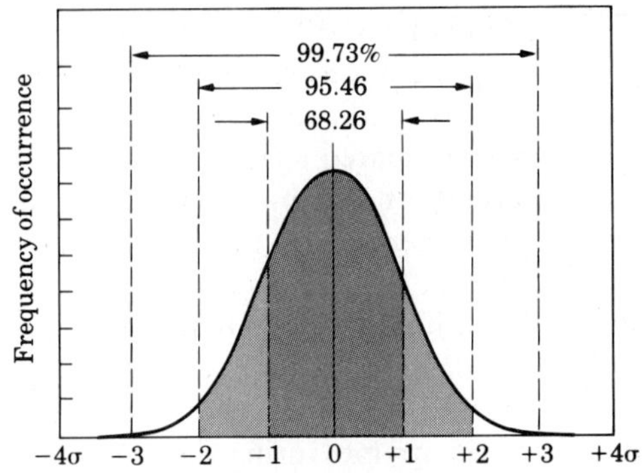

FIGURE 36.7
A normal distribution curve indicating areas within each range of standard deviation. Note: The greater the range, the higher the percentage of parts that fall within it.

has two important features. First, it shows that most part diameters tend to cluster around an *average* value (*arithmetic mean*). This average is usually designated as $\bar{x}$ and is calculated from the expression

$$\bar{x} = \frac{x_1 + x_2 + x_3 + \cdots + x_n}{n}, \tag{36.1}$$

where the numerator is the sum of all measured values (diameters), and n is the number of measurements (number of shafts).

The second feature of this curve is its width, indicating the *dispersion* of the diameters measured. The wider the curve, the greater the dispersion. The difference between the largest value and the smallest value is called the *range*, R:

$$R = x_{\text{max}} - x_{\text{min}}. \tag{36.2}$$

Dispersion is estimated by the **standard deviation**, which is generally denoted as σ and is obtained from the expression

$$\sigma = \sqrt{\frac{(x_1 - \bar{x})^2 + (x_2 - \bar{x})^2 + (x_3 - \bar{x})^2 + \cdots + (x_n - \bar{x})^2}{n - 1}}, \qquad (36.3)$$

where x is the measured value for each part. Note from the numerator in Eq. (36.3) that as the curve widens, the standard deviation becomes greater. Also note that σ has the unit of linear dimension. In comparing Eqs. (36.2) and (36.3), we note that the range R is a simpler and more convenient measure of dispersion.

Since we know the number of turned parts that fall within each group, we can calculate the percentage of the total population represented by each group. Thus Fig. 36.7 shows that the diameters of 99.73 percent of the turned shafts fall within the range of $\pm 3\sigma$, 95.46 percent within $\pm 2\sigma$, and 68.26 percent within $\pm 1\sigma$. Thus only 0.27 percent fall outside the $\pm 3\sigma$ range.

36.8 Statistical Process Control

If the number of parts that do not meet set standards (defective parts) increases during a production run, we must be able to determine the cause (incoming materials, machine controls, degradation of metalworking fluids, operator boredom, or others) and take appropriate action. Although this statement at first appears to be self-evident, it was only in the early 1950s that a systematic statistical approach was developed to guide operators in manufacturing plants.

This approach advises the operator to take certain measures and when to take them in order to avoid producing further defective parts. Known as **statistical process control** (SPC), this technique consists of several elements: control charts and setting control limits, capabilities of the particular manufacturing process, and characteristics of the machinery involved.

36.8.1 Control charts

The frequency distribution curve in Fig. 36.6 shows a range of shaft diameters being produced that may fall beyond the predetermined design tolerance range. Figure 36.8 shows the same bell-shaped curve, which now includes the specified tolerances for the diameter of turned shafts.

Control charts graphically represent the variations of a process over a period of time. They consist of data plotted during production, and typically, there are two plots. The quantity $\bar{x}$ (Fig. 36.9a) is the average for each subset of samples taken and inspected, say, a subset consisting of 5 parts. A sample size of between 2 and 10 parts is sufficiently accurate, provided that sample size is held constant throughout the inspection.

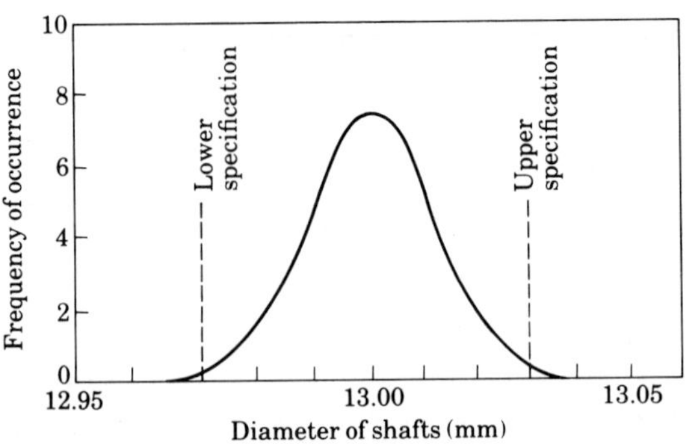

FIGURE 36.8
Frequency distribution curve, showing lower and upper specification limits.

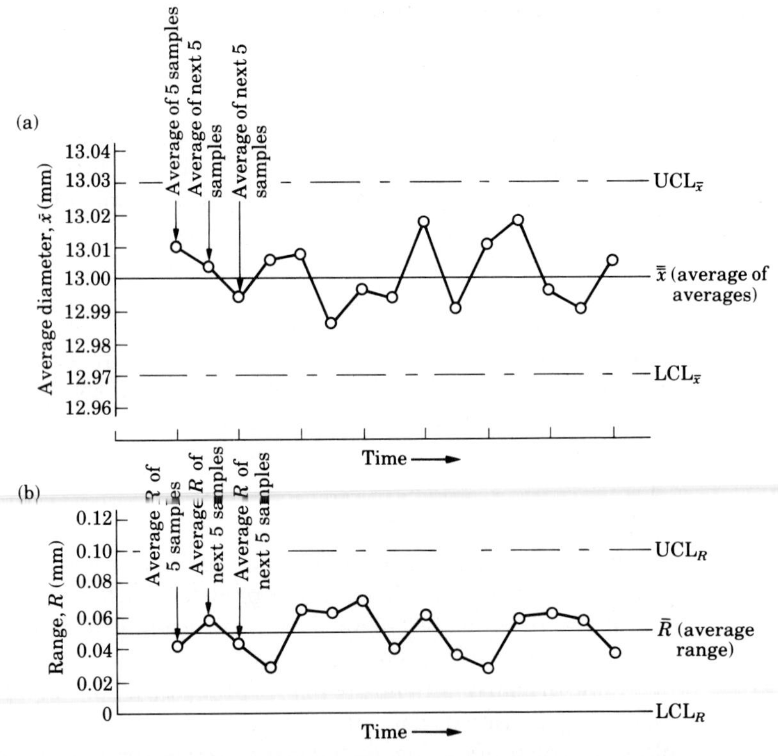

FIGURE 36.9
Control charts used in statistical quality control. The process shown is in statistical control because all points fall within the lower and upper control limits. In this illustration sample size is 5 and the number of samples is 15.

TABLE 36.2
CONSTANTS FOR CONTROL CHARTS

SAMPLE SIZE	A_2	D_4	D_3	d_2
2	1.880	3.267	0	1.128
3	1.023	2.575	0	1.693
4	0.729	2.282	0	2.059
5	0.577	2.115	0	2.326
6	0.483	2.004	0	2.534
7	0.419	1.924	0.078	2.704
8	0.373	1.864	0.136	2.847
9	0.337	1.816	0.184	2.970
10	0.308	1.777	0.223	3.078
12	0.266	1.716	0.284	3.258
15	0.223	1.652	0.348	3.472
20	0.180	1.586	0.414	3.735

The frequency of sampling depends on the nature of the process. Some processes may require continuous sampling, whereas others may require only one sample per day. Quality-control analysts are best qualified to determine this frequency for a particular situation. Since the measurements in Fig. 36.9(a) are made consecutively, the abscissa of these control charts also represents time. The solid horizontal line in this figure is the *average of averages* (*grand average*), denoted as $\bar{\bar{x}}$, and represents the population mean. The upper and lower horizontal broken lines in these control charts indicate the control limits for the process.

The **control limits** are set on these charts according to statistical-control formulas designed to keep actual production within the usually acceptable $\pm 3\sigma$ range. Thus for $\bar{x}$

$$\textbf{Upper control limit } (\mathrm{UCL}_{\bar{x}}) = \bar{x} + 3\sigma = \bar{\bar{x}} + A_2\bar{R} \tag{36.4}$$

and

$$\textbf{Lower control limit } (\mathrm{LCL}_{\bar{x}}) = \bar{x} - 3\sigma = \bar{\bar{x}} - A_2\bar{R}, \tag{36.5}$$

where A_2 is read from Table 36.2 and $\bar{R}$ is the average of R values.

These limits are calculated on the basis of historical production capability of the equipment itself. They are not generally associated with either design tolerance specifications or dimensions. They indicate the limits within which a certain percentage of measured values are normally expected to fall because of the inherent variations of the process itself, upon which the limits are based. The major goal of statistical process control is to improve the manufacturing process with the aid of control charts to eliminate assignable causes. The control chart continually indicates progress in this area.

The second control chart (Fig. 36.9b) shows the range R in each subset of samples. The solid horizontal line represents the average of R values in the lot, denoted as $\bar{R}$, and is a measure of the variability of the samples. The upper and lower control limits for R are obtained from the equations

$$\mathrm{UCL}_R = D_4\bar{R} \tag{36.6}$$

and

$$\text{LCL}_R = D_3 \bar{R}, \tag{36.7}$$

where the constants D_4 and D_3 are obtained from Table 36.2. This table also includes the constant d_2, which is used in estimating the standard deviation from the equation

$$\sigma = \frac{\bar{R}}{d_2}. \tag{36.8}$$

When the curve of a control chart is like that shown in Fig. 36.9(a), we say that the process is "in good statistical control." In other words, there is no clear discernible trend in the pattern of the curve, the points (measured values) are random with time, and they do not exceed the control limits. However, you can see that curves such as those shown in Figs. 36.10(a), (b), and (c) indicate certain trends.

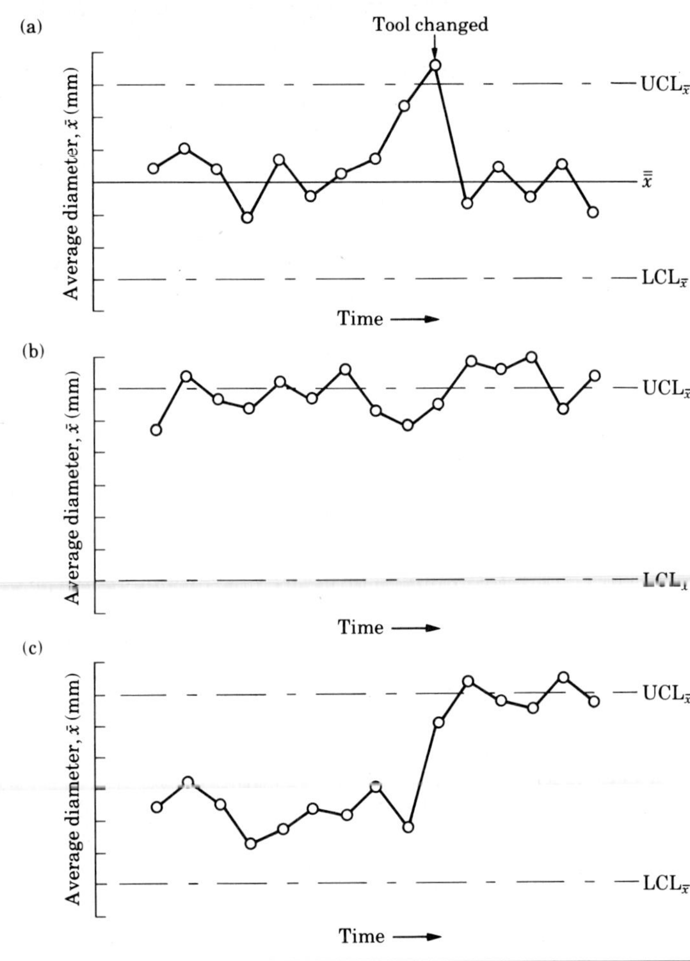

FIGURE 36.10 Control charts. (a) Process begins to become out of control because of factors such as tool wear. The tool is changed and the process is then in statistical control. (b) Process parameters are not set properly; thus all parts are around the upper control limit. (c) Process becomes out of control because of factors such as a sudden change in the properties of the incoming material.

For example, note in the middle of curve (a) that the diameter of the shafts increases with time. The reason for this increase may be a change in one of the process variables, such as wear of the cutting tool. If, as in curve (b), the trend is toward consistently larger diameters, hovering around the upper control limit, it could mean that the tool settings on the lathe may not be correct and, as a result, the parts being turned are consistently too large. Curve (c) shows two distinct trends that may be caused by factors such as a change in the properties of the incoming material or a change in the performance of the cutting fluid (for example, its degradation). These situations place the process "out of control."

Analyzing patterns and trends in control charts requires considerable experience in order to identify the specific cause(s) of an out-of-control situation. Such causes may be those changes outlined at the beginning of Section 36.7. Overcontrol of the manufacturing process—that is, setting upper and lower control limits too close to each other (smaller standard deviation range)—is a further reason for out-of-control situations. This is the reason for calculating control limits on process capability rather than on a potentially inapplicable statistic.

36.8.2 Process capability

Process capability is defined as the limits within which individual measurement values resulting from a particular manufacturing process would normally be expected to fall for random variations only. Thus process capability tells us that the process can produce parts within certain limits of precision. Since a manufacturing process involves materials, machinery, and operators, each can be analyzed individually to identify a problem when process capabilities do not meet part specifications.

• Example: Calculation of control limits and standard deviation.

The data in the accompanying table show length measurements (in.) taken on a machined workpiece. Sample size is 5 and the sample number is 10; thus the total number of parts measured is 50. The quantity $\bar{x}$ is the average of five measurements in each sample. We first calculate the average of averages $\bar{\bar{x}}$,

$$\bar{\bar{x}} = \frac{44.296}{10} = 4.430 \text{ in.},$$

and the average of R values,

$$\bar{R} = \frac{1.03}{10} = 0.103 \text{ in.}$$

Since the sample size is 5, we determine from Table 36.2 the following constants: $A_2 = 0.577$, $D_4 = 2.115$, and $D_3 = 0$. We can now calculate the control limits using

Eqs. (36.4)–(36.7). Thus, for averages we have

$$\mathrm{UCL}_x = 4.430 + (0.577)(0.103) = 4.489 \text{ in.},$$

$$\mathrm{LCL}_x = 4.430 - (0.577)(0.103) = 4.371 \text{ in.},$$

and for ranges we have

$$\mathrm{UCL}_R = (2.115)(0.103) = 0.218 \text{ in.},$$

$$\mathrm{LCL}_R = (0)(0.103) = 0 \text{ in.}$$

We can calculate the standard deviation using Eq. (36.8) and a value of $d_2 = 2.326$. Thus,

$$\sigma = \frac{0.103}{2.326} = 0.044 \text{ in.}$$

SAMPLE NUMBER	x_1	x_2	x_3	x_4	x_5	$\bar{x}$	R
1	4.46	4.40	4.44	4.46	4.43	4.438	0.06
2	4.45	4.43	4.47	4.39	4.40	4.428	0.08
3	4.38	4.48	4.42	4.42	4.35	4.410	0.13
4	4.42	4.44	4.53	4.49	4.35	4.446	0.18
5	4.42	4.45	4.43	4.44	4.41	4.430	0.04
6	4.44	4.45	4.44	4.39	4.40	4.424	0.06
7	4.39	4.41	4.42	4.46	4.47	4.430	0.08
8	4.45	4.41	4.43	4.41	4.50	4.440	0.09
9	4.44	4.46	4.30	4.38	4.49	4.414	0.19
10	4.42	4.43	4.37	4.47	4.49	4.436	0.12

• Example: Maintaining accuracy in boring using control charts.

The workpiece shown in the accompanying figure is made of gray cast iron and is bored to the tolerances indicated (5.5125/5.5115 in.). These parts were bored on a chucking machine. Each of the 18 points plotted on the vertical axis of the control chart represents the average of bore diameter measurements made on four parts (sample size). The horizontal broken lines at +0.0005 in. and −0.0005 in. represent upper and lower specified limits, respectively. The solid line $\bar{\bar{x}} = 0.00017$ in. is the estimate of the process capability based on a study of several bored samples on the machine. The upper and lower control limits are then calculated from $\bar{\bar{x}}$. We note that samples 4–9 show a definite trend toward undersize bored holes. If the operation had been continued without any changes, the successive bored holes very likely would have been out of tolerance. To avoid this situation (out of control), the boring tools were reset toward the upper control limit before parts in sample 10 and the rest were bored. *Source:* ASM International.

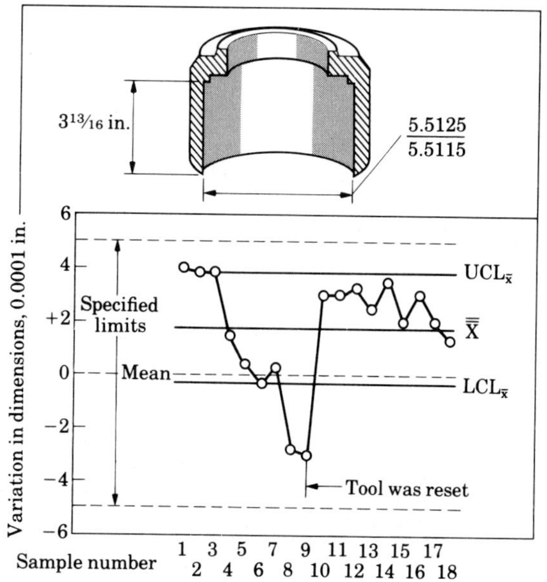

36.9 Acceptance Sampling and Control

Acceptance sampling consists of taking only a few random samples from a lot and inspecting them for the purpose of judging whether the entire lot is acceptable or should be rejected or reworked. Developed in the 1920s and used extensively during World War II for military hardware (MIL STD 105), this statistical technique is widely used and valuable. It is particularly useful for inspecting high production rate parts, where 100 percent inspection would be too costly.

A variety of acceptance sampling plans have been prepared for both military and national standards, based on an acceptable, predetermined, and limiting percentage of nonconforming parts in the sample. If this percentage is exceeded, the entire lot is rejected, or it is reworked if economically feasible. Note that the actual number of samples (not percentages of the lot that are in the sample) can be significant in acceptance sampling. The greater the number of samples taken from a lot, the greater will be the percentage of nonconforming parts and the lower the probability of acceptance of the lot. **Probability** is defined as the relative occurrence of an event. The probability of acceptance is obtained from various operating characteristic curves, one example of which is shown in Fig. 36.11.

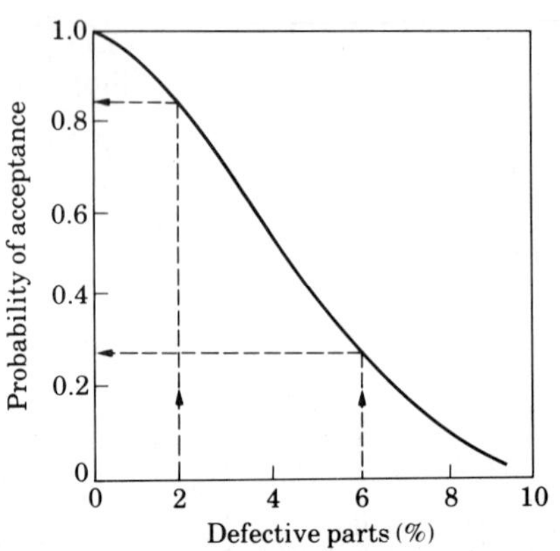

FIGURE 36.11
A typical operating-characteristic curve used in acceptance sampling. The higher the percentage of defective parts, the lower is the probability of their being accepted by the consumer. There are several methods of obtaining these curves.

The **acceptance quality level** (AQL) is commonly defined as the level at which there is a 95 percent probability of acceptance of the lot. This percentage would indicate to the manufacturer that 5 percent of the parts in the lot may be rejected by the consumer (called *the producer's risk*). Likewise, the consumer knows that 95 percent of the parts are acceptable (called *the consumer's risk*).

Lots that do not meet the desired quality standards can be salvaged by the manufacturer by a secondary rectifying inspection. In this method, a 100 percent inspection is made of a rejected lot, and the defective parts are removed. This time-consuming and costly process is an incentive for the manufacturer to control the production process.

Acceptance sampling requires less time and fewer inspections than do other sampling methods. Consequently, inspection of the parts can be more detailed. Keep in mind, however, that automated inspection techniques are being developed rapidly so that 100 percent inspection of all parts is indeed feasible and can be economical.

36.10 Total Quality Control and the Quality Circle

The **total quality control** (TQC) concept is a management system emphasizing the fact that quality must be designed and built into a product. Defect *prevention*, rather than defect detection, is the major goal. Total quality control is a systems approach in that both management and workers make an integrated effort to manufacture high-quality products consistently. All tasks concerning quality improvements and responsibilities in the organization are clearly identified. The TQC concept also requires 100 percent inspection of parts, usually by automated inspection systems. No defective parts are allowed to continue through the production line.

A related concept that has gained wide acceptance is the **quality circle**. This activity consists of regular meetings by groups of workers who discuss how to improve and maintain product quality at all stages of the manufacturing process. Worker involvement and responsibility are emphasized. Comprehensive training is provided so that the worker can become capable of analyzing statistical data, identifying causes of poor quality, and taking immediate action to correct the situation. Putting this concept into practice recognizes the importance of quality assurance as a major company-wide management policy, affecting all personnel and all aspects of production.

36.11 Quality Engineering as a Philosophy

Many of the quality-control concepts and methods that we have described have been put into larger perspective by certain experts in quality control. Notable among these experts are W. E. Deming and G. Taguchi, whose philosophies of quality and product cost have had a major impact on modern manufacturing. We outline their philosophies of quality engineering in this section.

36.11.1 Deming methods

During World War II, Deming and several others developed new methods of statistical process control in manufacturing plants for wartime industry. The need for statistical control arose from the recognition that there were variations in the performance of machines and people and the quality and dimensions of raw materials. Their efforts, however, involved not only statistical methods of analysis, but a new way of looking at manufacturing operations to improve quality and lower costs. Recognizing the fact that manufacturing organizations are systems of management, workers, machines, and products, Deming's basic ideas are:

- Define management's commitment to product quality and productivity. Making profits is essential, but it is not the organization's only purpose. Emphasize the continuity and consistency of the organization's reason for being.
- Recognize that high quality does not necessarily mean high cost. Making a defect-free part actually costs less.
- Avoid traditional adversarial relationships between management and workers. Encourage open communication among all groups in the

organization and problem solving through teamwork. Break down barriers in communication between various departments in the organization.

- Require that managers clearly identify those problems that are caused by the workers and those that are caused by the system. Make continued efforts to identify problems in the system and find ways to solve them.
- Recognize that workers know where potential improvements are possible. In addition to performing their jobs, workers are intelligent and capable of creative ideas and providing insight.
- Recognize pride of workmanship and provide the techniques and tools to enable workers to improve their performance. Avoid slogans, posters, numerical goals, and production quotas.
- Do not allow commonly accepted levels of defective materials, delays, and defective parts. Reduce the number of suppliers and purchase materials on a statistical basis, not price.
- Use modern statistical methods and teach them to the workers to enable them to identify problems and improve quality and productivity.
- Institute training programs for advancing the education of employees, allowing them to keep abreast of new developments in materials, processes, and technologies.

Note that Deming places great emphasis on communication, direct worker involvement, and education in statistics, as well as in modern manufacturing technology. His ideas have been widely accepted in Japan but only in some segments of the U.S. manufacturing community.

36.11.2 Taguchi methods

In the Taguchi methods, high quality and low costs are achieved by combining engineering and statistical methods to optimize product design and manufacturing processes. Loss of quality is defined as the financial loss to society after the product is shipped, with the following results:

a) Poor quality leads to customer dissatisfaction.
b) Costs are involved in servicing and repairing defective products, some in the field.
c) The manufacturer's credibility is diminished in the marketplace.
d) The manufacturer eventually loses its share of the market.

The Taguchi methods of quality engineering emphasize the importance of:

- Enhancing cross functional team interaction. In this interaction, design engineers and process or manufacturing engineers communicate with each other in a common language. They quantify the relationships between design requirements and the manufacturing process.
- Implementing experimental design, in which the factors involved in a process or operation and their interactions are studied simultaneously.

In *experimental design*, the effects of controllable and uncontrollable variables on the product are identified. This approach minimizes variations in product dimensions and properties, bringing the mean to the desired level. The methods used for experimental design are complex, involving the use of fractional factorial design and orthogonal arrays, which reduce the number of experiments required. These methods are also capable of identifying the effect on the product of variables that cannot be controlled (called *noise*), such as changes in environmental conditions.

The use of these methods allows rapid identification of the controlling variables and determination of the best method of process control. These variables are then controlled, without the need for costly new equipment or major modifications to existing equipment. For example, variables affecting tolerances in machining a particular component can be readily identified, and the correct cutting speed, feed, cutting tool, and cutting fluids can be specified.

• Example: Increasing quality without increasing the cost of a product.

A manufacturer of clay tiles notices that, because of temperature variations in the kiln used to fire the tiles, excessive scrap was being produced, adversely affecting the company's profits. The manufacturer first considered purchasing new kilns having better temperature controls. This solution, however, would require a major capital investment. A study was then undertaken to determine whether modifications could be made in the composition of the tile so that it became less sensitive to temperature fluctuations during firing. Based on factorial design of experiments, in which the factors involved in a process and their interactions are studied simultaneously, it was found that increasing the lime content of the tiles made them less sensitive to temperature variations during firing. This modification, which was also the low-cost alternative, was implemented, reducing the scrap substantially and improving tile quality.

36.12 Reliability

We know that, eventually, all products fail in some manner or other. Automobile tires become smooth, motors burn out, hot-water heaters begin to leak, and machinery stops functioning properly. **Reliability** is defined as the probability that a product will perform its intended function, in the manner expected, for a specified period of time. The more critical the application of a particular product is, the higher its reliability should be. Thus the reliability of an aircraft jet engine or a medical instrument should be much higher than that for a kitchen faucet or a mechanical pencil. From the topics that we have discussed in this chapter, you can see that as the quality of each component of a product increases, so too does the reliability of the final product. Reliability, of course, also depends on whether a product is properly used and maintained.

The expected reliability of a product depends on the nature of the product. For example, for an ordinary steel chain, the reliability of each link in the chain is important. Similarly, the reliability of each gear in a machine's gear train is important. This condition is known as *series reliability.* On the other hand, for a steel cable consisting of many individual wires, the reliability of each wire is not as critical. This condition is known as *parallel reliability.* The parallel reliability concept is important in the design of backup systems, which permit a product to continue functioning in the event one of its components fails. Electrical or hydraulic systems in an aircraft, for example, are backed by mechanical systems, which are called *redundant systems.*

Predicting reliability has become an important science and involves complex mathematical relationships. The importance of predicting the reliability of the critical components of civilian or military aircraft is obvious. The reliability of an automated and computer-control high-speed production line, with all its complex mechanical and electronic components, is also important, as its failure can result in major economic losses to the manufacturer.

SUMMARY

Several nondestructive and destructive testing techniques, each having its own applications, advantages, and limitations, are available for inspection of completed parts and products. The traditional approach has been to inspect the part or product after it is manufactured and to accept a certain number of defective parts. The trend now is toward on-line, 100 percent inspection of all parts and products being manufactured.

Quality must be built into products. Quality assurance is concerned with various aspects of production, such as design, manufacturing, and assembly, and inspection at each step of production for conformance to specifications. Statistical quality control and process control have become indispensable to modern manufacturing. They are particularly important for interchangeable parts and in reducing manufacturing costs.

Implementation of total quality control and quality circles are among the important recent developments for the prevention of defects in manufacturing. These activities should have the full support of management. All personnel from design to final assembly and inspection should have an active role in it, as clearly emphasized in the Deming and Taguchi methods of quality engineering.

TRENDS

- The trend of on-line automated inspection has become well established and will continue to be an essential element in manufacturing operations.
- The present goal in quality assurance is to manufacture parts and products that are 100 percent acceptable. Total quality control will continue to be a crucial manufacturing activity.
- Sensors based on a variety of principles are being developed for on-line inspection for all aspects of quality. Real-time process control of quality is being studied and implemented extensively.

KEY TERMS

Acceptance quality level
Acceptance sampling
Automated inspection
Control charts
Control limits
Distribution
Frequency distribution
Lot size
Lower control limit
Method of attributes
Method of variables
Nondestructive testing
Normal distribution curve
Population
Probability
Process capability
Quality
Quality assurance
Quality circle
Random sampling
Reliability
Sample size
Standard deviation
Statistical process control
Statistical quality control
Statistics
Total quality control
Upper control limit

BIBLIOGRAPHY

Aft, L.S., *Fundamentals of Industrial Quality Control.* Reading, Mass.: Addison-Wesley, 1986.

Anderson, D.M., *Design for Manufacturability, Optimizing Cost, Quality, and Time-to-Market.* Lafayette, Calif.: CIM Press, 1991.

Besterfield, D.H., *Quality Control,* 2d ed. Englewood Cliffs, N.J.: Prentice-Hall, 1986.

Buck, O. and S.M. Wolf, *Nondestructive Evaluation: Application to Materials Processing.* Metals Park, Ohio: American Society for Metals, 1984.

Clements, R.B., *The Handbook of Statistical Methods in Manufacturing.* Englewood Cliffs, N.J.: Prentice-Hall, 1991.

Deming, W.E., *Out of the Crisis.* Cambridge, Mass.: MIT Press, 1986.

Dhillon, B.S., *Quality Control, Reliability, and Engineering Design.* New York: Marcel Dekker, 1985.

Enrick, N.L., *Quality, Reliability, and Process Improvement*, 8th ed. New York: Industrial Press, 1985.

Feigenbaum, A.V., *Total Quality Control*, 3d ed., revised. New York: McGraw-Hill, 1991.

Grant, E.I., and R.S. Leavenworth, *Statistical Quality Control*, 6th ed. New York: McGraw-Hill, 1987.

Hansen, B.L., and P.M. Ghare, *Quality Control and Application*. Englewood Cliffs, N.J.: Prentice-Hall, 1987.

Hart, M., and R. Hart, *Quantitative Methods for Quality and Productivity Improvement*. Milwaukee: The American Society for Quality Control, 1989.

Juran, J., *Quality Control Handbook*. New York: McGraw-Hill, 1979.

——, and F.M. Gryna, Jr., *Quality Planning and Analysis*, 2d ed. New York: McGraw-Hill, 1980.

Kane, V., *Defect Prevention: Use of Simple Statistical Tools*. New York: Marcel Dekker, 1989.

Lester, R.H., N.L. Enrick, and H.E. Mottley, Jr., *Quality Control for Profit*, 2d ed. New York: Marcel Dekker, 1985.

Lewis, E.E., *Introduction to Reliability Engineering*. New York: Wiley, 1987.

Metals Handbook, 9th ed., Vol. 17: *Nondestructive Evaluation and Quality Control*. Metals Park, Ohio: ASM International, 1989.

Montgomery, D.C., *Introduction to Statistical Quality Control*. New York: Wiley, 1985.

Nondestructive Testing Handbook, Vol. 1: Leak Testing. R.C. McMaster (ed.), 1982; *Vol. 2: Liquid Penetrant Tests*. R.C. McMaster (ed.), 1982; *Vol. 3: Radiography and Radiation Testing*. L.E. Bryant (ed.), 1985. Metals Park, Ohio: American Society for Metals.

Oakland, J.S., *Statistical Process Control: A Practical Guide*. New York: Wiley, 1986.

Pyzdek, T., *What Every Engineer Should Know About Quality Control*. Milwaukee: The American Society for Quality Control, 1989.

Robinson, S.L., and R.K. Miller, *Automated Inspection and Quality Assurance*. New York: Marcel Dekker, 1989.

Ross, P., *Taguchi Techniques for Quality Engineering*. New York: McGraw-Hill, 1988.

Simmons, D.A., *Practical Quality Control*. Reading, Mass.: Addison-Wesley, 1979.

Taguchi, G., *Introduction to Quality Engineering*. Lanham, Maryland: UNIPUB/Kraus International, 1986.

Tool and Manufacturing Engineering Handbook, Vol. 4: Assembly, Testing, and Quality Control. Dearborn, Mich.: Society of Manufacturing Engineers, 1986.

Wadsworth, H.M., *Handbook of Statistical Methods for Engineers and Scientists*. New York: McGraw-Hill, 1990.

REVIEW QUESTIONS

36.1 Explain the difference between destructive and nondestructive testing techniques.

36.2 Describe the basic features of nondestructive testing techniques that use electrical sources of energy.

36.3 Identify the nondestructive techniques that are capable of detecting internal flaws and those that detect external flaws only.

36.4 Why is a developing agent used in the liquid-penetrant technique?

36.5 If the metal particles in magnetic-particle inspection are the same color as the workpiece itself, how would you go about producing a color contrast to detect flaws?
36.6 How is the depth of a flaw measured in ultrasonic testing?
36.7 Give examples from your own experiences in which the acoustic-impact technique of flaw detection has been useful.
36.8 What are the limitations of radiographic techniques?
36.9 State the differences between radiography and fluoroscope techniques.
36.10 Cite applications of thermal inspection techniques in engineering practice.
36.11 How are large workpieces accommodated in acoustic holography?
36.12 Give several examples of destructive techniques.
36.13 What are the advantages of automated inspection? Why is it becoming an important part of manufacturing engineering?
36.14 Explain the difference between in-process and postprocess inspection of manufactured parts. What trends are there in such inspection?
36.15 What are the functions of sensors in inspection?
36.16 Describe your understanding of good quality and poor quality in a product.
36.17 Explain why major efforts are now being made to build quality into products.
36.18 Why must we continue to make inspections once we have made a product that we know is of acceptable quality?
36.19 Name several material and process variables that can influence product quality.
36.20 What are chance variations?
36.21 Describe the terms sample size, random sampling, population, and lot size.
36.22 Explain the difference between method of variables and method of attributes.
36.23 What is a normal distribution curve?
36.24 Define standard deviation. Why is it important?
36.25 Describe what is meant by statistical process control.
36.26 Why are control charts made? How are they used?
36.27 What do control limits indicate?
36.28 Define process capability. How is it used?
36.29 What is acceptance sampling? Why was it developed?
36.30 Explain the difference between producer's risk and consumer's risk.
36.31 Explain your understanding of the quality circle. Why is it an important concept in manufacturing engineering?
36.32 Describe the basic concepts of the (a) Deming methods and (b) Taguchi methods.
36.33 What is the difference between series and parallel reliability?

QUESTIONS AND PROBLEMS

36.34 Which of the nondestructive inspection techniques are suitable for nonmetallic materials?
36.35 Describe situations in which nondestructive testing techniques are unavoidable.
36.36 Should products be designed and built for a certain expected life? Explain.
36.37 Comment on the ideas of Deming and Taguchi. What aspects of their concepts would be difficult to implement in a typical manufacturing facility? Why?
36.38 What is the consequence of setting lower and upper specifications closer to the peak of the curve in Fig. 36.8?
36.39 Identify several factors that can cause a process to become out of control.

36.40 Give examples that produce curves similar to Figs. 36.10(a) and (b).

36.41 Why is reliability important in manufacturing engineering? Give several examples.

36.42 Assume that in the example problem in Section 36.8 the number of samples was 5 instead of 10. Using the top half of the data in the table, recalculate control limits and the standard deviation. Compare your observations to the results from using 10 samples.

36.43 Calculate the control limits for averages and ranges for the following: number of samples = 4, $\bar{\bar{x}} = 70$, $\bar{R} = 7$.

36.44 Calculate the control limits for the following: number of samples = 3, $\bar{\bar{x}} = 36.5$, $UCL_R = 4.75$.

36.45 In an inspection with sample size 7 and a sample number of 50, it was found that the average range was 12 and the average of averages was 75. Calculate the control limits for averages and ranges.

36.46 Determine the control limits for the data shown in the table below.

x_1	x_2	x_3	x_4
0.55	0.60	0.57	0.55
0.59	0.55	0.60	0.58
0.55	0.50	0.55	0.51
0.54	0.57	0.50	0.50
0.58	0.58	0.60	0.56
0.60	0.61	0.55	0.61

36.47 The average of averages of a number of samples of size 9 was determined to be 125. The average range was 17.82 and the standard deviation was 6. The following measurements were taken in a sample: 120, 132, 124, 130, 118, 132, 135, 121, and 127. Is the process in control?

36.48 Would it be desirable to incorporate some of the nondestructive inspection techniques into metalworking machinery? Give a specific example and make a sketch of such a machine.

36.49 Give examples of the acoustic-impact inspection technique, in addition to those given in the text.

36.50 According to one electronics manufacturer, the goal in making computer chips is now 6σ. What does this mean? What percentage of the parts can be defective for this condition? What problems could be encountered in reaching this goal?

37

Human-Factors Engineering, Safety, and Product Liability

37.1 Introduction

Understanding the interaction of human beings with machines and the workplace environment is essential for the proper design of machinery and the development of safe and efficient working conditions in manufacturing plants. Regardless of the level of automation, workers are always involved in one or more aspects of production, including the efficient use, maintenance, and repair of machinery, tooling, and equipment.

Human-factors engineering, also called human engineering, is concerned with all aspects of human–machine interactions. An essentially synonymous term is *ergonomics*, based on the Greek words *ergo*, meaning work, and *nomics*, meaning management. Human-factors engineering has two major goals: (1) to maximize the quality and efficiency of work; and (2) to maximize human values such as operator safety, comfort, and satisfaction and to minimize fatigue and stress. These goals occasionally conflict, requiring design engineers to rely on their judgment in developing a reasonably safe machine or workplace.

In this chapter we describe those aspects of human-factors engineering that are particularly relevant to manufacturing operations. We also outline the basic design considerations for the safe interaction of human beings with machinery and the workplace environment. Finally, we discuss the increasingly important topic of *product liability*, which considers the legal responsibilities of the manufacturer and the user of a product in the event of bodily injury or physical and financial losses caused by the malfunction, misuse, or failure of a product.

37.2 Human-Factors Engineering

Human-factors engineering deals with applying the knowledge gained from human physiology and psychology concerning the characteristics and capabilities of the human body and mind. These include such factors as height, weight, vision, hearing, posture, strength, age, intelligence, educational level, dexterity, and reaction time. Numerous statistical data relating to these characteristics are available.

37.2.1 Workstations

Machinery and related equipment should be designed so that they can be operated with undue strain on the worker (for comfort), and with minimum unnecessary and wasted motion (for efficient operation). Muscle-strain injuries are common in the workplace. Such injuries can be avoided through application of proper lifting procedures and the use of mechanical devices when necessary. Wasted motion can be minimized, for example, by arranging sequential controls in series.

Controls and displays should be placed in appropriate locations and should not interfere with each other. They should be identified and marked clearly so that they can be understood and operated without confusion. Local and background lighting of machinery should be adequate to minimize eye strain, and glare from improper lighting or reflective surfaces should be minimized. The correct type of information display depends on the application. For example, numerical counters are preferable to all other display methods when precise values of static information are desired. However, for rapidly changing information, a pointer moving along a fixed scale is preferable to a counter because of the limited time available for reading values on the scale.

Workstations should be arranged to provide efficient movement of people, parts, and products. Sufficient space should be provided for storage, access, maintenance, and material-handling equipment when necessary.

37.2.2 Noise

Noise can significantly affect operator performance and health. Sustained exposure to high noise levels can cause permanent deafness. Excessive noise also interferes with communication and can lead to misunderstood messages. Noise intensity, measured in decibels by a sound meter, varies widely depending on the type of equipment involved (Fig. 37.1). Although they are somewhat controversial, permissible noise levels have been established by OSHA (Fig. 37.2). Note that noise level and operator exposure time are inversely related. High levels of noise emission by machines may not be objectionable in itself; it is worker exposure to noise that is important.

Although all machines emit noise, major sources of noise typically are air-actuated mechanisms, gears, material-handling equipment, chutes for conveying goods, impacting-type machinery, hydraulic pumps, motors, fans, and blowers. Noise levels can be reduced by muffling equipment, totally enclosing machinery, lining housings and chutes, and modifying machinery components. In areas of excessive noise, personal protective equipment such as ear plugs or ear muffs can reduce worker exposure to noise. Regular maintenance and replacement of worn components, such as gears and bearings, are also effective means of reducing noise levels. Machinery and floor vibration not only emits noise, but also is detrimental to the dimensional accuracy and surface finish of parts produced.

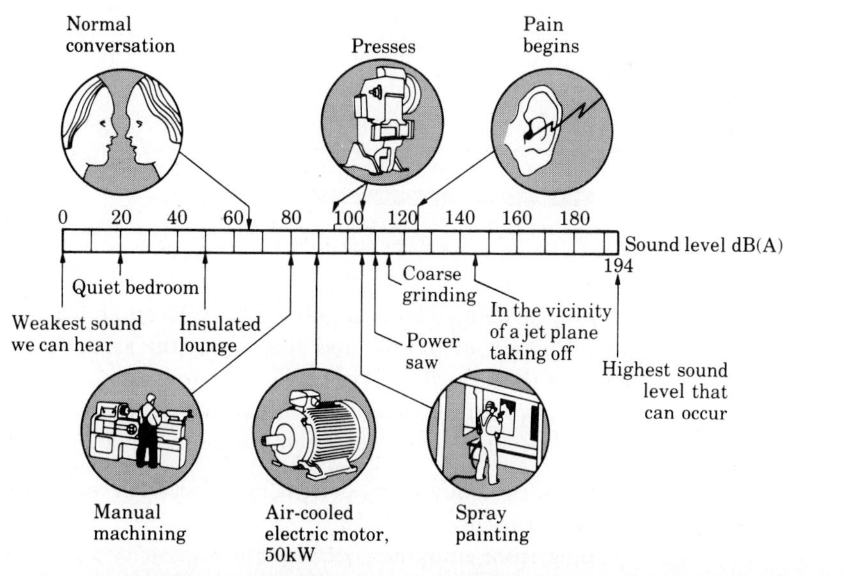

FIGURE 37.1
Typical sound levels for various operations. *Source:* Occupational Safety and Health Administration.

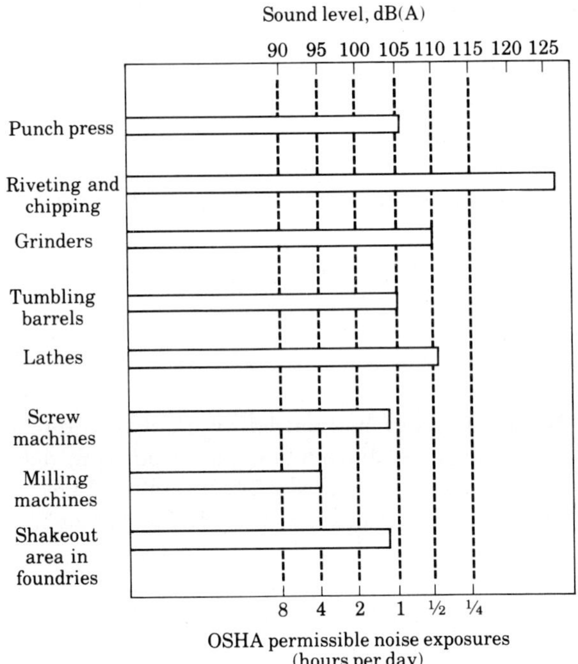

FIGURE 37.2 Sound levels of various types of machinery at operator's position and permissible noise exposure in terms of hours per day. Note the inverse relationship between noise level and permissible time. *Source:* Occupational Safety and Health Administration.

37.3 Safety

Safety may be defined as a judgment of the acceptability of danger, where **danger** is the combination of hazard and risk. Thus the safety of a machine or workplace depends on the hazard and risk involved with machine operation. **Hazard** is defined as an injury producer, while **risk** is defined as the likelihood (probability) that an injury will occur. The causes of employee injury are varied and include:

a) Parts of the body being caught in or between machine components.
b) Being struck by an object.
c) Falling from equipment or structures.
d) Slipping or tripping on walking or working surfaces.
e) Explosions and fires.
f) Exposure to dangerous levels of electricity.

g) Exposure to extremes of temperature and burns.
h) Exposure to and ingestion of toxic chemicals.
i) Excessive physical strain.

Employers are responsible for providing a reasonably safe workplace. Various safety and health standards have been promulgated in the United States, most notably by the American National Standards Institute. These standards have been the basis for many governmental regulations adopted by OSHA. Safety literature is available from the National Safety Council (NSC) and similar organizations, both in the United States and in other industrialized countries.

By their very nature, manufacturing operations involve safety hazards of varying degrees. Nevertheless, a reasonably safe working environment is imperative. Some hazards are open, and the means of controlling them are obvious, while others are hidden and require additional precautions. Safety professionals and organizations have defined the following priorities for eliminating hazards in the workplace:

1. Eliminate the hazard through machine design.
2. Apply safeguarding technology.
3. Use warning signs and labels.
4. Train and instruct the worker.
5. Prescribe personal protective equipment.

37.3.1 Safeguarding

Certain common machinery **safeguarding methods** have been developed. We describe them briefly below.

Barrier guards. When properly designed and maintained, *barrier guards* prevent operator exposure to common injury producers such as nip points and pinch points. They may be fixed, adjustable, or self-adjusting. Barrier guards also identify hazardous areas of machinery and, in some applications, prevent projectiles (such as broken pieces of a grinding wheel) from being thrown from the machine. Mechanical, electrical, hydraulic, and optical *interlocks* have been applied to prevent machine operation unless barrier guards are in place (Fig. 37.3). Unless they are bypassed or are unreliable, interlocks increase the effectiveness of barrier guards in most applications.

Safety devices. Passive and active *safety devices* help reduce the risk or severity of injury. These devices include pull-back mechanisms for the operator's hands, seat belts, dead-man controls, and presence-sensing devices (Fig. 37.4).

Lockouts. An operator can physically separate—or lock out—a machine from its power source, making powered operations impossible. *Lockout* procedures call for operators or maintenance personnel to utilize a padlock or similar means to prevent other personnel from reconnecting the power source.

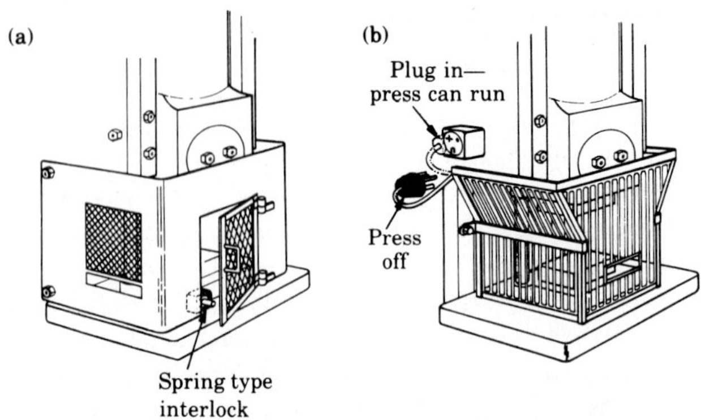

FIGURE 37.3
Barrier guards: (a) spring-type interlock shuts off power to machine when guard door is opened; (b) guard can only be removed by removing the plug, which then shuts off power to machine. *Source:* Triodyne, Inc.

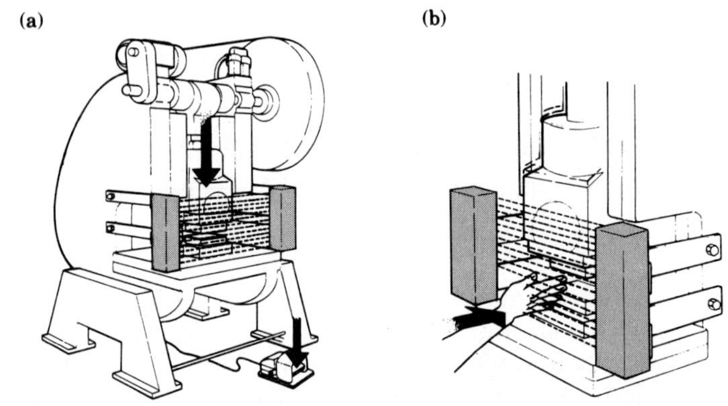

FIGURE 37.4
(a) Presence-sensing device, with light beams forming a curtain across the zone of operation. (b) Breaking the curtain of light beams by operator's hands sets brake on machine and disconnects clutch. *Source:* Triodyne, Inc.

Warnings. Posted signs, signals, and instructions warn the worker about hidden hazards, danger-control methods, or nonobvious consequences. Although inconsistent in actual use, the format and size of *warnings* have been prescribed by industry standards and governmental codes.

Personal protective equipment. Goggles, face shields, ear plugs, helmets, respirators, and protective clothing such as gloves and aprons are types of *personal protective equipment* that reduce worker exposure to hazards.

Approximately 85 percent of the accidents involving machinery are caused by the operator, 5 percent are caused by mechanical failure, and the remainder are caused by other factors. Employers are required to retrofit old machinery with guards. Manufacturers must include adequate guarding and safety devices on new machinery. Furthermore, employers who emphasize training are much more successful in reducing accidents through hazard recognition techniques and the development of safer work practices. Now that literally all machines are guarded, experience shows that machine designers and builders cannot influence safety to the extent that machine users can.

37.4 Environmental Considerations

The possible adverse effects of industrial activities and related operations on the health of individual workers as well as on our environment is, by now, well recognized by manufacturing industries, the public at large, and local, state, and federal governments. Environmental conditions regarding air, water, and land quality, and their control, are important aspects of all manufacturing activities, as is the conservation of natural resources. An understanding and appreciation of our ecological system is an integral part of this activity, including potential global climate change, stratospheric ozone depletion, loss of the earth's biological diversity, as well as less global issues such as oil spills, nuclear power-plant accidents, hazardous waste sites, the spread of pesticides, and leaking underground storage tanks.

Although there are many factors involved in environmental considerations, those that are impacted primarily by manufacturing activities can be summarized as follows: (a) pollutants from industrial plants such as solid and liquid waste, wastewater, and discharges from various facilities such as heat-treating plants, and (b) atmospheric pollutants, such as air pollution resulting from foundries and the use of coolants, lubricants, and various fluids in manufacturing operations. Over the past 20 years, the United States Congress has enacted various legislation that regulates a wide variety of industrial practices to improve our environment. We briefly outline the major regulations below, as they relate to manufacturing operations.

The Occupational Safety and Health Act (OSHA) concerns itself with unsafe conditions in the workplace. It encourages both employers and employees to reduce hazards in the workplace, including exposure to chemical and toxic substances, and implements existing or new safety and health programs. The National Institute for Occupational Safety and Health (NIOSH) was established by this Act. It conducts research, experiments, and demonstrations to identify toxic materials, determines safe exposure levels, and develops methodology for establishing health hazards.

The Environmental Protection Agency (EPA), established by the Executive Branch of the U.S. government, is charged with several responsibilities concerning toxic substances, clean air, water pollution, safe drinking water, and solid waste. The Toxic Substances Control Act regulates the manufacture, processing, distribution, use, and disposal of chemicals that present an unreasonable risk to humans or the environment. The Clean Air Act has the responsibility to set national air-quality standards, through controlling emissions from stationary as well as mobile sources, including, for example, emission of volatile components in lubricants. The Federal Water Pollution Act, also known as the Clean Water Act, has the responsibility to enforce the limiting of direct or indirect discharge of pollutants into navigable waters. The Safe Drinking Water Act provides for the protection of drinking water supplies and systems from contamination by toxic wastes, through the establishment of national drinking water quality standards.

Solid waste is controlled through two statutes. One is the Resource Conservation and Recovery Act, which requires the EPA to establish standards and regulations for hazardous waste generation and its storage, transportation, treatment, and disposal. Wastes are considered hazardous if they exhibit one or more of the following characteristics: ignitability (fire or explosion), corrosivity (damage to containers, thus allowing escape of wastes), or toxicity. The other statute is the Comprehensive Environmental Response, Compensation and Liability Act, which regulates existing hazardous waste sites, including unauthorized discharges and spills.

It is apparent from this brief review that manufacturing engineers and the management of manufacturing enterprises have great responsibility in protecting our environment. Although significant costs can be involved in implementing a large number of regulations that are mandated by local, state, and federal governments, in the long run it is in the interest of all of us to ensure the maintenance of a healthy and safe environment in which to live and work.

37.5 Product Liability

In the United States alone, at least 10 million people (4 percent of the population) suffer injuries on the job each year. At least 10,000 injuries are fatal (although this number has been slowly declining), and 30 percent of the injuries are classified as severe. As a result, millions of workdays are lost, with direct and indirect costs estimated at more than \$30 billion per year.

Assume that you have just purchased a hammer with a wooden handle. After using it for a few months, the handle breaks and the head of the hammer hits you, causing serious injury to your hand and requiring surgery. Further assume that the injury leaves your hand permanently deformed, and as a result, you cannot continue in your former job, which provided you with a good income. What recourse should you have to recover for damages to your hand, time lost from work, and its effects on your future employment and livelihood?

Until about one hundred years ago, the prominent legal theory applicable to machinery manufacturers was based on the Roman doctrine of *caveat emptor*, meaning *let the buyer beware*. If an injury resulted from the use of an unreasonably dangerous product, the injured party had virtually no legal recourse for obtaining compensation for the injury. In about 1900, the magnitude of the U.S. economy called for a different legal theory to bind manufacturers to their products. Since then, judicial decisions have gradually changed product liability doctrines.

37.5.1 Negligence

Under the legal theory of **negligence**, a party is liable for damages if it fails to act as a reasonable and prudent party would under like or similar circumstances.

For negligence theory to apply, the injured party, or plaintiff, must demonstrate two conditions: (1) that the standard of care was violated by the accused party, or defendant; and (2) that this violation was the proximate cause of the accident. The plaintiff must also demonstrate no contributary negligence in causing his or her own misfortune.

Several states rely on the concept of *comparative negligence*, or comparative fault, when deciding to what degree each party is responsible for an accident. For example, in comparative negligence, a jury may find that a plaintiff was 50 percent responsible for the accident and reduce the monetary reward by that amount. The manufacturer and distributor of the product involved may each be held 25 percent liable and thus share equally in the plaintiff's reduced compensation.

37.5.2 Strict liability

Under the legal theory of **strict liability**, the plaintiff must prove that:

- The product contained a defect that rendered it unreasonably dangerous (such as a bolt in a lawnmower that had a crack in it).
- The defect existed at the time the product left the defendant's hands (the manufacturer used a cracked bolt).
- The defect was a proximate cause of the injury (the crack propagated during use of the product by the plaintiff, and the bolt broke, injuring the plaintiff) and thus the product was unreasonably dangerous.

Under strict-liability laws, the actions of the plaintiff are irrelevant (the lawnmower was used on rough terrain). Whether the plaintiff acted as a reasonable and prudent party has no bearing on a claim based on strict liability. The emphasis is on the product, and a defense based on the contributory negligence of the plaintiff is invalid in most jurisdictions. Furthermore, manufacturers are required to anticipate reasonably foreseeable misuses of that product by the consumer (dropping the lawnmower on the pavement while removing it from the trunk of an automobile).

37.5.3 Defects

The definition of a defect has evolved over time. We may now define a **defect** as a fault, flaw, or irregularity that causes weakness, failure, or inadequacy in the form and function of a product that presents a risk of injury. A defect may result from product design, material selection, and/or manufacturing or production error. However, not all products that present risk of injury are defective. A sharp knife is not a defective product and is not unreasonably dangerous because the sharpness of the blade is essential for the intended use of the knife. The risk of injury is outweighed by the usefulness of the knife, which is made possible by its sharpness. Furthermore, the training that each of us receives at an early age in handling a knife helps to make this product reasonably safe.

Legal tests of whether a product is defective generally involve the following factors.

- Usefulness and desirability of the product.
- Availability of safer alternative products or work methods.
- Likelihood of injury and its probable seriousness.
- Obviousness of the hazard.
- Common knowledge and normal public expectation of the danger involved in the use of the product.
- Avoidability of injury by ordinary care in the use of the product.
- Ability to eliminate the hazard without seriously impairing the usefulness of the product.
- State-of-the-art in the particular industry at the time of product manufacture.
- Cost of making the product safer.
- Consumer willingness to pay for a higher priced but safer product.
- Bargaining power of the manufacturer as contrasted to that of the consumer.

Warnings and instructions. A product can be considered to be defective if warnings that would have prevented the accident are not provided with the product. Warning theories are complex, not all warnings improve operator safety, and warnings are not a substitute for training and instructions. Relevant instructions for use and maintenance should be presented in a clear and readable manner, affixed directly to the product. Warnings should clearly instruct the user about what to do and what not to do in using the product safely. Details should be presented in instruction, operating, and maintenance manuals that accompany the product.

• Example: Press builder liable for failure to warn.

An employee of a company was injured when his leg was caught in a hydraulic press. The press had a safety bar designed to prevent an operator from coming into contact with the pinch point created by the moving and stationary members of the press. At the time of the accident, however, the safety bar was not attached. The employee filed suit against the press manufacturer, claiming that the company had a duty to warn about the dangers of using the press without the safety bar. During the trial, the manufacturer claimed that (a) the injury was due to the failure of the employer to make sure that the safety bar was attached, and that (b) it had no duty to warn because, if properly attached and maintained, the bar would have prevented the accident. The court ruled that the manufacturer is liable because it should have realized that the safety bar had to be removed in order to service the machine. Since the manufacturer should have realized that the bar might not be properly replaced after servicing the machine, it should have placed a warning on the machine advising operators of the need for reinstalling the safety bar.

Example: Guarding of multipurpose press brake.

An employee was injured while working on a press brake (see Fig. 16.23f). As he reached into the press to remove a part that had become stuck in the die, the press ram came down and severed three fingers. The employee sued the press manufacturer, claiming that the press was defectively designed because it had been sold without guards to keep his hand out of the point of operation. The court ruled that the press manufacturer was not responsible for the accident because (a) it had sold the press to the employee's company without dies, and (b) under these circumstances the manufacturer had no duty to equip the press brake with all the various safety devices that might be required for safe operation.

37.5.4 Designing and manufacturing safe products

Over time and on the basis of accumulated experience, certain guidelines have been established for designing and manufacturing safe products. It is essential that product safety be viewed in terms of its design, manufacture, distribution, and ultimate use. Product safety should be the collective responsibility of all parties concerned, including design, materials, and manufacturing engineers, machine operators, supervisors, inspectors, shipping personnel, and management. Product safety and loss-control programs should be implemented through committees representing all departments and having the full support of management. The basic guidelines for designing and manufacturing safe products are:

a) Product design concepts should anticipate obvious and possible dangers in product use and foreseeable misuse as well as the injuries that failure of one or more components could cause.
b) All those concerned with the designing and manufacturing of the product should be fully aware of current industry and government standards and regulations concerning that product. The product should at least meet those standards, even though the standards may not fully anticipate all possibilities in the use and misuse of the product. However, complying with these standards does not necessarily protect the manufacturer from product liability.

c) All stages of production (raw materials, manufacturing processes, assembly procedures, and inspection techniques) should be monitored carefully and continuously to ensure that no defective components pass from the production line. Major emphasis should be placed on quality assurance and 100 percent inspection of all parts.
d) Warning labels and instruction manuals should be prepared with great care, and all relevant aspects of possible use and foreseeable misuse of the product should be addressed.
e) It is essential to keep complete records of the design, manufacturing, testing, and quality control of all product components and their assembly, with dates and appropriate identification numbers.

SUMMARY

Because technology ultimately serves humanity, the well-being of human beings as operators of machinery and as users of products is an integral aspect of all manufacturing activities. Efficient and safe operations can be planned and conducted in the workplace through comprehensive studies of man–machine interactions. Designing properly the physical environment, including air quality, lighting, and noise, can increase job performance.

Safety should be the concern of all who are involved in design and production. Unsafe operation of machinery and products is a major source of worker and user injury and financial loss. A variety of safeguarding systems has been developed and should be used. Product liability has become an important aspect of manufacturing and using products and has a major impact on the economics of manufacturing operations.

TRENDS

- Human engineering and ergonomics continue to be an essential element in designing and using manufacturing machinery and equipment, particularly in view of advances made in automation, requiring a different employee role.
- Safety will continue to be the responsibility of all parties concerned with manufacturing and using both consumer and industrial products.
- Product liability will continue to be an important element in design and manufacturing. The upward trend in the number of product liability cases, the high cost of awards by courts, the cost of defending a case, and the cost of product liability insurance have become a major concern.
- Limitations on injury claims and awards and national standardization are being considered by state legislatures and the U.S. Congress.
- The comparative fault theory is becoming accepted in an increasing number of states in the United States.

KEY TERMS

Danger	Negligence	Safeguarding methods
Defect	Product liability	Safety
Hazard	Risk	Strict liability
Human-factors engineering		

BIBLIOGRAPHY

Accident Prevention Manual for Industrial Operations. Chicago: National Safety Council (revised periodically).

Best's Safety Directory, 25th ed., Vol. II, 1985, revised periodically.

Brown, S., (ed.), *The Product Liability Handbook: Prevention, Risk, Consequence and Forensics of Product Failure.* New York: Van Nostrand Reinhold, 1991.

Clark, T.S., and E.N. Corlett, *The Ergonomics of Workspaces and Machines: A Design Manual.* London: Taylor and Francis, 1984.

Colangelo, V.J., and P.A. Thornton, *Engineering Aspects of Product Liability.* Metals Park, Ohio: American Society for Metals, 1981.

Environmental Law Handbook, 10th ed. Rockville, Maryland: Government Institutes, Inc., 1989.

Faulkner, L.L., *Handbook of Industrial Noise Control.* New York: Industrial Press, 1976.

Hammer, W., *Handbook of System and Product Safety.* Englewood Cliffs, N.J.: Prentice Hall, 1972.

———, *Occupational Safety Management and Engineering.* Englewood Cliffs, N.J.: Prentice Hall, 1985.

Heinrich, H.W., D. Petersen, and N. Roos, *Industrial Accident Prevention*, 5th ed. New York: McGraw-Hill, 1980.

Kolb, J., and S.S. Ross, *Product Safety and Liability—A Desk Reference.* New York: McGraw-Hill, 1980.

Sanders, M.S., and E.J. McCormick, *Human Factors in Engineering and Design.* New York: McGraw-Hill, 1987.

Schoemacher, B., *Engineers and the Law—An Overview.* New York: Van Nostrand Reinhold, 1986.

Thorpe, J.F., and W.H. Middendorf, *What Every Engineer Should Know About Product Liability.* New York: Marcel Dekker, 1979.

Weinstein, A.A., A.D. Twerski, H.R. Piehler, and W.A. Donaher, *Product Liability and the Reasonably Safe Product.* New York: Wiley, 1978.

Woodson, W.E., *Human Factors Design Handbook.* New York: McGraw-Hill, 1981.

REVIEW QUESTIONS

37.1 Explain your understanding of what human-factors engineering involves. Why is it important?

37.2 What are the important features of a workstation so far as the worker is concerned?
37.3 Inspect some of the instruments with which you are familiar, and explain how you would modify them to make them easier to use and less susceptible to reading errors.
37.4 How may environmental conditions affect the performance of workers?
37.5 Why is the noise level in a manufacturing plant important? Name several possible effects it has on workers.
37.6 Describe your understanding of what safety means.
37.7 Explain what is meant by the terms hazard and risk. Give several examples of each from your own experience.
37.8 List the typical causes of injury to workers. How would you go about avoiding them?
37.9 Name some unsafe situations or practices in your own home. Why are they unsafe?
37.10 Describe common methods of safeguarding of machinery.
37.11 What is the basic principle of interlocks?
37.12 List some of the equipment used for personal protection of workers. Do they have any disadvantages?
37.13 Why is it that machine builders cannot influence the level of safety to the extent that machine users can?
37.14 Explain the concept of caveat emptor. Why was it commonly accepted until the 1900s?
37.15 What trends in product liability took place after 1900?
37.16 Explain what is meant by negligence. What is comparative negligence?
37.17 List the proofs to be presented by the plaintiff according to strict-liability law.
37.18 What is the legal definition of a defect? What are the legal tests of whether a product is defective?
37.19 Do you think that a pair of sharp scissors is a defective product? Explain.
37.20 List the basic guidelines for designing and manufacturing safe products.

QUESTIONS AND PROBLEMS

37.21 In Fig. 37.1, is the difference in the noise level between 70 and 80 dB the same as between 150 and 160 dB? Explain.
37.22 Explain why product liability has become such an important factor in manufacturing. Why has it lagged in most other countries?
37.23 Can some of the safeguarding devices themselves on machines be dangerous? Give some examples and explain.
37.24 Decribe some of the machines that you have come across or used. Explain any features that make them unsafe. Propose some design changes to make them safer.
37.25 Should monetary awards be limited in product liability cases? What are the pros and cons of such limits? Should they depend on the type of accident?
37.26 The monthly *American Machinist* trade magazine has a page in each issue that gives summaries of court cases on product liability. Make a review of the latest issues of this magazine and describe your observations. As an executive of a manufacturing firm, what policies would you implement to avoid such litigation?
37.27 Describe your own thoughts regarding the environmental considerations described in Section 37.4.
37.28 In Section 37.3.1, we outlined certain personal protective equipment. Do you think such equipment can itself present hazards to the worker? Give specific examples.

37.29 Most modern machines are now built with ergonomic considerations in mind, whereas older machines didn't generally have these features. Describe your opinions concerning retrofitting older machines to improve their safe and comfortable use.

37.30 Assume that you are working as a design and manufacturing engineer in a company. You would like to implement the various topics described in this chapter; however, they are costly and the management is reluctant to approve them to the same extent that you would like to see. How would you go about convincing your supervisors of the merits of your approach?

PART VIII

Manufacturing in a Competitive Environment

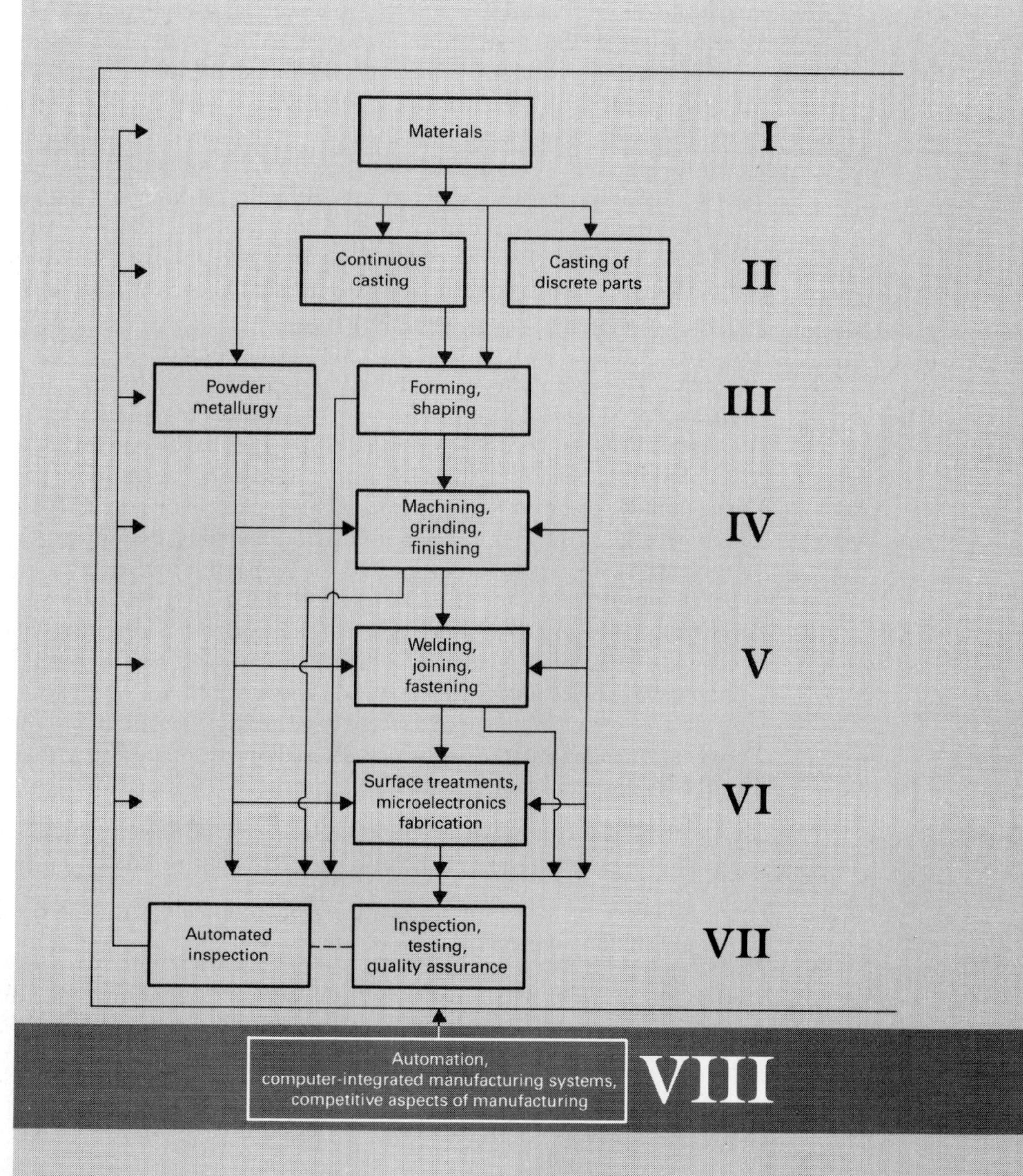

• INTRODUCTION •

The various processes and related operations you have learned about in this course constitute the important elements of manufacturing from an engineering point of view. Understanding these processes is fundamental. However, even more important is that you acquire a strong sense of the *context* of the individual subjects, that is, an understanding of how these subjects interrelate.

We have presented a schematic drawing at the beginning of each of the eight parts of this text, but we have not previously drawn your attention to it. Take a few moments now to study it and check your understanding of how the various subjects in this text interrelate. You will find this exercise useful even if you have not studied every part of the text. Note how the subjects in Part VIII (automation of manufacturing processes, computer-integrated manufacturing systems, and competitive aspects of manufacturing) encompass all the other topics that we have presented. We constructed the schematic in this manner to suggest that you should view these technical subjects in the context of the *competitive* forces that control the choice and implementation of various technologies in manufacturing products. We have regularly discussed and referred to many of the aspects of these forces—especially manufacturing cost, product quality, and the use of automation—and we hope that you have become sensitive to the significance of these factors as they relate to various manufacturing operations.

Advances in automating manufacturing processes, which we describe in Chapter 38, have been driven by several competitive forces, such as the need to improve productivity and product quality and decrease manufacturing costs. With potential benefits as fundamental as these, it is hardly surprising that automation has been implemented so widely in manufacturing operations. Beginning with the development of numerical control of machine tools in the 1950s, automation has become a key factor in several industries, enabling manufacturers to reverse the problems of escalating costs and/or declining product quality and thus enhance their competitive positions. As these technologies evolve, the number of applications expands rapidly. Although automation offers exciting benefits, we also recognize the potential disadvantages, limitations, and improper application of these technologies and their impact on the work force.

In this chapter we survey several automation technologies and focus on their application in manufacturing processes and operations. Among the subjects covered and their potential benefits are:

- Numerical control of machine tools, which allows flexible operation, improves productivity, and ensures consistent product quality.
- Adaptive control of manufacturing operations, optimizing production rates and minimizing costs.
- Material-handling systems and industrial robots, including automated guided vehicles, to increase operating efficiency and reduce labor costs.

- Sensor technology for automated measuring and monitoring of manufacturing operations and inspecting products for quality and dimensional accuracy.
- Flexible fixturing of workpieces as an aid in automated and computer-controlled manufacturing systems.
- Automated assembly to reduce manufacturing costs.

Much of the progress in automating manufacturing facilities is closely related to our ability to treat manufacturing operations as a *system.* As we successfully implement a systems approach to manufacturing, we can *integrate* various functions and activities that had previously been separate entities. In this way we can optimize the entire manufacturing process, rather than having to limit ourselves to considering isolated technical and human elements. In Chapter 39 we describe the integration of manufacturing operations and discuss the most promising technologies for accomplishing integration. We describe several topics that have been so widely discussed and are now so familiar that they are commonly referred to in terms of acronyms:

- Computer-aided design (CAD).
- Computer-aided manufacturing (CAM).
- Computer-integrated manufacturing (CIM).
- Computer-aided process planning (CAPP).
- Group technology (GT).
- Cellular manufacturing (CM).
- Flexible manufacturing systems (FMS).
- Just-in-time production (JIT).
- Manufacturing automation protocol (MAP).
- Artificial intelligence (AI).

Each of these technologies holds great promise for improving productivity, product quality, flexibility of operation, cost of manufacturing, and other performance criteria. The prospect of their total integration into a single system has profound implications for the future of manufacturing. We conclude the chapter with a discussion of such a vision: *the factory of the future.*

In the final chapter, we focus on three critical aspects of competition:

- Manufacturing economics, emphasizing the costs involved.
- Factors influencing selection of materials for products.
- Factors influencing selection of a manufacturing process.

In this chapter we highlight the major considerations within these areas and include a number of examples that illustrate the key factors involved. This discussion exposes you to the major factors involved in evaluating manufacturing alternatives, helping you develop an ability to identify and assess them.

We began this text with a General Introduction. In it we listed several goals of manufacturing:

- Fully meeting design and service requirements and product specifications.
- Finding the most economical methods of production.
- Building quality into the product at each stage of design and manufacturing.
- Ensuring that manufacturing processes are flexible enough to respond to changing market demand and product needs, both in variety and quantity.
- Viewing manufacturing as a system in order to integrate its parts and thus improve performance.
- Continually striving for higher levels of productivity.

A careful review of these goals will provide you with the proper context for studying the topics and examples presented in Part VIII.

38

Automation of Manufacturing Processes

38.1 Introduction

Until about four decades ago, most manufacturing operations were carried out on traditional machinery, such as lathes, milling machines, and presses, which lacked flexibility and required considerable skilled labor. Each time a different product was manufactured, the machinery had to be retooled, and the movement of materials had to be rearranged. The development of new products and parts with complex shapes required numerous trial-and-error attempts by the operator to set the proper processing parameters on the machine. Furthermore, because of human involvement, making parts that were exactly alike was difficult.

These circumstances meant that processing methods were generally inefficient and that labor costs were a significant portion of overall production costs. The need for reducing the labor share of product cost gradually became apparent, as was the need to improve the efficiency and flexibility of manufacturing operations. This need was particularly significant in terms of increased competition, both nationally and from other industrialized countries.

Productivity also became a major concern. Defined as the optimum use of all resources—materials, energy, capital, labor, and technology—or as output per employee per hour, **productivity** basically measures operating efficiency. With rapid advances in the science and technology of manufacturing and their gradual implementation, the efficiency of manufacturing operations began to improve and the percentage of total cost represented by labor costs declined.

How can productivity be improved? Mechanization of machinery and operations had, by and large, reached its peak by the 1940s. **Mechanization** runs a process or operation with the use of various mechanical, hydraulic, pneumatic, or electrical devices. Take, for example, the use of a simple can opener. Opening a thousand cans by hand would take a long time, would require much physical effort, and would be tedious. You would soon lose interest, and your efficiency would drop off. On the other hand, using an electric can opener—which is a mechanical device—takes less time and effort, but the job is still tedious, and your efficiency is still likely to drop after a while. Note that, in mechanized systems, the operator still directly controls the process, and must check each step of the machine's performance. If a tool breaks during machining, if parts overheat during heat treatment, if surface finish begins to deteriorate during grinding, and if dimensional tolerances become too large in metal forming, the operator has to intervene and change one or more process parameters.

The next step in improving the efficiency of manufacturing operations was **automation**, from the Greek word *automatos*, meaning self-acting. The word automation was coined in the mid-1940s by the U.S. automobile industry to indicate automatic handling of parts between production machines, together with their continuous processing at the machines. During the past three decades, major advances and breakthroughs in the types and extent of automation have occurred. These important developments were made possible largely through rapid advances in the capacity and sophistication of *control systems* and *computers*.

In this chapter, we follow the outline shown in Fig. 38.1, first reviewing the history and principles of automation and how they help us *integrate* various operations and activities in a manufacturing plant to improve productivity. We then introduce the important concept of control of machines and systems through *numerical control* and *adaptive control* techniques. An essential element in manufacturing is the movement of raw materials and parts in various stages of completion throughout the plant. We present the basic concepts of *material handling*, and how they have been developed into various systems, including the use of *industrial robots*, to improve their efficiency. We then introduce the subject of *sensor technology*, which is an essential element in the control and optimization of machinery, processes, and systems.

Next we describe the concept and applications of *flexible fixturing* and *assembly operations*. These methods enable us to take full advantage of advanced manufacturing technologies, particularly those of flexible manufacturing systems. Discussion of these topics concludes our review of automation of manufacturing processes. We will then be ready to consider major developments in integrated manufacturing systems and their impact on all aspects of manufacturing operations in Chapter 39.

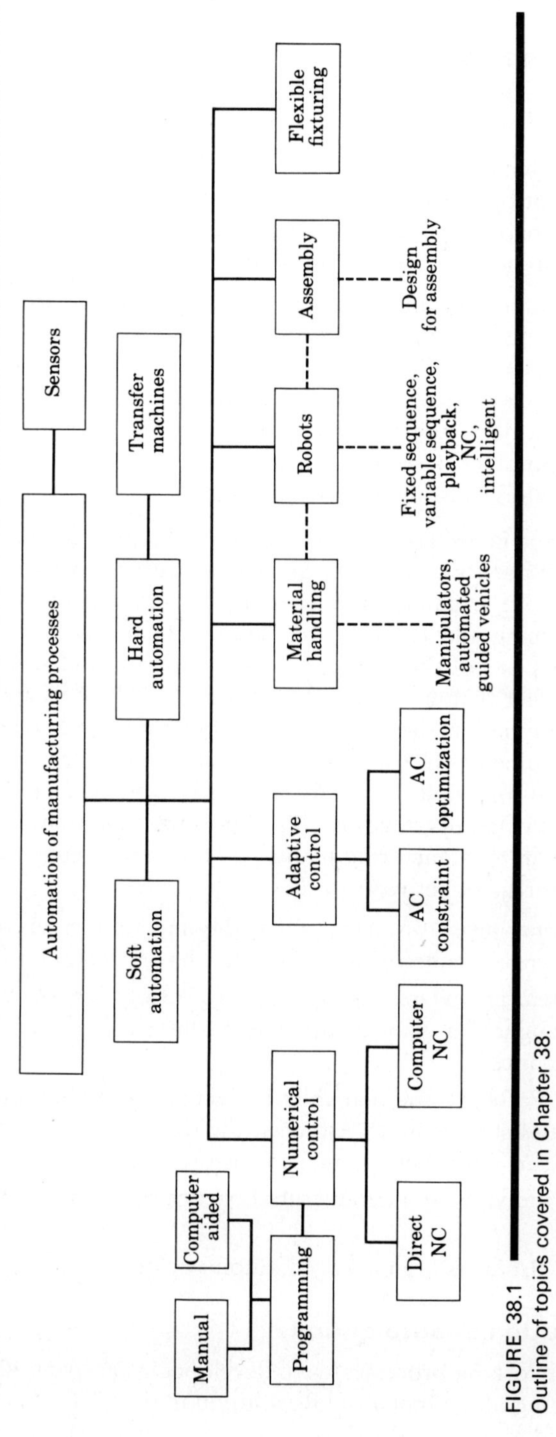

FIGURE 38.1 Outline of topics covered in Chapter 38.

38.2 Automation

Automation can generally be defined as the process of following a predetermined sequence of operations with little or no human labor, using specialized equipment and devices that perform and control manufacturing processes. The meaning and concept of automation has been variously interpreted as follows:

a) Semiautomatic or automatic material handling, workpiece loading and unloading in machines and workholding fixtures, and use of other labor-saving devices.
b) Automatic cycle control of machines and equipment, including use of mechanical devices, numerical control of machines, and use of computers.
c) Complete computer-based control of all aspects of manufacturing operations from raw materials to the finished product.

As we describe in Section 38.8 and Chapter 39, automation, in its full sense, is achieved through the use of a variety of devices, sensors, actuators, techniques, and equipment that are capable of observing the manufacturing process, making decisions concerning the changes that should be made in the operation, and controlling all aspects of the operation. Automation is and will continue to be an *evolutionary*, rather than a revolutionary, concept. All of us are familiar with the evolution of automation, beginning with hand tools and hand-operated simple machines, continuing to mechanized processes and machines, and finally moving to higher levels of automation. In the rest of this chapter and in Chapter 39, we give several examples of recent stages in this important evolution.

Automation in manufacturing plants has been implemented successfully in the following basic areas of activity:

- *Manufacturing processes.* Machining, forging, cold extrusion, and grinding operations are examples of processes that have been automated extensively.
- *Material handling.* Materials and parts in various stages of completion are moved throughout a plant by computer-controlled equipment without human guidance.
- *Inspection.* Parts are automatically inspected for quality, dimensional accuracy, and surface finish, either at the time of manufacturing (in-process inspection); or after they are made (postprocess inspection).
- *Assembly.* Individually manufactured parts are automatically assembled into a product.
- *Packaging.* Products are packaged automatically.

38.2.1 Evolution of automation

Although metalworking processes were developed as early as 4000 B.C., it was not until the beginning of the Industrial Revolution in the 1750s that automation began to be introduced in the production of goods. As you can see in Table 1 in the

General Introduction, machine tools, such as turret lathes, automatic screw machines, and automatic bottle-making equipment, were developed in the late 1890s and early 1900s. Mass-production techniques and transfer machines were developed in the 1920s. These machines had fixed automatic mechanisms and were designed to produce specific products. These developments were best represented by the automobile industry, which produced passenger cars at a high production rate and low cost.

The major breakthrough in automation began with numerical control (NC) of machine tools in the late 1940s. Since this historic development, rapid progress has been made in automating many aspects of manufacturing (Table 38.1). These

TABLE 38.1
DEVELOPMENTS IN THE HISTORY OF AUTOMATION OF MANUFACTURING PROCESSES

DATE	DEVELOPMENT
1500–1600	Water power for metalworking; rolling mills for coinage strips.
1600–1700	Hand lathe for wood; mechanical calculator.
1700–1800	Boring, turning, and screw cutting lathe; drill press.
1800–1900	Copying lathe; turret lathe; universal milling machine; advanced mechanical calculators.
1808	Sheet-metal cards with punched holes for automatic control of weaving patterns in looms.
1863	Automatic piano player (Pianola).
1900–1920	Geared lathe; automatic screw machine; automatic bottlemaking machine.
1920	First use of the word *robot.*
1920–1940	Transfer machines; mass production.
1940	First electronic computing machine.
1943	First digital electronic computer.
1945	First use of the word *automation*.
1948	Invention of the transistor.
1952	First prototype numerical-control machine tool.
1954	Development of the symbolic language APT (Automatically Programmed Tool); adaptive control.
1957	Commercially available NC machine tools.
1959	Integrated circuits; first use of the term *group technology*.
1960s	Industrial robots.
1965	Large-scale integrated circuits.
1968	Programmable logic controllers.
1970	First integrated manufacturing system; spot welding of automobile bodies with robots.
1970s	Microprocessors; minicomputer controlled robot; flexible manufacturing systems; group technology.
1980s	Artificial intelligence; intelligent robots; smart sensors; untended manufacturing cells.
1990s	Integrated manufacturing systems; just-in-time production; intelligent and sensor-based machines; telecommunications and global manufacturing networks.

involve the introduction of computers into automation, computerized numerical control (CNC), adaptive control, industrial robots, and computer-integrated manufacturing (CIM) systems, including computer-aided design and computer-aided manufacturing (CAD/CAM).

38.2.2 Goals and applications of automation

Automation has several primary goals:

- Integrate various aspects of manufacturing operations, so as to improve product quality and uniformity, minimize cycle times and effort, and thus reduce labor costs.
- Improve productivity by reducing manufacturing costs through better control of production. Parts are loaded, fed, and unloaded on machines more efficiently. Machines are used more effectively and production is organized more efficiently.
- Reduce human involvement, boredom, and possibilities of human error.
- Reduce workpiece damage caused by manual handling of parts.
- Raise the level of safety for personnel, especially under hazardous working conditions.
- Economize on floor space in the manufacturing plant by arranging machines, material movement, and related equipment more efficiently.

Automation and production quantity. Production quantity is crucial in determining the type of machinery and equipment—and the level of automation—required to produce parts economically. Before we proceed further, let's define some basic production terms. *Total production quantity* is defined as the total number of parts to be made. This quantity can be produced in individual batches of various *lot sizes*. Lot size greatly influences the economics of production, as we describe in Chapter 39. *Production rate* is defined as the number of parts produced per unit time, such as per day, month, or year. The approximate and generally accepted ranges of production volume are shown in Table 38.2 for some typical applications. As you might expect, *experimental* or *prototype* products represent the lowest volume.

TABLE 38.2
APPROXIMATE ANNUAL VOLUME OF PRODUCTION

TYPE OF PRODUCTION	NUMBER PRODUCED	TYPICAL PRODUCTS
Experimental or prototype	1–10	All
Piece or small batch	10–5000	Aircraft, special machinery, dies
Batch or high volume	5000–100,000	Trucks, agricultural machinery, jet engines, diesel engines
Mass production	100,000 and over	Automobiles, appliances, fasteners

Small quantities per year (Fig. 38.2) can be manufactured in *job shops* using various standard general-purpose machine tools (*stand alone* machines) or machining centers. These operations have high part variety, meaning that different parts can be produced in a short time without extensive changes in production operations and tooling. However, in job shops machinery generally requires skilled labor to operate, production quantity and rate are low, and hence cost per part can be high (Fig. 38.3). Production of parts involving a large labor component is known as *labor intensive*.

Piece-part production usually involves very small quantities and is suitable for job shops. The majority of piece-part production is in lot sizes of 50 or less. *Small-batch* production quantities usually range from 10 to 100, and general-purpose machines and machining centers with various computer controls are used. *Batch* production usually involves lot sizes between 100 and 5000 and utilizes machinery similar to that used for small-batch production but with specially designed fixtures for higher production rates.

Mass production involves quantities of 100,000 and over, and requires special-purpose machinery, called *dedicated machines*, and automated equipment for transferring materials and parts. Although the machinery, equipment, and specialized tooling are expensive, labor skills required and labor costs are relatively low because of the high level of automation. However, these production systems are organized for a specific product and, consequently, lack flexibility. Thus, as you can see in Fig. 38.2, production flexibility and productive capacity are inversely related. Most manufacturing facilities operate with a variety of machines in combination.

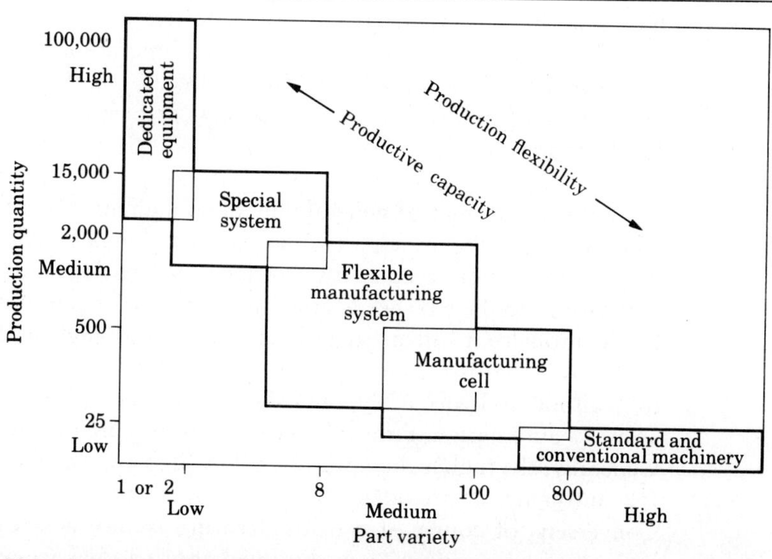

FIGURE 38.2 Capabilities of production systems and machines for various production quantities and part variety. See also Chapter 39.

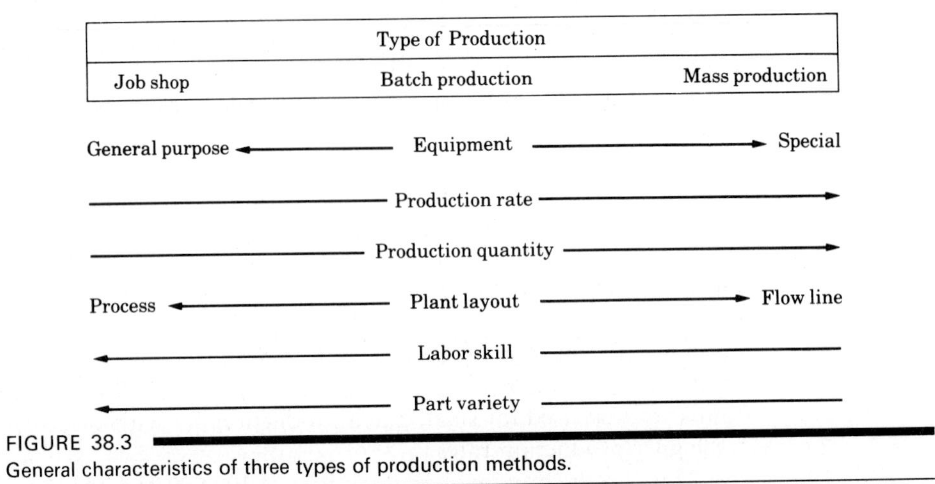

FIGURE 38.3
General characteristics of three types of production methods.

Applications of automation. Automation can be applied to manufacturing all types of goods, from raw materials to finished products, and in all types of production from job shops to large manufacturing facilities. The decision to automate a new or existing production facility requires the following additional considerations:

a) Type of product manufactured.
b) Quantity and rate of production required.
c) The particular phase of manufacturing operation to be automated.
d) Level of skill in the available workforce.
e) Reliability and maintenance problems associated with automated systems.
f) Economics.

Because automation generally involves high initial cost of equipment, as well as a knowledge of the principles of operation and maintenance required, a decision on its implementation, even at low levels of automation, must involve a careful study with regard to the true needs of an organization. It is not unusual for a company to begin automation with great enthusiasm and with high across-the-board goals, only to discover that its economic benefits were largely perceived rather than real, and that automation was, in the final assessment, not cost-effective. In many situations, *selective* automation, rather than total automation, of a facility would be desirable. Generally, the higher the level of skill available in the workforce, the less the need for automation, provided there is a sufficient number of workers available. Conversely, of course, if a manufacturing facility is already automated, the skill level required is lower. As we outlined above and as we will see in the rest of Part VIII, there are numerous important and complex issues involved in the decision-making process for automation.

38.2.3 Hard automation

In **hard,** or *fixed-position,* **automation,** the production machines are designed to basically produce a standardized product, such as engine blocks, valves, gears, and spindles. Although product size and processing parameters (such as speed, feed, and depth of cut) can be changed, these machines are specialized. They lack flexibility and cannot be modified to any significant extent to accommodate products that have widely different shapes and dimensions (see Group Technology, Section 39.8). Because these machines are expensive to design and construct, their economic use requires mass production of parts in very large quantities.

Machines used in hard-automation applications are usually built on the *building-block,* or *modular,* principle. They are generally called *transfer machines,* and consist of the following two major components: powerhead production units and transfer mechanisms.

Powerhead production units. Consisting of a frame or bed, electric driving motors, gearboxes, and tool spindles (Fig. 38.4), *powerhead production units* are self-contained. The components are commercially available in various standard sizes and capacities. They can easily be regrouped for producing a different part and thus have certain adaptability and flexibility. Transfer machines consist of two or more powerhead units, which can be arranged on the shop floor in linear, circular, or U patterns. The weight and shape of parts influence the arrangement selected. The arrangement is also important for continuity of operation in the event of tool failure or machine breakdown in one or more of the units. Buffer storage features are incorporated in these machines to permit continued operation.

Transfer mechanisms. *Transfer mechanisms* are used to move the workpiece from one station to another in the machine—or from one transfer machine to another—to enable various operations to be performed on the part. Workpieces are

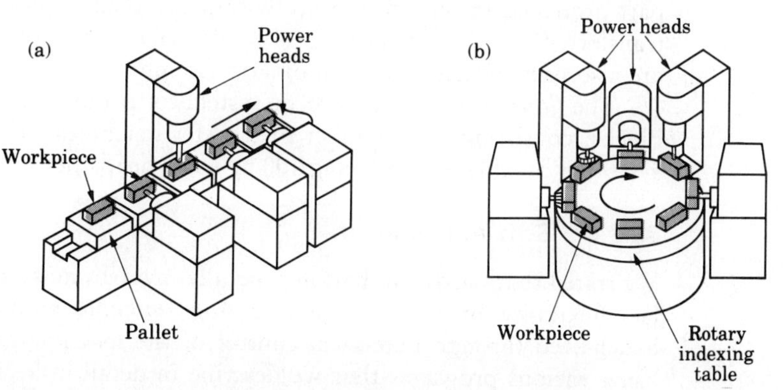

FIGURE 38.4 Arrangement of powerhead units in (a) straight and (b) circular patterns.

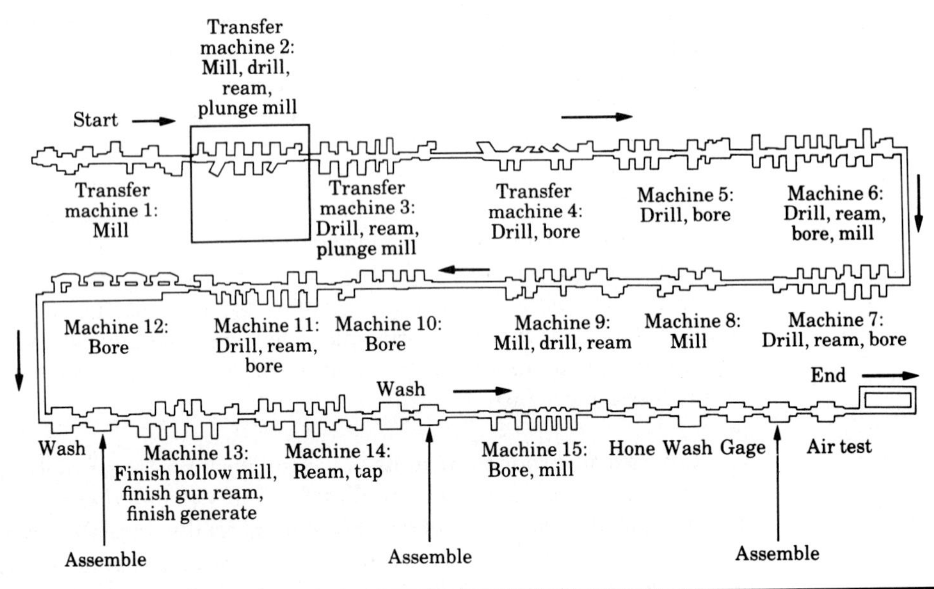

FIGURE 38.5
A large transfer line for producing engine blocks and cylinder heads. *Source:* Ford Motor Company and *American Machinist,* July 1985.

transferred by several methods: (1) rails along which the parts, usually placed on pallets, are pushed or pulled by various mechanisms (Fig. 38.4a); (2) rotary indexing tables (Fig. 38.4b); and (3) overhead conveyors. Transfer of parts from station to station is usually controlled by sensors and other devices. Tools on transfer machines can be changed easily using toolholders with quick-change features. These machines may be equipped with various automatic gaging and inspection systems. These systems are utilized between operations to ensure that the dimensions of a part produced in one station are within acceptable tolerances before that part is transferred to the next station. As we describe in Section 38.10, transfer machines are also used extensively in automatic assembly.

The *flow lines* for a very large system for producing cylinder heads for engine blocks, consisting of a number of transfer machines, is shown in Fig. 38.5. This system is capable of producing 100 cylinder heads per hour.

38.2.4 Soft automation

We stated that hard automation generally involves mass-production machines that lack flexibility. In **soft,** or *flexible* or *programmable,* **automation,** greater flexibility is achieved through numerical control of the machine and its various functions, using various programs that we describe in detail in Sections 38.3 and 38.4. Soft automation is an important development, because the machine can be easily and readily reprogrammed to produce a part that has a different shape or dimensions than the one just produced. Because of this capability, soft automation can produce parts with complex shapes. Further advances in flexible automation, with extensive use of modern computers, has led to the development of *flexible manufacturing systems* (see Section 39.10), with high levels of efficiency and productivity.

38.2.5 Programmable controllers

The control of manufacturing processes in a proper sequence, involving groups of machines and various material-handling equipment and accessories, has traditionally been done by timers, switches, relays, counters, and similar hardwired devices based on mechanical, electromechanical, and pneumatic principles. Beginning in 1968, **programmable logic controllers** (PLC; also called PC, but not to be confused with personal computer) were introduced to replace these devices. A programmable logic controller has been defined by the National Electrical Manufacturers Association (NEMA) as "a digitally operating electronic apparatus which uses a programmable memory for the internal storage of instructions for implementing specific functions such as logic, sequencing, timing, counting, and arithmetic to control, through digital or analog input/output modules, various types of machines or processes." The digital computer, which is used to control the functions of a programmable controller, is considered to be within this scope.

Because relay control panels can now be eliminated and PLCs can be reprogrammed and take less space, programmable controllers have become widely adopted in manufacturing systems and operations. Their basic functions are on–off, motion, sequential operations, and feedback control. In addition to their use in manufacturing process control, they are also used in system control with high-speed digital processing and communication capabilities. These controllers perform reliably in industrial environments and improve the overall efficiency of the operation.

38.3 Numerical Control

Numerical control (NC) is a method of controlling the movements of machine components by directly inserting coded instructions in the form of numerical data (numbers and letters) into the system. The system automatically interprets these data and converts it to output signals. These signals, in turn, control various machine components, such as turning spindles on and off, changing tools, moving the workpiece or the tools along specific paths, and turning cutting fluids on and off.

In order to appreciate the importance of numerical control of machines, let's briefly review how a process such as machining has been carried out traditionally.

After studying the working drawings of a part, the operator sets up the appropriate process parameters (such as cutting speed, feed, depth of cut, cutting fluid, and so on), determines the sequence of operations to be performed, clamps the workpiece in a workholding device such as a chuck or collet, and proceeds to make the part. Depending on part shape and the dimensional accuracy specified, this approach usually requires skilled operators. Furthermore, the machining procedure followed may depend on the particular operator, and because of the possibilities of human error, the parts produced by the same operator may not all be identical. Part quality may thus depend on the particular operator or even the same operator on different days or different hours of the day. Because of our increased concern with product quality and reducing manufacturing costs, such variability and its effects on product quality are no longer acceptable. This situation can be eliminated by numerical control of the machining operation.

We can illustrate the importance of numerical control by the following example. Assume that holes have to be drilled on a part in the positions shown in Fig. 38.6. In the traditional manual method of machining this part, the operator positions the drill with respect to the workpiece, using as reference points any of the three methods shown. The operator then proceeds to drill these holes. Let's assume that 100 parts, having exactly the same shape and dimensional accuracy, have to be drilled. Obviously, this operation is going to be tedious because the operator has to go through the same motions again and again. Moreover, the probability is high that, for various reasons, some of the parts machined will be different from others. Let's further assume that during this production run, the order for these parts is changed, so that 10 of the parts now require holes in different positions. The machinist now has to reset the machine, which will be time consuming and subject to error. Such operations can be performed easily by numerical control machines that are capable of producing parts repeatedly and accurately and of handling different parts by simply loading different part programs.

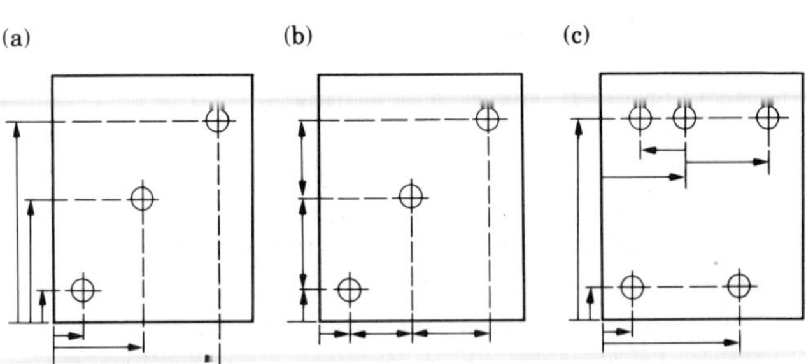

FIGURE 38.6
Positions of drilled holes in a workpiece. Three methods of measurements are shown: (a) absolute dimensioning, referenced from one point at the lower left of the part; (b) incremental dimensioning, made sequentially from one hole to another; and (c) mixed dimensioning, a combination of both methods.

In numerical control, data concerning all aspects of the machining operation, such as locations, speeds, feeds, and cutting fluid, are stored on magnetic tape, cassette, floppy or hard disks, or paper or plastic (Mylar, a thermoplastic polyester). Originally data were stored on punched 25-mm (1-in.) wide paper or plastic tape, although magnetic storage devices are now coming into increased use. The concept of NC control is that specific information can be relayed from these storage devices to the machine tool's control panel. On the basis of input information, relays and other devices (called *hard-wired controls*) can be actuated to obtain a desired machine setup. This method eliminates the need for manual setting of machine positions and tool paths or the use of templates and other mechanical guides and devices. Complex operations, such as turning a part having various contours and die sinking in a milling machine, can be carried out.

Numerical control has had a major impact on all aspects of manufacturing operations. It is a widely applied technology, particularly in the following areas:

a) Machining centers.
b) Milling, turning, boring, drilling, and grinding.
c) Electrical-discharge, laser-beam, and electron-beam machining.
d) Water-jet cutting.
e) Punching and nibbling.
f) Pipe bending and metal spinning.
g) Spot welding and other welding and cutting operations.
h) Assembly operations.

Numerical control machines are now used extensively in small- and medium-quantity production (typically 500 parts or less) of a wide variety of parts in small shops and large manufacturing facilities. Older machines can be retrofitted with numerical control.

38.3.1 Historical background

The basic concept of numerical control apparently was implemented in the early 1800s, when punched holes in sheet metal cards were used to automatically control weaving machines. Needles were activated by sensing the presence or absence of a hole in the card. This invention was followed by automatic piano players (Pianola), in which the keys were activated by air flowing through holes punched in a perforated roll of paper.

The principle of numerically controlling the movements of machine tools was conceived in the 1940s by J. Parsons in his attempt to machine complex helicopter blades. The first prototype NC machine was built in 1952 at the Massachusetts Institute of Technology. It was a vertical-spindle, two-axis copy milling machine, retrofitted with servomechanisms (see Section 38.3.4), and the machining operations performed consisted of end milling and face milling on an aluminum plate. The numerical data to be punched into the paper tapes were generated by a digital computer, which was being developed at the same time at MIT. The experiments successfully machined parts accurately and repeatedly without operator intervention. On the basis of this success, the machine-tool industry began building and marketing NC machine tools, the latest developments in which are machining centers, the principles of which we described in Section 24.2.

38.3.2 Advantages and limitations

Numerical control has the following advantages over conventional methods of machine control:

- Flexibility of operation and ability to produce complex shapes with good dimensional accuracy, repeatability, reduced scrap loss, and high production rates, productivity, and product quality.
- Tooling costs are reduced, since templates and other fixtures are not required.
- Machine adjustments are easy to make with minicomputers and digital readouts.
- More operations can be performed with each setup, and less lead time for setup and machining is required compared to conventional methods. Design changes are facilitated, and inventory is reduced.
- Programs can be prepared rapidly and can be recalled at any time utilizing microprocessors. Less paperwork is involved.
- Faster prototype production is possible.
- Required operator skill is less, and the operator has more time to attend to other tasks in the work area.

The major limitations of NC are the relatively high initial cost of the equipment and the need for programming and special maintenance, requiring trained personnel. Because NC machines are complex systems, breakdowns can be very costly, so preventive maintenance is essential. However, these limitations are often easily outweighed by the overall economic advantages of NC.

38.3.3 Computer numerical control

In the next step in the development of numerical control, the control hardware mounted on the NC machine was converted to local computer control with software. Two types of computerized systems were developed: direct numerical control and computer numerical control.

In **direct numerical control** (DNC), as originally conceived and developed in the 1960s, several machines are directly controlled step by step by a central main frame computer. In this system, the operator has access to the central computer through a remote terminal. Thus handling tapes and the need for computers on each machine are eliminated. With DNC, the status of all machines in a manufacturing facility could be monitored and assessed from the central computer. However, DNC had the crucial disadvantage that if the computer went down, all the machines became inoperative.

A more recent definition of DNC (now meaning *distributed* numerical control) includes the use of a central computer serving as the control system over a number of individual computer numerical control machines with onboard minicomputers. This system provides large memory and computational capabilities, thus offering flexibility, while overcoming the previous disadvantage of DNC.

Computer numerical control (CNC) is a system in which a minicomputer or microprocessor is an integral part of the control panel of a machine or equipment (onboard computer). The part program may be prepared at a remote site by the programmer, and may incorporate information obtained from drafting software packages and machining simulations to ensure that the part program is bug free. However, the machine operator can now easily and manually program onboard computers. The operator can modify the programs directly, prepare programs for different parts, and store the programs. Because of the availability of small computers with large memory, microprocessors, and program editing capabilities, CNC systems are widely used today. The importance of the availability of low-cost, programmable controllers in the successful implementation of CNC in manufacturing plants should be recognized.

The advantages of CNC over conventional NC systems are:

- Increased flexibility. The machine can produce a certain part, followed by other parts with different shapes and at reduced cost.
- Greater accuracy because of the higher sampling rate and speed of computers.
- More versatility. Editing and debugging programs, reprogramming, and plotting and printing part shape are simpler.

38.3.4 Principles of NC machines

The basic elements and operation of a typical NC machine are outlined in Fig. 38.7. The functional elements in numerical control and the components involved are:

a) *Data input.* Numerical information is read and stored in the tape reader or in computer memory.
b) *Data processing.* The programs are read into the machine control unit for processing.
c) *Data output.* This information is translated into commands, typically pulsed commands to the servomotor (Fig. 38.8). The servomotor moves the table on which the workpiece is placed to specific positions, through linear or rotary movements, by means of stepping motors, leadscrews, and other devices.

Types of control circuits. An NC machine can be controlled through two types of circuits: open loop and closed loop. In the **open-loop** system (Fig. 38.8a), the signals are given to the servomotor by the processor, but the movements and final destinations of the worktable are not checked for accuracy. The **closed-loop** system (Fig. 38.8b) is equipped with various transducers, sensors, and counters that measure the position of the table accurately. Through *feedback* control, the position

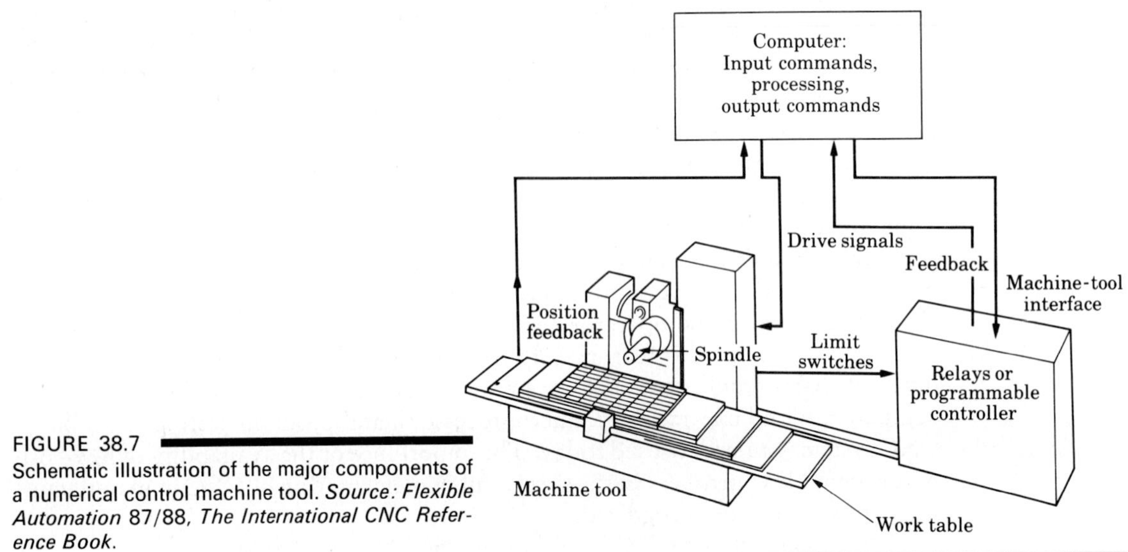

FIGURE 38.7 Schematic illustration of the major components of a numerical control machine tool. *Source: Flexible Automation* 87/88, *The International CNC Reference Book*.

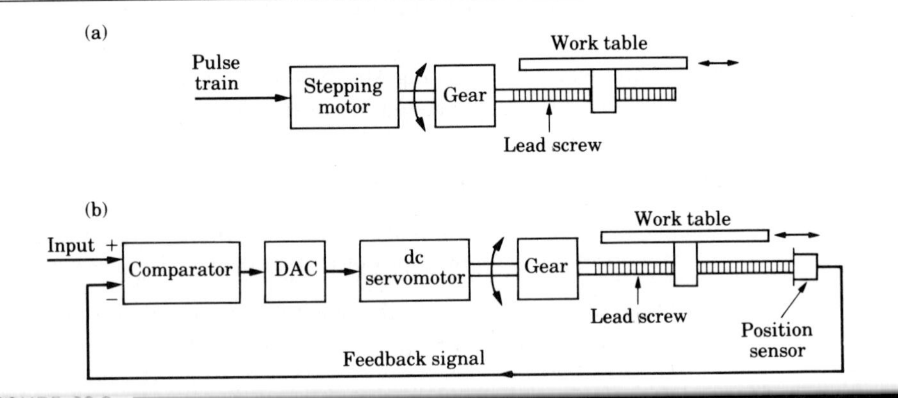

FIGURE 38.8 Schematic illustration of the components of (a) open-loop and (b) closed-loop control systems for a numerical control machine. DAC means digital-to-analog converter.

of the table is compared against the signal. Table movements terminate when the proper coordinates are reached. The closed loop system is more complicated and more expensive than the open-loop system.

Position measurement in NC machines can be accomplished through direct or indirect methods. In direct measuring systems, a sensing device reads a graduated scale on the machine table or slide for linear movement (Fig. 38.9a). This system is the most accurate because the scale is built into the machine, and backlash (the play

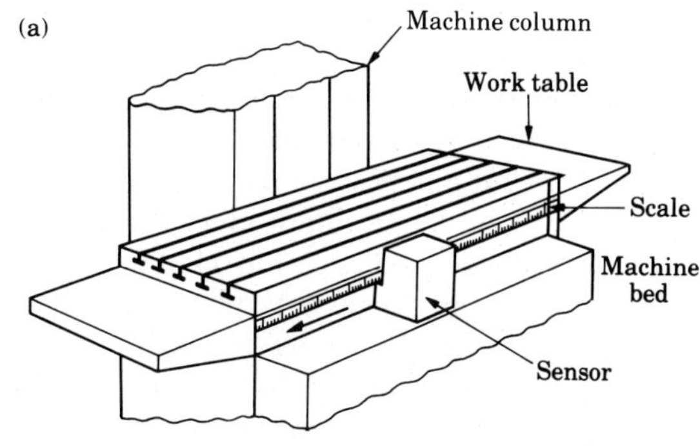

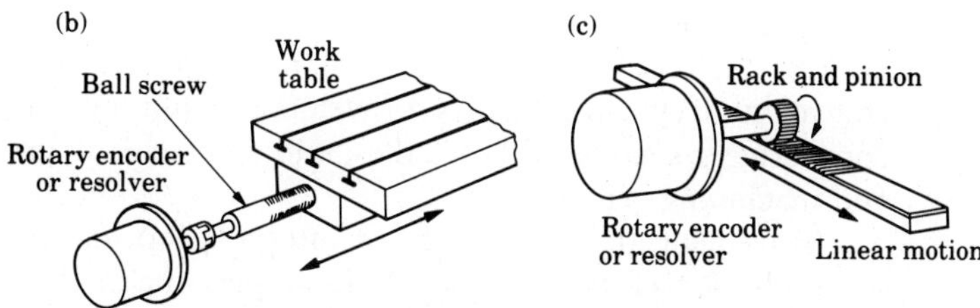

FIGURE 38.9
(a) Direct measurement of linear displacement of a machine tool worktable. (b) and (c) Indirect measurement methods (see also Fig. 38.10).

between two adjacent mating gear teeth) in the mechanisms is not significant. In indirect measuring systems, rotary encoders or resolvers (Figs. 38.9b and c) convert rotary movement to translation movement. In this system, backlash can significantly affect measurement accuracy. Position feedback mechanisms use various sensors, based mainly on magnetic and photoelectric principles (Fig. 38.10).

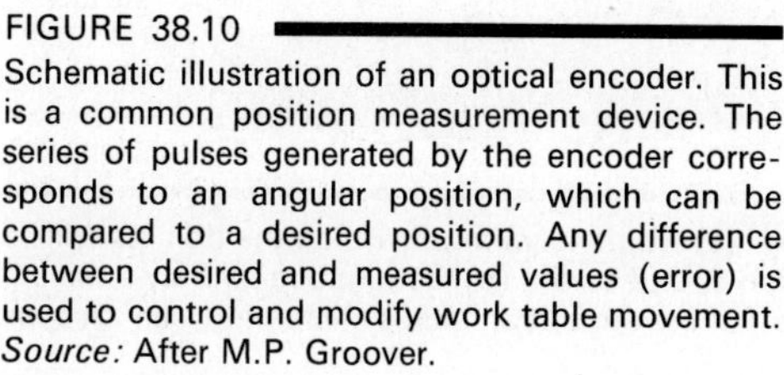

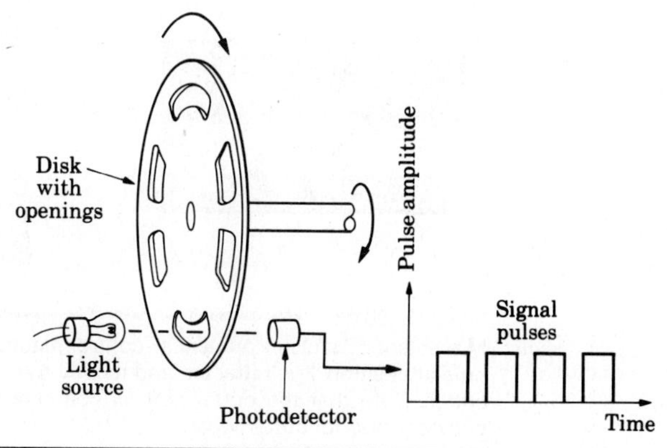

FIGURE 38.10
Schematic illustration of an optical encoder. This is a common position measurement device. The series of pulses generated by the encoder corresponds to an angular position, which can be compared to a desired position. Any difference between desired and measured values (error) is used to control and modify work table movement. *Source:* After M.P. Groover.

38.3.5 Types of control systems

There are two basic types of control systems in numerical control: point-to-point and contouring. In the *point-to-point* system, also called *positioning*, each axis of the machine is driven separately by leadscrews and, depending on the type of operation, at different velocities. The machine moves initially at maximum velocity in order to reduce nonproductive time but decelerates as the tool reaches its numerically defined position. Thus in an operation such as drilling or punching, the positioning and cutting take place sequentially (Fig. 38.11a). After the hole is drilled or punched, the tool retracts, moves rapidly to another position, and repeats the operation. The path followed from one position to another is important in only one respect: The time required should be minimized for efficiency. Point-to-point systems are used mainly in drilling, punching, and straight milling operations.

In the *contouring* system, also known as the *continuous path* system, positioning and cutting operations are both along controlled paths but at different velocities. Because the tool cuts as it travels along a prescribed path (Fig. 38.11b), accurate control and synchronization of velocities and movements are important. The contouring system is used on lathes, milling machines, grinders, welding machinery, and machining centers.

Movement along the path, or **interpolation**, occurs incrementally, by one of several basic methods (Fig. 38.12). Examples of actual paths in drilling, boring, and milling operations are shown in Fig. 38.13. In all interpolations, the path controlled is that of the center of rotation of the tool. Compensation for different tools, different diameter tools, or tool wear during machining, can be made in the NC program.

In *linear* interpolation, the tool moves in a straight line from start to end in two or three axes. Theoretically, all types of profiles can be produced by this method by

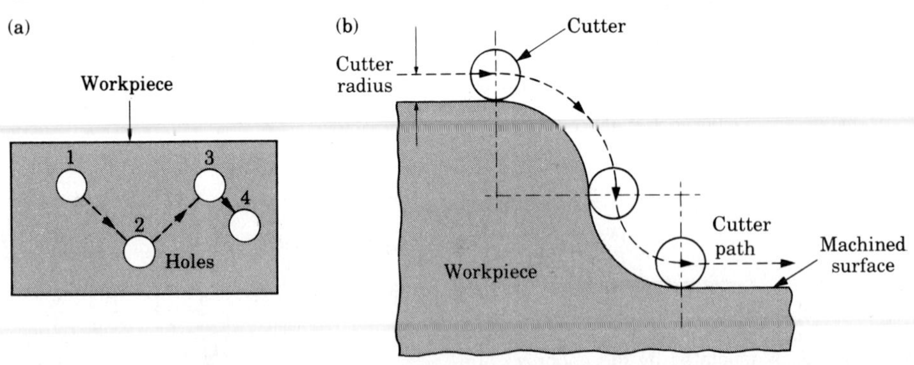

FIGURE 38.11
Movement of tools in numerical-control machining. (a) Point-to-point, in which the drill bit drills a hole at position 1, is retracted and moved to position 2, and so on. (b) Continuous path by a milling cutter. Note that the cutter path is compensated for by the cutter radius. This path can also be compensated for cutter wear.

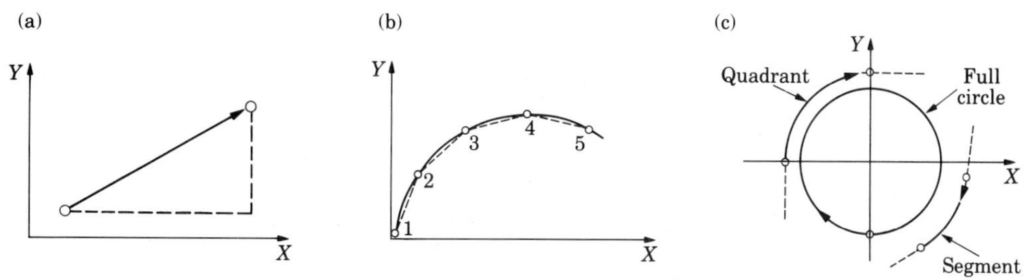

FIGURE 38.12
Types of interpolation: (a) linear, (b) continuous path approximated by incremental straight lines, and (c) circular.

making the increments between the points small. However, a large amount of data has to be processed in order to do so. In *circular* interpolation, the input required for the path are the coordinates of the end points, the coordinates of the center of the circle and its radius, and the direction of the tool along the arc. In *parabolic* and *cubic* interpolation, the path is approximated by curves using higher order mathematical equations. This method is effective in 4-axis or 5-axis machines and is particularly useful in die sinking operations for automotive bodies.

These interpolations are also used for industrial robot movements, which we discuss in Section 38.7.

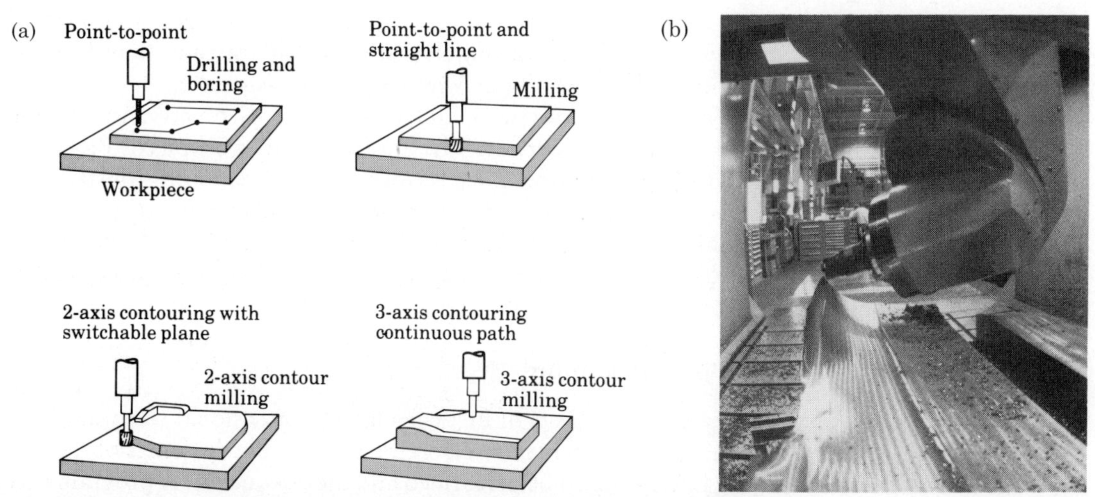

FIGURE 38.13
(a) Schematic illustration of drilling, boring, and milling with various paths. (b) Machining a sculptured surface on a 5-axis numerical control machine tool. *Source:* Courtesy of The Ingersoll Milling Machine Company.

38.3.6 Accuracy in numerical control

Positioning accuracy in NC machines is defined by how accurately the machine can be positioned to a certain coordinate system. An NC machine usually has a positioning accuracy of at least $\pm 3\ \mu m$ (0.0001 in.). **Repeatability**, defined as the closeness of agreement of repeated position movements under the same operating conditions of the machine, is usually around $\pm 8\ \mu m$ (0.0003 in.). **Resolution**, defined as the smallest increment of motion of the machine components, is usually about 2.5 μm (0.0001 in.).

The *stiffness* of the machine tool (see Section 24.3) and *backlash* in its gear drives and leadscrews are important for accuracy. Backlash can be eliminated with special backlash take-up circuits, whereby the tool always approaches a particular position on the workpiece from the same direction, as should be done in traditional machining operations. Rapid response to command signals requires that friction and inertia be minimized, say, by reducing the mass of moving components of the machine.

38.4 Programming for Numerical Control

A program for numerical control consists of a sequence of directions that causes an NC machine to carry out a certain operation, machining being the most commonly used process. **Programming for NC** may be done by an internal programming department, on the shop floor, or purchased from an outside source. Also, programming may be done manually or with computer assistance.

The program contains instructions and commands. Geometric instructions pertain to relative movements between the tool and the workpiece. Processing instructions pertain to spindle speeds, feeds, tools, and so on. Travel instructions pertain to the type of interpolation and slow or rapid movements of the tool or worktable. Switching commands pertain to on/off position for coolant supplies, spindle rotation, direction of spindle rotation, tool changes, workpiece feeding, clamping, and so on.

38.4.1 Manual programming

Manual part programming consists of first calculating dimensional relationships of the tool, workpiece, and work table, based on the engineering drawings of the part, and manufacturing operations to be performed and their sequence. A program sheet is then prepared, which consists of the necessary information to carry out the operation, such as cutting tools, spindle speeds, feeds, depth of cut, cutting fluids, power, and tool or workpiece relative positions and movements. On the basis of this information, the part program is prepared. Usually a paper tape is first prepared for trying out and debugging the program. Depending on how often it is to be used, the tape may be made of more durable Mylar.

Manual programming can be done by someone knowledgeable about the particular process and able to understand, read, and change part programs. Because they are familiar with machine tools and process capabilities, skilled machinists can do manual programming with some training in programming. However, the work is tedious, time consuming, and uneconomical—and is used mostly in simple point-to-point applications.

38.4.2 Computer-aided programming

Computer-aided part programming involves special symbolic programming languages that determine the coordinate points of corners, edges, and surfaces of the part (Fig. 38.14). *Programming language* is the means of communicating with the computer and involves the use of symbolic characters. The programmer describes the component to be processed in this language, and the computer converts it to commands for the NC machine. Several languages having various features and applications are commercially available. The first language that used English-like statements was developed in the late 1950s and is called APT (for Automatically Programmed Tools). This language, in its various expanded forms, is still the most widely used for both point-to-point and continuous-path programming.

Computer-aided part programming has the following significant advantages over manual methods:

- Use of relatively easy to use symbolic language. Several programs have been developed: ADAPT, IFAPT, MINIAPT, UNIAPT, AUTOSPOT, AUTOMAP, AUTOPROMPT, CAMPI, SPLIT, and COMPACT II.
- Reduced programming time. Programming is capable of accommodating a large amount of data concerning machine characteristics and process variables, such as power, speeds, feed, tool shape, compensation for tool shape changes, tool wear, deflections, and coolant use.
- Reduced possibility of human error, which can occur in manual programming.
- Ability to view a machining sequence on the screen for debugging.
- Capability of simple changeover of machining sequence or from machine to machine.
- Lower cost because less time is required for programming.

The use of programming languages or *compilers* not only results in greater part quality for the reasons mentioned above, but also allows for more rapid development of machining instructions. In addition, simulations can be run on remote computer terminals to ensure that the program functions as intended. This method prevents unnecessary occupation of expensive machinery for debugging procedures.

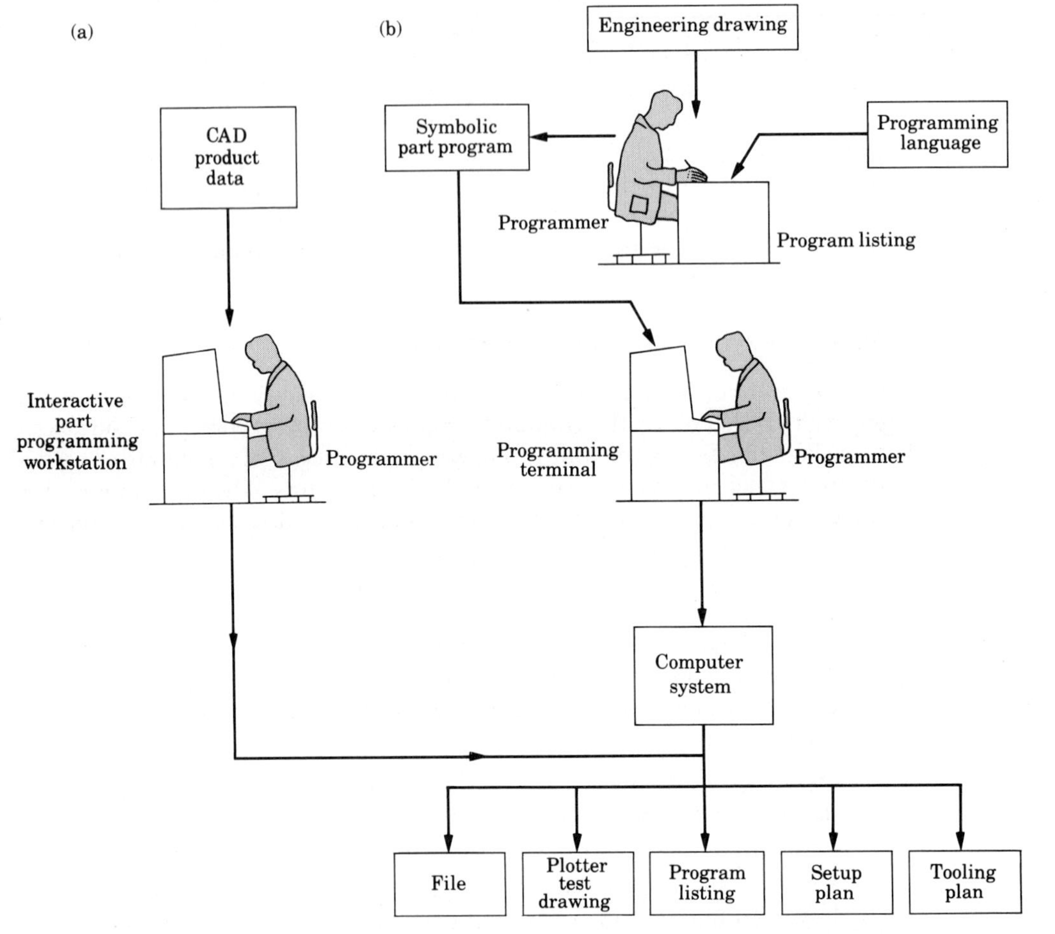

FIGURE 38.14
Outline of programming methods for numerical control. *Source: Flexible Automation* 87/88, *The International CNC Reference Book.*

Selection of a particular NC programming language depends on the following factors:

a) Level of expertise of the personnel in the manufacturing facility.
b) Complexity of the part.
c) Type of equipment and computers available.
d) Time and costs involved in programming.

Because numerical control involves the insertion of data concerning workpiece materials and processing parameters, programming must be done by operators or programmers who are knowledgeable about the relevant aspects of the manufacturing processes being used. Before production begins, programs should be verified, either by viewing a simulation of the process on a CRT screen or by making the part from an inexpensive material, such as aluminum, wood, or plastic, rather than the material specified for the finished part.

• Example: NC programming for machining.

The accompanying figure shows a 2-in. square and 0.75-in. thick workpiece that is machined around its circumference with an end mill to a depth of 0.125 in. Before the NC program is run, the origin of the axes is set at the lower left corner of the part (near D). The cutter has a diameter of 0.250 in. and starts at a position of

$$X = -1.500 \text{ in.}, Y = 0, Z = 1.500 \text{ in.}$$

The approach position of the cutter is shown in the figure. The NC program and its codes are given below. After rapid travel to within $-1.500 + 1.200 = 0.300$ in. of the part, the cutter moves along the following paths: (a) It travels in the $+X$ direction in linear interpolation, (b) machines the radius at A by counterclockwise circular interpolation, (c) travels in the $+Y$ direction, (d) machines the radius at B, (e) travels in the $-X$ direction, (f) machines the radius at C, (g) travels in the $-Y$ direction, (h) machines the radius at D, and (i) travels in the $-X$ direction to its starting position. Note in the program that the sums of all X and Y movements, respectively, are zero, indicating that the tool has returned to its starting position. Codes: G91—incremental value programming; M03—cutter spindle on, clockwise rotation; G00—rapid traverse; G01—linear interpolation; G03—circular interpolation, counterclockwise; M30—program end.

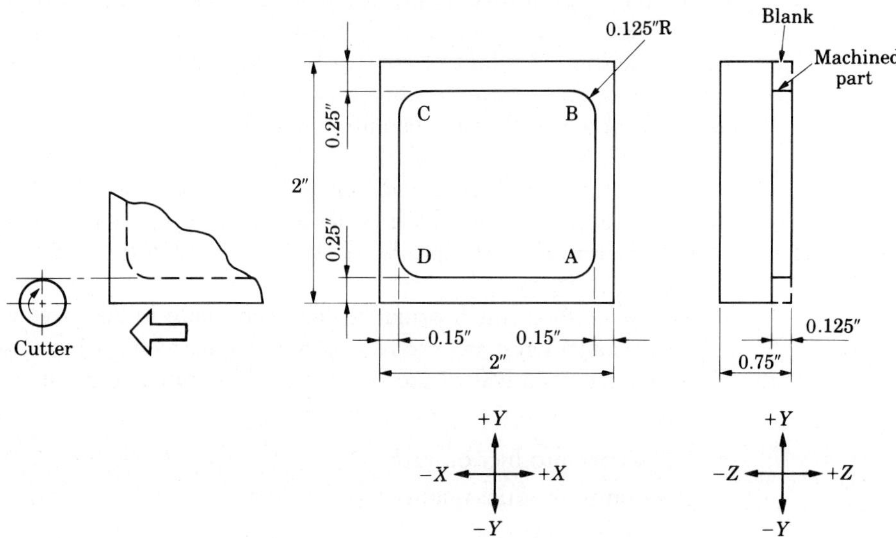

N	G (M)	X	Y	Z	
N00	G91				Set incremental programming mode.
N01	M03				Start cutter spindle.
N02	G00	0	0	−1625	Move cutter to depth of 0.125″.
N03	G00	1200	0	0	Rapid move to 0.300″ away from part.
N04	G01	2025	0	0	Cut until cutter reaches start of radius A.
N05	G03	125	125		Cut radius A using relative endpoints.
N06	G01	0	1250	0	Cut until cutter reaches start of radius B.
N07	G03	−125	125		Cut radius B using relative endpoints.
N08	G01	−1450	0	0	Cut until cutter reaches start of radius C.
N09	G03	−125	−125		Cut radius C using relative endpoints.
N10	G01	0	−1250	0	Cut until cutter reaches start of radius D.
N11	G03	125	−125		Cut radius D using relative endpoints.
N12	G00	0	0	1625	Rapid rise of cutter to original z-position.
N13	G00	−1775	0	0	Rapid movement of cutter to original x-position.
N14	M30				End of program.

Source: EMCO.

38.5 Adaptive Control

In **adaptive control** (AC), the operating parameters in an operation automatically adapt themselves to conform to new circumstances, such as changes in the dynamics of the particular process and any disturbances that may arise. It will be readily recognized that this approach is basically a feedback system. Consequently, the exact definition of adaptive control has escaped numerous attempts, although it is understood that adaptive control systems are inherently nonlinear. It is now generally accepted that a constant-gain feedback system is not adaptive control, the term *gain* being defined as the ratio of output to input in an amplifier.

Initial research in adaptive control began in the early 1950s and concerned the design of autopilots for high-performance aircraft, which operate over a wide range of altitudes and speeds. It was observed during tests that constant-gain feedback control systems would work well under some operating conditions but not others. Since then much progress has been made in adaptive control and several adaptive-control systems are now commercially available for applications such as ship steering, chemical-reactor control, rolling mills, and medical technology.

In manufacturing engineering, specifically, the purposes of adaptive control are:

- Optimize production rate.
- Optimize product quality.
- Minimize cost.

Although adaptive control has been used widely in continuous processing in the chemical industry and oil refineries for some time, its successful application to machining, grinding, forming, and other manufacturing processes is relatively recent. Application of AC in manufacturing operations is particularly important in situations where workpiece dimensions and quality is not uniform, such as a poor casting or an improperly heat-treated part.

Adaptive control is a logical extension of computer numerical control systems. As we described in Section 38.4, the part programmer sets the processing parameters, on the basis of existing knowledge of the material and data on the particular manufacturing process. In CNC machines, these parameters are held constant during a particular process cycle. In AC, on the other hand, the system is capable of automatic adjustments during processing through closed-loop feedback control (Fig. 38.15).

38.5.1 Principles and applications of adaptive control

The basic functions common to adaptive control systems are to:

a) Identify unknown parameters or measure performance, using sensors that detect parameters such as force, torque, vibration, and temperature.
b) Decide on a control strategy. Based on preset upper and lower thresholds, the system determines whether a particular measurement is exceeding the index of performance.
c) Modify the process parameters through input to the controller, such as changing the speed of operation, feed in machining, and so on.

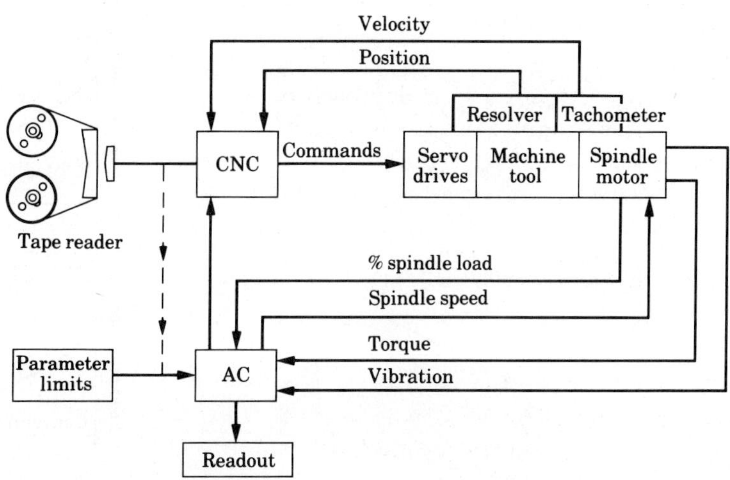

FIGURE 38.15 Schematic illustration of the application of adaptive control (AC) for a turning operation. The system monitors parameters such as cutting force, torque, and vibrations, and if excessive, modifies process variables such as feed and depth of cut to bring them to acceptable levels. *Source: Flexible Automation* 87/88, *The International CNC Reference Book.*

You may recognize that human reactions to occurrences in everyday life already contain adaptive control. When you drive a car on a rough road, you know how to steer to avoid potholes by visually and continuously observing the condition of the road, or your body feels the car's movements and vibrations. You then react by changing direction and speed to minimize the effects of the rough road on your car and to increase the comfort of the ride.

In an operation such as turning on a lathe, the adaptive control system senses real-time cutting forces, torque, temperature, tool wear rate, tool chipping or fracture, and surface finish of the workpiece. The system converts this information into commands that modify the process parameters on the machine tool to hold them constant (or within certain limits) or to optimize the cutting operation.

Those systems that place a constraint on a process variable (such as forces, torque, or temperature) are called *adaptive control constraint* (ACC) systems. Thus if the thrust force, the cutting force, and hence the torque increase excessively—because of a hard region in a cast workpiece, say—the adaptive control system changes the speed and/or feed to lower the cutting forces to acceptable levels (Fig. 38.16). Without adaptive control or direct operator intervention as in traditional machining operations, high forces may cause tools to chip or break and the workpiece to deflect excessively, thus losing dimensional accuracy. Systems that optimize an operation are called *adaptive control optimization* (ACO) systems. Optimization may involve maximizing material removal rate between tool changes

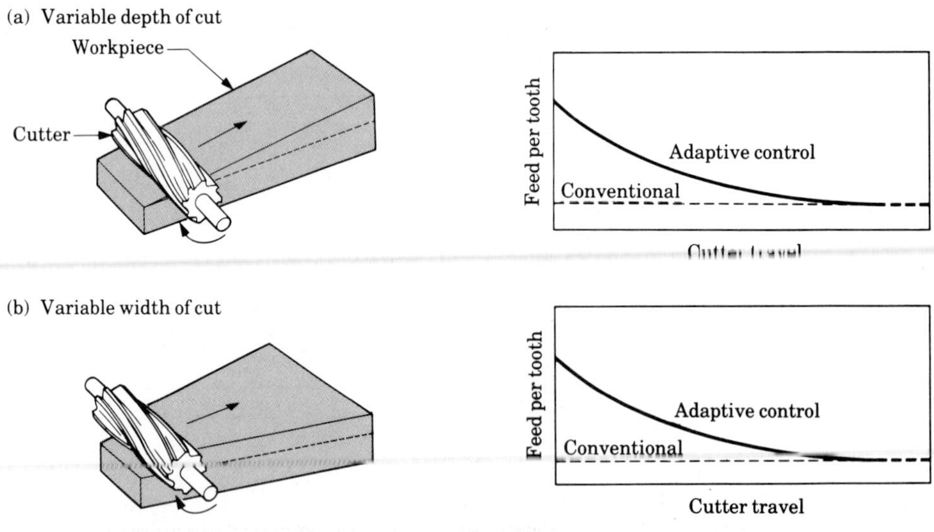

FIGURE 38.16 An example of adaptive control in milling. As the depth or width of cut increases, the cutting forces and torque increase. The system senses this increase and automatically reduces the feed to avoid excessive forces or tool breakage and to maintain cutting efficiency. *Source:* Y. Koren.

or resharpening or improving surface finish. Most systems are currently based on ACC because development and implementation of ACO is complex.

Response time must be short for AC to be effective, particularly in high-speed operations. Assume, for example, that a turning operation is being performed on a lathe at a spindle speed of 1000 rpm, and the tool suddenly breaks, adversely affecting the surface finish and dimensional accuracy of the part. In order for the AC system to be effective, the sensing system must respond within a very short time; otherwise the damage to the workpiece will be extensive.

For adaptive control to be effective in manufacturing operations, quantitative relationships must be known and stored in the computer software as mathematical models. For example, if tool wear rate in a machining operation is excessive, the computer must be able to know how much of a change (and whether increase or decrease) in speed and/or feed is necessary to reduce the wear rate to an acceptable level. The system should also be able to compensate for dimensional changes on the workpiece caused by tool wear and temperature rise (Fig. 38.17).

If the operation is grinding, for example, the computer software must contain quantitative relationships among process variables (wheel and work speeds, feed, type of wheel) and parameters such as wheel wear, dulling of abrasive grains, grinding forces, temperature, surface finish, and part deflections. Similarly, for bending of a sheet in a V-die, data on the dependence of springback on punch travel and other material and process variables must be stored in the computer software. Because of the many factors involved, mathematical equations for such quantitative relationships in manufacturing processes are difficult to establish. Compared to the other parameters involved, forces and torque have been found to be the easiest to monitor by AC. Various solid-state power controls are available commercially, in which power is displayed or interfaced with data acquisition systems. Coupled with CNC, adaptive control is potentially a powerful tool in optimizing manufacturing operations.

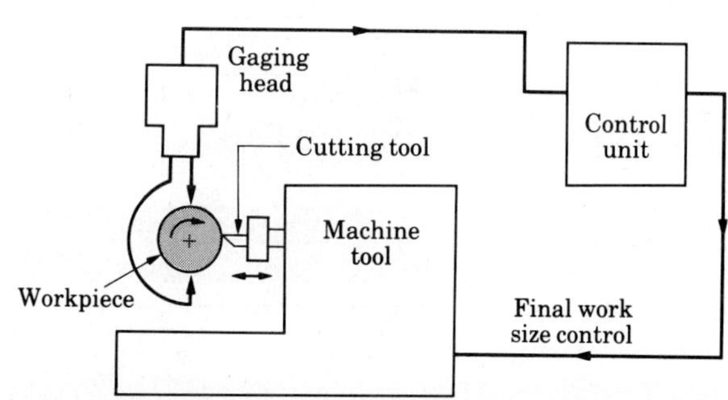

FIGURE 38.17

In-process inspection of workpiece diameter in a turning operation. The system automatically adjusts the cutting-tool position to produce the correct diameter.

38.6 Material Handling and Movement

During a typical manufacturing operation, raw materials and parts are moved from storage to machines, from machine to machine, from inspection to assembly and inventory, and finally to shipment. Workpieces are loaded on machines (a forging is mounted on a milling machine bed, or sheet metal is fed into a press for stamping), parts are removed from one machine and loaded on another (a machined forging is subsequently ground), and parts are inspected for flaws and dimensional accuracy prior to being assembled into a finished product. Similarly, tools, molds, dies, and various other equipment and fixtures are also moved in manufacturing plants. Cutting tools are mounted on lathes, dies are placed in presses or hammers, grinding wheels are mounted on spindles, and parts are mounted on special fixtures for dimensional measurement and inspection.

These materials must be moved either manually or by some mechanical means, and time is required to transport them from one location to another. Thus we may define **material handling** as the functions and systems associated with the transportation, storage, and control of materials and parts in the total manufacturing cycle of a product. The total time required for manufacturing depends on part size and shape and the set of operations required. Idle time and the time required for transporting materials can constitute the majority of the time consumed (Fig. 38.18). Because handling materials adds cost but not value to a product, it should be reduced as much as possible.

Plant layout is an important aspect of the flow of materials and components throughout the manufacturing cycle. As we note in our discussion of flexible manufacturing systems in Chapter 39, the arrangement of production machinery

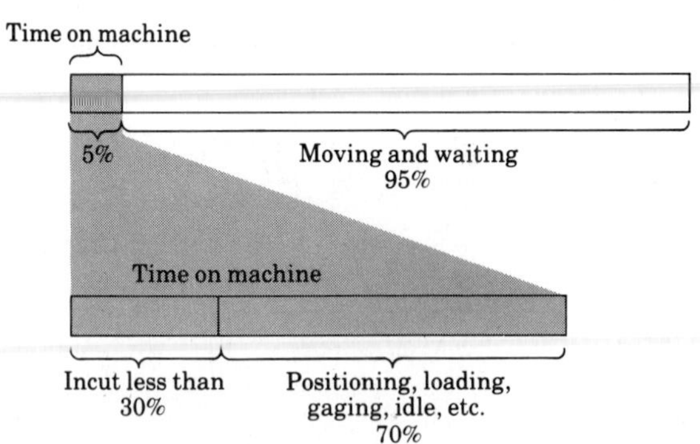

FIGURE 38.18 Time spent on various phases in a typical machining operation. Note that very little time is spent on actual machining. *Source:* Cincinnati Milacron, Inc.

and material-handling equipment should be orderly and efficient. The time and distances required for moving raw materials and parts should be minimized, and storage areas and service centers should be organized accordingly. For parts requiring multiple operations, equipment should be grouped around the operator or the industrial robot (see Cellular Manufacturing, Section 39.9).

Material handling should therefore be an integral part of planning, implementing, and controlling manufacturing operations. Furthermore, material handling should be repeatable and predictable. Here we define repeatability as the closeness of agreement of repeated positions under the same conditions to the same location. Consider, for example, what happens if a part or workpiece is loaded improperly in a forging die or in the chuck of a lathe. The consequences of such action may well be broken dies and tools, improperly made parts, or parts that are out of tolerance. This action can also present safety hazards and possibly cause injury to the operator, as well as to other nearby personnel.

38.6.1 Methods of material handling

Several factors have to be considered in choosing a suitable material-handling method for a particular manufacturing operation:

a) Shape, weight, and characteristics of parts.
b) Types of movement and distances involved and the position and orientation of parts during movement and at their final destination.
c) Conditions of the path along which parts are to be transported.
d) Degree of automation and control desired and integration with other systems and equipment.
e) Operator skill required.
f) Economic considerations.

For small-batch manufacturing operations, raw materials and parts can be handled and transported by hand, but this method is generally costly. Moreover, because it involves human beings, this practice can be unpredictable and unreliable and can be unsafe to the operator, depending on the weight and shape of the parts to be moved and environmental factors, such as heat and smoke in foundries and forging plants. In automated manufacturing plants, computer controlled material and parts flow is being rapidly implemented. These changes have improved repeatability and lowered labor costs.

38.6.2 Equipment

Various types of equipment can be used to move materials, such as conveyors, rollers, self-powered monorails, carts, forklift trucks, and various mechanical, electrical, magnetic, pneumatic, and hydraulic devices and manipulators. **Manipulators** are designed to be controlled directly by the operator, or they are automated for repeated operations, such as loading and unloading parts from machine tools, presses, and furnaces. Manipulators are capable of gripping and moving heavy parts and orienting them as required between manufacturing and assembly operations.

Machines are often used in a sequence where workpieces are transferred directly from machine to machine. Machinery with the capability to convey parts without the use of additional material-handling schemes are called *integral transfer devices*.

Industrial robots, specially designed pallets, and **automated guided vehicles (AGVs)** are used extensively in flexible manufacturing systems to move parts and orient them as required (Fig. 38.19). Thus flexible material handling and movement with real-time control has become an integral part of modern manufacturing.

Automated guided vehicles, which are the latest development in material movement in plants, operate automatically along pathways with in-floor wiring or tapes for optical scanning and without any operator intervention. This transport system has great flexibility and is capable of random delivery to different workstations. It optimizes the movement of materials and parts in cases of congestion

(a)

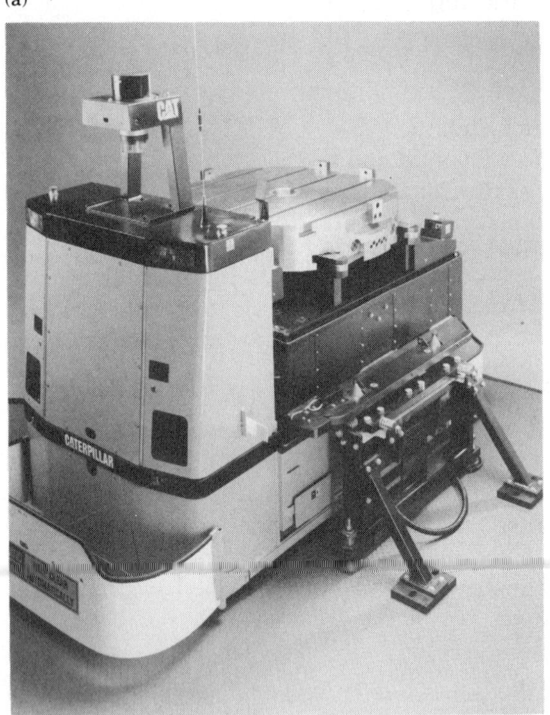

(b)

FIGURE 38.19

(a) A self-guided vehicle (Caterpillar Model SGC-M) carrying a machining pallet. The vehicle is aligned next to a stand on the floor. Instead of following a wire or stripe path on the factory floor, this vehicle calculates its own path and automatically corrects for any deviations. *Source:* Courtesy of Caterpillar Industrial, Inc. (b) A rectangular-coordinate checkerboard floor of uniform and contrasting colors serving as a navigational system of AGVs in an Apple Computer plant. An onboard computer stores information on positions, dimensions, and functions of various elements in the layout such as machines, workstations, storage racks, loading stations, and walls. The routes for specific functions are determined automatically. *Source:* Courtesy of Frog Systems and *Manufacturing Engineering Magazine*, Society of Manufacturing Engineers.

around workstations, machine breakdown (downtime), or failure of part of the system. The movement of AGVs are planned so that they interface with *automated storage/retrieval systems* (AS/RS). The latter utilizes warehouse spaces efficiently and reduces manpower.

Coding systems have been developed to locate and identify parts throughout the manufacturing system and to transfer them to their appropriate stations. *Bar codes* are the most widely used and the least costly. They are printed on labels, attached to the parts themselves, and read by fixed or portable code readers using light pens. Other identification systems are based on acoustic waves, optical character recognition, and machine vision (see Section 38.8).

38.7 Industrial Robots

The word **robot** was coined in 1920 by the Czech author K. Čapek in his play *R.U.R.* (Rossum's Universal Robots), and is derived from the word *robota*, meaning worker. An *industrial robot* has been defined as a reprogrammable multifunctional manipulator, designed to move materials, parts, tools, or other specialized devices by means of variable programmed motions and to perform a variety of other tasks. In a broader context, the term robot also includes manipulators that are activated directly by an operator.

More generally, an industrial robot has been described by the International Standards Organization (ISO) as follows: A machine formed by a mechanism including several degrees of freedom, often having the appearance of one or several arms ending in a wrist capable of holding a tool, a workpiece, or an inspection device. In particular, its control unit must use a memorizing device and it may sometimes use sensing or adaptation appliances to take into account environment and circumstances. These multipurpose machines are generally designed to carry out a repetitive function and can be adapted to other functions.

Introduced in the early 1960s, the first industrial robots were used in hazardous operations, such as handling toxic and radioactive materials and loading and unloading hot workpieces from furnaces and handling them in foundries. Some rule-of-thumb applications for robots are the three D's (dull, dirty, and dangerous, including demeaning but necessary tasks), and the three H's (hot, heavy, and hazardous). From their early uses for worker protection and safety in manufacturing plants, industrial robots have been further developed to improve productivity, increase product quality, and reduce labor costs. Computer-controlled robots were commercialized in the early 1970s, with the first robot controlled by a minicomputer appearing in 1974.

38.7.1 Components

While learning about robot components and capabilities, you might simultaneously observe the flexibility and capability of diverse movements of your arm, wrist, hand,

and fingers in reaching for and grabbing an object from a shelf or in operating your car. We describe below the basic components of an industrial robot (Fig. 38.20).

Manipulator. Also called *arm and wrist*, the *manipulator* is a mechanical unit that provides motions (trajectories) similar to that of a human arm and hand. The end of the wrist can reach a point in space with a specific orientation. There are three degrees of freedom each in linear and rotational movements, respectively. Manipulation is carried out using mechanical devices, such as linkages, gears, and various joints.

End effector. The end of the wrist in a robot is equipped with an *end-effector*, also called *end-of-arm tooling*. Depending on the type of operation, conventional end effectors are equipped with:

a) Grippers, hooks, scoops, electromagnets, vacuum cups, and adhesive fingers, for materials handling (Fig. 38.21).
b) Spray gun for painting.
c) Attachments for spot and arc welding and arc cutting.

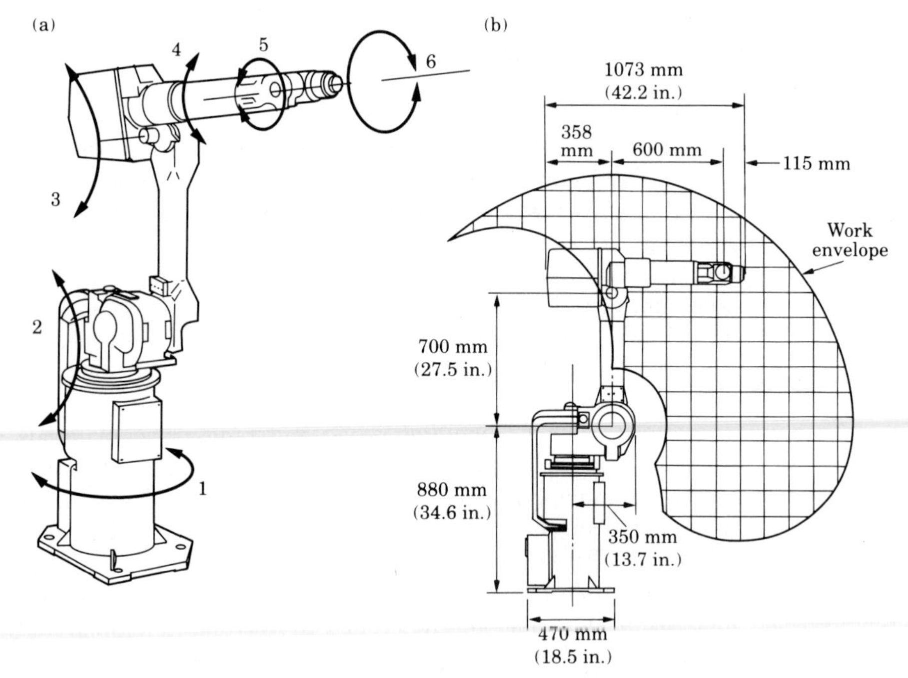

FIGURE 38.20
(a) Schematic illustration of a six-axes S-10 GMF robot. The payload at the wrist is 10 kg and repeatability is $\pm$ 0.2 mm (0.008 in.). The robot has mechanical brakes on all axes, which are coupled directly. (b) The work envelope of the robot, as viewed from the side. *Source:* GMFanuc Robotics Corporation.

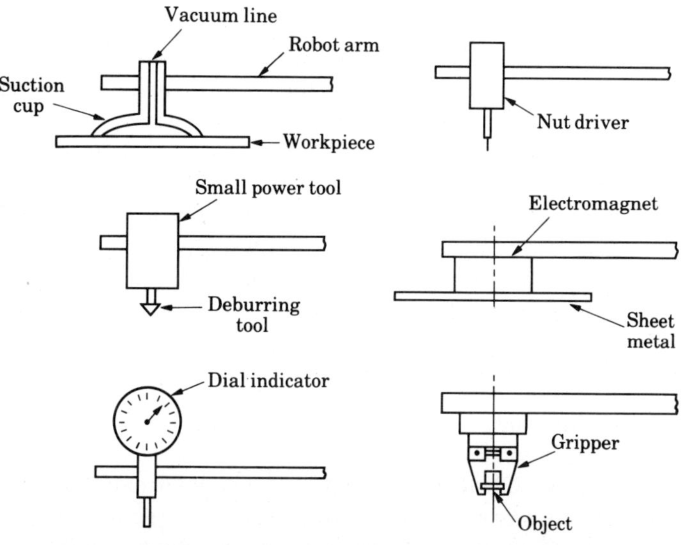

FIGURE 38.21
Various devices and tools attached to end effectors to perform a variety of operations.

d) Power tools, such as drills, nut drivers, and burrs.
e) Measuring instruments, such as dial indicators.

End effectors are generally custom made to meet special handling requirements. Mechanical grippers are the most commonly used and are equipped with either two or more fingers. The selection of an appropriate end effector for a specific application depends on factors such as the payload, environment, reliability, and cost.

Power supply. Each motion of the manipulator, in linear and rotational axes, is controlled and regulated by independent actuators, using electric, pneumatic, or hydraulic power supplies. Each source of energy and the motors involved have their own characteristics, advantages, and limitations.

Control system. Also known as the *controller*, the *control system* is the communications and information processing system that gives commands for the movements of the robot. It is the brain of the robot and stores data to initiate and terminate movements of the manipulator. It interfaces with computers and other equipment, such as manufacturing cells or assembly systems.

Feedback devices, such as transducers, are an important part of the control system. They transmit information to the control system on the position of various robot joints and linkages. In closed-loop control, the system automatically measures the degree to which the robot's movements conform to the desired response. It then utilizes this feedback to drive the system into conformance. Hence, the system has a self-correcting capability.

Robots with a fixed set of motions have open-loop control. In this system commands are given and the robot arm goes through its motions, but—unlike feedback in closed-loop systems—accuracy of the movements is not monitored. The system does not have a self-correcting capability.

As in numerical control machines, the types of control in industrial robots are point-to-point and continuous-path (see Section 38.3). Depending on the particular task, the positioning repeatability required may be as small as 0.050 mm (0.002 in.), as in assembly operations for electronic printed circuitry. Specialized robots can reach such accuracy, although most robots are unable to do so.

38.7.2 Classification

Robots may be classified by basic type: (a) cartesian or rectilinear, (b) cylindrical, (c) spherical or polar, and (d) articulated, or revolute, jointed, or anthropomorphic (Fig. 38.22). Robots may be attached permanently to the floor of a manufacturing plant, may move along overhead rails (*gantry robot*), or may be equipped with wheels to move along the factory floor (*mobile robot*). However, a broader classification of robots currently in use is most helpful for our purposes here. We use it to describe below various types of robots.

Fixed- and variable-sequence robots. The *fixed-sequence robot*, also called a *pick-and-place robot*, is programmed for a specific sequence of operations. Its movements are from point to point, and the cycle is repeated continuously. These robots are simple and relatively inexpensive. The *variable-sequence robot* can be programmed for a specific sequence of operations but can be reprogrammed to perform another sequence of operations.

Playback robot. An operator leads, or walks, the *playback robot* and its end effector through the desired path. In other words, the operator teaches the robot by showing it what to do. The robot memorizes and records the path and sequence of

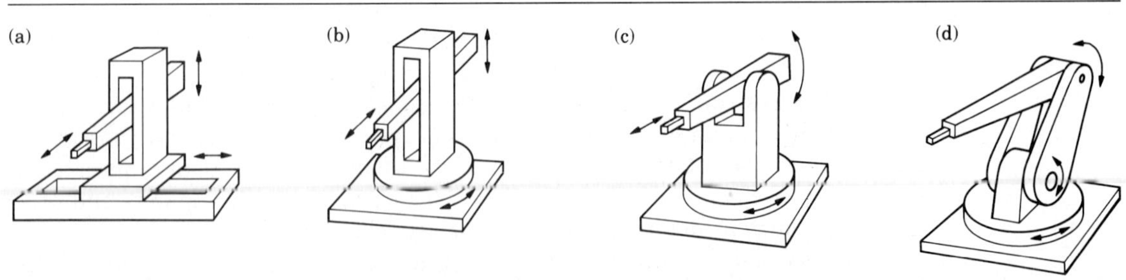

FIGURE 38.22
Types of industrial robots: (a) cartesian (rectilinear), (b) cylindrical, (c) spherical (polar), (d) articulated (revolute, jointed, or anthropomorphic).

motions and can repeat them continuously without any further action or guidance by the operator. Another method, the *teach pendant*, utilizes hand-held button boxes that are connected to the control panel and used to control and guide the robot and its tooling through the work to be performed. These movements are then registered in the memory of the controller and automatically reenacted by the robot.

Numerically controlled robot. The *numerically controlled robot* is programmed and operated much like a numerically controlled machine. The robot is servocontrolled by digital data, and its sequence of movements can be changed with relative ease. As in NC machines, there are two basic types of controls: point-to-point and continuous path. Point-to-point robots are easy to program and have higher load-carrying capacity and *work envelope* (the maximum extent or reach of the robot hand or working tool in all directions, also called working envelope; Fig. 38.23). Continuous path robots have greater precision than point-to-point robots, but have lower load-carrying capacity. Some advanced robots have a complex system of path control, enabling high speed movements with great accuracy.

Intelligent (sensory) robot. The *intelligent*, or *sensory*, *robot*, is capable of performing some of the functions and tasks carried out by human beings. The intelligent robot is equipped with a variety of sensors with visual (computer vision) and tactile (touching) capabilities (see Section 38.8). Much like humans, the robot observes and evaluates the immediate environment and proximity to other objects,

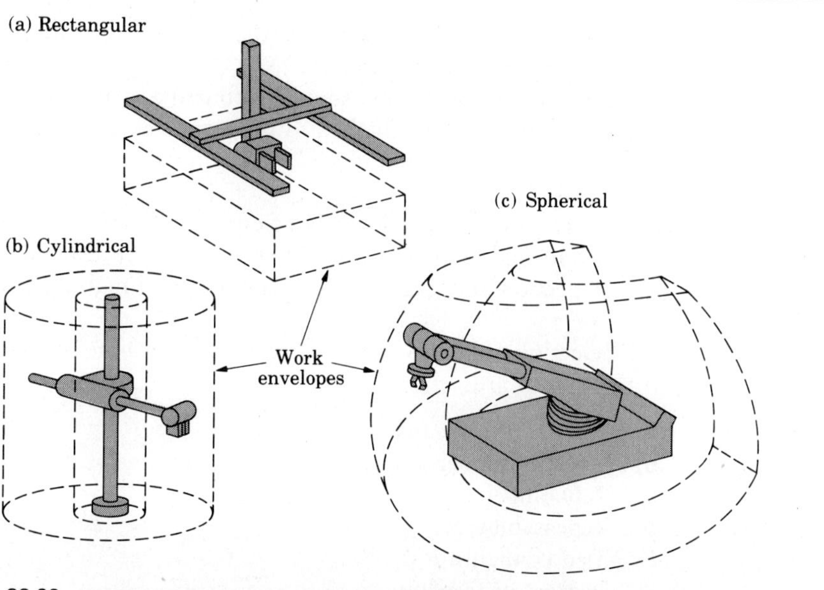

FIGURE 38.23
Work envelopes for three types of robots. The choice depends on the particular application.

machinery, etc., by perception and pattern recognition. It then makes appropriate decisions for the next movement and proceeds with them. Because its operation is so complex, powerful computers are required to control this type of robot.

Significant developments are taking place in intelligent robots so that the robot will behave more and more like humans, performing tasks such as moving among a variety of machines and equipment on the shop floor and avoiding collisions; recognizing, picking, and gripping properly the correct raw material or workpiece in progress; transporting it to a machine for processing or inspection; and assembling the components into a final product. It can be appreciated that in such tasks, the accuracy and repeatability of the robot's movements are important considerations, as are the economic benefits to be gained.

38.7.3 Applications and selection of robots

Major applications of industrial robots include the following.

a) Material handling, loading, unloading, and transferring workpieces in manufacturing operations. Examples are casting and molding, in which molten metal, raw materials, lubricants, and parts in various stages of completion are handled without operator interference. In heat treating, parts are loaded and unloaded from furnaces and quench baths. In forming operations, parts are loaded and unloaded from presses and various other types of metalworking machinery. All these operations can be performed reliably and repeatedly with robots, thereby improving quality and reducing scrap losses.
b) Spot welding automobile and truck bodies, producing welds of good quality (Fig. 38.24). Robots perform other, similar operations, such as arc welding, arc cutting, and riveting.
c) Machining operations, such as deburring, grinding, and polishing, with appropriate tools attached to their end effectors.
d) Applying adhesives and sealants, as in the automobile frame shown in Fig. 38.25.
e) Spray painting, particularly of complex shapes, and cleaning operations.
f) Automated assembly (Fig. 38.26; see also Section 38.10).
g) Inspection and gaging in various stages of manufacture.

Selection of robots. Factors that influence the selection of robots in manufacturing plants are:

a) Load-carrying capacity.
b) Speed of movement.
c) Reliability.
d) Repeatability.
e) Arm configuration.
f) Degrees of freedom.

FIGURE 38.24
Spot welding automobile bodies with industrial robots. *Source:* Courtesy of Cincinnati Milacron, Inc.

FIGURE 38.25
Sealing joints of an automobile body with an industrial robot. *Source:* Courtesy of Cincinnati Milacron, Inc.

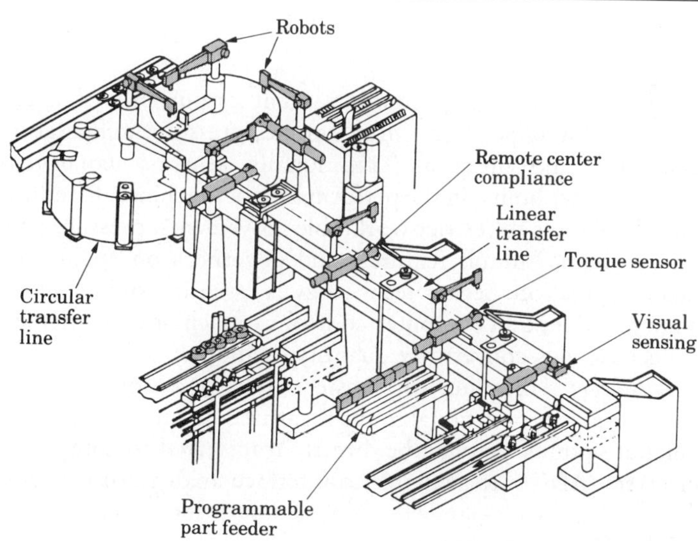

FIGURE 38.26
Automated assembly operations using industrial robots and circular and linear transfer lines. *Source: Computers in America*, January 1984.

g) Control system.
h) Program memory.
i) Work envelope.

Economics. In addition to technical factors, cost and benefit considerations are significant aspects of robot selection and their use. The increasing availability, reliability, and reduced costs of sophisticated intelligent robots are having a major economic impact on manufacturing operations and gradually replacing human labor. Whereas hourly wages are steadily rising, particularly in industrial nations, the cost of robot operation per hour has increased more slowly.

Robot safety. Depending on the size of the robot's work envelope, its speed, and its proximity to humans, safety in a robot environment is an important consideration. Particularly important are programmers and maintenance personnel who are in direct interaction with robots. In addition, the movements of the robot with respect to other machinery requires a high level of reliability in order to avoid collisions and serious damage, and its material-handling activities require proper securing of raw materials and parts in the robot gripper at various stages in the production line. We discuss further details on safety in Chapter 37.

38.8 Sensor Technology

A **sensor** is a device that produces a signal for purposes of detecting or measuring a property, such as position, force, torque, pressure, temperature, humidity, speed, acceleration, and vibration. Traditionally, sensors, actuators, and switches have been used to set limits on the performance of machines. Familiar examples are stops on machine tools to restrict worktable movements, pressure and temperature gages with automatic shutoff features, and governors on engines to prevent excessive speed of operation. Sensor technology is essential to data acquisition, monitoring, communication, and computer control of machines and systems.

Because they convert one quantity to another, sensors are also often referred to as *transducers*, meaning to transfer. *Analog* sensors produce a signal, such as voltage, that is proportional to the measured quantity. *Digital* sensors have numeric or digital outputs that can be directly transferred to computers. Analog-to-digital converters (ADC) are available to interface analog sensors with computers.

38.8.1 Classification

Sensors that are of interest in manufacturing may be classified generally as:

a) *Mechanical*—for measuring quantities such as position, shape, velocity, force, torque, pressure, vibration, strain, and mass.

b) *Electrical*—for measuring voltage, current, charge, and conductivity.
c) *Magnetic*—for measuring magnetic field, flux, and permeability.
d) *Thermal*—for measuring temperature, flux, conductivity, and specific heat.
e) *Others*—such as acoustic, ultrasonic, chemical, optical, radiation, lasers, and fiber optics.

Depending on its application, a sensor may consist of metallic, nonmetallic, organic, or inorganic materials and fluids, gases, plasmas, or semiconductors. Using the special characteristics of these materials, sensors convert the quantity or property measured to analog or digital output. An ordinary mercury thermometer, for example, is based on the principle of thermal expansion of mercury. Similarly, a machine part or a physical obstruction or barrier in a space can be detected by breaking the beam of light sensed by a photoelectric cell (see Fig. 37.4). A proximity sensor, which senses and measures the distance between it and an object or a moving member of a machine, can be based on acoustics, magnetics, capacitance, or optics. Other actuators physically contact the object and, usually by electromechanical means, take appropriate action.

Sensors are essential to the control of advanced and intelligent robots. Sensors are being developed with capabilities that resemble those of human beings (*smart sensors*).

Tactile sensing. *Tactile sensing* is the continuous sensing of variable contact forces, commonly by an array of sensors. Such a system is capable of performing within an arbitrary three-dimensional space. Fragile parts, such as glass bottles and electronic devices, can be handled by robots with *compliant*, or *smart*, end effectors. These effectors can sense the force applied to the object being handled, using piezoelectric devices, strain gages, magnetic induction, ultrasonics, and optical systems of fiber optics and light-emitting diodes. Tactile sensors capable of measuring and controlling gripping forces and moments in three axes are available commercially (Fig. 38.27).

The force sensed is monitored and controlled through closed-loop feedback devices. Compliant grippers with force feedback and sensory perception can be complicated, require powerful computers, and hence are costly. Anthropomorphic end effectors are being designed to simulate the human hand and fingers, with the capability of sensing touch, force, movement, and pattern. The ideal tactile sensor must also sense slip, a capability of human fingers and hand that we tend to take for granted, yet is so important in the use of robots.

Visual sensing (machine vision or computer vision). In *visual sensing*, cameras optically sense the presence and shape of the object (Fig. 38.28). There are two basic systems of machine vision: linear array and matrix array. In *linear array*, only one dimension is sensed, such as the presence of an object or some feature on its surface. *Matrix arrays* sense up to three dimensions and are, for example, capable of detecting a properly inserted component in a printed circuit or a properly made solder joint. When used in automated inspection systems, they can also detect cracks and flaws.

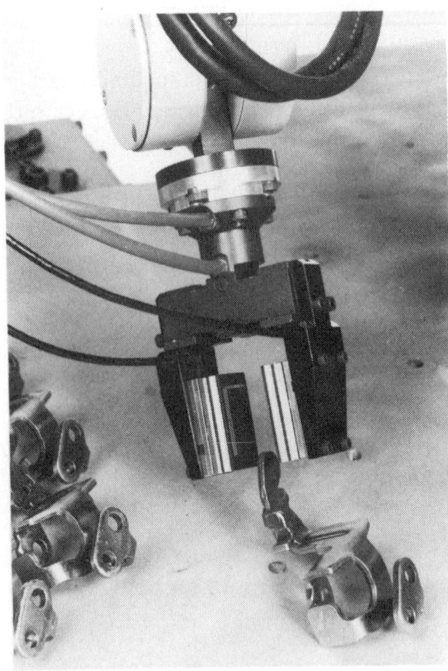

FIGURE 38.27
A robot gripper with tactile sensors. *Source:* Courtesy of Lord Corporation.

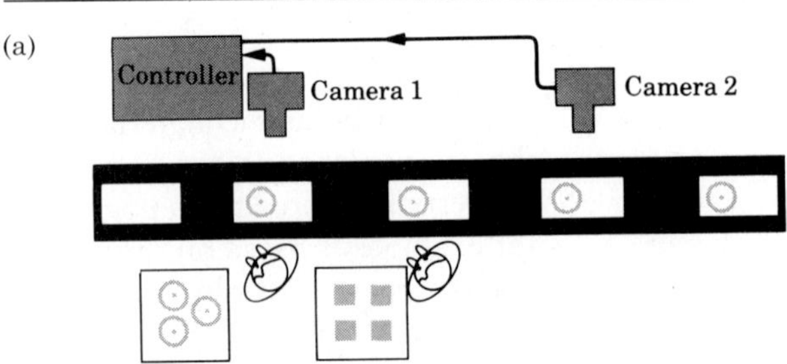

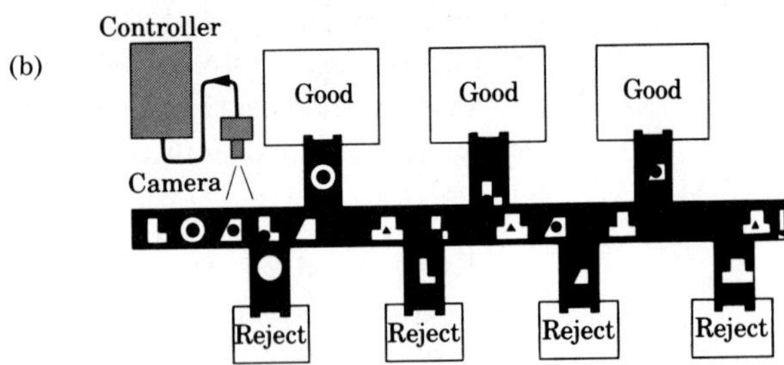

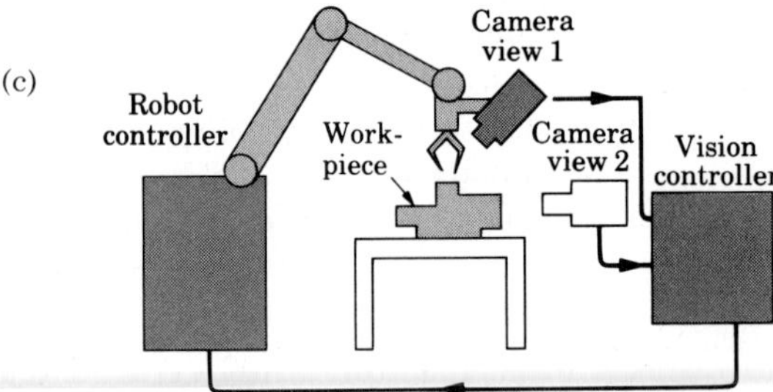

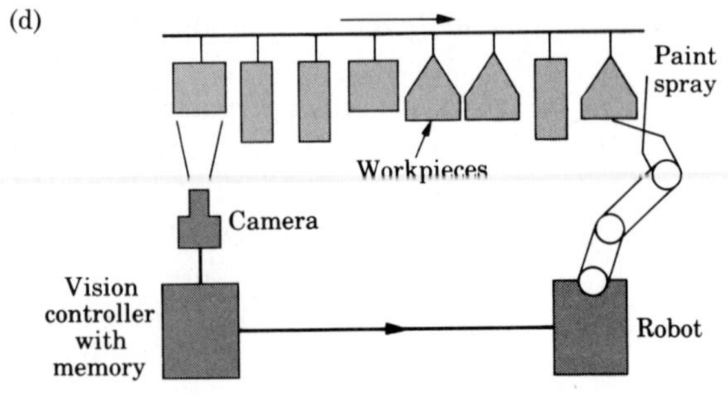

FIGURE 38.28
Examples of machine-vision application. (a) On-line inspection of parts. (b) Identification of parts with various shapes, inspection, and rejection of defective parts. (c) Use of cameras to provide positional input to a robot relative to the work-piece. (d) Painting parts with different shapes with input from a camera. The system's memory allows the robot to correctly identify the particular shape to be painted and to proceed with the correct movements of a paint spray attached to the end effector. *Source: Manufacturing Engineering,* November 1982.

In visual sensing, a microprocessor processes the image, usually in less than one second. The image is measured, and the measurements are digitized (*image recognition*). Machine vision is capable of on-line identification and inspection of parts—and rejection of defective parts. Several applications of machine vision in manufacturing are shown in Fig. 38.28. With visual sensing capabilities, end effectors are able to pick up parts and grip them in proper orientation and location. However, picking parts from a bin has proven to be a difficult task, because of the random orientation of parts in close proximity to each other.

The selection of a sensor for a particular application depends on factors such as the quantity to be measured or sensed, the sensor's interaction with other components in the system, expected service life, level of sophistication, difficulties associated with the sensor's use, power source, and cost. Another important consideration is the environment in which the sensors are to be used. Rugged sensors are being developed to withstand extremes of temperature, shock and vibration, humidity, corrosion, dust and various contaminants, fluids, electromagnetic radiation, and other interferences.

38.8.2 Sensor fusion

Although there is no clear definition of the term **sensor fusion**, it is generally understood that it basically involves the integration of multiple sensors in a manner in which the individual data from each of the sensors (force, vibration, temperature, dimensions, etc.) are combined to provide a higher level of information. It has been suggested that a common example of sensor fusion is someone drinking from a cup of hot tea or coffee. Although we take such an everyday event for granted, it can readily be seen that this process involves data input from the person's eyes, hands, lips, and tongue. There is real-time monitoring, through various senses, of relative movement and positions, as well as temperature. Thus, if the fluid is too hot, the hand movement of the cup toward the lip is controlled accordingly. The earliest applications of sensor fusion have been in (a) robot movement control and (b) missile flight tracking and similar military applications, primarily because these activities involve movements that mimic human behavior.

An example of sensor fusion is a machining operation in which a set of different but integrated sensors monitor (a) dimensions and surface finish of the workpiece, (b) cutting-tool forces, vibrations, and wear, (c) temperatures in various regions of the tool–workpiece system, and (d) spindle power. An important aspect in sensor fusion is sensor validation, whereby the failure of a sensor is detected so that the control system has high reliability. Thus redundant data from different sensors are essential. It can be seen that receiving, integration, and processing of all data from various sensors can be a complex problem.

Among new developments are **smart sensors**, which have the capability to perform a logic function, two-way communication, and make decisions and take appropriate actions. The necessary input and knowledge required to make a decision can be built into a smart sensor, whereby a chip with sensors can be programmed to turn a machine tool off when, for example, a cutting tool fails.

Likewise, a smart sensor can stop a mobile robot or a robot arm from accidentally coming in contact with an object or people by sensing quantities such as distance, heat, and noise.

Although to date there have been limited applications of sensor fusion in manufacturing operations, much research is being conducted in this area, with a potentially major impact on manufacturing in the near future. With advances in sensor size, quality, and technology, as well as developments in computer-control systems, artificial intelligence, expert systems, and neural networks (see Chapter 39), sensor fusion is bcoming practical and available at relatively low cost.

Example: Inspection of tie rod assembly with machine vision.

The accompanying figure shows an inner tie rod assembly for an automobile and its various features, such as the number of threads, the rod length, and the ball seating, that are inspected by machine vision. These features were previously inspected using mechanical probes. However, certain features such as the swage angle could not be inspected reliably because of imprecise fixturing or bent probes. A computer-controlled machine vision system was installed in the manufacturing plant and integrated with the assembly workstations. It was capable of analyzing 14 features on 15 families of inner tie rods in a period ranging from 1.5 s to 2.9 s, depending on the number of features to be inspected and the part model. The system is capable of inspecting 1500 tie rods per hour. The swage angle is measured with backlighting, and the software is programmed to reject parts if the angular tolerance of the swage angle is greater than $\pm 3°$. If this angle (measured from the longitudinal axis) is too small, the rod may pull free, and if too large, the rod will not swivel freely in its housing. *Source:* Automatix, Inc., and *Manufacturing Engineering*, February 1984.

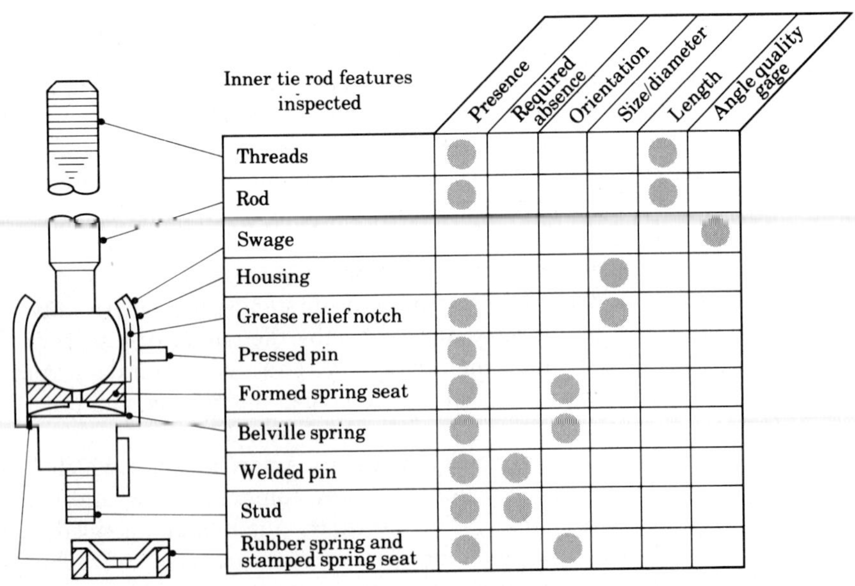

Inner tie rod features inspected	Presence	Required absence	Orientation	Size/diameter	Length	Angle quality gage
Threads	●				●	
Rod	●				●	
Swage						●
Housing				●		
Grease relief notch	●			●		
Pressed pin	●					
Formed spring seat	●		●			
Belville spring	●		●			
Welded pin	●	●				
Stud	●	●				
Rubber spring and stamped spring seat	●		●			

38.9 Flexible Fixturing

In previous chapters, we have described several workholding devices, such as chucks, collets, mandrels, and various fixtures, many of which are operated manually. Others are designed and operated at various levels of mechanization and automation, such as power chucks driven by mechanical, hydraulic, or electrical means. Workholding devices have certain ranges of capacity: Collets can accommodate round bars within a certain range of diameters; four-jaw chucks can accommodate square or prismatic workpieces having certain dimensions; and other devices and fixtures are made for specific workpiece shapes and dimensions. In manufacturing operations, the words clamp, jig, and fixture are often used interchangeably and sometimes in pairs, as in jigs and fixtures. *Clamps* are simple multifunctional devices, whereas *fixtures* are generally designed for specific purposes, with on and off features and shapes that are usually replicas of workpieces. *Jigs* have various reference surfaces and points for accurate alignment of parts and tools and are widely used in mass production.

The emergence of flexible manufacturing systems, which we describe in Chapter 39, has necessitated the use of workholding devices and fixtures that have a certain built-in flexibility. Generally called **flexible fixturing,** these devices are capable of quickly accommodating a range of part shapes and dimensions, without the necessity of changing or extensively adjusting the fixtures or requiring operator intervention, both of which would adversely affect productivity. A schematic illustration of a flexible fixturing system is shown in Fig. 38.29. The strain gage attached to the clamp senses the magnitude of the clamping force, and the system adjusts this force to keep the workpiece securely clamped on the work table.

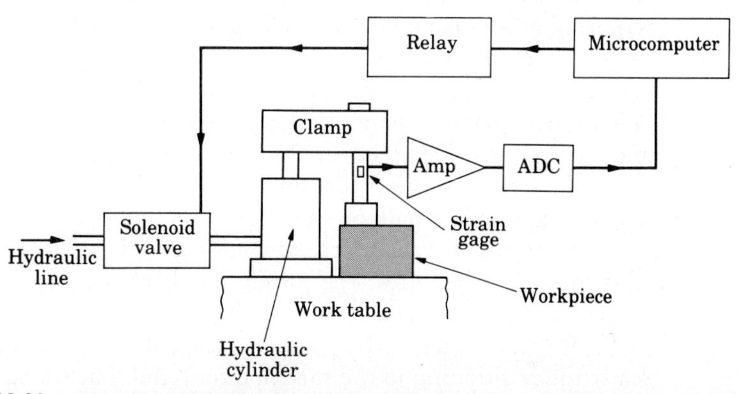

FIGURE 38.29

Schematic illustration of a flexible fixturing setup. The clamping force is sensed by the strain gage and the system automatically adjusts this force. *Source:* P. K. Wright.

38.9.1 Design considerations for flexible fixturing

Proper design of flexible workholding devices and fixtures is essential to the operation of advanced manufacturing systems. These devices should position the workpiece automatically and accurately and maintain its location precisely and with sufficient clamping force during the manufacturing operation. Fixtures should accommodate parts repeatedly in the same position and must have sufficient stiffness to resist without excessive deflection the forces developed. Fixtures and clamps should have low profiles so as to avoid colliding with tools and dies. Collision avoidance is also an important factor in programming tool paths.

Flexible fixturing must meet additional requirements in order to function properly in flexible manufacturing systems. Because of the increased efficiency of machining centers, machining time is very short. Consequently, to reduce cycle times the time required to load and unload parts should be minimal. The presence of loose chips between the locating surfaces of the workpiece and the fixture can be a serious problem. This type of situation does not exist in manual operations because the operator sees to it that these surfaces are cleaned. As with conventional workholding devices, the presence of even a single chip can result in an inaccurately produced part. Chips are most likely to be present in machining where cutting fluids are used and chips tend to stick to the wet surfaces.

A flexible fixture should accommodate parts made by various processes (such as casting, forming, powder metallurgy) and with dimensions and surface features that vary from part to part. The clamping force, in turn, must be estimated and applied properly. Significant progress is being made in the design of modular flexible fixtures.

• Example: Flexible fixturing of turbine blades.

A company manufactured 6000 different types of forged turbine blades, each requiring finish machining (by milling) of an airfoil contour. Because of the variety of blades made, the need for a flexible fixturing system became obvious. The accompanying figure shows a prototype design consisting of a pair of comformable clamps, octagonal in shape and hinged so that they can be opened to accept a blade and then closed. The base of the unit allows placing the clamps in any position along the length of the blade, and is thus able to accommodate blades of various lengths. The lower half of each clamp is equipped with a set of pneumatically operated plungers that are free to conform to the concave surface of the blade as it rests in the cradle. After the profile has been set, the plungers are mechanically locked in place. A high-strength belt is wrapped over the convex surface of the blade, forcing it against the plungers. The belt is then tightened and the assembly is ready for positioning in the machine tool and machining. The octagonal shape of the clamps allows positioning of the blade in different orientations. *Source:* M. R. Cutkosky, E. Kurokawa, and P. K. Wright.

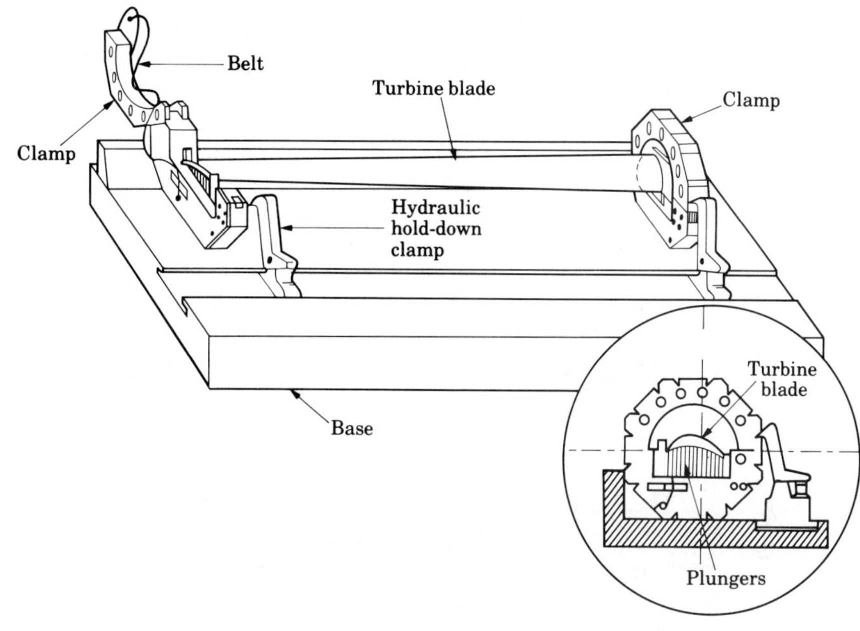

38.10 Automated Assembly

The individual parts and components made by various manufacturing processes are *assembled* into finished products by various methods. Some products are simple, having only two or three components to assemble, which can be done with relative ease, such as an ordinary pencil with an eraser, a frying pan with a wooden handle, or a beverage can. Most products, however, consist of many parts and their assembly requires considerable care and planning.

Traditionally, assembly has involved much manual work, contributing significantly to cost. The total assembly operation is usually broken into individual assembly operations, with an operator assigned to carry out each operation. Assembly costs are typically 25–50 percent of the total cost of manufacturing, with the percentage of workers involved in assembly operations ranging from 20 to 60 percent. In the electronics industries, some 40–60 percent of total wages are paid to assembly workers.

As production costs and quantities of products to be assembled increased, the necessity for automated assembly became obvious. Beginning with the hand assembly of muskets in the late 1700s and early 1800s with interchangeable parts, assembly methods have been vastly improved over the years. The first instance of large-scale modern assembly was the assembly of flywheel magnetos for the Model T Ford automobile, which eventually led to mass production of the automobile.

The choice of an assembly method and system depends on the required production rate and total quantities, product market life, labor availability, and most important, cost. *Automated assembly* can effectively minimize overall product cost. Although it requires a significant capital investment, automated machinery may be adapted with relative ease to permit assembly of a new product, thus becoming economical in the long run.

38.10.1 Assembly methods

You have seen that parts are manufactured with certain tolerances. Taking roller bearings as an example, you know that although they have the same nominal dimensions, some rollers in a lot are smaller than others by a small amount. Likewise, some bearing races are smaller than others in the lot.

There are two methods of assembly for such high-volume products: random assembly and selective assembly. In *random assembly*, parts are put together by selecting them randomly from the lots. In *selective assembly*, the rollers and races are segregated by groups of sizes, from smallest to largest. The parts are then selected to mate properly. Thus the smallest diameter rollers are mated with inner races having the largest outside diameter and with outer races having the smallest inside diameters.

38.10.2 Assembly systems

There are three basic types of assembly systems: synchronous, nonsynchronous, and continuous. In *synchronous systems*, also called *indexing*, individual parts and components are supplied and assembled at a constant rate at fixed individual stations. The rate of movement is based on the station that takes the longest time to complete its portion of the assembly. This system is used primarily for high-volume, high-speed assembly of small components. Transfer systems move the partially assembled parts from workstation to workstation by various mechanical means, as well as robots. Two typical transfer systems (*rotary* indexing and *in-line* indexing) are shown in Fig. 38.30. These systems can operate in either a fully automatic or a semiautomatic mode. However, a breakdown of one station can shut down the whole assembly operation. The part feeders supply the individual parts to be assembled and place them on other components, which are secured on work carriers or fixtures. The feeders move the individual parts by vibratory or other means through delivery chutes and ensure their proper orientation by various ingenious means (Fig. 38.31). Properly orientating parts and avoiding jamming are essential in all automated assembly operations.

In *nonsynchronous systems*, each station operates independently, and any imbalance is accommodated in storage (buffer) between stations. Thus if there are sufficient parts in the buffer, that particular station need not operate. Furthermore, if one station becomes inoperative for some reason, the assembly line continues to

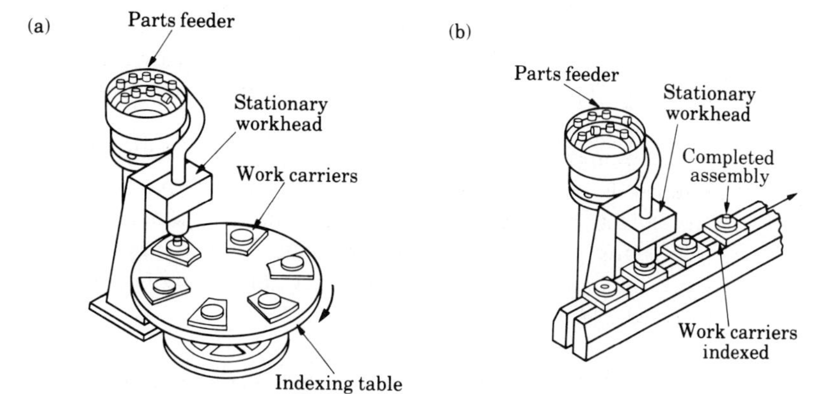

FIGURE 38.30 Transfer systems for automated assembly: (a) rotary indexing machine and (b) in-line indexing machine. *Source: G. Boothroyd.*

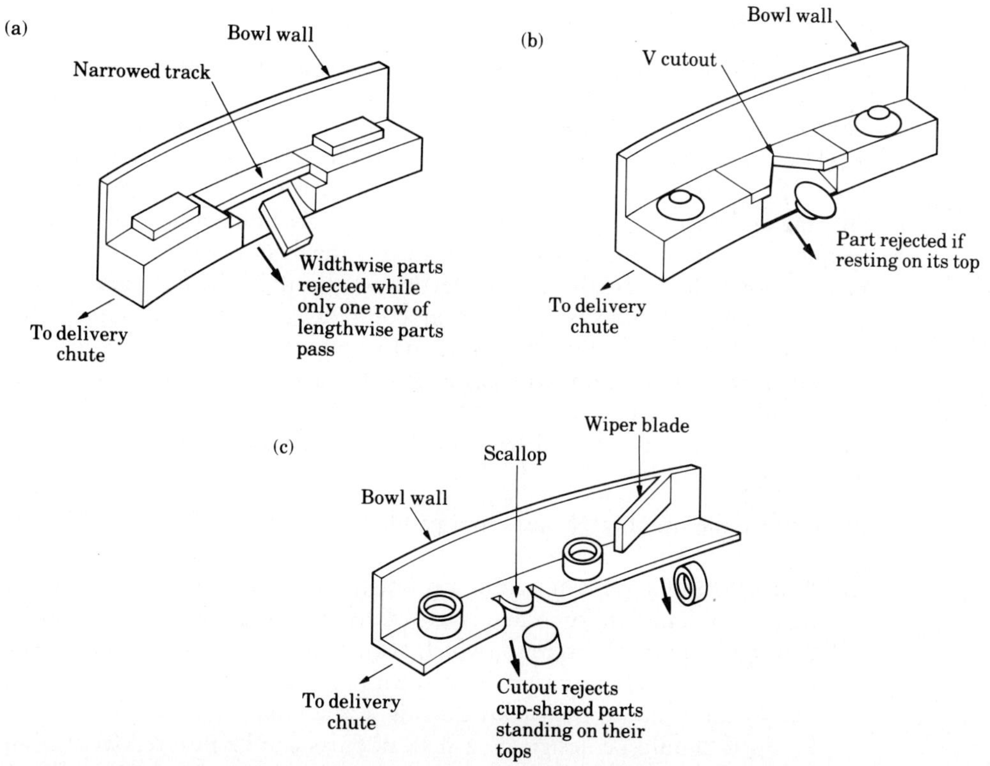

FIGURE 38.31 Various guides for proper orientation of parts for automated assembly. *Source:* G. Boothroyd.

operate until all the parts in the buffer have been used. Nonsynchronous systems are suitable for large components with many parts to be assembled and for types of assembly in which the times required for individual assembly operations vary widely. The speed of operation in this system is slower than the synchronous system.

In *continuous systems*, the product is assembled while moving at a constant speed on pallets or similar work carriers. The parts to be assembled are brought to the product on various workheads, and their movements are synchronized with the constant movement of the product. Typical applications of this system are in bottling and packaging plants and mass-production lines for automobiles and appliances.

Although these systems are set up for a certain product line, they can be modified for increased flexibility in order to assemble product lines that change with market demand. Such **flexible assembly systems (FAS)** utilize computer controls, interchangeable and programmable workheads and feeding devices, coded pallets, and automated guiding devices.

38.10.3 Design for assembly

Although the functions of a product and its design for manufacturing have been matters of considerable interest for some time, only recently has *design for assembly* (DFA) attracted special attention, particularly design for automated assembly. The need for reducing labor costs has necessitated greater use of automated assembly. In spite of the use of sophisticated mechanisms, robots, and computer controls, aligning and placing a simple square peg in a square hole with small clearances can be difficult in automated assembly. Because of the dexterity of the human hand and fingers—and their capability for feedback through various senses—assembly workers can manually assemble even complex parts without much difficulty. However, in hand assembly, the worker uses both hands, thus greatly increasing the flexibility of the operation, whereas automated assembly usually involves just one workhead.

Certain guidelines have been established as an aid in the design of parts for automated assembly:

a) Reduce the number of parts in a product so that fewer steps and lower costs are involved in its assembly.
b) When fasteners are required in a product, automated assembly with robots is capable of handling snap fits, rivets, adhesives, and welds better and faster than bolts, nuts, and screws, which require rotary motions. One exception is the use of screws on plastic or wooden parts, and sheet metals, since the screws can produce their own threads in the hole.
c) Products should be designed so that all parts can be inserted from a single direction, preferably vertically to take advantage of gravity. Assembly in two or more directions can present problems in automation.

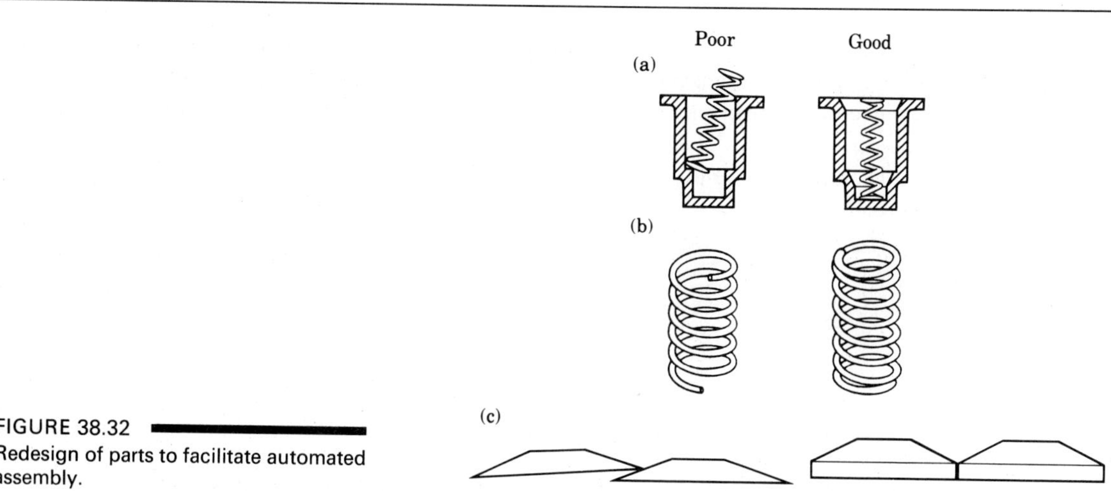

FIGURE 38.32 Redesign of parts to facilitate automated assembly.

d) Products should be designed, or existing products redesigned, so that there are no physical obstructions to the free movement of components during assembly (Fig. 38.32). Sharp external or internal corners should be replaced with chamfers, tapers, or radii.
e) Different orientation of parts to be assembled is a significant problem in automated assembly. Consequently, product designs should have a high degree of symmetry, either in shape or weight distribution, to help in their orientation during assembly.

SUMMARY

There are several levels of automation, from simple automation of machines to unmanned manufacturing cells to—ultimately—the factory of the future. Automation has been implemented successfully in manufacturing processes, material handling, inspection, assembly, and packaging. Production quantity and rate are important factors in determining the economical levels of automation. True automation began with the numerical control of machines, which has the capability of flexibility of operation, lower cost, and ease of making different parts with lower operator skill. Manufacturing operations are further optimized, both in quality and cost, by adaptive control techniques, which continuously monitor the operation and make necessary adjustments in process parameters.

Great advances have been made in material handling, particularly with the implementation of industrial robots and automated guided vehicles. The role of sensors is crucial in the implementation of these technologies, and a wide variety of sensors based on various principles have been developed and installed. Other advances include flexible fixturing and automated assembly techniques that reduce the need for worker intervention and that lower manufacturing costs. The efficient and economic implementation of these techniques requires that design for assembly be recognized as an important factor in the total design and manufacturing process.

TRENDS

- Because of high equipment costs, economic analysis of various aspects of automation in manufacturing plants and the use of industrial robots are becoming an important area of study.
- Together with extensive use of computers and rapid advances in sophisticated software, new concepts of manufacturing operations are being developed, including adaptive control of manufacturing systems and the use of multiple sensors.
- The trends in sensors are toward miniaturization, for compatibility with logic systems, and less emphasis on electromechanical systems. New developments include sensor fusion and smart sensors, which are capable of microcomputer-based calibration, computation, and decision making.

KEY TERMS

Adaptive control
Automated guided vehicles
Automation
Closed-loop control
Computer numerical control
Computer vision
Direct numerical control
Flexible assembly systems
Flexible fixturing
Hard automation
Interpolation
Machine vision
Manipulators
Material handling
Mechanization
Numerical control
Open-loop control
Productivity
Programmable logic controller
Programming for NC
Repeatability
Resolution
Robot
Sensor fusion
Sensors
Smart sensors
Soft automation

BIBLIOGRAPHY

Astrom, K.J., and B. Wittenmark, *Adaptive Control.* Reading, Mass.: Addison-Wesley, 1989.

Ballard, D.H., and C.M. Brown, *Computer Vision.* Englewood Cliffs, N.J.: Prentice-Hall, 1982.

Basics of Material Handling. Pittsburgh: The Material Handling Institute, 1981.

Boothroyd, G., C. Poli, and L.E. Murch, *Automatic Assembly.* New York: Marcel Dekker, 1982.

Castleberry, G.A., *The AGV Handbook.* Port Washington, Wis.: AGV Decisions, 1991.

Chang, C.-W., and M.A. Melkanoff, *NC Machine Programming and Software Design.* Englewood Cliffs, N.J.: Prentice-Hall, 1989.

Craig, J.J., *Introduction to Robotics: Mechanics and Control.* Reading, Mass.: Addison-Wesley, 1985.

Critchlow, A.J., *Introduction to Robotics.* New York: Macmillan, 1985.

Engleberger, J.F., *Robotics in Service.* Cambridge, Mass.: MIT Press, 1989.

Gevarter, W.B., *Intelligent Machines: An Introductory Perspective of Artificial Intelligence and Robotics.* Englewood Cliffs, N.J.: Prentice-Hall, 1985.

Groover, M.P., *Automation, Production Systems, and Computer-Integrated Manufacturing.* Englewood Cliffs, N.J.: Prentice-Hall, 1987.

——, M. Weiss, R.N. Nagel, and N.G. Odrey, *Industrial Robotics: Technology, Programming, and Applications.* New York: McGraw-Hill, 1986.

Harrington, J., Jr., *Understanding the Manufacturing Process.* New York: Marcel Dekker, 1984.

Holzbock, W.G., *Robotic Technology: Principles and Practice.* New York: Van Nostrand Reinhold, 1986.

Koren, Y., *Computer Control of Manufacturing Systems.* New York: McGraw-Hill, 1983.

——, *Robotics for Engineers.* New York: McGraw-Hill, 1985.

Lee, M.H., *Intelligent Robotics.* New York: Halsted Press, 1989.

McDonald, A.C., *Robot Technology: Theory, Design, and Applications.* Englewood Cliffs, N.J.: Prentice-Hall, 1986.

Miller, R.K., *Industrial Robot Handbook.* New York: Van Nostrand Reinhold, 1989.

NC/CIM Guidebook, published annually by *Modern Machine Shop*, Cincinnati, Ohio.

Owen, T., *Assembly with Robots.* Englewood Cliffs, N.J.: Prentice-Hall, 1985.

Pessen, D., *Industrial Automation*, New York: Wiley, 1989.

Ranky, P.G., and C.Y. Ho, *Robot Modelling, Control and Applications with Software.* New York: Springer-Verlag, 1985.

Rapello, R.G., *Essentials of Numerical Control.* Englewood Cliffs, N.J.: Prentice-Hall, 1986.

Rembold, U. (ed.), *Robot Technology and Applications.* New York: Marcel Dekker, 1990.

Rosheim, M.E., *Robot Wrist Actuators.* New York: Wiley, 1989.

Sandler, B.-Z., *Robotics: Designing the Mechanisms for Automated Machinery.* Englewood Cliffs, N.J.: 1991.

Seames, W., *Computer Numerical Control: Concepts and Programming*, 2d ed. Albany, N.Y.: Delmar, 1990.

Sharon, D., J. Harstein, and G. Yantian, *Robotics and Automated Manufacturing.* Aulander, N.C.: Pittman, 1989.

Snyder, W.E., *Industrial Robots: Computer Interfacing and Control.* Englewood Cliffs, N.J.: Prentice-Hall, 1985.

Tool and Manufacturing Engineers Handbook, 4th ed., Vol. 1: *Machining*. Dearborn, Mich.: Society of Manufacturing Engineers, 1983.

Tool and Manufacturing Engineers Handbook, 4th ed., Vol. 4: *Assembly, Testing, and Quality Control*. Dearborn, Mich.: Society of Manufacturing Engineers, 1986.

Ulsoy, A.G., and W.R. DeVries, *Microcomputer Applications in Manufacturing*. New York: Wiley, 1988.

Villers, P., and J.M. Brady, *The Robotic Handbook*. Menlo Park, California: Benjamin-Cummings, 1987.

Zeldman, M., *What Every Engineer Should Know About Robots*. New York: Marcel Dekker, 1984.

REVIEW QUESTIONS

38.1 Describe the differences between mechanization and automation. Give several specific examples for each.

38.2 Why is automation generally regarded as evolutionary rather than revolutionary?

38.3 Are there activities in manufacturing operations that cannot be automated? Explain.

38.4 Why should the level of automation in a manufacturing facility depend on production quantity and production rate?

38.5 Explain the difference between hard and soft automation. Why are they so called?

38.6 Describe the principle of numerical control of machines. What factors led to the need and development of numerical control? Name typical applications.

38.7 Discuss the advantages and limitations of NC. When would you not apply NC?

38.8 Explain the differences between direct numerical control and computer numerical control. What are their relative advantages?

38.9 Describe open-loop and closed-loop control circuits.

38.10 Discuss the types of interpolation in NC control systems.

38.11 What are the advantages of computer-aided NC programming?

38.12 Describe the principle and purposes of adaptive control. Give some examples of present applications and others that you think can be implemented.

38.13 What is the difference between AC constraint and AC optimization? Give examples.

38.14 What factors have led to the development of automated guided vehicles? Do they have any disadvantages? Explain.

38.15 List and discuss the factors that should be considered in choosing a suitable material-handling system.

38.16 Describe the features of an industrial robot. Why are these features necessary?

38.17 Explain why sensors have become so crucial in the development of automated manufacturing systems.

38.18 Discuss the principles of various types of sensors and give two applications for each type.

38.19 Why is there a need for flexible fixturing for holding workpieces? Are there any disadvantages? Explain.

38.20 Discuss why assembly costs are significant.

38.21 Explain the characteristics of different types of assembly systems.

38.22 Describe the concept of design for assembly. Why has it become an important factor in manufacturing?

38.23 Is it possible to have partial automation in assembly? Explain.

QUESTIONS AND PROBLEMS

38.24 Giving specific examples, discuss your observations concerning Fig. 38.2.

38.25 What are the relative advantages and limitations of the two arrangements for power heads shown in Fig. 38.4?

38.26 Discuss methods of on-line gaging workpiece diameters in turning operations other than that shown in Fig. 38.17. Explain the relative advantages and limitations.

38.27 Comment on the accuracies of hole positions in the three different methods shown in Fig. 38.6.

38.28 What is the function of the digital-to-analog converter (DAC) shown in Fig. 38.8(b)?

38.29 Is drilling the only application for the point-to-point system shown in Fig. 38.11(a)? Explain.

38.30 Give an example of a forming operation that is suitable for adaptive control similar to that shown in Fig. 38.16.

38.31 Comment on your observations and implications concerning Fig. 38.18.

38.32 Design two different systems of mechanical grippers.

38.33 Describe possible applications for industrial robots not discussed in this chapter.

38.34 What determines the number of robots in an automated assembly line such as that shown in Fig. 38.26?

38.35 Describe situations in which the shape and size of the work envelope of a robot (Fig. 38.23) can be critical.

38.36 Explain the functions of each of the components of the robot shown in Fig. 38.20.

38.37 Give some applications for the systems shown in Fig. 38.28(a) and (c).

38.38 Based on a system similar to that shown in Fig. 38.29, design a flexible fixturing setup for a lathe chuck.

38.39 Add others to the examples shown in Fig. 38.32.

38.40 Give examples of products that are suitable for the type of production shown in Fig. 38.3.

38.41 Give examples where tactile sensors would not be suitable. Explain why.

38.42 Give examples where machine vision cannot be applied properly and reliably. Explain why.

38.43 Think of a simple product to be made by end milling and prepare an NC program for it, similar to that shown in the example in Section 38.4. Explain why you chose those particular cutter paths. If this part is redesigned so that it now includes a threaded hole, how would you machine it? What type of machine would you recommend that still has NC features?

38.44 Explain the difference between an automated guided vehicle and a self-guided vehicle.

38.45 Choose one machine each from Parts II, III, and IV, and design a system in which sensor fusion can be used effectively. How would you convince a prospective customer of the merits of such a system? Would it be cost-effective?

38.46 Same as Problem 38.45, but for a flexible fixturing system.

38.47 It has been commonly acknowledged that, at their early stages of development and implementation, the usefulness and cost effectiveness of industrial robots have been overestimated. What reasons can you think of to explain this situation?

38.48 Describe the type of manufacturing operations (see Fig. 38.2) that are likely to make the best use of a machining center. Comment on the influence of product quantity and part variety.

38.49 Give a specific example each where (a) an open-loop and (b) a closed-loop control system would be desirable.

38.50 Think of a product and design a transfer line similar to that shown in Fig. 38.5. Specify the types and number of machines required.

39

Computer-Integrated Manufacturing Systems

39.1 Introduction

In describing manufacturing processes and operations in this text, we discussed each of them primarily as an individual activity with its own capabilities, applications, advantages, and limitations. On several occasions, we described the implementation and benefits of mechanization, automation, and computer control of various stages of manufacturing operations. In this chapter we focus our attention on the *integration* of manufacturing activities. Integration means that manufacturing processes, operations, and their management are treated as a *system*, allowing complete control of the manufacturing facility. In this way, productivity, product quality, reliability, and cost can be optimized.

Beginning with the 1960s, certain trends developed that had a major impact on manufacturing:

- International competition increased rapidly.

- Market conditions fluctuated widely.
- Customers demanded high-quality, low-cost products.
- Product variety increased substantially.
- Product life cycles became shorter.
- Labor costs increased significantly.
- Computer-controlled equipment became available.

These trends led to recognition of the need for an integrated approach to manufacturing. In *computer-integrated manufacturing*, the traditionally separate functions of research and development, design, production, assembly, inspection, and quality control are all linked. Consequently, integration requires that quantitative relationships among product design, materials, manufacturing process and equipment capabilities, and related activities be well understood. Thus changes in material requirements, product types, and market demand can be accommodated. The goal: to find the optimum method of producing high-quality, low-cost products efficiently.

In Chapter 38 we described the principles of automation, its benefits, and how it has been gradually introduced into manufacturing facilities, progressing from mechanization to numerical control machines, adaptive control, automated material handling, industrial robots, and automated assembly. We discussed the role and importance of sensors and various devices in automation. In spite of their benefits, however, these advances cannot by themselves be responsive to changing market demand. Why? Because the elements of automation exist in isolation; consequently, their performance can be optimized relative only to their individual tasks. However, within the context of a *system*, all operations become interdependent elements.

Therefore, machines, tooling, and manufacturing operations must have a certain built-in flexibility to respond to change and ensure *on-time delivery* of products to the customer. You can understand the significance of on-time product delivery by noting your own dissatisfaction when you do not receive something on its promised date. In industry, failure of on-time delivery can upset production plans and schedules and, consequently, have a significant economic impact. In a highly competitive environment, failure of on-time product delivery can cost a company its *competitive edge* because the customer will simply change suppliers.

We emphasized the importance of product quality in Chapter 36 and the necessity for total commitment of a company's management to total quality control. We stressed the importance of direct worker involvement and pride in quality control. Recall our statements that *quality must be built into the product*, that high quality does not necessarily mean higher cost, and marketing poor-quality products can indeed be very costly to the manufacturer. By integrating manufacturing functions, high quality is far more attainable. For example, by integrating design and manufacturing, design goals—such as function, durability, and cost—can be traced through various manufacturing alternatives during the design stage.

In the following sections, we describe major advances in integrating all aspects of manufacturing that have led to *computer-integrated manufacturing systems*.

Among these advances are:

- Computer-aided design.
- Computer-aided manufacturing.
- Rapid prototyping.
- Computer-aided process planning.
- Group technology.
- Cellular manufacturing.
- Flexible manufacturing systems.
- Just-in-time production.
- Artificial intelligence.

We present several examples to illustrate the major impact of these activities on the technology and economics of manufacturing. Because some of these advances are still in the development stage, some controversy exists among engineers and corporate management regarding their merits and economic justification. Thus understanding the characteristics, applications, potentials, advantages, and limitations of these technological developments and sound evaluations of their economic benefits are essential. Manufacturing engineers should be ready to implement new technologies as they become viable economic alternatives.

We conclude this chapter with a discussion and overview of the *factory of the future*. The prediction is that manufacturing operations in the future will be performed with little human labor; hence these facilities are also being called *unmanned factories*. Because of its potential impact on the workforce, the social implications of this trend are also discussed.

39.2 Manufacturing Systems

Although the word *system* is derived from the Greek *systema*, meaning to combine, it has come to mean an arrangement of physical entities characterized by identifiable and quantifiable interacting parameters. Because manufacturing entails a large number of interdependent activities consisting of distinct entities such as materials, tools, machines, power, and human beings, it should properly be regarded as a system. It is, in fact, a complex system because it is comprised of many diverse physical and human elements (see Fig. 2 in the General Introduction), some of which are difficult to predict and control, such as raw-material prices, market changes, and human behavior and performance.

Ideally, we should be able to represent a system by *mathematical* and *physical models*, which show us the nature and extent of interdependence. In a manufacturing system, a change or disturbance anywhere in the system requires that the system adjust itself in order to continue functioning efficiently. For example, if the supply of a particular raw material is reduced and hence its cost is increased because of, say, geopolitical reasons, alternative materials must be selected. This selection should be made only after careful consideration as to the effect of this change on product quality, production rate, and costs.

Similarly, if demand for a product is such that its shape, size, or capacity fluctuates randomly and rapidly, the system must be able to produce the modified product on short lead time and without the need for major capital investment in machinery and tooling. Although the labor share of product cost has been decreasing steadily over the years (in some cases it is now less than 10 percent of total product cost), the system must also be capable of absorbing some of the cost if worker pay escalates.

Modeling such a complex system can be difficult because we lack reliable data on many of the variables and cannot easily predict and control some of these variables. For example, machine characteristics, performance, and response to external disturbances cannot be modeled precisely, raw-material costs are difficult to predict accurately, and human behavior and performance are difficult to model.

In spite of these difficulties, considerable progress has been made in modeling and simulating manufacturing systems. In this way the effects of disturbances, such as product demand, materials changes, and machine performance, can be predicted with reasonable accuracy.

39.3 Computer-Integrated Manufacturing

The various levels of automation in manufacturing operations we described in Chapter 38 can be extended further by including information processing functions utilizing an extensive network of interactive computers. The result is **computer-integrated manufacturing** (**CIM**), which is a broad term describing the computerized integration of *all* aspects of design, planning, manufacturing, distribution, and management.

Computer-integrated manufacturing is a *methodology* and a *goal*, rather than an assemblage of equipment and computers. The technology for implementing CIM is well understood and available. However, because CIM ideally involves the total operation of a company, it must be comprehensive and have an extensive database. Consequently, if implemented all at once, CIM can be prohibitively expensive, particularly for small and medium-size companies.

Considerable experience and training is required in different areas of CIM and at all levels within a company if CIM is to operate effectively. Implementation of CIM in existing plants may begin with modules in various phases of the operation. For new manufacturing plants, comprehensive, long-range strategic planning covering all phases of the operation is necessary to take full advantage of CIM.

Thus such plans should include availability of resources; organizational mission, goals, and culture; existing and possible future technology; and degree of integration required.

Computer-integrated manufacturing systems consist of subsystems that are integrated into a whole. These subsystems consist of business planning and support, product design, manufacturing process planning, process control, shop floor monitoring systems, and process automation. The subsystems are designed, developed, and applied in such a manner that the output of one subsystem serves as the input of another subsystem. Organizationally, these subsystems are usually divided into business planning and business execution functions, respectively. *Planning* functions include activities such as forecasting, scheduling, material-requirements planning, invoicing, and accounting. *Execution* functions include production and process control, material handling, testing, and inspection.

The effectiveness of CIM depends greatly on the presence of a large-scale, *integrated communications system* involving computers, machines, and their controls. Major problems have arisen in factory communication because of the difficulty of interfacing different types of computers purchased from different vendors by the company at various times. As we describe in Section 39.12, the trend is strongly toward standardization to make communications equipment compatible.

The benefits of CIM include:

- Responsive to shorter product life cycles and changing market demand.
- Emphasis on product quality and its uniformity through better process control.
- Better use of materials, machinery, and personnel and reduction of inventory, thus improving productivity.
- Better control of production and management of the total manufacturing operation, resulting in lower product cost.

39.3.1 Database

An efficient computer-integrated manufacturing system requires a single database, which is shared by the entire manufacturing organization. A **database**, or *databank*, consists of up-to-date, detailed, and accurate data relating to products, designs, machines, processes, materials, production, finances, purchasing, sales, marketing, and inventory. This vast array of data is stored in computer memory and recalled or modified as necessary, either by individuals in the organization or by the CIM system itself in controlling various aspects of design and production.

A database generally consists of the following items, some of which are classified as technical and others as nontechnical:

a) Product data, such as part shape, dimensions, and specifications.
b) Production data, such as the manufacturing processes involved in making parts and products.
c) Operational data, such as scheduling, lot sizes, and assembly requirements.

d) Resources data, such as capital, machines, equipment, tooling, and personnel, and their capabilities.

Databases are built by individuals and various sensors in the machinery and equipment used in production. Data from the latter are collected automatically by a *data acquisition system* (DAS), such as number of parts being produced per unit of time, their dimensional accuracy, surface finish, weight, and so on, at specified rates of sampling. The components of DAS include minicomputers, microprocessors, transducers, and analog-to-digital converters (ADCs). Data acquisition systems are also capable of analyzing data and transferring them to other computers for purposes such as statistical analysis, data presentation, and forecasting product demand.

Several factors are important in the use and implementation of databases. They should be timely, accurate, easily assessible, easily shared, and user friendly. In the event that something goes wrong with the data, correct data should be recovered and restored. Because it is used for many purposes and by many people, the database must be flexible and responsive to the needs of different users. Companies must protect data against tampering or unauthorized use. CIM systems can be accessed by designers, manufacturing engineers, process planners, financial officers, and the management of the company by using appropriate access codes.

39.4 Computer-Aided Design

In traditional engineering design practice, a draftsperson typically uses a drafting board, drawing equipment, templates, and various handbooks and reference sources—a slow, often tedious, and not always accurate process. Activities such as generating and revising engineering drawings, modifying designs, and analyzing and optimizing designs are done with limited technological assistance. Furthermore, the inherently iterative, trial-and-error nature of the traditional design process usually leads to a lengthy product design phase.

Consider, for example, the design of a lawnmower in which the capacity of the motor is to be increased. The new motor has larger dimensions, thus requiring changes in the mower's design and method of mounting the motor. Such a change can involve significant effort, as the assembly and detail drawings have to be modified, new calculations have to be made, and the design changes have to be checked. Then revised drawings and manufacturing plans have to be prepared.

Computer-aided design (**CAD**) involves the use of computers to create design drawings (see Fig. 6 in the General Introduction). Computer-aided design is usually associated with *interactive computer graphics*, known as a *CAD system*. Computer-aided design systems are powerful tools and are used in mechanical design and geometric modeling of products and components. They simplify finite-element analysis of stresses, strains, deflections, and temperature distribution in structures and load-bearing members and generation, storage, and retrieval of NC data. These systems are also used extensively in the design of integrated circuits and other electronic devices.

In CAD, the drawing board is replaced by electronic input and output devices, an electronic plotter, and a data tablet (Fig. 39.1), which is divided into sections. Each section represents a mathematically defined geometric function, such as coordinate points, line, plane, circle, or cylinder, called *menu items*. Sitting in front of a graphics workstation, the user can generate parts of a drawing by touching the desired menu item with a stylus (electronic pencil) and entering the information on the keyboard. Other systems include the use of a trackball or a "mouse." Manipulation and modification of a design in CAD is done much more rapidly and accurately than is possible using traditional design methods. The design is continuously displayed on the cathode-ray tube (graphics screen). The final design is drawn automatically on a plotter that is interfaced with the computer.

39.4.1 Advantages of CAD systems

When using a CAD system, the designer can conceptualize the object to be designed more easily on the graphics screen and consider alternative designs or modify a particular design quickly to meet the necessary design requirements or changes. The designer can then subject the design to a variety of engineering analyses and identify

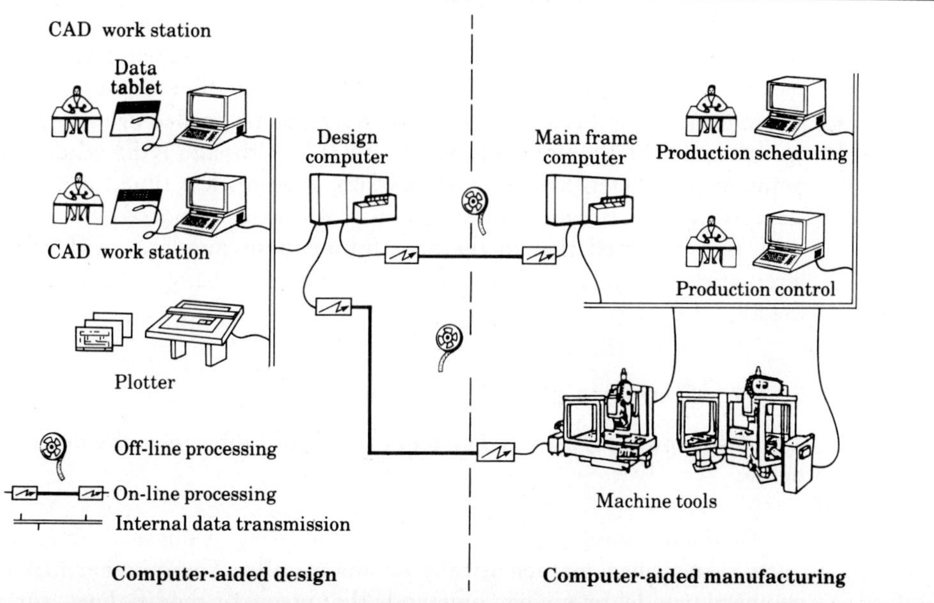

FIGURE 39.1
Information flow chart in CAD/CAM application. *Source: Flexible Automation, 87/88, The International CNC Reference Book.*

potential problems, such as excessive loads and deflection. The speed and accuracy of such analyses far surpasses traditional methods.

The CAD system generates working drawings for the product and its components quickly and accurately. These drawings generally are of higher quality and consistency than those produced by traditional manual drafting. They can be reproduced any number of times and at different levels of reduction and enlargement. In addition to the design's geometric and dimensional features, other information, such as a list of materials, specifications, and manufacturing instructions, are stored in the CAD database. Using such information, the designer can analyze the economics of various designs.

An important feature of CAD in machining operations is its capability to describe tool path in operations such as NC turning, milling, and drilling. The programmer provides instructions (programs) to the CAD system to automatically determine and optimize the tool path. With CAD's graphics capabilities, the designer can display and visually check the tool path for possible tool collisions with clamps, fixtures, or other interferences. The tool path can be modified at any time to accommodate other part shapes to be machined. Computer-aided design systems are also capable of coding and classifying parts into groups that have similar shapes (see Section 39.8), using alphanumeric coding.

Because of the availability of a wide variety of CAD systems with different characteristics supplied by different vendors, proper communication and exchange of data between these systems has become a significant problem (see also Section 39.12). The need for a single neutral format for better compatibility is presently filled largely with the *Initial Graphics Exchange Specification* (IGES). Thus vendors need provide only translators for their own systems to preprocess the data into the neutral format. Although IGES was originally developed for two-dimensional systems, some progress has been made for three-dimensional systems. A more recent development is a solid-model based standard, called *Product Data Exchange Specification* (PDES), which is based on IGES. Although IGES is adequate for most requirements, PDES has less memory size, requires less time for execution, and is less error-prone. Because of the existence of various standards in other countries as well, it is expected that in the near future these standards will culminate in an international standard, called *Standard for the Exchange of Product Model Data* (STEP).

39.4.2 Elements of CAD systems

The design process in a CAD system consists of four stages. We describe each stage below.

Geometric modeling. In *geometric modeling*, a physical object or any of its parts is described mathematically or analytically. The designer first constructs a geometric model by giving commands that create or modify lines, surfaces, solids, dimensions, and text that together are an accurate and complete two- or three-dimensional representation of the object. The results of these commands are displayed and can be moved around on the screen, and any section desired can be magnified to view details. These data are digital and are stored in the database contained in computer memory.

The models can be presented in three different ways. In *line* representation (*wire frame*; Fig. 39.2), all edges are visible as solid lines. This image can be ambiguous, particularly for complex shapes. However, various colors are generally used for different parts of the object, thus making the object easier to visualize. In the *surface* model, all visible surfaces are shown in the model, and in the *solid* model, all surfaces are shown but the data describe the interior volume.

The three types of wire-frame representations are 2-D, $2\frac{1}{2}$D, and 3-D. The 2-D image shows the profile of the object or part. The $2\frac{1}{2}$-D image can be obtained by translational sweep, that is, moving the 2-D object along the z axis. For round objects, a $2\frac{1}{2}$-D model can be generated by simply rotating a 2-D model around its axis.

Solid models can be constructed from "swept volumes" (Figs. 39.2b and c) or by the techniques shown in Fig. 39.3. In *boundary representation* (B-rep), surfaces are combined together to develop a solid model (Fig. 39.3a). In *constructive solid geometry* (CSG), simple shapes such as spheres, cubes, blocks, cylinders, and cones (called *primitives of solids*) are combined to develop a solid model (Fig. 39.3b). Programs are available whereby the user selects any combination of these primitives and their sizes, and combines them into the desired solid model. Although solid models have advantages such as ease of design analysis (see below) and preparation for manufacturing the part, they require more memory and processing time than the wire-frame and surface models shown in Fig. 39.2.

Design analysis and optimization. After the design's geometric features have been determined, the design is subjected to an engineering analysis. This phase may consist of analyzing stresses, strains, deflections, vibrations, heat transfer, temperature distribution, casting, solidification, or tolerance. Various sophisticated software packages having capabilities to compute these quantities accurately and

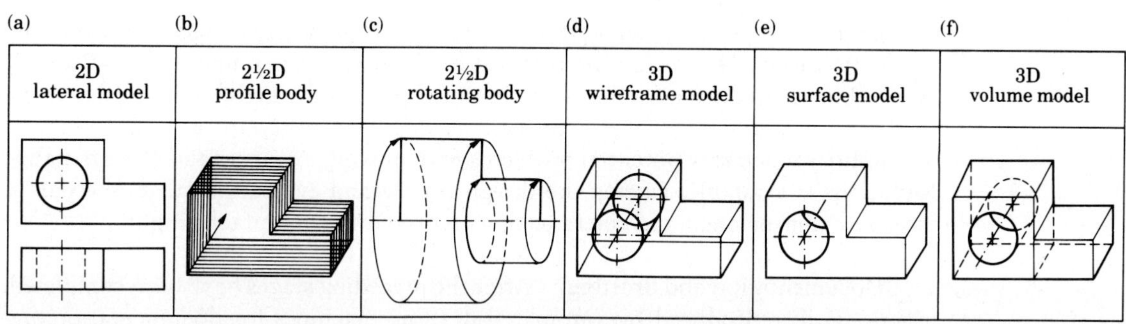

FIGURE 39.2
Types of modeling for CAD. *Source: Flexible Automation, 87/88, The International CNC Reference Book.*

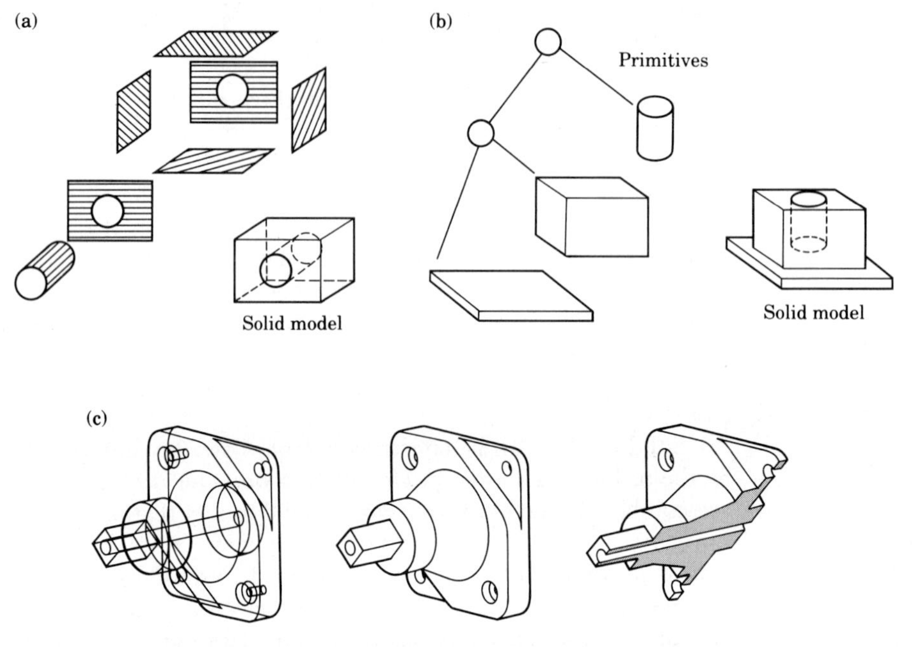

FIGURE 39.3
(a) Boundary representation of solids, showing the enclosing surfaces of the solid model and the generated solid model. (b) A solid model represented as compositions of solid primitives. (c) Three representations of the same part by CAD. *Source:* P. Ranky.

rapidly are available. Because of the relative ease with which such analyses can now be made, designers are increasingly willing to thoroughly analyze a design before it moves on to production. Experiments and field measurements may be necessary to determine the effects of loads, temperature, and other variables on the designed components.

Design review and evaluation. An important design stage is review and evaluation to check for any interference between various components in order to avoid difficulties during assembly or use of the part, and whether moving members, such as linkages, are going to operate as intended. Software having animation capabilities to identify potential problems with moving members and other dynamic situations is available. During the design review and evaluation stage, the part is precisely dimensioned and toleranced, as required for manufacturing it.

Documentation and drafting. After the preceding stages have been completed, the design is reproduced by automated drafting machines for documentation and reference. Detail and working drawings are also developed and printed. The CAD system is also capable of developing and drafting sectional views of the part, scaling the drawings, and performing transformations to present various views of the part. Although in CAD systems much of the design process is carried out on workstations connected to a main frame computer, smaller systems are being developed for use on personal computers (PCs).

39.5 Computer-Aided Manufacturing

Computer-aided manufacturing (CAM) is the use of computers and computer technology to assist in all phases of manufacturing a product, including process and production planning, machining, scheduling, management, and quality control. Computer-aided manufacturing encompasses many of the technologies that we have described in Chapter 38 and this chapter. Because of the benefits, computer-aided design and computer-aided manufacturing are sometimes combined into CAD/CAM systems. This combination allows the transfer of information from design into the planning for manufacture of a product without the need to manually reenter the part geometry. The database developed in CAD is stored and processed further by CAM into the necessary data and instructions for operating and controlling production machinery, material-handling equipment, and automated testing and inspection for product quality.

The emergence of CAD/CAM has had a major impact on manufacturing by standardizing product development and reducing design effort, tryout, and prototype work, thus resulting in significantly reduced costs and improved productivity. Typical applications of CAD/CAM include:

a) Programming for NC, CNC, and industrial robots.
b) Design of dies and molds for casting, in which, for example, shrinkage allowances are preprogrammed.
c) Dies for metalworking operations, such as complex dies for sheet forming and progressive dies for stamping.
d) Design of tools and fixtures and EDM electrodes.
e) Quality control and inspection, such as coordinate-measuring machines.
f) Process planning and scheduling.
g) Plant layout.

• Example: Application of CAD/CAM to make a mold.

The metal mold used for injection molding of plastic blender jars is made on a CNC milling machine using a ball-nosed end mill, as illustrated in the accompanying figure. The jar is about 230 mm (9 in.) high and has a maximum diameter of 127 mm (5 in.). First, a 3-D wireframe model of the mold is made, as seen in the photograph on the left, and viewed and inspected for various geometric features.

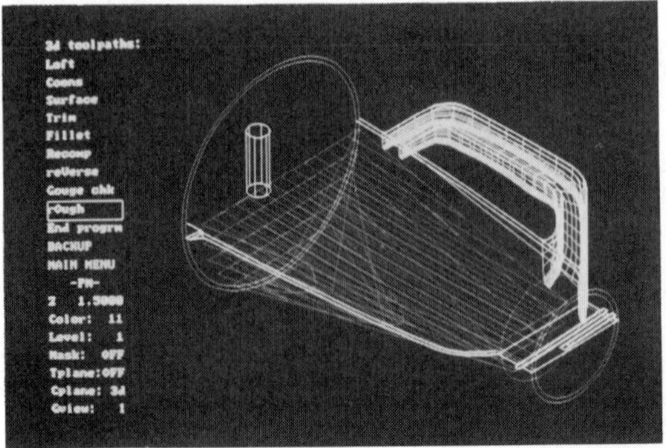

Next, an offset is added to each surface to account for the nose radius of the end mill, thus determining the cutter path, that is, the center of the machine spindle. The NC programming software then executes this program on the CNC milling machine, producing the die cavity with proper dimensions and accuracy. Electrical-discharge machining can be used to make this mold; however, it was found that EDM was about twice as expensive as machining the mold and had less dimensional accuracy. *Source*: Mold Threads Inc., and *Manufacturing Engineering Magazine*, March 1991, Society of Manufacturing Engineers.

39.6 Rapid Prototyping

An important recent advance in manufacturing is **rapid prototyping**, a process by which a solid physical model of a part is made directly from a three-dimensional CAD drawing. Also called *desktop manufacturing* or *free-form fabrication* and developed in the mid-1980s, rapid prototyping entails several different techniques that allow making a prototype, that is, a first full-scale model of a product (see Fig. 6 in the General Introduction). In order to appreciate the importance and economic impact of rapid prototyping, let us consider a design that is in its conceptual stage. First, through a three-dimensional CAD system, the design is viewed in its entirety and at different angles on the cathode-ray tube. Before that particular product is made, a prototype is manufactured and studied thoroughly from esthetic, technical, and functional aspects, using materials such as plastics or metals.

Making a prototype has traditionally involved actual manufacturing processes using a variety of tooling and machines, and usually taking anywhere from weeks to months, depending on part complexity. Rapid prototyping reduces this time significantly, as well as the cost, by using various consolidation processes such as resin curing, sintering, deposition, and solidification techniques. Generally used for prototype production, these techniques are being developed further so that they can also be used for low-volume production. Although the two most developed systems are stereolithography and selective laser sintering, we describe below the characteristics of all major processes, many of which are still under development and whose full potential has not yet been assessed.

39.6.1 Stereolithography

The **stereolithography** process is based on the principle of curing (hardening) a photopolymer into a specific shape (Fig. 39.4). A vat, containing a mechanism whereby a platform can be lowered and raised vertically, is filled with a photocurable liquid acrylate polymer. The liquid is a mixture of acrylic monomers, oligomers (polymer intermediates), and a photoinitiator. When the platform is at its highest position, the layer of liquid above it is shallow. A laser, generating an ultraviolet beam, is then focused along a selected surface area of the photopolymer at surface *a* and moved in the *x-y* direction. The beam cures that portion of the photopolymer, say, a ring-shaped portion, producing a solid body. The platform is then lowered enough to cover the cured polymer with another layer of liquid polymer, and the sequence is repeated. In Fig. 39.4, the process is repeated until level *b* is reached. Hence, thus far we have a cylindrical part with a constant wall thickness. Note that the platform is now lowered by a vertical distance *ab*.

At level *b*, the *x-y* movements of the beam are wider so that we now have a flange-shaped piece that is being produced over the previously formed part. After the proper thickness of the liquid has been cured, the process is repeated, producing another cylindrical section between levels *b* and *c*. Note that the surrounding liquid polymer is still fluid because it has not been exposed to the ultraviolet beam, and that the part has been produced from the bottom up in individual "slices." The unused portion of the liquid polymer can be used again to make another part or

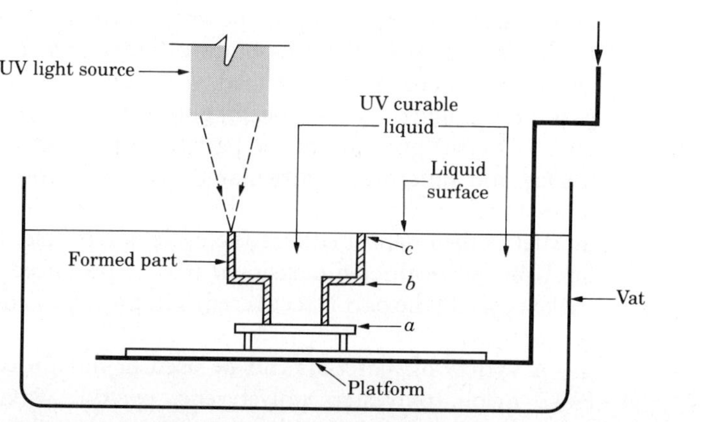

FIGURE 39.4 Schematic illustration of the stereolithography process. *Source:* Ultra Violet Products, Inc.

another prototype. Note that the word stereolithography, as used to describe this process, comes from the fact that the movements are three-dimensional and the process is similar to lithography, in which the image to be printed on a flat surface is ink-receptive and the blank areas are ink-repellent. After completion, the part is removed from the platform, blotted, cleaned ultrasonically and with an alcohol bath, and subjected to a final curing cycle.

The CAD system software "slices" the 3-D model for subsequent incremental curing cycles. By controlling the movements of the beam and the platform through a servo-control system, a variety of parts can be formed by this process. Total cycle times range from a few hours to a day. Progress is being made toward improvements in (a) accuracy and dimensional stability of the prototypes produced, (b) less expensive liquid modeling materials, (c) CAD interfaces to transfer geometrical data to model-making systems, and (d) strength so that the prototypes made by this process can be truly considered prototypes and models in the traditional sense. Sloping surfaces tend to be rough because of the manner in which layer-by-layer curing takes place. Depending on capacity, the cost of these machines is in the range of \$100,000 to \$400,000, while the cost of the liquid polymer is on the order of \$300 per gallon. Maximum part size is 0.5 m × 0.5 m × 0.6 m (20 in. × 20 in × 24 in.). Automotive, aerospace, electronics, and medical industries are among others using stereolithography as a rapid and inexpensive method of producing prototypes, and thus reducing product development cycle times. One major application is in the area of making prototype molds and dies for casting and injection molding.

39.6.2 Selective laser sintering

This is a process based on sintering (heating without melting to produce a coherent mass) of metallic or nonmetallic powders selectively into an individual object. The basic elements in this process are shown in Fig. 39.5. The bottom of the processing chamber is equipped with two cylinders: (a) the powder-feed cylinder, which is raised incrementally to supply powder to the part-build cylinder through a roller mechanism, and (b) the part-build cylinder, which is lowered incrementally and where the sintered part is formed. A thin layer of powder is first deposited in the part-build cylinder. A laser beam, guided by a process control computer and based on a 3-D CAD program of the part to be produced, is then focused on that layer, tracing and sintering a particular cross-section into a solid mass. The powder in other areas remains loose, and they support the sintered portion. Another layer of powder is then deposited and the cycle is repeated. Successive cycles are repeated until the entire three-dimensional part is produced. The loose particles are then shaken off and the part is recovered. The part does not require further curing, unless it is ceramic.

A variety of materials can be used in this process, including polymers (ABS, PVC, nylon, polyester, polystyrene, epoxy), wax, metals, and ceramics with appropriate binders. Its applications are similar to those for stereolithography. The cost of these machines is in the range of \$200,000 to \$450,000.

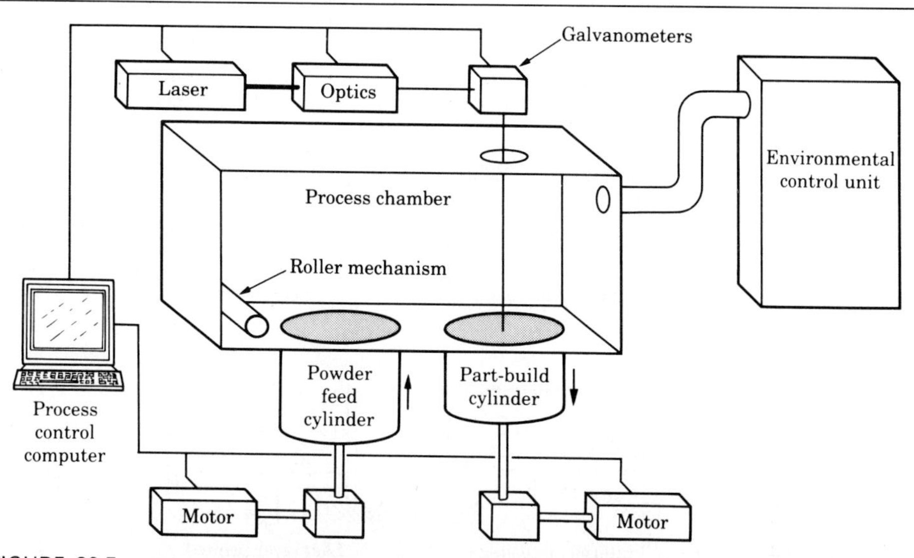

FIGURE 39.5
Schematic illustration of the selective laser sintering process. *Source:* C. Deckard and P. F. McClure.

39.6.3 Other methods of rapid prototyping

Several other methods of rapid prototyping, with a price range between \$50,000 and \$500,000, are in various stages of development as described below. All use a CAD system interfaced with the machine.

Three-dimensional printing. This process deposits powdered materials in thin layers and selectively binds the powder by ink-jet printing of an inorganic binder material (such as colloidal silica), without the use of a laser (Fig. 39.6). A piston supporting the powder bed is lowered incrementally, and with each step a layer is deposited and joined by the binder. Powder materials used are aluminum oxide, silicon carbide, silica, and zirconia.

Ballistic particle manufacturing. In this process streams of material, such as plastics, ceramics, metals, and wax, are ejected using an ink-jet mechanism through a small orifice at a surface (target). The mechanism uses a piezoelectric pump, which operates when an electric charge is applied, generating a shock wave that propels 50-μm droplets of wax at a rate of 10,000 per second. The operation is repeated in a manner similar to other processes described above to form the part, with layers of wax deposited on top of each other. The ink-jet head is guided by a three-axis robot.

Photochemical machining. This process is similar to stereolithography but uses two laser beams intersecting each other to form the part. One beam moves in

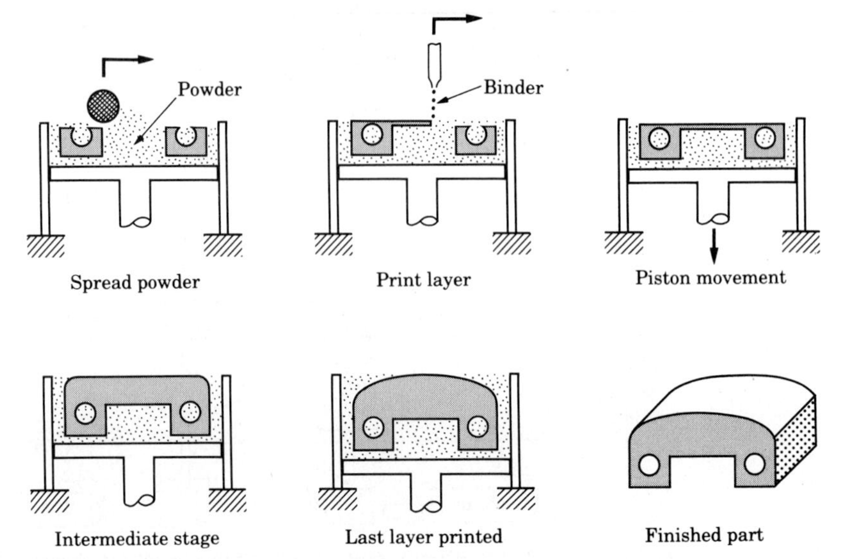

FIGURE 39.6
Schematic illustration of the three-dimensional printing process. *Source:* E. Sachs and M. Cima.

the *x-y* plane and the other in the *x-z* plane; thus part production is more versatile and no longer done in layers as in other processes.

Optical fabrication. Similar to stereolithography, this system uses a visible light argon-ion laser, and the part is built on a stationary platform.

Fused deposition modeling. In this process a thermoplastic filament, similar to wire, is fed through a heated extruding head. The filament, made of materials such as wax or nylon, melts at a temperature just above its solidification state prior to its deposition on a platform to produce the part. The prototype made requires no further curing.

Solid base curing. In this process the part is formed by multiple layers. The photopolymer liquid in each layer is covered with a photomask and cured in a few seconds by a strong ultraviolet lamp. The unexposed liquid polymer is then removed and the voids are filled with molten wax to support the next layer. This sequence is repeated until the entire part is formed.

Laminated object manufacturing. This process is based on the principle of layers of sheet, such as metal foil, paper, or plastic, bonded together in a stack. Beginning with the top sheet, a properly adjusted laser cuts each sheet successively, one at a time, to a particular profile. The unused portions are discarded; thus this process removes material rather than consolidating it as done in other rapid prototyping processes. Sheet thickness may be as small as 0.05–0.12 mm (0.002–0.005 in.).

Example: Rapid prototyping in the automotive industry.

It has been estimated that the cost of prototyping new automobile components can run to hundreds of millions of dollars annually. Furthermore, some parts may require a number of design iterations before being finalized and could take at least a year to make. Among various large manufacturing organizations, Ford Motor Company has been one of the early users of rapid prototyping technology. In a specific example, Ford engineers used stereolithography (Section 39.6.1) to make a plastic prototype of a new rocker arm, which took about 24 hours to make. In contrast, a conventionally made prototype would have taken about three months to produce and would cost about $30,000 more. The plastic prototype enabled the engineers to make design modifications within a matter of hours and finalize the design.

In another example, the engineers made a model of an experimental manifold with stereolithography. The model was then used in tests to increase air flow and optimize engine power. It was estimated that traditional prototyping would have taken 83 days longer to make and would cost $12,000 more. *Source*: *Manufacturing Engineering*, June 1991.

39.7 Computer-Aided Process Planning

In order for a manufacturing operation to be efficient, all of its diverse activities must be planned. *Process planning* is concerned with selecting methods of production, tooling, fixtures and machinery, sequence of operations, and assembly. It is traditionally done by process planners. The sequence of processes and operations to be performed, machines to be used, standard time for each operation, and similar information are documented on a *routing sheet* (Fig. 39.7). This task is highly labor intensive and time consuming and relies heavily on the experience of the process planner.

Computer-aided process planning (CAPP) accomplishes this complex task by viewing the total operation as an integrated system, so that the individual operations and steps involved in making each part are coordinated with others and are performed efficiently and reliably. Computer-aided process planning is an important link between CAD and CAM. Although CAPP requires extensive software and good coordination with CAD/CAM, as well as other aspects of integrated manufacturing systems discussed in the rest of this chapter, it is a powerful tool for efficiently planning and scheduling manufacturing operations. It is particularly effective in small-volume, high-variety parts production involving machining, forming, and assembly operations.

ROUTING SHEET

CUSTOMER'S NAME: Midwest Valve Co. PART NAME: Valve body

QUANTITY: 15 PART NO.: 302

Operation no.	Description of operation	Machine
10	Inspect forging, check hardness	Rockwell tester
20	Rough machine flanges	Lathe No. 5
30	Finish machine flanges	Lathe No. 5
40	Bore and counter bore hole	Boring mill No. 1
50	Turn internal grooves	Boring mill No. 1
60	Drill and tap holes	Drill press No. 2
70	Grind flange end faces	Grinder No. 2
80	Grind bore	Internal grinder No. 1
90	Clean	Vapor degreaser
100	Inspect	Ultrasonic tester

FIGURE 39.7
An example of a simple routing sheet. These sheets may include additional information on materials, tooling, estimated times for each operation, processing parameters such as cutting speeds, feeds, etc., and other information and are called *operation sheets*. The routing sheet travels with the parts from operation to operation.

39.7.1 Elements of CAPP systems

There are two types of computer-aided process planning systems: *variant* and *generative* process planning.

In the **variant system,** the computer files contain a standard process plan for the part to be manufactured. The search for a standard plan is made in the database by a code number for the part based on its shape and manufacturing characteristics (see Group Technology, Section 39.8). The standard plan is then retrieved, displayed for review, and printed as a route sheet. The process plan includes information such as type of tools and machines to be used, the sequence of manufacturing operations to be performed, speeds, feeds, time required for each sequence, and so on. Minor modifications of an existing process plan, which are usually necessary, can also be made. If the standard plan for a particular part is not in computer files, one that is close to it with a similar code number and an existing route sheet is retrieved. If a route sheet does not exist, one is made for the new part and stored in computer memory.

In the **generative system**, a process plan is automatically generated on the basis of the same logical procedures that would be followed by a traditional process planner in making that particular part. Such a system is complex because it must contain comprehensive and detailed knowledge of part shape, dimensions, process capabilities, selection of manufacturing methods, machinery, and tools required, and the sequence of operations to be performed. (We discuss these capabilities of computers, known as *expert systems*, in Section 39.13.) The generative system is thus capable of creating a new plan instead of using and modifying existing plans, as in the variant system. Although presently used less commonly than the variant system, this system has advantages such as (a) flexibility and consistency for process planning for new parts and (b) higher overall planning quality because of the capability of the decision logic in the system to optimize the planning and utilize up-to-date manufacturing technology.

Process planning capabilities of computers can be integrated in the planning and control of the production systems. These activities are a subsystem of computer-integrated manufacturing that we described in Section 39.3. Several functions, such as capacity planning for plants to meet production schedules, control of inventory, purchasing, and production scheduling, can be performed.

39.7.2 Advantages of CAPP systems

The advantages of CAPP systems over traditional process planning methods include standardization of process plans, thus improving the productivity of process planners, reducing lead times, reducing planning costs, and improving the consistency of product quality and reliability. Process plans can be prepared for parts having similar shapes and features, and they can be retrieved easily to produce new parts. Process plans can be modified to suit specific needs.

Route sheets can be prepared more quickly. Compared to traditionally hand-written route sheets, computer printouts are neater and much more legible. Other functions, such as cost estimating and work standards, can be incorporated into CAPP.

39.7.3 Material-requirements planning and manufacturing resource planning

Computer-based systems for managing inventories and delivery schedules of raw materials and tools are called **material-requirements planning** (MRP). This activity, sometimes regarded as a method of inventory control, involves keeping complete records of inventories of materials, supplies, parts in various stages of production, orders, purchasing, and scheduling. Several files of data are usually involved in a master production schedule.

A further development is **manufacturing resource planning** (MRP-II), which controls all aspects of manufacturing planning through feedback. Although the system is complex, MRP-II is capable of final production scheduling, monitoring actual performance and output, and comparing them against the master production schedule.

39.8 Group Technology

Group technology (GT) is a manufacturing concept that seeks to take advantage of the *design* and *processing similarities* among parts. This concept was first developed in Europe at about the turn of the century. It involved categorizing parts and recording them manually in card files or catalogs; designs were retrieved manually, as needed. This concept began to evolve further in the 1950s, and the term *group technology* was first used in 1959. However, not until the use of interactive minicomputers became widespread in the 1970s did the use of GT grow significantly.

In large manufacturing facilities, tens of thousands of different parts are made and placed in inventory. Although many of these parts have certain similarities in shape and method of manufacture, each part was traditionally viewed as a separate entity and produced in batches, involving considerable duplication and redundancy. The characteristic of similar parts (Fig. 39.8) suggests that benefits can be obtained by *classifying* and *coding* these parts into families. Results of surveys in manufacturing plants have repeatedly shown the prevalence of similar parts. One survey found that 90 percent of 3000 parts made by a company fell into five major families of parts. Such surveys consist of breaking each product into its components and identifying similar parts. For example, a pump can be broken into the basic components of motor, housing, shaft, seals, and flanges. In spite of the variety of pumps manufactured, each of these components is basically the same in terms of

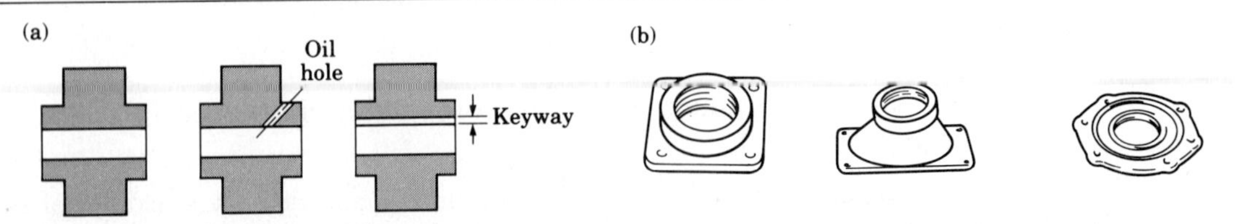

FIGURE 39.8
(a) Grouping parts according to their geometric similarities. (b) Grouping parts according to their manufacturing similarities. *Source:* Organization for Industrial Research.

design and manufacturing methods. Consequently, all shafts can be placed in one family, unless there are major differences in shape, materials, and production method.

This approach becomes more attractive in view of consumer demand for an ever-greater variety of products in smaller quantities. Under these conditions, maintaining high efficiency in batch operations is difficult. Thus overall manufacturing efficiency is hurt because nearly 75 percent of manufacturing today is batch production. Manufacturing surveys have also shown that designs are not necessarily optimized, either for product manufacture or assembly. Thus questions have been raised, for example, about why a particular part should have so many different sizes of fastener.

The traditional product flow in a batch manufacturing operation is shown in Fig. 39.9. Note that machines of the same type are arranged in groups, that is, groups of lathes, of milling machines, of drill presses, and of grinders. In such a layout (*functional layout*), there is usually considerable random movement, as shown by the arrows that indicate movement of materials and parts. Such an arrangement is not efficient because it wastes time and effort. The machines in *cellular manufacturing* are arranged in a more efficient product flow line (*group layout*; Fig. 39.9b on p. 1188). We describe further details in Section 39.9.

39.8.1 Advantages of group technology

The advantages of group technology include standardization of part design and minimization of design duplication. New part designs can be developed using similar previous designs, thus saving a significant amount of time and effort. The designer can quickly determine whether a similar part already exists in the computer files.

Data reflecting the experience of the designer and manufacturing process planner are stored in the database. Thus a new and less experienced engineer can quickly benefit from that experience by retrieving any of the previous designs and process plans. Costs can be estimated more easily, and relevant statistics on materials, processes, number of parts produced, and other factors can be more easily obtained.

Process plans are standardized and scheduled more efficiently, orders are grouped for more efficient production, and machine utilization is improved. Setup times are reduced, and parts are produced more efficiently with better and more consistent product quality. Similar tools, fixtures, and machinery are shared in the production of a family of parts. Programming for NC is more automatic.

With the implementation of CAD/CAM, cellular manufacturing, and CIM, group technology is capable of improving productivity and reducing costs in batch production to approach those of mass production. Potential savings in each of the various design and manufacturing phases can range from 5 percent to 75 percent.

39.8.2 Classification and coding of parts

Parts are identified and grouped into families by **classification and coding** (C/C) **systems**. Classification and coding of parts is a complex and critical first step in GT.

(a)

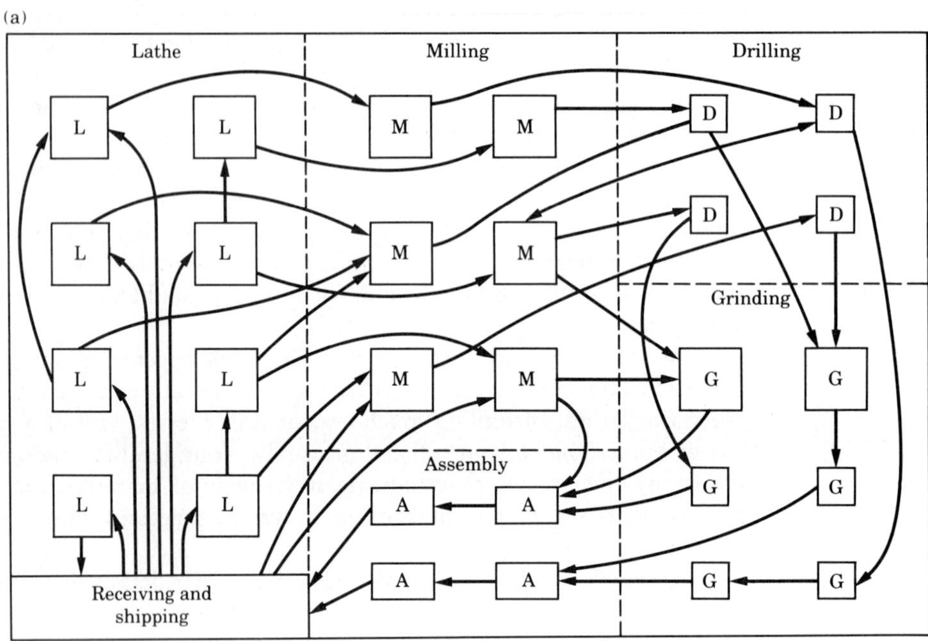

(b)

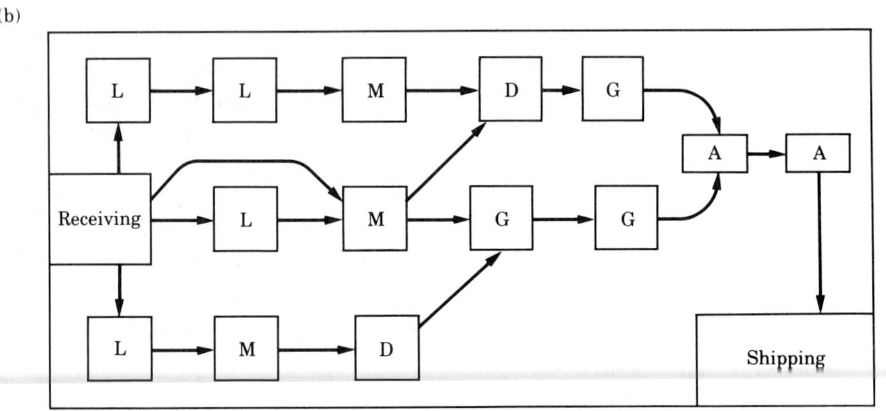

FIGURE 39.9

(a) Functional layout of machine tools in a traditional plant. Arrows indicate the flow of materials and parts in various stages of completion. (b) Group-technology (cellular) layout. Legend: L = lathe, M = milling machine, D = drilling, G = grinding, A = assembly. *Source:* M. P. Groover.

It may be done according to the part's design attributes or manufacturing attributes (see Fig. 39.8).

Design attributes pertain to similarities in geometric features and consist of the following:

a) External and internal shapes and dimensions.
b) Aspect ratio (length-to-width or length-to-diameter).

c) Tolerances specified.
d) Surface finish specified.
e) Part function.

Manufacturing attributes pertain to similarities in the methods and sequence of manufacturing operations performed on the part. As we have stated several times, selection of a manufacturing process or processes depends on many factors, among which are the shape, dimensions, and other geometric features of the part. Consequently, manufacturing and design attributes are interrelated. Manufacturing attributes consist of the following:

a) Primary processes.
b) Secondary and finishing processes.
c) Tolerances and surface finish.
d) Sequence of operations performed.
e) Tools, dies, fixtures, and machinery.
f) Production quantity and production rate.

From these lists, you can see that coding can be time consuming and requires considerable experience in designing and manufacturing products in general. In its simplest form, coding can be done by viewing the shape of the parts in a generic way and classifying them. Examples are parts with rotational symmetry, parts that are rectilinear in shape, and parts that have large surface-to-thickness ratios. Parts being reviewed and classified should be representative of the company's product lines. A more thorough method is to review all of the data and drawings concerning the design *and* production of parts.

Parts may also be classified by studying the production flow of parts during their manufacturing cycle (called *production flow analysis*, or PFA). Recall from Section 39.7 that route sheets clearly show process plans and operations to be performed. One drawback to PFA is that a particular route sheet does not necessarily indicate that the total operation is optimized. In fact, depending on the experience of the process planner, route sheets for manufacturing the same part can be quite different. We have previously noted the benefits of computer-aided process planning in avoiding such problems.

39.8.3 Coding systems

Coding can be based on a company's own system or one of several classification and coding systems that are available commercially. Because of widely varying product lines and organizational needs, none of the C/C systems has been universally adopted. Whether the system was developed in-house or was purchased, it should be compatible with the company's other systems, such as NC machinery and CAPP systems.

The code structure for part families consists of numbers, letters, or a combination of both. Each specific part or component of a product is assigned a code, which may pertain to design attributes only (generally less than 12 digits) or manufactur-

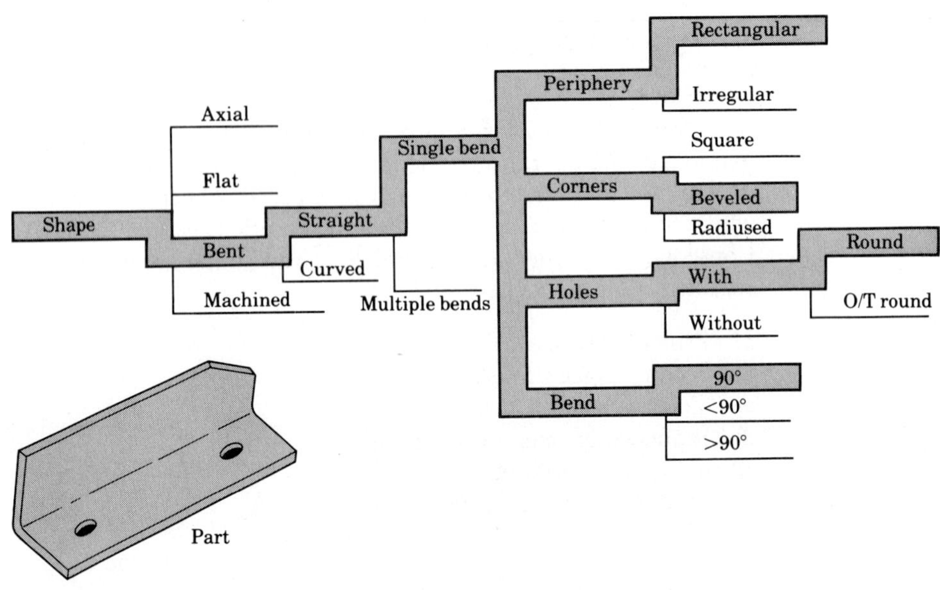

FIGURE 39.10
Logic-tree classification for a sheet-metal bracket. *Source:* G. W. Millar, *Manufacturing Engineering*, November 1985.

ing attributes only, although most advanced systems include both, using as many as 30 digits.

The three basic levels of coding vary in degree of complexity. They are:

- *Hierarchical coding.* In this code, also called *monocode*, the interpretation of each succeeding digit depends on the value of the preceding digit. Each symbol amplifies the information contained in the preceding digit, so a digit in the code cannot be interpreted alone. The advantage of this system is that a short code can contain a large amount of information. However, this method is difficult to apply in a computerized system.
- *Polycodes.* Each digit in this code, also known as *chain type*, has its own interpretation, which does not depend on the preceding digit. This structure tends to be relatively long, but allows identification of specific part attributes and is well suited to computer implementation.
- *Decision-tree coding.* This system, also called *hybrid codes*, is the most advanced and combines both design and manufacturing attributes (Fig. 39.10).

Two major industrial coding systems are the Opitz and the MultiClass system. The Opitz system, developed in the 1960s in West Germany by H. Opitz (1905–1977), was the first comprehensive coding system presented. The basic code consists of nine digits (12345 6789) representing design and manufacturing data (Fig. 39.11). Four additional codes (ABCD) may be used to identify the type and sequence of production operations. This system has two drawbacks: (1) It is

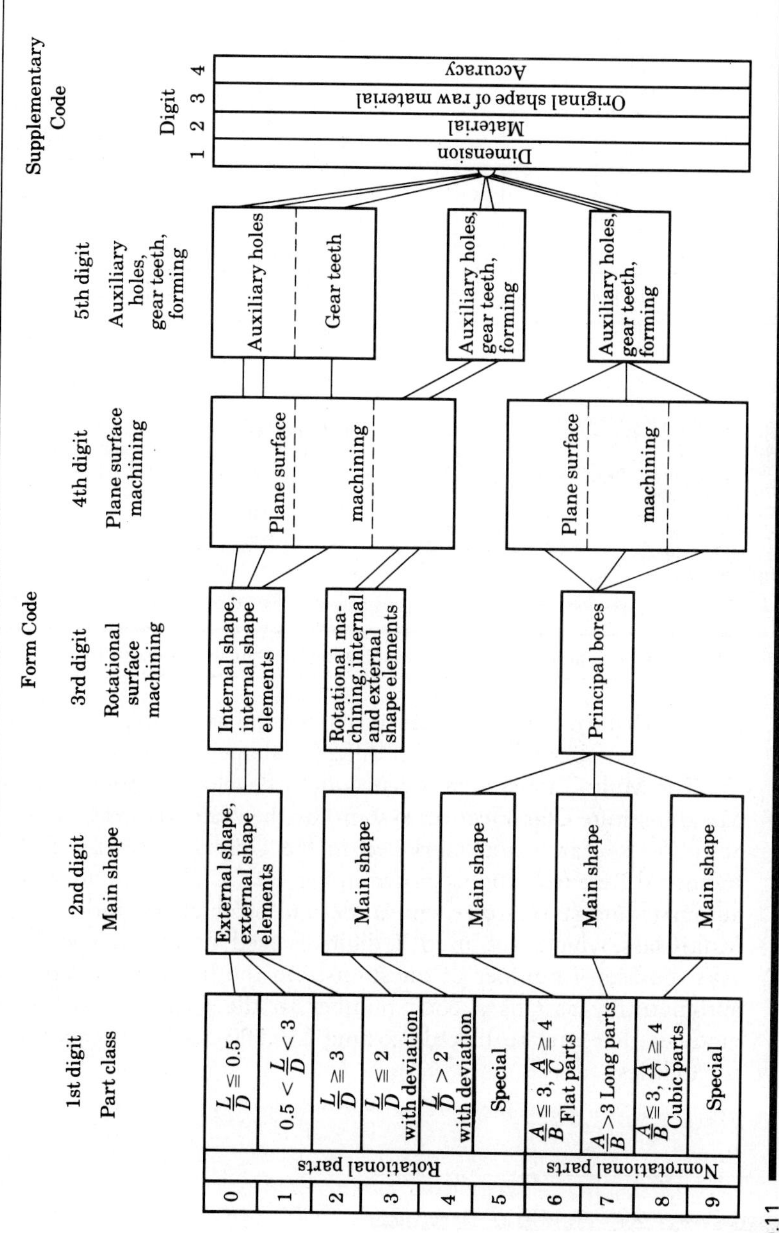

FIGURE 39.11
Classification and coding system according to Opitz, consisting of 5 digits and a supplementary code of 4 digits.

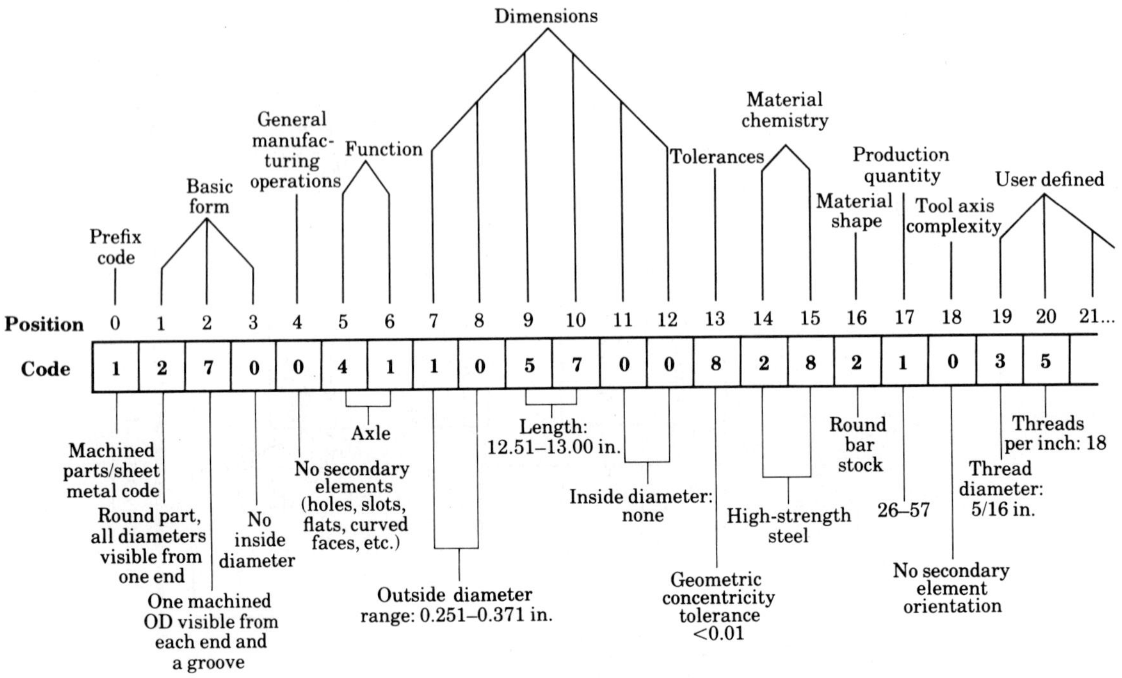

FIGURE 39.12
Typical MultiClass code for a machined part. *Source:* Organization for Industrial Research.

possible to have different codes for parts that have similar manufacturing attributes, and (2) a number of parts with different shapes can have the same code.

The MultiClass system was originally developed under the name MICLASS (for Metal Institute Classification System) by the Netherlands Organization for Applied Scientific Research and marketed in the United States by The Organization for Industrial Research and is shown in Fig. 39.12. This system was developed to help automate and standardize several design, production, and management functions. MultiClass, which uses up to 30 digits, is used interactively with a computer, which asks the user a number of questions. On the basis of the answers, the computer automatically assigns a code number to the part. The software is available in modules that can be linked, costing $50,000–$500,000 each, depending on their capabilities.

39.9 Cellular Manufacturing

The concept of group technology can be implemented effectively in **cellular manufacturing**, consisting of one or more manufacturing cells. A *manufacturing cell* is a small unit of one to several workstations within a manufacturing system. The workstations usually contain one (single machine cell) or more machines (group machine cell), each performing a different operation on the part. The machines can be modified, retooled, and regrouped for different product lines within the same family of parts. Manufacturing cells are particularly effective in producing families of parts that have relatively constant demand.

Cellular manufacturing thus far has been utilized primarily in machining and sheet-metal forming operations. The machine tools commonly used in manufacturing cells are lathes, machining centers, milling machines, drills, and grinders. For sheet forming, equipment consists of shearing, punching, bending, and other forming machines. This equipment may be special-purpose machines or CNC machines. Cellular manufacturing has some degree of automatic control for:

a) Loading and unloading raw materials and workpieces at workstations.
b) Changing tools at workstations.
c) Transferring workpieces and tools between workstations.
d) Scheduling and controlling the total operation in the cell.

Central to these activities is a material-handling system to transfer materials and parts among workstations. In attended (manned) machining cells, materials can be moved and transferred manually by the operator, unless the parts are too heavy or the movements are hazardous, or by an industrial robot located centrally in the cell. Automated inspection and testing equipment can also be a part of this cell.

Manned manufacturing cells are typically attended by one operator, who oversees all the machines in a cell. Because of the variety of machines and processes involved, the operator becomes multifunctional and is not subjected to the tedium involved with working on the same machine. Consequently, productivity is improved, which is an important feature of manned cells.

39.9.1 Manufacturing cell design

Because of the unique features of manufacturing cells, their design and implementation in traditional plants require reorganization of the plant and rearrangement of existing product flow lines. The machines may be arranged along a line, in a U-shape, in an L-shape, or in a loop. For a group-machine cell where materials are handled by the operator, the U-shaped arrangement is convenient and efficient because the operator can easily reach the various machines. For mechanized material handling, the linear arrangement and loop layout are more efficient. Selecting the best machine and material-handling equipment arrangement also depends on factors such as production rate and type of product and its shape, size, and weight.

39.9.2 Flexible manufacturing cells

In the introduction to this chapter and in Chapter 38, we stressed that, in view of rapid changes in market demand and the need for more product variety in smaller quantity, flexibility of manufacturing operations is highly desirable. Manufacturing

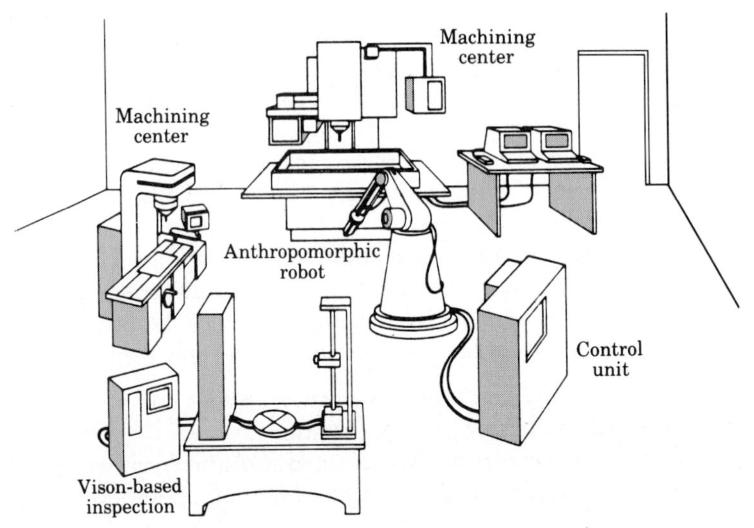

FIGURE 39.13 Schematic view of a flexible manufacturing cell, showing two machine tools, an automated part inspection system, and a central robot serving these machines. *Source:* P. K. Wright.

cells can be made flexible by using CNC machines and machining centers, with industrial robots or other mechanized systems for handling materials. An example of **flexible manufacturing cells** (FMC) for machining operations is shown in Fig. 39.13. In another flexible manufacturing cell, turbine blades are made by closed-die forging (Section 14.3). A robot places cylindrical billets into a furnace, and then transfers them to the forging hammer for forming the blade. A second robot transfers the forged blades to a cropper to trim the flash. The blades are then inspected by machine vision for dimensional accuracy.

Flexible manufacturing cells are usually unattended (unmanned), and their design and operation are more exacting than for other cells. The selection of machines and robots, including the types and capacities of end effectors and their control systems, are important to the proper functioning of the FMC. Significant changes in demand for part families should be considered during design to ensure that the equipment involved has the proper capacity and flexibility.

As in other flexible manufacturing systems (which we describe in Section 39.10), the cost of flexible cells is high. Proper maintenance of tools and machinery is essential, as are two or three shift operations of the cells for economic reasons.

• Example: Manufacturing cells in a small machine shop.

The following is an actual example of the application of the concept of manufacturing cells in a small shop. Company A has only 10 employees and 11 milling machines and machining centers (turning). These machines are set up in cells, such as a milling cell and a turning cell. The machines in the cells are arranged to allow an operator to machine a part in the most efficient and precise manner. Each cell allows the operator to monitor the performance of the machines in the cell. During the past 14 years, over 1200 different product lots have been produced, with quantities ranging from one part to as many as 35,000 parts of the same design. The parts are monitored as they are produced. Each employee in the shop is involved in the programming and the running of the machines, as well as the in-process inspection of parts.

39.10 Flexible Manufacturing Systems

A **flexible manufacturing system (FMS)** integrates all major elements of manufacturing that we have described into a highly automated system. First utilized in the late 1960s, FMS consists of a number of manufacturing cells, each containing an industrial robot serving several CNC machines and an automated material-handling system, all interfaced with a central computer. Different computer instructions for the manufacturing process can be downloaded for each successive part passing through the workstation. This system is highly automated and is capable of optimizing each step of the total manufacturing operation. These steps may involve one or more processes and operations, such as machining, grinding, cutting, forming, powder metallurgy, heat treating, and finishing, as well as handling raw materials, inspection, and assembly. The most common applications of FMS to date have been in machining and assembly operations.

Flexible manufacturing systems represent the highest level of efficiency, sophistication, and productivity that has been achieved in manufacturing plants (Fig. 39.14). The flexibility of FMS is such that it can handle a *variety of part configurations* and produce them *in any order*. Hence FMS can be regarded as a system that combines the benefits of two systems: (1) the highly productive but inflexible transfer lines (see Section 38.2.3); and (2) job-shop production, which can produce large product variety on stand-alone machines but is inefficient. Table 39.1 shows the relative characteristics of transfer lines and FMS. Note that in FMS the time required for changeover to a different part is very short. The quick response to product and market-demand variations is a major attribute of FMS. A variety of FMS technology is available from machine-tool manufacturers.

Compared to conventional systems, the benefits of FMS are:

a) Parts can be produced randomly and in batch sizes as small as one and at lower unit cost.
b) Direct labor and inventories are reduced, with savings estimated at 80–90 percent over conventional systems.

(a)

(b)

KEY:

1 Four CNC machining centers

2 Four tool-interchange stations, one per machine, for tool storage chain delivery via computer-controlled cart

3 Cart maintenance station; coolant monitoring and maintenance area

4 Parts wash station, automatic handling

5 Automatic workchanger (10 pallets) for online pallet queue

6 One inspection module—horizontal type coordinate measuring machine

7 Three queue stations for tool delivery chains

8 Tool delivery chain load/unload station

9 Four part load/unload stations

10 Pallet/fixture build station

11 Control center, computer room (elevated)

12 Centralized chip/coolant collection/recovery system (---- flume path)

13 Three computer-controlled carts, with wire-guided path

↷ Cart turnaround station (up to 360° around its own axis)

A Vestibule: construction schedule; general system layout in model form; production capabilities.

B Catwalk: fixture buildup station; part load/unload stations; CNC machining centers with 2-pallet automatic workchangers; tool delivery chain load/unload stations; queue stations for tool delivery chains.

C Control center: FMS control (DEC PDP 11/44 with peripherals); material-handling control (PDP 11/24); and computerized coordinate measuring machine (control PDP 11/24).

D Automatic workchanger pallet queue; automatic parts wash station; cart maintenance area; automatic coolant maintenance area.

FIGURE 39.14
(a) A general view of a flexible manufacturing system, showing several machine tools and an automated guided vehicle carrying a workpiece on a pallet. (b) Identification of components and machinery in the flexible manufacturing system shown in (a). *Source:* Courtesy of Cincinnati Milacron, Inc.

TABLE 39.1
COMPARISON OF THE CHARACTERISTICS OF TRANSFER LINES AND FLEXIBLE-MANUFACTURING SYSTEMS

CHARACTERISTIC	TRANSFER LINE	FMS
Types of parts made	Generally few	Infinite
Lot size	>100	1–50
Part changing time	$\frac{1}{2}$ to 8 hr	1 min
Tool change	Manual	Automatic
Adaptive control	Difficult	Available
Inventory	High	Low
Production during breakdown	None	Partial
Efficiency	60–70%	85%
Justification for capital expenditure	Simple	Difficult

c) Machine utilization is as high as 90 percent, thus improving productivity.
d) Shorter lead times are required for product changes.
e) Production is more reliable because the system is self-correcting and product quality is uniform.
f) Work-in-progress inventories are reduced.

39.10.1 Elements of FMS

The basic elements of a flexible manufacturing system are workstations, automated handling and transport of materials and parts, and control systems. The types of machines in workstations depend on the type of production. For machining operations, they usually consist of a variety of three- to five-axis machining centers, CNC lathes, milling machines, drill presses, and grinders. Also included are various other pieces of equipment for automated inspection (including coordinate-measuring machines), assembly, and cleaning. Other types of operations include sheet metal forming, punching and shearing, and forging, which include heating furnaces, forging machines, trimming presses, heat-treating facilities, and cleaning equipment. The workstations in FMS are arranged to yield the greatest efficiency in production, with an orderly flow of materials, parts, and products through the system.

Because of FMS flexibility, the material-handling, storage, and retrieval systems are very important. Material handling is controlled by a central computer and performed by industrial robots, conveyors, cranes, and automated guided vehicles

(Fig. 38.19). The system should be capable of transporting raw materials, blanks, and parts in various stages of completion to any machine, in random order, at any time. Prismatic parts are usually moved on specially designed *pallets*. Parts with rotational symmetry, such as those for turning operations, are usually moved by robots and mechanical devices.

The computer control system of FMS is its brains and includes various software and hardware. It controls the machinery and equipment in workstations and for transporting raw materials, blanks, and parts in various stages of completion from machine to machine. It also stores data and provides communication terminals that display the data visually.

39.10.2 Scheduling

Because FMS involves a major capital investment, efficient machine utilization is essential; that is, machines should not stand idle. Consequently, proper scheduling and process planning are crucial. Scheduling for FMS is *dynamic*, unlike that in job shops where a relatively rigid schedule is followed to perform a set of operations during a certain period of time. The scheduling system for FMS clearly specifies the types of operation to be performed on each part and identifies the machines or manufacturing cells to be used. Dynamic scheduling is capable of responding to quick changes in product type and hence is responsive to real-time decisions. Because of the flexibility in FMS, no setup time is wasted in switching manufacturing operations since the system is capable of performing different operations in different orders on different machines. However, the characteristics, performance, and reliability of each unit in the system must be checked to ensure that as parts move from workstation to workstation they are of acceptable quality and dimensional accuracy.

39.10.3 Economic justification of FMS

FMS installations are very costly, typically starting at well over $1 million. Consequently, a thorough cost–benefit analysis must be conducted before a final decision is made. This analysis should include factors such as costs of capital, energy, materials, and labor, expected markets for the products to be manufactured, and anticipated fluctuations in market demand and product type. An additional factor is the time and effort required for installing and debugging the system. Typically, an FMS system can take 2–5 years to install and at least 6 months to debug. Although FMS requires few, if any, machine operators, the personnel involved with the total operation must be trained and highly skilled. These personnel include manufacturing engineers, maintenance engineers, and computer programmers.

As we indicated in Fig. 38.2, the most effective FMS applications have been in medium-volume batch production. High-volume, low-variety parts production is best obtained from transfer machines (dedicated equipment). Low-volume, high-variety parts production can best be done on conventional standard machinery, with or without numerical control, or by machining centers. When a variety of parts is to be produced, FMS is suitable for production volumes of 15,000–35,000 aggregate parts per year. For individual parts of the same configuration, production may reach 100,000 units per year.

Example: Flexible manufacturing systems in large vs. small companies.

Because of the advantages of FMS technology, many manufacturers have long considered implementing a large-scale system in their facilities. After detailed review, however, and on the basis of the experience of a number of companies, most manufacturers have decided on a smaller, simpler, modular, and less expensive system that is more cost-effective. These systems include flexible manufacturing cells (whose cost would be on the order of a few hundred thousand dollars) and even stand-alone machining centers and various CNC machine tools that are easier to control than FMSs. There is now a general feeling that when FMS became an established alternative, the expectations were high, due partly to the promises of overly enthusiastic vendors. In some cases, extensive computerization has led to much confusion and inefficiency in company operations. Particularly for smaller companies, important considerations include not only the fact that large capital investment and major hardware and software acquisitions are necessary, but also that efficient operation of a large FMS requires extensive training of personnel.

On the other hand, there are several examples of successful and economically viable implementation of FMSs in a number of large companies. The accompanying table shows the results of a recent survey of 20 operating systems in the United States, indicating the improvements gained over prior methods. Some systems are now capable of economically producing lot sizes of one part. In spite of the large cost, the system has paid for itself in a number of companies. *Source for data:* Frost & Sullivan, Inc.

	PRIOR METHOD	FMS	IMPROVEMENT (%)	RANGE OF IMPROVEMENT FOR 20 SYSTEMS (%)
MACHINE TOOLS	29	9	70	60–90
DIRECT LABOR	70	16	77	50–88
MACHINE EFFICIENCY	20%	70%	50	15–90
PROCESSING TIME				
• *DAYS*	18.6	4.2	77	30–90
• *NUMBER OF OPERATIONS*	15	8	47	
FLOOR SPACE	1500 m^2	500 m^2	66	30–80
PRODUCT COST	$2000	$1000	50	25–75

39.11 Just-in-Time Production

The **just-in-time (JIT) production** concept was developed in Japan to eliminate waste of materials, machines, capital, manpower, and inventory throughout the manufacturing system. The JIT concept has the following goals:

- Purchase supplies just in time to be used.
- Produce parts just in time to be made into subassemblies.
- Produce subassemblies just in time to be assembled into finished products.
- Produce and deliver finished products just in time to be sold.

In traditional manufacturing, parts are made in batches, placed in inventory, and used whenever necessary. This approach is known as a *push system*, meaning that parts are made according to a schedule and are in inventory to be used if and when they are needed. Just-in-time is a *pull system*, meaning that parts are produced to order, and production is matched with demand for final assembly of products. There are no stockpiles, with the ideal production quantity being one (*zero inventory*, *stockless production*, *demand scheduling*). Moreover, parts are inspected by the worker as they are manufactured and are used within a short period of time. In this way, the worker maintains continuous production control, immediately identifying and correcting defective parts. The worker takes pride in product quality. Unnecessary motions in stockpiling and retrieving parts from storage are eliminated.

Implementation of the JIT concept requires that all aspects of manufacturing operations be carefully reviewed and monitored so that all nonvalue-adding operations and use of resources are eliminated. This approach emphasizes pride and dedication in producing high-quality products, eliminating idle resources, and joint problem solving by teamwork among workers, engineers, and management to quickly solve any problems that occur during production or assembly.

The ability to detect production problems in producing parts just in time has been likened to the level of water (representing the inventory levels) in a lake covering a bed of boulders. The boulders represent production problems. When the water level is high (high inventories), the boulders are not exposed. On the other hand, when the level is low (low inventories), the boulders are exposed; they can be found, observed, and removed. This analogy indicates that high inventory levels can mask quality and production problems involving parts that are stockpiled. Delayed inspection, especially, causes faulty parts production.

The just-in-time concept also includes delivery of supplies and parts from outside sources, thus reducing in-plant inventory. Suppliers are expected to deliver pre-inspected goods as they are needed for production, often on a daily basis. This requires reliable suppliers and close cooperation and trust between the company and its vendors. The JIT concept of purchasing and delivery is a significant departure from traditional purchasing of supplies from one or more vendors, whereby deliveries are made, with lead times of weeks or months, in larger quantities and stored in inventory. Although a buffer is built into the traditional system, it tends to create large inventory levels and to make quality control of incoming supplies difficult.

39.11.1 Kanban

The JIT concept was first demonstrated on a large scale in 1953 at the Toyota Motor Company, under the name **kanban**, meaning visible record. These records usually consist of two types of cards (kanbans). One is the *production card*, which authorizes the production of one container or cart of identical, specified parts at a workstation. The other card is the *conveyance* or *move card*, which authorizes the transfer of one container or cart of parts from that workstation to the workstation where the parts will be used. The cards contain information on the type of part, place of issue, part number, and the number of items in the container or cart. Recently, these cards have been replaced by bar-coded plastic tags and other devices. The number of containers in circulation at any time is completely controlled and can be scheduled as desired for maximum efficiency.

39.11.2 Benefits of JIT

The advantages of JIT are:

- Low inventory carrying costs.
- Fast detection of defects in production or in delivery of supplies and hence low scrap loss.
- Reduced inspection and rework of parts.
- High-quality parts at low cost.

Implementation of just-in-time production has resulted in reductions of 20–40 percent in product cost, 60–80 percent in inventory, up to 90 percent in rejection rates, 90 percent in lead times, 50 percent in scrap, rework, and warranty costs, and increases of 30–50 percent in direct labor productivity and 60 percent in indirect labor productivity.

39.12 Manufacturing Communications Networks

In Chapter 38 and this chapter, we have described the use of computers in all phases of product design, manufacture, assembly, and inspection. We have pointed out that, because of its many benefits, total integration of diverse manufacturing activities is a major goal of modern manufacturing. In order to maintain a high level of coordination and efficiency of operation in integrated manufacturing, an extensive, high-speed, and interactive *communications network* is required.

A significant advance in communications technology is the **local area network (LAN)**. In this hardware and software system, logically related groups of machines and equipment communicate with each other. Recall the importance and efficiency of grouping related machines and activities by such means as group technology and manufacturing cells using robots and various other material-handling equipment. A LAN ties these groups to each other, bringing different phases of manufacturing into a unified operation.

A local area network can be quite large and complex, linking hundreds or even thousands of machines and devices in several buildings. Various network layouts (Fig. 39.15) of copper cables or fiber optics are used over distances ranging from a

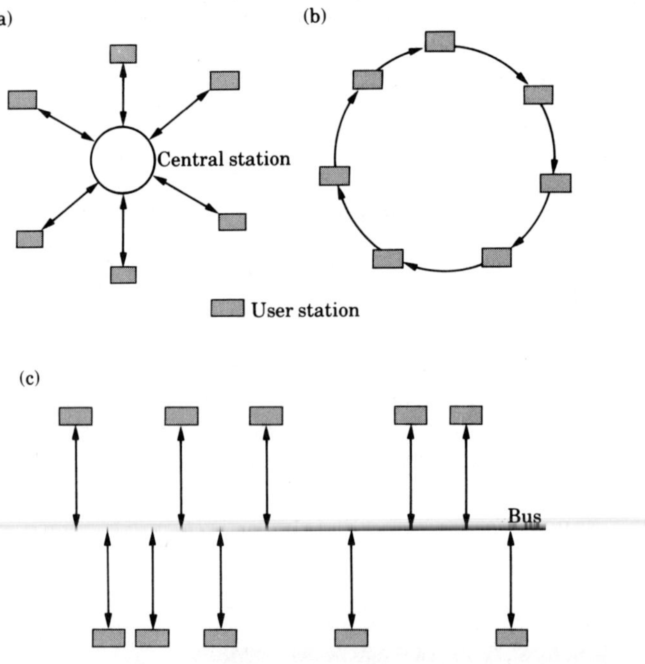

FIGURE 39.15 Three basic types of topology for local area network (LAN). (a) The *star* topology is suitable for situations that are not subject to frequent configuration changes. All messages pass through a central station. Telephone systems in office buildings usually have this type of topology. (b) In the *ring* topology all individual user stations are connected in a continuous ring. The message is forwarded from one station to the next until it reaches its assigned destination. Although the wiring is relatively simple, the failure of one station shuts down the entire network. (c) In the *bus* topology all stations have independent access to the bus. This system is reliable and is easier to service. Because its arrangement is similar to the layout of machines in a factory, its installation is relatively easy and can be rearranged when machines are rearranged.

few meters to as much as 30 km (20 mi). For larger distances, **wide area networks (WAN)** are used.

39.12.1 Communications standard

Typically, a manufacturing cell is built with machines and equipment purchased from one vendor, another cell with machines from another vendor, and yet another from another vendor. Thus a variety of programmable devices are involved, driven by several computers and microprocessors purchased at various times from different vendors and having various capacities and levels of sophistication. Each cell's computers have their own specifications and proprietary standards and cannot communicate beyond the cell with others, unless equipped with custom-built interfaces. This situation has created "islands of automation," and in some cases up to 50 percent of the cost of automation was related to difficulties in communication between manufacturing cells and other parts of the organization.

The existence of automated cells that could only function independently from each other—and without a common base for information transfer—led to the need for a *communications standard* to improve communications and the efficiency of computer-integrated manufacturing. The first step toward standardization began with formation of a task force at the General Motors Corporation in 1980. Its charter was to "identify communications standards which provide for multi-vendor data communications in the factory environment." After considerable effort, and on the basis of existing national and international standards, a set of communications standards known as **manufacturing automation protocol (MAP)** was developed.

Because of the diverse interests involved, implementation of MAP is a gradual process. However, its capabilities and effectiveness were demonstrated in 1984 by the successful interconnection of devices from a number of vendors. As a result, the importance of a worldwide communications standard is now recognized, and vendors are designing their products in compliance with these standards.

Because it has been accepted worldwide, MAP has adopted the International Organization for Standardization (ISO)/Open System Interconnect (OSI) reference model. The ISO/OSI model has a hierarchical structure in which communication between two users is divided into seven layers (Fig. 39.16). Each layer has a special task such as mechanical and electronic means of data transmission, error detection and correction, correct message transmission, controlling the dialog between users, translating messages into a common syntax, and ensuring that the data transferred is understood. The operation of this system is complex. Basically, each message or data from user A is transmitted to user B sequentially through successive layers. Additional messages are added to the original message as it travels from layer 7 to layer 1. The complete message (packet) is then transmitted through the communications medium to user B, through layer 1 to layer 7. The transmission takes place through coaxial cable, fiber-optic cable, microwave, and similar devices.

Communication protocols are now being extended to office automation as well, with the development of *technical and office protocol* (TOP). In this way total communication (MAP/TOP) is being established among the factory floor and offices at all levels of an organization.

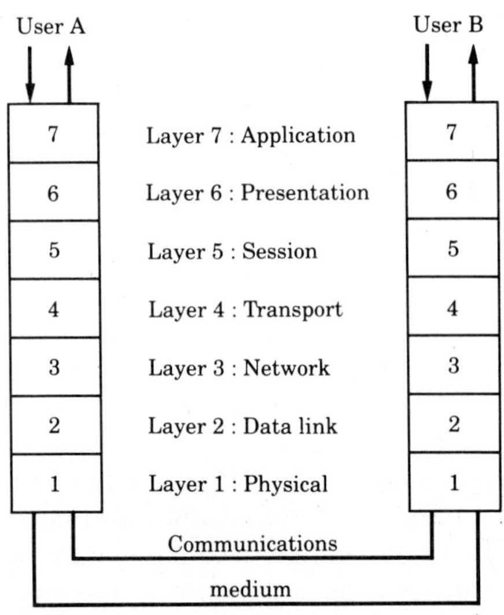

FIGURE 39.16
The ISO/OSI model for manufacturing automation protocol (MAP).

39.13 Artificial Intelligence

Artificial intelligence (AI) is that part of computer science concerned with systems that exhibit the characteristics usually associated with intelligence in human behavior such as learning, reasoning, problem solving, understanding language, and so on. The goal of AI is to simulate human behavior on the computer. The art of bringing relevant principles and tools of AI to difficult application problems is known as *knowledge engineering*.

Artificial intelligence is likely to profoundly affect design, automation, and overall economics of manufacturing operations, due in large part to the advances in memory expansion, that is, VLSI chip design and their decreasing costs (Fig. 39.17). Artificial intelligence packages costing on the order of a few thousand dollars have been developed, some of which can be run on personal computers. Thus AI has become accessible to office desks and shop floors.

39.13.1 Elements of artificial intelligence

In general, artificial-intelligence applications in manufacturing encompass the following activities: expert systems, natural language, and machine (computer) vision. We describe each of these elements below.

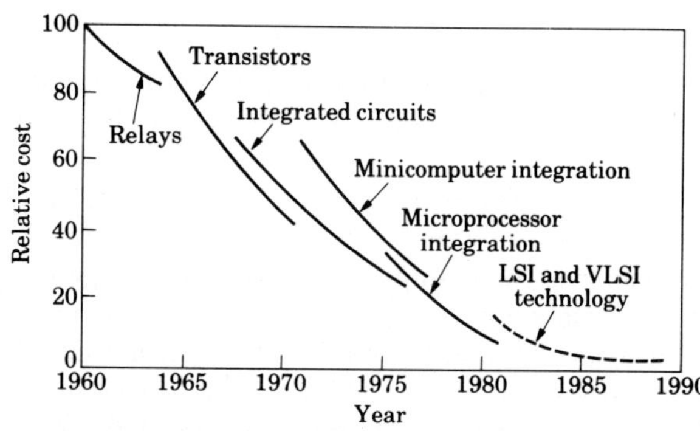

FIGURE 39.17 Relative cost of electronic and computer components. *Source: Flexible Automation,* 87/88, *The International CNC Reference Book.*

Expert systems. An **expert system** (ES), also called a *knowledge-based system*, can be defined as an intelligent computer program that has the capability to solve difficult real-life problems using a *knowledge base* and *inference* procedures (Fig. 39.18). The goal of an expert system is the capability to perform an intellectually demanding task as well as human experts would. The knowledge required to perform this task is called the *domain* of the expert system. Expert systems utilize a knowledge base containing facts, data, definitions, and assumptions. They also have the capacity for a heuristic approach, that is, making good judgments by discovery

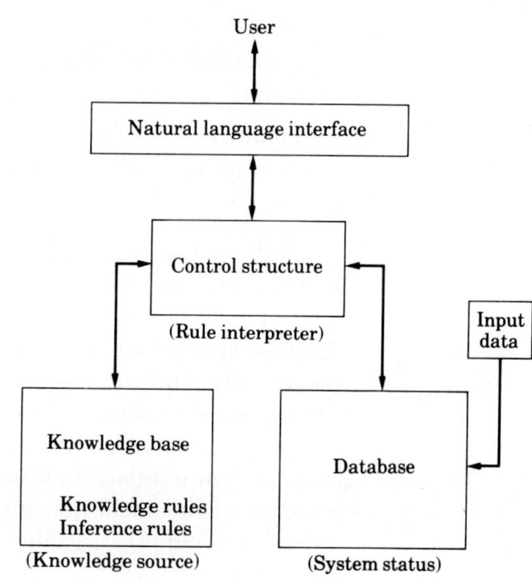

FIGURE 39.18 Basic structure of an expert system. The knowledge base consists of knowledge rules (general information about the problem) and inference rules (the way conclusions are reached). The results are communicated to the user through the natural language interface. *Source:* K. W. Goff, *Mechanical Engineering,* October 1985.

and revelation, and making good guesses, just as an expert would. The knowledge base is expressed in computer codes usually in the form of *if–then* rules, with a series of questions, and the mechanism to use these rules to solve problems is called an *inference engine*. Expert systems can also communicate with other computer-software packages.

To construct expert systems for solving the complex design and manufacturing problems encountered, one needs both a large amount of knowledge and some mechanisms for manipulating that knowledge to create solutions. Because of the difficulty involved in accurately modeling the many years of experience of an expert or a team of experts, and the complex inductive reasoning and decision-making capabilities of humans (including the capacity to learn from mistakes), developing knowledge-based systems requires considerable time and effort.

Expert systems work on a real-time basis, and their short reaction times provide rapid responses to problems. The most common programming languages are LISP and PROLOG, although some work is also being done with C++, and other languages can also be used. A more recent development is expert system software *shells* or *environments*, also called *framework systems*. These software packages are essentially expert-system outlines that allow a person to write specific applications to suit special needs. Writing these programs requires considerable experience and time.

Several expert systems have been developed and used since the early 1970s, utilizing computers with various capacities, for specialized applications such as:

- Missile guidance systems.
- Problem diagnosis in locomotive engines and various other types of machines and equipment and determination of corrective actions.
- Seismic hazard assessment.
- Modeling and simulation of production facilities.
- Computer-aided design, process planning, and production scheduling.
- Management of investment portfolios.
- Financial planning.
- Management of a company's manufacturing strategy.

Natural language processing. Traditionally, obtaining information from a database in computer memory has required utilization of computer programs that translate natural language to machine language. Natural language interfaces with database systems are now in various stages of development. These systems allow a user to obtain information by entering English language commands in the form of simple, typed queries.

Software shells are available and are used in applications such as scheduling material flow in manufacturing and analyzing information in databases. Significant progress is being made on computers that will have speech synthesis and recognition (*voice recognition*) capabilities, thus eliminating the need to type commands on keyboards.

Machine vision. We described the basic features of machine vision in Section 38.8. Computers and software for artificial intelligence can be combined with cameras and other optical sensors. These machines can then perform operations such as inspecting, identifying, and sorting parts and guiding robots (intelligent robots; Fig. 39.19), which would otherwise require human intervention.

Neural networks. Although the goal of AI is the simulation of human intelligence, a profound difference exists in the fundamentals of the thought processes between computers and the human brain. Human thinking involves very complex interactions of neurons in the brain. Each neuron contacts many other neurons, so that very large numbers of neurons coordinate and interact during the thought process. The human brain has about 100 billion linked neurons and more than 1000 times that many connections; hence **neural networks** are also known as *connectionist* architecture or *connectionism.* Computer thinking typically involves the interaction of the processor bit with a single bit of RAM. Many bits can be involved in a decision, but not simultaneously or interactively unless costly nested looping schemes are utilized.

The structure of neural networks makes them nonalgorithmic, massively parallel computing devices in which many inputs are handled simultaneously. However, unlike parallel computers, neural networks do not require complex programming to break up a problem into a form suitable for parallel processing. Thus since they can classify data and make weighted decisions without firm rules, they are particularly useful for ill-defined problems, or where the data are fuzzy, incomplete or contradictory, or problems that require a large number of rules. Other features of neural networks include the capacity (a) to be trained to store, process, and retrieve information, (b) to self-organize and learn to produce a desired goal, and (c) to find good, quick but approximate solutions to highly complex problems.

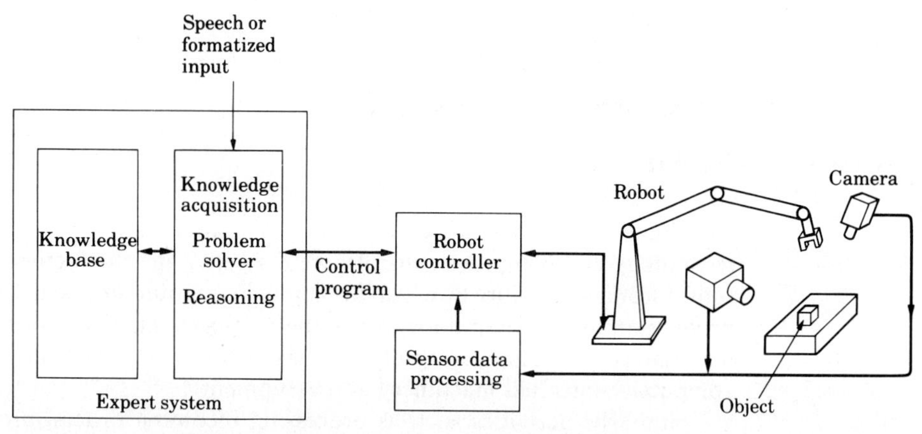

FIGURE 39.19 Expert system as applied to an industrial robot guided by machine vision.

Among various applications of neural networks relevant to manufacturing are sensor fusion and signal processing, monitoring and process control, process-control simulation and modeling, diagnostics of mechanical equipment, and material handling. There is a great amount of current research taking place concerning this subject and its various applications. Many believe that only through advances in neural network approaches will true artificial intelligence evolve. This field is still very controversial, however, and its ultimate effect on AI will be determined only over a period of time.

• Example: Application of expert systems in TV picture-tube making.

Color TV picture tubes are formed in a mold and then annealed in large furnaces to relieve residual stresses. The furnace temperature, annealing time, cooling rate, and exit temperature are important variables, but difficult to predict and control for each type and size of picture tube. The temperature in each furnace is controlled by gas burners, and dampers, louvers, and ports are used to regulate the air flow. These parameters are usually determined on the basis of many years of experience by the company's expert.

It was decided that this task could be taken over by an expert system, based on three knowledge-based systems. The first system diagnoses tube breakage and its causes. On the basis of the experience and input of the company's expert, the second system simulates the annealing process and relates the effect of various parametric changes on furnace temperature. The third system expands on the expert's knowledge by including a planner that, through intelligent trial and error, determines the necessary parametric changes to be made when new situations arise. The planner is capable of checking the validity of its guesses and adapts its strategy accordingly. All three systems are in the LISP language. *Source:* Corning Glass Works and *Manufacturing Engineering*, March 1988.

39.14 Factory of the Future

On the basis of the advances made to date in all aspects of manufacturing technology and computer controls, we may envisage the **factory of the future** as a fully automated facility in which human beings would not be directly involved with production on the shop floor (hence the term *untended factories*). All manufacturing, material handling, assembly, and inspection would be done by automated and computer-controlled machinery and equipment.

Similarly, activities such as processing incoming orders, production planning and scheduling, cost accounting, and various decision-making processes (usually performed by management) would also be done automatically by computers. The role of human beings would be confined to activities such as supervising, maintaining (especially preventive maintenance), and upgrading machines and equipment; shipping and receiving supplies and finished products; providing security for the plant facilities; and programming, upgrading, and monitoring computer programs, and monitoring, maintaining, and upgrading hardware.

Industries such as some food, petroleum, and chemical already operate automatically with little human intervention. These are continuous processes and, unlike piece part manufacturing, are easier to automate fully. Even so, the direct involvement of fewer people in manufacturing products is already apparent: Surveys show that only 10–15 percent of the workforce is directly involved in production. Most of the workforce is involved in gathering and processing information.

Virtually untended manufacturing cells already make products such as engine blocks, axles, and housings for clutches and air compressors. For large-scale, flexible manufacturing systems, however, highly trained and skilled personnel will always be needed to plan, maintain, and oversee operations.

The reliability of machines, control systems, and power supply is crucial to full factory automation. A local or general breakdown in machinery, computers, power, or communications networks will, without rapid human intervention, cripple production. The computer-integrated factory of the future should be capable of automatically rerouting materials and production flows to other machines and to the control of other computers in case of such emergencies.

In the implementation of advanced concepts of manufacturing, with their attendant impact on the operation of the factory of the future, certain developments and improvements are bound to continue at a rapid pace:

- Common communications networks, software, and standards for every aspect of manufacturing technology, from product design to manufacturing.
- Reduction in the size and efficiency of knowledge-rich machine controllers and their integration into machines, rather than as separate units.
- Availability of off-the-shelf control hardware.
- Reliable voice-recognition capabilities so that machines can understand and implement spoken messages.
- Sensor fusion, with rugged sensors, and feedback methods for monitoring and diagnostics of all aspects of manufacturing operations, augmented by developments in expert systems technology.
- Cost-effective system integration at all levels of an organization.

39.14.1 Impact on the workforce

An important consideration is the nature and extent of the fully automated factory's impact on employment, with all its social ramifications. The fear of unemployment persists despite strong evidence that advances in technology create more jobs—or at least maintain the same number—rather than eliminating them. Although projections indicate that there will be a decline in the number of machine tool operators and tool and die workers, there will be major increases in the number of computer service technicians and maintenance electricians.

Thus the generally low-skilled, direct labor force engaged in traditional manufacturing will shift to an indirect labor force, with special training or retraining required in areas such as computer programming, information processing, CAD/CAM, and other high-technology tasks that are essential parts of computer-integrated manufacturing. Development of more user-friendly computer software is making retraining of the workforce much easier.

Opinions about the impact of unmanned factories diverge widely. Consequently, predicting the nature of future manufacturing strategies with any certainty is difficult. Although economic considerations and trade-offs are crucial, companies that do not install computer-integrated operations are in jeopardy. It is now widely recognized that, in a highly competitive marketplace, rapid adaptability is essential to the survival of a manufacturing organization.

39.14.2 Status of advanced manufacturing technologies and strategies

The technological developments we have described in Chapters 38 and 39, as well as in other chapters of this text, are constantly being analyzed by the senior management of manufacturing industries with a view to their near- and long-term economic impact on their operations. Over the past few years much has been written, discussed, and argued concerning the relative manufacturing technologies and industrial strengths of the United States, Japan, Europe, South Korea, Taiwan, and so on, particularly in regard to the automotive, electronics, and computer industries.

We should point out that there are various complex issues involved vis-á-vis the characteristics of each industrial nation, its past, its social structure and standard of living, the nature, educational level, and loyalty of the workforce, the relationships among management, labor, and government, interactions among companies and research universities and organizations, productivity and effective utilization of capital equipment, the missions and operational philosophies of the management of industrial and business organizations, stockholder attitudes, and the challenges that a nation has faced and will be facing in the future in the highly competitive international marketplace.

The term *world class* is being used more and more frequently to describe the quality and level of manufacturing activities, signifying the fact that products made must meet high international standards and be acceptable worldwide. It has also been pointed out that world class is not so much a fixed target for a manufacturing company or a country to reach, but that it is a moving target and that it acquires a higher and higher level as time passes. Thus manufacturing organizations must be aware of this moving target and plan and execute their programs accordingly.

Although it is beyond the scope of this book to discuss management and investment strategies and decision-making processes in manufacturing industries, certain trends have been noted regarding the adoption of advanced manufacturing technologies. Several surveys of U.S. manufacturers have recently been conducted regarding this matter. We outline below the results of some recent surveys.

In a survey of manufacturers conducted in 1989, the *adoption status* of advanced manufacturing technologies was ranked as follows (*Source*: Ernst & Young):

TECHNOLOGY	IN PLACE	IN PROGRESS	CONSIDERING	NO PLANS
NC programming	15%	16%	36%	33%
CAD/CAM	26	30	30	14
Robotics	15	11	30	43
Electronic data interchange	10	20	40	30
Process controllers	16	25	36	23
Automated data collection	11	19	43	27
Computer-aided inspection	10	16	43	13
CAPP	6	8	41	45
MRP II	10	15	33	42
AI/Expert systems	4	6	35	55

In another survey of senior manufacturing, operations, engineering, and technology managers, published in 1990, the incentives (drivers) involved in *competitiveness* and *investment*, respectively, were ranked as follows, in decreasing order of perceived importance (*Source*: Harbor Research, Inc.):

Competitiveness:

Higher product quality.

Lower product cost.

Faster time to market.

More rapid delivery.

Greater manufacturing flexibility.

Investment:

Lower manufacturing costs.

Higher product quality.

Increased human resource productivity.

Increased capital resource productivity.

Using latest technology.

Greater manufacturing flexibility.

Faster time to market.

Thus, in order to be competitive, it is more important to improve product quality and reduce cost than, say, to improve delivery time of the product to customers. Also, a company would be more motivated to increase its investment if as a result, manufacturing costs are lowered and product quality is increased. Faster time to market the product is not an important incentive for new investment.

In a 1990 survey, manufacturing managers and executives were asked to state the CIM *strategies* that they would emphasize over the next few years. The responses are tabulated below as percentages (*Source*: *Industry Week*):

Bar coding/automated data collection	62%
CAD	57
CAM	57
MRP or MRP II	56
Computer-supported SPC	51
Company-wide networks for information sharing	47
Linking CAD to CAM	41
CNC	37
Computer-aided engineering	36
Electronic data interchange	33
Area or plant networks for equipment control/feedback	32
Automated testing/inspection	31
CAPP	28
Robotics	22
FMS	22

In the same survey, the perceived *benefits* of CIM were listed as follows, in decreasing order:

- Lower manufacturing costs.
- Higher product quality.
- Better production control.
- Faster responsiveness.
- Reduced inventories.
- Higher flexibility.
- Suitable for small-lot production.

In response to a question regarding *obstacles* to rapid adoption of CIM technology by U.S. manufacturing companies, the following reasons were given, again in decreasing order:

- Lack of in-house expertise.
- Top management does not grasp benefits of CIM.
- Inadequate planning or lack of vision.

- Inadequate cost-justification methods.
- Unavailable funds.
- Fear of poor implementation.
- No need for CIM.

There have been significant differences in perceptions and interpretations in the various trends observed regarding these issues in various countries, particularly in comparing the United States, Japan, and Europe, and much controversy still exists. For example, it has been determined that, in 1989, only 23 percent of the machine tools consumed in the United States were equipped with CNC, whereas in Europe and Japan this figure was 80 percent. Referring to the table below (*Source*: *American Machinist*) and in view of the importance of machine tools as the backbone of manufacturing, it has been suggested that the manufacturing infrastructure in the United States requires considerable rebuilding.

ESTIMATED 1990 PER-CAPITA MACHINE-TOOL CONSUMPTION

Switzerland	$200
West Germany	96
Singapore	70
Japan	61
Italy	54
Sweden	48
France	45
Taiwan	37
Canada	34
South Korea	33
United Kingdom	31
Spain	27
Soviet Union	20
United States	17
Mexico	3
China	1
India	1

It has also been pointed out that in the implementation of advanced manufacturing technology, U.S. firms rely heavily on vendor recommendations for specific hardware and software needed for manufacturing their product lines. However, proper communication between the parties involved has not always been achieved and expectations have not always been fulfilled. On the other hand, Japanese firms view product design and manufacturing systems as an integral part of their activities, and rely heavily on their own engineers and in-house expertise in the design of their integrated manufacturing systems. The specific equipment and systems decided on and needed are then purchased from suppliers.

In comparing the United States and Japan, it has been repeatedly stated that the management of U.S. firms, as well as stockholders, are generally more concerned with almost immediate payback on any expenditure and with short-term benefits and quarterly and annual reports, rather than long-term goals and benefits. In contrast, Japanese firms expect payback over a much longer period of time, and management and stockholders are interested in the long-term growth of the company, rather than in a year's return. Germany appears to be somewhere between these two attitudes.

The constant emphasis on worker participation, preventive maintenance, and especially product quality in Japanese manufacturing firms is well known, as is their determination to eliminate waste whether it is in materials, machinery, time, work-in-progress inventory, or human resources. There are strong indications and proof that many U.S. firms are, with innovative approaches, making significant progress in this regard, even though these efforts are a somewhat belated reaction to external challenges rather than a direct action on their part.

The facts and thoughts we presented above are among the most important factors for serious consideration by the management of any manufacturing organization as well as all manufacturing engineers in the coming years. The priorities set and the decisions taken will, by and large, determine not only the nature of the factory *of* the future, but as one author has pointed out, the factory *with* a future.

SUMMARY

Computers are having a major impact on all aspects of manufacturing processes and operations, as also evidenced by the extensive bibliography at the end of this chapter. Integrated manufacturing systems are being implemented to various degrees to optimize operations, reduce costs, and improve product quality and productivity. Computer-integrated manufacturing has become the important means of improving productivity, responding to changing market demand, and better controlling manufacturing and management functions. With extensive use of computers and rapid developments in sophisticated software, designs and their analysis and simulation are more detailed and thorough.

New developments in manufacturing operations, such as group technology, cellular manufacturing, and flexible manufacturing systems, are contributing significantly to improved productivity. Artificial intelligence is likely to open new opportunities in all aspects of manufacturing engineering. The factory of the future will extend these developments further, while their benefits and economic justification continue to be debated at all levels of management and the workforce.

TRENDS

- The trend in manufacturing is for greater product variety, shorter product life cycles, and increased emphasis on product quality at low cost.
- Utilization of computers in all phases of manufacturing will continue to grow rapidly in order to increase productivity and reduce delivery lead times and costs.
- The support of management and all others concerned in an organization will continue to be an essential element in the successful adoption of new technologies.
- Because of the high costs involved, a thorough analysis of costs and benefits related to implementation of various aspects of integrated manufacturing systems will continue to be an important area of study.
- Operators in the factory of the future are likely to perform tasks such as supervision, upgrading computers and computer programs, equipment maintenance, plant security, and shipping and receiving supplies and finished products.

KEY TERMS

Artificial intelligence
Attributes
Cellular manufacturing
Classification and coding systems
Computer-aided design
Computer-aided manufacturing
Computer-aided process planning
Computer-integrated manufacturing
Database
Expert systems
Factory of the future
Flexible manufacturing cells
Flexible manufacturing system
Generative system
Group technology
Just-in-time production
Kanban
Local area network
Manufacturing automation protocol
Manufacturing resource planning
Material-requirements planning
Neural networks
Rapid prototyping
Stereolithography
Variant system
Wide area network

BIBLIOGRAPHY

Anderson, D.M., *Design for Manufacturability, Optimizing Cost, Quality, and Time-to-Market*. Lafayette, Calif.: CIM Press, 1991.

Artificial Intelligence Handbook. Vol. 1: *Principles*; Vol. 2: *Applications*. Research Triangle, N.C.: Instrument Society of America, 1989.

Bedworth, D.D., M.R. Henderson, and P.M. Wolf, *Computer-Integrated Design and Manufacturing*. New York: McGraw-Hill, 1991.

Black, J.T., *The Design of the Factory with a Future*. New York: McGraw-Hill, 1991.

Bonetto, R., *Flexible Manufacturing Systems in Practice*. New York: Hemisphere, 1988.

Chang, T.-C., *Expert Process Planning for Manufacturing*. Reading, Mass.: Addison-Wesley, 1990.

——, and R.A. Wysk, *An Introduction to Automated Process Planning Systems*. Englewood Cliffs, N.J.: Prentice-Hall, 1985.

——, R.A. Wysk, and H.-P. Wang, *Computer-Aided Manufacturing*. Englewood Cliffs, N.J.: Prentice-Hall, 1991.

Corbett, J., M. Dooner, J. Meleka, and C. Pym, *Design for Manufacture: Strategies, Principles and Techniques*. Reading, Mass.: Addison-Wesley, 1991.

Dwyer, J., and A. Ioannou, *MAP and TOP: Advanced Manufacturing Communications*. New York: Wiley, 1988.

Ettlie, J.E., and H.W. Stoll, *Managing the Design-Manufacturing Process*. New York: McGraw-Hill, 1990.

Foston, A.L., C.L. Smith, and T. Au, *Fundamentals of Computer-Integrated Manufacturing*. Englewood Cliffs, N.J.: Prentice-Hall, 1991.

Galbiati, L.J., *Machine Vision and Digital Image Processing Fundamentals*. Englewood Cliffs, N.J.: Prentice-Hall, 1990.

Groover, M.P., *Automation, Production Systems, and Computer-Integrated Manufacturing*. Englewood Cliffs, N.J.: Prentice-Hall, 1987.

——, and E.W. Zimmers, Jr., *CAD/CAM: Computer-Aided Design and Manufacturing*. Englewood Cliffs, N.J.: Prentice-Hall, 1984.

Ham, I., K. Hitomi, and T. Yoshida, *Group Technology: Applications to Production Management*. Boston: Kluwer-Nijhoff, 1985.

Harmon, P., and D. King, *Expert Systems*. New York: Wiley, 1985.

Hernandez, A., *Just-In-Time Manufacturing: A Practical Approach*. Englewood Cliffs, N.J.: Prentice-Hall, 1989.

Hirano, H., *JIT Factory Revolution: A Pictorial Guide to Factory Design of the Future*. Cambridge, Mass.: Productivity Press, 1988.

Hordeski, M., *Computer Integrated Manufacturing: Techniques and Applications*. Blue Ridge Summit, Pa.: TAB, 1988.

Kaewert, J.W., and J.M. Frost, *Developing Expert Systems for Manufacturing: A Case Study Approach*. New York: McGraw-Hill, 1990.

Koenig, D.T., *Computer-Integrated Manufacturing: Theory and Practice*. Bristol, Pa.: Hemisphere, 1990.

Koren, Y., *Computer Control of Manufacturing Systems*. New York: McGraw-Hill, 1983.

Kumara, S., R. Kashap, and A. Soyster, *Artificial Intelligence: Manufacturing Theory and Practice*. Norcross, Ga.: Institute of Industrial Engineers, 1989.

Kusiak, A., *Intelligent Manufacturing Systems*. Englewood Cliffs, N.J.: Prentice-Hall, 1990.

Lee, M.H., *Intelligent Robotics*. New York: Halsted Press, 1989.

Luggen, W.W., *Flexible Manufacturing Cells and Systems*. Englewood Cliffs, N.J.: Prentice-Hall, 1991.

Maus, R., and J. Keyes (eds.), *Handbook of Expert Systems in Manufacturing*. New York: McGraw-Hill, 1991.

Milner, D.A., and V. Vasiliou, *Computer-Aided Engineering for Manufacture*. New York: McGraw-Hill, 1987.

Mitchell, F.H., Jr., *CIM Systems: An Introduction to Computer-Integrated Manufacturing*. Englewood Cliffs, N.J.: Prentice-Hall, 1991.

NC/CIM Guidebook, published annually by *Modern Machine Shop*, Cincinnati, Ohio.
Nevins, J.L., and D.E. Whitney (eds.), *Concurrent Design of Products and Processes: A Strategy for the Next Generation in Manufacturing*. New York: McGraw-Hill, 1989.
Nolen, J., *Computer-Automated Process Planning for World-Class Manufacturing*. New York: Marcel Dekker, 1989.
Parsaye, K., and M. Chignell, *Expert Systems for Experts*. New York: Wiley, 1988.
Pimentel, J.R., *Communications Networks for Manufacturing*. Englewood Cliffs, N.J.: Prentice-Hall, 1990.
Ranky, P.G., *Computer Integrated Manufacturing: An Introduction with Case Studies*. Englewood Cliffs, N.J.: Prentice-Hall, 1986.
——, *Computer Integrated Manufacturing*. Englewood Cliffs, N.J.: Prentice-Hall, 1989.
——, *The Design and Operation of FMS*. New York: North-Holland, 1983.
Rodd, M.G., and D. Farzin, *Communication Systems for Industrial Automation*. Englewood Cliffs, N.J.: Prentice-Hall, 1990.
Sandgren, E., *Design Optimization*. Reading, Mass.: Addison-Wesley, 1991.
Schonberger, R.J., *Japanese Manufacturing Techniques*. New York: Free Press, 1982.
——, *World Class Manufacturing*. New York: Free Press, 1986.
Suh, N.P., *The Principles of Design*. New York: Oxford University Press, 1990.
Suzaki, K., *The New Manufacturing Challenge*. New York: Free Press, 1987.
Talavage, J., and R.G. Hannam, *Flexible Manufacturing Systems in Practice: Applications, Design, and Simulation*. New York: Marcel Dekker, 1988.
Taylor, D., *Computer Aided Design*. Reading, Mass.: Addison-Wesley, 1991.
Teicholz, E. (ed.), *CAD/CAM Handbook*. New York: McGraw-Hill, 1985.
Tool and Manufacturing Engineers Handbook, 4th ed., Vol. 5: *Manufacturing Management*. Dearborn, Mich.: Society of Manufacturing Engineers, 1987.
Ulsoy, A.G., and W.R. DeVries, *Microcomputer Applications in Manufacturing*. New York: Wiley, 1988.
Valliere, D., *Computer-Aided Design in Manufacturing*. Englewood Cliffs, N.J.: Prentice-Hall, 1990.
Wright, P.K., and D.A. Bourne, *Manufacturing Intelligence*. Reading, Mass.: Addison-Wesley, 1988.
Zeid, I., *CAD/CAM Theory and Practice*. New York: McGraw-Hill, 1991.
Zuech, N., *Applying Machine Vision*. New York: Wiley, 1988.

REVIEW QUESTIONS

39.1 What are the trends in product characteristics that had a major impact on manufacturing?

39.2 In what ways have computers impacted on manufacturing?

39.3 What advantages are there in viewing manufacturing as a system? What are the components of a manufacturing system?

39.4 What is meant by "modeling"? What purposes does it have? Give two specific examples.

39.5 Discuss the benefits of computer-integrated manufacturing operations.

39.6 What is a database? Why is it necessary? Why should the management of a company have access to databases?

39.7 Explain how a CAD system operates.

39.8 What are the advantages of CAD systems over traditional methods of design? Are there any limitations?
39.9 Give three specific applications for CAD/CAM.
39.10 Describe the purposes of process planning. How are computers used in such planning?
39.11 Explain the features of two types of CAPP systems.
39.12 Describe the features of a route sheet. Why is it necessary?
39.13 What is group technology? Why was it developed? Explain its advantages.
39.14 Explain the difference between design attributes and manufacturing attributes. Why are their classification and coding essential in group technology?
39.15 Describe the features of various coding systems.
39.16 What is a manufacturing cell? Why was it developed?
39.17 Is there a minimum to the number of machines in a manufacturing cell? Explain.
39.18 Describe the principle of flexible manufacturing systems. Why do they require major capital investment?
39.19 Why is a flexible manufacturing system capable of producing a wide range of lot sizes?
39.20 What are the benefits of just-in-time production? Why is it called a pull system? What is a kanban?
39.21 Explain the function of a local area network.
39.22 What are the advantages of a communications standard?

QUESTIONS AND PROBLEMS

39.23 Describe the elements of artificial intelligence. Why is machine vision a part of it?
39.24 Why are expert systems called "expert"? Comment on their capabilities.
39.25 Explain why humans will still be needed in the factory of the future.
39.26 Explain in detail the principle of computer-aided manufacturing to an older worker in a manufacturing facility who is not familiar with computers.
39.27 Discuss your observations concerning the contents of Table 39.1.
39.28 Explain the various subsystems of a CIM system.
39.29 Give examples of primitives of solids other than those shown in Figs. 39.3(a) and (b).
39.30 Review various parts discussed in this text and group them in a manner similar to those shown in Fig. 39.8.
39.31 Explain the logic behind the arrangements shown in Fig. 39.9b.
39.32 Think of a product and make a logic-tree chart similar to that shown in Fig. 39.10.
39.33 Think of a commonly used product and design a manufacturing cell for making it, describing the features of the machines and equipment involved.
39.34 Describe your observations concerning Fig. 39.14(b).
39.35 What should be the characteristics of the guidance system for automated guided vehicles?
39.36 Give examples in manufacturing engineering in which artificial intelligence could be effective.
39.37 Describe your opinions concerning voice recognition capabilities of future machines and controls.
39.38 Would machining centers be suitable for just-in-time–type production? Explain.
39.39 In Section 39.6, we described several techniques for rapid prototyping. On the basis of the topics covered in this text, do you think there can be other methods of rapid prototyping? Describe your ideas, explaining the advantages and limitations of your method over others already described in this chapter.

39.40 Surveys have indicated that 95 percent of all parts made in the United States are produced in lots of 50 or less. Comment on this observation and describe your thoughts regarding the implementation of the technologies outlined in Chapters 38 and 39.

39.41 Assume that you are asked to rewrite Section 39.14 on the Factory of the Future. Outline your thoughts regarding this topic and present them in a brief paper.

39.42 Comment on the statistical data given in Section 39.14.2.

39.43 Give an example of (a) push and (b) pull systems to describe the fundamental difference between the two methods.

39.44 It has been suggested by some that ultimately artificial intelligence systems will be able to replace the human brain. Do you agree? Explain.

39.45 Assume that you own a manufacturing company and you are aware that you have not taken full advantage of the technological advances in manufacturing, but would like to do so and have the necessary capital. Describe how you would go about analyzing your company's needs and how you would plan to implement this technology. Consider technological as well as human aspects.

39.46 Think of a simple product and make a route sheet similar to that shown in Fig. 39.7. If the same part is given to another person, is it likely that the route sheet developed will be different? Why?

39.47 Give a specific example each where the (a) variant system and the (b) generative system of CAPP are desirable.

39.48 In Problem 38.50, assume that the type of product and/or its size will change within two years. How will this affect your transfer line design? Which of the technologies described in this chapter would you consider, and why?

39.49 We stated that neural networks are particularly useful where the problems are ill defined and the data are fuzzy. Give examples in manufacturing where this is the case.

39.50 With specific examples, describe your own thoughts concerning the state of manufacturing in the United States as compared to other industrialized nations.

40

Competitive Aspects and Economics of Manufacturing

40.1 Introduction

Manufacturing high-quality products at the lowest possible cost requires an understanding of the often complex relationships among many factors. You have seen that product design, selection of materials, and manufacturing processes are interrelated. Designs are often modified to improve product performance, take advantage of the characteristics of new materials, and make manufacturing and assembly easier, thus reducing costs. Because of the wide variety of materials and processes available today, the task of producing a high-quality product by selecting the best materials and processes while minimizing costs has become a major challenge. Meeting this challenge requires not only a thorough knowledge of the characteristics of materials and processes, but also innovative and creative approaches to design and manufacturing technology.

The cost of a product often determines its marketability and customer satisfaction. Consequently, each production step must be analyzed with a view toward ensuring an optimal economic outcome (Fig. 40.1). In this chapter we

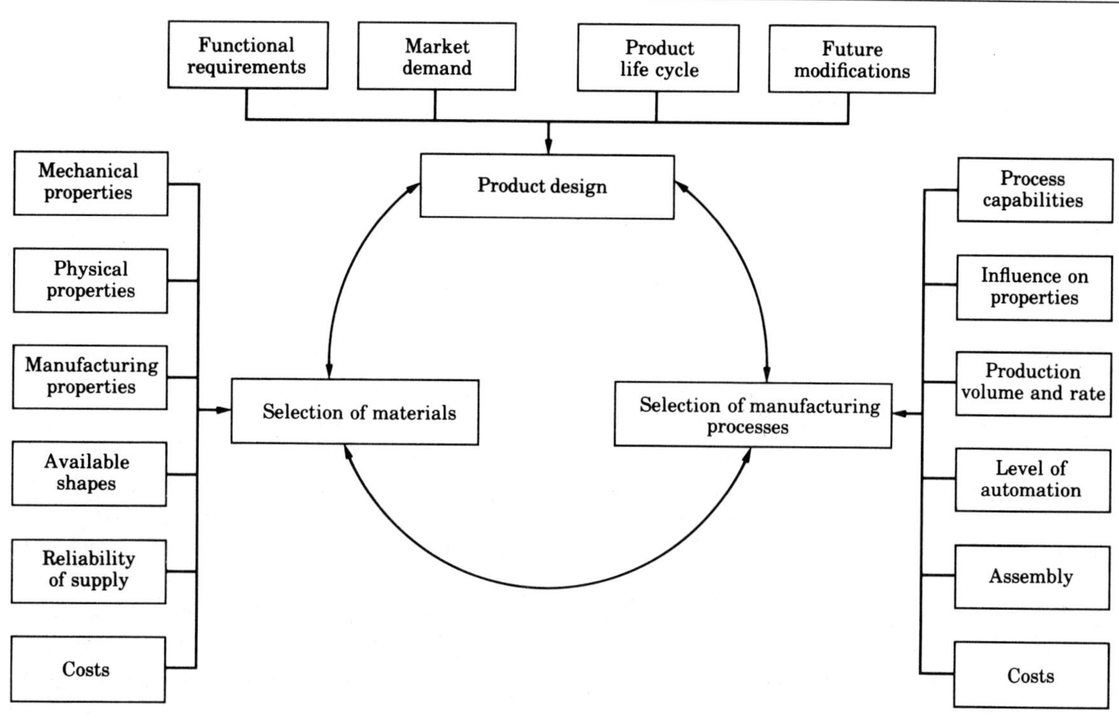

FIGURE 40.1
Interrelationships among various factors influencing product cost.

consider the following approaches to minimize costs while fulfilling design, quality, and service requirements:

a) Selection of materials.
b) Product design and quantity of materials.
c) Substitution of materials.
d) Selection of manufacturing processes.
e) Process capabilities.
f) Manufacturing costs.
g) Cost reduction.

Although we have described the economics of various manufacturing processes, we take a broader view in this chapter and summarize the important overall manufacturing cost factors. We look at cost reduction methods, particularly value engineering, which is a powerful tool that can be used to evaluate the cost of each manufacturing step in terms of its share of the value added to the product. Finally, we list a number of questions that manufacturing engineers and other members of a manufacturing organization must ask concerning the design and manufacture of a product.

40.2 Selection of Materials

In selecting materials for a product, we must have a clear understanding of the *functional* requirements for each of its individual components. We described the general criteria for selecting materials in the General Introduction. We discuss now further details of selecting materials.

40.2.1 Mechanical, physical, and chemical properties

Mechanical properties are strength, toughness, ductility, stiffness, hardness, and resistance to fatigue, creep, and impact. Physical properties include density, melting point, specific heat, thermal and electrical conductivity, thermal expansion, and magnetic properties. Chemical properties with which we are primarily concerned are oxidation and corrosion resistance. We described these properties and their relevance to product design and manufacturing in various chapters in Part I.

Selection of materials is now easier and faster with availability of computer software, which are in a sense computerized handbooks and thus have greater accessibility. However, to facilitate the selection of materials, as well as other parameters described below, expert-system software (smart databases) are now being developed. With proper input of design and functional requirements, these systems are capable of selecting appropriate materials for a particular application just as an expert or a team of experts would.

40.2.2 Shapes of commercially available materials

When selecting materials, we need to know the shapes and sizes of materials that are commercially available (Table 40.1) in order to avoid additional processing, unless it is economically justified. Materials are available in various forms: casting, extrusion, forging, bar, plate, sheet, foil, rod, wire, or powder. Purchasing materials in shapes that generally require the least additional processing means that we also have to consider characteristics such as surface quality, tolerances, and straightness. The better and more consistent these characteristics are, the less additional processing is required.

For example, if you want to produce simple shafts with good dimensional accuracy and surface finish, you should purchase round bars that are turned and centerless ground to the dimensions needed. Unless your facilities are capable of making round bars economically, it is cheaper to purchase them. On the other hand, if you have to make a stepped shaft with different diameters along its length, you should purchase a round bar (with a diameter equal to the largest diameter of the stepped shaft) and turn it on a lathe. For stock that have wide tolerances, or are warped or out of round, you must order larger sizes to ensure proper finished product size.

Each manufacturing process produces parts or stock that have their own shape, surface finish, and tolerance characteristics. Hot-rolled or hot-drawn products, for

TABLE 40.1
COMMERCIALLY AVAILABLE FORMS OF MATERIALS

MATERIAL	AVAILABLE AS
Aluminum	P, F, B, T, W, S, I
Copper and brass	P, f, B, T, W, s, I
Magnesium	P, B, T, w, S, I
Steels and stainless steels	P, B, T, W, S, I
Precious metals	P, F, B, t, W, I
Zinc	P, F, B, W, I
Plastics	P, f, B, T, w
Elastomers	P, b, T
Ceramics (alumina)	p, B, T, s
Glass	P, B, T, W, s
Graphite	P, B, T, W, s

Note: P, plate or sheet; F, foil; B, bar; T, tubing; W, wire; S, structural shapes; I, ingots for casting. Lowercase letter indicates limited availability.

example, have coarser surface finishes and greater tolerances than cold-rolled or cold-drawn products. The wall thickness of welded tubing is more uniform than that of seamless tubing. Extrusions have greater cross-sectional tolerance than those made by roll forming. Turned bars have coarser surface finish than ground bars. You can generally find information on these characteristics in catalogs furnished by suppliers of these materials.

40.2.3 Manufacturing properties

Manufacturing properties are typically castability, formability, machinability, weldability, and hardenability by heat treatment. Because materials have to be formed, shaped, machined, ground, fabricated, and heat treated into individual parts and components of specific shapes and dimensions, these properties are crucial to proper selection of materials.

Recall that the quality of a material can greatly influence its manufacturing properties. A rod or bar with a longitudinal seam will develop cracks during upsetting and heading operations. Bars with internal defects and inclusions will crack during seamless-tube production. Porous cast workpieces will produce poor surface finish when machined. Blanks that are heat treated nonuniformly or bars that are not stress relieved will distort during finishing operations, such as grinding.

Similarly, incoming stock that has variations in composition and microstructure cannot be heat treated or machined properly. Sheet-metal stock with variations in its cold-worked conditions will spring back differently during bending and other forming operations. If prelubricated sheet-metal blanks are supplied with nonuniform lubricant thickness, their formability, surface finish, and overall quality will be adversely affected.

• Example: Effect of workpiece hardness on cost in drilling.

Gear blanks, forged from 8617 alloy steel and having a hardness range of 149–156 HB, required drilling a hole 75 mm (3 in.) in diameter in the hub. The blanks were drilled with a standard helix drill. However, after only 10 pieces the drill began to gall, temperature increase was excessive, the drill became dull, and the holes had a rough surface finish. In order to improve machinability and reduce galling (see Section 20.9), the hardness of the gear blanks was increased to the range of 217–241 HB by heating to 840 °C (1540 °F) and quenching in oil. When the blanks were drilled at this hardness, galling was reduced, the surface finish was improved, drill life increased to 50 pieces, and the cost of drilling was reduced by 80 percent. *Source:* ASM International.

40.2.4 Reliability of supply

In the General Introduction, we pointed out geopolitical factors that can affect the supply of strategic materials. Other factors such as strikes, shortages, and reluctance of suppliers to produce materials in a particular shape, quality, or quantity also affect reliability of supply. Even though availability of materials may not be a problem for a country as a whole, it can be a problem for certain industries or because of the location of a manufacturing plant.

40.2.5 Costs of materials and processing

Because of its processing history, the *unit cost* of a material (cost per unit weight or volume) depends not only on the material itself (see Table 6.1), but also on its shape, size, and condition. Because more operations are involved in its production, the unit cost of thin wire, for example, is higher than that of a round bar made of the same material. The unit cost of metal foil is higher than that of metal plate. Similarly, powder metals are more expensive than bulk metals. The cost of materials generally decreases as the purchase quantity increases, much like packaged food products you purchase at supermarkets: the larger the quantity per package, the lower the cost per unit weight.

The cost of a particular material is subject to fluctuations, which may be caused by factors as simple as supply and demand or as complex as geopolitics. If a product is no longer cost competitive in the marketplace, alternative and less costly materials should be selected. For example, the copper shortage in the 1940s caused the U.S. government to mint pennies from zinc-plated steel. Similarly, when the price of copper increased substantially during the 1960s, electrical wiring for residential use was, for a time, made from aluminum (see the example in Section 3.7).

If scrap is produced during manufacturing, as in sheet-metal fabricating, forging, and machining, the value of the scrap is deducted from the material's cost to obtain net material cost. The value of the scrap depends on the type of metal and the demand for it. The value is usually 10–40 percent of the original cost of the material. Typical percentages of scrap produced in selected manufacturing processes are given in Table 40.2.

TABLE 40.2
APPROXIMATE AMOUNT OF SCRAP PRODUCED IN VARIOUS MANUFACTURING PROCESSES

PROCESS	SCRAP (%)
Machining	10–60
Hot closed-die forging	20–25
Sheet-metal forming	10–25
Cold or hot extrusion, forging	15
Permanent-mold casting	10
Powder metallurgy	5

40.3 Product Design and Quantity of Materials

With high production rates and less labor, the cost of materials becomes a significant portion of product cost. Although the cost of materials cannot be reduced below a certain level, efforts can be made to reduce the amount of material involved in components that are to be mass produced. Since the overall shape of the part is usually optimized during the design and prototype stages, further reductions in the amount of material used can be obtained only by reducing the part's thickness. To do so requires selection of materials having high strength-to-weight or stiffness-to-weight ratios (see Section 3.2).

Many engineering designs have traditionally been done intuitively, usually based on experience and empirical information. Although the components in some products may be underdesigned, parts are generally overdesigned (called *irrational design*), presumably with safety in mind. New techniques such as finite-element analysis, minimum-weight design, design optimization, and computer-aided design and manufacturing have greatly facilitated design analysis and optimization and product simulation.

Implementing new design techniques and minimizing the amount of materials utilized can, however, present significant problems in manufacturing. Producing parts with thin cross-sections can be difficult, such as welding, casting, forging and sheet-stretching operations on thin sections. In forging, thin parts require high forces (see Eq. 14.1). In welding, they tend to distort. In casting, thin sections present difficulties in mold-cavity filling and in maintaining dimensional accuracy and good surface finish. In sheet-forming operations, formability may be reduced (see Section 16.4).

Conversely, utilizing parts with thick sections can slow production in processes such as casting and injection molding because of the length of time required for cooling and removing the part from the mold. Abrupt changes in part cross-section, such as from thick to thin, contribute to problems and may lead to defective parts.

40.4 Substitution of Materials

There is hardly a product on the market today for which substitution of materials has not played a major role in helping companies maintain their domestic and international competitive positions. Automobile and aircraft manufacturing are examples of major industries where substitution of materials is an important ongoing activity. A similar trend is evident in sporting goods and other consumer-product areas.

Although new products continually appear on the market, the majority of both design and manufacturing activities is concerned with improving existing products. Product improvements are resulting from substitution of materials, as well as implementation of new or improved processing techniques, better control of processing parameters, and plant automation.

There are a number of reasons for substituting materials in existing products, including:

a) Reduction in costs of materials and processing.
b) Ease of manufacturing.
c) Improvements in performance, such as reduction in weight (Fig. 40.2) and better resistance to wear, fatigue, corrosion, and other characteristics.
d) Increase in stiffness-to-weight and strength-to-weight ratios.
e) Simplification of assembly and installation and for conversion to automated parts assembly.
f) Ease of maintenance and repair.
g) Unreliable domestic and overseas supply of materials.
h) Legislation and regulations prohibiting the use of certain materials in products for environmental control.

• Example: Aluminum vs. steel beverage cans.

The market for beverages cans in the United States is estimated at $4 billion a year and offers an example of how material substitution can have a major economic impact. For many years, beverage cans were made of steel in three pieces: the top, the bottom, and the seam-welded cylindrical body. Since the development of the aluminum can in 1958, however, aluminum has gained an increasingly larger share of the market, so that 95 percent of the 78 billion beverage cans produced today are

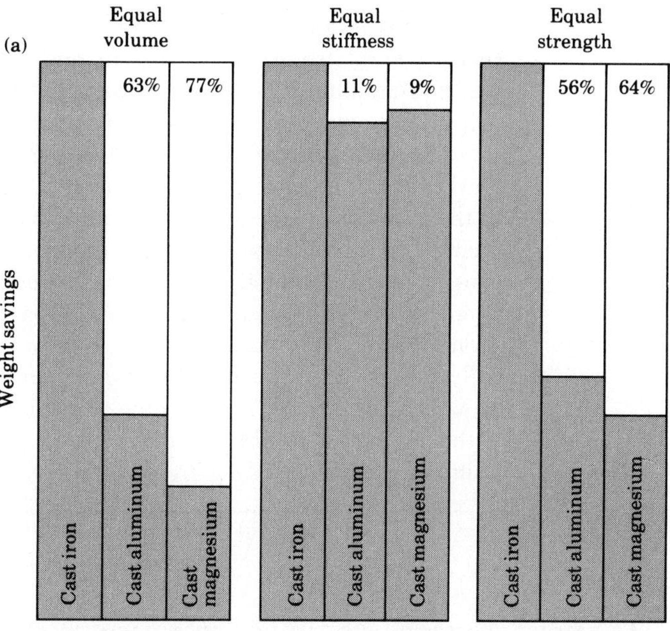

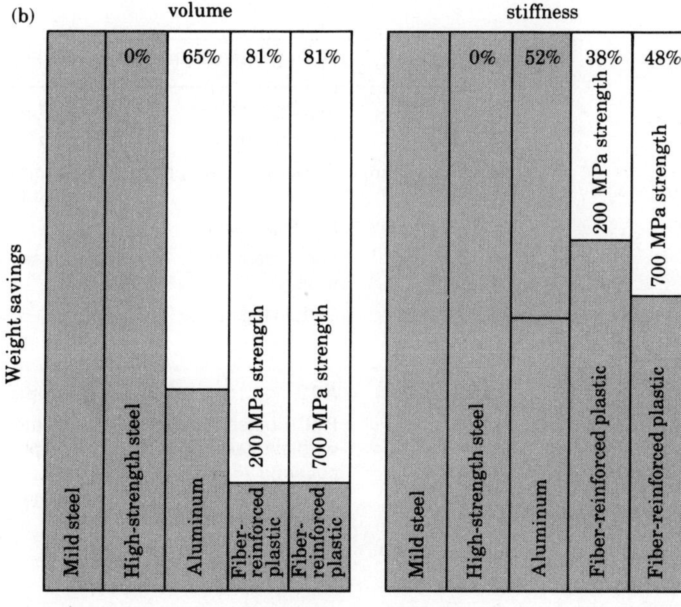

FIGURE 40.2 Weight reduction by material substitution: (a) cast metals and (b) wrought metals and reinforced plastics. *Sources:* W. D. Compton and N. A. Gjostein, *Scientific American*, and C. L. Magee, Ford Motor Company.

made of aluminum. The two-piece aluminum cans (top and body) have such advantages as lightness, recyclability, and the lack of seams that were unsightly and rust-prone in steel cans. Because of the value of aluminum, it is cost effective to recycle aluminum cans—another attractive feature. There are now about 10,000 can collection points, and more than 50 percent of the cans produced are recycled.

The steel industry is now making great strides to capture a larger share of this market. Two-piece steel cans (tin plate) have now been developed, although the top of the can is still made of aluminum because of the lower effort involved in opening the tops. Because the cost of aluminum has increased at a higher rate than steel, steel cans have the potential of becoming competitive. Furthermore, steel cans can be magnetically separated from other garbage and can be recycled, although the economics of recycling is still a matter of dispute. The steel industry is planning collection centers for cans in large cities for recycling. Some brewing companies are now showing interest in switching to steel cans. It is estimated that the cost saving with steel cans is on the order of $5 to $7 per thousand cans, or an estimated $500 million a year.

• Example: Material substitution in a vacuum cleaner.

The materials used in a household (Hoover) vacuum cleaner are shown in the accompanying figure. As shown in the table, some of these materials were different in older models of this product. The table indicates the reasons for the changes made. *Source: Metal Progress.*

PART	FORMER MATERIAL	NEW MATERIAL	REASON FOR CHANGE
Bottom plate	Steel stampings	Die-cast aluminum	Easier servicing
Wheels	Phenolic resin	Medium-density polyethylene	Less noise
Wheel mounting	Screw-machine parts	Preassembled with a cold-headed steel shaft	Lower cost, simpler to replace
Switch toggle	Phenolic resin	ABS	Tougher
Handle tube	Lock-seam steel tubing	Electric seam-welded tubing	Lower cost, better dimensional control
Motor hood	Cellulose acetate	ABS	Better heat and moisture resistance, no warpage
Rug nozzle	ABS	High-impact styrene	Lower cost
Hose	PVC-coated wire with single-ply PVC covering	PVC-coated wire with double-ply covering separated by a nylon reinforcement	More durable, lower cost
Bellows, cord insulation, bumper strips	Rubber	PVC	Lower cost, bettter aging, better choice of colors, less marking

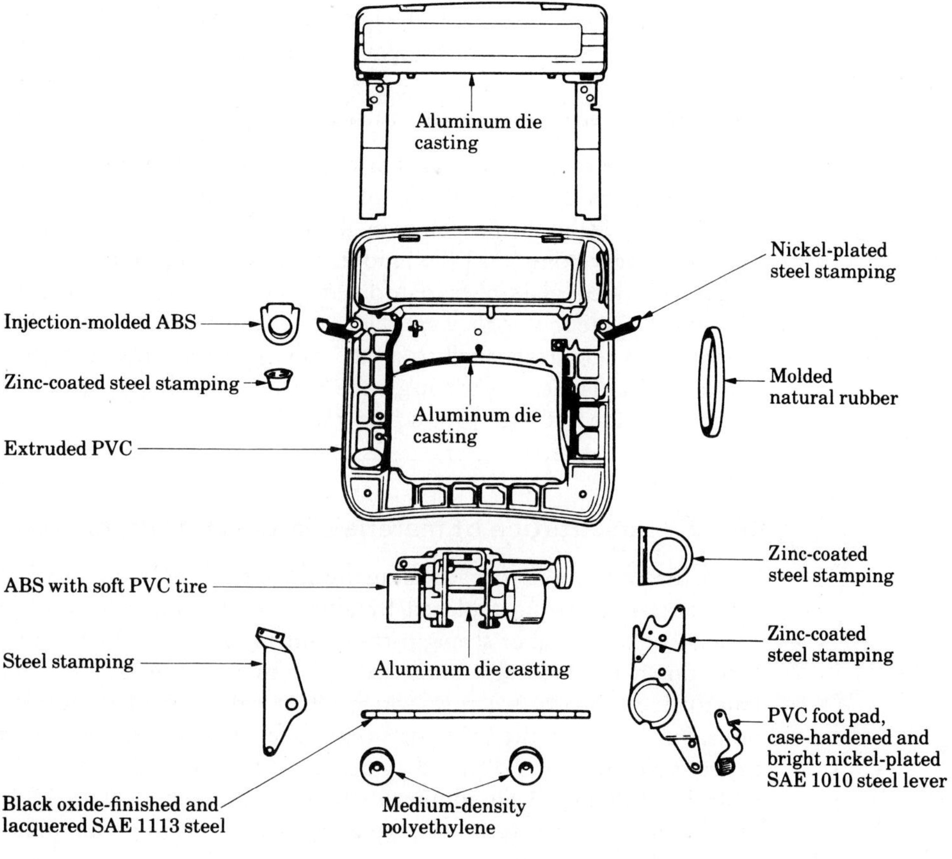

40.4.1 Substitution of materials in the automobile industry

The automobile is a good example of effective substitution of materials in order to achieve one or more of the objectives just listed. Several examples are:

a) All or part of the metal body replaced by plastic or reinforced plastic parts.
b) Metal bumpers, gears, pumps, fuel tanks, housings, covers, clamps, and many other similar components replaced by plastic substitutes.
c) Metal engine components replaced by ceramic and reinforced plastic parts.
d) Composite-material driveshafts used in place of all-metal driveshafts.
e) Various other substitutions, such as cast-iron to cast-aluminum engine blocks, forged to cast crankshafts, and forged to cast, powder-metallurgy, or composite-material connecting rods.

The automobile industry is a major consumer of both metallic and nonmetallic materials. Consequently, there is constant competition among suppliers, particularly in the steel, aluminum, and plastics industries. Automotive industry engineers and management are constantly investigating the advantages and limitations of these principal materials with regard to their applications and costs.

• Example: Material substitution in automobile engines.

Four components in the Ferrari Formula I engine that were once made of steel and specialty alloys are now made of composite materials. The oil pump, turbo impeller, and turbo inlet are now made of injection-molded carbon-fiber reinforced polyetheretherketone (PEEK), and the water-pump impeller is now made of injection-molded glass-fiber reinforced polysulfone (see Chapters 7, 9, and 18). These composite materials were selected for the characteristics of high strength-to-weight ratio and resistance to wear, oils, greases, and corrosion. Furthermore, unlike steels, plastics allow for greater design flexibility, and complex parts can be injection-molded at high production rates.

40.4.2 Substitution of materials in the aircraft industry

In the aircraft and aerospace industries, conventional aluminum alloys (2000 and 7000 series) are being replaced by aluminum–lithium alloys and titanium alloys because of their higher strength-to-weight ratios. Forged parts are being replaced with powder-metallurgy parts that are manufactured with better control of impurities and microstructure, which enhance the overall economics of the substitutions. Furthermore, the P/M parts require less machining and produce less scrap of expensive materials. Advanced composite materials and honeycomb structures are replacing traditional aluminum airframe components (Figs. 40.3 and 40.4), and metal-matrix composites are replacing some of the aluminum and titanium structural components.

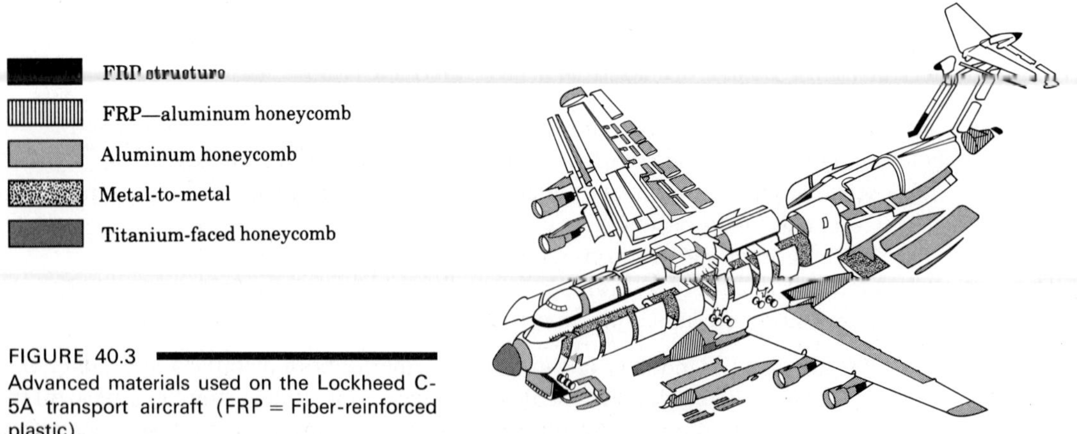

FIGURE 40.3 Advanced materials used on the Lockheed C-5A transport aircraft (FRP = Fiber-reinforced plastic).

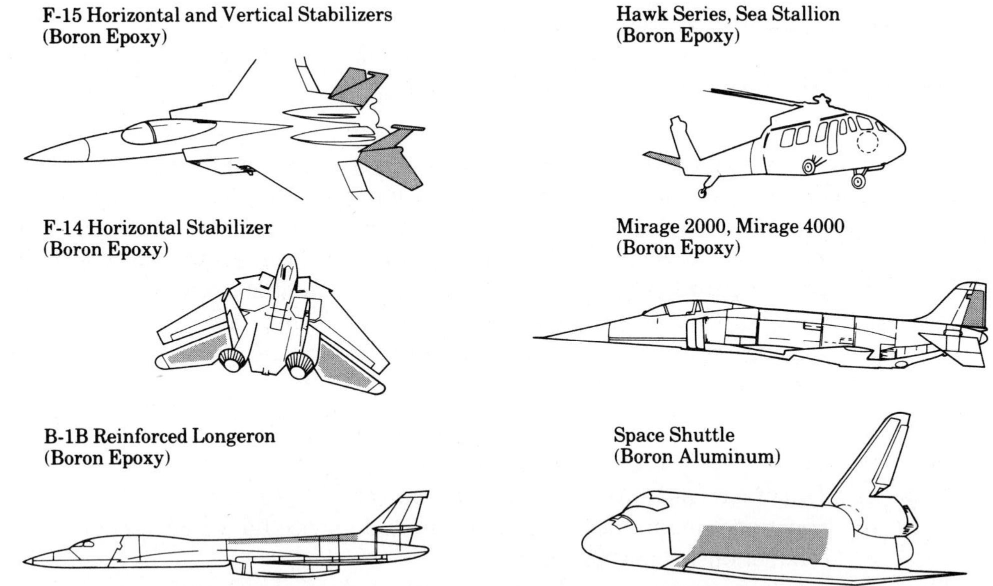

FIGURE 40.4
Boron epoxy and boron aluminum composite materials in aerospace applications. *Source:* AVCO Specialty Materials Division.

- **Example: Material changes from C-5A to C-5B military cargo aircraft.**

The accompanying table shows the changes made in materials for various components of the new aircraft and the reasons for the changes. *Source:* H. B. Allison, Lockheed-Georgia.

ITEM	C-5A MATERIAL	C-5B MATERIAL	REASON FOR CHANGE
Wing panels	7075-T6511	7175-T73511	Durability
Main frame			
Forgings	7075-F	7049-01	Stress corrosion resistance
Machined frames	7075-T6	7049-T73	
Frame straps	7075-T6 plate	7050-T7651 plate	
Fuselage skin	7079-T6	7475-T61	Material availability
Fuselage underfloor end fittings	7075-T6 forging	7049-T73 forging	Stress corrosion resistance
Wing/pylon attach fitting	4340 alloy steel	PH13-8Mo	Corrosion prevention
Aft ramp lock hooks	D6-AC	PH13-8Mo	Corrosion prevention
Hydraulic lines	AM350 stainless steel	21-6-9 stainless steel	Improved field repair
Fuselage failsafe straps	6Al-4V titanium	7475-T61 aluminum	Titanium strap disbond

40.5 Selection of Manufacturing Processes

In this section we describe the importance of proper manufacturing process and machinery selection, and how the selection process relates to characteristics of materials, tolerances, surface finishes obtained, and cost. As we have pointed out, many traditional manufacturing processes have been automated and are being computer controlled to optimize the processes. Computerization is also effectively increasing product reliability and quality and reducing labor costs.

The choice of a manufacturing process is dictated by various considerations, such as:

a) Characteristics and properties of the workpiece material (Table 40.3).
b) Shape, size, and thickness of the part.
c) Tolerances and surface finish requirements.
d) Functional requirements for the expected service life of the component.
e) Production volume (quantity).
f) Level of automation required to meet production volume and rate.
g) Costs involved in individual and combined aspects of the manufacturing operation.

You have seen that some materials can be processed at room temperature, but others require elevated temperatures, which means additional furnaces and appropriate tooling. Some materials are soft and ductile, whereas others are hard, brittle, and abrasive, thus requiring special processing techniques and tool and die materials. Different materials have different manufacturing characteristics, such as castability, forgeability, workability, machinability, and weldability. Few materials have the same favorable characteristics in all categories. For example, a material that is castable or forgeable may later present problems in machining, grinding, or finishing operations that may be required in order to produce a product with acceptable surface finish and dimensional accuracy.

• Example: Process selection in making a part.

Assume that you are asked to make the part shown in the figure on p. 1234. You should first determine the part's function, the type of loads and environment to which it is to be subjected, the tolerances and surface finish required, and so on. For the sake of discussion, let's assume that the part is round, 125 mm (5 in.) long and that the large and small diameters are 38 mm and 25 mm (1.5 in. and 1.0 in.), respectively. Let's further assume that, for functional requirements such as stiffness, hardness, and resistance to elevated temperatures, this part should be made of metal.

Which manufacturing process would you choose, and how would you organize the production facilities to manufacture a cost-competitive, high-quality product?

TABLE 40.3

COMMONLY USED MATERIALS IN VARIOUS MANUFACTURING PROCESSES

TYPE OF PART \ MATERIAL	IRON	CARBON STEEL	ALLOY STEEL	STAINLESS STEEL	TOOL STEEL	ALUMINUM ALLOYS	COPPER ALLOYS	MAGNESIUM ALLOYS	NICKEL ALLOYS	ZINC ALLOYS	TIN ALLOYS	LEAD	TITANIUM	PRECIOUS METALS
Extrusions	—	○	○	○	—	●	●	●	○	○	○	○	○	—
Metal stampings	—	●	●	○	—	●	●	○	○	○	—	—	—	●
Metal spinnings	—	●	○	●	—	●	●	○	●	○	○	○	—	—
Cold-headed parts	—	●	○	○	—	●	●	—	○	—	—	○	—	—
Impact extrusions	—	●	○	—	—	●	●	●	○	●	●	●	—	—
Swaged and bent tubing	—	●	●	●	—	●	●	○	●	○	○	—	○	—
Roll-formed sections	—	●	●	●	—	●	●	—	—	●	—	—	—	—
Powder-metal parts	●	○	○	○	○	○	●	—	○	—	—	—	○	—
Forgings	—	●	●	●	○	●	●	●	○	—	—	—	○	—
Screw-machine parts	○	●	○	○	—	●	●	○	○	○	—	—	○	—
Electrical-discharge-machined parts	—	○	○	○	●	○	○	—	○	—	—	—	○	—
Electrochemically machined parts	—	○	●	○	●	—	○	—	●	—	—	—	●	—
Chemically machined parts	—	●	○	●	○	●	●	●	○	—	—	—	○	—
Sand-mold castings	●	●	●	●	○	●	●	●	●	○	○	○	—	—
Permanent-mold castings	●	○	—	—	—	●	●	●	○	○	○	○	—	—
Ceramic-mold castings	●	●	●	●	●	○	●	○	●	○	—	—	—	—
Plaster-mold castings	—	—	—	—	—	●	●	○	—	●	○	○	—	—
Centrifugal castings	●	●	●	—	—	●	●	—	●	—	—	—	—	—
Investment castings	—	●	●	●	●	●	●	○	●	—	—	—	—	○
Die castings	—	—	○	○	○	●	○	○	—	●	○	○	—	—

Note: ●, frequently processed with this method; ○, sometimes processed with this method; —, seldom or never processed with this method.
Source: After J. G. Bralla.

We have stated several times that parts should be produced at or near their final shape (net or near-net shape manufacturing). This approach eliminates much secondary processing, such as machining, grinding, and other finishing operations, reduces total manufacturing time, and thus lowers the cost.

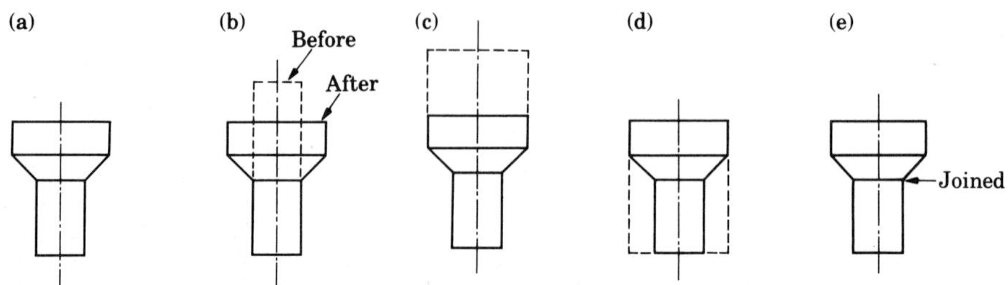

This part is relatively simple and could be suitably manufactured by different methods. You can readily see that the part can be made by (a) casting and powder metallurgy, (b) upsetting, (c) extrusion, (d) machining, or (e) joining two pieces together. For net-shape processing, the two logical processes are casting and powder metallurgy, each process having its own characteristics and various tooling and labor costs. This part can also be made by cold, warm, or hot forming. One method is by upsetting a 25-mm (1-in.) diameter round bar in a suitable die to form the larger end. The partial extrusion of a 38-mm (1.5-in.) diameter bar to reduce the diameter to 25 mm is another possibility. Each of these processes shapes the material without any waste.

This part can also be made by machining a 38-mm diameter bar stock to obtain the 25-mm diameter section. However, machining will take much longer than forming, and some material will be wasted as metal chips. On the other hand, unlike net-shape processes, which require special dies, machining does not require special tooling, and the operation can easily be carried out on a lathe. Note also that you can make this part in two pieces and then join them by welding, brazing, or adhesive bonding.

Each process has its own characteristics, production rate, surface finish, dimensional accuracy, and final product properties. Note that all these processes require materials in different shapes, ranging from metal powders to round bars. Because of the various operations involved in producing the materials, costs depend not only on the type of material (ingot, powder, drawn rod, extrusion), but also on its size and shape. Table 5.1 shows that, per unit weight, square bars are more expensive than round bars, and cold-rolled plate and sheet are more expensive than hot-rolled plate and sheet. Also, hot-rolled bars are much less expensive than powders of the same metal.

By now, you should realize that even a simple part such as this requires considerable thought before you can finally select a process and initiate a manufacturing plan. In addition to technical requirements, process selection also depends on factors such as required production quantity and rate, as we describe in the next section. Summarizing our discussion, we can say that if only a few parts are needed, machining this part is the economical method. However, as the production quantity increases, producing this part by a heading operation or by cold extrusion would be the proper choice. Method (e) would be the most appropriate choice if the top and bottom portions of this part were made of different metals.

40.6 Process Capabilities

We have shown that all manufacturing processes have certain advantages and limitations. Casting and injection molding, for example, can generally produce more complex shapes than can forging and powder metallurgy. The reason? Because the molten metal or plastic is capable of filling complex mold cavities. On the other hand, forgings generally can be made into complex shapes by additional machining and finishing operations, and have toughnesses that are generally superior to castings and powder metallurgy products.

Recall that the shape of a product may be such that it should be fabricated from several parts, then joined with fasteners or by brazing, welding, or adhesive bonding techniques. The reverse may be true for other products, whereby manufacturing them in one piece might be more economical because of significant assembly-operation costs.

Other factors that you have to consider in process selection are the minimum section size or dimensions that can be satisfactorily produced (Fig. 40.5). For example, very thin sections can be obtained by cold rolling, but processes like sand casting or forging prohibit forming such thin sections.

40.6.1 Tolerances and surface finish

Tolerances and surface finishes produced are important aspects of manufacturing. They have a particular bearing on subsequent assembly operations, proper operation of various machines and instruments, and consistency of overall appearance. The ranges of surface finishes and tolerances obtained in manufacturing processes are given in various chapters and summarized in Figs. 40.6 and 40.7. You have seen that in order to obtain finer surface finishes and closer tolerances, additional finishing operations, better control of processing parameters, and the use of higher quality equipment may be required.

The closer the tolerance required, the higher the cost of manufacturing will be (Fig. 40.8); the finer the surface finish required, the longer manufacturing will take, increasing the cost (Fig. 40.9). In machining aircraft structural members made of titanium alloys, for example, as much as 60 percent of the cost of machining the part is consumed in the final machining pass in order to hold proper tolerances and

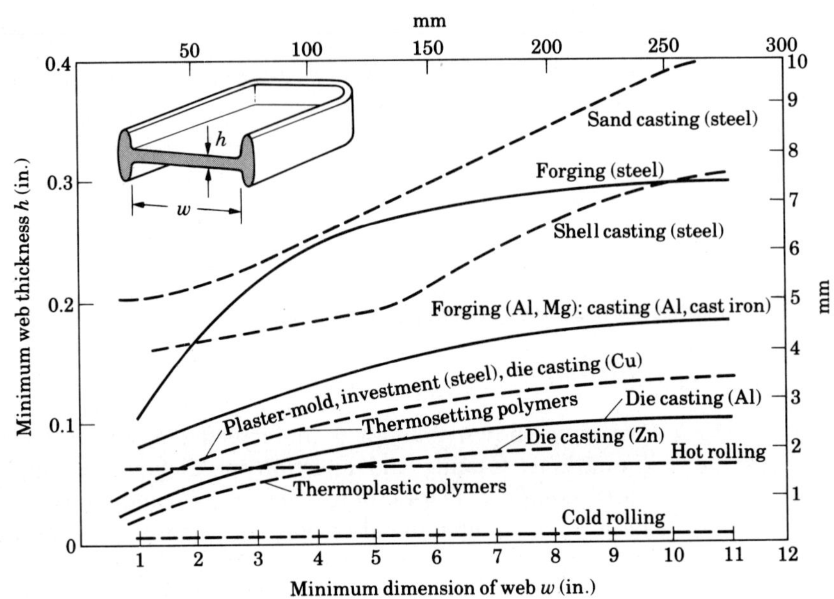

FIGURE 40.5 Process capabilities for minimum part dimensions. *Source:* J. A. Schey, *Introduction to Manufacturing Processes,* 2d ed. New York: McGraw-Hill, 1987.

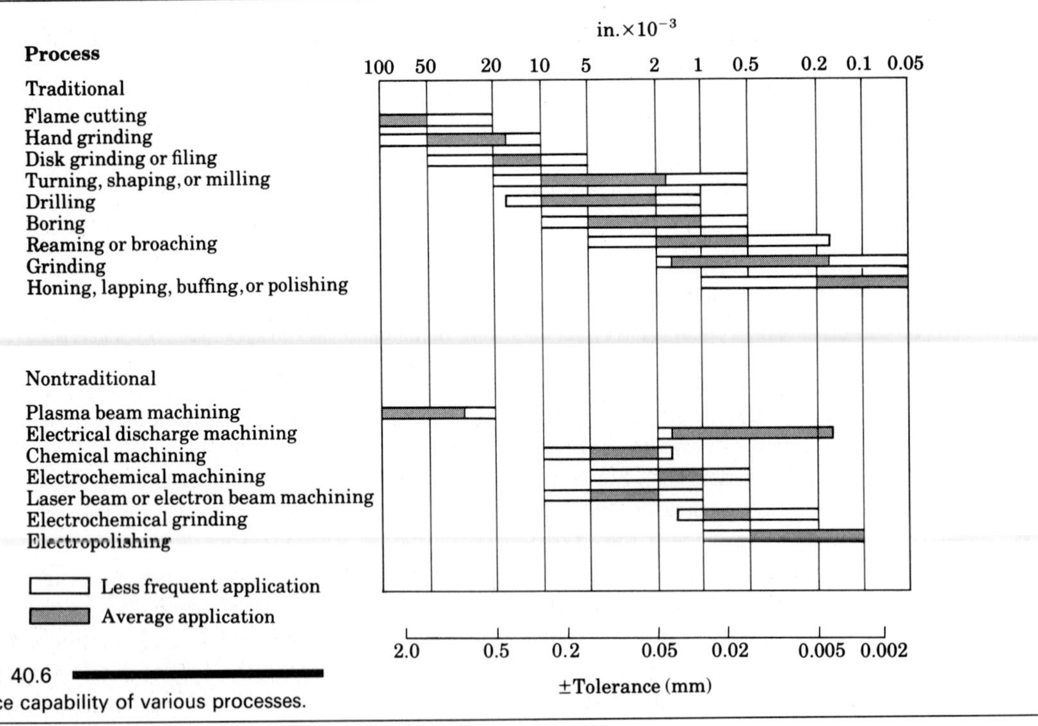

FIGURE 40.6 Tolerance capability of various processes.

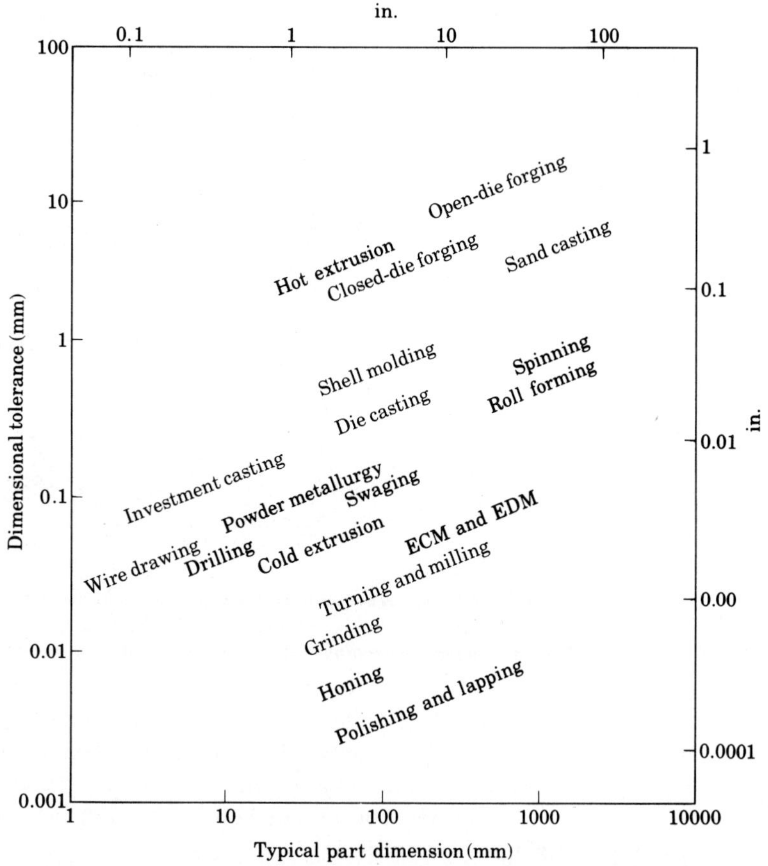

FIGURE 40.7 Tolerances as a function of component size for various manufacturing processes. Note that, because many factors are involved, there is a broad range for tolerances. See also Fig. 22.15.

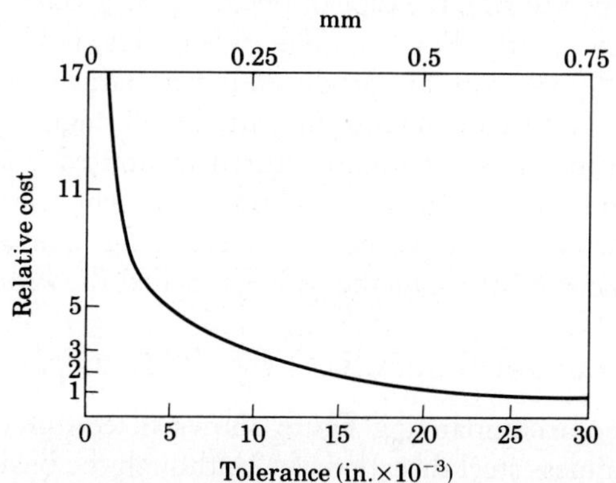

FIGURE 40.8 Relationship between relative cost and tolerance.

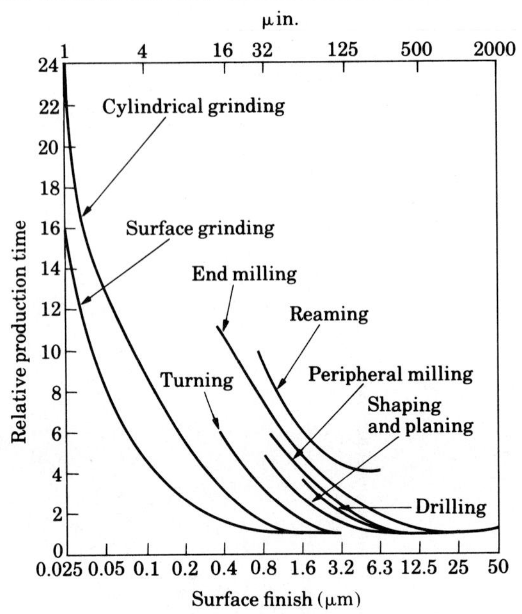

FIGURE 40.9
Relative production time as a function of surface finish produced by various manufacturing methods. *Source: American Machinist.* See also Fig. 25.34.

surface finishes. Unless specifically required with proper technical and economic justification, parts should be made with as coarse a surface finish and with as wide a tolerance as functionally and aesthetically acceptable. Thus the importance of interaction and continuous communication between designer and manufacturing engineer is obvious.

40.6.2 Production volume

Depending on the type of product, the **production volume**, or *quantity* (lot size), can vary greatly. For example, paper clips, bolts, washers, spark plugs, bearings, and ball point pens are produced in very large quantities. On the other hand, jet engines for large commercial aircraft, diesel engines for locomotives, and propellers for ocean liners are manufactured in limited quantities. Production quantity plays a significant role in process and equipment selection. In fact, an entire manufacturing discipline is devoted to determining mathematically the optimum production quantity for a specific process, called the *economic order quantity*.

• Example: Economical quantities for different production methods.

The accompanying figure shows three different methods for making a type 347 stainless steel threaded cap. Although the basic shape of the cap was the same, each method required a different design. The original order for this part was 100 caps. After a brief analysis, it was obvious that the most economical and quickest method of production was machining the cap from a solid round bar stock, as shown in (a). When the order was increased to 1000 caps, it was determined that a two-piece assembly might be more economical. The threaded part was machined from a smaller bar stock and the other piece was made of 0.0625-in. thick sheet, stamped and pressed on the machined part, as shown in (b). It was found, however, that the cost per piece was not significantly different from that of method (a).

The order was increased to 5000 caps, which called for a further review of the cap design. It was determined that an economical means of producing this cap in larger quantities was to make it in one piece, by first press forming (drawing) a 0.093-in. thick sheet and then machining the threads. The sheet had to be thicker than in method (b) because of the required threading. Because of the smaller amount of material removed by machining and the speed with which press forming could be done, both material and labor costs in this method were lower than those for the other two methods. As a result, the manufacturing cost per cap was reduced by nearly 50 percent. *Source:* ASM International.

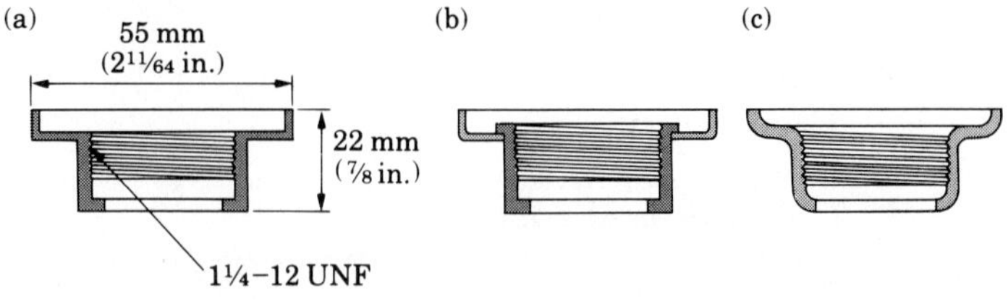

40.6.3 Production rate

A significant factor in manufacturing process selection is the **production rate**, defined as the number of pieces to be produced per unit of time, such as per hour, month, or year. Processes such as powder metallurgy, die casting, deep drawing, and roll forming are high-production-rate operations. On the other hand, sand casting, conventional and electrochemical machining, spinning, superplastic forming, adhesive and diffusion bonding, and processing of reinforced plastics are relatively slow operations. These rates can be increased by automation or the use of multiple machines. However, a slower production rate does not necessarily mean that the manufacturing process is inherently uneconomical.

40.6.4 Lead time

The choice of a process is greatly influenced by the time required to start production, or **lead time**. Processes such as forging, extrusion, die casting, roll forming, and sheet-metal forming may require extensive—and expensive—dies and tooling. In contrast, most machining and grinding processes have an inherent flexibility, and utilize tooling that can be adapted to many requirements in a relatively short time. Recall also our discussion of machining centers, flexible manufacturing cells, and flexible manufacturing systems, which are capable of responding quickly and effectively to product changes both in type and quantity.

40.7 Manufacturing Costs

We have shown in various chapters that economics is one of the most important aspects of any manufacturing operation. In this section we describe the various costs involved in manufacturing a product in order to identify those factors that can help minimize them, while maintaining quality.

Designing and manufacturing a product according to certain specifications and to meet service requirements is only one aspect of production. In order for the product to be successfully marketed, its cost must be competitive with similar products in the marketplace. Although different methods are used to account for product cost, the total cost consists of the following basic items:

a) Materials costs.
b) Tooling costs.
c) Labor costs.
d) Fixed costs.
e) Capital costs.

40.7.1 Material costs

We described **material costs** in Section 40.2. We also have given some cost data in various tables. Thus further discussion of material costs is not necessary.

40.7.2 Tooling costs

Tooling costs are the costs involved in making the tools, dies, molds, patterns, and special jigs and fixtures necessary for manufacturing a product or component. The tooling cost is greatly influenced by the production process selected. For example, if a part is to be made by casting, the tooling cost for die casting is higher than that for sand casting. Similarly, the tooling cost in machining or grinding is much lower than that for powder metallurgy, forging, or extrusion. In machining operations, the choice of cutting tool materials can be significant. For example, carbide tools are more expensive than high-speed steel tools, but tool life is longer.

If a part is to be manufactured by spinning, the tooling cost for conventional spinning is much lower than that for power spinning. Tooling for rubber forming processes is less expensive than that for male-and-female die sets used for drawing and stamping of sheet metals. High tooling costs, on the other hand, can be justified for high-volume production of a single item. As we have stated previously, the expected life of tools and dies, and their obsolescence because of product changes, are also important considerations.

40.7.3 Labor costs

Labor costs are generally divided into direct and indirect costs. The *direct labor cost* is for the labor directly involved in manufacturing the part (productive labor). This cost includes all labor from the time materials are first handled to the time the product is finished. This time is generally referred to as *floor-to-floor time*. For example, a machine operator picks up a round bar from a bin, machines it to the shape of a threaded rod, and returns it to another bin. The direct labor cost is calculated by multiplying the labor rate (hourly wage, including benefits) by the time that the worker spends producing the part.

The time required for producing a part depends not only on required part size, shape, and dimensional accuracy, but also on the workpiece material. We discussed, for example, cutting speeds in Part IV and pointed out that the cutting speeds for high-temperature alloys are lower than those for aluminum or plain-carbon steels. Consequently, the cost of machining certain aerospace materials is much higher than that for the more common alloys.

Indirect labor costs are those that are involved in servicing the total manufacturing operation. This cost is comprised of activities such as supervision, repair, maintenance, quality control, engineering, research, sales, and also generally includes the cost of office staff. Because these costs do not contribute directly to the production of finished parts or are not chargeable to a specific product, they are referred to as *overhead*, or the *burden rate*, which is charged proportionally to all products. The personnel involved are categorized as nonproductive labor.

40.7.4 Fixed costs

Fixed costs include the costs of power, fuel, taxes on real estate, rent, and insurance, and capital, including depreciation and interest. The company would have to pay these costs regardless of whether it made a particular product. Thus fixed costs are not sensitive to production volume.

Capital costs represent the capital investment in land, buildings, machinery, and equipment and represent major expenses for most manufacturing operations (Table 40.4). Note the wide range of prices in each category and the fact that some machines cost a million dollars and more. Interestingly, the cost of traditional machine tools, per unit weight, is historically roughly equivalent to the cost of steak per pound. This cost can easily double for machinery having computer controls.

Let's assume that a company has decided to begin manufacturing a variety of valves. A new plant has to be built, or an old plant remodeled, with all the necessary machinery, support equipment, and facilities. To cast and machine the valve bodies, melting furnaces, casting equipment, machine tools, quality control equipment, and other related equipment and machinery of various sizes and types have to be purchased. If valve designs are to be changed often, or if product lines vary greatly, the manufacturing equipment must have sufficient flexibility to accommodate these requirements. Machining centers and flexible manufacturing systems are especially suitable for this purpose. The equipment and machinery we have just listed are capital cost items and, as you can appreciate, require major investment.

In view of generally high equipment costs, particularly those involving flexible manufacturing systems, high production rates are often required to justify large expenditures and hold product unit cost at a competitive level. Lower unit costs can

TABLE 40.4
APPROXIMATE RANGES OF MACHINERY BASE PRICES

TYPE OF MACHINERY	PRICE RANGE ($000)
Broaching	10–300
Drilling	10–100
Electrical discharge	30–150
Electromagnetic and electrohydraulic	50–150
Gear shaping	100–200
Grinding	
Cylindrical	40–150
Surface	20–100
Headers	100–150
Injection molding	30–150
Boring	
Jig	50–150
Horizontal boring mill	100–400
Flexible manufacturing system	>1000
Lathe	10–100
Single- and multi-spindle automatic	30–250
Vertical turret	100–400
Machining center	100–1000
Mechanical press	20–250
Milling	10–250
Robots	20–200
Roll forming	5–100
Rubber forming	50–500
Stretch forming	400–>1000
Transfer machines	100–>1000
Welding	
Electron beam	200–1000
Spot	10–50
Ultrasonic	50–200

Note: Prices vary considerably, depending on size, capacity, options, and level of automation and computer controls.

be achieved by continuous production, involving round-the-clock operation, so long as demand warrants. Proper equipment maintenance is essential to ensure high productivity. Any breakdown of machinery (*downtime*) can be very expensive, ranging from a few hundred dollars to thousands of dollars per hour.

40.7.5 Relative costs

The costs we have described are interrelated, with **relative costs** depending on many factors. Consequently, the unit cost of the product can vary widely. For example, some parts may be made from expensive materials but require very little processing, such as minting gold coins. Thus the cost of materials relative to direct labor costs is high. On the other hand, some products may require several production steps to process relatively inexpensive materials, such as carbon steels. An electric motor, for example, is made of relatively inexpensive materials, yet many different manufacturing operations are involved in making the housing, rotor, bearings, brushes, and other components of the motor. In such cases, assembly operations can become a significant portion of the overall cost.

An approximate but typical breakdown of relative costs in today's manufacturing is as follows:

Design	5%
Material	50
Direct labor	15
Overhead	30

Note the rather small contribution to cost by direct labor, due largely to automation, and the even smaller contribution of design. However, design, in its comprehensive sense of design for manufacture and assembly, including simultaneous engineering, generally has the largest influence on the quality and success of a product in the marketplace.

Cost reductions can be achieved by careful analysis of all the costs incurred in each step in manufacturing a product, with methods that are described in detail in some of the references in the Bibliography at the end of this chapter. We have emphasized opportunities for cost reduction in various chapters throughout this text. Among these are simplifying part design and the number of subassemblies required, specifying broader tolerances, using less expensive materials, investigating alternative methods of manufacturing, and using more efficient machines and equipment.

The introduction of automation and up-to-date technology in a manufacturing facility is an obvious means of reducing costs. These, however, must be considered with care and after thorough cost-benefit analyses, with reliable data input, and due consideration of technical as well as human factors. Implementation of high technology, which can be quite expensive, should be made only after full assessment of the more obvious cost factors.

Undoubtedly you have noted that, over a period of time, the costs of products such as calculators, computers, and digital watches have decreased, whereas the costs of other products such as automobiles, aircraft, houses, and books have gone up. These differences result from the rates of change in various costs over time, including labor, machinery, materials, market demands, domestic and international competition, and worldwide economic trends (demand, exchange rates, and tariffs), and the inevitable impact of computers on all aspects of manufacturing.

• Example: Cost comparison in making gears by machining and powder metallurgy.

The oil pump gear for a truck is made of steel and requires a tensile strength of 700 MPa (100,000 psi) and hardness of 260–302 HB or equivalent. The gear is 40 mm (1.56 in.) in diameter and 50 mm (2 in.) long, and has a semicircular keyway. Full dimensions and tolerances are shown in the accompanying figure. This gear can be produced by machining (hobbing; Section 23.9) or by powder-metallurgy techniques (Chapter 17). The table below shows cost comparisons between the two processes, using a cost base of 100 for a hobbed gear. The analysis indicates that it is much more economical to produce this gear by P/M techniques. Note that the

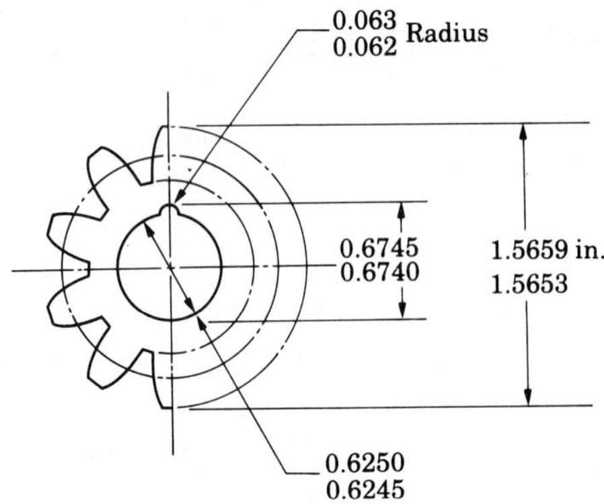

hobbing operation constitutes the largest portion of the total cost of making the gear by machining. In the P/M method, the largest cost is the cost of the metal powder. *Source:* S. S. McGee and F. K. Burgess, *International Journal of Powder Metallurgy and Powder Technology*, October 1976.

HOBBED GEAR	PERCENT OF TOTAL	P/M GEAR	PERCENT OF HOBBED TOTAL
Material: SAE 1045, including 2% setup scrap and 46% chips	15.10	Material: MPIF FC-0208-S (7.0 g/cm^3 density) 5% scrap	9.97
Operations:		Operations:	
Bar chuck, cut off and bore	8.49	Compact (100-ton press)	2.37
Broach keyway	3.17	Sinter	2.56
Hob teeth	47.50	Harden	1.92
Harden	1.92	Grind ends perpendicular to pitch diameter	5.93
Grind ends perpendicular to pitch diameter	5.93	Deburr	0.53
Deburr	0.53	Inspect	0.26
Inspect	0.26	Perishable tools and gages, per piece	8.19
Perishable tools and gages, per piece	17.10		31.73
	100.00		

• Example: Design changes and cost reduction.

The original design for an aluminum bellcrank assembly is shown in the accompanying figure (left). This part was made by die casting. Quality was inconsistent, however, and there were excessive dimensional variations and porosity in critical regions and problems in finish machining it. Consequently, scrap was significant. Alternative design and material were considered (right), consisting of a fabricated stamped-steel crank with a separate round hub, which was pressed into a hole. This design had the same load-deflection characteristics, but its cost was about 32 percent less than the die-cast bellcrank. *Source*: *Metals Handbook*, 9th ed., Vol. 3, ASM International.

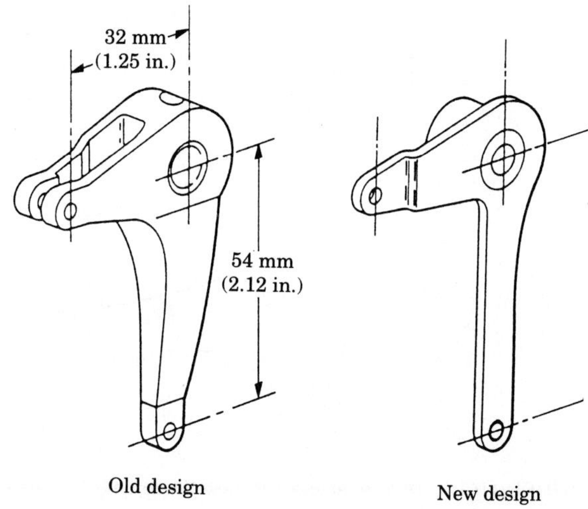

40.7.6 Manufacturing costs and production volume

One of the most significant factors in manufacturing costs is production volume. Large production volumes require high production rates. High production rates require the use of mass-production techniques, involving special machinery and proportionately less labor and plants operating on two or three shifts per day. On the other hand, small production volumes usually mean larger direct labor involvement.

Figure 40.10 shows a downward trend in the unit-cost curves as production volume increases. This result reflects proportionately lower tooling and capital costs per piece. Once the tooling is made, a greater number of produced parts will lower the cost of tooling per part. Note in Fig. 40.11 that at low production volumes, material costs are low relative to direct labor costs, whereas at high volumes the opposite is true.

As we described in Section 38.2, *small-batch production* is usually done on general-purpose machines, such as lathes, milling machines, and hydraulic presses. The equipment is versatile, and parts with different shapes and sizes can be produced by appropriate changes in the tooling. However, direct labor costs are high because these machines are usually operated by skilled labor.

For larger quantities (*medium-batch production*), these same general-purpose machines can be equipped with various jigs and fixtures, or they can be computer

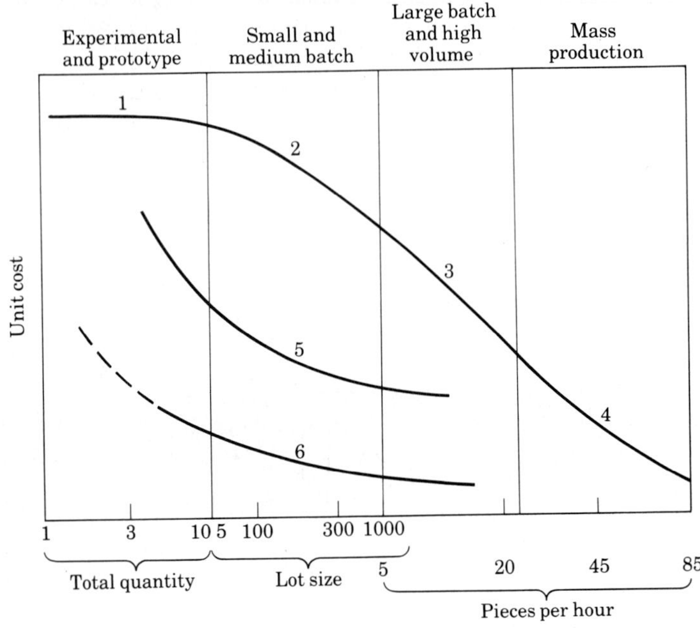

FIGURE 40.10 Unit cost of product as a function of production method and quantity. 1 = tool-room machinery; 2 = general-purpose machine tools; 3 = special-purpose machines; 4 = automatic transfer lines; 5 = NC and CNC machine tools; and 6 = computer-integrated manufacturing systems. *Source:* Cincinnati Milacron, Inc.

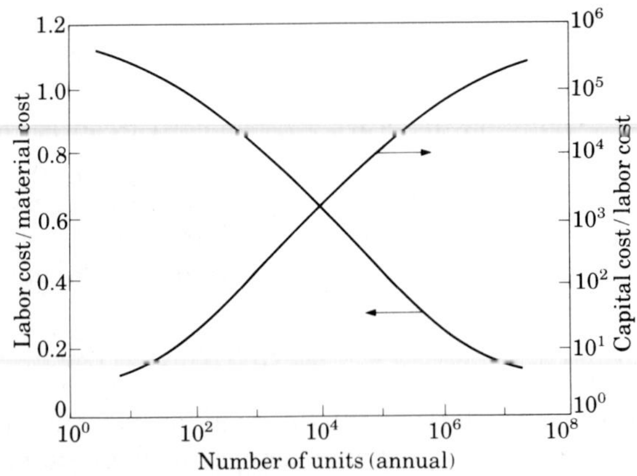

FIGURE 40.11 Relative cost of labor, materials, and capital as a function of annual production volume. Note the difference in the scales for the two ordinates in the figure. *Source:* After N. P. Suh.

controlled. To reduce labor costs further, machining centers and flexible manufacturing systems have been developed. For quantities of 100,000 and higher, the machines involved are designed for specific purposes (*dedicated machines*) and perform a variety of specific operations with very little labor involved.

• Example: Effect of quantity on relative costs.

The aircraft roller support shown in the accompanying figure is made of 8630 steel. It can be manufactured either by machining it from a rectangular bar stock or by finish machining from an investment-cast part. Whereas machining requires standard tooling, investment casting requires special tooling, the cost of which can be high. However, the special tooling cost per piece decreases as the number of parts made increases. On the other hand, machining takes considerably longer time and wastes a great deal of material, whereas casting does not. In the final analysis, we find that for smaller quantities machining is more economical, but as the quantity increases, investment casting-finish machining becomes more economical, as shown in the figure. It can be estimated that for a quantity of, say, 1000 parts, cost reduction is about 50 percent. *Source*: *Metals Handbook*, 9th ed., Vol. 3, ASM International.

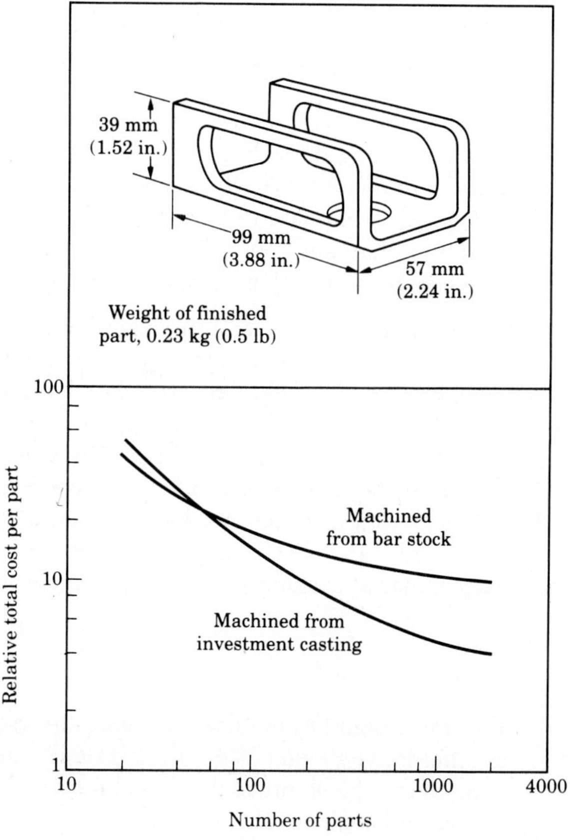

40.8 Value Engineering

We have identified several areas of activity in manufacturing where cost reduction is possible. As we stated in the General Introduction, manufacturing adds value to materials as they become discrete products and are marketed. Because this value is added in individual steps during creation of the product, utilization of value engineering (value analysis, value control, and value management) is important.

Value engineering is a system that evaluates each step in design, materials, processes, and operations so as to manufacture a product that performs its intended functions and has the lowest possible cost. A monetary value is established for each of two product attributes: (a) use value, reflecting the functions of the product; and (b) esteem or prestige value, reflecting the attractiveness of the product that makes its ownership desirable. The *value* of a product is then defined as the ratio of product function and performance to the cost of the product. Thus the goal of value engineering is to obtain maximum performance per unit cost.

Value analysis generally consists of five phases:

a) *Information* phase to gather facts and determine costs.
b) *Analysis* phase to define functions and identify problem areas and opportunities.
c) *Creativity* phase to seek ideas to respond to problems and opportunities, without judging the values of these ideas.
d) *Evaluation* phase to select ideas to be developed and the costs involved.
e) *Implementation* phase to present facts, costs, and values to management, develop a plan, and motivate positive action to obtain a commitment of necessary resources to accomplish the task.

Value engineering is an important and all-encompassing, interdisciplinary activity. It is usually coordinated by a value engineer and conducted jointly by designers, engineers, quality control, purchasing, and marketing personnel, and managers. In order for value engineering to be effective, it requires the full support of the company's top management. Implementation of value engineering in manufacturing facilities has resulted in benefits such as significant cost savings, reduced lead times, better product quality and performance, reduced product weight and size, and shorter manufacturing times. To properly assess the value of each step in manufacturing a product, several groups of questions have to be asked.

I. Product design

- Can the design be simplified without adversely affecting its functions? Have all alternative designs been investigated? Can unnecessary features or some components be eliminated or combined with others? Can the design be made lighter and smaller?

- Are the dimensional tolerances and surface finish specified necessary? Can they be relaxed?
- Will the product be difficult to assemble and take apart for maintenance? Is the use of fasteners minimized?
- Does each part have to be manufactured in the plant? Are some parts commercially available as standard products from outside sources?

II. Materials

- Do the materials selected have properties that are much above minimum requirements and specifications?
- Can some materials be replaced by others that are cheaper?
- Do the materials selected have the proper manufacturing characteristics?
- Are the materials to be ordered available in standard sizes, dimensions, surface finish, and tolerances?
- Is material supply reliable? Are there likely to be significant price fluctuations?

III. Manufacturing processes

- Have all alternative manufacturing processes been investigated?
- Are the methods chosen economical for the type of material, shape to be produced, and desired production rate? Can requirements for tolerances, surface finish, and product quality be met consistently?
- Can the part be formed and shaped to final dimensions rather than made by material-removal processes? Are machining, secondary processes, and finishing operations required? Are they necessary?
- Is tooling available in the plant? Can it be purchased as standard items?
- Is scrap produced? If so, what is the value of the scrap?
- Are processing parameters optimized? Have all the automation and computer-control possibilities been explored for all phases of the manufacturing operation? Can group technology be implemented for parts with similar geometric and manufacturing attributes?
- Are inspection techniques and quality control being implemented properly?

SUMMARY

Competitive aspects of production and costs are among the most significant considerations in manufacturing. Regardless of how well a product meets design specifications and quality standards, it must also meet economic criteria in order to be competitive in the domestic and international marketplace. The total cost of a product includes several costs, such as costs of materials, tooling, capital, labor, and overhead. Materials costs can be reduced through careful selection so that the least costly material can be identified and selected, while maintaining design and service requirements, functions, and specifications for good product quality.

Substitution of materials, modification of designs, and relaxing of tolerance and surface finish requirements are important methods of cost reduction. Certain guidelines for designing products for economic production have been established. Although labor costs are becoming a small percentage of product costs, they can be reduced further through the use of automated and computer-controlled machinery. Automation requires significant capital expenditures; however, a well-planned production facility can significantly improve productivity.

TRENDS

- The highly competitive nature of domestic and international markets will continue to challenge manufacturing engineers to reduce product costs. Recent studies show major potentials for cost reduction in materials and overhead costs.
- Material substitution and design modifications are subjects that continue to be studied closely and implemented at increasing rates. Various computer software packages are available to assist in the selection process and optimization of manufacturing operations.
- Value engineering continues to be a powerful tool in reducing costs while improving product function and performance.

KEY TERMS

Fixed costs	Material costs	Relative costs
Labor costs	Production rate	Tooling costs
Lead time	Production volume	Value engineering

BIBLIOGRAPHY

Anderson, D.M., *Design for Manufacturability, Optimizing Cost, Quality, and Time-to-Market*. Lafayette, Calif.: CIM Press, 1991.

Bralla, J.G. (ed.), *Handbook of Product Design for Manufacturing—A Practical Guide for Low-Cost Production*. New York: McGraw-Hill, 1986.

Corbett, J., M. Donner, J. Meleka, and C. Pym, *Design for Manufacture: Strategies, Principles and Techniques*. Reading, Mass.: Addison-Wesley, 1991.

Dieter, G.E., *Engineering Design*, 2d ed. New York: McGraw-Hill, 1991.

Enrick, N.L., and H.E. Mottley, *Manufacturing Analysis for Productivity and Quality/Cost Enhancement*, 2d ed. New York: Industrial Press, 1983.

Ettlie, J.E., and H.W. Stoll, *Managing the Design-Manufacturing Process*. New York: McGraw-Hill, 1990.

Malstrom, E.M., *What Every Engineer Should Know About Manufacturing Cost Estimating*. New York: Marcel Dekker, 1981.

—— (ed.), *Manufacturing Cost Engineering Handbook*. New York: Marcel Dekker, 1984.

Manufacturing Engineering Management, 4th ed., *Vol. 5: Tool and Manufacturing Engineers Handbook*. Dearborn, Mich.: Society of Manufacturing Engineers, 1988.

Michaels, J.V., *Design to Cost*. New York: Wiley, 1989.

Ostwald, P.F., *Cost Estimating*, 2d ed. Englewood Cliffs, N.J.: Prentice-Hall, 1984.

Pugh, S., *Total Design: Integrated Methods for Successful Product Engineering*. Reading, Mass.: Addison-Wesley, 1991.

Sandgren, E., *Design Optimization*. Reading, Mass.: Addison-Wesley, 1991.

Suh, N.P., *The Principles of Design*. New York: Oxford University Press, 1990.

Trucks, H.E., and G. Lewis (eds.), *Designing for Economical Production*, 2d ed. Dearborn, Mich.: Society of Manufacturing Engineers, 1987.

Winchell, W., *Realistic Cost Estimating for Manufacturing*, 2d ed. Dearborn, Mich.: Society of Manufacturing Engineers, 1989.

REVIEW QUESTIONS

40.1 List and describe the major considerations involved in selecting materials for products.

40.2 Why is a knowledge of available shapes of materials important? Give five specific examples.

40.3 Is it always desirable to purchase stock that is close to the final dimensions of a part to be manufactured? Explain and give some examples.

40.4 Describe manufacturing properties of materials. Give three examples demonstrating the importance of this information.

40.5 What course of action would you take if the supply of a material selected for a product line becomes unreliable?

40.6 Give three examples demonstrating the relationship between costs of materials and their processing.

40.7 Describe the potential problems involved in reducing the quantity of materials in products.

40.8 Why is material substitution an important aspect of manufacturing engineering? Give five examples from your own experience or observations.

40.9 Why has material substitution been particularly critical in the automotive and aerospace industries?

40.10 Discuss your thoughts concerning the replacement of aluminum beverage cans with steel cans.

40.11 What factors are involved in the selection of manufacturing processes? Explain why they are important.

40.12 What is meant by process capabilities? Select four different and specific manufacturing processes and describe their capabilities.
40.13 Why is production volume significant in process selection?
40.14 Give three examples of processes that have capabilities for high production rate and three that have low capabilities. Why should a company use a process that is inherently slow?
40.15 Discuss the disadvantages of long lead times in production.
40.16 What is meant by economical quantity?
40.17 Describe the various costs involved in manufacturing. Explain how you could reduce each of these costs.
40.18 What is relative cost? Name several processes with widely different relative costs.
40.19 Describe the characteristics of machines suitable for different production volumes.
40.20 What is value engineering? What are its benefits?
40.21 What is meant by tradeoff? Why is it important in manufacturing?

QUESTIONS AND PROBLEMS

40.22 Explain why the larger the quantity per package of food products, the lower the cost per unit weight.
40.23 Explain why the value of the scrap produced in a manufacturing process depends on the type of material.
40.24 Comment on the magnitude and range of scrap shown in Table 40.2.
40.25 Describe the difference between the cost and the price of a product. Do you think customers are willing to pay a higher price for a product that has good styling and design, even though these do not add significantly to the function and performance of the product? Explain with appropriate examples.
40.26 Describe your observations concerning the information in Table 40.3.
40.27 Explain how the high cost of some of the machinery listed in Table 40.4 can be justified.
40.28 Other than the size of the machine, what factors are involved in the range of prices in each machine category shown in Table 40.4?
40.29 Give several examples of applications of the information given in Fig. 40.2.
40.30 Explain the reasons for the relative positions of the curves shown in Fig. 40.5.
40.31 What factors are involved in the shape of the curve shown in Fig. 40.8?
40.32 Make suggestions as to how to reduce the dependence of production time on surface finish, as shown in Fig. 40.9.
40.33 How do you explain the trends in Fig. 40.10?
40.34 Why does the abscissa in Fig. 40.10 change from quantity or lot size to production rate?
40.35 Discuss your observations concerning Fig. 40.11.
40.36 Select three different products and make a survey of the change in their prices over the past ten years. Discuss the reasons for the changes in their prices.
40.37 Discuss the tradeoffs involved in selecting one of the two materials for each application listed. Also discuss typical conditions to which these products are subjected in their normal use.

a) Steel vs. plastic paper clips

b) Forged vs. cast crankshafts
c) Forged vs. powder-metallurgy connecting rod
d) Plastic vs. sheet metal light-switch plates
e) Glass vs. metal water pitchers
f) Sheet metal vs. cast hubcaps
g) Steel vs. copper nails
h) Wood vs. metal handle for a hammer
i) Sheet metal vs. reinforced plastic chairs

40.38 Discuss the manufacturing process or processes that are suitable for making the products listed in Question 40.37. Explain whether they would need additional operations such as coating, plating, heat treating, and finishing. If so, make recommendations and give the reasons for them.

40.39 In Fig. 2.1(a) we described the shape of a typical tensile-test specimen with round cross-section. Assuming that the starting material (stock) is a round rod and only one specimen is needed, discuss the processes and the machinery by which the specimen can be made, including their relative advantages and limitations. Describe how the process you selected can be changed for economical production as the number of specimens to be made increases.

40.40 The figure below shows the components of an electric hand drill. Select five components from this product and explain in detail (a) their functions, (b) suitable materials, and (c) processes by which these components can be made economically.

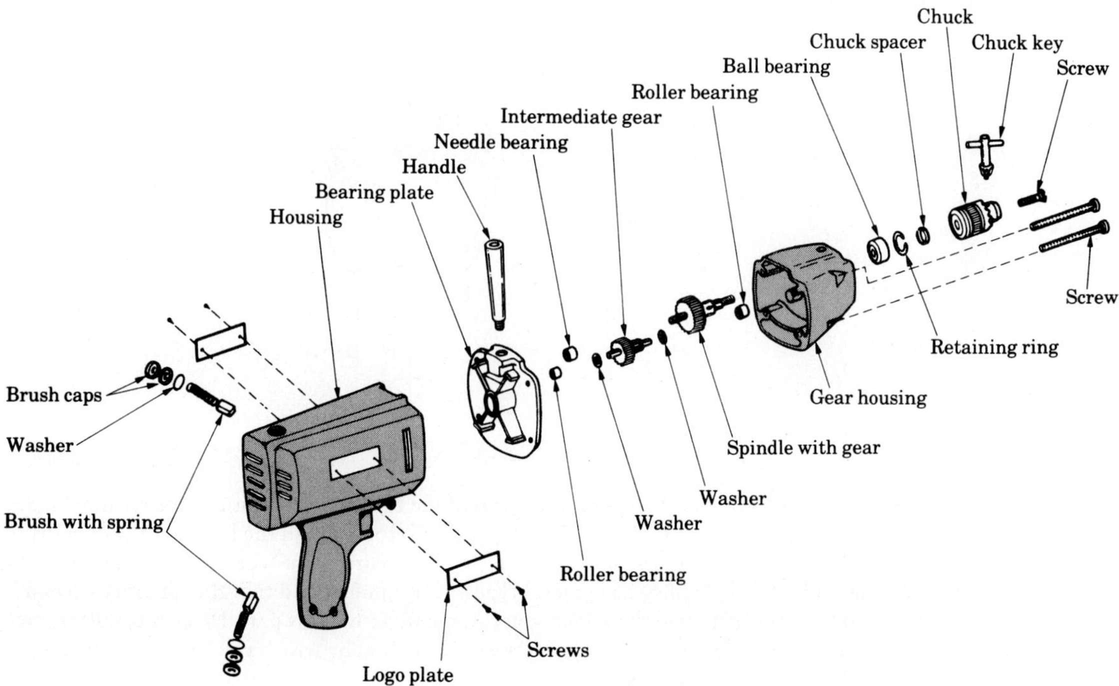

40.41 The figure below shows a sheet metal part made of steel. Consider either press-brake forming or contour roll forming as the alternative processes to form this part. Discuss how this part can be formed by either of the processes, the design and materials for tooling for each process, and how your selection of a process may be changed as (a) the length of the part increases and (b) the number of parts required increases.

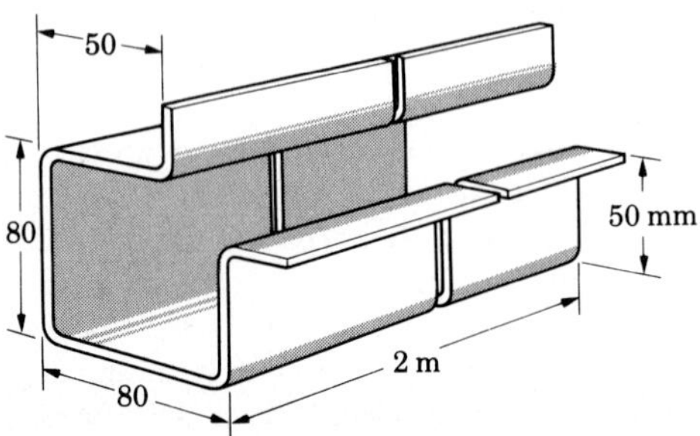

40.42 The part shown in the figure below is a carbon-steel segment gear. The small hole is for clamping the part on a spindle with a bolt and nut. When at first the demand was low, this part was made by machining from bar stock. As the demand increased, the process was changed to extrusion. (a) Describe the procedure for machining this part from bar stock and the processes and machines you would recommend. (b) Explain how this part can be extruded. (c) Would you produce the slot at the bottom during extrusion by proper die design, or would you prefer to slit it after the part is extruded? How would you produce the slot on a production basis? Explain the relative advantages and limitations.

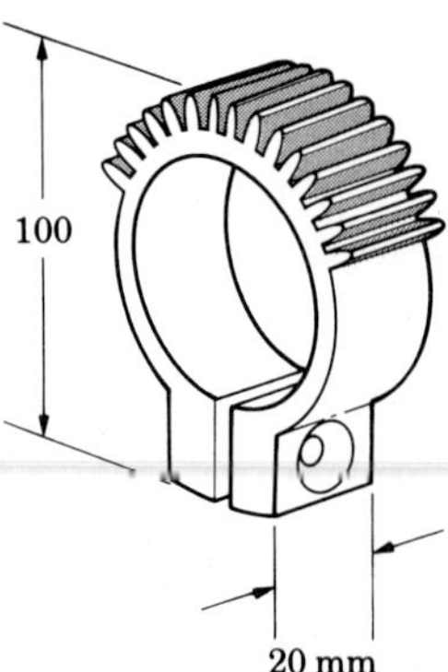

40.43 The part shown in the figure at the top of the following page is a cable terminal made of 304 stainless steel. The cable is slipped into the hole on the left, and the cylindrical end of the part is swaged over the cable, securing the cable. Assuming that the manufacturing facility has several types of equipment and that the quantity needed is 500, suggest processes by which you can manufacture this part. Discuss details of each process recommended. Is swaging the only method by which you can attach the cable? Discuss other alternatives.

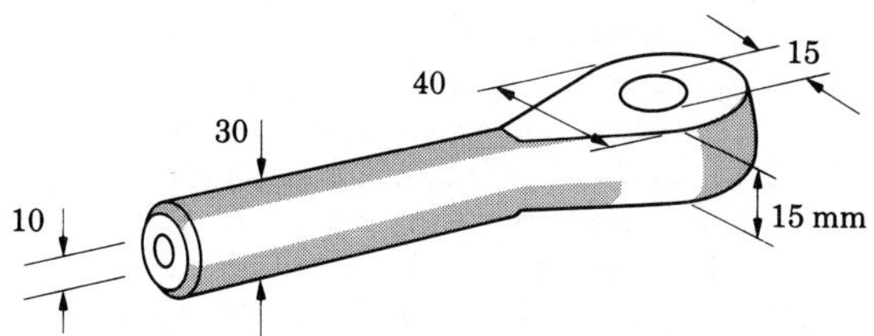

40.44 Several methods can be used in making the metal part shown in the figure below. List these methods, and for each method, explain (a) the details concerning the equipment needed, (b) estimated production time per part, and (c) amount of scrap produced.

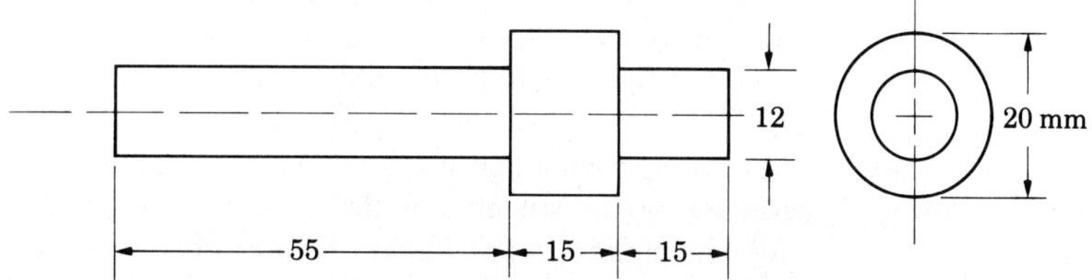

40.45 Review the manufacturing processes for the part in Question 40.44 and explain which process would produce a part that (a) may not have sufficient fatigue strength, (b) may not resist high temperatures, (c) would require additional finishing operations, and (d) would require the largest equipment.

40.46 Assume that the part shown in Question 40.44 is made of (a) thermoplastic and (b) thermosetting plastic. Explain which processes would be suitable for each of these materials.

40.47 In Table 40.1 we have listed several materials and their commercially available shapes. By contacting suppliers of materials, complete this list to include the following materials: (a) titanium, (b) superalloys, (c) lead, (d) tungsten, and (e) amorphous metals.

40.48 Select three products commonly found in your home. State your opinions as to (a) what materials were used in the product and why, and (b) how they are made and why those particular processes were used.

40.49 Inspect the components under the hood of your automobile. Identify three parts each that have been produced to (a) net-shape and (b) near-net-shape condition. Comment on the design and production aspects of these parts and how the manufacturer achieved the net-shape condition.

40.50 Discuss comprehensively the factors that influence the choice between the following pairs of processes.

a) Sand casting vs. die casting of a fractional electric-motor housing
b) Machining vs. forming a large gear
c) Forged vs. powder-metallurgy gear
d) Frying pan made of a casting vs. stamped sheet metal
e) Outdoor furniture made of aluminum tubing vs. cast iron
f) Welded vs. cast machine tool structures
g) Thread rolled vs. machined bolt for high strength application
h) Thermoformed plastic vs. molded thermoset fan blade for household use

40.51 Comment on the differences, if any, between the designs, materials, and processing and assembly methods used for making products such as hand tools and ladders for professional use vs. consumer use.

40.52 Other than casting, which processes could be used singly or in combination in making the automotive parts shown in Fig. 11.2 and the hydraulic-turbine runner in Fig. 11.11? Would they be economical? Explain in terms of production volume.

40.53 Discuss production and assembly methods that can be used to build the structures for the different presses shown in Fig. 16.51.

40.54 The capabilities of machining processes are outlined in Fig. 22.1. Inspect the various shapes produced and suggest alternative processes. Comment on the properties of materials that would influence your suggestions.

40.55 The double-flare V-groove weld shown in Fig. 27.1(a) is used to weld two solid cylindrical bars along their length. Assuming that the material is carbon steel, comment on whether you would still use this process if the diameter of the bars is 2 mm (0.080 in.) each. If not, which processes could be used to make it? If the material is changed to copper, would your answer be different? Explain.

40.56 The accompanying figure is a schematic illustration of a human-powered submarine designed by the students at the University of South Florida. It is prepared for an underwater race at 15 to 20 ft of water off Florida's east coast. Make a preliminary recommendation for the types of materials to be used for the shell of this submarine and the manufacturing methods to be employed. Give reasons for your recommendations. Assume that (a) only one submarine will be built and (b) speed, overall performance, and cost-effectiveness are among the important judging characteristics for this race.

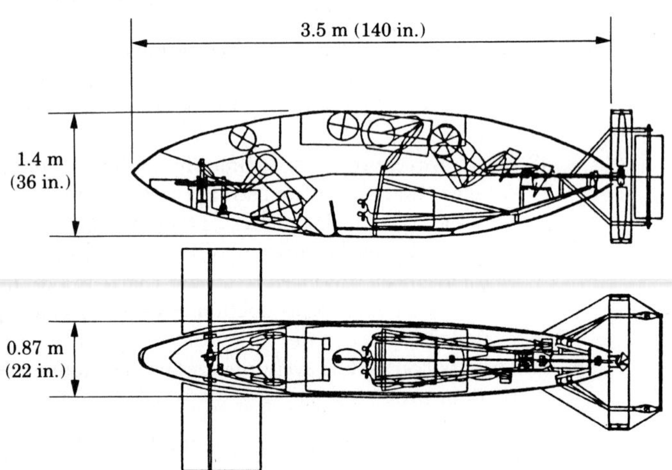

40.57 If the quantity in Problem 40.56 is increased to, say, 5 or 10, would your answers be different? Explain.

40.58 The illustrations below show two different designs for a fuel-pump lever arm. Assuming that the longest dimension is about 50 mm (2 in.), suggest several methods of manufacturing these two designs and the machinery and tooling needed. Also comment on the relative costs involved.

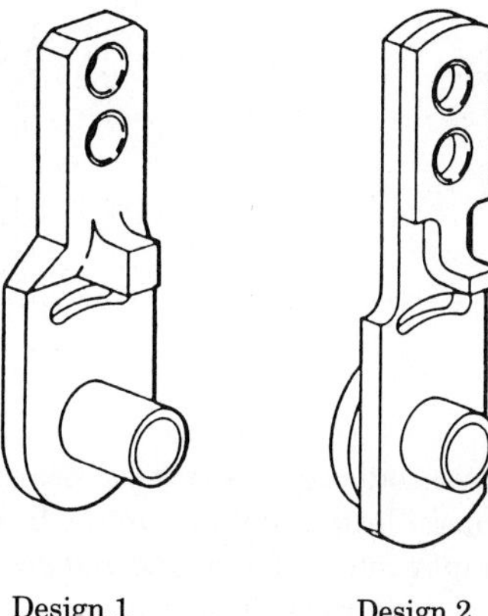

40.59 The illustrations below show a steel mounting bracket that can be manufactured by either (a) casting or (b) stamping of sheet metal. Describe the manufacturing sequence and processes involved in making each of these parts. Also comment on the relative costs involved. Can you suggest other methods of manufacturing this part? Explain.

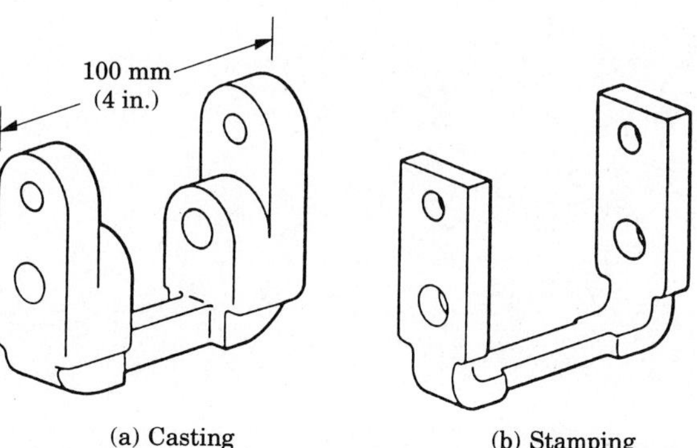

40.60 If the dimensions of the parts in Problems 40.58 and 40.59 were (a) ten times and (b) one hundred times larger, how different would your answers be?

40.61 The illustration at the top of the following page is that of an outdoor gas grill with three burners. Describe the materials and processes you would use in manufacturing the major components of this product, including racks, gas tank, burners, lid, and so on. Comment on the reasons for your selections. *Source for illustration*: Weber-Stephen Products Co.

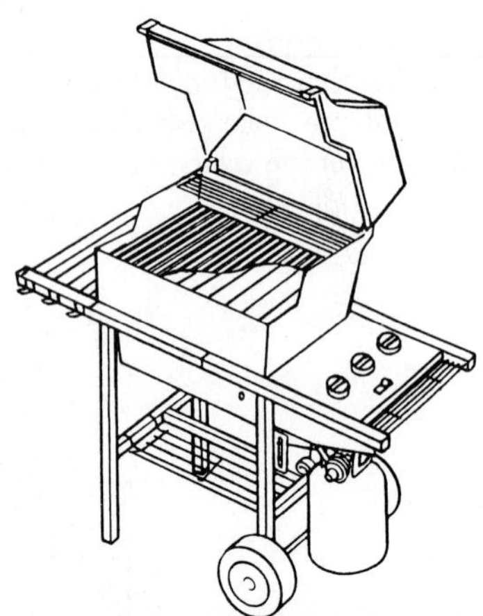

40.62 We introduced this text by showing a photograph of a jet engine (see Fig. 1 in the General Introduction). On the basis of the topics covered in this text, select any three components in this engine and describe the materials and processes that you would use in making them in quantities of, say, 1000. Remember that these parts have to be manufactured at minimum cost while maintaining quality, product integrity, and reliability.

Abbreviations

AA Arithmetical average; The Aluminum Association
AAAI American Association for Artificial Intelligence
AACE American Association of Cost Engineers
AAES American Association of Engineering Societies
AAR Association of American Railroads
ABS Acrylonitrile-butadiene styrene
AC Adaptive control; air cooled
ACC Adaptive control constraint
ACerS American Ceramic Society
ACI Alloy Casting Institute
ACM Association for Computing Machinery
ACO Adaptive control optimization
ACQWL American Center for Quality of Work Life
ACS American Chemical Society
ACT Acceptance criteria for testing
A/D Analog-to-digital converter
ADC Analog-to-digital converter
ADCCP Advanced data communication control procedures
ADCI American Die Casting Institute
ADDRG American Deep Drawing Research Group
ADI Austempered ductile iron
ADP Automatic data processing
ADS Automated design system
AEC Architecture, engineering, and construction
AEI Associated Electrical Industries
AES Abrasive Engineering Society
AFIT American Fabricating Institute of Technology
AFM Abrasive-flow machining
AFOSR Air Force Office of Scientific Research
AFP/SME Association for Finishing Processes of SME
AFS American Foundrymen's Society
AGDM Agent directory manager
AGMA American Gear Manufacturers Association
AGV Automated guided vehicle
AI Artificial intelligence

AIA Aerospace Industries Association
AIAA American Institute of Aeronautics and Astronautics
AID Auto-interactive design
AIIE American Institute of Industrial Engineers
AIME American Institute of Mining, Metallurgical and Petroleum Engineers
AIMI Automatic Identification Manufacturers Incorporated
AIMS Advanced integrated-manufacturing system
AISC American Institute of Steel Construction
AISI American Iron and Steel Institute
AJM Abrasive-jet machining
AK Aluminum killed
ALGOL Algorithmic language
ALU Arithmetic/logic unit
AML A manufacturing language
AMMRC Army Materials and Mechanics Research Center
AMRF Advanced Manufacturing Research Facility
AMS Aerospace Material Specification
AMSA American Metal Stamping Association
AMTDA American Machine Tool Distributors Association
ANSI American National Standards Institute
AOD Argon–oxygen decarburization
AP Array processor
APAS Adaptable-programmable assembly system
APC American Productivity Center
API American Petroleum Institute
APL A programming language
APMA American Productivity Management Association
APMI American Powder Metallurgy Institute
APT Automatically programmed tools
AQL Acceptable quality level
AROM Alterable read-only memory
ASCE American Society of Civil Engineers
ASCIE American Standard Code for Information Exchange
ASCII American Standard Code for Information Interchange
ASEE American Society for Engineering Education
ASHRAE American Society of Heating, Refrigerating and Air-Conditioning Engineers
ASLE American Society of Lubrication Engineers (now STLE, Society of Tribologists and Lubrication Engineers)
ASM American Society for Metals (now ASM International)
ASM Analytic solid modeling
ASME American Society of Mechanical Engineers
ASNT American Society for Nondestructive Testing
ASPE American Society for Precision Engineering
ASQC American Society for Quality Control
AS/RS Automatic storage and retrieval system
ASTM American Society for Testing and Materials
ATC Automatic tool changer
ATE Automatic test equipment
AUTOSTOP Automatic system for positioning of tools
AVA Automated Vision Association
AWC Automatic work changer
AWG American Wire Gage
AWS American Welding Society

BARE Boolean assessment of rules and equation
BASIC Beginner's All-purpose Symbolic Instruction Code
bcc Body-centered cubic
BCD Binary-coded decimal
BCRIA British Cast Iron Research Association
BDC Bottom dead center
BEST/SME Burr, Edge and Surface Conditioning Technology Division of SME
BHMA British Hard Metal Association
BHN Brinell hardness number (obsolete; see HB)
BHRA British Hydromechanics Research Association
BISRA British Iron and Steel Research Association
BLU Basic length unit
BMC Bulk-molding compound
BOM Bill of materials
BR Back rake angle
B-rep Boundary representation
BRU Basic resolution unit
BS British Standards Institute
BSC Binary synchronous communication
BTR Behind-tape reader
BTU British thermal unit
BUE Built-up edge
BWG Birmingham wire gage

C A high-level programming language
CAD Computer-aided design; computer-aided drafting
CADAM Computer-aided design and manufacturing
CAE Computer-aided engineering

CAI Computer-aided instruction
CAK Computer-aided kinematics
CAL Conversional algebraic language
CAM Computer-aided manufacturing
CAME Computer-assisted mechanical engineering
CAM-I Computer Aided Manufacturing—International, Inc.
CAP Consolidation by atmospheric pressure
CAPP Computer-aided process planning
CAQC Computer-aided quality control
CASA/SME Computer and Automated Systems Association of SME
CAT Computer-aided testing
CATI Computer-aided testing and inspection
CBEMA Computer and Business Equipment Manufacturing Association
CBN Cubic boron nitride
C/C Coding/classification
CCD Charge-coupled device
CCPA Cemented Carbide Producers Association
CDA Copper Development Association
CDC Cutter diameter compensation
CFRP Carbon-fiber reinforced plastic
CGS Centimeter-gram-second
CHM Chemical machining
CID Charge imaging device
CIM Computer-integrated manufacturing; Canadian Institute of Mining and Metallurgy
CIMA Construction Industry Manufacturers Association
CIMS Computer-integrated manufacturing system
CIP Cold isostatic pressing
CIRP International Institution for Production Engineering Research
CL Cutter location; center line
CLIPS A language-intelligent production system
CLU Control loops unit
CMC Ceramic-matrix composite
CMM Coordinate measuring machine
CMOS Complementary metal-oxide semiconductor
CMPM Computer-managed parts manufacture
CMS Cellular manufacturing system
CNC Computer numerical control
COBOL Common Business Oriented Language
COD Crack opening displacement; cost on delivery
CoG/SME Composites Group of SME
COM Computer output to microfilm
CP Continuous path
CPM Critical path method
CP/M Control program for microcomputers
CPS Characters per second
CPU Central processing unit
CR Cold rolled
CRP Capacity resource planning
CRT Cathode-ray tube
CSA Canadian Standards Association
CSG Constructive solid geometry
CSIRO Commonwealth Scientific and Industrial Research Organisation
CSME Canadian Society of Mechanical Engineers
CT Continuous transformation
CVD Chemical vapor deposition
CW Cold welding
CWQC Company-wide quality control

DA Design automation
D/A Digital-to-analog converter
DAC Digital-to-analog converter
DARPA Defense Advanced Research Projects Agency
DAS Dendrite-arm spacing
DBMS Data-base management system
DBTT Ductile-to-brittle transition temperature
DC Device coordinates
DCE Data communications equipment
DCEN Direct current electrode negative
DCEP Direct current electrode positive
DCL Depth-of-cut line
DCTL Direct-coupled transistor logic
DDA Digital differential analyzer
DDAS Direct data acquisition system
DDC Direct digital control
DEU Data entry unit
DFW Diffusion welding (bonding)
D&I Drawn and ironed
DIF Document interchange format
DIN Deutsches Institut für Normung (German Institute for Standards; German Industrial Standard)
DIP Dual in-line package
DMA Direct memory access
DMIC Defense Metals Information Center
DMIS Dimensional measurement interface specification
DNC Direct numerical control; distributed numerical control
DOM Drawn over mandrel
DOS Disk operating system
DP Degree of polymerization; diametral pitch; data processing

DPH Diamond pyramid hardness number, Vickers (obsolete; see HV)
DPI Dots per inch
DPRO Digital position readout
DPU Data processing unit
DQ Drawing quality
DRC Design rules checking
DRO Digital readout
DRS Data reduction subsystem
DSA Data structure architecture
DT Dynamic tear (test)
DTIC Defense Technical Information Center
DVST Direct-view storage tube
DWT Drop-weight test
DX Data transfer
DXF Data exchange format

EAROM Electrically alterable read-only memory
EBCDIC Extended binary-coded decimal interchange code
EBM Electron-beam machining
EBW Electron-beam welding
ECD Electrochemical deburring
ECEA End cutting-edge angle
ECG Electrochemical grinding
ECL Emitter-coupled logic
ECM Electrochemical machining; error-correcting memory
EDC Extended data comparison
EDG Electrical-discharge grinding
EDM Electrical-discharge machining
EDP Electronic data processing
EDX Energy dispersion x-ray
EHD Elastohydrodynamic (lubrication)
EIA Electronics Industries Association
ELC Expulsion limit controller
ELPO Electrophoretic primer
EMI Electromagnetic interface
EM/SME Electronic Manufacturing Group of SME
EOB End of block
EOC Economic order quantity
EOP End of program
EOT End of tape
EP Extreme pressure
EPA Environmental Protection Agency
EPROM Erasable programmable read-only memory
EROM Erasable read-only memory
ES Expert system
ESW Electroslag welding
ETP Electrolytic tough pitch (copper)
EXW Explosion welding

FAS Flexible assembly systems
FC Furnace cooled
FCAW Flux-cored arc welding
fcc Face-centered cubic
FDC Factory data collection
FEA Finite-element analysis
FEM Finite-element method
FF Flip-flop
FIA Forging Industries Association
FIFO First in, first out
FIS Flexible inspection system
FLC Forming limit curve
FLD Forming limit diagram
FMA Fabricating Manufacturers Association
FMC Flexible manufacturing cell
FMS Flexible manufacturing system; Federation of Materials Societies
FOB Freight on board
FOF Factory of the future
FORTRAN Formula Translation
FOW Forge welding
FPU Floating-point unit
FRN Feed rate number
FRP Fiberglass-reinforced plastic
FRW Friction welding
FW Flash welding

GFRP Graphite-fiber reinforced plastic; Glass-fiber reinforced plastic
GINO Graphics input and output
GMAW Gas metal-arc welding; metal inert-gas
GNP Gross national product
GPS General problem solver
GRN Glass-reinforced nylon
GT Group technology
GTAW Gas tungsten-arc welding; tungsten inert-gas
GU Graphical unit
GWI Grinding Wheel Institute

HAZ Heat-affected zone
HB Brinell hardness number
hcp Hexagonal close-packed
HDLC High-level data link control protocol
HDPE High-density polyethylene
HEM Hostile environment machines

HERF High-energy-rate forming
HGVS Human-guided vehicle system
HIP Hot isostatic pressing
HIPS High-impact polystyrene
HK Knoop hardness number
HMC High-strength molding compound
HMS High-modulus strength
HOL High-order language
HR Rockwell hardness number, including scales such as HRA, HRB, HRC, etc; hot rolled
HSLA High-strength low-alloy (steels)
HSS High-speed steel
HV Vickers hardness number

IACS International Annealed Copper Standard (for electrical conductivity)
IAMS Institute of Advanced Manufacturing Sciences
IAQC International Association of Quality Circles
IC Integrated circuit
ICAM Integrated computer-aided manufacturing
ICFG International Cold Forging Group
ICG Interactive computer graphics
IDDRG International Deep Drawing Research Group
IDS Integrated design support
IEEE Institute of Electrical and Electronic Engineers
IFI Industrial Fasteners Institute
IGC Institute for Graphic Communication
IGES Initial graphics exchange specification
IGS Interactive graphics system
IIE Institute of Industrial Engineers
IKBS Intelligent knowledge-based systems
I/M Ingot metallurgy
IMACS International Association for Mathematics and Computers in Simulation
IME Institution of Mechanical Engineers
IMV Industrial machine vision
INFAC International Flexible Automation Center
I/O Input/output
IOCS Input/output control system
IR Infrared; industrial robot
ISA Instrument Society of America
ISI Iron and Steel Institute
ISO International Standards Organization
IT Isothermal transformation
ITS International temperature scale
IU Image understanding
IVD Ion-vapor deposition
IVP Initial value problem

JIC Joint Industry Conference
JIRA Japanese Industrial Robot Association
JIS Japanese Industrial Standard
JIT Just-in-time
JSAE Japan Society of Automotive Engineers
JSME Japan Society of Mechanical Engineers
JSPE Japan Society of Precision Engineering
JSTPE Japan Society for Technology of Plasticity

KBS Knowledge-based system
KE Knowledge engineer
KHN Knoop hardness number (obsolete; see HK)
KR Knowledge representation

LAN Local area network
LASER Light Amplification by Stimulated Emission of Radiation
LBM Laser-beam machining
LBW Laser-beam welding
LC/SME Lasers' Committee of SME
LCCC Leadless ceramic chip carrier
LCD Liquid crystal display
LCL Lower control limit
LDH Limiting dome height
LDPE Low-density polyethylene
LDR Limiting drawing ratio
LED Light-emitting diode
LIM Liquid injection molding
LIS Large interactive surface
LISP List processing language
LMC Low-pressure molding compound; least material condition
LNG Liquefied natural gas
LPG Liquefied petroleum gas
LS Limit switch
LSD Least significant digit
LSI Large-scale integration
LTPD Lot tolerance percent defective
LUT Lookup table
LVDT Linearly variable differential transformer

MAP Manufacturing Automation Protocol
MBM Magnetic-bubble memory
MCD Machine control data
MCIC Metals and Ceramic Information Center
MCR Master control relay
MCU Machine control unit

MDC Machinability Data Center
MDI Manual data input
MDNA Machinery Dealers National Association
MEEF Manufacturing Engineering Education Foundation
MFP Master family planning
MHI Material Handling Institute
MIC Management (manufacturing) information system
MICR Magnetic ink character recognition
MIG Metal inert-gas; gas metal-arc welding
MIL Military; one thousandth of an inch
MIS Management information system
MLC Multilayer chip capacitor
MMC Maximum material condition; metal-matrix composite
MMCIAC Metal Matrix Composites Information Analysis Center
MOD Module
MODEM Modulator/demodulator
MOS Metal-oxide semiconductor
MPC Metal Properties Council; Manufacturing Productivity Center
MPIF Metal Powder Industries Federation
MP/M Multiprogramming control program for microprocessor
MPTA Metal Powder Technology Association
MRP Material requirements planning
MRP II Manufacturing resources planning
MSI Medium-scale integration
MTBF Mean time between failures
MTDR Machine Tool Design and Research (Conference)
MTIAC Manufacturing Technology Information Analysis Center
MTM Methods time data
MTS Multipoint terminal software
MUM Methodology for unmanned manufacturing
MVA/SME Machine Vision Association of SME
MWD Molecular-weight distribution

NACE National Association of Corrosion Engineers
NAE National Academy of Engineering
NAMRC North American Manufacturing Research Conference
NAMRI North American Manufacturing Research Institution
NAMTAC National Association of Management and Technical Assistance Centers
NAS National Academy of Sciences; National Aerospace Standards
NAS-NRC National Academy of Sciences—National Research Council
NASA National Aeronautics and Space Administration
NBS National Bureau of Standards (now **NIST**)
NC Numerical control
NCGA National Computer Graphics Association
NDC Normalized device coordinates
NDE Nondestructive evaluation
NDI Nondestructive inspection
NDT Nondestructive testing; nil ductility transition
NDTT Nil ductility transition temperature
NEC National Electrical Code
NEMA National Electrical Manufacturers Association
NFPA National Fire Protection Association
NIDA National Industrial Distributor Associations
NIOSH National Institute for Occupational Safety and Health
NIST National Institute of Standards and Technology
NLGI National Lubricating Grease Institute
NLI Natural language interface
NLP Natural language processing
NLU Natural language understanding
NMTBA National Machine Tool Builders' Association
NRC National Research Council; Nuclear Regulatory Commission
NSC National Safety Council
NSF National Science Foundation
NTIAC Nondestructive Testing Information Analysis Center
NTIS National Technical Information Service
NTMA National Tooling and Machining Association

OAW Oxyacetylene welding
OBI Open-back inclinable
OC Operating characteristic
OCR Optical character recognition
OEM Original equipment manufacturer
OFC Oxyfuel gas cutting
OFHC Oxygen-free, high-conductivity (for copper)
OFW Oxyfuel welding
OHW Oxyhydrogen welding
OMR Optical mark recognition

ONR Office of Naval Research
OQ Oil quenched
ORSA Operations Research Society of America
OS Operating system
OSHA Occupational Safety and Health Administration
OSI Open systems interconnection

PA Polyamide
PAC Plasma-arc cutting
PAL Phase-alternating line
PAM Plasma-arc machining
PAW Plasma-arc welding
PC Programmable controller; personal computer; printed circuit; polycarbonate
PCB Printed circuit board
PCD Polycrystalline diamond
PD Pitch diameter
PDES Product data exchange specification
PDL Programmable data logger
PE Proportional error; polyethylene
PERA Production Engineering Research Association
PEW Percussion welding
P/F Precision forging
PH Precipitation hardenable
PHD Plastohydrodynamic (lubrication)
PI Proportional integrating; polyimide
PID Proportional integral derivative
PLC Programmable logic controller
PLCC Plastic-leaded chip carrier
P/M Powder metallurgy
PMMA Polymethylmethacrylate
PP Polypropylene
PREP Plasma rotating electrode process
PROLOG Programming in logic
PROM Programmable read-only memory
PS Polystyrene; polysulfone
PSZ Partially-stabilized zirconia
PTFE Polytetrafluoroethylene (Teflon)
PTP Point-to-point
PVC Polyvinyl chloride
PVD Physical vapor deposition
PWM Pulse-width modulation
PZT Lead zirconate titinate

QA Quality assurance
QC Quality control
QCI Quality Circle Institute
QMS Quality management system

RA Reduction in area
RAC Reliability Analysis Center
RAM Random-access memory
RBS Rule-based system
RD Rolling direction
RDLP R&D limited partnership
RE Rare earth
REP Rotating electrode process
RF Radio frequency
RIA Robotic Industries Association
RIM Reaction-injection molding
RI/SME Robotics International of SME
RJE Remote job entry
ROC Rapid omnidirectional compaction
ROI Return on investment; region of interest
ROM Read-only memory
ROS Read-only storage
RPW Projection welding (resistance)
RPY Roll-pitch-yaw
RRIM Reinforced reaction-injection molding
RSEW Seam welding (resistance)
RSM Rapidly solidified materials
RSP Rapid solidification process
RSPD Rapid solidification plasma deposition
RSW Spot welding (resistance)
RT Room temperature
RTM Resin-transfer molding
RW Resistance welding
R/W Read/write
RWM Read/write memory

SAE Society of Automotive Engineers
SAMPE Society for the Advancement of Material and Process Engineering
SAN Styrene acrylonitrile
SAP Sintered aluminum powder; structural analysis program
SAW Submerged-arc welding
SBQ Salt-bath quench
SCAD Super computer-aided design
SCAE Society for Computer-Aided Engineering
SCARA Selective compliance assembly robot arm
SCEA Side cutting-edge angle
SCR Silicon-controlled rectifier
SCSI Small computer system interface
SCTE Society of Carbide and Tool Engineers
SEM Scanning electron microscope
SESA Society for Experimental Stress Analysis
SFC Shop-floor control (system)

SFSA Steel Founders' Society of America
SG Seeded gel
SI Système International d'Unités
SIAM Society of Industrial and Applied Mathematics
SIC Standard Industrial Classification
SLF Slip-line field
SMAW Shielded metal-arc welding
SMC Sheet-molding compound
SME Society of Manufacturing Engineers
SMED Single minute exchange of die
SPC Statistical process control
SPE Society of Plastics Engineers
SPF/DB Superplastic forming/diffusion bonding
SPI Society of the Plastics Industry
SPICE Simulation program for integrated circuits engineering
SQC Statistical quality control
SR Side rake angle
SSA Simplex search algorithm
SSI Small-scale integration
SSU Saybolt Seconds Universal
SSW Solid-state welding
STA Solution treated and aged
STEP Standard for the Exchange of Product Data
STLE Society of Tribologists and Lubrication Engineers
STQ Solution treated and quenched
SW Stud welding
SWG Steel-wire gage

TCM Thermochemical machining
TCP Tool center point
TCP/IP Transmission control protocol/Internet protocol
TCS Terminal control system
TD Transverse direction
TDC Top dead center
TED Thermal energy deburring
TEM Transmission electron microscope
TFE Tetrafluoroethylene
TIG Tungsten inert-gas; gas tungsten-arc welding
TIR Total indicator reading
TMC Thick molding compound
TMS The Metallurgical Society of AIME
TOP Technical and Office Protocol
TPA Target point align
TQC Total quality control
TRIP Transformation-induced plasticity
TRS Transverse rupture strength
TTL Transistor-transistor logic
TTS Technology Transfer Society
TTT Time temperature transformation
TZM Titanium-zirconium-molybdenum alloy

UBT Upper-bound technique
UCL Upper control limit
UDK User-defined key
UHM Ultra-high modulus
UHMWPE Ultrahigh-molecular-weight polyethylene
UL Underwriters' Laboratories
ULSI Ultra large-scale integration
UM Ultrasonic machining
UMC Unmanned machining cell
UNC Unified coarse thread
UNF Unified fine thread
UNS Unified Numbering System
USM Ultrasonic machining
USW Ultrasonic welding
UTS Ultimate tensile strength
UV Ultraviolet

VA Value analysis
VDI Verein Deutscher Ingenieure; virtual device interface
VDU Visual display unit
VHN Vickers hardness number (obsolete; see HV)
VHSIC Very high speed integrated circuit
VLSI Very large scale integration
VMM Variable mission manufacturing
VNC Voice programming for NC
VRP View reference point
VTL Vertical toolchanger lathe

WAN Wide area network
WC Tungsten carbide
WFS Wet flexural strength
WIP Work in progress
WQ Water quenched
WRC Welding Research Council

YAG Yttrium aluminum garnet

ZD Zero defect

Index

LIST OF EXAMPLES IN TEXT